حضرة صاحب الجلالة السلطان قابوس بن سعيد المعظم

His Majesty Sultan Qaboos bin Said

كلمة معالى سعيد بن أحمد الشعرى
وزير النفط والمعادن ، الرئيس الفخرى للندوة

أصحاب السمو والمعالى
ضيوفنا الكرام

يطيب لى أن أرحب بكم أجمل ترحيب ، فى هذا التجمع العلمى الكبير ، ويشرفنى أن أحييكم فى هذه الندوه العلميه ، التى عرفت طريقها الى النور ، بفضل قائد البلاد المفدى حضرة صاحب الجلالة السلطان قابوس بن سعيد المعظم وتوجيهاته السامية ، بتشجيع البحث العلمى فى كل ميدان ، لما له من فوائد كبيرة ، كأساس لفهم كوكبنا هذا وما تكمن فى باطنه من ثروات تعود بالخير والرخاء على البشريه جمعاء .

أصحاب السمو والمعالى
المشاركون الكرام

لقد ظلت السلطنة لسنوات عديدة تجتذب الكثير من الباحثين والدارسين من كل أنحاء العالم ، لاجراء البحوث العلمية على صخورها المتميزة والتعرف على نشأة مصادرها المعدنية ، التى تساهم فى تطور علوم الأرض .. واننا فى عُمان ننتظر بلهفة كبيرة حصيلة مناقشاتكم ، ونتائج دراساتكم القيمة واثقين انه سيكون لها اكبر الأثر فى تنمية ثرواتنا المعدنية .

اننى إذ اتقدم ببالغ الشكر والتقدير ، لصاحب السمو السيد فهد بن محمود آل سعيد لتكرمه بإفتتاح هذه الندوة نيابة عن صاحب الجلاله السلطان المعظم ، فانه لايفوتنى ان اشكر كل الذين ساهموا فى إحياء فكرة هذه الندوة ، كما أوجه شكرى الى جامعة السلطان قابوس ، هذا الصرح العلمى ، الذى تعتمد عليه البلاد فى تخريج الاجيال القادمة المؤهلة ، من أجل مستقبل مشرق . كما أتقدم بالشكر لمنظمة اليونسكو ، ولكل من ساهم بآرائه وأبحاثه فى هذه الندوة ، وأخص بالشكر وزارة الدفاع وشرطة عمان السلطانية و وزاره الاعلام على تعاونهم لانجاح هذه الندوة .

والسلام عليكم ورحمة الله وبركاته ...

سعيد بن احمد بن سعيد الشنفري
وزير النفط والمعادن
Said bin Ahmed bin Said Al Shanfari
Minister of Petroleum and Minerals

The Speech of H.E. Said Bin Ahmed Al Shanfari, Minister of Petroleum and Minerals, Patron of the Symposium

Your Royal Highness Sayed Fahad Bin Mahmoud al Said,

Your Royal Highness Members of the Honourable Royal Family,

Your Excellencies Ministers,

Your Excellencies,

Our Esteemed Guests,

It gives me great pleasure to welcome you in this scientific gathering in Oman which has been made possible through the directives of H.M. Sultan Qaboos Bin Said to encourage research and to better understand our planet and its resources, in hopes of bringing prosperity to all mankind.

Your highness, Excellencies and Esteemed participants,

For many years, the Sultanate has been attracting the attention of researchers and scientists from all Countries. Their interest in interpreting the evolution of the earth's crust and its mineral resources shall be enhanced by studying ancient analogues which will induce human ability to find more resources and contribute to the advancement of earth sciences. We, in Oman look forward with keen interest to the outcome of your discussions and the results of your studies, which shall play a major role in the development of our resources.

Finally, I wish to extend my deep thanks and appreciation to His Royal Highness Sayed Fahad Bin Mahmoud for the opening of this Symposium on behalf of H.M. The Sultan. I wish to thank also all those who initiated the idea of holding this conference. I extend my thanks and appreciation also to Qaboos University, the great scientific edifying centre, on which the country is depending on to educate the qualified graduates for a better future. I extend my thanks as well to the UNESCO and all those who have contributed with their thoughts and research to this Symposium. I wish to thank the Ministry of Defence, Royal Oman Police and the Ministry of Information for their cooperation to the success of this Symposium.

May Almighty God extend His Blessings on you and enable you to carry out your honourable scientific mission and achieve your ambitious objectives.

بفضل التوجيهات الساميه لحضره صاحب الجلاله السلطان قابوس بن سعيد المعظم وحرص حكومته الرشيده على تشجيع البحث العلمى والنهوض بهذا الوطن الحبيب وتنويع مصادر دخله القومى فقد حرصت وزاره النفط والمعادن منذ تأسيسها قبل عقدين من الزمن ، على تنميه وإستغلال ما يكمن فى باطن هذه الارض المعطاء من ثروات معدنيه .

وما هذه الندوه الدوليه الهامه التى إستجابت لها المؤسسات العلميه والجامعات من كل أنحاء العالم الآ اشاره واضحه ودليلا قاطعا على تلك الجهود الحثيثه المستمره التى بذلتها وزاره النفط والمعادن فى مجال تنميه الثروات المعدنيه وتنشيط قطاعى الاستخراج المعدنى والتعدين .

وتأمل الوزاره ان تستمر هذه الجهود ويزدهر البحث العلمى وتزداد فيه المشاركه العمانيه الفعاله .

والله ولى التوفيق ...

محمد بن حسين بن قاسم
مدير عام المديريه العامه للمعادن

In accordance with the directives of H.M. Sultan Qaboos Bin Said, to encourage scientific research in all fields and in conformity with the endeavour of his rightly guided government to promote this beloved country, boost and diversify its economy, the Ministry of Petroleum & Minerals has since its foundation, about two decades ago, strived for developing and exploiting profitably the country's mineral wealth.

This International Symposium on Ophiolite which received a great response from Universites and Scientific Organizations from all parts of the world, is the best indication for the continuous effort made by this Ministry to develop and stimulate the mineral sector in the Sultanate of Oman.

The Ministry hopes that the research efforts would continue and that the Omani participation would grow.

MOHAMMED BIN HUSSAIN BIN KASSIM
Director General of Minerals

شكـــــــر

نتوجه بخالص الشكر وعميق الامتنان الى كل أولئك الذين ساهموا في إنجاح هذه الندوة ونخص بالشكر المؤسسات والمنظمات التاليه :-

منظمة اليونســــكو
جامعة السلطان قابوس
ديوان البلاط السلطاني
وزارة الدفـــــاع
شرطة عمان السلطانيه
دائرة العلاقات العامه بوزارة النفط والمعادن
شركة تنمية نفط عمان

Acknowledgement

The help and cooperation of the following Institutions and Organizations is gratefully acknowledged:

UNESCO
Sultan Qaboos University
Diwan of Royal Court
Ministry of Defence
Royal Oman Police
Department of Public Relations of the Ministry of Petroleum & Minerals
Petroleum Development Oman

OPHIOLITE GENESIS AND EVOLUTION OF THE OCEANIC LITHOSPHERE

Petrology and Structural Geology

VOLUME 5

Series Editor:

ADOLPHE NICOLAS

Department of Earth Sciences,
University of Montpellier, France

The titles published in this series are listed at the end of this volume.

Ophiolite Genesis and Evolution of the Oceanic Lithosphere

Proceedings of the Ophiolite Conference, held in Muscat, Oman, 7–18 January 1990

Edited by

TJ. PETERS, A. NICOLAS and R.G. COLEMAN

KLUWER ACADEMIC PUBLISHERS

DORDRECHT / BOSTON / LONDON

Library of Congress Cataloging-in-Publication Data

Ophiolite Conference (1990 : Muscat, Oman)
Ophiolite genesis and evolution of the oceanic lithosphere : proceedings of the Ophiolite Conference : held in Muscat, 7-18 January 1990 / edited by Tj. Peters, A. Nicolas, R. Coleman.
p. cm. -- (Petrology and structural geology ; v. 5)
Includes bibliographical references and index.
ISBN 0-7923-1176-0 (HB : acid-free paper)
1. Ophiolites--Oman--Congresses. 2. Submarine geology--Congresses. 3. Geology--Oman--Congresses. I. Peters, Tjerk. II. Nicolas, A. (Adolphe), 1936- . III. Coleman, Robert Griffin, 1923- . IV. Title. V. Series.
QE462.06062 1990
552'.3--dc20 91-10740

ISBN 0-7923-1176-0

Published by Kluwer Academic Publishers,
P.O. Box 17, 3300 AA Dordrecht, The Netherlands.

Kluwer Academic Publishers incorporates the publishing programmes of
D. Reidel, Martinus Nijhoff, Dr W. Junk and MTP Press.

Sold and distributed in the U.S.A. and Canada
by Kluwer Academic Publishers,
101 Philip Drive, Norwell, MA 02061, U.S.A.

In all other countries except for the Sultanate of Oman, sold and distributed
by Kluwer Academic Publishers Group,
P.O. Box 322, 3300 AH Dordrecht, The Netherlands.

On behalf of the scientific community, the editors thank the Ministry of Petroleum and Minerals for a financial contribution which has lowered the list price of this publication and thereby afforded its widest possible distribution.

Cover page: Pangea 200 M.a. ago and location of more recent ophiolites from: "Les Montagnes sous la Mer", A. Nicolas, 1990, BRGM.

Printed on acid-free paper

Printed in the Netherlands

Table of Contents

Introduction

Ophiolites are key sources of information regarding the genesis and evolution of oceanic lithosphere. Over the past decades, the geological study of ophiolites has provided a wealth of insight into lithospheric processes and has proved to be an indispensible prerequisite to interpreting geophysical and other investigations of the crust underlying recent oceans. The Oman Ophiolite offers the most complete and structurally undisturbed sections of the oceanic crust in vast, clean exposures. It is, therefore, most fortunate for the scientific community that Mhd. Kassim, Director General of Minerals and Dr. Hilal Al Azri, Director of the Geological Survey, took upon themselves the task of organizing in Oman an international meeting on ophiolites.

Having planned for an attendance of only 100 to 150 persons, the logistics of the organizing committee were put to a severe test by the 300 participants who eventually arrived from 27 countries. The 14 field trips, most of which were conducted twice, provided the participants with an excellent introduction to the geology of Oman, the ophiolite sequence, and related phenomena. Good exposures, competent excursion leaders, excellent organization aided by the armed forces, and the hospitality of the local population, combined to make the excursions a total success. The participants were thus well tuned in for the following Ophiolite Symposium held on the newly built Campus of the Sultan Qaboos University. Of the 160 scientific contributions, some 80 were presented orally and the remainder as posters. The hospitality of our Omani hosts was crowned by his Majesty Sultan Qaboos' invitation for all the participants to visit his palace. The resulting party evolved into a scene from "1001 nights" as the long awaited rain drenched and soothed the participants and ophiolites in the Muscat area, at the very moment when his Majesty Sultan Qaboos appeared.

The symposium brought together marine geophysicists and other investigators of recent oceanic lithosphere, and field geologists working on ophiolite belts. The ophiolite outcrops provided a unique environment to discuss the models for the structure of the oceanic crust and mantle developed by these two disciplines. The field evidence for the products of various processes, such as hydrothermal convection cells, could be readily observed and checked by the scientists. Arguments initiated in the field, including the implications of laboratory work, were discussed further in oral and poster presentations during the symposium.

The present Proceedings Volume represents a wide spectrum of topics covered in the Symposium and is divided into the following chapters: Processes at Spreading Centers, Magmatic Ores, Hydrothermalism, Tectonics of Emplacement and Metamorphism, Palaeogeographic Setting of the Oman Ophiolite, Palaeoenvironment of other Ophiolites, and Mapping Ophiolites. The first chapter is subdivided into processes occurring either in the crust or in the mantle. The scientific problems and solutions discussed in these main chapters are briefly outlined below.

In treating oceanic processes, the volume reflects on some of the highlights of the field excursions and conference, most notably the nature of magma chambers, wehrlite intrusions, mantle diapirism and ridge hydrothermalism. Recent models based on field studies in Oman have shed light on problems regarding the size of magma chambers below ridges, but on this topic the marine scientists and ophiolite geologists remain divided. Resolution of this controversy clearly requires more data from both sides and a better knowledge of the physical state of the plutonic material that fills a magma chamber. For example, the intense magmatic deformation in layered gabbros points to a fairly consolidated medium. Further studies of mantle diapirs, which are well exposed in Oman, are likely to play a major role in clarifying our understanding of ridge processes. The abundant wehrlite intrusions seem to have been emplaced near the edge of the magma chamber, with a possible origin in the transition zone.

The enigma as to why certain ophiolite complexes are mineralized and others are not is now partly understood in the light of the variable evolutionary paths, and in particular spreading rates, of oceanic lithosphere. Another problem that is well illustrated by recent discoveries of black smokers in Oman is the dynamics of the hydrothermal cells that cool accreting ridges. Fluids have strongly affected the mineralogy of the crustal section of the Oman ophiolite. For example, they have lead to the formation of the copper ores that were mined as early as 3000 B.C.

The palaeogeographic setting of the ophiolites of Oman and other regions is crucial to the questions of how, where and why the ophiolites were formed, and of how they adopted their present positions. One of the most fundamental questions of all remains open: Why was the Oman ophiolite thrust onto the Arabian Continent rather than subducted beneath Eurasia? There is general agreement that the Oman ophiolite formed in the Neotethys – an ocean which opened during the Permian along a NW-SE trending rift zone. However, it is not known if a carbonate platform separated the Arabian Continental Shelf and the adjacent Hawasina Basin from the oceanic trough in which the ophiolite formed. Detailed basin analyses have been started to help resolve this dilemma.

The metamorphic sole at the base of the Semail Ophiolite formed as the

Semail Nappe made its way towards the Arabian platform, bulldozing and overriding the oceanic sediments in front of it. This remarkable event has suggested a model for high T – low P ("ironing") metamorphism. Although promising, the model is at present too simple to account for the complex history of the metamorphic rocks below the Semail Ophiolite, as revealed by recent investigations.

The evidence from other ophiolites in the Tethyan ocean and from earlier periods, illustrates their diverse compositions and tectonic environments. It is to be expected that new geophysical campaigns on land and in the ocean will discover examples of geodynamic environments where ophiolites are presently being formed.

The excellent exposures in the Oman mountains and their coverage by the geologic maps of Glennie et al., the Open University, and the B.R.G.M., make them an excellent location to test satellite images for mapping purposes. The 1:100'000 scale geologic maps now available for the entire mountain range will provide future investigators with a base to map certain key areas in more detail, and to produce special topical maps.

This Proceedings Volume underscores the significance of the Oman Ophiolite as a type locality for the oceanic lithosphere. Given the importance of the questions yet to be answered, the Oman Ophiolite will undoubtedly continue to attract geologists of various specializations and viewpoints who seek to understand its composition, structure and genesis.

Tj. Peters, A. Nicolas and R.G. Coleman

For a comprehensive introduction to the geology of the Oman mountains the reader is referred to the following publications:

Glennie, K.W., Boeuf, M.G.A., Hughes Clarke, M.W., Moody-Stuart, M., Pilaar, W.F.H. and Reinhardt, B.M., 1974. Geology of the Oman Mountains. Verh. Kon. Ned. Geol. Mijnb. Gen.

Journal of Geophysical Research, 1981. 86, B4.

Lippard, S.J., Shelton, A.W. and Gass, I., 1986. The Ophiolite of Northern Oman. Mem. Geol. Soc. London, 11.

Tectonophysics, 1988, p. 51.

Robertson, A.H.F., Searle, M.P. and Ries, A.C., 1990. The Geology and Tectonics of the Oman Region. Geol. Soc. Spec. Publ, p. 49.

Part I
Processes at Spreading Centers
A) Crust

Ridge Crest Magma Chambers: A Review of Results from Marine Seismic Experiments at the East Pacific Rise

ROBERT S. DETRICK
Graduate School of Oceanography, University of Rhode Island, Kingston, RI 02881

Abstract

Recent seismic studies along the northern East Pacific Rise have documented the existence of a thin, narrow crustal magma body that is significantly smaller than the magma chambers incorporated into many earlier ridge crest geological models. The predominately molten part of the chamber is only 1–2 km wide and less than a kilometer thick, although it can extend as a nearly continuous feature for distances of several kilometers to several tens of kilometers along the ridge crest. This thin, sill-like body of melt is surrounded by a much wider zone of anomalously low seismic velocities that is interpreted as ranging from a partially molten crystal mush to the solidified (but still hot) plutonic rocks of the lower oceanic crust. This magma-sill model of a mid-ocean ridge magma chamber has important implications for the petrological and geochemical variability of mid-ocean ridge basalts and the origin of the thick cumulate sections found in ophiolites.

Introduction

Most current models for the generation of oceanic crust require the presence, at least intermittently, of a shallow zone of molten rock within the crust at oceanic spreading centers. These crustal magma 'chambers' provide the crucible within which much of the chemical fractionation of mantle-derived, parental magmas is thought to occur. They are also believed to be the source of most eruptive activity, and provide the heat which drives hydrothermal circulation. As recently as the early 1980's few good geophysical constraints existed on the crustal structure of mid-ocean ridges allowing crustal magma bodies with a variety of different shapes and sizes to be incorporated into ridge crest geological models (e.g. Cann, 1974; Dewey and Kidd, 1977; Bryan and Moore, 1977; Pallister and Hopson, 1981; Lewis, 1983; Nicolas et al., 1988). In many of these models a relatively wide (up to 20 km across)

Tj. Peters et al. (Eds), Ophiolite Genesis and Evolution of the Oceanic Lithosphere, 7–20.

and thick (up to 4 km high) body of melt was envisioned at ridge crests, at least along faster spreading ridges like the East Pacific Rise.

Over the past decade three important seismic experiments have been carried out, all along the northern East Pacific Rise (EPR) between 9°N and 13°N, that have fundamentally changed this picture of ridge crest magma chambers: the 1982 MAGMA experiment located near 13°N (McClain et al., 1985; Burnett et al., 1989; Caress et al., in press); a 1985 two-ship multichannel seismic (MCS) survey of the EPR between 9°N and 13°N (Detrick et al., 1987; Mutter et al., 1988; Harding et al., 1989; Vera et al., 1990), and a 1988 MIT/WHOI OBS seismic tomography experiment located on the EPR near 9°30′N (Toomey et al., 1990). Taken together, these studies have provided strong support for the existence of crustal magma chambers at the East Pacific Rise. However, the size of the magma bodies documented in these studies is much smaller than previously assumed.

In this paper we briefly review the new constraints these experiments provide on the existence, dimensions and along-axis continuity of crustal magma chambers at the East Pacific Rise. We then present a crustal magma chamber model based on these results and briefly discuss its implications for crustal accretion models.

Evidence for a Crustal Seismic LVZ at the EPR

The existence of a crustal seismic low velocity zone (LVZ) at the East Pacific Rise has been demonstrated by numerous conventional seismic refraction experiments dating back to the mid-1970's (Orcutt et al., 1975; Rosendahl et al., 1976; Lewis and Garmany, 1982; McClain et al, 1985). The evidence for this LVZ is based on a dramatic attenuation of arrivals for lines shot along the ridge crest, and travel time delays for ray paths passing beneath the ridge on lines shot across the ridge axis. The principal uncertainty in these studies concern the width of the LVZ and its internal velocity structure. Estimates range from a width of 20 km (Rosendahl et al., 1976) to < 0.5 km (Lewis and Garmany, 1982) depending on the whether the LVZ is assumed to be entirely molten (Vp $\sim 3\ \mathrm{km\ s^{-1}}$) or largely solidified (Vp $\sim 6\ \mathrm{km\ s^{-1}}$).

Two-ship multichannel seismic (Harding et al., 1989; Vera et al., 1990) and tomographic investigations (Burnett et al., 1989; Caress et al., in press; Toomey et al., 1990) recently carried out along the northern East Pacific Rise have placed much better constraints on the width and internal velocity structure of this LVZ. Figure 1 shows the 2–D structure of the axial LVZ at 9°30′N based on two-ship expanding spread profile solutions and cross-axis reflection profiles published by Vera et al. (1990). Similar seismic structures have been reported across the EPR at 13°N using two-ship MCS data (Harding et al., 1989) and at both 13°N and 9°30′N using OBS tomographic techniques (Burnett et al., 1989; Caress et al., in press; Toomey et al., 1990).

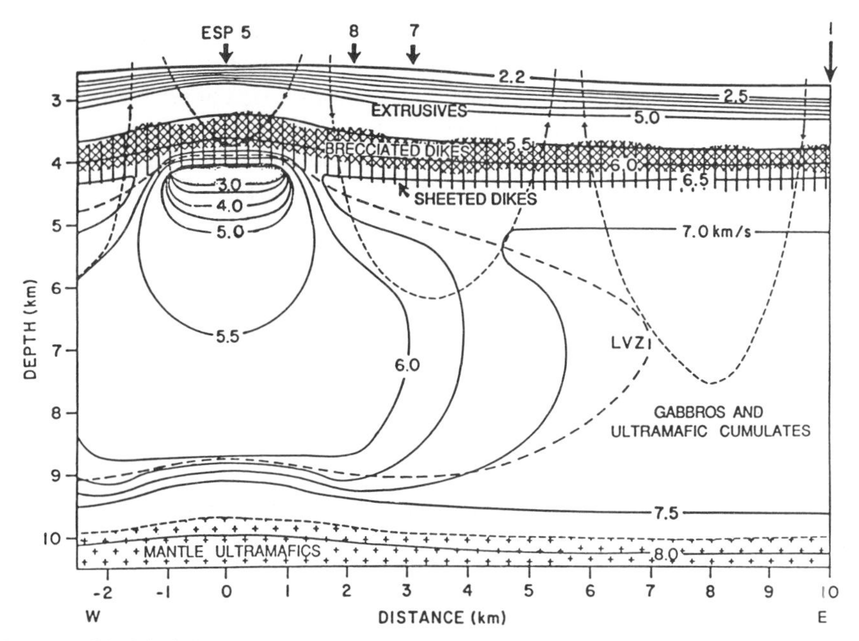

Figure 1. Model of the crustal structure of the East Pacific Rise near 9°30′N from Vera et al. (1990). Isovelocity contours are based on ESP solutions and cross-axis reflection lines. The axial LVZ has a half-width of up to 6 km or more. Seismic velocities within most of the LVZ are within 1 km s^{-1} of normal layer 3 seismic velocities. The lowest velocities (< 5 km s^{-1}) are confined to a narrow (< 2 km wide), relatively thin (< 1 km thick) sill-like body in the mid-crust beneath the rise axis.

All of these studies demonstrate that the EPR is associated with a large crustal LVZ that extends from 1–2 km below the sea floor at the ridge crest to the base of the crust. The LVZ has a somewhat trapezoidal shape in cross-section with a flat, narrow (~1 km half-width) top beneath the rise axis that deepens systematically beneath the ridge flanks. At its base the LVZ has a half-width of up to 6 km or more. The seismic velocities within most of this broad LVZ are within 1–2 km s^{-1} of normal layer 3 seismic velocities. The lowest velocities (<5 km s^{-1}) are confined to a narrow (<2 km wide), relatively thin (<1 km thick) mid-crustal zone beneath the rise axis.

Tomographic imaging of the EPR near 9°30′N indicates that this crustal LVZ is an axially continuous feature, although its width and internal velocity structure vary significantly along the ridge crest (Toomey et al., 1990). At 1–3 km depth the lowest velocities occur near the center of a small, morphologically defined ridge segment with higher velocities near the ends of the segment. The LVZ can also be offset slightly from the topographic rise crest. Toomey et al. (1990) report that near 9°30′N the LVZ at depths of 1–3 km is offset about 1 km to the west of the bathymetric rise crest. Immediately south of this area, seismic reflection data indicate a similar, but more pronounced westward offset in the location of a shallow intracrustal reflector identified with the top of the LVZ (Mutter et al., 1988).

Laboratory measurements show that compressional wave velocities of igneous rocks decrease from $\sim 6\,km\,s^{-1}$ at 0°C, to 4–5 $km\,s^{-1}$ at the onset of melting, to $<3\,km\,s^{-1}$ for basaltic melts (Murase and McBirney, 1973; Manghnani et al., 1986). Thus the relatively high ($>5\,km\,s^{-1}$) compressional wave velocities that characterize most of the seismic LVZ at the EPR suggests the bulk of the LVZ is at subsolidus temperatures. Any largely molten crustal magma body at the EPR must have dimensions considerably smaller than the width of the LVZ. This distinction between a relatively broad, largely solidified seismic LVZ, and a much narrower molten magma chamber is a very important one which was not always made in earlier interpretations of ridge crest seismic data.

Evidence for a Crustal Magma Chamber at the EPR

Geophysically, we define a magma chamber as a predominately molten body of any size within the crust. It may be entirely liquid, but it can include a partially molten crystal mush. Seismically it will be characterized by low compressional wave velocities (<3–$4\,km\,s^{-1}$) and zero or near-zero shear wave velocities. By this definition recent tomographic studies at the EPR (Burnett et al., 1989; Caress et al., in press; Toomey et al., 1990) have not resolved the existence of a crustal magma chamber. Although the tomographic inversions indicate a zone of low crustal velocities beneath the rise axis, the velocities are not low enough to *require* the presence of a molten body. Given the 1–2 km spatial resolution of these tomographic inversions in the mid-to-lower crust, these results place an effective upper bound on the possible dimensions of a magma chamber at the EPR.

The most direct geophysical evidence for the existence of a crustal magma chamber at the EPR is the high-amplitude, sub-horizontal upper crustal reflector reported by Herron et al. (1978; 1980), Hale et al. (1982) and Detrick et al. (1987) on CDP reflection profiles shot across and along the rise axis. This relatively flat lying event, which can be unequivocally tied to the top of the broader seismic LVZ, typically lies 1–2 km below the sea floor at the ridge crest (Figure 2). The amplitude of this reflector at vertical incidence, its phase reversal relative to the sea floor reflection at some locations, and its amplitude-offset behavior require a substantial drop in P and S wave velocities below this horizon that are consistent with the interpretation of this event as a reflection from the roof of an axial magma chamber (Harding et al., 1989; Vera et al., 1990).

In an effort to constrain the values of V_p and V_s in the magma chamber Harding et al. (1989) and Vera et al. (1990) have carefully modeled the variation in amplitude of this event as a function of shot-receiver offset (Figure 3). At 9°30′N Vera et al. (1990) found a V_p of $3\,km\,s^{-1}$ and V_s of

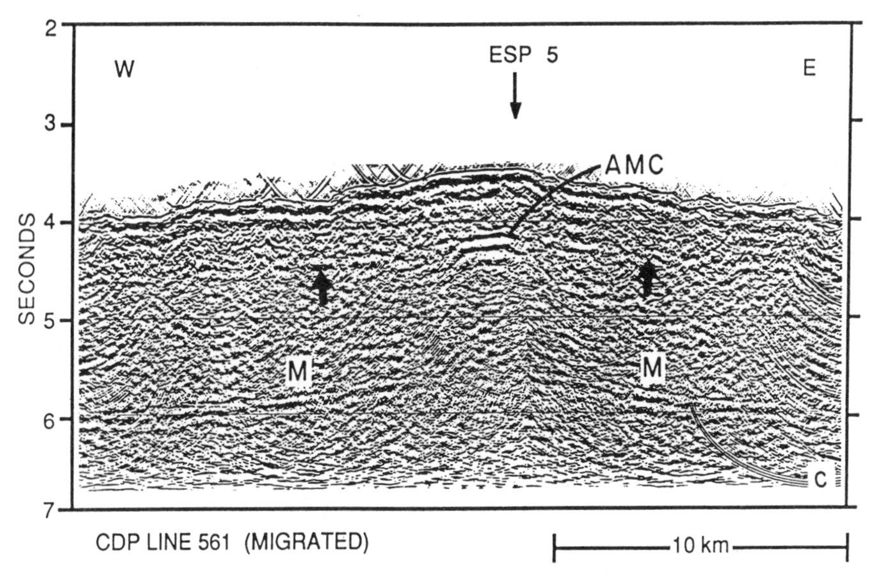

Figure 2. CDP Line 561 across the East Pacific Rise near 9°30′N showing the high amplitude reflection from the roof of the axial magma chamber (AMC) beneath the rise axis and Moho reflections (M) beneath the ridge flanks. This profile has been F-K migrated. (from Vera et al., 1990).

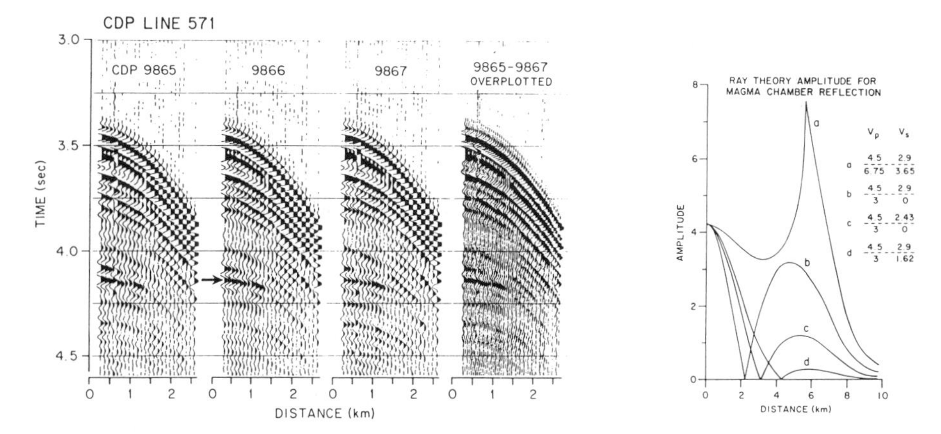

Figure 3. Selected CDP gathers from Line 571 near 9°30′N on the East Pacific Rise are shown on the left. The magma chamber reflection is indicated by the arrow. The amplitude of this event decreases rapidly at offsets greater than 1 km. The ray theory amplitude vs offset models shown on the right indicate that this requires a substantial drop in P and S wave velocities across this interface. Model B, which best fits the observed data, indicates the presence of liquid magma. (From Vera et al., 1990.)

zero best fit the amplitude and phase patterns seen in the original data, a result consistent with the presence of liquid magma beneath this interface. However, at 13°N Harding et al. (1989) infer somewhat higher compressional wave velocities and non-zero shear wave velocities that indicate a partially molten body. These differences may explain the weaker vertical incidence reflections from the chamber roof at 13°N compared to 9°N, suggesting significant along-axis variability in the composition of the magma chamber.

Dimensions of the Magma Chamber

The depth to the top of the magma chamber at the EPR is well-constrained by the two-way travel time of the magma chamber reflection and the seismic velocity structure of the crustal lid derived from ESP refraction profiles. Between 9°N and 13°N the roof of the magma chamber varies in depth between 1.2 and 2.4 km below the sea floor, with an average depth of about 1.5 km. At the ridge axis, the top of the axial LVZ approximately coincides with the layer 2/3 boundary; off-axis the top of the LVZ occurs at progressively deeper levels within seismic layer 3 (Harding et al., 1989; Vera et al., 1990).

The principal constraint on the width of the crustal magma chamber at the EPR comes from the width of the reflection from the chamber roof and coincident seismic refraction data shot along and parallel to the rise axis. On migrated stacked sections, like the profile shown in Figure 2, the magma chamber event is typically less than 3–4 km in width (Detrick et al., 1987). However, Kent et al. (1990) have noted that some of this apparent width is due to undermigration of diffracted energy generated from the edges of what is an even smaller body of melt. The presence of these diffractions require a sharp contrast in acoustic impedance. This contrast most likely originates from an abrupt transition from melt to rock at the edges of the chamber. By forward modeling of these diffraction hyperbola Kent et al. (1990) have shown that the distance between the best fitting point diffractors, and by inference the width of the predominately molten part of the magma chamber, is only 800–1200 m. This very narrow magma chamber width is consistent with coincident ESP refraction profiles which show that even only 2 km from the rise axis no evidence exists for the extremely low compressional wave velocities that would be associated with a crustal magma body (Harding et al., 1989; Vera et al., 1990). Seismic reflection data from the Valu Fa Ridge in the Lau Basin also indicate a similarly narrow axial magma body associated with that back-arc spreading center (Morton and Sleep, 1985; Collier and Sinha, 1990).

The height of the magma chamber at the EPR is not as well constrained by available seismic data as either its depth or width. Seismic velocity models of the EPR derived from tomographic studies provide the best constraints on the root structure of the magma chamber and indicate the largest velocity perturbations from a normal crustal velocity structure (>1 km s^{-1}) occur in the mid-crust between depths of 2–3 km (Burnett et al., 1989; Caress et al., in press; Toomey et al., 1990). Since basaltic melt is expected to have compressional wave velocities 2–3 km s^{-1} lower than normal oceanic crust, these results place an upper bound on the thickness of a predominately molten crustal magma body of about 1 km. Kent et al. (1990) have derived a lower bound on this thickness by reflectivity modeling of the interference effects between a wavelet reflecting off the top and bottom of a thin layer of melt as its thickness decreases. A layer thickness of ~ 10–50 m is required to explain the lack of a distinct basal reflection in the observed data. However, the absence of this basal reflection can also be explained by a gradual increase in velocities across the lower boundary due to a transition from melt to a crystal mush. In this case the thickness of the molten layer cannot be constrained by modeling reflections from the roof of the chamber. Thus, the best estimate that we can presently make of the thickness of the predominantly molten part of the magma chamber at the EPR is that it is a few tens of meters to a kilometer thick. This result is consistent with gravity data from the EPR which do not require the presence of a magma chamber of any appreciable size (Madsen et al., 1990).

Along-axis Variability of the Magma Chamber

Recent seismic experiments provide the first constraints on along-axis variations in the crustal magma chamber at the EPR. Remarkably, despite its narrow width, reflections from the roof of the magma chamber can be traced as a nearly continuous event for distances of several kilometers to several tens of kilometers along the rise axis (Detrick et al., 1987; Figure 4). The magma chamber event is frequently continuous across many smaller ridge axis discontinuities (e.g. Devals), although gaps in the reflector are present at larger ridge axis offsets including OSC's. Mutter et al. (1988), for example, have mapped the occurrence of the magma chamber reflection around the large 9°03′N OSC and shown that this OSC marks a major disruption in the continuity of the axial magma chamber, although the broader axial LVZ may simply bend to follow the ridge. In contrast, they find the magma chamber is continuous across the 9°06′N Deval although it is offset significantly from the location of the topographic rise crest. Kim and Orcutt (1989) have shown that the gaps in the axial reflector are not artifacts caused by increased topographic scattering in the vicinity of ridge axis discontinuities, although in some cases they may be caused by the ship sliding off the top of the narrow axial magma chamber.

Macdonald and Fox (1988) have noted that along-strike variations in the axial magma chamber are associated with subtle changes in the morphology of the ridge. Where a magma chamber is present the ridge often has a broad, inflated cross-section with a well-defined summit graben. Where the magma

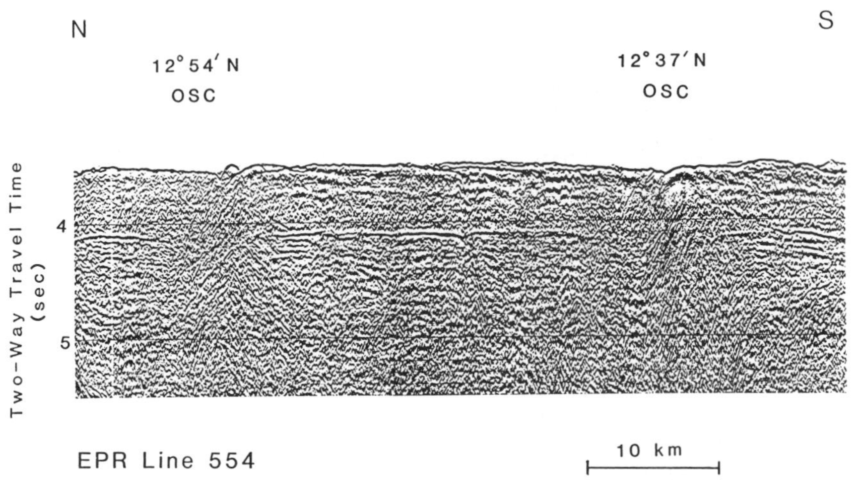

Figure 4. CDP Line 554 along the East Pacific Rise from 12°37′N to 12°54′N (from Detrick et al., 1987). The high-amplitude, mid-crustal reflection 1–2 km below the sea floor is a reflection from the top of what is inferred to be the top of an axial magma chamber. Note the reflector is continuous for a distance of at least 20 km along the ridge axis, but disappears in the vicinity of both OSC's.

chamber reflection is absent the ridge becomes narrower and steeper in cross-section and the axial summit graben frequently disappears. There is some correlation between the depth of the magma chamber (that is, the thickness of the crustal lid above the magma chamber) and the depth of the rise axis. In general, the shallowest parts of the rise axis are associated with the thinnest crustal lids, while the magma chamber reflection is deeper, more discontinuous and, in some cases, absent altogether where the rise axis is deep (Detrick et al., 1987). Altogether, the magma chamber reflection is present along about 61% of the EPR between 9°N and 13°30′N (Detrick et al., 1987), providing the best present estimate of the occurrence of crustal magma chambers at a regional scale along a fast spreading mid-ocean ridge.

Magma Chamber Model

Figure 5 shows an interpretative model of an EPR magma chamber based on the recent seismic results described above. The predominantly molten part of the magma chamber is confined to a narrow (1–2 km wide), sill-like body located about 1.5 km below the sea floor that may be anywhere from a few tens of meters to a kilometer in thickness. The large acoustic impedance contrast across the relatively flat roof of this reservoir explains the bright intra-crustal reflector seen at the EPR. This molten layer is shown grading downward into a partially solidified crystal mush zone which marks the transition from the (mostly liquid) chamber interior to the largely solidified (but still hot) surrounding rock. Within this mush layer the liquid fraction

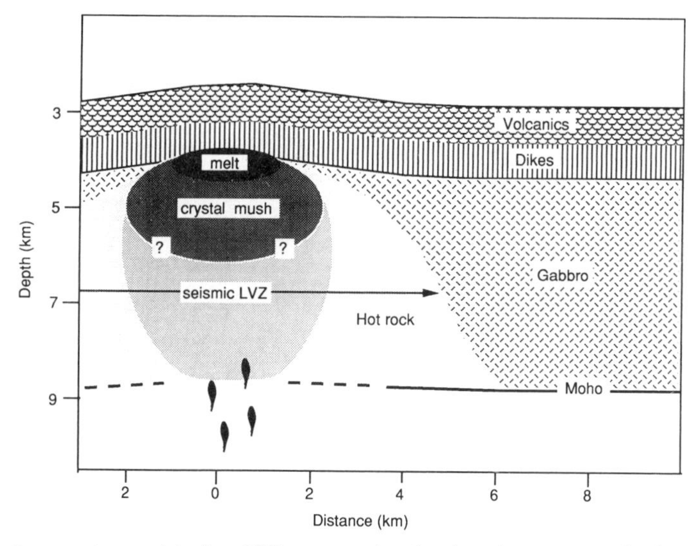

Figure 5. Interpretive model of an EPR magma chamber based on recent seismic results. The essential elements of this model are a narrow, sill-like body of melt 1–2 km below the sea floor that grades downward into a partially solidified crystal mush zone which marks the transition from the (mostly liquid) chamber interior to the largely solidified (but still hot) surrounding rock. The bulk of the axial LVZ is inferred to be composed of the slowly cooling cumulate rocks of layer 3.

presumably varies from close to 100% at the top to at most a few percent of intercumulus liquid at its base. The bulk of the broad axial LVZ is inferred to be composed of the slowly cooling cumulate rocks of layer 3. Moho reflections can be traced on seismic reflection profiles to within 2–3 km of the rise axis indicating a complete crustal section, including a well-defined crust/mantle boundary, form within about 50,000 years of crustal emplacement.

This model differs in some significant ways from many earlier models of ridge crest magma chambers (Cann, 1974; Dewey and Kidd, 1977; Bryan and Moore, 1977; Pallister and Hopson, 1981; Nicolas et al., 1988). First, the size of the predominately molten part of the magma chamber is quite small. Not only is the molten reservoir extremely narrow, but it also is relatively thin and essentially confined to the mid-crust. This is fundamentally different than many ophiolite-based crustal accretion models (e.g. Pallister and Hopson, 1981) which have assumed relatively wide (up to 20 km across) molten bodies with heights equivalent to the thickness of the plutonic section (3–5 km). It also is in direct conflict with some recent thermal models of the EPR (e.g. Wilson et al., 1988), which although incorporating a relatively narrow magma chamber, require a much larger reservoir of convecting magma. With a width of 1–2 km and a thickness on the order of 10^2 m the volume of melt in the model shown in Figure 5 is only 0.1–0.2 km^2 per km of ridge length. Kent et al. (1990) have noted that this volume is approximately equivalent to the volume of extrusives produced during a single eruption assuming an eruption interval of 50–600 years (Macdonald, 1982) and the spreading rate of 55–60 mm/yr for this section of the EPR (Klitgord and Mammerickx, 1982).

Another important aspect of this model is the explicit recognition of the existence of a crystal mush zone at the margins of the magma chamber. Recent views on how crystallization proceeds in magma chambers recognize the magma chamber margins as the most likely sites where crystallization occurs (McBirney and Noyes, 1979; Irvine et al., 1983; Brandeis and Jaupart, 1986). Thus petrologically and geochemically, the "magma chamber" will consist of *both* the well-mixed, convecting, liquid interior part of the chamber and a bordering, partially solidified mush zone that progressively moves through the chamber as solidification proceeds (Langmuir, 1989). The shape and volume of this mush zone are not well-constrained geophysically, primarily because of a poor understanding of how seismic velocity varies as a function of partial melt fraction. The main geophysical constraint on the extent of this mush zone comes from the relatively small reduction in compressional wave velocity (<1.0 km s^{-1}) associated with the bulk of the LVZ at the EPR. These near-normal velocities indicate subsolidus temperatures, or at most a few percent of intercumulus melt, in much of the LVZ thus confining a crystal mush layer with a high melt fraction ($>30\%$) to a relatively small volume in the mid-crust beneath the rise axis. This is not in accord with the magma chamber model proposed by Nicolas et al. (1988) which includes a large triangular-shaped, mush filled chamber with dimensions comparable to the width of the entire LVZ.

Discussion

The narrow, sill-like magma chamber body documented geophysically at the EPR is hard to reconcile with traditional ideas of fractional crystallization and crystal settling in a large molten magma chamber to form the cumulate gabbros of layer 3 (e.g Bryan and Moore, 1977; Pallister and Hopson, 1981). In these models the thickness of the cumulate sequence (>3 km) reflects the height of the magma chamber in which it formed. Recent geophysical data from the EPR clearly preclude a molten body of these dimensions. An alternative to this crystal settling model is *in situ* crystallization in a solidification zone at the margins of a small molten chamber (e.g. McBirney and Noyes, 1979). In this case an extensive thickness of cumulates develops from progressive crystallization and subsidence of the floor of a relatively thin ($\sim$100 m high) chamber (Dewey and Kidd, 1977; Browning, 1984). This model not only satisfies geophysical observations which require a volumetrically small crustal magma chamber, but it is more consistent with recent laboratory and theoretical results on how crystallization is likely to occur in magma chambers (Huppert and Sparks, 1980; Huppert and Turner, 1981; Brandeis and Jaupart, 1986).

Ophiolite studies have provided strong evidence for large viscous (magmatic) flow structures (shear bands, isoclinal folds, boudinage structures) within the layered gabbro section (Nicolas, 1989). These structures, which have been ascribed to a mechanical coupling between partially molten gabbros and subsolidus plastic flow in the underlying mantle (Nicolas, 1989), are difficult to reconcile with the model shown in Figure 5 if the bulk of the LVZ is completely solidified since the limit between plastic and viscous flow for an isotropic solid is usually estimated at about 0.35–0.40 melt/solid fraction (Usselman and Hodge, 1978). It is possible that this viscous/plastic flow threshold is drastically lowered for strongly anisotropic solids like layered gabbros (A. Nicolas, pers. comm., 1990), but there is not yet experimental evidence to support this hypothesis. It thus appears that in order to reconcile the relatively high seismic velocities found in the axial LVZ at the EPR with ophiolite evidence for viscous, sub-horizontal flow structures in the gabbroic section, the "hot rock" domain shown in Figure 5 must contain enough melt to behave rheologically as a viscous body, but a small enough melt fraction to respond as a solid at seismic wavelengths (few hundred meters).

The sill-like magma chamber body shown in Figure 5 also has important implications for the petrologic and geochemical variability of mid-ocean ridge basalts and the origin of the thick cumulate sections found in ophiolites. Extensive rock sampling along several different sections of the EPR have revealed variations in basalt geochemistry on an even finer scale than the tectonic segmentation of the ridge defined by transforms and OSC's (Langmuir et al., 1986; Sinton et al., 1988; 1991) This result is in apparent conflict with seismic data which indicate the existence of crustal magma chambers that are relatively continuous along-strike between major ridge axis discontinuities (Detrick et al., 1987). The existence of a narrow, thin molten layer, and variations in the relative proportion of melt to mush within the chamber, may resolve this dilemma (Macdonald, 1989; Kent et al., 1990). The lateral scale over which magma can mix in a sill-like body will be short, especially if a significant fraction of the chamber is filled with a viscous crystal mush that will not vigorously convect. Thus a continuous chamber may retain chemically distinct magmas until the time of eruption, and the observed geochemical segmentation along the EPR may primarily reflect the lateral scale of mixing within the chamber rather than the longer wavelength tectonic segmentation of the ridge (Macdonald, 1989).

The principal uncertainties in the model shown in Figure 5 center around the root structure of the magma chamber and the relationship between the observed velocity anomalies and the fraction of partial melt that might be present. How large is the partially solidified crystal mush zone surrounding the magma sill? How is this upper level magma body replenished? Is there a narrow stalk of partial melt feeding this body or is it replenished by episodic magma diapirs? Is each ridge segment centrally fed or is the magma chamber replenished at multiple points along a given ridge segment? Addressing these questions is really a two-fold problem. First it will require improved resolution of the seismic velocity structure within the LVZ and the crustal structure of the rise axis. With improvements in experimental design and tomographic inversion methods, including the use of waveform amplitude information, it should be feasible to obtain a nominal spatial resolution of 500 m in these models. Of equal importance, however, will be obtaining better experimental data on the effect of small fractions of partial melt on the seismic velocity and rheological properties of crustal rocks. This type of information, which is extremely limited at the present time, is essential in order to translate these seismic results into geological models of magma chamber processes.

Conclusions

Recent seismic studies along the northern East Pacific Rise have documented the existence of a thin, narrow crustal magma body that is significantly smaller than the magma chambers incorporated into many earlier ridge crest geological models. The predominately molten part of the chamber is only 1–2 km wide and less than a kilometer thick although it can extend as a nearly continuous feature for distances of several kilometers to several tens of kilometers along the ridge crest. This thin, sill-like body of melt is surrounded by a much wider zone of anomalously low seismic velocities that is interpreted as ranging from a partially solidified crystal mush to the solidified (but still hot) plutonic rocks of the lower oceanic crust. This magma-sill model for ridge crest magma chambers has important implications for the petrological and geochemical variability of mid-ocean ridge basalts and the origin of the thick cumulate sections found in ophiolites.

References

Brandeis, G. and C. Jaupart, 1986. On the interaction between convection and crystallization in cooling magma chambers, Earth. Planet. Sci. Lett., 86: 345–361.

Browning, P., 1984. Cryptic variation within the cumulate sequence of the Oman ophiolite: magma chamber depth and petrological implications. In: I.G. Gass, S.J. Lippard and A.W. Shelton (Ed.), Ophiolites and Oceanic Lithosphere, Geol. Soc. London Spec. Publ., 13: 71–82.

Bryan, W.B. and J.G. Moore, 1977. Compositional variations of young basalts in the Mid-Atlantic Ridge rift valley near lat. 36°49'N, Geol. Soc. Am. Bull., 88: 556–570.

Burnett, M.S., Caress, D.W. and J.A. Orcutt, 1989. Tomographic image of the magma chamber at 12°50'N on the East Pacific Rise, Nature, 339: 206–208.

Cann, J.R., 1974. A model for oceanic crustal structure developed, Geophys. J.R. astron. Soc., 39: 169–187.

Caress, D., Burnett, M. and J. Orcutt, in press. Tomographic image of the axial low velocity zone at 12°50'N on the East Pacific Rise, J. Geophys. Res.

Collier, J. and M. Sinha, 1990. Seismic images of a magma chamber beneath the Lau Basin back-arc spreading centre, Nature, 346: 646–648.

Detrick, R.S., Buhl, P., Vera, E., Mutter, J., Orcutt, J., Madsen, J. and T. Brocher, 1987. Multichannel seismic imaging of a crustal magma chamber along the East Pacific Rise, Nature, 326: 35–41.

Dewey, J.F. and W.S.F. Kidd, 1977. Geometry of plate accretion, Geol. Soc. Soc. Am., 88: 960–9687.

Hale, L.D., Morton, C.J. and N.H. Sleep, 1982. Reinterpretation of seismic reflection data over the East Pacific Rise, J. Geophys. Res., 87: 7707–7717.

Harding, A.J., Orcutt, J., Kappus, M., Vera, E., Mutter, J., Buhl, P., Detrick, R. and T. Brocher, 1989. The structure of young oceanic crust at 13°N on the East Pacific Rise from Expanding Spread Profiles, J. Geophys. Res., 94: 12,163–12,196.

Herron, T.J., Ludwig, W.J., Stoffa, P.L., Kan, T.K., and P. Buhl, 1978. Structure of the East Pacific Rise from multichannel seismic reflection data, J. Geophys. Res., 83: 798–804.

Herron, T.J., Stoffa, P.L. and P. Buhl, 1980. Magma chamber and mantle reflections – East Pacific Rise, Geophys. Res. Lett., 7: 989–992.

Huppert, H.E. and R.S.J. Sparks, 1980. The fluid dynamics of a basaltic magma chamber replenished by influx of hot, dense ultrabasic magma, Contrib. Mineral. Petrol., 75: 279–289.

Huppert, H.E. and J.S. Turner, 1981. A laboratory model of replenished magma chamber, Earth Planet. Sci. Lett., 54: 144–152.

Irvine, T.N., Keith, D.W. and S.G. Todd, 1983. The J-M platinum-palladium reef on the Stillwater complex, Montana: II. Origin by double-diffusive convective magma mixing and implications for the Bushveld complex, Bull. Soc. Econ. Geol., 78: 1287–1334.

Kent, G.M., Harding, A.J. and J.A. Orcutt, 1990. Evidence for a smaller magma chamber beneath the East Pacific Rise at 9°30′N, Nature, 344: 650–653.

Kim, I. and J.A. Orcutt, 1989. Effects of disrupted axial morphology on the amplitude strength of the crustal magma chamber reflection at the East Pacific Rise, EOS, (Trans. Am. Geophys. Union), 70: 1317.

Klitgord, K.D. and J. Mammerickx, 1982. Northern East Pacific Rise: Magnetic anomaly and bathymetric framework, J. Geophys. Res., 87: 6725–6750.

Langmuir, C.H., 1989. Geochemical consequences of in situ crystallization, Nature, 340: 199–205.

Langmuir, C.H., Bender, J.F. and R. Batiza, 1986. Petrologic and tectonic segmentation of the East Pacific Rise, 5°30′-14°30′N, Nature, 322: 422–426.

Lewis, B.T.R., 1983. The process of formation of ocean crust, Science, 220: 151–157.

Lewis, B.T.R. and J.D. Garmany, 1982. Constraints on the structure of the East Pacific Rise from seismic refraction data, J. Geophys. Res., 87: 8,417–8,425.

Macdonald, K.C., 1989. Anatomy of the magma reservoir, Nature, 339: 178–179.

Macdonald, K.C., 1982. Mid-ocean ridges: Fine scale tectonic, volcanic and hydrothermal processes within the plate boundary zone, Ann. Rev. Earth Planet. Sci., 10: 155–190.

Macdonald, K.C. and P.J. Fox, 1988. The axial summit graben and cross-sectional shape of the East Pacific Rise as indicators of axial magma chambers and recent volcanic eruptions, Earth Planet. Sci. Lett., 88: 119–131.

Madsen, J.A., Detrick, R.S., Mutter, J.C., Buhl, P. and J.C. Orcutt, 1990. A two- and three-dimensional analysis of gravity anomalies associated with the East Pacific Rise at 9°N and 13°N, J. Geophys. Res., 95: 4967–4987.

Manghnani, M.H., Sato, H. and C.S. Rai, 1986. Ultrasonic velocity and attenuation measurements on basalt melts to 1500°C: Role of composition and structure in the viscoelastic properties, J. Geophys. Res., 91: 9333–9342.

McBirney, A.R. and R.M. Noyes, 1979. Crystallization and layering of the Skaergaard intrusion, J. Petrol., 20: 487–554.

McClain, J.S., Orcutt, J.A. and M. Burnett, 1985. The East Pacific Rise in Cross-Section: A Seismic Model, J. Geophys. Res., 90: 8627–8640.

Morton, J.L. and N. Sleep, 1985. Seismic reflections from a Lau Basin magma chamber, in Scholl, D.W. and Vallier, T.L., eds., Geology and offshore resources of Pacific island arcs - Tonga region, Circum-Pacific Council for Energy and Mineral Resources Earth Science Series, v. 2: Houston Texas, 441–453.

Murase, T. and A. McBirney, 1973. Properties of Some Common Igneous Rocks and Their Melts at High Temperatures, Geol. Soc. Amer. Bull., 84: 3563–3592.

Mutter, J.C., Barth, G.A., Buhl, P., Detrick, R.S., Orcutt, J. and A. Harding, 1988. Magma distribution across ridge-axis discontinuities on the East Pacific Rise from multichannel seismic images, Nature, 336: 156–158.

Nicolas, A., 1989. Structures of Ophiolites and Dynamics of Oceanic Lithosphere. Kluwer, Dordrecht, 367 pp.

Nicolas, A., Reuber, I. and K. Benn, 1988. A new magma chamber model based on structural studies in the Oman ophiolite, Tectonophysics, 151: 87–105.

Orcutt, J.A., Kennett, B.L.N, Dorman, L.M and W. Prothero, 1975. A low velocity zone underlying a fast spreading ridge crest, Nature, 256: 475–476.

Pallister, J.S. and C.A. Hopson, 1981. Samail ophiolite plutonic suite: field relations, phase variation, cryptic variation and layering and a model of a spreading ridge magma chamber, J. Geophys. Res., 86: 2593–2644.

Rosendahl, B.R., Raitt, R.W., Dorman, L.M., Bibee, L.D., Hussong, D.M. and G.H. Sutton, 1976. Evolution of oceanic crust: 1. A physical model of the East Pacific Rise crest derived from seismic refraction data, J. Geophys. Res., 81: 5294–5304.

Sinton, J.M., Smaglik, S.M., and J.J. Mahoney, 1988. Along-axis magmatic variations at super-fast spreading: East Pacific Rise, 13–23°S, EOS (Trans. Am. Geophys. Union), 69: 1473.

Sinton, J.M., Smaglik, S.M., Mahoney, J.J. and K.C. Macdonald, 1991. Magmatic processes at superfast spreading oceanic ridges: Glass compositional variations along the East Pacific Rise 13°–23°S, J. Geophys. Res., 96: 6133–6155.

Toomey, D.R., Purdy, G.M., Solomon, S.C. and W.S.D. Wilcock, 1990. The three-dimensional seismic velocity structure of the East Pacific Rise near latitude 9°30′N, Nature, 347: 639–645.

Vera, E.E., Buhl, P., Mutter, J.C., Harding, A.J., Orcutt, J.A. and R.S. Detrick, 1990. The structure of 0–0.2 My old oceanic crust at 9°N in the East Pacific Rise from expanded spread profiles, J. Geophys. Res., 95: 15,529–15,556.

Usselman, T.M. and D.S. Hodge, 1978. Thermal central of low pressure fractionation processes, J. Volcanol. geotherm. Res., 4: 265–281.

Wilson, D.S., Clague, D.A., Sleep, N.H. and J. Morton, 1988. A model for narrow, steady state magma chambers on fast spreading ridges, J. Geophys. Res., 93: 11,974–11,984.

Accommodation Zones and Transfer Faults: Integral Components of Mid-Atlantic Ridge Extensional Systems

JEFFREY A. KARSON
Department of Geology, Duke University, Durham, NC 27706

Abstract

The style of tectonic extension varies rapidly along the Mid-Atlantic Ridge primarily as a consequence of along-axis variations in magma budget and episodic magmatism. Such variations require various types of strike-slip and oblique-slip faults or fault zones in order to accommodate variations in extensional strain at different scales. These types of accommodation zones and smaller scale transfer faults are well known in continental extensional terranes and are an important aspect of the extensional geometry along slow-spreading ridges and in ophiolites.

Introduction

Oceanic plate separation with a magma deficit results in tectonic extension of the axial lithosphere along slow-spreading mid-ocean ridge plate boundaries. Lithospheric necking (Tapponier and Francheteau, 1978) may result in relatively minor faulting and fissuring (e.g., Macdonald and Luyendyk, 1977; Macdonald, 1986) or brittle deformation affecting the full thickness of the crust (Harper, 1984; Toomey et al., 1985; 1988; Karson et al., 1987; Karson, 1990; White et al., 1990). The extent to which faulting affects the axial lithosphere is determined by the "magma budget" defined as the time-averaged rate of magmatic construction per unit of plate separation and by the thermal structure of the axial lithosphere. A low magma budget and episodic magmatism characterize much of the Mid-Atlantic Ridge (MAR) resulting in substantial extensional faulting in the median valley. The style of tectonic extension across the median valley is still poorly known, however, a number of general statements can be made at this time.

1. The range of styles of extensional faulting observed on the MAR span the same spectrum of geometries known from continental extensional terranes (e.g., Wernicke and Burchfiel, 1982). These include both steeply and gently dipping normal faults, listric normal faults and combinations of

Tj. Peters et al. (Eds), Ophiolite Genesis and Evolution of the Oceanic Lithosphere, 21–37.

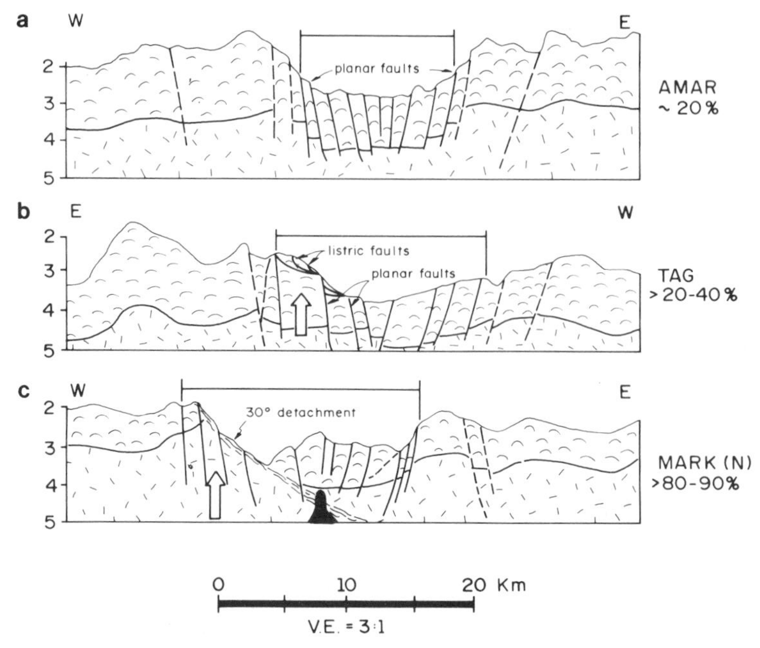

Figure 1. Summary of extensional fault systems known for the Mid-Atlantic Ridge with estimates of extension in the last 1 Ma (right). a) Nonrotational planar normal faults (e.g., AMAR Area-Macdonald, 1986); b) Listric normal faults (e.g., TAG Area-Karson and Rona, 1990); c) Low-angle normal (detachment) fault (e.g., northern MARK Area-Karson et al., 1987). Ornaments: lumpy pattern – basaltic pillow lavas; random dashes – gabbroic rocks; bold lines – faults (solid – active, broken – inactive); squiggles – shear zone; black – gabbroic intrusion. Bar indicates approximate region of active faulting.

these three types (Figure 1). Other fault geometries with conjugate fault pairs suggested from morphological data such as bathymetry and side-scan sonar (e.g., Harrison and Steiltjes, 1977) have not yet been observed.

2. The magnitude of tectonic extension experienced over the past 1 Ma, corresponding approximately to the age of the median valley, is typically between 10 and 20% but locally reaches more than 100% (Karson and Winters, 1987; Karson, 1990).

3. The style of faulting that has been observed from near-bottom surveys using submersibles and camera sleds is variable along the ridge axis sometimes changing rapidly over only a few kilometers. This is a reflection of both magmatic and tectonic segmentation of the ridge axis (Karson et al., 1987; Zonenshain et al., 1989; Sempere et al., 1990).

In order for axial lithosphere to extend under these conditions, both normal faults and related strike-slip or oblique-slip faults must slip in concert to accommodate a general strain. Such fault assemblages may be referred to as "extensional systems" analogous to thrust systems of compressional terranes (Boyer and Elliot, 1982). It is a kinematic requirement that multiple fault systems (fault plane and slip-direction pairs) function together as the lithosphere extends. Similar strain accommodation arguments have been made from the perspective of crystallographic slip-systems referred to as the von Mises criterion (e.g., Paterson, 1969) and from mesoscopic shear zones (Mitra, 1979). Such three-dimensional geometric compatibility is a basic condition of any consideration of crustal strain.

One of the neglected aspects of sea floor spreading along slow-spreading ridges is the role of strike-slip and oblique-slip faults in the evolution of the median valley morphology and structure. These structures are so closely spaced along spreading centers (on the order of 10 km) that they are also likely to be important in the evolution of many ophiolite complexes.

A Hierarchy of Faults at High Angles to Spreading Centers

Strike-slip and oblique-slip faults that are parts of the extensional systems of the MAR occur at different scales. The largest of these, transform fault zones, are strike-slip portions of plate boundaries. At an intermediate scale, features referred to as accommodation zones are necessary to link differently spreading ridge segments with little or no axial offset. The smallest examples, transfer faults, are minor offsets or jogs in individual fault strands and may be only metres to hundreds of meters in length. Transform faults are discussed here only briefly in order to clearly separate them from the smaller scale examples.

Transform Faults and Related Features

The largest scale strike-slip features are transform faults which have been extensively discussed in the marine geological literature (Wilson, 1965; Fox and Gallo, 1986; 1988). These are large-scale strike-slip fault zones typically tens of hundreds of kilometers in length, that link offset spreading center segments. The portions of the fault zone between the offset ridge axes, transform faults, are active plate boundaries; whereas, the portions beyond the ridge-transform intersections, non-transform segments of fracture zones, are intra-plate contacts between two segments of lithosphere of different age (Karson and Dewey, 1978). The older of the two segments contains a highly deformed and metamorphosed edge that once passed through the transform fault. The younger side was welded to the older at the ridge-transform intersection and never passed through the transform region. With simple, symmetrical spreading, the ridge offset is constant and halves of magnetic anomalies will be offset by this same length across the non-transform portion of the fracture zone (Figure 2a). Spreading is not always symmetrical across transform faults leading to lengthening or shortening of the offset and complex magnetic anomaly patterns (Schouten and Klitgord, 1982).

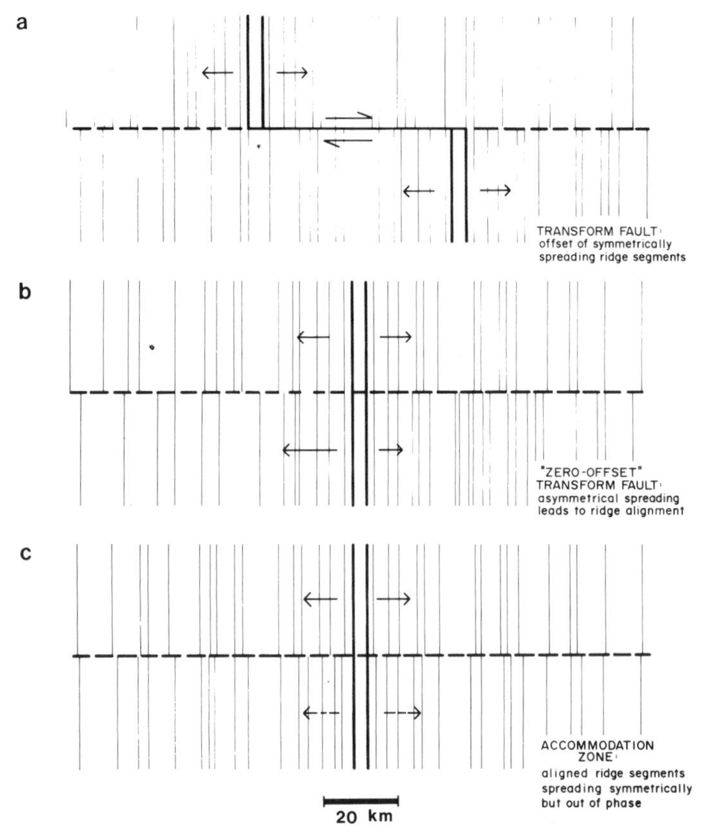

Figure 2. Schematic representation of major along-axis discontinuities linking discrete spreading cells. a) Transform fault; b) Zero-offset transform fault; c) Accommodation zone. Note differences in magnetic anomaly patterns developed through time for these different types of discontinuities.

Related features are Zero-Offset Transforms (Figure 2b) which result from asymmetric spreading on adjacent ridge segments to the point that the transform offset becomes negligable or even reversed (Schouten and White, 1980). Judging from magnetic anomaly patterns in the Central North Atlantic, these types of offsets are only ephemeral features that revert to significant transform offsets over a few millions of years (Schouten and White, 1980).

Both transforms and zero-offset transforms are lithosphere plate boundaries and as such must cut the full thickness of the plates. They separate spreading segments with distinct and independent magmatic and tectonic processes. They produce strong geophysical discontinuities in the oceanic lithosphere and are marked by obvious basement bathymetric depressions reflecting crustal density anomalies (Detrick and Purdy, 1980; Schouten and

White, 1980). These are distinct from along-axis discontinuities that will be referred to below as "accommodation zones" which link adjacent spreading cells and may show persistent magnetic anomaly mismatches along non-transforms (Figure 2c) with no consistent lateral offset.

Accommodation Zones

Accommodation zones are regions of complex strike-slip and/or oblique-slip faulting that link independently extending rift segments. They are well known from continental rifts (Bosworth, 1985; Rosendahl et al., 1986; Ebinger, 1988) and are well developed in regions of both amagmatic and magmatic rifting. In the Turkana Region of the East African Rift, tectonic and magmatic segmentation of the rift occurs on a scale of 40–50 km, similar to that along the MAR (Dunkleman et al., 1988; 1989; Karson and Curtis, 1989). In Turkana, individual half-graben with axial volcanic centers are linked end-to-end along the rift by oblique-slip accommodation zones. These faults permit the individual rift segments to extend with different styles (different asymmetry of half-graben in this case), at different rates and with different contributions from magmatic construction. Accommodation zones that separate discrete segments of the MAR are kinematically equivalent to those of continental rifts, but with some interesting geometric variations that result from rapid along-axis discontinuities in the magnitude of extension controlled by variations in magma budget and magmatic episodicity.

Accommodation zones are different from transform fault zones in several respects. They may superficially resemble zero-offset transforms but will not be linked to systematic anomaly offsets along flow-lines into older crust. They also appear to lack the strong gravity signatures typical of transforms (Lin et al., 1989). Accommodation zones exist where the axial lithosphere is mechanically continuous across the spreading center and do not necessarily cut through the lithosphere. As parts of the lithospheric extensional system they may merge downward into the plastic lower crust or upper mantle in regions of distributed strain. These regions may be steep or inclined, such as deeper detachment surfaces. Ultimately these faults and shear zones must communicate with the underlying asthenosphere, however, accommodation zones are probaby not accurate reflections of the offsets in the bases of plates. In fact, accommodation zones may define the seafloor intersection of detachment surfaces dipping parallel to the ridge axis. Unlike transforms, accommodation zones need not create significant purturbations of the underlying upper mantle.

Perhaps the most fundamental difference between transform faults and accommodation zones is nature of the extensional regions they connect. Transform faults on the MAR link regions of *crustal accretion*, whereas, accommodation zones link regions of *mechanical extension*. For short time scales or short transforms there may be some overlap in the use of these terms.

Below, a few basic types of accommodation zones are outlined. Some of these have been identified on the MAR and examples are given. Other geometries are expected to exist based on the occurrence of different normal fault geometries. Here only simple combinations of symmetrical and asymmetrical rift structures are considered. This is by no means a complete list and the reader will be able to combine some of the types of ridge segments discussed to create other possibilities.

Accommodation Zones Linking Two Symmetrical Segments

Many segments of the MAR are roughly symmetrical in map view and cross section (Phillips and Flemming, 1978; Kong et al., 1988; Sempere et al., 1990). The best known of these areas are the Narrowgate and AMAR Rifts segments of the FAMOUS Area, MAR at 37 degrees N (Macdonald and Luyendyk, 1977; Crane and Ballard, 1981; Stakes et al., 1984). The Narrowgate Rift is about 20 km in length and has an hourglass form in plan view. The floor of the median valley varies from a width of about 6 km near the transforms at either end of the segment to as little as 1–2 km at its narrowest central point. The axial region is dominated by very young (<10,000 yrs) volcanoes. In contrast, the AMAR Rift, located just across Transform B (15 km lateral offset) to the south has a broad U-shaped median valley with a broad, flat floor that is 8–10 km wide. Faults in both of these segments are considered to be steep and planar.

One type of accommodation zone to consider is a fault zone that joins two such symmetrical, but different ridge segments (Figure 3a). Such a structure would occur if two ridge segments like those described above were joined end-to-end. This could be the result of the two segments passing through a similar cyclic history (Crane and Ballard, 1981), but being out of phase with one another. Alternatively, the segments may perennially spread with different styles due to different magma budgets. Although planar, non-rotational faults are considered here, rift segments dominated by planar, rotational or listric faults could produce generally similar features. In any case, the active faults of the accommodation zone would extend between the outermost active rift faults on either side.

Accommodation Zones Linking Two Asymmetrical Rifts

Both multibeam bathymetry surveys (Kong et al., 1988; Sempere et al., 1990) and detailed near-bottom studies (Karson et al., 1987; Eberhart et al., 1989) have shown that asymmetrical rift segments are also very common on the MAR. Therefore the juxtaposition of pairs of asymmetrical rifts or symmetrical/asymmetrical rift pairs should be expected. In asymmetrical rifts, two different rift styles have been described. In both cases, there is marked asymmetry across the median valley in the relief of the walls, the geometry of faulting, the magnitude of tectonic extension and the types of

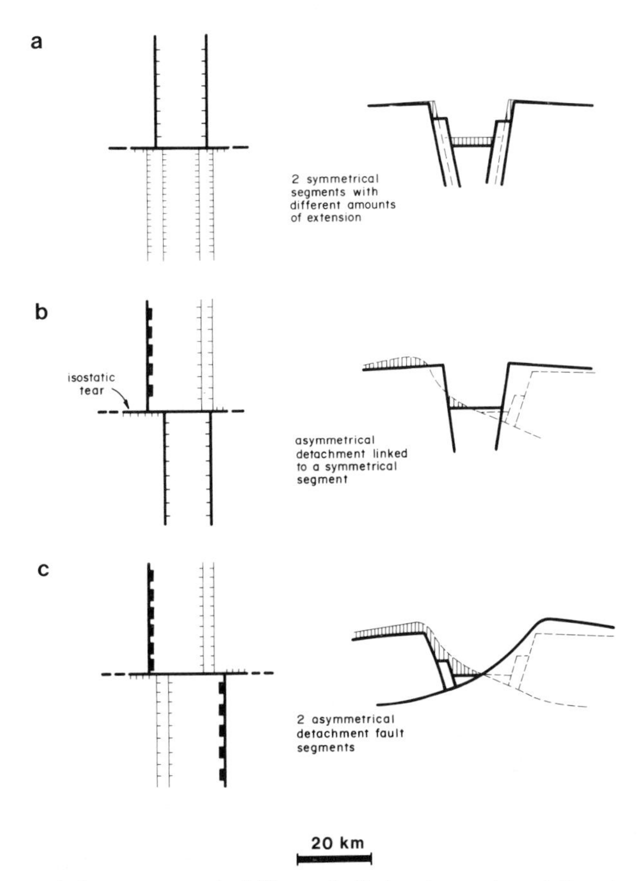

Figure 3. Accommodation zones required kinematically by along-axis variations in extensional style. Three simple cases are shown with plan views (left) and cross sections along accommodation zones (right). Lined pattern shows extent of fault scarps at a high angle to the ridge axis. a) Two differently extended but symmetrical rift segments; b) Asymmetrical/symmetrical rift pair; c) Two asymmetrical rifts with detachment faults. Note the predicted "isostatic tears" developed along the accommodation zones due to differential isostatic uplift of juxtaposed hanging wall and footwall blocks. Although only steep accommodation zones are shown here, low-angle geometries are also possible. Scale bar is 20 km.

lithologic exposures (Karson, 1990). In general, it appears that asymmetrical rifts have experienced more tectonic extension than the symmetrical ones described above; however, this statement is based upon a very small number of examples that have been studied in detail by near-bottom geological surveys. The best known examples are described very briefly below.

One highly asymmetrical rift segment occurs in the spreading cell just

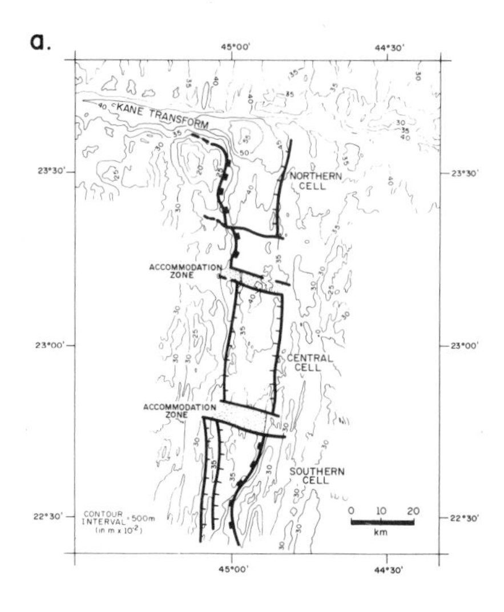

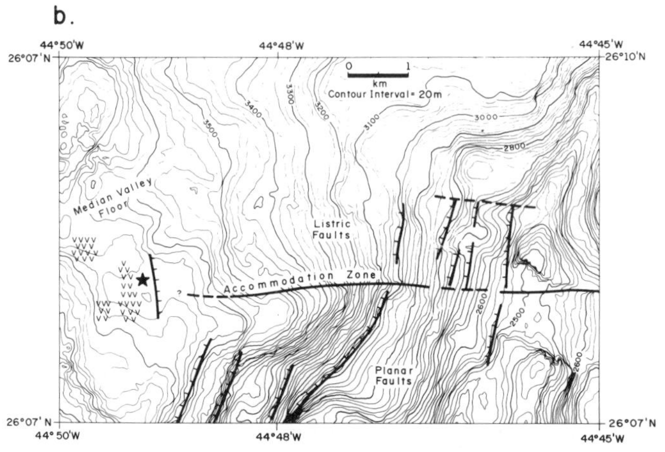

Figure 4. Highly schematic maps of accommodation zones on the MAR. a) MARK Area-two asymmetrically extended areas (north and south) linked to a central symmetrically spreading cell; b) TAG Area-two asymmetrically extended rift segments facing the same direction. Ornament: random dashes – gabbroic and ultramafic rocks; bold lines – major faults (tick marks on hanging wall of normal faults, boxes on upper plate of detachment faults), star – major hydrothermal sites.

south of the Kane Transform (Figure 4a) at 24 degrees N on the MAR (MARK Area). The eastern median valley wall is a simple block-faulted pillow lava terrane like the walls of the FAMOUS rifts. In contrast, the western wall is almost 1500 m higher and significantly steeper than the eastern wall (Karson and Dick, 1983; Karson et al., 1987; Karson, 1990). It is essentially a single major crustal shear zone that dips about 30 degrees to the ENE. Variably deformed gabbroic rocks crop out over much of this surface which is 8–10 km in width. This area is interpreted as a major detachment fault similar to those of highly extended terranes exposed in the metamorphic core complexes of the Basin and Range Province of the western U.S.A. (Wernicke et al., 1981; Miller et al., 1983; Davis and Lister, 1988). This highly extended oceanic terrane has experienced at least 80% horizontal extension. Areas to the south exposing serpentinized peridotites probably experienced even more extension. Detailed observations in this area show that this deformation is not the result of transform fault deformation, but rather the product of 0.5 to 1 Ma of tectonic extension without significant magmatic construction (Karson, 1989; 1990).

In the southern part of the MARK Area, about 100 km south of the Kane Transform, another highly asymmetrical segment occurs (Figure 4a). In this area the *eastern* wall is much higher and steeper. Microseismic studies show that active faulting extends into the mantle and defines a 45 degree west-dipping normal fault zones (Toomey et al., 1985; 1988). No geological data are yet available from this area, but the asymmetrical spreading style is well defined by the morphology and seismicity.

Detailed studies at the TAG Area at 26 degrees N (Figure 4b) also document a very asymmetrical rift valley structure (Eberhart et al., 1989; Zonenshain et al., 1989; Karson and Rona, 1990). In this case, a block-faulted western wall with only basaltic rocks exposed faces a much higher, steeper eastern wall with outcrops of lavas, dikes and minor gabbros. Fault patterns on this wall vary along strike from planar, nonrotational normal faults with large offsets to the south (Zonenshain et al., 1989) to listric normal faults and related structures to the north (Karson and Rona, 1990). In the northern area, listric faults sole out to very shallow dips (<30 degrees) at structural depths of only 300–500 m and may be related to hanging wall deformation above a gently west-dipping detachment fault. Late high-angle normal faults have disrupted the entire area creating a very complexly faulted terrane (Karson and Rona, 1990).

These asymmetrical rift segments are not unlike the half-graben structures developed along continental rifts and similar accommodation zone linkages should be expected. Two basic geometric parameters will dictate the kinematics of the accommodation zones: the sense of asymmetry and the amount of along-axis overlap of the rift border faults (e.g., Rosendahl et al., 1986). Where the polarity is opposed across the accommodation zone and overlap is very small, a strike-slip fault zone will exist at shallow levels where the footwalls of the major rift faults are in contact. Hanging wall blocks will have various oblique-slip relative motions. This is most likely for opposing detachment faults (Figure 3c). Such structures are well known from both continental margins (Lister et al., 1986) and highly extended continental terranes (e.g., Wernicke et al., 1988). Opposing asymmetrical rift structures with different styles, for example a detachment fault and a segment with asymmetrical planar faulting, will have dominantly oblique-slip faults in the intervening accommodation zones.

For accommodation zones joining asymmetrical rifts of the same polarity with minimal overlap, oblique-slip faults will dominate. An excellent example occurs in the TAG Area (Figure 4a) where a steeply dipping accommodation zone separates two asymmetrical rift segments that both have more intensely faulted eastern walls, but have very different fault styles as described above. Another example occurs in the northern part of the MARK Area where two eastward dipping detachment faults appear to have experienced different amounts of extension. To the north, only gabbroic rocks are exposed but to the south another detachment fault exposes only serpentinites. It is suggested that a low-angle oblique-slip fault zone separates these two regions.

Overlap of rift segments with asymmetrical forms is apparent in bathymetric maps of the MAR (Figure 4a, Brown et al., 1990; Sempere et al., 1990). Such overlapping rift structures can produce apparently symmetrical structures in some cross sections (Rosendahl et al., 1986). Similar features in the East African Rift do not have a consistent orientation with respect to the inferred plate separation direction. Therefore, these types of accommodation zones on the MAR should not be used as relative plate motion indicators.

Accommodation Zones Linking Symmetric/Asymmetric Rift Pairs

A third possible configuration involves the juxtaposition of symmetrical and asymmetrical rift segments (Figure 3b). Symmetrical segments like those described above from the FAMOUS Area are in some places found only a few kilometers along strike from regions of very asymmetrical extension. These accommodation zones will have oblique-slip movements that vary in detail across the fault zone depending upon which footwall and hanging wall blocks are in contact at a given point.

Two examples of this type of accommodation zone are present in the MARK Area (Figure 4b). They occur at the southern and northern ends of a symmetrically spreading cell in the central part of the area. Detachment structures occur to the north and south of this region, thus requiring accommodation zones. The northern accommodation zone has been described as a "cell boundary zone" (Karson et al., 1987; Brown and Karson, 1988). Similar features are also recognized in morphological data from slow-spreading ridges and given a variety of names including zero- or small-offset transforms (Schouten and White, 1980; Schouten and Klitgord, 1982), discordant zones (Grindlay et al., 1988) and shear zones (Bickell et al., 1989).

In the preceding sections, accommodation zones have been discussed primarily as steeply dipping interfaces between contrasting rift segments, following studies of similar structures in continental rifts. Accommodation zones, however, need not have steep dips. Especially in regions where detachment surfaces with complex ramp and flat forms interact, gently dipping accommodation zones with very subtle surface expressions may occur. These are analogous to large-scale lateral ramps and duplex terminations in thrust systems (Boyer and Elliot, 1982). Such structures are implied by gently inclined crustal seismic reflections seen in marine multichannel seismic reflection lines oriented along isochrons (Mutter et al., 1989; White et al., 1990).

Transfer Faults and Other Small-Scale Offsets

Examples of strike-slip and oblique-slip faults at a smaller scale than those of accommodation zones or transform faults also occur along the MAR. These are individual fault strands or fault zones that are short relative to the normal faults they connect. These are transform faults and other minor offsets along normal faults that permit along-strike variations in the shape or path of a normal fault. They should have net slips that are comparable to those of the adjacent normal faults. Unlike the larger-scale examples given above, these faults are unlikely to extend off-axis.

Transfer faults are oblique to right-angle jogs or offsets in normal faults (Figure 5a) that have the same slip direction as the adjacent normal faults (Gibbs, 1984). They are simply parts of a normal fault surface that are differently oriented from the main dip-slip fault surface. Ideally they do not extend beyond their intersections with the normal fault segments that they connect.

Transfer faults are completely analogous to lateral ramps along a gently inclined thrust or detachment fault surface. They link normal faults with the same sense of slip or polarity. Some authors have extended the use of this term to faults linking normal faults with opposing slip directions as in Figure 3c (e.g., Lister et al., 1986) that have very different fault kinematics and that in this discussion are a class of accommodation zone.

A variety of normal fault styles have been observed in extensional terranes and it is possible that different fault surfaces with similar slip directions will be kinematically linked along the walls of the median valley of the MAR. Some of these are shown in Figure 5. It should be noted that only the examples in Figures 5a and 5e strictly meet Gibb's (1984) definition of a transfer fault. The others shown are similar kinematically to the accommodation zones with the same facing directions. These are included in this discussion of transfer faults mainly because of their similar small scales.

Several of these types of structures have been observed along major normal fault zones of the MAR. The planar/listric (Figure 5b) and listric/listric (Figure 5d) types are found in the TAG Area (Figure 4b – Karson and Rona, in press). Lateral ramps and minor tear faults have been found

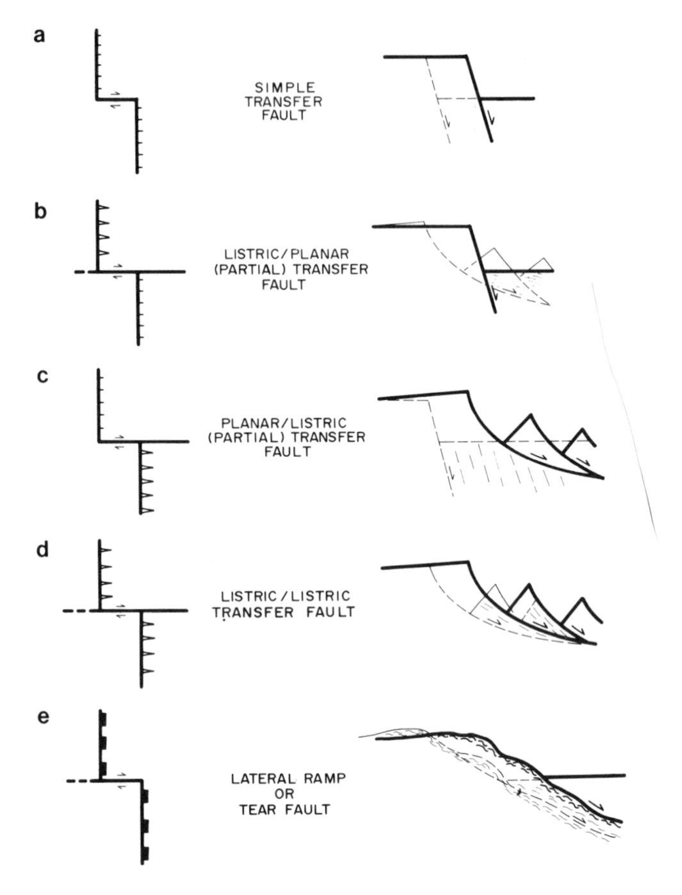

Figure 5. Transfer faults and major fault-bounding tilt domains. Plan views (left) and cross sections along transfer faults (right). Fine lines show relative slip directions of adjacent blocks. Near block in bold line; far block in fine dashed line. a) simple transfer fault: slip-line for transfer fault is parallel to that of the normal faults; b) Listric faults adjacent to the footwall of a planar fault; c) Planar faults adjacent to the footwall of a listric faulted block; d) Adjacent blocks bounded by listric normal faults; e) Lateral ramp or tear fault along a detachment surface. Note only a and e are true transfer faults in a kinematic sense; the others are more akin to small-scale accommodation zones. Although only steep transfer faults are shown, low-angle geometries are also possible. CH3 Figure captions

along the detachment surface on the western wall of the northern part of the MARK Area (Figure 4a – Karson, 1989). Where major rift faults terminate along-strike within an individual spreading cell, transfer faults are likely. Alternatively, rift faults may terminate in regions of distributed fracturing.

Discussion

Although accommodation zones and transfer faults of various types are a kinematic requirement of the type of tectonic extension that occurs across most of the MAR, little attention has been paid to them up to this point. One reason for this is that they have a very subtle bathymetric expression and are not readily seen even in Sea Beam multibeam bathymetric charts with 20 m contour intervals. This is probably due in part to the low relief produced along such fault zones because of their dominant strike-slip and oblique-slip fault movements. Lack of lateral continuity of transfer faults and their occurrence in very steep terranes of the median valley walls make them difficult to image with side-scan sonar systems and they may be removed from processed bathymetric maps by contouring programs designed to "smooth" rough terrane and connect offset bathymetry with smooth curves rather than sharp offsets.

Both accommodation zones and transfer faults are most easily detected in patterns of bathymetric lineaments. Owing to the lack of vertical fault motions in the central parts of accommodation zones, they are unlikely to be clearly defined within the median valley of the MAR. Instead, they are probably easier to spot from discontinuities along the median valley walls which may lead to small (few kilometers) offsets of the bathymetric axis of the rift valley. Along rift shoulders, distinctive normal faults trending parallel to the accommodation zones may form due to along-strike differences in the isostatic uplift of the median valley walls. This would be the result of different amounts of tectonic extension and unloading of the footwall of major fault zones between adjacent spreading cells. These "isostatic tears" would probably overprint accommodation zones resulting in a highly variable fault-slip history along the strike of the fault zone. Reverse faults or scissors faults should be expected in these areas.

It is unlikely that fault-slip studies of such fault zones will be possible in oceanic crust, however, such studies can be conducted in ophiolites. Workers in ophiolite terranes should be aware that these types of structures exist and that much can be added to the understanding of extensional processes in slow-spreading environments by the documentation of the geometric and kinematic development of such fault zones.

At present, it is unclear how long accommodation zones may persist at a given point along a spreading axis. Studies to date have concentrated on the median valley which was formed in the last 1 Ma. Future off-axis surveys will help to show if accommodation zones bounding discrete spreading cells are stable in time or if they are periodically truncated near the ridge axis by major ridge-parallel normal faults or by major magmatic constructional features (Karson and Rona, 1990).

In addition to their kinematic significance, accommodation zones and transfer faults may play an important role in the localization of hydrothermal venting and perhaps even magmatism along the MAR. It has long been recognized that massive sulfide ore bodies tend to occur at the intersections of major fault zones or lineaments (e.g., Franklin et al., 1981; Schiffman et al., 1987). It is becoming increasingly clear that many sea floor hydrothermal systems on intermediate- to slow-spreading ridges are located along fault zones (McConachy et al., 1986; Mevel et al., 1989) and often along lineament or fault intersections (Rona and Clague, 1989; Rona et al., 1990; Karson and Rona, 1990). Since fault zones provide permeable pathways for the outflow of hydrothermal fluids driven by magmatic heat, fault intersections should provide regions of especially high permeability that might be reactivated by movements on either fault. Accommodation zones and transfer faults form a kinematic basis for the interpretation of such fault-intersection controlled deposits.

Concluding Remarks

Despite their recognized importance as major structures in continental rift systems and their role in the kinematics of lithospheric extension, oblique- and strike-slip faults below the scale of transforms have not been given serious consideration in oceanic spreading environments. These are integral components of all lithospheric extensional systems and in order to fully understand extension on the MAR, marine geologists and geophysicists must develop techniques to better image these features. They are documented by near-bottom surveys using submersibles and deep-towed camera systems, however, they generally fall below the resolution of most swath mapping techniques. Future studies will help to develop correlations between prominent bathymetric features and the finer scale aspects of extensional systems of the sea floor. It is likely that the observed morphology of the MAR and other slow-spreading ridges may be dominantly the reflection of along-axis variations in tectonic extension rather than variations in magmatic construction as implied by many existing models.

Even though only a small portion of the MAR has been surveyed in sufficient detail to resolve accommodation zones, transfer faults and related features, they appear to be common features. If similar crust is preserved in ophiolites, these types of faults and fault zones should be anticipated. In the last decade there has been a tendency for all strike-slip faults in ophiolites to be interpreted as transform-related features. This is not necessarily the case and consideration should be given to the other types of strike-slip fault described in this paper.

Acknowledgments

This work was supported by NSF grant OCE 8693817 and ONR Grant 313–4005. The author thanks Dr. A Nicolas for his thoughtful review of the manuscript.

References

Bickell, J.D., Macdonald, K.C., Miller, S.P., Lonsdale, P.F. and Becker, K., 1986. Tectonics of the Nereus Deep, Red Sea: a deep-tow investigation of a site of initial rifting. Marine Geophysical Researches, 8: 131–148.

Brown, H.S., Sempere, J.-C., Schouten, H. and Purdy, G.M., 1990. Morphology and tectonics of non-transform offsets along the Mid-Atlantic Ridge between 24°N and 30°40'N. EOS, American Geophysical Union, 70: 1306.

Brown, J.R. and Karson, J.A., 1988. Variations in axial processes on the Mid-Atlantic Ridge: The median valley of the MARK Area. Marine Geophysical Researches, 10: 109–138.

Bosworth, W., 1985. Geometry of propagating continental rifts. Nature, 316: 625–627.

Boyer, S.E. and Elliot, D., 1982. Thrust systems. American Association of Petroleum Geologists Bulletin, 66: 1196–1230.

Crane, K. and Ballard, R.D., 1981. Volcanics and structure of the FAMOUS Narrowgate Rift: Evidence for cyclic evolution. Journal of Geophysical Research, 86: 5112–5124.

Davis, G.A. and Lister, G.S., 1988. Detachment faulting in continental extension: Perspectives from the southwestern U.S. Cordillera. Geological Society of America, Special Paper 218: 133–159.

Dunkleman, T.J., Karson, J.A. and Rosendahl, B.R., 1988. Structure of the Turkana Rift, Kenya. Geology, 16: 258–261.

Dunkleman, T.J., Rosendahl, B.R. and Karson, J.A., 1989. Structure and stratigraphy of the Turkana Rift from seismic reflection data. Journal of African Earth Sciences, 8: 489–510.

Eberhart, G.L., Rona, P.A. and Honnorez, J., 1988. Geologic controls of hydrothermal activity in the Mid-Atlantic Ridge. Marine Geophysical Researches, 10: 233–259.

Ebinger, C.J., 1989. Geometric and kinematic development of border faults and accommodation zones, Kivu-Rusizi Rift, Africa. Tectonics, 8: 117–134.

Fox, P.J. and Gallo, D.G., 1986. The geology of North Atlantic transform plate boundaries and their aseismic extensions. In: P.R. Vogt and B.E. Tucholke (Editors), The Geology of North America, The Western North Atlantic Region. Geological Society of America, Boulder, CO, pp. 157–172.

Fox, P.J. and Gallo, D.G., 1988. Transforms of the eastern central Pacific. In: E.L. Winterer, D.M. Hussong and R.W. Decker (Editors), The Geology of North America, The Eastern Pacific Ocean and Hawaii. Geological Society of America, Boulder, CO, pp. 111–124.

Franklin, J.M., Lydon, J.W. and Sangster, D.F., 1981. Volcanic associated massive sulphide deposits. In: B.J. Skinner (Editor), 75th Anniversary Volume, Economic Geology, 455–627.

Gibbs, A.D., 1984. Structural evolution of extensional basin margins. Journal of the Geological Society of London, 141: 609–620.

Grindlay, N.R., Fox, P.J. and Macdonald, K.C., 1988. Tectonic characteristics of ridge axis discordant zones in the South Atlantic. EOS, 69: 1425.

Harper, G.A., 1984. Tectonics of slow-spreading mid-ocean ridges and consequences of variable depth to the brittle/ductile transition. Tectonics, 4: 393–409.

Harrison, C.J.A. and Steiltjes, L., 1977. Faulting within the median valley. Tectonophysics, 38: 137–144.

Karson, J.A., 1989. Geometry and kinematics of extensional tectonics on the Mid-Atlantic

Ridge: Analogs from continental rifts. Geological Society of America, Abstracts with Programs, 21: A205.

Karson, J.A., 1990. Seafloor spreading on the Mid-Atlantic Ridge: Implications for the structure of ophiolites and oceanic lithosphere produced in slow-spreading environments. In: J. Malpas, E.M. Moores, A. Panayioutou and C. Xenophontos (Editors), Proceedings of the Symposium "TROODOS 1987", Cyprus Geological Survey Dept., Nicosia, pp. 547–555.

Karson, J.A. and Curtis, P.C., 1989. Tectonic and magmatic processes in the Eastern Branch of the East African Rift and implications for magmatically active continental rifts. Journal of African Earth Sciences, 8: 431–453.

Karson, J.A. and Dewey, J.F., 1978. Coastal Complex, western Newfoundland: an early Ordovician Oceanic fracture zone. Geological Society of America Bulletin, 89: 1037–1049.

Karson, J.A. and Dick, H.J.B., 1983. Tectonics of ridge-transform intersections at the Kane Fracture Zone. Marine Geophysical Researches, 6: 51–98.

Karson, J.A. and Rona, P.A., 1990. Block-tilting, transfer faults and structural control of magmatic and hydrothermal processes in the TAG area, Mid-Atlantic Ridge 26°N. Geological Society of America Bulletin, 102: 1635–1645.

Karson, J.A. and Winters, A.T., 1987. Tectonic extension on the Mid-Atlantic Ridge. EOS, 68: 1508.

Kong, L.S., Detrick, R.S., Fox, P.J., Mayer, L.A. and Ryan, W.B.F., 1988. The morphology and tectonics of the MARK Area from Sea Beam and Sea MARC I observations (Mid-Atlantic Ridge 23°N). Marine Geophysical Researches, 10: 59–90.

Lin, J., Purdy, G.M., Schouten, H., Sempere, J.-C. and Zervas, C., 1990. Evidence from gravity data for focused magmatic accretion along the Mid-Atlantic Ridge. Nature, 334: 627–632.

Lister, G.S., Etheridge, M.A. and Symonds, P.A., 1986. Detachment faulting and the evolution of passive continental margins. Geology, 14: 246–250.

Macdonald, K.C., 1986. The crest of the Mid-Atlantic Ridge: Models for crustal generation processes and tectonics. In: P.R. Vogt and B.E. Tucholke (Editors), The Geology of North America, The Western Atlantic Region, Geological Society of America, Boulder, CO, pp. 54–58.

Macdonald, K.C. and Luyendyk, B.P., 1977. Deep-tow studies of the structure of the Mid-Atlantic Ridge crest near 37°N (FAMOUS). Geological Society of America Bulletin, 88: 621–636.

McConachy, T.F., Ballard, R.D., Mottl, M.H., and von Herzen, R.P., 1986. Geologic form and setting of a hydrothermal vent field at lat. 10°56'N, East Pacific Rise. Geology, 14: 295–298.

Mevel, C., Auzende, J.-M., Cannat, M., Donval, J.-P., Dubois, J., Fouquet, Y., Gente, P., Grimand, D., Karson, J.A., Segonzac, M. and Stievenard, M., 1989. La ride du Snake Pit (dorsal medio-Atlantique, 23°22′N): resultats preliminaires de la campagne HYDROSNAKE. Compe Rendus Academie Sciences Paris, 308 II: 545–552.

Miller, E.L., Gans, P.B. and Garing, J., 1983. The Snake Range decollement- an exhumed mid-Tertiary ductile-brittle transition. Tectonics, 2: 239–263.

Mitra, G., 1979. Ductile deformation zones in Blue Ridge basement rocks and estimates of finite strain. Geological Society of America Bulletin, 90: 935–951.

Mutter, J.C., Buhl, P., Detrick, R.S. and Morris, E., 1989. Seismic images of dipping events along the axial valley of the Mid-Atlantic Ridge–MARK Area: Implications for the structure of slow-spreading ridges. EOS, 70: 1326.

Patterson, M.S., 1969. Ductility of rocks. In: A.S. Argon (Editor), Physics of Strength and Plasticity, MIT Press, Cambridge, MA, pp. 377–392.

Phillips, J.D. and Fleming, H.S., 1978. Multi-beam sonar study of the Mid-Atlantic Ridge Rift Valley, 36°-37°N. Geological Society of America, Map and Chart Series MC-19.

Rona, P.A. and Clague, D.A., 1989. Geologic controls on hydrothermal discharge on the northern Gorda Ridge. Geology, 17: 1097–1101.

Rona, P.A., Denlinger, R.P., Fisk, M.R., Howard, K.J., Taghou, G.L., Klitgord, K.D.,

McClain, J.S., McMurray, G.R. and Wiltshire, J.C., 1990. Major off-axis hydrothermal activity on the northern Gorda Ridge. Geology, 18: 493–496.

Rosendahl, B.R., Reynolds, D.J., Lorber, P.M., Burgess, C.F., McGill, J., Scott, D., Lambiase, J.J. and Derksen, S.J., 1986. Structural expressions of rifting: lessons from Lake Tanganyika, Africa. In: L.E. Frostick, R.W. Renaut, I. Reid and J.-J. Tiercelin (Editors), Sedimentation in the African Rifts, Geological Society of London, Special Publication 25, pp. 29–43.

Sempere, J.-C. and Purdy, G.M. and Schouten, H., 1990. Segmentation of the Mid-Atlantic Ridge between 24° and 30°40′N. Nature, 344: 427–431.

Schouten, H. and Klitgord, K.D., 1982. The memory of the accerting plate boundary and the continuity of fracture zones. Earth and Planetary Science Letters, 59: 255–266.

Schouten, H. and White, R.S., 1980. Zero offset fracture zones. Geology, 8: 175–179.

Schiffman, P., Smith, B.M., Varga, R.J. and Moores, E.M., 1987.

Geometry, conditions and timing of off-axis hydrothermal metamorphism and ore-deposition in the Solea graben. Nature, 325: 423–425.

Stakes, D.S., Shervais, J.W. and Hopson, C.A., 1984. The volcano tectonic cycle of the FAMOUS and AMAR Rift Valleys, Mid-Atlantic Ridge (36°47′N): Evidence from basalt glass and phanocryst compositional variations for a steady-state magma chamber beneath the valley mid-sections, AMAR 3. Journal of Geophysical Research, 89: 6995–7028.

Tapponier, P. and Francheteau, J., 1978. Necking of the lithosphere and mechanics of slowly accreting plate boundaries. Journal of Geophysical Research, 83: 3955–3970.

Toomey, D.R., Solomon, S.C., Purdy, G.M. and Murray, M.H., 1985. Microearthquakes beneath the median valley of the Mid-Atlantic Ridge near 23°N: Hypocenters and focal mechanisms. Journal of Geophysical Research, 90: 5443–5458.

Toomey, D.R., Solomon, S.C. and Purdy, G.M., 1988. Microearthquakes beneath the median valley of the Mid-Atlantic Ridge near 23°N: Tomography and tectonics. Journal of Geophysical Research, 93: 9093–9112.

Wernicke, B., 1981. Low-angle normal faults in the Basin and Range Province-Nappe tectonics in an extending orogen. Nature, 291: 645–648.

Wernicke, B.P., Axen, G.J. and Snow, J.K., 1988. Basin and Range extensional tectonics near the latitude of Las Vegas, Nevada. Geological Society of America Bulletin, 100: 1738–1757.

Wernicke, B. and Burchfiel, B.C., 1982. Modes of extensional tectonics. Journal of Structural Geology, 4: 105–115.

White, R.S., Detrick, R.S., Mutter, J.C., Buhl, P., Minshull, T.A. and Morris, E., 1990. New seismic images of oceanic crustal structure. Geology, 18: 462–465.

Wilson, J.T., 1965. A new class of faults and their bearing on continental drift. Nature, 207: 343–347.

Zonenshain, L.P., Kazmin, M.I., Lisitsin, A.P., Bogdanov, Y.A. and Baranov, B.V., 1989. Tectonics of the Mid-Atlantic Ridge Rift Valley between the TAG and MARK areas (26°–24°N): Evidence for vertical tectonism. Tectonophysics, 159: 1–23.

Rooting of the Sheeted Dike Complex in the Oman Ophiolite

A. NICOLAS AND F. BOUDIER*

* *Laboratoire de Tectonophysique, URA 1370 CNRS, Université Montpellier II, Pl. Bataillon – 34060 Montpellier, France*

Abstract

The root zone of the sheeted dike complex representing a thin zone (hundred meters thick) of extreme thermal gradient (~5°C/m) is regarded as a thermal boundary between the convective magma chamber system below, and the main convective hydrothermal circuit which closes above, at the base of this root zone. The root zone of the sheeted dike complex is located at the top of the high level foliated gabbro unit, where the foliation steepens, and where the first diabase dikes appears. It is a complex zone characterized by mutual intrusions of microgabbros dikes (that we call protodikes) with brownish microgranular contacts against the gabbro matrix. Upward, viscous flow in the protodikes and in the reheated enclosing gabbros generate a diffuse transition to the sheeted complex. Protodike margins stretched in the enclosing flowing doleritic gabbros form a complicated network which can be depicted thanks to microstructural analysis. Later diabase dikes cross-cut the section. These relationships are obscured by the hydrothermal circulation which has generated, in particular, isotropic amphibole gabbro veins. These veins tend to propagate horizontally; they may be interpreted as the downward closure of the main hydrothermal convective circuit.

Introduction

In ophiolites where a magma chamber has been identified, the zone located between the estimated roof of the magma chamber and the base of the sheeted dike complex, in which this unit is rooting, is of special interest because, within it, the temperature drops by some 550°C (Nicolas, 1989). This is deduced from the 400°C temperature at the base of the sheeted dikes (Nehlig and Juteau, 1988) and from a temperature at the top of the magma chamber around 950°C, bracketed within 1050°C and 850°C, respectively by the wet basalt solidus (Mysen and Boettcher, 1974) and the wet tonalite

Tj. Peters et al. (Eds), Ophiolite Genesis and Evolution of the Oceanic Lithosphere, 39–54.

solidus (Wyllie, 1980). Whatever the precise figure, this corresponds to an extreme thermal gradient (~5°C/m) considering that this root zone may not exceed one hundred meters, as shown below. It has been regarded as a thermal boundary layer between a convective magma chamber and the main convective hydrothermal circuit closing at the base of the sheeted dike unit (Nehlig, 1989; Nicolas, 1989). A new hydrothermal convection regime above the water supercritical conditions (temperatures in excess of 700°C) may have occurred in this zone (Delaney et al., 1987; Kelly and Delaney, 1987) and below (Gregory and Taylor, 1981). Data on this root zone are therefore critical in view of modeling crustal accretion at oceanic ridge. Unfortunately, only few detailed descriptions of this zone have been published so far (Allen, 1975, in Cyprus; Rosencrantz, 1983, in Bay of Islands; Pedersen, 1986 in Norway; Pallister, 1981; Pallister and Hopson, 1981; Smewing, 1981; Browning, 1982 and Rothery, 1983; all the latter authors in the Oman ophiolite). From these descriptions and the authors' experience, the Oman ophiolite seems best suited for the kind of observations which are looked for here. As illustrated by the Karmoy ophiolite in Norway (Pedersen, 1986), in many ophiolites and also locally in Oman (Pallister, 1981) the rooting of the sheeted dikes is commonly obliterated by successive gabbro intrusions and hydrous mobilisations, possibly related to environments of multiple magma chambers.

Revisiting the problem of the rooting of the sheeted dike complex in Oman after Rothery's (1983) detailed study is justified by the numerous new observations made, since, in this critical zone. We have also reinterpreted the structures of the plutonic section of this ophiolite showing, in particular, that the gabbros have been subjected to an intense magmatic flow that is responsible for the development of magmatic foliations and lineations, and for the tectonic transposition of the layering, that is the rotation and stretching of the layering into the foliation plane (Nicolas et al., 1988a and b). Magmatic foliation and layering, which are parallel to the Moho at the base of the crustal section, progressively steepen up section, a conclusion already reached in Oman by Rothery (1983). A close inspection of the high level gabbros has shown that they have a magmatic foliation which tends to become parallel to the attitude of the diabase dikes of the overlying sheeted dike unit. It is within these gabbros that we locate the root zone of the sheeted dike unit and not within isotropic gabbros, as admitted before (Dewey and Kidd, 1977; Harper, 1984; Juteau et al., 1988).

The object of this paper is to describe the root zone of the sheeted dike unit in the Oman ophiolite on the basis of new observations in the field and of a reinterpretation of the structures of the high level gabbros and of their origin within a magma chamber shaped with a narrow roof (Nicolas et al., 1988a). We wish also to speculate on the origin of this thermal boundary layer characterized by a very sharp thermal gradient between two systems, the magma chamber and the sheeted dike complex, which are cooled by magmatic and hydrothermal convection, respectively.

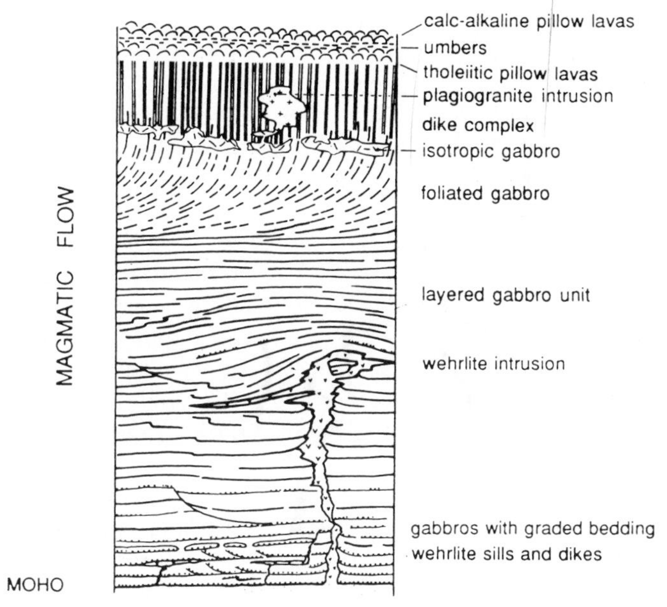

Figure 1. Synthetic structural log of the crustal section showing the steepening of the upper gabbro foliation in the root zone of the sheeted dike complex.

High Level Gabbros and the Roof of Magma Chambers

Previous observations and conclusions about the high level gabbros and their origin (Nicolas, 1989) should be recalled now. These gabbros have also been called 'isotropic', 'massive', 'vari-textured', 'foliated' or 'laminated', and 'plated' gabbros. We will mainly refer at them as foliated gabbros. As shown in Figure 1, they overlie the layered gabbros. The transition between layered and high level gabbros, is smooth, being marked by a progressive loss of the layered character. Both formations are well foliated and lineated (Figure 2A); commonly, layered gabbros grade into gneissic gabbros, before reaching the foliated gabbros. The layering of basal gabbros is parallel to the Moho and to the plastic foliation in the underlying peridotites. The foliation in these layered and foliated gabbros is of magmatic origin, and strain is caused by mechanical coupling with the fast flowing underlying mantle peridotites (Nicolas et al., 1988a). Upsection, the steepening of foliation of these layered gabbros with reference to the Moho surface is documented by the observation that the angle between gabbro layering and diabase dike swarms which locally intrude these gabbros, is considerably smaller than 90° (Browning, 1982; Rothery, 1983; Nicolas, 1989). This rotation becomes important within the foliated gabbros, as illustrated by the map of the Samrah area in the Semail massif (Figure 3). This new magmatic flow orientation has been attributed to viscous flow along the cooling walls of the magma chamber

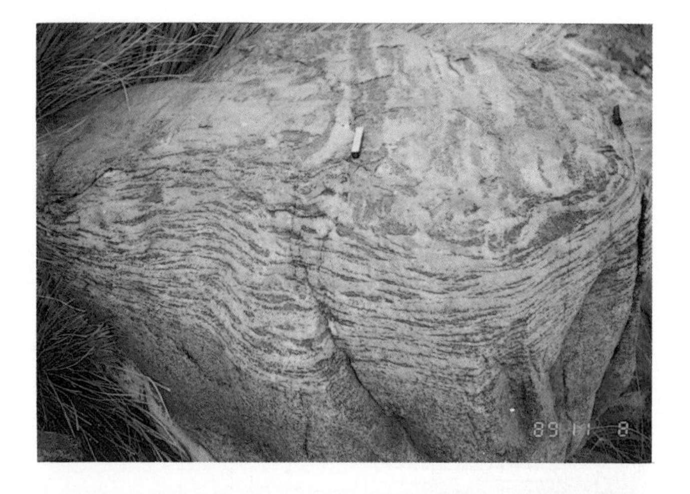

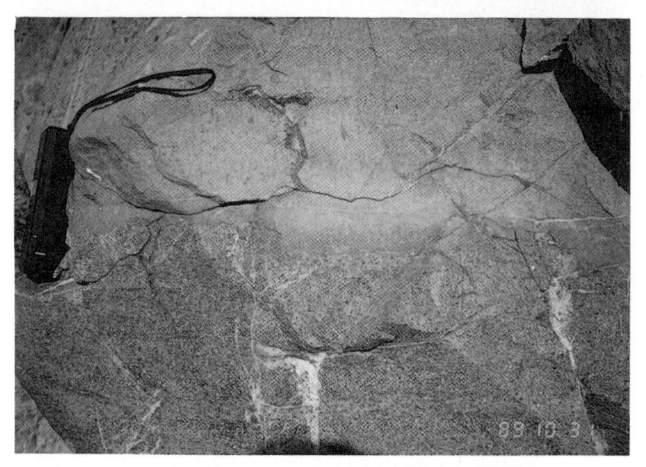

Fig. 2(A) and 2(B).

(Figure 4). In this sense, the high level gabbros merit the name of 'plated gabbros'.

When approaching the base of the sheeted dike unit several observations can be made in many sections of the Oman ophiolite. The foliation becomes progressively weaker and, commonly, very difficult to pick up in the field, thus explaining why such gabbros have been described as 'isotropic' or 'massive'. Some vague layering, often anorthositic, can be observed, parallel to the foliation.

- Structurally, the foliation tends to become parallel to the dikes of the sheeted dike unit in the root zone of this unit. Locally, large and varying angles are however measured between the foliated gabbros and the diabase dikes. The mineral lineations also become steeper on average.

Fig. 2(C), (D) and (E).

Figure 2. Photographs of field structures.
A) magmatic foliation and lineation in gneissic gabbro (W. Gideah);
B) hornfelsic margin of a protodike in contact with a doleritic gabbro from the protodike root zone (W. Wukabah);
C) mutually intrusive protodikes (Ibra road located by a star in Figure 3);
D) late diabase dike intrusive in the protodikes of Figure 2C (Ibra road);
E) and F) fragments of protodikes stretched in flowing doleritic gabbros from the protodike root zone (E,W. Falah; F, W. Rajmi);
G) isotropic amphibole-recrystallized veins in doleritic gabbro of the root zone, locally truncated by a diabase-plagiogranite magmatic breccia (W. Rajmi).

- Hydrous recrystallization of these gabbros also increases progressively within two main habits: swarms of millimeter-sized veins of dark green amphiboles and centimeter-sized veins of quartz-epidote on the one hand, and decimeter-to meter-sized patchy bands of amphibole gabbro and diorite with a fine-grained to pegmatitic isotropic texture, on the other hand (Figure 2G). These patches may be magmatic intrusions but detailed obser-

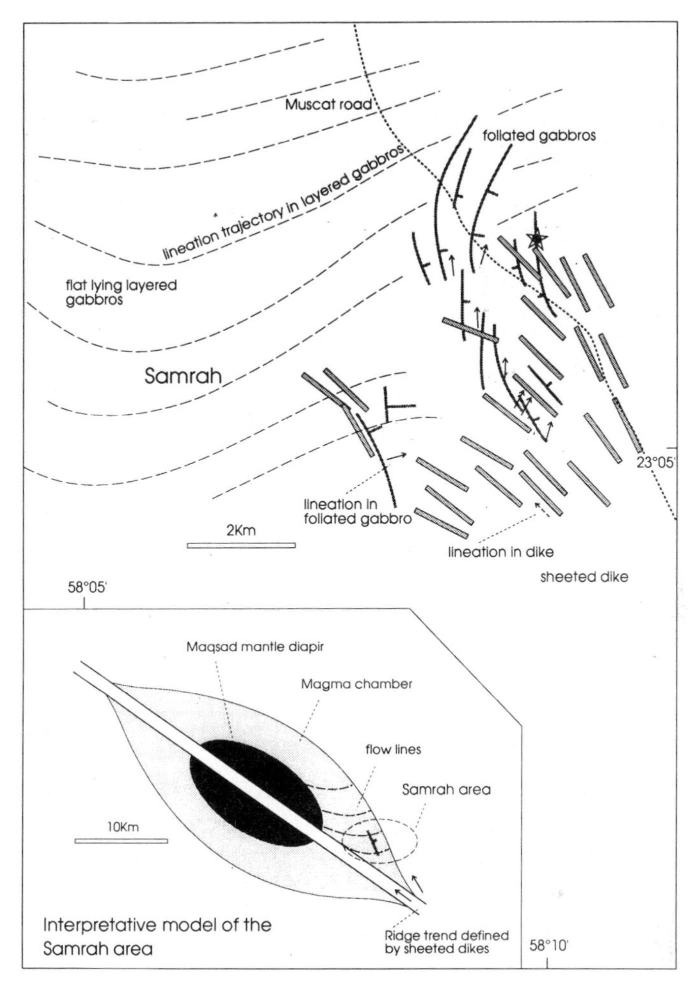

Figure 3. Structural map of the Samrah area (Semail massif) exposing the root zone of the sheeted dike complex (star locates outcrops described in Figure 2). Geometrical relationship between the foliation in high level gabbros and the sheeted dikes trend is assumed to indicate the ridge direction; this relationship suggests that this area is derived from the SE end of a ridge magma chamber as shown by the interpretative sketch, locating the Samrah area with respect to the Maqsad diapir (Nicolas et al., 1988b).

vations show that they are commonly local recrystallizations from the foliated gabbro, admittedly caused by water intrusion. Whitish plagioclase-quartz-epidote segregations are common within the patches. They seem to feed some plagiogranitic dikelets intruding the foliated gabbro (Figure 2G). This material could result from the wet anatexis of the still hot gabbros (Payne and Strong, 1979; Gerlach et al., 1981; Pedersen and Malpas, 1985). The foliated gabbro itself becomes hydrated and its microtexture changes as described below.

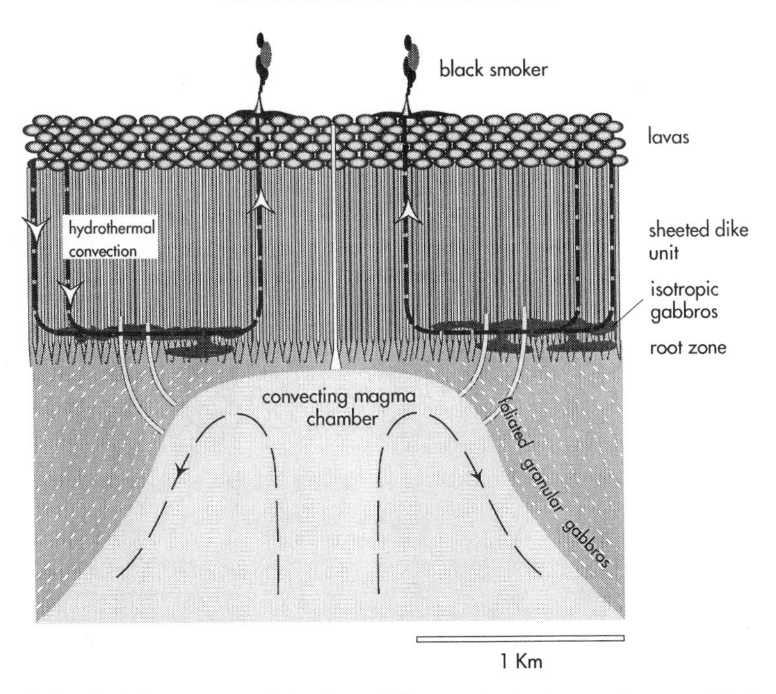

Figure 4. Sketch of the root zone of the sheeted dike complex in the tent-shape model of magma chamber (Nicolas et al., 1988a).

- The dioritic dikelets and epidote veins are, on average, parallel to the diabase dikes of the overlying sheeted dike unit (Figure 5). Although individual measurements of the attitude of the patchy isotropic gabbros are very approximate, it is probably significant that, averaged on 188 patches, they tend to be flat-lying (Figure 5).

Root Zone in the Field

At the top of the foliated gabbro unit and at the level where the first diabase dikes belonging to the sheeted dike unit appear, the field structures are generally obscure. This results from the large amount of hydrous recrystallization, and/or intrusion, of isotropic gabbros, diorites and plagiogranites. The foliation which is thus mainly imprinted into amphibole-rich doleritic gabbros, is difficult to identify. Within such gabbros, we have observed the rooting of the overlying sheeted dikes in several localities in Oman, including the Ibra road in the Semail massif (Figure 3), the Wadi Falah and Huffi area in the Rustaq-Haylayn massif, Wadi Rajmi in Northern Oman and the Masari area in the United Arab Emirates where the zone of rooting of the sheeted dikes is exposed over a considerable area; to this list which is evidently not

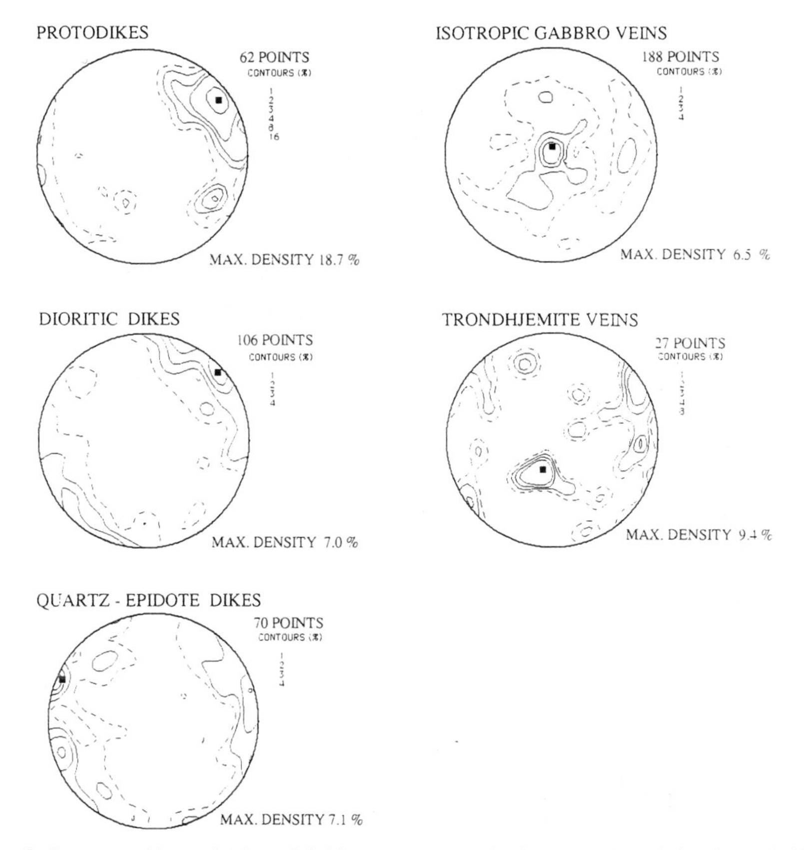

Figure 5. Stereographic projection of field measurements in the root zone of the sheeted dike complex. The compiled data come from throughout the Oman ophiolite where the overall sheeted dike trend is NNW-SSE. Lower hemisphere projection, geographical reference system, counting area: 1% of the hemisphere.

exhaustive should probably be added Rothery's (1983) observations in the Maydan syncline of the Wuqbah massif.

The most significant feature in these localities is the presence of microgabbro dikes with brownish and microgranular contacts against the gabbro matrix (Figure 2B, C). Smaller dikes, 10–50 cm across, are entirely microgranular; larger ones, in the meter range or above, retain this microgranular aspect at their margins but grade inside into a doleritic and variously foliated facies (Figure 6C), ranging in grain size from coarse diabase to gabbro. Because of the clear grading into diabase dikes we call these coarser and granular dikes, *diabase protodikes*. Lenses of the brownish protodike margins can be found isolated within the foliated doleritic gabbro and elongated parallel to its foliation (Figure 2E and F). They are interpreted as pieces of a protodike margin which has been detached and stretched into the foliation during flow. Combining field evidence from the various localities where protodikes have

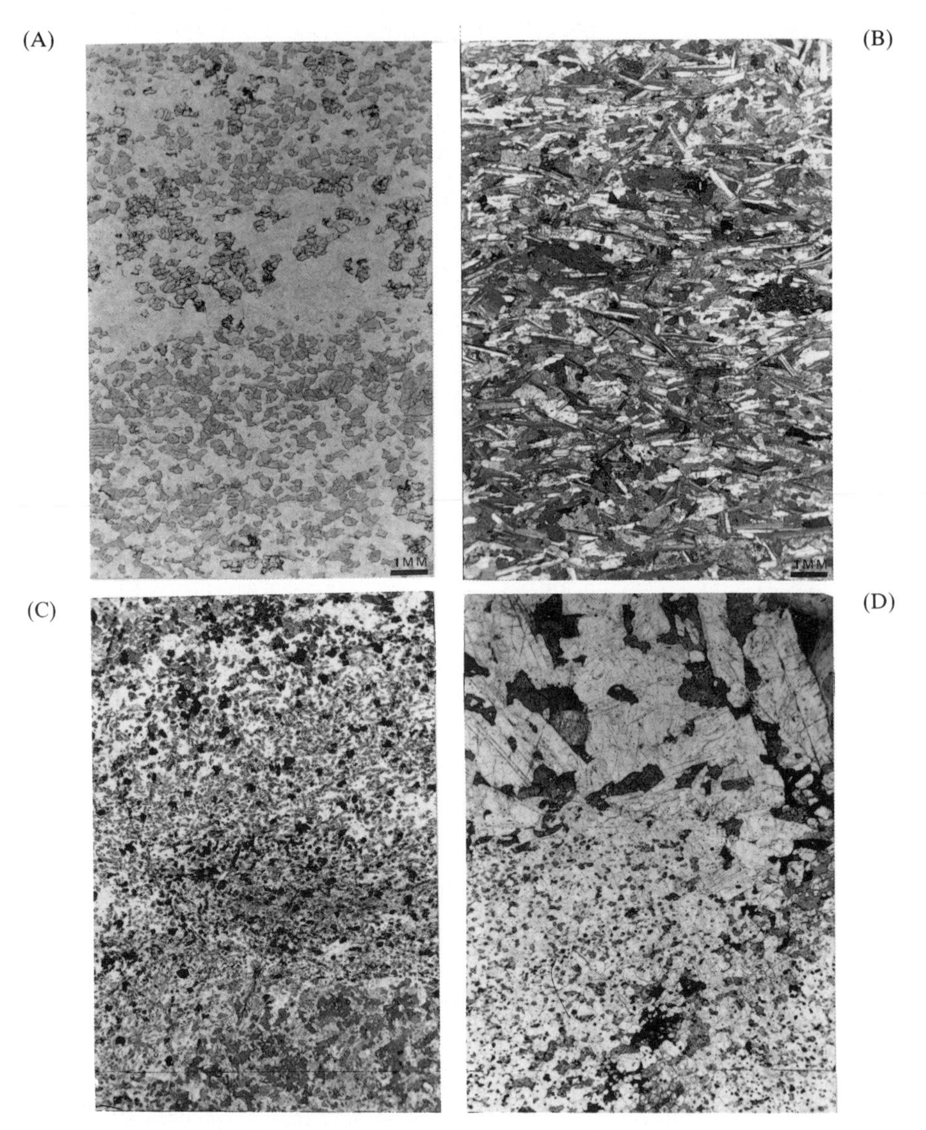

Figure 6. Microstructures in thin sections.

A) Granular texture in foliated high level gabbros: this texture is very similar to that of the layered gabbros which are in continuity with the high level gabbros. It is dominantly composed of plagioclase, usually bytownite, and diopside. Olivine is fairly common, orthopyroxene, exceptional. Plagioclase and diopside form a regular foam texture, indicative of a high degree of grain boundary mobility, requiring very high temperatures of equilibration. Plagioclase is generally flattened with a strong preferred orientation. Olivine appeared late in the crystallizing sequence forming a chain network, which has been stretched in well foliated gabbros. The absence of plastic strain in crystals and the nature of the minerals fabrics led to the conclusion that this foliation resulted from a large magmatic flow with no, or insignificant contribution of plastic flow (Nicolas et al., 1988a; Benn and Allard, 1989).

been observed, sketches of their setting can be drawn (Figure 7) and a sequence of events, proposed.

- In the lowest horizon where they have been observed such as Huffi in the Rustaq-Haylayn massif, protodikes cut a granular foliation in olivine-pyroxene gabbros (Figure 6A) which is parallel to a faint layering, locally preserved. These gabbros and their foliation are in continuity with the underlying layered gabbros. Therefore, two foliations in gabbros can be distinguished (Figure 7): this early granular, pyroxene foliation as a matrix and, inside the protodikes, the doleritic amphibole foliation (Figure 6B). The early foliation may be contorted even at the scale of a few meters. Attitudes of lineations associated with the two foliations are also different.
- Above, diabase dikes become important, say 50%. In contrast with typical dikes of the sheeted dike unit, they have no chilled margins and their matrix, where it is not replaced by isotropic amphibole gabbro-diorite, is constituted by the doleritic, amphibole gabbro. These sheeted dikes are not necessarily parallel to the gabbo foliation, as seen along the Ibra road (Figure 3). Brownish margins of protodikes and isolated lenses of such margins of protodikes are still recognized within this foliation.
- Upsection (Figure 7), isotropic, amphibole gabbros-diorites develop increasingly at the expense of all foliated gabbros. They can become intrusive, cutting the protodikes and early diabase dikes, also brecciating them (Figure 2C), but in turn they can be transacted by late diabase dikes. The foliated gabbros and protodikes are also cut by plagiogranite veins. Plagiogranites form, together with diabases, some spectacular breccias (Fig-

---

B) Doleritic texture in foliated gabbros: the high level gabbros closer to the sheeted dike rooting zone are commonly hydrated with brown to green amphiboles developed around clinopyroxenes during and after the magmatic stage. The texture under the microscope becomes accordingly doleritic: the plagioclase crystallizes in rectangular laths which constitute a tight lattice within an amphibole-rich matrix. The tabular plagioclase have an outer rim extending into the matrix which is strongly zoned and more acidic in composition, a change possibly correlated with hornblende coming in on the liquidus (Hopson et al., 1981). In contrast with the preceeding granular texture, this suggests a fast crystallization ending a comparatively low temperatures in the presence of water. The plagioclase laths can be strongly oriented, defining a foliation and a lineation ascribed to magmatic flow.

C) Microgranular texture in protodikes margins: the border facies of protodikes has a brownish hornfelsic facies which, under the microscope (central part of the photograph), reveals a microgranular texture with a 300 μm pavement of plagioclase and clinopyroxene; olivine and orthopyroxene are more common than in the high level gabbros. Opaques, mainly ilmenites are remarkably abundant, forming 30 to 50 μm round-shaped grains which crystallized at the same time as the other minerals. The contact with the intruded gabbro is straight cut and the grain size is reduced compared to inside the dike; in this direction, the microgranular texture coarsens, becomes foliated, tends to be progressively hydrated and replaced by the foliated, doleritic texture.

D) Secondary static gabbro-diorite recristallizations locally developed at the expense of all these facies, here the microgranular texture of a protodike margin.

ure 2G). The rounded and lobate contours of diabase fragments and mutual intrusive contacts suggest that the breccias derive from the unmixing of granitic and basaltic magmas.

Thickness of the Root Zone

It is first necessary to exclude from this discussion the case of sheeted dikes cutting at any level into the crust and even in the peridotite transition zone, as seen for instance in the area covering the eastern and western terminations of the Haylayn-Rustaq and Nakhl massifs. New crust was created there possibly as the result of a propagating rift. In the common situation where the layered and foliated gabbros are only exceptionally cut by diabase dikes, the thickness of the root zone of the sheeted dike unit, seems to be around 100m in Oman (Juteau et al., 1988), but may be less (Pallister and Hopson, 1981; Rothery, 1983; Lippard et al., 1986); it is between several tens of meters to a couple of hundred meters in Bay of Islands (Dewey and Kidd, 1977; Rosencrantz, 1983). It is difficult to propose a precise figure. First, the limits of this zone are not clear cut. One has to decide what is the fraction of diabase dikes with respect to the gabbros screens defining the base of the sheeted dikes unit; actually, this is not the most critical problem because, once these dikes appear in significant number, it is usually a question of tens of meters before reaching the sheeted dikes unit with one hundred percent dikes, as shown by Rothery (1983) in the Maydan syncline area. If we look for the disappearance of the protodikes downsection to define the lower limit of the rooting zone, the difficulty is greater. We have seen that a doleritic foliation is associated with the central part of these dikes, but it is difficult in the field to distinguish it clearly from the granular foliation of the high level gabbros and it is possible that microgranular gabbros can, independently, recrystallize into doleritic gabbros where they have been in contact with water at high temperature. The observation of brownish margins of protodikes, constitute a better criterion. Using it suggests that only seldom, in Oman, protodikes root much below the level where the first diabase dikes appear. Rothery (1983) also observed that the thickness and depth of the root zone seemed to vary laterally. This may be the result of a diachronous generation of protodikes and diabase dikes as will be envisaged in the discussion.

Discussion

Let us summarize the main observations made in the zone of protodikes before addressing the questions of how they originate, how their foliation relates to that of high level gabbros crystallized in the magma chamber and, eventually, how the roof of the magma chamber could look.

As defined by earlier workers, the root zone of the sheeted dike unit is limited upward by the disappearance of gabbro screens and downward by

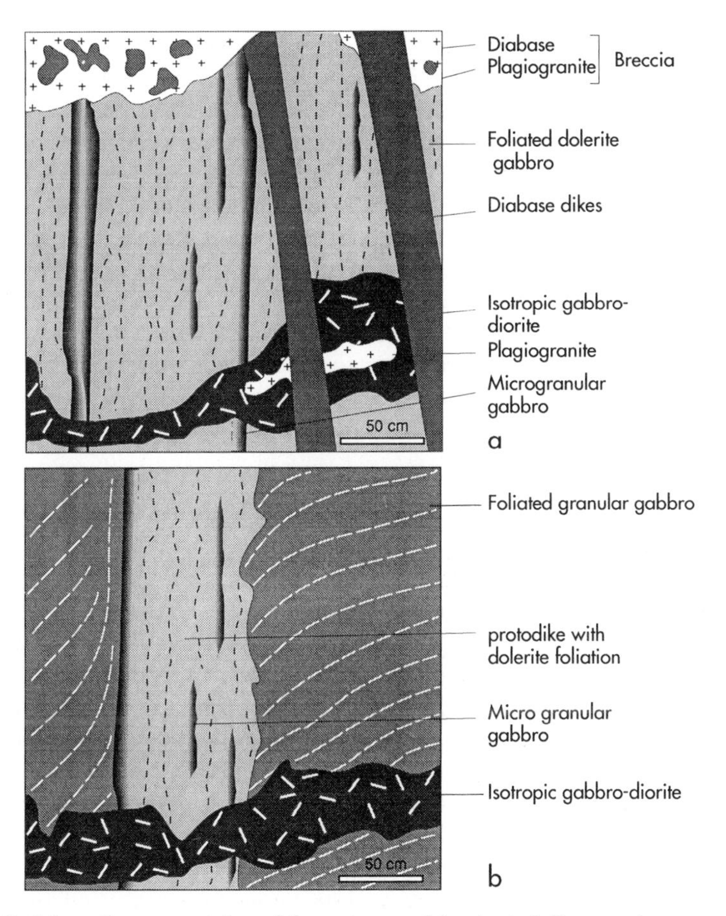

Figure 7. Schematic representation of the root zone of the sheeted dike complex: a, upper level; b, lower level.

that of diabase dikes, say less than 10% diabase dikes with respect to high level gabbro. It ranges from a few tens of meters to, at most a few hundreds of meters. Here, we define its lower limit as that of the protodikes, extending it downward only slightly. In Rosencrantz' (1983) detailed description of this root zone in Bay of Islands, our limit would coincide with his diabase dike 'keels'. The protodikes are basaltic dikes presenting cooled margins of brownish, microgranular gabbro. They grade inwards into a foliated, doleritic gabbro or coarse diabase, depending on the depth below the sheeted dikes. These dikes and their foliation cut the granular foliation of the high level gabbro, but these relations are obscured by the pervasive recrystallization of all gabbros and protodikes into amphibole, isotropic, gabbros-diorites and their intrusion by these isotropic gabbros as well as by plagiogranites and late diabase dikes (Figure 7).

Flow Structures in Protodikes

The continuity between the cooled microgranular margins and the doleritic inner foliation in the protodikes suggests that this foliation is induced by the flow of basaltic melt inside the dike. The coarse inner foliation and the microgranular border facies, compared to the grain size of diabases and their chilled margins respectively, indicate that the protodikes were emplaced in an environment at higher temperature than the diabase dikes of the sheeted dike unit. The presence of slivers of microgranular margin facies inside the doleritic foliated gabbros may be highly significant. The dike, emplaced in an environment which is already close to the melting point may be able to heat it up to the point of plastic or even viscous yielding. Thus, the protodike may be able to partly digest its own border and to incorporate fragments of it in this flowing melt.

Roof of Magma Chamber

The indication that, close to their base, protodikes are able to soften or, possibly remelt their surrounding suggests that the limit between the still partially molten formations at the top of the magma chamber and protodikes may be wavy. The magmatic flow in the granular, pyroxene gabbros from the top of the magma chamber is frozen, and plated as a magmatic foliation, along the irregular roof and the steep walls of the chamber during the drag, either imposed by the coupling with lower levels or caused by thermal convection (Figure 4). When a new protodike pierces the roof of the chamber, it cuts, deforms and possibly remelts on a small scale this granular foliation, as suggested by field observations.

The occurrence of diabase dikes cutting in various proportions the root zone and oriented closely parallel to the protodikes, may be related to the drifting of the root zone away from the magma chamber axis. During this drifting, the temperature falls rapidly and if melt is injected from the side of the magma chamber, it will freeze as diabase dikes within the root zone. Some of these later diabase dikes may feed the V2 volcanism which immediately follows the main V1 or Geotimes volcanism, as has been proposed by Beurrier (1987) on the basis of their geochemical affinities.

Concluding Remarks

The root zone of the sheeted dike unit described in this paper reveals a remarkable gradation between, above, the diabase dikes and, below, the foliated, granular gabbros frozen at the top of the magma chamber. The root zone is primarily composed of foliated, doleritic gabbros, here called protodikes. When a protodike intrudes the root zone, its thermal effect can locally remobilize the foliated gabbros plated at the roof of the magma chamber. This roof may thus be irregular-shaped at small scale, being composed of flaring chimneys at various stages of consolidation.

The root zone is at most, a few hundred meters thick, which is extraordinarily thin, considering that the temperature drops across it by some 500°C; this may correspond to the sharpest thermal gradient for any natural steady state system. Several authors have proposed that the main hydrothermal system, taking advantage of fissures parallel to the diabase dikes, closes at the base of the sheeted dike unit (Stern and Elthon, 1979; Nehlig and Juteau, 1988). Consequently, this zone may be looked at as a thermal boundary layer between two actively convecting systems: the magma chamber, below, and the main hydrothermal system above. The root zone is more isotropic than the sheeted dike unit, although it is largely constituted by foliated rocks. Water penetration would be more difficult than above, in the sheeted complex as suggested by the few quartz-epidote veins compared to above, where they have a density comparable to that of diabase dikes (Nehlig, 1989). Apart from millimeter-sized swarms of dark-green amphibole veins, the other evidence of water permeation is the occurrence of patchy zones of gabbro-diorites which develop at the expense of the foliated gabbros, generating locally a wet anatexis with the possibility of subsequent intrusion into surrounding rocks. The attitude of these hydrated patches is generally flat and we are inclined to interpret them as a diffuse downward closure of the active hydrothermal circuit which cools the sheeted dike unit located above so efficiently. The comparative inefficiency of this hydrothermal circulation in the root zone would explain the extremely sharp thermal gradient characterizing this zone.

Acknowledgments

This work has been improved by C. MacLeod's, D.A. Rothery's and C. Mével's reviews. It has been partly supported by the ATP 'Géodynamique' and DBT of the CNRS-INSU (contribution DBT 303).

References

Allen, C.R., 1975. The petrology of a portion of the Troodos plutonic complex, Cyprus. Ph.D., Univ. of Cambridge (unpubl.).

Benn, K. and Allard, B, 1989. Preferred mineral orientations related to magmatic flow in ophiolite layered gabbros. J. Petrology, 30: 925–946.

Beurrier, M., 1987. Géologie de la nappe ophiolitique de Samail dans les parties orientale et centrale de l'Oman. Thèse Doc. Etat, Paris 6, 406 p.

Browning, P., 1982. The petrology, geochemistry and structure of the plutonic rocks of the Oman ophiolite. Ph.D. The Open University, 404 p.

Delaney, J.R., Mogk, D.W. and Mottl, M.J., 1987. Quartz-cemented breccias from the Mid-Atlantic Ridge: samples of a high salinity hydrothermal upflow zone. J. Geophys. Res., 92: 9175–9192.

Dewey, J.F. and Kidd, W.S.F., 1977. Geometry of plate accretion. Geol. Soc. Amer. Bull., 88: 960–968.

Gerlach D.C., Avé Lallemant, H.G. and Leeman, W.P., 1981. An island arc origin for the

Canyon Mountain Ophiolite Complex, Eastern Oregon, U.S.A. Earth Planet. Sci. Lett., 53: 255–265.

Gregory, R.T. and Taylor, H.P., 1981. An oxygen isotope profile in a section of Cretaceous oceanic crust, Samail Ophiolite, Oman: evidence for $\delta^{18}0$ buffering of the oceans by deep (> 5 km) seawater-hydrothermal circulation at mid-ocean ridges. J. Geophys. Res., 86: 2737–2755.

Harper, G.D., 1984. The Josephine ophiolite, northwestern California. Geol. Soc. Am. Bull., 95: 1009–1026.

Hopson, C.A., Coleman, R.G., Gregory, R.T., Pallister, J.S. and Bailey, E.H., 1981. Geologic section through the Samail ophiolite and associated rocks along a Muscat-Ibra transect. J. Geophys. Res., 86: 2527–2544.

Juteau, T., Ernewein, E., Reuber, I., Whitechurch H. and Dahl, R., 1988. Duality of magmatism in the plutonic sequence of the Sumail nappe. Tectonophysics, 151: 107–135.

Kelly, D.S. and Delaney, J.R., 1987. Two-phase separation and fracturing in mid-ocean ridge gabbros at temperatures greater than 700°C. Earth Planet. Sci. Lett., 83: 53–66.

Lippard, S.J., Shelton, A.W. and Gass, I.G., 1986. The ophiolite of Northern Oman. Geol. Soc. London Mem., 11, 178 p.

Mysen, B.O. and Boettcher, A.L., 1974. Melting of a hydrous mantle. I: Phase relations of natural peridotite at high pressures and temperatures with controlled activities of water, carbon dioxide, and hydrogen. J. Petrol., 16: 520–548.

Nehlig, P., 1989. Etude d'un système hydrothermal océanique fossile: l'ophiolite de Semail (Oman). Thèse Doc. Univ. Brest, 308 p.

Nehlig P. and Juteau, T., 1988. Flow porosities, permeabilities and preliminary data on fluid inclusions and fossil thermal gradients in the crustal sequence of the Sumail ophiolite (Oman). Tectonophysics, 151: 199–221.

Nicolas, A., 1989. Structures of ophiolites and dynamics of oceanic lithosphere. Kluwer Acad. Publ., 367 p.

Nicolas, A., Reuber, I. and Benn, K., 1988a. A new magma chamber model based on structural studies in the Oman ophiolite. Tectonophysics, 151: 87–105.

Nicolas, A., Ceuleneer, G. and Boudier, F., 1988b. Mantle flow patterns and magma chambers at ocean ridges: evidence from Oman Ophiolite. Marine Geophys. Res., 9: 293–310.

Pallister, J.S., 1981. Structure of the sheeted dike complex of the Samail ophiolite near Ibra, Oman. J. Geophys. Res., 86: 2661–2672.

Pallister, J.S. and Hopson, C.A., 1981. Samail ophiolite plutonic suite: field relations, phase variation, cryptic variation and layering, and a model of a spreading ridge magma chamber. J. Geophys. Res., 86: 2593–2644.

Payne J.G. and Strong, D.F., 1979. Origin of Twillingate trondhjemite, North-Central Newfoundland: partial melting in the roots of an island arc. In: 'Trondhjemites, dacites, and related rocks', F. Barker ed., Elsevier, Amsterdam, 489–516.

Pedersen, R.B., 1986. The nature and significance of magma chamber margins in ophiolites: examples from the Norwegian Caledonides. Earth Planet. Sci. Lett., 77: 100–112.

Pedersen, R.B. and Malpas, J., 1985. The origin of oceanic plagiogranites from the Karmoy ophiolite, Western Norway. Contr. Mineral. Petrol., 88: 36–52.

Rosencrantz, E., 1983. The structure of sheeted dikes and associated rocks in North Arm Massif, Bay of Islands Ophiolite Complex, and the intrusive process at oceanic spreading centers. Canad. J. Earth. Sci., 20: 787–801.

Rothery, D.A., 1983. The base of a sheeted dyke complex, Oman ophiolite: implications for magma chambers at oceanic spreading axes. J. Geol. Soc., London, 140: 287–296.

Smewing, J.D., 1981. Mixing characteristics and compositional differences in mantle-derived melts beneath spreading axes: evidence from cyclically layered rocks in the ophiolite of North Oman. J. Geophys. Res., 86: 2645–2660.

Stern, C. and Elthon, D., 1979. Vertical variations in the effects of hydrothermal metamorphism in Chilean ophiolites: their implications for ocean floor metamorphism. Tectonophysics, 55: 179–213.

Wyllie, P.J., 1980. The origin of kimberlite. J. Geophys. Res., 85: 6902–6919.

Diabase Dikes Emplacement in the Oman Ophiolite: A Magnetic Fabric Study with Reference to Geochemistry

P. ROCHETTE[1], L. JENATTON[1], C. DUPUY[2], F. BOUDIER[2] and I. REUBER[2]
[1]*LGIT, Observatoire de Grenoble, IRIGM BP 53X 38041 Grenoble Cedex, France*
[2]*Université Scientifique et Technique du Languedoc, 34095 Montpellier Cedex, France*

Abstract

The anisotropy of magnetic susceptibility, coupled with geochemical identification of magma sources, has been investigated in 360 samples from 67 basaltic dikes of the sheeted dike complex of the Oman ophiolite. Two thirds of the analysed dikes have MORB affinities, while the others, previously related to island arc setting, show REE patterns and trace elements evolution more in agreement with an off-axis ridge magmatism and an interaction between magma and ultramafic country rocks. Only half of the dikes yield primary flow fabrics with a minimum magnetic axis K3 close to dike pole and a maximum axis K1 parallel to flow line. Secondary fabrics, mainly with K1 parallel to dike pole, are more likely to occur in thick ($\geq$1.5 m) dikes with high susceptibility and MORB composition. Flow lines appear quite dispersed with an overall tendency toward vertical flow. Dikes of MORB type composition show a distinct obliquity of their flow plane relative to dike margins, suggesting that the dikes were horizontally sheared at the end of emplacement. This shear could be a signature of oblique spreading during the accretion of the Omani oceanic crust.

Introduction

With the aim of understanding the processes leading to the accretion of the oceanic crust, in general, and of the Oman ophiolite in particular, structural mapping and microstructural studies have been extensively performed in Oman, leading to detailed knowledge of the deformation of the peridotite (e.g. Boudier et al., 1985; Nicolas et al., 1988a) and more recently of the gabbro layers (Nicolas et al., 1988b). With the discovery of mantle diapirs, a major aim of further studies became the mapping of such diapirs and the identification of their influence on the crustal structures, with in view the understanding of the segmentation of oceanic rifts (Nicolas et al., 1988a; Sempéré et al., 1990). In this respect the study of basaltic flow lines in the

Tj. Peters et al. (Eds), Ophiolite Genesis and Evolution of the Oceanic Lithosphere, 55–82.

sheeted dike complex may be potentially useful to detect an underlying diapir as flow lines are expected to diverge from the diapiric center which delivers the main part of the magma extracted from the mantle (Rabinowicz et al., 1984). Thus, a systematic mapping of flow structures in the diabase dikes throughout the ophiolite belt was undertaken. However, the study of flow structures appeared to be very difficult using microstructural techniques due to the very weak preferred orientation and fine grain size in those dikes.

Therefore the possibility to use the technique of the anisotropy of magnetic susceptibility (AMS) to study the emplacement fabric of the dikes has been tested in the Oman ophiolite. The measurement of AMS provides a susceptibility ellipsoid whose principal axes are labelled K1 > K2 > K3. Numerous studies have established that this ellipsoid is an image of the preferred orientation of magnetic minerals in the rock (Hrouda, 1982). The advantages of this technique are mainly a very good sensitivity to weak orientation and the rapidity of measurements allowing statistical and cartographic studies. In basaltic rocks, the magnetic minerals are essentially titanomagnetite or magnetite grains in multidomain state, i.e. with grain size larger than a few m, whose maximum susceptibility is defined by the long axis of the grain. It has been shown in basaltic dikes that early crystallized acicular grains are usually oriented within the magmatic flow plane and along the magmatic flow line (e.g. Shelley, 1985); thus the magnetic expression of this flow fabric should be a maximum axis (K1) parallel to flow line and a minimum axis (K3) perpendicular to the flow plane, on the average parallel the dike margins. Already published AMS studies of basaltic dikes strongly support this model (Hrouda, 1985), in particular the extensive study of Hawaian dikes of Knight and Walker (1988). These authors have also shown that the imbrication angle between K1 and the dike plane allows to infer the sense of the flow. On the other hand, Khan (1962) and Ellwood (1978) have occasionally found other relationships such as K2 parallel to flow line, in agreement with Jeffrey's model of laminar flow (1922). Park et al. (1988) have reviewed the processes that may lead to other types of fabric acquired after emplacement. Such processes include the effect of cooling stress resulting in K1 axes perpendicular to dike plane (Dawson and Hargraves, 1985), subsequent regional deformation and late stress effects.

Sampling and Geological Setting

The sheeted dike complex, typical of the oceanic crust, is well represented along the Semail ophiolite with very good exposures unaffected by obduction related deformation (Figure 1). This late stage is only visible through large scale folds of moderate dip (Pallister, 1981) and recorded in the basal sole of the ophiolite and along strike slip shear zones (Boudier et al., 1985, 1989).

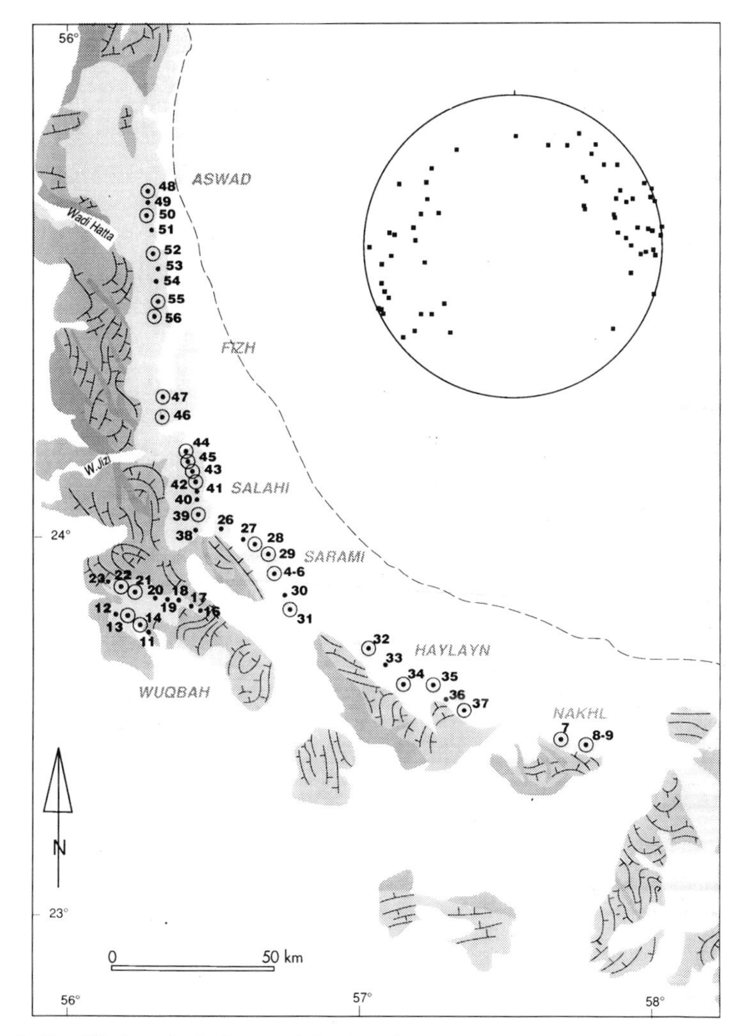

Figure 1. Simplified geological map of the Semail ophiolite (dark gray: peridotites; light gray: crustal sequence) with sampling sites, circled when studied geochemically. General foliation is indicated in the peridotites.The inserted stereoplot shows the sampled dike poles in geographic coordinates. Sites 1–3 come from Ibra zone (SE corner).

The stereoplot in Figure 1 illustrates the tendency for subvertical NNW striking dike planes; dispersion of the dike planes in our sampling is mainly due to original local variations of the extensional stresses.

The dikes are variously affected by ridge metamorphism and hydrothermalism in greenschist or amphibolite facies. As our study was the first attempt to use AMS in such an environment, a complete coverage of the ophiolite was undertaken resulting in 360 oriented cores drilled in 67 dikes from 56 localities, with usually at least 5 samples per dike (Figure 1 and Table 1).

Table 1. Data on all dikes classified according to fabric types and site number (see Fig. 1 for location); when more than one dike has been obtained in a site each dike is labelled with a letter A, B, C; N is sample number per dike, when less samples are used for mean directions dike name is shown with *; Km with its standard deviation is in 10-3 Sl; thickness in m; D3, I3 and E3 are declination, inclination and confidence angles for the average minimum direction K3, except for intermediate fabrics where intermediate directions are reported; D1, I1 and E1 refers to maximum directions K1 in all cases; Group label corresponds to geochemical type (see Table 2).

	N	Km	SD	L	F	Thickness	D3	I3	E3	D1	I1	E1	Group
Normal													
5A	7	0.84	0.76	1.008	1.002	0.32	278	21	39,25	148	59	29,23	
5B	4	32.00	3.60	1.016	1.032	0.63	100	4	40,21	6	45	60,14	II
6A	5	35.80	22.70	1.009	1.010	1.30	73	8	26,15	167	31	32,16	I
6B	4	35.10	9.40	1.006	1.012	2.50	91	24	39,15	208	45	44,21	
8A	3	0.91	0.33	1.022	1.053	1.60	236	9	56,37	142	23	43,14	I
8B	3	6.00	4.70	1.003	1.052	2.00	266	11	41,3	4	35	75,21	
9A	4	17.40	20.30	1.005	1.017	0.70	251	1	31,3	342	44	53,13	
9B	8	7.40	6.50	1.004	1.026	0.65	279	3	6,2	12	43	14,4	I
9C	4	0.95	0.55	1.001	1.006	0.20	101	4	18,10	193	24	62,13	
12A*	5	7.30	26.90	1.000	1.002	2.50	248	22	82,5	348	87	76,64	
13A	4	0.72	0.19	1.003	1.004	0.35	280	2	25,3	185	68	52,5	
14	5	1.47	3.19	1.001	1.003	0.80	277	33	44,15	150	43	29,7	III
15A	5	21.00	23.60	1.000	1.016	1.10	93	2	19,9	183	19	82,12	
16	5	11.70	27.90	1.003	1.014	1.20	288	7	25,5	190	51	43,11	
17	5	59.40	10.40	1.009	1.037	0.76	109	4	16,4	18	4	9,4	
18	5	7.90	11.80	1.014	1.014	1.00	105	15	24,17	271	75	28,17	
20*	6	5.80	8.10	1.006	1.060	1.50	100	11	44,15	200	41	34,14	
22B	4	57.50	15.20	1.006	1.009	0.50	68	14	46,10	338	0	44,12	
24*	5	6.30	11.60	1.005	1.006	1.50	96	15	74,55	226	67	80,72	
28*	5	4.10	4.50	1.004	1.015	0.70	297	25	31,4	133	64	40,20	II
29*	5	16.30	7.10	1.007	1.011	0.50	121	6	22,14	223	62	35,13	I
33	6	36.50	9.60	1.009	1.021	0.40	257	9	13,8	164	16	20,6	
34	7	0.57	0.15	1.002	1.003	0.80	76	2	11,6	173	76	13,3	II
38	5	38.10	16.20	1.012	1.024	1.00	100	13	16,6	196	25	18,6	
40	5	23.30	4.20	1.006	1.001	1.40	93	21	46,12	349	27	1,6	
43	5	39.20	19.20	1.004	1.022	1.20	111	10	22,11	359	65	49,12	I
44	5	64.90	40.60	1.014	1.010	1.10	112	10	29,7	235	72	12,6	I

45	5	31.40	14.10	1.002	1.003	1.20	279	8	56,22	10	8	26,12	I
46	5	11.10	12.00	1.004	1.003	1.20	268	4	35,12	36	83	24,5	II
55	5	18.10	5.40	1.006	1.032	2.50	282	0	14,9	12	57	22,10	III
56	4	26.80	8.80	1.006	1.028	3.00	268	11	13,8	58	77	68,13	III
Intermediate													
3	4	22.70	18.10	1.005	1.004	1.50	82	5	46,10	189	72	48,10	
7A	4	11.20	18.80	1.011	1.009	2.20	253	4	33,11	159	46	33,4	
7B	4	14.30	12.20	1.014	1.006	1.10	239	19	68,6	352	49	68,13	I
23	5	14.50	9.10	1.008	1.013	1.50	80	5	30,10	173	34	27,11	
27	6	25.10	11.80	1.005	1.004	1.20	79	12	48,13	174	22	17,8	
42	5	18.10	7.60	1.010	1.031	3.20	74	4	12,8	339	51	27,6	I
49	5	22.70	16.40	1.004	1.014	1.50	265	0	69,13	174	63	69,7	
54A*	5	33.30	13.00	1.003	1.020	1.30	97	33	34,8	191	7	62,14	
Reverse													
2	6	17.090	6.10	1.004	1.001	0.60	173	7	24,8	81	14	26,17	I
4	8	111.00	51.00	1.009	1.005	3.00	356	15	27,9	259	23	23,6	I
8C	5	0.76	0.12	1.012	1.023	2.00	207	48	23,16	298	2	59,7	II
10A*	4	66.20	20.00	1.005	1.010	1.50	350	14	47,35	253	27	41,16	
25	5	16.80	5.70	1.006	1.003	1.50	169	32	40,11	271	18	28,11	
26	5	69.60	30.00	1.010	1.017	1.50	8	23	21,13	101	8	39,17	
35	5	105.90	64.00	1.022	1.017	2.00	355	0	30,10	85	9	36,5	I
36	5	6.00	4.00	1.008	1.000	1.50	14	39	87,14	113	11	33,14	
39	5	2.70	5.90	1.010	1.003	1.60	161	11	57,19	256	24	19,8	II
41	5	16.30	12.10	1.012	1.004	2.00	357	61	26,13	97	5	19,12	
48A	5	9.40	7.20	1.007	1.012	1.70	166	35	15,6	262	9	10,5	I
48B	3	14.20	4.10	1.004	1.022	0.60	163	40	53,6	272	21	71,38	
51*	5	7.80	2.90	1.003	1.013	3.00	182	19	66,22	83	22	66,45	
52	5	31.90	33.00	1.009	1.016	1.70	168	23	27,12	258	0	27,5	I
53	5	9.30	4.00	1.001	1.008	0.60	177	26	19,7	84	6	66,17	
54B	3	39.50	6.50	1.006	1.011	2.50	7	24	55,23	107	21	61,39	

Table 1. (Continued)

	N	Km	SD	L	F	Thickness	D3	I3	E3	D1	I1	E1	Group
Others													
1	14	10.10	5.90	1.003	1.013	0.80	5	6	5,	102	50	19,	
11A	4	15.90	4.80	1.007	1.002	1.50							
11B	3	0.49	0.04	1.001	1.001	0.50							
13B	3	0.49	0.10	1.001	1.001	0.35							III
19	5	5.40	1.70	1.006	1.003	2.00							
21	5	4.40	1.90	1.002	1.004	1.00	45	23	21,5	161	42	27,6	I
22A	4	96.50	9.60	1.010	1.016	0.50	38	17	45,5	143	41	43,6	I
30	5	19.80	5.70	1.004	1.009	1.30							
31	5	17.70	10.40	1.004	1.005	1.80							I
32	5	15.30	3.40	1.002	1.004	1.60							I
37	5	9.90	9.50	1.002	1.006	3.00	326	9	44,4	94	75	67,,	II
50	5	12.30	10.40	1.009	1.009	0.70	216	0	24,15	306	60	35,,,	I

On the outcrop, selection criteria for a given dike were the presence and integrity of chilled margins, the concordance of the dike plane with the general strike and dip of the outcrop, and its limited fracturing and epidotitization. Weathering was usually negligible due to sampling in wadi's beds. Complete cross sections through the dikes were often realized and the distance of samples to the margin carefully noted, together with the dike thickness. Dike plane and samples were precisely oriented using solar and magnetic compass, revealing that local magnetic declination varies from 0 to 10° W.

Magnetic Measurements

General features and fabric types

The anisotropy of magnetic susceptibility in low field was measured on standard 25 mm cylindrical samples with the high sensitivity ($5.\ 10^{-8}$ SI) Czech induction bridge KLY-2. Maximum and minimum susceptibility directions (K1 and K3) of each sample are derived from a set of 15 measurements in different orientations; magnetic axes are described by declination and inclination. A coordinate system related to the dike is chosen to be able to compare data from every dike regardless of their field position: on the stereoplot N is the dike strike and vertical is the dike plane. Tilt corrections used to transfer from geographic coordinates due to non verticality of the actual dike plane, are usually small and do not exceed 30 degrees (see Figure 1). The measurement also provides mean susceptibility $Km = (K1 + K2 + K3)/3$ in SI unit and various anisotropy ratios: $P = K/K3$, $L = K1/K2$ and $F = K2/K3$, respectively total anisotropy degree, lineation and foliation parameters. As in other studies of basaltic dikes, mean susceptibility is generally very high (up to 10^{-1} SI, equivalent to 5–10% of magnetite) but anisotropy ratios are quite low: $L \leq 1.02$ and $F \leq 1.05$ (Table 1). Nevertheless the magnetic directions appear to be significant down to anisotropy ratio of 1.002. The significance of the magnetic fabric at the dike scale is tested using the tensorial mean statistics of Jelinek (1978). It provides mean directions for K1 and K3 with ellipses of confidence, together with L and F of the mean tensor. The pairs of confidence angles E3, E1 reported in Table 1 are the half-axes of the confidence ellipses on K3, K1 axes. They largely overestimate the dispersion when number of samples is less than 6, as exemplified by dikes 9A and 9B from the same site: the actual grouping of directions is equally good for both but confidence angles are much better for dike 9A with 8 samples.

Within a single dike, the magnetic directions are usually well grouped or homogeneously dispersed in a plane. However in some dikes, marked with a star on Table 1, one or two samples fall well away from the grouping and were thus rejected in the statistics. In others, no clustering appears at all; in

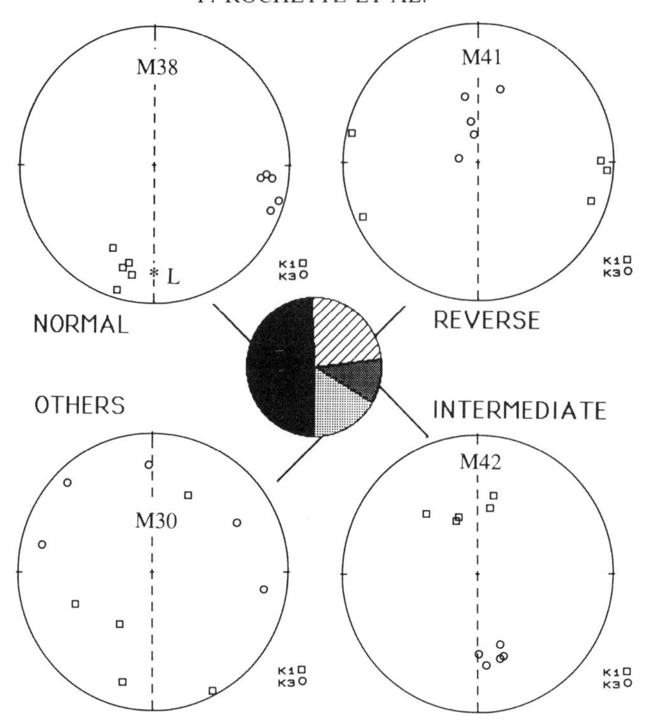

Figure 2. Examples of the 4 fabric types encountered in the dikes, with their respective frequency in a pie diagram. Magnetic anisotropy data (maximum and minimum susceptibility axes, respectively squares and circles) are plotted in "dike" coordinate, i.e. dike plane (dashed line) tilted to the vertical and rotated to a NS strike; L indicates flow line measured on the field.

such a case no mean directions are reported and the mean L and F are derived from arithmetical mean.

Such an analysis at the single dike scale leads to define various fabric types according to the orientation of the mean magnetic axes with respect to the dike pole, i.e. the E-W axis:

- almost half of the dikes (Figure 2) yield minimum axes clustered close to E-W as expected for a flow fabric; they will therefore be referred as normal type (N);
- the reverse type (R) corresponds to a maximum axis grouped at the dike pole and concerns 24% of the dikes;
- in a few examples, K3 and K1 axes are well grouped but within the dike plane, indicating that the intermediate axis is parallel to dike pole (I type);
- finally the remaining 12 dikes show either completely dispersed directions, as exemplified in Figure 2, or well grouped directions with an angular deviation from dike pole larger than 35 degree (5 dikes), or a mixture of normal, reverse and even intermediate types. These will be grouped in "others" (O) type.

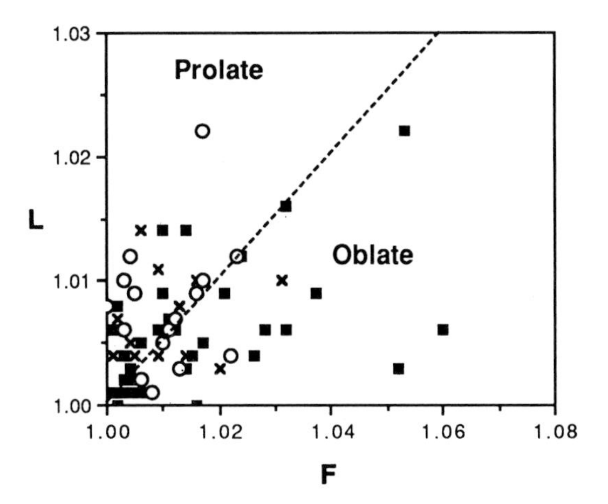

Figure 3. Mean dike values of anisotropy ratio L versus F for normal (black squares), reverse (open circles) and intermediate + others (crosses) fabric types.

The identification of N type as the primary flow fabric is confirmed by the coincidence of K1 mean direction with flow lineation in the few cases where it was visible on the outcrop. In the example of Figure 2, the flow line (L) was estimated from the elongation of elliptical inclusions while a faint preferred orientation of amphibole or plagioclase rods were detected in some other outcrops. Before interpreting further the N type in terms of emplacement fabric the fact that about half of the sampling failed to reveal this expected fabric has to be explained.

Secondary fabrics appear generally more isotropic than the primary N type: in Figure 3 L and F mean dike values are on the average smaller for R, I, O types than for N type. It is also found that large F values are only observed in normal type and that the magnetic fabric shape is more linear for secondary fabric types: an oblate ellipsoid ($L \leq F$) is observed in 88% of the dikes in the normal type and only 64% in the other types.

As R type is the more abundant after N and because at least part of I and O cases may derive from mixing of R and N types, the origin of reverse fabrics will be mainly investigated. Apart for being more linear, reverse fabrics are characterized by a good grouping of maximum directions with an average within 1° of the E–W axis, and minimum directions mainly within 30° of the horizontal (Figure 4).

Magnetic Mineralogy

The Omani sheeted dike complex has been already studied by Luyendick et al. (1982) and Shelton (1984) in terms of magnetic mineralogy and paleomagnetism. They described a very complex picture where the primary signal,

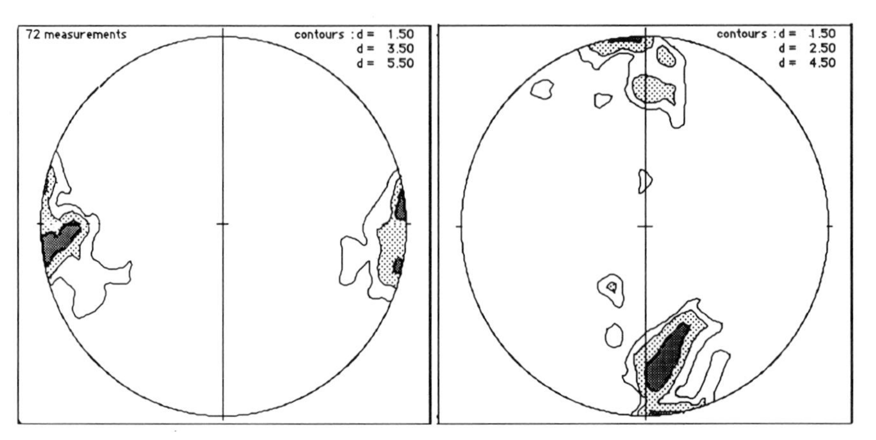

Figure 4. Density contours of individual maximum and minimum susceptibility directions for all samples from dikes of reverse fabric type. Solid line correspond to dike plane. The stereoplots are obtained using a program designed by Pecher (1989).

carried by magmatic titanomagnetites, was strongly affected by later events including recrystallisation, hydrothermalism or later alteration resulting in the formation of new magnetic phases. Such processes are likely to reset the primary N fabric. The other possibility for a mineralogical origin of abnormal fabrics is if very fine grains (single domains) are the carrier of AMS, as discussed by Potter and Stephenson (1988) and Rochette (1988). In such a case the long axis of the grain becomes the minimum magnetic axis; therefore a normal petrofabric results in a reverse magnetic fabric.

A magnetic mineralogy study has been performed. Only a brief report of the complex results will be given here. In each sample two or three magnetic minerals among the following list are usually detected: titanomagnetite, pure magnetite, maghemite, hematite and pyrrhotite. Due to their much larger susceptibility, only the first three (cubic spinel minerals) can contribute to the AMS signal. If pure magnetite or maghemite are in situ replacements of primary titanomagnetite (by exsolution in the magnetite case) or if their shape mimics the one of plagioclase crystals for example, the resulting magnetic fabric will have the same orientation as the primary fabric. If not the secondary fabric may be isotropic, i.e. only diluting the primary fabric, or it may be guided by cooling stresses, microfractures opening during hydrothermalism, or the regional stress field as suggested by Park et al. (1988). However, neither the nature of the magnetic minerals nor their grain size (as defined by Lowrie-Fuller tests) appeared to show clear-cut correlation with fabric type.

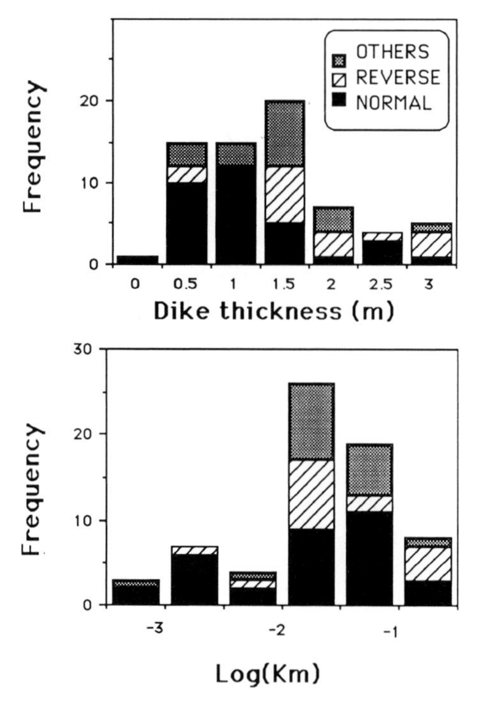

Figure 5. Histograms of mean susceptibility (Km) and dike thickness for normal (black), reverse (stippled) and other (gray) fabric types.

Origin of the Secondary Fabric Types

On the other hand, fabric type is found to be correlated with mean susceptibility and with dike thickness: secondary fabrics (R, I and O) are more abundant for larger values of these parameters (Figure 5).

Large values of susceptibility are not directly related to iron amount but to the magnetite (or related spinels) amount, and also to grain size: larger grain size corresponds to larger susceptibility. A possible explanation for the correlation with fabric type is that the abundance of (large) magnetite grains is related to the intensity of hydrothermalism and recrystallization in the rock. These are the processes invoked for secondary fabric.

77% of the dikes thinner than 1.5 m are normal, compared to only 28% in the thicker ones. The fact that the primary flow fabric is much more easily lost in thick dikes could have different reasons:

- the flow can be slower and less regular, thus leading to a poorly defined fabric more easy to overprint;
- due to larger heat capacity the magma could remain fluid, inside the chilled margins, after the end of the flow; then local convection, gravity settling and cooling stresses could reorient the magnetic grains.

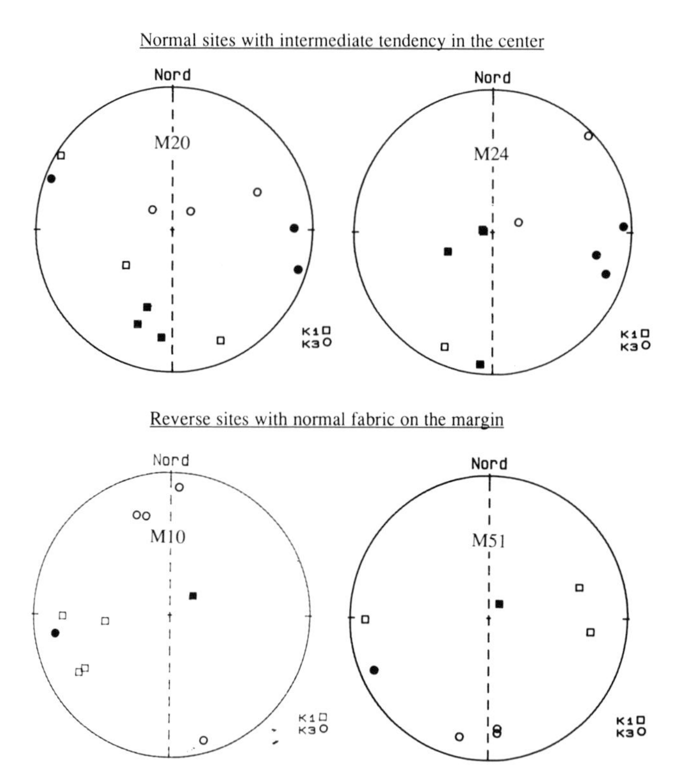

Figure 6. Examples of individual dikes showing a control of fabric type with distance to the chilled margin. Black symbols correspond to samples within 10 cm of the dike margin.

- the hydrothermal activity may be more important within large dikes: in the field the green color typical of epidotitization is more often observed in large dikes.

These interpretations are supported by the observations from several dikes showing mixed fabric types (Figure 6). Systematically the samples within 10 cm from the border are normal while reverse or intermediate behaviors appear when proceeding toward the center. In fact the margins experience conditions typical of much smaller dikes, either in terms of early freezing of the fabric or larger velocity gradient or smaller impact of hydrothermalism due to finer grain size.

It is likely that multiple sources of secondary fabrics are present, and that part of the complexity of the magnetic dataset is due the overall weakness of the preferred orientation and to mixing of fabrics of different origin. Therefore completely solving the problem would require detailed study of each dike. Reverse dikes yield relatively well grouped near horizontal minimum directions, i.e. along the ridge axis (Figure 4). This feature suggests that the fabric is ruled at the crustal scale possibly by the stress regime. It

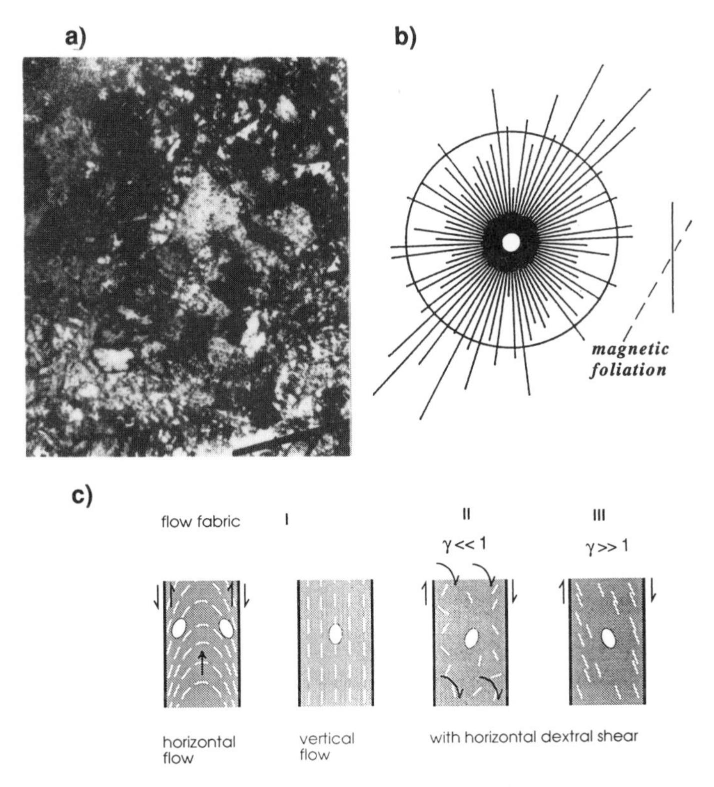

Figure 7. (a) Microphotograph of a thin section in reverse dike M41 cut in the plane containing K1(dashed line) and K3 (solid line, equal to the dike plane trace with length corresponding to 1 mm). (b) Rose diagram of plagioclase fabric in the plane perpendicular to magnetic foliation and dike plane from normal dike M43; dike plane is vertical. (c) Sketches of fabric acquisition model during flow and subsequent dextral shear. The ellipse represents the magnetic anisotropy.

would lead to K1 parallel to maximum extension and K3 parallel to minimum extension, as in a solid state deformation. Whether this causal relationship is actually due to a late emplacement deformation of the magmatic mush or if it is acquired through the geometry of temperature gradient and hydrothermal activity remain to be established although the latter explanation is more likely. Indeed fluid migration, driven by temperature gradient, should proceed from inside to outside, perpendicular to the dike plane, i.e. along K1, but also to a lesser extent vertically within the dike, i.e. roughly along K2 (see Nehlig, 1990). A preliminary microscopic study (Figure 7a) shows that in reverse type dike M41 opaque grains occur as clusters of euhedral grains and that the elongation of these clusters tentatively represents the fluid migration direction. As a cluster of spherical magnetite grains exhibits a maximum susceptibility along its elongation, such a microstructural process could lead to the observed reverse fabric.

Interpretation of Primary Flow Fabrics

1. *Lineations*

In primary flow fabrics the magnetic lineation, i.e. the grouping of K1 axes, represents the magma flow direction. This was observed for example by Knight and Walker (1988) and is directly confirmed by our observations in a few cases where lineation is visible (M5A, M9A, M38, M43). Therefore magnetic fabric measurements appear to be the best way to undertake a systematic study of dike flow lines in the ophiolite as these lines are not directly visible in a large number of dikes. However our sampling is a bit scarce due to the presence of secondary fabrics and to the fact that lineations are not always as well defined as in Figure 2 (see confidence angles in Table 1). Thus only a preliminary interpretation in terms of dip of the flow line within the dike plane may be put forward. In a hot spot situation with a single magma feeder, flow lines in a large area are expected to be well grouped near the horizontal, as exemplified by the Hawaian case (Knight and Walker, 1988). On the contrary an ideally cylindrical mid-oceanic magmatic chamber should produce consistently vertical flow lines in the sheeted dike complex. The present dataset on maximum directions for all samples of normal dikes (Figure 8a) shows a large scatter with a larger density along the vertical line. This way of examining the data may be misleading because it includes dispersion that is due to measurement uncertainties often quite high in those almost isotropic rocks. Considering mean dike lineation, as done in Figure 8b) may be more significant. The tendency towards subvertical directions still subsists but a large number of lineation appears also near the horizontal. Despite the problem of the variable confidence that can be put on those directions, it appears clearly that the flow lines geometry correspond to an intermediate situation between the hot spot situation and the cylindrical case, in agreement with present models of the Omani accretion involving a series of distinct feeders at the top of mantle diapirs along the ridge.

2. *Foliations*

By definition of the normal type, the minimum susceptibility directions K3 are concentrated near the E-W axis. However a distinct asymmetry appears with a maximum density showing a clockwise rotation with respect to E–W. This obliquity appears through the dataset as a whole but can be observed also in each dike as exemplified by the N type example in Figure 2 with a clear clockwise obliquity, which corresponds in Table 1 to D3 larger than 90 or 270. Some dikes have also anticlockwise or not significant obliquity (Table 1). The reality of this effect may be questioned on a single dike by considering that the precision on dike plane measurement is quite poor because this plane is often irregular, inducing a mean flow oblique with respect to the margin where structural measurement was done. Another likely source of obliquity could be the imbrication on the margins due to flow sense (Shelley,

NORMAL FABRICS

MAXIMUM DIRECTIONS K1 MINIMUM DIRECTIONS K3

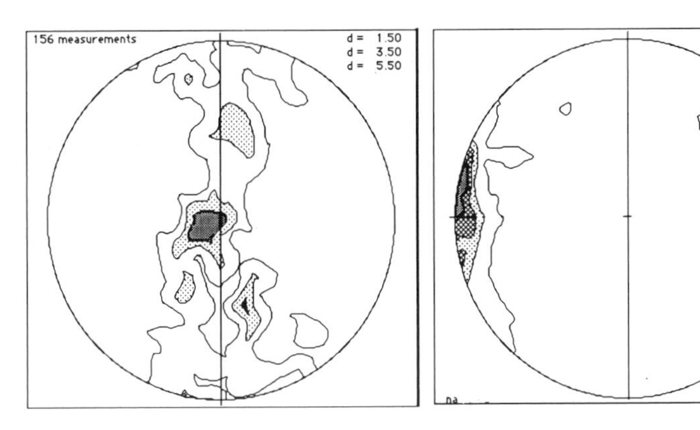

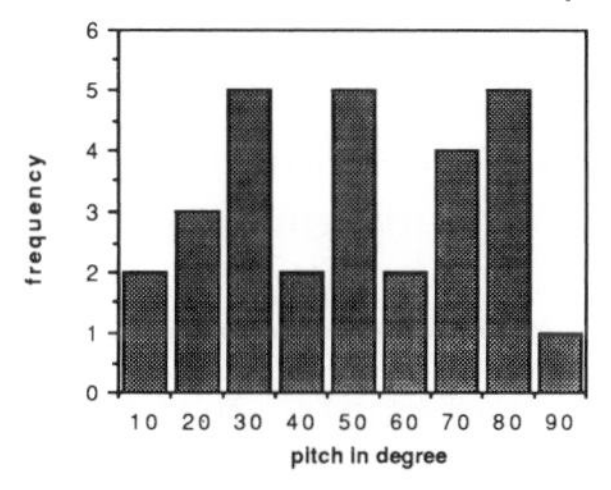

Figure 8. Density contours of individual maximum and minimum susceptibility directions for all samples of normal fabric, together with histogram of mean dike lineation dip.

1985; Knight and Walker, 1988). However this imbrication should result in a symmetric pattern with respect to dike center (Figure 7c first sketch). Therefore, considering the nature of the sampling and the repeated character of this obliquity in a number of independent dikes from separated outcrops, lead to the conclusion that the only explanation for the observed obliquity or loss of symmetry with respect to dike center, seems to be a horizontal strike slip parallel to the dike margins. The corresponding shear in the dike should occur while the rock is still viscous because no evidence of solid state deformation (such as deformed minerals or schistosity) is visible in the outcrop or in thin sections. Moreover the plagioclase platelets also show a plane of preferred orientation oblique on the dike margins and parallel to the magnetic foliation plane (Figure 7b). Another evidence that the postulated shear occurred very early is that the secondary reverse fabric exhibits no sign of obliquity (Figure 4).

Once the existence of this late emplacement shear on the dike margins is admitted, the sense of shear has to be inferred. However the relationship between fabric obliquity and sense of simple shear is quite a debated matter (e.g. Blumenfeld and Bouchez, 1988). The first problem would be to establish if magnetite grains or aggregates behave as rigid or deformable markers, what mean aspect ratio they have, and if they are isolated in a viscous matrix or mechanically interacting. The first two questions cannot be solved directly without a thorough microfabric study beyond the scope of the present paper.

However the fact that magnetite grains are smaller than the plagioclase grains and yield a similar fabric suggests that they share the same features: rigid, elongated and interacting markers. On the other hand this deformation is superimposed on the magmatic flow. From field observations it can be deduced that displacements on the dike margins are quite moderate, i.e. of the same order as dike thickness corresponding to $\gamma \leq 1$. The shear due to the flow is probably much larger, except right in the center. Therefore the horizontal strike slip movement can have an effect only when flow is almost stopped and on an already well oriented fabric. In the most common case of vertical flow this primary fabric will be strictly parallel to the margins when considering the horizontal plane (case I in Figure 7c). A moderate dextral shear applied to such a fabric, concerning rigid elongated markers, should rotate all grains clockwise and thus result in a clockwise obliquity of the fabric (case II) provided that shear is not large enough to produce imbrication, i.e. counterclockwise obliquity (case III).

The same dextral shear acting on markers with the same viscosity as their matrix and initial fabric parallel to the margin would also produce counterclockwise obliquity. Actual viscosity contrasts during obliquity acquisition are very difficult to evaluate so that the sense of shear indicated by our data remains debated. Although dextral sense is tentatively favored, sinistral shear is also possible and a direct confirmation on dikes where strike slip is identified on the field is strongly needed.

These interpretations could also be tested by looking at the fabric geometry within the dike. Although the number of samples per dike is small and that fabric features are very variable in details, two interesting further evidences of shear can be highlighted. Obliquity is often larger away from the margins, and a difference in lineation dip is sometimes observed between W and E margins. The W margin lineation appears deviated towards the N compared to the E margin in 6 dikes while 2 dikes yield the reverse deviation. In the case of an upward flow this deviation corresponds to a dextral shear if lineation is considered to record the magma direction with respect to a fixed margin. However such a difference between margins may also be linked to a ridge polarity, i.e. colder margin away from the ridge invoked by some authors (Kidd and Cann, 1974) but this postulated temperature gradient appears to be cancelled by hydrothermalism .

Finally it is worth pointing out that the selection criteria for N type was a mean K3 within 35 degree of the dike pole. However the postulated shear may have rotate K3 out of these limits in some cases classified in I or O type. A possible candidate could be M37 (O type) where the subvertical mean K1 coincides with the flow line estimated on the field.

Geochemical Characteristics

Previous studies have shown that the basaltic dikes of the Oman ophiolite include two types of composition (Alabaster et al., 1982; Beurrier et al., 1989): the dominant type is tholeiitic and has been related to the Geotimes volcanic sequence; the second type differs chemically and shows a calc-alkaline tendency. In addition, Beurrier et al. (1989) have suggested on the basis of field observations a chronological evolution from the first to the second type. Thus the chemical characterization of the rocks considered in the present study is essential in order to tentatively constrain a relative chronology of emplacement at the scale of the sampling. Thirty specimens have been chosen (Figure 1) out of the magnetic sampling. These specimens have been analyzed for major and trace elements (Table 2).

Presentation of the Data

All the analyzed dikes (Table 2) have basaltic compositions with [Mg] ratio bracketed between 0.4 and 0.7, indicating that they have suffered variable low P fractionning. In addition, the large dispersion of Na and K content records the low grade metamorphic alteration.

Selected elements converge to individualize three groups of rocks having distinct geochemical features. For instance, on Ti versus Mg diagram (Figure 9), group I which gathers most of the analyzed samples is characterized by the highest TiO2 content which increases with differentiation, up to [Mg] = 0.45, then tends to decrease. This differentiation trend, as well as the [Ti/V] ratio comprised between 23 and 45, is typical of Mid Ocean Ridge Basalts (MORB). Groups II and III have lower TiO2 content, which remains constant during differentiation. In addition [Ti/V] ratios are between 15 and 21 for group II, between 10 and 15 for group III. These features are not typical of MORB, and more likely found in Island Arc Tholeiites (IAT).

K and Na, several incompatible elements are affected by secondary hydrothermal alteration. This is indicated on Figure 10a) where Sr is plotted versus Ce. Dispersion of the data accounts for the secondary mobilization of Sr. The following other elements have been also affected: K, Rb, Na, Ba, U, Cu and Au. Conversely Hf (Fig 10b) as well as Th are not affected. Indeed Mc Culloch et al. (1981) have shown that the REE are not affected by hydrothermal alteration.

The REE patterns normalized to chondrites have been reported for different samples of each of the previously delineated groups (Figure 11a). In group I, the convex REE pattern is marked by a depletion in LREE. The ratio [La/Yb]N between 0.7 and 0.9 supports an origin from a depleted upper mantle source. The increase of REE during differentiation is accompanied by a slight increase of the [La/Yb]N ratio and a marked anomaly of Eu in the most differentiated specimens. It suggests that the variation recorded

Table 2. Major and trace elements amounts respectively in % and ppm analysed for 30 selected dikes classified from N to S in the ophiolite. Magma type is indicated above site name.

Group	I	I	III	I	III	II	I	II	I	I	I	I	I	I
Ref.	M50	M48	M55	M52	M56	M46	M47	M39	M44	M43	M45	M42	M21	M22
SiO_2	50.00	52.60	52.60	52.21	58.78	50.28	53.56	49.70	52.98	48.73	51.05	49.50	55.11	54.92
Al_2O_3	14.53	15.00	15.05	15.27	14.70	15.26	14.38	14.86	14.63	15.46	14.33	14.59	14.38	14.21
Fe_2O_3	9.37	9.50	8.97	11.79	9.08	9.26	14.40	8.37	11.65	10.25	11.14	12.05	10.91	12.50
MnO	0.19	0.12	0.15	0.22	0.15	0.11	0.09	0.15	0.15	0.18	0.21	0.21	0.17	0.15
MgO	7.80	5.92	6.83	5.25	4.70	6.34	4.57	8.88	4.56	8.61	6.28	6.55	5.50	4.10
CaO	9.16	8.00	6.95	6.68	4.13	9.12	2.21	9.90	5.62	8.77	6.65	10.25	4.52	5.57
Na_2O	3.78	4.20	4.28	4.55	3.95	3.90	5.70	3.07	5.55	3.00	5.27	3.35	5.35	4.02
K_2O	0.25	0.19	1.15	0.21	0.80	0.19	0.11	0.08	0.21	0.06	0.07	0.09	0.41	0.27
TiO_2	1.06	1.14	0.50	1.38	0.50	0.84	1.61	0.80	1.95	1.30	1.59	1.56	1.00	1.42
L.O.I	2.80	2.23	3.10	2.00	3.13	3.70	3.00	3.40	2.07	2.53	2.52	1.22	2.43	2.25
Σ	98,94	98.90	99.58	99.56	99.92	99.00	99.63	99.21	99.37	98.89	99.11	99.37	99.78	99.41
[Mg]	0.646	0.579	0.626	0.498	0.533	0.598	0.412	0.699	0.465	0.649	0.554	0.549	0.528	0.421
Rb	2	1	10	2	6	3	1	1	2	3	2	3	3	2
Sr	179	192	143	157	128	274	87	188	154	188	112	167	172	190
V	273	287	366	323	293	248	360	229	339	228	318	312	331	333
La	2.00	2.90	0.49	3.00	0.92	1.90	3.60	1.20	4.60	3.50	3.60	4.20	1.40	2.80
Ce	6.60	9.20	1.20	9.00	2.20	5.90	10.10	3.80	14.20	9.70	10.20	12.10	4.00	8.80
Nd	6.20	8.10	1.30	9.50	3.00	4.40	8.70	3.90	12.10	8.80	8.10	10.70	4.50	7.60
Sm	2.35	3.20	0.58	3.35	0.96	1.93	3.29	1.66	4.19	3.14	3.12	3.76	1.82	2.78
Eu	0.93	1.11	0.26	1.22	0.37	0.74	1.16	0.71	1.32	1.09	1.22	1.37	0.65	1.07
Tb	0.72	0.85	0.28	1.02	0.40	0.57	0.91	0.50	1.25	0.81	0.89	0.95	0.59	0.89
Yb	2.36	3.16	1.31	3.53	2.05	1.97	3.18	1.85	4.17	2.97	2.93	3.83	2.75	3.36
Lu	0.41	0.50	0.23	0.59	0.35	0.33	0.51	0.30	0.67	0.47	0.47	0.58	0.44	0.54
Th	0.09	0.30	0.12	0.18	0.22	0.23	0.33	0.13	0.33	0.22	0.31		0.24	0.18
Hf	1.70	2.20	0.40	2.40	0.80	1.40	2.30	1.06	3.30	2.30	2.20	2.60	1.30	2.10

Table 2. Part 2.

Group	III	III	II	I	I	I	II	I	II	II	I	I	I	I	I	I
Ref	M13	M14	M28	M29	M31	M4	M5	M6	M37	M34	M32	M35	M8	M7	M9	M2
SiO_2	50.28	49.30	50.66	50.86	54.17	53.95	59.77	54.53	49.90	52.10	51.25	54.25	49.50	52.98	51.05	52.80
Al_2O_3	14.40	14.96	15.72	14.90	14.14	14.00	14.36	14.78	15.00	16.45	15.18	15.28	14.84	14.61	15.36	14.14
Fe_2O_3	8.07	8.16	8.62	11.57	10.71	12.90	9.18	10.81	8.05	8.87	10.00	10.75	8.22	9.75	10.84	8.06
MnO	0.09	0.12	0.14	0.18	0.18	0.13	0.07	0.06	0.14	0.13	0.12	0.12	0.14	0.15	0.16	0.14
MgO	8.25	8.05	6.75	5.53	3.76	4.16	3.23	4.15	8.28	6.93	6.15	3.85	9.98	6.44	5.67	5.51
CaO	11.87	11.79	10.18	9.89	5.52	4.53	4.41	6.25	11.27	9.08	7.17	4.60	12.22	9.17	8.44	4.85
Na_2O	2.25	2.37	4.08	3:68	6.34	5.74	4.23	4.58	2.71	1.46	4.10	6.35	2.10	3.40	3.83	5.20
K_2O	0.08	0.10	0.07	0.24	0.10	0.19	0.14	0.37	0.08	0.05	0.38	0.48	0.07	0.06	0.07	0.10
TiO_2	0.52	0.61	0.87	1.66	1.72	1.67	0.73	1.53	0.86	0.72	1.21	1.41	0.78	0.97	1.31	1.34
L.O.I.	3.61	3.66	2.07	0.59	2.32	1.51	2.97	2.32	2.77	3.35	3.47	1.87	1.24	1.61	2.30	7.75
Σ	99.42	99.12	99.16	99.10	98.96	98.78	99.09	99.38	99.06	99.14	99.03	98.96	99.09	99.14	99.03	99.89
[Mg]	0.691	0.682	0.634	0.518	0.437	0.418	0.435	0.46	0.693	0.631	0.573	0.443	0.73	0.595	0.536	0.59
Rb	1	1	1	1	1	2	2	3	1	1	2	3	2	2	2	1
Sr	177	207	190	179	137	90	234	184	187	150	268	283	140	175	160	153
V	230	229	234	319	279	244	246	335	214	265	235	263	218	271	300	253
La	0.60	0.80	2.20	4.30	6.80	7.90	1.30	4.10	1.60	1.30	3.30	4.40	1.80	2.40	2.90	3.40
Ce	1.60	2.60	6.80	12.90	19.20	22.90	3.90	11.80	5.60	3.70	9.70	13.30	5.40	6.40	8.20	9.50
Nd	1.90	3.00	5.80	11.10	17.70	17.00	3.80	9.50	4.70	3.40	8.50	11.60	4.70	5.40	7.10	7.40
Sm	0.97	1.24	2.17	4.11	5.08	5.66	1.54	3.43	1.92	1.38	3.05	4.23	1.65	2.05	2.70	2.65
Eu	0.39	0.51	0.83	1.33	1.78	1.61	0.78	1.34	0.74	0.59	1.10	1.35	0.65	0.79	0.98	1.19
Tb	0.32	0.39	0.58	1.12	1.14	1.45	0.55	0.93	0.52	0.41	0.80	1.05	0.46	0.60	0.70	0.75
Yb	1.44	1.86	2.01	3.83	4.57	5.92	2.45	3.63	1.91	1.52	2.83	4.18	1.64	2.27	2.79	2.72
Lu	0.23	0.29	0.32	0.62	0.73	0.92	0.37	0.54	0.30	0.27	0.45	0.66	0.26	0.37	0.45	0.42
Th	0.06				0.46	0.32	0.13		0.08			0.29		0.17	0.27	0.36
Hf	0.60	0.80	1.50	2.90	3.80	5.20	1.40	2.60	1.30	1.00	2.10	3.20	1.10	1.40	1.70	2.30

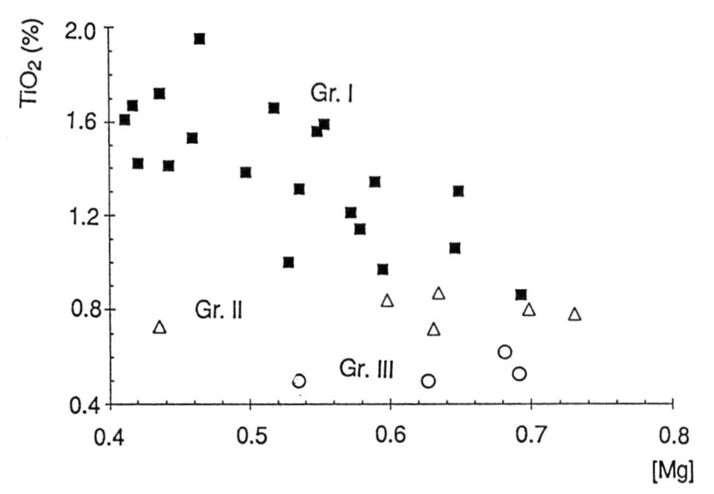

Figure 9. TiO2 versus [Mg] ratio in analysed diabase samples (square: group I; triangles: group II; circles : group III).

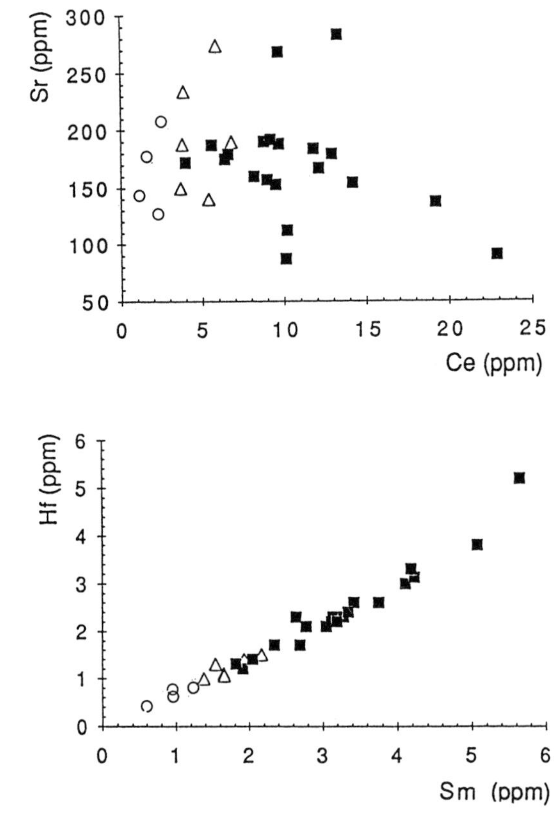

Figure 10. Sr versus Ce and Hf versus Sm in the analyzed diabase samples.

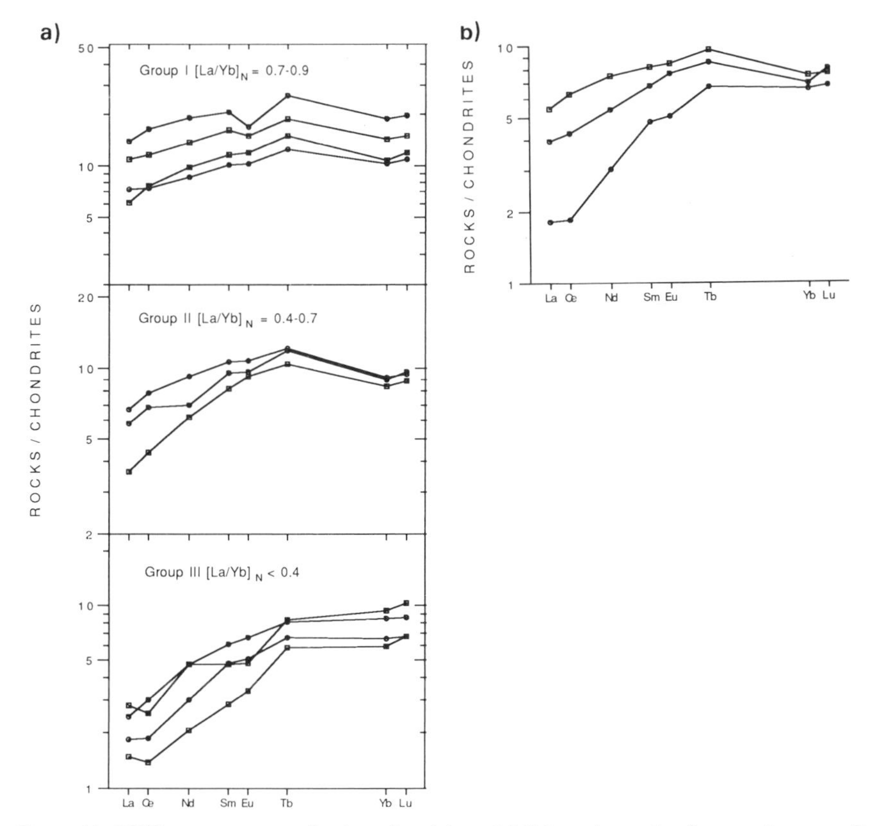

Figure 11. REE patterns normalized to chondrites; (a) Selected samples from each group; (b) Most primitive samples of each group.

in this group are dominantly controlled by clinopyroxene and plagioclase fractionation. This group has MORB characteristics, with REE content similar to South East Indian ridge (Michard et al., 1986).

In groups II and III (Figure 11a), the REE patterns are characterized by a more marked depletion of LREE; the [La/Yb]N ratios are also lower than in group I. These low REE values are not observed in MORBs. They are occasionally found in IAT, but do not represent typical pattern of IAT.

As Ti and LREE, the incompatible elements Th and Hf decrease from group I to group III. The correlation of these elements is shown on Figure 12. Hf/Th ratios cluster around 10 in the three groups. This value is different from that encountered in IAT and in boninites from Cyprus. It is slightly below the Hf/Th ratio characterizing N-type MORBs, but lies in the range of Hf/Th ratios of Indian Ocean MORB (Dosso et al., 1988).

Interpretation

The three groups chemically recognized in the sheeted dike complex compare with the overlying lava flows units (Beurrier et al., 1989): the tholeiitic dikes

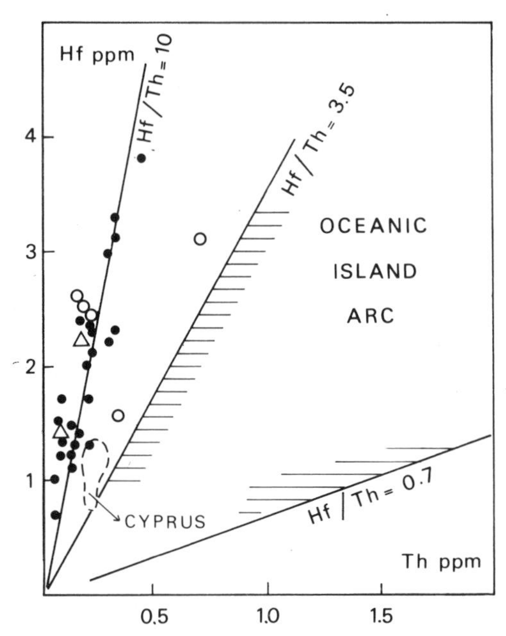

Figure 12. Hf versus Th for the analysed diabase samples (dots), compared with IAT and Cyprus domains (after Rautenschlein et al., 1985), N-type MORB (triangles after Viereck et al., 1989) and Indian Ocean MORB (open circles after Dosso et al., 1988).

of group I are similar to the Geotimes lava flows whereas the basaltic rocks of group II and III resemble respectively to Alley and Lasail units. This suggests a genetic link between the sheeted dike complex and the overlying flow units. According to the sequence of lava flow eruption, it may be also postulated that the dike of group I are emplaced first and were followed by the dike of group II and III.

The presence of magmas with MORB affinities associated with magmas having island arc affinities are not typical of oceanic ridge. According to Pearce (1981) such magmatic association suggests that, at least, part of the Oman ophiolite might have formed in a marginal or a back arc basin. In fact this interpretation, based on the comparison with basalts of well known geological setting is more or less in conflict with geological and structural interpretations (Boudier et al., 1988). This is true for the Oman ophiolite as well as for several other ophiolitic complexes (Nicolas, 1989).

In the present study, the island arc affinities of some basaltic dikes is inferred from the depletion of the most incompatible trace elements and from the relatively low content of Ti and its constancy during the differentiation. In the absence of a proven subduction, which process may generate basalt with MORB or island arc affinities ?

In order to appreciate this problem, the REE pattern of the most primitive sample in each group are reported on Figure 11b). This figure shows a progressive decrease of REE content from the dikes of group I to group III. The decrease is accompanied by a constancy of several incompatible trace elements ratios (HP/Th = 10 ± 2; Th/La = 0.08 ± 0.01; Ti/Tb = 9700 ± 100). Two possible explanations could be proposed for this geochemical evolution. The first one is a multi-stage melting process in which the residue left from an earlier melting event is available for further basaltic liquid extraction as the residue continues its diapiric rise. However, such alternative is contradicted by Nd isotopic values which tend to decrease from group I to group III (Dupré and Michard, pers. comm.).

The remaining explanation involves interaction between basaltic liquid and enclosing ultramafic rocks within the transition zone below the ridge, a process which have been clearly evidenced on the field (Nicolas et al., 1988a). Such an interaction may generate calc-alkali basalts from a tholeiitic magma (Kelemen et al., 1990). Preliminary investigations indicate that most of the geochemical features described in this study for basaltic dikes with island arc affinities may be explained by such a model (work in progress). But isotopic variations imply that tholeiitic magma and harzburgite have distinct Nd isotopic ratio.

Discussion

The interest of putting together geochemical and fabric data is that both are independently related to crustal and local scale conditions of magma emplacement. Three clearcut correlations appear when comparing the magnetic features of group I (typical MORB) and groups II-III (calk-alkaline affinities) dikes:

1) low susceptibility values appear essentially in groups II-III dikes (Table 1);
2) MORB dikes exhibit much less primary fabrics: normal fabrics are found in about 2/3 of groups II-III dikes and in only 1/3 of group I dikes;
3) when separating K3 directions of normal dikes according to geochemical affinity, it appears that the obliquity is highlighted in group I dikes and disappears in groups II-III (Figure 13).

Group II-III have consistently lower Fe_2O_3 amounts (<9.5% while all but two group I dikes yield values >9.5%) and their larger amounts of Al_2O_3 and CaO tend to decrease the free oxide amount, thus accounting for their smaller mean susceptibilities. This brings a new light on the correlation between susceptibility and fabric type. Are primary fabrics more abundant in low susceptibility dikes because they are more likely to belong to groups II-III, or the contrary? There is another possible explanation for relationship 2): if groups II-III dikes have been emplaced in a later event than the accretion related MORB dikes, then they should have intruded a colder

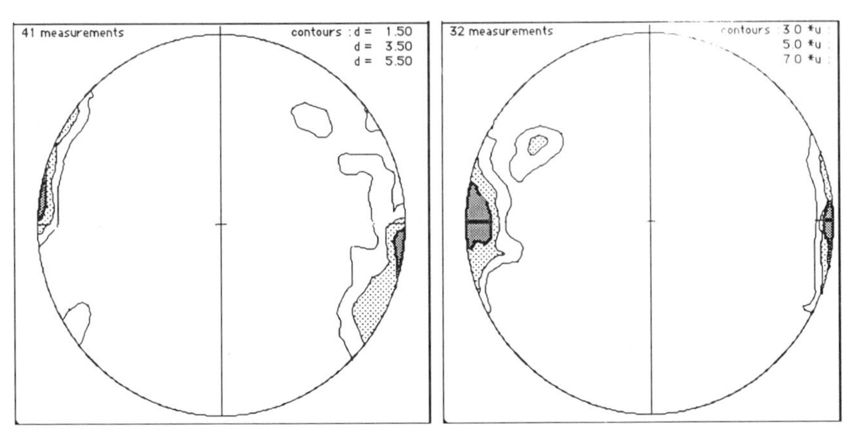

Figure 13. Density contours in dike coordinates of minimum axes for normal dike separated according to geochemical characters: (a) group I dikes; (b) group II-III dikes.

crust, thus suffering less hydrothermalism and transformations leading to secondary fabrics. On the other hand how does the absence of obliquity in groups II-III dikes can be integrated in this two stages model?

For that purpose the geodynamic framework has to be discussed. The presence of a strike-slip movement almost synchronous of the extension resulting in dike emplacement can be interpreted in terms of oblique spreading, i.e. a relative divergence of the oceanic plate not perpendicular to the ridge axis. Such a situation does not belong to orthodox plate tectonics as it is less stable than a ridge perpendicular to extension plus a transform and implies some kind of discontinuous kinematics. However it has been recently identified in the incipient rift of Asal (Tapponier et al., 1990) where it is interpreted in terms of overlapping rifts, and in the midatlantic ridge N of Iceland (Dauteuil et al., 1990), a place where a change of spreading direction has still not been accommodated by rotation and segmentation of the ridge between new transforms, probably due to a slow spreading rate of 18 mm/a. So it appears that an oblique spreading can be stable for more than 1 Ma and that the Oman ophiolite could have been accreted in such a ridge. In fact oblique spreading has already been suggested in Oman on the basis of the late plutonic structures (Reuber, 1988) and of the discovery of major ridge parallel shear zones of high temperature in the mantle (Boudier et al., 1985, 1989).

The Oman ophiolite is believed to have been accreted just a few Ma before the beginning of obduction (Coleman, 1981; Montigny et al., 1988). It is tempting to relate the oblique spreading to the beginning of kinematic changes leading to obduction, or to a special setting of the Omani ridge needed to produce that quite unusual process leading to ophiolitic nappes.

The early and large clockwise rotation of the ophiolite tentatively suggested by paleomagnetic data (Thomas et al., 1988) is in good agreement with dextral shear on the ridge, as well as the ridge parallel early directions of transport of the nappe recorded in the mantle (Boudier et al., 1985). Moreover detailed analysis of accretion and detachment related flow directions in the mantle of Fizh and Hilti massifs (Figure 8 of Boudier et al. 1988) is consistent with a clockwise rotation of the extension direction. Although these geometric reconstructions need more detailed investigations, the present study suggest a clockwise progressive reorientation of the flow in the sheeted dikes, coupled with a continuous change of their chemistry. It emphasizes the point that group II-III dikes, previously quoted as having island arc affinity, have in fact been emplaced during the accretion history. Their geochemistry suggest a basaltic magma drawn from an already depleted mantle and that would have interacted with mantellic rocks such as the harzburgite. A genetic relationship with werhlitic intrusions is also suggested. The source of this differentiated magma is probably located at a larger distance from ridge axis than MORB type intrusion which would be concentrated right at the top of the magmatic chamber. This place is also the weakest part of the crust, with practically a null shear strength: it will concentrate any shear stress applied to the nascent oceanic crust. Therefore it can be imagined that, while the same shear stress is acting a group I dike is more likely to experience a strike slip movement than a Group II-III dike emplaced a few km away from the axis in an already rigidified crust.

Conclusions

This first study of the fabric of ophiolitic dikes using the magnetic anisotropy technique, coupled with geochemical identification of magma sources, shows quite promising results and provides a guide for further investigations:

- primary flow fabric are often reset by secondary processes more likely to occur in thick (>1.5 m) dikes with MORB affinities and high susceptibility.
- primary fabrics provide magmatic flow direction, and possibly sense, in the absence of any other microstructural markers;
- in the Oman ophiolite dike flow lines are quite dispersed with an overall tendency toward vertical flow;
- dikes of MORB type composition show a distinct obliquity of their flow plane on dike margins, typical of a late emplacement horizontal shear. This shear is related to an oblique spreading during the accretion of the Oman oceanic crust.

Commonly encountered differentiated magmas previously related to island arc setting have geochemical features (mainly REE patterns and trace elements trends) more in agreement with a pure accretionary origin with either multistage melting or interaction of the magma with enclosing ultramafic rocks. Intrusion of this magma may be simultaneous to the MORB magma intrusion but taking place at a larger distance from ridge axis, thus corresponding, on the average, to a slightly later emplacement in a given crustal section.

Future investigations should more firmly establish shear sense (here tentatively assigned to be dextral), collect more flow line data to be able to correlate them with the ophiolite segmentation pattern, as defined by mantle diapiric intrusions. The association of oblique spreading with obduction could be a coincidence or a general feature that could be tested in other ophiolites.

Acknowledgements

We are pleased to acknowledge A. Nicolas for discussions and final review, B. Dupré and A. Michard-Vitrac for providing unpublished isotopic data, and F. Hrouda for constructive review. The project has been supported by INSU (DBT program Géodynamique, contribution # 197).

References

Alabaster, T., Pearce, J.A. and Malpas, J., 1982. The volcanic stratigraphy and petrogenesis of the Oman ophiolite Complex. Contrib. Mineral. Petrol., 82: 168–183.

Beurrier, M., Ohnenstetter, M., Cabanis, B., Lescuyer, J.L., Tegyey, M. and Le Métour, J., 1989. Géochimie des filons doléritiques et des roches volcaniques ophiolitiques de la nappe de Semail: contraintes sur leur origine géotectonique au Crétacé Supérieur. Bull. Soc. Géol. Fr., 8: 205–219.

Boudier, F., Bouchez, J.L., Nicolas, A., Cannat, M., Ceuleneer, G., Misseri M. and Montigny R., 1985. Kinematics of oceanic thrusting in the Oman ophiolite: model of plate convergence. Earth Planet. Sci. Lett., 75: 215–222.

Boudier, F., Ceuleneer, G. and Nicolas, A., 1988. Shear zones, thrusts and related magmatism in the Oman ophiolite: initiation of thrusting on an oceanic ridge. Tectonophysics, 151: 275–296.

Boudier, F., Nicolas, A. and Ceuleneer, G., 1989. De l'accretion océanique à la convergence. Le cas de l'ophiolite d'Oman. Bull. Soc. Géol. France, 8: 221–230.

Coleman, R.G., 1981. Tectonic setting for ophiolite obduction in Oman. J. Geophys. Res., 86: 2497–2508.

Dauteuil, O., Brun, J.P., Avedik, F. and Geli, L., 1990. Structures of the oblique rifting in the Mohns ridge (Norwegian Sea), C.R. Acad. Sci. Paris, 311: 357–363.

Dawson, E.M. and Hargraves, R.B., 1985. Anisotropy of magnetic susceptibility as an indicator of magma flow directions in diabase dikes. EOS Trans. AGU, 66: 251.

Dosso, L., Bougault, H., Beuzart, P., Calvez, J.Y. and Joron, J.L., 1988. The geochemical structure of the South East Indian Ridge. Earth Planet. Sci. Lett., 88: 47–59.

Ellwood, B.B., 1978. Flow and emplacement directions determined for selected magmatic bodies using anisotropy of magnetic susceptibility measurements. Earth Planet. Sci. Lett., 41: 254–264.

Hrouda, F., 1982. Magnetic anisotropy of rocks and its application in geology and geophysics. Geophys. Surv., 5: 37–82.

Hrouda, F., 1985. The magnetic fabric of the Brno massif. Sbor. geol. ved, 19: 89–112.

Jeffrey, G.H., 1922. The motion of ellipsoidal particles emersed in a viscous fluid. Proc. R. Soc. London ser. A, 102: 161–179.

Jelinek, V., 1978. Statistical processing of anisotropy of magnetic susceptibility measured on group of specimens. Studia Geophys. Geodet., 22: 50–62.

Kelemen, P.B., Joyce, D.B., Webster, J.D. and Holloway, J.R., 1990. Reaction between ultramafic rock and fractionating basaltic magma – II – Experimental investigation of reaction between olivine tholeiite and harzburgite at 1150°-1050°C and 5 kb. J. Petrol., 31: 99–134.

Khan, M.A., 1962. The anisotropy of magnetic susceptibility of some igneous and metamorphic rocks, J. Geophys. Res., 67: 2873–2885.

Kidd, R.G.W. and Cann, J.R., 1974. Chilling statistics indicate on ocean floor spreading origin for the Troodos complex, Cyprus. Earth Planet. Sci. Lett., 24: 151–155.

Knight, M.D. and Walker, G.P., 1988. Magma flow directions in dikes of the Koolau complex, Oahu, determined from magnetic fabric studies. J. Geophys. Res., 93: 4301–4319.

Luyendick, B.P., Laws, B.R., Day, R. and Collinson, T.B., 1982. Paleomagnetism of the Samail ophiolite, Oman 1. The sheeted dike complex at Ibra. J. Geophys. Res., 87: 10883–10902.

Mc Culloch, M.T., Gregory, R.T., Wasserburg, G.J. and Taylor, H.P., 1981. Sm-Nd, Rb-Sr and O^{18}/O^{16} isotopic systematics in an oceanic crustal section: evidence from the Samail ophiolite. J. Geophys. Res., 86: 2721–2735.

Michard, A., Montigny, R. and Schlich, R., 1986. Geochemistry of the mantle beneath the Rodriguez Triple junction and the South-East Indian Ridge. Earth Planet. Sci. Lett., 78: 104–114.

Montigny, R., Le Mer, O., Thuizat, R. and Withechurch, H., 1988. K-Ar and $^{40}Ar/^{39}Ar$ study of metamorphic rocks asociated with the Oman ophiolite: tectonic implications. Tectonophysics, 151: 345–262.

Nehlig, P., 1989. Etude d'un système hydrothermal fossile: l'ophiolite de Semail (Oman). Thèse Doc. Univ. Brest, 308 p.

Nicolas, A., 1989. Structures of ophiolites and dynamics of oceanic lithosphere. Kluwer ed., London, 367 pp.

Nicolas, A., Ceuleneer, G., Boudier, F. and Misseri, M., 1988a. Structural mapping in the Oman ophiolites: mantle diapirism along an oceanic ridge. Tectonophysics, 151: 27–56.

Nicolas, A., Reuber, I. and Benn, K., 1988. A new magma chamber model based on structural studies in the Oman ophiolite. Tectonophysics, 151: 87–105.

Pallister, J.S., 1981. Structure of the sheeted dike complex of the Samail ophiolite near Ibra, Oman. J. Geophys. Res., 86: 2661–2672.

Park, J.K., Tanczyk, E. and Desbarats, A., 1988. Magnetic fabric and its significance in the 1400 Ma Mealy diabase dikes of Labrador, Canada, J. Geophys. Res. 93: 4301–4319.

Pearce, J.A., Alabaster, T., Shelton, A.W. and Searle, M.P., 1981. The Oman ophiolite as a Cretaceous arc-basin complex: evidences and implications. Phil. Trans. R. Soc. London, 300: 299–317.

Pecher, A., 1989. SchmidtMac, a program to display and analyse directionnal data. Computer & Geosciences, 15: 1315–1326.

Potter, D.K. and Stephenson, A., 1988. Single-domain particles in rocks and magnetic fabric analysis, Geophys. Res. Lett. 15: 1097–1100.

Rautenschlein, M., Jenner, G.A., Hertogen, J., Hofmann, A.W., Kerrich, R., Schmincke, H.U. and White, W.M., 1985. Isotopic and trace element composition of volcanic glasses from the Akaki Canyon, Cyprus, implication for the origin of the Troodos ophiolite. Earth Planet. Sci. Lett., 75: 369–383.

Reuber, I., 1988. Complexity of the crustal sequence in northern Oman ophiolite (Fizh and southern Aswad block): the effect of early slicing ? Tectonophysics. 151: 137–165.

Rochette, P., 1988. Inverse magnetic fabric in carbonate bearing rocks, Earth planet. Sci. Lett., 90: 229–237.

Sempéré, J.C., Purdy, G.M. and Schouten, S., 1990. Segmentation of the Mid-Atlantic Ridge between 24°N and 30°40'N. Nature, 344: 427–431.

Shelton, A.W., 1984. Geophysical studies on the northern Oman ophiolite. Ph. D. Thesis, Open Univ. Milton Keynes, 353 pp.

Shelley, D., 1985. Determining paleo-flow directions from groundmass fabrics in the Lyttelton radial dikes. J. Volcanol Geotherm. Res., 25: 69–79.

Tapponnier, P., Armijo, R., Manighettti, I. and Courtillot, V., 1990. Bookshelf faulting and horizontal block rotations between overlapping rifts in southern Afar. Geophys. Res. Lett., 17: 1–4.

Thomas, V., Pozzi, J.P. and Nicolas, A., 1988. Paleomagnetic results from Oman ophiolites related to their emplacement. Tectonophysics, 151: 297–322.

Viereck, L.G., Flower, M.F.J., Hertogen, J., Schmincke, H.U. and Jenner, G.A., 1989. The genesis and significance of N-MORB sub-types. Contrib. Mineral. Petrol., 102: 112–126.

Geometry and Flow Pattern of the Plutonic Sequence of the Salahi Massif (Northern Oman Ophiolite) – A Key to Decipher Successive Magmatic Events

I. REUBER
Laboratoire de Tectonophysique, U.S.T.L., Montpellier, France;
Current address: Institut Dolomieu, 15, rue M. Gignoux, 38031 Grenoble, France

Abstract

The model of a large steady state magma chamber for the crystallization of ophiolitic cumulates encounters a number of geophysical problems, and has not been evidenced in present day oceans. Detailed mapping in several areas of the Oman ophiolite has shown a complex structure for the plutonic sequence that is more likely the product of superposed magmatic events, than of fractional crystallization in one huge magma chamber.

In the Salahi massif (northern Oman), though known for its regular, flat lying cumulate sequence, it can be shown that:

- the tectonites-cumulates transition (MOHO) is an irregular, domed surface;
- distinct layers are commonly sills;
- layers marked by grain size variations may be due to recrystallisation;
- magmatic lineations may be dispersed on the same outcrop from one layer to the next;
- steepening of layers is a very early event;
- gabbros with a dominantly linear fabric in near to vertical position appear to trace feeder zones, that are in relation with wehrlites rooted in dunite domes.

Consequently successive cycles of magma generation and intrusion can be traced in the studied area.

These structures can be related to the progressive inflation of the transition zone, that consequently becomes unstable and intrudes upwards into more or less solidified pre-existing plutonics. Large amounts of magma appear to be injected as sills, that give the sequence its distinctly layered aspect.

Introduction

The traditional concept of a large (2 × 10 km wide) steady state magmatic chamber, as deduced from petrographic evidence (Greenbaum 1972, Parrot

Tj. Peters et al. (Eds), Ophiolite Genesis and Evolution of the Oceanic Lithosphere, 83–103.

and Ricou 1976, Pallister and Hopson 1981) for the crystallisation of ophiolitic cumulates has been questioned since a long time (Lister 1983, Nicolas et al. 1988b). Seismic data from actual oceanic ridges reveal the existence of magma chambers only exceptionally and for fast spreading ridges (Hale et al. 1982, McClain et al. 1985, Detrick et al. 1987) and constrain its width to a maximum of 2 to 6 km. For slower spreading rates, thermal modeling predicts the existence of intermittent magma chambers only, if any at all (Kusznir 1980, Lewis 1983, Lister 1977, 1983).

In the Oman ophiolite, some areas of the plutonic sequence now are mapped in detail and their complexity has been shown especially in the northern and central massifs as Fizh (Smewing 1980, Reuber 1988) and Haylayn (Juteau et al. 1988a, Beurrier 1987, Reuber et al. in press). These areas are interpreted in terms of special ridge configurations, while the near to horizontally exposed magmatic sequences of Wadi Tayin and Semail massifs in the southern part are described as representatives of large magmatic chambers(Pallister and Hopson 1981, Nicolas et al. 1990).

Among the northern massifs, the Salahi block appeared to have similar characteristics, showing a regular mantle flow pattern, regular sheeted dikes, and predominantly near to horizontal cumulates. Detailed re-mapping of the petrographic facies and of the magmatic flow structures, however, revealed a much more complex geometry of the magmatic sequence. Already Ernewein (1987, Ernewein et al. 1986, 1988) noticed the abundance of wehrlites cross-cutting the layered gabbros and related them to early obduction. This study shows that not only wehrlites, but equally gabbros and troctolites intrude pre-existing magmatics and the feeder zones of some of these have been traced in the field.

General Description of the Salahi-Hilti Massif

The Salahi massif stretches for more than 25 km south of Wadi Jizzi and west of the town of Sohar (Fig. 1). It is characterized by an exceptionally thick and complete volcanic sequence in its northern part (Alabaster et al. 1982, Pflumio 1988, Pflumio et al. submitted), that hosts the copper mineralisations of Lasail mine. The sheeted dikes show a relative uniform orientation, varying however between N 160 E and N 30 E with a dip between W 60 and E 70. No cross-cutting relationships have been observed yet. The orientations seem to vary gradually or across fault zones. NW-SE trending faults are of importance throughout the massif affecting the upper crustal sequence and also lower parts of the section in the north. An early oceanic initiation of the lineaments is probable, as the major fault zones are accompanied by large wehrlite bodies and plagiogranitic plugs. Further south, the plutonics are regularly exposed over a width of 8 km in average between the mantle peridotites in the West and the sheeted dike complex in the East (Fig. 2). The mean dip of the whole nappe, in accordance with other massifs

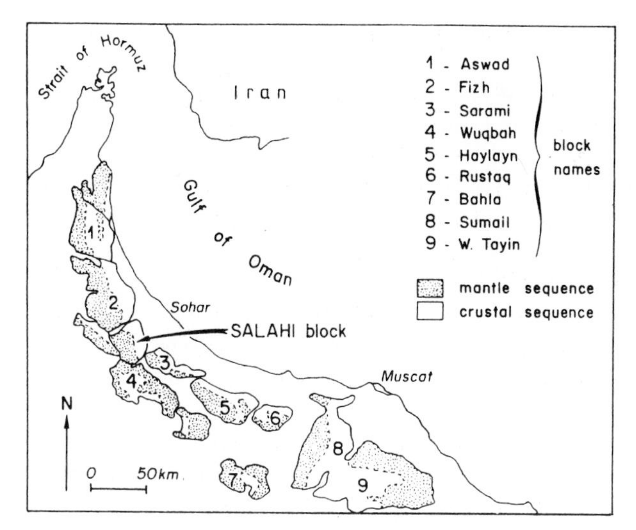

Figure 1. Location of the Salahi massif in the North-Oman ophiolite.

in the north is about 20° east, which results in an average thickness of the plutonic sequence of about 3 km. Layering commonly is flat, dipping generally less than 30° east or west, but layerings at high angle exist as well. This plutonic sequence is the main topic of the paper.

The Mantle Section

Plastic flow in the mantle resulted in a regular foliation dipping gently eastward associated with a generally east-west striking lineation. General eastward shear of the upper part is interrupted in a zone of shear inversion just below the Moho, as equally observed elsewhere in the Oman ophiolite (Ceuleneer 1986). In Wadi Hilti, only the uppermost part of the sequence is rich in gabbroic and pyroxenitic dikes, the harzburgites of Wadi Sudum are remarkably poor in dikes, but in the southern extremity of the massif the abundance of magmatic dikes increases again (Ceuleneer 1986). The major transition from mantle harzburgites to gabbros occurs along a gently eastward dipping surface in the western part of the mapped area (Fig. 2). The transition between harzburgites and gabbros is rapid in Wadi Hilti and further south in Wadi Sudum, whereas towards the north, in Wadi Salahi, a huge pile (more than 800 m thickness) of transitional dunites, with variable amount of interstitial clinopyroxene-plagioclase impregnation, and gabbro dikes and lenses is developed. East of the main peridotite-gabbro transition, recurrences of peridotites (mostly dunites, but occasionally also of harzburgites) are frequent throughout the sequence (Fig. 2, 8), as described below.

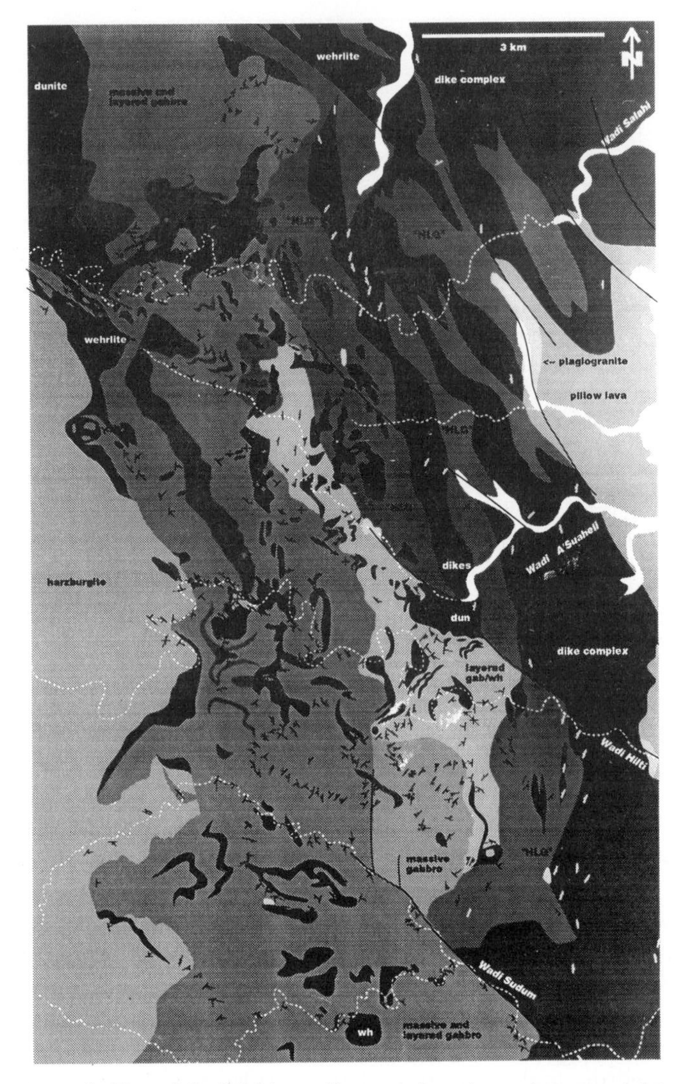

Figure 2. Map of the Salahi massif – crustal section and structural data.

The Wehrlites

Wehrlite bodies occur at any size and shape. Huge bodies are generally rooted in, and completely transitional with impregnated and 'dry' dunites (Fig. 2), as also common elsewhere (Reuber 1988, Juteau et al. 1988). They are abundant in the north, not only in relation with the fault zones. They are not always intrusive into layered gabbro, but multiple structural relations exist as equally observed in other places of the Oman ophiolite. All intermediate relations between wehrlites intrusive into layered gabbro, containing angular enclaves of layered gabbro (Fig. 3A) to wehrlitic inclusions in gabbro (Fig. 3H) can be encountered. Lobed contacts and reactive limits point to the coexistence of two highly viscous liquids (Fig. 3B, C). Different relations can be observed at a very small scale (Fig. 3E): gabbro sills intruding wehrlites just 5 m above lobed contacts between gabbro and wehrlite (Fig. 3F), and images of mixing or de-mixing between the two magmas that resemble graded bedding (Fig. 3G). Important viscous deformation stretches out any of these images to parallelise them into progressively regular bands (Fig. 3D).

The Gabbros

At a first view, the major part of the area can be mapped as gabbroic cumulates (Gass et al. 1980). At a second view, the abundance of intrusive wehrlites, and the occurrence of occasionally intrusive gabbros becomes evident (Ernewein 1987, Ernewein et al 1987, 1988). Only detailed observations give an idea of the multitude of successive magmatic events and of the complexity and variability of relations between gabbros, wehrlites and intermediate facies. Many of these structures occur at a scale that is impossible to represent on the map, thus the most important ones are described below.

Petrographically the gabbros range from troctolite to ferrogabbro, with normal cpx-gabbro being the far most abundant facies. Olivine, clinopyroxene and plagioclase are cumulus minerals. A small percentage of late interstitial orthopyroxene is common, and occasionally opx constitutes large oikocrysts. The grain size is generally of a few millimeters (1×3 mm for typical cpx or plag cumulus crystals), it may be larger in troctolite (5 mm approx.) and even more in the recrystallized pegmatitic parts occurring throughout the sequence (see below). The gabbros become near to microcrystalline in the root zone of the sheeted dikes, where they are occasionally rich in Fe-Ti oxides.

The nature and aspect of layering is quite variable. Graded beds, interpreted as images of gravity settling (Nicolas 1989) are rare, and restricted to the vicinity of the major peridotite gabbro transition, i.e. the basal tens of meters in the cumulates. These and all other structures are commonly transposed by strong magmatic flow. Other types of layering are more common:

1) *Layers with sharp limits, and often contrasting intensity of preferred mineral orientation*: Alternating layers of gabbro and wehrlite (Fig. 4A), or of gabbro in mela-gabbro-pyroxenite are well visible at long distance (Fig. 4B, 3E). In both cases, these layers can be identified as sills of one facies intruding the other (cf. Bedard et al. 1988). Boudinage of the more resistant layers (either wehrlite or gabbro, Nicolas et al. 1988b) and rare evidence of interconnecting dikes support this idea (Fig. 4C). Occasionally, important variations in the orientation of lineations from one layer

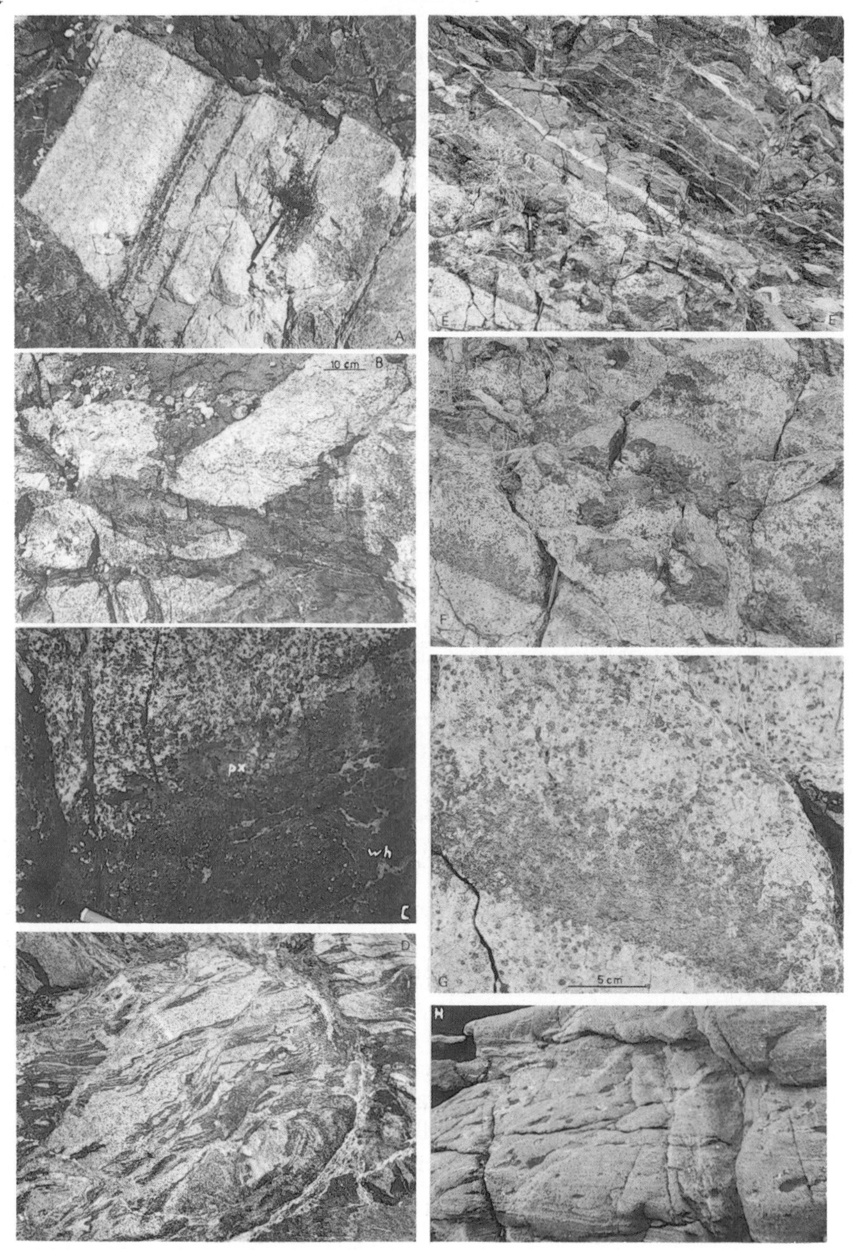

Figure 3. Relations between gabbro and wehrlites: (a) Angular block of layered gabbro included in wehrlite, Wadi Salahi; (b) Lobed contacts between gabbro and wehrlite indicating coexistance of the two magmas, Wadi Salahi; (c) Recrystallisation at the limit between gabbro and wehrlite, Wadi Salahi; (d) Strongly deformed magmatic melange of wehrlite and gabbro, Wadi Khabiyat in southern Fizh block; (e) Gabbro sills in wehrlite, overlying zone of magmatic melange between wehrlite and gabbro; north of Wadi Hatta; (f) Detail of the lower part of (e), showing lobed contacts between wehrlite and gabbro overlying zone of mixing that resembles a graded layering, shown in detail in (g). (h) Rounded enclaves of wehrlite in well deformed gabbro, Wadi Hilti.

Figure 4. Different types of layering: (a) Distinct layers of alternating gabbro and wehrlite – Wadi Sudum, cf. stereogram fig. 7b; (b) Distinct layers (sills) of gabbro in altered gabbro-pyroxenite – Wadi Hilti-S, interconnected by small dikes and cross-cut by dikes of pegmatitic hornblende-gabbro; (c) Interconnectd sills of leucogabbro in darker gabbro – Wadi Hilti-S; (d) Layers defined by grain size variations in the massive layered gabbro (Wadi Hilti) and in the fine grained gabbro (e) – Wadi A'Suaheli; (f) plagioclase-enriched schlieren in troctolite defining a "proto"-layering, Wadi Hilti/Wadi Sudum massive gabbro.

to the next in such assemblages (Fig. 7B) are easily explained if these represent sills.

2) *Layers consisting of grain size variations with gradual limits* (*Fig.* 4*D*,*E*) are, on the contrary, often poorly visible, and occur in coarse grained as well as in microcrystalline facies. Such layering is likely to result from

fluid (magma of hydrothermal) percolation of the still permeable rock during crystallization (cf. Robins et al. 1987).
3) Occasionally *plagioclase enriched schlieren* constitute an irregular layering, which can be explained as segregation of interstitial liquid (Fig. 4F) during compaction (Ceuleneer et al. in prep.)

Magmatic flow in the gabbros is indicated by the predominant orientation of acicular primary crystals as pyroxene and plagioclase (Nicolas et al. 1988b, Benn & Allard 1989). The first investigations along upper Wadi Hilti and lower Wadi A'Suaheli (Nicolas et al. 1988b), showed the predominance of east-westerly orientated lineations on mainly flat lying lamination planes. Their parallelism with corresponding mantle structures was interpreted as a coupling between mantle flow and overlying crust. Changes in the dip of layering/lamination and consequently also of the lineation were interpreted as late.

The more detailed data now available show in many places a more complex picture besides the common east-west orientations:

- Lineations may be extremely dispersed even on the same outcrop on successive parallel or progressively steepened layers (Fig. 7B, A).
- As high angle layers are intruded by low angle ones their steepening must be an early event (Fig. 5D, 6B).
- Some high angle structures, especially the line dominated gabbros are most likely in their original position, thus indicating the existence of vertical magmatic flow, well independent from the large scale plastic flow of the mantle.

Description of Particular Zones of the Crustal Sequence

Chosen areas of complex magmatic structures are described from north to south, as these may represent individual magmatic feeder zones:

A – Massive Gabbros of North-Salahi

The mountains north of Wadi Salahi are essentially composed of massive gabbros (Fig. 2). A weak layering is relatively steep in the central part of the area, and the magmatic lineation is dip parallel. In some places the linear fabric is stronger than the planar one. Towards the north and the east (possibly as well towards the west), the layering flattens. The mineral orientation is strong in the near to vertical orientations, but poor, absent or dispersed in the flat layers. This might point to an origin by magmatic flow for the first structures, by compaction for the latter ones. No rotation has been observed, but intermediate orientations that might result from magmatic flow lamination partly reoriented by flattening during compaction.

B – Dunite 'Upwelling' and Wehrlite-Gabbro Chaos of Salahi

The thick transitional dunites of Wadi Salahi contain numerous plagioclase-pyroxenite impregnations and occasionally gabbro lenses. A huge body of dunites occurs about 1.5 km east of their main transition into gabbros (Fig. 2, 8). This dunite-wehrlite contains angular enclaves of layered gabbro (Fig. 3A), whereas further east, at the edge of the massive gabbro, lobed contacts and pyroxene-rich reaction zones indicate coexistence of wehrlitic and gabbroic magma (Fig. 3B, C). The transition between these dunites and the massive gabbros upwards and towards the north is completely chaotic (Fig. 5A), witnessing multiple and alternating intrusions of gabbroic, intermediate and ultramafic magma, (Fig. 5B).

C – Couloir of Sub-Vertical Layered Gabbros Between Wadi Salahi and Wadi A'Suaheli

A zone of north-south striking,vertical, well layered gabbros can be followed in the upper Wadi A'Suaheli. The layering is of the distinct gabbro-wehrlite type that suggests a dike or sill assemblage. The lineation is systematically vertical or nearly so (>60°). These vertical layers are topped (cut magmatically) by horizontally layered gabbro with dynamic structures (Fig. 5D, 6B) in the northern and southern end of the couloir. Oblique crosscutting gabbros (Fig. 5C, 6A),vertical massive gabbros parallel to the main couloir (Fig. 6C), brecciating other layered gabbro, accompany the couloir in the West and the East.

D – Dunite 'Upwelling' of Lower Wadi A'Suaheli and Associated Structures

A huge body (about 1 km across) of dunite occurs south of lower Wadi A'Suaheli, in faulted contact with the dike complex. A near to vertical foliation and a dip-parallel lineation are occasionally visible in these dunites. They pass upwards gradually into wehrlites, which are rich in gabbro enclaves, lenses and dikes. The doming of these dunites and following intrusion of wehrlite bodies may be responsible for the steepening of the distinctly layered gabbros observed towards the west in Wadi A'Suaheli (Fig. 7, 7A). Steep massive gabbros just south of the dunites, are observed to bend into flatter ones towards the east and the west, in a divergent way that is impossible to obtain by late folding of a pre-existing layered sequence (Fig. 9).

E – Dunite-Harzburgite Domes of Upper Wadi Hilti

Several smaller peridotite occurrences are observed along the bottom of Wadi Hilti and along a southern confluent (Fig. 2, 7). The existence of foliated harzburgites in the central part of these dunites indicates that they are domes of mantle material. The dunites grade into wehrlites, which either are intrusive into layered gabbro (predominantly towards the east), or grade into concordant layered troctolite and gabbro by progressive appearance of troctolitic layers upsection, observed southward on top of the ridge.

Figure 5. Complex relations between the different facies: (a) chaos of gabbro in wehrlite – Wadi Salahi, dunite dome; (b) Wehrlite (2) intruding and enclaving layered gabbro and troctolite (1,1′), cut by layered gabbro (3), same area as (a); (c) Flat lying gabbro sill cutting oblique 'in situ' gabbroic layers within wehrlite – Wadi Hilti/Wadi Sudum massive facies. (d) Flat gabbro layers with slump cutting gabbro layers at high angles – Wadi Salahi-S.

Figure 6. Complex relations between the different facies, drawing after outcrop photos: (a) cross-cutting oblique layers of gabbros, with patches of recrystallisation along some contacts, upper Wadi A'Suaheli; (b) flat gabbro layers cutting gabbro layers at high angle – Wadi A'Suaheli; (c) magmatic breccia or gabbro feeder couloir with enclaves of older gabbro and some recrystallization patches – Wadi A'Suaheli.

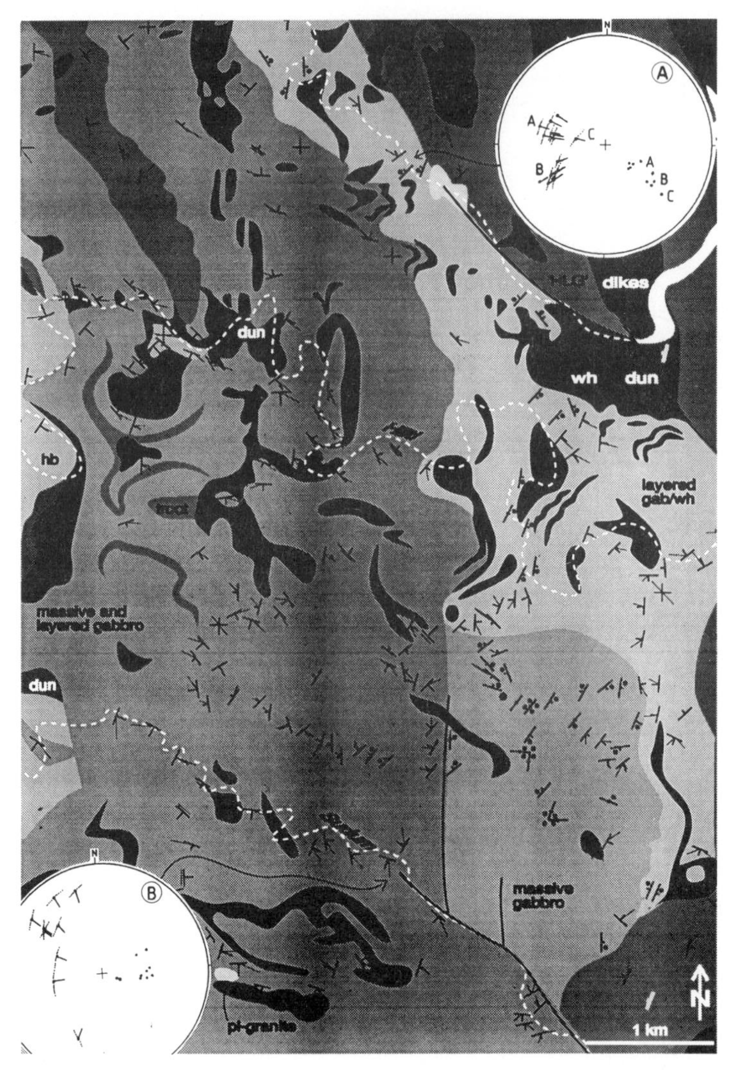

Figure 7. Detailed structural map of the area between Wadi Sudum and Wadi Hilti, and stereograms of distinctly layered gabbro-wehrlites showing the dispersion of the lineation between superposed layers (A, B).

F – Linear Vertical Gabbros between Wadi Hilti and Wadi Sudum

Another occurrence of massive gabbros very similar to the ones north of Wadi Salahi exists in the southern part of the massif.The central part of the occurrence is characterized by a two pyroxene gabbro, in which a strong vertical line is the dominant structural feature (Fig. 7, 10C). It grades into wehrlite with near to horizontal troctolite layers with gradual limits and poor mineral orientation (Fig. 4F, Fig. 10A, B). The vertical flow is not seen to bend into a horizontal position as further north (Fig. 9), but the horizontal

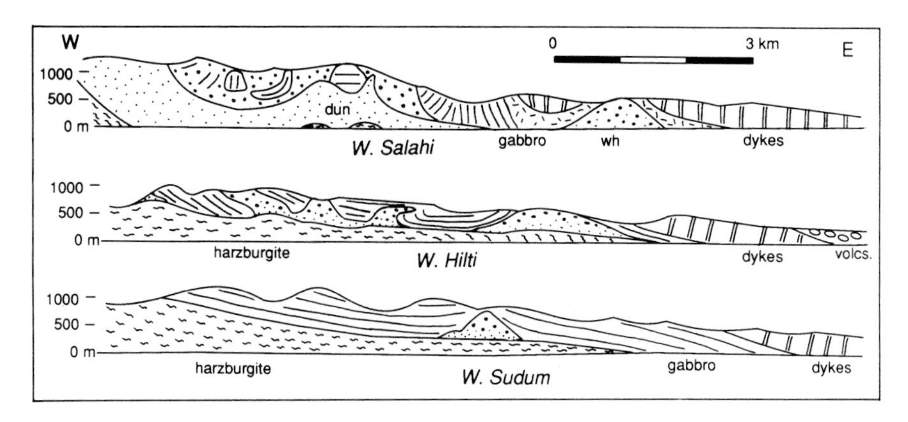

Figure 8. Serial cross-sections of the Hilti massif.

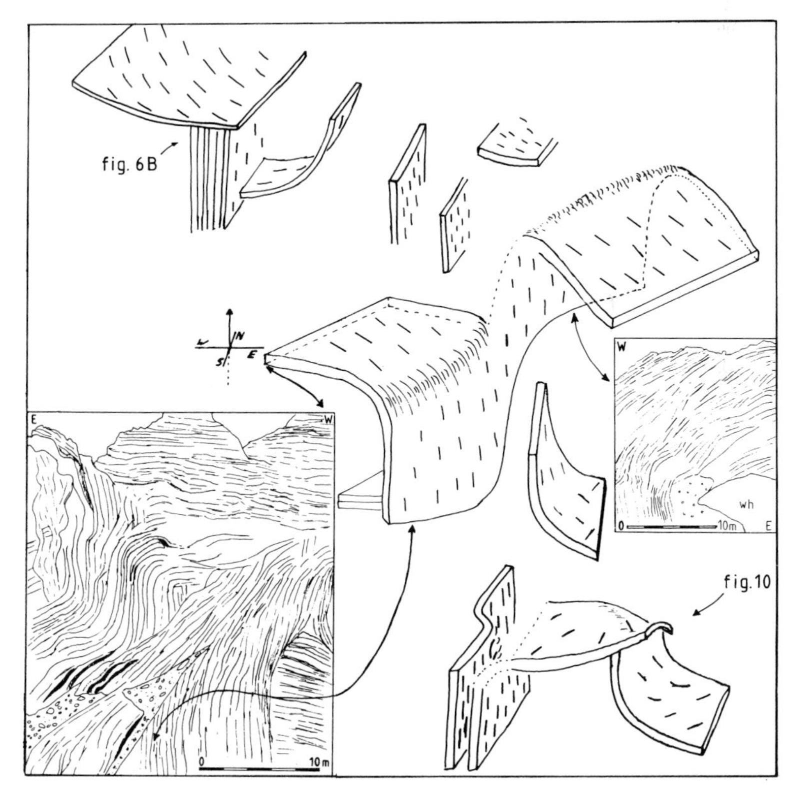

Figure 9. Three dimensional sketch of the bends and mutual relations between gabbro layers around Wadi Hilti – Wadi A'Suaheli (cf. carte fig. 7), with drawings after photo of bends.

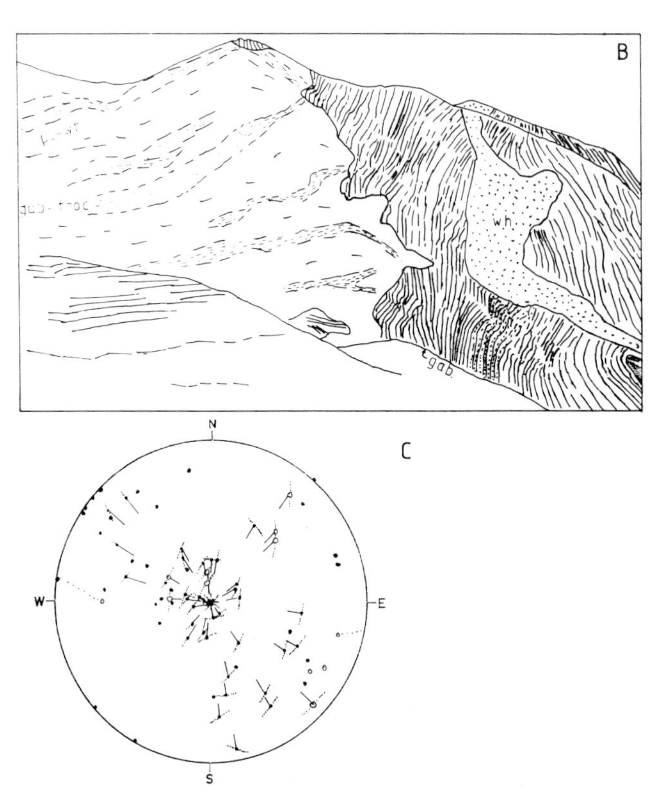

Figure 10. Eastern intrusive contact of the wehrlite – troctolite – massive gabbro complex with verticalized layered gabbro between Wadi Sudum and Wadi Hilti, (a) photo, (b) corresponding drawing, (c) stereogram showing structures of the massive facies (black dots) and adjacent layered facies (open circles).

layers appear to be of completely different origin, i.e. compaction. Eastward adjacent layers are steepened, and a large scale lobed contact witnesses its intrusive nature (Fig. 10A, B). The western contact is difficult to trace accurately due to numerous interfingering sill type layers.

Summary of the Observations

These observations do not fit into the idea of a regularly fed steady state magma chamber developing a near to horizontal cumulate layering and overlying the mantle peridotite with a flat Moho contact, on the contrary:

- the Moho contact is an irregular domed surface, with domes corresponding in general to thicker transition zones;
- gabbro layers are often sills in ultramafics, and wehrlite layers are sills in gabbros, equally gabbro layers in gabbro can be sills which is only less obvious to see;
- magmatic lineations may be dispersed on the same outcrop from one layer to the next supporting the idea of sill assemblages for the layered cumulates;
- steep orientations of layering are of early origin, and occasionally trace vertical magmatic flow;
- coarse grained and vary-textured isotropic gabbros do not only occur at the base of the dike complex, but throughout the sequence, where they are product of recrystallisation due to magma or hydrothermal fluid percolation;
- consecutive intrusions of magmatic bodies can be mapped under favorite conditions, their size is not more than few km across and about one km in height.

Comparison with other Areas of the Oman Ophiolite

Similar observations have been made throughout the Oman ophiolite. Especially the variation of the importance of transition zones shows a more or less regular alternation between huge transition zones and rapid peridotite gabbro contacts at a scale of approximately 10–20 km on average throughout the nappe (Reuber 1990). These variations are overprinted by smaller ones at a pluri-kilometric scale as observed not only in upper Wadi Hilti.

In the largest transition zones, comparable to the one in upper Wadi Salahi, sills, dikes and lenses of layered gabbros occur as above the Maqsad diapir (Nicolas et al. 1990), in the SE-Haylayn peridotite dome (Reuber et al. in press), in Western Nakhl (Reuber 1990), in Fizh block, Reuber 1988), etc. They may be called mega-sills or micro-magma chambers. An internal layering has been observed even in metre wide dikes. Their orientation is flat as in the Maqsad diapir or rather steep as in SE- Haylayn, western Nakhl and northern Fizh block. Such large transition zones generally grade into intrusive wehrlites (Southern Fizh block Reuber 1988, SE-Haylayn – Reuber et al. in press), commonly associated with coarse grained gabbros and pyroxenites representing typical recrystallized facies due to percolation.

For many of these complex structures, special accretionary environments have been proposed: Fizh – leaky transform (Smewing 1980), overlapping spreading center (Reuber 1988) or propagating rift (McLeod & Reuber 1990), and SE-Haylayn – off axis diapir + propagating rift (Reuber et al. in press). The compression during early obduction along the very young ridge as proposed by Coleman (1981) and Boudier et al. (1985) certainly helps to mess up the plutonic series that are barely crystallized. This major change from extension to compression is probably preceded by a change in ridge orientation, obviously resulting in complex accretionary patterns (Beurrier 1990, Reuber 1990).

However, also normal oceanic ridges show complicated structures indicating alternating magmatic and tectonic extension (Juteau et al. 1988b, Karson et al. 1987, Karson & Winters 1987). An explanation of the observed structures in the frame of regular oceanic spreading can be imagined (Reuber in press).

Interpretation

The exposed data clearly indicate that several magmatic events succeeded each other and overlap in space (Fig. 9). Different reactions of new-coming magma with pre-existing host rock can be deduced. The dynamics and reactions of rising melt – crystal mush assemblages with host rock – crystal mush or magma encountered depend largely on the volume rising upwards, its temperature and resulting viscosity, in comparison to the same characteristics of the encountered host material. General temperature gradients and the stress field may equally be of importance. Accordingly the following interactions are expected:

A) *The magma meets with magma* – no structures of the classical case of refilling a magma chamber are fossilized (Huppart & Sparks 1980, Huppart et al. 1983, 1986, Campbell & Turner 1986).

B) *The magma meets with a crystal mush of considerably higher viscosity* – the melange of the two magmas is difficult and mechanical mixing results in lobed contacts (Fig. 3C, F). The formation of sills due to differences in viscosity and density is probably favoured in a dynamic situation.

C) *The magma meets with a crystal frame and interstitial liquid* – percolation of the frame with new hot magma results in exceptionally coarse facies without dynamic structures, also commonly located at the limit between the intrusion and the host rock (Fig. 6A, C).

D) *The magma meets with consolidated rock* – Different interactions can result according to the volumes, dynamics and heat transfers involved:

D1) *Hydraulic fracturing* resulting in anastomosing breccia, in vertical dike swarms, perpendicular to extension, or in sills fracturing parallel to a pre-existing horizontal fabric are abundant where the host rock reacts brittle, the magma is relatively fluid and has overpressure.

D2) *Bending* of the host rock layers (Fig. 9, 10) appears to be produced by the largest and more viscous intrusive magma volumes, while the host gabbro was still hot enough to allow folding.

D3) *Re-melting* may occur especially when the magma meets with hydrated rock, that has a lowered point of partial melting. Melange of the melts may be an efficient way to introduce water into the magma that may change its petrographic characteristics such as crystallisation sequence, and thus simulate a different magma source (Lippard et al. 1986). Segregating melt that intrudes upwards may form plagiogranite dikes and plugs including mafic xenoliths.

The feeder areas in the Salahi massif can be traced by dunite domes and/or vertical structures. Only for the youngest events, the original volumes fed by the same feeder channel can approximately be delimited: The Hilti massif displays two main centres dominated by massive gabbros with characteristic steep linear fabrics, located in the North and in the South, approximately 15 km apart (Fig. 1, 7). The dominant vertical line in their central part strongly suggests vertical magmatic flow as expected in a magmatic feeder zone. The poor flat layering in the main body may be due to compaction following intrusion (Fig. 10). Along the eastern contact of the Hilti-Sudum massive gabbros, the pre-existing layered gabbros and wehrlites were bent into a vertical position during upward intrusion of this magmatic body and recrystallization facies is abundant. Along the western contact interfingering sills within older gabbros make it impossible to accurately delimit the extension of the magmatic body. The north-Salahi massive gabbros are bordered by recrystallized facies in the East, they pass into the chaotic zone towards the South that is rooted westward in transitional dunites.

The different types of contacts may be related to differences in the structural levels presently exposed. The western contacts represent 1 to 1.5 km lower levels than the ones exposed along the eastern contacts of Hilti-Sudum, and northern Salahi massive gabbros respectively. These height differences correspond nearly to the lowermost and uppermost lateral contacts of each magma body.

The concept of successive magma chambers does not necessarily imply intermittent melt generation, which in any case cannot be excluded. In the most idealised case of steady state regular magma supply from lower mantle zones in an amount corresponding to a fast spreading rate, a steady state magma chamber still has poor chances to be maintained because of important hydrothermal cooling (Lister 1983, Nehlig 1989). Furthermore, the thickness

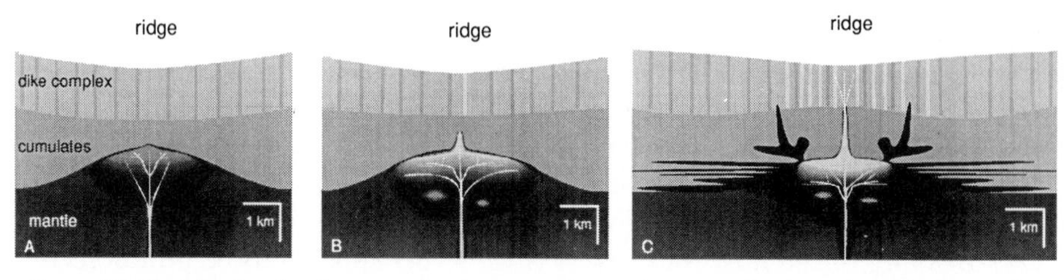

Figure 11. Sketch of cycle of progressive impregnation and consecutive intrusion of magma; the magma supply is shown to cease, as the precedingly affected transition zone moves away due to spreading; the vertical scale of wehrlite sills is exaggerated for reasons of presentation.

variations of the transition zones indicate an alternating cyclic evolution: absorption of magma would be followed by instability of this zone, its upward bulging and expulsion (Fig. 11). The extracted crystal mush would leave a dry dunite behind, and separate itself into xenocryst rich wehrlite sills injecting into the lower parts of the crustal sequence, and a more gabbroic melt crystallising in a small magma chamber. Taking into account the observed dimensions (1 km height, few km wide), magma supply and cooling rate, the lifetime of such magma chamber can be calculated as a few 10 000 years (Reuber in press).

The Two Main Differences with the Classical Concept are:

- Layering is to a small extent only the product of gravity settling or another primary process in a static magma chambers, formed horizontally, but to a larger extent the product of sill and dike injection, strong magmatic flow or percolation-recrystallisation.
- The total thickness of the plutonic sequence is progressively accreted by sill injections and considerably greater than the thickness of the magma chamber.
- The magma chamber is not only fed by melt generated in depth in the conditions of generating true MORB-type magma, but large amounts of supplementary melting of already depleted harzburgite modify the petrological characteristics of the sequence (Fisk 1986, Kelemen 1990, Kelemen et al. 1990). This variation is most important for the initial part of each cycle and the most ultramafic magmas. Later the composition may gradually shift back to MORB-type magma. Supplementary water naturally enters the system by melting of host rock hydrated due to hydrothermalism.

The future will show, whether the more and more non-MORB lavas dredged at ocean ridges give a more complex picture for normally working oceanic ridges, or whether the effect of deceleration of spreading before change to compressional regime is a necessary condition to create a petrographically and structurally more complicated oceanic crust as observed in Oman.

Conclusions

Based on the evidence of numerous mutually intrusive magmatic bodies observed in an area previously reported for its regularity of the cumulate sequence in the northern Oman ophiolite, a model for the generation of these magma bodies by inflation of the transition zone has been proposed: Tholeitic magma generated in depth, infiltrates the uppermost mantle and induces supplementary partial melting until this zone becomes unstable and bulges upwards. The feeder zone of these magma bodies could be traced by massive gabbros showing a vertical line dominated fabric. Large amounts of magma intrude sideways into pre-existing more or less solidified cumulates as sills to create the most frequent type of layering and to increase the thickness of the plutonic sequence. Simple calculations (Reuber in press) show that the times involved for inflation of the transition zone and crystallization of the magma bodies are in the order of a few 1000 to 10 000 years each, for a spreading rate of 10 cm/year.

Acknowledgements

I greatly acknowledge stimulating discussions with F.Boudier, M. Cannat, G. Ceuleneer, T. Juteau, and A. Nicolas. I am thankful to C. Lister for his review, to G. Ceuleneer for repeatedly reading my manuscript, to E. Ball for drawing the maps of Figures 2 and 7, and to C. Nevado for numerous thin sections. Logistic support from the Omani Ministry for Petroleum and Minerals, financial support from various projects of French CNRS and PNEHO are gratefully acknowledged. It is always a pleasure to remember the field work in Salahi massif together with M. Ernewein, C. Pflumio, P. Nehlig, and the great hospitality of the local people throughout the studied area.

References

Alabaster, T., Pearce, J.A. and Malpas, J., 1982. The volcanic stratigraphy and petrogenesis of the Oman ophiolite complex. Contr. Mineral. Petrol., 81: 168–183.

Bédard, J.H., Sparks, R.S.J., Renner, R., Cheadle, M.J. and Hallworth, M.A., 1988. Peridotite sills and metasomatic gabbros in the eastern layered series of the Rhum complex. J. Geol. Soc. London, 145: 207–224.

Benn, K. and Allard, B., 1989. Preferred mineral orientations related to magmatic flow in ophiolite layered gabbros. J. Petrol., 1989: 925–946.

Beurrier, M., 1987. Géologie de la nappe ophiolitique de Samail dans les parties orientale et centrale de l'Oman. Thèse Doc. Etat, Paris 6, 406p.

Beurrier, M., 1990. The ophiolites of the central Oman mountains: an example of propagating rift into former oceanic crust. Symp. ophiol. genesis and evol. oc. lithosphere, Muscat-Oman, abstract.

Boudier, F., Bouchez, J.L., Nicolas, A., Cannat, M., Ceuleneer, G., Misseri, M. and Montigny,

R., 1985. Kinematics of oceanic thrusting in the Oman ophiolite: model of plate convergence. Earth and Planet. Sci. Lett., 75: 215–221.
Campbell, I.H. and Turner, J.S., 1986. The influence of viscosities on fountains in magma chambers. J. Petrol. 27: 1–30.
Ceuleneer, G., 1986. Structure des ophiolites d'Oman: flux mantellaire sous un centre d'expansion océanique et charriage à la dorsale. Thèse Doc. Univ. Nantes, 217 p.
Coleman, R.G., 1981. Tectonic setting for ophiolite obduction in Oman. J. geophys. Res., 86: 2497–2508.
Detrick, R.S., Bulh, P., Vera, E., Mutter, J., Orcutt, J., Madsen, J. and Brocher, T., 1987. Multichannel seismic imaging of a crustal magma chamber along the East Pacific Rise. Nature, 326: 35–41.
Ernewein, M., 1987. Histoire magmatique d'un ségment de croûte océanique téthysienne: pétrologie de la séquence plutonique du massif ophiolitique de Salahi (nappe de Semail, Oman). Thèse Doc. Univ. Strasbourg, 205 p.
Ernewein, M. and Whitechurch, H., 1986. Les intrusions ultrabasiques de la séquence crustale de l'ophiolite d'Oman: un évènement témoin de l'extinction d'une zone d'accrétion océanique? C.R. Acad. Sci. 303: 379–384.
Ernewein, M., Pflumio, C. and Whitechurch, H., 1988. The death of accretion zone as evidenced by the magmatic history of the Sumail ophiolite (Oman). Tectonophysics, 247–274.
Fisk, M.R., 1986. Basalt magma interaction with harzburgite and the formation of high-magnesium andesites. Geophys. Res. Lett., 13: 467–470.
Gass, I.G. et al., 1980. The Open University Oman Project – map 2.
Greenbaum, D., 1972. Magmatic processes at ocean ridges: evidence from the Troodos massif, Cyprus. Nature, 238: 18–21.
Hale, L.D., Morton, C.J. and Sleep, N.H., 1982. Reinterpretation of seismic reflection data over the East Pacific rise. J. geophys. Res., 87: 7707–7717.
Huppart, H.E. and Sparks, R.S.J., 1980. The fluid dynamics of basaltic magma replenished by influx of hot dense ultrabasic magma. Contrib. Mineral. Petrol., 75: 279–289.
Huppart, H.E., Sparks, R.S.J. and Turner, J.S.,1983. Laboratory investigations of viscous effects in replenished magma chambers. Earth PLanet. Sci. Lett. 65: 377–381.
Huppart, H.E., Sparks, R.S.J., Whitehead, J.A. and Hallworth, M.A., 1986. Replenishment of magma chambers by light inputs. J. Geophys. Res. 91: 6113–6122.
Juteau, T., Ernewein, M., Reuber, I., Whitechurch, H. and Dahl, R., 1988a. Duality of magmatism in the plutonic sequence of the Sumail nappe, Oman. Tectonophysics, 151: 107–135.
Juteau, T., Cannat, M. and Lagabrielle, Y., 1988b. Peridotites in the upper oceanic crust away from transform zones: a comparison of the results of previous DSDP and ODP legs. Proc. ODP, Init. Repts, pp. 106–109.
Karson, J.A. and Winters, A.T., 1987. Tectonic extension on the Mid-Atlantic Ridge. EOS, 68: 1508.
Karson, J.A., Thompson, G., Humphris, S.E., Edmond, J.M., Bryan, W.B., Brown, J.R., Winters, A.T., Pockalny, R.A., Casey, J.F., Campbell, A.C., Klinkhammer, G., Palmer, M.R., Kinzler, R.J. and Sulanowska, M.M., 1987. Along-axis variations in seafloor spreading in the MARK area. Nature, 328: 681–685.
Kelemen, P.B., 1990. Reaction between ultramafic rock and fractionating basaltic magma I. Phase relations, the origin of calc-alkaline magma series, and the formation of discordant dunite. J. Petrology, 31: 51–98.
Kelemen, P.B., Joyce, D.B., Webster, J.D. and Holloway, J.R., 1990. Reaction between ultramafic rock and fractionating basaltic magma II. Experimental investigation of reaction between olivine tholeiite and harzburgite at 1150–1050°C and 5 kb. J. Petrology, 31: 99–134.
Kusznir, N.J., 1980. Thermal evolution of the oceanic crust; its dependence on spreading rate and effect on crustal structure. Geophys. J. r. astron. Soc., 61: 167–181.
Lewis, B.T.R., 1983. The process of formation of ocean crust. Science, 220: 151–157.
Lippard, S.J., Shelton, A.W. and Gass, I.G., 1986. The ophiolite of Northern Oman. Geol. Soc. London Mem., 11, 178 p.

Lister, C.R.B., 1977. Qualitative models of spreading center processes, including hydrothermal penetration. Tectonophysics, 37: 203–218.

Lister, C.R.B., 1983. On the intermittency and crystallization mechanisms of sub-seafloor magma chambers. Geophys. J. r. Astron. Soc., 73: 351–365.

Lister, C.R.B., (in press). The upward migration of self-convecting magma bodies.

McClain, J.S., Orcutt, J.A., and Burnett, M., 1985. The East Pacific Rise in cross section: a seismic model. J. geophys. Res., 90: 8627–8639.

McLeod, C.J. and Reuber, I., 1990. Complex rift geometry in the northern Oman ophiolite: a possible overlapping spreading center. Symp. ophiol genesis and evol. oc. lithosphere, Muscat-Oman (abstract).

Nicolas, A., 1989. Structures of ophiolites and dynamics of oceanic lithosphere. Kluwer Academic Publishers, 666p.

Nicolas, A., Ceuleneer, G., Boudier, F. and Misseri, M., 1988a. A structural mapping in the Oman ophiolites: mantle diapirism along an oceanic ridge. Tectonophysics, 151: 27–56.

Nicolas, A., Reuber, I. and Benn, K., 1988b. A new magma chamber model based on structural studies in the Oman ophiolite. Tectonophysics, 151: 87–105.

Nicolas, A., Pallister, J. and Ceuleneer, G., 1990. The ophiolite of southern Oman mountains, mantle diapir and transform zone in the Sumail and Wadi Tayin massifs. Symp. ophiol genesis and evol. oc. lithosphere, Muscat-Oman, Excusion C, guidebook, 35p.

Nehlig, P., 1989. Etude d'un système hydrothermal océanique fossile: l'ophiolite de Semail, Oman.Thèse Doc. Univ. Brest, 313 p.

Pallister, J.S. and Hopson, C.A., 1981. Samail ophiolite plutonic suite: field relations, phase variation, cryptic variation and layering, and a model of a spreading ridge magma chamber. J. geophys. Res., 86: 2593–2644.

Parrot, J.F. and Ricou, L.E., 1976. Evolution des assemblages ophiolitiques au cours de l'expansion océanique. Cahiers ORSTOM Sér. Geol., 7: 49–68.

Parsons I. (ed.) 1987. Origins of Ignous Layering, NATO ASI series 196, 666p.

Pflumio, C., 1988. Histoire magmatique et hydrothermale du Bloc de Salahi: Implications sur l'origine et l'évolution de l'ophiolite de Semail (Oman). Thèse Doc. Ecole des Mines, Paris, 243 p.

Pflumio, C., Michard, A. and Juteau, T. (in prep.): Petrology and Geochemistryof the volcanic sequence of the Salahi block (Northern Oman): Implications for the origin and evolution of the Sumail ophiolite.

Rabinowicz, M., Nicolas, A. and Vigneresse, J.L., 1984. A rolling mill effect in asthenospheric beneath oceanic spreading centers. Earth Planet Sci. Lett., 67: 97–108.

Rabinowicz, M., Ceuleneer, G. and Nicolas, A., 1987. Melt segregation and flow in mantle diapirs below spreading centers: evidence from the Oman ophiolites. J. geophys. Res., 92: 3475–3486.

Robins B., Haukvik, L. and Jansen, S., 1987. The organization and internal structures of cyclic units in the Honningsvag intrusive suite, north-Norway: implications for intrusive mechanisms, double diffusive convection, and pore magma infiltration. In: Parsons I. (ed.) Origins of Ignous Layering, NATO ASI series 196: 287–312.

Reuber, I., 1988. Complexity of the crustal sequence in northern Oman ophiolite (Fizh and southern Aswad block): the effect of early slicing ? Tectonophysics, 151: 137–165.

Reuber, I., 1990. Geométrie et dynamique de l'accrétion dans les ophiolites téthysiennes: Himalaya du Ladakh, Oman, Turquie. Thèse d'état, Univ. de Brest, pp. 389–612.

Reuber, I. (in press). Diapiric magma intrusions in the plutonic sequence of the Oman ophiolite traced by the geometry and flow pattern of the cumulates. Proc. Symposium on Diapirism, Bandarabbas-Iran.

Reuber, I., Nehlig, P. and Juteau, T. (in press). Axial segmentation at a fossil oceanic spreading center: off-axis mantle diapir and advancing ridge tip in the Haylayn block (Semail nappe, Oman). J. Geodynamics

Smewing, J.D., 1980. An upper Cretaceous ridge-transform intersection in the Oman ophiolite. In: Proc. Inter. Ophiolite Symp., Geol. Surv. Dept. Cyprus, pp. 407–413.

Part I
Processes at Spreading Centers
B) Mantle

The Upward Migration of Self-Convecting Magma Bodies

C.R.B. LISTER
School of Oceanography, WB-10, University of Washington, Seattle, Washington 98195, U.S.A.

Abstract

A magma body existing in an upper mantle containing a small percentage of partial melt should be subject to internal convection. The presence of partial melt means that the matrix follows closely the temperature gradient versus pressure, or depth, of the mantle solidus. This gradient is several times the adiabatic gradient of a mixed magma body, and the difference is available to drive convection. A magma body rising by Stokes drift in a viscous upper mantle at 0.1 m y^{-1} would contain 11 km^3. If thermal convection were dominant, the body would elongate vertically and migrate upward at 18 m y^{-1} by melting material off the top and crystallising it at the bottom. However, in a chemically inhomogeneous system, the process of dissolving an unlike solid off the top is limited by chemical diffusion and the convection behaves as chemically driven. For olivine dissolving into dry basalt the upward velocity drops to 0.16 m y^{-1}, taking 0.7 My to traverse 100 km.

Larger bodies move more rapidly than smaller ones and would tend to sweep them up, as well as concentrating incompatible elements into the melt. The difference between ridge-crest upwelling, where mantle matrix rises to near the surface as well as does the magma, and island 'hot spot' volcanism, where the crystals are irreversibly left behind, can account for the differences in basalt composition. The peculiar sequence and timing of eruptions on the Hawaiian Islands is readily explained by magma self-convection and thermal diffusion in the lower lithosphere. While the mantle may be quite heterogeneous, there is no need to postulate complex sequential melting events, or different geochemical reservoirs that persist over geologic time.

Introduction

Somewhere in the upper mantle, basaltic magmas separate from their parent ultramafic rocks. Suggested mechanisms range from steady flow through intergranular pores (Stolper et al., 1981; Richter and McKenzie, 1984) to

Tj. Peters et al. (Eds), Ophiolite Genesis and Evolution of the Oceanic Lithosphere, 107–123.

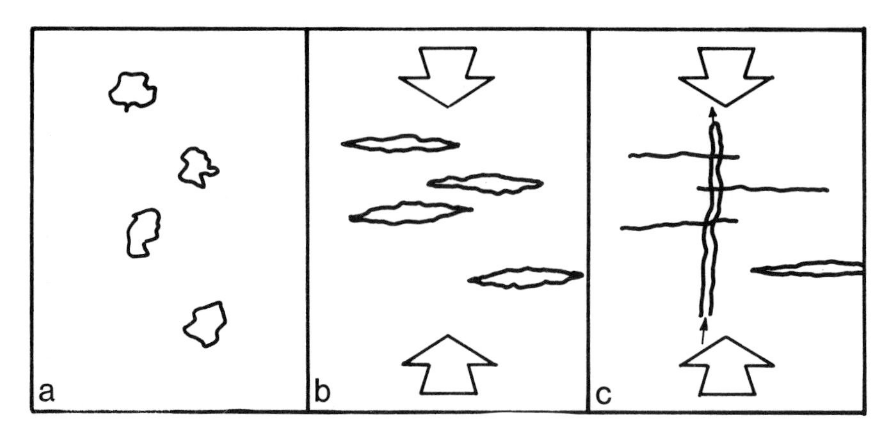

Figure 1. A schematic of the sequence of magma separation from a partially melted rock. (a) Melt develops at the sites of garnet crystals and remains in centimetre-sized blobs. (b) Deviatoric strain squashes these blobs into thin discs perpendicular to the compression axis. (c) A crack opens parallel to the principal compression and the differential pressure drains the discs into the crack.

vein and dyke-draining models of varying complexity (Maaloe, 1981; Nicolas, 1986; Sleep, 1988). In the first case, actual segregation can only occur where, paradoxically, the temperature drops and the rock has no permeability, due to the vanishing melt content; in the second case no mechanism has been advanced for the *initiation* (as opposed to the growth) of veinlets. However, Maaloe (1981) points out that melting, in a spinel-lherzolite, can only start where grains of four different minerals meet at a corner, and Nicolas (1986) gives examples of protogranular spinel-lherzolites where former melt pockets were of centimetre dimensions. If the latter rock is deformed, such equi-dimensional pockets would become elongated perpendicular to the direction of compression, not, as in Sleep (1988), perpendicular to the direction of maximum tension.

The transition from the centimetre scale of veinlets to the metre scale of dykes is a difficult one, but one can note that veinlets shaped by deformation must contain magma at a positive pressure differential, whereas cracks parallel to the direction of compression contain a pressure *less* than that in the rock (Sleep, 1988). Any inhomogeneity that can initiate a crack in the latter direction, large enough to intersect pressurised veinlets, will cause them to drain and the fracture to grow (Figure 1). Nicolas (1986) believes that episodic dyke events reach directly to the surface and feed a crustal magma chamber. Dyke-like propagation in a plastic matrix is possible only at small scales and along the axis of compression. The latter is vertical near the axis of a plume impinging on crust or lithosphere (Figure 2a). In the more generalised mantle diapir of Figure 2b, magma pockets in the circumferential shear zone drain outward, thus denuding the diapir of some of its buoyancy. At the apex of the diapir, however, magma can collect by gravitational flow into a pocket.

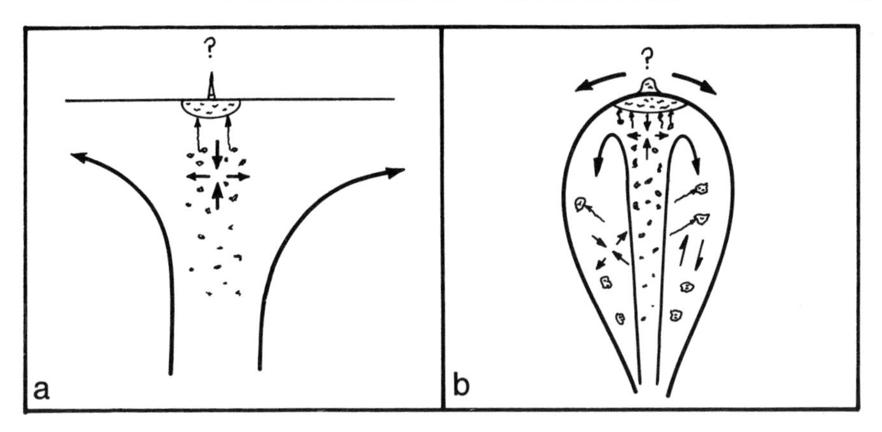

Figure 2. (a) When a mantle diapir impinges on a more solid layer, the deviatoric strain favours the vertical separation of magma from the matrix through veins (Figure 1c) aided by gravity. (b) A generalised mantle plume, far from a viscosity boundary, can segregate efficiently only at the apex. The shear at the sides generates veins that would lead magma out of the plume.

Here one must consider what happens at larger scales. A matrix containing partial melt is plastic rather than brittle and the growth of elongated dykes occurs when that is specifically favoured by the local stress field. As soon as buoyancy forces in the magma exceed the deviatoric stress in the matrix, the dynamics of the flow tend towards those of fluids. It is, at first, not obvious why scale is important, and why magma under high buoyancy pressure does not simply fracture the rock and propagate a dyke. Increasing scale increases magma pressure and also increases the velocity of propagation of a body by plastic flow (see below). Most plastic materials obey power-laws of creep, which means that the creep rate increases rapidly above some critical level of stress; conversely, the stress rises little above this critical value no matter how high the creep rate. Crack propagation occurs due to leverage at the tip from the stress in the material, and therefore, for pressure-fracturing, is likely to be limited in velocity by the yield process in a plastic material (this subject has not, to my knowledge, been investigated, and is far too complex to be treated further here). If the crack propagation rate is, indeed, limited by plasticity, then at a large-enough scale, propagation by other mechanisms is more important, and dykes do not form. That this occurs in nature is confirmed by the roundness of volcanic plugs and the shape of granite batholiths; while for deeper levels, zoned ultramafic complexes have shapes more consistent with magma-body processes than with dyking (Murray, 1972). When a low-viscosity fluid rises in a matrix of high-viscosity fluid, the bubbles become spherical, not elongated (Batchelor, 1957). The scale at which the transition occurs can be estimated as follows. Take diapirs of 10 km dimension (Nicolas et al., 1988) moving at $0.1\,\mathrm{m\,y^{-1}}$. The strain rate is of order $3 \times 10^{-13}\,\mathrm{s^{-1}}$, and if the matrix viscosity is 10^{18} Pa-s (Cooper and Kohlstedt, 1984) the stress is 3×10^5 Pa. A comparable differential pressure is produced by a magma column 6m high if the density difference between it and the matrix is 0.5 Mg m^{-3}. For magma bodies of height significantly greater than this, buoyancy and fluid mechanics dominate over local stress. Thus, the rise of magma bodies capable of adding significantly to a magma chamber, triggering an eruption, or filling a ridge-crest dyke, is dominated by buoyancy and fluid-dynamic processes.

Rates of Rise of Magma Bubbles

The rate of rise of a very-fluid bubble in a much more viscous matrix is essentially independent of the viscosity of the thin fluid and is

$$v = \frac{1}{3}\frac{a^2 g}{\nu}\left(1 - \frac{\rho^1}{\rho}\right) = 5.2 \times 10^{-8} a^2 = 0.1 \text{ m y}^{-1} \quad (1)$$

(Batchelor, 1967). Here a is the sphere radius in m, g is gravity, ρ^1 is the magma density (2.8Mgm^{-3}), ρ is the matrix density (3.3 Mg m^{-3}), and ν is the kinematic viscosity ($3 \times 10^{14} m^2 s^{-1}$). To reach a velocity equal to that chosen as representative of a diapiric upwelling (10 cm y^{-1}) requires the sphere to be 1.4 km in radius, or 11 km^3 in volume. At this size, the magma could separate from the generalised mantle plume of Figure 2b and make its own way toward the surface.

Could such a magma bubble break through the lithosphere to the surface of a mature oceanic plate? Lister (1986) has shown that differential thermal stresses in the cold and elastic part of the lithosphere make that region highly susceptible to vertical dyking. However, there is a wide transition zone from the solidus temperature of upper mantle material, about 1420 K, to the rigidus where it behaves elastically, about 800 K. The rock deforms by creep processes in this zone, but the effective viscosity rises by many orders of magnitude from the 10^{18} Pa-s of a matrix containing partial melt. There is thus no simple mechanical way for the magma bubble to penetrate: the rock is too stiff for the plastic-flow ascent velocity to be significant, but not brittle enough to crack rapidly and sustain a propagating dyke. Spence and Turcotte (1985) have shown that magma must move rapidly in a narrow dyke not to be frozen by cooling from the walls, and, as discussed above, any other mechanism offering faster propagation should dominate over dyking.

The Melting-Through Method

If a fluid is hotter than the melting-point of a solid, it can melt its way through: this is the well-known principle of metal-cutting by an oxy-acetylene torch. That this is at least part of the emplacement method of granitoid batholiths became clear to this author on a stroll along the beach at Capetown in 1964. The section proceeds from low-grade sedimentary rock through a very high-grade metamorphosed version of the same material, to melted material. Further into the batholith there are stoped blocks of the sedimentary rock, preserving the strata, in the melt, and finally a more uniform igneous rock of a somewhat different composition. The overlying sedimentary strata of Table Mountain are flat-lying and undeformed, but these may or may not predate the emplacement of the batholiths.

The case of a basalt pocket in an ultramafic matrix is different. At the microscopic scale, and in the absence of a temperature gradient, the pockets are in equilibrium (Nicolas, 1986). If there is a temperature gradient, ultramafic minerals are more soluble in the liquid at the high temperature side of a pocket than at the low one. The concentration gradient causes diffusion of the mineral across the liquid, and the pocket moves towards the high temperature side (Lescher and Walker, 1982).

At the kilometre scale, gravity becomes important. Not only is there a buoyancy force trying to move the magma bubble upward, but, even in a stable asthenosphere, there are temperature gradients in both the magma and the matrix. If the magma is mixed, it has an adiabatic temperature gradient in it:

$$\left(\frac{\partial T}{\partial P}\right)_S = \frac{\alpha T}{\rho^1 c_p} \simeq 25\ \mathrm{K\ GPa^{-1}} \qquad (2)$$

where α is the expansion coefficient of the melt ($5 \times 10^{-5}\ \mathrm{K^{-1}}$), T is absolute temperature ($\sim$1700–2000 K), and c_p is the specific heat at constant pressure ($1300\mathrm{j\ kg^{-1}\ K^{-1}}$). If the mantle material is in equilibrium with partial melt, its temperature gradient must be that of the solidus (from Takahashi, 1986).

$$\left(\frac{\partial T}{\partial P}\right)_M = 130\ \mathrm{K\ GPa^{-1}}. \qquad (3)$$

The gradient of equilibrium temperature between matrix materials and liquid is substantially larger than the adiabatic gradient in the basalt, so that, if the top of the magma chamber were in equilibrium with the fluid, the bottom would be too hot and would drive convection in the magma (Figure 3). In a first look at this convection, the super-adiabatic temperature gradient ($\beta = 100\ \mathrm{K\ GPa^{-1}} = 3\ \mathrm{K\ km^{-1}}$) can be entered into the formula for the Rayleigh number

$$\mathcal{A} = \frac{\alpha g \beta h^4}{\kappa^1 \nu^1} = 830\ h^4 = 6.8 \times 10^{16} \qquad (4)$$

where g is gravity ($10\ \mathrm{m\ s^{-2}}$), h is the height of the bubble (3,000 m here), κ^1 is the thermal diffusivity of basalt magma ($5 \times 10^{-7}\ \mathrm{m^2\ s^{-1}}$) and ν^1 its kinematic viscosity ($.0036\ \mathrm{m^2\ s^{-1}}$). Convection in magma of tholeiitic viscosity thus begins when the body has a height of 2 m or more, and is extremely vigorous in a bubble 3 km high (Lister, 1983).

Convection transports heat according to the relation

$$Q = k\beta N \quad W\,m^{-2} \tag{5}$$

where k is the thermal conductivity of liquid basalt (about $2\,W\,m^{-1}\,K^{-1}$), and, by experiment N, the Nusselt number has been found to be approximately

$$N \cong 0.062\,\mathcal{A}^{1/3} \tag{6}$$

in very vigorous convection in a fluid of moderate Prandtl number (Katsaros et al., 1977). Combining equations 4, 5, and 6 gives a value for the heat flux:

$$Q = 0.0035\,h^{4/3} = 151\,W\,m^{-2} \tag{7}$$

Finally, this heat flux is applied to melting off, or dissolving, ultramafic minerals at the top of the bubble, and recrystallising them at the base. The upward migration of the magma bubble occurs at a velocity

$$v = \frac{Q}{\rho L} = 0.062\frac{k\beta}{\rho L}\left(\frac{\alpha g\beta}{\kappa^{1}\nu^{1}}\right)^{1/3} h^{4/3} = 6.7\times10^{-5}\,h^{4/3} = 2.8\,m\,y^{-1} \tag{8}$$

which is much faster than the Stokes velocity of the same bubble even in an asthenosphere of low viscosity. Here L is the latent heat of solution of the ultramafic rock in the basaltic melt (6.7×10^{5} J kg^{-1}) (Fukuyama, 1985).

Dissolution of dissimilar minerals, such as of olivine into basalt, has several consequences. One is that the convective drive is increased, because olivine is a dense component of the melt (Lister, 1983). The other is that, once the system becomes chemically inhomogeneous, considerations such as the solubility of a mineral in a melt near the liquidus become important, and chemical diffusion may become a rate-limiting parameter.

Limiting Phenomena

If the matrix (mantle) and magma (basalt) were of the same composition, then the process at the top boundary would be one of congruent melting. For most substances, this process is fast, and the familiar ice-and-water bath does not deviate much from 0°C unless melting and freezing rates are high. A value for the characteristic rate of solution for diopside into its own melt is $78\,m\,y^{-1}\,{}^{\circ}C^{-1}$ (Kuo and Kirkpatrick, 1985). While those authors' measured rates for enstatite and olivine were lower by an order of magnitude, they were not for true congruent melting, and so diffusion was probably limiting (see below). The disequilibrium required for $2.8\,m\,y^{-1}$, .04°C, is a small fraction of the available drive, 9°C, and therefore not significant. Because melting rate should be proportional to $(\beta h)^{4/3}$, the fraction used up by dissolution disequilibrium would increase slowly with further increase in magma-chamber height, but should not be significant for any reasonable size of chamber in the earth.

When a mineral such as olivine dissolves into a fluid such as a basalt near its liquidus, there is a large difference between the temperature of melting of the solid and the temperature of the dissolution: 620°C at atmospheric pressure, dropping to 410°C at 40 kb (Hyndman, 1985, Figs. 4–14 and 4-8). To increase the proportion of olivine in the solution requires an increase in temperature, according to the slope of the liquidus versus composition. Conversely, 1°C temperature rise produces an increase in olivine concentration of about .001 at atmospheric pressure and .0016 at 40 kb. Since the entire drive for convection, even for a magma chamber several km high, is only a few degrees, it is clear that the liquid in contact with the olivine crystals cannot dissolve very much of them. Any increased concentration near the olivine surface needs to diffuse away before more can dissolve.

Since chemical diffusion rates are much slower than those for heat, one could imagine another extreme scenario where the liquid was sufficiently stirred to be adiabatic in temperature and the solute/solution disequilibrium at the boundaries drove chemical convection. The concentration of the dense phase would be higher at the top of the chamber, where the temperature is high relative to the liquidus slope, than near the bottom. Now the effective expansion coefficient for the changing mixture would be the difference in density between the dissolved solid and that of the liquid as a whole, divided by the latter. For olivine and basalt the values are 2.80 (2.93 with 10% Fayalite by volume) and 2.60 Mg m^{-3} (Lister, 1983), leading to an α' of .077 (or 0.13, per unit concentration). The super-adiabatic temperature gradient β must be multiplied by the inverse liquidus slope c_T to obtain the concentration disequilibrium difference between the bottom and the top of the chamber. Finally, if the thermal diffusivity κ is replaced by the chemical diffusivity γ, then the same method of convective analysis can be applied:

$$\mathcal{A}_{\text{CHEM}} = \frac{\alpha' g \beta c_T h^4}{\gamma \nu} \tag{9}$$

and

$$v \simeq 0.062\left(\frac{\alpha' g}{\nu}\right)^{1/3} \gamma^{2/3} (\beta c_T h)^{4/3} = 7.9 (\beta c_T h)^{4/3} \text{ m y}^{-1} \tag{10}$$

if γ is a representative value for olivine in a basaltic liquid, 6×10^{-10} m^2 s^{-1}. It has been assumed that a chemical Nusselt number can be calculated in the same way as the thermal one: this may not be correct, since the chemical Prandtl number ν/γ is much higher than the thermal one. Conservatively, both the chemical transport relation to Rayleigh number and the critical Rayleigh number for onset of chemical convection should be verified by experiment.

The critical height for onset of chemical convection, if the downward increase in temperature is limited to the adiabatic gradient for stirred magma, is only 20 cm. However, transports are quite low because of the low diffusivity, and the velocity of upward melting of a body 3 km high is, if $\beta h = 9°C$ and $c_T = .0016$, only .03 m y^{-1}. This is 100 times slower than the thermal calculation given, and therefore chemical diffusion should be the operating limiting parameter.

Shape of Upward-Convecting Magma Bubbles

It will be noticed that equation (1) contains the radius of a spherical magma bubble, while equations (8) and (10) contain only the height. In the numerical estimation h was made equal to a. However, there is no reason why a self-convecting magma body should have the shape of a bubble. If a cupola developed on the magma chamber, convection would be even more vigorous into the increased height, and so the cupola would grow at the expense of the broad surface. The magma body should continually elongate in the vertical direction until it becomes narrow enough to impede the vigour of the convection. There is no experimental data on this for high Rayleigh-number convection, so I will estimate the limiting height to diameter ratio at 10:1. For a comparable volume of magma to the 3 km-diameter sphere, 11 km^3, the elongated magma "worm" would be 12 km high, and its velocity, based on pure chemical convection of olivine, would be 0.16 m y^{-1}.

The comparison between the two modes of transport is shown graphically in Figure 4. The velocities are shown by arrows at the same scale, and the thermal diffusion aureoles $\sqrt{\kappa t}$ are sketched in. If the magma and the country rock are at different temperatures, this is the distance over which there would be substantial thermal exchange. The total trail diameters are 2 km for the rising Stokes bubble and 4.4 km for the worm, not very different. However, for the former the *chemical* exchange aureole would be exceedingly thin (0.7 m) as the stream lines are smooth and the chemical diffusivity is 10^7 times smaller than the thermal diffusivity in rock (Misener, 1974). For the worm, the exchange is complete over the fluid diameter of 1.2 km, and the trail consists mostly of cumulates. I will propose (below) that this chemical exchange could be a contributor to the compositional differences between ocean-island and ridge-crest eruptives.

Difference Between the Top and Bottom Boundaries

Observable field contacts between magmas and country rock are generally sharp on the centimeter scale. Relating to the stability of the upper boundary, I have personally observed stoping of a metamorphosed sedimentary rock into a granite batholith on the beach at Capetown, but this was on the meter scale expected for flaws and joints in the pre-existing sedimentary rock. For

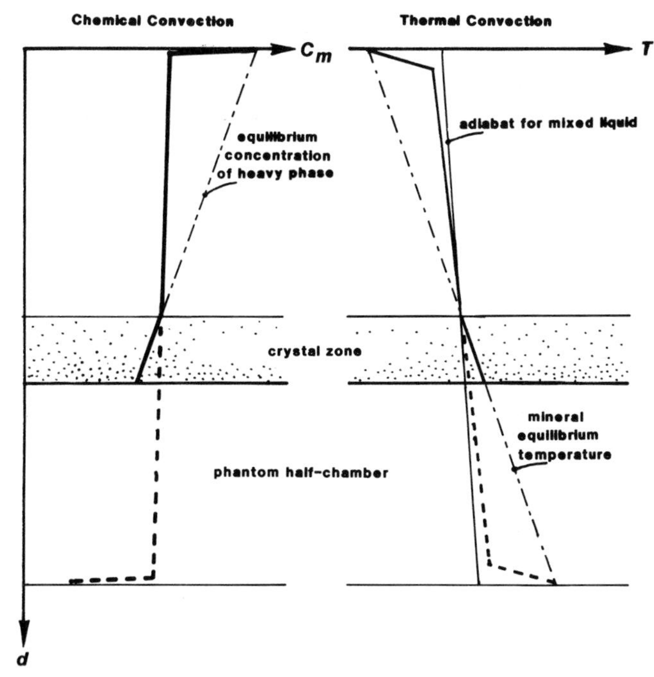

Figure 3. The concentration-drive for chemical convection (left) compared to the thermal drive for standard convection (right). Both are developed by the steeper change with pressure of the crystal-liquid equilibrium temperature than of the adiabatic temperature. The inverse liquidus slope of increasing concentration in the crystal of interest connects the two. Boundary-layer thickness for chemical convection is much smaller than for thermal convection: both are greatly exaggerated to be shown. The crystal-settling bottom boundary is a much more efficient diffuser than a sharp solid/liquid interface and the conventional two-boundary chamber with the same convective flux would be almost twice as high (see text).

well-annealed mantle, or slowly-cooled lithosphere, there is no reason to expect such flaws or such macro-stoping. Micro-stoping could occur for a mineral mixture, where one component constituting at least 35% by volume of the matrix dissolves more rapidly than the other. The appropriate advance rate of "melting" would then be that of the faster-dissolving component. For example, diopside has a ten-times greater inverse liquidus slope in basalt than olivine, and this change, appearing to the 4/3 power in (10) dominates over its smaller α'. Also, diffusion rates are decreased more strongly by the temperature-depression due to the presence of water than they are enhanced by the presence of the volatile, so the advance rate of a magma body is dependent on the water content of the magma as well as the composition of the country rock. This is of interest in relation to the known eruptive sequence at hot-spot volcanoes like Hawaii, especially as the advance rate of chemically-controlled convection matches well with the observed time scale between different magmas of order million years (see below).

The lower boundary is one of settling crystals, as observed in numerous

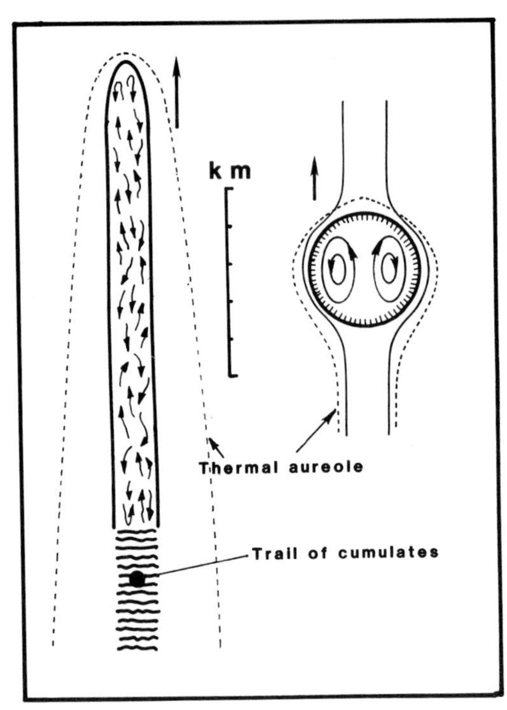

Figure 4. Comparison between upward migration of a self-convecting magma worm and a sphere of the same volume (11 km^3) rising by Stokes drift. Thermal-diffusion layers ($\sqrt{\kappa t}$) are shown by the dashed outlines, scaled for a diffusivity of 10^{-6} m^2 s^{-1}. Worm velocity 0.16 m y^{-1}; sphere drift 0.1 m y^{-1} in material of viscosity 10^{18} Pa-s (10^{19} poise). Arrows show upward advance in 10,000 y.

exposures of cumulate gabbro, both of layered intrusions and of ophiolite suites (e.g., McCallum et al., 1980). These crystals settle slowly (~0.5 mm s^{-1}) so that, although they may have a low concentration in the liquid (e.g., Lister, 1983), there can be a substantial number of them in the lowest few hundred meters of the chamber. This means that the area available for diffusion onto their surfaces is much greater than the area at a sharp boundary, and crystallisation at the bottom is unlikely to be rate limiting. (It should be noted that the settling itself does not assist diffusion, as the Reynolds number is of order .0003). Such an efficient boundary can be modeled by doubling the equivalent height of the magma chamber above the crystallisation zone for the purpose of calculating the transport rate (Figure 3), since the calculation assumes two diffusive boundary layers.

Boundary Layers

At a sharp solid boundary, both heat and chemical flux are transported entirely by diffusion. The rate of transport is defined as being increased by the factor N, the Nusselt number, and from the geometry of Figure 3 one can quickly deduce that the boundary-layer thickness must be $h/2N$. For the thermal convection case, this is already small: with parameters as before and a chamber height of 3 km it would be only 13 cm thick. Under conditions of chemical convection, it is still thinner: about 1 mm! This justifies the assertion above, that a crystal-settling region is a relatively efficient boundary – it would not be the case if the natural boundary layer were comparable in thickness to the zone of crystallisation.

There is another boundary thickness that is important when chemical convection is dominant. Some heat is transported by the chemical process, but the slow rate of advance of the boundary allows a thick thermal boundary layer to develop in the rock. Its thickness is κ/v, in a $1/e$ sense, and calculates to be of order 1 km. Magma bodies of less than 1 km height should not disturb the thermal state of their surroundings appreciably, but should continue to convectively diffuse upwards as along as the general gradient of temperature against depth is less steep than that of crystal-liquid equilibrium.

Mixed Convection

The lack of a large thermal transport associated with chemical convection means that upward movement of magma bodies might be confined to regions with a small geothermal gradient. These conditions are most closely met in upwelling plumes, whose gradient should be solid-mineral adiabatic (even lower than magma-adiabatic). Another region should be the asthenosphere, whose scale and viscosity imply a very low temperature difference across it:

$$\Delta T_c = \frac{1700\ \kappa\nu}{\alpha g\ h^3} \simeq 0.4°C \tag{11}$$

is the superadiabatic temperature difference for convective onset in a region 150 km thick with viscosity $\nu = 3 \times 10^{18}$ Stokes, thermal diffusivity $\kappa = .01$ $cm^2\ s^{-1}$, and expansion coefficient $\alpha = 4 \times 10^{-5}°C^{-1}$. Conversely, the lower lithosphere, with temperature increasing downwards relatively steeply, should be a zone where magma pockets of all sizes are stable.

How, then, can magma bodies penetrate the lithosphere to produce, for example, the hot-spot volcanism of ocean island chains? Penetration through dykes should be confined to the brittle upper lithosphere (e.g., Lister, 1986, and below) and must occur in times measured in hours to preserve magma liquidity (e.g., Spence and Turcotte, 1985). A non-convecting magma body is stabilised by the greater dissolution of the heavy phase at the base than at the top. However, if it is already convecting, a cold top surface would enhance the thermal convection that would be occurring anyway if the crystal dissolution were not the rate-limiting process. Therefore, the body can convect thermally to maintain the temperature of the top boundary above the

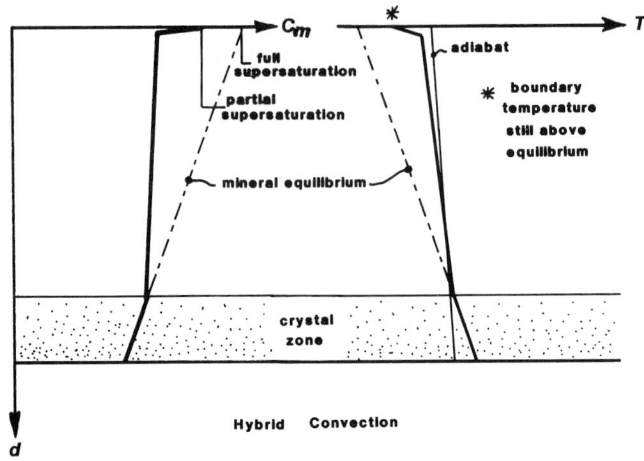

Figure 5. Conditions in a convecting magma chamber where the magnitude of heat loss by conduction at the top, Q_{SOLID}, lies between the maximum thermal flux transported by pure chemical convection Q_{CHEM} and the maximum available from thermal convection Q_{THERM}. If $Q_{SOLID} > Q_{THERM}$, the top boundary could grow downward by freezing, while if $Q_{SOLID} \ll Q_{CHEM}$, the chamber remains closely adiabatic. A Q_{SOLID} of order Q_{CHEM} must drop the temperature at the upper boundary and initiate thermal convection. There is no stationary equilibrium of $Q_{SOLID} = Q_{CHEM}$, since the latter requires dissolution to be taking place for it to exist.

liquidus, and continue to dissolve material more slowly at the top, and advance upwards.

The mechanism is explained in Figure 5. If chemical convection is dominant, the temperature in the magma conforms closely to the adiabat, leaving a substantial temperature rise at the top boundary from the state of mineral equilibrium. This temperature rise is what drives the chemical dissolution, producing the chemical convection. If the heat-flux loss at the top boundary is less than the maximum available from *thermal* convection driven by the disequilibrium between top and bottom of the chamber, then the level of thermal convection needed to provide it does not use up the full disequilibrium temperature difference. The remainder is a disequilibrium temperature available to drive dissolution. Thus the magma body continues to advance, but at a reduced velocity.

Penetration of Cold Lithosphere

Under old ocean floor and under stable continental cratons, the lithosphere grows by conductive cooling to thicknesses of order 100 km. The temperature profile will have approximately an error-function form prior to the first disturbance by magma (Lister, 1977). According to Lister (1986) the rigidus of mantle rock is reached when it cools to about 800 K (T_2), and we will be generous and let it cool from only 1600 K (T_1), the last temperature at which the magma is likely to be in equilibrium with the rock. If the temperature of the cooled surface is about 300 K (T_0), the rigidus defines a position on the error-function curve:

$$\frac{T_2 - T_0}{T_1 - T_0} \simeq 0.3846 = \mathrm{erf}(y_0); \quad y_0 = 0.3553. \tag{12}$$

The heat required per unit area to bring all the material deeper than where $T = T_2$ up to T_1 is then

$$\begin{aligned} Q^1 &= \int_{y_0}^{\infty} \rho c_\rho T_1 \,\mathrm{erfc}\left(\frac{z}{2\sqrt{\kappa t}}\right) dz \\ &= 2\sqrt{\kappa t}\, \rho c_\rho T_1 \left\{\frac{1}{\sqrt{\pi}} \exp(-y_0^2) - y_0 \,\mathrm{erfc}\, y_0\right\} \\ &= 0.72\sqrt{\kappa t}\, \rho c_\rho T_1. \end{aligned} \tag{13}$$

This heat can be supplied by a column of magma of height h when

$$Q^1 = h\rho^1 L \quad \text{(area ratio)} \tag{14}$$

where the 'area ratio' is the ratio of magma-worm cross-section to the area that includes lateral thermal diffusion. Allowing for the radial decay in temperature, this ratio is about 5, so that

$$h = 3.6\sqrt{\kappa t}\,\frac{\rho c_p T_1}{\rho^1 L} = 600 \text{ km @ } t = 100 \text{ My}. \tag{15}$$

Since magma columns are unlikely to reach a height of 600 km in the earth, the same spot in the lithosphere must be repeatedly attacked by magma bodies before one can penetrate to the level of dyke propagation.

Zone Refining: PLUME versus MORB

When a basaltic magma body melts its way upward through an ultramafic matrix it should become progressively more enriched in the 'incompatible' elements that fractionate out of olivine and pyroxene crystals. At depth, garnet crystals should also form at the crystallising base of the magma body. Where a mantle plume rises all the way to the surface, what was residuum at depth is brought up to be re-equilibrated with melt as magma bodies pass through it when it is near the surface. Thus, at whatever depth melting is

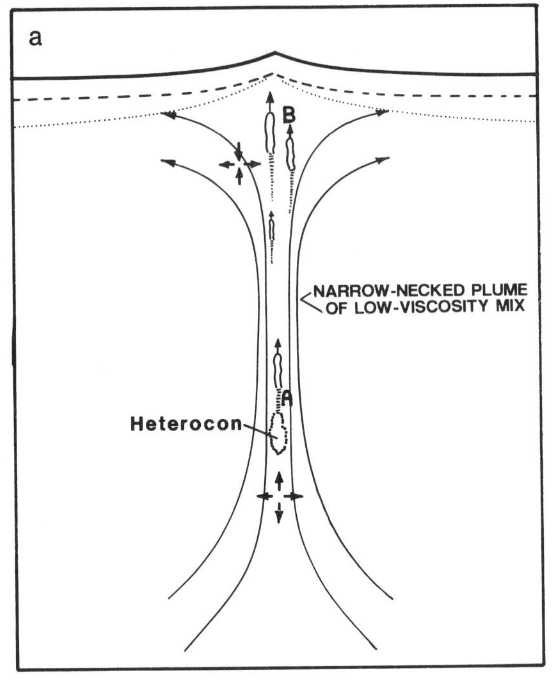

Fig. 6(a)

initiated and magma bodies separate, the chemical signature of the final magma is that of shallow melting (Figure 6a).

When magma separates from an asthenospheric plume at the base of the lithosphere, the garnet crystals are left behind at that point. Moreover, penetration of the lower lithosphere by melting leaves *all* of the residuum behind, and, however complex the process may be in detail, causes substantial enrichment of any magma that finally penetrates to the surface (Figure 6b).

Anderson (1981) summarises the differences between ocean-island basalts (PLUME) and mid-ocean-ridge basalts (MORB). The elements Yb and Y are relatively enriched in MORB: the explanation of Figure 6a is that garnet crystals, a residuum at depth and enriched in Yb and Y, are brought up by the mantle plume to be remelted by passage of magma bodies near the surface. Conversely, the elements most enriched in PLUME, K, Rb, Ba, La, V have the strongest partition coefficents out of the main lithospheric minerals. The scenario of Figure 6b causes this quite naturally. There is no need to postulate separate source regions for the two types of erupted basalt, or to argue about what the composition of 'primary' magmas might be (O'Hara, 1982). 'Primary' here is where magma can separate enough from the matrix to begin moving up by self-convection. Its composition may be very different from what finally erupts.

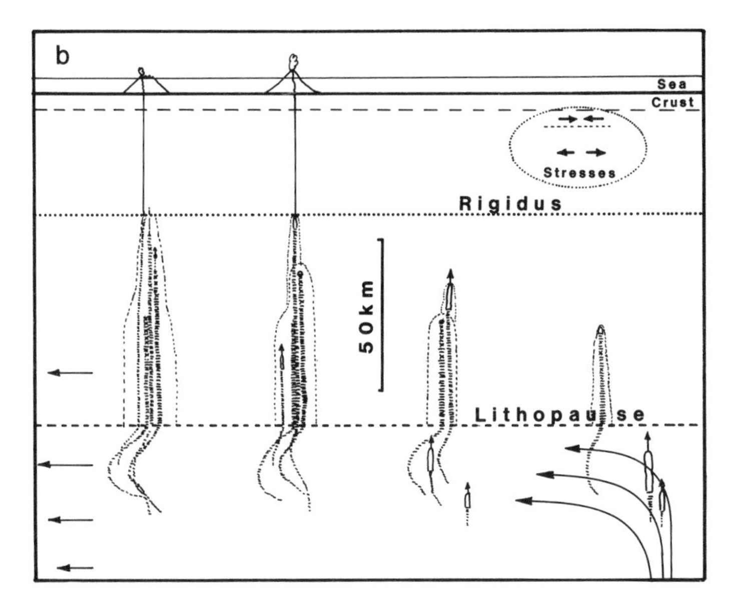

Figure 6. (a) Even if substantial separation occurs at depth (through melting of enriched inhomogeneities, e.g., Sleep, 1984), the garnet-enriched trails A would be brought up by the general flow to be remelted at B. Thus, the final mean magma signature would be that of shallow melting. (b) Penetration of a mature lithosphere by repeated attack from self-convecting magma bodies. Thermal-diffusion aureoles are shown growing with time (light dashed outlines). Dashed line: mechanical base of lithosphere; dotted line: position of rigidus; double arrows: differential thermal stresses in elastic layer; large arrows: velocity of plate, and flow in asthenosphere.

Chromitites, Hawaiian Volcanic Suites, and Conclusions

This paper has described a physical phenomenon, magma self-convection, that should occur if magma bodies exist in the upper mantle. The question is whether there is any evidence left in the rocks of the cumulate-like trail of magma passage, after further plastic deformation and annealing. One piece of evidence, perhaps not convincing by itself, is the common presence of chromite bodies in the peridotite underlying the magmatic section of an ophiolite (e.g., a few observed by Nicolas et al., 1988, Figure 4). Many of these bodies occur in the dunite sequence at the base of the cumulate gabbros. This is in accord with the idea that precipitation from the magma occurs at a particular stage in its evolution by crystallisation, as has long been known (Wagner, 1924). In some cases, upward propagating magma bodies can reach this level of evolution deeper in the ridge-crest diapir, and the chromite lenses so formed escape redispersion by the passage of new magma of less evolved composition. The presence of even a few chromite bodies below the petrologic Moho show that magmatic activity, including crystallisation, did occur in the upper mantle.

The puzzle of the peculiar sequence of Hawaiian eruptions, from alkalic-to-tholeiitic-to-alkalic-to-nephelinitic (Feigenson, 1986) is very elegantly solved by Figure 6b. Passage of a magma worm through fresh lithosphere causes extreme concentration of incompatible elements, but passage through the trails of previous worms causes less and less. Later, after the large magma bodies have either congealed in the lithosphere, or burst through, come smaller ones, originally smaller and thus slower, and also more enriched because of the greater ratio of crystal flux to magma volume.

Last come the volatile-rich and low-solidus residues of bodies that did not quite congeal, and that can continue slowly upward through the part of the lithosphere warmed by thermal diffusion from previous passages, delivering the nephelinites. It has already been mentioned that the presence of volatiles reduces the temperature of the magma body and therefore also the diffusion rate of chemical inhomogeneities. A slowing of convective transport by a factor of three would give the right order of magnitude for the delay in arrival of the nephelinites on Oahu.

It is clear that the thermal and geometric complexity of the situation would allow one to match almost any detailed eruptive scenario, and the elegance lies primarily in the absence of the need to mix various different mantle sources of additionally varying partial melt ratio (Feigenson, 1986). Of course, elegance is no guarantee of correctness, but I am persuaded by the power of two simple and understood physical processes,self-convection and thermal diffusion, to replace very complex scenarios of undefined physics. The upper mantle may well be heterogeneous on various scales, but there is no need for peculiarly separate geochemical reservoirs to be lurking below, unmixed over geologic time, waiting to supply the various different kinds of surface lavas.

Acknowledgements

This paper was conceived with the support of National Science Foundation grants OCE-861464 and OCE-8710059. The author thanks especially the University of Otago, New Zealand, for providing a William Evans Visiting Professorship that made the gestation and early writing possible. This is University of Washington School of Oceanography Contribution No. 1868.

References

Anderson, D.L., 1981. Hotspots, basalts and the evolution of the mantle., Science, 213: 82–89.

Batchelor, G.K., 1967. An introduction to fluid mechanics. Cambridge Univ. Press, p. 236.

Cooper, R.F. and Kohlstedt, D.L., 1984. Solution-precipitation-enhanced diffusional creep of partially-molten olivine-basalt aggregates during hot-pressing., Tectonophys., 107: 207–233.

Feigenson, M.D., 1986. Constraints on the origin of Hawaiian lavas., J. Geophys. Res., 91: 9383–9393.

Fukuyama, H., 1985. Heat of fusion of basaltic magma., Earth Plan. Sci. Lett., 73: 407–414.
Hyndman, D.W., 1985. Petrology of igneous and metamorphic rocks, 2nd Ed., McGraw-Hill, New York, 786 pp.
Katsaros, K.B., Liu, W.T., Businger, J.A. and Tillman, J.E., 1977. Heat transport and thermal structure in the interfacial boundary layer measured in an open tank of water in turbulent free convection., J. Fluid Mech., 83: 311–335.
Kuo, L.-C. and Kirkpatrick, R.J., 1985. Kinetics of dissolution in the system diopside-forsterite-silica., Am. J. Sci., 285: 51–90.
Lescher, C.E. and Walker, D., 1988. Cumulate maturation and melt migration in a temperature gradient., J. Geophys. Res., 93: 10295–10311.
Lister, C.R.B., 1977. Estimators for heat flow and deep rock properties based on boundary-layer theory., Tectonophys., 37: 157–171.
Lister, C.R.B., 1983. On the intermittency and crystallisation mechanisms of sub-sea-floor magma chambers., Geophys. J. Roy Astr. Soc., 73: 351–366.
Lister, C.R.B., 1986. Differential thermal stresses in the earth., Geophys. J. Roy. Astr. Soc., 86: 319–330.
Maaloe, S., 1981. Magma accumulation in the ascending mantle., J. Geol. Soc. London, 138: 223–236.
McCallum, I.S., Raedecke, L.D. and Mathez, E.A., 1980. Investigations of the Stillwater complex: part 1, stratigraphy and structure of the banded zone., Am. J. Sci., 280–A: 59–87.
Misener, D.J., 1974. Cationic diffusion in olivine 1400°C and 35 kbar, in: Hoffman, A.W. (Ed.), Geochemical Transport and Kinetics, Carnegie Inst. of Wash., pp. 117–128.
Murray, C.G., 1972. Zoned ultramafic complexes of the Alaskan type: feeder pipes of Andesitic volcanoes, in:, Geol. Soc. Amer. Mem., 132 (Hess vol.): 313–335.
Nicolas, A., 1986. A melt extraction model based on structural studies in mantle peridotites., J.Petrol., 27: 999–1022.
Nicolas, A., Boudier, F. and Cueleneer, G., 1988. Mantle flow patterns and magma chambers at ocean ridges: evidence from the Oman ophiolite., Mar. Geophys. Res., 9: 293–310.
O'Hara, M.J., 1982. MORB–a mohole misbegotten., Trans. Amer. Geophys. U., 63: 537.
Richter, F.M. and McKenzie, D.P., 1984. Dynamical models for melt segregation from a deformable matrix., J. Geol., 92: 729–740.
Sleep, N.H., 1984. Tapping of magmas from ubiquitous mantle heterogeneities: an alternative to mantle plumes?, J. Geophys. Res., 89: 10029–10041.
Sleep, N.H., 1988. Tapping of melt by veins and dikes., J. Geophys. Res., 93: 10255–10272.
Spence, D.A. and Turcotte, D.L., 1985. Magma-driven propagation of cracks., J. Geophys. Res., 90: 575–580.
Stolper, E., Walker, D., Hager, B.H. and Hayes, J.F., 1981. Melt segregation from partially-molten source regions: the importance of melt density and source-region size., J. Geophys. Res., 86: 6261–6271.
Takahashi, E., 1986. Melting of a dry peridotite KLB-1 up to 14 GPa: implications on the origin of peridotitic upper mantle., J. Geophys. Res., 91: 9367–9382.
Wagner, P.A., 1924. On magmatic nickel deposits of the Bushveldt igneous complex, Mem. 21, Geol. Surv. Un. Sou. Africa.

Melt Migration and Depletion – Regeneration Processes in Upper Mantle of Continental and Ocean Rift Zones

N.L. DOBRETSOV and I.V. ASHCHEPKOV
Institute of Geology and Geophysics, Siberian Division of the Academy of Sciences of the USSR, Novosibirsk, USSR

Abstract

Tectonized harzburgites constituting the ultramafic basement of inter-arc and back-arc basins and forming the lower part of corresponding ophiolites provide textural and mineralogical evidences that some veins and other local mineral segregations resulted from crystallization from infiltrated basic-ultrabasic melt and the reaction of that melt with harzburgites. Veins of dunite are found by reaction. Veins of pyroxenite and gabbro may be formed by disequilibrium crystallization from infiltrated melts during the general cooling. These vein assemblages occur in many ophiolites including Upper Precambrian and Palaeozoic ophiolites of the Urals and southern Siberia.

The evolution of the upper mantle beneath Baikal Rift Zone (BRZ) in the Cenozoic was estimated based on the studies of a great number of mantle xenoliths from the basaltic lavas and tuffs of three stages of volcanic activity (about 30 m.y., 18–9 m.y. and 5–2 m.y.). The xenoliths of the first stage are composed of Mg- and Cr-rich minerals whereas those of the second and the third stages are enriched in Al, Fe, Ti, Na. Magma segregation was deepest at the early stage, shifted upwards at the second stage and went down again at the third stage. These changes may be correlated with the compositions of lavas and the scale of volcanic activity. The geotherms plotted using garnet-bearing assemblages show that the temperatures for the same depth levels increased by about 100–150°C from the first to the last stage. We suppose that the heating and alteration of mantle was started by the intrusion of the picritic magma at the depth of 100–120 km. Green pyroxenites crystallized from picritic melts while black pyroxenites and amphibole-phlogopite cumulites were formed from basaltic liquids. A direct evidence of infiltration of such melts is glass-bearing veins in the mantle lherzolites. The composition of these glasses from the Cenozoic mantle xenoliths of BRZ includes olivine-rich to felsic differentiated liquids.

The composition of rocks and minerals show the same similarities and differences between the processes in oceanic and continental upper mantle. The main differences can be related to composition of migrating melt and

Tj. Peters et al. (Eds), Ophiolite Genesis and Evolution of the Oceanic Lithosphere, 125–146.

great depth of processes in continental upper mantle. In both cases the interactions of mantle peridotites with basic-ultrabasic melts and evolution of melts due to those reactions from picritic to basaltic liquids were followed by mantle diapiric rise. The texture-temperature relationships and distribution of different mantle peridotites reflect the local zoning in the mantle formed as a result of mantle diapirism. Such a process of melt migration and compositional change in the mantle has been referred to as "mantle metasomatism" (Manzies and Hawkesworth, 1987) or "paratexis" (Dobretsov, 1980). It might result in progressive depletion of the mantle as shown in the ophiolite harzburgite mantle or in the regeneration of the mantle at the regressive stage. A good example of the regeneration of the mantle is in BRZ.

Introduction

The origin of most veins and mineral segregations in the upper mantle samples both from oceanic and continental rift has, commonly, been explained by fluid-rock reactions broadly termed mantle metasomatism (Savelyeva, 1987; Whilshere et al., 1980; Lloyd and Bailey, 1975). Recently many authors have come to the conclusion that the main factor producing such veins is infiltration of intergranular melt into the rocks of the upper mantle (Dobretsov, 1981; Melyachovetskii et al., 1986; Francis, 1987; Edgar et al., 1989, etc.).

The vein stockworks of dunites and pyroxenites in harzburgites are an example of such a process. Whereas the origin of large lenses of dunites usually associated with pyroxenite veins is a special problem (Efimov et al., 1977), the dunite veins have been interpreted (Dobretsov et al., 1977) as traces of the infiltration of ultrabasic melts through solid harzburgites rather than as metasomatites or as a result of melting of peridotites "in situ". At a progressive stage they "wash out" all the components excluding residual olivine of extremely magnesian composition. The dunite veins and lenses are later followed by pyroxenite veins and after that by gabbros. A similar schema appears to be favoured in interpretation for such veins in the Oman ophiolites (Boudier, Nicolas, 1988; Juteau et al., 1990 etc.). The origin of dunite, pyroxenite and gabbro veins in harzburgites can be related to the repeated migration of melt through depleted harzburgite mantle. Such a situation is likely typical of marginal seas and inter-arc basins.

At the same time deep-seated xenoliths in kimberlites and alkaline basalt from the continental lithosphere mantle have generated a great debate on the role of the melt filtration. In this case the most direct evidence of such a process is the occurrence of veinlets with glass in mantle peridotites (Edgar et al., 1989; Dobretsov, Ashchepkov, 1990).

We shall try to summarise our observations in ancient (Riphean – Lower Palaeozoic) ophiolites of the Urals and southern Siberia and to compare them

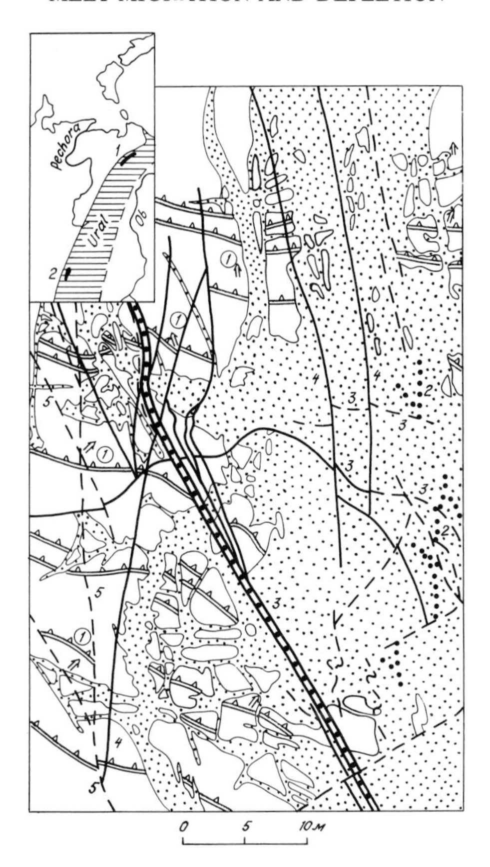

Figure 1. Map of the Voikar massif, Hayla river, showing the relationships between metamorphic harzburgites with early bending and lineation (1) and dunites (dotted area), veins of chromites (2), early enstatites and clinopyroxenites (3), late dunites with websterite rims (4) and latest clinopyroxenites and gabbro (5) (modified from Savelyeva,1987). In the upper left corner the location of Voikar (1) and Kraka (2) massifs in the Urals are shown.

with new data obtained from studying the Baikal rift zone. This comparison is intended to evaluate the role of melt migration in the process of evolution (depletion and regeneration) of the upper mantle in the oceanic and continental rift zones.

Supply for Magmatic Chamber in Ophiolites

Our discussion is based on our studies of ophiolites of the Ilchir belt in the southern Siberia, Voikar belt in the Polar Urals (Dobretsov et al., 1979; Dobretsov et al., 1985) and G.A. Savelyeva's work on Voikar belt and Kraka lherzolite massifs in the South Urals (Savelyeva, 1987). The location of the Uralian belts is shown in Figure 1 and that of the Ilchir belt near Baikal is

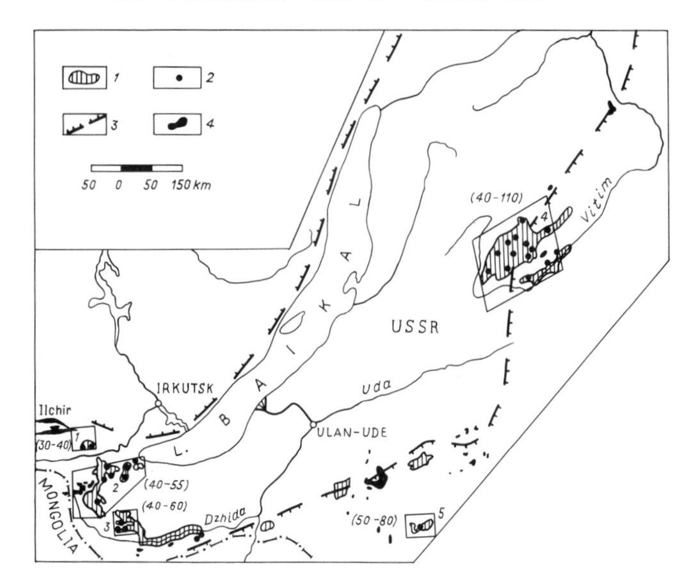

Figure 2. Geological sketch map showing the distribution of Cenozoic basalts (1); location of mantle inclusions (2) in Miocene and in Oligocene; in Pliocene basalts and ophiolite ultramafic bodies (4) in the Baikal rift zone boundaries of which (3) are given according to geophysical data (Krylov, 1984). The basaltic fields studied in detail are: 1. Tunka, 2. Khamar-Daban, 3. Dzhida, 4. Vitim, 5. Tchikoy (the intervals in km of xenoliths trapping are given in parenthesis).

shown in Figure 2. These ophiolites are described in detail in the cited works. We are interested only in the mantle part of the ophiolite section, therefore we include a brief summary of the mantle rocks in these sections.

The Voikar massif in the Polar Urals is one of the largest ophiolitic massifs in the USSR; it is more than 200 km long and up to 15–20 km in width. The mantle section is predominantly harzburgite with small relict bodies of lherzolites which are boudinated in the foliation plane marked by banding of harzburgites. The harzburgites are characterised by an abundance of dunite bodies and multiphase veins of dunite, pyroxenite and gabbro (Figure 1). As a whole, the ultrabasic part of the Voikar massif is petrographically similar to the southern sector of Semail ophiolites in Oman (Boudier, Nicolas, 1988; Nicolas, 1989).

The Ilchir belt (Figure 2) looks like peridotites of the central and northern sector of Oman (Juteau et al., 1990; Reuber, 1989). There are both a dunite-harzburgite complex similar to the Voikar massif and a wehrlite-pyroxenite complex cut by numerous veins of clinopyroxenites and gabbros. The wehrlite-pyroxenite complex is more usual for the northern branch of the Ilchir belt.

In the Kraka massifs (Figure 1), spinel and spinel-plagioclase in the lherzolites are predominant. Harzburgites form thin restite (together with dunites) lenses in lherzolites in the upper part of the section. Pyroxenite and dunite veins in lherzolite and harzburgite are rare and thin. Pyroxenites and wehrlites are the transitional zone between dunite-harzburgite rim and hornblende gabbro. Deformation structures of upper harzburgite and pyroxenite-wehrlite zones are inconformable with respect to the folding and the plastic flow structures in lherzolites. Because of lherzolite abundance and geochemical features this section resembles the possible mantle of oceanic type (Savelyeva, 1987).

The interrelations of the numerous and various veins in the Voikar belt are shown in Figure 1. Veins in the Ilchir belt show similar relationships excluding the enstatite and websterite veins. The earliest formations are microveinlets and pencil-like segregations of two pyroxenes, secondary olivine and spinel: these are most common in the transitional harzburgite-lherzolite rocks. They are nearly normal to the early banding of the harzburgites and often parallel to early lineation in harzburgites marked by spinel (see Figure 1). In our opinion, such aggregates are relics of the melt filtrated slowly through a solid substrate. New dunite veins cut these first generation veins. Pipe-like bodies up to 100–200 m in diameter (in Figure 1 the marginal part of such a pipe is presented) are oriented nearly normal to the harzburgite banding and surrounded with the vein stockwork of various veins. Chromite veins and cross-cutting veins of enstatite and clinopyroxenite occur in the centre of these bodies (Figure 1). The presence of enstatite veins is typical of the Voikar massif, the veins being rather deformed as well. These vein assemblages are related to progressive stages of filtration of picritic melt trough harzburgite followed by "washing-out" of all easily melted components to form reaction-restite dunites.

In the Voikar and Ilchir massifs, the websterite and wehrlite veins rich in clinopyroxene and without dunite rims form at the late regressive stage. They are followed by veins of clinopyroxenites and, finally, by gabbros including orthopyroxene-anorthite veins and zonal olivine – two-pyroxene – diopside gabbro. As in the Oman ophiolites, late veins of pyroxenite and gabbro are concentrated predominantly in the upper part of the harzburgites near the layered gabbros and gabbro-norites. This transitional zone has a vein-brecciated structure and has the highest vein concentration. The late websterites, wehrlites and clinopyroxenites are distinguished in many sections by lower mg′ (0.87–0.82) as well as by higher CaO, TiO_2 (0.12–0.25%) and Al_2O_3 compared to the earliest websterite and enstatite veins. Zonal gabbro and gabbro-norite veins have higher and more varied mg′ (0.84–0.70) and Al_2O_3 (6–2 wt.%) (Figure 3).

In contrast to the harzburgite-dominated section just discussed, the western zone of the Voikar belt and the northern branch of the Ilchir belt are characterised by thick pyroxenite-wehrlite complexes (Savelyeva, 1987; Dobretsov et al., 1977, 1985). The upper part of the ultrabasic sequence is composed of the impregnated dunites overlain by interbedded dunite, clinopyroxenite, wehrlite and then bedded wehrlite – pyroxenite – amphibole gabbro. This complex forms lens-like or dome-like structures and is replaced along its margins by gabbro-amphibolites. Around these marginal transition

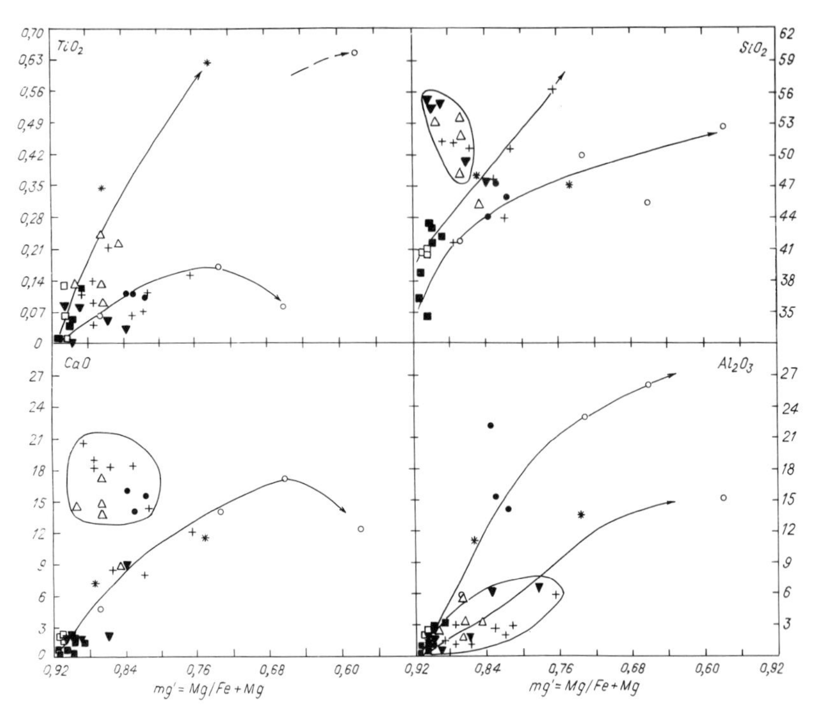

Figure 3. Composition of the metamorphic peridotites and veins in the typical ophiolites of the USSR (Voikar, Ilchir, Kraka). Black squares – dunites and harzburgites; open squares – lherzolites; black triangles – enstatitites; open triangles – wehrlites; crosses – clinopyroxenites, black dots – websterites, open dots – gabbro-norites; stars – average compositions of ophiolites and initial liquids of abyssal tholeiites from Table 1. Lines show trends of magmatic differentiation.

zones veins of massive wehrlite and pegmatitic gabbro are common. This sequence of rocks is very similar to dunite-wehrlite domes described in Central-Northern Oman (Reuber, 1988; Juteau et al., 1990).

The compositions of vein rocks and some typical minerals from the above ophiolites are summarised in Table 1 and Figures 3, 4. Most peridotites, some pyroxenites and all gabbro and gabbro-norite veins form a single trend of magmatic differentiation (Figure 3). This trend corresponds to the points of theoretical initial melts calculated for ophiolites and initial picritic liquid for abyssal tholeiite (Table 1). The composition of initial picritic melt was estimated from the distribution coefficient between olivine and liquid (Irvine, 1977), and possible composition of the initial liquid in Ilchir ophiolites was calculated as an average-weighted value from the composition of cumulates and dikes + basalts compared with upper gabbro (Dobretsov, 1980). Two trends are observed for TiO_2 and SiO_2. Plagioclase lherzolites, wehrlite and late clinopyroxenite have higher values of TiO_2, SiO_2 (Figure 3) and Na, Ba, Sr, Zr and Y (sample c-31 in Table 1) than harzburgites, websterites, orthopyroxenites and some clinopyroxenites that are distinguished also for

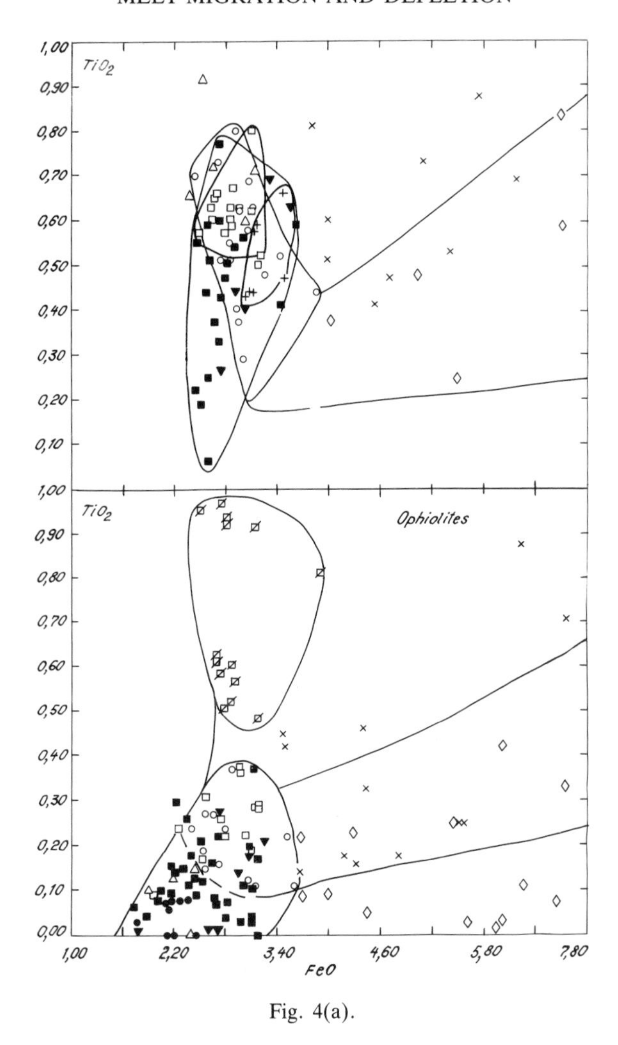

Fig. 4(a).

high contents of CaO and lower contents of Al_2O_3, Sr, and Ba, at similar Mg/Mg + Fe as in peridotites or magnesian gabbros. Along with the geological nature of these veins this is consistent with a reactional character of those rocks.

The mineral compositions confirm the regularities which are based on the bulk rock composition. This will be discussed below by comparison with minerals of the continental mantle xenoliths (Table 2, Figure 4).

According to Dobretsov (1980), Dobretsov and Zonenshain (1989), Quick and Gregory (1990) there are several types of ophiolites with distinctive magmatic supply systems. The first type is typical of mid-oceanic ridges and probably of the Kraka lherzolite massifs. The mantle substrate in this type is composed of lherzolite with an upper dunite-harzburgite zone of restricted

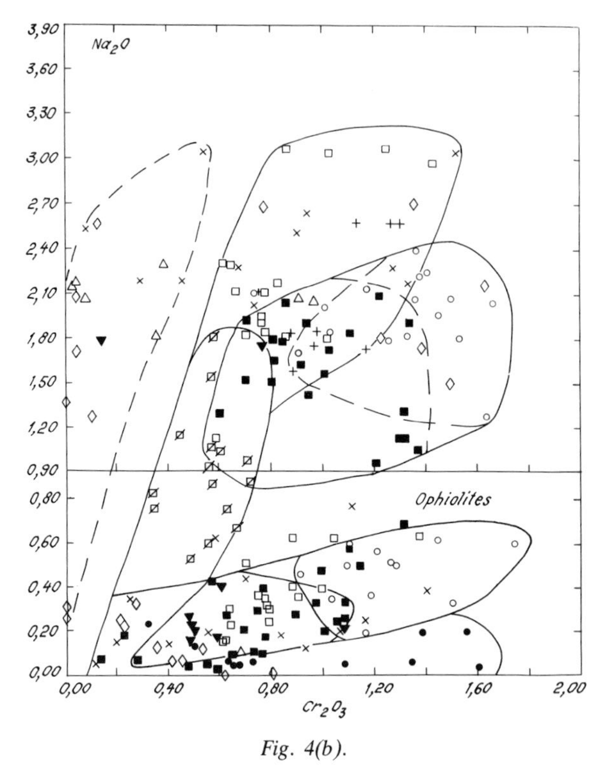

Fig. 4(b).

Figure 4. The compositions of the mantle peridotite clinopyroxenes from ophiolites (lower parts) and for deep- seated inclusion from Baikal rift (upper parts). Ophiolite rocks: diamonds – ophiolite pyroxenites and gabbronorites; bended crosses – wehrlites; crossed squares – Newfoundland lherzolites; open squares – Pacific ocean peridotites; black squares – Urals harzburgites and lherzolites; black dots – Tibet harzburgites; open dots – Urals plagioclase lherzolites; black triangles – Voikar pyroxenites; open triangles – Kraka pyroxenites.

Xenoliths from Baikal rift: black squares – dry spinel lherzolites; open dots – garnet lherzolites; open triangles – garnet websterites; open squares – amphibole lherzolites; crosses – phlogopite lherzolites; diamonds – cumulative pyroxenites; bended crosses – hydrous reactional mantle peridotites, modified by reactions.

thickness (up to 1.0–1.5 km). Gabbro, pyroxenite and dunite veins are rare and thin; these veins reflect fusion "in situ" and/or slow migration of melt.

Ophiolites with a system of dunite bodies and veins and deformed/undeformed multistage pyroxenite and gabbro veins define a second type of the magmatic supply system. This is typical of the thick harzburgite metamorphic peridotites similar to those of the Voikar, Ilchir and Oman ophiolites. These are interpreted to be possible fragments of back-arc or inter-arc basement, or other oceanic crust with multistage evolved and depleted upper mantle. In the second type, the earlier bedding in harzburgite and pencil-like aggregates or veinlets of the Sp-Cpx-Opx-Ol association may be due to the slow migration of the intergranular melt. This stage is similar to the veins in the

Table 1. Selected analyses of rocks and minerals from Ilchir and Volkar ophiolites.

	Ilchir vein pyroxenites and gabbro-pyroxenites						Ilchir peridotite and vein minerals				Voikar vein pyroxenites and gabbro-norites								
Components	1290	c-26a	1358	c-31	2560	2560a	1290 sp	1290 opx	2560 cpx	2560a cpx	211–2	215–3	255–33	11–8	14–16	14–4	212–5	Initial picrite melt (2)	Average[3] composition of ophiolites
SiO_2	54.46	49.66	48.15	52.26	47.40	46.04	0.10	56.08	53.67	53.68	55.41	51.95	45.78	53.60	42.00	43.56	45.41	47.70	46.40
TiO_2	0.02	0.05	0.13	0.15	0.11	0.10	–	–	0.20	n.d.	0.08	0.09	0.22	0.24	0.06	0.13	0.08	0.63	0.69
Al_2O_3	0.30	1.61	5.83	8.77	15.44	14.19	16.17	0.73	0.66	0.78	1.80	1.70	3.09	3.21	5.70	20.85	26.02	13.72	12.5
Cr_2O_3	–	–	–	–	–	–	55.06	0.44	0.22	0.24	0.32	0.28	0.46	0.30	0.16	0.03	0.08	0.13	–
Fe_2O_3	0.06	n.d.	n.d.	n.d.	0.81	0.09	0.26	–	–	–	0.93	2.11	1.86	3.06	1.48	1.88	1.99	0.37	–
FeO	7.1	8.6	5.93	8.53	3.63	4.90	17.39	7.61	5.39	5.25	5.27	3.95	8.06	3.09	8.04	8.76	2.54	9.17	11.9
MnO	0.18	0.15	0.13	0.20	0.11	0.11	0.17	0.41	0.12	0.28	0.17	0.17	0.18	0.14	0.18	0.13	0.07	0.16	0.18
MgO	33.50	29.42	21.42	15.61	11.82	12.35	11.24	32.79	17.06	16.87	32.70	21.09	29.68	21.05	34.32	12.04	4.73	15.80	16.55
CaO	0.60	2.17	13.78	12.02	13.97	15.52	–	1.39	21.95	22.12	2.01	17.23	8.69	14.75	4.67	11.19	17.01	10.14	10.6
Na_2O	0.43	0.27	0.53	0.80	1.75	1.14	–	–	0.22	0.33	0.11	0.29	0.36	0.38	0.19	1.06	1.35	2.07	1.18
K_2O	0.05	0.06	0.07	0.05	0.28	0.02	–	–	n.d.	n.d.	0.17	0.06	0.08	0.08	0.01	0.17	0.10	0.12	0.02
P_2O_5	0.05	0.01	0.05	0.04	n.d.	n.d.	–	–	–	–	0.03	0.02	0.02	0.09	0.04	0.09	0.01	0.05	0.06
CO_2			0.45	0.40			–	–	–	–	0.09	0.10	–	–	0.30	–	–	–	–
H_2O	1.80	7.57	2.77	1.83	4.76	5.02	–	–	–	–	0.43	0.54	1.45	0.32	2.24	0.16	0.95	–	–
Total	99.75	99.57	99.54	100.66	100.08	99.48	100.39	99.45	99.49	99.55	99.67	99.57	99.93	100.32	99.39	100.05	100.34	99.99	100.08
Sr	5	5	7	168	270	230	–	–	–	–	–	–	–	–	–	–	–		
Ba	9	7	7	15	45	60	–	–	–	–	–	–	–	–	–	–	–		
Cr	770	2300	1600	700	360	340	–	2000	1000	1200	1300	48	–	–	–	–	1500		
Co	36	34	64	40	35	37	210	40	45	49	32	23	–	–	–	–	55		
Ni	380	430	490	68	120	130	2000	490	220	205	230	20	–	–	–	–	700		
Ca	120	7	7	7	200	51	–	–	–	–	–	–	–	–	–	–	–		
Zn	230	–	68	n.d.	37	41	–	–	–	–	–	–	–	–	–	–	–		
V	50	–	122	171	50	50	–	–	–	–	–	–	–	–	–	–	–		
Zr	–	–	7.8	16	27														
Y	–	–	5.6	4.6	8.3	8.2													

[1] m′ = MgO/MgO + FeO, mol %
[2] Irvine, 1977
[3] Dobretsov, 1981
Compositions of rocks and minerals from (Dobretsov et al., 1985; Savelyeva, 1985).

first type of ophiolite but was longer or repeated to result in a large thickness of harzburgites, abundant multistage veins of dunite, pyroxenite and gabbro.

A third type of ophiolite exhibits dunite-wehrlite domes or lenses. These developed later than the veins characteristic of the second ophiolite type and may be connected with filtration of more alkaline melt of possibly deepest origin (Central Oman, West Voikar, North Ilchir ophiolites). Vein series in this case are represented by massive wehrlite, pyroxenite and pegmatitic gabbro which cut the lower gabbro of the previous stage. In the metamorphic peridotites we can see not only veins but also impregnated dunites and peridotites.

Composition and proportion of cumulate rocks, dikes and basalts correlates in general with the type of magmatic supply systems.

Veins and Reacted Rocks in the Baikal Rift Zone and the Regeneration of the Continental Mantle

The Baikal Rift Zone is an example of the evolution of the mantle caused by uprising of deep-seated melts. Three major stages of its evolution have been revealed by study of various mantle xenoliths from several areas of young basaltic volcanism (see Figure 2). The first stage (about 25 m.y.) was found only in Vitim plateau and was marked by an eruption of picritic tuffs carrying abundant garnet and spinel lherzolite inclusions. During the second stage (18–9 m.y.) a flood basalt plateau having overall thickness of about 500 m was erupted. Xenolith-bearing undersaturated alkaline basalts prevailed at the third stage, 5–2 m.y. (Dobretsov et al., 1989, Ashchepkov et al., 1990).

We have studied the mineral and bulk rock compositions of mantle xenoliths and their host lavas in four volcanic areas of the Baikal rift zone (BRZ) – Tunka (axial rift valley of BRZ), Khamar-Daban range (60 km from the axis), Dzhida river basin (120 km) and Vitim plateau (250 km from the axis (Figure 2)

Xenoliths of mantle peridotites are found in every volcanic area of the Baikal rift zone (Figure 2). Their petrography and composition vary within large intervals (Ashchepkov, 1990). The set and depth of xenolith capture depends on the tectonic position of the volcanic region relative to the depth of the asthenosphere (Ashchepkov et al., 1988; Dobretsov et al., 1989). Garnet-bearing lherzolites (up to 30 kbars) were found in a great distance from the axis of the rift in the Vitim area. Among the xenoliths from Dzhida only garnet-bearing cumulates are found. In the central part of Khamar-Daban only spinel-bearing assemblages were distributed, and only various low-pressure cumulates and rare depleted spinel lherzolites occur in basaltic lavas in the central part of the rift valley in Tunka.

Xenoliths of the first stage, are composed of Mg- and Cr-rich minerals and porphyroblastic garnets surrounded by olivine-rich zone. The rims of garnets around some spinel grains in other samples suggest the long and complex metamorphic history for these mantle rocks. Some of these xenoliths define a depleted mantle with model Sm-Nd age near 1.6–2.0 b.y. (Dobretsov et al., 1989). Besides garnet lherzolites the nodules of hot magnesian websterites and fragments of them were found at the first stage.

The second, Miocene, stage is characterised by low-temperature enriched Al and Na-rich mineral aggregates showing plastic deformations and kelyphite aggregates substituting garnet. At the beginning of this stages in some locations there are also high-temperature websterites and enstatites with zones and fragments of porphyroblastic harzburgites. They seem to be the veins with contact zones that are similar to dunite rims around pyroxenite veins in ophiolites. There are traces of melt filtration in the form of cross veinlets having lherzolite composition with minerals slightly enriched in Al, Fe, Ti, Na in common spinel lherzolites of this stage. Low-temperature veins of clinopyroxenite seem to be typical of the upper frontal part of the mantle diapir whereas more high-temperature veins of pyroxenites with dunite-harzburgite rims and relict garnet are typical of its deeper part, or of a later stage of melt migration. But large garnet occurs only in the deep-seated websterite veins of the first and third stages. As a whole, the veins of second stage vary from low-chrome olivine websterites with $mg' = 88–85\%$ to garnet websterites and clinopyroxenites with $mg' = 85–80\%$.

The xenoliths of the third stage exhibit some differences compared to other xenolith suites. At first they are the most enriched in Al, Fe, Ti, Na. This refers both to the typical green xenoliths (lherzolites, websterites) and abundant black clinopyroxenites ($mg' = 80–68\%$), with an Al-, Ti-, Na-rich augite as a principal mineral. Similar black pyroxenites occur but very rarely at stages I and II. Less common lherzolite and websterite xenoliths with hydrous minerals (phlogopite and amphibole) occur among the stage III xenoliths especially in Dzhida volcanic region (Melyakhovetskii et al., 1986). The phlogopite- and amphibole-bearing rocks also have veinlets of glass, the significance of which we shall discuss below.

Basalts of all three stages contain megacrysts of clinopyroxene and feldspar and additionally garnet, amphibole and phlogopite at the third stage. Megacrysts and minerals of black pyroxenites and amphibole-bearing pyroxenites are of similar composition. Both these rocks and megacrysts may be cumulative rocks or products of interaction with filtrated melt.

We suggest the general changes in the rift volcanism are linked with the evolution of the large diapir comparable to the anomalous mantle in BRZ and some local diapirs for instance in Vitim plateau (Dobretsov and Ashchepkov, 1989), believed to exist there.

The mass balance between the cumulative and green lherzolitic groups of inclusions are believed to be connected with the permeability of the mantle for melts. For example the volcanoes located at the crossing of Trance Khamar-Daban zone with rift faults contain a lot of cumulative high-temperature pyroxenites.

Compositions of the above groups of xenoliths and typical minerals are summarised in Table 2 and Figs. 4–6. Compositions of depleted lherzolites, garnet lherzolites and some lherzolites of the stages I and II as well as garnet websterites form a trend at Figs. 5 and 6 close to ophiolite trends (Figure 3). The glass compositions described below fall into the same trends. The second trend is formed by spinel lherzolites, phlogopite- amphibole-bearing rocks and black pyroxenites. This trend is generally similar to the alkaline wehrlite-pyroxenite trend in ophiolites, although somewhat different from it.

Many interrelations in these mantle xenoliths are difficult to interpret because of their small size and fragmentation of many xenoliths. However, the xenoliths are little altered and evidence of melt filtration and interaction with mantle rocks is well documented. These samples with evidence of melt filtration are subdivided into two groups: 1) high-temperature veins and contact zones near them; 2) veinlets with glass, amphibole or phlogopite, secondary olivine, clinopyroxene, spinel.

In complex xenoliths the host minerals are gradually enriched in Fe, Ti, Al towards the contacts of the veins. In some cases pyroxene compositions suggest a rise of temperatures in this direction based on data calculated from different geothermometers. The most useful geothermometer was that of Bertrand-Mercier (1985). The thickness of the contact zone depends on the temperatures of contacts and the type of vein. Near the most high-temperature orthopyroxenite veins ($T \geq 1200°C$, $m' = 80–82\%$) the compositions of host minerals varied slightly but inside veins these compositions rapidly changed. At the contacts of low-temperature essentially amphibole-phlogopite veinlets ($T = 900–1500°C$, $m' = 65–75\%$) the changes in mineral composition are noticeable only within 4–6 cm. The filtration and crystallization of melts can produce a range of cumulates and associated, impregnated or porphyroblastic, peridotites due to the interaction of different melts with the mantle rocks. Besides diffusion interaction, phase breakdown (especially garnet) during heating is possible. At decreasing temperatures, the crystallization of veins occurs including veinlets with glass.

Glasses have been found in the lherzolites most enriched in Fe, Ti, Al, K and Na. They are observed in interstices in the form of films enveloping olivine and clinopyroxene grains or in the form of more regular veinlets with amphibole (or phlogopite), secondary olivine, clinopyroxene and spinel. One such veinlet from specimen 17/1 (region of the Vitim river) is shown in Figure 7 and the glass and mineral compositions of the host lherzolite are shown in Table 3. Olivine and clinopyroxene here are more magnesian than the primary minerals of lherzolite. Clinopyroxene also contains less Si, more Ti and Ca. Spinel is poorer in magnesium and contains much more Cr. The mg′ and calcium content of the glass (gl1 in Table 3) are close to the composition of the earliest picrites on the Vitim plateau but glass is markedly richer in Al, Si and Na (i.e. in albite or jadeite components). The disappearance of orthopyroxene and tendency in the composition of glass (enrichment

Table 2. The representative analyses of mantle rocks and minerals from BRZ.

	Rocks							Minerals														
Compo-	358–	302–	315–	315–	303–	313–		17–44				17–75			302–12		313–10					
nents	–1	–12	–46	–11	–11	–54		cpx	ol	opx	cpx	sp	opx	cpx	sp	opx	opx	cpx	ol	gr		
SiO_2	44.8	43.06	50.6	50.2	42.68	42.44	SiO_2	52.28	40.57	55.15	51.91	–	54.75	52.03	0.08	54.70	55.88	52.77	40.76	42.98		
TiO	0.05	0.29	0.52	0.95	0.99	0.53	TiO_2	0.67	–	0.16	0.44	0.28	0.12	0.81	0.55	0.19	0.08	0.52	–	0.19		
Al_2O_3	2.0	1.89	8.0	16.8	14.56	5.59	Al_2O	5.36	0.03	3.75	5.35	51.48	3.84	5.74	48.80	4.69	2.85	4.98	–	21.21		
Fe_2O_3	1.86	1.56	2.44	4.17	5.28	0.39	Cr_2O_3	0.78	–	0.36	0.90	11.10	0.33	0.86	13.05	0.36	0.55	1.38	–	0.93		
FeO	6.33	7.82	5.54	5.42	6.80	7.99	FeO	2.20	10.28	6.36	2.90	13.28	6.29	5.32	15.06	6.55	6.35	3.31	9.76	8.12		
MgO	44.0	43.04	18.6	7.6	13.78	38.39	MgO	0.07	0.19	0.10	0.07	0.15	0.14	0.10	0.19	0.14	0.10	0.10	–	0.28		
CaO	0.91	1.59	12.5	9.5	12.97	2.54	CaO	14.61	49.0	33.20	15.92	23.97	33.71	15.57	22.49	32.03	32.99	15.94	49.33	20.87		
Na_2O	0.06	0.35	1.10	3.4	1.04	1.41	Na_2O	22.05	–	0.71	21.04	–	0.57	18.32	–	1.04	0.84	18.37	0.11	5.11		
K_2O	n.d.	0.19	0.04	0.68	0.43	0.17	K_2O	1.70	–	0.05	1.41	–	0.08	0.88	–	0.12	0.15	1.96	–	0.08		
P_2O_5	0.03	n.d.	0.01	0.09	0.04	n.d.	–	–	–	–	–	–	–	–	–	–	–	–	–	–		
CO_2	0.07	n.d.	0.08	0.12	0.12	n.d.	–	–	–	–	–	–	–	–	–	–	–	–	–	–		
H_2O	0.06	0.57	0.29	0.35	0.44	0.79	–	–	–	–	–	–	–	–	–	–	–	–	–	–		
Total	100.17	100.49	99.72	99.27	99.13	100.39	Total	99.62	100.07	99.79	99.93	100.24	99.78	99.63	100.22	99.82	99.77	99.23	99.97	99.77		
Ni	1884	2200	450	44	390	1800		400	1800	1000	350	2000	2000	–	–	–	1000	380	2500	100		
Co	90	68	46	30	60	561		25	120	80	26	250	120	–	–	–	100	25	120	50		
V	41	74	42	108	170	47		260	2	72	180	500	2	–	–	–	120	250	2	63		
Zn	46	–	299	220	220	50		19	60	37	23	430	67	–	–	–	56	26	65	22		
Sc	7.7	–	22	29	–	–		80	2	18	80	3	2	–	–	–	5	35	2	120		
Y	0.5	–	6.4	23	–	–	Cu	14	8.2	11	14	18	12	–	–	–	11	7.7	9	7.7		
La	–	–	2	6	–	–																
Sr	3.2	4.3	94	555	–	–																
Ba	2	30	42	280	–	–																

The analyses of rocks and microprobe minerals from (Ashchepkov, 1990).

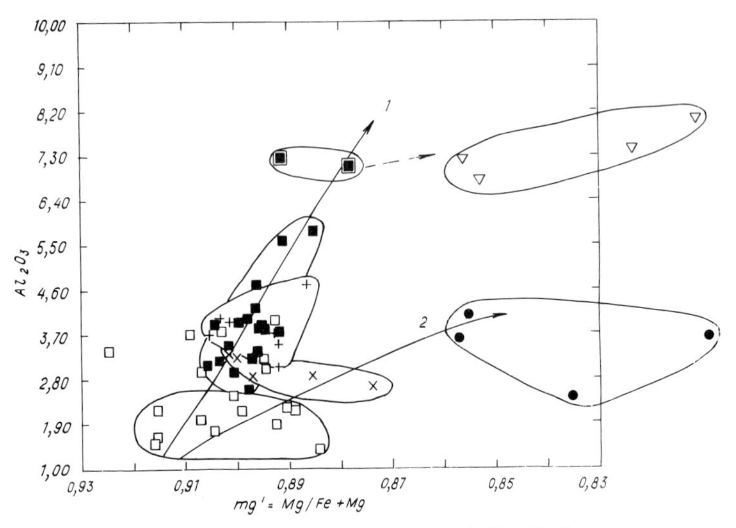

Figure 5. Compositions of deep-seated inclusions from the Baikal rift. Black squares – garnet lherzolites; double squares – garnet websterites; open squares – spinel lherzolites; crosses – amphibole lherzolites; bended crosses – phlogopite lherzolites; triangles – green cumulative pyroxenites; black dots – reactional mantle peridotites. Lines 1 and 2 show two ways of differentiation.

in SiO_2 and alkali content, decreasing in MgO and FeO) are similar to the experimental glass composition (gl9 in Table 3) produced by the interaction of melt with mantle peridotites and assimilation of orthopyroxene (Fisk, 1986). Similar results are obtained by numerical calculations (Kelemen, 1986).

Other glasses (gl2–gl4 from specimen 302–17) are richer in K_2O and poorer in magnesium (mg′ = 64–59%). Their TiO_2 contents are close to the above picrites. These felsic glasses are similar to those in mantle xenoliths from Eifel, Germany (Edgar et al., 1989). Glasses in basalts (mg′ = 0.37–0.28) and minerals in them are markedly different in magnesium content and other parameters, so glasses in veinlets in lherzolites can be neither the result of basalt liquid intrusion, nor the product of their interaction near surface (Kelemen, 1986). The veinlet composition calculated from the observed modal composition (20% Ol, 10% Cpx, 1% Sp, 4% (hornblende + phlogopite)) is richer in magnesium (mg′ = 86%) and Na, Si than picrite from the Vitim plateau at a similar Al, Ca, K content. In the diagram (Figure 6) based on the ratio mg′-SiO_2 and mg′-Al_2O_3 the calculated vein composition falls on the trend phlogopite-amphibole rocks – pyroxenites – black pyroxenites.

The primary melt was probably transitional between the composition of picrite and reactional veinlets with glass. Its interaction with mantle peridotites led to a change of composition and regeneration of the mantle and the appearance of a new melt similar at first stage to the Vitim plateau picrite

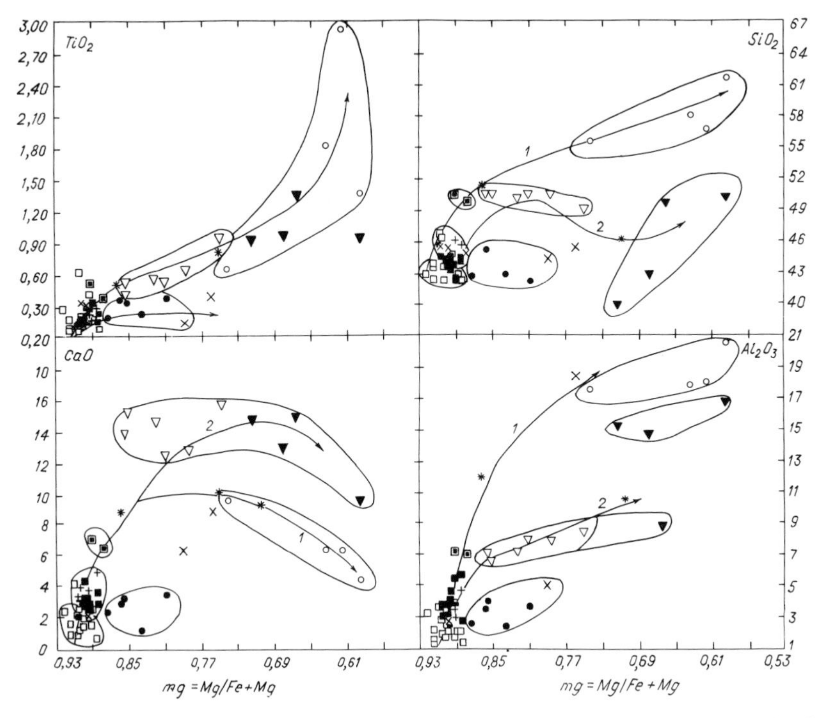

Figure 6. Compositions of the deep-seated inclusions from the Baikal rift and trends of the mantle melt differentiation (lines 1 and 2). Signs are the same as in Figure 4 and additionally: black triangles – black cumulative pyroxenites; open dots – intergranular glasses in peridotite xenoliths; stars – average composition of the intergranular veins and the composition of host picrite tuff from Table 3. Broken lines are trends of magmatic differentiation in ophiolites from Figure 3.

and than to the plateau basalts. This process began 25–27 m.y. after the intrusion of the first stage picrites.

The level of magma segregation was the deepest at the early stage, it shifted upwards at the second stage and went down again at the third stage, which may be correlated with the composition of magma and the scale of volcanic activity. The geotherms plotted using garnet-bearing assemblages (Nickel, Green, 1985; Bertrand, Mercier, 1985) show that the temperatures for the same depth levels increased by about 150°C from the first to the last stage (Figure 8). Since the geotherm of the first stage has an inflection point ("K" in Figure 8) due to the presence of high-temperature (>1250°C) green websterites and tectonized lherzolites we suppose that the heating of the mantle was caused by the intrusion of the picritic magma at a depth of 100–120 km. Sm-Nd isochrone ages for earlier peridotites range from 47 to 27 m.y. (Dobretsov et al., 1989). The Sm-Nd isotopic data were obtained in Mainz, BRG (Ionov and Jagoutz, 1988). The younger values (27 m.y.) can be in-

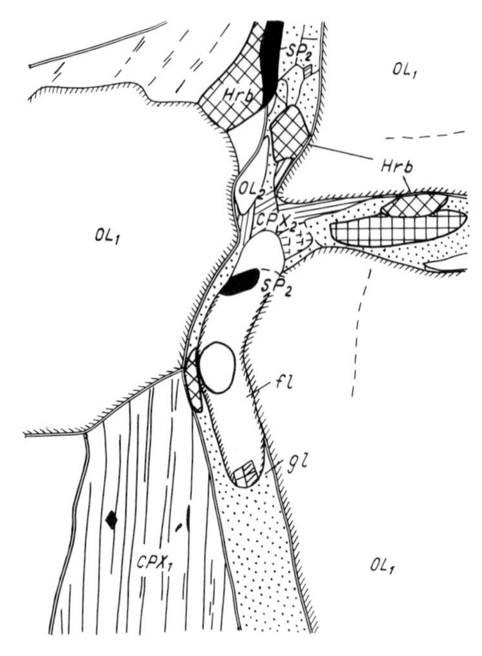

Figure 7. The scheme of the thin section of a lherzolite inclusion containing veinlets of intergranular glass (gl, dotted area) and small secondary mineral grains of amphibole (Hrb), spinel (Sp2, black), olivine (Ol2, white) and clinopyroxene (Cpx2) between coarse-grained primary mineral grains. Some gaseous blebs and fluid inclusions (fl) in glass are also shown. Length of bar is 100 mm.

terpreted as the eruption age of the host picrite-basalts and the older one may show a earlier stage of mantle alteration.

The heating of the lithospheric mantle accompanied by garnet (O'Neill, 1981) and amphibole breakdown, reaction of mantle peridotites with picritic magmas, migration of intergranular liquids and evolution of the picritic melts to form basaltic ones were followed by mantle uplift. This diapir stopped near the Moho and was the source for plateau basalts of the second stage. Zonal distribution of the average mantle xenolith temperatures at the Vitim plateau (Ashchepkov, 1990) supports the real existence of some local diapir developed from the large diapir elongated along the rift zone. When it cooled down, tectonic activization in Pliocene resulted in eruption of alkaline basalts from deeper levels of the altered mantle. The basalts of the third stage were originated from a residual melt and reflect the cooling and cessation of deep melt migration. They contain xenoliths of the previous stage of mantle regeneration.

Comparison and Discussion

We have examined the trends and composition of mantle rocks in ophiolitic and continental xenoliths. In conclusion, it will be useful to compare the

Table 3. Composition of minerals and glasses from glass-bearing lherzolites.

	Primary lherzolite minerals				Minerals in glass veins 17/1				17/1		Glasses in lherzolites				Glasses in basalts				Experimental
Components	ol	opx	cpx	sp	ol1	ol2	cpx	sp	veins[a] in average	picrite	gl1	gl2	gl3	gl4	gl5	gl6	gl7	gl8	gl9
SiO_2	40.85	56.6	54.82	0.76	41.1	41.1	50.6	1.49	51.7	44.93	55.76	56.74	58.0	61.7	52.82	53.46	51.1	50.0	53.5
TiO_2	0.00	0.11	0.18	0.12	0.02	0.00	0.42	0.21	0.50	1.88	0.64	2.96	1.83	1.37	0.09	0.60	2.22	2.34	1.31
Al_2O_3	0.00	3.04	4.58	45.60	0.13	0.09	4.43	20.1	12.0	10.04	17.6	18.0	17.8	20.5	16.33	16.45	16.0	14.9	14.65
Cr_2O_3	0.02	0.33	2.12	20.65	0.09	0.08	2.98	45.9	0.80	2.14	0.05	0.07	0.00	0.00	0.00	0.00	0.00	0.00	0.07
FeO	9.38	6.5	2.76	13.41	7.00	7.2	2.06	16.34	4.7	11.02	4.21	3.35	3.0	2.75	10.68	10.43	9.77	14.0	8.50
MnO	0.16	0.15	0.11	0.22	0.11	0.18	0.07	0.46	0.08	0.24	0.06	0.05	0.06	0.05	0.01	0.01	0.11	0.11	0.15
MgO	50.0	34.0	15.23	18.33	51.1	49.3	15.77	14.2	16.5	17.68	6.82	3.01	2.91	2.27	3.11	2.31	2.98	4.62	7.45
CaO	0.07	0.31	18.11	0.00	0.28	0.29	22.60	0.0	8.9	9.51	9.54	6.36	6.39	4.46	9.15	10.15	8.78	8.12	11.27
Na_2O	0.0	0.09	2.60	0.0	0.0	0.0	0.65	0.0	2.93	0.72	4.49	4.21	4.47	1.92	3.79	4.04	3.37	2.14	3.06
K_2O	0.00	0.00	0.00	0.00	0.00	0.00	0.01	0.00	1.49	1.34	1.68	5.89	5.10	4.00	1.26	1.79	4.67	3.73	0.37
H_2O[b]	–	–	–	–	–		–	–	0.40	0.10	<1.0	–	0.50	1.00	2.00	1.00	1.00	1.00	0.7
Total	99.48	100.13	99.87	99.04	99.83	98.24	99.59	98.70	100.00	100.8	100.9	101.10	100.01	100.02	99.24	100.24	100.00	100.56	100.03
mg′[c]					0.82	0.84			0.86	0.71	0.74	0.64	0.63	0.59	0.34	0.28	0.35	0.37	0.60

[a] 65% gl1 + 20% (ol1 + ol2) + 10% cpx + 1% sp + 4% phl
[b] H_2O contents are theoretical
[c] m′ = MgO/MgO + FeO, mol%

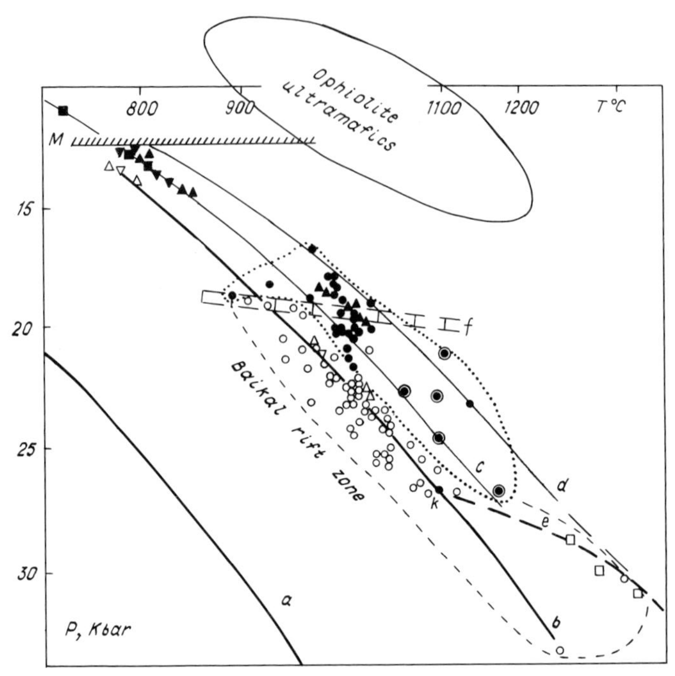

Figure 8. Thermal evolution of the Vitim plateau mantle reconstructed by the P-T – estimates for deep-seated < xenoliths (see text for explanation). Open marks – xenoliths from picrite-basalts of I stage; black marks – xenoliths from Pliocene hawaiites of III stage. Dots – garnet lherzolites; double circles – the same enriched in Fe; triangles – spinel lherzolites; and depleted spinel lherzolites; squares black pyroxenite cumulates; and green pyroxenite cumulates; 9) standard geotherms: a) shield, b) oceanic; 10) geotherm by xenoliths from Oligocene picrite-basalts; k,e = inflection of the geotherm formed by the picrite melt migration; dotted field of T-P estimates by xenoliths from Pliocene basalts; f = the line of garnet-spinel transition; M = Moho boundary. Broken line shows field of T-P estimates by xenoliths from Oligocene picrite-basalts.

mineral compositions of the two suites and to show general similarities of melt migration processes in the oceanic and continental mantle. For this comparison we use more widely cited data (from Newfoundland and Tibet ophiolite, Mongolia and Pyrenees xenoliths etc.)

Clinopyroxene is the most abundant mineral in the veins and host peridotites from ophiolites and upper continental mantle. When comparing the clinopyroxene composition from different ophiolitic rocks with those from deep-seated xenoliths of BRZ (Tables 1–3 and Figure 4) the greatest similarity appeared in the components (FeO, MgO, TiO_2, Cr_2O_3) which most strongly depend on the rock (or melt) composition. With the trend of changes of the composition of peridotites, reaction rocks and veins (see Figure 3, 5 and 6) being similar, it is obvious that the composition ranges and trends of the change of these components in clinopyroxenes will be similar as well.

Thus the ranges of changing FeO and TiO_2 and their general trend of differentiation are similar both for peridotites of ophiolites and deep-seated xenoliths and for their clinopyroxenes (Figure 4A). The clinopyroxenes most poor in titanium and iron are those from harzburgites and spinel lherzolites (Voikar, Ilchir, Tibet etc.). The highest TiO_2 clinopyroxenes (0.5–0.95% TiO_2) are from lherzolites and wehrlites of Newfoundland (Mercier et al., 1984) and correspond to titanium-rich (0.5–0.9% TiO_2) pyroxenes from amphibole-bearing peridotites and veins. A higher iron content (3–7% FeO) is characteristic for pyroxenes from veins in the mantle xenoliths. There are two levels of TiO_2 content in such xenoliths: a higher one for amphibole-phlogopite pyroxenites and a lower one for "dry" pyroxenites. Two similar ranges in TiO_2 contents are typical of clinopyroxenes from two stage veins in ophiolites, but there is a third level of the titanium-poorest clinopyroxenes (up to 0.15–.0.25% TiO_2) at the same content of FeO (3–7% FeO). Some pyroxenes from early veins fall into this range. They are common in or near dunite bodies in the Voikar and Ilchir massifs.

Similar ranges are seen for Cr_2O_3 (Figure 4B), but contents of Na_2O, Al_2O_3 and SiO_2 related to the crystallization conditions are sharply different. The pyroxenes from deep-seated xenoliths were crystallized at higher temperatures and especially pressures, therefore, they are richer in jadeite (0.9–3.0% Na_2O). In ophiolite pyroxenes, the contents of Na_2O 0.9–1.8% are reached only in the Newfoundland ophiolites and in those from the western metamorphic part of the Voikar belt. Most ophiolites pyroxene contain less than 0.85% Na_2O and the Cr_2O_3 content is similar to the groups under comparison. Thus, the Cr-rich clinopyroxenes from garnet peridotites (0.9–1.75% Cr_2O_3) have their analogues from Kraka plagioclase lherzolites and some pyroxenites (0.9–1.80% Cr_2O_3, see Figure 4B). The pyroxenes from xenoliths of spinel lherzolites are poorer in Cr (0.4–1.40% Cr_2O_3) and similar to pyroxenes from the Kraka spinel lherzolites and harzburgites of various massifs (0.2–1.0% Cr_2O_3). The trend of increasing Na_2O with a slight increase in Cr_2O_3 is fixed for clinopyroxenes from Newfoundland and some rocks of the Voikar belt. Amphibole- and phlogopite bearing xenoliths display a similar trend. In the deep-seated xenoliths there is a Cr-poor (to 0.3% Cr_2O_3) but Na-rich (1.2–3.0% Na_2O) group of pyroxenes from amphibole-phlogopite veinlets and some websterites, which are close to pyroxenes of eclogitic paragenesis.

Thus, mantle lherzolites can be subdivided into altered (enriched or depleted) and primary groups in both cases of oceanic and continental mantle. But dunites and depleted rocks are more abundant in the oceanic mantle (Figs. 5, 6), whereas hydrous and LREE-rich lherzolites and cumulates can be found in the continental mantle. This difference is due to thermal conditions of melt migration which are determined by the scale and intensity of the mantle convection. A large volume of melts ascending over a long time from beneath the mid-ocean ridges and repeated migration in back-arc or inter-arc basement leads to a strong depletion of the mantle. Pyroxene geothermometry and solid solution structures demonstrate that the temperatures of the ocean mantle often exceed 1300°C (Dobretsov et al., 1977; Nicolas 1989). These types of pyroxenes are rarely found in the cool lithosphere mantle under the continents. This extensive melting might occur at the earlier stage of continental mantle origin but is usually not preserved during later stages. Alkali melt migration and interaction of magma with wall rock mantle peridotites tends to enrich continental rift peridotites. A shallow depth of the origin for ophiolite peridotites is indicated by the low jadeite-component content in their pyroxenes. These differences are connected with the regime of filtration.

A process of melt migration in the mantle and related depletion-regeneration of mantle was called "mantle metasomatism" (Lloyd and Baily, 1975; Menzies and Hawkesworth, 1987; Efimov et al., 1978; Whilshire et al., 1980; Fabries et al., 1989), mantle-melt interaction (Fisk,1989; Francis, 1987) or "paratexis" (Dobretsov, 1980; Dobretsov et al., 1985). The latter term differs from the partial melting of mantle (or anatexis in the continental crust). It means extensive interaction between melts and host mantle rocks with compositional alteration of both melts and mantle. It might result in the depletion in repeated stages as in the ophiolite harzburgite mantle and in regeneration of the mantle on a progressive stage as in the Baikal rift zone (Ashchepkov, 1990) or in the Pyrenees (Azambre, Fabries, 1989).

Acknowledgements

We are greatly indebted to S.H. Bloomer, J. Sinton and P.B. Kelemen for thoughtfully and carefully reviewing the manuscript and for their constructive remarks. G.A. Savelyeva supplied the geological map and some additional information on the Voikar and Kraka massifs. Discussions with R.G. Coleman helped focus our ideas.

References

Ashchepkov, I.V., 1990. Deep-seated inclusions at Baikal rift. Novosibirsk. Nauka, 210 p. (in Russian).

Ashchepkov, I.V., Dobretsov, N.L. and Kalmanovich, M.A., 1988. Garnet lherzolites in alkaline picrites and basanites from Vitim plateau – Doklady Akademii Nauk SSSR, 302, 2: 417–420. (in Russian).

Azambre, B. and Fabrie, J., 1989. Mesozoic evolution of upper mantle beneath eastern Pyrenees: evidence from xenoliths from Triassic and Cretaceous alkaline volcanics of eastern Corbieres (France), Tectonophysics., 170, 3/4: 213–230.

Bertrand, P. and Mercier, J.J.-C., 1985. Mutual solubility of coexisting ortho- and clinopyroxene: toward an absolute geothermometer for natural systems?, Earth Planet. Sci. Lett., 86: 109–120.

Boudier, F. and Nicolas, A. (Eds.), 1988. The ophiolites of Oman, Tectonophysics, 151: 255.

Dobretsov, N.L., 1980. Introduction to global petrology. Novosibirsk. Nauka, 280 p. (in Russian).

Dobretsov, N.L., Ashchepkov, I.V. and Ionov, D.A., 1989. Evolution of the upper mantle and basaltic volcanism in Baikal rift zone. In: Crystalline crust in space and in time (magmatism). Moscow, Nauka, pp. 5–20. (in Russian).

Dobretsov, N.L., Kazak, A.P. and Moldovantsev, Yu. E, et al., 1977. Petrology and metamorphism of ancient ophiolites. Novosibirsk. Nauka, 221 p. (in Russian).

Dobretsov, N.L., Konnikov, E.G., Medvedev, V.N. and Sklyarov, E.V., 1985. Ophiolites and olistostroms of East Sayan. In: Riphean-Paleozoic ophiolites of North Eurasia, edited by N.L. Dobretsov. Novosibirsk, Nauka, p. 34–58 (in Russian with English abstracts).

Dobretsov, N.L. and Zonenshain, L.P., 1989. The evolution of pre-Mesozoic ophiolites in Northern Eurasia: A comparative review. Chem. Geol., 77: 323–330.

Edgar, A.D., Lloyd, F.E., Forthys, D.M. and Barnret, R.L., 1989. Origin of glass in upper mantle xenoliths from the quarternary volcanic of Gees, West Eifel, Germany, Contrib. Miner. Petrol., 103, 3: 277–286.

Efimov, A.A., Lennykh, V.I., Puchkov, V.N., Savelyev, A.A., Savelyeva, G.N. and Yaseva, R.L., 1978. Guidebook for excursion "Ophiolites of Polar Urals". Moscow, Geological Institute, 165 p. (in Russian).

Fabries, J, Bodinier, J.L., Dupuy, C., Lorand, J.P. and Benkerrou, C., 1989. Evidence for modal metasomatism in the orogenic spinel lherzolite body from Caussou (Northern-East Pyrenees, France), J. Pertrol., 30, 1: 199–228.

Fisk, M.R., 1986. Basalt magma interaction with harzburgite and the formation of high-magnesium andesites, Geophys. Res. Lett., 13, 5: 467–470.

Francis, D., 1987. Mantle-melt interaction recorded in spinel lherzolites from the Alligator Lake volcanic complex, Yukon, Canada, J. Petrol., 28: 569–598.

Fujii, T. and Scarf, C.M., 1985. Composition of liquid coexisting with spinel lherzolites at 10 kbar and the genesis of the MORB, Contrib. Mineral. Petrol., 90: 18–28.

Jagoutz, E., Palme, N., Buddenhausen, N. et al., 1979. The abundances of major, minor and trace elements in the Earth mantle as derived from primitive ultramafic nodules. 11 Lunar Sci. Conf., p 10.

Juteau, T., Beurrier, M. and Nehlig, P., 1990. Symposium on ophiolite genesis and evolution of oceanic lithosphere. Excursion D.

Irvine, T.N., 1977. Definition of primitive liquid composition for basic magmas. Carnegiee Inst. Wash. Year Book 76: 459–461.

Kelemen, P.B., 1986. Assimilation of ultramafic rock in subduction related magmatic areas, Journal of Geology, 94: 829–843.

Krylov, S.V., 1984. Combining of the methods of seismology of explosions and earthquakes for investigations of deep structure of Baikal rift. In: Regional complex geophysical explorations of crust and upper mantle. Moscow, Radio and Communication, pp. 80–87. (in Russian).

Lloyd, F.E. and Bailey, d.K., 1975. Light element metasomatism of the continental mantle: the evidence and consequences. Phys. Chem. Earth, 9: 275–280.

Melyakhovetskii, A.A., Ashchepkov, I.V. and Dobretsov, N.L., 1986. Amphibole- and phlogopite-bearing deep-seated inclusions of Bartoy volcanoes (Baikal rift zone), Doklady Akademii Nauk SSSR, 286, 5: 1215–1219 (in Russian).

Menzies, M.A. and Hawkesworth, C.J. (eds), 1987. Mantle metasomatism. London: Academic Press, 422 pp.

Mercier, J.C-C., Benoit, V. and Gerardea, J., 1984. Equilibrium state of diopside-bearing harzburgites from ophiolites: geobarometric and geodynamic implications, Contrib. Mineral. Petrol., 85, 3: 391–404.

Nickel, K.G. and Green, D.H., 1985. Empirical geothermobarometry for garnet peridotites and implication for the nature of lithosphere, kimberlites and diamonds, Earth Planet. Sci. Lett., 73: 158–170.

Nicolas, A., 1986. A melt extraction model based on structural studies in mantle peridotites, J. Petrol., 27, 4: 999–1022.

Nicolas, A. and Dupay, A., 1984. Origin of ophiolitic and oceanic lherzolites, Tectonophysics, 110: 177–187.

Quick, J.E. and Gregory, R.T., 1990. Magmathermal alteration of the upper mantle: evidence from the comparative anatomy of the Samal, Trinity and Darb Zubayan ophiolites, Abstr. Symposium of ophiolite genesis and evolution of oceanic lithosphere. UNESCO - Sultan Qaboos Univ., Muscat.

Reuber, I., 1988. Complexity of the crustal sequence in the Northern Oman ophiolite (Fizh and southern Aswad blocks): the effect of early slicing?, Tectonophysics, 151: 137–165.

Savelyeva, G.N., 1987. Gabbro - ultrabasic assemblages of the Urals ophiolites and their analogues in modern oceanic crust. Moscow, Nauka, 245 p. (in Russian).

Sobolev, V.S., Dobretsov, N.L. and Sobolev, N.V., (Eds) 1975. Deep-seated xenoliths and upper mantle. Novosibirsk, Nauka, 271 p. (in Russian).

Whilshire, H.G., Pike, N.J.E., Meyer, C.E., Schwarzman, E.C., 1980. Amphibole-rich veins in lherzolite xenoliths Dish Hill and Deadman Lake California, Am. J. Sci., 280–A: 576.

Evidence for a Paleo-Spreading Center in the Oman Ophiolite: Mantle Structures in the Maqsad Area

GEORGES CEULENEER
Laboratoire de Tectonophysique, Université de Montpellier II, 34095 Montpellier, France
(Present address: GRGS – UPR 234, OMP, Av. Ed. Belin, 31400 Toulouse, France)

Abstract

Mantle diapirs about ten kilometers in size have been recognized in the Oman ophiolite. A detailed study of the best preserved of these diapirs, cropping out in the Maqsad area, has been undertaken in order to understand melt migration processes in the asthenosphere beneath oceanic spreading centers. New results include the accurate location of the paleo-spreading axis related to the Maqsad diapir. It is evidenced by a 1–2 km wide dunitic corridor striking parallel to the sheeted dyke complex, bearing witness, in its structure and petrology, to pervasive soaking of the mantle by a basaltic melt; most of the upwelling flow is channelled at shallow depth within and beneath this horizon; away from it, the horizontal asthenospheric flow displays a clear divergent pattern. The zone of diverging flow extend 10–15 km on both flanks of the presumed paleo-ridge axis. The relations between the asthenospheric flow trajectories and a large mylonitic shear zone suggest that the Maqsad diapir corresponds to a late spreading event largely contemporaneous with early emplacement tectonics. Dykes facies and intrusion pattern at the periphery of the zone of diverging flow suggest that it ascended through a previously accreted lithospheric segment and that mantle diapirism beneath spreading centers is unsteady.

This reconstitution of a frozen ridge segment provides a logical structural framework for the interpretation of melt extraction structures from the mantle. The upwelling zone is made of homogeneous harzburgites (20–30% opx) which are interpreted as the residue left after partial melting and melt extraction al greater (>10 km) but unknown depth in the diapir. Basaltic melt relics trapped there are particularly scarce pointing to the efficiency of melt extraction processes at great depth. The occurrence in the upwelling of solid/melt reaction structures reflects intense circulation of a hot basaltic melt equilibrated at depth. Abundant melt extraction structures are observed in the dunitic corridor at the very top of the upwelling. They point to the formation of a very low viscosity crystal mush horizon. Most of the basaltic melt delivered into the overlying magma chamber is produced by the com-

Tj. Peters et al. (Eds), Ophiolite Genesis and Evolution of the Oceanic Lithosphere, 147–173.

paction of this horizon. Focusing the magmatic activity at ridge axis might be accounted for by the coupling between plastic flow and melt migration leading to the formation of a narrow crystal mush horizon at the top of mantle upwellings. The zone of diverging flow is devoid of melt extraction structures but abundantly intruded by gabbro dykes. This zonation points to significant temperature decrease away from the upwelling. The progressive deformation of the crystal mush·horizon away from the axial zone is correlated to the increase of the curvature radius of the mantle flow lines at the top of the upwelling. This evolution likely reflects a progressive increase in the effective viscosity of the uppermost mantle away from the ridge axis.

1. Introduction

Mantle flow patterns and melt extraction processes related to the formation of the oceanic crust are poorly constrained by geophysical data along present-day spreading centers. Fragments of fossil oceanic crust and uppermost mantle are exposed in the ophiolite massifs. The peridotites from the mantle section of ophiolites, when unaffected by obduction-related deformation, have kept a record of solid-state mantle flow in their structures induced by high temperature and low deviatoric stress deformation. Mapping these structures gives an accurate image of the asthenospheric flow geometry beneath the ridge of origin, provided that the ophiolite can be restored in a structural framework related to the paleo-spreading direction. In the same way, the distribution of basaltic components trapped within the peridotites allow to constrain melt migration processes beneath oceanic ridges. (An exhaustive presentation of the method of structural analysis of mantle peridotites can be found in Nicolas, 1989).

The Oman ophiolite, thanks to its size, good preservation and perfect outcrop conditions, is probably the best place in the world to study the deep processes taking place beneath oceanic ridges (Coleman, 1981). Structural mapping in Oman have revealed a few areas of paleo-vertical mantle flow about 10 Km in extension; at the top of these vertical pipes, the plastic flow structures rotate to the horizontal in a narrow transition zone a few hundred meters thick and diverge in every directions, mainly along the paleo-ridge axis. This flow pattern has been interpreted as the footprint left by mantle diapirs comparable in size to those invoked to account for the small-scale segmentation of present-day ridges (Ceuleneer and Nicolas, 1985; Rabinowicz et al., 1987; Ceuleneer et al., 1988; Nicolas et al., 1988a,b).

Given the likely complexity of mantle diapirism and related magmatic processes, our descriptions were based on a rather low density of field measurements and observations (a kilometric network was adopted for systematic mapping of the Oman ophiolite). During the winter 1988–89, I have revisited the diapiric structure recognized in the Maqsad area. This choice was dictated by the exceptional exposure, in Maqsad, of the mantle/crust transition zone whose importance for melt extraction processes was stressed by Rabinowicz et al. (1987). This detailed mapping resulted in the accurate localization of the paleo-spreading axis related to the Maqsad diapir. This result is important in that it provides a logical structural framework for the interpretation of melt extraction structures in mantle peridotites (dyke injection patterns, dyke composition, distribution of porous flow structures, . . .) and for the interpretation of the structures mapped in the overlying crustal section in terms of magma chamber processes (Nicolas et al., 1988b,c). Data treatment is in progress; the complete data set and maps will be published later. In this short paper, I summarize the main evidences for the existence of a paleo-spreading center frozen in the Maqsad area together with some field observations relevant to melt extraction processes beneath oceanic ridges. These results were presented during the field trip C of the Oman ophiolite symposium.

2. General Structure

Obducted during Maestrichtian times, the Oman ophiolite is now exposed as a crescent-shaped nappe 500 km in length and 50–100 km in width. It has been slightly dismembered to form about twelve massifs during post-obduction tectonic events (Fig. 1) (Glennie et al., 1974). The considerable thickness of the nappe precluded tight folding of the ophiolite during or after its obduction. The Maqsad area belongs to the Sumail massif which crops out on the southeastern slope of the Jabal Akhdar (Fig. 1). Compared to most of the Oman massifs, it has not been strongly affected by mylonitic deformation related to intra-oceanic thrusting (Ceuleneer, 1986; Boudier et al., 1988). Emplacement-related mylonites are restricted to the northwestern edge of the massif, where the peridotites are in contact with the metamorphic sole, and to a northwest trending dextral mylonitic shear zone cutting the massif from Muqhbariah to Saymah (Fig. 2) (Ceuleneer, 1986; Beurrier, 1988).

The Sumail massif comprises a nearly complete ophiolitic sequence from the sheeted dyke complex down to the mantle peridotites. On a regional scale, the lithological contacts dip moderately (<15°) to the East, an attitude which is reflected in the preferred orientation of the magmatic layering at the base of the crustal section (Fig. 3b). The sheeted dyke complex strikes NW–SE (Fig. 3a); this is the assumed orientation of the paleo-ridge axis in the Sumail massif. At some places, the paleo-Moho is disrupted by normal faults inducing vertical displacements ranging in amplitude from less than one meter to several tens of meters. The development of a mylonitic foliation bearing a down the dip lineation is exceptionally observed in the peridotites cropping out in the vicinity of these faults. They are always associated with alteration minerals formed by high-temperature hydrothermal circulation

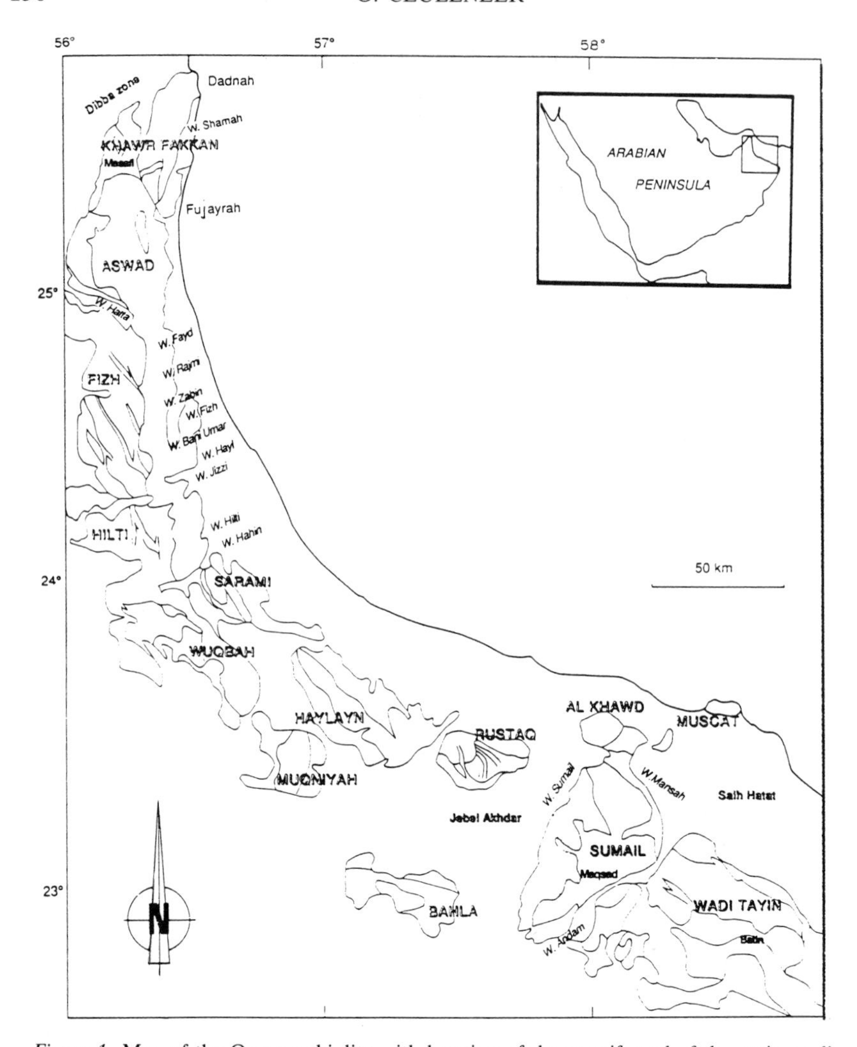

Figure 1. Map of the Oman ophiolite with location of the massifs and of the main wadis.

(recrystallization of the gabbros into assemblages of pistacite, zoisite, chlorite, . . .). Most of the copper occurrences in mantle peridotites lie along these fault zones (Al Azri and Beurrier, this volume). Very locally, they induce block rotation of up to a few tens of degrees as evidenced on Figure 3b where the poles of magmatic layering in the crustal section define a girdle around a direction close to the strike of the paleo-ridge. Although further investigations are necessary to clarify the origin of these structures, it is probable that they formed early in the tectonic history of the massif, possibly during the oceanic spreading stage (Beurrier, 1988).

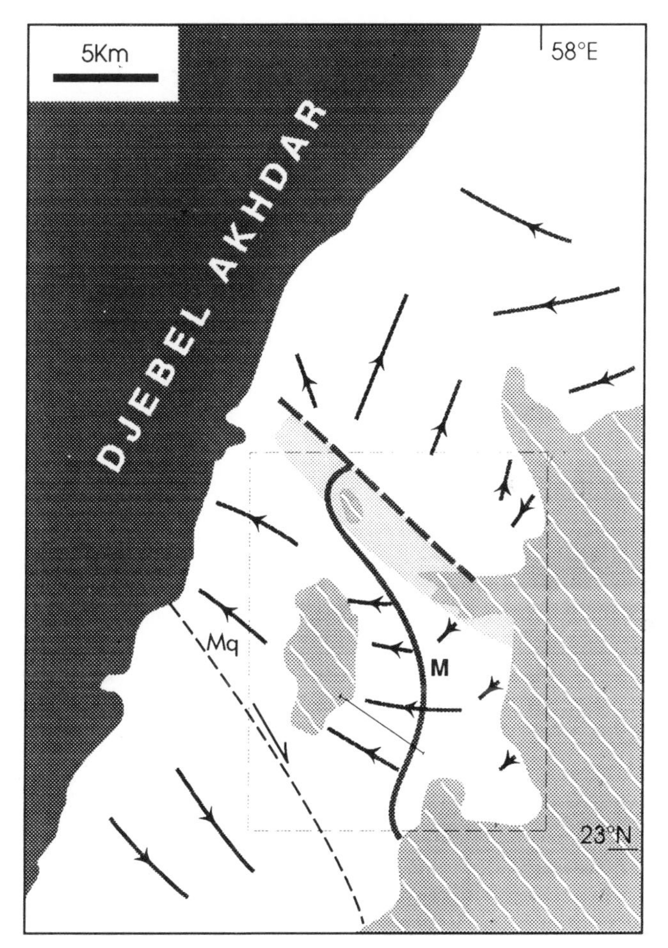

Figure 2. Simplified tectonic sketch of the Sumail peridotites showing the area of vertical mantle flow structures delimited by a thick line, the orientation and shear senses of the diverging flow from the diapir and the zone of undeformed dunites (pale grey). Mq: the Muqbariah shear zone; M: Maqsad oasis. Intermediate gray with white stripes: crustal section. Dark grey: mesozoic limestones. Dotted square: field of Figure 6. Section refer to Figure 6.

3. Mantle Flow Structures

Vertical Flow

The Maqsad area is characterized by foliations and lineations at a high angle to the paleo-Moho (Figs. 2 and 4). At shallow paleo-depths, the zone of vertical flow is restricted to a narrow band 1–2 km in width striking parallel to the sheeted dyke complex. When deeper levels of the mantle section are considered, the zone of vertical flow opens out toward the SE, reaching a maximum width of 8 km at the place where the mantle outcrops are concealed

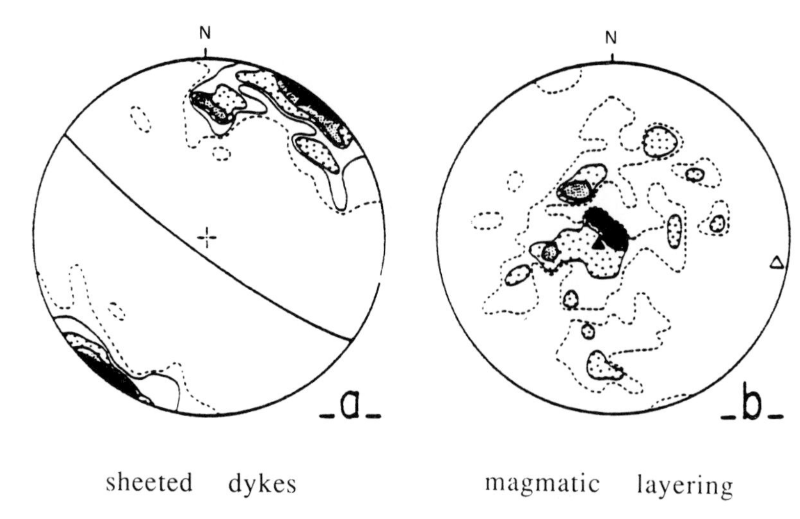

Figure 3. Orientation of the crustal structures in the Sumail massif. Equal area projection, lower hemisphere. (a) Wadi Andam sheeted dyke complex; 177 measurements; contours at 1.1, 2.2, 4.5, 8.4 and 15%; preferred dyke orientation:125 SW 82. (b) Magmatic layering at the base of the crustal section; 80 measurements; contours at 1.25, 2.50, 3.75 and 5.00%; preferred layering orientation: 164 E 08 (▲); girdle axis (△): 100 E 06.

by the crustal section. The azimuth of the vertical foliations range from WNW–ESE to NNE–SSW with a NW–SE preferred orientation, parallel to the paleo-ridge axis (Fig. 4a). Their westward preferred dip of 75° (Fig. 4a) and the regional southeastward tilt of the massif are consistent with a sub-vertical orientation of these flow structures at the time they were frozen. These foliation planes bear sub-vertical lineations (average pitch of 90°) (Fig. 4b). In the center of the vertical conduit, plastic deformation, although clearly identifiable, is not very pronounced. This moderate finite strain is illustrated by the poor olivine lattice fabric and by its strong obliquity with the shape fabric (Fig. 5a). The shear senses deduced from this obliquity are very consistent through all the zone of vertical flow (Fig. 6): an upward relative movement along the northeastern side of the foliation planes is recorded.

The northwest termination of the vertical conduit is exposed in the massif: it is characterized by sub-horizontal flow lines striking parallel to the sheeted dyke complex. This zone is not marked by any change in the foliation planes azimuth, which remains NW–SE; however, their dip is there poorly defined, and the strain facies is strongly linear. Where the foliation dip is found sub-horizontal, shear senses indicate northwestward divergence of the flow, i.e. away from the main vertical conduit (Fig. 6).

Rotation of the Vertical Flow to the Horizontal

In the narrow corridor where the vertical flow reaches shallow paleo-depths, no rotation of the flow can be observed as the plastic deformation structures

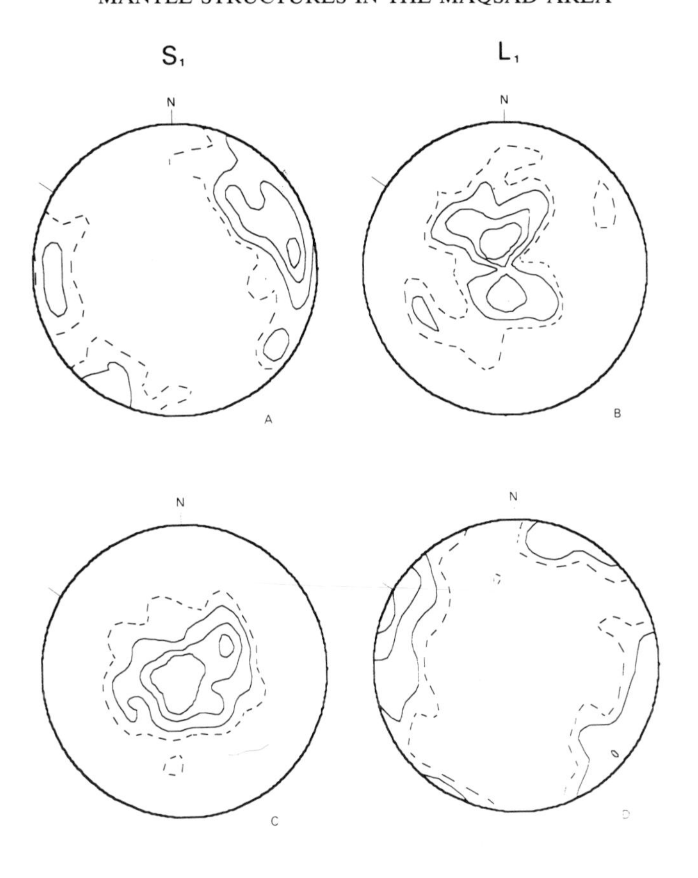

HT° structures in mantle peridotites

Figure 4. High temperature foliations and lineations in mantle peridotites of the Maqsad area. Equal area projection, lower hemisphere. Contours: 1, 2, 4 and 6%. (a) Foliations in the area bounded by the thick line in Figure 2; 91 measurements. (b) Associated lineations. (c) Foliations in the zone of diverging flow; 135 measurements. (d) Associated lineations. Strips: azimuth of the sheeted dyke complex.

become more and more tenuous and finally vanish in a dunitic zone about 300 m thick at the transition between the harzburgites and the first gabbroic cumulates. The absence of plastic deformation in these dunites is corroborated by their largely random lattice fabric (Fig. 5b). The occurrence of undeformed dunites is exceptional in Oman. In the Maqsad area, they are restricted to a narrow band (1–2 km in width) elongated along a NW–SE direction cropping out on a distance of about 15 km (Fig. 2). It has to be stressed that, in spite of their lack of plastic strain fabric, individual olivine grains constituting these dunites present commonly sub-grain boundaries diagnostic of hot working in asthenospheric conditions. Accordingly, a mantle origin is ascribed to this dunitic transition zone. Its origin is to be

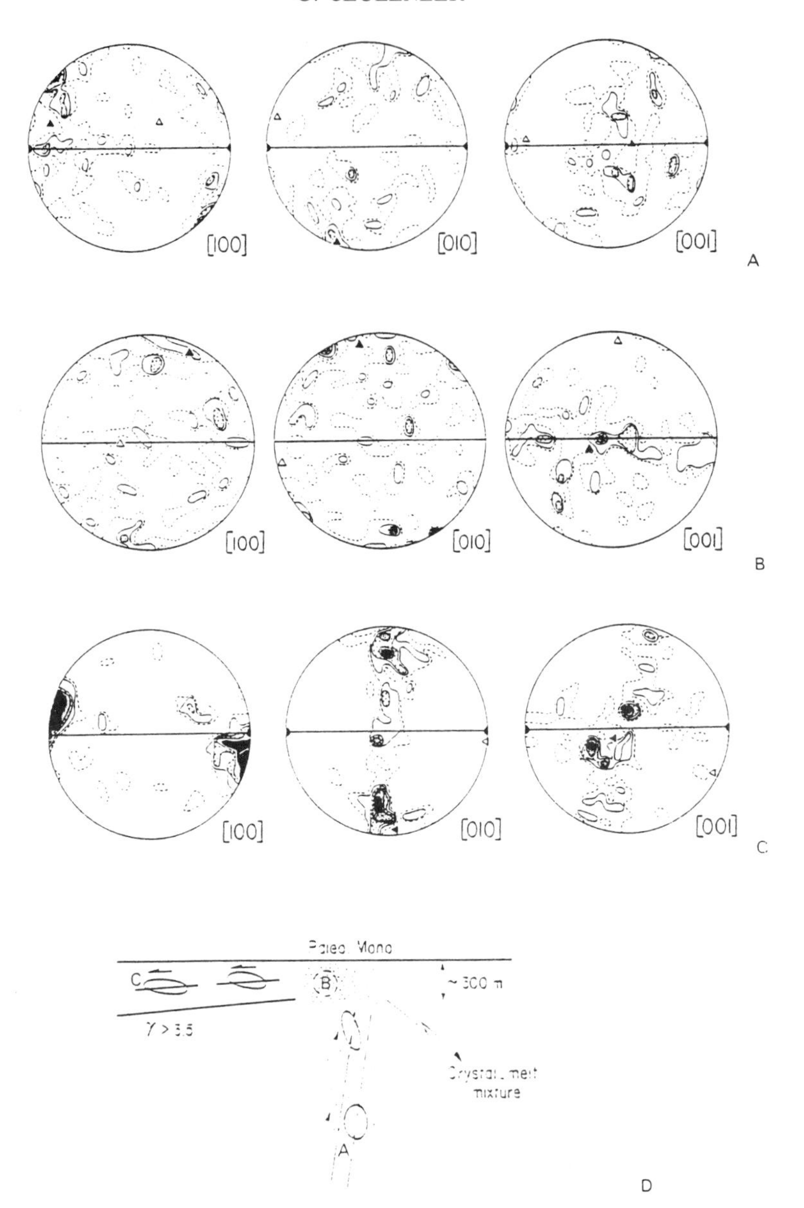

Figure 5. Olivine preferred orientations in the Maqsad area. Equal area projection on the lower hemisphere; 100 measurements; contours at 1, 2, 3, 4 and 10%; horizontal line: foliation and/or banding trace; dot: lineation. (a) harzburgite in the zone of vertical flow. (b) dunite at the top of the zone of vertical flow. (c) harzburgite in the zone of diverging flow. (d) Interpretative sketch in terms of finite strain (After Ceuleneer and Nicolas, 1985 and Rabinowicz et al., 1987).

Figure 6. Shear senses in the Maqsad area (location, see Figure 2). Arrow: relative direction of movement of the upper block in area of weakly dipping foliations; open arrow: westward movement; full arrow: eastward movement. Stripe: trace of the foliation plane in area of sub-vertical lineations with + indicating the upwelling block. Dotted area: undeformed dunites.

looked for in the compaction of a crystal/melt mixture (Rabinowicz et al., 1987). Observations relevant to this process are presented in section 4.

Away from this area, in the diverging zone of the diapir, the rotation of the vertical flow structures becomes more and more progressive. At a distance of several kilometers from the band of undeformed dunites, it initiates in the harzburgites at a paleo-depth of about 2 km. The progressive rotation of both the foliation planes and lineations is well illustrated on the stereonet of Figure 7. On this figure, it appears also that the rotation of the foliation from a vertical to a horizontal attitude is correlated with a rotation of its azimuth from the NW–SE value associated to the vertical dips to a SW–NE azimuth when the foliation becomes sub-horizontal. This evolution, which has been put clearly in evidence on the southwestern flank of the diapir, just reflects the southeastward regional tilt of the massif. Such azimuthal rotations were previously misinterpreted in terms of curved flow plane trajectories in the diapir.

On the northeastern flank of the zone of vertical flow, however, the

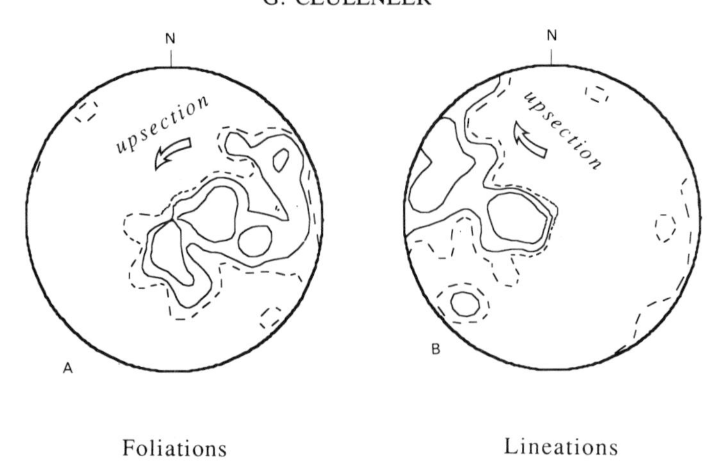

Figure 7. Progressive rotation of the plastic flow structures in section located on Figure 2. Equal area projection on the lower hemisphere; 50 measurements; contours at 2, 4, 8 and 12%.

transition between the vertical and horizontal flow has not been put in evidence. This can be ascribed partly to a wide serpentinized fracture bounding to the northeast the zone of undeformed dunites and precluding detailed observations in the critical zone. However, as it is correlated with some other asymmetrical features of the diapir, it is also possible that the rotation was originally more abrupt there.

Diverging Flow

The zone of horizontal flow is characterized by intense plastic strain shown by the development of a strong lattice fabric slightly oblique to the shape fabric (Fig. 5c). Aspect ratios of antinodular chromite ore indicate that finite strains reaching 3.5 are recorded in the transition zone within several kilometers of the zone of undeformed dunites (Fig. 8a, b). This high plastic strain affects both the harzburgites and the dunitic transition zone and is correlated with a marked reduction of the transition zone thickness which grades from a few hundred meters to a few tens of meters on a horizontal distance of 5 to 10 kilometers.

Two groups of flow directions have been mapped in the zone of horizontal foliations (Figs. 4c, d): one strikes NNE, i.e. sub-parallel to the presumed spreading direction, the other ranges in azimuth from E–W to NW–SE, i.e. sub-parallel to the paleo-ridge axis. NNE striking horizontal lineations are almost entirely restricted to the northeastern flank of the diapir (Figs. 2, 6). Associated shear senses (Fig. 6) indicate relative northeastward movement of the uppermost parts of the asthenosphere, i.e. away from the zone of vertical flow. A few tens of meters below the Moho, a shear sense inversion is observed (Fig. 6), suggesting mechanical coupling between the diverging flow and the already crystallized crust (Rabinowicz et al., 1984).

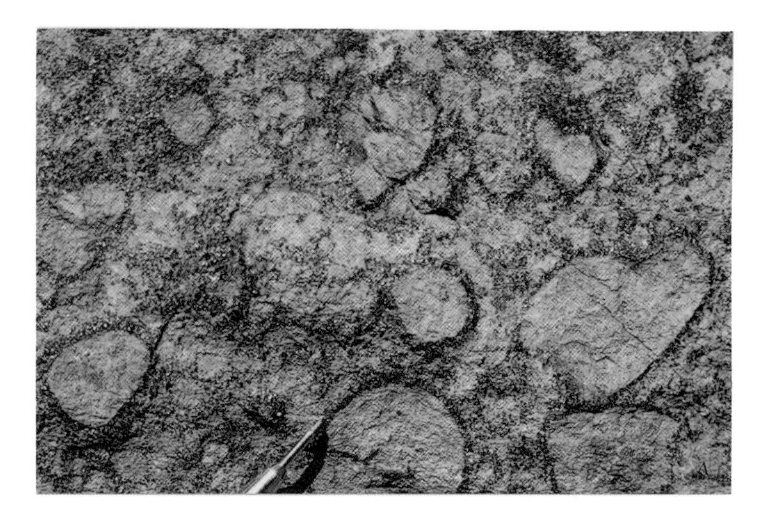

Figure 8. Magmatic structures in the Maqsad area. (**a**) Undeformed antinodular chromite ore texture characteristic of the zone of undeformed dunites. (**b**) Strongly deformed antinodular ore texture characteristic of the zone of diverging flow. Black stripe: 2 cm.

Close to the zone of undeformed dunites and of vertical flow, the lineations have an E–W preferred orientation. To the west of this zone, shear senses are very consistent and point to relative westward flow of the upper parts of the mantle (Fig. 6). Moving away from the band of undeformed dunites, on the western flank of the upwelling zone, the mantle flow adopts progressively a NW–SE azimuth (Figs. 2, 6).

Extent of the Diverging Flow

On the eastern flank of the Maqsad diapir, NE trending flow directions are mapped to a distance of about 10 km away from the zone of undeformed dunites. The northern part of the Sumail massif has recorded a horizontal westward flow direction (Fig. 2). The intersection between the westward and northeastward flow at the northeastern end of the diverging zone is characterized by an abundance of gabbro sills intruding the harzburgites (Fig. 8c). These sills range in thickness from several decimeters to a few meters and constitute about 20% in volume of the mantle section there. Most of these gabbro sills display a pegmatitic texture, with decimetric plagioclase and clinopyroxene crystals (Fig. 8d). Clinopyroxenes are commonly rimmed by brown amphiboles. Such dykes have not been recognized elsewhere in the Sumail massif and are rather uncommon in the Oman peridotites. This area is also cross-cut by abundant NE-trending diabase dykes. Locally, a group of about 10 dykes displaying self intrusion relationships has been found, forming some kind of small "sheeted dyke complex" rooted in the harzburgites.

On the western flank of the diapir, the diverging flow probably extended to the Muqbariah shear zone, according to the lack of continuity of the asthenospheric flow patterns on both sides of this structure. Unfortunately, the original contact relationships between the westward diverging flow and the adjacent mantle section are concealed by the shear zone.

4. Melt Extraction Structures

Vertical Pipe

The zone of vertical flow is constituted of homogeneous harzburgites relatively rich in orthopyroxene (20–30%) and frequently containing trace amounts of residual diopside associated with enstatite crystals. Such harzburgites are interpreted as the residue left after basaltic melt extraction at greater depth in the diapir (e.g. Dick, 1977; Nicolas, 1986). Structures related to melt migration are very uncommon in the vertical pipe. The most frequently observed ones are centimeter to decimeter-thick discordant dunitic bands underlined at some places by small and discontinuous gabbroic pockets or by individual crystals of plagioclase or clinopyroxene. These probably crystallized from a basaltic melt trapped in closing cracks on its way to the surface; the dunites are interpreted as the reaction product between this magma and the peridotitic wall rock following a well documented mechanism (Quick, 1981; Berger and Vannier, 1984; Kelemen, 1990). However, the zone of vertical flow is almost entirely devoid of well-developed mafic dykes. Apart from the occurrence of residual dunites, from which the dissolution of enstatite is inferred, petrographic evidence for disequilibrium between the

Figure 8. (**c**) Thick gabbro sills at the northeastern extremity of the zone of diverging flow from the Maqsad diapir. (**d**) Detail on these sills showing the pegmatitic texture. Length of the ruler: 10 cm.

upwelling peridotites and the melt migrating through them are not very common. In the zone of vertical flow, at a depth of about 2 km beneath the paleo-Moho, a mafic dyke has been found with a troctolitic composition (about 70% plagioclase and 30% olivine). Olivine crystals have a clearly deformed texture, contrasting with the magmatic texture of the plagioclase crystals (Fig. 8e). Two characteristics point to the desequilibrium between

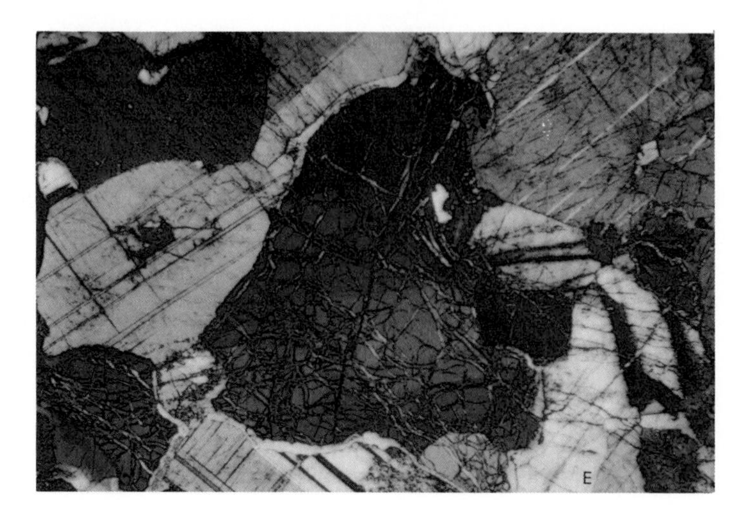

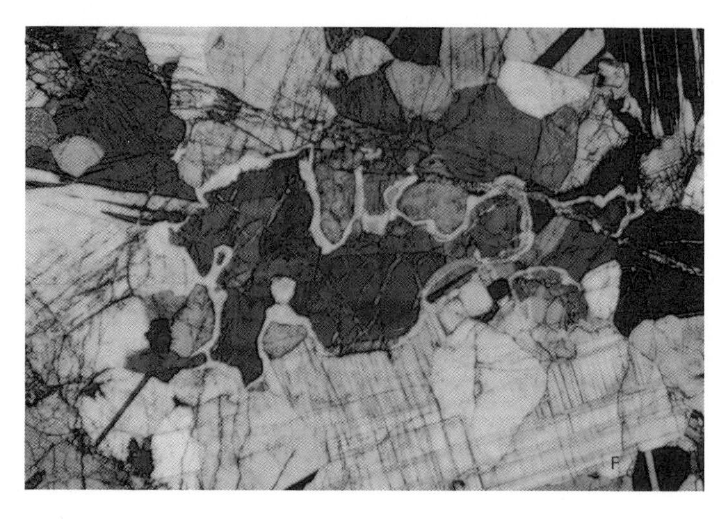

Figure 8. (**e**) Olivine crystal (4 mm in size) from a troctolite dyke. Subgrain boundaries are characteristic of a mantle origin. The plagioclase has an undeformed texture. Zone of vertical flow within the Maqsad diapir. (**f**) Same troctolite dyke. Detail on a strongly corroded mantle olivine (6 mm in size) rimmed with orthopyroxene.

the residual mantle and the migrating melt: the convex inward grain boundaries of olivine grains, conferring to some of them an amoeboid shape, and rims of orthopyroxene more or less developed around olivine grains (Fig. 8f). Higher in the section, the olivine xenocrysts become less abundant and the mafic dyke grades in composition to gabbro as clinopyroxene joins plagioclase as a cumulus phase.

Transition Zone

At the top of the zone of mantle upwelling, the transition zone between the harzburgites and the first layered gabbros is exceptionally thick (200–300 m) compared to its average thickness in Oman (a few meters to a few tens of meters). Dunites (99% olivine, 1% chromian spinel) constitute about 90% of its volume. The remaining 10% of the transition zone is constituted by dunites with variable percentages of interstitial plagioclase, clinopyroxene and, more exceptionally, Ti-rich chromian spinel locally forming ore concentrations. Gabbro dykes and sills are locally abundant. Brown amphibole has been observed but is very uncommon. The transition zone in Maqsad is completely devoid of orthopyroxene. These various lithologies do not define large scale layering. However their small-scale structure is interesting to analyze because they are relevant to melt extraction mechanism from a crystal/melt mixture. Different stages of this process can be observed in Maqsad.

At some places interstitial clinopyroxene and plagioclase, interpreted as the relicts of the trapped melt, are homogeneously distributed within the dunites (Fig. 8g). However, such homogeneous impregnation was apparently not a stable configuration: evidence of melt segregation is particularly abundant in the transition zone. Porous flow occurs but, as far as it can be inferred from the present distribution of interstitial clinopyroxene and plagioclase, it was not an efficient melt migration mechanism on distances greater than several decimeters, exceptionally a few meters. Percolating melt pools in centimetric to metric pockets where the increasing melt/matrix ratio leads to the complete desorganization of the solid framework (Fig. 8h). This critical melt fraction, corresponding to the second percolation threshold, is estimated to be about 25% (modal percentage of clinopyroxene and plagioclase at which the olivine grains are no longer adjacent).

Further melt segregation from the transition zone involves the opening of cracks frequently connected to these magma pockets. The rooting of subvertical veinlets (millimeter to centimeter-thick) draining the basaltic melt from the impregnated bodies is illustrated on Figure 8i. When crack thickness exceeds about 10 cm, the crystal mush itself, and not only the interstitial melt, can be mobilized and intrude the fracture (Fig. 8j). Compaction of this wehrlitic-troctolitic "magma" occurs within the dykes. Melt first segregates along the walls of the fracture; this is evidenced by the progressive increase of clinopyroxene and plagioclase concentration from the center to the wall of the fracture, where gabbroic composition are first achieved. At a height ranging from a few meters to a few tens of meters above the roots of the dykes, all the olivine crystals transported by the melt have settled and the dyke composition has evolved to an olivine-free gabbro. These gabbro dykes, rooted in the impregnated dunites, can be followed up to the Moho (Fig. 8k). This vertical compositional gradient within the dykes might also reflect a dying system; it is not excluded that, when the magma was actively flowing

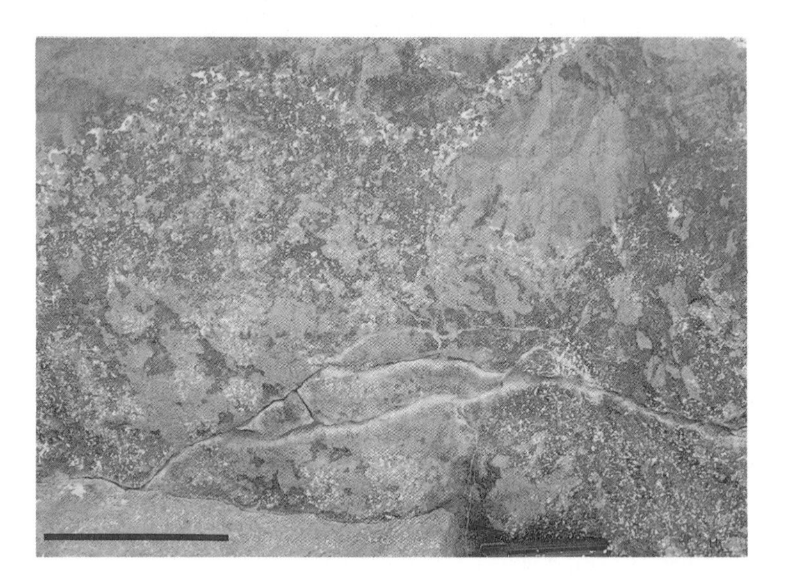

Figure 8. (**g**) Homogeneous impregnation in the zone of undeformed dunite at the top of the upwelling zone. Note the clear interstitial nature of the clinopyroxene and plagioclase crystals (white grains). Ruler: 10 mm. (**h**) Pooling of the melt into decimetric impregnation pockets in the zone of undeformed dunites at the top of the upwelling. Black stripe: 5 cm.

within the fractures, most of the olivine crystals carried by the melt were transported up to crustal levels and discharged within the magma chamber.

The same kind of compaction phenomena within vertical cracks has been observed in a decametric chromitite dyke, with the coarse chromite nodules settling at the base of the dyke, and the fine-grained disseminated chromite ore being trapped at its top, with a regular decrease in the size of the

Figure 8. (**i**) Rooting of a gabbroic dykelet on an impregnated pocket. Zone of undeformed dunites at the top of the upwelling.

Figure 8. (**k**) Group of gabbroic dykes (arrow) rooted in the zone of impregnated dunites and feeding the overlying crustal section. M: Moho. (**l**) Gabbro sill rooted on a magma pocket. Zone of undeformed dunites at the top of the upwelling. Black stripe: 10 cm.

chromite nodules in between. The centimetric size of chromitite nodules points to magma ascent velocities of at least a few m/s (Lago et al., 1982). The great concentration of chromite reflects the high volume of melt having circulated in such fractures. These cracks were probably the main draining system of the mantle diapir.

Lateral melt injection from a magma pocket in sub-horizontal fracture planes is also frequently observed in the transition zone (Fig. 8l). The melt expelled in that way remains trapped within the mantle section where it crystallizes as gabbro sills. As the top of the transition zone is approached, these sills become more and more abundant and may have lateral extension of several hundred meters. The alternance of closely spaced gabbroic sills with compacted dunitic horizons mimics a magmatic layering with which it is currently assimilated.

Diverging Zone

The compaction of the transition zone continues away from the zone of undeformed dunites, now enhanced by the very large plastic strain which leads to the direct injection of wehrlitic bodies into the barely crystallized crustal section (Benn et al., 1988; Reuber, 1988). The rare relicts of melt remaining trapped within the dunites are progressively deformed and, a few kilometers away from the zone of undeformed dunites, form a concordant layering. No more porous flow structures are observed.

Contrasting with the upwelling zone, the diverging zone of the diapir is intensely cross cut by gabbroic dykes displaying clear intrusive relationships with the harzburgites (harrisitic growth on the walls of the dikes, lack of reaction aureole in the wall rock, . . .). All textural varieties are observed, from coarse-grained gabbros to microgabbros with chilled margins and, finally, to diabases. Orientation patterns of these dykes, together with the dykes measured in the upwelling and in the transition zone, are presented in the next section.

Dyke Injection Pattern

The goal of this section is to illustrate, with a few examples, the structural control exerted by mantle flow on the dyke injection geometry in the Maqsad diapir, an observation which, in turn, is consistent with the hypothesis that these dykes are contemporaneous with the ascent of the diapir. A more advanced discussion of their significance in terms of melt extraction mechanism from a mantle upwelling would require petrological studies.

In the zone of vertical flow, the rare mafic dykes and the discordant dunitic bands clearly strike perpendicular to the flow lineation and have a preferred sub-vertical dip. In the example chosen (Fig. 9a), the lineation strikes parallel to the sheeted dyke complex, resulting in dykes normal to the azimuth of the sheeted dyke complex. At the top of the upwelling, in the undeformed transition zone, the subvertical fractures rooting in the impregnated dunites do not show clear preferred azimuth, apart from a slight concentration parallel to the sheeted dyke complex (Fig. 9b). The same pattern is observed for gabbro dykes whose rooting relationships have not been demonstrated (Fig. 9c). In the diverging zone, the dykes tend to be oriented either parallel to the sheeted dyke complex or normal to the mantle

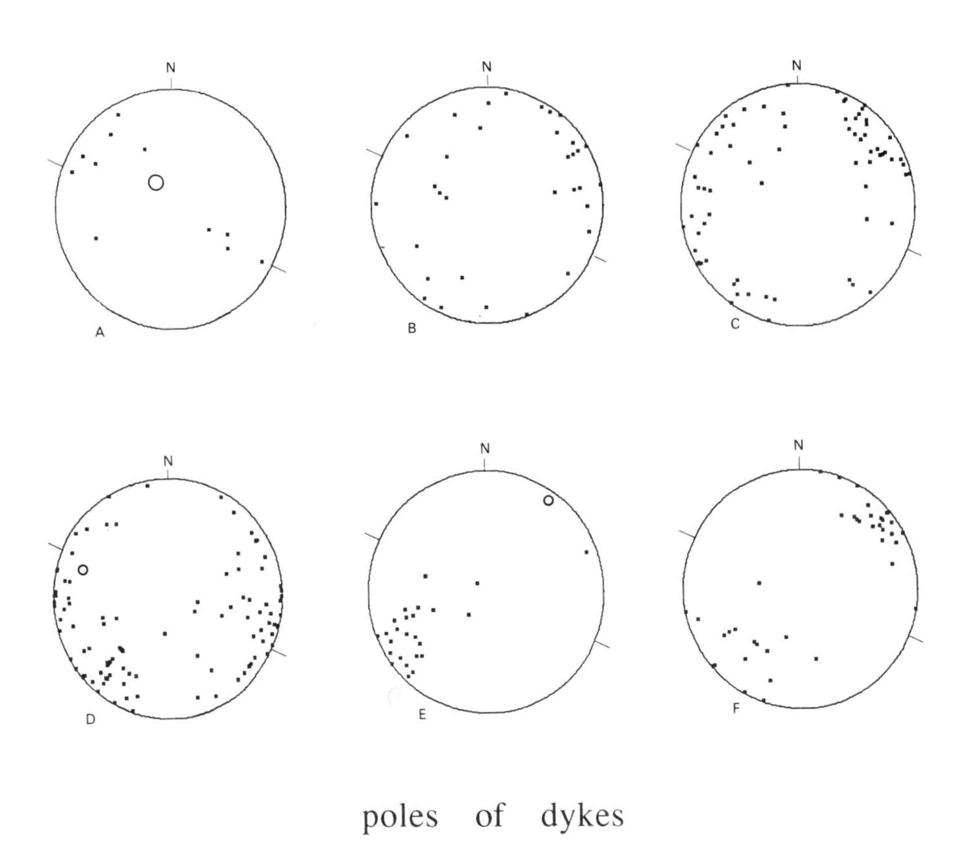

Figure 9. Orientation of gabbroic dykes in mantle peridotites from the Maqsad area. Squares: poles of dykes; open circle: local orientation of the plastic lineation in the peridotites; bars: preferred azimuth of the Wadi Andam sheeted dyke complex. (a) *In situ* dykes and discontinuous dykelets in the zone of vertical flow (11 measurements). (b) Gabbro dykes rooted in the impregnated dunites of the transition zone; 34 measurements; dykes with a dip <30°, which are very frequent there, are not included for clarity. (c) Gabbro dykes intrusive (no evidence of rooting relationships) in the dunites of the transition zone; 80 measurements. (d) Gabbro dykes intrusive in the southwestern diverging zone of the diapir; 116 measurements. (e) Gabbro dykes intrusive in the northeastern diverging zone of the diapir; 38 measurements. (f) diabase and microgabbro dykes intrusive in the diapir; 42 measurement.

flow direction. This pattern is clearly illustrated on the southwestern diverging zone of the diapir, where the flow direction is parallel to the sheeted dyke complex (Fig. 9d). Accordingly, in the northeastern diverging zone of the diapir, where the flow direction is perpendicular to the strike of the sheeted dyke complex, all the gabbro dykes have the same preferred orientation (Fig. 9e). Finally, the diabase dykes intruding the mantle section through all the Maqsad area display a clear preferred orientation parallel to the sheeted dyke complex (Fig. 9f).

5. Discussion

Position of the Paleo-Spreading Axis

Vertical lineations related to asthenospheric flow were frozen in the Maqsad area. Relatively broad at depth (several kilometers), the upwelling narrows at Moho level where it is restricted to a 1–2 km wide corridor parallel to the strike of the paleo-ridge. There, it is capped with a thick dunitic zone whose structure and composition reveal the transit of a considerable volume of basaltic melt. In the zone of vertical flow, the shear senses, reflecting local flow rate gradients at the time when the plastic strain structures were frozen, can be interpreted in terms of faster upwelling beneath the dunitic corridor than away from it. In the present interpretation, this corridor is though to mark the position of the paleo-spreading axis. Moving away on both sides of the upwelling zone, the dunites and underlying harzburgites are progressively affected by intense horizontal plastic flow. The most convincing evidence for the juxtaposition of the dunitic corridor with the paleo-spreading axis is probably given by the shear senses recorded in these zones of horizontal flow. They indicate very consistently westward divergence on its southwestern side and northeastward divergence on its northeastern side.

A Late Spreading Event Contemporaneous with Early Emplacement Tectonics

The total extent of the diverging zone away from the Maqsad diapir is about 25 km measured perpendicularly to the strike of the paleo-ridge. The presumed paleo-ridge axis is located right in the middle of this zone (Figs. 2, 10c). The mantle peridotites on both sides of the diverging zone were likely accreted, before the initiation of the Maqsad diapir, from spreading cells located to the west and/or to the east along the Oman ridge, as suggested by their E–W asthenospheric flow direction (Figs. 2, 10). Abundant pegmatitic sills observed in deep levels (a few km below Moho) of the mantle section, at the intersection between the northeast diverging zone and the adjacent peridotites, could be related to the first gabbroic melts extracted from the top of the Maqsad diapir as it ascended through a previously

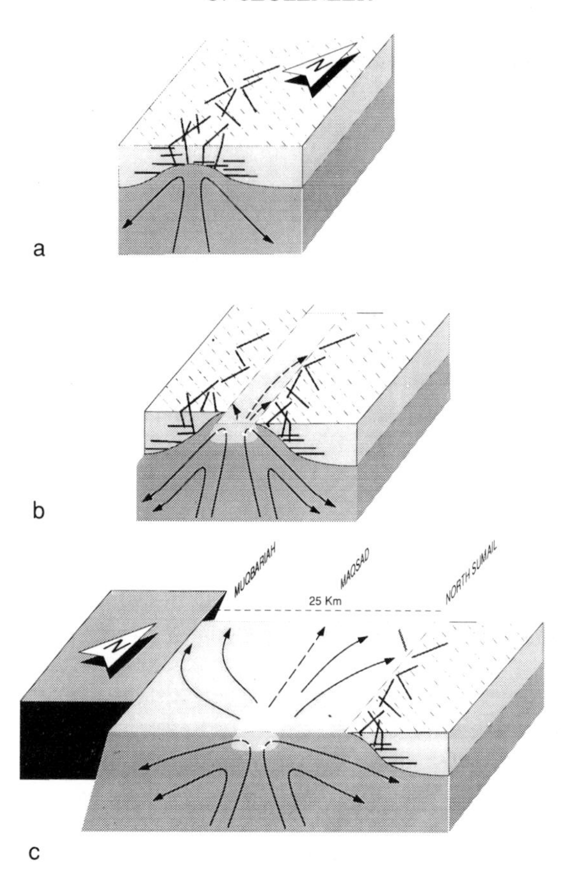

Figure 10. Sketch for the opening of the Maqsad area. (a) The Maqsad diapir rises through a previously accreted segment of the Oman paleoridge (pale grey) having recorded an E–W flow direction (dotted lines). Gabbro sills and diabase dykes (thick stripes) are emplaced at this stage. (b) The diapir has pierced the lithosphere. The Muqbariah shear zone was probably already active at this stage. (c) Expansion stops after the creation of a 25 km wide lithospheric segment, resulting in the frozen flow pattern observed now. White: the zone of undeformed dunites.

accreted lithospheric segment (Fig. 10a). If the origin of these pegmatitic sills is actually related to some kind of "underplating" process, it would mean that mantle diapirism beneath spreading centers is not steady. On a particular ridge segment, the ascent of two successive diapirs would be separated by a period of reduced magmatic activity and of progressive cooling of the uppermost mantle. Minimum periodicities of the order of several 10^5 years seem to be in good agreement with the Oman data considering (i) that before the rise of the Maqsad diapir the mantle significantly cooled down to a depth of at least 2 km, and (ii) that 25 km of new lithosphere was accreted from the Maqsad diapir before it was sampled by the obduction process.

On the southwestern diverging zone of the Maqsad diapir, flow lines rotate progressively into parallelism with the Muqbariah mylonitic shear zone (Figs. 2, 4, 10c). The sense of this rotation is consistent with dextral movements along the Muqbariah shear zone deduced from petrofabric analysis of the mylonites. This rotation occurred as the diverging mantle was still very ductile, as no low temperature microstructure overprint them except at the contact with the mylonites. The Muqbariah shear zone is related to the early movements of oceanic thrusting of the ophiolite. This suggests that the Maqsad diapir was largely contemporaneous with the first detachment of the Oman ophiolite and that the accretion of 25 kilometers of lithosphere from this upwelling corresponds to the last activity of the Oman paleo-ridge. This deduction is consistent with the general emplacement model of the Oman ophiolite at an oceanic spreading center, supported by structural and geochronological data. (Boudier and Coleman, 1981; Ceuleneer, 1986; Boudier et al., 1985, 1988; Montigny et al., 1988).

Plastic Flow, Melt Extraction, and Concentration of Magma Discharge Beneath Oceanic Spreading Axis

The general distribution of melt extraction structures in the Maqsad area supports the structural analysis presented above. Basaltic melt relics trapped within the upwelling zone are particularly scarce, pointing to the efficiency of melt extraction processes in the ascending part of a diapir. The occurrence there of solid/melt reaction structures reflects intense circulation of a hot basaltic melt equilibrated at greater depth. Among them are abundant dm-thick dunitic bands of large vertical extension from which it is inferred that melt transport in the diapir is focused in zones of restricted extension, probably cracks, and does not involve large-scale porous flow. Contrasting with the central part of the upwelling, the zone of diverging flow of the diapir is almost entirely devoid of reaction structures but is abundantly intruded by gabbro dykes. This zonation points to progressive temperature decrease away from the main upwelling zone of the diapir. These "off-axis" mafic dykes were extracted from the upwelling at an unknown depth ranging from the depth where partial melting initiates (probably around 80 km) to several kilometers beneath the Moho, the maximum depth accessible to the observation in Oman. Their petrological study should provide some clues to melt migration mechanism at depth within a mantle upwelling.

Most of the "active" melt extraction structures preserved in the Maqsad area are observed in the undeformed dunitic transition zone at the top of the upwelling. Locally the high melt fractions trapped in these dunites suggest that the second percolation threshold was exceeded. This is confirmed by the local "magmatic" behaviour of this crystal mush, like its injection in cracks. Typical compaction distances deduced from the spacing of melt-enriched pockets in the impregnated dunites range from a several centimeters to about one meter; these values are compatible with viscosities of the order of 10^{10} to 10^{12} Pa.s (Scott and Stevenson, 1986; Sleep, 1988) intermediate between the viscosities of basaltic melt and of solid mantle. The transition zone is essentially (about 90%) composed of dunites devoid of interstital phases precipitated from a trapped basaltic magma. However, the regional absence of plastic strain fabric in the dunites at the top of the upwelling, even where there is no more evidence of magmatic impregnation, suggests that the entire transition zone has been pervasively soaked with a basaltic melt. Although the impregnation leading to the fabric destruction was not necessarily synchronous at the scale of the entire diapir, indirect structural evidence suggest that, at one stage of its evolution, it was more important than what is observed today. A viscosity drop of at least four orders of magnitude at the top of the diapir has been invoked to account for the sharp rotation of the plastic flow lines beneath the Moho (Rabinowicz et al., 1987). In that model, we suggested that the origin of the low viscosities had to be due to a catastrophic increase of the melt/rock ratio induced by a discontinuity in the pressure field at the top of the diapir. Such low viscosities are compatible with the one deduced from the compaction structures observed in the transition zone. The sharpness of the rotation of the asthenospheric flow structures has been shown to be correlated with the deformation state of the transition zone: where this one is thick and undeformed, no rotation is observed, which means that it occurred within a material where no plastic deformation could be recorded, most probably a crystal mush. Where the transition zone is strongly laminated and plastically deformed, precluding large interstitial melt fractions, the rotation of the plastic flow structures is more progressive. This observation demonstrates the relationship between the impregnation (i.e. viscosity) and the plastic flow geometry. It also suggests pervasive impregnation of the transition zone at the scale of the entire diapir. In steady-state regime, a dynamical equilibrium was probably maintained between the melt pooling in the transition zone and the melt extracted from it to feed the magma chamber. The undeformed structure of the transition zone is probably inherited from this stage. However, its present composition is likely inherited from a period when it was still compacting but no longer supplied with melt. This is consistent with the scarcity, in Maqsad, of structures indicative of feeding of the transition zone with magma batches from below and, on the other hand, with the abundance of frozen compaction structures relevant to the drainage of magma toward the overlying magma chamber. Accordingly, some of the petrological characteristics of the Maqsad diapir are possibly attributable to a dying system.

Magmatic activity at most present-day oceanic ridges is concentrated in a narrow neo-volcanic zone ranging in width from a few hundred meters to a few kilometers (e.g. Macdonald, 1982). Attempts have been made to account for this remarkable feature by a peculiar melt extraction pattern, involving either large scale porous flow or dyke injection, or a combination of these two mechanisms (e.g. Spiegelman and McKenzie, 1987; Phipps-Morgan, 1987; Scott and Stevenson, 1989). In Maqsad, where a wide spectrum of melt/solid ratios can be observed, the present distribution of interstitial melt relics suggests that porous flow was not an efficient melt migration mechanism on distances exceeding a few meters, whatever the melt fraction. On the other hand, nothing in the dyke injection pattern indicates that crack orientation in the diapir could induce magma concentration beneath the ridge. Inside the upwelling, the dykes may be oriented at high angle to the ridge axis, depending on the asthenospheric flow direction. In the crystal mush, from where the melt is drained toward the surface, no preferred dyke orientation is observed at all. As the mantle flows away from the axial zone and progressively cools, the dyke injection pattern becomes more and more determined by the tensile stress related to the spreading direction; no clear evidence of melt injection toward the ridge axis has been found there. The structure of the Maqsad diapir suggests that the concentration of magmatic activity beneath the spreading axis is attributable to a feedback process between mantle solid-state flow and melt extraction following mechanisms discussed by Rabinowicz et al (1984, 1987), Scott and Stevenson (1989) and Buck and Su (1989). This mechanism originates from concentration of asthenospheric flow in low viscosity zones, i.e. where the melt/rock ratio is maximum. The feedback consists of the fact that such channelling of the solid-state flow enhances melt delivery in the zones already enriched in magma. In Maqsad, this zone of flow concentration defines a narrow corridor at the top of the mantle upwelling elongated parallel to the paleo-ridge axis. This horizon, from where most of the melt feeding the magma chamber is extracted, is comparable in width with the neo-volcanic zone along present-day spreading centers (about 2 Km). Further concentration of the eruptive activity is to be looked for in the geometry of the magma chamber itself (Nicolas et al., 1988 b and c).

Acknowledgments

This work was made possible thanks to facilities in Oman provided by the Ministry of Petroleum and Minerals. I am very indebted to Dr. Al Azri, director of the Oman Geological Survey, for the constant support and interest he brought to my work in Oman, to Michel Beurrier and his team for providing usefull facilities in Mascate, to professor A. Nicolas who introduced me to the study of ophiolites, to Pierre Genthon for help in field work, to Peter Kelemen for a constructive review of the manuscript, to all of them and to Michel Rabinowicz and Marc Leblanc for stimulating discussions. Financial support was provided by the Centre National de la Recherche Scientifique. The is a contribution DBT/INSU No. 300 'Dynamicque Globale'.

References

Benn, K., Nicolas, A. and Reuber, I., 1988. Mantle-crust transition zone and origin of wehrlitic magmas: evidence from the Oman ophiolite. Tectonophysics, 151: 75–85.

Berger, E.T. and Vannier, M., 1984. Les dunites en enclaves dans les basaltes alcalins des iles oceaniques: approche petrologique. Bull. Mineral., 107: 649–663.

Beurrier, M., 1988. Géologie de la nappe ophiolitique de Samail dans les parties orientale et centrale des Montagnes d'Oman. Documents du BRGM, 128, 412 p.

Boudier, F. and Coleman, R.G., 1981. Cross section through the peridotites in the Samail ophiolite, Southeastern Oman. J. Geophys. Res., 86: 2573–2592.

Boudier, F., Bouchez, J.L., Nicolas, A., Cannat, M., Ceuleneer, G., Misseri, M. and Montigny, A., 1985. Kinematics of oceanic thrusting in the Oman ophiolite. Model of plate convergence. Earth Planet. Sci. Lett., 75: 215–222.

Boudier, F., Ceuleneer, G. and Nicolas, A., 1988. Shear zones, thrusts, and related magmatism in the Oman ophiolite: initiation of thrusting on an oceanic ridge. Tectonophysics, 151: 275–296.

Buck, W.R. and Su, W., 1989. Focused mantle upwelling below Mid-Ocean ridges due to feedback between viscosity and melting. Geophys. Res. Lett., 16: 641–644.

Ceuleneer, G., 1986. Structure des ophiolites d'Oman: flux mantellaire sous un centre d'expansion océanique et charriage à la dorsale. These Doc. Univ. Nantes, 338 p.

Ceuleneer, G. and Nicolas, A., 1985. Structures in podiform chromite from the Maqsad district (Sumail ophiolite, Oman). Mineralium Depos., 20: 177–185.

Ceuleneer, G., Nicolas, A. and Boudier, F., 1988. Mantle flow patterns at an oceanic spreading centre: the Oman peridotites record. Tectonophysics, 151: 1–26.

Coleman, R.G., 1981. Tectonic setting for ophiolite obduction in Oman. J. Geophys. Res., 86: 2497–2508.

Dick, H.J.B. Evidence of partial melting in the Josephine peridotite. Oregon Dept. Geol. Mineral. Ind., 96: 59–62.

Glennie, K.W., Boeuf, M.G.A., Hughes-Clark, M.W., Moody-Stuart, M., Pilaar, W.F.H. and Reinhardt, B.M., 1974. Geology of the Oman Mountains, Verh. K. Ned. Geol. Mijnbouwkd Genoot, 31, 423 p.

Kelemen, P.B., (in press), Reaction between ultramafic rock and fractionation basaltic magma 1. Phase relations, the origin of calc-alkaline magma series, and the formation of discordant dunite. J. Petrol., in press.

Lago, B., Rabinowicz, M. and Nicolas, A., Podiform chromite ore bodies: a genetic model. J. Petrol., 23: 103–125.

Macdonald, K.C., 1982. Mid-ocean ridges: fine scale tectonic, volcanic, and hydrothermal processes within the plate boundary zone. Ann. Rev. Earth Planet. Sci., 10: 155–190.

McKenzie, D.P., 1984. The generation and compaction of partially molten rock. J. Petrol., 25: 713–765.

Montigny, R., Le Mer, O., Thuizat, R. and Whitechurch, H., 1988. K-Ar and 40Ar/39Ar study of metamorphic rocks associated with the Oman ophiolite: tectonic implications. Tectonophysics, 151: 345–362.

Nicolas, A., 1986. Melt extraction model based on structural studies in mantle peridotites. J. Petrol., 27: 999–1022.

Nicolas, A., 1989. Structures of ophiolites and dynamics of oceanic lithosphere. Kluwer Acad. Publ., 367 p.

Nicolas, A., Boudier, F. and Ceuleneer, G., 1988. Mantle flow patterns and magma chambers at ocean ridges: evidence from Oman ophiolite. Marine Geophys. Res., 9: 293–310.

Nicolas, A., Ceuleneer, G., Boudier, F. and Misseri, M., 1988a. A structural mapping in the Oman ophiolites: mantle diapirism along an oceanic ridge. Tectonophysics, 151: 27–56.

Nicolas, A., Reuber, I. and Benn, K., 1988b. A new magma chamber model based on structural studies in the Oman ophiolite. Tectonophysics, 151: 87–105.

Phipps Morgan, J., 1987. Melt migration beneath mid-ocean spreading centers. Geophys. Res. Lett., 14;: 1238–1241.

Quick, J.E., 1981. Petrology and petrogenesis of the Trinity peridotite, an upper mantle diapir in the eastern Klamath Mountains, northern California. J. Geophys. Res., 86: 11837–11863.

Rabinowicz, M., Ceuleneer, G. and Nicolas, A., 1987. Melt segregation and flow in mantle diapirs below spreading centers: evidence from the Oman ophiolites. J. Geophys. Res., 92: 3475–3486.

Rabinowicz, M., Nicolas, A. and Vigneresse, J.L., 1984. A rolling mill effect in asthenospheric beneath oceanic spreading centers. Earth Planet. Sci. Lett., 67: 97–108.

Reuber, I., 1988. Complexity of the crustal sequence in northern Oman ophiolitc (Fizh and southern Aswad block): the effect of early slicing? Tectonophysics, 151: 137–165.

Scott, D.R. and Stevenson, D.J., 1986. Magma ascent by porous flow. J. Geophys. Rcs., 91: 9283–9296.

Scott, D.R. and Stevenson, D.J., 1989. A self-consistent model of melting, magma migration and buoyancy-driven circulation beneath mid-ocean ridges. J. Geophys. Res., 94: 2973–2988.

Sleep, N.H., 1988. Tapping of melt by veins and dykes. J. Geophys. Res., 93: 10255–10272.

Spiegelman, M. and McKenzie, D., 1987. Simple 2–D models for melt extraction at mid-ocean ridges and island arcs. Earth Planet. Sci. Lett., 83: 137–152.

Structural and Petrological Features of Peridotite Intrusions from the Troodos Ophiolite, Cyprus

ROGER LAURENT, CLAUDE DION and YVES THIBAULT
Département de Géologie, Université Laval, Ste-Foy, Québec, Canada G1K 7P4

Abstract

Strings of intrusive peridotite cut through the layered ultramafic cumulates and gabbroic rocks of the Troodos plutonic core. The massive and undeformed peridotite has a typical poikilitic texture with olivine (Fo 84–90) and spinel grains (Cr^* 59–68) enclosed by oikocrysts of magnesian augite (En51 Wo43 Fs6), orthopyroxene (En 83–87) and calcic plagioclase (An 92–93). Late interstitial pargasitic hornblende is locally abundant.

Bulk chemistry and mineralogy of the poikilitic peridotite are close to those of undepleted mantle lherzolite suggesting that the intrusive rock is the product of extensive melting of a fertile mantle source. The parental magma consisted of a mixture of basaltic melt and mantle residual solid phases such as olivine and spinel. It is postulated that such a magmatic mixture can only be collected in mantle diapirs reaching the base of ophiolite crusts.

Introduction

About twenty years ago as the result of the development of plate tectonics concepts, the Troodos ophiolite of Cyprus was recognized as a well-preserved fragment of late Cretaceous Tethyan oceanic crust (Moores and Vine, 1971). Since then, and because of its controversial origin, the ophiolite has been the focus of intensive research, the most recent one being that undertaken by the International Crustal Drilling Project. Our studies of the plutonic rocks of Mount Olympus are part of this program. Detailed field mapping at the 1:5,000 scale has allowed us to document the presence of poorly-known intrusive bodies of peridotite emplaced into the layered ultramafic and mafic rocks of the ophiolite crustal sequence. The purpose of this paper is to give a comprehensive description of these peridotite intrusions and evaluate their significance. The data were collected in the area of the

Tj. Peters et al. (Eds), Ophiolite Genesis and Evolution of the Oceanic Lithosphere, 175–194.

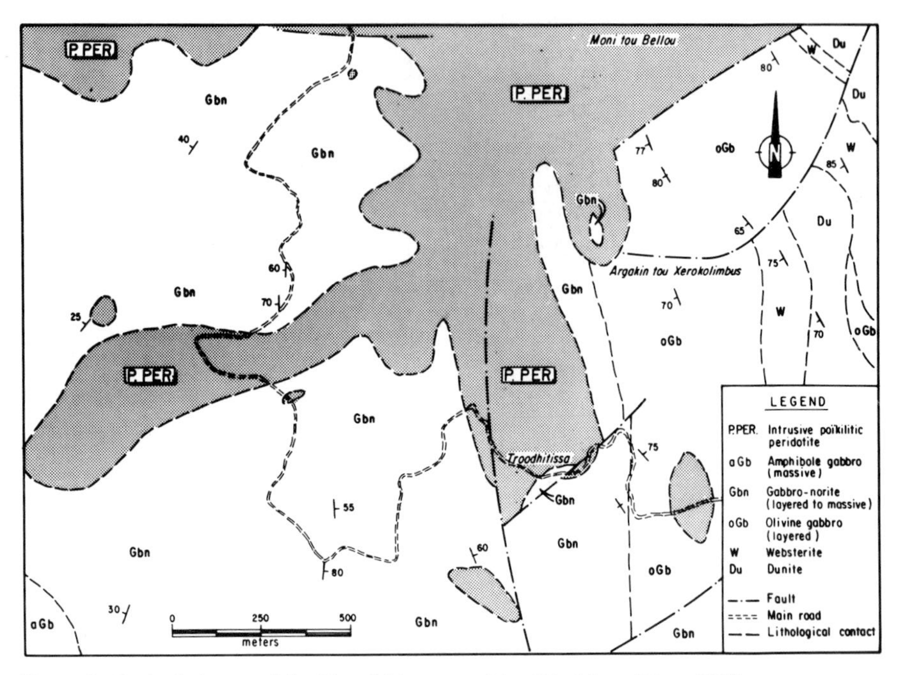

Figure 1. Geological map of the Troodhitissa area (simplified from Dion, 1987).

Troodhitissa monastery on the southwestern flank of Mount Olympus in the plutonic core of the Troodos Massif (Figure 1).

Field Relations

The basement of the Troodos ophiolite consists of harzburgite tectonites associated with chromite-bearing dunites. These highly residual mantle rocks form the core of the plutonic complex (Wilson, 1959; George, 1978). Sequences of layered and folded plutonic rocks of magmatic origin lie in fault contact against the harzburgitic core on all sides. The unconformity between the residual harzburgite and the overlying rocks of magmatic origin is usually recognized as the local *Moho*. The Troodhitissa area is structurally situated right above this unconformity.

Multiple intrusions of distinct age within the plutonic assemblage of magmatic origin have been described by Allen (1975), Benn and Laurent (1987) and Malpas et al. (1989). The plutonic rocks have been divided into two main sequences: (1) an early stratiform cumulate suite, and (2) a late intrusive suite. Near Troodhitissa the early stratiform suite is composed of layered cumulate dunite and wehrlite, layered olivine gabbro and layered to massive gabbro-norite. Vari-textured amphibole gabbro and irregular bodies of plagiogranite occur higher up in the sequence. Regional strike of layering is

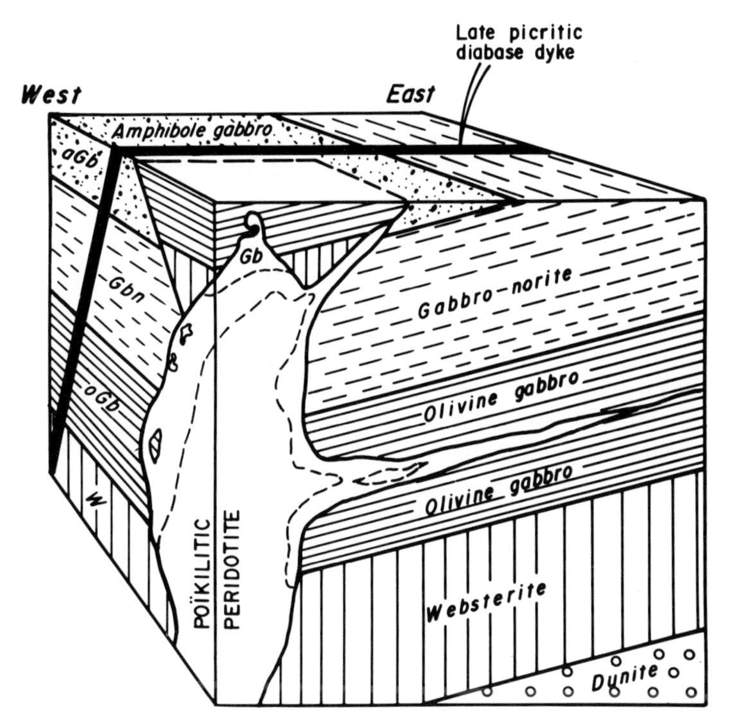

Figure 2. Structural sketch of the relationships between the peridotite intrusion and the early stratiform cumulate suite (after rotation of the units back to their assumed original position).

approximately to the north; dips vary from subvertical to less than 30° to the west. In contrast, the late plutonic suite consists of intrusive bodies of massive peridotite, from a few meters to more than 2 kilometers in size, that display a characteristic poikilitic texture. The large peridotite intrusion, which extends on more than 2 km across the map area (Figure 1), is zoned, varying in composition from dunite to olivine gabbro. The gabbroic fraction represents less than 3% of the whole body.

The intrusive peridotites occupy about 25% of the Troodhitissa area; they are concentrated near the unconformity at the base of the early stratiform suite. This unconformity has probably captured and channeled the peridotitic magmas rising from the mantle. Above the unconformity, positions of peridotite intrusions are controlled by NE-SW and E-W faults and their irregular shapes by the structure of the country-rock. The peridotite has cut its way up through the early suite host rocks. "En bloc" elevation of the overlying gabbroic rocks can be locally observed above the peridotite bodies indicating that the country rock was uplifted along vertical faults. The geometrical relations are illustrated in Figure 2 representing the assumed position of the units at the time of intrusion, the main elongation of the peridotite intrusion defining the vertical direction. The intrusive peridotites cut across the layering of the country-rock or may follow the bedding or other weakness zones of the layered sequence forming irregular sills of poikilitic peridotite. No evidence of chilled margins is found at intrusive contacts, but layered dunite and wehrlite hosts are frequently impregnated by clinopyroxene within several meters around their contact with the massive intrusive peridotite while gabbroic rocks are recrystallized as olivine gabbro-norite.

Petrography

Intrusive peridotites are massive but compositionally zoned. Their base consists of dunitic and wehrlitic rocks which grade progressively upward into the main body of plagioclase lherzolite. Minor amounts of olivine melagabbro are intermixed with the lherzolite near the top of intrusions and, along contacts, some massive gabbros occur discontinuously. The average plagioclase lherzolite is medium to coarse grained with a poikilitic texture. Olivine (>40%) and chromite (<3%) are included into large oikocrysts of clinopyroxene (5–35%) and orthopyroxene (0–15%). Plagioclase (0–25%) and prismatic amphibole (0–20%) are major interstitial phases. Olivine in pyroxene oikocrysts occurs as 1 to 5 mm anhedral crystals containing few chromite inclusions. Most olivine grains seem to be undeformed though some crystals display kink bands and wavy extinctions. However, observations in these grains of high-temperature plastic deformation microstructures (Benn and Laurent, 1987) suggest that significant amounts of olivine are mantle xenocrysts. Chromite is present as euhedral to anhedral, fine grained (0.05 to 0.25 mm) crystals which are intergranular to olivine or included in the oikocrysts. Clinopyroxene forms 3 to 50 mm anhedral and homogeneous oikocrysts without compositional zoning or exsolutions. With few exceptions, the crystals are not deformed. A late magmatic hornblende rims clinopyroxenes which can also be replaced by acicular actinolite of metamorphic origin. Orthopyroxene is texturally identical to the clinopyroxene and contains inclusions of olivine, chromite and anhedral clinopyroxene. Undeformed anorthitic plagioclase occupies interstices between the clinopyroxene oikocrysts.

The crystallization order of the poikilitic peridotite is first olivine and chromite, later followed successively by clinopyroxene, orthopyroxene, plagioclase and hornblende. Significant fractions of olivine and anhedral chromite are xenocrysts that were carried by the magma (crystal mush) and which have reacted with it during the ascent and cooling. Some of the chromite could also represent a reaction product, formed at the time of emplacement, between the magma and assimilated early liquidus phases and wall rock minerals (Bowen, 1928). The olivine melagabbros intermixed with the lherzolite in the upper part of the intrusion are wholly recrystallized xenoliths of the partly assimilated layered country-rock. Along the contact of peridotite intrusions then occurs locally a massive olivine gabbro-norite a few meters thick. The same rock occurs also as segregated pockets within the poikilitic peridotite. This gabbro-norite consists of coarse and variously resorbed olivine grains within a fine grained granoblastic matrix of plagioclase, clinopyroxene, orthopyroxene and amphibole. We interpret this gabbro-norite as a product of the mixing of the melted host rock with the intrusive magma. These observations suggest that the country rock was nearly completely melted in small amount and partially melted in large amount along the contact with the peridotite intrusion. Both host rock and intrusive peridotite were heated to high temperatures and then cooled slowly together, preventing the development of chilled margins in the peridotite. Slow cooling was necessary in order to sustain the conditions of crystallization required to produce the typical poikilitic texture of the peridotite.

Bulk Chemistry

Table 1 presents 12 analyses of poikilitic peridotites, 5 analyses of associated gabbros, the respective average composition of both groups (P.PER. and P.GB) and, for comparison, 1 analysis of residual harzburgite (KB 4032) from the Troodos mantle core (methods of analysis are described in Laurent and Hebert, 1989). The chemistry of poikilitic peridotites is characterized by high MgO and low CaO and Al_2O_3 contents compared to their associated gabbros which, on Figure 3, plot in a cluster centered along the trend defined by the host cumulate rocks of the Cyprus deep drill Cy-4 (Laurent and Hebert, 1989; Laurent, 1990). Compositions intermediate between the gabbro group and the poikilitic peridotites are not found in significant amounts.

The bulk composition of the average plagioclase lherzolite (P.PER., Table 1) is then compared to estimated fertile mantle compositions. Mantle peridotites are mainly composed of SiO_2, MgO and FeO. The proportion of these three major components is similar in mantle peridotites and our average plagioclase lherzolite. Furthermore, oxides such as Al_2O_3, CaO and FeO which tend to be concentrated in the liquid during the melting of peridotite are in equal proportions in our plagioclase lherzolite, the Zabargad spinel lherzolite (Bonatti et al., 1986) and pyrolite composition (Ringwood, 1975). The average composition of 384 spinel lherzolites (Maaløe and Aoki, 1977) is lower in CaO and Al_2O_3 than P.PER. The Troodos harzburgite (KB 4032, Table 1) has the most refractory composition of the group. Similar relations are given by the respective contents of Cr and Ni. P.PER. has concentrations of Cr and Ni very close to those of Ringwood's pyrolite. In all cases, the average composition of the ultramafic cumulate rocks from Cy-4 (Laurent and Hebert, 1989) plot away from P.PER. and other mantle compositions. The poikilitic plagioclase lherzolite of the late Troodos intrusions clearly is not the product of a remobilization of the basal cumulates. On the contrary, similarity in bulk composition of P.PER. and fertile mantle peridotites suggests a common origin. The chemical evidence implies the possibility that the late intrusions derive from a magma consisting of extensively melted fertile peridotite with addition of basaltic magma.

Table 1. Chemical analyses of poikilitic peridotites and associated gabbros, and of one tectonite harzburgite, from southwestern Troodos.

	1	2	3	4	5	6	7	8	9	10
Ox wt%	CD8522	CD8526	CD8533	CD8577	CD8584	CD8603	CD8608	YT8502	KB4078	KB4084
SiO_2	43.23	45.73	44.33	42.36	41.58	42.95	45.18	45.79	41.20	44.75
TiO_2	0.08	0.05	0.11	0.04	0.03	0.06	0.05	0.11	0.01	0.05
Al_2O_3	6.46	3.91	5.10	2.31	2.20	2.53	6.07	7.18	0.65	4.86
Fe_2O_3	4.53	4.27	4.80	5.55	5.23	4.99	4.31	0.97	1.77	1.73
FeO	5.53	5.30	4.83	4.56	4.79	5.05	4.95	7.66	8.14	9.27
MnO	0.16	0.15	0.16	0.16	0.15	0.14	0.15	0.15	0.17	0.18
MgO	35.15	34.05	35.96	43.26	44.12	42.48	33.80	30.02	46.98	34.43
CaO	4.48	6.36	4.49	1.67	1.83	1.79	5.49	7.56	1.05	4.55
Na_2O	0.22	0.10	0.10	0.01	0.01	0.00	0.00	0.46	0.01	0.02
K_2O	0.02	0.02	0.03	0.03	0.04	0.01	0.00	0.01	0.01	0.01
P_2O_5	0.13	0.05	0.09	0.04	0.01	0.00	0.00	0.09	0.01	0.16
Ni (ppm)	1133	1225	1283	2130	1651	2560	1400	828	1440	1040
Cr (ppm)	2203	3021	2385	2720	3501	1763	1880	1595	118	200
V (ppm)	74	92	83	26	47	72	80	109	nd	nd
Co (ppm)	54	117	nd	120	nd	119	112	nd	nd	nd
Zr (ppm)	7	8	6	1	6	5	8	8	nd	nd
FeO*	9.61	9.14	9.16	9.55	9.49	9.55	8.83	8.53	9.73	10.83
FeO*/MgO	0.27	0.27	0.25	0.22	0.22	0.22	0.26	0.28	0.21	0.31
Mg#	88.1	88.3	89.0	90.5	90.6	90.2	88.7	86.6	90.0	85.6

Table 1. (continued)

Ox wt%	11 KB4150	12 KB8535	13 P. PER	14 CD8552	15 CD8553	16 CD8605	17 YT8504	18 KB4092	19 P. GB.	20 KB4032
SiO_2	44.84	47.67	44.13	46.96	45.57	48.82	52.56	52.37	49.25	44.35
TiO_2	0.04	0.08	0.06	0.10	0.16	0.11	0.28	0.09	0.15	0.01
Al_2O_3	7.18	3.93	4.36	14.50	6.89	11.09	10.01	6.26	9.75	0.75
Fe_2O_3	1.66	1.69	3.46	1.44	2.77	0.89	1.02	1.66	1.56	nd
FeO	7.78	10.95	6.57	3.94	6.97	5.00	5.55	2.87	4.87	8.54
MnO	0.16	0.23	0.16	0.11	0.17	0.13	0.16	0.13	0.14	0.13
MgO	30.21	29.27	36.64	17.36	30.73	19.36	13.58	17.42	19.69	45.18
CaO	7.96	6.05	4.44	14.91	6.25	14.20	15.97	18.80	14.03	0.90
Na_2O	0.08	0.11	0.11	0.49	0.36	0.27	0.68	0.27	0.41	0.01
K_2O	0.01	0.01	0.02	0.07	0.02	0.02	0.04	0.03	0.04	0.01
P_2O_5	0.09	0.02	0.07	0.12	0.11	0.12	0.15	0.10	0.12	0.12
Ni (ppm)	1040	680	1368	306	931	346	138	16	347	2220
Cr (ppm)	164	88	1637	1398	3354	2200	1041	70	1613	260
V (ppm)	nd	32	68	75	134	76	215	24	104	nd
Co (ppm)	nd	nd	104	nd	nd	72	nd	nd	72	nd
Zr (ppm)	nd	nd	6	5	9	1	13	nd	7	nd
FeO*	9.27	12.47	9.68	5.24	9.46	5.80	6.47	4.36	6.27	8.54
FeO*/MgO	0.31	0.43	0.27	0.30	0.31	0.30	0.48	0.25	0.33	0.19
Mg#	85.9	81.3	87.9	86.4	86.2	86.1	79.6	88.8	85.4	90.5

All samples are from the Troodhitissa-Caledonian Falls area. Samples CD are from Dion (1987), YT from Thibault (1987) and KB from Benn (1986). Numbers 1 to 12 are poikilitic peridotites. Number 13 (P.PER.) is the average composition of samples 1 to 12. Numbers 14 to 18 are gabbros associated with the poikilitic peridotites. Number 19 (P. GB.) is the average composition of samples 14 to 18. Number 20 (KB 4032) is a tectonite harzburgite. Major oxides have been normalized to 100%.

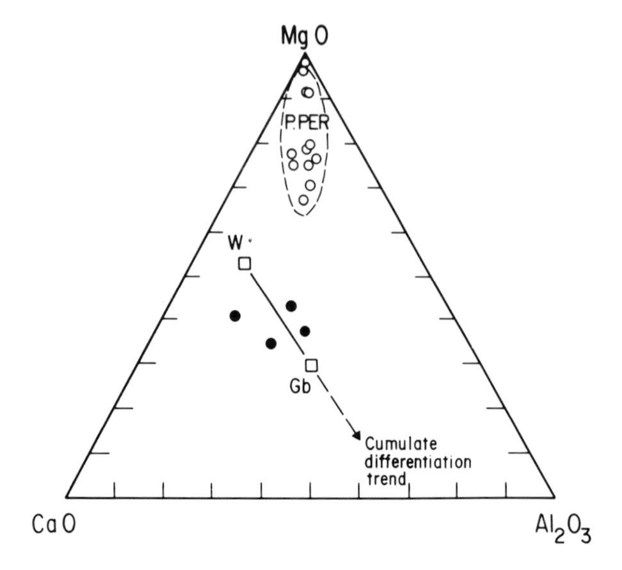

Figure 3. MgO-CaO-Al_2O_3 bulk chemistry diagram for the poikilitic peridotites (open circles) and associated gabbros (full circles). W and Gb are respectively average compositions of the ultramafic cumulates (W) and gabbroic rocks (Gb) from the deep drill Cy-4.

Mineral Chemistry

Representative average compositions of the main mineral components of the poikilitic plagioclase lherzolite are given in Tables 2 to 5. They were collected on a wavelength dispersive ARL electron microprobe at Laval University (conditions and methods are described in Laurent and Hebert, 1989). Compositional variations of the main minerals are limited to a narrow range as shown by Figure 4 summarizing the data for pyroxenes, olivine and plagioclase. Olivine and clinopyroxene are highly magnesian, with $Mg^{\#}$ values (100 Mg : Mg + Fe) ranging from 84 to 92. Orthopyroxene and amphibole, late comers in the crystallization sequence, have $Mg^{\#}$ values of 86 to 92, while plagioclase is wholly anorthitic (An 92–93). Homogeneity in composition of the main minerals reflects the limited range of differentiation within the peridotite intrusion. Only chromian spinels by comparison show relatively large variations with $Mg^{\#}$ ranging from 40 to 55 and $Cr^{\#}$ (100 Cr : Cr + Al) from 59 to 68. The Na_2O and TiO_2 contents of late magmatic amphiboles also vary significantly. These features suggest that the spinels result from a more complex evolution than most main minerals, and that contents of Na_2O and TiO_2 were heterogeneously distributed within the residual melts, having reacted with pyroxenes to form the amphiboles.

$Mg^{\#}$ values and NiO contents of olivine are positively correlated (Figure 5A). Mole percents of forsterite vary from 89.5 to 84 while NiO contents are between 0.32 and 0.14 wt% (equivalent to 2,500–1,100 ppm Ni). These data from the Troodhitissa area (Dion, 1987; Thibault, 1987) confirm those

Table 2. Microprobe analyses of olivine from the troodos poikilitic peridotite.

Ox wt%	CD8522	CD8526	CD8533	CD8553	CD8577	CD8584	CD8603	CD8605	CD8608	YT8502
SiO_2	39.60	41.06	40.48	38.81	39.20	41.15	39.80	39.02	40.83	39.66
FeO	12.04	12.50	11.53	13.73	11.05	10.46	9.96	15.05	11.32	15.09
MnO	0.21	0.23	0.22	0.30	0.21	0.16	0.16	0.25	0.20	0.26
NiO	0.17	0.20	0.20	0.14	0.32	0.25	0.27	0.16	0.22	0.19
MgO	47.66	46.20	46.92	45.35	49.29	48.45	48.32	45.04	46.93	44.25
CaO	–	–	–	–	0.06	0.05	–	–	–	–
Sum	99.68	100.20	99.35	98.34	100.12	100.52	98.51	99.51	99.49	99.45
O = 4.0										
Si	0.9864	1.0152	1.0066	0.9886	0.9705	1.0059	0.9932	0.9875	1.0119	1.002
Fe^{2+}	0.2506	0.2585	0.2396	0.2924	0.2286	0.2138	0.2078	0.3184	0.2345	0.319
Mn	0.0044	0.0049	0.0046	0.0065	0.0043	0.0032	0.0034	0.0053	0.0042	0.005
Ni	0.0033	0.0040	0.0041	0.0029	0.0064	0.0049	0.0054	0.0033	0.0043	0.003
Mg	1.7688	1.7019	1.7386	1.7211	1.8181	1.7649	1.7969	1.6981	1.7330	1.666
Ca	–	–	–	–	0.0015	0.0013	–	–	–	0.000
Sum	3.0136	2.9844	2.9934	3.0114	3.0295	2.9941	3.0068	3.0125	2.9881	2.997
Fo	87.40	86.60	87.69	85.21	88.64	89.05	89.48	83.99	87.89	83.70
Mg#	87.59	86.82	87.89	85.48	88.67	89.05	89.64	84.2		
1	88.08	83.94								

Fo = Forsterite (mol %); Mg# = 100 Mg/(Mg + Fe).

Table 3. Microprobe analyses of clinopyroxene from the Troodos poikilitic peridotite.

Ox wt%	CD8522	CD8526	CD8533	CD8553	CD8577	CD8584	CD8603	CD8605	CD8608	YT8502
SiO_2	53.06	53.12	52.33	52.74	50.16	52.46	53.39	53.09	54.58	52.34
Al_2O_3	2.61	2.12	3.13	2.85	2.66	2.37	2.01	2.65	1.84	2.35
TiO_2	0.14	0.04	0.10	0.11	0.07	0.05	0.12	0.15	0.07	0.08
Cr_2O_3	0.68	0.68	0.78	0.63	0.52	0.66	0.41	0.60	0.61	0.71
FeO	3.79	3.46	3.61	4.16	3.06	3.29	2.66	4.51	3.72	3.55
MnO	0.14	0.16	0.14	0.17	0.16	0.10	0.09	0.13	0.12	0.14
MgO	18.05	17.20	19.10	18.12	18.05	17.83	16.68	17.82	17.80	18.49
CaO	21.69	22.61	20.97	21.33	24.17	21.37	23.07	20.93	20.96	21.44
Na_2O	0.11	0.09	0.06	0.19	0.34	0.20	0.16	0.15	0.10	0.08
K_2O	–	–	–	–	0.05	–	–	–	–	0.00
Sum	100.25	99.46	100.23	100.30	99.24	98.34	98.56	100.03	99.80	99.18
O = 6.0										
Si	1.9266	1.9459	1.8985	1.9168	1.8627	1.9376	1.9653	1.9330	1.9785	1.9990
Al IV	0.0734	0.0541	0.1015	0.0832	0.1164	0.0624	0.0347	0.0670	0.0215	0.079
Al VI	0.0381	0.0371	0.0320	0.0391	0.0000	0.0407	0.0526	0.0467	0.0571	0.022
Ti	0.0037	0.0012	0.0028	0.0031	0.0020	0.0013	0.0032	0.0040	0.0019	0.002
Cr	0.0194	0.0196	0.0224	0.0182	0.0153	0.0193	0.0118	0.0174	0.0175	0.020
Fe^{2+}	0.1150	0.1058	0.1093	0.1265	0.0950	0.1016	0.0817	0.1373	0.1127	0.109
Mn	0.0043	0.0050	0.0044	0.0051	0.0050	0.0032	0.0026	0.0041	0.0037	0.004
Mg	0.9762	0.9385	1.0324	0.9811	0.9987	0.9813	0.9149	0.9666	0.9614	1.011
Ca	0.8435	0.8871	0.8152	0.8302	0.9612	0.8455	0.9094	0.8161	0.8137	0.843
Na	0.0075	0.0064	0.0045	0.0132	0.0245	0.0143	0.0114	0.0108	0.0070	0.005
K	–	–	–	–	0.0024	–	–	–	–	–
Sum	4.0079	4.007	4.0230	4.0164	4.0830	4.0071	3.9877	4.0029	3.9750	4.018
Mg#	89.45	89.87	90.45	88.58	91.31	90.63	91.81	87.56	89.51	90.27
En	50.34	48.47	52.64.	50.50	48.60	50.88	47.94	50.24	50.83	51.39
Fs	6.15	5.72	5.80	6.77	4.62	5.27	4.44	7.35	6.15	5.76
Wo	43.50	45.81	41.56	42.73	46.78	43.85	47.63	42.41	43.02	42.85

Mg# = 100 Mg/(Mg + Fe); En = Enstatite; Fs = Ferrosilite; Wo = Wollastonite (mol%).

Table 4. Microprobe analyses of orthopyroxene from the Troodos poikilitic peridotite.

Ox wt%	CD8522	CD8553	CD8605	YT8502
SiO_2	56.04	56.17	56.10	54.84
Al_2O_3	1.93	1.80	1.59	1.60
TiO_2	0.05	0.00	0.06	0.27
Cr_2O_3	0.38	0.37	0.26	0.27
FeO*	7.41	7.60	8.75	9.33
MnO	0.21	0.23	0.24	0.27
MgO	33.80	32.83	31.67	32.07
CaO	1.51	2.23	2.04	1.26
Na_2O	0.01	0.02	0.00	0.01
Sum	101.35	101.23	100.71	99.92
O = 6.0				
Si	1.9282	1.9389	1.9536	1.9307
Al IV	0.0718	0.0611	0.0464	0.0664
Al VI	0.0065	0.0121	0.0188	0.0000
Ti	0.0014	0.0000	0.0014	0.0071
Cr	0.0102	0.0100	0.0072	0.0075
Fe^{2+}	0.2131	0.2192	0.2547	0.2747
Mn	0.0062	0.0067	0.0071	0.0081
Mg	1.7325	1.6886	1.6433	1.6827
Ca	0.0557	0.0823	0.0762	0.0475
Na	0.0009	0.0013	0.0000	0.0007
Sum	4.0266	4.0202	4.0087	4.0254
Mg#	89.05	88.51	86.58	85.97
En	86.30	84.57	82.94	83.59
Fs	10.92	11.31	13.21	14.05
Wo	2.78	4.12	3.84	2.36

Mg# = 100 Mg/(Mg + Fe); En = Enstatite, Fs = Ferrosilite; Wo = Wollastonite (mol %).

from the nearby Caledonian Falls area (Benn, 1986; Benn and Laurent, 1987) and show that olivine from the peridotite intrusions is systematically enriched in NiO compared to olivine from the host layered cumulate rocks, a point which has been questioned by Malpas et al. (1989, p. 47). Olivine of the poikilitic plagioclase peridotite has NiO contents similar to olivine from mantle peridotites though it is on the average more fayalitic than the latter (Basaltic Volcanism Study Project, 1981). Like olivine, orthopyroxene is also richer in FeO than orthopyroxenes from typical mantle peridotites, its enstatite content varying from 87 to 83 mole %. This difference is due in part to its late appearance in the crystallization order and to re-equilibration reactions with clinopyroxene at subsolidus temperatures. The main effect of re-equilibration, as illustrated in Figure 4, is to decrease the $Mg^{\#}$ value of the orthopyroxene losing MgO to the clinopyroxene and gaining FeO (Huebner, 1980). On the other hand, its wollastonite content, less than 4 mole %, and its Al_2O_3 (1.5 to 2.0 wt %), Cr_2O_3 (0.26 to 0.38 wt %) and TiO2 (0.0 to 0.27 wt %) contents are similar to orthopyroxenes of many mantle peridotites.

Compositions of clinopyroxene plotted in the pyroxene quadrilateral (Figure 4) define two groups falling respectively in the augite and diopside fields,

Table 5. Microprobe analyses of spinels from the Troodos poikilitic peridotite.

Ox wt%	CD8522	CD8526	CD8533	CD8553	CD8577	CD8584	CD8603	CD8608	YT8502
Al_2O_3	16.92	20.33	16.84	19.76	16.09	19.40	16.16	17.12	19.86
TiO_2	0.39	0.33	1.10	0.13	0.57	0.31	0.19	0.51	0.37
Cr_2O_3	47.39	45.20	46.00	43.57	49.07	48.44	51.35	44.22	45.49
FeO	25.20	25.48	27.88	25.52	24.28	21.43	21.73	28.90	22.45
MnO	0.13	0.13	0.14	0.13	0.15	0.12	0.14	0.14	0.12
MgO	10.83	9.61	9.11	10.31	10.09	11.37	11.25	8.48	11.93
ZnO	0.09	0.08	–	0.09	0.16	0.08	–	–	–
V_2O_3	0.17	–	–	0.22	–	–	–	–	–
Sum	101.13	101.15	101.07	99.72	100.41	101.16	100.81	99.37	100.22
O = 24.0									
Al	5.0273	6.0021	5.0590	5.8905	4.8484	5.6767	4.7968	5.2253	5.834
Ti	0.0734	0.0614	0.2146	0.0241	0.1097	0.0588	0.0353	0.0990	0.069
Cr	9.4384	8.9634	9.2713	8.7106	9.9302	9.5453	10.3017	9.0877	8.961
Fe^{3+}	1.3524	0.9117	1.2405	1.3065	1.0019	0.6604	0.8310	1.4889	1.069
Fe^{2+}	3.9667	4.4318	4.7353	4.0992	4.2063	3.8001	3.7684	4.7878	3.609
Mn	0.0270	0.0271	0.0296	0.0279	0.0330	0.0253	0.0295	0.0316	0.025
Mg	4.0621	3.5873	3.4498	3.8795	3.8398	4.2178	4.2373	3.2797	4.430
Zn	0.0175	0.0153	–	0.0176	0.0306	0.0155	–	–	–
V	0.0351	0.0000	–	0.0441	–	–	–	–	–
Sum	24.0	24.0	24.0	24.0	24.0	24.0	24.0	24.0	24.0
Cr#	65.1	59.90	64.69	59.65	67.21	62.72	68.16	63.48	60.57
Mg#	50.60	44.74	42.31	48.60	47.70	52.60	52.93	40.66	55.11
$Y(Fe^3)$	8.55	5.74	8.02	8.22	6.35	4.16	5.22	9.42	6.74

Cr# = 100 Cr/(Al + Cr); Mg# = 100 Mg/(Mg + Fe); $Y(Fe^3) = 100\ Fe^3/(Fe^3 + Al + Cr)$. Fe^2 and Fe^3 according to stochiometry.

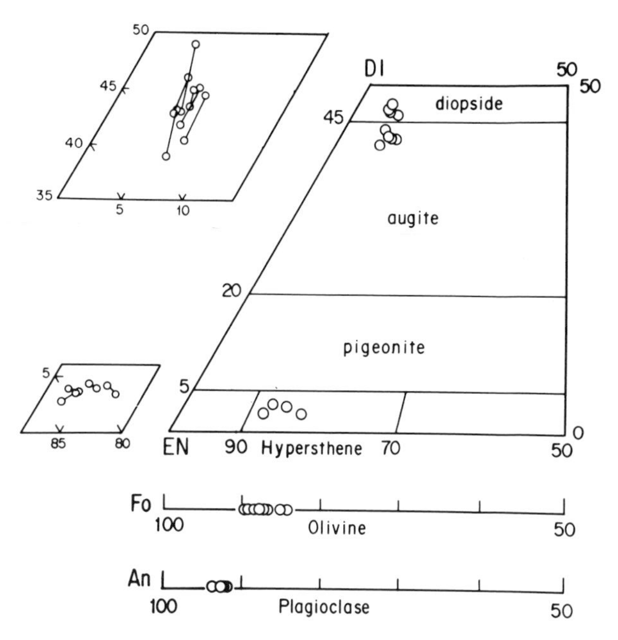

Figure 4. Mineral chemistry of pyroxenes, olivine and plagioclase from the poikilitic peridotite.

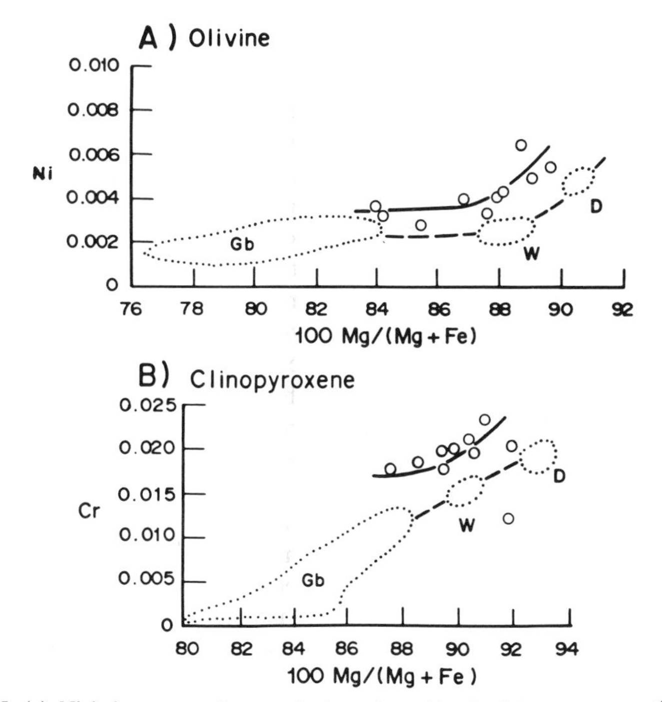

Figure 5. (a) Nickel contents (in atomic formula unit) of olivine versus $Mg^{\#}$; (b) Chrome contents (in atomic formula unit) of clinopyroxene versus $Mg^{\#}$. Open circles are for the poikilitic peridotite samples. The interrupted line joining D (dunite), W (wehrlite) and Gb (gabbro) defines the compositional trend of olivine and clinopyroxene from the early stratiform suite.

according to the new nomenclature of Morimoto (1989). In the enlargement of the diopside corner, tie-lines connect extreme compositions of clinopyroxene measured in the same sample. Ranges in composition vary from less than 1 to more than 7 mole % wollastonite. We assume that these variations are mainly due to subsolidus re-equilibration (Fleet, 1974). The average composition of the high-temperature (solidus) clinopyroxene is a magnesian augite (En51 Wo43 Fs6) with a Mg# value close to 90. Subsolidus reactions have given the clinopyroxene a diopsidic composition (En48 Wo48 Fs4) and a Mg# value of about 92. The major effect of re-equilibration to lower temperatures for the clinopyroxene was a significant enrichment in CaO and a slight increase of its MgO/FeO ratio. Clinopyroxene compositions plotting on tie-lines indicate an evolution from the high-temperature solidus to compositions of progressively lower subsolidus temperatures. Clinopyroxene contains higher concentrations of Al_2O_3, Cr_2O_3 and TiO_2 than do coexisting orthopyroxenes. Its contents of Cr_2O_3 (0.41–0.78 wt %) are similar to those of clinopyroxenes from spinel lherzolite xenoliths while its Al_2O_3 (2.0 to 3.0 wt %) and TiO_2 (0.05 to 0.14 wt %) contents are similar to those of clinopyroxenes from garnet lherzolite xenoliths (Basaltic Volcanism Study Project, 1981). Only its concentration of Na_2O is anomalously low (0.06 to 0.34 wt %) which, to some extent, may be due to a reaction with amphibole and to subsolidus re-equilibration. Mg# values of clinopyroxene are positively correlated to its Cr_2O_3 content (Figure 5B). Again, these new data confirm those of Benn (1986) and Benn and Laurent (1987) and show that clinopyroxene from the peridotite intrusions is enriched in Cr_2O_3 compared to clinopyroxenes of equivalent Mg# value from the host layered cumulate rocks. Late magmatic amphibole rimming clinopyroxene is a pargasitic and edenitic hornblende. Frequently the amphibole is zoned with a crystal core rich in Al_2O_3, Cr_2O_3, TiO_2 and Na_2O. Composition of this hornblende also is quite distinct from that of the calcic amphiboles, depleted in all these components, of the host layered cumulate rocks (Laurent and Hebert, 1989; Hebert and Laurent,1990). A graph of the Mg/Fe divalent cation ratios of clinopyroxene versus coexisting olivine (Figure 6A) shows that Mg and Fe are linearly distributed between these minerals. The average partition coefficient calculated for this equilibrium is 1.30 ± 0.03. If the scattering observed for some of the samples was caused by subsolidus re-equilibration, as it has been discussed before, we may conclude that most olivine and clinopyroxene of the peridotite intrusions first crystallized in mutual equilibrium. The application of the geothermometer of Lindsley and Anderson (1983) provides another method to check the relation between clinopyroxene and orthopyroxene. Analyses representative of their respective high-temperature solidus compositions were plotted in the pyroxene quadrilateral (Figure 6B) according to the conventions described by Lindsley and Anderson. The isotherms drawn for this geothermometer are based on a confining pressure lower than 2 kbar (200 MPa) (Lindsley, 1983). This condition applies to the peridotite intrusions since they were emplaced at a depth of less than 7 km in the

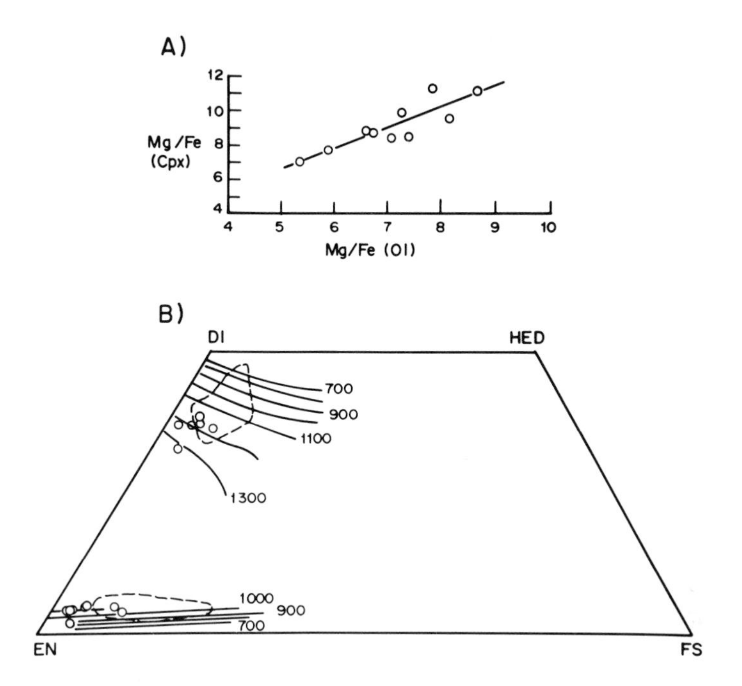

Figure 6. (a) Mg/Fe ratios of coexisting clinopyroxene and olivine in poikilitic peridotites. B) Pyroxene quadrilateral with isotherms for the geothermometer of Lindsley and Anderson (1983). Open circles are pyroxenes from the poikilitic peridotites. Compositional fields of pyroxenes from the early stratiform suite are indicated by interrupted lines.

gabbroic crust. In Figure 6B, the distribution of the analyses with respect to the isotherms suggests that temperatures of crystallization for clinopyroxene are close to 1200°C and that orthopyroxene has crystallized at lower temperature around 1100°C or less. Considering the limitations of the method (Lindsley, 1983), these figures are, at best, broad estimates of crystallization temperatures. Nevertheless, they are realistic magmatic temperatures and agree with the relative order of crystallization observed between the two types of pyroxene.

Spinels from the poikilitic plagioclase peridotite are plotted in Figure 7. They are clearly distinguished from spinels of the host Troodos cumulate rocks and from spinels of mid-ocean ridge peridotites (MOR) by their higher $Cr^{\#}$-ratio. The compositional range of these spinels is intermediate between the field of the MOR spinels and the field of boninite series rocks as defined by Dick and Bullen (1984). Experimental fusion of mantle peridotites demonstrates that $Cr^{\#}$ and $Mg^{\#}$ in spinel increase with the percentage of fusion (Dick et al., 1984). This evidence indicates that Al_2O_3 and $FeO + Fe_2O_3$ are partitioned into the liquid whereas Cr_2O_3 and MgO are concentrated in the spinel. Reversely, the first spinels formed in the melt during crystallization can be expected to have high $Cr^{\#}$ and $Mg^{\#}$ values. Spinels of the peridotite intrusions have these characteristics as shown in Figure 7. The analyses

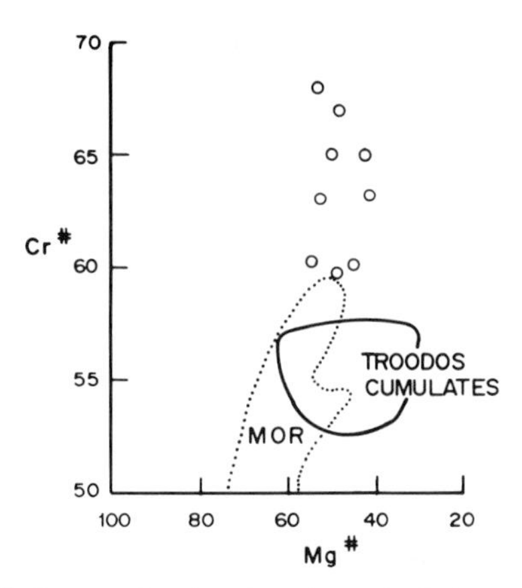

Figure 7. Cr$^{\#}$ versus Mg$^{\#}$ diagram for the chromian spinels from the plagioclase peridotites (open circles). Field of Troodos cumulates defined according to our data. Field of mid-ocean ridge peridotites (MOR) from Dick and Bullen (1984).

selected are mainly from the center of zoned spinel grains. Cr$^{\#}$ and Mg$^{\#}$ values decrease towards their margins indicating that the spinel has reacted with the melt or later, at subsolidus temperatures, with its olivine or pyroxene hosts. Subsolidus re-equilibration of the spinel through Mg-Fe exchange with the host silicate would decrease its Mg$^{\#}$ value and proportionally increase the Mg$^{\#}$ value of the silicate (Cameron, 1975). This effect explains the lack of a strong positive correlation between Cr$^{\#}$ and Mg$^{\#}$ in the spinels analyzed. If the first-formed spinels have the highest Cr$^{\#}$ value, they should also have the highest Mg$^{\#}$, but this is not clearly observable here. We must assume that the Mg$^{\#}$ value has been modified by subsolidus reactions with the host silicates. Along with amphibole these spinels are the only mineral components of the peridotite that are chemically distinct from the mineral assemblage of common mantle peridotites. This is not surprising in view of the fact that these minerals are very sensitive petrogenetic indicators of the conditions that prevail at the time of their crystallization (Irvine, 1965).

Discussion and Conclusion

The first petrological problem concerning the peridotite intrusions is the significance of their composition. What is the source of this rock and what are the relative roles of partial melting and crystal accumulation in producing it? We have shown that the bulk chemistry of the poikilitic plagioclase lherzolite matches undepleted mantle compositions implying the possibility that this rock derives from the extensive melting of a fertile mantle source. The mineral chemistry, with the exception of spinels, also broadly matches the composition of the major minerals from mantle peridotites, thus confirm-

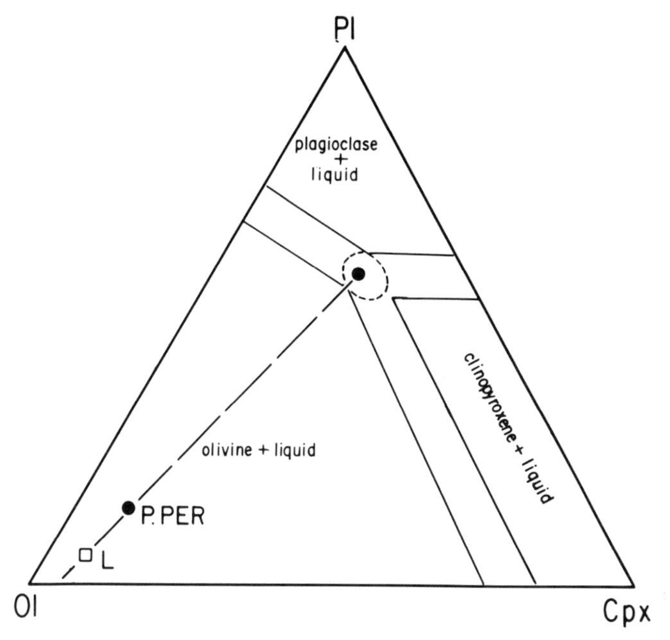

Figure 8. Simplified basalt system (Cox et al., 1979). P.PER. is the average composition of the poikilitic peridotite, L is an average mantle spinel lherzolite (Maaløe and Aoki, 1977).

ing the possibility of a mantle origin. In contrast, a cumulate origin for this rock is most unlikely because the minerals which would have fractionated from a basaltic or picritic melt to produce the lherzolitic composition could not have the chemistry of the actual mineral components. They would have more evolved characteristics such as lower $Mg^{\#}$ values, lesser contents of NiO and Cr_2O_3 and higher concentrations of less compatible elements. If we accept that the poikilitic plagioclase lherzolite (P.PER.) was somehow extracted from the mantle, the next problem is to determine whether or not its composition reflects without fractionation that of the source. On Figure 8, which shows the simplified basalt system "olivine-clinopyroxene-plagiocla-se" of Cox et al. (1979), we have plotted the composition of P.PER. and that of an average spinel lherzolite L (Maaløe and Aoki, 1977). Melting of P.PER. at the eutectic of the three components system would yield about 18% more of basaltic liquid than L. In other words, the composition of the poikilitic plagioclase lherzolite is 18% richer in basaltic components than the spinel lherzolite. For the time being we have no evidence allowing us to make a choice between two alternative possibilities. Either P.PER. represents the complete melting of highly fertile mantle or P.PER. results from the addition of a partial melt of basaltic composition to an extensively melted source similar in composition to the spinel lherzolite L. The possibility of contamination of a magma of composition L by assimilation of about 18% of crustal rocks to produce the composition of P.PER. is for thermodynamic reasons far less probable.

A preliminary conclusion of our analysis is that the parental magma was a mixture of variable proportions of basaltic melt and residual solid phases such as olivine and spinel. The texture of the rock is indicative of a system containing at least 50% of melt (oikocrysts and interstitial phases) at the time of emplacement into the present setting. The crystal mush or crystal-liquid suspension was highly viscous allowing it to carry crystals and xenoliths. This dense magma, which could not rise easily through the light upper crust, pooled massively at the base of the crust. Some of the host rocks were melted along contacts and cooling and solidification occurred slowly. During the emplacement the spinels, as first suggested by Bowen (1928), may have been partly remelt several times and may have grown each time a little richer in the high-melting Cr_2O_3 component before capture by a pyroxene oikocryst. Zoning of the intrusion presumably resulted from the process of filter pressing. The buoyant liquid in the lower part of the intrusion rised upwards through the crystal mush column towards the lower pressure zones of the upper part of the intrusion, thus generating the petrographic zoning observed in the peridotite body.

The origin of the peridotitic magma is a challenging problem. In the present state of knowledge it can only be treated in a speculative manner. The formation of a crystal mush of this composition requires unusual conditions and its emplacement into the crust implies that the melt was unable to segregate from its crystalline matrix. Benn and Laurent (1987) assumed that the crystal mush originated by compaction of a zone of melt concentration located below a spreading center. The suprasubduction zone environment of formation of the Troodos ophiolite postulated by most recent workers suggests an alternative hypothesis. In this context of active tectonics we may consider the possible role of diapirs generated in the deep and hot mantle and undergoing partial melting as they rise adiabatically. If the rate of rise is fast, the melts formed may congregate instead of being fractionally removed during the ascent. The presence of magmatic amphiboles in the poikilitic peridotite shows that the magma was hydrated. Introduction of water from the subduction zone environment into the upper melting zone of a rising diapir would still increase the percentage of fusion, and melt segregation could be inhibited by extensive deformation of the environment. In some cases, these conditions, if sustained, may have allowed the diapir to reach the base of the crust through a thinned lithosphere, where it released crystal mushes that resulted in the intrusion of poikilitic peridotites into the crust. In Cyprus, the harzburgitic mantle core of the Troodos has intruded and deformed the crustal rocks of the early stratiform plutonic suite. The mantle dome of the Troodos is evidence to support the view that mantle diapirs may penetrate the base of (marginal basin ?) ophiolitic crust sequences.

Acknowledgements

Field and laboratory work for this paper were supported by the Natural Sciences and Engineering Research Council of Canada. We thank C. Xenophontos and the Geological Survey of Cyprus for help with logistics, J. Malpas, T. Calon and S. Dunsworth for stimulating discussion, and K. Benn for his active participation to our work. The manuscript has benefited from the constructive comments of G. Ceuleneer and the editing of J. Riva.

References

Allen, C.R., 1975. The petrology of a portion of the Troodos plutonic complex, Cyprus. Ph.D. thesis, Cambridge University, Cambridge, 161 p.

Basaltic Volcanism Study Project, 1981. Basaltic volcanism on the terrestrial planets. Pergamon Press, Inc., New York, 1286 p.

Benn, K., 1986. Petrology of the Troodos plutonic complex in the Caledonian Falls area, Cyprus. M.Sc. thesis, Laval University, Quebec, 223 p.

Benn, K. and Laurent, R., 1987. Intrusive suite documented in the Troodos ophiolite plutonic complex, Cyprus. Geology, 15: 821–824.

Bonatti, E., Ottonello, G. and Hamlyn, P.R., 1986. Peridotites from the island of Zabargad (St. John), Red Sea: petrology and geochemistry. Journal of Geophysical Research, 91: 599–632.

Bowen, N.L., 1928. The evolution of the igneous rocks. Dover Publ. Inc., New York, 334 p.

Cameron, E.N., 1975. Postcumulus and subsolidus equilibration of chromite and coexisting silicates in the Eastern Bushveld Complex. Geochemica et Cosmochemica Acta, 39: 1021–1033.

Cox, K.G., Bell, J.D. and Pankhurst R.J., 1979. The interpretation of igneous rocks. George Allen & Unwin ltd., London, 450 p.

Dick, H.J.B. and Bullen, T., 1984. Chromian spinel as a petrogenetic indicator in abyssal and alpine-type peridotites and spatially associated lavas. Contributions to Mineralogy and Petrology, 86: 54–76.

Dick, H.J.B., Fisher, R.L. and Bryan, W.B., 1984. Mineralogic variability of the uppermost mantle along mid-ocean ridges. Earth Planetary Sciences Letters, 69: 88–106.

Dion, C., 1987. Géologie de la région de Troodhitissa, complexe plutonique du Troodos, Chypre. M.Sc. thesis, Laval University, Quebec, 280 p.

Fleet, M.E., 1974. Partition of major and minor elements and equilibration in coexisting pyroxenes. Contributions to Mineralogy and Petrology, 44, 259–274.

George, R.P.Jr., 1978. Structural petrology of the Olympus ultramafic complex in the Troodos ophiolite, Cyprus. Geological Society of America Bulletin, 89, 845–865.

Hebert, R. and Laurent, R., 1990. Mineral chemistry of the plutonic section of the Troodos ophiolite: new constraints for genesis of arc-related ophiolites. In: A. Panayiotou (ed), Ophiolites and oceanic lithosphere, Proceedings Symposium Troodos 87, Geological Survey Department, Cyprus: 149–163.

Huebner, J.S., 1980. Pyroxene phase equilibria at low pressure. In: C.T. Prewitt (ed), Pyroxenes, Reviews in Mineralogy, Mineralogical Society of America, 7, 213–288.

Irvine, T.N., 1965. Chromian spinel as a petrogenetic indicator; Part 1, Theory. Canadian Journal of Earth Sciences, 2, 648–671.

Laurent, R., 1990. Parental magma and crystal fractionation modelling of the Cy-4 plutonic rocks, Troodos ophiolite, Cyprus. In: A. Panayiotou (ed), Ophiolites and oceanic lithosphere, Proceedings Symposium Troodos 87, Geological Survey Department, Cyprus: 139–148.

Laurent, R. and Hebert, R., 1989. Petrological features of gabbroic and ultramafic rocks from deep drill Cy-4, Cyprus. In: I.L. Gibson, J. Malpas, P.T. Robinson and C. Xenophontos (ed), Cyprus Crustal Study Project: Initial Report, Hole Cy-4, Geological Survey of Canada, Paper 88–9, 115–145.

Lindsley, D.H., 1983. Pyroxene thermometry. American Mineralogist, 68, 477–493.

Lindsley, D.H. and Anderson, D.J., 1983. A two pyroxene thermometer. Proceedings of the 13th Lunar and Planetary Science Conference, Part 2. Journal of Geophysical Research, 88 suppl., A887–A906.

Maaløe, S. and Aoki, K., 1977. The major element composition of the upper mantle estimated from the composition of lherzolites. Contributions to Mineralogy and Petrology, 63, 161–173.

Malpas, J., Brace, T. and Dunsworth, S.M., 1989. Structural and petrologic relationships of the Cy-4 drill hole of the Cyprus Crustal Study Project. In: I.L. Gibson, J. Malpas, P.T. Robinson and C. Xenophontos (Eds.), Cyprus Crustal Study Project: Initial Report, Hole Cy-4, Geological Survey of Canada, Paper 88–9,39–67.

Moores, E.M. and Vine, F.J., 1971. The Troodos massif, Cyprus, and other ophiolites as oceanic crust: evaluation and implications. Philosophical Transactions of the Royal Society, London, A-268, 443–466.

Morimoto, N., 1989. Nomenclature of pyroxenes. Canadian Mineralogist, 27, 143–156.

Ringwood, A.E., 1975. Composition and petrology of the earth's mantle. McGraw-Hill, Inc., New York, 618 p.

Thibault, Y., 1987. Géologie et pétrologie des gabbros et des dykes de la région de Phini, complexe ophiolitique de Troodos, Chypre. M.Sc. thesis, Laval University, Quebec, 225 p.

Wilson, R.A.M., 1959. The geology of the Xeros-Troodos area. Geological Survey Department, Cyprus, Memoir 1, 135 p.

Evolutional History of the Uppermost Mantle of an Arc System: Petrology of the Horoman Peridotite Massif, Japan

NATSUKO TAKAHASHI
Department of Earth Sciences, Faculty of Science, Kanazawa University, Kanazawa, Ishikawa 920, Japan

Abstract

The Horoman peridotite massif, Hidaka belt, Hokkaido, northern Japan, is composed of three peridotite suites each of which has different age and origin. The first is the Main Harzburgite–Lherzolite Suite (MHLS), which occupies most of the Horoman massif. The MHLS forms a layered complex of harzburgite, spinel lherzolite and plagioclase lherzolite with gradual lithological boundaries. Characteristics of mineral chemistry of the MHLS show a residual origin after arc magma extraction. The second is the Banded Dunite–Harzburgite Suite (BDHS), which is characterized by a conspicuous layering of dunite, harzburgite and olivine orthopyroxenite with sharp boundaries. The BDHS peridotites probably make exotic blocks within the MHLS. The BDHS peridotites are cumulates from some high-Mg magma (i.e. high magnesian andesite or boninite), deduced from mineralogical characteristics; high Cr/(Cr + Al) atomic ratio of spinel, high Fo content and low Ni content of olivine. They were formed in a fore-arc setting. The third is the Spinel-rich Dunite–Wehrlite Suite (SDWS), which always occurs as layers in the harzburgite of the MHLS. The SDWS is mainly composed of dunite enriched with chromian spinel and sometimes with clinopyroxene. The mineral chemistry indicates that the SDWS is of cumulus origin from the magma which had been released from (or coexisted with) the surrounding MHLS harzburgite. The Horoman peridotite massif is, therefore, a fragment of arc mantle which had evolved from a fore-arc mantle. Tectonics and geology of Hokkaido indicate that the Horoman massif had been derived from the southwestern margin of the Okhotsk (or North America) plate.

1. Introduction

Chemical characteristics of magmatism in a region are closely related to the tectonic setting, and their temporal variation, if any, can be usually detected by a chemical variation of "frozen magma" on the surface of the earth. The

Tj. Peters et al. (Eds), Ophiolite Genesis and Evolution of the Oceanic Lithosphere, 195–205.

temporal variation of the magmatism, must be closely correlated with that of petrological characteristics of the underlying mantle materials. We could detect, therefore, histories of igneous activity recorded in the upper mantle within alpine-type peridotite massifs which are mantle slices exposed on the earth's surface.

The Horoman peridotite massif, in the Hidaka belt, Hokkaido, northern Japan, is composed of quite various kinds of ultramafic and mafic rocks, which had been produced by time-integrated igneous processes in the upper mantle. The Horoman peridotite massif, therefore, will provide a particularly good field for an investigation to reveal the successive igneous processes recorded in the upper mantle.

2. Geological Background

The Horoman peridotite massif is situated in the southern end of the Hidaka metamorphic belt in Hokkaido, Northern Japan. The massif is approximately 8 × 10 Km in plan and is in a fault contact with surrounding various metamorphic and igneous rocks, such as biotite gneiss, schistose amphibolite and gabbroic rocks. This massif belongs to the Main Zone of the Hidaka belt (Komatsu et al., 1986). The Main Zone is arcuate and convex westward, and is interpreted to be a crust-mantle section of an island arc (Komatsu et al., 1986). The bottom of the zone is westward, partly faced to the Western Zone, another crust-mantle section, and therefore, the Horoman massif constitutes the lowest part of the Main Zone section (Fig. 1). The Main Zone is interpreted to be composed of tectonic slices derived from a southwestern margin of the North America (or Okhotsk) plate thrusted up by collision with the Eurasia plate in the late Miocene to Pliocene (Komatsu et al., 1987).

The Horoman massif is mainly composed of various kind of peridotites. Small amounts of gabbroic rocks are present, showing conspicuous layered structures with peridotites especially in the upper part of the massif. This massif has a total thickness of approximately 3000 m (i.e. Niida, 1984). The rocks are well exposed and the degree of serpentinization is very low.

3. Petrology of the Horoman Peridotite Massif; Distinction of Three Kinds of Peridotite Suite

The southern part of the massif was mainly investigated, and the lithological map of this area is shown in Fig. 2. It is possible to classify the peridotitic rocks into three suites in terms of the mode of occurrence, the modal composition (Fig. 3) and the mineral chemistry (e.g., Fig. 4). Selected microprobe analyses are listed in Table 1.

The first is the Main Harzburgite–Lherzolite Suite (MHLS), which consists

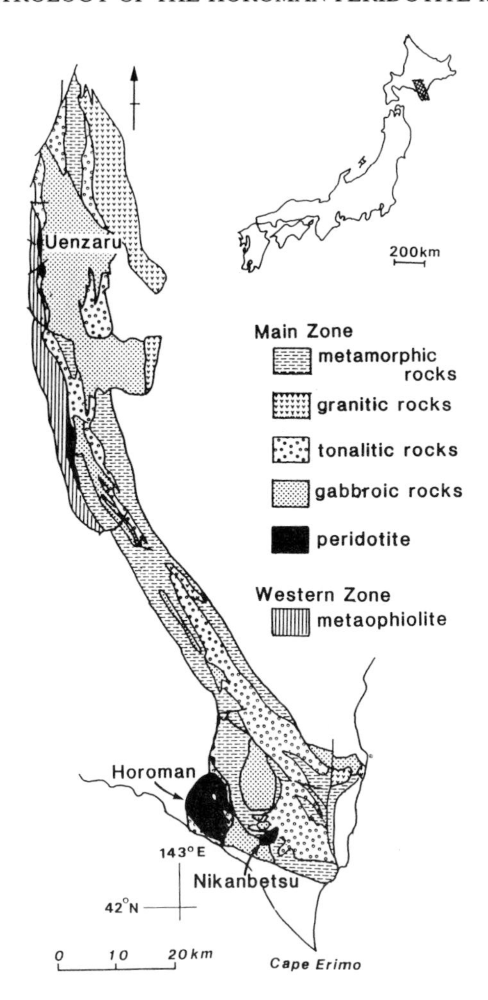

Fig. 1. Index map showing the locality of the Horoman peridotite massif in the Hindaka Main Zone (after Komatsu et al., 1986), northern Japan.

of a layered complex of harzburgite, spinel lherzolite and plagioclase lherzolite (Fig. 3). It occupies the most part of the massif (Fig. 2). Most of the "dunite" reported by previous workers (e.g. Niida, 1984; Obata and Nagahara, 1987) is "harzburgite", because it is usually contain more than 5 volume % of orthopyroxene.

The second is the Banded Dunite–Harzburgite Suite (BDHS). This suite is composed of depleted dunite, harzburgite and olivine orthopyroxenite which are generally free of clinopyroxene with rare exceptions (Fig. 3). They show conspicuous compositional layering with sharp boundaries; individual layers range from a few centimeters to a few meters in thickness. The BDHS rocks occur in the MHLS rocks but have the ill continuity along the strike (Fig. 2). Perhaps they make blocks in the MHLS rocks.

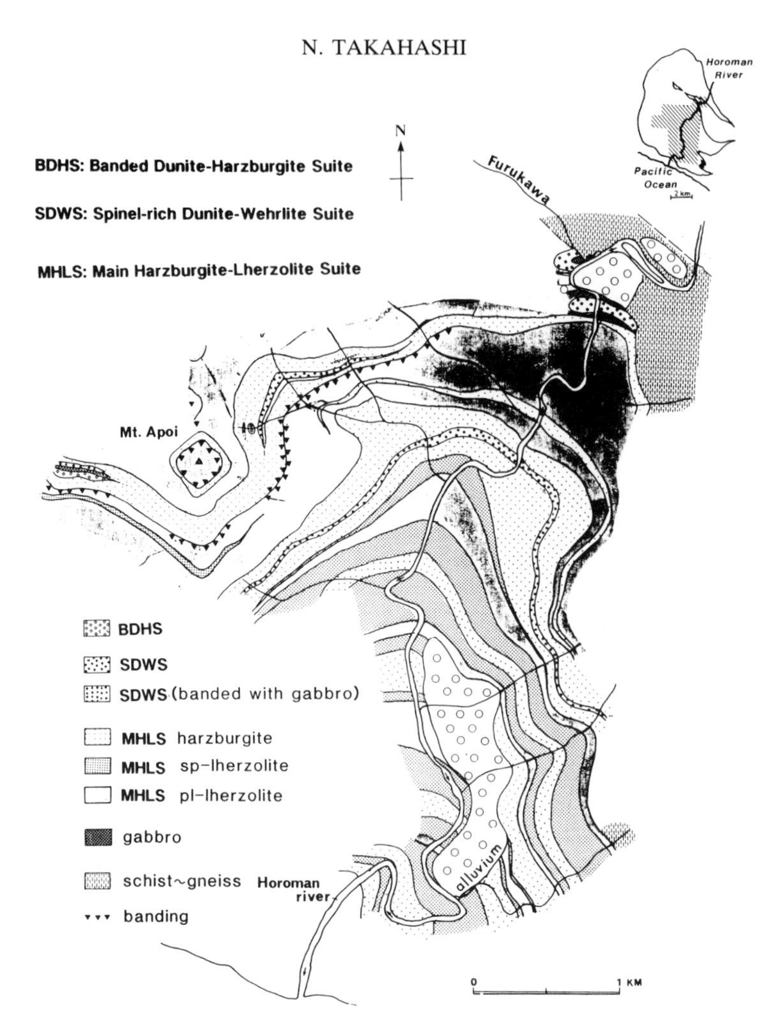

Fig. 2. Lithological map of the southern part of the Horoman peridotite massif.

The third is the Spinel-rich Dunite–Wehrlite Suite (SDWS), which is composed of clinopyroxene-bearing dunite, with local wehrlitic portions (Fig. 3). The SDWS rocks usually contain abundant euhedral chromian spinel (up to 5 volume %). The SDWS rocks always occur as layers (maximum thickness is 16 m) *in the MHLS harzburgite*, the most depleted member of the MHLS (Fig. 2).

3.1. *MHLS Peridotites*

The MHLS peridotites show a gradual change in lithology; hazburgite gradually changes into plagioclase lherzolite through spinel lherzolite in terms both of the mode and of the mineral chemistry (Figs. 3 and 4). The mineral chemistry is excellently correlated with the modal composition; the Fo content of olivine and the Cr/(Cr + Al) atomic ratio of coarse discrete spinel

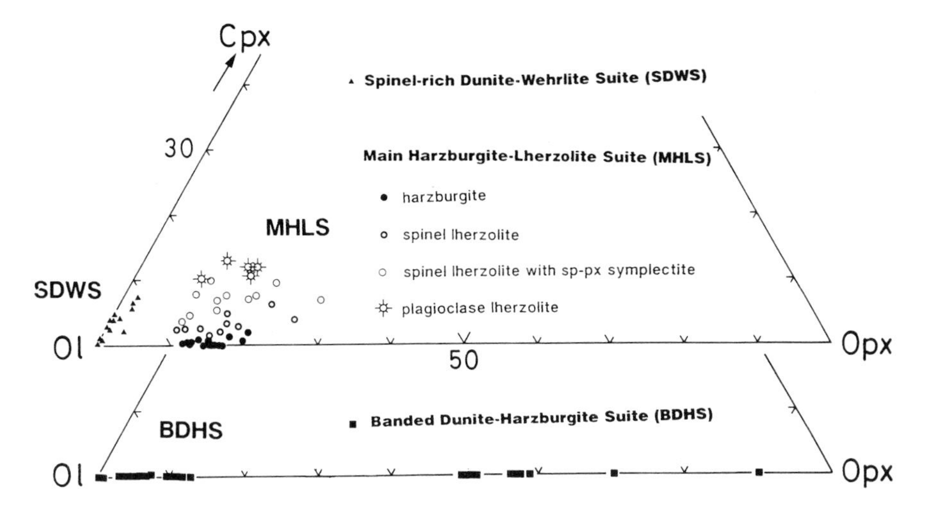

Fig. 3. Modal composition of the Horoman peridotites. Note that harzburgite and lherzolite in this paper are defined on the basis of the ratio of clinopyroxene to total pyroxenes (Cpx/(Cpx + Opx) < 0.1 for harzburgite).

decrease with an increase of the volume of total pyroxenes and the volume ratio of clinopyroxene to total pyroxenes (Figs. 3 and 4). Two-pyroxene spinel symplectites, which are remnants of decomposed garnets (Takahashi and Arai, 1989), occur mainly in the spinel lherzolite with a particular chemical range (Figs. 3 and 4). The Fo content of olivine ranges from 92.8 to 90.8 in harzburgite, 91.8 to 90.0 in spinel lherzolite and 90.0 to 88.8 in plagioclase lherzolite (Fig. 4). The Cr/(Cr + Al) atomic ratio of coarse discrete spinel core ranges from 0.68 to 0.45 in harzburgite, 0.46 to 0.18 in spinel lherzolite, and 0.17 to 0.07 in plagioclase lherzolite, positively correlated with the Fo content of coexisting olivine (Fig. 4). The whole MHLS peridotites are enclosed in the olivine-spinel mantle array of Arai (1987) (Fig. 4). NiO content in olivine is positively correlated with the Fo content; the MHLS peridotites are almost included in the mantle olivine array of Takahashi et al. (1987) (Fig. 5). The fine-grained chromian spinel associated with plagioclase in plagioclase lherzolite has the Cr/(Cr + Al) atomic ratio around 0.30, more Cr-enriched than the large discrete one. The An content of plagioclase is around 60.

3.2. *BDHS Peridotites*

The BDHS peridotites show conspicuous compositional layerings with sharp boundaries. The BDHS rocks are strongly deformed as compared with the nearby MHLS and the SDWS; spinel is flattened along the foliation. Thin parallel fractures, which are vertical to the lineation plane and filled by olivine, sometimes cut the flattened spinel. Coarse olivine (approximately 5 cm in diameter) is strongly deformed and is partly recrystallized into aggre-

Table 1. Selected microprobe analyses of minerals in the Horoman peridotites. Ol, olivine. Opx, orthopyroxene. Cpx, clinopyroxene. Sp, spinel (coarse discrete). Pl, plagioclase. 1 to 3, MHLS harzburgite. 4 to 8, MHLS plagioclase lherzolite (analysed at the Chemical Analysis Center, University of Tsukuba). 9 to 10, BDHS dunite. 11 to 12, BDHS harzburgite. 13 to 15, SDWS dunite (analysed at University of Tokyo). Analyses were done at the Geological Survey of Japan, if not otherwise mentioned.

	MHLS								BDHS				SDWS		
	Harzburgite BCD10			Pl-Lherzolite 85070717a					Dunite FR08b		Harzburgite AP01p		Dunite HD41		
	1 Ol	2 Opx	3 Sp	4 Ol	5 Opx	6 Cpx	7 Sp	8 Pl	9 Ol	10 Sp	11 Ol	12 Opx	13 Ol	14 Cpx	15 Sp
SiO_2	41.050	57.262	0	40.62	55.57	51.86	0	52.51	41.832	0	41.244	57.518	40.626	52.386	0.019
TiO_2	0.032	0.040	0.027	0	0.13	0.40	0.11	–	0	0.185	0	0.011	0	0.222	0.222
Al_2O_3	0.009	1.510	20.143	0.01	4.25	4.88	55.22	30.63	0	5.783	0	0.963	0	4.345	32.360
Cr_2O_3	0.013	0.424	47.946	0	0.46	1.19	11.14	0	0	58.323	0.010	0.523	0.007	1.231	30.098
FeO*	7.782	5.710	15.843	9.66	6.49	3.07	15.28	0.18	6.497	27.260	0.905	4.990	9.967	2.561	21.376
NiO	0.429	0.067	0.014	0.32	0.07	0.11	–	–	0.380	0.041	0.276	0.096	0.252	0.089	0.263
MnO	0.101	0.071	0.233	0.19	0.19	0.04	0.08	–	0.082	0.300	0.049	0.144	0.160	0	0.255
MgO	51.150	34.850	12.576	49.41	32.59	15.63	19.55	–	52.163	6.757	50.968	35.395	49.198	15.184	14.685
CaO	0.022	0.733	0.011	0.02	0.79	21.45	–	12.58	0.012	0	0.007	0.545	0.202	24.370	0
Na_2O	0.033	0.022	0	–	0.03	0.75	–	4.44	0.007	0	0.008	0.034	0.006	0.863	0
K_2O	0	0	0	–	0	0	–	0.02	0	0	0	0	0	0	0
Total	100.621	100.689	100.021	100.23	100.57	99.38	101.38	100.36	100.973	98.649	100.467	100.219	100.418	101.251	99.278
O	4	6	4	4	6	6	4	8	4	4	4	6	4	6	0.001
Si	0.993	1.959	0	0.996	1.914	1.900	0	2.372	1.000	0	0.998	1.971	0.994	1.893	0.118
Al	0	0.061	0.747	0	0.173	0.211	1.702	1.630	0	0.240	0	0.039	0	0.185	005
Ti	0.001	0.001	0.001	0	0.004	0.011	0.002	–	0	0.005	0	0	0	0.006	0.698
Cr	0	0.012	1.177	0	0.013	0.035	0.230	–	0	1.636	0	0.014	0	0.035	0.536
Fe*	0.157	0.163	0.501	0.198	0.187	0.094	0.334	0.007	0.130	0.809	0.160	0.143	0.204	0.774	0.006
Ni	0.008	0.002	0	006	0.002	0.003	–	–	0.007	0.001	0.005	0.003	0.005	0.003	0.006
Mn	0.002	0.002	0.006	0.004	0.005	0.001	0.001	–	0.003	0.009	0.001	0.004	0.003	0	0.642
Mg	1.844	1.777	0.589	1.806	1.673	0.854	0.762	–	1.860	0.357	1.838	1.808	1.794	0.818	0
Ca	0.001	0.277	0	0.001	0.029	0.842	–	0.609	0	0	0	0.020	0.005	0.944	0
Na	0.002	0.002	0	–	0.002	0.054	–	0.389	0	0	0	0.002	0	0.061	0
K	0	0	0	–	0	0	–	0.001	0	0	0	0	0	0	
Total	3.008	4.006	3.021	3.011	4..002	4.005	3.031	5.008	3.000	3.057	3.002	4.004	3.005	4.022	4.012

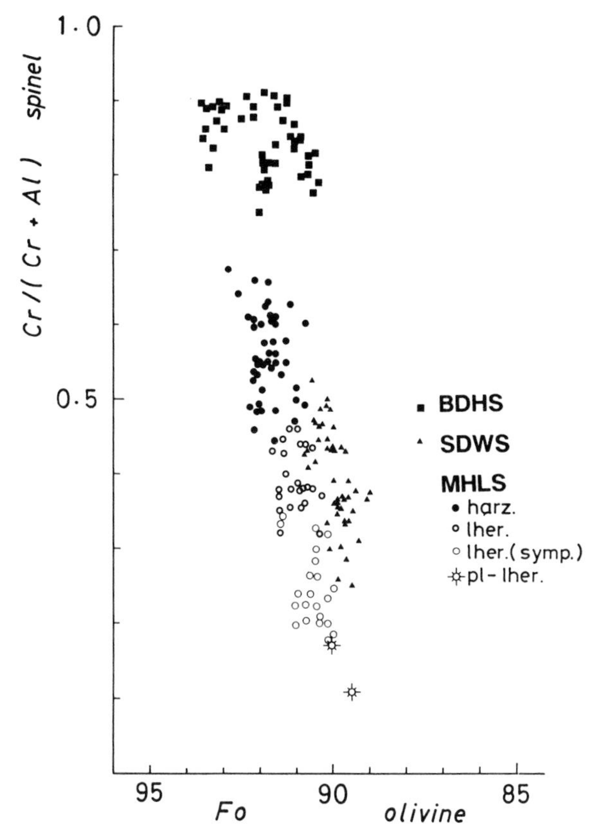

Fig. 4. Relationships between Fo content of olivine and Cr/(Cr + Al) atomic ratio of coarse discrete spinel in the Horoman peridotites.

gates of neoblasts along subgrain boundaries. The contact between the BDHS and the MHLS can be observed at a western ridge of Mt. Apoi (Apoi dake), one of the peaks of this area. Towards the contact with the BDHS, the MHLS harzburgite tends to become slightly depleted in orthopyroxene. However, the contact is sharp with a modal gap of orthopyroxene (<4%). The modal gap is coincident with a chemical gap (especially NiO content of olivine). BDHS peridotites contain extremely Cr-rich spinel (Cr/(Cr + A1) = 0.73 – 0.92) and magnesian olivine (Fo = 91.5 – 94.0). There is a clear chemical gap between the BDHS and the MHLS (Fig. 4). The NiO content of olivine shows a marked decrease as the Fo content decreases (Fig. 5), which can be explained by the olivine fractionation model of Takahashi et al. (1987). The Fe^{3+} content of spinel tends to increase as the Fo content of olivine decreases.

It is doubtless that BDHS peridotites are early cumulates from high magnesian andesites or related magma which have olivine and spinel phenocrysts

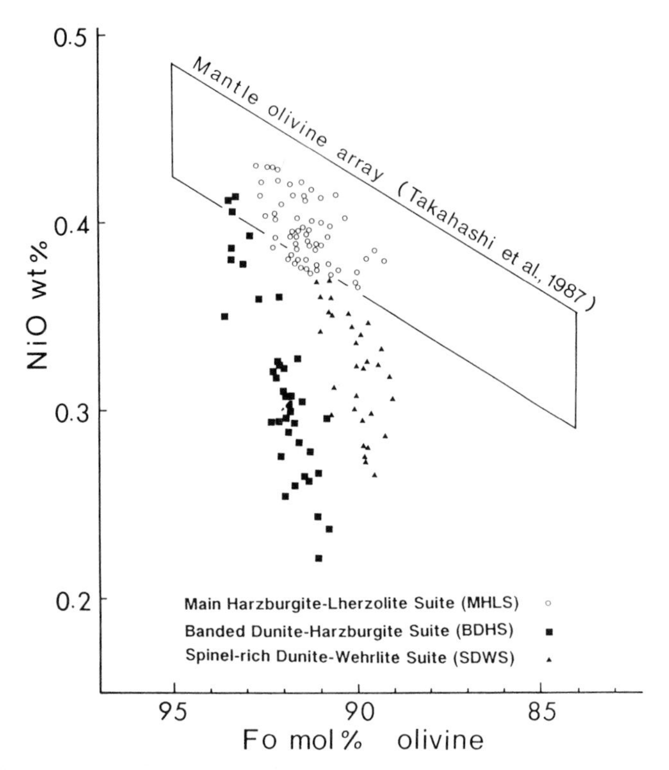

Fig. 5. Relationship between NiO content and Fo content of olivine in the Horoman peridotites. Note that the NiO contents in SDWS and BDHS olivine show a rapid decrease with a slight decrease of the Fo content, which indicates a cumulus origin.

similar in composition to the constituents of the BDHS peridotites (Arai & Takahashi, 1986).

3.3. *SDWS Peridotites*

The SDWS peridotites always occur as layers in the MHLS harzburgite. The modal composition of the MHLS harzburgite changes rapidly near the boundaries within several centimeters; towards the contact with the BDWS, the modal amount of orthopyroxene in harzburgite decreases at the boundary, the volume of spinel suddenly increases. The chromian spinel in the SDWS peridotites has a shape of spindle with a long axis parallel to the lineation of the rock. The SDWS peridotites contain Ti-bearing, relatively Al-rich spinel ($TiO_2 = 0.12 - 0.87$ wt%, $Cr/(Cr + Al) = 0.24 - 0.56$) (Fig. 6) and Fe-rich olivine ($Fo = 88.8 - 91.2$) relative to the surrounding MHLS harzburgite ($Fo = 91.5 - 92.4$). The spinel is rich in Fe^{3+} and the olivine is low in Ni (Fig. 5). The mineral chemistry of the SDWS and the surrounding MHLS harzburgite changes systematically. Across the contact, the Fo mol %

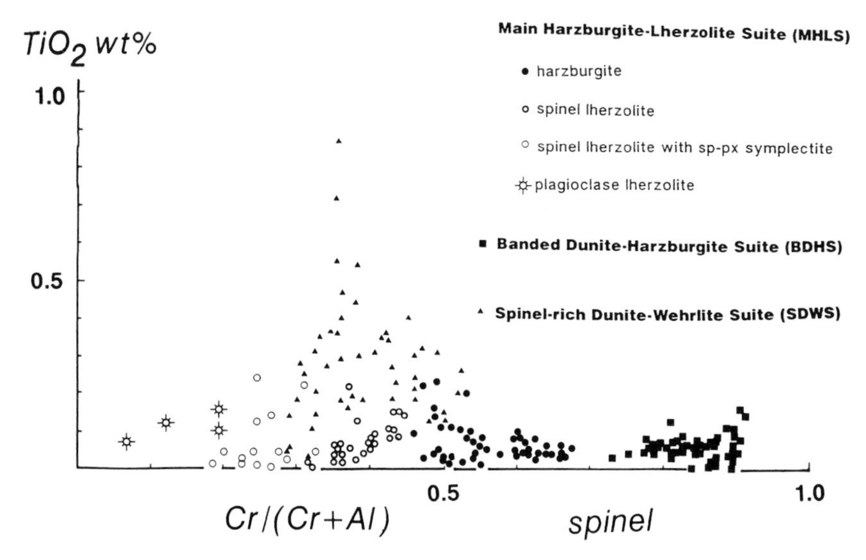

Fig. 6. Relationships between TiO_2 content and Cr/(Cr + Al) atomic ratio of coarse discrete spinel core.

of olivine and Cr/(Cr + Al) atomic ratio of spinel gradually decrease from the MHLS harzburgite (Fo = 92, Cr/(Cr + Al) = 0.62) to the SDWS (Fo = 90.2, Cr/(Cr + A1) = 0.39) in several tens centimeters. But NiO content of olivine and TiO_2 content of spinel have a clear chemical gap at the contact. The chemical variations in the main part of the SDWS layer are slight relative to those in the marginal part in contact with MHLS.

4. History of the Horoman Massif; A Disscusion

The field relations and mineral chemistry indicate that the BDHS peridotites are exotic blocks in the MHLS peridotites, whereas the SDWS peridotites are cumulates from the melt coexisted the MHLS harzburgite. Therefore, three peridotite suites, the BDHS, the MHLS, the SDWS had been formed in this order, which is consistent with the difference in the degree of deformation.

The first stage, that is the formation of the BDHS, is the activity of some high magnesian andesite (or related magmas), which could produce the BDHS peridotites as early cumulates (Arai and Takahashi, 1986). It is possible that the mantle of this stage had belonged to a fore-arc setting.

The second stage, the main stage, is a diapiric intrusion of the MHLS and subsequent formation of the SDWS. It is possible that the diapir had been derived from the garnet peridotite field (>20 Kb) evidenced by the preservation of the garnet remnants (two-pyroxene spinel symplectites) (Takahashi and Arai, 1989). Plagioclase in plagioclase lherzolite is interpreted to be a subsolidus reaction product from aluminous two-pyroxene spinel symplectite in the latest uppermost mantle condition. The MHLS diapir had attained to the uppermost mantle and had captured the pre-existing BDHS peridotite blocks.

Obata and Nagahara (1987) ascribed the lithological change of the MHLS to a mixing of melt and residue in various proportions. However, their interpretaion is not sufficiently convincing for the origins of the two-pyroxene spinel symplectite and the SDWS.

In the MHLS peridotites Cr/(Cr + Al) ratio of spinel and the NiO content of coexisting olivine systematically vary with the Fo content of olivine in the way that they are of residual origin after various degree of partial melting.

The MHLS diapir had been probably related to a genesis of some island arc magma because of their mineral chemistry (Arai, 1989). In this stage, the Horoman mantle had belonged to an arc stage proper.

The mode of occurence of the SDWS strongly indicates that the formation of the SDWS was closely related to that of the MHLS harzburgite. Several possibilites on the process for formation of the SDWS include 1) the segregated melt was gathered near the center of the MHLS harzburgite and precipitated the SDWS as cumulates, 2) the exotic melt which intruded into the garnet-bearing lherzolite (present plagioclase lherzolite and symplectite-bearing spinel lherzolite) formed the harzburgite as a reaction product between the melt and the lherzolite and the SDWS as cumulates from the modified melt. A further study will be needed to unravel the genesis of the SDWS.

In the third stage, the alkali metasomatism possibly related to a latent alkali basaltic magmatism was widespread in the Horoman massif (Arai and Takahashi, 1989); secondary-textured phlogopite and amphibole occur in all peridotite suites. This kind of alkali metasomatism is widely observed in the peridotite massifs of the Hidaka belt (Takahashi et al., 1989). It is interpreted that the alkali magmatism had been performed along a shear zone near the boundary between the Eurasia and the North America plates just before the collision.

In conclusion the Horoman peridotite massif is a slice of the upper mantle which had experienced a fore-arc setting followed by an arc setting proper at the western margin of the Okhotsk plate. Alkali metasomatism had been slightly modified the massif at the latest stage.

Acknowledgements

The writer wishes to express her hearty appreciation to Prof. Shoji Arai, Kanazawa University, for his guidance, encouragement and discussion. She is also grateful to Prof. Eiichi Takahashi, Tokyo Institute of Technology, for his discussion, encouragement and supply of unpublished data, to Prof.

Masaaki Obata, Kumamoto University, and to Dr. Kazuhito Ozawa, University of Tokyo, for their discussions and encouragements. She is much indebted to Mr. H. Yoshida, University of Tokyo, Mr. N. Nishida, University of Tsukuba, Dr. M. Aoki and Dr. T. Urabe, Geological Survey of Japan, for their help for microprobe analysis.

References

Arai, S., 1987. An estimation of the least depleted spinel peridotite on the basis of olivine-spinel mantle array, Neues Jb. Miner. Mh., 8: 347–354.

Arai, S., 1989. Origin of ophiolitic peridotites. Jour. Geogr., (Tokyo), 98–3: 45–54. (in Japanese)

Arai, S. and Takahashi, N., 1986. Petrographical notes on deep-seated and related rocks (4) Highly refractory peridotites from Horoman ultramafic complex, Hokkaido, Japan, Ann. Rep. Inst. Geosci. Univ. Tsukuba, 12: 76–78.

Arai, S. and Takahashi, N., 1989. Formation and compositional variation of phlogopites in the Horoman peridotite complex, Hokkaido, northern Japan: implications for origin and fractionation of metasomatic fluids in the upper mantle, Contrb. Mineral. Petrol., 101: 165–175.

Komatsu, M., Miyashita, S. and Arita, K., 1986. Composition and structure of the Hidaka metamorphic belt, Hokkaido – historical review and present status –. Monogr. Assoc. Collab., Japan: 31,189–203. (in Japanease with English abstract)

Niida, K., 1984. Petrology of the Horoman ultramaphic rocks, Jour. Fac. Sci. Hokkaido Univ., Ser. IV, 21: 61–81.

Obata, M. and Nagahara, N., 1987. Layering of alpine-type peridotite and the seggregation of partial melt in the upper mantle, Jour. Geophys. Res. 92: 3467–3474.

Takahashi, E., Uto, K. and Schilling, J.G., 1987. Primary magma compositions and Mg/Fe ratios of their mantle residues along Mid Atlantic Ridge 29°N to 73°N, Technical Report of ISEI, Okayama Univ., Ser. A, 9: 1–14.

Takahashi, N. and Arai, S., 1989. Textual and chemical features of chromian spinel-pyroxene symplectites in the Horoman peridotites, Hokkaido, Japan, Sci. Rep. Inst. Geosci. Univ. Tsukuba. Sec. B, 10: 45–55

Takahashi, N., Arai, S. and Murota, Y., 1989. Alkali metasomatism in peridotite complexes from the Hidaka belt, Hokkaido, northern Japan, Jour. Geol. Soc. Japan, 95: 311–329. (in Japanese with English abstract)

Part II
Magmatic Ores

Cu-Ni-PGE Magmatic Sulfide Ores and their Host Layered Gabbros in the Haymiliyah Fossil Magma Chamber (Haylayn Block, Semail Ophiolite Nappe, Oman)

M. LACHIZE[1], J.P. LORAND[2] and T. JUTEAU[1]

[1] *Groupement de Recherches "Genèse et Evolution des Domaines Océaniques" et Unité de Recherche Associée au CNRS no. 1278, Univ. Bretagne Occidentale, 6 Avenue le Gorgeu, 29287 Brest Cedex, France*

[2] *AVH Fellow, Pr. G. Friedrich, Inst. für Mineralogie und Lagerstättenlehre, R.W.T.H. Aachen, Technical University of Aachen, Wüllverstr. 2, D-5100 Aachen, Germany*

Abstract

A disseminated sulfide-rich zone has been discovered in the gabbroic magma chamber of the Wadi Haymiliyah (Haylyn Block), located in the upper part of the main coarse-grained layered gabbro unit, immediately below a thick, laminated and fine-grained noritic unit. In this 150 m thick zone, thin, fine-grained, concordant layers of noritic gabbro first appear in the layered gabbros, and contain most of the disseminated sulfides. The cumulates are weakly affected by hydrothermal alteration and hence, magmatic textures are well preserved. At the base of the section, the cumulates comprise olivine gabbros, gabbros and two-pyroxene gabbros. In the uppermost two-thirds of the sulfide-rich zone, two-pyroxene gabbros become predominant and grade locally into noritic gabbros and norites. This zone constitutes a transitional unit between an open-system fractionation represented by the main layered gabbro unit, and a closed-system fractionation corresponding to the planar-laminated noritic gabbros.

At the thin section scale, the sulfide content ranges from <1% to 30%. From the bottom to the top, the relative proportion of sulfides increases in parallel with the abundance of orthopyroxene. When fine-grained two-pyroxene gabbros and coarse-grained gabbros are in contact, the sulfides are preferentially concentrated in the two-pyroxene gabbros. Sulfides are observed as spherical inclusions in plagioclase, clinopyroxene and orthopyroxene crystals, as well as intercumulus grains. "Net-textured" sulfides are observed in the layers richest in sulfides. Textural criteria evoke the segregation of an immiscible sulfide liquid which separated from a basaltic melt, prior to the first appearance of orthopyroxene as a main cumulus phase.

Three sulfide associations were observed: 1) Pyrrhotite with pentlandite and chalcopyrite characterize the base of the sulfide-rich zone. 2) Massive pyrite may be associated with chalcopyrite and pyrrhotite, accounting for more than 60% of the sulfide paragenesis. 3) In hydrothermally altered

Tj. Peters et al. (Eds), Ophiolite Genesis and Evolution of the Oceanic Lithosphere, 209–229.

samples, pyrrhotite is pseudomorphosed into marcassite $\pm$ pyrite ("bird eyes" textures). This transformation may be related to low-temperature off-axis hydrothermal circulations. Pentlandite and chalcopyrite occur in two habits, as blocky grains adjacent to pyrrhotite, or as exsolution bodies within pyrrhotite. PGE analyses in one of the layers richest in sulfides yield (in ppb): Os < 8, Ir < 3, Ru 32, Rh 4, Pt 37, Pd 130 and Au 150. The chondrite normalized PGE pattern has a positive slope, with a Pd/Ir ratio > 30 and Au/Ir ≈ 160 (normalized to chondrites) typical of magmatic concentrations. Textural and mineralogical criteria are in agreement, supporting a magmatic origin for these sulfide concentrations. These sulfides are contemporaneous to the fractionation of orthopyroxene, and were precipitated as immiscible liquid droplets from an already evolved basaltic magma. The absence of sulfides in coarse-grained gabbros suggests that the MORB-type magma which has periodically replenished the Haymiliyah magma chamber was sulfur-undersaturated. Sulfur saturation is due to an increase of volatile content in the already evolved magma.

Introduction

PGE-Cu-Ni sulfide mineralizations are common in continental igneous layered intrusions (Naldrett et al., 1979; Naldrett, 1981). These mineralizations are composed of four main sulfides: monoclinic pyrrhotite + pentlandite + chalcopyrite + pyrite occurring in order of decreasing abundance (Craig and Kullerud, 1969; Duke and Naldrett, 1976; Von Grunewaldt, 1979). Current popular hypotheses proposed to explain these ores favour segregation of immiscible sulfide liquids from sulfur-saturated basaltic magmas (Haughton et al., 1974; Duke, 1979; Irvine et al., 1983).

In ophiolite complexes, massive sulfide deposits are found to occur preferentially within the pillow lava section, but they are clearly related to volcanogenic and hydrothermal circulation at mid-oceanic spreading centres (Silitoe, 1973; Upadhyay and Strong, 1973). By contrast, cumulative rocks are generally lacking in magmatic sulfide ores (Naldrett, 1973). So far, the only one occurrence of magmatic ores is that which has been described in feldspathic cumulates in the Bay-of-Islands ophiolite complex, Newfoundland (Lydon and Richardson, 1988). The other deposits of restricted size which are located in the transition zone separating the cumulative rocks from the underlying tectonized mantle-derived harzburgites, have been interpreted as hydrothermal (e.g. Economou and Naldrett, 1984; Foose et al., 1985; Talhammer et al., 1986). In the present paper, we describe sulfide-rich noritic gabbro layers associated with the layered gabbros of the Haymiliyah fossil magma chamber (Haylayn Block, ophiolite of Oman). We present unambiguous mineralogical and geochemical evidence for a magmatic origin of these sulfides. Accordingly these ores would represent the second example of magmatic sulfide mineralization discovered in the layered gabbros of the oceanic lithosphere.

Geological Setting

The Sumail ophiolite outcrops over more than 30,000 km^2 along the length of the Oman Mountains. It forms a huge thrust sheet of Cretaceous oceanic lithosphere over 600 km long and 150 km wide which was broken up during its emplacement into about a dozen major tectonic blocks of unequal size (Glennie et al., 1974) (Fig. 1). The Wadi Haymiliyah area is located in the central part of the Haylayn block (central part of the Oman Mountains). There, the harzburgite tectonites make up the basal and major part of the ophiolite pile. Their basal contact with the Hawasina volcanosedimentary nappes is marked by discontinuous outcrops of metamorphic rocks (Searle and Malpas, 1980). The plutonic sequence is well developed and reaches up to 5 km in thickness (Fig. 2). It is intruded by numerous discordant and concordant ultramafic bodies and sills composed of wehrlites and dunites. These rocks have been interpreted as late intrusions arising from a second-stage melting of mantle material (Ernewein and Whitechurch, 1986; Juteau et al., 1988 a and b; Benn et al., 1988).

At the base, the plutonic sequence starts with 2000 m of layered gabbros which display adcumulus texture. The basal layered gabbros, 40 m thick, are composed of olivine leucogabbros including olivine-rich troctolite layers, in which the cumulus phases crystallize in the following order: chrome spinel, olivine, clinopyroxene and plagioclase. This crystallization order is different from that described in the N-MORBs and may be related either to tholeiites enriched in volatiles or to MORB-type magmas with a high Si/Al ratio. Orthopyroxene appears as a post-cumulus phase only (Dahl, 1984; Juteau et al., 1988a). The intermediate layered gabbros show internal modal variations from troctolite or plagioclase wehrlite at the bottom to gabbro and anorthosite at the top. Upward, noritic gabbros become abundant; in the main layered gabbro unit, these rocks occur at the top of cyclic units several metres in thickness which start with plagioclase wehrlite and troctolite.

Above the lower gabbroic sequence, fine-grained massive and homogeneous noritic gabbros become predominant over a thickness of 1000 m. These noritic gabbros are composed of plagioclase, clinopyroxene and orthopyroxene adcumulates ($<5\%$ intercumulus material). They are characterized by planar-laminated structure due to the orientation of pyroxene and feldspar crystals along a lamination plane. This unit contains numerous recurrences of coarse-grained gabbros.

The planar-laminated noritic gabbros are cross-cut by anastomosing sills of dioritic to plagiogranitic composition and hydrothermal veins. Upsection, the latter become more abundant, so that the planar-laminated noritic gabbros are brecciated over the top 500 m of their thickness. This magmatic

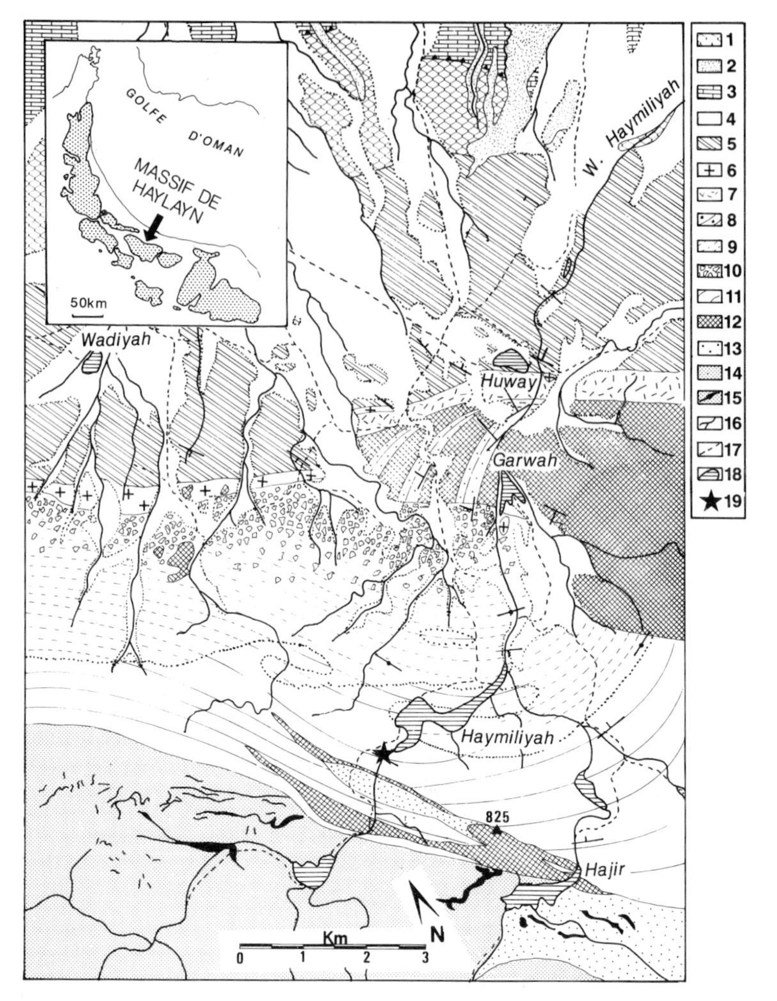

Figure 1. Detailed geological map of the Wadi Haymiliyah area, after Nehlig and Juteau (1988). 1 = gravels and sands; 2 = Quaternary terraces; 3 = sediments of the Hawasina nappes; 4 = extrusive lavas; 5 = sheeted dyke complex, with the indication of the strike and dip of the dykes; 6 = plagiogranites; 7 = high-level isotropic gabbros; 8 = upper layered cumulate gabbros, intruded by numerous wehrlitic sills (not shown); 9 = fine-grained, planar-laminated noritic gabbros; 10 = magmatic breccias; 11 = lower layered gabbroic sequence, with indication of the strike and dip of the magmatic layering; 12 = wehrlitic intrusive bodies; 13 = dunitic intrusive bodies; 14 = mantle harzburgites; 15 = granites; 16 = wadis; 17 = path roads; 18 = oases; 19 = sulfide-rich zone. The approximate limits between layered gabbros and planar-laminated noritic gabbros are outlined by dotted lines.

breccia horizon is overlain by olivine clinopyroxene layered gabbros in which the internal layering is totally discordant on the underlying planar-laminated structure. The upper layered cumulates culminate with several tens of metres of plastically deformed laminated noritic gabbros.

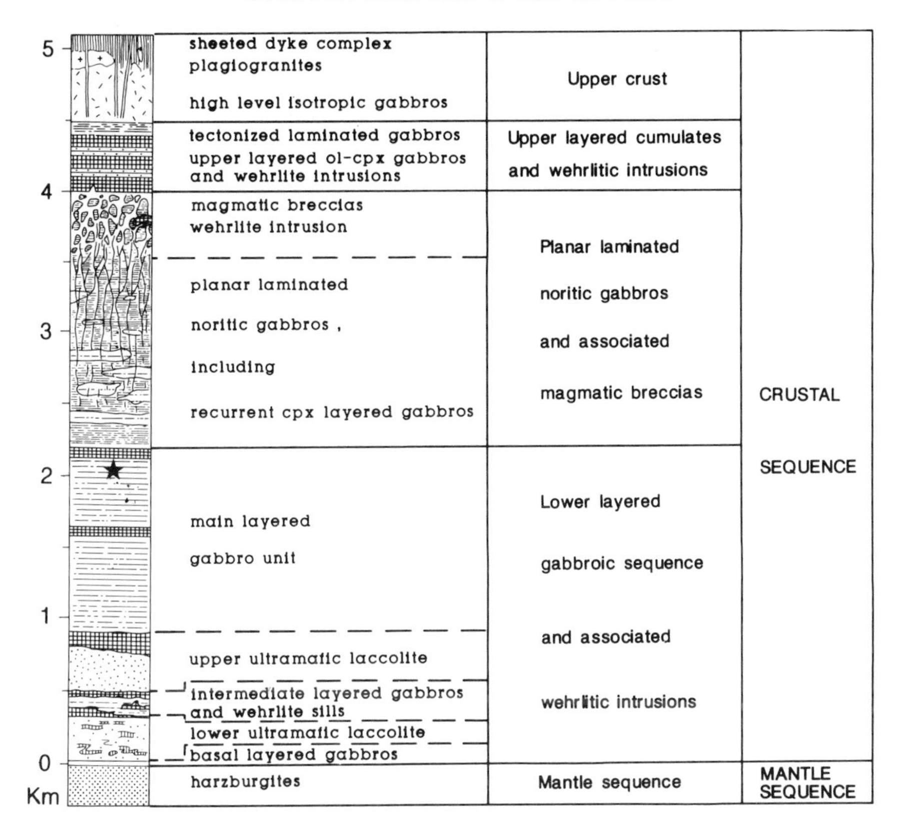

Figure 2. Stratigraphic column of the Wadi Haymiliyah section, after Juteau et al. (1988).The whole plutonic sequence (lower crust) has been rotated in order to restore the petrological Moho to a horizontal attitude. The star represents the sulfide-rich zone.

The uppermost gabbros consist of a 1500 m thick unit of high-level isotropic gabbros. The transition into the overlying sheeted dyke complex is marked by a 100 m thick zone in which dykes become more abundant up section. The dyke complex is 500 m thick and consists of mutually intrusive subparallel doleritic dykes having variable thicknesses.

Phase Composition and Magmatic Evolution of the Whole Magma Chamber

The composition of the four main cumulus phases varies sympathetically (Dahl, 1984). Mafic indices for olivine, orthopyroxene and clinopyroxene are rather constant throughout the layered gabbros whatever their structural height in the cumulate pile (Fo 84–88; En 80–88; Mg Cpx 85–92). By contrast, mafic indices drop significantly in the planar-laminated noritic gabbros which are enriched in Fe (Fo 74–85; En 70–76; Mg Cpx 72–82). The An content of plagioclase ranges between 5 and 90. The highest values (55–90) are found in the layered gabbros, whereas plagioclase in the planar-laminated noritic gabbros is Na-enriched, sometimes reaching An5–Ab95 in composition.

Mineralogical and chemical features (metre-sized mineral-graded layers showing depositional features, zig-zag patterns of cryptic variations, organization in cyclic units and the nearly constant An and Mg ratios) suggest that the layered gabbros separated from MORB-type magma through an open-system fractionation along the floor and the walls of a steady state magma chamber (Jackson et al., 1975; Juteau and Whitechurch, 1980; Pallister and Hopson, 1981; Juteau et al., 1988). In this model, the main layered sequence would represent in-situ bottom crystallization, and the upper layered gabbros, roof crystallization, respectively. By contrast, the planar-laminated noritic gabbros would correspond to a closed-system fractionation (Juteau et al., 1988b). According to these authors, the magma chamber was cut off from its source supply at this stage of its evolution.

Petrography of the Sulfide-Rich Zone

The sulfide-rich zone (about 150 m thick) is located at the top of the main layered gabbro unit just below the planar-laminated noritic gabbros. This zone is composed of thin layers of fine-grained noritic gabbros (average grain size <1 mm) interbedded within coarse-grained gabbros. Locally, both rock types are involved in viscous flow deformation pattern, in spite of the fact that they cross-cut each other at shallow angles ($<20°$).

The cumulates are weakly affected by hydrothermal alteration. Magmatic minerals are sometimes replaced by secondary minerals. Olivine fracture planes are filled with serpentine + magnetite, plagioclase is replaced by secondary minerals while clinopyroxene is uralitized and orthopyroxene replaced by bastite-like minerals. The magmatic parageneses are also locally replaced along veins and veinlets by green amphibole, prehnite, epidote and calcite (Nehlig and Juteau, 1988). These alterations are far from being complete, so that the cumulative magmatic textures remain well preserved.

The sulfide-rich zone starts with the uppermost cyclic unit of the main layered gabbro unit (samples HM 16, HM90 and HM88; Fig. 3). This cyclic unit consists of a lowermost troctolitic layer, an intermediate layer of olivine gabbro and culminates with a coarse-grained gabbro. The olivine gabbro layers contain minor amounts of chrome spinel and orthopyroxene, both minerals being cumulus phases. Above the sample HM 74, the cumulate pile is composed of three main rock types: olivine gabbro, gabbro and two-pyroxene gabbro. The coarse-grained gabbros show local modal evolution towards nearly pure clinopyroxenite just below the planar-laminated noritic gabbros. In the uppermost two-thirds of the studied zone, there is an increasing proportion of two-pyroxene gabbros which alternate with gabbros and olivine gabbros (Fig. 3). In parallel, the relative proportion of olivine decreases in the olivine gabbros, this decrease is balanced by an increase in the amount of orthopyroxene. When present (samples HM 76, HM 78, HM 81 . . .), olivine is anhedral, displays lobate grain boundaries and tends to be interstitial, forming networks in-between clinopyroxene and plagioclase. Moreover, there is evidence that this mineral is followed by orthopyroxene in the crystallization order of the cumulus phases: orthopyroxene crystallizes as rims around large crystals of olivine. Olivine is generally corroded and, in this case, always occurs as inclusions in large subhedral orthopyroxene. Relic olivine has been observed in the samples HM 75, HM 21 and HM 96A and may be oxidized into orthopyroxene + magnetite symplectite-like intergrowths (Jonhston and Stout, 1984).

The two-pyroxene gabbros are characterized by varying Opx/Cpx modal ratios. Hence, they grade locally into noritic gabbros and norites. Orthopyroxene tends to be nearly spherical or subhedral while clinopyroxene occurs as anhedral crystals.

The cumulates are characterized by the crystallization of cumulus Fe-Ti oxides (magnetite and ilmenite) and abundant interstitial brown hornblende (Fig. 3).

Mineral Chemistry

The evolution of the phase chemistry in the lower half of the sulfide-rich zone is in continuity with those in the underlying main layered gabbro unit. The An content of plagioclase and the Mg ratio of mafic cumulus phases vary sympathetically. Both decrease only slightly, suggesting that the magma chamber was still operating as an open-system with regular replenishment by fresh magma. On the contrary, the bulk chemical evolution over the zone located between the samples HM 10 and HM 94 is similar to that described in the overlying planar-laminated noritic gabbros. The highest Mg ratio and An content correspond to recurrent layers of coarse-grained clinopyroxenites while the two-pyroxene gabbros and the noritic gabbros are noticeably enriched in Fe. The iron-enrichment trend is accompanied by crystallization of cumulus Fe-Ti oxides, interstitial brown hornblende and by oxidation of olivine. These mineralogical and chemical features reflect an increase in the volatile and Ti contents (Kushiro, 1979), suggesting that the magma chamber started to evolve as a closed-system during segregation of the two-pyroxene gabbros.The tooth-shaped chemical trends at this level (Fig. 3) may be explained by progressively dying episodic replenishments of less evolved magma. To summarize, because of its structural position and petrological characteristics, the sulfide-rich zone clearly constitutes a transitional unit between the main layered gabbro unit and the planar-laminated noritic gabbros.

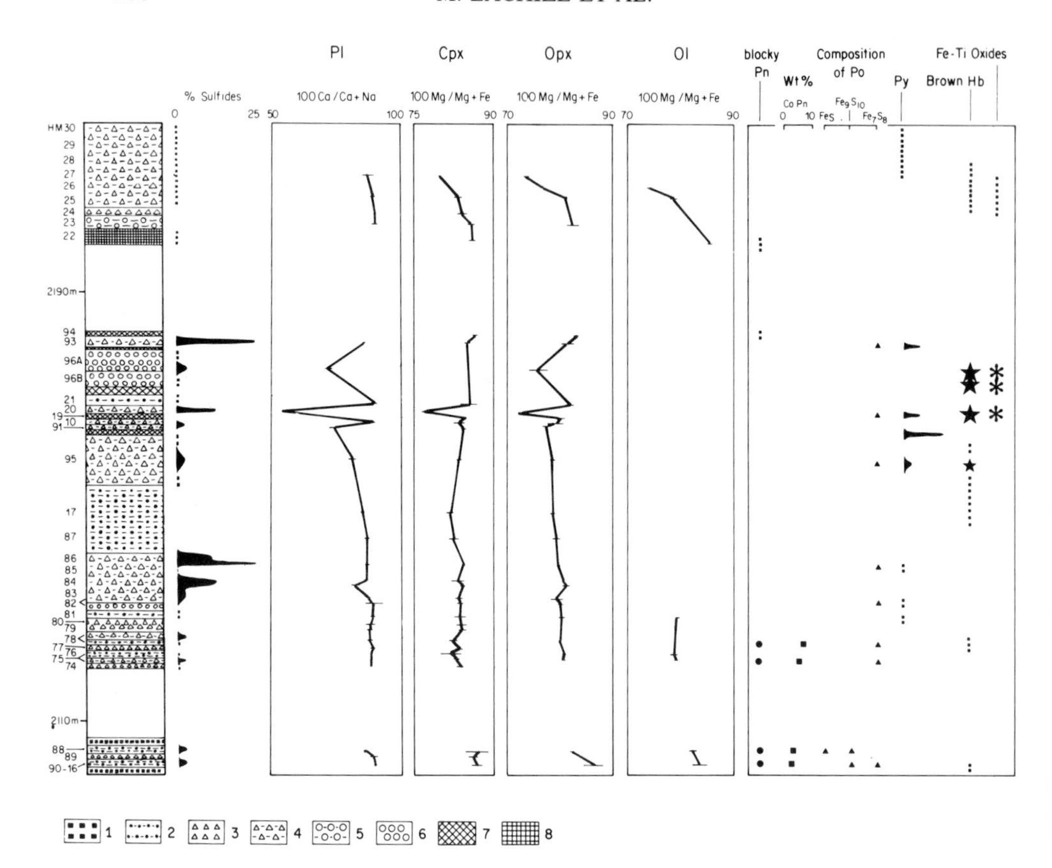

Figure 3. Overall chemical evolution of the main silicates together with mineralogy and chemistry of sulfide phases in the sulfide-rich zone of the Wadi Haymiliyah section. 1 = troctolite; 2 = olivine gabbro; 3 = gabbro; 4 = two-pyroxene gabbro; 5 = noritic gabbro; 6 = norite; 7 = clinopyroxenite; 8 = plagioclase wehrlite. Pn = pentlandite; Po = pyrrhotite; Py = pyrite; Hb = magmatic hornblende. Samples are indicated to the left of the lithologic column.

Abundance and Distribution of Sulfides

The sulfide content ranges between <1% to 30% by volume. The coarse-grained gabbros are almost totally devoid of sulfides, or contain two or three chalcopyrite grains per polished thin section. By contrast, sulfide enrichments are noticed in olivine gabbros and in the fine-grained two-pyroxene gabbros (Fig. 3). From the bottom to the top of the sulfide-rich zone, the relative proportion of sulfides increases in parallel with the abundance of orthopyroxene. The olivine gabbros HM 90 and HM 88 from the cyclic unit of layered gabbro contain 2–3% disseminated sulfides. The highest sulfide contents are found in two-pyroxene gabbros (samples HM 86 and HM 93, 30% and 17% by volume respectively). On the outcrop scale, these sulfide-rich samples form layers clearly discordant to the plane of magmatic layering of the host gabbros, moreover, they may exhibit cross-bedding and viscous flow. On the thin section scale, when fine-grained two-pyroxene gabbros and coarse-grained gabbros are in contact, the sulfides are preferentially concentrated in the two-pyroxene gabbros, but their abundance decreases sharply in the gabbros. This close relationship between sulfides and orthopyroxene is similar to that described in intracontinental layered gabbros (e.g. Hulbert and Von Grunewaldt, 1982). However, it does not hold true for Fe-Ti rich norites (e.g. sample HM 96) and laminated noritic gabbros. In the former rock type, the relative proportion of sulfides decreases significantly in cm-thick oxide-rich layers as previously described by Page (1971) in the Stillwater layered complex (U.S.A.). Likewise, the sulfides disappear almost totally from the planar-laminated noritic gabbros overlying the sample HM 22 (Fig. 3).

The sulfides are found to occur in two textural settings: as inclusions in silicates and as interstitial grains. The sulfide inclusions have been observed within orthopyroxene, clinopyroxene and plagioclase, but are conspicuously lacking in olivine. When included in cumulus euhedral orthopyroxene, the rounded olivine oikocrysts may be surrounded by a thin sulfide rim. In the other cumulus phases, the inclusions are rounded droplet-like bodies ranging in diametre from 10 to 100μm (Fig. 4a). These inclusions are randomly distributed between orthopyroxene, clinopyroxene and plagioclase but their average abundance increases slightly in the sulfide-rich layers HM 93 and HM 86A. There, up to five or six inclusions may be observed in a single clinopyroxene crystal. On the contrary, the planar-aminated noritic gabbros are devoid of sulfide inclusions.

Shape and size of interstitial grains vary according to the sulfide content of the host rock. In samples containing less than about 10% sulfides, the interstitial sulfide grains occur as discrete anhedral bodies isolated from each other in the silicate matrix (Fig. 4b). These grains average 200 μm in diametre, with maximum dimensions up to 500 μm × 2 mm. They are located at the triple junctions of silicate crystals and display curvilinear grain margins which are typically molded over the adjacent magmatic silicates (Fig. 4b). In the more massive ores, individual interstitial grains may be in optical continuity over distances of several mm. Isolated and rounded silicates may be partially or totally engulfed in a sulfide matrix (Fig. 4c), therefore giving rise to a "net-textured" ore (Ewers and Hudson, 1972; Naldrett, 1973). This peculiar texture, coupled with the numerous sulfide droplets included in the cumulus silicates obviously evokes segregation of an immiscible sulfide liquid (Naldrett, 1973; Usselmann et al., 1979). Field and textural evidence indicates that this liquid separated from the basaltic melt parent to the olivine gabbros and two-pyroxene gabbros, immediately after olivine but prior to the first appearance of orthopyroxene as main cumulus phase. The external morphology of the sulfide blebs interstitial to the silicates suggests that they formed as immiscible sulfide droplets which did not solidify until much of the engulfing silicates had crystallized (Duke and Naldrett, 1976).

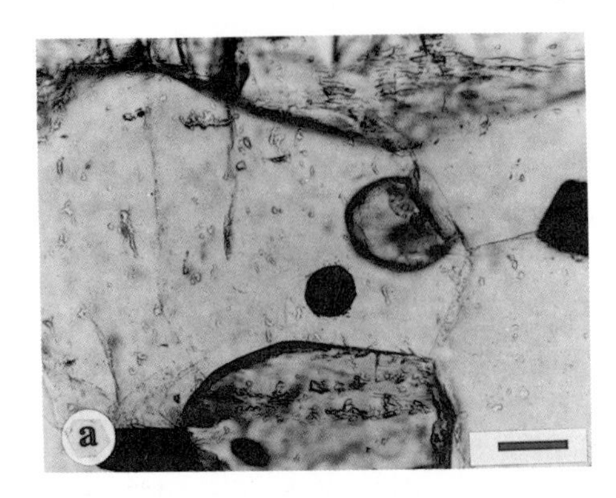

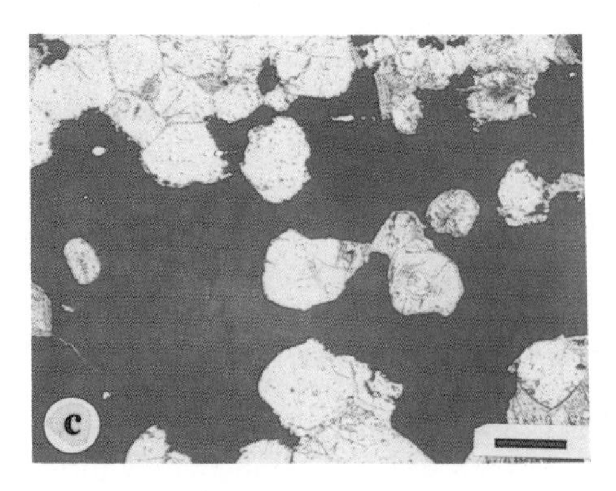

Figure 4. Textural relationships between sulfides and silicates. (a) Sulfide inclusion (dark) in plagioclase. Plane polarized transmitted light; scale bar = 100 μm. (b) Disseminated interstitial sulfides showing curvilinear grain boundaries. Plane polarized transmitted light; scale bar = 500 μm. (c) Net textured ore showing silicates partially or fully entrapped within sulfide grains. Plane polarized transmitted light; scale bar = 200 μm.

Analytical Method

The sulfides have been investigated on polished thin section using both transmitted and reflected light microscopy. Spot analyses of sulfide phases were performed with an automated CAMEBAX electron microprobe (M.N.H.N.: Muséum National d'Histoire Naturelle, Paris) using 15 kV acceleration voltage, 10 nA current beam and 6s/peak, 6 s/background counting times. Ni, Cu and Co were analysed with pure metal standards while Fe and S were determined using a natural pyrite. Whole-rock analyses of S, Cu, Ni, Co, Zn and noble metals (Au and Platinum-group elements, PGE) were obtained for samples HM 95 and HM 93. Sample HM 95 contains about 2 wt.% disseminated sulfides, whereas sample HM 93 is richer in sulfides (17%) and displays net-textured sulfides. Sulfur analyses were performed by the barium sulfate method (M.N.H.N.) while transition metals were analysed by AAS (Brest). The six PGE and Au were determined by NiS fire assay preconcentration followed by ICP/MS analyses (X Ray Assay laboratory, Canada). Detection limits are 5–8 ppb for Os, 2–3 ppb for Ir, 1 ppb for Ru, 2 ppb for Rh, 2 ppb for Pt, 1 ppb for Pd and 2 ppb for Au. Precisions are about 15% for Os, Ir, Ru, Rh and Pd and 25% for Pd (Talkington and Waltkinson, 1986). The results are listed in Table 1.

Sulfide Mineralogy

The sulfides are composed of four predominant phases: pyrrhotite, chalcopyrite, pentlandite and pyrite. This paragenesis is considered to be typical of magmatic sulfide ores and results from subsolidus decomposition of high-temperature monosulfide solutions (Craig and kullerud, 1969; Kullerud et al., 1969). The respective proportions of each sulfide displays a base-to-top gradual evolution which parallels that of the host rocks (Fig. 3). The olivine gabbros of the basal cyclic unit and the two-pyroxene gabbros located below the sample HM 79 are characterized by pyrite-free, three phase assemblages. Moreover, the pyrrhotite is composed of two-phase intergrowths between FeS and Fe_9S_{10} or $Fe_9S_{10} + Fe_7S_8$ in the lowermost olivine gabbros, but its composition grades upward into a uniform one-phase Fe_7S_8-type pyrrhotite. Pentlandite and chalcopyrite occur in two habits, as blocky grains adjacent to pyrrhotite or as exsolution bodies within this sulfide (Fig. 5a). The flame-like exsolution bodies of pentlandite in pyrrhotite may form at temperatures as low as 150°C by subsolidus diffusion of Ni in solid solution in the pyrrhotite (Kelly and Vaughan, 1983). The blocky pentlandite display Fe/Ni atomic ratios close to 1, in accordance with sub-solidus re-equilibration in the presence of pyrrhotite in the sulfide paragenesis (Misra and Fleet, 1973). The most significant chemical variations of the blocky pentlandite are related to Co content which increases regularly from the lowermost olivine gabbro HM 90 (1 wt.%) to the two-pyroxene gabbro HM 78 (≈ 7 wt.%) (Fig. 3). This

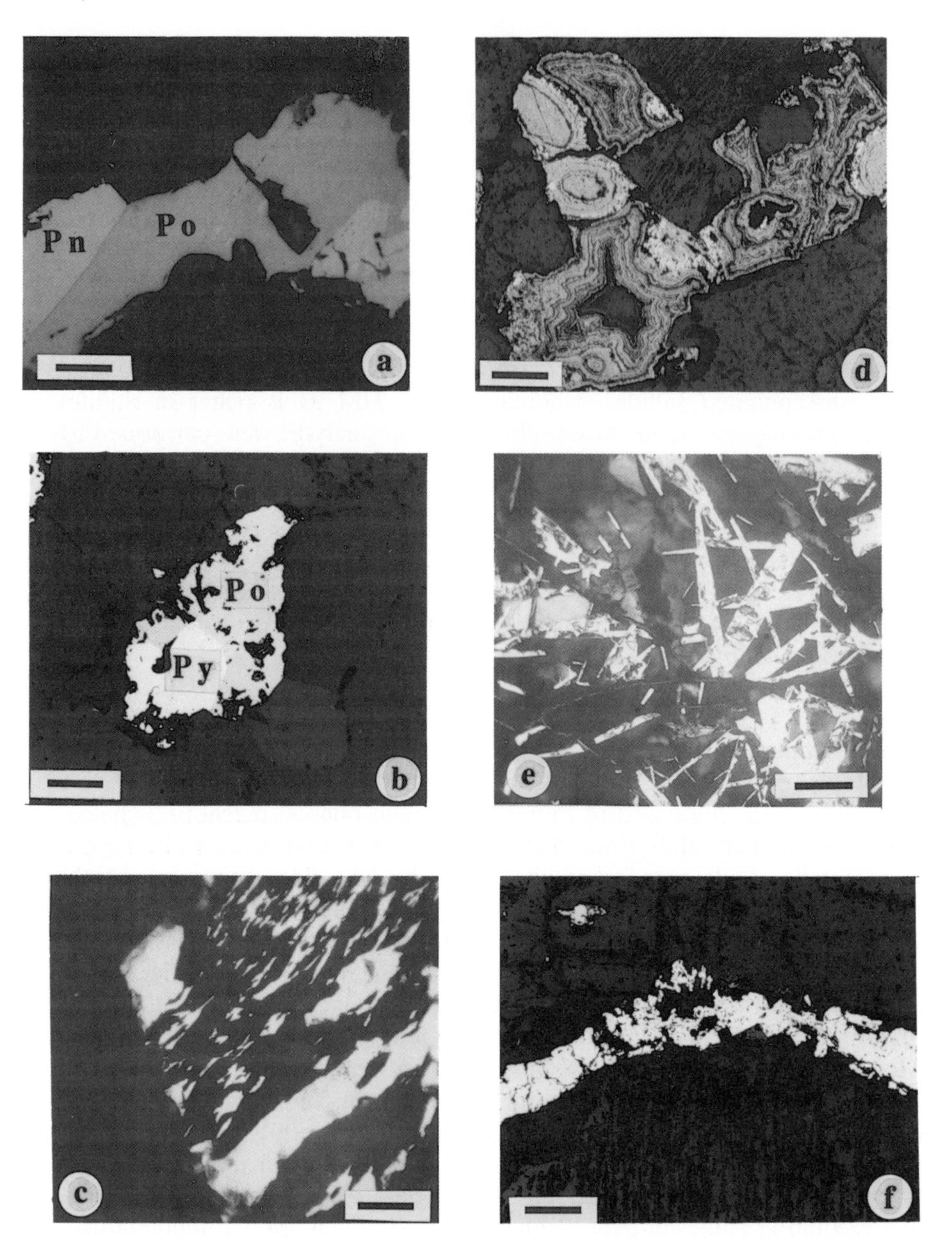

Figure 5. (a) Blocky pentlandite (Pn) adjacent to pyrrhotite (Po). Plane polarized reflected light; scale bar = 100 μm. (b) Primary pyrite (Py) associated with Po. Plane polarized reflected light. Scale bar = 100 μm. (c) Magmatic sulfide grains penetrated by laths of hydrous silicate. Plane polarized reflected light; scale bar = 50 μm. (d) Pyrrhotite pseudomorphed by hydrothermal marcassite and pyrite forming "bird-eyes" textures. Plane polarized reflected light; scale bar = 100 μm. (e) coarse laths of secondary pyrrhotite intergrown with serpentinized olivine. Plane polarized reflected light; scale bar = 50μm. (f) Hydrothermal fracture plane filled with euhedrally shaped secondary pyrite. Plane polarized reflected light; scale bar = 50 μm.

increasing Co content is negatively correlated with the relative abundance of blocky pentlandite which decreases progressively in parallel with the disappearance of olivine from the two-pyroxene gabbros. Chalcopyrite may form discrete blocky grains isolated from pyrrhotite and pentlandite, with dimensions up to 3 × 2 mm. This is especially the case for the olivine gabbros HM 90 and HM 76 which are chalcopyrite-rich. The peculiar behaviour of blocky chalcopyrite is usually attributed to the lower solidification temperature of Cu-enriched sulfide liquids with respect to the monosulfide solid solution precursor of pentlandite + pyrrhotite assemblages (Kullerud et al., 1969; Pasteris, 1984; Lorand, 1987).

Blocky pentlandite totally disappears while pyrite starts to crystallize above the sample HM 79 (Fig. 3). Primary pyrite occurs as massive Co-rich grains (e.g. Ewers and Hudson, 1972) almost constantly associated with pyrrhotite and chalcopyrite (Fig. 5 b). Pyrrhotite has a uniform composition of monoclinic-type pyrrhotite (Fe_7S_8; Kissin and Scott, 1982) which contains sparse exsolutions of pentlandite. Chalcopyrite is always subordinate. The most obvious mineralogical change in this zone is that involving pyrite (Fig. 3). This sulfide is scarce from the sample HM 80 to the sample HM 95 and was never encountered as inclusions in silicates. Its abundance markedly increases in the uppermost sulfide-rich type layers located above the sample HM 95; moreover, pyrite is locally included together with chalcopyrite and pyrrhotite in the cumulus silicates. This increase in pyrite contents is coeval with the breakdown of relict olivine into orthopyroxene and magnetite and with the precipitation of Fe-Ti oxides and brown hornblende (Fig. 3).

In the pyrite-rich samples, modal compositions of interstitial sulfides vary greatly over distances of a few hundreds of μm. Pyrite, chalcopyrite and pyrrhotite may form discrete monomineralic grains. When the three sulfides are associated, pyrite contains numerous inclusions of pyrrhotite and chalcopyrite. According to experimental phase diagrams in the Cu-Fe-S ternary system, pyrite, chalcopyrite and monoclinic-type pyrrhotite can coexist stably at temperature only below 334°C (Cabri, 1973).

Hydrothermal Alteration

The whole 200 m thick sulfide-rich zone displays signs of low-temperature hydrothermal alteration whose intensity varies greatly irrespective of the height in the lithological column. The incipient alteration is marked by recrystallization of magmatic sulfide grains which are penetrated by coarse laths of hydrous silicates and lose their smooth curvilinear grain boundaries (Fig. 5c). This stage affects principally chalcopyrite which is highly mobile in hydrothermal fluids (Crerar and Barnes, 1976). Unlike chalcopyrite, pyrrhotite underwent in-situ replacement by concentrically-zoned Fe sulfides (pyrite, marcassite and probably melnikovite) locally developing "bird-eyes" textures (Fig. 5d) (Mac Farlane and Mossman, 1981).

In addition to remobilization and replacement of magmatic sulfides, hydro-

Table 1. S, Cu, Ni, Co, Zn and noble metal contents of the samples HM 95 and HM 93.

	HM 95	HM 93
Wt. %		
S	0.67	5.82
ppm		
Co	41	320
Ni	171	2050
Cu	35	3310
Zn	26	17
ppb		
Os	<5	<8
Ir	<2	<3
Ru	34	32
Rh	<2	4
Pt	<2	130
Pd	4	150
Au	2	
(Pd/Ir)N		42.5
(Au/Ir)N		160
Cu/Cu+Ni		0.61
Pt/Pt+Pd		0.2

thermal alteration has also introduced external sulfur into the magmatic sequence along veins and veinlet networks. Serpentine pseudomorphs after olivine are intergrown with secondary pyrrhotite ranging in composition between FeS and Fe_7S_8 (Fig. 5e). Hydrothermal fracture planes in pyroxenes and plagioclase are filled with chains and veinlets of Co-free pyrite developing sometimes euhedral shapes (Fig. 5f).

Abundance of Transition and Precious Metals

In accordance with their chalcophile properties, the Ni, Co and Cu contents are positively correlated with sulfur and increase significantly from the sample HM 95 to the sample HM 93 (Table 1). Os, Ir and Ru contents are independent of sulfur. In both samples, Os and Ir contents are below the detection limit of the analytical method while the Ru contents are nearly identical. By contrast, Rh, Pt, Pd and Au are much more concentrated in the sulfide-rich sample HM 93. These data are consistent with the now widely held hypothesis of two groups of PGE, the first one (Os, Ir and Ru) displaying strongly siderophile properties, and the second one (Rh, Pt and Pd) exhibiting chalcophile tendencies (Crockett, 1979; Naldrett et al., 1979; Naldrett, 1981; Barnes et al., 1985). The analytical results for PGE contents have been first normalized to 100% sulfides and then normalized to chondrite. The sample HM 93 is characterized by a steep positive chondrite-normalized PGE

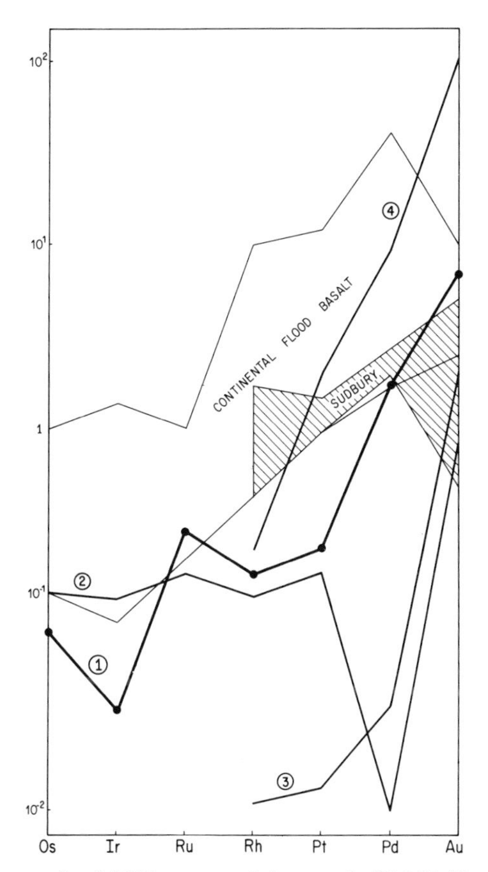

Figure 6. Chondrite-normalized PGE pattern of the sample HM 93 (1). Other curves (2,3,4) shown for comparison are from ophiolitic complexes (2 = magnetite-sulfide hydrothermal ore from Eretria, Tsangli area (Greece); Economou and Naldrett, 1984; 3 = hydrothermal As-rich sulfide mineralization from Pevkos, Cyprus; Foose et al., 1985; 4 = magmatic sulfide-rich ore associated with cumulus pyroxenites; Southwestern Oregon (USA); Foose et al., 1985). Fields of continental flood basalts and of Sudbury are taken from Naldrett (1981).

pattern with a jagged shape between Os and Ru. (Fig. 6). Jagged patterns are generally found in magmatic sulfides which underwent hydrothermal remobilization (Cornelius et al., 1987; A.J. Naldrett, personal communication to J.P. Lorand, 1987). However, hydrothermal alteration should have remobilized Pd and Au rather than the strongly siderophilic PGE which are considered to be inert in most geological fluids (Keays et al., 1982; Mountain and Wood, 1988). This jagged pattern is more likely to be due to the poor accurracy of Os and Ir analyses at tenors close to the detection limits of the analytical method.

The Pd and Au contents of the sample HM 93 are of the same order of magnitude as in the Sudbury-type sulfide ores (Naldrett et al., 1979, 1982) and in continental flood basalts (Naldrett, 1981; Barnes et al. 1985; Lorand,

1989). The steep positively trending pattern is nearly identical to that of magmatic sulfide ores associated with cumulus pyroxenites in ophiolites from southwestern Oregon, USA (Foose et al., 1985) (Fig. 6). $(Au/Ir)_N$ ratio higher than 100 are also characteristic of immiscible sulfide globules separated from MORB (Peach and Mathez, 1986).

Discussion and Conclusion

Textural and field evidence indicates that the sulfides separated as immiscible liquid droplets from the silicate melt parent to the olivine gabbros and the two-pyroxene gabbros. This interpretation is further supported by the mineralogical evolution in sulfide assemblages and the PGE data. From the lowermost olivine gabbros to the uppermost sulfide-rich layers, the progressive disappearance of pentlandite, coupled with the increasing modal proportions of monoclinic pyrrhotite and pyrite are signs of a progressively decreasing Ni/Fe ratio and an increasing sulfur fugacity. This evolution parallels the massive precipitation of orthopyroxene instead of olivine, and increasing oxygen fugacity which is marked by precipitation of magnetite and Fe-Ti oxides. Similar trends have been reported in continental layered intrusions (Chamberlain, 1967; Duke and Naldrett, 1976; Duke, 1979; Ripley, 1979; Von Grunewaldt, 1979). These trends are considered to strongly control the bulk sulfide compositions during fractional crystallization of the melt. In addition, the sample HM 93 plots in the field of tholeiitic-related magmatic sulfide deposits in the Pt/(Pt + Pd) vs Cu/(Cu + Ni) diagram of Naldrett (1981) (Fig. 7). These mineralogical and compositional features clearly distinguish the Haymiliyah sulfide-rich samples from hydrothermal sulfide ores associated with serpentinized chromite bodies at Eretria (Othris; Greece)(Economou and Naldrett, 1984) or located in the serpentinized transition zone at Pevkos (Limassol Forest, Cyprus) (Foose et al., 1985; Talhamer et al., 1986). Although Au enriched, these ores display very different PGE patterns characterized by a pronounced depletion in Pd relative to Au (Fig. 6). Both contain predominant troilite-type pyrrhotite instead of the monoclinic-type observed in the Haymiliyah chamber. Finally, the Pevkos ore is characterized by abundant Ni arsenides (Foose et al., 1985) which have never been observed in our samples.

According to microscopic and textural criteria, the sulfide melt separated shortly after olivine, but prior to the massive precipitation of orthopyroxene despite close association with the initial precipitation of orthopyroxene. Another way to qualitatively determine the period at which immiscibility occurred is to examine the Cu/Cu + Ni and Pd/Ir ratios of sulfide-rich layers. Under sulfur-undersaturated conditions, the fractional crystallization process is expected to deplete the residual mafic melt in Ni and siderophilic PGE (Os, Ir) while increasing the contents of incompatible PGE (Pd, Pt) and of Au and Cu (Duke, 1979; Naldrett and Duke, 1980; Barnes et al., 1985). The

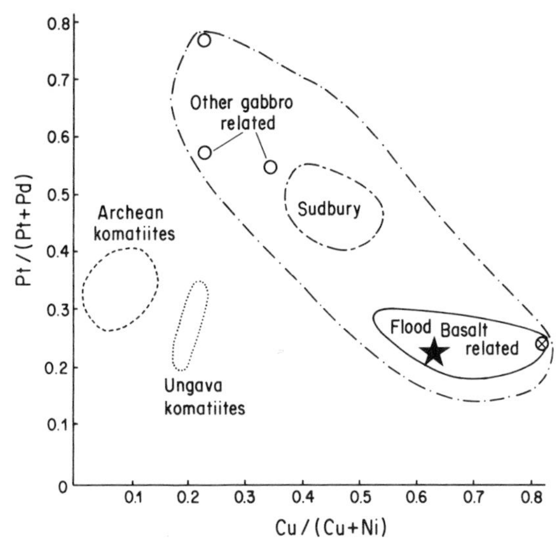

Figure 7. Plot of HM 93 (star) in a Pt/(Pt + Pd) versus Cu/(Cu + Ni) diagram (Naldrett, 1981).

high Cu/Cu + Ni, $(Pd/Ir)_N$ and $(Au/Ir)_N$ ratios (0.6, 42 and 160, respectively; Table 1) are therefore strongly consistent with immiscibility intervening late after the differentiation of the main gabbro unit. Clearly, the absence of sulfide in these rocks, even when they are interbedded with sulfide-rich two-pyroxene gabbros and noritic gabbros, suggests that the MORB-type magma which has periodically replenished the Haymiliyah magma chamber was sulfur-undersaturated. This conclusion seems to be at odds with the assumption that MORBs are sulfur-saturated in their mantle source region (e.g. Hamlyn and Keays, 1986 and reference therein). We suggest that the sulfur deficiency in the initial MORB-type magma of the Haymiliyah magma chamber arose from sulfide segregation during its extraction from the oceanic upper mantle, a process which is now well documented in the mantle sequence of the Oman ophiolites (Lorand, 1988).

Many studies have shown the sulfide-rich ores of the layered intrusions could result from a mixing process between two parental magmas of different compositions (Irvine et al, 1983; Hoatson and Keays, 1989; Naldrett, 1989). This model cannot account for mineralogical and chemical data of the sulfide-rich zone which show an evolution of the Haymiliyah magma chamber towards a progressively closed system, at the level of sulfide precipitation. The most likely hypothesis to explain sulfur saturation is an increasing volatile content in the already evolved melt parent to the two-pyroxene gabbros when the magma chamber started to evolve as a closed system (e.g. Von Grunewaldt, 1979). This role of a volatile component receives further support from the relative abundance of interstitial brown hornblende in the sulfide-rich zone. Experimental data of Haughton et al. (1974) and Buchanan and Nolan (1979) have shown that for oxygen fugacity conditions below the NNO solid buffer curve, the solubility of sulfur in mafic melts is controlled by the following exchange reaction

$$FeO_{(m)} + \frac{1}{2} S_{2(v)} \rightleftharpoons FeS_{(m)} + \frac{1}{2} O_{2(v)} \tag{1}$$

(m = basaltic melt; v = vapor)

At constant FeO, an increase of fS_2 due to accumulation of volatiles increases the FeS content of the basaltic melt until the sulfur saturation level is reached.

The increase of oxygen fugacity which produced precipitation of Fe-Ti oxides would have normally enhanced sulfide liquid immiscibility because it reduces the solubility of sulfur in the basaltic melt by driving the equation (1) to the left. The drop of sulfide content in the Fe-Ti oxide-rich noritic layers suggests that the oxygen effect was not prevailing. Obviously, it has been counterbalanced by the increasing FeO content in these rocks which markedly increase the solubility of sulfur in basaltic melts (Haughton et al., 1974; Buchanan and Nolan, 1979). It may be inferred from the rapid succession of sulfide-poor and sulfide-rich samples over a thickness of a few metres that small scale fluctuations of FeO and oxygen fugacity intervened during sulfide precipitation, in accordance with our hypothesis of closed system fractional crystallization.

The lack of any sulfide-rich layers in the whole mass of planar-laminated noritic gabbros above the 200–m thick sulfide-rich zone is also noteworthy. This suggests that the conditions suitable for sulfur saturation and sulfide precipitation were maintained only over a restricted period of the evolution of the Haymiliyah magma chamber. More analytical work is needed to clarify this assumption, but it may be adequately explained by a sulfur depletion in the resident magma after separation of the sulfide-rich zone and by the overall iron-enrichment trend which characterizes the planar-laminated noritic gabbros.

Acknowledgments

We are indebted to the Oman Ministry of Petroleum and Minerals, represented by M.H. Kassim and to the BRGM team in Oman for their technical support for the field studies.Financial support was provided by the French CNRS (Unité de Recherche Associée no. 736 "Minéralogie des roches profondes et des météorites", MNHN, Paris, and Unité de Recherche Associée no. 1278, UBO, Brest) and INSU grant (no. 90 DBT 5.14 to J.P. Lorand; paper no. 189). We also thank M. Guiraud for reading earlier draft of the manuscript. The authors are very grateful to Dr. S.J. Edwards for helpful reviews for the manuscript.

References

Anders, E. and Ebihara, M., 1982. Solar-system abundances of the elements., Geochim. Cosmochim. Acta, 46: 2363–2380.
Barnes, S.J., Naldrett, A.J. and Gorton, M.P., 1985. The origin of the fractionation of Pt group elements in terrestrial magmas., Chem. Geology, 53 (3–4): 303- 323.
Benn, K., and Nicolas, A., 1988. Mantle-crust transition zone and origin of wehrlitic magmas: Evidence from the Oman ophiolite. in: "The ophiolites of Oman". F. Boudier and A. Nicolas, (Eds.), Tectonophysics, 151: 75–86.
Buchanan, D.L. and Nolan, J., 1979. Solubility of sulfur and sulfide immiscibility in synthetic tholeiitic melts and their relevance to Bushveld complex rocks., Can. Miner., 17: 483–494.
Cabri, L.J., 1973. New data on phase relations in the Cu-Fe-S system., Econ. Geol., 68: 443–454.
Chamberlain, J.A., 1967. Sulfides in the Muskox intrusion., Canad. J. Earth. Sci., 4: 105–153.
Cornelius, M., Stumpfl, E.F., Gee, D. and Prochaska W., 1987. Platinum group elements in mafic-ultramafic gneiss terrain, Western Australia., Mineral. Petrol., 36: 247–265.
Craig, J.R. and Kullerud, G., 1969. Phase relations in the Cu-Fe-Ni-S system and their applications to magmatic ore deposits, "Magmatic Ore Deposits", Econ. Geol. Monog. 4: 343–358.
Crerar, D.A. and Barnes, H.L., 1976. Solubilities of chalcopyrite and chalcocite assemblages in hydrothermal solutions at 200–390°C., Econ. Geol., 71: 772–794.
Crockett, J.H., 1979. Platinum-group elements in mafic and ultramafic rocks: a survey., Canad. Mineral., 17: 391–402.
Dahl, R., 1984. Etude géométrique, pétrologique et géochimique de la séquence crustale de l'ophiolite d'Oman, massif de Rustaq (bloc de Haylayn). Un modèle tridimensionnel de zone d'accrétion. Thèse Univ. Clermont-Ferrand, France, 264p.
Duke, J.M. and Naldrett, A.J., 1976. Sulfide mineralogy of the Main Irruptive, Sudbury, Ontario., Can. Mineral., 14: 450–461.
Duke, J.M., 1979. Computer simulation of the fractionation of olivine and molten sulfide from mafic and ultramafic magmas., Can. Miner., 17: 507–514.
Economou, M. and Naldrett, A.J., 1984. Sulfides associated with podiform chromite at Tsangli, Eretria, Greece. Mineral. Deposit., 19: 289–297.
Ernewein, M. and Whitechurch, H., 1986. Les intrusions ultrabasiques de la séquence crustale de l'ophiolite d'Oman: un évènement témoin de l'extinction d'une zone d'accrétion océanique?, C.R. Acad. Sci., Paris, 303, II: 379–384.
Ewers, W.L. and Hudson, D.R., 1972. An interpretative study of a nickel-iron sulfide ore intersection, Lunon Shoot, Kambalda, Western Australia., Econ Geol, 76: 1075–1092.
Foose, M.P., Economou, M. and Panayiotou, A., 1985. Compositional and mineralogic constraints on the genesis of ophiolite-hosted nickel mineralization in the Pevkos area, Limassol Forest, Cyprus., Miner. Deposita, 20 (4): 234–240.
Glennie, K.W., Boeuf, M.G.A., Highes-Clarke, M.W., Moody-Stuart, M., Pilaar, W.F.H. and Reinhardt, B.M., 1974. Geology of the Oman Mountains, Part one (Text), Part two (Tables and Illustration), Part Three (enclosures) Kon. Nederlands Geol. Mijb. Gen. Ver. Verh., 31, 423p.
Hamlyn, P. and Keays, R.R., 1986. Sulfur saturation and second-stage melts: application to the Bushveld platinum metal deposits., Econ. Geol., 81: 1431–1445.

Haughton, D.R., Roeder, P.L. and Skinner, B.J., 1974. Solubility of sulfur in mafic magmas., Econ. Geol., 69: 451–467.

Hoatson, D.M. and Keays, R.R., 1989. Formation of the Platiniferous sulfide horizons by crystal fractionation and magma mixing in the Munni Munni Layered Intrusion, West Pilbara Block, Western Australia., Econ. Geol, 84 (7): 1775–1804.

Hulbert, L.J. and Von Gruenewaldt, G., 1982. Ni, Cu and Pt mineralization in the Lower Zone of the Bushveld complex, south of Potgietersrus., Econ. Geology, 77 (6): 1296–1306.

Irvine, T.N., Keith, D.W. and Todd, S.G., 1983. The J.M. Platinum-Palladium Reef of the Stillwater Complex, Montana, II: Origin by double diffusive convective magma mixing and implications for the Bushveld complex., Econ. Geol., 78: 1287–1348.

Jackson, E.D., Green, H.W., II, and Moore, E.M., 1975. The Vourinos ophiolite, Greece: cyclic units of lineated cumulates overlying harzburgite tectonite., Geol. Soc. Amer. Bull., 86: 390–398.

Johnston, A.D. and Stout, J.H., 1984. Development of Opx Fe/ Mg ferrite symplectite by continuous olivine oxidation., Contr. Miner. Pet., 88 (1–2): 196–202.

Juteau, T. and Whitechurch, H., 1980. The magmatic cumulates of Antalya (Turkey): Evidence of multiple intrusions in an ophiolitic magma chamber. In: A. Panayiotou, (Ed.), Ophiolites, Proc. Int. Ophiolite Symp. (Cyprus, 1979). Geol. Surv. Dep. Cyprus, Nicosia, pp. 377–391.

Juteau, T., Beurrier, M., Dahl, R. and Nehlig, P., 1988a. Segmentation at a fossil spreading axis: the plutonic sequence of the Wadi Haymiliyah area (Haylayn block, Samail ophiolite nappe, Oman).in: "The ophiolites of Oman". F. Boudier and A. Nicolas, (Eds.), Tectonophysics, 151: 167–197.

Juteau, T., Ernewein, M., Reuber, I., Whitechurch, H. and Dahl, R., 1988b. Duality of magmatism in the plutonic sequence of the Sumail Nappe, Oman. In: "The ophiolites of Oman". F. Boudier and A. Nicolas, (Eds.), Tectonophysics, 151: 107–135.

Keays, R.R., Nickel, D.H., Groves, D.I. and Mac Goldrick, P.J., 1982. Iridium and palladium as discriminants of volcanic-exhalative, hydrothermal, and magmatic nickel sulfide mineralizations., Econ. Geol., 77: 1535–1547.

Kelly, D.P. and Vaughan, D.J., 1983. Pyrrhotine-pentlandite ore textures: a mechanistic approach., Mineral. Mag., 47 (4), 453–463.

Kissin, S.A., Scott, S.D., 1982. Phase relations involving pyrrhotite below 350°C., Econ. Geol., 77: 1739–1755.

Kullerud, G., Yund, R.A. and Moh, G., 1969. Phase relations in the Cu-Fe-S, Cu-Ni-S and Fe-Ni-S systems. in "Magmatic ore deposits"., Econ. Geol. Monogr., 4: 323–343.

Kushiro, I., 1979. Fractional crystallization of basaltic magma. In: The evolution of igneous rocks, Fiftieth Anniversary Perspectives., H.S. Yoder J.R. (ed), pp. 171–204.

Lorand, J.P., 1987. Caractères minéralogiques et chimiques généraux des microphases du système Cu-Fe-Ni-S dans les roches du manteau suprieur: exemples d'hétérogénéités en domaine sub-continental., Bull. Soc. géol. France, (8), tIII, no. 4: 643–657.

Lorand, J.P., 1988. Cu-Fe-Ni sulfide assemblages of tectonite peridotites from the Maqsad district, Sumail ophiolite, southern Oman: implications for the origin of the sulfide component in the oceanic upper-mantle. In: "The ophiolites of Oman". F. Boudier and A. Nicolas, (Eds.), Tectonophysics, 151: 57–74.

Lorand, J.P., 1989. Sulfide petrology of spinel and garnet pyroxenite layers from mantle-derived spinel peridotite massifs of Ariège (Northeastern Pyrenees, France)., J. Petrol., 30: 987–1015. Lydon, J.W. and Richardson, D.G., 1988. Distribution of PGE in sulphides of the Bay of Islands ophiolite complex, Newfounlands. In: Geo-Platinum 87 (Prichard et al. (Eds)), Elsevier, pp. 251–252.

Macfarlane, N.D. and Mossman, D.J., 1981. The opaque minerals and economic geology of the Nemeiben ultramafic complex, Saskatchewan., Canada. Miner. Deposita, 16: 409–425.

Mason, B., 1971. Elemental abundances in meteorites. New York, Gordon and Brache Sci. Publ, 555p.

Misra, K. and Fleet, M.E., 1973. The chemical composition of synthetic and natural pentlandite assemblages., Econ. Geol., 68: 518–539.

Mountain, B.W. and Wood, S.A., 1988. Chemical controls on the solubility, transport and deposition of platinum and palladium in hydrothermal solutions: a thermodynamic approach., Econ. Geol, 83: 492–511.
Naldrett, A.J., 1973. Nickel sulfide deposits. Their classification and genesis with special emphasis on deposits of volcanic associations., Can. Inst. Met. Trans., 76: 183–201.
Naldrett, A.J., 1981. Platinum-group element deposits: Canadian Inst. Mining Metallurgy Spec., 23: 1286–1295.
Naldrett, A.J., 1989. Magmatic sulfide deposits. Oxford Monogr. Geol. Geophys., 14, 186p.
Naldrett, A.J. and Cabri, L., 1976. Ultramafic and related mafic rocks: their classification and genesis with special reference to the concentration of nickel-sulfides and platinum group elements., Econ. Geol. Lancaster, 71: 1131–1158.
Naldrett, A.J. and Duke, J.M., 1980. Platinum metals in magmatic sulfide ores., Science, 208: 1417–1424.
Naldrett, A.J., Hoffman, E.L., Green, A.H., Chou, C.L., Naldrett, S.R. and Alcock, R.A., 1979. The composition of Ni sulfide ores with particular reference to their content of PGE and Au., Can. Mineral., 17 (2): 403–415.
Naldrett, A.J., Innes, D.G., Sowa, J. and Gorton, M.P., 1982. Compositional variations within and between five Sudbury ore deposits., Econ. Geology, 77 (6): 1519–1534.
Nehlig, P. and Juteau, T., 1988. Flow porosities, permeabilities and preliminary data on fluid inclusions and fossil thermal gradients in the crustal sequence of the Sumail ophiolite (Oman). In: "The ophiolites of Oman". F. Boudier and A. Nicolas, (Eds.), Tectonophysics, 151: 199–221.
Page, N.J., 1971. Sulfide minerals in the G. H. Chromitite zone of the Stillwater complex Montana. U.S. Geol. Survey Prof. Paper., 694, 20p.
Pallister, J.S. and Hopson, C.A., 1981. Samail ophiolite plutonic suite: field relations, phase variations cryptic variations and layering, and a model of a spreading ridge magma chamber., J. Geophys. Res., 86: 2593–2644.
Pasteris, J.D., 1984. Further interpretation of the Cu-Fe-Ni sulfide mineralization in the Duluth complex, Northeastern Minnesota., Can. Mineral., 22: 39–54.
Peach, C.L. and Mathez, E.A., 1986. Gold and iridium in sulphides from submarine basalt glasses. E.O.S. (abstract), 410.
Ripley, E.M., 1979. Sulfide petrology of basal chilled margins in layered sills of the Archean Deer Lake Complex, Minnesota., Contr. Miner. Petrol., 69: 345–354.
Searle, M.P. and Malpas, J., 1980. Structure and metamorphism of rocks beneath the Semail ophiolite of Oman and their significance in the ophiolite obduction., Trans. R. Soc. Edinb. Earth Sci., 71: 247–262.
Sillitoe, R.H., 1973. Formation of volcanogenic massive sulphide deposits., Econ. Geol., 68: 1321–1336.
Sun, S.S., 1982. Chemical composition and origin of the earth 's primitive mantle., Geochim. Cosmochim. Acta, 46: 179–192.
Talhammer, O., Stumpfl, E.F. and Panayiotou, A., 1986. Postmagmatic, hydrothermal origin of sulfide and arsenide mineralizations at Limassol Forest, Cyprus., Mineral. Deposita, 21: 95–105.
Talkington, R.W. and Watkinson, D.M., 1986. Whole rock platinum-group element trends in chromite-rich rocks in ophiolitic and stratiform igneous complexes. In: Metallogeny of basic and ultrabasic rocks. Proceedings IMM conference, Edinburgh, 9–12 April 1985, Inst. Min. Metall., London, Ed. Gallagher et al., pp. 427–440.
Upadhyay, H.D. and Strong, D.F., 1973. Geological setting of the Betts Cove copper deposits, Newfoundland: An example of ophiolite suite mineralization., Econ. Geol., 68: 161–168
Usselman, T.M., Hodge, D.S., Naldrett, A.J. and Campbell, I.H., 1979. Physical constraint on the characteristics of Ni sulfide ores in ultramafic lavas., Can. Mineral., 17 (2): 361–371.
Von Gruenewaldt, G., 1979. A review of some recent concepts of the Bushveld complex with particular reference to the sulfide mineralization. Can. Mineral. 17(2): 233–256.

Platinum-Group Elements and Gold in Ophiolitic Complexes: Distribution and Fractionation from Mantle to Oceanic Floor

MARC LEBLANC
C.N.R.S., Centre Géologique et Géophysique, Université Montpellier II, 34095 Montpellier Cedex 2 (France)

Abstract

This review paper attempts to discuss the distribution of noble metals in ophiolitic complexes and suggests new targets for exploration.

Platinum-Group Elements (PGE) are nearly unfractionated in the less depleted mantle rocks (lherzolites). Residual dunites are clearly empoverished in Pt, Pd, and Au which were collected by silicate melts during partial melting. The chromite grains of the podiform chromitites, which crystallized early in magmatic conduits in the mantle, include Os-Ir-Ru minerals. Podiform chromitites are PGE-poor (100–500 ppb) or PGE-rich (up to 1 ppm), the highest PGE contents (>ppm) are often related to the presence of interstitial sulphides. In the crustal section PGE are fractionated among the magmatic rocks (Pd/Ir ratio increases from dunite to gabbro). The very low PGE content of the basaltic lavas involves a PGE segregation in a magmatic sulphide phase before emplacement at surface. PGE-rich sulphides are locally present at the base of the crustal section (in chromite-rich dunite or in pyroxenite) or among the upper gabbros. Hydrothermal processes related to serpentinization concentrate mainly gold (1–10 ppm) in sulphide-rich serpentinites, or in carbonatized (listwaenite) or silicified serpentinites where the noble metals are in sulphides and/or arsenides. Convective hydrothermal systems strongly leached the magmatic pile, then precipitated on the ocean floor (i) Massive Sulphide Deposits (Cyprus-type MSD) which may be gold-rich (>1ppm); and (ii) metalliferous sediments displaying slight enrichments in Pt-Pd and Au. Up to 1 ppm Pt may be found in oceanic Mn-nodules. Lastly, weathering and erosion should concentrate PGE in laterites and placers.

PGE are an economically important group of elements and new targets should be explored in ophiolites: (1) chromitite pods (Os-Ir-Ru) which may be also Ru-Pt-Pd-rich when sulphides are present; (2) sulphide-rich serpentinites or carbonatized-serpentinites (listwaenites) mainly for gold; (3) sulphide-bearing magmatic rocks (Ru-Pt-Pd); (4) exhalative metalliferous sediments; and (5) laterites and placers.

Tj. Peters et al. (Eds), Ophiolite Genesis and Evolution of the Oceanic Lithosphere, 231–260.

I. Introduction

The most important primary economic deposits of the Platinum-Group Elements (PGE) are hosted by stratiform mafic-ultramafic magmatic complexes, as Bushveld and Stillwater complexes, and by magmatic sulphide deposits. Although the ophiolite suite is one of the more common mode of occurence of ultramafic rocks, economic deposits of PGE in ophiolites are apparently insignificant. Almost all reported PGE concentrations in ophiolites are associated with the chromitite pods, which are generally small-sized bodies. Very little is known about the distribution of PGE within a single ophiolite. In the extensive literature on ophiolite, information about the noble metals is fragmentary, except a few works like this of Oshin and Crocket (1982) on the Thetford ophiolite. The more abundant data concern the podiform chromitites. This present study is mainly a review paper and includes preliminary results of a current work on the Bou Azzer ophiolite, Morocco(Leblanc and Fischer, 1990), and the Semail ophiolite, Oman. The present paper attempts (i) to document the abundance levels of noble metals in the ophiolite-forming rocks, (ii) to discuss their distribution and fractionation trends, and (iii) to examine their petrogenetic and economic implications.

II. Preliminary Considerations

A. PGE Geochemistry

The platinum-Group Elements represent a coherent group of siderophile elements. They include poorly soluble elements (Ir, Os, and Ru) and more soluble elements (Rh, Pt, and Pd) in basaltic melts (Amossé et al., 1987). Consequently Ir, Os, and Ru tend to be early concentrated in the most refractory cumulates; whereas Rh, Pt, and Pd behave apparently as incompatible elements being progressively concentrated in the liquid. Partition coefficients for all PGE between immiscible silicate and sulphide liquids are of the order of 10^3 to 10^5 (Campbell and Barnes, 1984). Thus, the fractionation and concentration of PGE in magmatic conditions will depend strongly (1) on the relative timing of sulphide saturation and crystal fractionation, and (2) on the behaviour of the sulphide phase (Barnes et al., 1988). Sulphur is an extremely mobile element during hydrous alteration of rocks. Consequently, some form of hydrothermal transport and reduction of PGE may also be expected (Mountain and Wood, 1988).

B. Analytical Data and Use of Chondrite-Normalized Curves (CN Curves)

In most rocks PGE and gold are present at very low concentration levels (ppb). Analytical techniques require a preconcentration process, generally a fire assay procedure by nickel sulphide collection. The most usual final measurement techniques include atomic absorption spectrometry, instrumental neutron activation analysis and inductively coupled plasma-atomic emission spectrometry. The usual detection limits are 1ppb. Naldrett et al., (1979) showed that chondrite normalized noble metals analysis plotted in order of decreasing melting point (Os, Ir, Ru, Rh, Pt, Pd, Au) display smooth CN curves. The range, the shape, and the slope (Pd/Ir ratio) of the CN curves are representative of PGE fractionation (Barnes et al., 1985). The conventional normalizing values used are the C_1 chondrite (Naldrett and Duke, 1980).

III. Mantle Section

A. Peridotites

Most mantle peridotites from ophiolitic complexes are more or less intensively serpentinized harzburgites and dunites. Very little is known about the PGE distribution in primitive mantle rocks except for mantle xenoliths and for the high-temperature intrusive lherzolite massifs such Ronda (Stockman, 1976) and Lherz (Begou et al.; 1989 Lorand, 1989). Lherzolites display a restricted range of PGE contents (30–60 ppb) and exhibits roughly chondritic ratios of noble metals, except for a slight gold enrichment (Fig. 1). The lherzolite field is nearly flat (Pd/Ir = 1–2), except for small Ru positive and Pt negative anomalies. It covers the range 0.05–0.15 times chondrite, and corresponds fairly well to the CN curve for the average mantle composition (Sun, 1982). The associated harzburgites and dunites are clearly depleted in Pt, Pd, and Au with respect to lherzolite (Fig. 1); the PGE content of dunite is <30 ppb. The few available data concerning the ophiolitic harzburgites and dunites are quite similar to that of the same rock-types from lherzolite massifs (Fominykh and Kostova, 1970; Stockmann, 1976; Oshin and Crocket, 1982). They exhibit slightly negatively sloped PGE patterns, although some harzburgites display a flat unfractionated PGE pattern. In contrast, the various types of the mafic layers from the lherzolite massifs (Stockman, 1976; Leblanc et al, 1990) exhibit positively sloped PGE patterns (Pd/Ir = 2 to 20), and show a strong gold enrichment. From lherzolite to harzburgite, then to dunite, the copper, gold, and sulphur contents decrease; whereas these elements are both enriched in mafic layers (Table I).

B. Chromites

Typical ophiolitic podiform chromitite (Page and Talkington, 1984) is characterized by Os, Ir, and Ru abundances in the range of 0.05 to 1.0 times chondritic values, and Pt and Pd abundances about 0.01 times chondritic. Thus, Pt and Pd are 10 to 100 times depleted in chromitite relative to Os, Ir and Ru. The resulting CN curves (Fig. 2) are consequently characterized by a steep negative slope from Ru to Pt (Ru/Pt = 2 to 25). The Os, Ir and

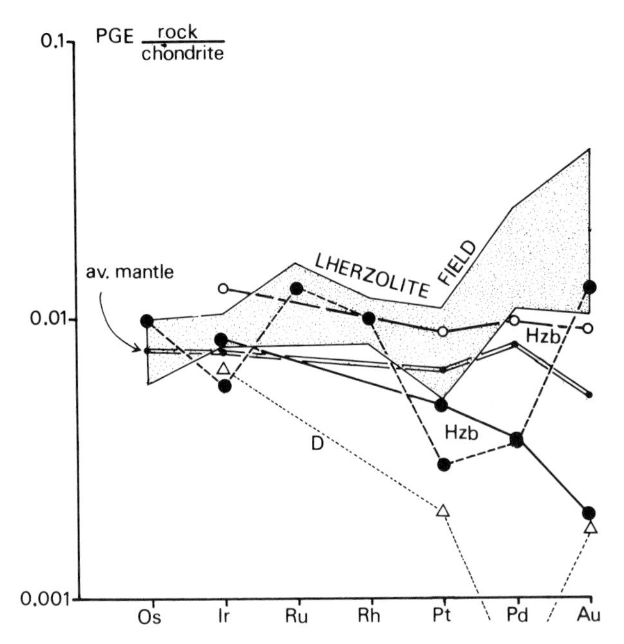

Figure 1. Chondrite normalized PGE diagram for the mantle rocks showing the field of lherzolite alpinotype massifs (Stockman, 1976; Begou, 1989; Lorand, 1989 and unpublished data from Gervilla and Leblanc, current work) and the average mantle composition (Sun, 1982). Individual trends for harzburgite (Hzb) and dunite (D) are from lherzolite massifs (black symbols) or from ophiolitic complexes (open symbols). The ophiolitic harzburgite is from the Josephine ophiolite (Stockman, 1976), and the ophiolitic dunite is from the Theteford ophiolite (Oshin and Crocket, 1982).

Ru portion of the CN curves is relatively flat, although the Os/Ir and Ir/Ru ratios are variable resulting from the relative proportion of laurite-erlichmanite (RuS_2 -OsS_2) and Ru-Os-Ir bearing alloy inclusions in chromite. Minerals that contain Pt, Pd, and Rh are generally absent in chromite, however Pt-Fe alloys, Pt arsenides and sulphides have been reported exceptionally as interstitial phases.

Considering the chromitites of different ophiolitic complexes, the PGE content and fractionation (Pd/Ir ratio) vary in a wide range (Fig. 2–3). Most of the ophiolitic podiform chromitites contain between 100 and 500 ppb PGE and have moderate to low Pd/Ir ratios (0.8 to 0.1). Although some ophiolites, such as the Massif du Sud of New Caledonia (Legendre, 1982; Page et al., 1982a) are characterized by PGE-rich chromitites (>750 ppb) which display a very low Pd/Ir ratio (<0.1) and a CN curve with a steeper negative slope (Fig. 4). The PGE content of chromitite in ophiolite seems to increase with decreasing Pd/Ir ratio (Fig. 3), that is to say that the high PGE contents result mainly from a concentration of Ir relatively to Pd. The variations in PGE content observed between the chromitites of different ophiolitic complexes (Fig. 2) are at first sight without clear connection with any other

Table I. Platinum group element and gold average concentrations (ppb) in ophiolitic rocks. Concentrations in sulphides are normalized to 100% sulphides.

Rock type	Os	Ir	Ru	Rh	Pt	Pd	Au	ΣPGE	Pd/Ir	S(ppm)	References
Mantle	4.02	4.15	n.d.	n.d.	6.74	4.49	0.79	<35.	1.08	n.d.	1
Spinel-lherzolite	6.7	4.0	6.1	2.0	6.7	4.5	1.5	30.0	1.12	150	2,3,4
Harzburgite	n.d.	3.2	n.d.	n.d.	10.	3.8	1.5	<30.	1.18	74	5
Dunite	3.2	4.0	4.5	1.0	1.5	0.5	0.2	14.7	0.12	30	4
Serpentinite	37.0	3.4	5.3	1.0	4.0	2.4	8.9	20.0	0.70	n.d.	6
Chromitite PGE-poor	447.0	25.9	55.2	5.2	10.2	10.9	6.1	144.4	0.42	n.d.	7,8,9,10
Chromitite PGE-rich		280.0	303.0	15.0	15.3	3.8	2.0	1064.1	0.013	n.d.	11,12,13
Ultramafic cumulates	3.0	2.3	4.8	1.7	17.3	12.5	1.9	41.6	5.4	n.d.	5,15,15,*
Mafic cumulates	0.05	0.04	0.18	0.14	4.9	3.0	1.0	8.31	75.0	n.d.	5,14,15,*
Magmatic sulphides	230.0	208.0	225.0	211.0	1865.04	4870.0	117.0	7609.0	23.4	x.10%	16
MORB	n.d.	0.02	n.d.	n.d.	n.d.	0.83	1.3	<3.	41.0	800	17
Boninite	n.d.	0.06	n.d.	n.d.	n.d.	15.0	1.9	<25.	250	55	17
Hydrothermal sulphides	27.7	9.2	9.2	9.2	9.2	27.7	527.0	92.0	3.0	x.10%	*
Metalliferous sediments	1.0	7.5	1.0	1.0	6.0	10.5	21.5	27.0	1.4	900	*

References: 1 = Sun (1982); 2 = Stockman (1976); 3 = Lorand (1989); 4 = unpublished data from (Gervilla and Leblanc); 5 = Oshin and Crocket (1982); 6 = Leblanc and Fischer (1989); 7 = Economou (1986); 8 = Page *et al.*, (1984); 9 = Prichard *et al.*, (1986); 10 = Talkington and Watkinson (1986); 11 = Page *et al.*, (1982[a]); 12 = Legendre (1982); 13 = Dupuy *et al.*, (1981); 14 = Page *et al.*, (1983); 15 = Becker and Agiorgitis (1978); 16 = Barnes *et al.*, (1988); 17 = Hamlyn *et al.*, (1985).
n.d. = no data available.

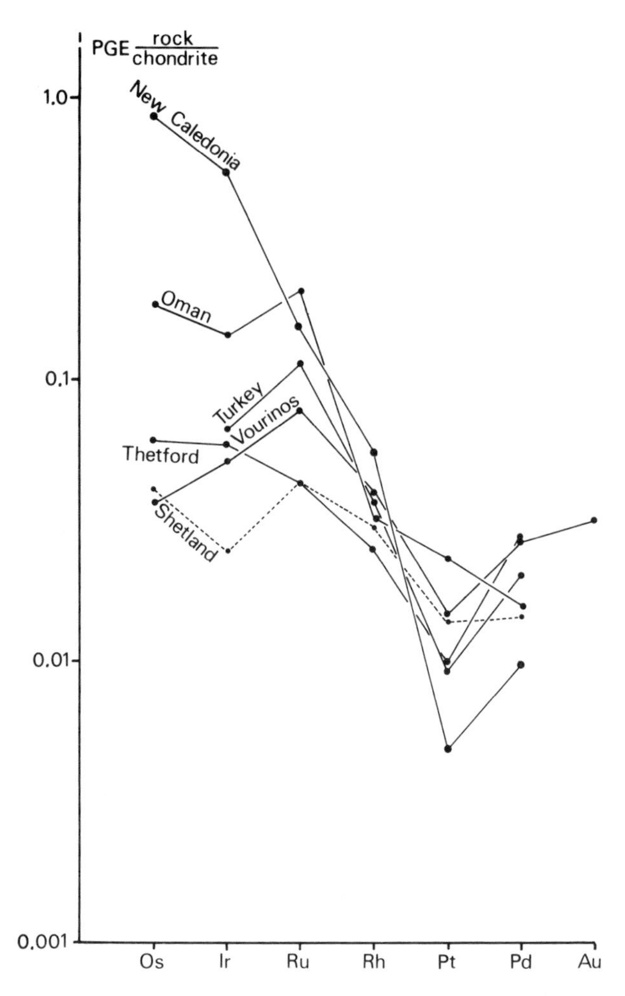

Figure 2. Chondrite normalized average PGE content for podiform chromitites from different ophiolitic complexes: New Caledonia (Legendre, 1982); Oman (Legendre, 1982): Turkey (Page et al., 1984); Vourinos, Greece (Economou, 1986); Thetford, Appalachians (Talkington and Watkinson, 1986); and average of normal chromitite pods (E and Quoys) from Shetland (Prichard et al., 1986).

characteristics of the ophiolite. The spreading-rate of the oceanic crust which regulates partial melting and magmatic flow in mantle (Boudier and Nicolas, 1985) should control the genesis of the chromitite bodies and their PGE content. Most of the PGE-poor chromitites are in low-spreading ophiolites, and some of the PGE-rich chromitites may be ascribed to fast-spreading ophiolites following the criterion of Boudier and Nicolas (1985). Age is not an important factor: ophiolites of Mesozoic, Paleozoic, or Upper Proterozoic ages may have chromitites with similar PGE contents (Fig. 5). Nevertheless a link between the PGE content and the chromite composition is suggested

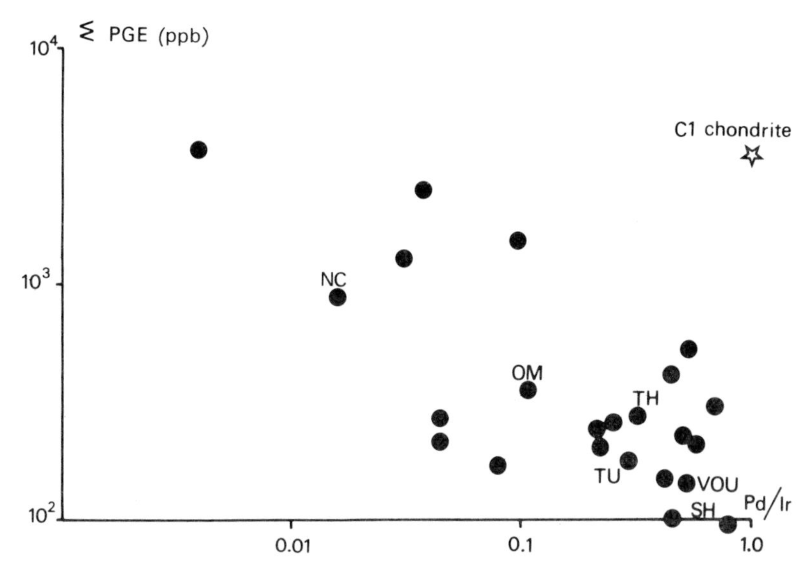

Figure 3. PGE contents versus Pd/Ir ratio diagram for podiform chromitite from various ophiolitic complexes. The abbreviations refer to the ophiolites of the Figure 2. Note that the anomalously PGE-rich Cliff chromitite, Shetland, is far away on the right side of the diagram (ΣPGE = 50,800 ppm; Pd/Ir = 12).

by Economou (1986) for chromitite samples from different ophiolites of Central Greece; the Cr-poor ores display lower PGE level compared to the Cr-rich ores (Fig. 4). Variations are also observed at the scale of a single ophiolite. For example, among the 40 chromitite occurrences of the Massif du Sud, New Caledonia, analysed by Page et al., (1982a), three distinct, although similar, patterns can be recognized (Fig. 4); the variations in PGE content, Pd/Ir ratio, and Ir relatively to Ru cannot be correlated neither with the spatial distribution of the orebodies, their shape and orientation, the ore type, nor the chromium content (Cassard, 1980). Only a possible negative correlation between the PGE content of the chromitite and the size of the orebodies has been found.

In the Ibra massif of the Semail ophiolite, Oman, a slight but distinct variation (Fig. 6) is observed from the deepest chromitites in the mantle section to those located along the transition zone, just below the gabbroic cumulates (Page et al.,1982b): the chromitites from the transition zone have slightly lower PGE content and display a slight Pd-enrichment (Pd/Ir = 0.2 against 0.10 for the mantle chromitites). In the Maqsad diapiric structure, described in the Semail ophiolite by Ceuleneer et al., (1988), there is a vertical chromitite dike which was not deformed by plastic flow. The chromitite ore displays well-preserved magmatic textures which are from top to bottom: dispersed euhedral chromite crystals, small skeletal cubic-shaped nodules and, massive oval nodules. Preliminary data on this chromitite orebody show (i) low PGE contents (≤100 ppb), in comparison with concordant

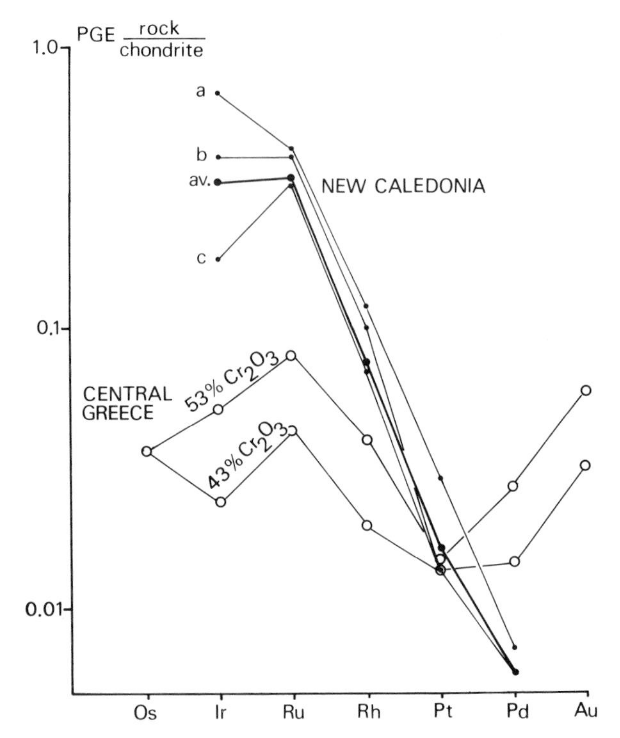

Figure 4. Chondrite normalized PGE patterns for PGE-rich chromitites from New Caledonia (Page et al., 1982a), and PGE-poor chromitites from Central Greece (Economou, 1986). Three slightly different patterns may be defined among the New Caledonian chromitites. Among the Greek chromitites the Cr-richer display higher PGE contents.

chromitite pods of Oman (Brown, 1982; Legendre, 1982; Page et al., 1982b); and (ii) a strong fractionation from top to bottom (Pd/Ir = 0.6 to 3.6). The corresponding PGE patterns are distinct from the average PGE pattern of the Oman chromitite (Fig. 6) and rather resemble PGE pattern of magmatic chromitite. Sulphide-silicate inclusions are present in chromite ore (Lorand, 1988). At the top of the Maqsad diapiric structure, the surrounding mantle rocks were strongly impregnated by percolating magmas resulting in a metasomatic enrichment of the sulphide component (Lorand, 1988).

Unusually high PGE values (>1 ppm), two orders of magnitude higher than those found generally in ophiolitic chromitites, have been recorded in Shetland (Prichard et al., 1986); Skyros, Greece (Wolf and Agioritis, 1978); White Hills, Newfoundland (Talkington and Watkinson, 1986); Pole Corral, Klamath Mountain (Moring et al., 1988); and also in Norwegian Caledonide ophiolites (Barnes et al., 1988). Most of these PGE-rich chromitites exhibit a classic PGE pattern for podiform chromitite in ophiolite (Fig. 7) with an extremely steep negative slope (Pd/Ir < 0.1). In contrast, the small chromitite body of Cliff, Shetland, shows a strong positive slope in its CN curve (Pd/Ir = 11–14). Although located in mantle harzburgite, among other chro-

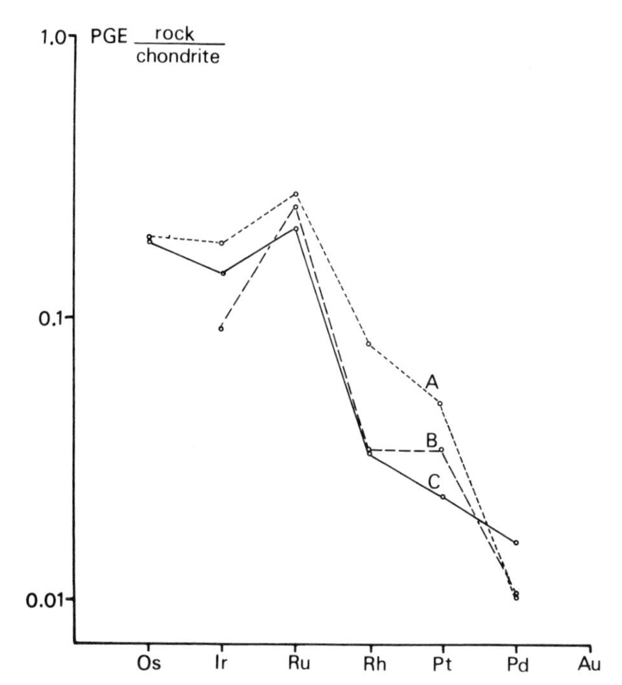

Figure 5. Chondrite normalized PGE content for podiform chromitites from ophiolites of different ages: (A) Upper Proterozoic (Arabia; Legendre, 1982); (B) Paleozoic (Newfoundland; Page and Talkington, 1984): and (C) Mesozoic (Oman; Legendre, 1982). Note the similarities of the corresponding patterns.

mitite bodies displaying classical PGE patterns (Fig. 2), Cliff display a positively sloped pattern (Fig. 7) which is much more typical of layered stratiform magmatic complexes, such as the Bushveld and Stillwater complexes. The majority of Platinum Group Minerals in Cliffs are Ru-Os-Ir minerals included in chromite; Rh-Pt-Pd minerals which are mainly arsenides and sulfo-arsenides are present in serpentinite interstitial to chromite grains, in association with Ni-Cu sulphides and Ni-arsenides (Prichard et al., 1986).

C. Serpentinite and Associated Rocks (Fig. 8)

Serpentinites exhibit flat PGE patterns (Pd/Ir = 0.7 to 1.5) and low PGE contents (20–25 ppb) which are closely similar to those of residual mantle peridotites (Fig. 8A). Their CN curves show a very slight Rh, Pt, and Pd depletion, and are characterized by a slight gold enrichment (3–9 ppb on average). Serpentinite comprises a secondary mineral assemblage (serpentine minerals, magnetite, accessory sulphides) and relict grains of chromite. Chrysotile (Fig. 8A) shows a strong Rh, Pt, Pd, and Au enrichment, and a Ir, Ru impoverishment resulting in a steep positively sloped CN curve (Pd/Ir $>$ 2); the PGE pattern of magnetite is characterized by a positive Ru

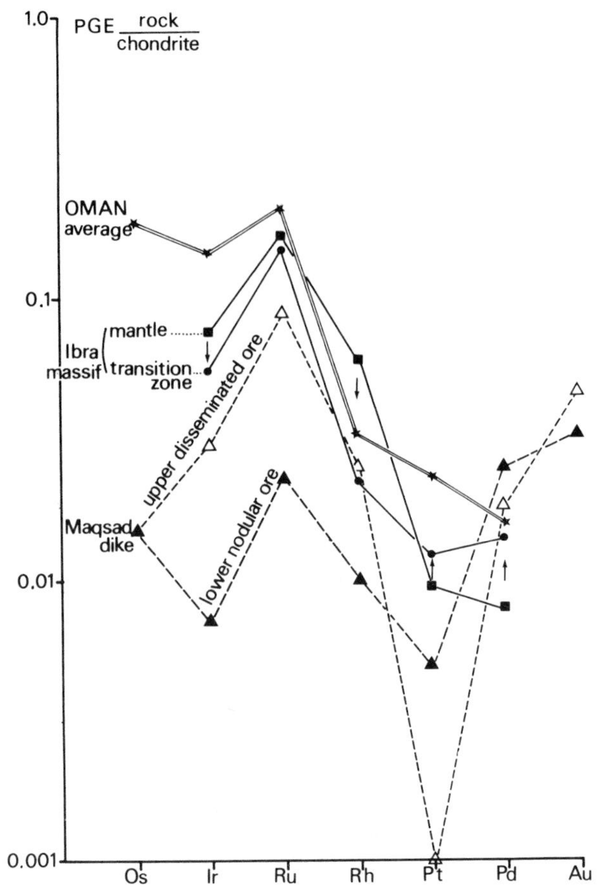

Figure 6. Chondrite normalized PGE content for the podiform chromitites of Oman. The average data are from Legendre (1982). The evolution from mantle to the transition zone in the Ibra massif is from Page et al.(1982b), and the PGE fractionation in the Maqsad dike is from unpublished preliminary data (current work).

anomaly and a slight Pt, Pd, and gold enrichment (Fischer et al., 1989). The accessory Cr-spinels from harzburgite and dunite (Oshin and Crocket, 1982; Cocherie et al., 1989) show PGE patterns respectively similar to that of harzburgite and dunite; their PGE content is five to ten times higher (Fig. 8B) suggesting that Cr-spinel is an important PGE-bearing mineral.

Sulphide-bearing serpentinites have been described in Eretria, Greece (Econonou and Naldrett, 1984), Limassol, Cyprus (Foose et al., 1985), Oregon (Foose, 1986) and Bou Azzer, Morocco (Leblanc and Fischer, 1989). They correspond either to particular horizons of sulphide-and chromite-bearing black serpentinite within the serpentinite massifs (Bou Azzer), or to mineralized zones along sheared borders of serpentinite massif (Limassol). The mineralization consists mainly of Fe-Ni-Cu sulphides. Their magmatic

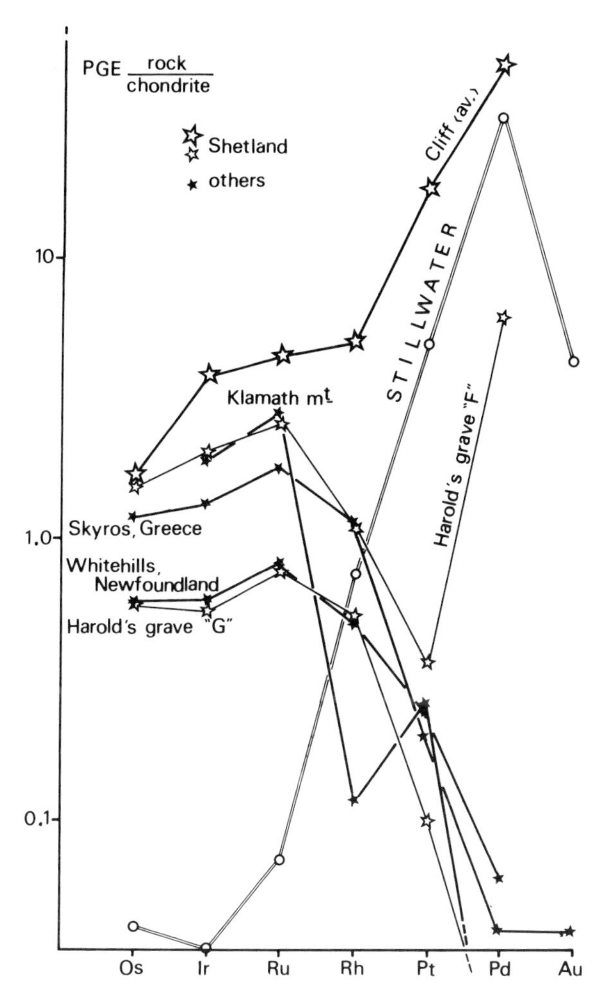

Figure 7. Chondrite normalized curves for PGE-rich podiform chromitites (compare with Figure 2). The trend of the platiniferous JM reef of the Stillwater magmatic complex is given for comparison (Barnes et al., 1985).

origin is uncertain as they were removed during serpentinization into a secondary assemblage comprising accessory arsenides and magnetite. PGE concentration in sulphides is 2 to 5 times average serpentinite (Fig. 8C). The CN PGE pattern of the Bou Azzer sulphides is jagged and exhibits positive Ru-Rh and negative Pt anomalies (Fig. 8C). This pattern is similar to that of sulphide-bearing komatiite, except for lower PGE contents and a gold enrichment for the sulphides of serpentinites (0.1–8 ppm Au). The highest gold contents (Limassol) are in arsenide-bearing samples. These high As and gold contents may indicate a significant hydrothermal contribution (Foose et al., 1985).

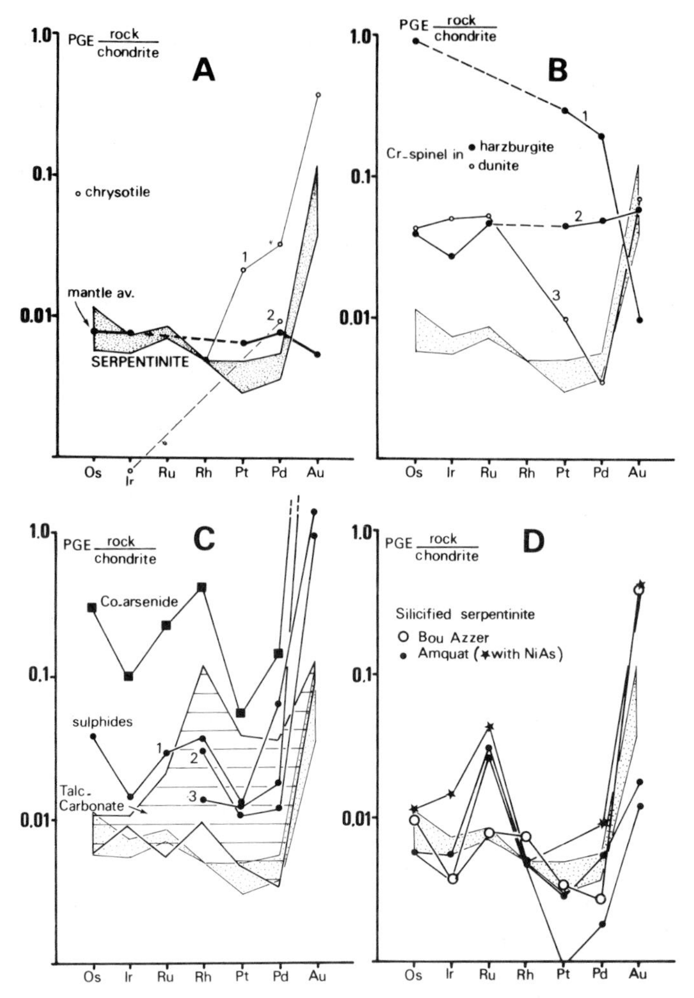

Figure 8. PGE chondrite normalized content in serpentinite and associated rocks. (a) Field of serpentinite rocks (Leblanc and Fisher, 1989) with reference to average mantle composition (Sun, 1982). Data for separated chrysotile are form (1) Fisher (1988), and (2) Oshin and Crocket (1982). (b) Separated Cr-spinels are 5 to 100 times richer, except for gold, than the serpentinized ultramafic rocks. (1) Oshin and Crocket (1982); (2) (3) Cocherie et al., 1989). (c) Fractionation of Rh, Pt, and Pd in the talc-carbonate rocks which are hydrothermally derived from serpentinite. Sulphides and arsenides in serpentinite show similar patterns characterized by negative Ir, Pt and positive Rh anomalies, and display a strong gold enrichment. (d) The silicified serpentinites from Bou Azzer, Morocco (Leblanc and Fisher, 1989) and Amquat, Oman (Stanger, 1985) are nearly similar to normal serpentinites, except for a Ru, and gold enrichment in the NiAs-bearing silicified serpentinite.

During hydrothermal processes, generally related to serpentinization, serpentinite may be progressively transformed in talc-carbonate-silica rocks. Talc-carbonate and carbonate rocks are richer than serpentinite in PGE and gold, 60–70 ppb and 15–20 ppb respectively. Their PGE patterns are positively sloped, although a negative Pt anomaly is often observed in carbonate rocks. In a CN diagram (Fig. 8C), the wide field of the talc-carbonate and carbonate rocks overlaps the narrower field of serpentinite. In carbonatized serpentinite (listwaenite) gold may be concentrated (0.1–10 ppm) in sulphide or arsenide bearing listwaenites (Buisson and Leblanc, 1987). Many examples of gold mineralization in listwaenite lenses have been reported throughout the world in serpentinite massifs from various ophiolitic complexes (Buisson and Leblanc, 1986). The cobalt-arsenide mineralization of Bou Azzer, Morocco (Leblanc and Billaud, 1982) is located in carbonate-quartz rocks resulting from the hydrothermal alteration of serpentinite. Extremely high gold and high PGE contents are found in the Bou Azzer cobalt ores: from 5 to 20 ppm and 40 to 2000 ppb respectively (Leblanc and Fisher, 1990). Precious metals are born by the cobalt arsenide which has an average content of 120 ppm of gold, and 580 ppb of PGE. The PGE pattern is flat (Pd/Ir = 1–2) and comprises positive Rh and negative Pt anomalies. The cobalt-ores display PGE pattern which are similar to that of the sulphide-rich serpentinites of Bou Azzer (Fig. 8C), suggesting a kind of genetic relationship.

Serpentinite may be strongly silicified along major tectonic contacts resulting in a jasper-like quartz-rock (birbirite) which still shows relict chromite grains and structures from serpentinite. Accessory sulphides, sulpho-arsenides or arsenides ares locally present. Birbirite has generally low PGE contents (30–60 ppb). The PGE patterns (Fig. 8D) are roughly flat (Pd/Ir = 1–2) and show positive Ru and negative Pt-Pd anomalies. Gold may be enriched in birbirite. Gold-bearing silicified serpentinites (Au $>$ 0.1 ppm) are recorded in New Caledonia (Jacob, 1985) and in Turkey (current work), along major faults or thrust planes. In the silicified serpentinite of New Caledonia, Jacob (1985) noted a clear positive correlation between Au and As, the highest average gold contents (2.5 ppm) are related to the presence of arsenopyrite. The birbirite of Amquat, Oman (Stanger, 1985) contains from 30 to 60 ppb PGE and from 2 to 70 ppb Au, the highest contents are in niccolite (NiAs)-bearing samples (current work).

The mercury-gold mineralizations found in various ophiolitic serpentinite massifs are generally ascribed to late hydrothermal circulations (Henderson, 1969). Among them, the important gold deposit of McLaughlin, California (Homestake Ltd, unpubl. reports), discovered in 1981 (30 Mt at 5ppm Au), is located in serpentinites which were silicified and carbonatized by post-obduction geothermal systems. At last the richest parts of the most famous gold lode in the world, e.g. the Mother Lode of California, are in carbonatized serpentinite wallrocks. No PGE data are avalaible for these important gold deposits.

IV. Crustal Section

The crustal section of ophiolites comprises mainly magmatic rocks including from bottom to top: layered mafic-ultramafic rocks(generally described as

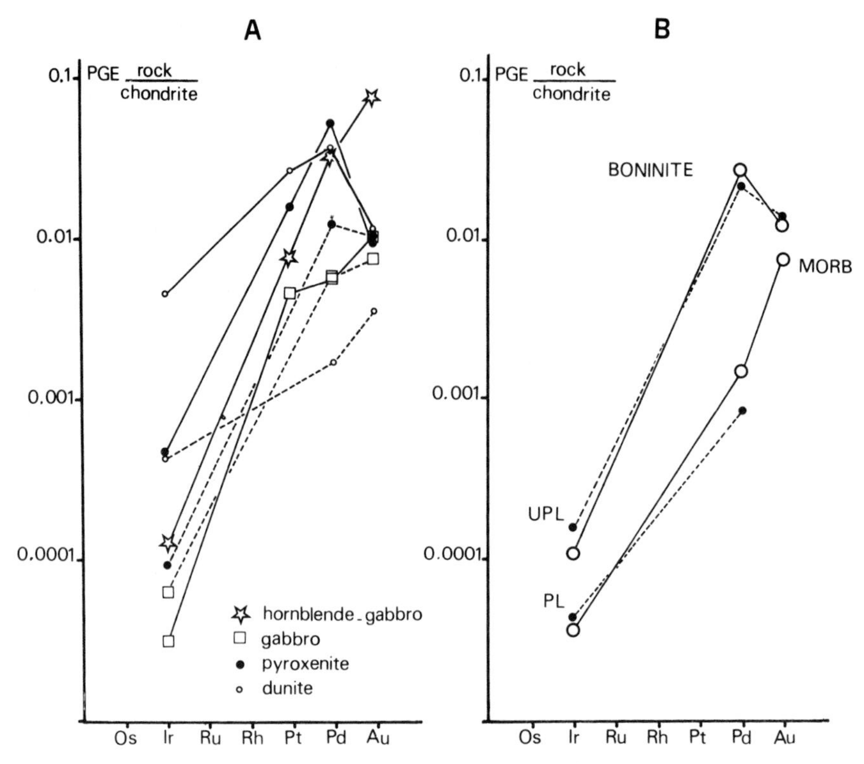

Figure 9. PGE chondrite normalized diagrams (A) for ophiolitic cumulate rocks from Thetford (Oshin and Crocket, 1982) and, in dashed lines, from Troodos (Becker and Agioritis, 1978); and (B) for Çmafic lavesÇ from Troodos, lower pillow-lavas (LPL) and upper pilow-lavas (UPL) (Becker and Agioritis, 1978; Hawlyn et al., 1989). Average values for boninites and Middle Oceanic Ridge Basalts are from Hamlyn et al. (1989).

cumulates), isotropic or pegmatoid gabbros, sheeted dike complexes, mafic lavas and metalliferous sediments

The nature and origin of the transition zone between crust and mantle is a debatable problem. Nicolas and Prinzhofer (1983) have shown from structural evidence that the dunite of the transition zone results from a complex interaction between a residual mantle rock and active percolating mafic magmas. Nevertheless as many authors describe the transition zone as the lower part of the cumulate pile, the belonging of the chromitite orebodies of the transition zone to the mantle section or to the crustal section is debatable.

A. *Ultramafic-Mafic Cumulates* (*Fig.* 9*A*)

1. *Magmatic differentiation from dunite to gabbro*

The variation in the PGE patterns within the ophiolitic cumulates has led Oshin and Crocket (1982) to suggest that the PGE are strongly fractionated among dunites, pyroxenites and gabbros. In contrast the fractionation of gold among cumulate rocks is very slight (Dupuy et al., 1981): gold contents are low in dunite (≤4 ppb) and decrease towards gabbros (≤1 ppb). PGE contents vary from 10 to 50 ppb, the highest values are in pyroxenite. Iridium contents drop from dunite to pyroxene-bearing cumulates, and palladium contents increase from dunite to gabbros through pyroxenite which exhibits the highest Pd values (15–30 ppb). Consequently the Pd/Ir ratio increases from dunite (4–8) to late gabbros (250–300). This evolution of Pd/Ir ratio is consistent with a depletion of Ir and an increase in Pd with continuing magmatic differentiation that is also clearly marked by a Ni decrease and a Cu enrichment (Dupuy et al., 1981). The PGE pattern for the cumulate dunite is positively sloped, this pattern is antithetic to that of the mantle refractory dunite which exhibits a negative slope (Fig. 2). Nevertheless some differences in PGE content and CN curves appear between the dunites, mainly in the transition zone between the mantle and crustal section where there is no unequivocal evidence for a magmatic or refractory origin of the dunite. In contrast, the CN curves for the pyroxenites and the gabbros display very steep positive slopes in a restricted range (Fig. 9A); they result mainly from a strong depletion in Ir contents which are in a factor of 10 to 100 under the values obtained for dunite. These PGE patterns are similar to those obtained from cumulate sequences of stratiform magmatic complexes (Page et al., 1982). Palladium and gold are enriched in pegmatoïd hornblende-gabbros (Oshin and Crocket, 1982), these elements were probably mobbilized with sulphur during late deuteric processes. In contrast, serpentinization of cumulate dunites has not significantly affected noble metal content (Oshin and Crocket, 1982).

2. *Chromitite Layers* (*Fig.* 10*A*)

Chromite is generally a minor phase in olivine cumulates; although chromite-rich dunites and chromitite layers are present in the ultramafic cumulates of some ophiolitic complexes. The Zambales ophiolite hosts along the mantle crust boundary important lenses of massive chromitite (Coto) which are capped by olivine-gabbro or plagioclase-dunite (Violette, 1980; Leblanc and Violette, 1983). Lenses of chromitite and layers of disseminated chromite have been recorded in the ultramafic cumulates of the Theteford ophiolite, Appalachians (Gauthier et al., 1989), the Leka ophiolite, Norway (Barnes et al., 1988) and the Kizildag ophiolite, Turkey (Ragoshay, 1986). In this last ophiolite, there is a 2 km-thick dunite pile (Kuzulyuksek area) comprising about 2000 chromitite layers.

Noble metal content and PGE pattern of the chromitite layers and lenses from the transition zone and the lower part of the magmatic pile (Fig. 10A) are similar to those of podiform chromitite. They are both characterized by a negative slope (Pd/Ir < 1). Among the chromitites of the Thetford ophiolite (Laurent and Kacira, 1987) there is no significative difference between the podiform chromitite bodies of the mantle harzburgite and the chromitite

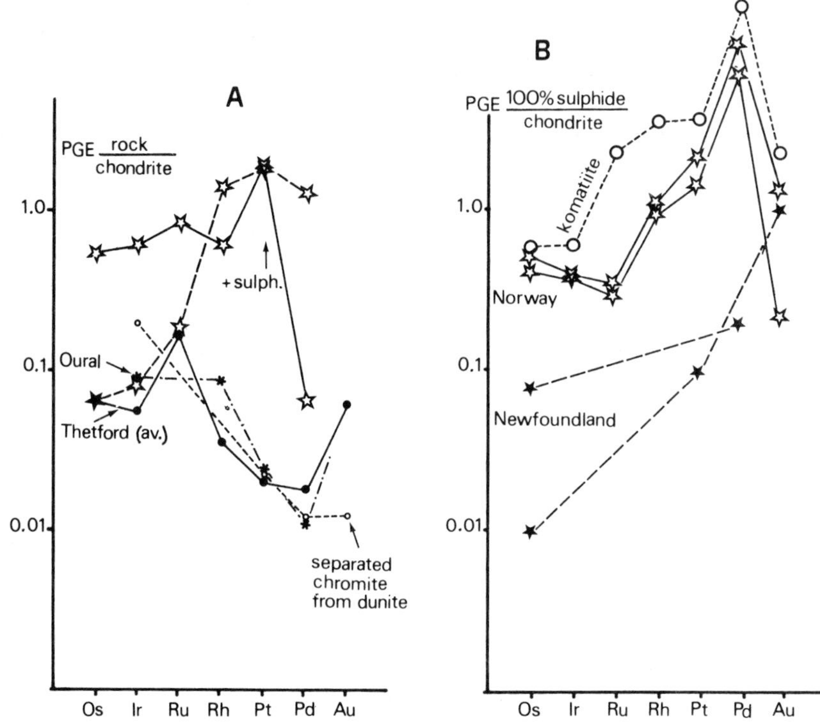

Figure 10. Chondrite normalized diagram for the noble metal contents (A) in chromitite, and (B) sulphide concentrations from the magmatic crustal section. (a) Chromitites from Thetford, Appalachians (Gauthier et al., 1989) and Oural (Page et al., 1982c). The patterns with open stars correspond to PGE-rich chromitites with interstitial sulphides from Thetford (Gauthier et al., 1989). The separated chromite is from a dunite cumulate of Thetford (Oshin and Crocket, 1982). (b) Sulphide concentrations (normalized to 100% sulphide) from Newfoundland (Lydon and Richardson, 1988), and Norway (Barnes et al., 1988). The reference trend for the sulphides of komatiite is from Barnes et al. (1985).

bodies of the ultramafic cumulate sequence. Furthermore similar characteristics are found in the accessory chromite grains separated from a cumulate dunite of Thetford (Oshin and Crocket, 1983). Nevertheless in the Voikar-Syninsky ophiolite, Polar Urals (Page et al., 1983), the chromitite layers are PGE-richer than the podiform chromitite.

Very high PGE contents (3–10 ppm) have been recorded in chromitite lenses from the basal part of the cumulate sequence in Acoje, Philippines (Bacuta et al., 1988); Leka, Norway (Barnes et al., 1988); Thetford, Appalachians (Gauthier et al., 1989). Platinum has been mined as by-product in Acoje, Philippines. In Thetford the PGE-rich chromitites which are located at the dunite-pyroxenite boundary, show a positive slope (Fig. 10A) which corresponds mainly to a strong Rh, Pt, and Pd enrichment. Nevertheless some other PGE-rich chromitites exhibit different PGE-pattern as for example the Leka chromitite which is Os, Ir, and Ru enriched. High sulphur contents, and visible base metal sulphides are present in these PGE-rich chromitites (Bacuta et al., 1988; Barnes et al., 1988) leading to suggest a PGE concentration in a sulphide phase. Inclusions of laurite, Os-Ir-Ru alloys and Fe-Ni sulphides are present in chromite, whereas complex Pt-Pd-Fe-Ni alloys, Fe-Ni-Cu sulphides and Rh-arsenides are located in the interstitial serpentinized silicate matrix (Barnes et al., 1988; Gauthier et al., 1989). A primary magmatic origin , superposed by later alteration related to serpentinization processes, was suggested by Bacuta et al. (1988) for the interstitial sulphides.

3. *Sulphide Mineralization* (*Fig.* 10*B*)

Sulphide mineralization within the cumulate section of ophiolites is uncommon. Disseminated Fe-Ni-Cu-sulphides are present in black dunite lenses in the lower part of the *ultramafic cumulates* of Acoje, Philippines (Paringit, 1975). These sulphides are associated with Pt-Pd-Te-Bi-S-As minerals that occur as a secondary assemblage related to serpentinization processes (Orberger et al., 1988). Analytical data concerning the noble metal content in this sulphide mineralization are not available in literature, but in the chromitite lenses found in the same zone and which comprise accessory sulphides (Bacuta et al., 1988) up to 6 ppm of Pt and 8 ppm of Pd have been reported. Sulphide concentration occurs in three occurrences within the *cumulate gabbros* of the Bay of Island ophiolite, Newfoundland (Lydon and Richardson, 1988). The primary sulphide assemblage is pyrrhotite -chalcopyrite -pentlandite with textures that vary from interstitial to massive sulphide; the Cu/Ni ratios fall within the range of magmatic sulphides (Rajamani and Naldret, 1978). Their PGE contents normalized to 100% sulphide are relatively low (100 ppb Pt and Pd each); the corresponding CN patterns are positively sloped and show a gold enrichment (Fig. 10B). In contrast, the magmatic sulphide mineralization in the sheeted dyke compex of the Karmoy ophiolite, Norway (Barnes et al, 1988), shows very high PGE contents, up to 9 ppm, and low gold contents. The corresponding PGE pattern resembles that of the sulphides of komatiites (Fig. 10B) which is marked by a strong Pd enrichment. Lastly, in many parts of the gabbros of the Troodos ophiolites, Cyprus, there are small sporadic occurrences of Fe-Cu-Co-Ni sulphide mineralization which is mainly in the form of sulphide veins associated with hydrothermal alteration of the surrounding gabbros (Constantinou, 1980). Native gold is present in chalcopyrite (50–200 ppm Au in chalcopyrite), but no PGE data are available for this sulphide mineralization.

B. Mafic Lavas and Exhalative Metalliferous Sediments

1. *Mafic Lavas* (*Fig.* 9*B*)

Because of the very few data available in literature for fresh ophiolitic mafic lavas, ophiolitic and Mid-Oceanic Ridge Basalts will be both considered there. They show a low concentration of noble metal (<20 ppb), and exhibit extremely steep patterns (Pd/Ir = 20–200). Platinum, palladium, and gold are the least depleted noble metals. Two types of lavas can be recognized: (i) MORB which are characterized by lower PGE contents and relatively less fractionated PGE (Pd/Ir = 20–50); (ii) Mg-rich basalts and boninites which show unusually high Pd contents and slightly higher Ir and Au contents compared with MORB, and that consequently exhibit more fractionated PGE patterns (Pd/Ir = 100–250). Sulphur contents in boninite are exceptionally low (<100 ppm) and are in marked contrast to the S content of 800 ppm for the MORB magmas which generally are S-saturated (Hawlyn et al., 1985). The PGE-poor nature of MORB magmas can thus be explained by the separation of an immiscible sulphide phase during the early stages of fractional crystallisation (Hawlyn et al., 1985). Immiscible Fe-Ni-sulphide melts are preserved as globules in submarine basaltic glasses; sulphide contains 44–210 ppb Ir and 1–4 ppm Au (Peach and Mathez, 1988). Equilibrium between the two melts in these basaltic glasses has been nearly approached allowing to estimate that distribution coefficients for Ir and Au between sulphide and silicate melts are both of the order of 10^4 (Peach and Mathez, 1988).

From a comparative study of unaltered and hydrothermally altered Hawaiian basalts, Crocket and Kabir (1988) show that Ir, Pd, and Au are two to three times concentrated in altered basalts suggesting that hydrothermal exhalative activity may efficiently transport noble metals.

2. *Massive Sulphide Deposits*

Convective hydrothermal circulations related to submarine volcanism can deposit, through a stockwork feeder zone, massive sulphide orebodies upon the oceanic floor, in different types of geodynamic setting (Hutchinson, 1973). The Cyprus type, e.g. the ophiolitic type, of massive sulphide deposits is characterised by higher copper grades (0.5–4%). The source-zone for copper is the basaltic pile from which copper was mobilised and transported by hydrothermal flows. Another characteristic of the ophiolitic sub-type of massive sulphide deposit is the presence of moderate gold grades. Most of them contain less than 1.5 ppm Au (Boyle, 1979). In contrast relatively high gold contents (2–5 ppm average) have been recorded in the massive sulphide deposits of Ergani and Küre, Turkey (Helke, 1964; Çatagay et al., 1980); Eastern Metals, Quebec Appalachians (Gauthier, 1985); Al Ajal and Rakah, Oman (Lescuyer et al., 1988). Native gold and electrum inclusions are present in copper sulphide (Ergani and Rakah orebodies). Gold and arsenic contents are positively correlated. In most of these gold-bearing massive sulphide deposits, a genetic relationship with ultramafic rocks may be suggested considering the presence of anomalous cobalt contents (0.1–0.5% Co) and of Cr-minerals. A more obvious evidence is found in the Ergani massive sulphide deposit which lies upon a serpentinite footwall (Bamba, 1874), and in the Eastern Metals orebody which is located along the silicified and carbonatized border of a serpentinite massif (Gauthier, 1985). Thus hydrothermal convective circulations through ultramafic rocks may result in limited mobilization and transportation of Co, Cr, and gold by hydrothermal fluids up to their precipitation with the base metal sulphides upon the oceanic floor. In present-day submarine oxide-sulphide deposits from Atlantis II Deep, Red Sea (Oudin, 1987), the average gold content is approximately 2 ppm. Electrum occurs as inclusions in sulphides, and gold precipitation is related to boiling fluids (Oudin, 1987). In sulphide mounds (Crocket, 1989) the highest gold contents (4 ppm) are present nearby the hot event or in the vicinity of serpentinite rocks.

Very little is known concerning the distribution of PGE in massive sulphide deposits. From one analysis of copper-sulphide stringers which cut the mafic lavas of the Bou Azzer ophiolite, Morocco (current work), the sulphides are PGE-poor (92 ppb) and gold-rich (530 ppb); their PGE pattern (Fig. 11) ressembles PGE patterns of sulphides in serpentinites.

3. *Exhalative Metalliferous Sediments* (*Fig*. 11)
Oxides and/or silicates predominate in the metallifous sediments deposited on the oceanic crust as the result of seawater-basalt interaction. Iron-rich sediments upon the Bou Azzer ophiolite, Morocco (Leblanc, 1981), and upon the Semail ophiolite, Oman, were analysed for PGE and gold (current work). They display low PGE contents (<30 ppb) but relatively high gold contents (21 ppb on average). Palladium (2–13 ppb), and at a lesser extent platinum (1–10 ppb) are the most abundant PGE. The CN curves show a positive slope from Pt to Pd, and gold. The Bou Azzer samples are characterized by negative Ir, positive Rh, and negative Pt anomalies; whereas the Oman samples have a definite positive Ir anomaly and a lower Pd/Ir ratio (1–2).

A very strong platinum concentration (510 ppb on average) has been recorded in the Pacific ferromanganese nodules and crusts (Stumpfl, 1986; Hein et al., 1988). The average Pd, Ru and Ir contents are respectively of 16, 3.5, and 2.5 ppb. Relative to the sea-water content the ferromanganese crusts contain 2.10^6 times more Pt. This strong Pt enrichment has been either ascribed to cosmic spherules incorporated into the ferromanganese crusts or to a selective coprecipitation of Pt with the Fe-Mn hydroxides (Halbach et al., 1989).

V. Subsequent Metamorphism and Weathering

A. Metamorphism

Gold can be expelled from ophiolitic ultramafic rocks during metamorphism, as suggested by Zappettini et al., (1983) for some Variscan gold-bearing veins of the Massif Central, France. In the Western Alps, many gold-bearing

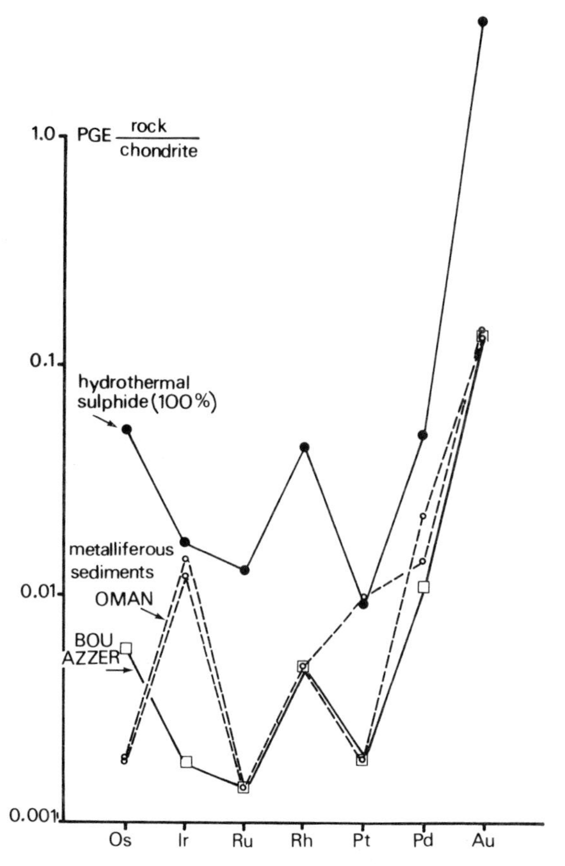

Figure 11. Chondrite normalized PGE contents for hydrothermal copper sulphides, and exhalative metalliferous sediments of the Bou Azzer (Morocco), and Oman ophiolites.

veins are located in ophiolitic formations (Brigo et al., 1977). At a smallest scale of observation, a gold enrichement (0.1 ppm) has been described in the carbonatized shell of ultramafic xenoliths in the anatexites of the French Massif Central (Leblanc and Didier, 1987). No data are available concerning the mobilization of PGE from ophiolitic rocks during metamorphism, although there is compelling evidence for transport and deposition of PGE in hydrothermal solutions (Stumpfl, 1986; Mountain and Wood, 1988).

B. Weathering

The PGE occur as native elements, sulphides, sulphosalts or arsenides. Under oxidizing conditions they are progressively fully reduced in the native state (Bowles, 1988). PGE and gold distribution in lateritic profiles indicates that the noble metals are more or less mobile in the supergene zone. Pd and Au are among the most mobile, but locally Ir and Rh may also be concentrated in the oxidation zone of PGE-bearing sulphide mineralizations (Plimer and Williams, 1988). In Cyprus, some gossans upon massive sulphide deposits have been worked for their gold content (Constantinou, 1980). Nevertheless no PGE concentration has been reported in the weathering and laterite profiles developed upon ophiolites, as for example in New Caledonia (G. Trolly, personal communication).

In contrast, many occurrences of Os-Ir-(Ru) and Pt-Fe alloys have been found in placers derived from ophiolitic complexes. In the North American Cordillera, many gold placers contain Os-Ir-Ru alloys. Most of the PGE produced in the western United States come from placers as a by-product of gold mining (Peterson, 1984). The main placers are located in the Klamath Mountains (Os-Ir > Pt-Fe), and along the foot-hills of the Sierra Nevada (Pt-Fe > Os-Ir). PGE-bearing placer from ophiolitic complexes have been also investigated in NW-China, Tasmania, Borneo and Papua-New Guinea (Mertie, 1969; Cabri, 1982). Most minerals occur as small rounded grains, but platinum-metal nuggets have been reported from the Trinity ophiolite, California (Snetsinger, 1971). There is evidence of reworking and reconcentration of PGE during transport from their ophiolitic source-rocks to placers (Burgath, 1988).

VI. Discussion

The noble metal distribution and fractionation observed in ophiolitic rocks (Fig. 12), from mantle to oceanic floor, may be ascribed to three different processes: (i) partial melting, (ii) magmatic fractionation, and (iii) hydrothermal alteration. These processes will be more or less efficient depending on the behaviour of the minerals that hold the PGE and gold.

A. Mineralogical Control

In ultramafic mantle rocks the PGE and gold are mainly held by Cr-spinel (Fig. 8B) and intergranular sulphides (Mitchell and Keays, 1981; Oshin and Crocket, 1982; Cocherie et al., 1989). In podiform chromitites, two different occurrences of Platinum-Group Minerals are present (Talkington et al., 1984; Legendre and Augé, 1986; Augé, 1987): (1) *inclusions in chromite* of almost exclusively Os-Ir-Ru alloys and sulphides which are commonly associated with small amounts of Fe-Ni-Cu sulphides, even in the same inclusion; (2) *interstitial assemblage* of PGM including Rh-Pt-Pd sulphides, sulpho-arsenides, and arsenides, and of Fe-Ni-Cu sulphides. The interstitial assemblage will be easily melted during partial melting, dissolved during hydrothermal alteration, and oxidized during weathering; whereas the PGE phases which are included within the chromite crystals will be more easily preserved.

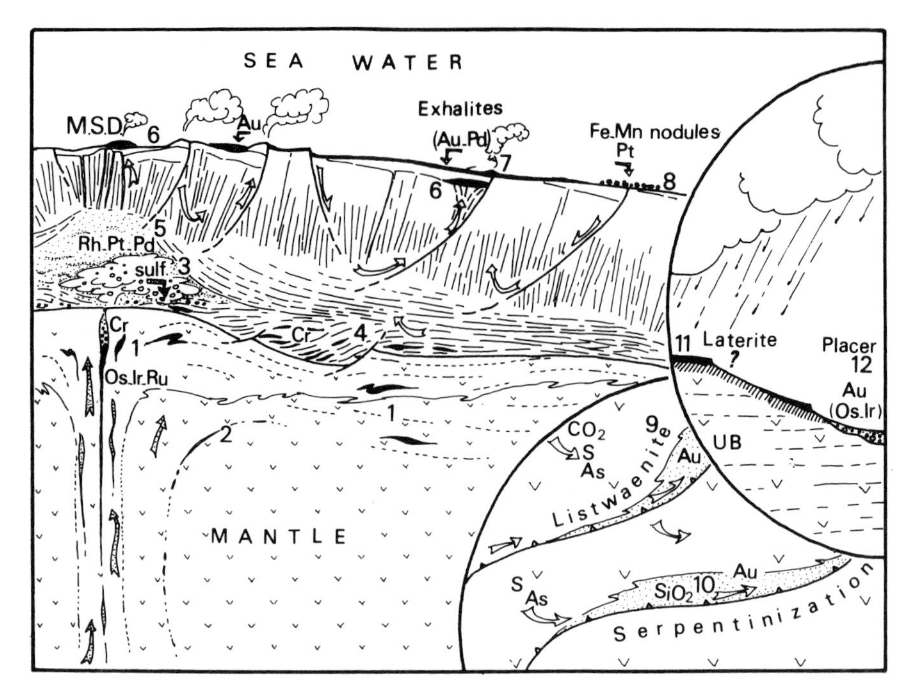

Figure 12. Prospective sketch for possible concentrations of noble metals in an ophiolitic complex: (1) chromitite pods and (2) sulphide-rich rocks in the mantle section; (3) magmatic sulphides in the lower cumulates, with stratiform chromitites (4), or at the top of the magmatic chamber (5); (6) hydrothermal Massive Sulphide Deposits (MSD), and (7) metalliferous sediments (exhalites), with the oceanic Fe-Mn nodules for reference (8); (9) carbonatized (listwaenite) and (10) silicified serpentinites along tectonic contacts; (11) laterite profiles on weathered ultramafic rocks, and placer deposits (12).

B. Partial Melting

The less depleted mantle rocks exhibit a nearly flat and unfractionated PGE pattern (Fig. 1). From lherzolite to dunite the PGE, Au, Cu, and S contents decrease (Table I). Palladium is strongly depleted whereas iridium remains nearly unchanged resulting in a strong variation in Pd/Ir ratio, from 2.8–1 to .05–.01. This evolution is consistent with the removal of a partial melt enriched in Pd, leaving a refractory residuum relatively enriched in Ir and depleted in Pd.

Palladium is mainly held by interstitial sulphide grains, while the Ir-minerals are mainly included within the Cr-spinel accessory crystals. At low degrees of partial melting the intergranular sulphides well be easily melted but the resulting sulphide liquid remains more or less in situ as a restite phase. At less 25% partial melting is required to dissolve all sulphide liquide in the silicate melt (Sun, 1982; Barnes et al., 1985). Differences in solubilities may also account for the PGE fractionation (Makovicky et al., 1986; Amosse et al., 1987). The nearly chondritic PGE pattern of some harzburgites can be attribed to the persistence of a residual sulphide phase or to re-equilibrations with percolating magmas. At more important degrees of partial melting the removal of a S-Cu-Pd-rich silicate melt leaves a S-Cu-Pd-depleted residual dunite.

C. Magmatic Fractionation

By using a mass balance equation it may be easily shown that the PGE content in primitive mantle rocks cannot be matched by the PGE content of magmatic rocks (lavas and cumulates) and ultramafic residue, assuming a reasonnable degree of partial melting (F = 30%). The main problem results from the very low PGE content of the basaltic lavas. They should have lost large amounts of PGE somewhere between their mantle source zone and their emplacement at surface, either during chromite fractionation (Hertogen et al., 1980), or sulphide segregation (Hawlin et al., 1985). The mass balance needs to involve the formation of chromitite bodies and sulphide concentrations in the following calculated proportions: 0.13 and 0.06% wt respectively. These proportions are roughly consistent with observed abundance of chromite and sulphide in ophiolites. The sulphide concentration of the Karmoy ophiolite, Norway (Barnes et al., 1988) could represent (Fig. 10B) a PGE-rich sulphide melt segregated from MORB-type basalts (Barnes et al., 1988).

At low sulphur fugacities and high-temperature (1200–1300°C), Os-Ir-Ru will concentrate into alloys, leaving an Os-Ir-Ru depleted melt. The chromite growing grains may include these minute Os-Ir-Ru euhedral crystals. The common association in chromite of Os-Ir-Ru alloys, Os-Ir-Ru sulphides, and Fe-Ni-Cu sulphides, suggests a possible collection of PGE from the magma by a sulphide phase. The concentration of PGE in solid solution in magmatic sulphides has been reported by numerous authors; they suggest (Naldrett et al., 1979) that the PGE will be collected from the silicate melt by droplets of an immiscible sulphide liquid segregated at high temperature from the magma when sulphur saturation is reached. Partition coefficients for the PGE between silicate and sulphide liquid are very high, from 10^3 to 10^5 (Campbell and Barnes, 1984). Almost all the PGE will be collected from the silicate melt, preserving the same Pd/Ir ratio in the sulphide phase (Barnes et al., 1985). For example, a basaltic liquid which has lost iridium by crystal fractionation and which is consequently Pd enriched, will give an Ir-poor and Pd-rich sulphide phase.

The PGE-rich chromitites could result from a PGE-rich magma generated by high degrees of partial melting from a PGE-enriched mantle source (metasomatism). Nevertheless the PGE-enrichment in chromite should result in a more simple way from a local concentration, during the chromite crystallization, of Os-Ir-Ru crystals and droplets of PGE-bearing sulphide liquid if sulphur saturation occured early. In a turbulent magmatic flow the Os-Ir-Ru crystals and the sulphide droplets can be mechanically concentrated, as suggested by A. Nicolas (pers. communication), following the ellutriation model of Lago et al. (1982) for the genesis of the chromitite pods. The recognition of abundant hydrous and Na-rich silicates included within chromite grains has led some workers to emphasize the influence of a fluid phase in the formation of chromite (Johan et al., 1983; Johan, 1986); such a phase should play an efficient role in PGE concentration.

The PGE are strongly fractionated among dunites, pyroxenites, and gabbros of the crustal section. The gabbros are PGE-depleted (<10 ppb) and display high Pd/Ir ratio (up to 120). The highest PGE contents are often related to accessory magmatic Fe-Cu-Ni sulphides (Lydon and Richardson, 1988), either in chromite-rich dunites just upon the transition zone, or in pyroxenite layers just below the gabbroic pile. A similar location at the base of the plagioclase-cumulus rocks is observed for the platiniferous horizons of the stratiform magmatic complexes.

For gold the mass balance calculation implies a metasomatic gold enrichment for ophiolitic mantle rocks or/and a gold extraction from the crustal section, probably by late hydrothermal processes.

D. Hydrothermal Processes

Serpentinization has not significantly changed the noble metal content of ultramafic rocks, except for a gold enrichment. In sulphide and/or arsenide bearing serpentinites the PGE content is 3 to 5 times higher, and gold is still more concentrated up to 100 times average serpentinite content. Gold is easily mobilized from the accessory interstitial sulphides and magnetite grains in serpentinite during low temperature (250°C) hydrothermal processes related to serpentinization; from the stability field of the gold-bearing minerals and associated gangue minerals, gold was probably transported as thiocomplexes (Buisson and Leblanc, 1986). Economic gold grades are present in carbonatized serpentinite (e.g. listwaenite), in sulphide and/or arsenide-rich zones (Buisson and Leblanc, 1987).

Gold enrichments are also found in the hydrothermal deposits of base metals massive sulphides, and in the metalliferous exhalative sediments upon the ophiolitic oceanic floor. Very little is known about the PGE contents in these exhalative deposits; but Pt and Pd may be mobilized and transported as chlorides in low-temperature hydrothermal fluids (Stumpfl, 1986).

VII. Conclusions

1. The Platinum-Group Elements and gold are strongly fractionated from their mantle source-rock during partial melting and subsequent magmatic differentiation. Consequently the various ophiolitic rock-types exhibit different noble metal contents and specific PGE chondrite normalized patterns.
2. The geochemical behaviour of the PGE and gold is strongly dependant

on the presence of sulphides which are the main noble metal bearing minerals. During partial melting the interstitial sulphides will be easily liquefied, then dissolved in the silicate melt. During magmatic differentiation Os-Ir-Ru alloys and sulphides will first crystallise, at low sulphur fugacity and high temperature. They can be included in growing chromite crystals leaving a Rh-Pt-Pd-Au enriched liquid; at higher sulphur fugacities an immiscible sulphide liquid can segregate and collect almost all noble metals, leaving a depleted silicate liquid.

3. Two main types of magmatic concentration of PGE are present in ophiolite: in chromitite and in sulphide deposits. Os, Ir, and Ru are strongly concentrated in chromitite, as mineral inclusions in chromite. A mechanical concentration of these PGM in a turbulent flowing magma should explain the PGE enrichment of some chromitite pods (>1 ppm PGE). Sulphide concentrations are present mainly in the lower ultramafic part of the crustal section; they are Rh, Pt, Pd, and gold-rich. These sulphide occurrences can be a new target for noble metal exploration in ophiolites.
4. Gold, and at a lesser extent Pd, and Pt, are easily mobilized by hydrothermal circulations. During serpentinization gold is concentrated up to economic grades mainly in carbonatized serpentinite (gold-bearing listwaenite) and in sulphide and/or arsenide rich-zones in serpentinite. Massif sulphide deposits in ophiolites are often characterized by relatively high gold contents; gold was probably extracted from ultramafic mantle rocks by hydrothermal convective systems. Lastly slight Pt, Pd, and gold enrichments are observed in metalliferous exhalative sediments. All sedimentary exhalative deposits are possible targuets for noble metals exploration.

Acknowledgments

I am endebted to G. Ceuleneer, C. Dupuy, and A. Nicolas for their helpfull comments and constructive discussion, and I thank the anonymous reviewers for their critical and pertinent comments.

References

Amossé, J., Fischer, W., Allibert, M., et Piboule, M., 1986. Méthode de dosage d'ultratraces de platine, palladium, rhodium et ordans les roches silicatées par spectrométrie d'absorption atomique électronique., Analysis, 14(1): 26–31.

Augé, T., 1987. Chromite deposits in the Northern Oman ophiolite: mineralogical contrainsts., Mineral. Deposita, 22: 1–10.

Bacuta, G.C., Gibbs, A.K., Kay, R.W., and Lipin, B.R., 1988. Platinum-Group Element abundance in chromite déposits of the Acoje. In: "Geo-Platinum 87", Prichard, H.M., Potts, P.J., Bowles, J.F.W., and Cribb, S.J., Edit., Elsevier Applied Science, Publ., London, U.K., pp. 387–382. Bamba, T., 1974. Copper deposits in the Ergani mining district, Southeastern Turkey. Mining Geology, 24: 389–400.

Barnes, S.J., Naldrett, A.J., and Gorton, M.P., 1985. The origin of the fractionation of platinum-group elements in terrestrial magams., Chem. Geol., 53: 303–323.

Barnes, S.J., Boyd, R., Korneliussen, A., Nilsson, L.P., Pedersen, R.B., and Robins, B., 1988. The use of normalization and metal ratios in discriminating between the effects of partial melting, crystal fractionation, and in: "Geoplatinum 87", Prichard, H.M., Potts, P.J., Bowles, J.F.W., and Cribb, S.J. Edit., Elsevier Applied Science, Publ., London, U.K., pp. 113–144.

Becker, E., and Agiorgitis, G., 1978. Iridium, osmium, and palladium distribution in rocks of the Troodos complex, Cyprius., Chem. Erde Bd., 37: 302–306.

Begou, P., Amossé, J., Fischer, W., et Piboule, M., 1989. Distribution des éléments du groupe du platine (PGE) dans les péridotites massives à spinelle de l'étang de Lherz (Ariège, France): résultats préliminaires., C. R. Acad. Sci. Paris, 309: 1177–1182.

Boudier, F., and Nicolas, A., 1985. Harzburgite and lherzolite subtypes in ophiolitic and oceanic environment., Earth Planet. Sci. Let., 76: 84–92.

Bowls, J.F.W., 1988. Further studies of the development of Platinum-Group Minerals in the laterites of Freetown layered complex, Sierra Leone. In: "Geo-Platinum 87", Prichard, H.M., Potts, P.J., Bowles, J.F.W., and Cribb, S.J., Edit., Elsevier Applied Science, Publ., London, U.K., pp. 273–280.

Boyle, R.W., 1979. The geochemistry of gold and its deposits. Geological survey of Canada Bulletin, 280, 584 p.

Brigo, L., Ferrario, A., and Zuffardi, P., 1977. Gold distribution in the Alps. In: "Metallogeny and plate tectonics in the NE Mediterranean".Jankovic, S., Edit., Jugoslovensko udruzenje Nauka; drustvo, Belgrad, Yugolsavia, pp. 429–438.

Brown, M.A., 1982. Chromite deposits and their ultramafic host rocks in the Oman ophiolite. Ph. D. Thesis, Open Univ., Milton Keynes, U.K., 264 p.

Buisson, G., and Leblanc, M., 1986. Gold bearing listwaenites (carbonatized ultramafic rocks) in ophiolite complexes. In: "Metallogeny of basic and ultrabasic rocks", M.J. Gallagher, R.A. Ixer, C.R. Neary and H.M. Prichard Edit., The institution of Mining and Metallurgy Publ., London, U.K., pp. 121–132

Buisson, G., and Leblanc, M., 1987. Gold in mantle peridotites from Upper Proterozoic ophiolites in Arabia, Mali, and Morocco., Econ. Geol., 82: 2091–2097.

Burgath, K.P., 1988. Platinum-Group Minerals in ophiolitic chromitites and alluvial placer deposits, Mineratus-Bobaris area, Southeast Kalimantan. In: "Geoplatinum 87", Prichard, H.M., Potts, P.J., Bowles, J.F.W., and Cribb, S.J., Edit., Elsevier Applied Science, Publ., London, U.K., pp. 383–404.

Cabri, L.J., 1982. Classification of platinum-group element deposits with reference to the Canadian Cordillera. In: "Precious metals in the Northern Cordillera", A.A. Levinson, Edit., The association of Exploration Geochemists Pub., Ontario, Canada, spec. publ., no. 10: 21–32.

Campbell, I.H., and Barnes, S.J., 1984. A model for the geochemistry of the platinum-group elements in magmatic sulphide deposits., Canad. Mineral., 22: 151–160.

Cassard, D., 1980. Structure et origine des gisements de chromite du Massif du Sud (Ophiolites de Nouvelle-Calédonie). Thèse Doct. 3° cycle, Univ. Nantes, France, 239 p.

Catagay, A., Pehlivanoglu, H., and Altun, Y., 1980. Kure piritli bakir yataklarini kobasl-altin mineralleri. Maden tetrik ve Arama Enstitusu Dergisi,Ankara, 93/94: 110–117.

Ceuleneer, G., Nicolas, A., and Boudier, F., 1988. Mantle flow patterns at an oceanic spreading centre: The Oman peridotites record.Tectonophysics, 151, 1–4: 1–26.

Cocherie, A., Augé, T., and Meyer, G., 1980. Geochemistry of the platinum-group elements in various types of spinels from the Vourinos ophiolitic complex, Greece., Chem. Geol., 77: 27–39.

Constantinoù, G., 1980. Metallogenesis associated with the Troodos ophiolite. In: "Ophiolites", Proceedings Int. Ophiolite Symp. 1979, A. Panayiotou Edit., Cyprus Geol. Survey Publ., pp. 663–674.

Crocket, J.H., and Kabir, A., 1988. PGE in Hawaiian basalt: implications of hydrothermal alteration on PGE mobility in volcanic fluids. In: "Geoplatinum 87", Prichard, H.M., Potts, P.J., Bowles, J.F.W., and Cribb, S.J., Edit., Elsevier Applied Science, Publ., London, U.K., pp. 259–260.

Dupuy, C., Dostal, J., and Leblanc, M., 1981. Distribution of copper and gold in ophiolites from New Caledonia. Canad. Mineral., 19: 225–232.

Economou, M.I., 1986. Platinum-Group Elements (PGE) in chromite and sulphide ores within the ultramafic zone of some Greek ophiolite complexes. In: "Metallogeny of basic and ultrabasic rocks", M.J. Gallagher, R.A. Ixer, C.R. Neary and H.M. Prichard Edit., The institution of Mining and Metallurgy Publ., London, U.K., pp. 441–454.

Economou, M.I., and Naldrett, A.J., 1984. Sulfides associated with podiform bodies of chromite at Tsangli, Eretria, Greece., Mineral. Deposita, 19: 289–297.

Fischer, W., Amossé, J., and Leblanc, M., 1988. PGE distribution in some ultramafic rocks and minerals from the Bou Azzer ophiolitic complex (Morocco). In: "Geoplatinum 87", Prichard, H.M., Potts, P.J., Bowles, J.F.W., and Cribb, S.J., Edit., Elsevier Applied Science, Publ., London, U.K., pp. 199–210.

Fominykh, V.G., and Kvostova, V.P., 1970. Platinum content of Ural dunites., Doklady Akad. Nauk S.S.S.R., 191: 443–445.

Foose, M.P., 1986. Setting of a magmatic sulfide occurrence in a dismembred ophiolite, southewestern Oregon. US Geol. Survey Bull., 1626–A, 21 p.

Foose, M.P., Economou, M., and Panayiotou, A., 1985. Compositional and mineralogic constraints on the genesis of ophiolite hosted nickel mineralization in the Pevkos area., Mineral. Deposita, 20: 234–240.

Gauthier, M., 1988. Synthèse métallogénique de l'Estrie et de la Beauce (secteur Sud). Ministère Energie Ressources, Quebec, Canada, rapport MB 85–20, 74 p.

Gauthier, M., Corriveau, L., Trottier, J., Laflamme, J.H.G., et Bergeron, M., 1989. Chromitites platinifères des complexes ophiolitiques de l'Estrie-Beauce, Appalaches du Sud du Québec. Mineral. Deposita, (in press).

Halbach, P., Kriete, C., Prause, B., and Puteanus, D., 1985. Mechanisms to explain the platinum concentration in ferromanganese seamount crusts., Chem. Geol., 76: 95–106.

Hamlyn, P.R., Keays, R.R., Cameron, W.E., Crawford, A.J., and Waldron, H. M., 1985. Precious metals in magnesian low-Ti lavas: implications for metallogenesis and sulfur saturation in primary magmas., Geoch. Cosmoch. Acta, 49: 1797–1911.

Hein, J.R., Schwab, W.C, and Davis, A.A., 1988. Cobalt- and platinum-rich ferromanganese crusts and associated substrate rocks from the Marshall Islands., Marine Geol., 78 (3–4): 255–283.

Helke, A., 1964. Die kupferlagerstatte ergani Maden in der Turkei., N. Jb. Min. Abh., Bd, 101 (H3): 233–270.

Henderson, F.B., 1969. Hydrothermal alteration and ore deposition in serpentinite-type mercury deposits., Econ. Geol., 64: 489–499.

Hertogen, J., Janssen, M.J., and Palme, M., 1980. Trace elements in ocean ridge basalt glasses: implications for fractionation during mantle evolution and petrogenesis., Geochim. Cosmochim. Acta, 44: 2125–2143.

Hulin, C.S., 1950. Results of study of Ni-Pt ores and concentrates: Acoje Mining Company, Philippine Island., Philippine Geologist, 4: 11–23.

Hutchinson, R.W., 1973. Volcanogenic sulfide deposits and their metallogenic significance., Economic Geology, 68: 1223–1246.

Jacob, M., 1985. Etude géologique, minéralogique et géochimique des anomalies en As, Sb, Au, et W et indices minéralisés liés aux fractures régionales de la cote Sud-Est de la Nouvelle-Calédonie. Thèse Doct. 3° cycle, Univ. Paul Sabatier, Toulouse, France, 167 p.

Johan, Z., 1986. Chromite deposits in the Massif du Sud ophiolite, New Caledonia; genetic considerations. In: "Chromites" UNESCO's IGCP-197 Project" Metallogeny of ophiolites", Edit. Theophrastus, Athens, pp. 311–339.

Johan, Z., Dunlop, H., Le Bel, L., Robert, J.L., and Volfinger, M., 1983. Origin of chromite deposits in ophiolitic complexes: evidence for a volatile and sodium-rich reducing fluid phase., Forstchr. Miner., 61: 105–107.

Lago, B., Rabinowicz, M., and Nicolas, A., 1982. Podiform chromite orebodies: a genetic model., Jour. Petrology, 23: 103–125.

Laurent, R., and Kacira, N., 1987. Chromite deposits in the Appalachian ophiolites. In: "Evolution of chromium ore fields", C.W. Stowe edit., Van Nostrand Reinhold Co., New York, pp. 169–193.

Leblanc, M., 1981. Late Proterozoic ophiolite of Bou-Azzer (Morocco): evidence for Pan-African plate tectonics.In: "Precambrian Plate Tectonics", Kroner A. (ed), Elsevier publ., Amsterdam, pp. 435–451.

Leblanc, M., and Billaud, P., 1982. Cobalt arsenide orebodies related to an Upper Proterozoic ophiolite: Bou Azzer (Morocco). Econ. Geol., 77: 162–175.

Leblanc, M., and Violette, J.F., 1983. Distribution of aluminum-rich and chromium-rich chromite pods in ophiolite peridotites., Econ. Geol., 78: 293–301.

Leblanc, M., and Didier, J., 1987. Enclaves ultrabasiques carbonatisées avec traces d'or dans les anatexites du Haut-Allier (France)., Bull. Mineral., 110: 359–371.

Leblanc, M., and Fischer, W., 1990. Gold and Platinum-Group Elements in cobalt-arsenide ores: hydrothermal concentration from a serpentinite source-rock (Bou Azzer, Morocco)., Mineral. Petrol., 42: 197–209.

Leblanc, M., Gervilla, F., and Jedwab, J., 1990. Noble metals segregation and fractionation in magmatic ores from Ronda and Beni Bousera lherzolite massifs (Spain, Morocco)., Mineral. Petrol., 42: 233–248.

Legendre, O., 1982. Mineralogie et géochimie des platinoïdes dans les chromitites ophiolitiques. Thèse Doct. 3° Cycle, Univ. Paris 6, 171 p. Legendre, O., and Augé, T., 1986. Mineralogy of platinum-group mineral inclusions in chromitites from different ophiolitic complexes. In: "Metallogeny of basic and ultrabasic rocks", Gallagher, M.J., Ixer, R.A., Neary, C.R., and Prichard, H.M., edit., The institution of Mining and Metallurgy, Publ., London, U.K., pp. 361–372.

Lescuyer, J.L., Oudin, E., and Beurrier, M., 1988. Reviews of the different types of mineralization related to the Oman ophiolitic volcanism. Proceedings of the Seventh Quadriennial IAGOD Symposium. Schweizerbart'sche Verlagsbuch handlung (Nägele u. Obermiller), Edit., Stuttgahrt, DFR, pp. 489–500.

Lorand, J.P., 1988. Fe-Ni-Cu sulfides in tectonite peridotites from the Maqsad district, Sumail ophiolite, southern Oman: implication for the origin of the sulfide component in the oceanic upper mantle., Tectonophysics, 151 (1/4): 57–74.

Lorand, J.P., 1989. Abundance and distribution of Cu-Fe-Ni sulfides, sulfur, copper and platinum-group elements in orogenic type spinel lherzolite massifs of Ariége (northeastern Pyrénées, France)., Earth Planet. Sci. Letters, 93: 50–64.

Lydon, J.W., and Richardson, D.G., 1988. Distribution of PGE in sulphides of the Bay of Islands ophiolite complex, Newfoundland. In: "Geo-platinum 87", Prichard, H.M., Potts, P.J., Bowles, J.F.W., and Cribb, S.J., Edit., Elsevier Applied Science Publ., London, U.K., pp. 251–252.

Makovicky, M., Makovicky, E., and Rose-Hansen, J., 1986. Experimental studies on the solubility and distribution of platinum group elements in base-metal sulphides in platinum deposits. In: "Metallogeny of basic and ultrabasic rocks", Gallagher, M.J., Ixer, R.A., Neary, C.R., and Prichard, H.M., Edit., The institution of Mining and Metallurgy Publ., London, U.K.

Mertie, J.B., 1969. Economic geology of the platinum metals., U.S. Geol. Survey. Prof. Paper, 630, 120 p.

Mitchell, R.H., and Keays, R.R., 1981. Abundance and distribution of gold, palladium and iridium in some spinel and garnet lherzolites. Implications for the nature and origin of precious metal-rich intergranular components in the upper mantle., Geochem. Cosmoch. Acta, 45: 2425–2442.

Moring, B.C., Page, N.J., and Oscarson, R.L., 1988. Platinum Group Element mineralogy of the Pole Coral podiform chromite deposit, Rattlesnake Creek terrane, Northern California. In: "Geoplatinum 87", Prichard, H.M., Potts, P.J., Bowles, J.F.W., and Cribb, S.J., Edit., Elsevier Applied Sciences, Publ., London, U.K., pp. 257–258.

Mountain, B.W., and Wood, S.A, 1988. Solubility and transport of platinum-group elements in hydrothermal solution: thermodynamic and physical chemical constraints. In: "Geoplatinum 87", Prichard, H.M., Potts, P.J., Bowles, J.F.W., and Cribb, S.J., Edit., Elsevier Applied Science, Publ.,London, U.K., pp. 57–82.

Naldrett, A.J., and Duke, J.M., 1986. Pt metals in magmatic sulfide ores; the occurrence of these metals is discussed in relation to the formation and importance of these ores., Science, 208: 1417–1484.

Nicolas, A., and Prinzhofer, A., 1983. Cumulative or residual origin for the transition zone in ophiolites: structural evidence., Jour. Petrology, 24: 188–206.

Oberger, M., Friedrich, G., and Woermann E., 1988. Platinum-group element mineralization in the ultramafic sequence of the Acoje ophiolite block, Zambales, Philippines. In: "Geoplatinum 87", Prichard, H.M., Potts, P.J., Bowles, J.F.W., and Cribb, S.J., Edit., Elsevier Applied Science, Publ., London, U.K., pp. 361–380.

Odin, E., 1987. Trace element and precious metal concentrations in East Pacific Rise, Cyprus and Red Sea submarine sulfide deposits.In: "Marine minerals", P.G. Teleki et al. Edit., D. Reidel Publishing Company, 349–362.

Oshin, I.O., and Crocket, J.H., 1982. Noble metals in Thetford Mines ophiolites, Quebec, Canada. Part I: distribution of gold, iridium, platinum, and palladium in the ultramafic and gabbroic rocks., Econ. Geol., 77: 1556–1570.

Page, N.J., Cassard, D., and Haffty, J., 1982a. Palladium, platinum, rhodium, ruthenium, and iridium in chromitites from the Massif du Sud and Tiebaghi Massif, New Caledonia., Econ. Geol., 77: 1571–1577.

Page, N.J., Pallister, J.S., Brown, M.A., Smewing, J.D., and Haffty, J., 1982b. Palladium, platinum, rhodium, iridium, and ruthenium in chromite-rich rocks from the Samail ophiolite, Oman., Canad. Mineral., 20: 537–548.

Page, N.J., Aruscavage, P.J., and Haffty, J., 1982c. Platinum Group Elements in rocks from the Voikar-Syninsky ophiolite complex, Polar Urals, U.S.S.R. Mineral. Deposita, 18: 443–455.

Page, N.J., Gruenewaldt, G., Haffty, J., and Aruscavage, P.J., 1983. Comparison of platinum, palladium, and rhodium distribution in some layered intrusions with special reference to the late differentiates (upper zones) of the Bushveld complex, South Africa., Econ. Geol., 77: 1405–1418.

Page, N.J., and Talkington, R.W., 1984. Palladium, platinum, rhodium, ruthenium and iridium in peridotites and chromitites from ophiolite complexes in Newfounfland., Canad. Mineral., 22: 137–149.

Page, N.J., Engin, T., Singer, D.A., and Haffty, J., 1984. Distribution of platinum-group elements in the Bati Kef chromite deposit, Guleman-Elazig area, Eastern Turkey., Econ. Geol., 79: 177–184.

Paringit, R.V., 1975. Nickel sulfide deposits and exploration works at Acoje mine, Zambales province, Philippines., Geol. Soc. Philippines Jour., 29: 16–27.

Peach, C.L., and Mathez, E.A., 1988. Gold and iridium in sulphides from sumarine basalt glasses. In: "Geo-Platinum 87", Prichard, H.M., Potts, P.J., Bowles, J.F.W., and Cribb, S.J.Edit., Elsevier Applied Science Publ., London, U.K., pp. 409–410.

Peterson, J.A., 1984. Metallogenic maps of the ophiolite belts of the Western United States., U.S. Geol. Survey, Map I-1505, 16 p.

Plimer, I.R., and Willians, P.A., 1988. New mechanism for the mobilization of the Platinumt-Group Elements in the supergen zone. In: "Geo-Platinum 87", Prichard, H.M., Potts, P.J., Bowles, J.F.W., and Cribb, S.J. Edit., Elsevier Applied Science Publ., London, U.K., pp. 83–92.

Prichard, H.M., Neary, C.R., and Potts, P.J., 1986. Platinum group minerals in the Shetland ophiolite. In: "Metallogeny of basic and ultrabasic rocks" Gallagher, M.J., Ixer, R.A., Neary, C.R., and Prichard, H.M. (Ed), The Institution of Mining and Metallurgy Publ., London, U.K., pp. 395–414.

Ragoshay, M., 1986. Les chromites et leurs gisements dans les complexes ophiolitiques de la chaîne du Taurus (Turquie), comparaison avec les gisements omanais. Thèse Doct. Univ. Louis Pasteur, Strasbourg, France, 206 p.

Rajamani, V., and Naldrett, A.J., 1978. Partitioning of Fe, Co, Ni and Cu between sulphide liquid and basaltic melts and the composition of Ni-Cu sulfide deposits., Econ. Geol., 73: 82–93.

Snetsinger, K.G., 1971. A platinum-metal nugget from Trinity Country, California., American Mineral., 56: 1101–1105.

Stanger, G., 1985. Silicified serpentinite in the Semail nappe of Oman., Lithos, 18: 13–22.

Stockman, H.W., 1982. Noble metals in the Ronda and Josephine peridotites. Ph D., thesis, Massachusett Institute of Technology, Cambridge, Mass., USA (unpubl) 180 p.

Stumpfl, E.F., 1986. Distribution, transport and concentration of platinum group elementsIn: "Metallogeny of basic and ultrabasic rocks", Gallagher, M.J., Ixer, R.A., Neary, C.R., and Prichard, H.M. (Ed), The Institution of Mining and Metallurgy Publ., London, U.K., pp. 379–394.

Sun, Shen-Su., 1982. Chemical composition and origin of the earth's primitive mantle., Geochim. Cosmochim. Acta, 46: 179–192.

Talkington, R.W., Watkinson, D.M., Whittaker, P.J., and Jones, P.C., 1984. Platinum-group minerals and other solid inclusions in chromite of ophiolite complexes: occurrence and petrological significance., Tschermaks Mineral. Petrol. Mitt., 32: 285–301.

Talkington, R.W., and Watkinson, D.M., 1986. Whole rock platinum-group element trends in chromite-rich rocks in ophiolitic and stratiform igneous complexes. In: "Metallogeny of basic and ultrabasic rocks", Gallagher, M.J., Ixer, R.A., Neary, C.R., and Prichard, H.M. Edit., The Institution of Mining and Metallurgy Publ., London, U.K., pp. 427–440.

Violette, J.F., 1980. Structure des ophiolites des Philippines (Zambales et Palawan) et de Chypre. Ecoulement asthénosphérique sous les zones d'expansion. Thèse Doct. 3° Cycle, Univ. Nantes, France, 152 p.

Wolf, R., and Agiorgitis, G., 1978. On an unusual platinum element enrichment in chromites from Skyros Island, Greece., News Jahrb. Mineral. Monatshefte, 1: 39–41.

Zappetini, E., Picot, P., and Sabourdy, G., 1983. Nouvelles données sur le gisement aurifère du Châtelet (Massif Central français), liaison génétique probable entre l'or et les roches ultrabasiques., C. R. Acad., Sci. Paris, 297: 351–354.

Chromite-Rich and Chromite-Poor Ophiolites: The Oman Case

A. NICOLAS* and H. AL AZRI**
* *Laboratoire de Tectonophysique, URA 1370 CNRS, Université Montpellier II, Pl. Bataillon – 34060 Montpellier, France*
***Ministry of Petroleum and Minerals – P.O. Box 551, Muscat, Oman*

Abstract

Chromite deposits in Oman belong dominantly to the concordant structural type, meaning that these have been intensely deformed by plastic flow and tectonically rotated to become parallel to the peridotite foliation. This occurred soon after their formation within the transition zone below the ridge of origin. Subconcordant and discordant pods are also present. The latter have preserved delicate magmatic structures showing that they have only been little deformed after their formation in melt-carrying conduits. Regionally, the chromite deposits have been dominantly found, so far, in restricted areas whereas large areas of Oman seem to be devoid of deposits. Maqsad, one of the largest chromite districts, was also an area of mantle diapirism below the ridge of origin. This association is well explained if it is considered that most of mantle melt feeding the crust at ridges is expected to be delivered through such diapirs. Although Oman is the largest and best exposed ophiolite in the world, it seems to be comparatively poor in chromite due to the spreading situation. In the Lherzolite Ophiolite Type (LOT), thought to be derived from slow spreading ridges, the chromite deposits are absent as a result of chromium being retained in mantle diopside during partial melting. The chromite deposits are restricted to the Harzburgite Ophiolite Type (HOT) in which chromium has passed into the melt. Although Oman belongs to the HOT group, it seems to have been a particularly fast spreading HOT. It is suggested that in such a situation the transition zone, which is the level where the chromite normally precipitates from the melt, due to temperature drop, could have remained too hot to allow for abundant chromite formation.

Introduction

Chromite deposits are a normal ingredient of ophiolite complexes and, with the Stratiform Complex deposits, they represent the main chromium re-

Tj. Peters et al. (Eds), Ophiolite Genesis and Evolution of the Oceanic Lithosphere, 261–274.

source. However, a closer look at the situation in ophiolites reveals that they are not equally distributed among ophiolite massifs. In some ophiolites, chromite occurrences have never been observed, contrasting with the majority of ophiolites where chromite is ubiquitously distributed, provided that is exposed the transition zone between layered gabbros and tectonic peridotites where chromite deposits are generally located. This discrepancy has been related to the nature of the ophiolites (Nicolas, 1986): chromite is generally absent in those ophiolites where the peridotite section is lherzolitic, belonging to the Lherzolite Ophiolite Type (LOT) and, on the contrary present in those where this section is harzburgitic, (belonging to the Harzburgite Ophiolite Type, HOT). Chromite occurrence is indeed one of the characteristics proposed by Boudier and Nicolas (1985) to distinguish LOT and HOT. As noted by Roberts (1988), with increasing melting, chromium contained initially into the spinel phase is retained by this phase which becomes a chromitite. We conclude that the source of chromite during partial melting is the diopside. Chromite deposits in LOT would be absent because in such peridotites the chromium hosted by clinopyroxene in fertile mantle lherzolite, is still largely retained by this mineral after the melting episode related to ophiolite generation. On the other hand in HOT peridotites, the clinopyroxene has been molten out with, as a result, chromium having been incorporated into the basaltic melt. This opens the possibility for the later segregation of chromium into chromite deposits during basalt crystallization.

With respect to this division, the Oman ophiolite represents a noteworthy exception. Although it belongs to the harzburgitic subtype and has the largest and best exposures for an ophiolite in the world, so far only a comparatively limited amount of chromite occurrences have been found. It is the object of this paper to present the first general synthesis on chromite occurrences in the Oman ophiolite and to discuss why this ophiolite makes an exception to the general rule exposed above.

Structure of Chromite Bodies and Mechanisms of Formation

Cassard et al. (1981) suggest that the chromite bodies form, within the transition zone of ophiolites, as discordant pods with respect to the attitude of the high temperature flow plane and foliation in the surrounding mantle peridotite; they are next tectonically transposed into subconcordant and concordant lenses by the very large plastic flow and related foliation recorded in the peridotites just below the Moho (Fig. 1). The discordant pods are themselves generated, locally, inside the dikes feeding the overlying crust with melt, provided 1) that chromite can precipitate from the melt, and 2) that a chromite trap is created along a segment of the dike. The first condition could be fulfilled in the transition zone just below the crustal layered gabbros where a temperature drop and an oxygen fugacity increase would allow for appearance of chromite on the basalt liquidus; the increase in oxygen fugacity

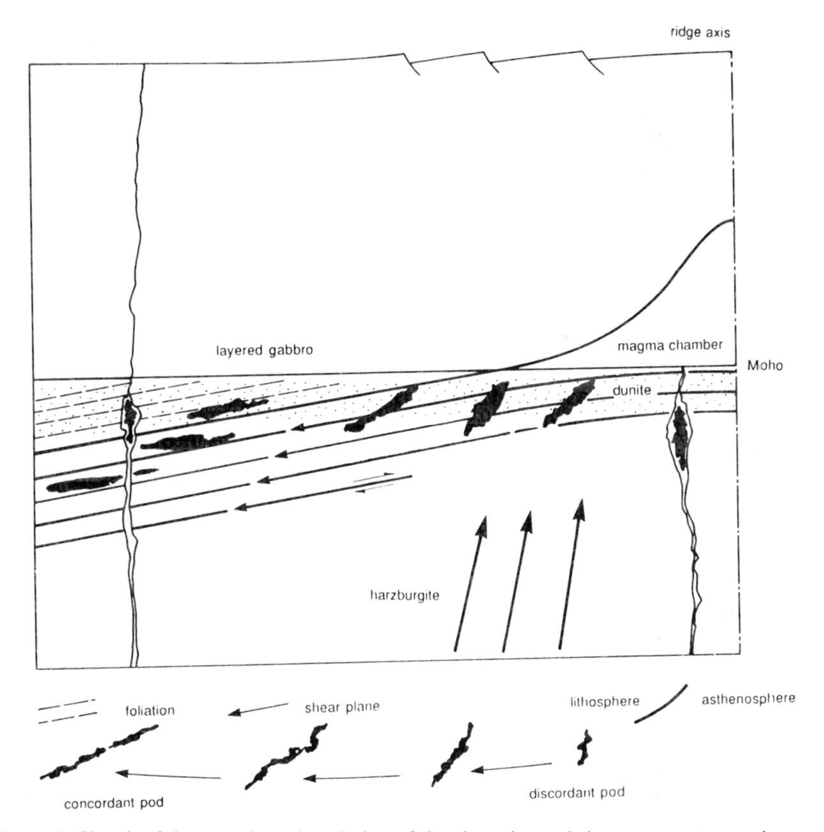

Figure 1. Sketch of the genesis and evolution of the chromite pods in uppermost oceanic mantle beneath an active spreading ridge (Cassard et al., 1981).

could possibly be related to a minor contamination by sea-water. A chromite trap may be created if an enlargement locally forms within the conduit, as suggested by a physical and numerical modelling (Lago et al., 1982).

A positive test for these ideas, is that magmatic growth and settling structures in chromite ore are observed only in discordant pods. Delicate textures such as nodular, orbicular, occluded, or chain textures (Thayer, 1969) are generally destroyed by the flow transposing the discordant pods into subconcordant and concordant lenses. In these deformed lenses, the ore is mainly disseminated or massive. Massive chromite, which is much stronger than the surrounding dunite, develops pull-apart structures with olivine and even dunite in-fillings (Fig. 3); the antinodular texture (rounded dunite inclusions in massive chromite inherited from the magmatic stage), is now strongly flattened. Only occasionally within larger lenses where the strain may have been less severe, nodular ore has been preserved. When such strain-induced features are taken into account it becomes clear that, in ophiolites, the stratiform chromite deposits generally regarded as produced by magmatic accumulation at the base of a magma chamber are indeed concordant deposits, with however a few exceptions (see Nicolas, 1989; Dogan Paktunc, 1990).

Chromite Deposits in the Oman Ophiolite

More than 200 deposits have been so far reported in Oman by the various groups which have been studying them (see Burgath et al., 1982; Augé, 1982; Augé and Roberts, 1982; Ceuleneer and Nicolas, 1985; Christiansen, 1985, 1986. Information can also be found in maps and specialized volumes devoted to the Oman ophiolite: 'Oman Ophiolite', *J. Geophys. Res.*, 1981; 'Mineral deposits of the Mountains of Northern Oman', *Sultanate of Oman*, Map scale 1/1000,000, 1985; 'The Ophiolite of Northern Oman', *Geol. Soc. London Mem.*, 1986; Geol. Map Sultanate Oman, scale 1/100,000, 1986; 'The Ophiolite of Oman', *Tectonophysics*, 1988). Figure 2 is a compilation of these different works, locating chromite occurrences within the ophiolite belt. We describe first their general features as examplified by a few typical deposits and discuss next their position within the ophiolite sequence and their location in map with respect to the still poorly known segmentation of the ophiolite.

Typical Features of Chromite Deposits in Oman

With the exception of a few uncommon deposits such as those described by Ceuleneer and Nicolas (1985) in the Maqsad area of the Semail massif, structural and deformational studies show that most Omanese chromite bodies belong to the concordant and subconcordant types as defined by Cassard et al. (1981). This means that they form lenses flattened parallel to the high temperature foliation of the surrounding dunites and harzburgites and elongated parallel to the spinel lineation measured in this foliation plane as illustrated by the Shamis 2 body in the Rajmi massif (Fig. 3). In this respect, we disagree with former interpretations which interpreted them as stratiform deposits (Brown, 1982; Augé, 1982). Discordant and subconcordant bodies are less common (Fig. 4).

In a number of cases, mainly in the western part of Wadi Tayin, the discordant pods have been folded by the high temperature plastic flow as shown by the geometrical relations of the folding with the foliation-lineation referential and by the fact that the same phase equilibrium conditions have been maintained during the folding episode (Fig. 5). This, however, has not resulted in a complete transposition (Fig. 5b). Finally, discordant bodies, whose shape is unrelated to the foliation attitudes in surrounding peridotites, are also observed (Fig. 6).

Discordant pods escape plastic flow when they have been emplaced either

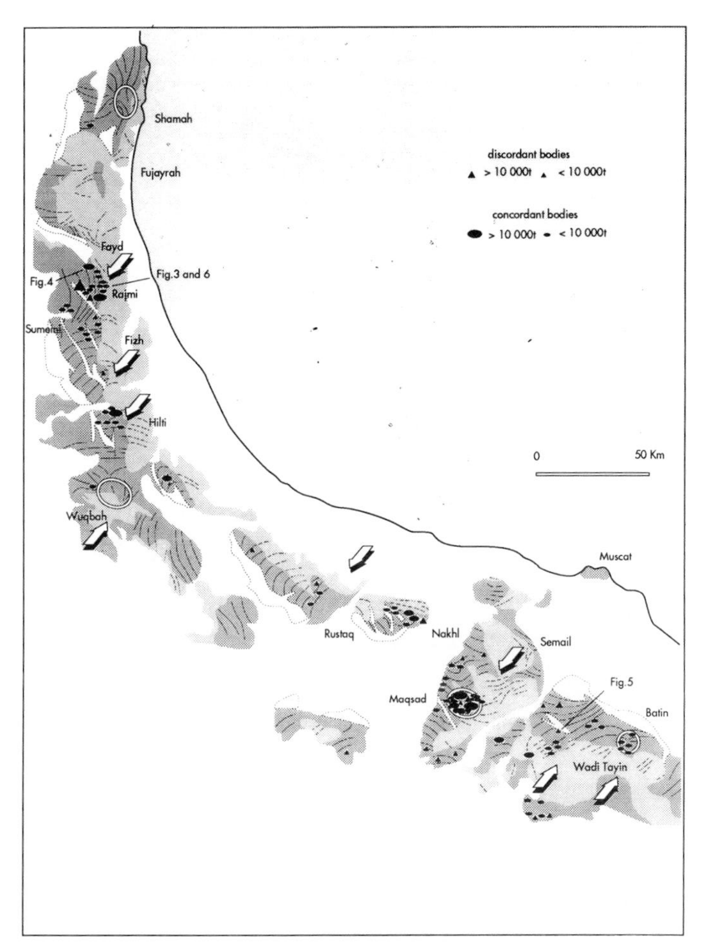

Figure 2. Location map of chromite deposits in the Oman ophiolite, superimposed on a simplified structural map of lineation trajectories in peridotites (thick continuous lines) and in gabbros (dashed lines), showing diapiric areas by the steepening and convergence of lineations the mantle. Petrological trails of presumed diapirs are located by arrows.

outside the deforming transition zone below the ridge of origin, for instance at the margin of a diapiric uprise, or within the central area of a diapir, now preserved in the ophiolite because it was frozen in situ. Such a special situation probably results from an early oceanic detachment having sliced an active spreading center (Nicolas, 1989), as illustrated by the Maqsad diapir (Ceuleneer et al., 1988). All observed pods in this diapir are concordant except three, the 'stratiform', the 'diffuse discordant' and the 'discordant nodular' pods described by Ceuleneer and Nicolas (1985) which are undeformed chromite bodies located close to the center of the diapir, in a zone of very weak plastic strain.

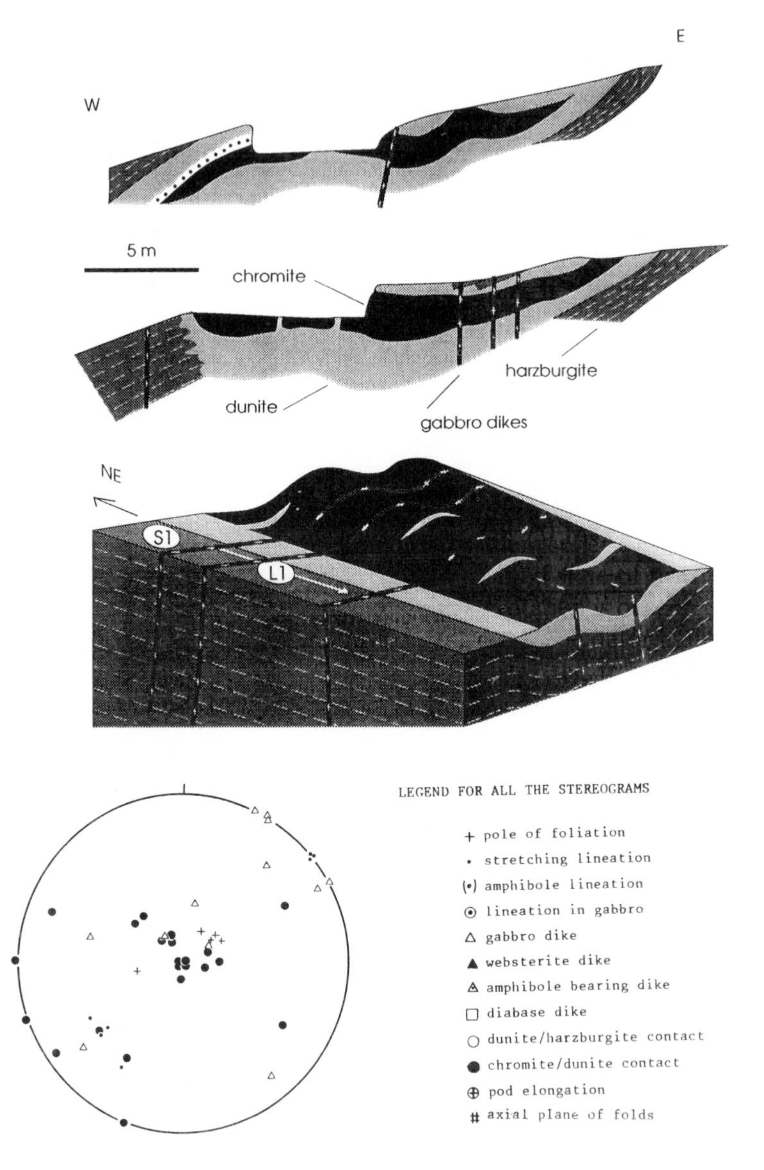

Figure 3. Concordant chromite lens (Shamis 2, Wadi Rajmi). The two cross sections from the northern extremity of the pod and the 3–D model show the flattening of the chromite lens parallel to the foliation in surrounding peridotites and its elongation parallel to the mineral lineation (elongation ratio of about 4/1); dunite veins perpendicular to this elongation indicate that the pull apart, visible at hand specimen scale in massive chromite ore, also stretches this ore at the outcrop scale. Stereographic projections of structures are represented in the lower hemisphere, geographical reference system. Planes are represented by their poles.

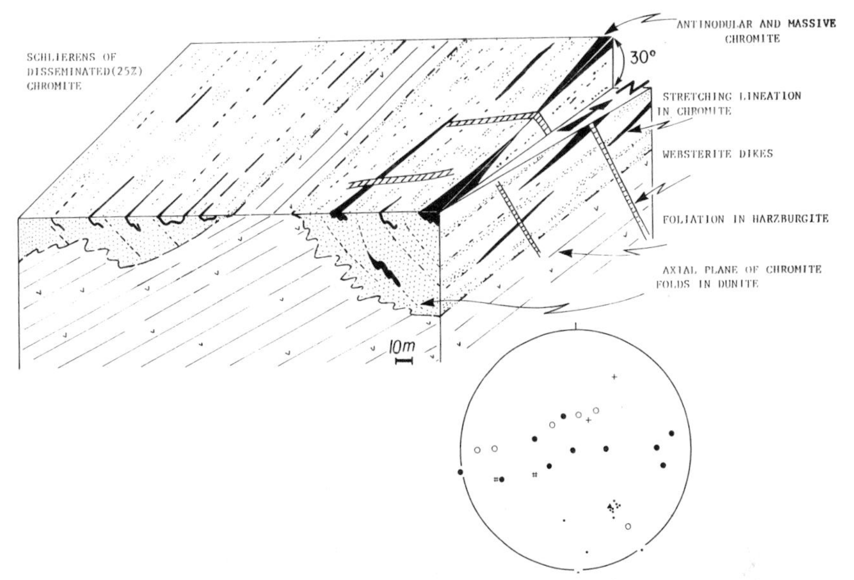

Figure 4. Subconcordant chromite lens and schlieren (85OA61 pod, Wadi Fayd). Note the elongation, as in Figure 3, and the folding of the chromite lenses parallel to the mineral lineation. Websterite dikes are emplaced at high angle to the mineral lineation, in the extension orientation.

Location of Chromite Bodies within the Oman Ophiolite

As observed in other ophiolites, the chromite occurrences in Oman are usually located within the transition zone at the top of the harzburgite section, just below the layered gabbro section. They occur either within the dunites which separate the harzburgites from the layered gabbros or more commonly, just below, in the irregular horizon where dunites root into harzburgites. However, a number of occurrences have been found deeper within the harzburgite section. This is the case of a large subconcordant body in Wadi Fayd (Fig. 4) located 2–3 km below the Moho, of several bodies in Wadi Rajmi, one or two kilometers below the Moho and even deeper, at the base of the peridotite section in the Sumeini window; also within the Wadi Tayin peridotite massif, several chromite bodies are located at some three kilometers or more below the Moho. In this area, the enclosing harzburgites have a steep foliation and a horizontal lineation over a 10 km thickness which have been interpreted as resulting from a transform zone activity (Nicolas et al., 1988).

In map (Fig. 2), chromite occurrences tend to concentrate in a few areas. This becomes more obvious if the size of the chromite deposits is taken into account. Clearly two districts contain most of the recognized chromite occurrences: the Maqsad area in the Semail massif and the Rajmi area in the northern massifs. Others areas comparatively rich in chromite are western

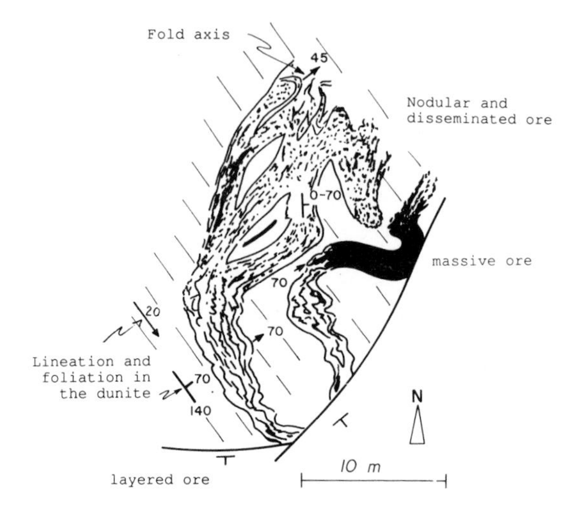

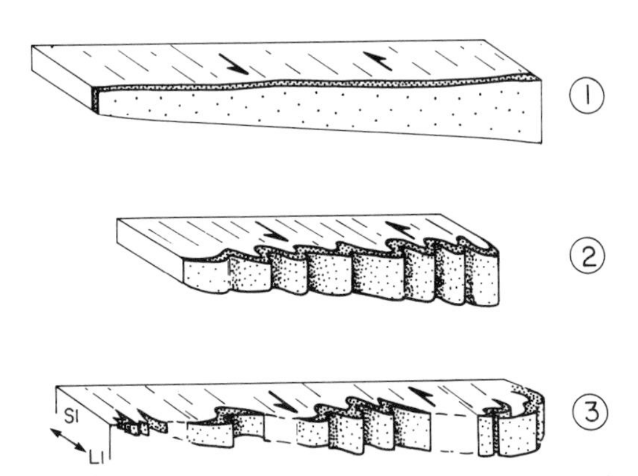

Figure 5. Folded discordant pod (81OM137, Wadi Tayin). (a) View in map of the folded pod: fold axes are at high angles to the mineral lineation in surrounding peridotites. (b) Model of formation of this type of pods, starting in 1) with a discordant chromite dike (arbitrary orientation), progressively folded (2) and stretched (3) by the high temperature plastic flow. The pod would be located at the right of sketch 3, which incorporates data on three other nearby pods (center and left part of the sketch).

Wadi Tayin, the eastern part of the Nakhl massif, the northern part of the Hilti massif and the Fizh area. Significantly, wide areas extending over a few tens of kilometers seem to be completely sterile.

The transition zone in areas where chromite bodies cluster is specially rich in dunites, gabbro dikes and impregnation structures. The crustal section above these areas is also rich in wehrlitic intrusions at the expense of layered

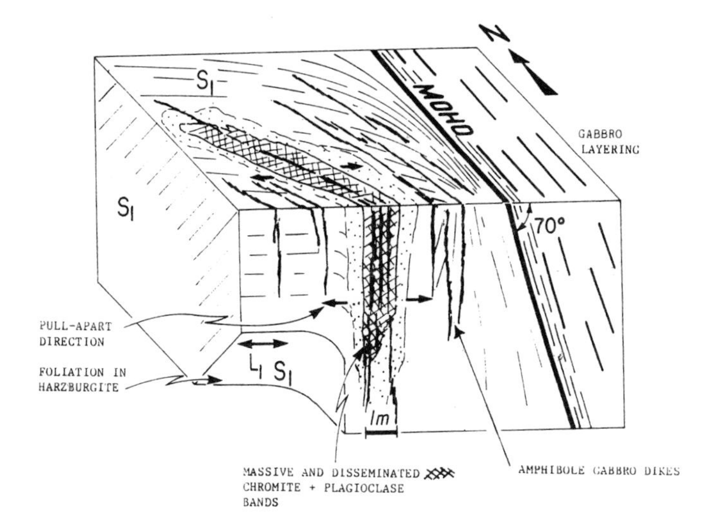

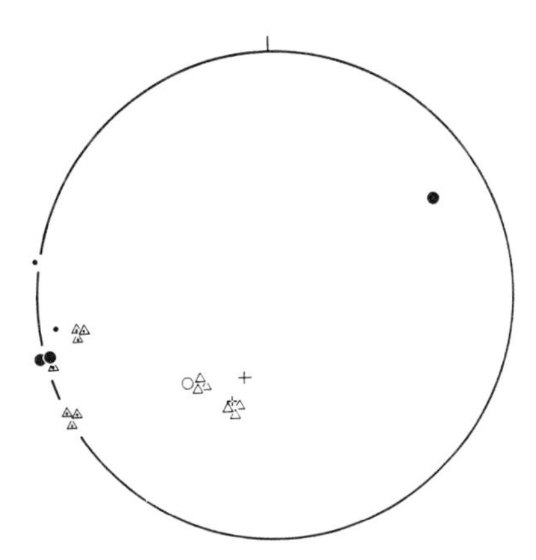

Figure 6. Discordant pod (85OA51, Wadi Rajmi). Located close to the vertical and sheared Moho of Wadi Rajmi, this pod is clearly discordant with respect to the surrounding high temperature foliation in harzburgites. It is however in an extension orientation with respect to the mineral lineation in the peridotite. This is also reflected by pull apart textures and injection of pargasite-gabbro dikes parallel to the chromite lens.

gabbros. Such features indicate a particularly active melt discharge from the mantle with vigourous crust-mantle interaction at Moho depth. They have been regarded as typical of an area of mantle diapirism below the ridge of origin (Rabinowicz et al., 1987; Nicolas et al., 1988). Finally, the Maqsad area, which is the richest chromite district in Oman, displays a frozen diapiric structure in the tectonic peridotites (Ceuleneer et al., 1988).

Discussion

In this discussion, we address two points regarding chromite occurrences in the Oman ophiolite. The first one deals with the preferred location of chromite deposits within the ophiolite and the second, with the question of why the Oman ophiolite is comparatively poor in chromite compared to other HOT ophiolites.

Chromite Occurrence and Mantle Diapirs

The coincidence of the chromite-rich district in Oman, the Maqsad area in the Semail massif, with a mapped diapiric structure in the peridotite section tends to confirm a previous conclusion (Nicolas and Violette, 1982), based on studies in other ophiolites, namely that the largest chromite deposits seem to be located at the top of mantle diapirs. Melt extracted from the mantle is dominantly channelled to the surface through such diapirs (Rabinowicz et al., 1984; Nicolas, 1986). It is assumed that most of the chromite formed by early crystallization from this pristine melt at the top of the diapirs.

However, other mapped diapirs in the Oman ophiolite show either little chromite, as Batin in the Wadi Tayin massif, and Wuqbah, or no detected chromite, as Shamah in the Northern United Arab Emirates (Fig. 2). In contrast, the Rajmi area, which is the second largest chromite district in Oman, does not display the typical steep and concentric mantle foliation of a diapir pattern, but a flat-lying mantle structure. The latter difficulty may be accounted for by recalling that, in ophiolites, a diapiric structure is preserved only if the ophiolite represents an active oceanic center; otherwise, the ophiolites derive from one flank of an oceanic ridge, being detached at some distance from that ridge. In this situation, the mantle flow issued from a diapir has been rotated to the horizontal before being frozen and all *structural* evidence has been destroyed. However, it could be expected that if a diapir has created a distinctive *petrologic* signature, this signature would not be destroyed by drifting away from the ridge and, consequently, that a petrological trail should make possible to locate diapirs along strike in the ophiolite. This has been tempted in Oman (Nicolas et al., 1988), considering the above-mentioned signs of a particular magmatic activity at the top of diapirs. Nine such areas have been spotted along the entire Oman ophiolite belt (Fig. 2); it is encouraging to check that the four mapped diapirs belong to such areas. The Rajmi, Hilti and western Wadi Tayin chromite districts are also located within such areas, suggesting that they are derived from once upstream diapirs. The Fizh and Rustaq-Nakhl chromite districts coincide only approximately with a petrological trail and finally, the Fujairah and southern Fizh trails show no association with a chromite district.

On the other hand, it is not well understood why three of the four mapped diapirs are chromite-poor. Shamah exposures in the United Arab Emirate are below the transition zone; in this case, chromite deposits, which usually

Table 1. Chromite ore reserves and/or production in the main producing ophiolites

Countries	Estimated reserves (or annual production) in thousands of metric tons of chromite ore	Source
Greece	Vourinos : 4000	Rassios and Vacondios, 1986
	Othris : 1000	Vacondios, 1986
Zambales	Acoje Mine : 1730	Bacuta, 1978
(Philippines)	Coto Mine : 6340	Bacuta, 1978
	Other districts : 1300	Hock and Friedrich, 1985
New Caledonia	Tiébaghi : 2700	Leblanc and Violette, 1983
	Other districts : 1000	Leblanc, 1987
Troodos	Kokinorotsos Mine : 800	Leblanc, 1987
(Cyprus)	total production to 1980 : 6000	Pantakis, 1980
Turkey	Guleman : 1500 (production to 1977)	Engin et al., 1987
	Total annual production : 500	Thayer and Lipin, 1978
Pakistan	Zhob Valley	Ahmad and Bilgrami, 1987
	Total production to 1980 : 1500	
Albania	Total annual production : 500	Thayer and Lipin, 1978
Urals	Total annual production : 1600	Thayer and Lipin, 1978

cluster in the transition zone would have been eroded. No explanation can be afforded for the cases of Wuqbah and Batin.

Why Comparatively Little Chromite Has Been Found in the Oman Ophiolite?

In the introduction of this paper, the point was made that chromite deposits in ophiolites are restricted to the harzburgite subtype (HOT). Though belonging to HOT, the Oman ophiolite is comparatively poor in this mineral resource, although considering the ophiolite belt size, chromite is potentially an important economic target. As immediatly visible in Figure 2, there are peridotite massifs extending over tens of kilometers where no chromite has been yet reported and, in the chromite-rich districts, the largest recognized deposits do not seem to exceed a few hundred thousand tons of chromite, compared to a few millions tons in the richest HOT ophiolites, such as those of Greece or Philippines (Table I). Oman would not be a single exception, as the large Bay of Islands ophiolite in Newfoundland seems to be also comparatively chromite-poor (Malpas and Strong, 1975).

Because the plutonic sequence in the Oman ophiolite is so thick and continuous it has been proposed (Nicolas, 1989) that this ophiolite was formed at a fast spreading ridge, where seismic reflection imaging has revealed over comparable distances along strike, a continuous magma chamber (Detrick et al., 1987), in contrast with slow spreading ridges where no magma chamber has been so far clearly documented. Another evidence is derived from the observation that the high temperature foliations in the peridotite section just below the crustal section are parallel to the surface bounding the

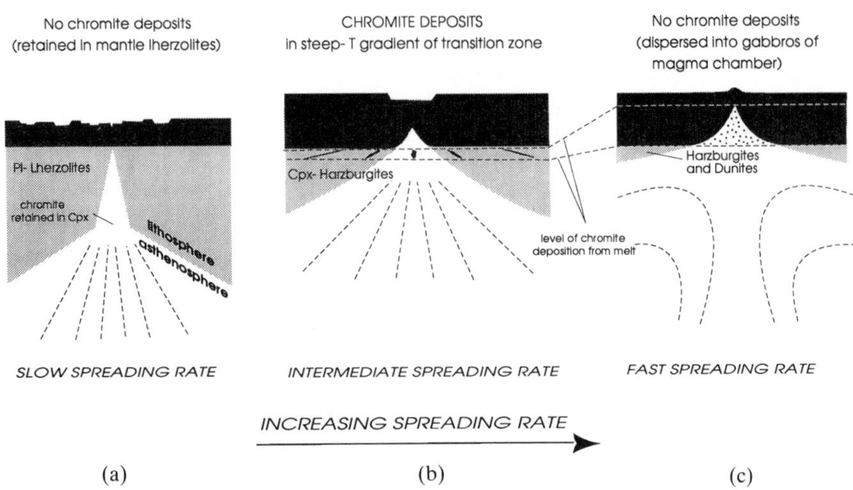

Figure 7. Model relating chromite deposits occurrence in LOT or HOT ophiolites to spreading rate at the ridge of origin.

two sections, regarded as the Moho. Modelling of mantle flow below a ridge and how it has accreted to the overlying lithosphere shows that the dip of the high temperature foliations can be equated with the dip of the lithosphere thickening away from the ridge; thus a flat-lying foliation means a flat-lying lithosphere, which itself correlates to a fast spreading situation, say faster than a few cm/y (Nicolas, 1989, p. 29).

On the other hand, it is widely recognized that chromite deposits cluster within the transition zone of HOT ophiolites. Above, we have explained this by the mantle cooling and the possible increase in oxygen fugacity affecting this zone below an oceanic ridge, two factors which would favour chromite and olivine crystallization from the rising melt (Fig. 7b). It is here speculated that, in the case of the Oman ophiolite, the spreading rate may have been faster than in most other ophiolites and that, consequently, the transition zone below the ridge would have been maintained at temperatures above the basalt liquidus temperature. If this is true, chromite would not precipitate within the dikes carrying melt through this zone, but within the magma chamber where it would remain scattered in the layered gabbros (Fig. 7c). Indeed, compared to other HOT ophiolites richer in chromite deposits, Oman seems to have thicker layered gabbros, suggesting that the magma chamber of origin could have been larger and more continuous in space and time; this, in turn, would suggest a faster spreading rate than in chromite-rich ophiolites.

A final remark deals with the observation that compared to other HOT ophiolites, chromite deposits in Oman seem to be less strongly clustered in the transition zone, with a number of pods observed several kilometers below the Moho (Fig. 2). Taking radically the model proposed above, it should be concluded that no chromite deposit can form during the steady-state functioning of the Oman ophiolite (Fig. 7c). This is obviously excessive, as it has been proposed that several chromite districts, such as Rajmi, derive from a steady-state functioning as suggested by their important deformation as concordant pods. It is however speculated that many chromite deposits may have formed during the waning stage of the spreading center activity, when, as a result of cooling, the liquidus temperature of basalt moved from a position within the magma chamber down into the transition zone and, in order to explain the deep chromite deposits, finally deeper into the harzburgite sequence. In this interpretation, the deepest chromite deposits would have formed from the last ridge activity of the Oman ophiolite. This is examplified by the Wadi Tayin and Sumeini discordant bodies.

Acknowledgements

F. Boudier is thanked for producing unpublished data on several chromite deposits and reviewing this paper. R. Laurent's and F.G. Christiansen's reviews are also acknowledged. This work has been possible through continuous support over years from the Ministry of Petroleum and Minerals of Oman and the Centre National de la Recherche Scientifique-Institut National des Sciences de l'Univers (DBT contribution 301).

References

Ahmad, Z. and Bilgrami, S.A., 1987. Chromite deposits and ophiolites of Pakistan, In: Evolution of chromium fields, C.S. Stowe (ed), Van Nostrand Reinhold Co., New-York, pp. 239–264.

Augé, T., 1982. Etude minéralogique et pétrographique de roches basiques et ultrabasiques du complexe ophiolitique du Nord Oman, Relations avec les chromitites comparaison avec deux complexes dArabie Saoudite, Thèse 3ème cycle, Univ. Orléans, Fr., 263 p.

Augé, T. and Roberts, S., 1982. Petrology and geochemistry of some chromitiferous bodies within the Oman ophiolite.; Ofioliti, 7: 133–154.

Bacuta, G.C., 1978. Geology of some Alpine-type chromite deposits in the Philippines: Manila, Philippines., Bur. Mines, unpub. Rep., 22 p.

Boudier, F. and Nicolas, A., 1985. Harzburgites and lherzolites subtypes in ophiolitic and oceanic environments., Earth Planet. Sci. Lett., 76: 84–92.

B.R.G.M./Oman, 1985. Mineral deposits of the mountains of Northern Oman, Map 1/1,000,000, Sultanate of Oman, (Ed.)

Brown, M.A., 1982. Chromite deposits and their ultramafic host rocks in the Oman ophiolite., Dep. Earth Sci. Open Univ., Ph. D. Thesis, 263 p.

Burgath, K.P., Mohr, M., Ramunlmair D. and Steiner, L., 1982. The chromite potential in North Central Oman (new discoveries). Federal Inst. Geosc. Nat. Res., Hannover, unpublished Rep., 89 p.

Cassard, D., Nicolas, A., Rabinowicz, M., Moutte, M., Leblanc, M. and Prinzhofer, A., 1981. Structural classification of chromite pods in southern New Caledonia., Econ. Geol., 76: 805–831.

Ceuleneer, G. and Nicolas, A., 1985, Structures in podiform chromite from the Maqsad district (Sumail ophiolite, Oman)., Mineral. Deposita, 20: 177–185.

Ceuleneer, G., Nicolas, A. and Boudier, F., 1988. Mantle flow patterns at an oceanic spreading centre: the Oman peridotites record., Tectonophysics, 151: 1–26.

Christiansen, F.G., 1985. Deformation fabric and microstructures in ophiolitic chromitites and host ultramafics., Sultanate of Oman, Geologische Rundschau, 74: 61–76.

Christiansen, F.G., 1986. Structures of ophiolitic chromite deposits, Thesis Lic. Scient. degree, Univ. Aarhus, Denmark.

Detrick, R.S., Bulh, P., Vera, E., Mutter, J., Orcutt, J., Madsen, J. and Brocher, T., 1987: Multi-channel seismic imaging of a crustal magma chamber along the East Pacific Rise., Nature, 326: 35–41.

Dogan Paktunc, A., 1990. Origin of podiform chromite deposits by multistage melting, melt segregation and magma mixing in the upper mante., Ore Geol. Rev., 5: 211–222.

Engin, T., Ozkoçak, O. and Artan, U., 1987. General geological setting and character of chromite deposits in Turkey. In: Evolution of chromium fields, Stowe CX.W. (ed.), New-York, Van Nostrand Reinhold Co. pp. 195–219.

Hock, M. and Friedrich, G., 1985. Structural features of ophiolitic chromitites in the Zambales Range, Luzon, Philippines., Mineral. Deposita, 20: 290–301.

Lago, B.L., Rabinowicz, M. and Nicolas, A., 1982. Podiform chromite ore bodies: a genetic model., J. Petrol., 23: 103–125.

Leblanc, M., 1987, Chromite in oceanic arc environments: New-Caledonia, In: Evolution of chromium ore fields, C.W. Stowe (ed.), Van Nostrand Reinhold Co, pp. 265–295.

Leblanc, M. and Violette, J.F., 1983. Distribution of aluminium-rich and chromium-rich chromite pods in ophiolite peridotites., Economic Geology, 78: 293–301.

Malpas, J. and Strong D.F., 1975. A comparison of chromite-spinels in ophiolites and mantle diapirs of Newfoundland., Geochim. Cosmochim Acta, 39: 1045–1060.

Nicolas, A., 1986. Structure and petrology of peridotites., Rev. Geophys., 24: 875–895.

Nicolas, A., 1989. Structures of Ophiolites and dynamics of oceanic lithosphere, Kluwer Ed., 367 p.

Nicolas, A. and Violette, J.F., 1982. Mantle flow at oceanic spreading centers: models derived from ophiolites., Tectonophysics, 81: 319–339.

Nicolas, A., Reuber, I. and Benn, K., 1988. A new magma chamber model based on structural studies in the Oman ophiolite., Tectonophysics, 151: 87–105.

Pantakis, M.T., 1980. Chromite mineralization associated with the Troodos ophiolite, Cyprus, International Symposium on Metallogeny of mafic and ultramafic complexes of the eastern Mediterranean and western Asian area and its comparison with similar metallogenic environments of the word, pp. 91–97. Athens: UNESCO IGCP Publication, 1, National Technical Univ., Athens.

Rabinowicz, M., Nicolas, A. and Vigneresse, J.L., 1984. A rolling mill effect in asthenospheric beneath oceanic spreading centers., Earth Planet. Sci. Lett., 67: 97–108.

Rabinowicz, M., Ceuleneer, G. and Nicolas, A., 1987. Melt segregation and flow in mantle diapirs below spreading centers: evidence from the Oman ophiolites., J. Geophys. Res., 92: 3475–3486.

Rassios, A. and Vacondios, I., 1986. Chromite mineralization and Mining at Vourinos., EEC Internal Rep., Athens, pp. 128–142.

Roberts S., 1988. Ophiolitic chromitite formation: a marginal basin phenomenon?, Econ. Geology, 83: 1034–1036.

Thayer, T.P., 1969. Gravity differenciation and magmatic re-emplacement of podiform chromite deposits., Econ. Geol. Monogr., 4: 132–146.

Thayer, T.P. and Lipin B.R., 1978. A geological analysis of world chromite production to the year 2000 A.D., Proceedings 107th Annual Meeting Council of Economics of AIME, pp. 143–146.

Vacondios, I., 1986. Chromite mine in Tsangli, Workshop on Greek ophiolites., Inst. Geol. Min. Expl., Athens, pp. 55–57.

Diamonds: the Oceanic Lithosphere connection with Special Reference to Beni Bousera, North Morocco

P.H. NIXON, D.G. PEARSON[1] and G.R. DAVIES[2]
Department of Earth Sciences, The University, Leeds, U.K.
[1]*Present address: Carnegie Institution of Washington, Dept. of Terrestrial Magnetism, 5241 Broad Branch Rd. NW, Washington D.C. 20015.*
[2]*Present address: Dept. of Geological Sciences, University of Michigan, 1006 C.C. Little Bldg, Ann Arbor, Michigan 48109–1063.*

Abstract

The two main diamond associations are either of eclogitic or peridotitic mineralogy, exemplified by inclusions in diamond or rocks in which diamond has formed, notably xenoliths in kimberlites. Outside cratonic areas where kimberlites typically occur, diamonds have been recorded in alpine peridotites and gravels associated with them. Graphite pseudomorphs after diamonds occur in the Beni Bousera peridotite massif (Kornprobst, 1969) within garnet pyroxenites or eclogites (Slodkevich, 1980b). They have cubic morphology and contain cpx inclusions of high pressure cubo-octahedral morphology similar to those in natural diamonds. They possess isotopically light $\delta^{13}C$ values. The garnet pyroxenites have similar mineralogy to diamond bearing eclogite xenoliths and have a wide oxygen isotope variation ($\delta^{18}O$ = 4.9 to 9.3%) – much greater than that of the associated peridotites.

They are believed to represent portions of hydrothermally altered subducted oceanic lithosphere that resided at great depth ($\gg$ diamond stability limit) before diapiric uprise and final cumulate segregation. The evolution of these rocks is compared with that of the sparsely diamondiferous alpine peridotites which probably have a simpler and shorter subduction history to relatively shallow depths.

Introduction

The existence of diamonds in kimberlite and lamproite volcanic intrusions (pipes) is well known. These diamonds formed long before eruption, during mantle events unrelated to the generation of the volcanic host rock (Kramers, 1979; Richardson et al., 1984; Jaques et al., 1989). For this reason, attention is increasingly directed at diamond-bearing mantle xenoliths and diamond inclusions (divided into eclogite and peridotite groups) as being more directly related to diamond genesis, rather than the kimberlite-lamproite hosts that

Tj. Peters et al. (Eds), Ophiolite Genesis and Evolution of the Oceanic Lithosphere, 275–289.

transported them to the surface (Boyd and Gurney, 1986). The origin of both eclogite and some peridotite xenoliths has been attributed to subduction.

Eclogites, not necessarily diamondiferous, from the Colorado Plateau kimberlites were regarded as subducted and metamorphosed ophiolites by Helmstaedt and Doig (1975). Sharp (1974) suggested a subducted oceanic plate origin for "some inclusions", including diamond in kimberlite.

Ringwood (1977) proposed that the depleted Mg, Cr-rich peridotitic assemblages in kimberlites (knorringite garnet, forsterite, enstatite and magnesiochromite) resulted from subduction of an oceanic cumulate assemblage and consequent depletion by partial melting followed by metamorphic re-equilibration. In order to explain the low Ca mineralogy of the diamond peridotite suite Schulze (1986) suggested removal of calcium during serpentinisation and subduction of oceanic lithosphere, a feature concomitant with the formation of rodingites. Serpentinites now represented in alpine-type and abyssal peridotites have similar major element proportions to the subcalcic garnet harzburgites and dunites that make up the diamondiferous peridotite xenolith lithology (Schulze, 1986).

Significantly, large tectonically emplaced outcrops of eclogitic and peridotitic rocks are found in the vicinity of many occurrences of unexplained sporadic alluvial diamonds, i.e. those far removed from cratonic areas or their Proterozoic margins where kimberlites or lamproites occur. These "anomalous" diamonds are commonly reported in Phanerozoic mobile belts or collision zones containing alpine type peridotites, e.g. the Urals, Armenia, Appalachians, north-eastern USSR and Tibet (Fig. 1; see reviews by Kaminskii, 1984; Dawson, 1983; and Nixon et al., 1986).

Many alpine peridotites (Thayer, 1960) have an obvious ophiolite connection and are composed predominantly of Cr spinel harzburgites. Others which lack this *obvious* connection contain a higher proportion of pyroxenite layers and may include lithologies with pyrope (-almandine) garnet. This group includes the peridotite massifs of the western Mediterranean; Ronda and Ojen (S. Spain), Lherz (S.W. France) and Beni Bousera, N. Morocco. However, of these only Beni Bousera has yielded unequivocal evidence of diamond association - in the form of octahedral graphite pseudomorphs* (described below).

Beni Bousera

The peridotite massif is well known from the work of Kornprobst and coworkers (Kornprobst and Vielzeuf, 1984, and references therein). It is emplaced within nappe sequences of the Rifean fold Belt which extend around the

*Since this article was written, octahedral graphite pseudomorphs have been found in pyroxenites in the Ronda massif, S. Spain (unpublished data, G.R. Davies, P.H. Nixon, M. Obata and D.G. Pearson).

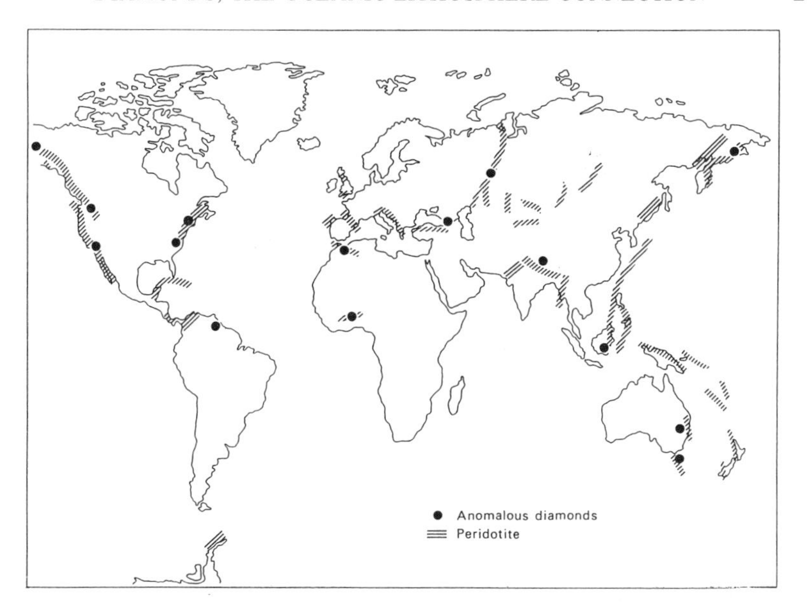

Figure 1. Distribution of anomalous diamonds and their proximity to alpine peridotites/ophiolites.

Alboran Sea and which flank other peridotite massifs, notably the Ronda within the contiguous Betic Fold Belt (Fig. 2). Adjacent to the Beni Bousera peridotite massif is the Filali unit composed of graphite-bearing garnet sillimanite granulites (kinzigites) grading out into lower-grade kyanite and andalusite gneisses (Kornprobst, 1974).

The predominant rock type within the Massif is variably altered spinel lherzolite defined on the basis of presence of clinopyroxene). Harzburgites (clinopyroxene absent) are rare and highly serpentinised. Garnet lherzolite and garnet spinel lherzolite also occur especially near to the western margin of the massif (Pearson, 1989). Dunitic lenses are reported by Kornprobst (1969, 1974) and Reuber et al. (1982). There is abundant petrographic evidence (exsolution features) for sub-solidus re-equilibration and high temperature crystal plastic deformation at some stage during the evolution and emplacement of the peridotites.

Pyroxenite layers with variable proportions of clinopyroxene, orthopyroxene, spinel and garnet occur within the peridotite (Fig. 3). Based on measured field sections these form up to 10% of a 174 m section locally but overall are estimated to form 1–3% of the complex (Pearson, 1989), in good agreement with the calculations of Allegre and Turcotte (1986). Since these layers include the garnet pyroxenites containing the graphite pseudomorphs after diamond (see below) their origin is crucial to the understanding of that of the diamond itself.

The pyroxenite layers are well exposed in the south western part of the complex especially in the Oued el Jouj area. They are typically < 5 cm thick

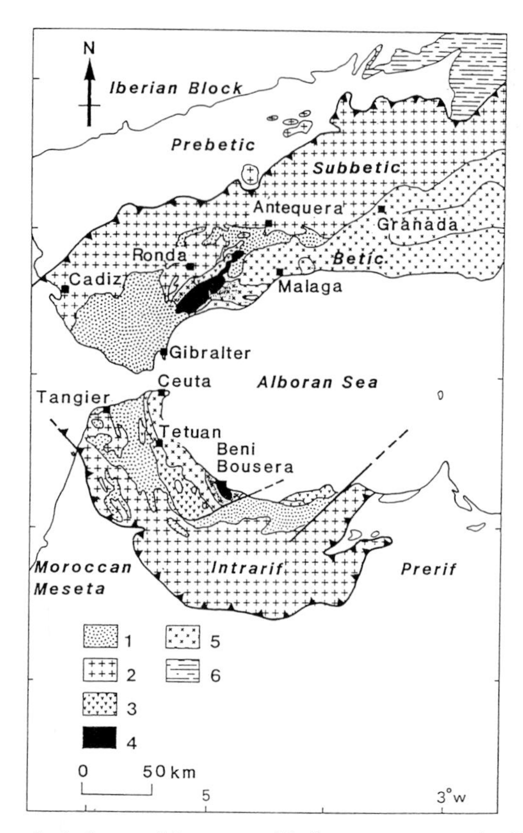

Figure 2. Regional geological map of the western Mediterranean area, after Bonini et al. (1973). Key: (1) Cretaceous and Tertiary flysch. (2) Mesozoic, Subbetic deep water carbonates and volcanics. (3) Drosdale, Mesozoic and Tertiary epicontinental seas. (4) Peridotite bodies, including the Ceuta outcrop. (5) Ghomaride/Malaguide sequence of late Paleozoic clastics. The outlined area within this outcrop represents Alpujarride (S. Spain) or Sebtide sequence (Morocco) which contains the Filali unit. (6) Late Cenozoic cover of the Prebetic zone and Meseta.

with parallel sides and are traceable over several tens of metres. The thicker bodies (>5 cm to 3 m) are lensoid; these include most of the garnet bearing types and all those with graphite pseudomorphs. Garnet is only found in the Al-augite group of pyroxenites and is absent from the volumetrically subordinate Cr-pyroxenite group. This group is not dealt with here; for details see Pearson (1989). The symmetrical mineralogical zonation in some layers (Fig. 3) suggests that pyroxenite magma intruded into peridotite and crystallised from the conduit walls inwards. There is some induced melting of the peridotite to produce dunite and harzburgite selvedges (Kornprobst, 1969; Pearson, 1989). The narrow zone of depletion (<10 cm) against the pyroxenite layers (*however thick these may be*) does not support earlier suggestions that the pyroxenites are derived from the host peridotites by partial melting.

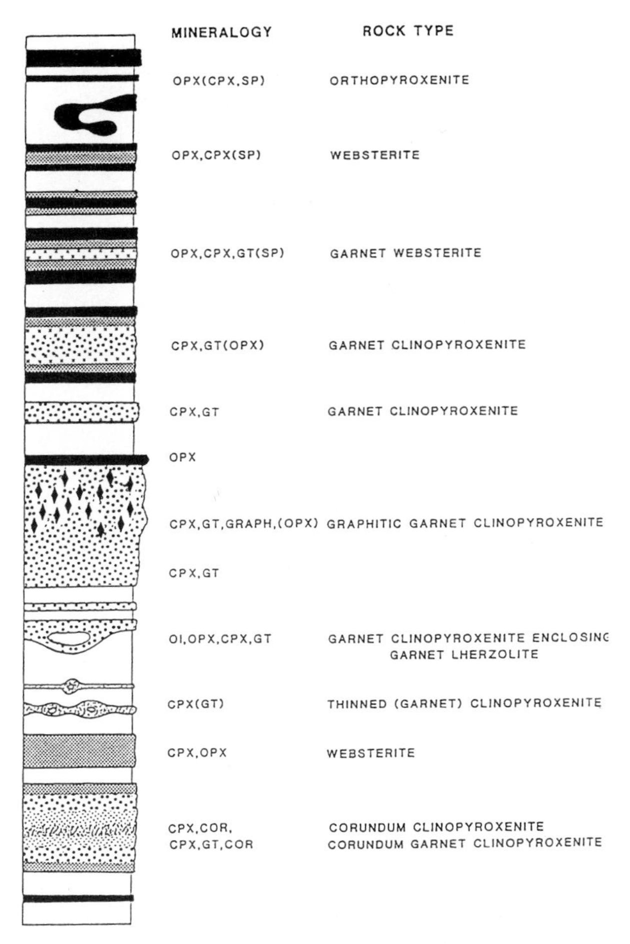

Figure 3. Representation of the types of pyroxenite layers within the Beni Bousera peridotite massif. There is no scale and no proportions of rock types is implied, the peridotite (white) being grossly under-represented except locally. The corundum clinopyroxenites are described by Kornprobst et al. (1982 and 1991). The garnet clinopyroxenites are the eclogites of some other writers (see text).

Most small scale igneous textures related to pyroxenite magma intrusion have been destroyed during subsequent layer parallel shearing and re-equilibration, but some relict small rounded apophyses locally intrude the peridotite. The linearity exemplified by most layer boundaries has probably resulted from shearing due to high strain developed between the rheologically distinct peridotite and pyroxenite. Highly stretched and thinned garnet clinopyroxenite layers within the host spinel peridotite grade, with intimate tectonic mixing, into the garnet lherzolites +/− spinel (Kornprobst, 1966). Pearson (1989) has argued on the basis of petrography and whole rock chemistry that

Figure 4. Graphite pseudomorph after diamond in garnet clinopyroxenite. The edge of the octahedron is 7 mm. Oued el Jouj, Beni Bousera.

the garnet lherzolites are a product of tectonic mixing *and* equilibration at high (mantle) temperatures.

Graphite Garnet Pyroxenites

Four garnet pyroxenite horizons are known to contain graphite octahedra (Fig. 4) which may comprise up to 15% of the rock (Slodkevich, 1980a). The graphite has not been observed to transgress from the pyroxenite layer into the peridotite nor has disseminated graphite been found in the adjacent peridotites (Pearson et al. 1989a). It is thus argued that the chances are remote of the graphite originating from the kinzigites surrounding the massif or from cross-cutting late-stage, low temperature Cu-Ni mineralised veins, known to contain graphite. The evidence that the graphite octahedra are pseudomorphs after diamond (Slodkevich, 1980a,b; Nixon et al., 1986) has been strengthened by the recent work of Pearson et al. (1989a, b; 1990; in prep.). It is summarised below:

1. All morphological forms of the graphite are exhibited by diamond. These forms are sharp-edged octahedra (Fig. 4) having edge lengths up to 12 mm, with or without rounded fibrous coats (the most abundant form); flattened/sheared octahedra (on (111) plane); rhombicuboctahedra; contact twins (macles) mostly with fibrous coats; irregular assemblages resembling framesite diamond (Pearson et al., 1989a).

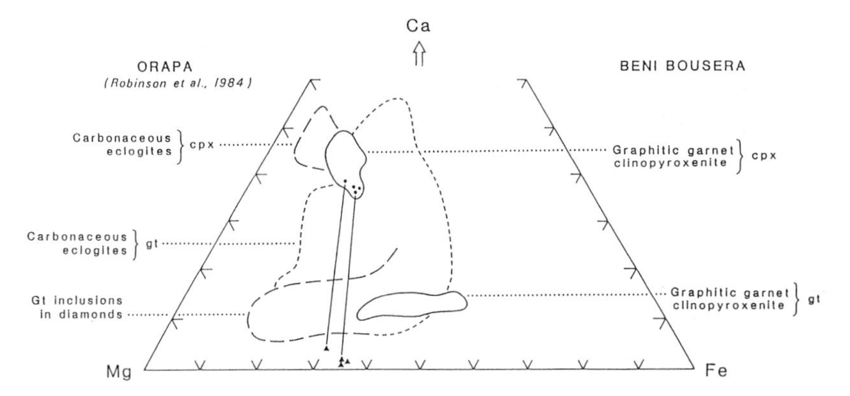

Figure 5. Fields of coexisiting garnet-clinopyroxene-orthopyroxene in graphite garnet pyroxenites/carbonaceous eclogites from Beni Bousera/Orapa kimberlite, Botswana, after Pearson et al. (1989a). Tie lines join coexisting cpx-opx. The field for eclogitic garnet inclusions in Orapa diamonds falls within that of the carbonaceous (graphite or diamond) eclogite garnets, Orapa data from Robinson et al. (1984).

2. X-ray diffraction and SEM evidence indicate that there is strong preferred orientation of graphite (0001) parallel to (111) of the octahedra, i.e. the same preferred orientation displayed by graphitised diamonds but *not* graphitised cubic minerals such as kamacite.
3. SEM reveals surface features (trigons) similar to those of diamond (Pearson et al., 1989a).
4. Raman spectra of the graphite are consistent with crystallisation at high temperature (>600°C) (Pearson and Pasteris, unpublished data).
5. Some clinopyroxene and garnet inclusions in the graphite show cubo-octahedral facetting as in natural diamonds (Slodkevich, 1980a; Pearson et al., 1989a).
6. The type of inclusions within the graphites, viz. clinopyroxene (abundant), garnet, pyrrhotite, and ilmenite (recorded by Slodkevich, 1980a) is the same as that of the eclogite-type of diamonds (Meyer, 1987). The compositions of clinopyroxenes and garnets likewise are close to those of the fields of diamond inclusions (eclogite type) and minerals within carbonaceous (graphite/diamond) eclogites (Robinson et al., 1984) (Fig. 5).

The clinopyroxenes of both the graphite inclusions and the host garnet pyroxenites at Beni Bousera range from aluminous sodic augite to omphacite but are generally too jadeite poor to warrant the host rock term eclogite. The clinopyroxene compositions reflect varying degrees of re-equilibration during which a significant, but variable amount of garnet (pyrope-almandine) has exsolved together with some plagioclase and orthopyroxene. Despite substantial re-equilibration, some garnets retain a slight Si excess and small amounts of Na_2O (0.05 wt%; but up to 0.2 wt% according to Slodkevich (1980a). Both features can be interpreted as a result of (relict?) high pressure solid solution of clinopyroxene in the garnet (Sobolev and Lavrent'ev, 1971; Akaogi and Akimoto, 1977) exemplified by the extreme values ($SiO_2 = 47$ wt%; $Na_2O = 1$ wt%) in the eclogitic garnet inclusions of diamonds in rapidly emplaced kimberlites (Moore and Gurney, 1985).

The close similarity between the Beni Bousera garnet pyroxenites and eclogite xenoliths is reflected also in their major element compositions which are picritic, except for lower total alkalis, with the graphitic garnet pyroxenites being generally less magnesian (MgO, 10.8–15.0 wt%) than those without graphite (11.8–19.5 wt%). Corundum garnet pyroxenites described by Kornprobst et al. (1991) are compositionally distinct from the majority of the pyroxenites in having higher Mg/Fe ratios and Al_2O_3 contents. Most of the pyroxenites have major element characteristics comparable to ultramafic and mafic cumulates from ophiolites (Pearson, 1989). The low bulk rock K_2O, P_2O_5, TiO_2 and Na_2O is also compatible with this view (Pearson, op.cit.).

The graphite garnet pyroxenites are notable is possessing low Ni (usually < 200 ppm) and Cr usually (<1000) both of which are poorly correlated with the wide range in Mg nos. (38–53) and showing shallower trends that the pyroxenite group as a whole. This and a similar lack of correlation between incompatible elements and Mg nos has led Pearson (1989) to propose that the pyroxenite suite was not derived from a single homogeneous source. This is further supported by the presence of both positive and negative Eu anomalies (Fig. 6). It is logical to explain these features in terms of plagioclase fractionation in a protolith such as a layered sequence of ophiolites (cf. Kornprobst et al., 1991) prior to subduction and high pressure equilibration/melting.

The extreme LREE depletion of the graphite-bearing garnet pyroxenites, indicative of several partial melt events noted by Nixon et al. (1986) has been confirmed $(Ce/Yb)_n = 0.0018–0.09$) and supports a complex origin for the pyroxenites (Loubet and Allegre, 1982). Other garnet and non garnetiferous pyroxenites show less extreme depletion but the data preclude them from being derived from the surrounding peridotites ($Ce/Yb_n = 0.228–1.840$) (Pearson, 1989) as does the presence of the Eu anomalies.

Carbon and Oxygen Isotopes

Graphite from the garnet pyroxenites is characterised by isotopically light $\delta^{13}C$ values ranging from -16.4 to -27.6% and mean of -22 to -23% (Slodkevich and Lobkov, 1983; Pearson et al., 1989b). These values could have resulted from an original biogenic component in subducted sediments (Nixon et al., 1986; Pearson et al., 1989b). Moreover the fact that some eclogite-type diamonds from kimberlites also have similar "light" carbon isotope signatures (Sobolev et al., 1979; Deines et al., 1984) could reflect a similar origin. However, most eclogite-type diamonds and practically all the peridotite-type diamonds from kimberlite, have been shown by the above workers to lie within generally accepted mantle values of −8 to −2% $\delta^{13}C$. Pearson et al., (1989b; in press) have discussed the possibility that the Beni Bousera massif has undergone large scale carbon exchange with the crustal kinzigites and graphite Cu-Ni-Fe mineralised veins ($\delta^{13}C = -17$ to -22%). They discounted this possibility on the grounds that the carbon is restricted to only four garnet pyroxenite layers and is absent from the immediately surrounding peridotites.

Oxygen isotope determinations have been carried out on peridotite and pyroxenite minerals from Beni Bousera (Pearson et al., 1991). Clinopyroxene values from spinel lherzolites range from $\delta^{18}O = 5.3$ to 6.0%, i.e. within the range of mantle clinopyroxenes determined by Kyser et al. (1981) and Javoy (1980). The pyroxenite clinopyroxenes show a much greater oxygen isotope variation than that of the peridotites with values ranging from 4.9 to 9.3% (Fig. 6). The values of two graphite bearing garnet pyroxenites are 5.2 and 7.5%. The small magnitude of the cpx-melt and gt-melt fractionation factors at mantle temperature (Kyser et al., 1981; Kyser, 1986) precludes the pyroxenites being derived from surrounding peridotites, even if substantial crystal fractionation took place.

The range in $\delta^{18}O$ is comparable to that observed in eclogite xenoliths from kimberlite (clinopyroxene range is 2.2 to 7.2; Macgregor and Manton, 1986). These authors have suggested that the eclogites represent subducted hydrothermally altered oceanic lithosphere. Garnet-clinopyroxene mineral pairs from graphitic and non graphitic garnet pyroxenites and clinopyroxene-orthopyroxene pairs from websterites are in isotopic equilibrium, plotting close to $\text{delta}_{\text{gt-cpx/cpx-opx}} = 0\%$ demonstrating that the range is not due to post emplacement low temperature open system processes.

The $\delta^{18}O$ range in the Beni Bousera clinopyroxenes is within that of Phanerozoic ophiolites such as Semail (Gregory and Taylor, 1981) and Macquarie Island (Cocker et al., 1982) see Figure 7. In view of the ophiolitic affinities of the Beni Bousera pyroxenites suggested on the grounds of major and minor element chemistry plus the high $^{87/86}Sr$ and low $^{143/144}Nd$ (Nixon et al., 1986; Pearson, 1989) the most obvious interpretation of the oxygen isotope data is that the pyroxenite suite originated from subducted hydrothermally altered oceanic lithosphere from which deep-seated melts were later generated - a conclusion supported by the clinopyroxenes having high $^{87/86}Sr$ and variable $^{206/204}Pb$ (Pearson et al., 1989b; 1990).

Diamonds and the Oceanic Lithosphere

The importance of the graphitised diamonds from Beni Bousera is that the host garnet pyroxenites can be equated with diamondiferous eclogite xenoliths found in kimberlites (and lamproites). Thus, the evolution of the pyroxenites from an *oceanic* origin through deep subduction is increasingly certain

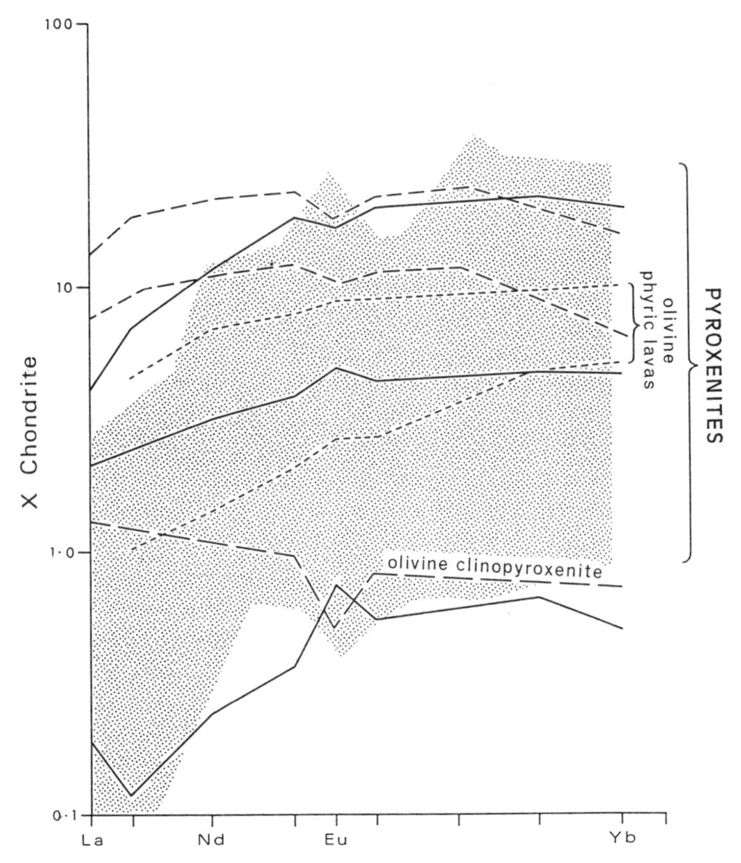

Figure 6. Chondrite normalised REE patterns for various units of ophiolites (gabbros, solid lines; dykes, dashed) compared to the REE range observed in the Beni Bousera pyroxenites (stipple). Ophiolite data from Coleman (1977) and Pallister and Knight (1981).

to apply to those diamondiferous eclogite bodies that reside near the base of the *continental* lithosphere that were intercepted and incorporated into kimberlites. Although, as noted above, these eclogites give Archean to L. Proterozoic ages – usually much older than the enclosing kimberlite, there seems no reason to suppose that younger eclogitic diamonds could not be found in kimberlites, depending on the pre-cratonic tectonic history. The same should apply to some peridotite diamonds if these are also of subducted origin.

Secondly, the Beni Bousera occurrence provides a link that might explain the distribution of anomalous diamonds (Fig. 1). Although many reports of solitary diamonds in alluvial gravels can be discounted, there are sufficient number, in some cases with illustrations (e.g. diamond in harzburgite, Koryak Mountains, USSR, Shilo et al., 1981) that the connection between alpine-peridotite and diamond can be believed. Examples of a connection with

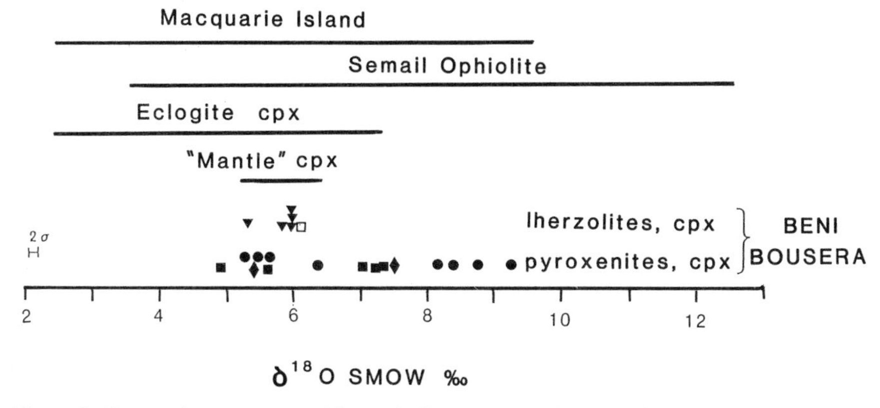

Figure 7. Oxygen isotope compositions of clinopyroxenes from Beni Bousera peridotites and pyroxenites (Pearson et al., 1989b; Pearson, 1989) compared with those from mantle peridotite and eclogite xenoliths (Kyser et al., 1981; R.S. Harman, pers. comm.; MacGregor and Manton, 1986). Range for Semail ophiolites (McCulloch et al., 1981) and Macquarie Island ophiolites (Cocker et al., 1982) also shown. Symbols: diamonds, graphite garnet pyroxenites; solid squares, garnet clinopyroxenites; open square garnet lherzolite; triangles, spinel lherzolites; circles websterites.

peridotite are more abundant than one with eclogite (garnet clinopyroxenite). Alpine peridotites in which diamond survives have clearly, evolved along a different PT path to Beni Bousera type pyroxenites although they could represent the depleted ultramafic component of the ophiolite succession. A comparison of both evolutionary paths is attempted in Figure 8.

In case B (Beni Bousera) hydrothermally altered oceanic crust with accumulated carbonaceous sediment and the underlying harzburgitic part of the slab is subducted to depths well within the diamond stability field, possibly to 400 km indicated by the deepest diamonds (on eclogitic mineral inclusion evidence; Moore and Gurney, 1985) or to the 670 km discontinuity (Ringwood, 1989; Pearson, 1989). Dehydration melting and thermal equilibration took place over a long period of time (Pearson, 1989). Later diapiric upwelling (Ringwood, 1989) induced decompression melting of former oceanic crust. High pressure garnet(majorite), diamond, and at shallower levels, pyroxene may be expected in the cumulate products veining the surrounding peridotite. With further asthenospheric uplift, diamond is graphitised and final melting at PT's representing intersection with a low volatile peridotite solidus gives rise to decoupling of trace element systematics followed by formation of the present cumulate assemblage (Pearson, 1989). The diapiric ascent and intrusion in the Western Mediterranean during the Neogene (Kornprobst and Vielzeuf, 1984) resulted from, or was a cause of major extensional rifting.

Peridotites containing graphite pseudomorphs after diamond do not appear to have been recorded. They do not occur at Beni Bousera in spite of

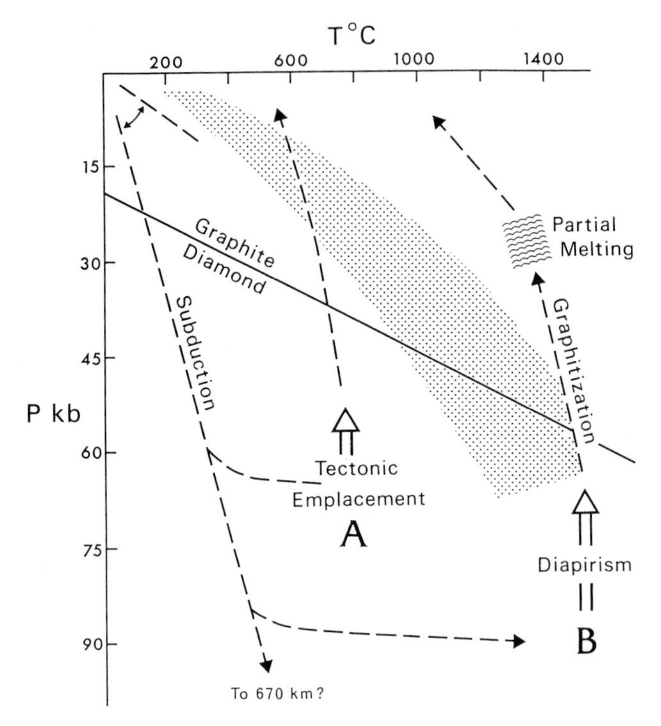

Figure 8. Portrayal of cold slab subduction and geotherms (stippled) ranging from continental to oceanic. Two scenarios are envisaged, both in which the slabs enter the diamond stability field. In B (Beni Bousera) oceanic lithosphere is deeply subducted with consequent melting and segregation of carbonaceous basaltic (ophiolitic) crust both during descent and ascent. The ultimate depth of subduction and temperatue equalisation with ambient mantle is unknown but could be 670 km (Ringwood, 1989). Graphitisation of diamonds within the eclogitised basaltic crust takes place during hot diapiric upwelling. In A (Tibet) rapid, relatively shallow subduction is followed by mainly tectonic rapid upthrust before temperature equalisation is attained. At these low temperatures the kinetic energy levels necessary to dissociate carbon atoms in the diamond structure to form graphite are not reached and diamonds persist metastably into the shallow crust.

the intimate tectonic mixing between peridotite and pyroxenite referred to above.

Peridotites containing diamond, exemplified by those in Tibet (Fang and Bai, 1981) are mineralogically distinct from the garnet clinopyroxenites of Beni Bousera nor are the high percentages of carbon present. They are part of extensive belts described as ophiolites (Girardeau and Mercier, 1988). Many are Phanerozoic. They are typically basalt-depleted harzburgites and dunites with high Cr magnesiochromite and thus are similar to the peridotite suite of diamond inclusions and xenoliths (Meyer, 1982). This is not entirely the case as knorringite garnet is rarely recorded, indeed lower pressure almandine is more common in the associated alluvial gravels, together with such minerals as corundum, platinoid metals, rutile, staurolite and florencite (Kaminskii, 1984).

To explain the absence of graphitisation of the diamonds requires rapid transportation out of the diamond stability field at low temperatures, below those that would produce the high activation energy levels necessary to initiate inversion to graphite. Few quantifiable data are available but it is reasoned that relatively cool tectonic emplacement rather than diapiric upwelling would be appropriate. The path A shown in Figure 8 illustrates subduction to shallower depth than in the case of B (Beni Bousera) with a shorter, less complicated evolution at lower temperatures.

To summarise, both eclogitic and peridotitic diamonds have originated within the Earth's oceanic lithosphere which was subducted to great depths either to be erupted to the surface by deep-seated continental volcanoes or to be emplaced tectonically in mobile belts.

Acknowledgements

The initiation of our studies on the Beni Bousera complex, and the graphite garnet pyroxenites in particular was made possible through the introduction of the rocks to PHN by Dr. V.V. Slodkevich in 1985. We gratefully acknowledge a review of this paper by Dr. J. Kornprobst.

References

Akaogi, M. and Akimoto, S., 1977. Pyrope-garnet solid solution equilibria in the systems $Mg_4Si_4O_{12}$ - $Mg_3Al_2Si_3O_{12}$ and $Fe_4Si_4O_{12}$ - $Fe_3Al_2Si_3O_{12}$ at high pressures and temperatures., Phys. Planet. Earth Interiors, 15: 90–106.

Allegre, C.J. and Turcotte, D.L., 1986. Implications of a two component marble-cake mantle., Nature, London. 323: 123–126.

Bonini, W.E., Loomis, T.P. and Robertson, J.D., 1973. Gravity anomalies, ultramafic intrusions and the tectonics of the region around the Straits of Gibralter., J. Geophys. Res., 78: 1372–1382.

Boyd, F.R. and Gurney. J.J., 1986. Diamonds and the African lithsophere., Science, 232: 472–477.

Cocker, J.D., Griffin, B.J. and Muehlenbachs, K., 1982. Oxygen and carbon isotope evidence for seawater hydrothermal alteration of the Macquarie Island ophiolite., Earth Planet. Sci. Lett., 61: 112–122.

Coleman, R.G., 1977. Ophiolites. Springer-Verlag, New York, 229 pp.

Dawson, J.B., 1983. New developments in diamond geology., Naturwissenschaften, 70: 586–593.

Deines, P., 1980. The carbon isotopic composition of diamonds: relationship to diamond shape, color, occurrence and vapor composition., Geochim. Cosmochim. Acta, 44: 943–961.

Fang, Chingson and Bai Wenji, 1981. The discovery of Alpine-type diamond bearing ultrabasic intrusion in Xizang (Tibet)., Int. Geol Rev., 27: 455–457 (in Chinese).

Girardeau, J. and Mercier, J-C.C., 1988. Petrology and texture of the ultramafic rocks of the

Xigaze ophiolite (Tibet): constraints for mantle structure beneath slow-spreading ridge., Tectonophysics, 147: 33–58.

Gregory, R.G. and Taylor, H.P., 1981. An oxygen isotope profile in a section of Cretaceous oceanic crust, Samail ophiolite, Oman: evidence for $\delta^{18}O$ buffering of oceans by deep (> 5 km) seawater hydrothermal alteration at mid ocean ridges., J. Geophys. Res., 86: 2737–2755.

Helmstaedt, H. and Doig, R., 1975. Eclogite nodules from kimberlite pipes of the Colorado Plateau – samples of subducted Franciscan-type oceanic lithosphere., Phys. Chem. Earth, 9: 95–111.

Jacques, A.L., Hall, A.E. et al., 1989. Composition of crystalline inclusions and C-isotopic composition of Argyle and Ellendale diamonds. In: Ross et al. (Eds), "Kimberlites and related rocks", Vol.2., Geol. Soc. Aust, Spec. Pub., 14, Blackwell: pp. 966–990.

Javoy, M., 1980. $^{18}O/^{16}O$ and D/H ratios in high temperature peridotites., Colloq. Int. C.N.R.S. 272: 279–287.

Kaminskii, F.V., 1984. The diamond content of non-kimberlitic eruptive rocks. Izd-vo Nedra, Moscow. 171 pp (in Russian).

Kornprobst, J., 1966. A propos des péridotites du massif des Beni Bousera (Rif Septentrional, Maroc)., Bull. Soc. Fr. Minéral. Cristallog., 89: 399–404.

Kornprobst, J., 1969. Le massif ultrabasique des Beni Bouchera (Rif Interne, Maroc)., Contrib. Mineral. Petrol., 23: 283–322.

Kornprobst, J., 1974. Contribution a l'etude petrographique et structurale de la zone interne du Rif (Maroc Septentrional)., Notes. Serv. Geol. Maroc. T251: 256 pp.

Kornprobst, J., Piboule, M. and Roux, L., 1982. Corundum bearing garnet pyroxenites at Beni Bousera (Morocco): An exceptionally Al-rich clinopyroxene from grospydites associated with ultramafic rocks., Terra Cognita, 2: 257–259.

Kornprobst, J. and Vielzeuf, D., 1984. Transcurrent crustal thinning: a mechanism for the uplift of deep continental crust/upper mantle associations. In: J. Kornporbst (ed) "Kimberlites II: The Mantle and Crust-Mantle Relations", Elsevier, Amsterdam, pp.347–355.

Kornprobst, J., Piboule, M., Roden, M. and Tabit, A., 1990. Corundum-bearing garnet clinopyroxenites at Beni Bousera (Morocco): original plagioclase-rich gabbros recrystallized at depth within the mantle?, J. Petrol., 31: 717–745.

Kramers, J.D., 1979. Lead, uranium, strontium, potassium and rubidium in inclusion-bearing diamonds and mantle derived xenoliths from Southern Africa. Earth Planet. Sci. Lett., 42: 58–70.

Kyser, T.K., 1986. Stable isotope variations in the mantle. In: J.W. Valley, H.P. Taylor and J.R. O'Neil (eds), "Stable Isotopes in High Temperature Geological Processes", Reviews in Mineralogy., Vol.16. Min. Soc. Am., pp.141–164.

Kyser, T.K., O'Neil, J.R. and Carmichael, I.S.E., 1981. Oxygen isotope thermometry of basic lavas and mantle nodules., Contrib. Mineral. Petrol., 77: 11–23.

Loubet, M.. and Allègre, C.J., 1982. Trace elements in orogenic lherzolites reveal the complex history of the upper mantle., Nature, 298: 809–814.

MacGregor, I.D. and Manton, W.I., 1986. Roberts Victor eclogites: an ancient oceanic crust., J. Geophys. Research., 91: 14063–14079.

McCulloch, M.T., Gregory, R.T., Wasserburg, G.J. and Taylor, H.P., 1980. A neodymium, strontium and oxygen isotopic study of the Cretaceous Samail ophiolite and implications for the petrogenesis and seawater-hydrothermal alteration of oceanic crust., Earth Planet. Sci. Lett., 46: 201–211.

Meyer, H.O.A., 1982. Mineral inclusions in natural diamond. International Geological Proceedings: Gemological Institute of America, Los Angeles, pp. 447–465.

Moore, R.O. and Gurney, J.J., 1985. Pyroxene solid solution in garnets included in diamond., Nature, London, 318: 533–555.

Nixon, P.H., Davies, G.R., Slodkevich, V.V. and Bergman, S.C., 1986. Graphite pseudomorphs after diamond in the eclogite-peridotite massif of Beni Bousera, Morocco, and a review of anomalous diamond occurrences., Fourth Internat. Kimberlite Conf., Perth, W. Australia, extend. abstr.: pp. 412–414.

Pallister, J.S. and Knight, R.J., 1981. Rare-earth element geochemistry of the Samail ophiolite near Ibra, Oman., J. Geophys. Res., 86: 2673–2697.

Pearson, D.G. 1989. The petrogenesis of pyroxenites containing octahedral graphite and associated mafic and ultramafic rocks of the Beni Bousera peridotite massif, N. Morocco., Ph.D. thesis, Dept. Earth Sci., Univ. Leeds, 413 pp.

Pearson, D.G., Davies, G.R., Nixon, P.H. and Milledge, H.J., 1989a. Graphitized diamonds from a peridoite massif in Morocco and implications for anomalous diamond occurrences., Nature, London. 338: 60–62.

Pearson, D.G., Davies, G.R., Nixon, P.H., 1989b. Graphite-bearing pyroxenites from Morocco: evidence of recycled oceanic lithosphere and the origin of E-type diamonds. Extended abstracts: Workshop on Diamonds 28th Internat., Geol. Cong. Washington, 83–86.

Pearson, D.G., Davies, G.R., Nixon, P.H., Greenwood, P.B. and Mattey, D.P., 1991. Oxygen isotope evidence for the origin of pyroxenites in the Beni Bousera peridotite massif, N. Morocco: derivation from subducted oceanic lithosphere. Earth and Planetary Sci. Lett., 102: 289–301.

Pearson, D.G., Davies, G.R., Nixon, P.H. and Mattey, D.P., 1991. A carbon isotope study of diamond facies pyroxenites and associated rocks from the Beni Bousera peridotite, N. Morocco. In Menzies, M.A., Dupuy, C. and Nicoles, A. (eds), "Orogenic Lherzolites and Mantle Processes", Spec. Pub., J. Pet. (in press).

Reuber, I., Michard, A., Chalcouan, A., Juteau, T. and Jermoumi, B., 1982. Structure and emplacement of the Alpine type peridoites from Beni Bousera. Rif Morocco: a polyphase tectonic interpretation., Tectonophysics., 82: 231–251.

Richardson, S.H. Gurney, J.J., Erlank A.J. and Harris, J.W., 1984. Origin of diamonds in old enriched mantle., Nature, London, 310: 198–201.

Ringwood, A.E., 1977. Synthesis of pyrope-knorringite solid solution series., Earth Planet. Sci. Lett., 36: 443–448.

Ringwood, A.E., 1989. Constitution and evolution of the mantle. In: Ross et al. (eds), "Kimberlites and Related Rocks", Vol.2, Geol. Soc. Aust. Spec. Pub. 14, Blackwell: 457–485.

Robinson, D.N., Gurney, J.J. and Shee, S.R., 1984. Diamond eclogite and graphite eclogite xenoliths from Orapa Botswana. In: J. Kornprobst (ed), "Kimberlites II, The Mantle and Crust-Mantle Relations", Elsevier: Amsterdam., pp. 11–24.

Schulze, D.J., 1986. Calcium anomalies in the mantle and a subducted metaserpentinite origin for diamonds., Nature, 319: 483–485.

Sharp, W.E., 1974. A plate tectonic origin for diamond-bearing kimberlites., Earth Planet. Sci. Lett., 21: 351–354.

Shilo, L.A., Kaminskii, F.V., Palandzhyan, S.M. et al., 1981. First diamond finds in alpine-type ultramafic rocks of northeastern USSR., Doklady Akad. Nauk SSSR, 241: 179–182.

Slodkevich, V.V., 1980a. Polycrystalline aggregates of octahdral graphite. Doklady, Akad. Nauk SSSR, 253: 194–196.

Slodkevich, V.V., 1980b. Graphite pseudomorphs after diamond., Internat. Geol. Rev. 25, No.5.

Slodkevich, V.V. and Lobkov, V.A., 1983. Isotopic composition of graphite from the mafic-ultramafic pluton at Beni Bousera, Morocco (in Russian)., Doklady Akad. Nauk SSSR, 272: 698–701.

Sobolev, N.V. and Lavrent'ev, Ju.G., 1971. Isomorphic sodium admixture in garnets formed at high pressures. Contrib. Mineral. Petrol. 31: 1–12.

Sobolev, N.V., Galimov, E.M. Ivanovskaya, I.N. and Yefimova, E.S., 1979. Isotopic composition of carbon of diamonds containing crystalline inclusions. (in Russian)., Doklady Akad. Nauk SSSR, 249: 1217–1220.

Thayer, T.P., 1960. Some critical differences between alpine-type and stratiform peridotite-gabbro complexes., Internat. Geol. Congr. 21st Sess. Copenhagen. 13: 247–259.

Part III
Hydrothermalism

Lithospheric Stretching and Hydrothermal Processes in Oceanic Gabbros from Slow-Spreading Ridges

CATHERINE MÉVEL[1] and MATHILDE CANNAT[2]
CNRS URA 736 Pétrologie-Magmatologie-Métallogénie Université P. et M. Curie, T26, 4 place Jussieu 75252 Paris cedex 05, France
[2]*GDR Genèse et Evolution des Domaines Océaniques UBO, 6 avenue Le Gorgeu, 29287 Brest cedex, France*

Abstract

Abundant oceanic gabbros created in slow-spreading ridges have been collected by dredging, drilling or with submersibles (Atlantic ocean crust, Mid-Cayman Rise, South West Indian Ridge). A review of published studies as well as work in progress show that these gabbros may be extensively metamorphosed and more or less deformed. Structural and petrological investigations suggest that shearing starts in the lower crust at very high temperature, before the complete solidification of the magma chamber. Continuing shearing allows seawater penetration and formation of synkinematic amphibole as temperature decreases. In the absence of ductile deformation, metamorphic reactions result from interaction between gabbros and a seawater-derived fluid phase circulating through a crack network.

We propose a model to explain the metamorphic and deformational characteristics of oceanic gabbros. We suggest that early lithospheric stretching beneath the ridge allows seawater penetration in the lower crust when it is still very hot, through permeability created by shear zones and associated synkinematic cracks. Therefore, hydration of the lower crust starts at high temperature (750°C), in contrast with a simple cracking front model in which hydration starts at temperature below 500°C. The amount of stretching may be related to the spreading rate through the magma budget.

Gabbroic series from ophiolite complexes may show either this early stretching and high temperature metamorphism associated with ductile shear zones (Western Alps ophiolites) or a crack network related to the cracking front and moderate temperature metamorphism (Haylayn massif, Oman ophiolite).

Introduction

Hydrothermal circulation of seawater below mid-ocean ridges is now a widely accepted process. It explains heat flow anomalies, and results in the metamor-

Tj. Peters et al. (Eds), Ophiolite Genesis and Evolution of the Oceanic Lithosphere, 293–312.

phism of crustal rocks at depth and in the formation of black smokers and associated metalliferous deposits on the seafloor.

Models for hydrothermal convective cells generally consider that hydrothermal circulation is mostly confined to the upper part of the oceanic crust: volcanic effusives, dike complex, and the gabbro/dike complex transition zone (Wolery and Sleep, 1976; Mottl, 1983; Fehn et al., 1983; Bowers and Taylor, 1985). In these models, interpillow spaces and joints parallel to dike margins are considered to be the main cause of crustal permeability.

Gabbroic rocks recovered from the seafloor, however, often display evidence for extensive interaction with hot seawater-derived fluids (see references infra). In such rocks hydrothermal reactions which occurred at temperatures lower than about 500°C may have resulted from seawater circulation down cracks created as a response to the thermal contraction of the cooling lithosphere, as it spread away from the ridge axis (Lister, 1974). Many of these plutonic rocks have, however, also experienced hydrothermal reactions which occurred at temperatures greater than 500°C. Such high temperature metamorphic reactions cannot be explained by Lister's (1974) thermal contraction and cracking front model; morover they suggest that early hydrothermal circulation near the ridge axis was not confined to the uppermost lavas and sheeted dikes crustal units.

Most oceanic gabbros were collected from slow-spreading ridges and fracture zones. In this paper, we will not consider the samples collected along fracture zones, where large scale strike-slip faults may be expected to provide access to seawater into the deepest crustal levels (Francis, 1981). Excluding fracture zones, it becomes clear that emplacement of lower crustal rocks in or near the seafloor, where they can be dredged, drilled, or sampled by submersibles, is characteristic of slow-spreading environments. Along the fast-spreading East Pacific Rise, away from fracture zones, plutonic rocks have only been recovered in peculiar tectonic settings such as the Mathematician Ridge failed rift (Vanko and Batiza, 1982), the Hess Deep propagator (Francheteau et al., 1990), or Pito Deep cruise of the R/V Sonne (Hekinian, personal communication). By contrast, gabbros have frequently been recovered along the slow-spreading Mid-Atlantic Ridge (see synthesis in Lagabrielle and Cannat, 1990), Mid-Caman Rise (CAYTROUGH, 1979), and the Indian Ridge (Engel and Fisher, 1975 ; Meyer et al., 1989).

Another characteristic of slow-spreading ridges is that the plutonic rocks recovered there are often plastically deformed. Textures vary from slightly deformed with abundant magmatic porphyroclasts in a recrystallized matrix, through gneissic, to truly mylonitic. This abundance of deformed samples is not consistent with the predictions of classical oceanic spreading models (i.e., Lachenbruch, 1973; Sleep, 1969), which assume that the new lithosphere formed at the ridge axis is then translated passively away from the ridge, following the diverging motion of the plates. By contrast, significant deformation of the new lithosphere near the ridge axis is consistent with more recent models of slow-spreading axial processes. These models predict that, because of low and discontinuous magma supply, the lithosphere at the ridge axis must be thick, and undergo tectonic extension (Tapponnier and Francheteau, 1978; Phipps Morgan et al., 1987; Karson, 1991; Cannat et al., 1991 a and b). So far evidence in support of these predictions has come from three locations along the slow-spreading Mid-Atlantic and Southwest Indian ridges:

1. the MARK area (Mid-Atlantic Ridge south of Kane Fracture Zone, 23°N), where normal faulting microearthquakes with focal depths down to 8 km occur right at the ridge axis (Toomey et al., 1988), indicating that the lithosphere there is indeed thick, and subjected to extensional deformation. In this same region, serpentinized mantle-derived peridotites and gabbros crop out in the Median Valley wall; these rocks are deformed by low to moderately (30–60°) dipping normal faults ((Dick et al., 1981; Karson and Dick, 1983; Karson et al., 1987; Mével et al., 1991);
2. the western intersection of the Mid-Atlantic Ridge with the Vema Fracture Zone, where seismic refraction experiments show the existence of low (30°) dipping normal faults extending at depths of 2 km below the ridge/transform intersection nodal basin (Bowen and White, 1986). These faults may be the deep equivalent of the normal faults observed in the MARK area gabbro outcrops;
3. the 500.7 m deep drill hole at Site 735 (Ocean Drilling Program, Leg 118) in the Southwest Indian Ocean, which provided gabbro samples stretched along low to moderately dipping ductile shear zones developed in granulite facies to amphibolite facies metamorphic conditions (Cannat, 1991; Cannat et al., 1991; Stakes et al., 1991). These high temperature metamorphic conditions indicate that the extensional deformation took place in the immediate vicinity of the Southwest Indian Ridge axis (Cannat et al., 1991; Dick et al., 1991).

In this paper, we review the metamorphic characteristics of gabbro samples collected along the slow-spreading Mid-Atlantic Ridge, Mid-Cayman Rise, and Southwest Indian Ridge (Figure 1). We analyse the relationships between these metamorphic characteristics and the deformational history of the gabbros. Based on these observations, we propose a model for hydrothermal circulation at slow-spreading ridges during periods of low magma supply and tectonic extension of the lithosphere. This model involves circulation of seawater at high temperatures into the deepest crustal levels, along low to moderately dipping normal faults and shear zones.

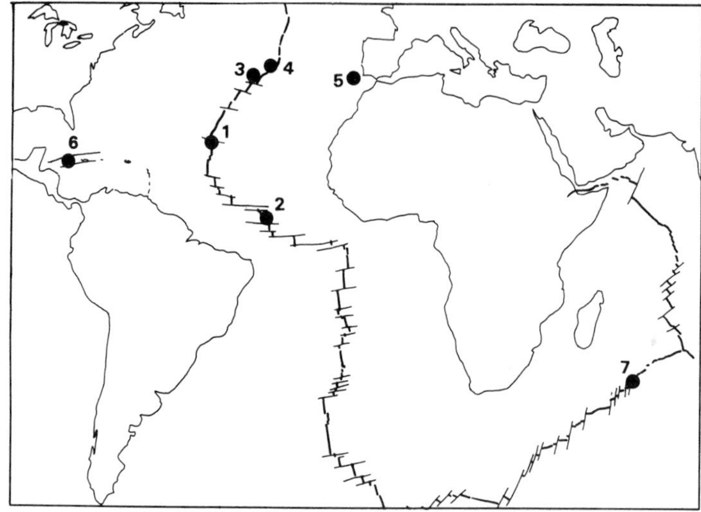

Figure 1. Location map of the areas where the gabbro suites referred to were collected (see text for references). 1 = MARK area; 2 = 6°N on the M.A.R.; 3 = DSDP Hole 334; 4 = DSDP Hole 556; 5 = Gorringe Bank; 6 Mid-Cayman Rise; 7 = ODP Hole 735B.

Location and Metamorphic Characteristics of the Gabbro Samples

Sample Collections (*Figure* 1)

In the MARK area (Mid-Atlantic Ridge/Kane fracture zone intersection, 23–24°N), gabbros crop out on the wall of the axial valley, on the eastern flank of the intersection massif (Dick et al., 1981; Karson and Dick, 1983; Mével et al., 1991). The gabbros are cut by normal fault surfaces, oriented parallel to the ridge axis and low- to moderately dipping toward the axis. Brief descriptions of gabbro samples collected with the submersibles ALVIN and NAUTILE are given respectively in Karson and Dick (1983) and Mével et al. (1991). Detailed petrological studies of the NAUTILE samples are in progress.

A suite of metagabbros together with greenschist facies metabasalts, serpentinites ans metaserpentinites was dredged at 6°N on the lower slopes of the East wall of the M.A.R. axial valley, and studied by Bonatti et al. (1975).

Two DSDP drill holes away from the M.A.R. axis penetrated gabbros at very shallow level. At site 334, gabbros occur under 59 m of basalts (DSDP Leg 37 Scientific Party, 1977). Their metamorphic history has been studied by Helmstaedt and Allen (1977). At site 556, gabbros occur as breccia elements together with mantle peridotites, under 96 m of basalts (DSDP Leg 82 Scientific Party, 1985). Their metamorphic history has been studied by Mével (1987).

Two other sets of samples come from the Gorringe bank, an anomalously high structure close to the Azores-Gibraltar line, off the coast of Portugal.

Gabbros drilled there during DSDP Leg 13 were studied by Honnorez and Fox (1972). Gabbros dredged on mount Gettysburg were studied by Prichard and Cann (1982). Gabbros collected with the submersible CYANA on both Gettysburg and Ormonde mounts (CYAGOR II Group, 1984) were studied by Mével (1988).

The Mid-Cayman Rise gabbros have been recovered with the Alvin in the walls of the axial rift valley (CAYTROUGH, 1979). The geology of the area is detailed in Stroup and Fox (1981). Microstructures of the gabbro suite have been studied by Malcolm (1981) and the metamorphic history by Ito and Anderson (1983).

The Southwest Indian ocean gabbros were drilled at ODP site 735B on a shallow flat terrace adjacent to the Atlantis II fracture zone (ODP Leg 118 scientific Party, 1989). The detailed magmatic stratigraphy of the gabbro section is given in Dick et al. (1991). Structure and metamorphism are discussed in Cannat et al. (1991a), Dick et al. (1991) and Stakes et al. (1991). Further petrological studies of metamorphic processes in this gabbros sequence are in progress.

Samples from the MARK area, Gorringe Bank, and Site 735 were recovered in the vicinity of a fracture zone, but the geometry of deformational structures in these samples (inferred from submersible observations or from the drill cores) is not compatible with strike-slip deformation in the transform fault: in the MARK area, normal faults cutting the gabbro outcrops are ridge-parallel and low to moderately dipping (Dick et al., 1981; Karson and Dick, 1983; Mével et al., 1991); in the Gorringe Bank, the gabbro foliation has a low dip (CYAGOR II Group, 1984); the ductile shear zones at Site 735 are low to moderately dipping, with down-dip stretching lineations (Shipboard Scientific Party, 1989; Cannat et al., 1991a and b). Therefore we infer that these deformational structures resulted from tectonic processes occurring at the ridge axis.

Metamorphic Characteristics.

All the gabbro collections considered in this paper are characterized by the coexistence of fresh rocks with deformed and undeformed metamorphic rocks (Figure 2). The main rock types are troctolites, pyroxene and olivine gabbros, gabbronorites, ferrogabbros. Late diorites, trondhjemites and plagiogranites may locally cross-cut the previous rock types, forming small dikelets or pockets. As many authors have pointed out, the nature and composition of metamorphic minerals in oceanic gabbros is strongly influenced, besides the temperature, by the nature and composition of the magmatic phases present in the rocks. The bulk rock composition of the gabbro samples varies only within a narrow range, corresponding to the differenciation trend of a tholeiitic liquid. The number of metamorphic species is rather restricted. Amphiboles and plagioclases are the most widespread secondary minerals. Other secondary phases include essentially pyroxenes, olivine, il-

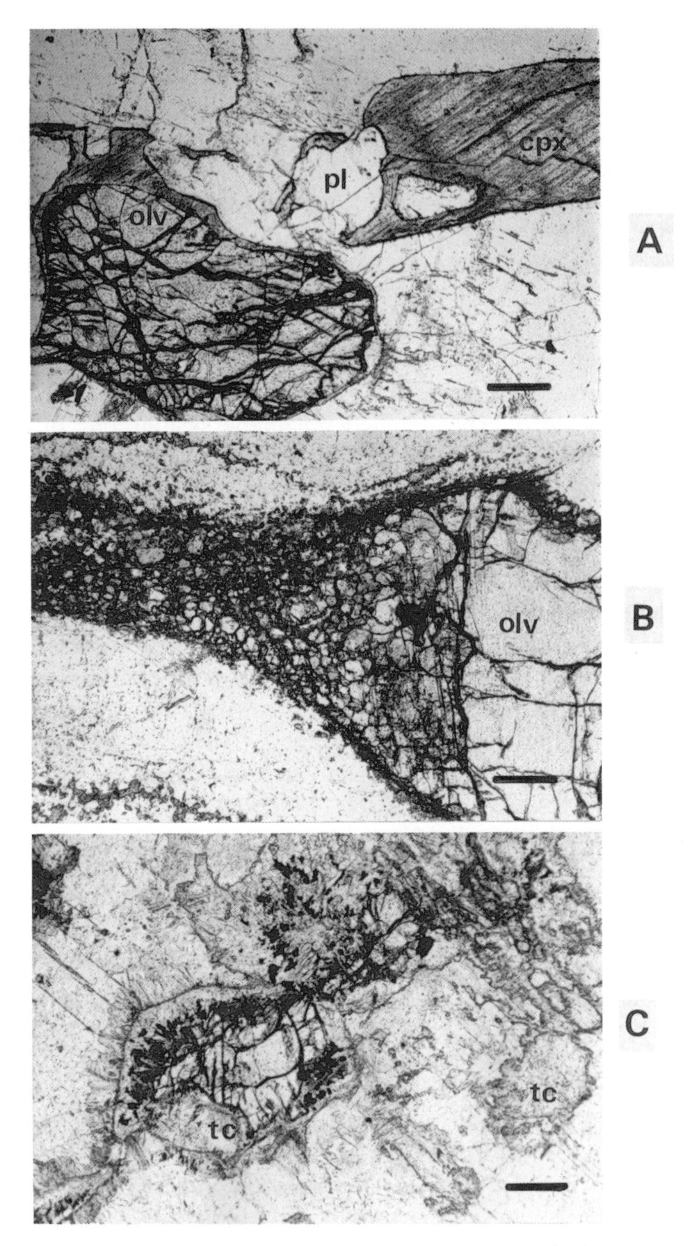

Figure 2. Microphotographs of textures coexisting in oceanic gabbro collections: example of Hole 735B. A = fresh olivine (olv) – clinopyroxene (cpx) and plagioclase (pl) gabbro (sample 735B 85–R6 #4). B = magmatic olivine partly recrystallized as granular grains in a ductile shear zone (sample 735B 8D-1 #4). C = Olivine partly pseudomorphosed by talc (tc) + magnetite (in black) in an undeformed metagabbro (sample 735B 87–R7 #6). Scale bars = 200 μm.

menite, epidote, talc, magnetite, quartz, chlorite, prehnite, and low temperature minerals such as zeolites, clay minerals, and calcite. In most of the rocks, metamorphic textures suggest a lack of complete metamorphic equilibrium: magmatic relicts are abundant, either as partially pseudomorphed or as porphyroclasts. Homogeneization is therefore incomplete. Textures often indicate sequential crystallization, evidenced by compositional zoning of secondary minerals, particularly amphiboles. This zoning may result either from progressive change in external conditions or from the evolution of the reacting fluid phase composition. It is therefore difficult to precisely assess temperature conditions. This estimation can be inferred from mineral assemblages, pyroxene pairs when present, amphibole and plagioclase compositions.

Distribution of Metamorphic Crystallizations in the Lower Crust

Samples dredged or collected by submersible do not provide continuous cross-sections of gabbro outcrops, and there is no way to check the geometrical relationships between the deformational structures observed in each sample. Drilling provides more continuous sets of samples but the degree of continuity depends on recovery rates. Drilled samples are also oriented with respect to the vertical. Therefore the relationships between rock types, metamorphic characteristics and structural evolution are more easily constrained in drilled samples than in samples dredged or collected by submersible. Because of exceptionally high recovery rates, ODP hole 735 (Southwest Indian Ocean) provides a unique opportunity to examine the distribution of metamorphic crystallizations within a gabbro section 500.7 m long. Metamorphosed gabbros from site 735 are often remarkably similar in texture and mineralogy to samples from the other localities examined in this paper. This similarity suggests that we may use the observations carried out at site 735 as guidelines to interpret the metamorphic characteristics of our other sets of samples. Figure 3 is a simplified structural and metamorphic log for the site 735 drill hole, adapted from Stakes et al. (1991) and Cannat et al., (1991). The Site 735 gabbroic section is divided into six lithologic units (ODP Leg 118 Scientific Party, 1989). Pockets or dikelets of late differentiated diorite or plagiogranite intrude the gabbros at various levels. The deformation column points out the heterogeneous distribution of ductile shear zones. Some portions of the hole are completely undeformed, mainly between core 57 and 76. The number of macroscopic veins per meter of core does not correlate with the degree of deformation. Mylonitic gabbros (Type V in the deformation column) have very few veins. Moderately deformed gabbros contain abundant veins, some of which are synkinematic as discussed later. But there are also undeformed intervals which contain abundant veins, like those around core 63 and 69. The alteration index in the last column is estimated from the volume proportion (measured by point counting) of secondary amphibole (Amp) replacing primary (magmatic) clinopyroxene

ODP site 735 B

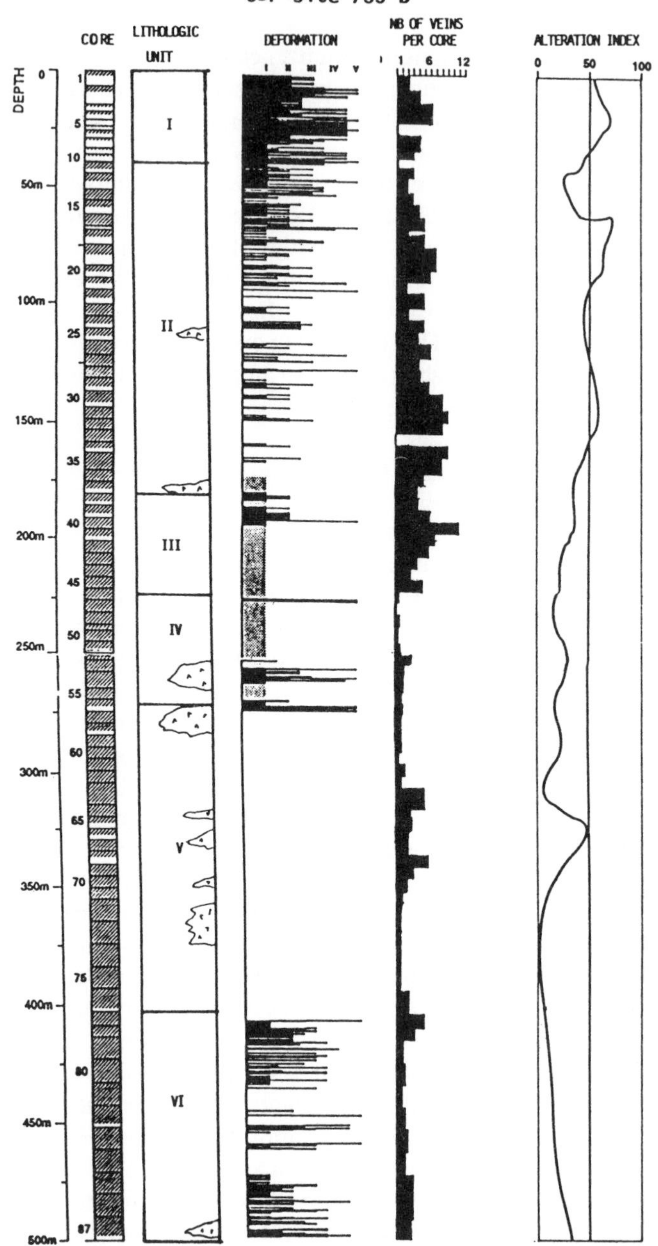

Figure 3. Lithostratigraphic log of Hole 735B, adapted from Stakes et al. (1991) and Cannat et al. (1991). Left hand columns: depth below seafloor and core numbers. Lithologic units: I = gabbronorite with minor olivine gabbro, II = olivine gabbro with intervals of oxide gabbro, III = olivine gabbro with disseminated oxides and olivine gabbro with intervals of oxide gabbro, IV = oxide gabbro, V = olivine gabbro, VI = olivine gabbro with intervals of oxide gabbro and troctolite. Deformation: I to V = deformation textures varying from weakly recrystallized and unfoliated to mylonitic. Number of veins per core counted on board. Alteration index: see text for explanation.

(Cpx). This alteration index is equal to Amp/(Amp + Cpx) $\times$ 100, where Cpx is the amount of relictual magmatic clinopyroxene. The alteration index is high in most shear zones and in intervals with numerous veins. In the absence of both shear zones and veins (for instance between cores 72 and 74), the alteration index falls close to zero. This indicates that the hydrothermal circulations which caused crystallization of secondary amphibole in the site 735 gabbros were related either to ductile deformation or to the development of a crack network. It also points out that there is no simple relationship between depth and metamorphic intensity or grade.

In the next two sections, we will describe metamorphic textures and mineral asssemblages characterizing first ductily deformed samples or zones and second undeformed samples or zones.

Ductile Deformation and Hydrothermal Metamorphism

Deformed gabbros from all sample collections display metamorphic crystallization at progressively decreasing temperatures and increasing water/rock ratio. This is evidenced by sequential paragenesis (amount of secondary amphibole increasing and secondary pyroxene decreasing) and by oxygen isotopes data (Stakes et al., 1991).

Early, highest temperature, and lowest water/rock ratio deformed assemblages display granular textures, made up of neoblasts with triple junctions (Figure 2B). These granular rocks occur in DSDP holes 334 (Helmstaedt and Allen, 1977) and 556 (Mével, 1987), in Gorringe bank (Prichard and Cann, 1982; Mével, 1988) and in ODP hole 735B (Cannat et al., 1991; Stakes et al., 1991) Depending on the primary mineralogy of the gabbros, the neoblasts consist of plagioclase, clinopyroxene, orthopyroxene, Ti-rich hornblende, and/or olivine. The proportion of Ti-rich hornblende neoblasts is always low, and their composition is similar to that of primary magmatic hornblende in the undeformed paragenesis (Ti-rich pargasites or pargasitic hornblende), although possibly a little lower in Al. Analyses of plagioclase neoblasts show exactly the same composition as the primary crystal, suggesting no exchange with a fluid phase. Coexisting orthopyroxene and clinopyroxene occur in some gabbronorites from DSDP Hole 334 (Helmstaedt and Allen, 1977) and ODP Hole 735B (Stakes et al., 1991). Equilibrium temperatures are estimated at 800°C for Hole 334 assemblages and varying from 850 to 900°C for Hole 735B assemblages. These temperatures are consistent with granulite facies metamorphic conditions.

When temperature decreases, percolation of an external fluid phase starts as evidenced by the development of abundant synkinematic amphibole in shear zones. The latter exist in all the rock collections mentioned. The fact that the fluid phase is seawater-derived is demonstrated by the oxygen isotope ratios (Ito and Clayton, 1983; Stakes et al., 1991).

The composition of these synkinematic amphiboles varies widely, as a

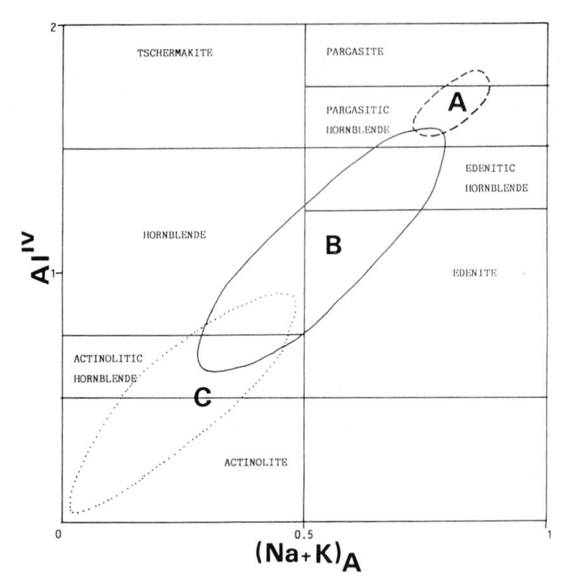

Figure 4. Compositional fields of secondary amphiboles from Hole 735B gabbros plotted in an Al versus (Na + K) diagram. A = brown amphiboles associated with granular pyroxene in early shear zones; B = green amphibole in shear zones and associated synkinematic veins; C = amphiboles in post-kinematic veins and pyroxene pseudomorphs.

function of the bulk rock composition (Fe/Mg), the fluid phase (particularly the Cl content), and also at a function of temperature (for the Si, Al and alkali content). Figure 4 shows the compositional fields of amphiboles from Hole 735B gabbros. Early amphiboles, associated with abundant secondary clinopyroxene, are typically pargasitic hornblendes, which may be either rich in titanium in ferrogabbros or poor in titanium in clinopyroxene or olivine gabbros. The disappearance of secondary clinopyroxene is accompanied by a progressive change of the amphibole composition which evolves to hornblende s.l. then actinolitic hornblende. In the meantime, the secondary plagioclase becomes more acidic, evolving toward andesine compositions (Stakes et al., 1991; Vanko and Stakes, 1991). Similar compositional evolution has been reported in the Mid-Cayman Trough gabbros (Ito and Anderson, 1983) and in the Gorringe bank gabbros (Mével, 1988).

The evolution of amphibole and plagioclase compositions, and the disappearance of secondary clinopyroxene suggest a progressive lowering of the temperature from upper amphibolite facies to greenschist/amphibolite transitional facies. The increasing proportion of amphibole suggests increasing water/rock ratio.

But the complete sequence does not exist everywhere: in Hole 556 gabbros, ductile deformation ceased just after high temperature recrystallization and no synkinematic hornblende has been observed (Mével, 1987). On the other hand deformation of the 6°N gabbros occurred at moderate tempera-

Figure 5. Microphotograph of synkinematic veins filled with amphibole. The veins cross-cut the plagioclase porphyroclasts but do not continue into the recrystallized matrix. Sample 735B 13–R1 #6. Scale bar = 200 μm.

ture, around 500°C or below and synkinematic amphibole is actinolitic hornblende and actinolite (Bonatti et al., 1975). In hole 334 gabbros, there is a gap between granulite facies assemblages and development of synkinematic amphibole which is actinolitic hornblende (Helmstaedt and Allen, 1977). In the Mid-Cayman Rise gabbros, most of the deformation occurred between 750 and 450°C (Ito and Anderson, 1983). These variations may be attributed to an incomplete sampling (compared to the almost continuous drill core at site 735B) or to local specific evolution.

Ductile shearing in the most highly deformed samples from Site 735 induced brittle failure and boudinage of nearby less deformed gabbros (Cannat, 1991; Cannat et al., 1991; Dick et al., 1991). This is evidenced by the texture shown in Figure 5: the cracks cross-cut relict magmatic plagioclase porphyroclasts but do not continue into the mylonitic matrix, demonstrating that it was still deforming when the cracks were created. The resulting synkinematic crack network was filled with hornblende compositionally similar to the hornblende which crystallized in the mylonitic foliation (Stakes et al., 1991). This crack network created a permeability allowing seawater circulation away from shear zones, and inducing pseudomorphic replacement of pyroxene, olivines, and plagioclase by secondary minerals. In the other sample sets, relationships between cracks and shear zones is difficult to precise because of the lack of sampling continuity. In Gorringe bank samples, however, similar synkinematic cracks are present.

Most plastically deformed gabbros exhibit high temperature (amphibolite facies or granulite facies) synkinematic assemblages. This probably results from the tendancy for ductile deformation to become more localized as

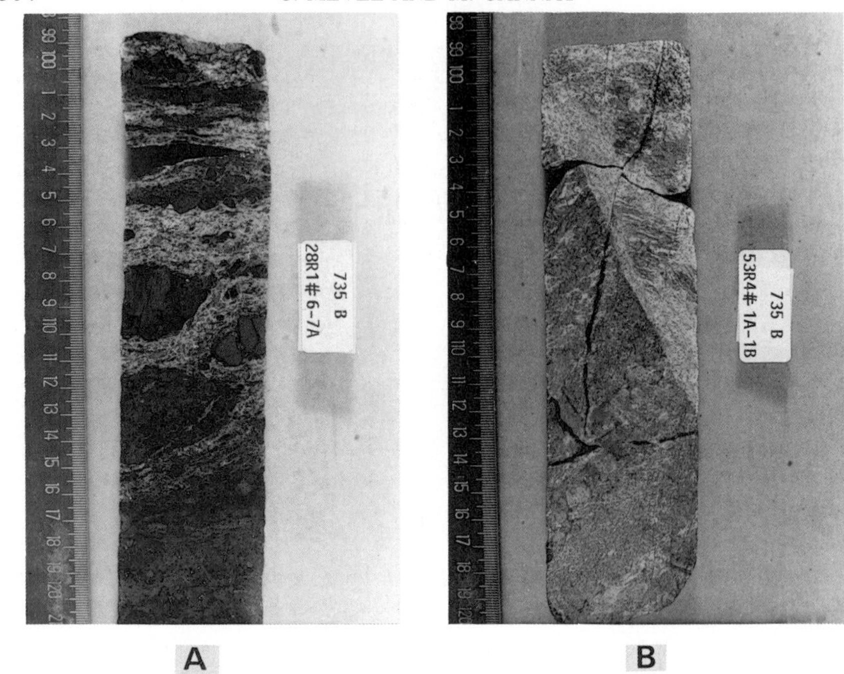

Figure 6. Textural relationship between late magmatic intrusions and deformation in Hole 735B. A = synkinematic late magmatic intrusion; the white trondhjemitic matrix of the magmatic breccia is clearly deformed (sample 735B 28–R1 #6 and 7. B = post kinematic late magmatic intrusion: the undeformed white trondhjemitic matrix encloses elements of deformed and foliated gabbro (sample 735B 53–R4 #1).

temperature decreases, and therefore as the yield strength of the rocks increases. This is well illustrated in site 735 gabbros, where shear zones formed under temperature conditions consistent with the development of metamorphic assemblages transitional between amphibolite and greenschists facies, are rare and thin (Cannat et al., 1991a).

Small pockets or dikelets (generally pluricentimetric) of differentiated rocks are quite common in the studied gabbro collections: they have been described in Hole 556 (Mével, 1987), Gorringe Bank (Mével, 1988), and in Hole 735B (Leg 118 Scientific Party, 1989; Stakes et al., 1991; Dick et al., 1991). They also occur in gabbros dredged from the Indian Ridge (Engel and Fisher, 1975). These rocks are made up of acidic plagioclase + quartz + biotite + zircon + sphene + apatite + diopside. They represent the product of the latest stage of magmatic crystallization of the gabbros. In the Gorringe bank sample set, acidic rocks are typically more deformed than the enclosing gabbros, as if they had concentrated shearing (Mével, 1988). In Hole 735B, late acidic rocks may be deformed in ductile shear zones (Figure 6A), but they more frequently postdate ductile deformation and associated hydration (Figure 6B) (Cannat et al., 1991; Dick et al., 1991). The injection of the late magmatic liquids in already deformed and hydrated rocks produced magmatic breccias with gneissic amphibolite elements in a cement of acidic rocks (Figure 6B). In Hole 556 gabbros, the only dioritic vein observed is undeformed and shows no relationship with the granular zones (Mével, 1987).

In summary, studies of deformed gabbros from The Mid-Atlantic, Mid-Cayman, and Southwest Indian slow-spreading ridges show that:

- ductile deformation may have started at temperature as high as 900°C, under mostly anhydrous conditions;
- continuing shearing as temperature decreased favored seawater penetration in synkinematic cracks, or along grain boundaries in shear zones; crystallization of synkinematic amphiboles occurs in the shear zones and the associated crack network;
- the last products of magmatic crystallization were often emplaced during and/or after shearing, when seawater penetration had already started.
- ductile deformation ceased when temperature dropped below 350–400°C.

Hydrothermal Metamorphism Under Static Conditions

Under static conditions, metamorphic reactions only result from the interaction of the gabbro with a fluid phase circulating through a crack network. The textures show that fluid circulation along grain boundaries is minor: the gabbro is commonly altered only within a few millimeters from the cracks. Farther away, the gabbros remain fresh.

Metamorphic reactions in the gabbros developed mostly during the initial period of opening of the cracks, at temperatures corresponding to amphibolite facies, or to the transitional facies between lower amphibolite and greenschist (temperatures between 500° and 350°C). When cracks cross-cut pyroxene crystals, the latter are replaced by amphibole; when they cross-cut olivine, the latter are replaced by an intergrowth of talc, magnetite, Ca and Fe-Mg amphiboles and sometimes chlorite (Figure 2C). Amphiboles vary from hornblende to actinolite (Mével, 1987; Stakes et al., 1991; Vanko and Stakes, 1991) (Figure 4). Plagioclase statically recrystallized along the edges of the cracks has andesine to oligoclase composition (Mével, 1987, 1988; Stakes et al., 1991; Vanko and Stakes, 1991). When cracks cross-cut shear zones, synkinematic minerals recrystallized to static assemblages (Figure 7).

Most cracks appear to have sealed quickly under high to moderate temperature conditions. They were filled with secondary minerals such as hornblende, actinolite, cpx, indicating temperatures of over 350°C. It is only under particular local conditions that they have been reopened and filled with lower temperature minerals. For instance, in the lower part of Hole 735B, hydrothermal breccias, likely related to the emplacement of late magmatic intrusions (diorites, trondhjemites), are associated with a crack network filled with low-temperature minerals such as chlorite, thompsonite,

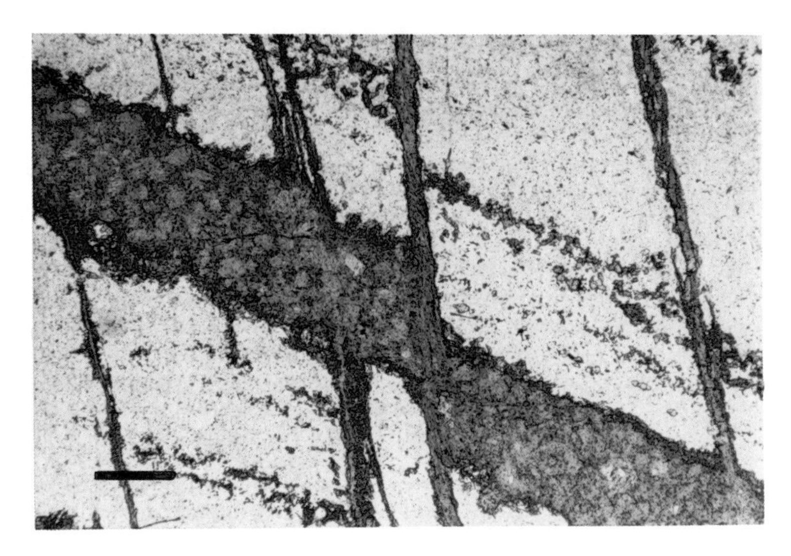

Figure 7. Microphotograph of postkinematic veins cross-cutting a foliated gabbro (samples 735B 13–R1 #6). In the foliation planes, synkinematic pyroxenes are replaced by static amphibole contemporaneous of the vein filling. Scale bar = 100 μm.

natrolite, prehnite (Stakes et al., 1991; Dick et al., 1991) indicative of greenschist to zeolite metamorphic facies. In the MARK area, a crack network parallel to low-dipping normal fault plane is consistently filled with greenschist facies minerals such as chlorite, epidote, albite, prehnite (Mével et al., 1991), overprinting higher temperature assemblages.

Discussion

The metamorphic and deformational characteristics of gabbro samples collected along the Mid-Atlantic, Mid-Cayman and Southwest Indian ridges suggest a model for the hydrothermal circulations in gabbros from slow spreading oceanic ridges. This model is summarized in the three cartoons of Figure 8, representing three successive cooling stage. Intentionally, there are no depth nor time scale in this figure.

Ductile shearing starts at very high temperature, when the gabbro section is still over 800°C and likely not completely solidified (Figure 8A). This is evidenced by the presence of a magmatic foliation in some portions of Hole 735B (Cannat et al., 1991; Dick et al., 1991). Shearing continues as temperature decreases (Figure 8B). Seawater starts to penetrate along grain boundaries in shear planes at temperatures as high as 700°C, evidenced by the synkinematic assemblages of pargasitic hornblende + clinopyroxene. Seawater circulation is also favored by the formation of a synkinematic crack network. The portions away from shear zones remain unaltered. This shear

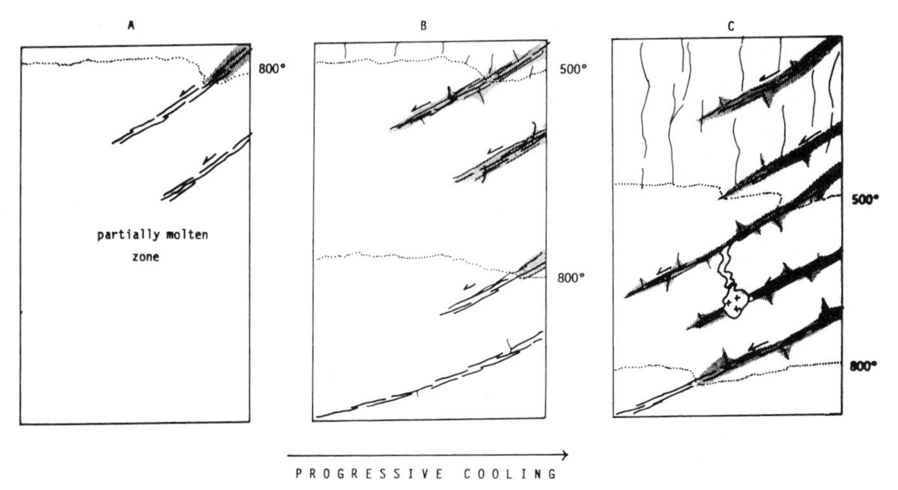

Figure 8. Cartoons summarizing three stages of the evolution of the lower crust as temperature decreases and stretching occurs (see text for explanation). Dotted areas: areas where seawater penetrates and reacts at high temperature, along shear planes and associated synkinematic cracks; subvertical thin lines = cracks related to the cracking front along which seawater penetrates at temperature lower than 500°C; dotted lines = isotherms. A late magmatic intrusion is figured with small crosses.

zone-related hydrothermal alteration is most efficient at temperatures over about 500°C. At lower temperatures, i.e. corresponding to the conditions of transition between amphibolite facies and greenschist facies, ductile deformation becomes very localized. The volume of gabbro affected by synkinematic hydrothermal metamorphism therefore becomes rapidly smaller. The latest differentiated magma are often emplaced at this stage, forming small pockets or dikelets. They may be deformed in the shear planes, in some cases concentrating the deformation. They can also postdate the ductile deformation and related seawater penetration.

Many gabbros also show evidence for the development of a late crack network after ductile deformation has ceased. This crack network crosscuts both fresh and undeformed rocks and metamorphosed and deformed rocks (Figure 8C). It creates a permeability allowing seawater circulation. The development of this post-ductile deformation crack network may be purely thermal, being related to Lister's (1974) cracking front which is efficient at temperatures below 500°C. Observation of the exceptionally continuous site 735 gabbro section suggests however that thermal contraction is not the leading cause for the crack network there. Cracks are not present everywhere. Their distribution (Stake et al., 1991; Dick et al., 1991) is not compatible with a simple cracking front model. The gabbros at site 735, cut by numerous shear zones, are rheologically heterogeneous and isotherms are disturbed by the penetration of seawater along fault planes. Therefore a regular vein network due to the progressive cooling of a homogeneous gabbro

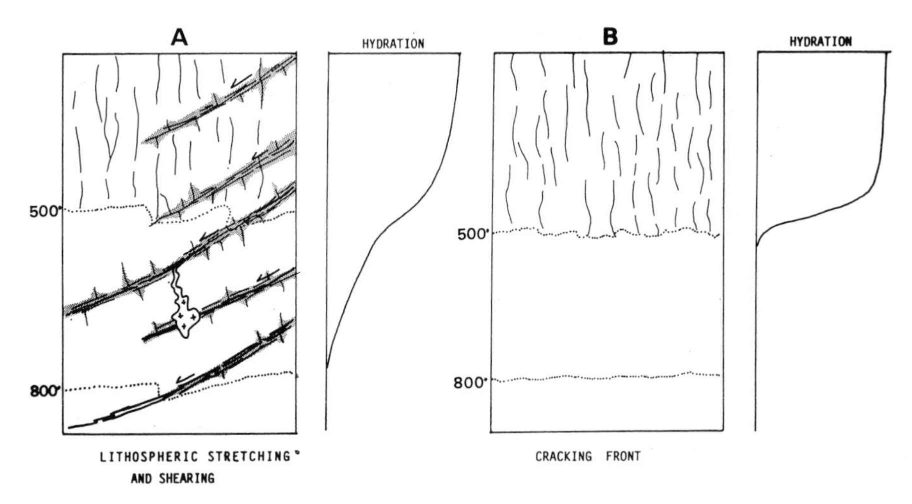

Figure 9. A = model for lower crust produced in slow spreading environment, where abundant early stretching occurs. Hydration of the gabbros starts at temperatures around 750°. B = cracking front model (after Lister, 1974). Hydration of the lower crust begins only when the cracks open, at temperature below 500°C. Symbols same as in Figure 8.

pile is not likely to form. Numerous cracks in the lower section of Hole 735B are actually inferred to result from hydraulic fracturing by late magmatic or hydrothermal fluids (Stakes et al., 1991). The persistance of tectonic activity may also systematically reopen the cracks. This is evidenced in the MARK area gabbros, where a dense crack network is parallel to the low-dipping fault surfaces (Mével et al., 1991) and not vertical, as expected from a cracking front model. High temperature hydrothermal brecciation is also present there (Delaney et al., 1987; Kelley and Delaney, 1987). Seismic experiments and submersible observations in the M.A.R. (Mutter, personal communication; Dick et al., 1981; Karson and Dick, 1983; Mével et al., 1991) indicate that the deep crust below the ridge is affected by low to moderately dipping normal faults. Such faults may induce brittle failure in the surrounding cooling gabbros, generating crack networks. Another possible effect of such faults may be to disrupt the original lithologic and metamorphic stratigraphy.

The main implication of our model is that early ductile shearing of the lithosphere, which we believe may be widespread along slow-spreading, magma starved, oceanic ridges, allows seawater to penetrate very early into the lower crust, producing hydrous metamorphic reactions at temperatures as high as 700°C (Figure 9A). It contrasts with a pure cracking front model, in which permeability is created only when temperature falls below 500°C (Figure 9B): i.e. farther away from the ridge (Lister, 1974). Such early and deep hydrothermal circulations may further accelerate the cooling of the lithosphere at the ridge axis, therefore increasing the effect of the reduced magma budget and allowing the lithosphere to thicken even more. Morphological studies of fast-spreading ridges also suggest the existence of alternating stages of magmatic and tectonic activity (Kappel and Ryan, 1986; Gente et al., 1986) and the occurrence of stretching at some stages cannot be completely ruled out. However it does not appear to be a common process leading to the exposure of deep crustal rocks in the axial valley as in slow spreading ridges. Presently we have only little information concerning the gabbros from fast-spreading ridges. Data about the metamorphism of the lower crust created at the fast-spreading East Pacific Rise and exposed in Hess Deep are not yet available. Gabbroic rocks collected from the Mathematician ridge include strongly deformed samples (Stakes and Vanko, 1986), but they occur in a failed rift, which is a peculiar tectonic environment. We think however that the amount of ductile shearing and therefore high temperature hydration of the lower crust may well be a function of the spreading rate.

The relationships between the fluids circulating in the lower crust through the shear zones and associated synkinematic cracks and the black smokers is not established yet. Since most models of convective cells producing the observed ridge axis hydrothermal vents imply interaction at temperature in the order of 400°C, there might be two different systems of seawater circulation. In any case, it should be pointed out that lithospheric stretching at slow-spreading ridges leads to the construction of an oceanic lithosphere which departs from widely accepted models implying a layered structure.

Comparison with Ophiolites

Since ophiolites are obducted fragments of oceanic crust, it is of great interest to compare the deformational and metamorphic characteristics in the gabbroic section of ophiolite sequences with the two models of Figure 9. Without attempting a general review of ophiolite complexes, we want to point out that the two situations corresponding to Figures 9A and 9B have been recorded in ophiolitic gabbros.

In the Western Alps, Le Chenaillet massif is a small ophiolitic body which is part of the Piemont zone. The ophiolite sequence includes serpentinized peridotites, gabbros and pillo-lavas; the contact between these different lithological units is interpreted as tectonic. The gabbroic section, about 500 m thick, is characterized by numerous high temperature ductile shear zones, formed under granulite to amphibolite facies conditions (Mével et al., 1978). The textures are very similar to those of Hole 735B gabbros, with granular dry assemblages and subsequent synkinematic amphibole. This early high temperature deformation and interaction with seawater occurred at temperature over 500°C, and before the complete crystallization of the magma chamber: outcrops of magmatic breccias diplay elements of deformed and recrystallized gabbros (flaser-gabbros) cemented by a matrix of diorite. This type of relationship is very similar to the situation illustrated in Figure 9A and interpreted as resulting from early stretching.

On the other hand, the Semail ophiolite, in Oman, is a very continuous ophiolite nappe, showing a complete sequence from mantle to extrusives, with a thick gabbro pile and an impressive dyke complex. Hydrothermal alteration of the gabbro section has been studied in the Haylayn massif (Nehlig and Juteau, 1988). The earliest evidence of hydrothermal alteration in these gabbros is related to a dense network of subparallel, millimiter thick, closely spaced amphibole veins, perpendicular to the layering. Amphibole compositions vary from actinolitic hornblende to actinolite suggesting interaction at temperatures below 500°C. As proposed by Nehlig and Juteau, the geometry of this crack network as well as the temperature of interaction are consistent with the cracking front model illustrated in Figure 9B.

The completely different pattern of alteration between the Chenaillet and Haylayn massifs may be related to the spreading rate of the ridge system in which they were formed. The Chenaillet massif would represent a slow-spreading oceanic crust and the Oman ophiolite may represent fragments of a fast-spreading ocean (Nicolas et al., 1988).

Acknowledgements

This work benefited of stimulating discussions with scientists onboard ships during cruises in slow-spreading environments (CYAGOR II, HYDROSNAKE, VEMANAUTE, ODP Leg 118), and after. We are particularly grateful to D.S. Stakes, J.A. Karson, D.A. Vanko, and J.C. Mutter. We also thank P. Nehlig for fruitful discussion about the Oman ophiolite. A. Nicolas and J. Honnorez gave contructive comments and contributed to improve the manuscript. This work was supported by CNRS-INSU through an "ASP ODP" and an IST "Geosciences Marines" grant.

References

Bonatti, E., Honnorez, J., Kirst, P. and Radicati, F., 1975. Metagabbros from the Mid-Atlantic Ridge at 06°N: contact hydrothermal dynamic metamorphism beneath the axial valley., J. Geol., 83: 61–78.

Bowen, A.N. and White, S., 1986. Deep-tow seimic profiles from the Vema transform and ridge/transform intersection, J. Geol. Soc. London, 143: 807–817.

Bower, T.S. and Taylor, H.P. Jr, 1985. An integrated chemical and stable isotope model of the origin of mid-ocean ridge hot spring system, J. Geophys. Res., 90, B14: 12583–12606.

Cannat, M., 1991. Plastic deformation at an oceanic spreading ridge: a microstructural study of the site 735 gabbros (Southwest Indian ocean). Proceedings of the ODP, Leg 118.

Cannat, M., Mével, C. and Stakes, D., 1991. Normal ductile shear zones at an oceanic spreading ridge: tectonic evolution of the site 735 gabbros, SW Indian Ocean, Proceedings of the ODP, Leg 118, part B.

Cannat, M., Mével, C. and Stakes, D., 1991. Stretching of the deep crust at the slow-spreading SW Indian Ridge. Tectonophysics.

CAYTROUGH, 1979. Geological and geophysical investigation of the Mid-Cayman Rise spreading center: intitial results and observations. In: Talwani, M., Harrison, C.G. and Hayes, D.E. (eds), Maurice Ewing series 2, Washington D.C, AGU, pp. 66–95.

CYAGOR II Group, 1984. Intraoceanic tectonism on the Gorringe Bank. In: I.G. Gan, S.J. Lippard and A.W. Shelton (eds), Opiolites and oceanic lithosphere, Blackwell, London, 113–120.

Delaney, J.R., Mogk, D.W. and Mottl, M.J., 1987. Quartz-cemented breccias from the Mid-Atlantic Ridge: samples of high salinity hydrothermal upflow zone J. Geophys. Res., 92: 9175–9192.

Dick, H.J.B., Thompson, G. and Bryan, W.B., 1981. Low-angle faulting and steady state emplacement of plutonic rocks at ridge-transform intersections, Eos, 62: 406.

Dick, H.J.B., Meyer, P.S., Bloomer, S., Kirby, S., Stakes, D.S. and Mawver, C., 1991. Lithostratigraphic evolution of an in-situ section of oceanic layer 3. Proceedings of ODP, vol. 118.

DSDP Leg 37 Scientific Party, 1977. Site 334, Initial Reports of the DSDP, 37: 201–238.

DSDP Leg 82 Scientific Party, 1985. Site 556. Initial Report of the DSDP, 82: 61–244.

Engel, C.G. and Fisher, R.L., 1975. Granitic to ultramafic rock complexes of the Indian ocean ridge system, western Indian ocean, Geol. Soc. Amer. Bullt., 86: 1553–1578.

Fehn, U., Green, K.E., Von Herzen, R.P. and Catthles, L.M., 1983. Numerical models for the hydrothermal field at the Galapagos spreading center, J. Geophys. Res., 88: 1033–1048.

Francheteau, J., Armijo, R., Cheminée, J.L., Hekinian, R., Lonsdale, P. and Blum, N., 1990. 1 Ma East Pacific Rise oceanic crust and uppermost mantle exposed in Hess Deep. Earth Plan. Sci. Letters, 101: 281–295.

Francis, T.J.G., 1981. Serpentinization faults and their role in the tectonics of slow-spreading ridges, J. Geophys. Res., 86(11): 616–622.

Gente, P., Auzende, J.M., Renard, V., Fouquet, Y. and Bideau, D., 1986. Detailed geological mapping by submersible of the East Pacific Rise axial graben near 13°N, Earth Planet. Sci. Letters, 78: 224–236.

Helmstaedt, H. and Allen, J.M., 1977. Metagabbronorites from DSDP Hole 334. An example of high temperature deformation and recrystallization near the Mid-Atlantic Ridge, Can. J. Earth Sci., 14: 886–898.

Honnorez, J. and Fox, P.J., 1972. Petrography of the Gorringe bank "basement". In: Ryan, W.B.F. and Hsü, H.J., Initial Reports of the Deep Sea Drilling Project, vol XIII: 747–752.

Ito, E. and Anderson, A.T. Jr, 1983. Submarine metamorphism of gabbros from the Mid-Cayman Rise: petrographic and mineralogic constraints on hydrothermal processes at slow-spreading ridges, Contrib. Mineral. Petrol., 82: 371–388.

Ito, E. and Clayton, R., 1983. Submarine metamorphism of gabbros from the mid-Cayman Rise: an oxygen isotopic study, J Geochim. Cosmochim. Acta, 47: 535–546.

Kappel, E.S. and Ryan, W.F.B., 1986. Volcanic episodicity and a non-steady state rift valley along northeast Pacific spreading centers: evidence from SeaMARC I, J. Geophy. Res., 91: 13925–13940.

Karson, J.A., 1991. Seafloor spreading on the Mid-Atlantic Ridge: implication for the structure of ophiolite and oceanic lithosphere produced in slow-spreading environments. In: J. Malpas, E.M. Moores, A. Panayiotou and C. Xenophontos (eds), Ophiolites: oceanic crust analogues, Geological Survey, Nicosia, Cyprus, 547–556.

Karson, J.A. and Dick, H.J.B., 1983. Tectonics of ridge-transform intersections at the Kane fracture zone., Mar. Geophys. Res., 6: 51–91.

Karson, J.A., Thompson, G., Humphris, S.E., Edmond, J.M., Bryan, W.B., Brown, J.R., Winters, A.T., Pockalny, R.A., Casey, J.F., Campbell, A.C., Klinkhammer, C., Palmer, M.R., Kinzler, R.J., and Sulanowska, M.M., 1987. Along axis variations in seafloor spreading in the MARK area Nature, 328: 681–685.

Kelley, D.S. and Delaney, J.R., 1987. Two phase separation and fracturing in mid-oceanic ridge gabbros at temperatures greater than 700°C, Earth Planet. Sci. Letters, 83: 53–66.

Lachenbruch, A.H., 1973. A simple mechanical model for oceanic spreading centers, J. Geophys. Res., 78: 3395–3417.

Lagabrielle, Y. and Cannat, M., 1990. Ancient alpine jurassic ocean floor resembles the modern central Atlantic basement. Geology, 18: 319–322.
Lister, C.R.B., 1974. On the penetration of water into hot rocks. Geophys, J.R. Astr. Soc., 39 465–509.
Malcolm, F.L., 1981. Microstructures of the Cayman trough gabbros, J. Geol., 89: 675–688.
Mével, C., 1987. Evolution of oceanic gabbros from DSDP leg 82: influence of the fluid phase on metamorphic crystallizations, Earth Planet. Sci. Letters, 83: 67–79.
Mével, 1988. Metamorphism of oceanic layer 3, Gorringe bank, Eastern Atlantic, Contrib. Mineral. Petrol., 100: 496–509.
Mével, C., Caby, R. and Kienast, J.R., 1978. Amphibolite facies conditions in the oceanic crust: example of amphibolitized flaser-gabbros and amphibolites from the Chenaillet ophiolite massif (Hautes Alpes, France), Earth Planet. Sci. Lett., 39: 98–108.
Mével, C., Cannat, M., Gente, P., Marion, E., Auzende, J.M. and Karson, J.A., 1991. Emplacement of deep crustal and mantle rocks on the west median valley wall of the MARK area (M.A.R., 23°N). Tectonophysics.
Meyer, P.S., Dick, H.J.B. and Thompson, G., 1989. Cumulate gabbros from the Southwest Indian Ridge, 54°S–7°16′E: implications for magmatic processes at a slow-spreading ridge.
Nicolas, A., Boudier, F. and Ceuleneer, G., 1988. Mantle flow patterns and magma chambers at ocean ridges. Evidence from the Oman ophiolite, Marine Geophys. Res., 9: 293–310.
Mottl, M.J., 1983. Metabasalts, hot springs and the structure of hydrothermal systems at mid-ocean ridges. Geol. Soc. Amer. Bull., 94: 161–180.
Nehlig, P. and Juteau, T., 1988. Flow porosities, permeabilities and preliminary data on fluid inclusions and fossil geothermal gradients in the crustal sequence of the Sumail ophiolite, Oman. Tectonophysics, 151, 199–221.
ODP Leg 118 Scientific Party, 1989. Proceedings of the ODP, volume 118.
Phipps-Morgan, J., Parmentier, E.M. and Lin, J., 1987. Mechanisms for the origin of Mid-Ocean Ridge axial topography: implication for the thermal and mechanical structure of accreting plate boundaries, J. Geophy. Res., 92: 12823–12836.
Prichard, H.M. and Cann, J.R., 1982. Petrology and mineralogy of dredged gabbros from Gettysburg bank, North Atlantic ocean., Contrib. Mineral. Petrol., 79: 46–55.
Sleep, N.H., 1969. Sensitivity of heat flow and gravity to the mechanism of seafloor spreading, J. Geophys. Res., 74: 542–549.
Stakes, D.S. and Vanko, D.A., 1986. Multistage alteration of gabbroic rocks from the failed Mathematician Ridge, Earth Planet. Sci. Letter, 79: 75–92.
Stakes, D.S., Mével, C., Cannat, M. and Chaput, T., 1991. Metamorphic stratigraphy of site 735B gabbros. Proceedings of ODP, Volume 118.
Stroup, J. and Fox, P.J., 1981. Geologic investigation in the Cayman trough: evidence for thin crust along the Mid-Cayman Rise, J. Geol., 89: 395–420.
Tapponnier, P. and Francheteau, J, 1978. Necking of the lithosphere and the mechanics of slowly accreting plate boundaries, J. Geophys. Res., 83(B8): 3955–3970.
Toomey, D.R., Solomon, S.C. and Purdy, G.M., 1988. Microearthquakes beneath the median valley of the Mid-Atlantic Ridge near 23°N: tomography and tectonics, J. Geophys. Res., 93 (B8): 9093–9112.
Vanko, D.A. and Batiza, R., 1982. Gabbroïc rock from the Mathematician Ridge failed rift, Nature, 300: 742–744.
Vanko, D.A. and Stake, D.S., 1991. Fluids in oceanic layer 3: evidence from veined rock, Hole 735B, South West Indian Ridge. Proceeding of the ODP, Volume 118.
Wolery, T.J. and Sleep, N.H., 1976. Hydrothermal circulation and geochemical flux at Mid-Ocean Ridges, J. Geol., 84: 249–275.

Evidences for Polyphased Oceanic Alteration of the Extrusive Sequence of the Semail Ophiolite from the Salahi Block (Northern Oman)

C. PFLUMIO
C.G.G.M., Ecole Nationale supérieure des Mines de Paris, 60 Bd. St. Michel, 75272 Paris, Cedex 06, France
Present address: Laboratoire de Géochimie et Métallogénie, UPMC, 4 place Jussieu, 75252 Paris Cedex 05, France

Abstract

The extrusive sequence of the Salahi block (northern Oman) consists of a well-developed dyke complex and thick volcanic member. The latter is composed of three events (V1, V2 and V3). Lenses of pelagic, often metalliferous sediments are present within the lava flows. The largest exposures of these sediments are observed at the interface of the different volcanic units. The V1–V2 contact is also the locus of the fault-controlled, Zuha sulphide prospect.

A strong metamorphic zonation overprints the extrusive sequence of the Salahi block (greenschist-facies assemblage in the dyke complex, prehnite-pumpellyite-facies assemblage in the lower part of the volcanic sequence, zeolite-facies and low-temperature assemblages in high stratigraphic level flows). Although the steep thermal gradients and static recrystallization suggest that the observed zonation had developed in response to seawater circulation, the dykes and lava flows alteration differs from that described in oceanic layer 2 in three respects: 1) by the pervasiveness of the recrystallization, 2) by the occurrence of a prehnite-pumpellyite-facies assemblage at the top of the first accretion-related volcanic event (V1), 3) by the relative scarcity of low-temperature minerals and the widespread development of phases uncommon in the modern oceanic crust (i.e. iron-rich pumpellyite) in the high stratigraphic level lavas (V2, V3).

Field observations and mineralogical study indicate that the Salahi block extrusive sequence has been subjected to three stages of hydrothermal circulation and to low-temperature oceanic alteration that were contemporaneous with the three magmatic events of the Semail complex. The superposition of alteration phases accounts for the peculiarities of the metamorphic zonation in the volcanic member of this ophiolite.

The origin of the Zuha mineralized zone is attributed to the hydrothermal phase activated by the second, off-axis, magmatic event. The Zuha prospect displays the characteristics of a hydrothermal discharge zone: sulphide-bearing lavas with a typical stockwork paragenesis (quartz, Fe-chlorite, rectorite,

Tj. Peters et al. (Eds), Ophiolite Genesis and Evolution of the Oceanic Lithosphere, 313–351.

titanite), appear in the uppermost V1 sequence below the gossans. This assemblage results from an interaction between V1 lavas and hot (250–300°C), relatively low pH, metal-loaded, upwelling fluids. After interaction with the volcanics and formation of the sulphide mineralization, the hydrothermal fluids were responsible for the formation of large, Mn-rich, sedimentary lenses at the top of the V1 unit in the vicinity of the mineralized zone.

Introduction

From the early 1970's up till now, accumulated observations during investigations of the sea bottom have established the existence of low-temperature/hydrothermal circulation through the oceanic crust. Sea-bottom temperatures and heat-flux anomalies are well-documented (Wolery and Sleep, 1976; Green et al., 1981). A large set of authigenic Fe-Mn oxide crusts (Bostrom and Peterson, 1966; Bonatti; 1983), metalliferous sediments (Sayles and Bischoff, 1973; Hoffert et al., 1979), and low- to high-grade altered oceanic basalts, diabases and gabbros (Cann, 1969; Humphris and Thompson, 1978a; Humphris et al., 1980; Ito and Anderson, 1983; Mevel, 1987) have been studied and such rocks are now currently interpreted as related to hydrothermal processes.

The discovery of active vents exhaling high temperature, metal-loaded fluids in the axial zones of medium- to fast-spreading mid-oceanic ridges (Francheteau et al., 1979) and the more recent observation of similar structures in other tectonic settings such as slow-spreading ridge axes (Detrick et al., 1986) or off-axis seamounts (Hekinian and Fouquet, 1985; Alt et al., 1987) have confirmed the occurrence of large-scale oceanic hydrothermal circulation. In parallel, experimental studies undertaken on seawater-basalt interaction have yielded secondary mineral assemblages and fluid compositions similar to the ones which are described in the oceanic crust (Mottl, 1983; Crovisier et al., 1983, Seyfried and Janecky, 1985). These experiments and the geochemical studies of both rocks and fluids samples have allowed much more insight into the physico-chemical conditions of alteration of the first kilometer of oceanic crust (Stakes and O'Neil, 1982; Kerridge et al., 1983; Bowers et al., 1985; Merlivat et al., 1987).

Furthermore, hydrothermal products such as metalliferous sediments, ochres and sulphide deposits have been observed in ophiolites of various ages (Robertson, 1976; Franklin et al., 1981). These complexes also exhibit evidences of hydrothermal alteration (recrystallization under low stress and high-temperature gradients) which is responsible for a metamorphic zonation similar to the one inferred in modern oceanic crust (Stern and Elthon, 1979; Gillis and Robinson, 1985; Thorette, 1986). However, ophiolitic rocks are often more deeply recrystallized than their oceanic equivalents as some have suffered one or more episodes of regional metamorphism following their oceanic history.

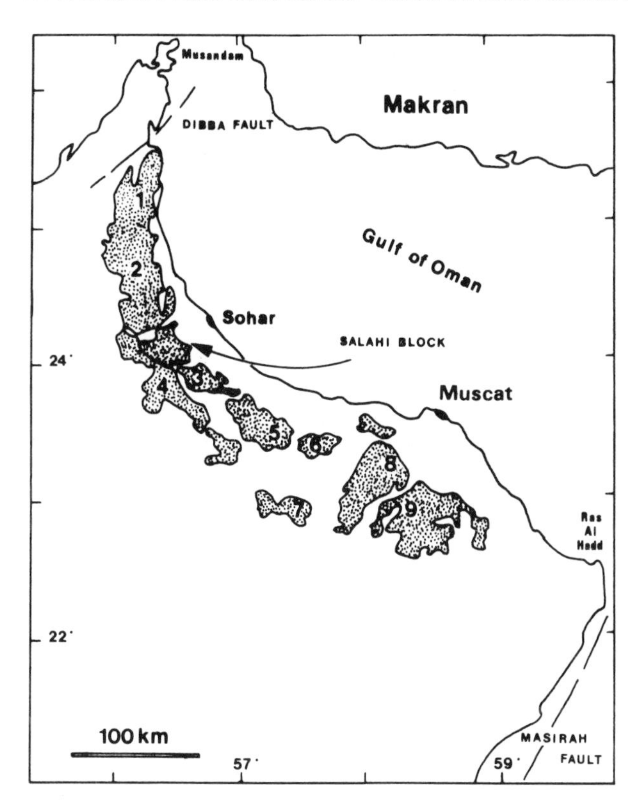

Figure 1. Location of the Salahi block in the Semail ophiolite belt. Blocks names: 1: Aswad; 2: Fizh; 3: Sarami; 4: Wuqbah; 5: Haylayn; 6: Rustaq; 7: Bahia; 8: Semail; 9: Wadi Tahin.

The Cretaceous Semail ophiolite in the Sultanate of Oman is the largest known piece of fossil oceanic lithosphere. This complex displays complete and well-preserved sections, reflecting the fact that no major, late-tectonic process (i.e. continental collision) has affected the eastern margin of the arabian plate. Furthermore, the Semail nappe is located at the top of the pile of allochtonous terrains which constitute the Oman mountains. Such geographic location and structural position explain why it has only registered the recrystallization events induced by oceanic processes. However, it is well-established that this ophiolite results from the superposition of several magmatic events (Pearce et al., 1981; Ernewein et al., 1988). Since hydrothermal circulation is principally driven by the heat released by magmatic processes, the Semail complex is likely to have been subjected to several stages of alteration and thus displays a different metamorphic zonation than that described in a classic mid-oceanic ridge crustal section.

This paper presents field and mineralogical evidence for such a polyphased hydrothermal history of the Semail ophiolite. The data have been collected in the extrusive sequence of the Salahi block which is located in the northern part of Oman (Fig. 1). A small, fault-controlled, sulphide prospect (Zuha) occurs within the lava flows of this tectonic block. The secondary assemblages which developed in the volcanics at the level of this mineralized zone, are also described and are compared with the assemblages which have been described in altered lavas from mining districts of Oman and Cyprus (Collinson, 1986; Richards et al., 1989). These data allow a reconstruction of the different oceanic alteration stages undergone by the ophiolite extrusive sequence; the propounded evolution is supported by geochemical data obtained on rocks from the Zuha prospect (Regba et al., this volume).

Geological Setting

The Semail ophiolite caps a complex of nappes which is composed mainly of pelagic and epicontinental sedimentary facies dated from Permian up to Upper Cretaceous. Several studies were focused on the different allochtonous slices and their results are gathered in special issues (Glennie et al., 1974; Coleman and Hopson, 1981; Lippard et al., 1986; Boudier and Nicolas, 1988). The ophiolite is separated from the underlying units by a metamorphic sole which is composed of intensely deformed lenses of oceanic crust and sediments and is characterized by an inverted, metamorphic gradient. Such a sole formed during the oceanic thrusting and obduction of the Semail complex (Alleman and Peters, 1972; Ghent and Stout, 1981; Lanphere, 1981).

During these thrusting events, the Semail complex was broken into several blocks which represent complete, non-deformed sections of oceanic lithosphere. The Salahi block comprising 800 km^2, is bounded by two major fault zones, the Wadi Jizi in the north and the Wadi Ahin in the south.

Geology of the Salahi Block

The Mantle and Plutonic Sequences

The Salahi block mantle peridotites are 4 to 7 km thick and consist mainly of harzburgites. These rocks are crosscut by scarce gabbro and pyroxenite dykes. The peridotites and to some extent the crosscutting dykes are affected by the two types of foliations which are commonly described in the ophiolite mantle sequence (Boudier and Coleman, 1981; Ceuleneer et al., 1988).

In the Salahi area, the plutonic sequence is only 3 to 3.5 km thick. The gabbroic cumulates represent the predominant rock-type of this sequence and are overlain by isotropic gabbros. This gabbroic series is crosscut by numerous mainly ultrabasic intrusions which belong to the second magmatic event of the ophiolite (Ernewein et al., 1988). Such intrusions are also common in other blocks (Juteau et al., 1988).

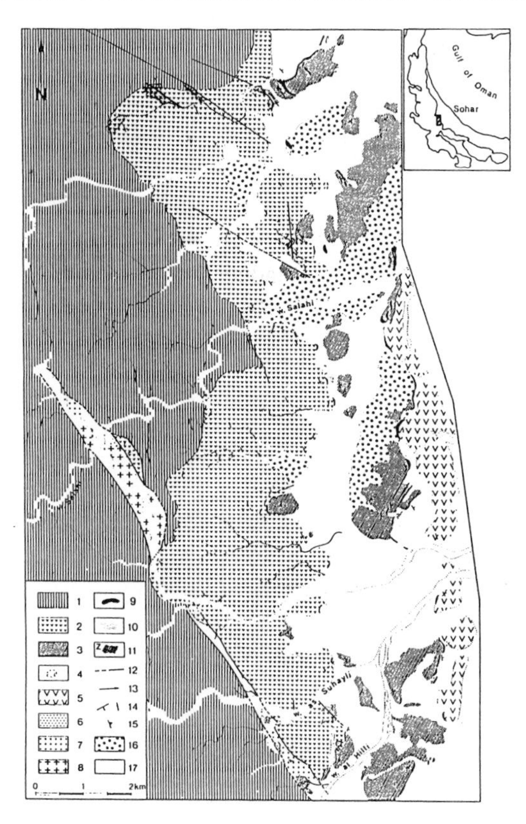

Figure 2. Geological map of the extrusive sequence of the Salahi block. 1: dyke complex; 2: V1 lavas; 3: V2 lavas; 4: basaltic picrites; 5: V3 lavas; 6: wehrlitic intrusions; 7: gabbroic intrusions; 8: plagiogranites; 9: metalliferous sediments; 10: oxidized stockworks; 11: Zuha prospect; 12: fault contact; 13: lava tube flow direction; 14: strike and dip of dykes and lava flows; 15: strike and dip of sills; ancient gravel terraces; 17: wadi gravels and present-day terraces.

The Extrusive Sequence

The extrusive sequence consists of a well-developed dyke complex and a thick volcanic member (Fig. 2). The main orientation of the dykes is 160, W60 whereas the lava flows exhibit an average 155, E35 orientation. The volcanic section comprises three main events called V1, V2 and V3 (Ernewein et al., 1988; Pflumio, 1988). Pelagic, often metalliferous, sediments can be observed within these lavas. The sediments largest exposures have been found at the contacts between the different volcanic units. Depending on their stratigraphic position, these sediments are either recrystallized in the prehnite-pumpellyite or in the zeolite facies (Karpoff et al., 1988), as are the surrounding lavas.

The dyke complex is 1600 meters thick. It consists mainly of dark green dyke swarms which are rooted in the gabbroic series. These dykes are cross-cut by a second generation of discreet irregularly-trending light-green dykes related to the second magmatic event of the ophiolite. Both kinds of dykes have suffered extensive hydrothermal alteration as shown by the frequent development of epidotized dykes (epidosites), and of veins or amygdules occupied by the assemblage epidote, quartz, sulphides (pyrite, chalcopyrite) throughout the entire sheeted-complex. Some hydrothermal veins can be organized in large scale networks which preferentially run at the dykes margins. Most of these veins are loaded with oxidized material (goethite, quartz) and likely represent weathered stockworks. The location of these networks at the dykes margins indicates that the fluids travelled preferentially in along-strike parallel planes (Nehlig and Juteau, 1988).

The dyke complex to volcanic sequence transition is sharp, taking place in only 2 to 3 meters. In this zone, the dykes become vesicular, start to display lobate features and hyaloclastites develop at their margin.

The first, axis, volcanic event, V1, lies over and was clearly fed by the dyke complex. It consists of 1000 to 1600 meters of aphyric, poorly vesicular, brownish pillow-lavas and massive flows. It has a tholeïtic affinity, high ϵ_{Nd} (around 8) and is considered as a mid-oceanic-ridge type of volcanism (Ernewein et al., 1988; Pflumio et al., 1990). According to isotopic and biostratigraphic dates, this first volcanic event occurred between early Cretaceous and early Cenomanian (McCulloch et al., 1981; Beurrier et al., 1987).

The second, off-axis, volcanic event, V2, is fed by sill complexes which are rooted in differentiated plutons (wehrlite, gabbro, plagiogranite). These intrusions crop out at the dyke complex – V1 interface along a N150 trending fault zone (Fig. 2). They are the higher-structural-level equivalent of the predominantly ultrabasic intrusions that crosscut the ophiolite gabbroic series.

In the vicinity of the intrusions, both V1 volcanics and dykes have been subjected to an intense hydrothermal alteration and are completely recrystallized as epidosite. The plagiogranite intrusions themselves display abundant aggregates of epidote in their matrix. The sills that feed the V2 volcanics, also exhibit a high degree of alteration and abundant oxidized cubes of pyrite can be seen in their groundmass. Lenses of pyrite-bearing, brown jasper (up to 2–3 meters long and .80 meter thick) are also present at the margins of many sills. These observations suggest that an important discharge of hydrothermal fluids and alteration episode occurred more or less simultaneously with the emplacement of the intrusions and sill complexes. Similar conclusions have been reached by Alabaster and Pearce (1985) on the basis of investigations in the Lasail copper mine area (southern part of the Fizh block, Fig. 1).

V2 is only separated from V1 by discontinuous lenses of metalliferous sediments which suggests that it was extruded a short time after the end of the axis volcanic event. This agrees with the biostratigraphic studies which have yielded Cenomanian-Turonian ages for the V2 episode (Tippit et al.,

1981), and with the conclusions reached on the temporal relationships between the plutonic events by Ernewein and Whitechurch (1986).

V2 is 1000 to 1300 meters thick and consists mainly of two volcanic facies (Pflumio et al., submitted). These lavas are characterized by a depletion in most of the incompatible elements (HFSI and REE) compared to V1, and by negative Ta anomalies; their ϵ_{Nd} are close to the V1 ones. These geochemical features suggest that the second off-axis event derived from a hydrated, second-stage melting of the same mantle which provided the first volcanic event.

The first facies appears at the base of the V2 sequence. It consists of primitive lavas which form up to 300-meters-thick discontinuous outcrops. These lavas are represented by small, green pillows and thin massive flows which are locally crosscut by dark brown, feeder sills. This volcanic facies was named the Lasail unit by Alabaster et al. (1980). In places where these lavas are absent from the bottom of the V2 sequence, thick metalliferous-sediment lenses occur at the V1–V2 interface.

The second facies consists of more evolved, vesicular pillow-lavas and massive flows. In the Salahi area, these flows represent the dominant V2 facies; they are equivalent to the Alley unit of Alabaster et al. (1980). These lavas have characteristic yellow and/or dark-green hues which are related to the presence of yellow Fe-pumpellyite and green celadonite. Pumpellyite-rich lavas locally form bright-yellow panels that crosscut the main flow orientation; such zones likely indicate fossil hydrothermal fluids channels. The vesicles of these lavas are up to 3 cm long and are most commonly filled with zeolites.

The last, within-plate, volcanic event, V3, consists of 200 meters of thick (up to 20 meters), columnar-jointed, massive flows. These lavas were extruded on top of the V2 sequence after a period of pelagic sedimentation and during the oceanic thrusting of the ophiolite (Ernewein et al., 1988). The V3 lavas are low ϵ_{Nd} (around 5), enriched tholeïtes. This event was named the Salahi unit by Pearce et al. (1981) because these authors only observed this unit in the Wadi Salahi area. Since then, an occurrence of this event has been reported from the Haylayn block in the central part of the ophiolite (Beurrier, 1987) suggesting that the V3 might have had a large lateral extension.

The contact between V2 and V3 is marked by the only continuous horizon of pelagic sediments in the Salahi block (Fig. 2). It consists of up to 30 meters of colorful deposits. .50 to 1 meter of purple, metalliferous layers similar to the ones commonly encountered as lenses within the V1 and V2 flows mark the base of this horizon. They are followed by 20 to 25 meters of cream- yellow, metal-poor, biogenic lime-mudstones and limestones which are capped by 3–5 meters reddish mudstones. This horizon is equivalent to the Suhaylah and part of the Zabyat formations which were first described further north at the level of the Wadi Jizi (Fleet and Robertson, 1980). Ages are only available on the Suhaylah formation and they indicate that it was

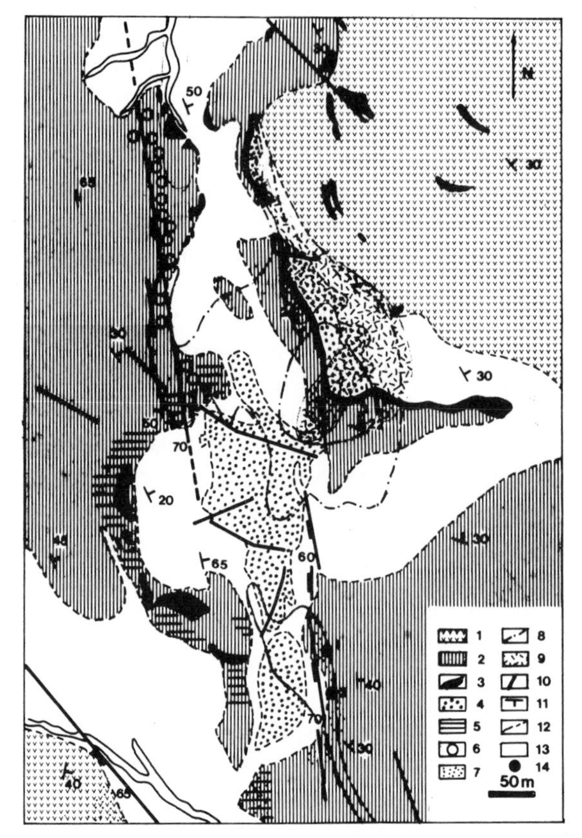

Figure 3. Geological map of the Zuha prospect (see caption 11 of fig 2 for location). 1: V2 lavas; 2: V1 lavas; 3: metalliferous sediments; 4: gossans; 5: jasper-bearing lavas; 6: epidosites; 7: lavas bearing oxidized hydrothermal veins; 8: 400 ppm Cu anomaly contour; 9: 400 ppm Zn anomaly contour; 10: major fault; 11: strike and dip with indication of the dip; 12: outcrop limit and geological contact; 13: wadi gravels; 14: 59–1 and 59–2 coreholes.

deposited by Coniacian-Santonian time (Tippit et al., 1981). A Campanian time for the extrusion of the V3 is therefore advanced in the absence of more precise dates.

The Zuha Sulphide Prospect

The Zuha sulphide prospect is located approximately one kilometer north of the Wadi Salahi (Fig. 2). A survey of the deposit reveals that its surface exposures (4 hematite-stained, siliceous gossans) lay at the interface of V1 and V2, on the eastern edge of a small graben (Fig. 2 and 3). The graben is bounded on both sides by V1 outcrops and is filled by V2 lavas. The western border fault is oriented N170 and the eastern one N145. On the surface, the structure of the sulphide deposit is also controlled by a N170 fault zone along which the gossans are aligned. N170 and N145 orientations are typical for the dyke complex in the Salahi area (Pflumio, 1988) and through the ophiolite (Pallister, 1981), and likely represent main conjugated directions which developed during the accretion.

In the area, west of the gossans, and in the fault zone, a great abundance of sulphide-bearing, red jasper is present around the V1 pillows or in sets of veins parallel to or crosscutting the flows. Along the N170 deposit-bounding faults, oxidized, stockwork-type veins are common, and the lavas are locally recrystallized to epidosite. A hydrothermal breccia which consists of angular fragments of altered lava (partially to totally epidotized), developed along the western bounding fault. These observations suggest that the western Zuha area was the locus of an intense hydrothermal circulation, and that the bounding faults channeled the mineralized fluids. The latter is clearly indicated by the Cu and Zn geochemical anomalies which are oriented N-S and are centred upon the faulted area (Fig. 3).

Most gossans sit on a .5 to 1.5 meter thick, horizon of purple, metalliferous sediments which extend up to the first outcrops of the overlying V2 lavas. In addition, the sediments can be traced laterally to the South-East where they crop out as thick lenses devoid of gossanous caps. Thick columnar-jointed flows lie over the gossans, whereas typical V2 vesicular lavas can be found above the distal lenses.

Both sites sediments have been analysed (Karpoff et al., 1988). The sediments which lie under the gossans exhibit the strongest enrichments in metals (Fe, Cu and Zn) of any of the samples analysed in the Salahi block. On the other hand, the distal lenses are characterized by a conspicuous manganese enrichment and low Fe, Cu and Zn contents, and are geochemically very similar to Mn-rich sediments sampled in the vicinity of high-temperature vents on East Pacific Rise (Germain-Fournier, 1986).

The Zuha prospect thus developed at the level of a ridge-parallel fault zone located on the eastern edge of a graben probably related to the accretion tectonics (Fig. 4). As a result, important discharge of hydrothermal fluids also occurred in the area west of the gossans. The mineralized deposit itself, probably lies within the uppermost V1 flows but the formations that constitute its apex crop out at the V1–V2 contact. These gossans and the underlying metalliferous sediments are the stratigraphic equivalents of Mn-rich, metalliferous-sediment lenses that preferentially develop in zones where the first primitive V2 flows are absent. This stratigraphic position suggests that the formation of Zuha and deposition of the Mn-rich sediments occurred at the beginning of the second volcanic event.

The Zuha sulphide deposit occupies a structural-stratigraphic position similar to the Lasail and Aarja massive sulphide deposits (Alabaster and Pearce, 1985; Haymon et al., 1989). Furthermore, strong analogies exist between the structural environment of Zuha and that of oceanic high-temperature discharge zones (Francheteau et al., 1979; Hekinian and Fouquet, 1985; Tufar et al., 1986).

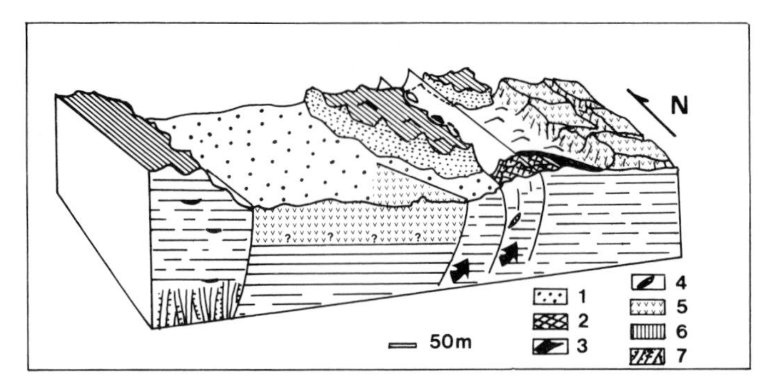

Figure 4. Schematic representation of the structural setting of the Zuha sulphide prospect. 1: wadi gravels; 2: Zuha main gossan; 3: metalliferous sediments; 4: sulphide deposit; 5: V2 flows; 6: V1 lavas; 7: dyke complex. The arrows indicate the ascending hydrothermal fluids.

Secondary Mineralogical Assemblages in the Extrusive Sequence

Dyke Complex

The most common secondary assemblage developed in the diabases of the dyke complex is quartz + albite + actinolite + chlorite + epidote + prehnite ± titanite ± sulphides (pyrite, chalcopyrite). This assemblage formed under greenschist- to prehnite-amphibolite-facies conditions (250–450°C, Liou et al., 1985). Actinolites are present as uralitization products after clinopyroxenes and are in many cases retrograded to chlorite (Table 1, Fig. 5). This phase exhibits compositions which indicate slight pargasite substitution. Some analyses display anomalous Al contents reflecting partially chloritized amphiboles. Compositions of amphiboles analysed in the isotropic gabbros have been reported on the same figure (Fig. 5; Pflumio, 1988). The absence of intermediate edenitic hornblende and edenite usually encountered in oceanic samples is striking. This seems to be a common aspect of Semail ophiolite gabbros (Nehlig and Juteau, 1988).

In the dykes, magmatic plagioclase is almost always albitized and chlorite, epidote, prehnite and quartz are commonly found in aggregates between the plagioclase laths. Epidote is also observed as small grains in the plagioclase laths, or in veins with or without prehnite but always accompanied by sulphides (pyrite and minor chalcopyrite). Titanite is a typical alteration product of titanomagnetite. Prehnite, epidote, quartz and titanite are the only phases which appear in the epidotized dykes.

The chlorites are pycnochlorites which exhibit a wide range of Fe/Fe + Mg ratios (.28 to .60, Fig. 6) and MnO contents less than .80 percent. No evolution of the Fe/Fe + Mg ratio or Al(IV) content of this phase has been observed through the dyke complex. The epidote pistacite-contents fall between .18 and .32 and the highest values were found in the vein epidotes or

Table 1. Selected amphibole compositions from the isotropic gabbros and dyke complex diabases. Pa: pargasite. Hb: hornblende, Act. Hb.: actinolitic hornblende. Act.: actinolite. Structural formulae calculated on the basis of 23 oxygens (lower table) and the sum Ca + Ti + Al + Cr + Fe + Mn + Mg + Ca normalised to 15 in order to calculate the Fe^{3+} content. Fe^{3+} = 46 − 46.N where N is the normalizing factor (after Laird and Albee, 1981), n.d. oxide not detected.

	AMPHIBOLES					
	CWS 2 Gabbro Hb.	CWS 2 Gabbro Act. Hb.	CWS 14 Gabbro Pa.	CWS 11 Dyke Hb.	OT 46 Dyke Act.	CP467 Dyke Act.
SiO_2	49.05	51.69	41.52	45.39	51.4	50.39
TiO_2	1.38	0.34	3.6	0.52	0.63	0.37
Al_2O_3	5.84	4.24	11.26	7.66	4.42	3.57
FeO	11.56	14.25	12.42	16.93	11.56	17.12
MnO	0.25	0.23	0.16	0.27	0.5	0.51
MgO	16.52	14.98	13.43	14.3	16.5	12.98
CaO	11.3	11.5	11.14	9.23	9.86	10.97
Na_2O	1.25	0.53	2.88	0.59	0.57	0.64
K_2O	0.28	0.03	0.2	0.07	0.05	0.06
Total	97.45	97.79	96.61	94.96	95.49	96.61
Si (T)	7.101	7.475	6.21	6.78	7.51	7.503
Ti (M1,2,3)	0.15	0.032	0.405	0.058	0.069	0.041
Al (T)	0.899	0.525	1.79	1.22	0.49	0.497
Al (M1,2,3)	0.097	0.198	0.195	0.129	0.272	0.129
Fe^{3+} (M1,2,3)	0.098	0.098	0	0.786	0	0.088
Fe^{2+} (M1,2,3)	1.09	1.430	1.406	0.843	1.066	1186
Fe^{2+} (M4)	0.214	0.187	0.146	0.406	0.347	0.183
Mn (M4)	0.031	0.028	0.02	0.034	0.062	0.064
Mg (M1,2,3)	3.564	3.229	2.994	3.184	3.593	2.88
Ca (M4)	1.755	1.785	1.788	1.48	1.546	1.753
Na (M4)	0	0	0.044	0	0.046	0
Na (A)	0.352	0.149	0.791	0.174	0.116	0.185
K (A)	0.052	0.006	0.038	0.014	0.009	0.011
Tot (M1,2,3)	5	5	5	5	5	5
Tot (M4)	2	2	2	2	2	2
Tot (A)	0.403	0.164	0.829	0.187	0.125	0.196
Total	15.403	15.154	15.829	15.187	15.125	15.196
Si	7.116	7.491	6.21	6.899	7.51	7.517
Ti	0.151	0.037	0.405	0.059	0.069	0.042
Al	0.999	0.724	1.985	1.372	0.761	0.628
Fe^{2+}	1.405	1.727	1.554	2.152	1.413	2.136
Mn	0.031	0.028	0.02	0.035	0.062	0.064
Mg	3.572	3.236	2.994	3.239	3.593	2.886
Ca	1.759	1.788	1.788	1.505	1.546	1.756
Na	0.352	0.149	0.835	0.174	0.161	0.185
K	0.052	0.006	0.038	0.014	0.009	0.011
Tot	15.436	15.187	15.829	15.449	15.125	15.225

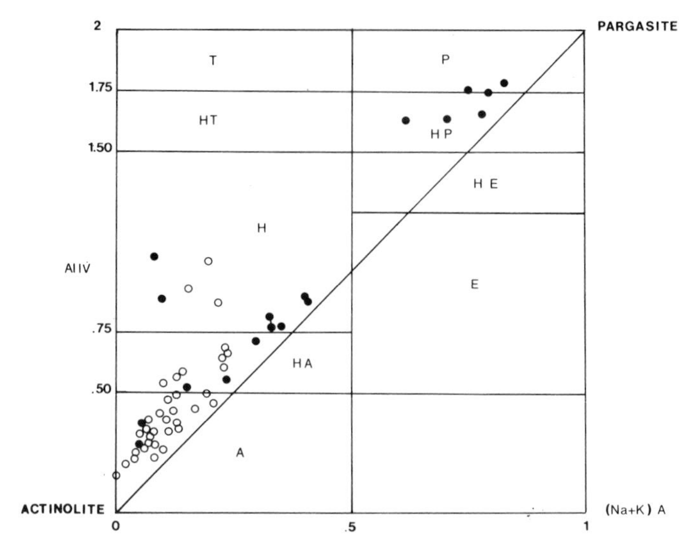

Figure 5. Compositions of amphiboles analysed in the dyke complex diabases (open circles) and in the isotropic gabbros (closed circles). A: actinolite; H.A.: actinolitic hornblende; H.: hornblende; H.T.: tremolitic hornblende; T: tremolite; P: pargasite; H.P.: pargasitic hornblende; H.E.: edenitic hornblende; E: edenite (after Leake, 1978).

in sulphides-bearing samples. There is no apparent evolution of the pistacite content with depth.

The prehnites exhibit Fe3 +/Fe3 + + AlVI ratios varying between .06 and .48. When only prehnites from the groundmass are considered, an increase in the Fe3 +/Fe3 + + AlVI ratio toward the top of the dyke complex is

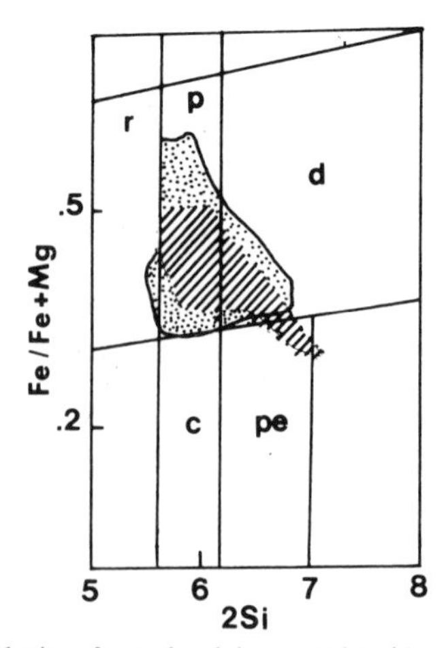

Figure 6. Compositions of chlorites from the dyke complex (dotted area) and V1 lavas (striped area). r, p, d, c and pe: ripidolite, pycnochlorite, diabantite, clinochlore and penninite composition fields (after Hey, 1954).

discernible. Such a progressive change in the composition of the prehnites can be interpreted as related to a decrease in temperature during fluid/rock interactions and/or an increase of the oxygen fugacity (Liou et al., 1983).

The second generation dykes display a lower temperature secondary mineral assemblage which consists of quartz + albite + chlorite + epidote + prehnite + laumontite ± titanite. The development of laumontite in the groundmass suggests that these rocks were recrystallized under prehnite-pumpellyite- to zeolite-facies conditions (150–300°C, Liou, 1971; Liou et al., 1985).

Volcanism 1

The secondary assemblage which developed in the V1 flows is quartz + albite + chlorite + epidote ± prehnite ± titanite ± sulphides (pyrite and scarce chalcopyrite). It formed under prehnite-pumpellyite-facies conditions (200–300°C, Liou et al., 1985).

Actinolites have only been observed in one sample, where they occur as aggregates in devitrified portions of a pillow-lava. Chlorite, epidote and quartz are common in the groundmass and in vesicles. Epidote is rarely an alteration product of the plagioclase. This phase is present in veins from where it can invade the spilitized lava. The epidotized lavas sampled at the level of the N170 Zuha bounding faults have been totally recrystallized into the assemblage epidote ± titanite ± pyrite, whereas the lavas sampled nearer the intrusions exhibit the assemblage epidote + prehnite + quartz ± titanite. Prehnite is only present in these samples. Pumpellyite has been encountered in two lavas where it is associated with epidote and quartz in vesicles.

The V1 chlorites are generally pycnochlorites which have a range of compositions similar to the dyke-complex chlorites (Fe/Fe + Mg from .29 to .58, MnO < 1.2%, Table 2, Fig. 6). The chlorites richest in MnO have been found in pillows or massive flows rims. The epidotes from groundmass, vesicles and veins exhibit pistacite contents from .21 to .30. These pistacite contents are slightly less than those of epidotes from the Zuha epidosites (.22 to .36), but fall in the same range as those of epidotes from the dykes. The V1 pumpellyites represent Al-rich varieties (Table 2).

Volcanism 2

The first primitive V2 flows usually exhibit the following secondary assemblage: albite + chlorite + prehnite + epidote + calcite ± titanite ± sulphides (pyrite + minor chalcopyrite). This assemblage formed under prehnite-pumpellyite-facies conditions. A similar secondary mineralogy without sulphides, has been observed in the thick columnar-jointed flows which are superjacent to the gossans.

Table 2. Selected chlorite (structural formulae calculated on the basis of 28 oxygens) and pumpellyite compositions (structural formulae normalized to 16 cations and cations repartition in the sites *W*, *X*, *Y*, *Z* according to Coombs et al. (1976)) from V1 lavas. n.d. not detected.

	Chlorites V1				Pumpellyites VI		
	CP 40 V1	CP 65 V1	CP 668 V1		CP 46 V1	CP 46 V1	CP 81 V1
SiO_2	27.57	26.6	29.18	SiO_2	34.68	35.06	35.86
Al_2O_3	17.26	17.38	15.82	TiO_2	0.02	0.1	0.03
FeO	25.05	26.3	21.62	Al_2O_3	18.01	17.7	22.33
MnO	0.49	0.55	1.21	Fe_2O_3	15.61	15.66	10.07
MgO	15.96	14.04	17.23	MnO	0.19	0.16	0.29
CaO	0.13	0.07	0.19	MgO	1.7	1.9	2.06
Na_2O	n.d.	0.02	n.d.	CaO	21.9	22	22.15
K_2O	0.04	0.06	0.06	Na_2O	0.03	0.02	0.05
Total	86.5	86.25	85.31	Total	92.14	92.6	92.89
				Si	5.91	5.95	5.92
				Al(Z)	0.09	0.05	0.06
Fe/Fe + Mg	0.468	0.495	0.413	Total Z	6	6	6
				Ti	0	0.01	0
				Al	3.53	3.49	4
Si	2.947	2.9	3.11	Fe	0.47	0.51	0
Al	2.174	2.216	1.987	Total Y	4	4	4
Fe2+	2.239	2.38	1.927				
Mn	0.044	0.05	0.109	Fe	1.53	1.49	1.25
Mg	2.542	2.425	2.737	Mg	0.43	0.46	0.51
Ca	0.014	0.008	0.022	Mn	0.03	0.02	0
Na	n.d.	0.004	n.d.	Al	0	0	0.26
K	0.005	0.008	0.008	Na	0.01	0	0
Total	9.968	9.996	9.9	Total X	2	1.99	2.02
				Ca	4	4	3.92
				Na	0	0	0.01
				Mn	0	0	0.04
				Total W	4	4	3.97

Actinolites have been observed in one sample coming from the vicinity of the intrusions. Chlorite is common in both groundmass and vesicles, and in places is accompanied by epidote. Prehnite is the dominant Ca-silicate in these lavas. It is locally abundant with calcite, in both vesicles and groundmass. Laumontite has been observed in one sample where it seals veins occupied by chlorite, prehnite, epidote and sulphides. Titanite is present as disseminated grains in the groundmass in association with chlorite.

The few analyzed chlorites are pycnochlorites characterized by a higher Mg content than in V1 lavas chlorites (Fig. 7, Table 3). This likely reflects the initial, more primitive geochemistry of these V2 lavas. The MnO content of these chlorites is also low (<.20 percent). Prehnites display Fe3 + /Fe3 + + Al(VI) ratios which range from .08 to .36, whereas the epidote-pistacite contents vary between .18 and .32. Both mineral phases compositions are close to the compositions in V1 volcanics and dyke complex.

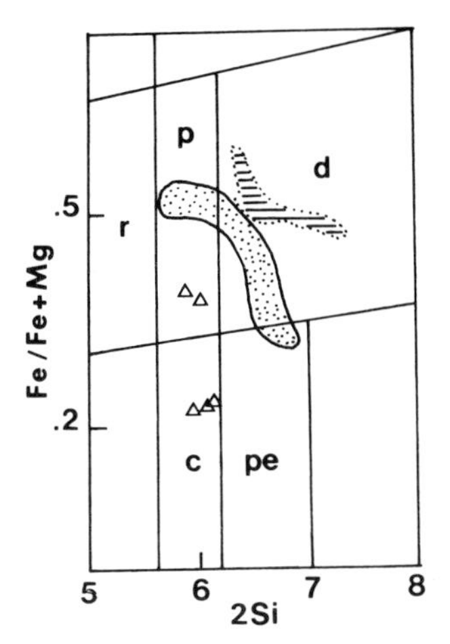

Figure 7. Compositions of chlorites from the primitive (triangles) and evolved V2 lavas (dotted area) and V3 volcanics (striped area). r, p, d, c and pe: ripidolite, pycnochlorite, diabantite, clinochlore and penninite composition fields (after Hey, 1954).

Table 3. Selected chlorite and celadonite compositions from V2 lavas. Structural formulae calculated on the basis of respectively 28 and 22 oxygens. Prim chlorite from primitive V2 lava. n.d. not detected.

	Chlorites V2				Celadonites V2			
	CP 758 V2 prim	CP 758 V2 prim	CP 89	CP 22	CP 495	CP 920	CP 92′	CP 173
SiO_2	30.99	30.91	31.28	30.51	54.8	56.12	55.25	55.45
TiO_2	n.d.	n.d.	n.d.	n.d.	0.05	0.05	0.31	0.21
Al_2O_3	18.07	17.59	13.27	14.56	2.19	5.81	4.37	7.52
FeO	13.1	13.33	19.58	21.2	19.54	14.64	16.34	15.48
MnO	0.17	0.2	0.23	0.17	0.07	n.d.	0.05	n.d.
MgO	25.13	24.6	20.86	17.31	6.07	6.84	6.21	4.34
CaO	0.37	0.21	0.24	0.51	0.07	0.22	0.06	0.32
Mg_2O	n.d.	n.d.	n.d.	n.d.	0.08	n.d.	0.02	0.1
K_2O	0.06	0.04	0.1	0.34	10.21	10.07	10.24	7.92
Total	87.89	86.88	85.32	84.6	93.08	93.75	92.86	91.34
Fe/Fe + Mg	0.226	0.233	0.336	0.407				
Si	3.035	3.065	3.322	3.255	4.138	4.069	4.098	4.082
Ti	n.d.	n.d.	n.d.	n.d.	0.002	0.002	0.017	0.011
Al	2.086	2.056	1.637	1.831	0.194	0.496	0.382	0.652
Fe	1.073	1.105	1.625	1.891	1.234	0.887	1.013	0.953
Mn	0.014	0.016	0.024	0.015	0.004	n.d.	0.003	n.d.
Mg	3.668	3.636	3.21	2.752	0.683	0.739	0.686	0.476
Ca	0.038	0.022	0.031	0.058	0.005	0.017	0.004	0.025
Na	n.d.	n.d.	n.d.	n.d.	0.011	n.d.	0.002	0.014
K	0.007	0.005	0.013	0.046	0.983	0.931	0.969	0.743
Total	9.924	9.908	9.865	9.851	7.258	7.145	7.178	6.959

The overlying, vesicular flows which constitute the remainder of the V2 sequence, exhibit a different secondary assemblage consisting of quartz + albite + chlorite (+ smectite, illite layers) + celadonite + Fe-pumpellyite + stilbite ± K-feldspar ± titanite ± calcite. These flows recrystallized at lower temperature than the underlying volcanics, under zeolite- to browstone-facies conditions (<200°C, Liou, 1971; Elthon, 1981).

Actinolite has been found in only a few samples as uralites. The scarce occurrence of this phase suggests that it is related to an early deuteric stage. Magmatic Ca-plagioclase is recrystallized to albite, and K-feldspar is also present (Table 4). Some plagioclase crystals may exhibit both fresh Ca-rich zones and Na and K-rich recrystallized areas. Chlorite and celadonite developed in both groundmass and vesicles and less often pseudomorphed plagioclases and clinopyroxenes. In vesicles, celadonite always occurs in a more peripheral position than chlorite. This mineral also developed in glassy portions.

In these lavas, pumpellyite with yellow to green pleochroism characteristic of the Fe-rich variety (Coombs, 1953) is common, whereas prehnite is very scarce. Fibrated pumpellyite occurs in vesicles whereas a more platy variety pseudomorphs the magmatic phases. This phase also occurs as a glass alteration product like celadonite. Stilbite is the characteristic Ca-zeolite of these upper-level V2 lavas (Table 4). It is always observed in the centre of vesicles and it is often accompanied by calcite. Small veinlets of calcite can also crosscut the zeolite crystals. Laumontite is present in a few samples. It is associated in the groundmass with chlorite or in veins with calcite. Mesolite has been observed once, at the periphery of vesicles filled by chlorite.

Chlorites are principally diabantites, except in the actinolite-bearing samples, where they are pycnochlorites (Table 3, Fig. 7). The diabantites often contain some calcium and potassium suggesting the presence of smectitic and illitic layers. Celadonites are characterized by low Fe-contents compared to typical oceanic celadonites (Table 3, Andrews, 1980; Laverne, 1987). Phyllosilicates with compositions intermediate between celadonite and chlorite occur toward the interior of vesicles suggesting a gradual evolution from one phase to the other as fluid composition and/or temperature were modified. The pumpellyites exhibit compositions which are characteristic of pumpellyites crystallizing under low-pressure, zeolite-facies conditions, and contrast with those of the pumpellyites from the V1 lavas (Table 5, Fig. 8, Glassley, 1975; Liou, 1979).

These observations show that a metamorphic zonation overprints the second volcanic event. In the first flows, prehnite-pumpellyite-facies assemblage is dominant whereas low-temperature and zeolite-facies mineral phases are characteristic of the higher-stratigraphic level lavas. Furthermore, the phase assemblage which developed in the lower V2 lavas is very similar to the one observed in the underlying V1 volcanics. The main difference is that the dominant Ca-silicate is prehnite in the V2 rocks and is epidote in the V1 flows. As for chlorite compositions, this feature is most probably related to

Table 4. Selected potassium feldspar (K-spar) laumontite (Laum.), stilbite (Stil.), and mesolite (Mes.) compositions from V2 lavas. Structural formulae calculated respectively on the basis of 8, 12, 18 and 30 oxygens. n.d. not detected.

	K-SPAR V2		
	CP 337	CP 337	CP 462
SiO_2	64.21	64.76	63.83
TiO_2	0.03	0.04	n.d.
Al_2O_3	19.54	19.16	17.67
FeO	0.02	n.d.	0.19
MnO	n.d.	n.d.	n.d.
MgO	n.d.	n.d.	n.d.
CaO	0.4	0.06	0.05
Na_2O	1.47	0.19	0.02
K_2O	14.08	16.24	16.54
Total	99.75	100.47	98.3
Ab	13.42	1.74	0.18
Or	84.57	97.95	99.56
An	2.02	0.304	0.25
Si	2.954	2.976	3.008
Ti	0.001	0.001	n.d.
Al	1.06	1.038	0.982
Fe^{2+}	0.001	n.d.	0.007
Mn	n.d.	n.d.	n.d.
Mg	n.d.	n.d.	n.d.
Ca	0.02	0.003	0.003
Na	0.131	0.017	0.002
K	0.827	0.952	0.995
Total	4.993	4.988	4.996

	ZEOLITES V2				
	CP 90 Laum	CP 83 SH1	CP 22 Stil	CP 83 Stil	CP 92 Mes.
SiO_2	50.45	59.71	57.3	58.3	41.76
Al_2O_3	20.53	15.74	18.47	18.87	30.15
Fe_2O_3	2.37	n.d.	0.08	0.12	1.91
MgO	1.35	n.d.	n.d.	n.d.	0.49
CaO	9.4	7.87	8.97	8.69	10.75
Na_2O	0.49	0.21	0.51	1.12	3.06
K_2O	0.64	0.1	0.07	0.34	1
Total	85.23	83.63	85.4	87.44	89.12
Si	3.96	6.89	6.55	6.53	5.09
Al	1.9	2.14	2.49	2.49	6.89
Fe^{3+}	n.d.	n.d.	0.01	0.01	0.29
Mg	n.d.	n.d.	n.d.	n.d.	0.14
Ca	0.79	0.98	1.1	1.04	2.24
Na	0.00	0.05	0.11	0.24	1.15
K	0.06	0.02	0.01	0.05	0.25
Total	7.09	10.07	10.27	10.37	19.03

Table 5. Selected pumpellyite compositions from the V2 lavas. Structural formulae calculated on the basis of 16 cations and cations repartition in the sites *W*, *X*, *Y*, *Z* according to Coombs et al. (1976). n.d. not detected.

	Pumpellyites V2				
	CP 83	CP 89	CP 89	CP 92	CP 175
SiO_2	35.73	34.56	34.15	35.20	34.86
TiO_2	0.08	0.15	0.30	0.41	0.37
Al_2O_3	14.03	12.83	12.89	13.80	17.97
Fe_2O_3	19.28	21.92	20.73	19.61	15.86
MnO	0.06	n.d.	n.d.	0.14	0.06
MgO	2.07	1.70	1.80	1.76	1.73
CaO	21.16	21.26	21.02	20.72	19.96
Na_2O	0.04	n.d.	0.11	0.05	0.11
Total	91.48	92.52	91.43	91.69	90.92
Si	6.06	6.04	6.04	6.16	6.04
Al (Z)	0.00	0.00	0.00	0.00	0.00
Z	6.06	6.04	6.04	6.16	6.04
Ti	0.02	0.02	0.04	0.05	0.05
Al	2.88	2.64	2.69	2.84	3.67
Fe	1.10	1.36	1.27	1.11	0.28
Y	4.00	4.00	4.00	4.00	4.00
Fe	1.44	1.52	1.49	1.47	1.89
Mg	0.54	0.44	0.47	0.46	0.45
Mn	0.01	n.d.	n.d.	0.02	0.00
Al	0.01	0.00	0.00	0.00	0.00
Na	0.00	0.00	0.02	0.00	0.00
X	2.00	1.78	1.95	1.95	2.34
Ca	3.98	3.95	4.04	3.55	3.70
Na	0.00	n.d.	0.00	0.00	0.02
Mn	0.00	0.00	0.00	0.00	0.00
W	3.96	3.98	4.04	3.88	3.72

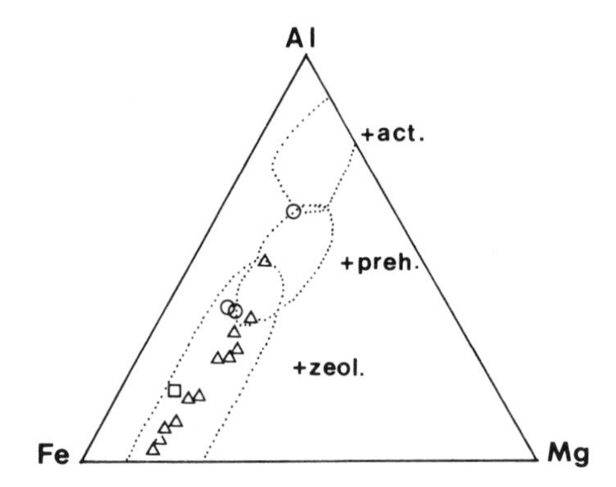

Figure 8. Compositions of pumpellyites from V1 (circles), evolved V2 (triangles) and V3 (square) lavas. Dotted fields: compositions of pumpellyite associated with actinolite, prehnite and with zeolites respectively from Coombs et al. (1976) and Liou (1979).

differences in rock rather than in fluid compositions, as at these temperatures, prehnite (which is lower in iron than epidote) is the calc-silicate expected to stabilize in an Mg-rich rock.

Volcanism 3

Although less pervasively developed, a secondary assemblage similar to the one described in the vesicular V2 lavas is present in the V3 flows. It consists of albite + chlorite (+ illite, smectite layers) ± celadonite ± Fe-pumpellyite- ± scolecite ± calcite ± titanite.

In these lavas, the magmatic plagioclase is mostly unaltered. Phyllosilicates are present as green to orange patches in between the plagioclase and clinopyroxene crystals. Celadonite has been observed in one sample. A yellow to green pumpellyite which is optically similar to the pumpellyite from the vesicular V2 lavas, developed in glassy portions. Zeolites also frequently occur as glass alteration products. Calcite appears in late veins or in the vesicles of the scarce pillow-lavas. Sphene is observed as an alteration product of Ti-oxide or in association with the phyllosilicates.

The phyllosilicates are chlorites with slightly higher silica, calcium and potassium contents than the chlorites from the high-stratigraphic level V2 lavas (Table 6, Fig. 7); these features reflect higher illite-smectite proportions. In addition, the orange colour and relatively high birefringence of these phases suggest that most of the iron is under the oxidized form. Microprobe analyses have provided a pumpellyite composition similar to those of the vesicular V2 lavas pumpellyites (Table 6, Fig. 8), and have allowed the identification of the zeolite scolecite.

Zuha Core Samples

Two coreholes, 59–1 and 59–2, were drilled at the foot of the major Zuha gossan on the eastern side of the main 400 ppm Cu anomaly (Fig. 3). Both holes are 91 meters long and were drilled in the V1 volcanics, perpendicular to the main flow orientation. They encountered only disseminated mineralization, mostly pyrite, in a volume less than 20 percent. Four lithologies have been distinguished in the cores (Fig. 9): weakly mineralized or unmineralized V1 lavas (Type I of Regba et al., this volume), mineralized lavas (Type II of Regba et al.), hematitized jasper veins, and lava breccias (type III of Regba et al.).

The weakly mineralized or unmineralized V1 lavas have the same secondary mineralogy as the V1 lavas outside the Zuha area. Nonetheless, the core samples are characterized by a greater degree of alteration as indicated by the scarce occurrence of fresh clinopyroxenes. Laumontite has been observed in some of these lavas cortex.

The mineralized lavas are characterized by a mineralogy dominated by

Table 6. Selected chlorite, scolecite and pumpelltyite and compositions from the V3 flows. Structural formulae for chlorites and pumpellyite calculated as in tables 3 and 4. Scolecite structural formulae calculated on the basis of 10 oxygens. n.d. not detected.

	CHLORITE V3				SCOLECITE V3	PUMPELLYITE V3	
	CP447	CP 447	CP 457	CP 458	CP 458		CP 458
SiO_2	32.34	34.22	29.44	30.65	46.05	SiO_2	34.7
TiO_2	n.d.	n.d.	0.2	n.d.	n.d.	TiO_2	n.d.
Al_2O_3	14.11	13.96	14.06	14.91	25.84	Al_2O_3	14.04
FeO	23.04	20.04	25.13	24.9	0.05	Fe_2O_3	20.61
MnO	13	.26	0.86	0.19	n.d.	MnO	n.d.
CaO	12.74	13.15	14.14	14.17	n.d.	MgO	1.25
Na_2O	45	.48	1.39	0.94	13.64	CrO	21.2
K_2O	.13	.08	0.08	0.03	0.26	Mn_2O	0.03
	1.02	.97	0.09	0.11	n.d.	Total	91.33
Total	83.96	83.09	85.39	85.85	85.84		
						Si	6.08
Fe/Fe + Mg	.503	.460	0.499	0.497		Al (Z)	0
						Z	6.08
Si	3.502	3.663	3.31	3.28	6.03		
Ti	n.d.	n.d.	0.02	n.d.	0	Ti	n.d.
Al	1.801	1.752	1.81	1.55	3.99	Al	2.9
Fe^{2+}	2.087	1.794	2.29	2.23	0.01	Fe	1.1
Mn	.011	0.023	0.08	0.02	n.d.	Y	4
Mg	2.056	2.097	2.3	2.25	n.d.		
Ca	0.052	0.055	0.16	0.11	1.92	Fe	1.62
Na	0.027	0.016	0.02	0.01	0.07	Mg	0.33
K	0.140	0.132	0.01	0.02	n.d.	Mn	n.d.
Total	9.680	9.535	9.89	9.79	12.01	Al	0
						Na	0.01
						X	1.96
						Ca	3.98
						Na	0
						Mn	0
						W	3.98

chlorite and quartz and the presence of up to 20 percent of disseminated sulfides. Quartz and chlorite are associated with rectorite and titanite. The common sulphides are pyrite and chalcopyrite. These rocks are devoid of albite as it has been replaced by rectorite and quartz. Rectorite occurs as transparent, needle-like crystals which polarize in the second order tints, and is present mainly on older plagioclase laths. The occurrence of this phase has been confirmed by R.X. analysis which detected the presence of a regular illite-smectite, interlayered phyllosilicate of the allevardite family. In the mineralized samples, the primary Ti-oxides have been totally pseudomorphosed by titanite. Titanite is also present in veins in association with quartz. As in unmineralized samples, pyrite is the most common disseminated sulphide. Pyrite and chalcopyrite also occur in veins, associated with chlorite and

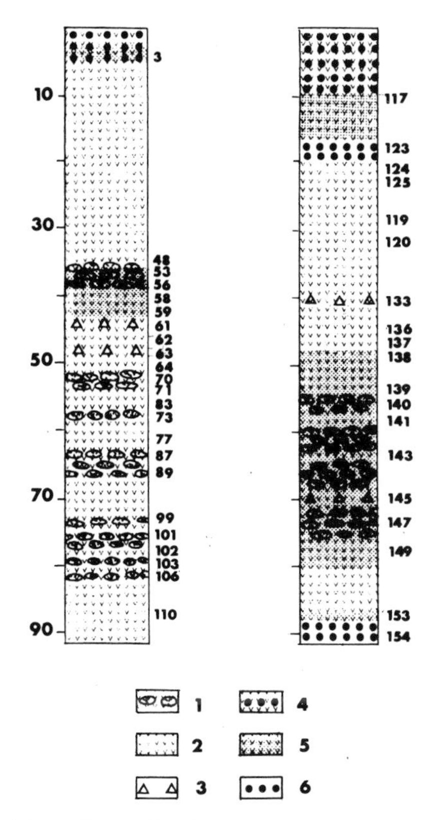

Figure 9. Distribution of the different lithologies and secondary assemblages encountered in the coreholes 59–1 and 59–2. Numbers refer to studied samples (see Regba et al., this volume). 1: V1 pillow-lavas with prehnite-pumpellyite-facies assemblage; 2: V1 lavas with intersertal texture and prehnite-pumpellyite-facies assemblage; 3: breccia; 4: silicified and ferruginous lava; 5: mineralized V1 lavas with quartz + Fe-chlorite + rectorite + titanite + sulphides assemblage; 6: jasper. Depth scale in meters.

quartz. Laumontite also is common alone in late veins or seals sulphides, chlorite and quartz-bearing veins.

The hematitized jasper veins consist of cryptocrystalline quartz pigmented by hematite, of pyrite and of minor chalcopyrite. Epidote commonly occurs either in the quartz groundmass or associated with sulphides in veinlets. In this case, the sulphides are late, relative to epidote, as in dyke complex veins. Laumontite has also been identified in these samples. The jasper veins are very similar to the red jasper rods which are observed in the field around the V1 pillow-lavas.

Lava breccias have been found in both cores at a depth of 42 to 47 meters. They consist of altered lava fragments in a matrix of smaller rock and mineral fragments. These breccias are macroscopically similar to the ones present along the Zuha western bounding fault, and very likely developed in relation to fracture zones. Two types of secondary assemblages have been identified

Table 7. Selected chlorite compositions from mineralized lavas in cores 59–1 and 59–2. Structural formulae calculated on the basis of 28 oxygens. n.d.: not detected.

	Chlorites mineralized samples							
	59–1–61	59–1–61	59–1–59	59–2–137	59–2–138	59–2–139	59–2–143	59–2–143
SiO_2	24.45	25.87	25.07	28.09	27.22	26.16	26.99	26.59
TiO_2	0.05	n.d.	0.02	0.1	n.d.	n.d.	0.06	n.d.
Al_2O_3	21.29	19.17	21.79	16.81	19.93	19.15	20.49	20.14
FeO	28.05	23.3	25.86	22.25	22.53	23.53	26.66	27.51
MnO	0.34	0.48	0.61	0.71	0.23	0.95	0.26	0.36
MgO	11.35	16.4	13.76	16.62	15.96	14.88	12.1	12.43
CaO	0.08	0.06	0.06	0.14	0.08	n.d.	0.09	0.18
Na_2O	n.d.	0.02	n.d.	n.d.	n.d.	0.03	0.05	0.05
K_2O	0.01	n.d.	0.07	0.11	0.04	n.d.	0.2	0.09
Total	85.61	85.33	87.27	84.83	85.99	84.7	86.9	87.34
Fe/Fe + Mg	0.581	0.443	0.513	0.428	0.442	0.47	0.552	0.553
Si	2.685	2.784	2.669	3.019	2.073	2.842	2.881	2.843
Ti	0.004	n.d.	0.001	0.008	n.d.	n.d.	0.004	n.d.
Al	2.756	2.431	2.734	2.129	2.48	2.452	2.578	2.538
Fe	2.579	2.097	2.302	2	1.989	2.157	2.38	2.46
Mn	0.031	0.043	0.055	0.064	0.02	0.087	0.023	0.031
Mg	1.859	2.63	2.185	2.662	2.511	2.409	1.925	1.981
Ca	0.009	0.006	0.006	0.016	0.009	n.d.	0.01	0.02
Na	n.d.	0.004	n.d.	n.d.	n.d.	0.006	0.01	0.01
K	0.001	n.d.	0.009	0.015	0.005	n.d.	0.027	0.012
Total	9.93	10	9.965	9.915	9.889	9.935	9.842	9.898

in these samples. A prehnite-pumpellyite-facies assemblage and locally pervasive epidote, alike in the breccias along the Zuha border fault and a secondary mineralogy similar to that found in the mineralized lavas are present. Sample 59–1–61 (Regba et al., example of type III) consists of quartz, chlorite, titanite and pyrite-rich fragments that are gathered in a pyrite-bearing chloritic cement. In another mineralized breccia, rectorite is present. Veins of laumontite are common in all breccias and crosscut both fragments and matrix. The rectorite-bearing breccia is cemented by laumontite.

Chlorites from the mineralized samples display pycnochloritic to ripidolitic compositions. Their compositional field partly overlaps the compositional field of the unmineralized samples chlorites from the vicinity of Zuha (Table 7, Fig. 10), but the mineralized-samples chlorites are slightly richer in Fe and Al. Their MnO content is quite variable, falling between .2 and 1.5 percent, and is apparently unrelated to the FeO content.

The compositions of these chlorites have been compared to the compositions of chlorites from greenschist facies, oceanic samples and from ophiolitic stockworks samples (Fig. 11). The chlorites from the Zuha mineralized samples are richer in Fe and Al than chlorites from oceanic spilitized samples

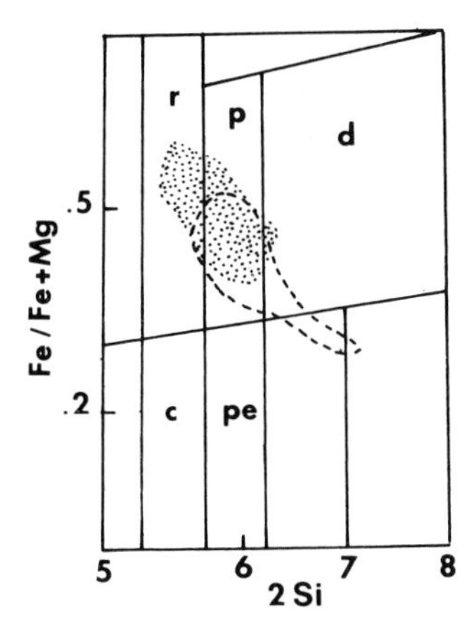

Figure 10. Compositions of chlorites from the Zuha mineralized V1 lavas (dotted area); dotted line: field of composition of chlorites from unmineralized V1 lavas. r, p, d, c, pe: ripidolite, pycnochlorite, diabantite, clinochlore and penninite composition fields (after Hey, 1954).

(Humphris and Thompson, 1978a), but they exhibit compositions similar to those of chlorites from mineralized breccias sampled in the Atlantic (Delaney et al., 1987). The chlorites from the unmineralized samples are richer in FeO than the chlorites from the oceanic spilitized samples, probably reflecting the initial composition of the lavas, as the V1 rocks are on the whole more differentiated than common MORB (Alabaster et al., 1982; Pflumio et al., submitted). The chlorites from the Zuha mineralized samples are less Fe-

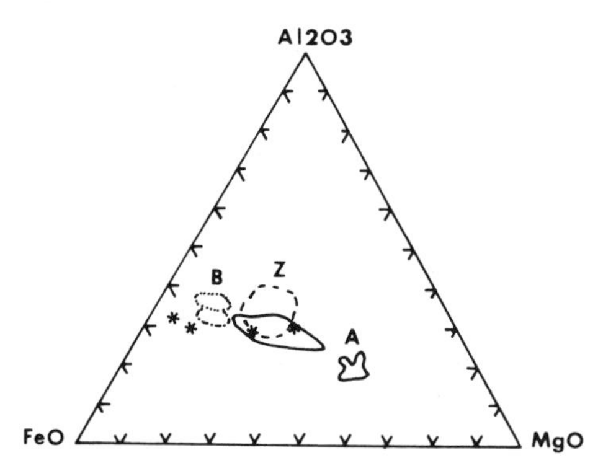

Figure 11. Distribution in the Al2O3 – FeO – MgO triangle of chlorite compositions from the Oman Bayda copper mine (B: dotted field = stockworks; dotted-strait line = mineralized breccia), Zuha mineralized V1 lavas and Salahi unmineralized V1 lavas (Z: respectively dotted and continuous contour line) and from Atlantic greenschist facies lavas (A) and mineralized breccias (stars).

rich than the chlorites from the mineralized, rectorite-bearing, stockwork and breccia samples from the Bayda copper mine in northern Oman (Collinson, 1986; Zuha and Bayda chlorites Fe/Fe + Mg respectively .38–.55 and .56–.79). They are on the other hand, less magnesian than the chlorites from the rectorite-bearing, alteration halos of the Kokkinopezoula sulphide deposit of Cyprus (Fe/Fe + Mg: .25–.40; Richards et al., 1989).

Crystallization temperature estimations have been made using the Cathelineau and Nieva (1985) geothermometer (Table 8). This geothermometer is based on the Al(IV) content of the chlorite. Chlorites of the mineralized samples exhibit the highest crystallization temperatures (235–295°C) and temperatures below 250°C are recorded for chlorites from the samples which are devoid of rectorite. In unmineralized samples from both inside and outside the Zuha mineralized zone, chlorites crystallized between 110 and 290°C and the highest temperatures are registered in the core-samples chlorites which crystallized near sulphide-bearing veinlets. These crystallization temperatures are higher than those of chlorites from vesicular V2 lavas and V3 flows (130–190°C); this agrees with the fact that the latter have been recrystallized in the zeolite facies. Crystallization temperatures of 220°C have been calculated for chlorites from the lower primitive V2 flows consistent with the prehnite-pumpellyite-facies assemblage of these lavas.

Table 8. Mean crystallization temperatures for chlorites from V1, V2 and V3 lavas and Zuha mineralized samples. Temperatures estimates are based on the Cathelineau and Nieva (1985)′ geothermometer, number between brackets is the number of analysed chlorites; in V2 volcanism, prim.: primitive flows, mass.: massive flows above Zuha. evol. evolved flows.

Unmineralized samples	Temperatures	Mineralized samples	Temperatures
59–1–48	124 ± 13(3)		
59–1–62	224 ± 1(3)		
59–1–83	211 ± 25(6)	59–1–3	234 ± 35(4)
59–1–73	268(2)	59–1–53	291 ± 3(3)
59–1–89	281 ± 8(2)	59–1–59	267 ± 1(2)
59–1–102	260 ± 4(5)	59–1–61	274 ± 13(11)
59–2–110	172(1)	59–2–138	254 ± 5(4)
59–2–124	231 ± 17(2)	59–2–139	261 ± 10(7)
59–2–119	240 ± 8(2)	59–2–143	268 ± 23(21)
59–2–120	234 ± 4(3)	59–2–149	256 ± 3(5)
59–2–133	246 ± 4(7)		
59–2–137	227 ± 8(16)		
V1		V1	
CP40	241(1)	OT95	223 ± 15(5)
Cp65	245 ± 10(7)	CP666	200 ± 9(3)
V3		V2	
CP 447	156 ± 24(5)	CO 759(m)	224 ± 10(3)
CP 457	164 ± 6(6)	CP 43 (mass)	191 ± 16(6)
CP 458	161 ± 13(2)	CP 89 (col.)	168 ± 6(3)

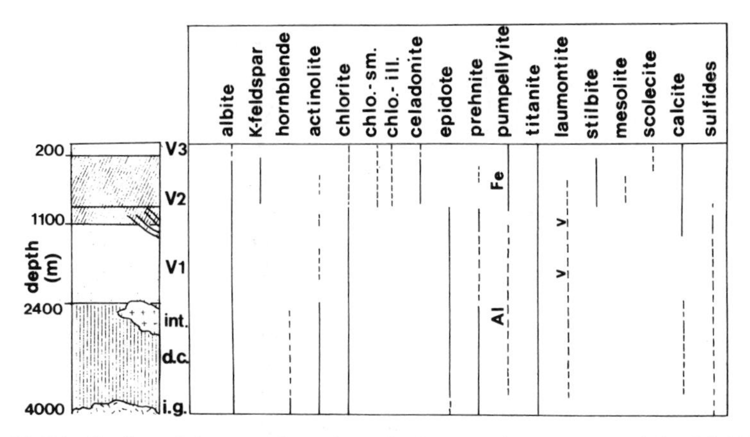

Figure 12. Distribution of the secondary phases in the extrusive sequence of the Salahi block. int = V2 related intrusive bodies; d.c. = dyke complex; i.g. = isotropic gabbros; Al, Fe = Al-rich versus Fe-rich; v = phase observed in veins.

Discussion

Origin of the Metamorphic Zonation in the Salahi Block Extrusive Sequence

The previous mineralogical descriptions point out a strong metamorphic zonation which overprints the extrusive sequence of the Salahi block. This zonation grades from greenschist facies in the dyke complex to brownstone facies in the uppermost lava flows; it is characterized by a minimum overlap between the different amphibole-, prehnite-epidote- and celadonite-pumpellyite-zeolite-bearing zones which coincide with the volcanic stratigraphy (Fig. 12 and 13). Furthermore, the recrystallization events which affected this sequence took place under static conditions as indicated by the lack of deformation features in the whole lava sequence.

Such a sharply defined zonation contrasts with that described from burial metamorphic terrains (Coombs et al., 1976; Kuniyoshi and Liou, 1976; Boles and Coombs, 1977). It is not either identical to the mineralogical zonation described in the modern oceanic layer 2 which has only undergone axial alteration episodes (low-temperature and hydrothermal) and off-axis aging (Humphris et al., 1980; Alt et al., 1986).

Nevertheless, the characteristic mineral assemblages and the sharpness of the different metamorphic zones in this ophiolitic extrusive sequence suggest that its recrystallization resulted from an interaction with seawater or seawater-derived fluids under high-temperature gradients. The oceanic origin of the interacting fluid is clearly demonstrated by the hydrogen isotopic compositions of altered rocks in this section (Regba et al., this volume) and the observed thermal gradient is approximately 100°C/km. D.S.D.P. hole 504B studies have provided the most complete descriptions of the alteration zones

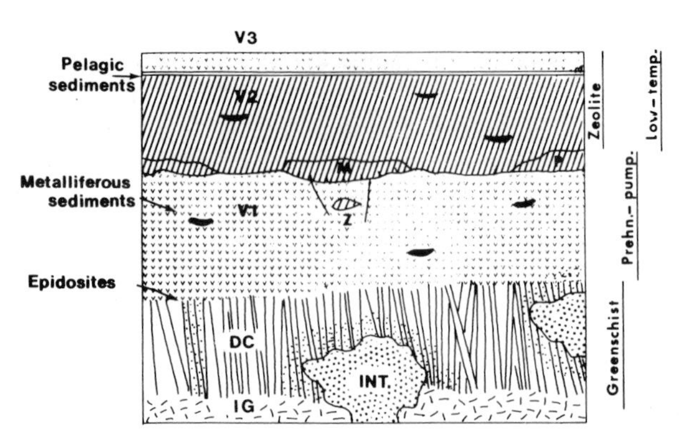

Figure 13. Summary of field observations and distribution of the metamorphic facies in the extrusive sequence of the Salahi block. IG = isotropic gabbros; DC = dyke complex; INT = V2–related intrusions; Z = Zuha sulphide deposit; M = V2 columnar-jointed flows; P = first, primitive V2 flows.

in the upper level of the modern oceanic crust (Fig. 14). The secondary mineral distribution in that hole is thus used in the following discussion as a reference for secondary-phase distribution in present-day oceanic extrusives.

The secondary assemblage which developed in the Salahi block dyke complex is similar to that described in leg 504B diabases. The two rock ensembles, however, differ in their degree of recrystallization. Indeed, the Salahi dykes are usually almost totally recrystallized, whereas in leg 504B dykes, occurrence of secondary phases is restricted to veins and to immediatly adjacent wall-rocks.

The V3 flows excepted, a higher degree of recrystallization also character-

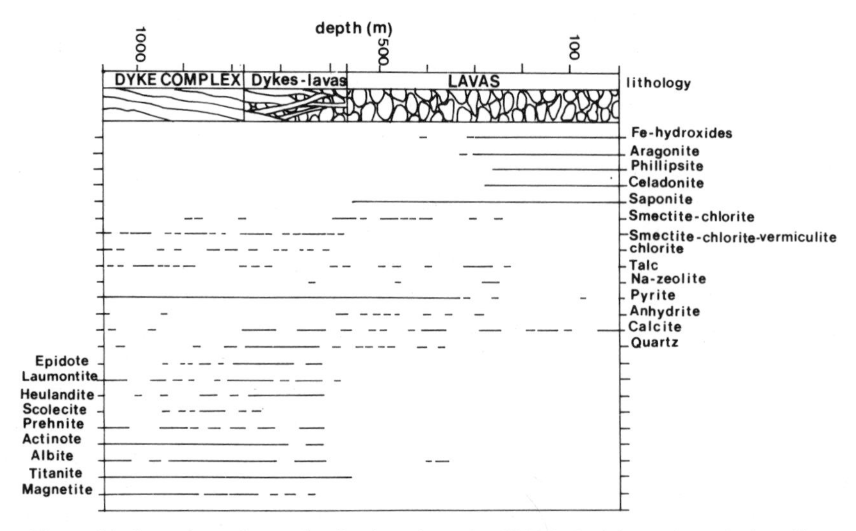

Figure 14. Secondary phases distribution along the 504B hole (after Alt et al. (1986)).

ises the Salahi block volcanics compared to hole 504B lavas. Besides, important mineralogical differences also exist. Most evidences of oceanic, low-temperature alteration are indeed missing (i.e. iddingsite, saponite, phillipsite, anhydrite). Further, this seems to be a general feature of the ophiolite extrusives as it has been emphasized by Alabaster and Pearce (1985) in the Lasail area. Only celadonite and K-feldspar in the vesicular V2 lavas and V3 flows can be interpreted as relicts of such a low-temperature alteration stage. In agreement with this is the occurrence of celadonite at the rims of the vesicles, suggesting that it is an early secondary product.

Additionally, a widespread development of Fe-rich pumpellyite as has been observed in the upper V2 lavas has never been described in modern oceanic volcanics. On the other hand, pumpellyite has been reported in association with celadonite and zeolites in the volcanics of the east Taïwan ophiolite (Liou, 1979) and in the volcanic member of the Del Puerto ophiolite (Ewarts and Schiffman, 1983). The similarity between the metamorphic zonation observed in the Salahi section and the one decribed in the Del Puerto extrusive sequence is striking. The same phase distribution is present with identical sharp transitions between the different mineral zones.

Alabaster and Pearce (1985) have reported almost constant secondary mineral compositions throughout the volcanic sequence. They have suggested, on the basis of these data, that the whole ophiolite extrusive section suffered a uniform, low-intensity, postmagmatic metamorphism which was superimposed on earlier hydrothermal events. This low intensity metamorphism was believed to have developed in a sealed, multipass hydrothermal system. These authors further argued that the development of greenschist-facies assemblage (i.e. represented by the phases albite and chlorite) throughout the lava sequence was related to prograde metamorphism, whereas the crystallization of stilbite, mesolite and clay minerals such as celadonite occurred by means of retrograde reactions as the hydrothermal system slowly cooled down. Overlying, locally thick, mèlange deposits were proposed as the appropriate cap for isolating the underlying lava sequence.

The data presented in this paper do not entirely support the model of Alabaster and Pearce (1985). Indeed, it has been shown that the mineral assemblage and the mineral compositions (in particular the chlorites composition which show increasing smectite/illite component upward) vary through the volcanic sequence. In addition, it is unlikely that phases such as celadonite are a product of retrograde reaction in a closed hydrothermal system as 1) celadonite typically crystallizes in open, oxidized environments (Wise and Eugster, 1964), 2) it commonly appears in the Salahi section at the vesicle rims or as a replacement of glass (and thus represents an early secondary product), and 3) it is also usually reported as early product in sediment-free oceanic crust (Laverne, 1987), as in the East Taïwan ophiolite which also does not display evidence of recrystallization in closed, multipass hydrothermal system (Liou, 1979). Further, the ubiquitous development of Fe-rich pumpellyite in the high-stratigraphic-level lavas suggests that this part of the extrusive sequence reacted with oxygenated, singlepass hydrothermal fluids, as proposed for the Del Puerto ophiolite volcanic section by Evarts and Schiffmann (1983).

On the other hand, the peculiar metamorphic zonation and the pervasiveness of the recrystallizations displayed by the Salahi block and likely by the whole ophiolite extrusives, can be accounted for in the context of only a polyphased hydrothermal system. In this model, each magmatic pulse in the ophiolite would activate a main hydrothermal phase which in the extrusive sequence, would superimpose on earlier low-temperature and hydrothermal alteration stages.

In addition to the mineralogical observations, field evidences have to be taken into consideration to reconstruct the alteration history of the upper part of the Salahi block. Both have been summarized on Fig. 13 which roughly corresponds to a N-S section through the mapped area.

First Magmatic Event Alteration Stage

No mineralogical evidence of the oceanic alteration associated with the accretion stage could be discerned in the V1 lavas. However, hydrothermal activity accompanied this event since metalliferous sediment lenses are found within the V1 flows. This hydrothermal phase is termed H1.

By analogy with modern oceanic crust, H1 may be in part responsible for the recrystallization of the dyke complex diabases in the greenschist facies. The pervasive recrystallization of this level, however, very likely occurred during a later hydrothermal phase as discussed below.

Second Magmatic Event Alteration Stage

The first volcanic event has been thoroughly recrystallized in the prehnite-pumpellyite facies. Such a high-temperature mineral assemblage only appears at the base of leg 504B volcanic section (Fig. 14). The pervasive development of this assemblage throughout the V1 sequence, including the sedimentary lenses, suggests that all these rocks have undergone a phase of recrystallization after the end of the first volcanic event. This phase had to be very pervasive in order to mineralogically homogenize the V1 pile. Such an homogenization is likely to have occurred in response to active hydrothermal convection.

The pervasive development of a similar prehnite-pumpellyite assemblage in the first V2 flows suggests that these lavas have been affected by the same hydrothermal phase as the V1 flows. This phase thus clearly postdates the accretion hydrothermal stage. It is termed H2.

H2 was most likely activated by the second magmatic event of the ophiolite. Indeed, several field evidences suggest that an important high-temperature-fluids discharge and alteration episode was induced by the high-stratigraphic-level V2–related intrusions (e.g. epidotized V1 lavas and diabases around these plutonic bodies, high degree of recrystallization of the V2–feeder sills, common occurrence of sulphide-bearing jasper lenses in between the feeder sills). Furthermore, since the second magmatic event occurred immediately after the first one and the V2–related plutons intruded a still-hot oceanic crust, not much heat was lost by conduction and a very efficient hydrothermal convection could develop in the surrounding crustal rocks during the early steps of the second magmatism (i.e. setting and first stages of differentiation of the intrusion, extrusion of the first primitive V2 flows).

The Zuha mineral zone and the manganiferous lenses that are located at the same stratigraphic level as the first V2 flows, probably resulted from this important hydrothermal convection. The absence of crosscutting veins bearing prehnite-pumpellyite-facies phases (epidote, prehnite) in the mineralized samples agrees with the fact that Zuha is not associated with the former accretion-related hydrothermal phase.

Evolved vesicular flows were extruded above earlier V2 flows and metalliferous sediments, whereas thick columnar-jointed flows were emplaced above Zuha. These lavas were progressively subjected to oceanic low-temperature alteration as evidenced only by the remaining celadonite and K-feldspar in the vesicular flows. In the meantime, the V1 sequence and earlier V2 flows were progressively heated up to prehnite-pumpellyite-facies conditions. This is a reasonable process to consider since the V1 volcanics and lower V2 lavas occupied a stratigraphic position relative to the intrusions (heat sources) which is similar to the position of the dyke complex with respect to the magma chambers in a classic oceanic section. In addition, the permeability and porositiy in the lower part of the volcanic section was subsequently diminished by the earlier, low-temperature crystallizations, allowing progressive heating. The upper vesicular flows apparently remained away from high heat-flow zones (i.e. discharge zones and intrusions) and as a result did not suffer prehnite-pumpellyite-facies recrystallization as did the columnar-jointed flows.

That hydrothermal activity further accompanied the construction of the V2 pile is evidenced by deposition of metalliferous sediments concomitantly with the extrusion of the upper lavas. However, this hydrothermal activity was less intense than at the beginning of the second volcanic event as it did not produce sulphide deposits.

When the second volcanic event ceased, the volcanic sequence was progressively buried under 30 m of metalliferous and pelagic sediments. The recrystallization of the second-generation of dykes under zeolite-facies conditions and the formation of late zeolite-bearing veins in the V1 lavas (at the level of Zuha) and in the V2 primitive flows likely occurred at this time when the intrusions cooled down, the volcanic sequence was progressively buried and the second phase of hydrothermal activity vanished. This alteration can be compared to off-axis aging described in modern oceanic crust which is also characterized by the appearance of zeolite veins (Alt et al., 1986).

Third Magmatic Event Alteration Stage

The occurrence of similar secondary phases in the high-level V2 lavas and in the V3 flows suggests that both lava sequences recrystallized under similar pressure and temperature conditions. Furthermore, the development of a zeolite-facies assemblage in the upper V2 was likely linked to an active fluid convection as indicated by the presence of pumpellyite-rich lava panels materialising hydrothermal fluid pathways.

The lack of evidence for a volcanic event postdating the V3 episode and the absence of indication that hydrothermal activity could be induced during oceanic thrusting suggest that the upper V2 lavas and (in part) the V3 flows, could be recrystallized during a hydrothermal phase activated by a third magmatic event which provided the V3 volcanism. This phase is termed H3.

The V3 volcanism represents a within-plate, seamount-type volcanic activity. Such a magmatism can induce active fluid convection and high-temperature alteration as shown by the formation of secondary products at temperatures higher than 250°C in basalts from off-axis seamounts in the Pacific ocean (Alt et al., 1987). In addition, since outcrops of V3 volcanism are also present in the central part of the ophiolite, this magmatism may have had a wider extent than inferred in previous works (Alabaster et al., 1982), and thus might be responsible for a hydrothermal activity which developed at the scale of the ophiolite.

The occurrence of a third volcanic event a few million years after the cessation of the V2 volcanism (i.e. from Turonian to Campanian), is therefore inferred to induce a third phase of hydrothermal activity responsible for the recrystallization of the V2 vesicular lavas under zeolite-facies conditions. The pervasive and homogeneous development of the zeolite-facies assemblage reflects the fact that the recrystallization took place in a permeable medium (i.e. in a pillowed section characterized by a high vesicularity) yet in relatively closed system (i.e. with a sedimentary layer and probably also some V3 flows capping the sequence).

The V3 flows were recrystallized mainly under low water/rock ratios (low degree of recrystallization; low Na and K uptake by the rocks (Pflumio, 1988)) but still under zeolite-facies conditions. Such an assemblage is uncommon in the upper lavas of oceanic layer 2. Its formation in the V3 lavas is likely related to the interaction of three factors: 1) the influence of the H3 hydrothermal phase, 2) the low water/rock ratio which inhibited the Mg and K uptake from seawater and therefore the formation of Mg and K-bearing low-temperature phases (e.g. saponite, phillipsite) and 3) the higher temperatures which prevailed in this section (composed mainly of thick massive flows).

The Zuha Stockworks

The mineralogy displayed by the Zuha mineralized lavas suggest that these rocks resulted from interaction with a silica-enriched (high modal quartz), metal-loaded (Fe-rich chlorites, occurrence of pyrite and chalcopyrite) fluid. In agreement with the mineralogical data, geochemical mass-balance calculations indicate a thorough increase in Fe, Cu, Zn, Li in the mineralized samples compared to the barren ones (Regba et al., this volume). On the other hand, the albite destabilization to an illite-smectite phase and the disappearance of all calc-silicates in the mineralized lavas reflect higher pH in the mineralizing fluids than in the fluids that gave rise to prehnite-pumpellyite-facies assemblage in the country rocks (Urabe et al., 1983). Finally, chlorites crystallization temperatures indicate that the mineralizing fluids were slightly hotter (250–300°C) than the ones that reacted with the country rocks (< 250°C) in agreement with the oxygen isotopic data (Regba et al., this volume).

Similar mineralogical changes are well-documented in stockworks or alteration halos of volcanogenic sulphide deposits (Date et al., 1983; Richards et al., 1989). Additional fluid inclusion studies indicate that these changes result from interaction between the lavas and a relatively low pH, high temperature, ore-bearing fluid, a hypothesis which is confirmed by thermochemical modeling (Pisutha-Arnond and Ohmoto, 1983; Cathles, 1983). Furthermore, compared to seawater, hydrothermal fluids sampled at the level of oceanic active vents display elemental variations which are similar to the elemental variations observed in unmineralized samples relative to mineralized samples (Regba et al., this volume).

The Zuha mineralized samples therefore represent V1 lavas which interacted with fluids similar to hydrothermal vent waters. These rocks display the same mineralogy as the stockworks of the Bayda mine (Northern Oman) and the alteration halo of the Kokkinopezoula deposit (Cyprus).

The continuity in composition between the chlorites of the unmineralized lavas and that of Zuha stockworks indicates that the stockworks were slightly recrystallized as the top of the V1 sequence was progressively heated to reach prehnite-pumpellyite facies-conditions. The cores 59–1 and 59–2 were drilled slightly outside of a main geochemical anomaly (Fig. 3) suggesting that the studied samples represent a more external part of the hydrothermal discharge zone than the Bayda stockworks. This would explain the relatively Fe-poor character of Zuha chlorites. Furthermore, fluid inclusions studies on Bayda stockworks indicate constant higher alteration temperatures (340°C; Nehlig and Haymon, 1987) than the temperatures inferred for Zuha samples on the basis on both chlorite and oxygen isotopic compositions (Regba et al., this volume).

Conclusion

The extrusive sequence of the Salahi block displays a well-developed dyke complex and a thick volcanic member. The latter is composed of three lava units (V1, V2 and V3). A strong metamorphic zonation overprints this sequence and grades from greenschist facies in the dyke complex to brownstone facies in the lavas. This zonation was induced by a succession of oceanic alteration stages (passive low-temperature and active hydrothermal phases) accompanying the construction of the Semail ophiolite (Fig. 15).

The first hydrothermal phase, H1, was induced by the first, accretion-related, magmatic event (Fig. 15–1). This magmatic event is represented by a main gabbroic series, a well-developed dyke complex and the V1 volcanic sequence. The direct evidence for H1 is the presence of metalliferous sedimentary lenses within the V1 flows. In the extrusive sequence, the mineralogical clues of this hydrothermal phase are now restricted to the dyke complex and are represented by greenschist-facies phases and discrete epidotized dykes. In the plutonic sequence, this axial circulation is indicated by the presence of actinolite-facies minerals which are well-developed in the isotropic gabbros and are restricted to or closely associated with veins in the underlying cumulates (Nehlig and Juteau, 1988). As the H1 phase induced the recrystallization of rocks at deep ophiolitic levels, the volcanic sequence was likely subjected to low-temperature alteration from which no mineralogical evidence remains.

A second off-axis magmatic event immediately followed the accretion. It consists of swarms of intrusive bodies which crosscut the former accretionary section up to the base of the V1 volcanics and fed a second volcanic unit, V2. During an early volcanic stage, the intrusions supplied the primitive V2 flows which locally developed to a thickness of 300 meters. The remainder of the V2 sequence consists mainly of highly vesicular, more evolved flows. This magmatic event activated a second hydrothermal phase, H2 (Fig. 15–2).

H2 induced the progressive recrystallization of the V1 lavas and interbedded sediments under prehnite-pumpellyite-facies conditions and the pervasive development of greenschist-facies assemblage in the dyke-complex diabases (Fig. 15–2A). The recrystallization of the lavas and dykes coincided with a strong enrichment of calcium in the fluids in relation to widespread development of albite (well-illustrated by the Na enrichment and the pronounced Ca-depletion of these rocks (Regba et al., this volume)). Such Ca-rich fluids when flowing upward, 1) either in the volcanic sequence along fault zones (e.g. Zuha area), 2) in the dyke complex along dykes or sets of dykes, or 3) in the vicinity of intrusions in response to an active convection, could induce the lavas and diabases epidotizations.

Experimental studies have shown that the epidotization process could be an efficient mechanism to lower the pH of a solution (Seyfried and Mottl, 1982) and that under low pH, metals are easily solubilized (Seyfried and Bischoff, 1977), a fact in agreement with the observation that epidosites are commonly depleted in copper and zinc (Richardson et al., 1987; Regba, unpublished data). However, the V1 lavas are also depleted in copper compared to MORB (Pflumio, 1988). The recrystallization of this volcanic level in the prehnite-pumpellyite facies can be at the origin of this depletion, as

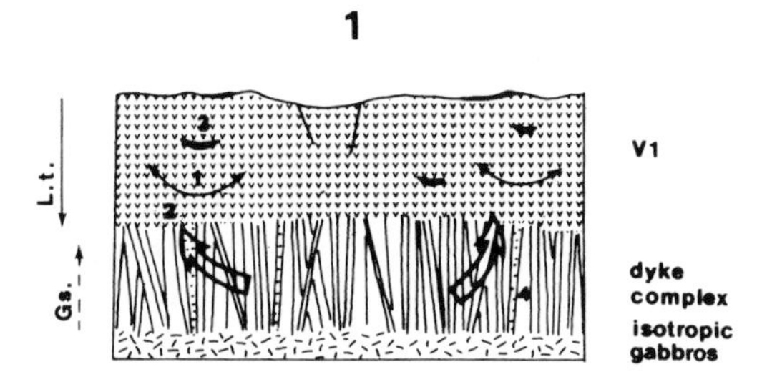

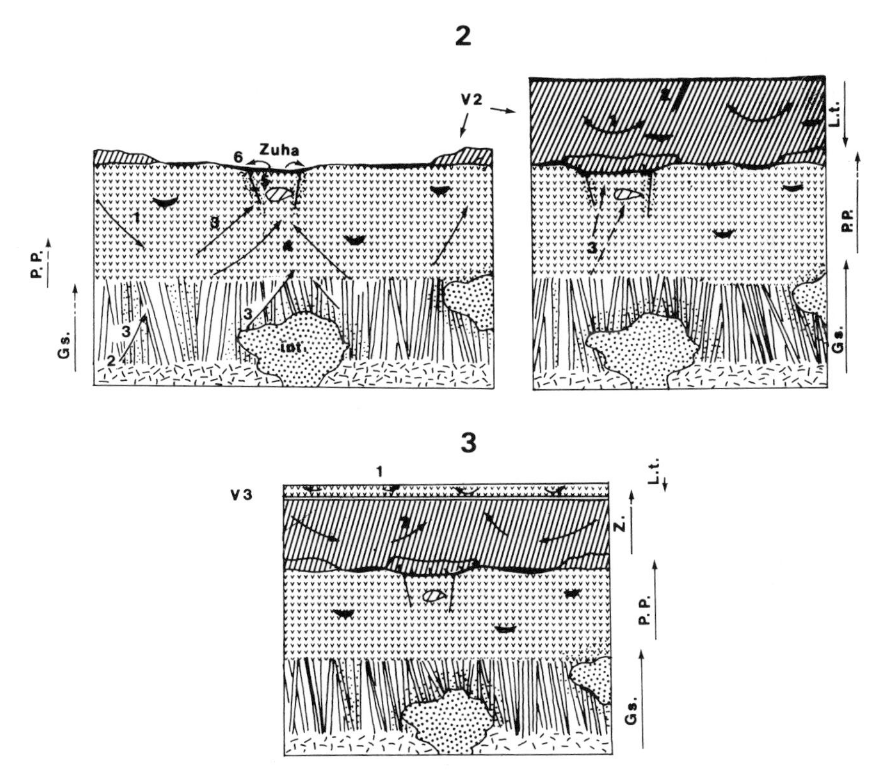

Figure 15. Sequence of oceanic alteration stages registered by the extrusive sequence of the Salahi block.

1) *First alteration stage.* (1) Low-temperature alteration of the V1 lavas. (2) Hydrothermal phase H1 (related to axis magmatism): deposition of metalliferous sediments within the V1 flows (3), and hydrothermal alteration of the dyke-complex diabases (development of greenschist facies assemblage and discreet epidotizations (4)).
2) *Second alteration stage.* Hydrothermal phase H2 (related to off-axis magmatism):

A. (1) progressive recrystallization of the V1 lavas and interbedded sediments under prehnite-pumpellyite-facies conditions, (2) pervasive recrystallization of the dyke-complex diabases under greenschist-facies conditions; (3) Ca-rich fluids discharging along fracture zones in the volcanic sequence, along sets of dykes or in the vicinity of intrusions and inducing epidotizations, (4) suggested by the studies on trace element mobility under similar temperatures and pressures (Humphris and Thompson, 1978b).

Therefore, such evolved fluids flowing out of the epidosite zones are inferred to induce the formation of high-temperature stockworks such as those underlying Zuha, by interaction with V1 lavas. A genetic relationship between the epidosites and the stockworks is further supported by their similar low $\delta^{18}O$ (Regba et al., this volume). A contribution of metal-bearing fluids from the country rocks to the mineral zone is also likely. The stockworks fed a small sulphide deposit which formed on the eastern side of a small graben and whose geometry was controlled by a N-S-trending fault zone. Both structures are related to the former accretionary tectonics.

Metalliferous sedimentary lenses formed contemporaneously at the top of the V1 sequence, either in close association with the deposit, or out of the mineralized zone in relation to local hydrothermal discharge. These sediments are similar to metalliferous deposits which develop at present in the neovolcanic zone of oceanic ridges, near active vents.

More differentiated V2 lavas were extruded onto the previous section with metalliferous sediments accumulating between them (Fig. 15–2B). These lavas suffered low-temperature alteration at the seawater-contact, while the whole V1 sequence and earlier V2 flows were progressively heated up to prehnite-pumpellyite-facies conditions. At the end of the second magmatic event, this extrusive section was buried under 30 meters of metalliferous and pelagic sediments. Lower-temperature circulation was responsible for the development of zeolite veins in high permeability zones such as the Zuha area.

A third hydrothermal phase, H3, occurred a few millions years after the end of V2. It was activated by the third magmatic event in the Semail ophiolite (Fig. 15–3) and induced the recrystallization of the V2 upper lavas and associated sediments under zeolite-facies conditions. The V3 volcanism is represented principally by thick massive flows which bear evidence of alteration under low water/rock ratios and zeolite-facies conditions. The development of such an assemblage at the top of the ophiolite volcanic

---

ascending low-pH, metal-rich and high-temperature fluids from the epidosites (dyke-complex- or intrusions-related), along fracture zones in the V1 sequence, (5) development of stockworks and mineralizations in V1 lavas at the level of main fracture zones, (6) deposition of metalliferous sediments at the top of the V1 sequence.

B. (1) low-temperature oceanic alteration of the vesicular upper V2 lavas. (2) Deposition of hydrothermal sediments within the flows and later over the volcanic sequence. (3) Last bursts of the H2 hydrothermal phase: late laumontite veins in the V1 lavas at the level of Zuha.

3) *Third alteration stage*. (1) Low-temperature oceanic alteration of the V3 flows. Hydrothermal phase H3 (related to within-plate magmatism): (2) pervasive recrystallization of the vesicular upper V2 flows and interbedded sediments under zeolite-facies conditions,. Int = Intrusion, Gs. = greenschist facies, P.P. = prehnite-pumpellyite facies, Z. = zeolite facies, L. t. = low temperature alteration. Straight line: pervasive development; dotted line: heterogeneous development.

sequence is interpreted as related mainly to the specific conditions of alteration in the V3 section compared to a more permeable, pillow-rich oceanic volcanic sequence.

Acknowledgements

I greatly acknowledge the Oman Ministry of Petroleum and Minerals represented by Mr. M. H. Kassim and Mr H. al Azri, for facilities during the field seasons. I benefited from fruitful discussions with C. Mevel, M. Besson and T. Juteau during the development of this study. I thank P. Agrinier, M. Regba and M. Loubet for their collaboration and G. Moutet for patience and effective support. Acknowledgments are due to Dr. T. Alabaster for reviewing the manuscript and to Dr. S.K. Mittwede for his constructive scientific comments and for improvements of the English version of the manuscript.

References

Alabaster, T., Pearce, J.A., Mallick, D.I.J. and Elboushi, I.M., 1980. The volcanic stratigraphy and location of massive sulfide deposits in the Oman ophiolite. In: A., Panayiotou (Ed), Ophiolites, Proceedings of the International Ophiolite Symposium, 1979, Cyprus., Nicosia Ministry of Agriculture Nat. Resources, Geol. survey dept.: 751–757.

Alabaster, T. and Pearce, J.A., 1985. The interrelationship between magmatic and ore-forming hydrothermal processes in the Oman ophiolite., Econ. Geol., 80: 1–16.

Alleman, F. and Peters, T., 1972. The ophiolite-radiolarite belt of the north-Oman mountains., Eclogae Geol. Helv., 65: 657–698.

Alt, J.C., Honnorez, J., Laverne, C. and Emmermann, N, R., 1986. Hydrothermal alteration of a 1 km section through the upper oceanic crust, DSDP hole 504B: the mineralogy, chemistry and evolution of seawater-basalt interactions., J. Geophys. Res., 91: 10309–10335.

Alt, J.C., Lonsdale, P., Haymon, R. and Muehlenbach K., 1987. Hydrothermal sulphide and oxide deposits on seamounts near 21°N, East Pacific Rise., Geol. Soc. Am. Bull., 98: 157–168.

Andrews, A.J., 1980. Saponite and celadonite in layer 2 basalts, D.S.D.P. Leg 37., Contrib. Mineral. Petrol., 73: 323–340.

Beurrier, M., 1987. Géologie de la nappe ophiolitique de Samail dans les parties orientales et centrale des montagnes d'Oman., Thèse Doc. ès Sci., Université Paris 6, France, document B.R.G.M., 128.

Beurrier, M., Bourdillon de Grissac, C., De Wever P and Lescuyer J.L., 1987. Biostratigraphie des radiolarites associées aux volcanites ophiolitiques de la nappe de Semail (Sultanat d'Oman): conséquences tectonogénétiques., C. R. Acad. Sci. Paris, 304, Ser. 2: 907–910.

Boles, J.R. and Coombs, D.S., 1977. Zeolite facies alteration of sandstones in the Soouthland syncline, New Zealand., Am. J. Sci., 277: 982–1012. Bonatti, E., 1983. Hydrothermal metal deposits from oceanic rifts: a classification. In: P.A. Rona et al. (Eds), Hydrothermal processes at seafloor spreading centers (NATO, Conf. Ser.), Plenum, New York and London: pp. 491–502.

Bostrom, K. and Peterson, M.N.A., 1966. Precipitates from hydrothermal exhalations on the East Pacific rise., Econ. Geol., 61: 1258–1265.

Boudier, F. and Coleman, R.G., 1981. Cross section through the peridotites in the Semail ophiolite, southeastern Oman mountains., J. Geophys. Res., 86: 2573–2592.

Boudier, F. and Nicolas, A., 1988. The ophiolites of Oman., Tectonophysics, 151, 1–4.

Bowers, T.C., Von Damm, K.L. and Edmond, J.M., 1985. Chemical evolution of mid-ocean ridge hot springs., Geochim. Cosmochim. Acta, 49: 2239–2252.

Cann, J.R., 1969. Spilites from the Carlsberg Ridge, Indian Ocean., J. Petrol., 10: 1–19.

Cathelineau, M. and Nieva, D., 1985. A chlorite solid solution geothermometer: the Los Azufres (Mexico) geothermal system., Contrib. Mineral. Petrol. 91: 235–244.

Cathles, L. M., 1983. An analysis of the hydrothermal system responsible for massive sulphide deposition in the Hokuroku basin of Japan., Econ. Geol. Mon., 5: 439–487.

Ceuleneer, G., Nicolas, A. and Boudier, F., 1988. Mantle flow pattern at an oceanic spreading center: the Oman peridotite record. In: F. Boudier and A. Nicolas (Eds), The ophiolites of Oman., Tectonophysics, 151: 1–26.

Coleman, R.G. and Hopson, C. A., 1981. The Oman ophiolite., J. Geoph. Res., 86: 2495–2782.

Collinson, T., 1986, Hydrothermal mineralization and basalt alteration in stockwork zones of the Bayda and Lasail massive sulphide deposits, Oman Ophiolite. M.A. Thesis, University of California, Santa Barbara.

Coombs, D.S., 1953. The pumpellyite mineral series., Mineral. Mag., 30: 113–135.

Coombs, D.S., Nakamura, Y. and Vuagnat, M., 1976. Pumpellyite-Actinolite facies Schists of the Taveyanne formation near Loèche, Valais, Switzerland., J. Petrol., 17: 440–471.

Crovisier, J.L., Thomassin, J.H., Juteau, T., Eberhardt, J.C., Touray, J.C. and Baillif, P., 1983. Experimental seawater-basaltic glass interaction at 50°C: study of the early developed phases by electron microscopy and X-Ray photoelectron spectrometry., Geochim. Cosmochim. Acta, 43: 377–387.

Date, J., Wanatabe, Y. and Saeki, Y., 1983. Zonal alteration around the Fukazawa Kuroko deposits, Akita prefecture, Northern Japan., Econ. Geol. Mon., 5: 365–386.

Delaney, J.R., Mogk, D.W. and Mottl, M.J., 1987. Quartz-cemented breccias from the Mid-Atlantic Ridge: samples of a high salinity hydrothermal upflow zone., J. Geophys. Res., 92: 9175–9192.

Detrick, R.S., Honnorez, J., Adamson, A.C., Garrett, W.B., Gillis, K.M., Humphris, S.E., Mevel, C., Meyer, P.S., Petersen, N., Rautenschlein, M., Shibata, T., Staudigel, H., Woolridge, A. and Yamamoto, K., 1986. Forages dans la dorsale médio-Atlantique: résultats préliminaires du Leg 106 du Joïdes resolution (Ocean Drilling Program)., C.R. Acad. Sci. Paris, 303, Ser. 2: 379–384.

Elthon, D., 1981. Metamorphism in oceanic spreading centers. In: C. Emiliani (Ed), The Sea, Vol 7, John Wiley: pp. 285–303.

Ernewein, M. and Whitechurch, H., 1986. Les intrusions ultrabasiques de la séquence crustale de l'ophiolite d'Oman: un evènement témoin de l'extinction d'une zone d'accrétion océanique?, C.R. Acad. Sci. Paris, 303, Ser. 2: 379–384.

Ernewein, M., Pflumio, C. and Whitechurch, H., 1988. The death of an accretion zone as evidenced by the magmatic history of the Semail ophiolite (Oman)., Tectonophysics, 151: 245–274.

Evarts, R.C. and Schiffman, P., 1983. Submarine hydrothermal metamorphism of the Del Puerto ophiolite, California., Am. J. Sci., 283: 289–340.

Fleet, A.J. and Robertson, A.H.F., 1980. Ocean-ridge metalliferous and pelagic sediments of the Semail nappe, Oman., J. Geol. Soc. London, 137: 403–422.

Germain-Fournier, B., 1986. Les sédiments métallifères océaniques actuels et anciens: caractérisation, comparaisons. Thèse Université Bretagne occidentale, Brest, France.

Francheteau, J., Needham, H.D., Choukroune, P., Juteau, T., Seguret, M., Ballard, R.D., Fox, P.J., Normak, W., Carranza, A., Cordoba, D., Guerrero, J. and Rangin, C., Bougault, H., Cambon, P. and Hekinian, R., 1979. Massive deep-sea sulphide ore discovered on the East Pacific Rise., Nature, 277: 523–528.

Franklin, J.M., Lydon, J.W. and Sangster, D.F., 1981. Volcanic-associated massive sulphide deposits., Econ. Geol. 75th. Anniv. Vol., pp. 485–627.

Ghent, E.D. and Stout, M.Z., 1981. Metamorphism at the base of the Samail ophiolite, southeastern Oman Mountains., J. Geophys. Res., 86: 2557–2571.

Glassley, W., 1975, Low variance phase relationships in a prehnite – pumpellyite facies terrain., Lithos, 8: 69–76.

Gillis, K. and Robinson, P.T., 1985. Low temperature alteration of the extrusive sequence, Troodos ophiolite, Cyprus., Can. Min., 23: 431–441.

Glennie, K.W., Boeuf, M.G.A., Hugues-Clark, M.W, Moody-Stuart, M., Pilaar, W.F.H. and Reinhardt, B.M., 1974. Geology of the Oman mountains., Kon. Ned. Geol. Mijbouwk Genoot. Vern., 31.

Green, K.E., Von Hertzen, R.P. and Williams, D.L., 1981. The Galapagos spreading center at 86° N: A detailed geothermal field study., J. Geophys. Res., 86: 979–986.

Haymon, R.M., Koski, R.A., and Abrams, M.J., 1989. Hydrothermal discharge zones beneath massive sulfide deposits mapped in the Oman ophiolite., Geology, 17: 531–535.

Hekinian, R. and Fouquet, Y., 1985. Volcanism and metallogenesis of axial and off-axial structures on the East-Pacific Rise near 13°N., Econ. Geol., 80 (2): 221–249.

Hey, M.H., 1954. A new review of the chlorites., Min. Mag., 30: 277–292.

Humphris, S.E. and Thompson, G., 1978a. Hydrothermal alteration of oceanic basalts by seawater., Geochim. Cosmochim. Acta, 42: 107–125.

Humphris, S.E. and Thompson, G., 1978b. Trace element mobility during hydrothermal alteration of oceanic basalts., Geochim. Cosmochim. Acta, 42: 127–136.

Humphris, S. E., Melson, W.G. and Thompson, R.N., 1980. Basalt weathering on the East Pacific rise and the Galapagos spreading center. Initial Reports of the Deep Sea Drilling Project, 54: 773–788. U.S. Government Printing Office, Washington, D.C.

Ito, E. and Anderson, A.T. Jr, 1983. Submarine metamorphism of gabbros from Mid-Cayman rise: petrographic and mineralogic constraints on hydrothermal processes at slow spreading ridges., Contrib. Mineral. Petrol., 82: 371–388.

Juteau, T., Ernewein, M., Reuber, I., Whitechurch, H. and Dahl, R., 1988. Duality of magmatism in the plutonic sequence of the Semail nappe, Oman., Tectonophysics, 151: 107–136.

Karpoff, A.M., Walter, A.V. and Pflumio, C., 1988. Metalliferous sediments within lava sequences of the Samail ophiolite (Oman): mineralogical and geochemical characterization, origin and evolution., Tectonophysics, 151: 223–246.

Kerridge, J.F., Haymon, R.M. and Kastner, M., 1983. Sulfur isotope systematics at the 21°N site, East Pacific Rise., Earth Planet. Sci. Lett., 66: 91–100.

Kuniyoshi, S. and Liou, J.G., 1976. Burial metamorphism of the Karmutsen volcanic rocks, northeastern Vancouver island, British Columbia., Am. J. Sci., 276: 1096–1119.

Laird, J. and Albee, A. L., 1981. High pressure metamorphism in mafic schists from northern Vermont., Am. J. Sci., 281: 97–126.

Lanphere, M.A., 1981. K-Ar ages of metamorphic rocks at the base of the Semail ophiolite, Oman., J. Geophys. Res., 86: 2777–2782.

Laverne, C., 1987. Les interactions basalte-fluides en domaine océanique. Minéralogie, pétrologie et géochimie d'un système hydrothermal: le puit 504B, Pacifique oriental. Thèse Doc. ès. Sci., Université Aix-Marseille, France.

Leake, B.E., 1978. Nomenclature of amphiboles., Am. Min., 63: 1023–1052.

Liou, J.G., 1971. Stilbite-Laumontite equilibrium., Contrib. Mineral. Petrol., 31: 171–177.

Liou, J.G., 1979. Zeolite facies metamorphism of basaltic rocks from the East Taiwan Ophiolite., Am. Mineral., 64: 1–14.

Liou, J.G., Maruyama, S. and Cho, M, 1985. Phase equilibria and mineral parageneses of metabasites in low-grade metamorphism., Mineral. Mag., 49: 321–333.

Liou, J.G., Hyung Shik K. and Maruyama, S., 1983. Prehnite-Epidote equilibria and their petrologic applications., J. Petrol., 24: 321–342.

Lippard, S.J., Shelton ,A.W. and Gass, I.G., 1986. The ophiolite of Northern Oman., Geol. Soc., London, Mem., 11.

Mc Culloch, M.T., Gregory, R.T., Wasserburg, G. J. and Taylor, H.P. Jr., 1981. Sm-Nd, Rb-Sr and O^{16}/O^{18} isotopic systematics in an oceanic crustal section: evidence from the Semail ophiolite., J. Geophys. Res., 86: 2721–2735.

Merlivat T, L., Pineau, F. and Javoy, M. 1987. Hydrothermal vent waters at 13°N on the East Pacific Rise: isotopic composition and gas concentration., Earth Planet. Sci. Let., 84: 100–108.

Mevel, C., 1987. Evolution of oceanic gabbros from D.S.D.P. Leg 82: influence of the fluid phase on metamorphic crystallizations., Earth Planet. Sci. Lett., 83: 67–79.

Mottl, M.J., 1983. Metabasalts, axial hot springs, and the structure of hydrothermal systems at mid-ocean ridges., Geol. Soc. Am. Bull., 94: 161–180.

Nehlig, P. and Haymon, R., 1987. Microthermometric study of fluid inclusions in a fossil ridge crest, hydrothermal discharge zone in the Bayda area (North Oman Ophiolite)., Eos Trans., A. G. U., 68: 1545.

Nehlig, P. and Juteau, T., 1988. Flow porosities, permeabilities and preliminary data on fluid inclusions and fossil thermal gradients in the crustal sequence of the Semail ophiolite (Oman)., Tectonophysics, V151: 199–221.

Pallister, J.S., 1981. Structure of the sheeted dyke complex of the Semail Ophiolite near Ibra., J. Geophys. Res., 86: 2661–2672.

Pearce, J. A., Alabaster, T., Shelton, A.W and Searle, M.P., 1981. The Oman ophiolite as an arc-basin complex: evidence and implications., Trans. R. Soc. London, 300: 299–317.

Pflumio, C., 1988. Histoire volcanique et hydrothermale du massif de Salahi: implications sur l'origine et l'évolution de l'ophiolite de Sémail (Oman). Thèse, Ecole Nat. Sup des Mines de Paris, France.

Pflumio, C., Michard, A., Whitechurch, H. and Juteau, T., 1990. Petrology of the extrusive sequence of the Salahi block (Northern Oman): implications for the origin and evolution of the Semail ophiolite. Symposium on Ophiolite Genesis and Evolution of the Oceanic Lithosphere, Muscat, Abstracts with program.

Pflumio, C., Juteau, T. and Michard, A., Petrology and geochemistry of the volcanic sequence of the Salahi block (Northern Oman): implications for the origin and evolution of the Semail ophiolite, submitted to Lithos, 1991.

Pisutha-Arnond, V. and Ohmoto, H., 1983. Thermal history and chemical and isotopic compositions of the ore forming fluids responsible for the Kuroko massive sulphide deposits in the Hokuroku district of Japan., Econ. Geol. Mon., 5: 523–558.

Regba, M., Agrinier, P., Pflumio, C., Loubet, M. A geochemical study of an oceanic, hydrothermal discharge zone: the Zuha sulphide prospect in the Salahi block (Semail ophiolite, Oman). This volume.

Richards, H.G., Cann, J.R. and Jensenius, J., 1989. Mineralogical and metasomatic zonation of alteration pipes of Cyprus sulphide deposits. J. Geophys. Res., 84: 91–115.

Richardson, C.J., Cann, J.R., Richards, H.G. and J.G. Cowan, 1987. Metal-depleted root zones of the Troodos ore-forming hydrothermal systems., Cyprus, Earth Planet. Sci. Let., 84: 243–253.

Robertson, A.H.F., 1976. Origins of ochres and umbers: evidences from Skouriotissa, Troodos massif, Cyprus., Trans. Inst. Min. Metall. Sec. B, Appl. Earth Sci.: 245–251.

Robertson, A.H.F. and Fleet, A.J., 1986. Geochemistry and paleo-oceanography of metalliferous and pelagic sediments from the late cretaceous Oman ophiolite., Mar. Petrol. Geol., 3: 315–337.

Sayles, F.L. and Bischoff, J.L., 1973. Ferromanganoan sediments in the equatorial East Pacific Rise., Earth Planet. Sci. Lett.: 19, 330–336.

Seyfried ,W.E. et Bischoff, J.L., 1977. Hydrothermal transport of heavy metals by seawater: the role of seawater/basalt ratio., Earth Planet. Sci. Lett., 34: 71–77.

Seyfried, W. E., and Mottl, M. J., 1982. Hydrothermal alteration of basalt by seawater under seawater dominated conditions., Geochim. Cosmochim. Acta, V46: 985–1002.

Seyfried, W. E. and Janecky, D.R., 1985. Heavy metal and sulfur transport during subcritical and supercritical hydrothermal alteration of basalt: influence of fluid pressure and basalt composition and crystallinity., Geochim. Cosmochim. Acta, 49: 2545–2560.

Stakes, D.S. and O'Neil, J.R., 1982. Mineralogy and stable isotope geochemistry of hydrothermally altered oceanic rocks., Earth Planet. Sci. Lett., 57: 285–304.

Stern, C. and Elthon, D., 1979. Vertical variations in the effects of hydrothermal metamorphism in Chilean ophiolites: Their implications for ocean floor metamorphism., Tectonophysics, 55: 179–213.

Thorette, J., 1986. Contribution à l'étude de l'hydrothermalisme océanique: exemple du district minéralisé de York Harbour (ophiolite de Blow-me-Down, Bay of Island, Terre -Neuve). Thèse, Ecole Nat. Sup. Mines Paris, France.

Tippit, P.R., Pessagno, E.A.Jr. and Smewing, J.D., 1981. The biostratigraphy of sediments in the volcanic unit of the Semail ophiolite., J. Geophys. Res., 86: 2756–2762.

Tuffar, W., Tuffar, E. AND Lange, J., 1986. Ore paragenesis of recent hydrothermal deposits at the Cocos-Nazca plate boundary (Galapagos rift) at 85°51' and 85°55'W: complex massive sulfide mineralizations and mineralized basalts., Geol. Rundschau, 75: 829–861.

Urabe, T. Scott, S.D. and Hattori, K., 1983. A comparison of footwall-rock alteration and geothermal systems beneath some Japanese and Canadian volcanogenic massive sulfide deposits., Econ. Geol. Mon., 5: 345–364.

Wise, W.S. and Eugster, H.P., 1964. Celadonites: synthesis, thermal stability and occurrences., Am. Min., 49: 1031–1083.

Wolery, T.J. and Sleep, T.J., 1976. Hydrothermal circulation and geochemical flux at mid-ocean ridges., J. Geol., 84: 249–275.

A Geochemical Study of a Fossil Oceanic Hydrothermal Discharge Zone in the Oman Ophiolite (Zuha Sulphide Prospect): Evidence for a Polyphased Hydrothermal History

M. REGBA[1], P. AGRINIER[2], C. PFLUMIO[3] and M. LOUBET[1]

[1] *Université Paul Sabatier, Laboratoire de Mineralogie, 38 rue des 36 Ponts, 31400 Toulouse*

[2] *Laboratoire de Géochimie des Isotopes Stables, Institut de Physique du Globe de Paris and U.E.R. des Sciences Physiques de la terre, Université Paris 7, 4 Place Jussieu, 75230 Paris, Cedex 05*

[3] *C.G.G.M., Ecole Nationale supérieure des Mines de Paris, 60 Boulevard Saint-Michel, 75072 Paris, Cedex 06*

Abstract

A geochemical study (of major and trace elements (including REE), as well as hydrogen, oxygen, neodymium and strontium isotopes) of the Zuha sulphide prospect in the Salahi block (northern part of the Sumail ophiolite, Oman) was focused on a pair of boreholes which were drilled at the foot of the main gossan, on the border of a 400 ppm Cu geochemical anomaly.

The geochemical data establish two main alteration styles (type I and type II) which developed in the V1 lavas around the Zuha prospect. Both types exhibit contrasting secondary mineralogy. Type I rocks display a classical prehnite-pumpellyite facies mineral assemblage (quartz, albite, chlorite, epidote, sphene), whereas type II rocks bear a typical stockwork alteration paragenesis (quartz, chlorite, rectorite, sphene, sulphides).

The diversity of chemical and isotopic compositions of the type I and type II alterations reflects distinct temperatures, water/rock ratios and fluid compositions, and suggests that these two types of alteration were formed in different parts of the same hydrothermal system. In this system, the type I mineral assemblage results from an interaction between V1 lavas and Na/Mg rich sea-water derived fluids in recharged zones, and the type II assemblage from an interaction between V1 lavas and upwelling hydrothermal fluids in a discharge zone, at higher temperature.

The δD compositions of these two types of alteration assemblages support the contention that the fluids responsible for these recrystallizations were sea-water derived. The oxygen isotope compositions indicate fluid rock interaction temperatures around 200°C for type I rocks and above 250°C for type II rocks, in agreement with mineralogical assemblage estimates.

Comparison of type I and type II rock compositions suggests that the fluids at the origin of the Zuha stockwerks were globally enriched in Fe, Si, Cu, and shows they were characterized by LREE and Eu enrichments. Thus, these fluids show strong analogies with modern, active hot springs.

Both type I and type II rocks however display $^{87}Sr/^{86}Sr$ ratios ($^{87}Sr/^{86}Sr$

Tj. Peters et al. (Eds), Ophiolite Genesis and Evolution of the Oceanic Lithosphere, 353–383.

0.705) which are higher than modern oceanic hydrothermal compositions. A polyphase hydrothermal history is advanced in order to explain these peculiar compositions. It is proposed that the hydrothermal system at the origin of the Zuha prospect developed in previously altered oceanic crust whose Sr isotopic composition was further shifted. These two alteration stages were successively activated by the first axial magmatic event (V1 type magmatism) and the second post accretion magmatic event (V2 type magmatism). The secondary mineralogical assemblages observed in the V1 lavas (prehnite, pumpellyite facies) and in the diabases from the dike complex (greenschist facies) were recrystallized during this second alteration stage.

Preliminary data obtained on epidote-rich dikes (episodites) and their host diabases in this Salahi block dike complex are also presented. The close similarities found between the $\delta^{18}O$ and $^{87}Sr/^{86}Sr$ compositions of some of these episodites and the type II stockwork rocks support the genetic relationship between these two rock facies. As advanced by several authors, these data support the contention that the epidosite zones represent fossil conduits of the hydrothermal, metal-bearing uprising solutions.

Introduction

It is now widely accepted that hydrothermal convective circulation associated with the formation of new oceanic crust takes place at mid-oceanic spreading centers, where it affects a large thickness of the oceanic crust. This hydrothermal process has also been observed in other geodynamic environments such as seamounts or back-arc areas (Horibe et al., 1986; Alt et al., 1987).

This process is important because it controls the heat loss from the earth (Wolery and Sleep, 1976; Sclater et al., 1981), the geochemical balances of the major, minor and trace elements in the ocean (Edmond et al., 1979a, b; Staudigel and Hart, 1983), has a significant effect on the chemistry, isotopic composition, mineralogy and physical properties of the oceanic crust (Humphris and Thompson, 1978a, b; Honnorez, 1981; Stakes and O'Neil, 1982; Mottl, 1983; Alt et al., 1986a, b) and is responsible for the formation of metal-rich deposits (Solomon and Walshe, 1979; Lalou, 1983; Bonatti 1983; Hekinian and Fouquet 1985; Alt et al., 1987, Rona, 1988).

Mineralogical and geochemical studies of modern oceanic crust sections as well as the analyses of oceanic vent hydrothermal solutions have contributed significantly to our understanding of hydrothermal alteration (Humphris and Thompson, 1978a, b; Edmond et al., 1979a, b; Staudigel et al., 1981a, b; Von Damn et al., 1985a, b; Lowell and Rona, 1985; Alt et al., 1986a, b). However, direct observation of this phenomenon is limited in the ocean and this prohibits the detailed analysis in space and time of the distribution of the alteration effects (i.e., up to now the deepest oceanic bore-hole has only reached the upper part of the sheeted dike complex (504 B, 1076 m), (Alt et al., 1986a)).

On the other hand, ophiolite complexes offer a 3D sampling of the oceanic crust for analysis. The study of this material has firmly established the existence of sea water/oceanic crust interaction (Spooner et al., 1974; Spooner et al., 1977; Mc Culloch et al., 1980; Lanphere and Coleman, 1981; Cocker et al., 1982; Schiffman et al., 1984), as well as the evolution of the intensity of this interaction with depth (Stern and Elton, 1979; Mc Culloch et al., 1981; Gregory and Taylor, 1981). Recent ophiolitic studies have emphasized the development of epidosite zones throughout the sheeted dike complex and their importance in the genesis of massive sulfide deposits (Richarson et al., 1987; Schiffman et al., 1987; Harper et al., 1988; Schiffman and Smith, 1988).

The Sumail ophiolite represents a piece of Tethyian oceanic lithosphere obducted at the end of the Cretaceous upon the Arabic margin. It is an excellent locus to study the effect of oceanic hydrothermalism. Indeed, this ophiolite can be considered as one of the best preserved piece of oceanic crust outcropping on land, without evidence of a later, superposed episode of regional metamorphism. This massif has already been the subject of numerous geological, metallogenical and geochemical studies (see Coleman and Hopson, 1981, Lippard et al., 1986 and Boudier and Nicolas, 1988).

This work is part of a French team project which was developed in cooperation with the Ministry of the Petroleum and Minerals of the Sultanate of Oman. It has been focused upon the Zuha sulphide prospect in the Salahi block (northern part of the Sumail ophiolite), which is considered to be a typical hydrothermal discharge zone (Pflumio, this issue). This work concerns two boreholes which were drilled at the foot of the Zuha major gossan and which cut across a stockwork zone.

This paper presents a set of geochemical data obtained on these core samples – including major, trace element (including REE), and hydrogen, oxygen, strontium and neodymium isotopic compositions.These combined geochemical tracers enable to get insight into the chemical and physical conditions (temperatures, water/rock ratios, fluid compositions) which governed the interactions between the hydrothermal solutions and the basaltic rocks, and into the evolution of this hydrothermal system.

Furthermore, preliminary Sr and O isotopic data on epidosites and surrounding diabases from the dike complex allow for the discussion of the relationship between these deeper level rocks and the stockwork zone.

Geological Setting

Magmatic Suites of the Salahi Block

The Zuha sulphide prospect is located in the Salahi block (northern part of the Sumail ophiolite, Oman) (Fig. 1). Integrated field, mineralogical and geochemical studies have allowed for the definition of a magmatic stratigra-

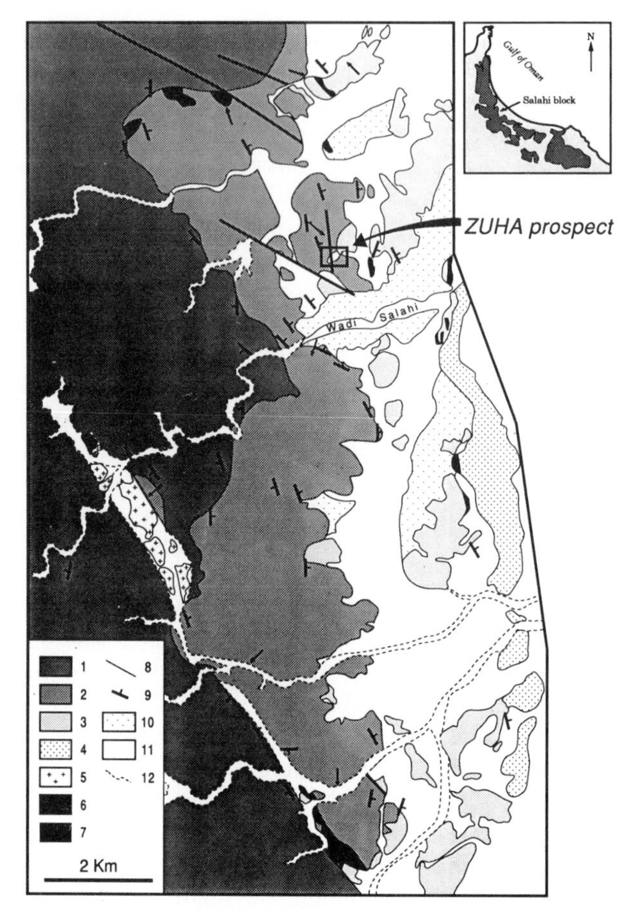

Figure 1. Geological map of the extrusive sequence of the Salahi block, 1: dike complex; 2: V1 lavas; 3: V2 lavas; 4: V3 lavas; 5: plagiogranites; 6: oxidized stockworks; 7: metalliferous sediments; 8: fault contact; 9: strike and dip of dikes and lavas; 10: ancient gravel terraces; 11: wadis gravels and present day terraces; 12: wadis.

phy in the area (Ernewein et al., 1988, Pflumio et al., submitted). Three magmatic events with tholeiitic affinities are distinguished:

- the V1 magmatism, genetically related to the main plutonic sequence and to the dike complex (previously Geotimes unit of Alabaster et al., 1982), is attributed to a spreading-axis volcanism.
- the V2 volcanism including the Lasail, Alley and clinopyroxenes units of Alabaster et al. (1982), is related to sets of plutonic intrusions and represents a slightly later, off-axis magmatism (Tippit et al., 1981; Ernewein et al., 1988).
- the V3 volcanism (Salahi unit of Alabaster et al., 1982), is considered to be a within-plate type volcanism (Ernewein et al., 1988).

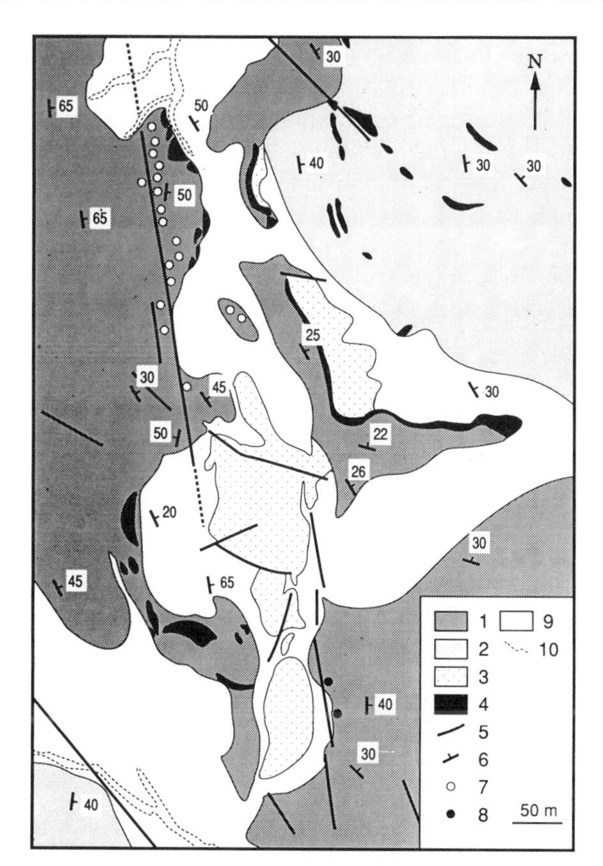

Figure 2. Geological map of the Zuha prospect (see caption 11 of fig. 2 for location). 1: V1 lavas; 2: V2 lavas; 3: gossans; 4: metalliferous sediments; 5: major fault; 6: strike and dip of dykes and lava flows; 7: epidosites; 8: 59–1 and 59–2 coreholes; 9: wadi gravels; 10: wadis.

The Zuha Sulphide Prospect

The volcanic sequence of the Sumail ophiolite encloses numerous massive sulphide deposits and disseminated sulphide mineralizations. In the field, most mineralized areas are characterized by the presence of a bright red, siliceous gossan. The massive ore bodies are located either between the V1 and the V2 sequences as in the Lasail and Arja mining districts or inside the V1 lavas sequence as at Bayda (Alabaster and Pearce, 1985; Haymon et al., 1989).

The Zuha sulphide prospect is located approximately 1 km north of the wadi Salahi (Fig. 1). Its surface exposures are four gossans which exactly lie at the interface of the V1 and V2 volcanisms (Fig. 2) (see Fig. 2 and 3 in Pflumio, this issue). The gossans are located along a N170 fault system, on the eastern side of a small graben which is presently occupied by some V2 flows. The N170 fault zone has been the loci of an important discharge of mineralized fluids as indicated by the N-S trending Cu and Zn geochemical anomaly contours that are superposed over this faulted area. As the dike complex also displays similar orientation at the scale of the ophiolite (Pallister, 1981; Pearce et al., 1981; Pflumio, 1988), the structural features which governed Zuha are most contemporaneous with the accretion stage.

Three core holes have been drilled in the V1 lavas at the foot of the major Zuha gossan. They were located on the border of the 400 ppm Cu anomaly contour and oriented perpendicular to the main directions of the lava flows. Only two holes, 59–1 and 59–2, have encountered disseminated mineralization (pyrite content of the lavas $< 20\%$).

Petrographical and Mineralogical Study

A history of the oceanic alteration of the extrusive sequence of the Salahi block has been proposed by Pflumio (this issue) based on mineralogical grounds (distribution and composition of the secondary minerals in the three volcanic events). The following is a brief description taken from this work.

The Secondary Assemblages in the Extrusive Sequence of the Salahi Block

The first volcanic episode, 1.3 km thick, consists of aphyric, red brown pillow lavas and massive flows. All the V1 lavas have been recrystallized under prehnite-pumpellyite facies conditions and these volcanics commonly exhibit the following secondary assemblage: quartz + albite + chlorite + epidote + sphene + pumpellyite + prehnite + pyrite + chalcopyrite. Conversely, the diabases of the dyke complex have been recrystallized under prehnite-greenschists or amphibolite facies conditions as indicated by the development of a secondary assemblage similar to the one previously detailed in V1 with in addition the occurrence of actinolite. Epidotised dikes (epidosites) are also quite commonly observed throughout the dike complex. These dikes are totally replaced by the assemblage epidote + prehnite + quartz + sphene. V1 lavas and dike complex diabases are also pervasively epidotised at the contact of the V2 related intrusions.

The V2 volcanism, 1 km thick, consists of two types of lavas. Primitive lavas corresponding to the Lasail unit of Alabaster et al., (1982) are common at the bottom of this sequence. More evolved, highly vesicular flows equivalent to the Alley unit of Alabaster et al., (1982) overlie the former lavas. They represent the common V2 facies in the studied area.

The primitive lavas exhibit the same secondary assemblage as V1 flows which is indicative of conditions of recrystallization under the prehnite-pumpellyite facies. Albite, prehnite, chlorite, epidote and sphene are the dominant secondary minerals. Laumontite has been observed in late veins. Pyrite and chalcopyrite develop in veins and they are associated with chlorite, prehnite and epidote. The secondary assemblage displayed by the evolved vesicular flows is the following: quartz + albite + Kfeldspar + celadonite + chlorite + Fe rich pumpellyite + stilbite + sphene + calcite. This assemblage implies that these lavas have suffered a low temperature oceanic alteration as well as a hydrothermal recrystallization under zeolite facies conditions.

The V3 volcanism consists of columnar-jointed lava flows only 200 m thick. These lavas are less recrystallized than the V1 and V2 flows They exhibit a secondary paragenesis (albite, chlorite, celadonite, iron rich pumpellyite, sphene and scolecite) which is similar to the one observed in the evolved, vesicular V2 flows. This indicates similar conditions of alteration (low temperature, zeolite facies).

Secondary Paragenesis Developed in the Zuha Sulphide Prospect Core Holes

Four lithological types have been distinguished in the cores: weakly mineralized or unmineralized lavas or type I samples, mineralized lavas are called type II samples, lava breccias which represent type III samples and hematitized jasper veins.

The weakly mineralized or unmineralized lava cores and rims display the same mineralogical association as V1 lavas sampled outside the prospect zone. Laumontite has been observed in late veins in the pillow lava rims.

The mineralized lavas (*Type II*) exhibit the following secondary assemblage: quartz + Fe-chlorite + rectorite + sphene + pyrite + chalcopyrite. This assemblage is similar to the one described in the stockwerks of the Oman Bayda copper mine (Collinson, 1986) and in the alteration halos of the Troodos Kokkinopezoulas sulphide deposit (Richards et al., 1989).

The disappearance of albite and occurrence of Fe-chlorite in these mineralized samples agree well with an interaction with low pH metal-rich fluids (Urabe et al., 1983).

Different types of breccias of hydrothermal origin displaying different secondary assemblages have been encountered in the drilled holes. The breccia type studied in the present paper is composed of lava fragments, which exhibit a secondary assemblage similar to the one of the mineralized samples': quartz + Fe-chlorite + sphene + pyrite + chalcopyrite. The matrix of this breccia is composed of quartz fragments, cubes of pyrite and chlorite. Veins of laumontite crosscut both fragments and matrix.

The hematitized jasper samples consist of quartz, hematite, pyrite and chalcopyrite. Laumontite is usually encountered either in the groundmass or in veins. Epidote often develops in the border of these jaspers. The core jaspers are very similar to the red jaspers which are observed in the field, in the V1 interpillow spaces and in V1pillow fractures.

In summary, two main mineralogical associations can be distinguished in the V1 flows at the level of the Zuha prospect: *a prehnite-pumpellyite facies assemblage* (type I) similar to the one developed in the entire V1 sequence, and a *stockwork type mineralogical assemblage* (*type II*). In addition, *late zeolite veins* are observed in both unmineralized and mineralized lavas in high permeability locations (i.e. pillow rims, interpillow spaces and breccias).

Analytical Techniques

Major elements (except Na and K) were measured by X-ray fluorescence (X.R.F.) analysis of fused powders on the SIEMENS X-ray spectrometer of the Toulouse University Petrological Laboratory (T.U.P.L.). Na and K were analyzed by flame spectrometry at the Toulouse University Mineralogical Laboratory (T.U.M.L). The total volatile elements were determined by loss on ignition (L.O.I) at 1000°C. Analytical precision under routine conditions is better than 2%.

Trace elements (excluding Li and REE) were also measured by X.R.F. spectrometry using pressed powder pellets. Matrix and instrumental effects were corrected by computation according to the method described by Bougault et al. (1977). Selected international rock standards were used for calibration. Analytical precision for trace elements other than Nb is within a few percent $<$ %). Nb precision is estimated at $\pm$ 1 p.p.m. for rocks whose Nb concentration is between 1 and 10 p.p.m.

Rare earth elements (R.E.E.), Rb and Sr analyses were performed by isotope dilution at T.U.M.L using a modified CAMECA 206 THN mass spectrometer and with a precision better than 2%. Chemical separation of the elements was carried out on a AG 50 W (200–400 mesh) cationic ion exchange column. Sr was separated from Ca using ammonium citrate as a complexing agent (Birck and Allègre 1978). The REE were divided into three fractions using the H.D.E.H.P (Di(2ethylhexyl)orthophosphoric acid, Richard et al. 1976).

Sr isotopic composition measurements were done on an automated Finnigan MAT 261 multicollector mass spectrometer at T.U.M.L. NBS 987 Standard was measured with a $^{87}Sr/^{86}Sr$ ratio of 0.71015 $\pm$ 0.0002 ($2\sigma/\sqrt{N}$). Correction of the mass discrimination effect is done by normalizing the $^{87}Sr/^{86}Sr$ ratio to a value of 0.1194. Analytical errors are better than $\pm$ 0.0002 ($2\sigma/\sqrt{N}$).

Chemical separation of Nd was done using the technique of Richard et al. (1976). The isotopic measurements were performed on an automated Finnigan MAT 261 multicollector mass spectrometer at T.U.M.L. The Nd was loaded on double Re filaments and measured as metal. Correction of the mass discrimination effect is done by normalizing the $^{146}Nd/^{14}4Nd$ ratio to a value of 0.721903.The average value of $^{143}Nd/^{144}Nd$ isotopic ratios for the Johnson Mattey standard was 0.511110 $\pm$ 0.000010 ($2\sigma/\sqrt{N}$). The relative external reproductibility of 10 samples is better than $\pm$ 0.002%.

Oxygen and hydrogen isotopic composition ($\delta^{18}O$ and δD) were measured at the "Laboratoire de Géochimie des Isotopes Stables" of the Université

Paris VII. Oxygen was extracted from silicate rocks by ClF_3 or BrF_5 (Borthwick and Harmon, 1982 and J.R. O'Neil. pers. com., 1985; Clayton and Mayeda, 1963) and reacted with carbon to give CO_2. Hydrogen, extracted mainly in the H_2O form by fusion in an induction furnace, was obtained by reduction on hot zinc metal (Coleman et al., 1982). The hydrogen yield was manometricaly measured for precise determination of water contents $[H_2O^+]$ (expressed in weight percent). CO_2 and H_2 were then analyzed with the mass spectrometer to obtain the $^{18}O/^{16}O$ and D/H ratios. The isotopic composition of a sample is given as:
$\delta_{sample} = (R_{sample}/R_{standard}-1) * 1000$ where R is $^{18}O/^{16}O$, or D/H and the standard is SMOW for oxygen and hydrogen. Reproducibility (s) of replicate oxygen and hydrogen analyses is about ±0.1 ‰ and ±1.5 ‰ respectively.

Major Elements, Trace Elements (Excluding REE), Nd Isotopic Compositions Geochemistry

Major and trace element data (excluding REE) for the cores 59–1 and 59–2 samples are given in Tables 1 and 2 respectively, with the composition of one core sample from hole 59–3 (59–3–166) (Table 2). The Sr and Nd isotopic data obtained on several type I and type II samples are reported in Table 3. In Figures 3a and 3b, major element compositions of core samples have been plotted according to their stratigraphical position.

The constancy of the ratio of some incompatible and immobile elements such as Y/Zr and Zr/Nb is a remarkable feature of these rocks (Table 1 and 2) and supports their consanguinity (i.e their belonging to a same magmatic suite which is V1). This conclusion is further demonstrated by the close similarity of the Nd isotopic compositions of all of these rocks (Table 3).

These core samples show Cr and Ni contents (from 0 to 95 ppm and from 0 to 16 ppm respectively) lower than expected values for mantle derived primary magmas (250 to 360 ppm) (Allegre and Minster, 1978; Bougault, 1980). In addition, they display dispersed TiO2 values (0.83 to 2.02%) and Ti/Zr ratios (which extend over a relatively large range (60 to 100)). These features do not result from secondary alteration effects as they concern elements considered to remain immobile during such phenomena (Pearce et al., 1973; Humphris and Thompson, 1978b). They are more likely related to differentiation processes of the magmas and more specifically to the crystallization of mafic phases (for Cr and Ni) and titaniferous phases (for Ti).

Compared to typical MORBs, the unmineralized core samples (type I) are characterized by a relatively uniform chemical composition with a notable enrichment in Na_2O and a depletion in CaO. The Na enrichment observed in these rocks results from an intense albitization of plagioclase, and suggests an alteration at water/rock ratios less than 10 (Mottl, 1983). The low water/rock ratios ($\approx$5) calculated below on the basis of $^{87}Sr/^{86}Sr$ isotopic compositions support this hypothesis (see below).

Table 1. Major element content in weitht % (X Ray Fluorescence analysis) of Zuha core samples from the respective holes 59-1 (Table 1a) and 59-2 (Table 1b) and 59-3 (sample 166, Table 1b). P.F.: Dehydration %.

Table 1a.

Samples	SiO_2	Al_2O_3	Fe_2O_3	MnO	MgO	CaO	Na_2O	K_2O	TiO_2	P_2O_5	P.F.	T
3	59.68	13.88	10.93	0.08	4.77	0.72	0.44	0.08	1.66	0.17	8.16	100.57
48	58.14	13.45	13.16	0.19	5.33	2.22	2.99	0.24	1.78	0.28	3.39	101.17
53	74.64	6.88	10.94	0.10	3.24	0.29	0.17	0.07	0.83	0.13	4.23	101.52
56	66.44	8.56	14.08	0.10	3.34	0.44	0.23	0.08	1.11	0.18	5.61	101.17
59	56.95	13.55	16.00	0.14	5.84	0.65	0.28	0.09	1.48	0.22	6.32	101.52
61	44.33	16.17	16.44	0.20	6.86	2.21	0.50	0.23	1.40	0.15	11.30	99.79
62	45.34	16.38	19.59	0.27	7.11	2.40	2.45	0.09	1.93	0.21	5.12	100.89
71	53.77	14.61	12.94	0.21	5.37	1.84	5.28	0.22	1.68	0.15	3.42	99.49
83	52.82	15.27	12.61	0.21	5.29	1.63	4.20	0.19	1.78	0.18	4.28	98.46
77	47.71	15.67	17.86	0.16	6.93	1.58	4.90	0.10	1.79	0.14	3.55	100.39
99	50.02	16.46	13.31	0.19	6.65	1.69	5.83	0.04	1.79	0.18	4.15	100.31
101	51.95	14.18	15.37	0.19	5.76	1.45	5.22	0.10	1.61	0.15	3.49	99.47
106	56.42	13.64	10.93	0.15	4.48	2.63	5.54	0.22	1.79	0.18	3.42	99.40

Table 1b.

Samples	SiO_2	Al_2O_3	Fe_2O_3	MnO	MgO	CaO	Na_2O	K_2O	TiO_2	P_2O_5	P.F.	T
133	56.03	13.66	v12.25	0.36	5.17	1.84	4.17	0.16	2.02	0.24	3.77	99.67
137	55.07	14.57	1.94	0.31	6.07	1.67	3.71	0.11	1.91	0.25	4.37	99.98
138	50.06	14.54	18.14	0.19	8.01	0.46	0.22	0.06	1.3	0.15	7.15	100.32
139A	44.84	15.23	18.84	0.31	10.61	0.49	0.18	0.16	1.86	0.25	7.55	100.32
140	55.73	12.90	15.76	0.14	6.02	0.57	0.29	0.09	1.56	0.23	6.23	99.52
141A	57.90	9.95	18.39	0.15	5.54	0.33	0.14	0.03	1.17	0.18	5.86	99.64
142	57.67	8.64	18.13	0.10	4.57	0.28	0.19	0.06	1.02	0.17	8.29	99.12
143A	63.29	9.08	14.92	0.13	4.94	0.42	0.20	0.08	1.02	0.17	5.68	99.93
147	54.29	12.46	14.89	0.20	&.44	0.80	0.21	0.07	1.31	0.21	6.97	99.85
154	91.65	1.72	4.23	0.02	0.34	0.35	0.06	0.05	0.08	–	2.33	100.83
166*	50.72	13.85	12.99	0.20	5.84	5.41	5.83	0.09	1.97	0.18	3.13	100.21

Table 2. Minor element content in p.p.m. (X Ray Fluorescence analysis) of core samples from the respective holes 59-1 (Table 2a), 59-2 (Table 2b) and 59-3 (sample 166, Table 2b).

Table 2a.

Samples	Type	V	Cr	Co	Ni	Cu	Zn	Sr	Y	Zr	Nb	Li	Y/Zr	Ti/Zr	Zr/Nb
3	II	154	76	–	6	75	142	199	52	146	5	6.8	0.36	68	29
48	I	195	3	13	–	–	107	51	66	159	5	9.4	0.41	67	31
53	II	110	11	14	–	6	53	17	27	78	5	9.4	0.35	64	19
56	II	147	12	13	–	27	51	32	38	94	4	15.3	0.41	71	23
59	II	204	6	–	–	–	87	55	54	118	5	8.5	0.45	75	23
61	Breccia	321	40	30	11	445	59	56	28	76	6	21.0	0.37	110	15
62	Breccia	357	13	45	3	–	93	70	41	106	4	8.9	0.39	109	26
71	I	353	86	22	8	87	95	59	44	106	4	7.7	0.42	95	26
83	I	341	94	28	11	85	12	40	54	127	3	5.7	0.42	84	42
77	I	359	4	10	11	127	57	35	42	102	2	4.0	0.41	105	34
99	I	298	20	31	16	68	67	28	52	105	5	3.2	0.49	102	26
101	I	365	90	25	13	87	11	26	44	91	3	7.2	0.48	106	30
106	I	378	91	22	9	43	13	67	51	124	5	5.7	0.42	86	24

Table 2b.

Samples	Type	V	Cr	Co	Ni	Cu	Zn	Sr	Y	Zr	Nb	Li	Y/Zr	Ti/Zr	Zr/Nb
133	I	179	46	14	–	17	12	32	72	191	7	12.8	0.38	63	27
137	I	192	–	14	1	–	220	37	71	176	8	8.1	0.40	65	22
138	II	312	61	12	–	–	110	40	31	88	4	8.5	0.35	91	22
139A	II	317	–	14	–	26	181	20	65	173	8	9.7	0.38	65	22
140	II	255	4	12	–	–	121	48	63	147	6	9.4	0.43	64	24
141A	II	205	8	11	–	106	63	14	42	108	6	8.9	0.39	65	18
142	II	156	–	18	–	729	58	9	36	88	5	9.7	0.41	69	20
143A	II	167	–	6	–	1262	163	23	47	98	4	7.2	0.48	62	24
147	II	204	–	4	–	698	129 16	20	61	159	6	14.6	0.38	50	26
154	JASPER	28	54	16	–	705	109	5	10	22	2	12.8	–	–	–
166*	III	438	–	37	13	4		44	49	117	7	4.1	–	–	–

Table 3. $^{87}Sr/^{86}Sr$ and $^{143}Nd/^{144}Nd$ compositions of the Zuha core samples. Index m and c refer respectively to the "measured" compositions and to the "computed initial compositions" 100 my ago. $\epsilon(0)$ = ϵNd of the rocks at present time; $\epsilon(T)$ = ϵNd of the rocks 100 my ago; W/R = Water rock ratios (by weight) for closed system are calculated according to the method described by McCulloch et al. (1981) assuming a Cretaceous sea water $^{87}Sr/^{86}Sr$ composition of 0.708 (Burke et al., 1985; Koepnick, 1985).

Samples	Nature	$^{87}Sr/^{86}Sr$ m	$^{87}Rb/^{86}Sr$	$^{87}Sr/^{86}Sr$ c	$^{143}Nd/^{144}Nd$m	$^{147}Sm/^{143}Nd$	$^{143}/^{144}Nd$c	$\epsilon(0)$	$\epsilon(T)$
59.3.166	Type I	0.705480 ± 20	0.03042	0.705440 ± 20	0.513083 ± 08	0.2040	0.512950 ± 08	8.69	8.60
59.2.133	Type I	0.705467 ± 24	0.08335	0.705350 ± 24	0.513056 ± 05	0.191	0.512933 ± 05	8.11	8.26
59.1.106	Type I	0.705077 ± 14							
59.2.138	Type II	0.705116 ± 24	0.03527	0.705070 ± 24	0.513103 ± 07	0.1678	0.512993 ± 07	9.07	9.44
59.2.147	Type II	0.705401 ± 26	0.07738	0.705290 ± 26	0.513057 ± 06	0.1680	0.512947 ± 06	8.97	8.55
59.1.59	Type II	0.704778 ± 24	0.03660	0.704730 ± 24					
59.1.53	Type II	0.704989 ± 26							
59.1.61	Breccia	0.707077 ± 20	0.21680	0.706770 ± 20					
59.2.154	Jasper	0.706580 ± 28	0.05784	0.706498 ± 28	0.513040 ± 08	0.1606	0.512935 ± 08	7.80	8.31
CP830	DySV	0.704619 ± 23	0.02420	0.704055 ± 23	0.513063 ± 07	0.1959	0.512935 ± 07	8.25	8.30
CPN831	Epidosite	0.703489 ± 14	0.00061	0.703488 ± 14	0.513111 ± 07	0.1978	0.512982 ± 07	9.19	9.22
CWS11	DySV	0.705984 ± 28	0.00825	0.705972 ± 28	0.513082 ± 07	0.2023	0.512950 ± 07	8.66	8.60
CWS17	Epidosite	0.704587 ± 35	0.00023	0.704587 ± 35	0.513056 ± 08	0.2092	0.512918 ± 08	8.15	7.99

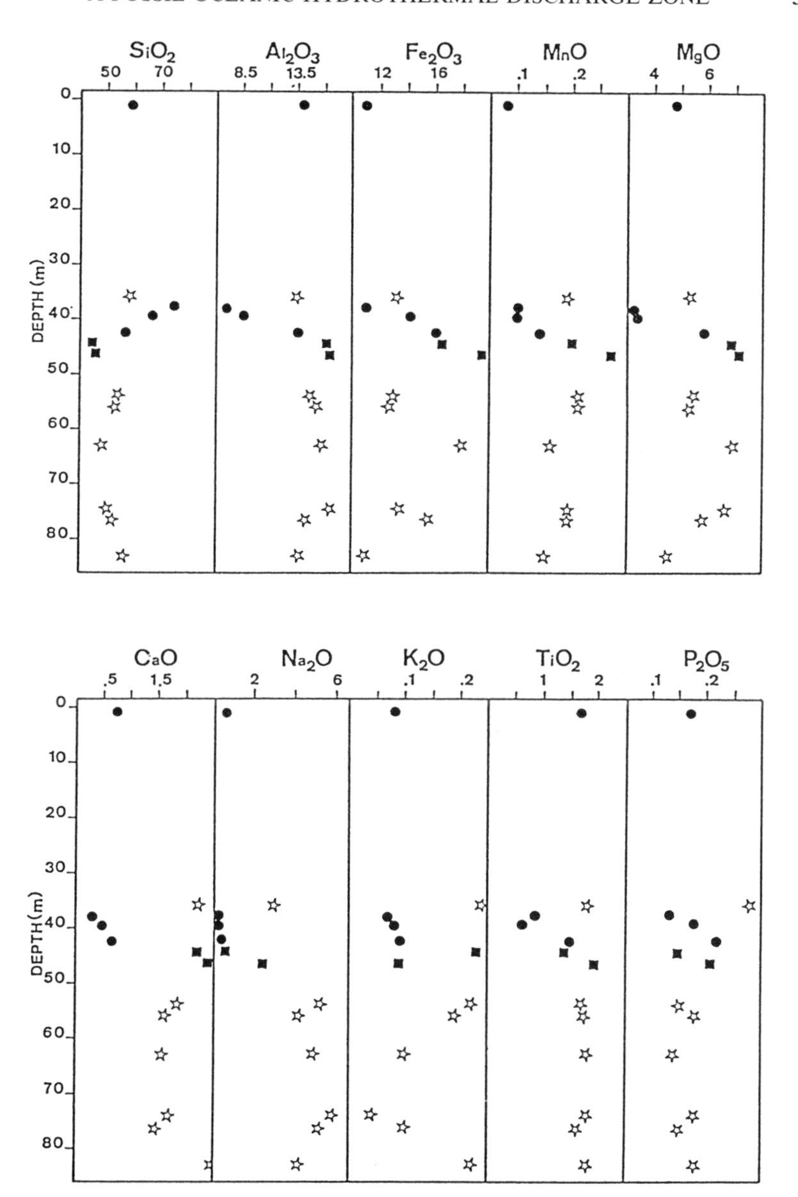

Figure 3a.

Figure 3. Major element composition of the core samples plotted in function of their stratigraphical position. Fig. 3a – Core 59-1. Fig. 3b – Core 59-2. Black dots = Type II rocks (mineralized). Open stars = Type I rocks (unmineralized). Black squares = breccia rocks. See text for type I and type II rocks definition.

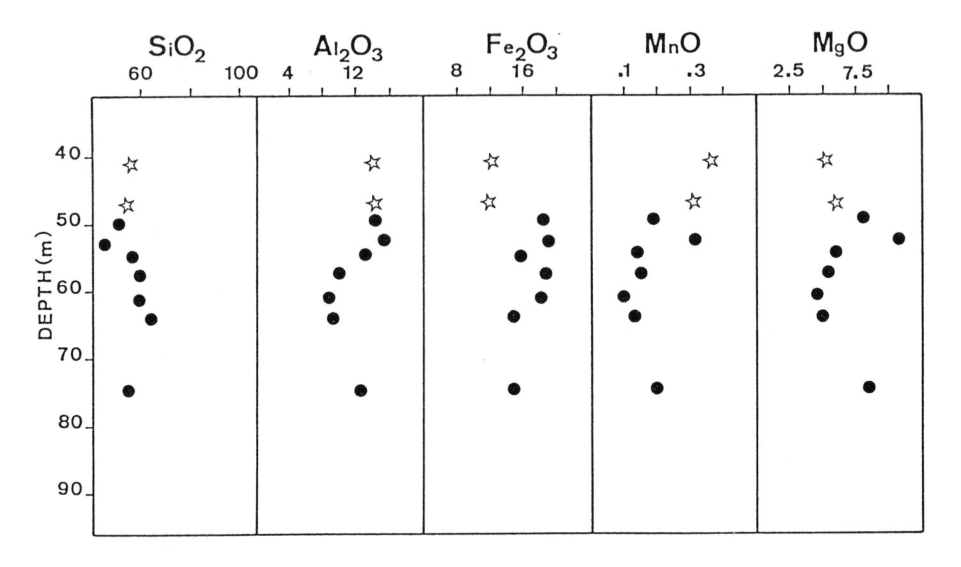
SiO2
60
100
Al2O3
4
12
Fe2O3
8
16
MnO
.1
.3
MgO
2.5
7.5
DEPTH (m)
40
50
60
70
80
90

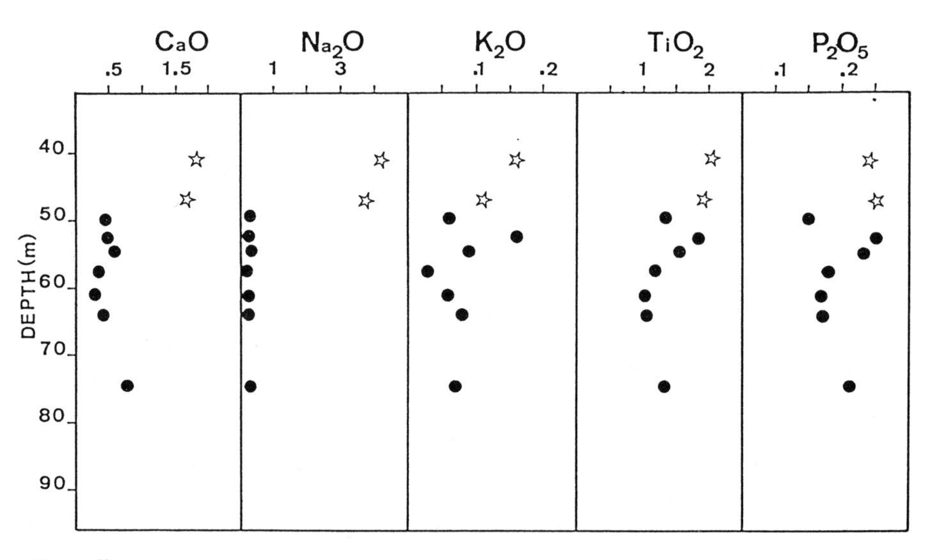
CaO
.5
1.5
Na2O
1
3
K2O
.1
.2
TiO2
1
2
P2O5
.1
.2
DEPTH (m)
40
50
60
70
80
90

Figure 3b.

Most type II mineralized samples are enriched in SiO2 compared to type I rocks. The type II rocks also exhibit an important enrichment in copper together with a depletion in Na_2O and CaO. These characteristics suggest an interaction with evolved hydrothermal fluids. Silicification, depletion in Na2O and CaO, and metal enrichments are commonly observed in stockwerks beneath ophiolitic massive sulphide deposits (Franklin et al., 1981; Collinson, 1986; Zierenberg et al., 1988).

The breccia core sample (59–1–61) is highly enriched in alkaline elements (K_2O = 0.23%, Rb = 4.26 ppm, Li = 21 ppm). A low temperature alteration effect can be suspected as alkali-element enrichments are commonly observed in low temperature altered oceanic rocks (Honnorez, 1981; Staudigel et al., 1979a, b; Hart and Staudigel, 1982).

In summary these first geochemical data: (a) support the consanguinity of the magmas at the origin of these different types of rocks (i.e., constancy of some incompatible element ratios, and Nd isotopic compositions). (b) demonstrate that some differentiation (crystallization) process operated on these magmas before their emplacement (i.e., variation in content of some compatible elements such as Cr, Ni, Ti), (c) relate some compositional characteristics, in particular the typeI/type II compositional, variations to the effects of hydrothermal alteration processes (Na, Ca, Si variations and Cu, Zn metals contents).

Type I/Type II Transformation: Mass Balance Calculations

As type I and type II rocks belong to the same V1 magmatic series, we can postulate that distinct conditions of alteration are responsible for their different mineralogical and geochemical characteristics. Several observations suggest that they formed in different parts of a single hydrothermal system with the type II rocks corresponding to the discharge zone (Pflumio, this issue). With the absence of V1 fresh rocks, a comparison via mass balance calculation of the type II rocks with the least altered type I rock has been made. The mass flux deduced from this calculation provides informations on the interaction processes which operated during the type II genesis, and in particular on the characteristics of the mineralized fluids which operated during this process.

Dilution or enrichment of elements due to the mass transfer do not allow a direct interpretation of the type I and type II data in terms of mass flux. The composition of type II rocks were thus recalculated considering that Zr remained immobile during interaction between V1 lavas and the upwelling hydrothermal fluid. This procedure is justified by the fact that in all the analyzed rocks, the Y/Zr and Nb/Zr ratios were almost constant (Tables 1 and 2). The comparatively least altered type I sample 3–166 lava (3.13% L.O.I.), was chosen as the reference sample for this calculation The mass fluxes for 100g of rock deduced from the difference between the corrected content and the reference sample content are reported in Table 4.

Table 4. Computed variations of compositions between type II rocks and a low altered type I reference rock (sample 3-166). These variations of compositions are interpreted in the text as resulting from an alteration of V1 lavas with upwelling hydrothermal fluids. The compositions of the type II rocks were recalculated considering the Zr remained immobile during the interaction process. The dilution coefficient is defined as the ratio between the Zr (reference sample)/Zr (sample).

Sample	59.2.139	59.2.140	59.2.141A	59.2.142	59.2.143	59.2.147
Dilution coefficient	0.70	0.80	1.08	1.33	1.20	0.74
SiO_2	−19.33	−6.15	11.8	26.00	25.20	−10.54
Al_2O_3	−3.20	−3.50	−3.10	−2.40	−2.90	−4.60
Fe_2O_3	0.20	−0.40	6.90	11.12	4.92	−1.97
MnO	0.02	−0.09	−0.04	−0.07	−0.04	−0.05
MgO	−4.71	−1.02	0.14	0.24	0.09	0.41
CaO	−5.07	−4.96	−5.05	−5.04	−4.91	−4.82
Na_2O	−5.71	−5.60	−5.68	−5.58	−5.59	−5.68
K_2O	0.02	−0.02	−0.06	−0.01	0.01	−0.04
Cu	14.20	–	111.50	966.00	1510.00	512.50
Zn	17.70	−12.20	−41.00	−31.90	87.00	−13.54
Li	2.76	3.44	5.60	8.90	4.60	6.80

The calculated fluxes underline a general enrichment in the metals Cu, Fe (in most samples), and Li; and depletion in Ca and Na. On the other hand, silica fluxes are more variable, although quartz is ubiquitous in these type II rocks. Three processes can be inferred to explain these chemical transfers.

a) The composition of the mineralizing fluid: the positive influx of Fe, Cu, and Li, and the ubiquitous presence of quartz suggests that the hydrothermal solutions reacting with the V1 lavas became saturated with respect to silica and metals. These characteristics are in agreement with the compositions of the fluids presently exiting from oceanic hydrothermal vents (EPR 13°N, 21°N, Galapagos, MAR, Edmond et al., 1982; Michard et al., 1984; Edmond et al., 1979, Campbell et al., 1988, Table 3). They also agree with the composition of the fluids experimentally derived from seawater-rock interactions at high temperatures (Seyfried and Janecky, 1985; Seyfried, 1987 and references therein).
b) The nature of the secondary phases in equilibrium with the upwelling mineralizing fluid: the destabilization of albite and Ca secondary phases (i.e epidote, prehnite) in favor of high-T clays (Fe-chlorite, and rectorite) resulting from interaction with low pH hydrothermal fluids explains the CaO, Na2O, Al2O3 decrease in the mineralized samples. The enrichment in K and Li also detected in some samples likely reflects the weight proportions of rectorite in these rocks.

   On the other hand, the variability in silica fluxes can be explained by the variable modal content of Fe-chlorite in the type II samples. This

Table 5. δD and $\delta^{18}O$ whole-rock compositions and water content of the samples.

Sample number	Nature	δ^{18} (%)	δD (%)	$(H^2O^+$	Minearology*
59-1-106	type I	10.2	−53	2.8	qz10ab70cl20
59-2-133	type I	9.1	−48	3.5	qz10ab60cl30
59-1-56	type II	3.5	−52	6.2	cl50qz30sp20
59-2-138	type II	2.9	−43	6.8	cl55qz25sp20
59-2-147	type II	3.5	−46	5.8	cl50qz30sp20
59-1-59	type II	3.9	−52	5.1	cl40qz30sp30
59-1-53	type II	6.4	−55	3.2	cl27qz43sp30
59-1-61	breccia	6.3	−49	6.0	ab50cl45ep5
59-1-62	breccia	5.4	−38	3.8	ab70cl20ep10
CP830	dySV	6.4	−55	3.4	a40cl30qz10
Cws11	dySV	8.2	−56	1.9	ab60an25cl15
CP831a	epidosite	4.9	−28	1.9	ep80qz20
Cws17	epidosite	3.9	−31	2.0	ep90qz10

Abbreviations: (H^2O^+) in Weight percent; n.a.; qz: quartz; ab: albite; an: anorthite; cl: chlorite; ep: epidote; sp: sphene; dy SV: dyke altered in greenschist facies.
* Approximative mineralogy according to petrographic study (Pflumio, 1987) and to the chemical composition. The number following the mineral name abbreviation indicates the weight of that mineral.

feature is supported by the negative MgO flux in most of the SiO2 depleted rocks.

c) Mixing processes: the slight positive influx in Mg is at first surprising since oceanic hot hydrothermal fluids display a deficit in this element. Such influx may be explained by a slight mixture of upwelling hydrothermal solution with cold seawater in the immediate vicinity of the Zuha mineralization zone. A seawater/hydrothermal fluid mixture ratio of 4% would be sufficient to explain this slight positive influx.

Hydrogen, Oxygen and Strontium Isotopic Compositions

Hydrogen Isotopic Compositions

The core samples display much higher water contents $[H_2O^+]$ (above 1.9 wgt.%, Table 5) than the fresh, mantle derived materials from the oceanic crust (i.e. MORBs). This confirms the importance of the hydration and alteration processes that have affected these rocks. The δD values (between −30 and −50) are in the range of those of similarly altered samples dredged or cored in the upper part of the oceanic crust (Javoy and Fouillac, 1979; Stakes and O'Neil, 1982, Kawahata et al., 1987). They agree well with the δD values expected from D/H fractionation factors between the secondary phases of the altered oceanic crust and seawater below 350°C (Bowers and Taylor, 1985; Graham et al., 1987). Therefore, on the D/H basis, we

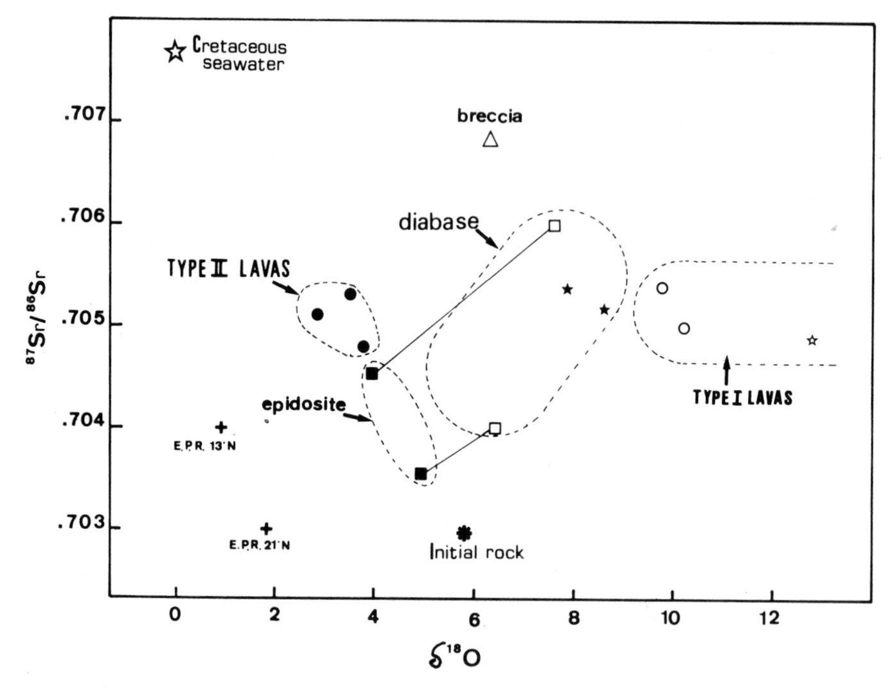

Figure 4. $^{87}Sr/^{86}Sr$ versus $\delta^{18}O$ composition of Zuha core samples and two epidosite zones and juxtaposed greenschist facies dikes of the Salahi dike complex at depth. Lines connect analyses of epidosites and host rocks. Open dots = Type I rocks (unmineralized); Black dots = Type II rocks (mineralized). Open squares = diabases; Black squares = epidosites. Open stars = basalts and black stars = diabases (from Ibra section (Mc Culloch et al., 1981)). Sr isotopic compositions of hydrothermal solutions are from Michard et al. (1981) (EPR13'N) and Albarède et al. (1980) (EPR21'N). Oxygen isotopic compositions of hydrothermal solutions are from Craig et al. (1980) (EPR21'N) and Merlivat et al. (1987) (EPR 13'N).

infer that the main fluids involved during the alteration of these samples from the upper part of the Sumail ophiolite were seawater derived fluids. Contribution of fluids which bear deep-seated hydrogen has been minor, interaction with such fluids should produce more D-depleted rocks ($-40 < \delta D < -90$) (Sheppard and Epstein, 1970; Kyser and O'Neil, 1982).

Strontium and Oxygen Isotopic Compositions

Strontium and oxygen isotopic systematics are powerful tools to get insight into the physical conditions of alteration (temperature, water/rock ratios, fluid pathways) (Taylor, 1977; Gregory and Taylor, 1981; McCulloch et al., 1981, Schiffman and Smith, 1988; Harper et al., 1988). Such measurements were undertaken on the core samples. The results are reported in Tables 3 and 5 , and plotted on Figure 4. Preliminary measurements of the Sr and O isotopic compositions of greenschist facies diabases and epidosites from the dike complex have also been presented in these tables and in Figure 4.

Table 6. Approximate isotopic temperatures (in degrees C) deduced from the oxygen isotope composition of the sample and the thermometric equations.

Sample number	Nature	−1*	0*	2*
59-1-106	type I	150	160	200
59-2-133	type I	150	170	210
59-1-56	type II		250	340
59-2-138	type II		260	360
59-2-147	type II		250	340
59-1-59	type II		240	330
59-1-53	type II		190	250
59-1-61	breccia	170	190	250
59-1-62	breccia	220	250	320
CP830	dySV		190	250
Cws11	dySV		190	240
CP831a	epidosite		180	270
Cws17	epidosite		190	300

* Calculated temperatures assuming $\delta^{18}O$ isotopic compositions of fluids. Mass balance considerations indicate that fluids with a low $\delta^{18}O$ (0 to −1)[1] are more likely to have been in equilibrium with ^{18}O enriched rocks relative to mantle composition ($\delta^{18}O \approx 5.5$)[1], while fluids with a high $\delta^{18}O$ (0 to 2)[1] are more likely to have been in equilibrium with ^{18}O depleted rocks.
Thermometric equations:
$10^3 \ln(\alpha\,^{18}O(qz\text{-}water)) = 3.10 * (10^6/T^2) - 3.3$ (Knauth and Epstein, 1976)
$10^3 \ln(\alpha\,^{18}O(ab\text{-}water)) = 2.91 * (10^6/T^2) - 3.4$ (O'Neil and Taylor, 1967)
$10^3 \ln(\alpha\,^{18}O(cl\text{-}water)) = 1.60 * (10^6/T^2) - 4.7$ (Wenner and Taylor, 1971)
$10^3 \ln(\alpha\,^{18}O(ep\text{-}water)) = 0.95 * (10^6/T^2) - 1.6$ (Matthews et al., 1983)
$10^3 \ln(\alpha\,^{18}O(sp\text{-}water)) = 0.90 * (10^6/T^2) - 1.6$ (Richter and Hoernes, 1988)
$10^3 \ln(\alpha\,^{18}O(an\text{-}water)) = 2.30 * (10^6/T^2) - 3.0$ (O'Neil and Taylor, 1967)

McCulloch et al. (1981) used a δ18O versus $^{87}Sr/^{86}Sr$ diagram to model the evolution of the Sr and O isotopic compositions that could be expected during seawater/oceanic crust interaction processes. The position of samples, in this diagram, reflects the effects of the temperature of equilibrium (which mainly determines the $d1^8O$) and the water/rock ratio (which determines the $^{87}Sr/^{86}Sr$ composition as well as the $d1^8O$).

Figure 4 underlines the strong contrast which exists between type I and type II sample $d^{18}O$ values. With respect to primitive mantle material, type I rocks display high $d^{18}O$ values ranging between 9 and 10 ‰. These values are comparable to those found in the upper members of ophiolitic complexes which are altered under zeolitic to lower greenschist facies conditions (Javoy, 1970; Spooner et al 1974; Gregory and Taylor, 1981; Cocker et al., 1982; Agrinier et al., 1988). These high values are easily explained by a seawater/-basalt interaction at moderate temperatures. Temperature ranges of 140 to 150°C, 150 to 170°C and 190 to 210°C are inferred with respectively an initial $\delta^{18}O$ of the fluids equal to −1,0 and 2‰ (Table 6). The higher temperature range is in good agreement with the crystallization temperatures of the chlorites calculated by Pflumio (this issue) which rely on the Al^{IV} content of this mineral.

The high $^{87}Sr/^{86}Sr$ ratios measured from these rocks (around 0.705) are compatible with an interaction of these rocks with Cretaceous sea water derived fluids (Cretaceous sea water $^{87}Sr/^{86}Sr$ close to 0.708 (Burke et al., 1982; Koepnick et al., 1985) at low water/rock ratios (around 5) (Tab. 3). Such isotopic ratios are in the range of ratios regularly encountered in rocks which are altered under similar conditions (Spooner, 1977; McCulloch et al., 1981).

Group II rocks display depleted $\delta^{18}O$ isotopic compositions, lower than 4.0‰, with the exception of the hyaloclastite 59–1–53 which displays a $\delta^{18}O$ of 6.4. However, this sample is located at the top of the mineralized zone of core 59–1 (Fig 3a) and its higher $\delta^{18}O$ may be related to the less pronounced influence of the hydrothermal alteration at this level; it can also be related to the peculiar lithology and mineralogy of this sample which is a breccia with a high quartz/chlorite ratio. Rocks with low $\delta^{18}O$ isotopic compositions similar to type II rocks are usually found in deep crustal levels (base of the sheeted dyke complex, gabbros) which have interacted with seawater derived fluids at high temperature (≥250°C) (Javoy, 1970; Gregory and Taylor, 1981; Cocker et al., 1982). Temperatures between 250°C and 360°C have been computed for a fluid with an initial $\delta^{18}O$ of +2 equivalent to the $\delta^{18}O$ values measured in the oceanic, hydrothermal fluids (Bowers and Taylor, 1985; Merlivat et al., 1987; Table 6). These temperatures are in good agreement with the alteration temperatures which were derived from the chlorite compositions (Pflumio, this issue). Similar low $\delta^{18}O$ values are also commonly found in the stockwerks of ophiolitic sulphide deposits (Spooner et al., 1977; Schiffman et al., 1987) and associated with Kuroko type mineralizations (Pisutha-Arnond and Ohmoto, 1983). Therefore, these data support the hypothesis that the group II rocks represent a part of a stockwork resulting from an interaction with hot, mineralizing fluids with characteristics equivalent to those sampled at the level of oceanic springs and to the solutions which give rise to massive sulfide deposits.

On the other hand, the $^{87}Sr/^{86}Sr$ isotopic ratios of type II rocks are rather high and equivalent to those of type I rocks (0.705). High temperature oceanic fluids usually display low 87Sr/86Sr isotopic ratios (0.703 for EPR 21°N, 0.704 for EPR 13°N) (Albarède et al., 1981; Michard et al., 1984) which reflect an equilibrium with deep-seated, unaltered, crustal rocks. The relatively high Sr isotopic ratios of the stockwork are quite remarkable and suggest that the fluids at the origin of the Zuha stockwerks have recorded a different hydrothermal history than the fluids sampled at the level of oceanic springs. This point will be more extensively discussed later on.

The breccia sample (59–1–61) differs from type II rocks. It displays a Sr enriched isotopic composition ($^{87}Sr/^{86}Sr = 0.7068$) close to the Cretaceous seawater's (Burke et al., 1985; Koepnick, 1985). This high value suggests that the alteration process took place under a high water/rock ratio ($W/R \approx 29$) (Tab. 3). Furthermore, the $\delta^{18}O$ value measured in this sample, although being lower than type I rocks ($\delta^{18}O \approx 6.0 \pm 0.5$‰), exceeds the

$\delta^{18}O$ values of type II samples. Alteration temperatures ranging between 190 to 250°C can be estimated for this sample (assuming $\delta^{18}O$ fluid = 0‰). These isotopic characteristics reflect the complex alteration history undergone by this breccia sample. Indeed, as shown by the mineralogy and previous geochemical data, this rock interacted with hot mineralizing fluids (development of a quartz + chlorite + sulfide mineralogical assemblage) and in a later stage was subjected to a lower temperature alteration (indicated by the occurence of laumontite veins, and by the high alkali content, and relatively high $\delta^{18}O$ composition of this sample).

The oxygen and Sr isotopic compositions of two epidosites and host greenschist facies dykes located in the Salahi block dyke complex have also been reported in Fig. 4.

The data indicate first the existence of large variations of $\delta^{18}O$ and $^{87}Sr/^{86}Sr$ compositions between the epidosites and the juxtaposed diabase dikes. These variations strongly support the existence of high lateral temperature and composition gradients in the fluids which flowed through the dikes. The dike complex structure constraining the pathway of the hydrothermal fluids should explain this feature (Nehlig and Juteau, 1988).

The epidosites from the dike complex exhibit lower $\delta^{18}O$ (≈4.4‰) than the $\delta^{18}O$ of the surrounding greenschist facies diabases (≈7.3‰) (Table 5). This agrees with the recrystallization of diabases into epidosites at temperatures above 240°C (Table 6).

In addition, greenschist facies dykes have lower δD (≈ −55‰) than epidotized dykes (≈ −30‰).This can be explained by the fact that Deuterium is preferentially incorporated into epidote relative to chlorite (Stakes and O'Neil, 1982; Graham et al; 1980 and 1987).

The data also show large variations of compositions between the two epidosite dike pairs. One of the epidosites displays $\delta^{18}O$ and $^{87}Sr/^{86}Sr$ compositions which are close to type II rocks ones although the other shows rather different compositions. These variations of compositions will be discussed later on.

Rare Earth Element Distributions

REE isotope dilution analyses are reported in Table 7 and illustrated with the classical chondrite normalized patterns in Figure 5.

Type I rocks display classical N-Morb type REE patterns which are flat from Lu to Nd and decrease from Nd to La. These patterns show rather high heavy REE contents (20 to 30 times chondrites) and exhibit negative Eu anomalies. They are similar to the profiles already obtained on V1 lavas in other parts of the Sumail ophiolite: Wadi Tayin (Pallister and Knight, 1981), Wadi Salahi (Ernewein et al., 1988 Pflumio, 1988) and Wadi Haymiliyah (Regba et al., submitted).

Type II rock REE patterns differ from those of type I. These patterns

Table 7. REE and Ba contents of the type I and type II rocks from Zuha bore samples.

Samples Nature	59.3.166 Type I	59.2.133 Type I	59.2.138 Type II	59.2.147 Type II	59.1.3 Type II	59.1.61 Breccia	59.2.154 Jasper
La	3.10	2.98	2.98	5.27	7.32	3.03	1.62
Ce	22.88	21.30	10.41	19.00	20.42	15.49	3.16
Nd	12.15	19.10	8.97	15.27	19.78	11.98	1.77
Sm	4.10	6.00	2.49	4.42	4.77	2.90	0.47
Eu	1.43	1.51	0.68	1.69	1.36	1.23	0.25
Gd	5.59	8.24	3.31	6.23	5.78	3.78	0.70
Dy	6.81	–	3.46	6.63	6.70	3.65	0.86
Er	–	–	–	–	–	2.45	–
Yb	4.33	6.20	2.13	3.83	4.38	2.45	0.39
Lu	0.70	1.16	0.39	0.54	0.62	0.35	0.12
Rb	0.49	1.00	0.50	0.50	0.56	4.26	0.22
Sr	46.59	34.70	41.00	18.69	179.40	56.83	11.00
Ba	–	6.10	11.21	4.78	8.95	8.44	5.89
Nd/Sm	2.96	3.18	3.60	3.46	4.15	4.13	3.77

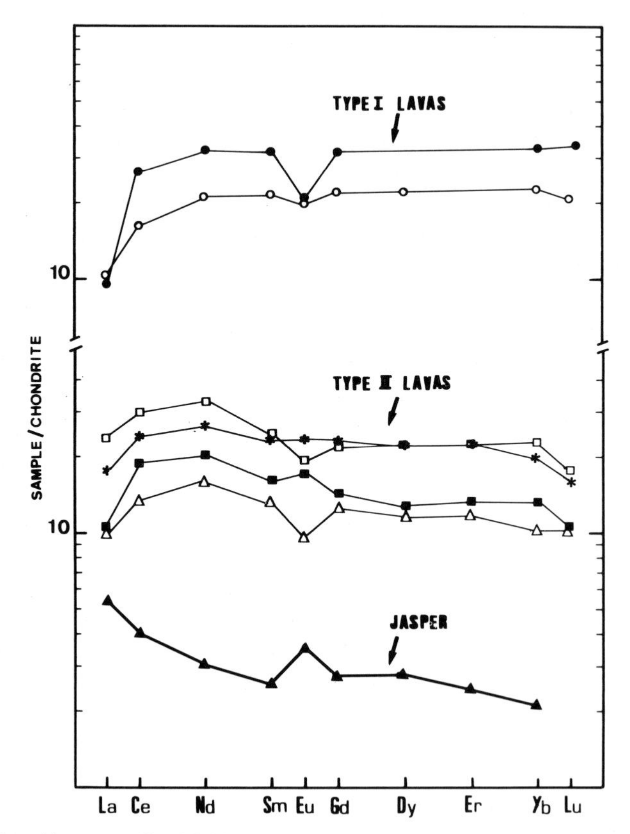

Figure 5. Chondrite normalized REE content of type I and type II lavas, breccia and jasper. Type I lavas: ● 59-3-166; ○ 59-2-133; Type II lavas: □ 59-1-3; ✻ 59-2-147; △ 59-2-138; Breccia: ■ 59-1-61; Jasper: ▲ 59-2- 154

display a light REE enrichment with respect to the heavy REE. Significantly La remains depleted with respect to Nd in these rocks and the Eu anomaly tends to disappear. The high Nd/Sm ratio stresses the relative light REE enrichment with respect to heavy REE (Tab. 7).

The REE pattern of the breccia sample (59–1–61) is similar to type II rock REE patterns (Fig. 5), with an enrichment in LREE with respect to the HREE. Such REE distribution further agree with the fact that this sample has interacted with the same hot, mineralizing fluids that gave rise to the type II samples.

As type I and type II rocks belong to the same V1 volcanic series, it is inferred that the relative light REE enrichment observed in type II rocks has a hydrothermal origin. The question now is whether this enrichment results from a rare earth fractionation between the hydrothermal fluids and the lavas during the interaction process or is a fluid characteristics. Mineralogically, type II rocks essentially differ from the type I samples by the abundance of quartz and sulfide and by the disappearance of sodic plagioclase in favour of high temperature phyllosilicates. According to Morgan and Wanddless (1980) and Rossman at al. (1987), quartz and sulfides only accomodate negligible quantities of rare earths and rather act as diluents for those elements. On the other hand, clay minerals may strongly trap rare earths. It has been shown that the exchange capacity of clay minerals is very important for alkaline elements (Berger et al., 1988). For REE, the available data concerning exchange reactions between clays and hydrothermal solutions (Berger, 1990), show the ability of these secondary phases at 40°C to become saturated in REE without fractionation of the LREE versus Sm. If we assume a similar behavior of this group of elements at high T (which remains to be proved), no internal fractionation should be expected.

These observations therefore suggest that the relative light REE and Eu enrichments in type II rocks reflect characteristics of the hydrothermal fluids. Such fluids would be analogous to oceanic hydrothermal fluids sampled from active chimneys in present day oceans (Michard et al., 1983; Campbell et al., 1988).

The core jasper (59–2–154) exhibits a REE pattern with characteristics which interestingly resemble those previously assigned to the hydrothermal fluids. It shows a light REE enrichment as well as a positive Eu anomaly. This pattern is also characterized by rather low REE contents (3 times chondrites) and the absence of a Ce anomaly. A number of arguments suggests that the REE distribution of this jasper directly reflects the REE distribution of the hydrothermal fluids with little or no fractionation:

a) As sea-water is characterized by a Ce negative anomaly (Hogdahl et al., 1968; De Baar et al, 1985), the absence of such an anomaly in this sample suggests an equilibrium with hydrothermal type fluids.
b) The similarity in Nd isotopic compositions of this sample ($\epsilon Nd = 8.31$)

(Table 3) and the V1 lavas further indicates that this hydrothermal fluid interacted with the basalts.

c) The REE distribution of this jasper (the pattern and the low REE content level) is similar to those of hydrothermal sediments S.S studied by Courtois and Treuil (1977) and Courtois (1981) in the Red Sea. These authors explained these patterns as reflecting a direct precipitation from hydrothermal solutions.

Discussion and Conclusions

Polyphased Hydrothermal History

The geochemical study carried out on the Zuha prospect allows for an assessment of the conditions for the development of the hydrothermal process in the upper part of this oceanic crust. The two main types of hydrothermal alterations affecting the V1 lavas, distinguished on mineralogical basis, could also be geochemically characterized. Type I rocks which display a prehnite-pumpellyite facies secondary assemblage are characterized by high $\delta^{18}O$ (>9%) and $^{87}Sr/^{86}Sr$ ratios (around 0.705) indicative of low water/rock ratios (around 5). Type II rocks which display a stockwerk-type assemblage are characterized by low $\delta^{18}O$ (<4) and $^{87}Sr/^{86}Sr$ ratios (around 0.705) similar to the Sr isotopic compositions of type I rocks.

The diversity of mineralogical, chemical and isotopic compositions displayed by the type I and type II rocks reflects distinct conditions of alteration (temperature, W/R ratios and fluid compositions) and suggests that these two types of rocks were formed in different parts of the same hydrothermal system. This contention is supported by numerous facts:

a) The δD values of these two types of rocks suggest that the fluids responsible for their recrystallization were seawater-derived.

b) Mineralogical assemblages and oxygen isotopic compositions indicate fluid-rock interaction temperatures below 200°C for type I rocks and above 250°C for type II rocks.

c) Alteration experiments show that under hydrothermal conditions, rock-seawater interactions are mainly characterized by Mg removal from seawater and formation of chlorite and Mg-rich clay minerals (Mottl, 1983 and references therein). Therefore, the intense albitization observed in type I rocks occurs probably in a recharge zone through the interaction between V1 lavas and seawater derived fluids with high Na/Mg ratios (Seewald, 1987).

d) Comparison of type I and type II rocks compositions indicates that the fluids at the origin of the Zuha stockwerks (type II) were uniformly silica, iron and copper enriched with REE distributions characterized by an enrichment in LREE compared to the HREE, a positive anomaly in Eu and a lack of anomaly in Ce. This suggests that type II rocks were formed by interaction between V1 lavas with discharged evolved fluids at high temperatures. These solutions thus display characteristics similar to those typically found in oceanic low pH hydrothermal fluids. Thermodynamic computations show that the chemistry of hot oceanic hydrothermal fluids is controlled by equilibrium with greenschist mineralogical assemblages (Bowers et al, 1988, Von Damm, 1988). The precipitation of sulphides results apparently from a cooling of hot hydrothermal solutions and/or a mixing with cold sea water (Janeckey and Seyfried, 1984, Janeckey and Shanks, 1988).

Some characteristics of both type I and type II rocks differ however from those described in modern oceanic crust (such as 504 B hole; (Alt et al., 1986; Kawahata et al., 1987):

a) the V1 lavas have been homogeneously altered at moderate temperatures in prehnite – pumpellyite facies (Pflumio, this issue);
b) the available data concerning the $^{87}Sr/^{86}Sr$ isotopic ratios of the V1 lavas in the Oman show that these lavas display a uniform isotopic composition (around 0,705) (this study; Mc Culloch et al., 1981; Lanphere, 1981);
c) the $^{87}Sr/^{86}Sr$ isotopic ratios displayed by the type II rocks (0.705) are higher than in modern oceanic hydrothermal fluids (0,703 – 0,704 (Albarede et al., 1981; Michard et al., 1984; Campbell et al., 1988);
d) the $^{87}Sr/^{86}Sr$ isotopic ratios displayed by the type II rocks (0.705) are relatively similar to those of type I rocks.

A polyphased hydrothermal history can explain all of these characteristics. The two first observations suggest that V1 unit has been homogeneously altered in a relatively closed system at low water / rock ratios. It is proposed that the hydrothermal system at the origin of the Zuha prospect developed in an oceanic crust already altered by a first event and whose Sr isotopic composition was yet shifted consecutively to this first event. Such a scenario is compatible with the polyphased magmatic history identified in the Oman ophiolite (Alabaster et al., 1980; Alabaster and Pearce, 1985; Pflumio, this issue). According to Pflumio, the first, axial, magmatic event of the Sumail ophiolite (V1 magmatism), during which the sheeted dyke complex and the main plutonic sequence formed, induces the first alteration stage. During this first magmatic event, the crust which is being constructed is progressively affected by the different types of oceanic alterations which are commonly described in present-day oceanic crust , i.e. low temperature alteration in the volcanic sequence and hydrothermal alteration in the dyke complex and lower volcanic sequence (Alt et al., 1986a). During this stage, the Sr isotopic composition of the extrusive sequence is progressively shifted from its original value of 0.703 to higher isotopic compositions by interaction with the seawater derived-fluids (Kawahata et al., 1971). The second alteration stage is activated by the second magmatic event of the ophiolite complex (illustrated by late intrusions and the V2 volcanism). The present mineralogical assemblages of the V1 lavas (prehnite-pumpellyite facies) and of the diabases of the dyke complex (pervasive greenschist facies) developed during this second alteration stage. The fluids which circulated during this second hydrothermal stage thus interacted at low water / rock ratios with the previously $^{87}Sr/^{86}Sr$ enriched rocks and therefore acquired higher $^{87}Sr/^{86}Sr$ isotopic compositions than hydrothermal fluids interacting with unaltered rocks such as those sampled at the level of modern ridge axis. The similarity of $^{87}Sr/^{86}Sr$ isotopic compositions between type I and type II rocks results from the subsequent exchange in the discharging zone at high water / rock ratios and during which the wall rocks acquired the $^{87}Sr/^{86}Sr$ isotopic composition of fluids.

Other consequence of this polyphased hydrothermal history concern the $^{18}O/^{16}O$ compositions. The distinct $\delta^{18}O$ oxygen isotopic compositions of type I and type II rocks result mainly from different temperatures of alteration (Table 6). However, a multistage hydrothermal alteration can generate ^{18}O enriched hydrothermal fluids. This enrichment occurs during the second alteration stage when the pristine sea water interacts, at relatively high temperature and low water/rock ratios, with rocks previously ^{18}O enriched during the first hydrothermal stage at low temperature. The fluids at the origin of type II rocks were apparently ^{18}O enriched (as modern oceanic hydrothermal solutions (Craig et al., 1980; Merlivat et al., 1987; Campbell et al., 1988)). In the Zuha prospect, this enrichment may partly result from such a polyphased hydrothermal history.

Diabase dike display $\delta^{18}O$ compositions ranging between 6.4 and 8.2, which are similar to the values obtained by Mc Culloch et al. (1981) and Gregory and Taylor (1981) in the sheeted dike complex. Gregory and Taylor (1981) explained the high $\delta^{18}O$ values in the dike complex as a result of a low temperature recharge and subsequent exchange cooling discharging fluids enriched in ^{18}O prior exchange with deeper section of the ophiolite at high temperature ($T > 300°C$). The ^{18}O enriched fluids generated during the second hydrothermal stage may also explain the high $\delta^{18}O$ compositions in the sheeted dike complex.

Finally, late, lower temperature circulations which are attributed to the last burst of this second hydrothermal phase (Pflumio, this issue), are also recorded in a highly permeable breccia sample (59–1–61), being characterized by relatively high $\delta^{18}O$ and $^{87}Sr/^{86}Sr$ compositions, and alkali enrichments.

Epidosite Zones and Mineralizing Fluids

It has been suggested by several authors that the hot and low pH mineralizing fluids could originate from deep seated epidote-rich levels, the epidosite zones representing fossil conduits for hydrothermal solutions (Seyfried and Janecky, 1985; Richardson et al; 1986; Schiffman et al; 1987 Schiffman and

Smith, 1988; Harper et al; 1988; Berndt et al; 1989). The evidences for such hypothesis are the abundance of epidosite zones under massive sulphide deposits (Richardson et al; 1986; Schiffman and Smith, 1988; Harper et al; 1988; Nehlig, 1989), the low metal content of these rocks (Richardson et al., 1986; Regba et al. in preparation), their low $\delta^{18}O$ values which reflect a high temperature alteration (Richardson et al., 1986; Schiffman and Smith, 1988; Harper et al., 1988; data from this study) and the results of water/rock experiments which demonstrate the ability of evolved fluids (Na, Ca, K, Cl rich) to precipitate Ca-silicates near supercritical conditions and to generate solutions similar to oceanic hydrothermal fluids (Seyfried and Janecky, 1985; Berndt et al., 1989).

The data presented in Fig. 4 show that there are large variations of isotopic compositions from one epidosite dike to another. Significantly one epidosite displays $\delta^{18}O$ and $^{87}Sr/^{86}Sr$ compositions close to type II rocks ones. This argues in favour of a genetic relationship between these two rock types. These variations of compositions may be indicative of different generations of epidosites. The epidosite with low $^{87}Sr/^{86}Sr$ ratio may have been formed during the first hydrothermal stage related to ridge magmatism. The epidosite with higher $^{87}Sr/^{86}Sr$ composition may have been formed during the second stage of hydrothermalism. A more detailed field study, for example trying to structurally differentiate the different orientations of epidosites dikes, could test this hypothesis.

Acknowledgments

We had fruitful discussions with P. Nehlig, T. Juteau, M. Rabinowitz and J. Schott. Some technical assistance of L. Briqueu and J. Lancelot from the Isotope Geochemical Laboratory of Montpellier University was highly appreciated. We thank D.S. Stakes and P. Schiffman for their useful comments. This work was supported by grants from A.T.P. Geologie et Geophysique des océans (code PIROCEAN, 81) and IFREMER (89.2.470–322).

References

Agrinier, P., Javoy, M. and Girardeau, J., 1988. Hydrothermal activity in a peculiar oceanic ridge: oxygen and hydrogen isotope evidence in the Xigaze ophiolite (Tibet, China)., Chemical Geology, 71: 313–335.

Alabaster, T., Pearce, J.A. and Malpas, J., 1982. The volcanic stratigraphy and petrogenesis of the Oman ophiolite complex., Contrib. Mineral. Petrol., 81: 168–183.

Alabaster, T. and Pearce, J.A., 1985. The interrelationship between magmatic and ore forming hydrothermal processes in the Oman ophiolite., Econ. Geology, 80: 1–16.

Allegre, C.J. and Minster, J.F., 1978. Quntitative models of trace elements behavior in magmatic processess., Earth. Planet. Sci. Lett., 38: 1–25.

Alt, J.C., Honnorez, J., Laverne, C. and Emmermann, R., 1986a. Hydrothermal alteration of a 1 km section through the upper oceanic crust, deep sea drilling project hole 504B: Mineralogy, chemistry and evolution of seawater-basalt reactions., J. Geophys. Res., 91: 10309–10335.

Alt, J.C., Muehlenbachs, K. and Honnorez J., 1986b. An oxygen isotopic profil through the upper kilometer of the oceanic crust, deep sea drilling project hole 504B., Earth. Planet. Sci. Lett., 80: 217–229.

Alt, J.C., Lonsdale, P., Haymon, R. and Muehlenbachs, K., 1987. Hydrothermal sulfide and oxide deposits on seamounts near 21°N, East Pacific Rise., Geol. Soc. Amer. Bull., 98: 157–168.

Berger, G., 1990. Distribution of trace elements brtween clays and zeolites and aqueous solutions similar to sea-water. Applied Geoch. In press.

Berndt, M.E., Seyfried, Jr W.E. and Beck, J.W., 1989. Hydrothermal alteration processes at Midocean Ridges: Experimental and theoretical constraints from Ca and Sr exchange reactions and Sr isotopic ratios., J. Geophys. Res., 93: 4573–4583.

Birck, J.L. and Allègre, C.J., 1978. Chronology and chemical history of parent body of basaltic chondrites studied by the ^{87}Rb-^{87}Sr method., Earth. Planet. Sci. Lett., 39: 37–51.

Bonatti, E., 1983. Hydrothermal metal deposits from the oceanic rifts: a classification. In "Hydrothermal processes at sea-floor spreading centers" Rona, P.A., Boström, K., Lambier, L. and Smith, Jr K.L. (Eds), Plenum Press New York, pp.491–502.

Borthwick, J. and Harmon, R.S., 1982. A note regarding ClF_3 as an alternative to BrF_5 for oxygen isotope analysis., Geochim. Cosmochim. Acta, 46: 1665–1668.

Boudier, F. and Nicolas, A., 1988. "The ophiolite of Oman.", Tectonophysics, Vol. 151.

Bougault, H. 1980. Contribution des éléments de transition à la compréhension de la genèse des basaltes océaniques. Thèse Doctorat d'Etat, Univ. Paris 7, France, 221 p.

Bougault, H., Cambon, P. and Toulhont, H., 1977. X-ray spectrometric analysis of trace elements in rocks correction for instrumental interferences., X-Ray Spectrom., 6 (2): 66–72.

Bowers, T.S., Von Damm, K.L. and Edmond, J.M., 1985. Chemical evolution of mid-ocean ridge hot springs., Geochim. Cosmochim. Acta, 49: 2239–2252.

Bowers, T.S. and Taylor, H.P. Jr., 1985. An integrated chemical and stable isotope model of the origin of midocean ridge hot spring system., J. Geophys. Res., 90: 12583–12606.

Bowers, T.S., Campbell, A.C., Measures, C.I., Spivack, A.J., Khadem, M. and Edmond, J.M., 1988. Chemical controls on the composition of vent fluids at 13°-11°N and 21°N, East Pacific Rise., J. Geophys. Res., 93: 4522–4536.

Burke, W.H., Denison, R.E., Heatherington, E.A., Koepnick, R.B., Nelson, H.F. and Otto, J.B., 1982. Variation of seawater $^{87}Sr/^{87}Sr$ throughtout Phanerozoic time., Geology, 10: 516–519.

Campbell, A.C., Palmer, M.R., Klinkhammer, G.P., Bowers, T.S., Edmond, J.M., Lawrence, J.R., Casey, J.F., Thompson, G., Humphris, S., Rona, P. and Karson, J.A., 1988. Chemistry of hot springs on the Mid-Atlantic Ridge., Nature, 335: 514–519.

Clayton, R.N. and Mayeda, T.K., 1963. The use of bromine pentafluoride in the extraction of oxygen from oxides and silicates for isotopic analysis., Geochim. Cosmochim. Acta, 27: 43–52.

Cocker, J.D., Griffin, B.J. and Muehlenbachs, K., 1982. Oxygen and carbon isotope evidence for seawater-hydrothermal alteration of the Macquarie Island Ophiolite., Earth. Planet. Sci. Lett., 61: 112–122.

Coleman, R.G. and Hopson, C.A., 1981. "Oman ophiolite.", J. Geophys. Res., Vol. 86.

Coleman, M.L., Shepherd, T.J., Durham, J.J., Rouse, J.E. and Moore, G.R., 1982. Reduction of water with zinc for hydrogen isotope analysis., Anal. Chem., 54: 993–995.

Collinson, T., 1986. Hydrothermal mineralization and basalt alteration in stockwork zones of the Bayda and Lasail massive sulphide deposits, Oman ophiolite. M.A. Thesis, University of California, Santa Barbara, 164 p.

Courtois, C. and Treuil, M., 1977. Distribution des terres rares et de quelques éléments en trace dans les sédiments récents des fosses de la Mer Rouge., Chemical Geology, 20: 57–72.

Courtois, C., 1981. Distribution des terres rares dans les dépôts hydrothermaux de la zone de Famous et de Galapagos., Comparaison avec les sédiments métallifères. Marine Geology, 39: 1–14.

De Baar, H.J.W., Bacon, M.P., Brewer, P.G. and Bruland, K.W., 1985. Rare earth elements in the Pacific and Atlantic oceans., Geochim. Cosmochim. Acta, 49: 1943–1959.

Edmond, J.M., Measures, C, Mc Duff, R.E., Chen, L.H., Collier, R., Grant, B., Gordon, L.I. and Corliss, J.B., 1979a. Ridge crest hydrothermal activity and the balances of the major and minor elements in the ocean, the Galapagos data., Earth. Planet. Sci. Lett., 46: 1–18.

Edmond, J.M., Measures, C, Mangum, B., Grant, B., Sclatter, F.R., Collier, R. and Hudson A., 1979b. On the formation of metal rich deposits at ridge crests., Earth. Planet. Sci. Lett., 46: 19–30.

Edmond, J.M., Von Damm, K.L., Mc Duff, R.E. and Measures, C., 1982. Chemistry of hot springs on the East Pacific Rise and their effluent dispersal., Nature, 297: 187–191.

Ernewein, M., 1987. Histoire magmatique d'un segment de croûte océanique téthysienne: pétrologie de la séquence plutonique du massif ophiolitique de Salahi (nappe de Semail, Oman). Thèse Univ. Louis Pasteur, Strasbourg, France, 205 p.

Ernewein, M., Pflumio, C. and Whitechurch, H., 1988. The death of accretion zone as evidenced by the magmatic history of the Semail ophiolite (Oman)., Tectonophysics, 151: 107–126.

Franklin, J.M., Lydon, J.W. and Sangster, D.F., 1981. Volcanic-associated massive deposits. Ec. Geology, 75th. Anniv., 485–627.

Graham, C.M., Sheppard, S.M.F. and Heaton, T.H.E., 1980. Experimental hydrogen isotope studies-I., Systematics of hydrogen isotope fractionation in the system epidote-H_2O, zoisite-H_2O and $Al(OH)H_2O$., Geochim. Cosmochim. Acta, 44: 353–364.

Graham, C.M., Viglino, J.A. and Harmon, R.S., 1987. Experimental study of hydrogen isotope exchange between aluminous chlorite and water and of hydrogen diffusion in chlorite., Amer. Mineral., 72: 566–579.

Gregory, R.T. and Taylor, H.P., 1982. An oxygen isotope profile in a section of Cretaceous oceanic crust, Semail ophiolite, Oman: Evidence of $\delta^{18}O$ buffering of the oceans by deep (>5 km) seawater – hydrothermal circulation at mid ocean ridges., J. Geophys. Res., 86: 2737–2755.

Harper, G.D., Bowman, J.R. and Kuhns, 1988. A field, chemical and stable isotope study of subseafloor metamorphism of the Josephine ophiolite, California – Oregon., J. Geophys. Res., 93: 4625–4656. Hart, S.R. and Staudigel, H., 1982. The control of alkalies and Uranium in seawater crust alteration., Earth. Planet. Sci. Lett., 58: 202–212.

Haymon, R.M., Kooki, R.A. and Abrams, M.J., 1988. Hydrothermal discharge zones beneath massive sulfide deposits napped in the Oman ophiolite., Geology, 17: 5315–3558.

Hekinian, R. and Fouquet, Y., 1985. Volcanism and metallogenesis of axial and off axial structures on the East Pacific Rise near 13°N., Econ. Geology, 80: 221–249.

Hogdahl, O.T., Melson, S. and Bowen, V., 1968. Neutron activation analysis of lanthanide elements in sea water., Adv. Chem. Ser., 73: 308–325.

Honnorez J., 1981. The aging of the oceanic crust at low temperature. In: The Sea, Ed. Emiliani C., Wiley and sons, 7, 525–587.

Horibe, Y., Kim, K.R. and Craig, H., 1986. Hydrothermal methane flumes in the Mariana back-arc spreading center., Nature, 324: 131–133.

Humphris, S.E. and Thompson, G., 1978a. Hydrothermal alteration of oceanic basalts by seawater., Geochim. Cosmochim. Acta, 42: 107–125.

Humphris, S.E. and Thompson, G., 1978b. Trace element mobility during hydrothermal alteration of oceanic basalts., Geochim. Cosmochim. Acta, 42: 127–136.

Janeckey, D.R. and Seyfried, Jr., W.E., 1984. Formation of massive sulfide deposits on oceanic crests: Incremental reaction models for mixing between hydrothermal solutions and seawater., Geochim. Cosmochim. Acta, 48: 2723–2738.

Janeckey, D.R. and Shanks, W.C. III, 1988. Computational modeling of chemical and sulfur isotopic reaction pricesses in seafloor hydrothermal systems: chimneys, massive sulfides, and subjacent alteration zones., Can. Mineral., 26: 805–825.

Javoy, M., 1970. Utilisation des isotopes de l'oxygène en magmatologie. Thèse d'éyay, Université Paris VII, 230 pp.

Javoy, M. and Fouillac, A.M., 1979. Stable isotope ratios in Deep Sea Drilling Project Leg 51 Basalts. Init. Report D.S.D.P., LI, LII, LIII, Washington (U.S. Gov. Print. Off.), pp. 1153–1157.

Kawahata, H., Kusakabe, M. and Kikuchi, Y., 1987. Strontium, oxygen and hydrogen isotope geochemistry of hydrothermally altered and weathered rocks in Deep Sea Drilling Project Hole 504B, Costa Rica Rift., Earth. Planet. Sci. Lett., 85: 343–355.

Koepnick, R.B., Burke, W.H., Denison, R.E., Heatherington, E.A., Nelson, H.F., Otto, J.B. and Waite, L.E., 1985. Construction of seawater $^{87}Sr/^{87}Sr$ curve for the Cenezoic and Cretaceous: supporting data., Chemical Geology (Isotope Geoscience Section 58: 55–81.

Knauth, L.P. and Epstein, S., 1976. Hydrogen and oxygen isotope ratios in nodular and bedded cherts., Geochim. Cosmochim. Acta, 40: 1095–1108. Kyser, T.K. and O'Neil, J.R., 1984. Hydrogen isotope systematics of submarine basalts., Geochim. Cosmochim. Acta , 48: 2123–2133.

Lalou, C., 1983. Genesis of ferromanganese deposits: Hydrothermal origin. In: "Hydrothermal process at sea floor spreading", Rona P.A., Boström L., Laubier K.L. and Smith Jr K.L. (Eds), Plenum Press, New York, pp. 503–534.

Lanphere, M.A. and Coleman, R.G., 1981. Sr isotopic tracer study of the Semail ophiolite, Oman., J. Geophys. Res., 86: 2709–2720.

Lippard, S.J., Shelton, A.W. and Gass, I.G., 1986. The ophiolite of Northern Oman. Memoir no. 11, The Geological Society, Blackwell Sci. Publi., Oxford, U.K.

Lowell, R.P. and Rona, P.A., 1985. Hydrothermal models for the generation of massive ore deposits., J. Geophys. Res., 90: 8769–8783.

Mc Culloch, M.T., Gregory, R.T., Wasserburg, G.J. and Taylor, Jr, H.P., 1981. Sm-Nd, Rb-Sr and $^{18}O/^{16}O$ isotopic systematics in oceanic crustal section: Evidence from the Samail ophiolite., J. Geophys. Res., 86: 2721–2735.

Matthews, A., Goldsmith, J.R. and Clayton, R.N., 1983. Oxygen isotope fractination between zoisite and water., Geochim. Cosmochim. Acta, 47: 645–654.

Merlivat, L., Pineau, F. and Javoy M., 1987. Hydrothermal vent waters at 13°N on the East Pacific Rise: isotopic composition and gas concentration.,, Earth. Planet. Sci. Lett., 84: 100–108.

Michard, A., Albarède, F., Michard, G., Minster, J.F. and Charlou, J.L., 1983. Rare earth elements and uranium in high temperature solutions from East Pacific Rise hydrothermal vent field (13°N)., Nature, 303: 795–797.

Michard, G., Albarède, F., Michard, A., Minster, J.F., Charlou, J.L. and Tan, N., 1984. Chemistry of solutions from the 13°N East Pacific Rise hydrothermal site., Earth. Planet. Sci. Lett., 67: 297–307.

Morgan, J.W. and Wandless, G.A., 1980. Rare earth element distribution in some hydrothermal minerals: evidence crystallographic control., Geochim. Cosmochim. Acta, 44: 973–980.

Mottl, M.J., 1983. Metabasalts, axial hot springs and the structures of hydrothermal at mid-ocean ridges., Geol. Soc. Am. Bull., 94: 161–180.

Nehlig, P., 1989. Etude d'un système hydrothermal océanique fossile: l'ophiolite de Semail (Oman). Thèse Univ. Bretagne Occidentale, Brest.

Nehlig, P. and Juteau, T., 1988. Flow porosities, permeabilities and preliminary data on fluid inclusions and fossil thermal gradients in the crustal sequence of the Semail ophiolite (Oman)., Tectonophysics, 151: 199–222.

O'Neil, J.R. and Taylor, Jr,. H.P. 1967. The oxygen isotope and cation exchange chemistry of feldspars. Am. Mineral., 52: 1414–1437. Pallister, J.S. and Knight, R.J., 1981. Rare Earth element geochemistry of the Samail ophiolite near Ibra, Oman., J. Geophys. Res., 86: 2673–2697.

Pallister, J.S., 1981. Structure of sheeted dike comlpex of the Samail ophiolite near Ibra, Oman., J. Geophys. Res., 86: 2661–2672.

Pearce, J.A. and Cann, J.R., 1973. Tectonic setting of basic volcanic rocks determined using trace element analysis., Earth Planet. Sci. Lett., 19: 290–300.

Pearce, J.A., Alabaster, T., Shelton, A.W. and Pearce M. P., 1981. The Oman ophiolite as a Cretaceous arc basin complex: evidence and implication., Phil. Trans. R. Soc. London, A-300, 299–317.

Pflumio, C., 1988. Histoire magmatique and hydrothermale du Bloc de Salahi: Implications sur l'origine et l'évolution de l'ophiolite de Semail (Oman). Nouvelle Thèse, Ecole des Mines, Paris, 243 p.

Pisutha-Arnong, V. and Ohmoto, H., 1983. Thermal histiry, and chemical and isotopic compositions of the ore- forming fluids responsible for the Kuroko massive deposits in the Hokuroku district of Japan. Econ. Geol. Monograph., 5: 523–558.

Richard, P., Schimuzu, N. and Allègre, C.J., 1976. $^{143}Nd/^{144}Nd$ a natural tracer: an application to oceanic basalts., Earth. Planet. Sci. Lett., 31: 269–278.

Richards, H.G. and Cann, J.R., 1989. Mineralogical and metasomatic zonation of the alteration pipes of Cyprus sulphide deposits. Econ. Geol., in press.

Richardson, C.J. Cann, J.R., Richards, H.G. and Cowan, J.G., 1987. Metal-depleted root zones of the Troodos ore-forming hydrothermal systems, Cyprus., Earth. Planet. Sci. Lett., 84: 243–253.

Richter, R. and Hoerness, S., 1988. The application of the increment method in comparison with experimentally derived and calculated O-isotope fractionations., Chem. Erde 48: 1–18.

Rona, P.A., 1988. Hydrothermal mineralization at oceanic ridges., Can. Mineral., 26: 431–465.

Rossman, G.R., Weiss, D. and Wasserburg, G.J., 1987. Rb, Sr, Nd and Sm concentrations in quartz., Geochim. Cosmochim. Acta, 51: 2325–2329.

Schiffman, P., Williams, A.E. and Evarts, R.C., 1984. Oxygen isotope evidence for submarine hydrothermal alteration of the Del Puerto ophiolite, California., Earth. Planet. Sci. Lett., 70: 207–220.

Schiffman, P., Smith, B.M., Varga, R.J. and Moores, E.M., 1987. Geometry, conditions and timing of off-axis hydrothermal metamorphism and ore-deposition in the Solea graben., Nature, 325: 423–425.

Schiffman, P. and Smith, B.M., 1988. Petrology and oxygen isotope geochemistry of a fossil seawater hydrothermal system within the Solea graben, Northern Troodos ophiolite, Cyprus., J. Geophys. Res., 93: 4612–4624.

Sclater, J.G., Parsons, B. and Janpart, C., 1981. Oceans and continents: similarities and differences in the mechanisms of heat losse., J. Geophys. Res., 86: 11535–11552.

Seewald, J.S., 1987. Na and Ca metasomatism during hydrothermal basalt alteration: An experimental and theoretical study. M.Sc. thesis, University of Minnesota, Minneapolis, Minnesota.

Seyfried, Jr, W.E. and Mottl, M.J., 1982. Hydrothermal alteration of basalt by seawater under seawater-dominated conditions., Geochim. Cosmochim. Acta, 46: 985–1002.

Seyfried, Jr, W.E. and Janeckey, D.R., 1985. Heavy metal and sulfur transport during subcritical and supercritical hydrothermal alteration of basalt: Influence of fluid pressure and basalt composition and crystallinity., Geochim. Cosmochim. Acta, 49: 2545–2560.

Seyfried, Jr, W.E., 1987. Experimental and theoretical constraints on hydrothermal alteration processes at Mid-Ocean ridges., Ann. Rev. Earth. Planet. Sci., 15: 317–335.

Sheppard, M.S.F. and Epstein, S., 1970. D/H and $^{18}O/^{16}O$ of mineral of possible mantle or lower crustal origin., Earth. Planet. Sci., 95: 232–239.

Spooner, E.T.C., Beckinsale, R.D., Fyfe, W.S. and Smewing, J.D., 1974. O^{18} enriched ophiolitic metabasic rocks from E., Liguria (Italy), Pindos (Greece) and Troodos (Cyprus). Contrib. Mineral. Petrol., 47: 41–62.

Spooner, E.T.C., Beckinsale, R.D., England, P.C. and Senior, A., 1977a. Hydratation, ^{18}O enrichment and oxidation during ocean floor hydrothermal metamorphism of ophiolitic metabasic rocks from E. Liguria, Italy., Geochim. Cosmochim. Acta, 41, 873–890.

Spooner, E.T.C., Chapman, H.J. and Smewing, J.D., 1977b. Strontium isotopic contamination and oxydation during ocean floor hydrothermal metamorphism of the ophiolitic rocks of the Troodos massif, Cyprus., Geochim. Cosmochim. Acta, 41, 873–890.

Solomon, M. and Walshe, J.L., 1979. The formation of massive deposits on the sea floor., Econ. Geol., 74,797–813.

Stakes, D.S. and O'Neil, J.R., 1982. Mineralogy and stable isotope geochemistry of hydrothermally altered oceanic rocks., Earth. Planet. Sci. Lett., 57, 285–304.

Staudigel, H., Muehlenbachs, K., Richardson, S.H. and Hart, S.R., 1981a. Agents of low temperature ocean crust alteration., Contrib. Mineral. Petrol., 77, 150–157.

Staudigel, H., Hart, S.R. and Richardson, S.H., 1981b. Alteration of the oceanic crust: Processes and timing., Earth. Planet. Sci. Lett., 52, 311–327.

Staudigel, H. and Hart, S.R., 1983. Alteration of basaltic glass: Mechanisms and significance for the oceanic crust seawater budget., Geochim. Cosmochim. Acta, 47, 337–350.

Stern, C. and Elthon, D., 1979. Vertical variations in the effects of hydrothermal metamorphism in Chilean ophiolites: their implications for ocean floor metamorphism., Tectonophysics, 55, 179–213.

Taylor, Jr H.P, 1977. Water/rock interactions and the origin of H_2O in granitic batholiths., J. Geo. Soc. London, 133, 509–558.

Tippit, P.R., Pessagno, Jr, E.A. and Smewing, J.D., 1981. The biostratigraphy of sediments in the volcanic unit of the Samail ophiolite., J. Geophys. Res., 86, 2756–2762.

Urabe, T., Scott, S.D. and Hattori, K., 1983. A comparison of footwall-rock alteration and geothermal systems beneath some japanese and canadian volcanogenic massive sulfide deposits., Econ. Geol. Monograph, 5: 345–364.

Von Damm, K.L., Edmond, J.M., Grant, B., Measures, C.I., Walden, B. and Weiss, R.F., 1985a. Chemistry of submarine hydrothermal solution at 21°north, East Pacific Rise., Geochim. Cosmochim. Acta, 49, 2197–2220.

Von Damm, K.L., Edmond, J.M., Measures, C.I. and Grant, B., 1985b. Chemistry of submarine hydrothermal solution at Guaymas Basin, Gulf of California., Geochim. Cosmochim. Acta, 49, 2221–2237.

Von Damm, K.L., 1988. Systematics of and Postulated controls on submarine hydrothermal solution chemistry., J. Geophys. Res., 93, 4551–4561.

Wenner, D.B. and Taylor, Jr. H.P., 1971. Temperatures of serpentinization of ultramafic rocks based on $^{18}O/^{16}O$ fractination between coexisting serpentine and magnetite., Contrib. Mineral. Petro., 32: 165–185.

Wolery, T.J. and Sleep, N.H., 1976. Hydrothermal circulation and geochemical flux at mid ocean ridges., J. Geol., 48: 249–275.

Zierenberg, R.A., Shanks III, W.C., Seyfried, Jr, W.E., Koski, R.A., and Striekler, M.D., 1988. Mieralization, alteration and hydrothermal metamorphism of the ophiolite-Hosted Turner- Albtight sulfide deposit, Southwestern Oregon., J. Geol., 93: 4657–4674.

Lead Isotope Geochemistry of Various Sulphide Deposits from the Oman Mountains

J.Y. CALVEZ and J.L. LESCUYER
Bureau de recherches géologiques et minières, B.P. 6009, 45060 Orleans Cedex 2, France

Abstract

A Pb isotope study was carried out on three main types of Cu(Au) hydrothermal mineralization in the allochthonous terrains of northern Oman: (1) the Al Ajal sulphide-hematite deposit of Late Permian age; (2) massive sulphide deposits emplaced during a major break in the Late Albian-Early Cenomanian volcanic activity; and (3) minor massive sulphides deposited during breaks in the Cenomanian-Turonian volcanic episode.

The Pb isotopic ratios of the Al Ajal sulphides are similar to those of the underlying tholeiite, indicating a basaltic source as the primary contributor of lead to the sulphide-hematite deposit.

The hydrothermal sulphide deposits of types 2 and 3, located at the top of the Samail ophiolite, have Pb isotope values which define two restricted and contrasting fields. The first, with the least radiogenic Pb, lies within the compositional fields defined by the values reported by Chen and Pallister (1981) for sulphides and rocks of the Samail ophiolitic suite, and by Gale et al. (1981) for metalliferous sediments from different localities in the Oman nappes. This field comprises samples from the massive sulphides and associated sediments deposited at the top of the V1 volcanites (Lescuyer et al., 1988) in the coastal area. Their isotopic composition confirms that Pb in the sulphides is primary and derived from the host basalts, a characteristic of many unsedimented ridges (Le Huray et al., 1988) which has been quoted as a distinguishing feature between the Oman and the Cyprus ophiolites.

The second, more radiogenic field incorporates the Cu-Au deposit of Rakah (type 2) and the gold-bearing sulphide deposits associated with the V2 volcanism (type 3). It also includes Pb data obtained by Chen and Pallister (1981) on serpentinized peridotites. Furthermore, it lies on an apparent mixing line between the least radiogenic field and that determined by lead in the sediments underlying the Oman nappes. This indicates that Pb in some massive sulphide deposits of northern Oman could be partly derived from a sedimentary source, suggesting possible subduction processes. This hypo-

Tj. Peters et al. (Eds), Ophiolite Genesis and Evolution of the Oceanic Lithosphere, 385–397.

thesis is supported by the geochemical characteristics of the volcanic rocks hosting the deposits.

Geological Setting of the Sulphide Deposits in Northern Oman

Sulphide deposits in the allochthonous terrains of northern Oman were emplaced at three different periods: Late Permian in the Hawasina nappes; and Late Albian – Early Cenomanian, and Cenomanian-Turonian in the Samail ophiolite (Lescuyer et al., 1988).

Late Permian Mineralization

The Al Ajal sulphide-hematite layer in the Hawasina nappes (Fig. 1) was deposited at the top of chloritized basaltic pillow lavas cut by stockwork ore. It is overlain by shale, siltstone and limestone of Late Permian age. The gold-bearing Cu-Zn-Ba deposit is characterized by alternating layers of massive iron sulphides (pyrite, minor pyrrhotite) and oxides (hematite) which show striking analogies with certain facies of the Red Sea mineralized muds (Oudin, 1983). The MORB composition of the footwall tholeitic basalt reflects a stage of aborted oceanization in the Hamrat Duru intracontinental basin which opened north of the unstable Arabian platform margin during the Late Permian (Bechennec et al., 1990). The turbidite succession and its volcanic basement were thrust over the platform during the Late Cretaceous, prior to emplacement of the overlying Semail ophiolitic nappe.

Late Albian – Early Cenomanian Mineralization

Main Characteristics

The major period of hydrothermal activity in the Samail oceanic basin occurred at the end of the first volcanic stage (V1). Radiolaria-bearing sediments, Late Albian-Early Cenomanian in age (Beurrier et al., 1987), which overlie the V1 basalts are locally enriched in pyrite, silica, and iron or manganese oxides. Contemporaneous deep fractures crossing the volcanic pile and the sheeted dyke complex down to the magmatic chamber (Fig.2) are usually parallel to the spreading axis (Haymon et al., 1989) and are marked by chloritization, pyritization, and silicification in the upper parts of the hydrothermal system. These paleofractures not only acted as pathways for the mineralizing fluids, but also delineated seafloor depressions where sulphides locally accumulated.

Massive ore in the sulphide deposits at the V1/V2 contact in the Samail nappe displays brecciated and collomorphous textures and is composed mainly of pyrite with small amounts of chalcopyrite, sphalerite, magnetite and marcasite (Ixer et al., 1984). Chalcopyrite-sphalerite-rich ore with "black

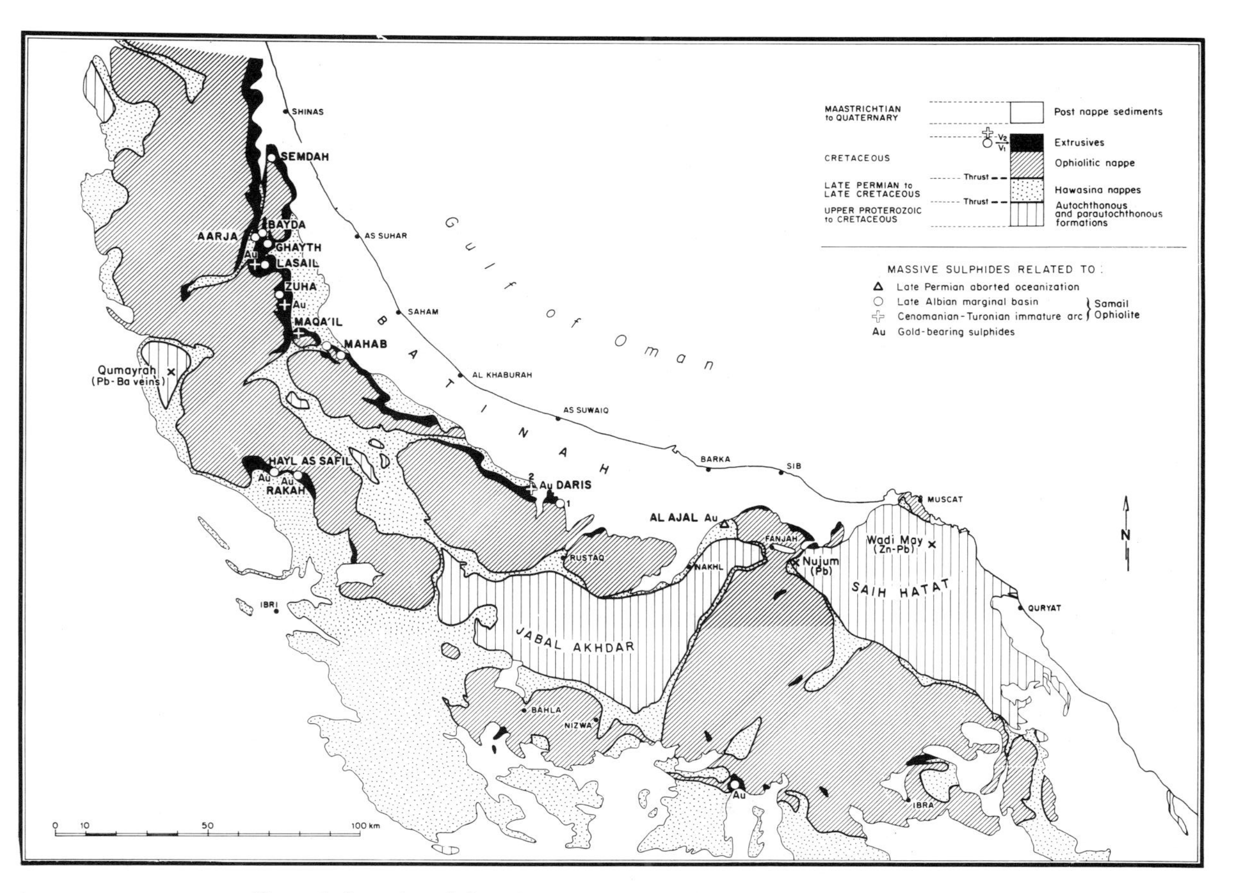

Figure 1. Location of the sulphide deposits in the northern Oman mountains.

smoker" chimney fragments marks the paleovents in the Rakah and Aarja deposits. Ghost textures of isocubanite, replaced by chalcopyrite, and of pyrrhotite and anhydrite, replaced by pyrite, have been observed in several deposits (Lescuyer et al., 1988). In the Rakah and Hayl as Safil deposits, several stages of ore deposition are marked by interbedded massive sulphide breccia and low-temperature silica-hematite deposits, locally containing hydrothermal fossil worm tubes.

Paleotectonic Setting

The 200–500–m-thick basalt and andesite (V1), which underly most of the massive sulphide deposits, were emplaced prior to 95Ma (Lower Cenomanian) along an accretion ridge system that was roughly parallel to the present north- to northwest-striking axis of the Samail ophiolite, as shown by the general strike of the underlying sheeted dykes. Geochemical characteristics of this magmatic stage (Beurrier et al., 1989) are similar to those of present-day mid-ocean or marginal basin basalt: $Ca/Al_2O_3 = 0.45$; $Ti/V = 20$ to 45; $Th/Ta = 1.44 \pm$. However, the low concentrations of Cr and Ni, as well as the relative enrichment of LIL-elements (Rb, Ba, Sr) of the V1 magmatism, as compared to typical MORB compositions, probably result from hydration of the mantle source in a possible back-arc setting. These data were obtained mainly by Beurrier et al. (1989) from samples of the crustal sequence exposed along the northeastern edge of the ophiolite (coastal zone).

Southwestwards, at a distance of about 100 km prior to the shortening of the ophiolitic nappe, the extrusive outliers located at the front of the Samail ophiolite, such as in the Rakah area (Figs. 1 and 2), display a high geochemical variability of the Late Albian V1 basalt (Masson, 1988): besides relatively normal MORB compositions, several samples show a strong Cr and Ni enrichment and low Ti/V ratios of about 10 which are typical of arc tholeiites. Although more detailed studies, including REE, Hf, Ta and Th analyses, are necessary to assess the precise paleotectonic setting of the southwestern part of the Samail oceanic crust, it is assumed that a back-arc spreading over a northeast-plunging subduction zone could explain the characteristics of the V1 magmatism (V1a in Fig. 2) in this area. Moreover, it should be noted that the massive sulphides deposited at the top of V1 in the inner zone of the Samail ophiolite are significantly enriched in precious metals (2 to 6g/t Au on average, with a maximum of 38g/t Au in a Cu-Zn-rich paleovent of the Rakah deposit) compared to the massive sulphide deposits of the coastal area, which contain less than 1g/t Au.

Cenomanian-Turonian Mineralization

Volcanic activity resumed soon after the Late Albian-Early Cenomanian break, with emission of basaltic to rhyolitic flows and domes (V2a), which were fed by swarms of dykes and sills, and related intrusives (trondhjemite, doleritic stock) along scattered submarine volcanic centres. Pelagic sedi-

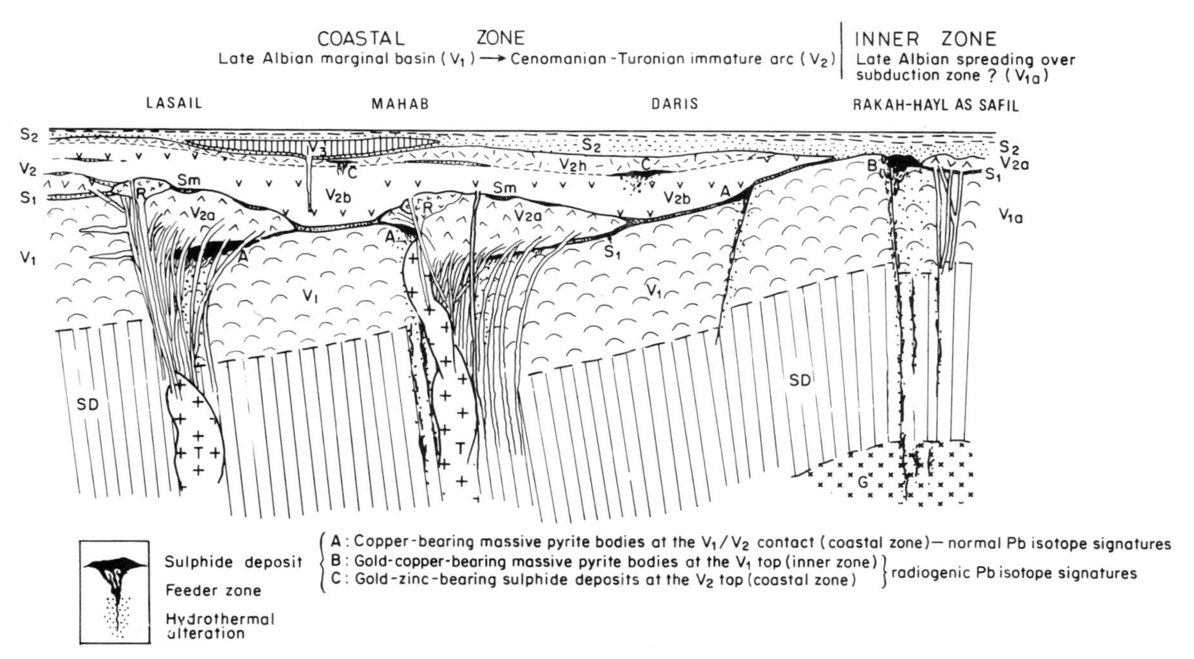

Figure 2. Schematic cross-section through the Cretaceous Samail oceanic crust – G: gabbro; SD: sheeted dykes; V1: lower volcanites from the coastal zone (Cr-Ni depleted and LIL-enriched MORB); V1a: Cr-enriched and Ti/V-depleted lower volcanites (inner zone); S1: Late Albian-Early Cenomanian pelagic and metalliferous sediments; V2: arc tholeiites (V2a: differentiated basalts, andesites and rhyolites [R] and related trondhjemites [T]; V2b: vesicular basaltic pillows; V2h: hyaloclastites); Sm: manganese-bearing pelagic sediments; V3: Salahi alkali basalts; S2: Turonian-Campanian pelagic sediments.

ments, locally enriched in manganese, were deposited during breaks in the volcanic activity. Towards the top, the V2 volcanic sequence is characterized in the coastal zone by extensive basaltic pillows (V2b) and hyaloclastites (V2h). Small Zn-Cu-Au-Ag-bearing siliceous stringers and sulphide deposits (Figs. 1 and 2) are associated with this last event in the Daris (Daris 2 deposit), Zuha (Salahi prospect) and Lasail (Gaddamah W prospect) areas.

The V2 extrusives, 0–1000 m thick, display geochemical characteristics of arc tholeiites (low Ti/V; high CaO/Al_2O_3 and Th/Ta) (Beurrier et al., 1989) and probably result from a subduction-related magmatism of Cenomanian to Turonian age (Beurrier et al., 1987).

Pelagic sediments and talus breccia, with local alkali basalt emissions (V3 on Fig. 2), were deposited at the top of the Samail extrusives from the Turonian to the Campanian, before the Late Campanian-Early Maastrichtian obduction of the ophiolite over the Arabian platform.

Lead Isotope Studies

Previous Studies

Several Pb isotope studies have already been made on the Oman ophiolites, metalliferous sediments and associated sulphide bodies. Chen and Pallister (1981) have shown that the Pb isotopic compositions of the majority of the ophiolitic rocks plot within the field of oceanic basalt close to the least radiogenic Pb of the present MORB (Figs. 3 and 4). This observation was confirmed by Hamelin et al. (1984) who consider, moreover, that the $^{208}Pb/^{204}Pb$ ratio of the Oman ophiolites, which is higher than that of MORB, is compatible with an origin in an inter-arc basin or in an immature arc. The isotopic compositions of the Lasail, Bayda and Aarja (Fe, Cu) sulphides, determined by Chen and Pallister (1981) (sulphides I on Fig. 3), are similar to those of the ophiolitic rocks, which suggested to the authors that the source of the Pb in the sulphides was mainly oceanic crust. The same authors have shown that the isotopic data obtained on serpentinized peridotite (not represented on Fig. 3, within the Cyprus ophiolites field on Fig. 4) would suggest that part of the serpentinization process could imply an addition of radiogenic Pb from an old continental crust or from pelagic sediments.

Gale et al. (1981) studied the isotopic composition of metalliferous (Fe and Mn oxides and hydroxides) sediments from the Lasail area, which are basal facies of a sedimentary accumulation on oceanic crust. They show that the representative points in the two Pb isotopic diagrams lie within the field determined by the underlying ophiolitic rocks (Fig. 3).

Hamelin et al. (1988), in their Pb isotope study of the Troodos massif of Cyprus, draw attention to the differences that exist between the isotopic signatures of the rocks, sulphides and sediments of the Troodos and Samail ophiolites. At Samail, they note the great similarity and homogeneity of the

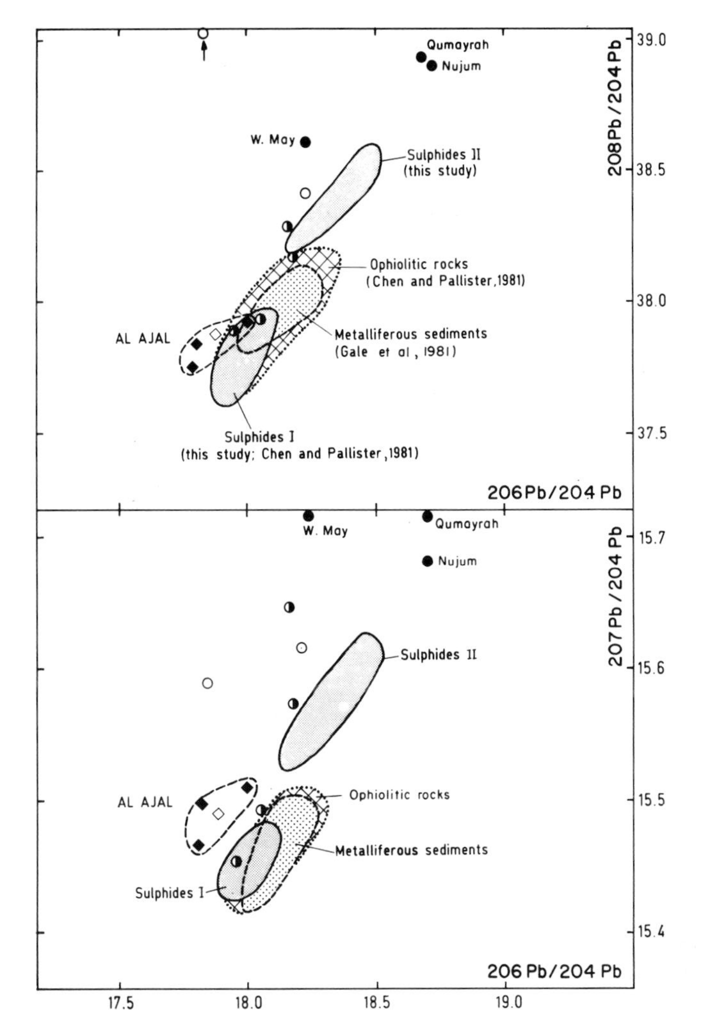

Figure 3. Lead isotope ratio diagrams showing the data fields of ophiolitic rocks, metalliferous sediments and sulphides from the Samail Ophiolite, and analytical points from Table 1 for galena (●), V_2 volcanic rocks (○), sediments (◑), and Late Permian Al Ajal sulphides (◆) and basalt (◇).

isotopic compositions of the sulphides, rocks and sediments which contrast with those of the Troodos where the isotopic compositions are heterogenous and more radiogenic. They conclude that these differences reflect different geodynamic situations: Troodos would be related to suprasubduction spreading, whereas Samail would be typical of normal oceanic crust. The involvement of pelagic sediments is one of the phenomena that they invoke to explain the isotopic variations between sulphides of different deposits and the more radiogenic character of the sulphides from certain mines in relation to the ophiolitic rocks.

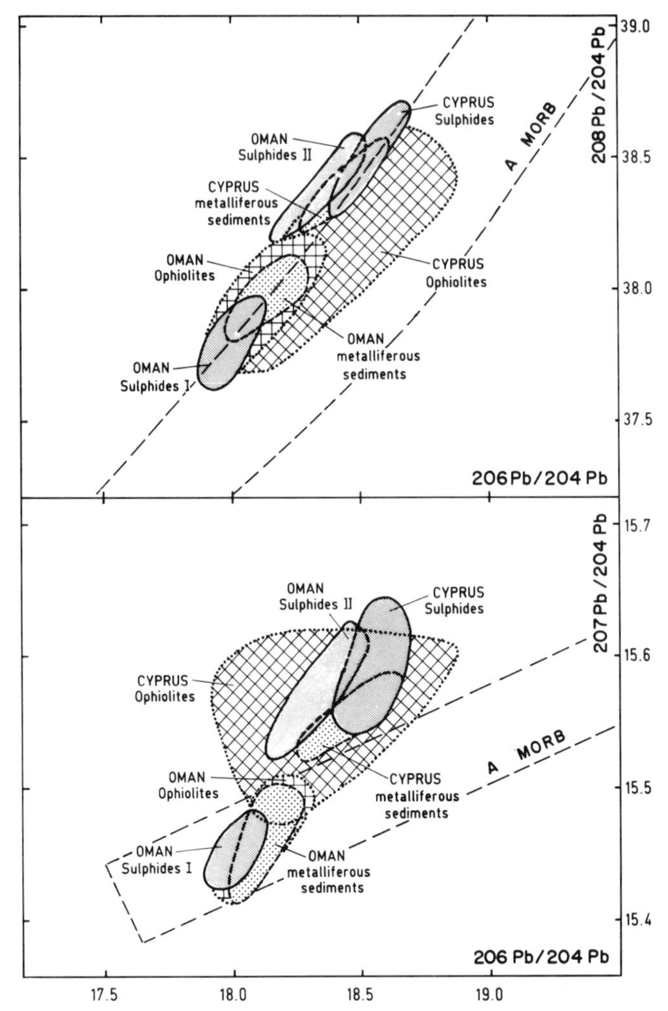

Figure 4. Lead isotope ratio diagrams showing the data fields of ophiolitic rocks, metalliferous sediments and sulphides from the Samail and Troodos ophiolites. Fields for the Troodos are drawn after data by Doe and Zartman (1979), Gale et al. (1981), Spooner and Gale (1982), Hamelin et al. (1984), Rautenschlein et al. (1985) and Hamelin et al. (1988). A MORB = Ancient (100Ma) Mid-Ocean Ridge Basalts field.

Present Study

Objective

The previous studies indicate a relatively simple isotopic regime for the rocks, sulphides and sediments of the Oman mountains. The objective of the present study was to see whether isotopic differences exist between the sulphide deposits associated with the two volcanic episodes (V1 and V2).

Theoretically the differences recorded at the geochemical level, and interpreted by Beurrier et al. (1989) as reflecting a geodynamic change, should also be seen at the isotopic level.

Analytical Techniques

Analyses were carried out at the Bureau de recherches géologiques et minières in Orléans, using the following preparations:

1. Galenas were dissolved in two drops of extra pure 7N.HBr to give a solution of $Pb(NO_3)_2$ with $7N.HNO_3$; 25–100 μl of this solution was extracted and gently dried.
2. Sulphides and gossan material were dissolved in a 1:1 mixture of 6N.HCl-$14N.HNO_3$, this procedure being repeated twice.
3. Volcanic rocks (fresh samples with no trace of hydrothermal alteration) and sediment samples were digested in a 28N.HF (3 ml) and $14N.HNO_3$ (1 ml) mixture; after drying the residue was dissolved in 2N.HCl (1 ml).

For the materials other than galena, the Pb was separated from the other elements by passing the solutions through an ion-exchange resin column in Cl form.

The experiments were carried out under a laminar flow-hood, with all reagents being redistillated using subboiling distillation techniques; the total blank was in the range of 0.5 ng.

For the analysis, approximately 300 ng of Pb in H_3PO_4 was loaded onto a rhenium single filament (40 micron) using a silica gel. The Pb isotope composition was measured with a Finnigan Mat 261 automated mass spectrometer, and a mass discrimination correction of 1.24 per mil per mass unit (estimated from NBS 982 reproducibility) was applied. The maximum errors of the ratios were usually in the range of 0.3–0.4 per mil per mass unit. The within-run error (2σ) is estimated to be lower than 0.25 per mil of mass unit, giving total absolute precision of about 0.01 for $^{206}Pb/^{204}Pb$ and $^{207}Pb/^{204}Pb$ and about 0.04 for $^{208}Pb/^{204}Pb$.

Main Results

The isotopic data, not corrected for in situ decay of U and Th, are given on Table 1 and shown on Figure 3; the analytical points are represented in relation to the fields of the ophiolitic rocks, the sulphides and the metalliferous sediments determined from existing data (Chen and Pallister, 1981; Gale et al., 1981). The Sulphides I field, in the two diagrams, was determined from values obtained by Chen and Pallister (1981) on the sulphide deposits of Lasail, Bayda and Aarja situated at the top of the V1 volcanics; it also includes the results determined during the present study on two samples from the Daris 1 sulphide body, one sample from the siliceous gossan at Zuha and one sample from the Hayl as Safil sulphide body – three deposits situated stratigraphically at the top of V1–type volcanites.

Table 1. Lead isotope ratios of galenas, sulphides, gossans, sediments and volcanic rocks from the northern Oman Mountains.

Sample No.	Locality	$^{207}Pb/^{206}Pb$	$^{206}Pb/^{204}Pb$	$^{208}Pb/^{204}Pb$
Galenas				
C 533	Wadi May (Late Permian)	15.734	18.244	38.603
JL 90-35	Qumayrah (Sumayni Gp)	15.719	18.696	38.930
Sulphides				
PS 12	Al Ajal (Late Permian)	15.468	17.810	37.766
PS 17	Al Ajal (Late Permian)	15.515	18.010	37.928
PS 18	Al Ajal (Late Permian)	15.501	17.819	37.876
PS 6	Rakah	15.608	18.513	38.524
PS 7	Rakah	15.593	18.480	38.501
PS 9	Rakah	15.620	18.451	38.504
PS 10	Rakah	15.607	18.461	38.551
CJ 57	Rakah	15.606	18.495	38.516
CJ 58	Rakah	15.602	18.476	38.503
PS 8	Rakah	15.608	18.500	38.526
PS 5	Daris 1	15.488	18.095	37.957
PS 1	Daris 2	15.547	18.379	38.355
PS 2	Daris 2	15.600	18.452	38.494
PS 4	Daris 1	15.443	18.047	37.884
PS 16	Hayl as Safil	15.461	18.114	37.956
CJ 55	Hayl as Safil	15.603	18.458	38.457
CU 73	Maqa'il (sulphide blocks in breccia)	15.544	18.223	38.258
Gossans				
CU 54	Gaddamah W (Lasail area)	15.533	18.177	38.200
CU 24	Zuha	15.451	18.034	37.761
Pelagic Sediments				
CU 13	Bayda	15.450	17.981	37.879
CU 09	Aarja	15.647	18.174	38.293
CU 07	Ghayth	15,492	18.070	37.949
CU 82	Semdah	15,574	18.200	38.163
Volcanic rocks				
AAl	Al Ajal (Late Permian)	15.504	17.985	37.943
JLL 147	Rakah area] V_{2a}	15.615	18.241	38.405
1844	Daris area] V_{2a}	15.587	17.860	39.729

The Sulphides II field, situated clearly above Sulphides I, is determined by samples from sulphide deposits and associated stockworks located either at the top of the V1 volcanites in the Rakah area (seven samples from the Rakah deposit and one from the Hayl as Safil deposit) or at the top of the V2 volcanites in the coastal zone (three samples of the Daris 2 and Gaddamah [immediately west of Lasail] mineralization and one sample of reworked massive sulphide in a talus breccia from the Maqa'il occurrence at the top of the ophiolitic pile). The definition of this field constitutes the original contribution of this study, because it shows that, contrary to the accepted isotopic range for the rocks and sulphides of the Samail ophiolitic nappe with a homogeneous composition within the field of ocean basalt, a greater diversity in isotopic signatures exists at the level of the (Pb-rich) sulphides, and that the source of the Pb in the mineralization cannot be only oceanic crust.

The pelagic sediments analyzed correspond to the V1/V2 contact and were collected close to laterally deposited sulphide bodies (Aarja, Bayda, Ghayth and Semdah). They show a very large range of composition, from only slightly radiogenic values, situated in the field of the metalliferous sediments (Fig. 3), to values close to those obtained for the sulphides of field II. It can be seen that these values define a narrow data field which passes through the data field of the metalliferous sediments analyzed by Gale et al. (1981); however, it barely passes through the data fields of Sulphides I and Sulphides II, which is especially evident on the $^{207}Pb/^{204}Pb$ v. $^{206}Pb/^{204}Pb$ diagram.

The two representative rocks of the V2a (differentiated tholeiitic) volcanism, collected from the coastal zone (Daris) and the Rakah area, show different isotopic compositions from those of the rocks of the lower ophiolitic sequence; in particular, they have higher $^{207}Pb/^{204}Pb$ ratios which are similar to those of the Sulphides II field. This is only an indicative trend. More detailed studies, involving corrections for in situ decay of U and Th, should be performed on V2 volcanic rocks in order to confirm it.

Two galenas were also analyzed; one collected in Wadi May from a Late Permian stratiform layer (with Pb-Zn) of the autochthonous series of the Saih Hatat region (south of Muscat), and the other from a Pb-Ba-F vein, probably of Tertiary age, which cuts the parautochthonous units of the Sumayni Group. The latter has an isotopic composition fairly close (Fig. 3) to that of a galena from Nujum (below the ophiolite sole, Fig. 1) analyzed by Stacey et al. (1980) and interpreted as reflecting an origin of the Pb from old evolved continental material. The former has an isotopic composition reflecting a similar origin, but with a lower $^{206}Pb/^{204}Pb$ ratio due to the difference in age.

The sulphides of the Late Permian stratiform Cu-Zn mineralization at Al Ajal and the footwall tholeiite have fairly close isotopic compositions situated slightly above the field of oceanic basalt; this conforms to the proposed origin (aborted oceanization) for this volcanism.

Discussion

The new data obtained on the sulphides of the mineralization associated with the V1 and V2 volcanism of the Samail ophiolitic pile, significantly broadens the previously established field of isotopic compositions towards more radiogenic values comparable with those of the sulphides from the Troodos (Cyprus) ophiolites. Thus the overall interpretation of the isotopic compositions of the sulphides of the Samail nappe mineralization as being restricted to typical MORB values (Hamelin et al., 1988) needs to be revised, as does the accepted isotopic distinction between the Samail and Troodos ophiolites and the geodynamic inferences that this implied.

In contrast to Cyprus, the spread of values ranges from uncontaminated MORB-type lead to evolved crustal lead. The sulphide leads associated with the V1 volcanism of the coastal zone are isotopically similar or close to those reported by Chen and Pallister (1981) and they plot in the MORB field. They indicate an origin from the oceanic crust, and are similar to those of sulphides from present non-sedimented mid-ocean ridges (Le Huray et al., 1988).

On the other hand, sulphides associated with the V2 volcanism and with the V1 volcanism of the inner zone of the Samail nappe contain a more radiogenic Pb with a non-MORB character. Such leads are found in sulphides of mineralization from sedimented mid-ocean ridges (Le Huray et al., 1988) or from island arcs. The isotopic signatures imply a derivation of the lead from sediments with no or only small basaltic components. The interpretation of the V2 volcanism as an immature arc magmatism above a subduction zone plunging to the northeast (Beurrier et al., 1989) is compatible with these Pb isotope data. The additional radiogenic component can be looked for in the galenas of the formations underlying the ophiolitic nappe or, more probably, in the sediments located in the V1–V2 interval which plot on a straight line in the Pb-Pb diagrams, indicating the mixture of two types of lead – the one with oceanic affinity, the other with continental affinity. The V2 volcanic rocks also show this radiogenic isotopic signature, which supports the hypothesis of contamination by continental material. Similarly, the radiogenic isotopic composition of the sulphides located at the top of the lower volcanites in the inner zone of the ophiolites (Rakah area) seems to confirm the existence of an early subduction process along the southwestern edge of the Samail oceanic crust.

From the metallogenic point of view, the presence of gold (2 to 6 g/t Au on average) associated with slightly anomalous As, Pb, Sb values in the sulphide deposits related to the V_2 volcanism and to the V_1 volcanism of the inner zone provides an independent argument to support the isotopic distinction between Sulphides I and Sulphides II of the Samail ophiolite.

References

Bechennec, F., Tegyey, M., Le Métour, J., Lescuyer, J.L., Rabu, D., and Milési, J.P., 1990. Igneous rocks in the Hawasina nappes and the Hajar Supergroup (Oman Mountains): their significance in birth and evolution of the composite extensional Arabian margin of the Eastern Thethys. Symposium on ophiolite genesis and evolution of oceanic lithosphere, Muscat.

Beurrier, M., Bourdillon-de-Grissac, C., De Wever, P., and Lescuyer, J.L., 1987. Biostratigraphie des radiolarites associées aux volcanites ophiolitiques de la nappe de Samail (Sultanat d'Oman): conséquences tectogénétiques. C.R. Acad. Sci. Paris, t.304, série II, no.15:907–910.

Beurrier, M., Ohnenstetter, M., Cabanis, B., Lescuyer, J.L., Tegyey,M., and Le Métour, J., 1989. Géochimie des filons doléritiques et des roches volcaniques ophiolitiques de la nappe de Semail: contraintes sur leur origine géotectonique au Crétacé supérieur., Bull. Soc. géol. Fr., (8), t.V., no.2:205–219.

Chen, J.H., and Pallister, J.S., 1981. Lead isotopic studies of the Semail ophiolite Oman., J. Geophys. Res., 92:11411–11415.

Gale, N.H., Spooner, E.T.C., and Potts, P.J., 1981. The lead and strontium isotope geochemistry of metalliferous sediments associated with Upper Cretaceous ophiolitic rocks in Cyprus, Syria and the Sultanate of Oman., Can. J. Earth Sci., 18: 1290–1302.

Hamelin, B., Dupré, B., and Allègre, C.J., 1984. The lead-isotope systematics of ophiolite complexes. Earth Planet. Sci. Letter, 67: 351–356.

Hamelin, B., Dupré, B., Brévart, D., and Allègre, C.J., 1988. Metallogenesis at paleo-spreading centres: Lead isotopes in sulfides, rocks and sediments from the Troodos Ophiolite (Cyprus). Chem. geol., 68: 229–238.

Haymon, R.M., Koski, R.A., and Abrams, M.J., 1989. Hydrothermal discharge zones beneath massive sulfide deposits mapped in the Oman ophiolite. Geology,17: 531–535.

Ixer, R.A., Alabaster, I., and Pearce, J.A., 1984. Ore petrography and geochemistry of massive sulphide deposits within the Semail Ophiolite, Oman., Inst. Min. Metall. Trans., 93, B114–B124.

Le Huray, A.P., Church, S.E., Koski, R.A., and Bouse, R.M., 1988. Pb isotopes in sulfides from mid-ocean ridge hydrothermal sites., Geology, 16: 362–365.

Lescuyer, J.L., Oudin, E., and Beurrier, M., 1988. Review of the different types of mineralization related to the Oman ophiolitic volcanism. Proceedings of the Seventh Quadriennal IAGOD Symposium, E.Schweizerbart'sche Verlagsbuchhandlung, Stuttgart: pp. 489–500.

Masson, J.M., 1988. Etude du métamorphisme hydrothermal océanique associé aux indices minéralisés en cuivre et zinc du secteur de Rakah, nappe ophiolitique de Semail (Oman). Diplôme d'Etudes Approfondies, GIS Océanologie et Géodynamique, Brest, 119p.

Oudin, E., 1983. Minéralogie de gisements et indices liés à des zones d'accrétion océaniques actuelles (Ride Est Pacifique et mer Rouge) et fossiles (Chypre)., Chron. rech. min., 470: 43–55.

Rautenschlein, M., Jenner, G.A., Hertogen, J., Hofmann, A.W., Kerrich,R., Schmincke, H.U., and White, W.M., 1985. Isotopic and trace element composition of volcanic glasses from the Akaki Canyon, Cyprus: implications for the origin of the Troodos ophiolite., Earth. Planet. Sci. Letter, 75: 369–383.

Spooner, E.T.C., and Gale, N.H., 1982. Pb isotopic composition of ophiolitic volcanogenic sulfide deposits, Troodos complex, Cyprus., Nature (London), 29: 239–242.

Stacey, J.S., Doe, B.R., Roberts, R.J., Delevaux, H.H., and Gramlish,J.W., 1980. A lead isotope study of mineralization in the Arabian Shield., Contrib. Mineral.Petrol., 74: 175–188.

Hydrothermal Metamorphism in Oceanic Crust from the Coast Range Ophiolite of California: Fluid-Rock Interaction in a Rifted Island Arc

PETER SCHIFFMAN[1], RUSSELL C. EVARTS[2], ALAN E. WILLIAMS[3], and WILLIAM J. PICKTHORN[2]
[1] *Dept. of Geology, University of California, Davis, California 95616*
[2] *U.S. Geological Survey, 345 Middlefield Road, Menlo Park, California 94025*
[3] *Dept. of Earth Sciences, University of California, Riverside, California 92521*

Abstract

Metamorphic-mineral parageneses and stable-isotope systematics in remnants of the southern portion of the California Coast Range ophiolite differ from those described in modern oceanic crust and many other ophiolites. Detailed studies of the Del Puerto and Point Sal ophiolite remnants reveal a style of submarine hydrothermal metamorphism characterized by (1) development within the extrusive section of a continuous low-P/T metamorphic facies series that includes zeolite-, prehnite-pumpellyite-, and greenschist-facies mineral assemblages; (2) *pervasive* alteration of the extrusive rocks; (3) absence of massive sulfide deposits; (4) large ^{18}O enrichments in metavolcanic rocks (up to 20‰ [SMOW] in the Del Puerto remnant); and (5) absence of corresponding ^{18}O depletions in associated plutonic rocks.

D/H whole-rock and mineral-separate data from the Del Puerto ophiolite remnant confirm that metamorphism occurred in the presence of a seawater-derived hydrothermal fluid. Oxygen-isotope analyses of primary minerals in Del Puerto remnant plutonic rocks indicate limited open-system behavior characterized by ^{18}O-enrichment, relative to primary magmatic values, in plagioclase. In the Coast Range ophiolite remnants, fluid inflow through a large volume of unidentified or missing rocks (possibly sedimentary cover, including tuffaceous cherts), must have enriched seawater in ^{18}O. Because the Coast Range ophiolite remnants did not form in a strongly extensional tectonic environment, major listric normal fault systems did not develop, and thus seawater penetration into the lower crust must have been limited. Consequently, isotopically shifted seawater produced strong ^{18}O enrichments in volcanic roof rocks during upflow, but did not significantly interact with the underlying plutonites.

The absence of massive sulfide deposits as well as the pervasive style of hydrothermal metamorphism and isotopic exchange in Coast Range ophiolite extrusive rocks implies that fluid flow was not strongly fracture controlled. The continuity of metamorphic zones within the extrusive section is more closely analogous to that characteristic of continental geothermal systems or

Tj. Peters et al. (Eds), Ophiolite Genesis and Evolution of the Oceanic Lithosphere, 399–425.

contact metamorphic aureoles as opposed to hydrothermal metamorphism in open ocean spreading centers. Exposure of sediment-capped, volcanic roof rocks to cooling plutons for relatively long periods of time allowed for more complete recrystallization and isotopic exchange, and development of zeolite- and prehnite-pumpellyite-facies assemblages, features that are typically absent from modern ocean crust.

Introduction

Although relatively little is known about the hydrothermal metamorphism of *in situ* oceanic crust, many features have been inferred from investigations of structurally coherent ophiolite complexes. In fact, current definitions and descriptions of ocean floor metamorphism (e.g., Mason; 1978) are based on observations from ophiolites, particularly from the better studied Tethyan complexes in Cyprus and Oman (e.g., Gass and Smewing, 1973; Lippard et al., 1986). The igneous geochemistry of most ophiolites differs significantly from that of crust formed at modern-day, open-ocean spreading centers (e.g., Robinson et al., 1983; Rautenschlein et al., 1985), but studies of secondary mineral assemblages have emphasized the similarities between the hydrothermal systems of oceanic crust and their inferred fossil analogues (e.g., Haymon and MacDonald, 1985; Oudin and Constantinou, 1984). The problem is one of perspective: our data base on modern, *in situ* oceanic crust is extremely limited-(i.e., only sites 504B and 735B have provided significant amounts of information on oceanic hydrothermal processes) and is restricted to one tectonic setting (oceanic crust formed at mid-oceanic spreading centers). Conversely, studies of fossil submarine hydrothermal systems are biased because we can only sample pieces of oceanic crust which were emplaced along continental margins. Many ophiolites apparently form because of spreading induced by oblique subduction and are later emplaced in response to basin closure during plate collisions (Moores, 1982). Ophiolites undoubtedly form in diverse tectonic environments (e.g., marginal basins, island arcs) which are not restricted to open-ocean spreading centers (Coleman, 1984). If rocks from these diverse tectonic settings have identifiable igneous geochemical fingerprints (e.g., Pearce and Cann, 1973), we can expect to observe analogous hydrothermal patterns as well.

This paper presents mineralogic and stable-isotope data on hydrothermally altered rocks from the California Coast Range ophiolite. We will show how the submarine hydrothermal features of this ophiolite are quite distinct and separable from those of both modern oceanic crust and other ophiolites. Our intention is to point out the variations in hydrothermal metamorphism that can occur within oceanic crust and to suggest how these may be related to specific physical and chemical conditions which are themselves indicative of distinct tectonic environments not restricted to open-ocean spreading centers.

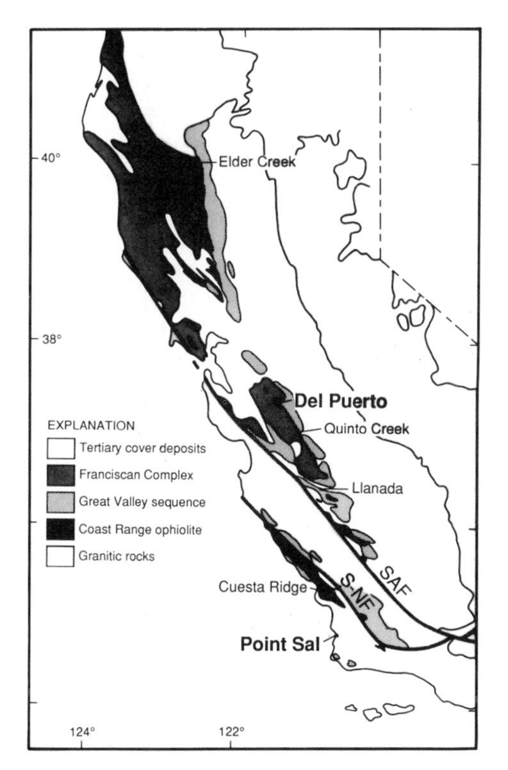

Figure 1. Simplified geologic map showing major litho-tectonic units and Coast Range ophiolite remnants described in the text. SAF, San Andreas fault; S–NF, Sur-Nacimiento fault zone. Modified after Shervais (1990).

The California Coast Range Ophiolite

The Late Jurassic Coast Range ophiolite of California is comprised of a series of discontinuous remnants that occur along an approximately 700–km-long segment of the northern and central Coast Ranges (Figure 1). The ophiolite remnants are depositionally overlain by Upper Jurassic sedimentary rocks of the Great Valley sequence and are tectonically underlain by Jurassic (and younger) rocks of the Franciscan Complex.

Although the Coast Range ophiolite was originally believed to represent crust formed at a mid-ocean spreading center (e.g., Bailey et al., 1970; Hopson et al., 1981), recent workers (e.g., Shervais, 1990) favor an island-arc setting for its remnants now present in the central Coast Ranges, including those at Del Puerto Canyon and Point Sal. The development of an island arc is likely because: (1) the volcanic members of some remnants (e.g., the Del Puerto, Llanada, and Quinto Creek remnants of Hopson et al., 1981) are conformably overlain by, or intercalated with turbiditic, volcaniclastic sedimentary rocks, including andesite-bearing conglomerates of non-ophiolitic origin (Evarts, 1977; Emerson, 1979; Robertson, 1989); (2) the volcanic member includes (the altered equivalents of) high Mg-basaltic andesites, andesites, dacites, and rhyodacites whose major and trace element geochemistry display strong island-arc characteristics (Shervais and Kimbrough, 1985); and (3) the associated plutonic sections include abundant silicic rocks (quartz diorite, tonalite, and trondhjemite). Furthermore, orthopyroxene- and hornblende-bearing cumulates containing highly calcic plagioclase, indicate calc-alkaline rather than tholeiitic affinities (Evarts, 1977; Hopson and Frano, 1977). The latter two petrologic and geochemical characteristics are consistent with a supra-subduction zone origin for the Coast Range ophiolite (Shervais, 1990). The presence of coarse-grained volcaniclastic sedimentary rocks argues for a close temporal and spatial relationship between ophiolite formation and active arc volcanism. The general absence of sheeted-dike complexes and presence of sill complexes (e.g., at Del Puerto Canyon: Evarts, 1977; Point Sal: Hopson et al., 1981) indicate that the tectonic environment in which the Coast Range ophiolite remnants were created was not characterized by pronounced extension such as recorded in the Tethyan ophiolites. Nevertheless, the creation of new oceanic crust implies at least incipient rifting of the island arc.

Metamorphic Petrology of the Coast Range Ophiolite

Although ophiolite complexes are invariably metamorphosed to some extent, it is essential to differentiate the effects of submarine hydrothermal metamorphism from those which can be attributed to burial, emplacement, or subsequent tectonism. The most unambiguous way to do this is by comparison of the inferred metamorphic gradient within the ophiolite pseudostratigraphy to that within bounding lithologic units. The sedimentary cover of the Great Valley sequence overlying the Coast Range ophiolite is as much as 15 km thick (Ingersoll, 1978). Although comprehensive, regional studies of the diagenesis of these sediments are lacking, albitized and laumontite- and/or prehnite-bearing sandstones occur at the base of the sequence (e.g., at Cache Creek, Dickinson et al., 1969; near Paskenta, Bailey and Jones, 1973). In the northern Coast Ranges near Paskenta, an 8.5–km-thick section of the Great Valley sequence has an inferred metamorphic gradient of 25°C/km (Suchecki and Land, 1983). Thus, burial metamorphic temperatures at the base of the sequence probably approached 200°C. Evarts and Schiffman (1983) pointed out however, that the inferred metamorphic gradient *within* the underlying Coast Range ophiolite remnant at Del Puerto is much greater (i.e., closer to 100°C/km) and can not be the product of burial metamorphism beneath the sediments. Rocks of the Coast Range ophiolite have been recrystallized in a style consistent with submarine hydrothermal metamorphism (Hopson et al., 1981), and there is no evidence for emplacement-related or subsequent dynamothermal metamorphism.

The metamorphic petrology of the Coast Range ophiolite has been studied

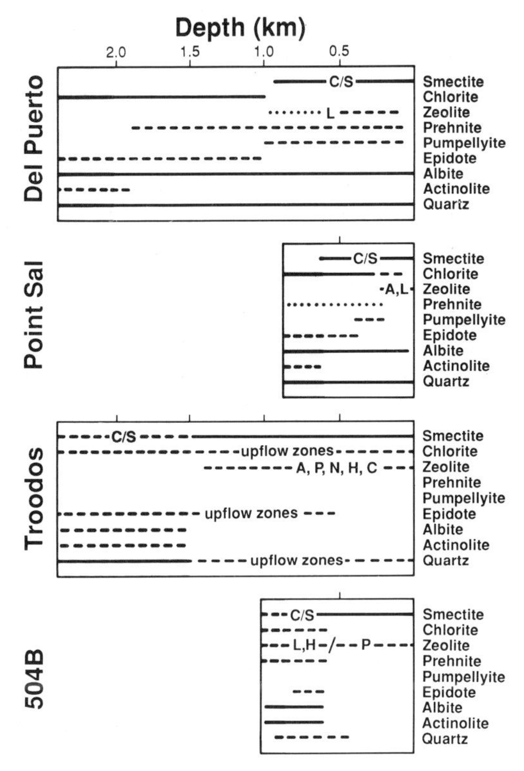

Figure 2. Metamorphic stratigraphy of the Del Puerto and Point Sal remnants of the Coast Range ophiolite, Troodos ophiolite, and DSDP Hole 504B in the Costa Rica rift. Diagram depicts alteration of volcanic (unpatterned) and dike/sill complex (stippled). Abbreviations: C/S = interstratified smectite/chlorite, A = analcime, L = laumontite, P = phillipsite, N = natrolite, H = heulandite, c = clinoptilolite. Data sources: Del Puerto (Evarts and Schiffman, 1983); Pt Sal (Bettison and Schiffman, 1988); Troodos (Gillis and Robinson, 1985; Schiffman and Smith, 1988); Hole504 B (Alt et al., 1986a).

in detail at two localities: Del Puerto (Evarts and Schiffman, 1983; Schiffman et al., 1984) and Point Sal (Schiffman et al, 1986; Bettison and Schiffman, 1988). The metamorphic petrology in these two remnants has characteristics that differ markedly from those of other well-studied, coherent ophiolite complexes (e.g., Troodos) or *in situ* modern, oceanic crust (i.e. 504B). Specifically, the Coast Range ophiolite remnants are characterized by: (1) continuous, low pressure/temperature (P/T) metamorphic facies (zeolite, prehnite-pumpellyite, and greenschist) series developed within extrusive and dike/sill members, and (2) volcanic sequences which exhibit pervasive alteration, including albitization, and localized Ca-metasomatism.

Metamorphic mineral zonations recorded within the Del Puerto and Point Sal remnants are shown in Figure 2. The following descriptions are abstracted from Evarts and Schiffman (1983), for the Del Puerto rocks, and Schiffman et al. (1986) and Bettison and Schiffman (1988), for the Point Sal rocks.

Del Puerto Ophiolite Remnant

In the Del Puerto ophiolite remnant, volcaniclastic sandstones of the Lotta Creek Formation (Raymond, 1973) rest on ophiolitic lavas and are locally intruded by quartz keratophyre sills which are petrographically similar to the most silicic lavas. These sandstones contain a wide assortment of zeolites (including analcime, heulandite, and laumontite), prehnite, pumpellyite, quartz, calcite, hematite and randomly interstratified chlorite/smectite, as well as unaltered igneous clinopyroxene, hornblende, Fe-Ti oxides and calcic plagioclase. The upper kilometer or so of the ophiolite volcanic section contains pumpellyite + albite + corrensite + quartz + hematite + titanite assemblages which locally include laumontite, prehnite, randomly interstratified chlorite/smectite, celadonite, calcite, and relict clinopyroxene. The lower part of the volcanic section is characterized by the appearance of epidote + discrete (i.e., non-interlayered) chlorite, at the expense of pumpellyite + interlayered chlorite/smectite; the pumpellyite-out, epidote-in, mineral zone boundary is subparallel to the ophiolite pseudostratigraphy. Actinolite appears as uralitic replacements of primary clinopyroxene near the base of the volcanic section. The upper levels of the plutonic complex, including the dike and sill complex, are generally characterized by greenschist- (or prehnite-actinolite-) facies assemblages consisting of actinolite + epidote + chlorite + albite + quartz +/− prehnite, and less common epidosites (i.e., granular, metasomatic rocks comprised of epidote + chlorite + quartz +/− albite +/− actinolite). Hydrothermal metamorphism of cumulates from the deeper parts of the plutonic complex is much more limited and localized. Cumulate gabbros may contain green to brown hornblende as replacements or rims on igneous pyroxene which may be of either late-magmatic or metamorphic origin. Mafic dikes which intrude these cumulates locally contain either low pressure amphibolite-facies assemblages characterized by actinolitic hornblende + calcic plagioclase +/− chlorite +/− magnetite, or greenschist-facies assemblages.

Point Sal Ophiolite Remnant

The metamorphic-mineral zonation within the Point Sal remnant is remarkably similar to that described above. The Point Sal section is overlain by thinly bedded tuffaceous cherts (Hopson et al., 1981; Robertson, 1989) which lack metamorphic index minerals. The upper basaltic lavas immediately below the cherts have assemblages containing pumpellyite, prehnite, laumontite, analcime, albite, randomly interstratified chlorite/smectite, celadonite, quartz, and calcite. Lavas lower in the section contain epidote + chlorite + rare andradite. Actinolite first appears in the dike complex, which is characterized by greenschist, prehnite-actinolite, or epidositic mineral assemblages. High level gabbroic rocks contain greenschist facies assemblages, which locally overprint earlier low pressure amphibolite facies assemblages.

Hydrothermal metamorphism within cumulate gabbros is restricted to veins and their alteration haloes; the vein minerals include prehnite, epidote, actinolite, and actinolitic hornblende. Away from these veins, the cumulate gabbros are petrographically fresh with the exception of serpentine which has replaced olivine and orthopyroxene, and brown hornblende which forms rims on pyroxenes.

Metasomatic Effects

Metasomatism of volcanic rocks, particularly well developed in the Del Puerto ophiolite remnant, is characterized by sodium enrichment, potassium depletion, and local redistribution of most other elements (Evarts, 1977; Evarts and Schiffman, 1983; Schiffman et al, 1984). Calcium enrichment is pronounced within structurally discordant metadomains that are locally developed within the volcanic and overlying volcaniclastic rocks. For example, meter-thick metadomains within zeolite-bearing volcaniclastic rocks cut bedding at high angles and display wholescale textural and mineralogical reconstitution to aggregates of coarsely crystalline prehnite, pumpellyite, calcite, and quartz; such features apparently reflect channelized upflow of Ca-rich fluids, presumably along high angle faults and fractures. Other Ca-rich metadomains are conformable, particularly the nearly monomineralic laumontite "mounds" which occur locally along the sediment-volcanic interface, and are believed to have formed by replacement of fine-grained tuffaceous basal sediments (Evarts and Schiffman, 1983). Within the volcanic section, pumpellyite- and epidote-rich metadomains, tens of meters in extent, are locally developed within mafic amygdaloidal flows and breccias. All primary phases in these metadomains are locally replaced by bimineralic Fe^{3+}-rich- pumpellyite + quartz assemblages (Evarts and Schiffman, 1983). Pumpellyite, prehnite, and epidote within the volcanic section and overlying sediments are generally ferric iron rich and co-exist with hematite. The hydrothermal fluids that altered these rocks were apparently quite oxidizing.

Seawater/rock interactions over large ranges in fluid/rock ratios generally produce Mg-enrichments in oceanic volcanic rocks (Seyfried, 1987). Quantifying the magnitude of these chemical fluxes during submarine hydrothermal alteration of Coast Range ophiolite volcanic rocks is impossible, because the absence of fresh glass precludes knowledge of the chemical composition of the protolith. However, it is possible to ascertain whether many of these rocks have experienced Mg-enrichments or depletions by evaluating deviations from equilibrium magmatic clinopyroxene/bulk rock Fe/Mg primary fractionations in clinopyroxene-phyric rocks. The method is applicable to sub-greenschist facies metsabasalts that contain a relatively small proportion (less than 5%) of unaltered relict clinopyroxene phenocrysts. Primary (i.e. magmatic) Fe/Mg fractionation factors for clinopyroxene/basaltic melt calibrated in a number of studies (Grove and Bence, 1977; Grove and Bryan, 1983; Nielsen and Dungan, 1983) indicate that the Fe/Mg distribution coef-

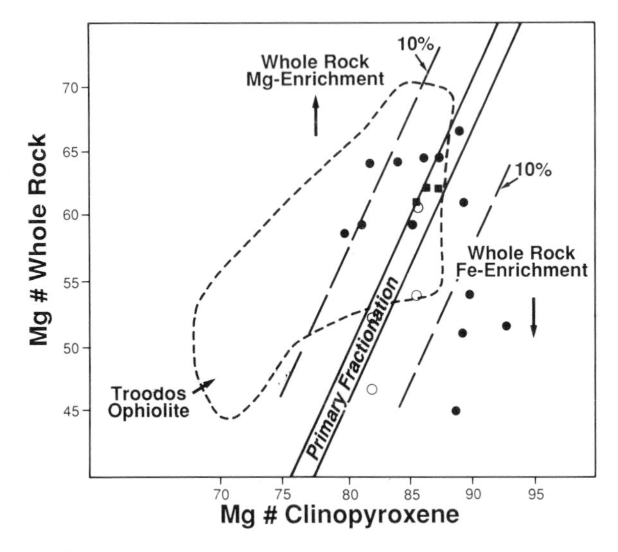

Figure 3. Mg# of clinopyroxene vs bulk rock in Point Sal (filled circles) and Del Puerto (open cirles) ophiolite volcanic members compared to data from the Troodos ophiolite (Schiffman and Smith, 1988). Filled squares represent clinopyroxene-fresh glass pairs from the Troodos ophiolite (Peter Thy, personal communication, 1989). Primary fractionation curve taken from Grove and Bence (1977).

ficients (K_d) are temperature dependent but relatively independent of rock composition (at least for basalts). For example, the Fe/Mg K_d's for coexisting clinopyroxene and glass in Troodos ophiolite lavas (Peter Thy, personal communication, 1989), are identical to those for quartz-normative lunar basalts (Grove and Bence, 1977). Thus, given the compositions of a fresh clinopyroxene phenocryst and its host rock, the experimentally determined K_d-values can be used to calculate a useful estimate of *original* melt Fe/Mg assuming that the bulk rock composition of sparsely-phyric lavas is a reasonable proxy for melt composition. The measured Fe/Mg ratio in the bulk rock then can be directly compared with that predicted from the cpx composition, in order to ascertain the nature of Mg (or Fe) metasomatism.

The Mg# (i.e. Mg/Fe + Mg) of clinopyroxene phenocrysts and their host rocks from the Point Sal and Del Puerto ophiolite remnants are presented in Figure 3. Of the seventeen samples analyzed, 5 plot within the range of equilibrium, magmatic fractionations between 975–1200°C (Grove and Bence, 1977), 5 exhibit Mg-enrichments (relative to Fe), and 7 exhibit Mg-depletions. For comparison, Figure 3 also shows the field of data for fifteen sub-greenschist facies volcanic and dike rocks from the Solea graben area of the Troodos ophiolite (Schiffman and Smith, 1988). The nearly uniform enrichments in Mg are consistent with the hypothesis that the Troodos metabasites were affected by alteration in a regime of seawater recharge (Schiffman and Smith, 1988). The variable Mg-enrichments and depletions exhibited by the Point Sal rocks must reflect a more complex alteration history, presumably encompassing both cold seawater recharge (producing Mg-enrichments) and hot, evolved seawater discharge (resulting in Mg-depletions).

Hydrothermal Mineralization

Massive sulfide mineralization has not been reported from the volcanic members of Coast Range ophiolite remnants, despite relatively good exposures of the volcanic and overlying sedimentary rocks. In the Del Puerto ophiolite remnant, hydrothermal mineralization is apparently limited to small manganiferous-oxide deposits and associated ferruginous chert of probable hydrothermal origin, which conformably overlie volcanic rocks (Evarts, 1978; Evarts and Schiffman, 1983). In the Point Sal ophiolite remnant, blocks of pillow lava and related breccias which are exposed as landslide debris (Location 15 of Hopson et al., 1975) are intensely epidotized and silicified, and locally contain appreciable disseminated pyrite.

Stable Isotope Geochemistry

The metamorphic mineral zonations described for the Coast Range ophiolite remnants clearly reflect recrystallization within a low P/T hydrothermal system. Stable isotope compositions of rocks and mineral separates help establish the nature of the hydrothermal fluid involved. Oxygen isotope analyses of whole rocks and mineral separates from the Del Puerto ophiolite remnant are consistent with a seawater origin for the hydrothermal fluids (Schiffman et al., 1984). Below, we present new oxygen isotopic data on whole rocks and mineral separates from the Del Puerto and Point Sal ophiolite remnants, as well as D/H data from the Del Puerto rocks. These data convincingly substantiate a submarine hydrothermal alteration of these rocks. Tables 1 and 2 contain combined oxygen and hydrogen isotopic data on the Del Puerto rocks and mineral separates. Table 3 contains oxygen isotopic data on the Point Sal rocks and minerals. Some of the oxygen isotopic data from the Del Puerto ophiolite remnant have been previously published (Schiffman et al., 1984) but are included for the sake of completeness.

Oxygen Isotopes

The oxygen isotope compositions of mafic extrusive rocks from the Coast Range ophiolite remnants are characterized by nearly ubiquitous enrichments in ^{18}O with respect to inferred magmatic values (Figure 4 and Table 4). The extent of ^{18}O-enrichment is inversely correlated with the grade of metamorphism for rocks of similar bulk composition (Tables 1 and 3; Figure 5). Epidote-free, zeolite- and prehnite-pumpellyite-facies mafic extrusive rocks have $\delta^{18}O$ between 8.2 and 19.6‰ (SMOW) and recrystallized at temperatures below about 225°C by analogy with modern geothermal systems (Bird

Table 1. Combined ^{18}O and D/H whole rock data for the Del Puerto ophiolite remnant.

Sample	Lithology	^{18}O	D/H*	Metamorphic grade
99–25 A	LCF Sandstone	16.9	−84.9	Zeolite
96–2	LCF Laum tuff	22.4	−69.5	Zeolite
78–3	LCF Sandstone	13.8	−57.7	Zeolite
95–28	LCF Conglomerate	11.9	−60.0	Zeolite
95–38	LCF Sandstone	14.7	−59.9	Zeolite
96–2 A	Basalt	14.3	−76.0	Pumpellyite
99–8	Basalt	19.6	−76.4	Pumpellyite
95–23	Basalt	12.1	−50.5	Pumpellyite
220–27 PM	Basalt	11.6	−95.6	Pumpellyite
99–9	Basalt	13.0	−79.0	Pumpellyite
220–14 A	Qtz Keratophyre	17.2	−71.1	Pumpellyite
220–12	Basalt	13.7	−75.5	Pumpellyite
95–12C	Keratophyre	12.0	−65.0	Pumpellyite
92–94	Basalt	13.4	−75.2	Pumpellyite
93–54	Basalt	8.3	−48.9	Epidote
93–79 C	Basalt	10.5	−48.1	Epidote
93–65	Keratophyre	8.1	−49.3	Epidote
93–38	Qtz Keratophyre	12.3	−60.8	Epidote
93–36	Basalt	9.4	−53.3	Epidote
93–72	Basalt	8.7	−70.3	Epidote
78–03	Epidosite Dike	8.4	−42.4	Epidote
65–21	Epidosite Dike	5.7	−38.2	Epidote
93–16 E	Hbl Tonalite	8.5	−54.6	Actinolite/Hbl
75–6	Mafic Diabase	6.0	−55.6	Actinolite
64–52	Serpentinite	5.6	−96.0	Lizardite
54–5	Serpentinite	6.9	−53.0	Antigorite

* In delta notation; per mil SMOW (also in Tables 2 and 3).
Note = whole rock oxygen isotopic data in Tables 1 and 2 are from Schiffman et al. (1984).
Abbreviation: LCF = Lotta Creek Formation, Laum = Laumontite, Qtz = Quartz, Hbl = Hornblende.

et al., 1984). Epidote-bearing mafic extrusive and dike rocks have $\delta^{18}O$ between 7.7 and 10.6‰. More silicic rocks (i.e., keratophyre and quartz keratophyre flows and dikes) are generally isotopically enriched with respect to mafic rocks of similar metamorphic grade (Tables 1 and 3). Epidosites are the only ^{18}O-depleted extrusive or intrusive rocks of the Coast Range ophiolite that we have analyzed with $\delta^{18}O$ values as low as 5.1‰. Petrographically fresh, cumulate gabbroic and related rocks have $\delta^{18}O$ values between 5.6 and 7.4‰; hornblende-bearing cumulates have values between 6.3 and 7.6‰ (Tables 2 and 3).

Table 2. Oxygen and hydrogen isotope data on Del Puerto mineral separates.

Sample	Lithology	W.R.	CPX	PLAG	OPX	HBL	OX	OL	Quartz	Epidote	Chlorite	PR/PM
DPV-04	Laumontite Vein											
95–38	LCF Conglomerate	14.7		12.8		8.8						11.8 PR
82–05	LCF Conglomerate	10.8										
220–14 A	Qtz Keratophyre	17.2		17.7		6.1 (−71.1)			22.8, 8.2			
220–27 CC	Spilite	13.4										
99–8	Spilite	19.6							24.2			11.8 PM
93–65	Qtz Keratophyre	8.1								10.8		
93–38	Qtz Keratophyre	12.3		13.8								
93–16 J	Qtz Diorite					5.7						
65–21	Epidosite Dike	5.7							9.2	3.1 (−36)	2.3	
65–33	Plagiogranite			12.1					10.2		3.8 (−43)	
65–38C	Plagiogranite			6.6		5.2			8.1		(−49)	
75–13	Hbl-Qtz Diorite	7.4		6.5		5.0			8.3			
76–6	Mafic Dike	6.0		6.5		6.0						
64–83 A2	Hbl Gabbro	7.6	6.1	7.5								
72–15	Gabbro Norite	7.4	5.8	6.7								
82–08 B	Gabbro Norite	5.9	4.9	6.0	5.4	4.4 (−70)	0.6					
75–20	Gabbro		5.8	6.8	6.0	5.4						
64–44 B	Hbl Gabbro Dike		5.7	7.7		4.8 (−47)		5.3				
64–56	Gabbro		5.2	7.5	5.6							
75–12	Gabbro		5.1	6.7	5.3							
64–49	Gabbro		5.6	9.8								
64–203	Wehrlite		6.8					4.4				
76–1	Peridotite		5.5	6.0								
64–157	Peridotite		5.6	6.7								

Abbreviations: W.R. = Whole Rock, CPX = Clinopyroxene, PLAG = Plagioclase, OPX = Orthopyroxene, HBL = Hornblende, OX = Fe-Ti Oxide, OL = Olivine, PR = Prehnite, PM = Pumpellytite, Qtz = Quartz, (Value) = D/H.

Table 3. Oxygen isotope data for the Point Sal Ophiolite remnant.

Sample	Lithology	Mineralogy	Whole Rock Mineral
Sub-Epidote-Grade Metavolcanics			
TC385	Tuffaceous Chert	QZ, IL	19.3
1A–13	Basalt	SM, CO, CH, AB	13.2
1A–14	Basalt	SM, CS, CH, CE, AB, QZ	17.7
1A–15	Basalt	SM, CS, CH, CE, AB	13.9
1A–21	Basalt	SM, CS, CH, QZ	15.6
1B–30	Basalt	CO, CH, AB	9.6
1B–42	Basalt	CO, CH, CE, AB	14.3
1B–47	Basalt	SM, CH, AB, QZ	9.5
1B–55	Basalt	CS, CO, CH, AB, QZ, PR	8.2
1B–56	Basalt	SM, CS, CH, PM, AB, QZ	13.5
1B–60	Basalt	SM, SC, CH, CE, AB, AN	13.4
L18–39	Basalt	CS?, PM, AB, QZ, CC	10.0
L18–41	Basalt	CS, CH, AB, PM, QZ, CC	9.6
L19–38	Basalt	CS?, CE, AB	13.7
L20–35	Basalt	SM, CS, CH, CE, AB, CC	13.5
L20–36	Keratophyre	SM, CO, CE, AB, QZ	16.1
Epidote-Grade Metavolcanics			
1B–8	Basalt	CH, AB	7.9
1B–21	Basalt	CH, AB	7.7
1B–23	Basalt	CH, AB, EP, QZ, CC, GT	7.7
L15–50	Keratophyre	CH?, EP, QZ	7.6
L18–44	Keratophyre	CH?, EP, AB, QZ, GT	8.8
Dikes			
L10–22 A	Epidosite	EP, ACT, CH, QZ, AB	10.0
L10–24	Plagiogranite	AB, QZ, PR, ACT, CH	10.1
L13–26	Epidosite	EP, CH, QZ, ACT	5.1
L14–52	Plagiogranite	AB, QZ, PR, EP, ACT	16.5
L16–47	Diorite	AB, ACT, EP, PR, QZ	13.1
L17–46	Diabase	CH, ACT, AB, EP	9.4
L18–42	Diabase	CH, AB	10.6
Plutonics			
L7–12	Olivine Gabbro	(OL, CPX, PL)	6.0
L8–16	Anorthosite	(PL, OL, CPX, HB)	6.3
L8–17	Noritic Dike	(OPX, PL)	5.6
L10–20	Olivine Gabbro	(OL, CPX, PL)	6.8
L13–27	Hornblende Gabbro	(HB, PL)	6.6
L3–07	Olivine Cpxite	(CPX) SERP	CPX = 5.4
L45–09	Olivine Cpxite	(CPX) SERP	CPX = 5.4

Abbreviations: SM = Smectite, CS = Chlorite/Smectite, CO = Corrensite, CH = Chlorite, AN = Analcime, PM = Pumpellyite, EP = Epidote, GT = Grandite, PR = Prehnite, ACT = Actinolite, HB = Hornblende, CE = Celadonite, QTZ = Quartz, AB = Albite, Cc = Calcite, OL = Olivine, CPX = Clinopyroxene, OPX = Orthopyroxene, PL = Plagioclase, SERP = Serpentine, IL = Illite, ? = Phyllosilicate identification not confirmed by XRD or Microprobe.

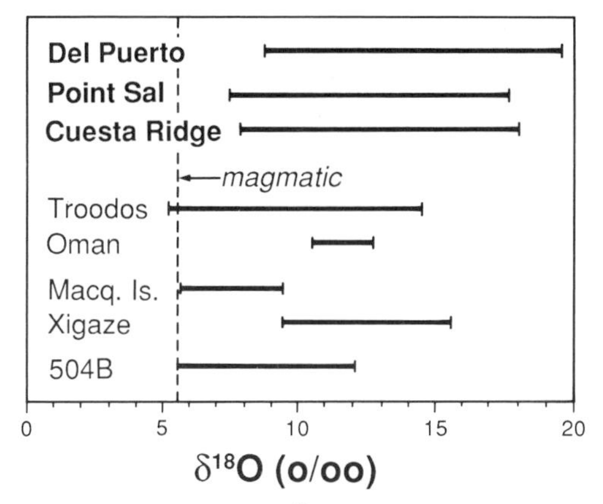

Figure 4. Comparison of range of whole rock $\delta^{18}O$ in mafic volcanic rocks from the Del Puerto, Point Sal, and Cuesta Ridge remnants of the Coast Range ophiolite versus other ophiolites and oceanic crust. Data sources: Coast Range ophiolite (this study; Schiffman et al., 1984; Magaritz and Taylor, 1976); Troodos (Schiffman and Smith, 1988; Heaton and Sheppard, 1977; Rautenschlein et al. 1985); Oman (Gregory and Taylor, 1984); Macquarie Island (Cocker et al., 1982); Xigaze ophiolite (Agrinier et al., 1988); Hole 504 B (Alt et al. 1986b).

Hydrogen isotopes

The hydrogen isotopic compositions of rocks from the Del Puerto ophiolite remnant (Table 1) are consistent with a submarine hydrothermal origin for their metamorphism. There is an inverse correlation of D/H with metamorphic grade, although it is not as strong as the corresponding inverse correlation of $\delta^{18}O$ with metamorphic grade. Zeolite- and pumpellyite-bearing metabasic rocks, which contain high modal proportions of interlayered chlorite/smectite, generally have the lowest δD (between −95 and −50‰ (SMOW)) values whereas epidote-bearing and higher grade metabasites, which contain discrete chlorite and (or) actinolite or actinolitic hornblende, are more D-enriched (with δD between −70 and −40‰). Aside from the epidote-free metabasites, which exhibit such marked enrichments in $\delta^{18}O$, metabasic volcanic and dike rocks from the Del Puerto ophiolite remnant are isotopically similar to those from the Troodos and Josephine ophiolites (Heaton and Sheppard, 1977; Harper et al, 1988), the Cuesta Ridge ophiolite remnant of the Coast Range ophiolite (Magaritz and Taylor, 1976), and modern oceanic crust (Stakes and O'Neil, 1982), as depicted in Figure 5. Hydrogen-isotope compositions of two chlorites, one epidote, and two amphiboles from Del Puerto dikes, high level intrusives, and cumulate rocks (Table 2) further support the whole rock data and demonstrate the deep penetration of seawater-derived hydrothermal fluids.

Table 4. Oxygen isotope composition of mafic rocks from ophiolites and oceanic crust.

	Coast range ophiolite							
	Del Puerto	Point Sal	Cuesta Ridge	Troodos	Oman	Macquarie Island	Xigase	504B
Extrusive Rocks	8.3–19.6	7.6–17.7	7.9–17.9	5.4–14.4	10.7–12.7	5.8–9.5	9.5–15.6	5.7–12.0
Dikes/Sills	5.7–8.4	5.1–10.6	8.8–9.9	2.8–13.6	2.9–11.3	4.0–8.8	4.4–8.9	3.6–6.5
Plutonic Rocks	5.8–7.6	5.6–6.8	6.1–8.3	5.3–6.5	2.4–6.4	3.6–5.7	4.8–11.5	
Source	1	This study	2	3, 4, 5	6, 7	8	9	10

(1) Schiffman et al. (1984); (2) Magaritz and Taylor (1976); (33 Rautenschlein et al. (1985); (4) Heaton and Sheppard (1977); (5) Schiffman and Smith (1988); (6) Gregory and Taylor (1981); (7) Stakes et al. (1984); (8) Cocker et al. (1982); (9) Agrinier et al., (1988); (10) Alt et al. (1986b).

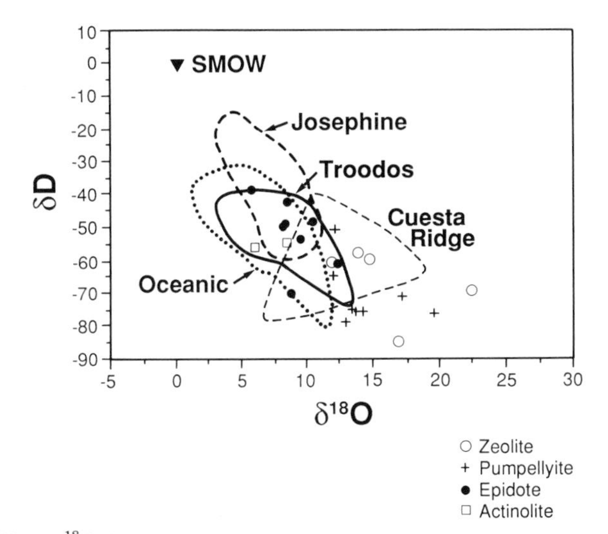

Figure 5. D/H vs $\delta^{18}O$ for rocks of the Del Puerto ophiolite remnant compared to modern oceanic crust (Stakes and O'Neil, 1982) and other ophiolites (Troodos ophiolite: Heaton and Sheppard, 1977; Josephine ophiolite: Harper et al., 1988; San Luis Obispo ophiolite (Cuesta Ridge ophiolite remnant): Magaritz and Taylor, 1976). Whole rock samples are plotted according to their metamorphic mineral grade.

Mineral separates from the plutonic rocks of the Coast Range ophiolite

Oxygen isotope compositions of mineral separates from Del Puerto plutonic rocks were determined to establish the systematics of fluid rock interaction within the deepest parts of the crustal section comprising the Coast Range ophiolite. Whole rock compositions of Del Puerto and Point Sal cumulate gabbroic rocks do not show ^{18}O depletions with respect to magmatic values that have been recorded in gabbros from the Oman (Gregory and Taylor, 1984; Stakes et al., 1984) and Macquarie Island (Cocker et al., 1982) ophiolites as well as from similar rocks dredged from the Mid Cayman Rise (Ito and Clayton, 1983). Furthermore, oxygen isotope analyses of mineral separates from Del Puerto plutonic rocks do not display ^{18}O-depleted plagioclase reported from these other localities (interpreted to reflect high temperature, low seawater/rock ratio interactions). Instead, eight plagioclase-clinopyroxene pairs from Del Puerto cumulate gabbros have $\delta^{18}O$ compositions either close to equilibrium magmatic values, or exhibit relative ^{18}O enrichment in the plagioclase (Figure 6a). Del Puerto gabbro plagioclase-cpx relations may indicate open system behavior (Gregory et al., 1989) at subsolidus temperatures exceeding the stability field of albite, such that plagioclase became ^{18}O-enriched, but not mineralogically transformed.

Quartz-plagioclase pairs from silicic intrusive rocks at Del Puerto yield similar results (Figure 6b): a hornblende quartz diorite and a plagiogranite

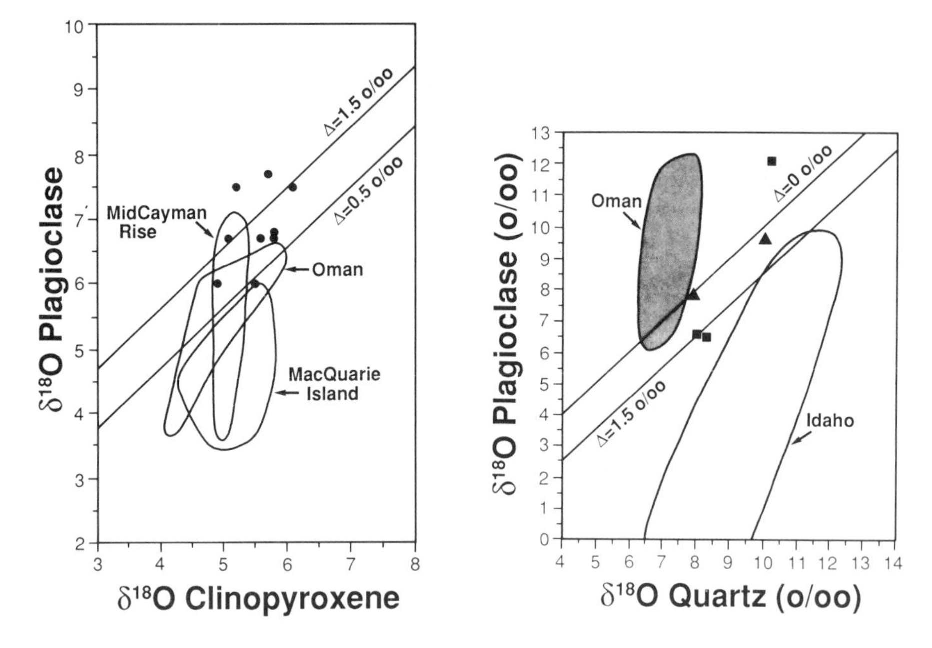

Figure 6. Delta-Delta plots for mineral separates from the Del Puerto remnant of the Coast Range ophiolite compared to data fields from other ophiolites and ocean crust. (a) clinopyroxene-plagioclase (Del Puerto data in filled circles) and (b) plagioclase-quartz (Del Puerto data in filled squares). Data sources: Del Puerto (this study, Schiffman et al., 1984); Oman (Gregory and Taylor, 1984; Stakes et al., 1984); Macquarie Island (Cocker et al., 1982); Mid-Cayman Rise (Ito and Clayton, 1983); Idaho Batholith (Gregory et al., 1989). The two filled triangular symbols in figure b are data for plagiogranites from the Cuesta Ridge ophiolite remnant (Magaritz and Taylor (1976).

retain essentially primary magmatic compositions and another plagiogranite contains relatively ^{18}O-enriched plagioclase. The absence of ^{18}O depletions in plagioclase from Del Puerto plutonic rocks indicates that they have not experienced significant high-temperature oxygen-isotope exchange with seawater, unless later, lower temperature interactions have overprinted this signature. Overprinting of high-temperature alteration by later, lower temperature interactions could account for the observed isotopic patterns, but there is no petrographic evidence of later overprinting (e.g., alteration of plagioclase to phases stable at low temperatures). Quartz-plagioclase pairs from Coast Range ophiolite remnants appear to show an open-system exchange trend essentially similar to those indicated in other hydrothermal systems (Figure 6b). This trend, however, is slightly offset (to higher quartz $\delta^{18}O$) from that of the Oman ophiolite (Gregory and Taylor, 1984; Stakes et al., 1984) although not by as great a degree as the Idaho meteoric hydrothermal system trend (Gregory et al., 1989).

Discussion

Hydrothermal Alteration in the Coast Range Ophiolite Compared to Other Ophiolites and Modern Oceanic Crust

The submarine hydrothermal metamorphism recorded in the Coast Range ophiolite remnants displays several mineralogic and geochemical characteristics that clearly differentiate the Coast Range ophiolite from other ophiolites and from all modern oceanic crust examined to date. Mineralogic data are summarized in Figure 2 and oxygen isotopic data are summarized in Figure 4–6 and Table 4.

1. *The Presence of "High-Temperature" Zeolite Mineral Paragenesis*
Coast Range ophiolite volcanic rocks contain heulandite, laumontite, and analcime. Other zeolites, such as phillipsite, natrolite, and clinoptilolite, that form at lower temperature and are common in the oceanic crust (Alt et al., 1986a, Gillis and Robinson, 1988) and in other ophiolites (Gillis and Robinson, 1985), are rare or absent.

2. *The Absence of Smectite*
A wide variety of smectites including saponite, nontronite, and beidellite are common products of low-temperature weathering of oceanic volcanic rocks (Alt and Honnorez, 1984; Gillis and Robinson, 1988). These smectites extend to considerable depth in the oceanic crust (e.g., within Hole 504B, Alt et al., 1986a) and within the volcanic sections of ophiolites such as Troodos (Gillis and Robinson, 1985; Schiffman and Smith, 1988) and Oman (Alabaster and Pearce, 1985). Discrete smectite has not been recognized within the Coast Range ophiolite. In both the Point Sal and Del Puerto ophiolite remnants, randomly and/or regularly interstratified chlorite/smectite occur in zeolite- and/or pumpellyite-bearing volcanic rocks; discrete chlorite occurs within epidote-bearing volcanic rocks. In the Troodos ophiolite, discrete chlorite and chlorite/smectite are virtually absent from volcanic rocks, except within regionally restricted upflow zones (Richards et al., 1989; Schiffman and Smith, 1988). In Hole 504B, these phases only occur at the base of the volcanic section, at a depth of approximately 1 km below the seafloor.

3. *The Presence of Pumpellyite*
Although pumpellyite is a mineral which commonly develops during the low grade metamorphism of basaltic rocks, it is relatively less common in ophiolites that have been unaffected by emplacement-related metamorphic overprinting, and even rarer in basaltic rocks dredged or drilled from the modern oceans (see review in Liou et al., 1987). In modern submarine environments, pumpellyite has been described from oceanic island sequences of volcanic/-volcaniclastic rocks that have undergone hydrothermal metamorphism in

response to intrusion of plutons (Staudigel and Schmincke, 1984; Rancon, 1985).

4. *Low P/T Metamorphic Facies Series Which Include Zeolite, Prehnite-Pumpellyite, and Greenschist Mineral Assemblages*
Studies of dredged and drilled samples suggest that metamorphic zonations within oceanic crust are apparently discontinuous, with markedly "stepped" thermal gradients between realms of seafloor weathering and greenschist alteration at deeper crustal levels (Mottl, 1983; Gillis and Robinson, 1988). The distribution of hydrothermal minerals within Hole 504B (Alt et al., 1986a), the only locality where *in situ* high temperature metamorphism within oceanic layer 2 has been well documented, supports these models. Moreover, metamorphic zonations described from most ophiolites (Coleman, 1984) include similar, discontinuous transitions between low-temperature alteration (characterized by smectites, celadonite, carbonates, and low-T zeolites) and greenschist facies metamorphism, without an intervening zone of prehnite-pumpellyite facies metamorphism. Prehnite-pumpellyite facies mineral assemblages have been described from the mafic extrusive rocks of many ophiolite complexes (e.g., the Josephine ophiolite (Harper et al., 1988, Zierenberg et al., 1988), the Pindos and East Liguria ophiolites (Spooner and Fyfe, 1973), and the Horokanai ophiolite (Ishizuka, 1985)). However, all these have undergone subsequent emplacement-related metamorphism which has variably obscured the original subseafloor metamorphic mineralogy (although which admittedly may include prehnite and pumpellyite as well). One notable exception is the volcanic section in the Salahi block of the Semail ophiolite, where the zeolite to prehnite-pumpellyite to greenschist facies transition has recently been documented (Pflumio, 1988). Similarly, young, metamorphosed volcanic/volcaniclastic sequences in the oceanic islands of La Palma (Staudigel and Schmincke, 1984) and Reunion (Rancon, 1985) contain zeolite to prehnite-pumpellyite to greenschist transitions akin to those of the Coast Range ophiolite remnants and the Salahi block of Semail.

5. *Albitization and Calcium Mobility Within the Extrusive Members*
In both the Del Puerto and Point Sal ophiolite remnants, the volcanic rocks retain no primary calcic plagioclase, and albitization is nearly ubiquitous even in zeolite and prehnite-pumpellyite facies rocks. In the modern oceanic crust and many ophiolites, albitization of primary plagioclase is generally restricted to greenschist facies rocks. The abundant veins of Ca-bearing minerals and local Ca-rich metadomains in the Del Puerto ophiolite remnant may be related to pervasive albitization (Na-Ca exchange) of calcic plagioclase in the volcanic rocks, which results in Ca-enrichment of the hydrothermal fluids (Evarts and Schiffman, 1983). Extreme Ca metasomatism in the form of epidotization of dike or extrusive rocks is becoming increasingly recognized within ophiolites (Richardson et al, 1987; Schiffman and Smith,

1988; Harper et al., 1988). Within the Coast Range ophiolite remnants, however, epidosites occur only sporadically in the deeper parts of the volcanic members and in the underlying dike-and-sill complexes. These epidosites generally contain some albite and actinolite, and mineralogically simple epidote + quartz rocks, common in the Troodos ophiolite (Schiffman and Smith, 1988), have not been described from Coast Range ophiolite remnants. Moreover, the $\delta^{18}O$ values of epidosites from the Coast Range ophiolite remnants (5.1 to 10.0‰; Tables 1 and 3) are significantly higher than those of the Troodos ophiolite (2.8 to 5.0‰; Schiffman et al., in press).

6. *Mg Enrichments Versus Depletions in Volcanic Rocks*
The hydrologic regimes of oceanic hydrothermal systems can be subdivided into zones of cold seawater recharge versus zones of hydrothermal discharge. Detailed study of the Solea graben in the northern Troodos ophiolite has shown that recharge in this system was regionally extensive within volcanic rocks, whereas discharge was restricted to discrete, near vertical, conduit-like zones which cross cut the dike complex and overlying volcanic rocks (Schiffman and Smith, 1988). Alteration of volcanic rocks in submarine recharge zones is characterized by Mg-enrichment (Mottl, 1983). In contrast, alteration in upflow regimes generally lacks Mg-enrichment, because the evolved hydrothermal fluid is Mg-depleted as a result of prior interaction with rock during the recharge cycle (Seyfried, 1987). The nearly ubiquitous Mg-enrichments recorded in volcanic and low-grade dike rocks from the Troodos ophiolite (Figure 3) are consistent with the hypothesis of regional scale downwelling of cold seawater within midoceanic spreading centers (Mottl, 1983). In the Point Sal ophiolite remnant, volcanic rocks exhibit a substantially different pattern of alteration, one characterized by domains of both Mg-enrichment and Mg-depletion. Thus, it would appear that fluid flow through the Point Sal crust was much more complex, and zones of recharge versus discharge are not as readily identifiable.

7. *Absence of Massive Sulfide Mineralization*
Massive sulfide deposition at modern, deep ocean spreading centers occurs where hydrothermal discharge is localized and focused, such as in black smoker fields (Haymon and MacDonald, 1985). Presumably similar hydrothermal processes occurred in Tethyan oceanic crust as well, because the massive sulfide deposits in the Oman and Troodos ophiolite are broadly analogous to modern examples (Koski, 1987). The absence of massive sulfide deposits within the Coast Range ophiolite remnants may reflect a fundamental difference in the nature of hydrothermal discharge in these systems. The Coast Range ophiolite hydrothermal systems lack evidence for intense high-temperature discharge such as the Fe-Cu-Zn sulfide deposits underlain by stockwork alteration zones typical of modern ocean spreading centers (Embley et al., 1988) and Tethyan ophiolites (Constantinou and Govett, 1973; Alabaster and Pearce, 1985). Instead, probable upflow zones are charac-

terized by low temperature Ca or Mn metasomatism as indicated by the laumontite "mounds", prehnite-pumpellyite veins, and Mn-oxide deposits of the Del Puerto ophiolite remnant.

8. *Extreme ^{18}O-Enrichments in Volcanic Rocks*

The mafic volcanic rocks of Coast Range ophiolite remnants exhibit the greatest ^{18}O enrichments recorded thus far in either ophiolites or modern oceanic crust (Figure 4). Basaltic rocks from the Del Puerto (Schiffman et al, 1984), Point Sal (this study), and Cuesta Ridge (Magaritz and Taylor, 1976) ophiolite remnants have maximum $\delta^{18}O$ values from 17.7 to 19.6‰ (Table 4), exceeding those from four other ophiolites (maximum $\delta^{18}O$ = 15.6‰ in Xiagze) and the modern oceans (maximum $\delta^{18}O$ = 12.0‰ in Hole 504B). The most ^{18}O-enriched basalts from these localities are those which have suffered the lowest grade of alteration (i.e., submarine weathering or zeolite facies metamorphism). The greater ^{18}O enrichments recorded in the Coast Range ophiolite rocks relative to rocks of similar grade and bulk composition elsewhere must reflect (1) interaction with isotopically "heavier" fluids, and/or (2) the more complete oxygen isotopic equilibration (or breakdown) of primary minerals such as Fe-Ti oxides, feldspars, and pyroxenes with the hydrothermal fluids. The latter mechanism is supported by the observation that primary plagioclases are ubiquitously recrystallized into albite (of high $\delta^{18}O$; see analyses in Table 2) in the Coast Range ophiolite volcanic rocks.

Schiffman et al. (1984) have previously favored isotopically heavy fluids to explain the ^{18}O enrichments found in the Del Puerto volcanic rocks. These fluids would be considerably ^{18}O enriched relative to seawater (0‰) and to modern day submarine hydrothermal vent fluids (−0.02 to +0.60‰; Merlivat et al., 1987; Shanks and Seyfried, 1987). Isotopically heavy (+2 to +4‰) fluids apparently reacted with diabase and gabbro to form epidosites in the Josephine ophiolite (Harper et al, 1988).

9. *Absence of ^{18}O-Depletions in Plutonic Rocks*

The origin of such isotopically heavy hydrothermal fluids is one major problem which remains. If the fluids were of seawater origin, as the hydrogen isotope data demonstrate, subsequent fluid-rock interactions can indeed enrich this seawater in ^{18}O, *but there must be a corresponding depletion of ^{18}O in plutonic rocks during this process*. In ophiolites, these ^{18}O-depleted rocks are typically found in underlying cumulate sequences (e.g. in the Semail ophiolite; Gregory and Taylor, 1984) which apparently underwent high temperature oxygen isotope exchange with seawater at low fluid/rock ratios. Schiffman et al. (1984) suggested that depleted gabbroic rocks in the Del Puerto ophiolite remnant were removed by faulting, inasmuch as the cumulate sequence in this remnant is incomplete (Evarts, 1977). However, the much more complete cumulate section of the Point Sal ophiolite remnant also lacks ^{18}O-depleted gabbroic rocks. Thus it is unlikely that the cumulate sections within the Coast Range ophiolite remnants have experienced the same kind of high temperature, low fluid/rock interactions that have been recorded in many other ophiolites and oceanic crust (Figure 6a). Oxygen isotopic data on mineral separates (Figures 6a and b) indicate open system exchange trends showing primarily feldspar ^{18}O enrichment in all samples from both volcanic and plutonic sections. Although the slightly heavy quartz values of the Coast Range ophiolite samples (Figure 6b) could indicate a slightly higher $\delta^{18}O$, intermediate composition source rock for these systems, this inference is not strongly supported by the plagioclase-pyroxene pair trends shown on Figure 6a. In any case, it appears that no ^{18}O depleted rocks are present in the preserved ophiolite.

Relationship Between Tectonic Setting and Hydrothermal Alteration: Fluid-Rock Interactions in a Rifted Arc Environment

It is clear that existing models for submarine metamorphism in deep, open-ocean spreading-centers (e.g., Mottl, 1983) cannot explain many of the mineralogic and isotopic characteristics of the Coast Range ophiolite. Most importantly, the pervasiveness of metamorphism, including zeolite and prehnite-pumpellyite mineral assemblages, and extent of ^{18}O-enrichments recorded in Coast Range ophiolite volcanic rocks, are features incompatible with modern day, open-ocean spreading-center environments.

The rifted island arc tectonic setting within which the Coast Range ophiolite formed is fundamentally distinct from that of deep ocean spreading centers in many ways, but foremost to this discussion are the following features: (1) the inherently slower spreading rates manifested by "non-sheeted" dike and sill complexes and anastomosing networks of relatively small, cross-cutting plutonic bodies (the Del Puerto remnant; Evarts, 1977); (2) the absence of well-developed, listric/normal fault systems characteristic of deep ocean extensional environments (Harper, 1985) and ophiolites formed in analogous slow-spreading environments (Varga and Moores, 1985); (3) the great compositional diversity of volcanic and plutonic rocks; and (4) the close spatial and temporal relationship of rift-related volcanism and arc-related volcaniclastic sedimentation, such as described from many Coast Range ophiolite remnants.

An ophiolite that develops slowly within a rifted arc may be buried beneath a volcaniclastic cover while magmatism is still active (such as in the Del Puerto ophiolite remnant). Magmatism responsible for the Coast Range ophiolite is interpreted to have persisted over a long period of time as indicated by the presence of two compositionally distinct volcanic members at Point Sal (Hopson and Frano, 1977) and by the presence of late-stage, cross-cutting plagiogranite intrusions within both the Point Sal and Del Puerto ophiolite remnants (Hopson et al., 1981; Evarts, 1977). Available ages for the Del Puerto rocks span a range of about 10 m.y. (Lanphere,

1971; Evarts, 1978; Hopson et al., 1981; Sharp and Evarts, 1982). Hydrothermal activity was probably episodic and related to distinct periods of magmatic intrusion. The final period of intrusive and associated hydrothermal activity was apparently extensive, because even the uppermost lavas in the Coast Range ophiolite remnants are pervasively metamorphosed. The pronounced metamorphic and isotopic effects in superjacent sedimentary rocks indicate that hydrothermal activity within the ophiolite remnants continued after the last pulse of volcanic extrusion.

The presence of the sediment cover, combined with the absence of a well developed, spreading-related, listric/normal fault system, probably exerted important controls on the nature of the thermal and chemical regime during hydrothermal metamorphism in the Coast Range ophiolite. The sediment cover would result in higher than ambient seafloor temperatures at the sediment/volcanic interface (<200°C, i.e. within the zeolite facies, by analogy with modern sediment-covered spreading centers such as in the Guyamas Basin (Einsele et al., 1980) or the northern Juan de Fuca Ridge (Davis et al., 1987) where sill-like bodies have intruded the sediment cap). Isotherms within volcanic rocks would be subhorizontal and evenly spaced because of the conductive nature of heat loss. The absence of listric/normal fault systems would limit vertical convective heat transport. Lateral fluid flow along beds of volcanic breccia would enhance development of subhorizontal as opposed to subvertical isotherms and promote pervasive metamorphism of the volcanic section beneath the sediment cover, a process believed to be responsible for the suppression of marine magmatic anomalies above modern sedimented spreading centers (Levi and Riddihough, 1986). The insulating effects of an impermeable sedimentary cap would result in longer-lived hydrothermal effects within the volcanic pile (Cathles, 1981). Cooling plutonic rocks would not experience high temperature hydrothermal activity (i.e., metamorphism or isotopic exchange), because seawater would have limited access to penetrate layer 3 in the absence of listric/normal fault-related ductile or brittle shear systems which apparently channelize hydrothermal fluids in deep ocean crust (Stakes, 1989).

The presence of sediment cover during metamorphism could have a significant effect on isotopic exchange within the volcanic pile. Inflow of seawater through heated biogenic and pelagic sediments could significantly enrich the fluids in ^{18}O (Fouillac and Javoy, 1988), and thus directly produce ^{18}O enrichments similar to those observed in the Coast Range ophiolite volcanic rocks. In such a situation, the ^{18}O source would be the sediment itself rather than deep, plutonic rocks as in other ophiolite settings (Gregory and Taylor, 1984). A single whole rock $\delta^{18}O$ value (19.3‰) for radiolarian chert overlying the Point Sal volcanic rocks implies considerable ^{18}O depletion from a typical pelagic sediment value of about 25–35‰ (Knauth and Epstein, 1975).

Although isotopically heavy sediments could produce fluid ^{18}O enrichment, mass balance considerations still require a large volume of such sediments in the recharge region of hydrothermal circulation. This causes several problems in the case of the Coast Range ophiolite. First, only thin interlayers of such pelagic sediment are found in the dominantly volcaniclastic sediment piles overlying Coast Range ophiolite remnants. Second, such fine-grained sediments often act as aquitards and might therefore provide only small quantities of ^{18}O enriched fluids. Third, large volumes of such sediments would have to be heated to temperatures at which significant isotopic exchange (enriching the fluid in ^{18}O) could occur. Such heating could not be produced by hydrothermal circulation itself since recharge areas would be regions of cooling. Heating resulting from burial metamorphism of a thick sediment section could, however, suffice if permeabilities remained high enough to permit the fluid flux necessary for hydrothermal circulation. Finally, as with any hypothesis requiring a large volume of isotopically depleted rock as the source for the enrichments observed in the Coast Range ophiolite, these depleted regions are not preserved or at least have not yet been identified.

Acknowledgements

The authors wish to thank a number of individuals for their contributions to our studies on hydrothermal processes in the Coast Range ophiolite: Cliff Hopson, for his gracious help in getting P. S. and A.E.W. started on the Point Sal work; Lori Bettison-Varga for her studies on Point Sal metavolcanic rocks; Randy Koski, Robert Zierenberg, Greg Harper, and Lori Bettison-Varga for stimulating discussions on submarine hydrothermal processes. The thorough and thoughtful reviews of G. Fruh-Green, J.G. Liou, Koski, and Zierenberg are greatly appreciated.

This work was supported by NSF grant EAR 83–16544 (to P.S. and A.E.W.)

References

Agrinier, P., Javoy, M. and Girardeau, J., 1988. Hydrothermal activity in a peculiar oceanic ridge: oxygen and hydrogen isotope evidence in the Xigaze Ophiolite (Tibet, China)., Chemical Geology, 71: 313–335.

Alabaster, T. and Pearce, J.A., 1985. The interrelationship between magmatic and ore-forming hydrothermal processes in the Oman Ophiolite., Economic Geology, 80: 1–16.

Alt, J.C. and Honnorez, J., 1984. Alteration of the upper oceanic crust, DSDP site 417: mineralogy and chemistry., Contributions to Mineralogy and Petrology, 87: 149–169.

Alt, J.C., Honnorez, J., Laverne, C. and Emmermann, R., 1986a., Hydrothermal alteration of a 1 km section through the upper oceanic crust, Deep Sea Drilling Project hole 504B: mineralogy, chemistry, and evolution of seawater-basalt interactions. Journal of Geophysical Research, 91: 10,309–10,335.

Alt, J.C., Muehlenbachs, K. and Honnorez, J., 1986b. An oxygen isotopic profile through the upper kilometer of the oceanic crust, DSDP Hole 504B., Earth and Planetary Science Letters, 80: 217–229.

Bailey, E.H., Blake, M.C. and Jones, D.L., 1970. On-land Mesozoic oceanic crust in California Coast Ranges., U.S. Geological Survey Prof. Paper, 700–C: C70–80.

Bailey, E.H. and Jones, D.L., 1973. Metamorphic facies indicated by vein minerals in basal beds of the Great Valley sequence, northern California. Journal of Research of the U.S. Geological Survey, 1: 383–385.

Bettison, L.A. and Schiffman, P., 1988. Compositional and structural variations of phyllosilicates from the Point Sal ophiolite, California., American Mineralogist, 73: 62–76.

Bird, D., Schiffman, P., Elders, W.A., Williams, A.E. and McDowell, S.D., 1984. Calc-silicate mineralization in active geothermal systems., Economic Geology, 79: 671–695.

Cathles, L.M., 1981. An analysis of the hydrothermal system responsible for the massive sulfide deposition in the Hokuroku basin of Japan., Economic Geology Monograph, 5: 439–487.

Cocker, J.D., Griffin, B.J. and Muehlenbachs, K., 1982. Oxygen and carbon isotope evidence for seawater-hydrothermal alteration of the Macquarie Island ophiolite., Earth and Planetary Science Letters, 61: 112–122.

Coleman, R.G., 1984. Preaccretion tectonics and metamorphism of ophiolites., Ofioliti, 9: 205–222.

Constantinou, G. and Govett, G.J.S., 1973. Geology, geochemistry, and genesis of Cyprus sulfide deposits., Economic Geology, 68: 843–858.

Davis, E.E., Goodfellow, W.D., Bornhold, B.D., Adshead, J., Blaise, B., Villinger, H. and Le Cheminant, G.M. 1987. Massive sulfide in a sedimented rift valley, northern Juan de fuca Ridge., Earth and Planetary Science. Letters, 86: 49–61.

Dickinson, W.R., Ojakangas, R.W. and Stewart, R.J., 1969. Burial metamorphism of the Late Mesozoic Great Valley Sequence, Cache Creek, California., Bulletin Geological Society of America, 80: 519–526.

Einsele, G., et al., 1980. Intrusion of basaltic sills into highly porous sediments, and resulting hydrothermal activity., Nature, 283: 441–445.

Embley, R.W., Jonasson, E.R., Perfit, M.R., Franklin, J.M., Tivey, M.A., Malahoff, A., Smith, M.F. and Francis. T.J.G., 1988. Submersible investigation of an extinct hydrothermal system on the Galapagos Ridge: sulfide mounds, stockwork zone, and differentiated lavas., Canadian Mineralogist, 26: 517–539.

Emerson, N.L., 1979. Lower Tithonian volcaniclastic rocks above the Llanada ophiolite, California: M. Sc. Thesis, Univ. of California, Santa Barbara, 65 pp.

Evarts, R.C., 1977. The geology and petrology of the Del Puerto ophiolite, Diablo Range, central California Coast Ranges. In: R.G. Coleman and W.P. Irwin (Eds), North American Ophiolites. Oregon Dept. Geology and Mineral Industries Bull., 95: 121–140.

Evarts, R.C., 1978, The Del Puerto ophiolite: Structural and petrologic evolution. Ph.D. dissertation, Stanford University, 409 pp.

Evarts, R.C. and Schiffman, P., 1983. Submarine hydrothermal metamorphism of the Del Puerto ophiolite. American Journal of Science, 283: 289–340.

Fouillac, A.M. and Javoy, M., 1988. Oxygen and hydrogen isotopes in the volcano-sedimentary complex of Huelva (Iberian Pyrite Belt): example of water circulation through a volcano-sedimentary sequence., Earth and Planetary Science Letters, 87: 473–484.

Gass, I.G. and Smewing, J.D., 1973. Intrusion, extrusion and metamorphism at constructive margins: evidence from the Troodos Massif, Cyprus., Nature, 242: 26–29.

Gillis, K.M. and Robinson, P.T., 1985. Low-temperature alteration of the extrusive sequence, Troodos Ophiolite, Cyprus., Canadian Mineralogist, 23: 431–441.

Gillis, K.M. and Robinson, P.T., 1988. Distribution of alteration zones in the upper oceanic crust., Geology, 16: 262–266.

Gregory, R.T., Criss, R.E. and Taylor, H.P., Jr., 1989. Oxygen isotope exchange kinetics of mineral pairs in closed and open systems: applications to problems of hydrothermal alteration of igneous rocks and Precambrian iron formations., Chemical Geology, 75: 1–42.

Gregory, R.T. and Taylor, H.P., Jr., 1984. An oxygen isotope profile in a section of Cretaceous oceanic crust, Samail ophiolite, Oman: Evidence for $\delta^{18}O$ buffering of the oceans by deep (> 5 km) seawater-hydrothermal circulation at mid-ocean ridges. Journal of Geophysical Research, 86: 2737–2755.

Grove, T.L. and Bence, A.E., 1977. Experimental study of pyroxene-liquid interaction in quartz-normative basalt 15597. Proceedings of the Lunar Science Conference, 8: 1549–1579.

Grove, T.L. and Bryan, W.B., 1983. Fractionation of pyroxene-phyric MORB at low pressure: an experimental study., Contributions to Mineralogy and Petrology, 84: 293–309.

Harper, G.D., 1985. Tectonics of slow-spreading mid-ocean ridges and consequences of a variable depth to the brittle/ductile transition., Tectonics, 4: 395–405.

Harper, G.D., Bowman, J.R. and Kuhns, R., 1988. A field, chemical, and stable isotope study of subseafloor metamorphism of the Josephine Ophiolite, California-Oregon., Journal of Geophysical Research, 93: 4625–4656.

Haymon, R.M. and Macdonald, K.C., 1985. The geology of deep-sea hot springs., American Scientist. 73: 441–449.

Heaton, T.H.E. and Sheppard, S.M.F., 1977. Hydrogen and oxygen isotope evidence for seawater-hydrothermal alteration and ore deposition, Troodos complex, Cyprus., Geological Society of London Special Publication 7: 42–57.

Hopson, C.A. and Frano, C.J., 1977. Igneous history of the Point Sal ophiolite, southern California. In: R.G. Coleman and W.P. Irwin (Eds), North American Ophiolites., Oregon Dept. Geology and Mineral Industries Bull., 95: 161–183.

Hopson, C.A., Frano, C.J., Pessagno, E.A., Jr. and Mattinson, J.M., 1975. Preliminary report and geologic guide to the Jurassic ophiolite near Point Sal, southern California coast. Geological Society of America 71st Annual Meeting Cordilleran Section, Field Trip Guide, 5, 36 pp.

Hopson, C.A., Mattinson, J.M. and Pessagno, E.A., 1981. Coast Range Ophiolite, western California. In: W.G. Ernst (Ed), The Geotectonic Development of California. Prentice-Hall, Englewood Cliffs, pp. 418–510.

Ingersoll, R.V., 1978. Petrofacies and petrologic evolution of the late Cretaceous fore-arc basin, northern and central California., Journal of Geology, 86: 335–352.

Ishizuka, H., 1985. Prograde metamorphism of the Horokanai Ophiolite in the Kamuikotan zone, Hokkaido, Japan., Journal of Petrology, 26: 391–417.

Ito, E. and Clayton, R.N., 1983. Submarine metamorphism of gabbros from the Mid-Cayman Rise: an oxygen isotope study., Geochimica et Cosmochimica Acta, 47: 535–546.

Knauth, L.P. and Epstein, S., 1975. Hydrogen and oxygen isotope ratios in silica from the JOIDES Deep Sea Drilling Project., Earth and Planetary Science Letters, 25: 1–10.

Koski, R.A., 1987. Sulfide deposits on the sea floor: geological model and resource perspectives based on studies in ophiolite sequences. In: P.G. Teleki et al., (Eds), Marine Minerals, Advances in Research and Resource Assessment. D Reidel, Dordrecht, pp. 301–306.

Lanphere, M.A., 1971. Age of the Mesozoic oceanic crust in the California Coast Ranges., Bulletin of the Geological Society of America. 82: 3209–3212.

Levi, S. and Riddihough, R., 1986. Why are marine magnetic anomalies suppressed over sedimented spreading centers?, Geology, 14: 651–654.

Liou, J.G., Maruyama, S. and Cho, M., 1987. Very low-grade metamorphism of volcanic and volcaniclastic rocks - mineral assemblages and mineral facies. In: M. Frey (Ed), Low Temperature Metamorphism. Blackie and Sons, Glasgow, pp. 59–113.

Lippard, S.J., Shelton, A.W. and Gass, I.G., 1986. The Ophiolite of northern Oman., Geological Society of London Memoir, 11, 178 pp.

Magaritz, M. and Taylor, H.P. Jr., 1976. Oxygen, hydrogen and carbon isotope studies of the Franciscan formation, Coast Ranges, California., Geochimica et Cosmochimica Acta, 40: 215–234.

Mason, R., 1978. Petrology of the metamorphic rocks. George Allen & Unwin, London.

Merlivat, L., Pineau, F. and Javoy, M., 1987. Hydrothermal vent waters at 13°N on the East Pacific Rise: isotopic and gas concentrations., Earth and Planetary Science Letters, 84; 100–108.

Moores, E.M., 1982. Origin and emplacement of ophiolites., Reviews of Geophysics and Space Physics, 20: 735–760.
Mottl, M.J., 1983. Metabasalts, axial hot springs, and the structure of hydrothermal systems at mid-ocean ridges., Geological Society of America Bulletin, 94: 161–180.
Nielsen, R.L. and Dungan, M.A., 1983. Low pressure mineral-melt equilibria in natural anhydrous mafic systems., Contributions to Mineralogy and Petrology 84: 310–326.
Oudin, E. and Constantinou, G., 1984. Black smoker chimney fragments in Cyprus sulfide deposits., Nature, 308: 349–353.
Pearce, J.A. and Cann, J.R., 1973. Tectonic setting of basic volcanic rocks determined using trace element analysis., Earth and Planetary Science Letters, 19: 290–300.
Pflumio, C., 1988. Histoire volcanique et hydrothermal du massif de Salahi: implications sur l'origine et l'evolution de l'ophiolite de Semail., Ph.D. dissertation, Ecole Nat. Sup. des Mines de Paris, 243 p. (in French).
Rancon, J.Ph., 1985. Hydrothermal history of Piton Des Neiges Volcano (Reunion Island, Indian Ocean)., Journal of Volcanology and Geothermal Research, 26: 297–315.
Rautenschlein, M., Jenner, G.A., Hertogen, J., Hofmann, A.W., Kerrich, R., Schmincke, H.-U. and White, W.M., 1985. Isotopic and trace element composition of volcanic glasses from the Akaki Canyon, Cyprus: implications for the origin of the Troodos ophiolite., Earth and Planetary Science Letters, 75: 369–383.
Raymon, L.A., 1973. Tesla-Ortigalita fault, Coast Range thrust fault, and Franciscan melange, northern Diablo Range, California., Geological Society of America Bulletin, 84: 3547–3562.
Richards, H.G., Cann, J.R., and Jensenius, J., 1989. Mineralogical zonation and metasomatism of the alteration pipes of Cyprus sulfide deposits., Economic Geology, 84: 91–115.
Richardson, C.J., Cann, J.R., Richards, H.G. and Cowan, J.G., 1987. Metal depleted root zones of the Troodos ore-forming hydrothermal systems, Cyprus., Earth and Planetary Science Letters, 84: 243–253.
Robertson, A.H.F., 1989. Paleogeography and tectonic setting of the Jurassic Coast Range ophiolite, central California: evidence from the extrusive rocks and the volcaniclastic sediment cover., Marine and Petroleum Geology, 6: 194–220.
Robinson, P.T., Melson, W.G., O'Hearn, T. and Schmincke, H.-U., 1983. Volcanic glass compositions of the Troodos ophiolite, Cyprus., Geology, 11: 400–404.
Schiffman, P., Bettison, L. and Williams, A., 1986. Hydrothermal metamorphism of the Point Sal remnant, California Coast Range ophiolite., Proceedings of the Fifth International Symposium on Water-Rock Interactions, 489–492.
Schiffman, P., Bettison, L.A. and Smith, B.M., 1990. Mineralogy and geochemistry of epidosites from the Solea graben, Troodos ophiolite, Cyprus. Troodos '87 Symposium Proceedings.
Schiffman, P. and Smith, B.M., 1988. Petrology and oxygen-isotope geochemistry of a fossil seawater hydrothermal system within the Solea graben, northern Troodos Ophiolite, Cyprus., Journal of Geophysical Research, 93: 4612–4624.
Schiffman, P., Williams, A.E. and Evarts, R.C., 1984. Oxygen isotope evidence for submarine hydrothermal alteration of the Del Puerto ophiolite, California., Earth and Planetary Science Letters, 70: 207–220.
Seyfried, W.E., Jr., 1987. Experimental and theoretical constraints on hydrothermal alteration processes at mid-ocean ridges., Annual Reviews of Earth and Planetary Sciences, 15: 317–335.
Shanks, W.C. III and Seyfried, W.E. Jr., 1987. Stable isotope studies of vent fluids and chimney minerals, Southern Juan de Fuca Ridge: sodium metasomatism and seawater sulfate reduction., Journal of Geophysical Research, 92: 11387–11399.
Sharp, W.D. and Evarts, R.C., 1982. New constraints on the environment of formation of the Coast Range ophiolite at Del Puerto Canyon, California., Geological Society of America, Abstract with Programs, 11: 127.
Shervais, J.W., 1990. Island arc and ocean crust ophiolites: contrasts in the petrology, geochemistry, and tectonic style of ophiolite assemblages in the California Coast Ranges. Troodos '87 Symposium.

Shervais, J.W. and Kimbrough, D.L., 1985. Geochemical evidence for the origin of the Coast Range ophiolite: a composite island arc-oceanic crust terrane in western California., Geology, 13: 35–38.

Spooner, E.T.C. and Fyfe, W.S., 1973. Sub-sea-floor metamorphism, heat and mass transfer., Contributions to Mineralogy and Petrology, 42: 287–304.

Stakes, D.S., 1989. Oxygen isotope and mineralogic profiles of oceanic layer 3: a comparison of the Semail ophiolite and ODP Site 735, SW Indian Ocean., EOS, 70: 1041.

Stakes, D.S. and O'Neil, J.R., 1982. Mineralogy and stable isotope geochemistry of hydrothermally altered oceanic rocks., Earth and Planetary Science Letters, 57: 285–304.

Stakes, D.S., Taylor, H.P. and Fisher, R.L., 1984. Oxygen isotope and geochemical characterization of hydrothermal alteration in ophiolite complexes and modern oceanic crust. In: I.G. Gass, S.I. Lippard, and A.W. Shelton (Eds), Ophiolites and Oceanic Lithosphere, Geological Society of London, Blackwell Scientific, Oxford, pp. 199–214.

Staudigel, H. and Schmincke, H.-U., 1984. The Pliocene Seamount series of La Palma/Canary Islands., Journal of Geophysical Research, 89: 11915–11215.

Suckecki, R.K. and Land, L.S., 1983. Isotopic geochemistry of burial metamorphosed volcanogenic sediments, Great Valley Sequence, Northern California., Geochimica et Cosmochimica Acta, 47: 1487–1499.

Varga, R.J. and Moores, E.M., 1985. Spreading structure of the Troodos ophiolite, Cyprus., Geology, 13: 846–850.

Zierenberg, R.A., Shanks, W.C. III, Seyfried, W.E. Jr., Koski, R.A. and Strickler, M.D., 1988. Mineralization, alteration, and hydrothermal metamorphism of the ophiolite-hosted Turner-Albright sulfide deposit, southwestern Oregon., Journal of Geophysical Research, 93, 4657–4674.

Sulfur Isotopic Profile Through the Troodos Ophiolite, Cyprus: Preliminary Results and Implications

JEFFREY C. ALT
The University of Michigan, Department of Geological Sciences, 1006 C.C. Little Building, Ann Arbor MI 48109–1063 USA

Abstract

The sulfide mineralogy, sulfur contents, and sulfur isotopic compositions were determined for a sect:on through the Troodos Ophiolite, Cyprus, as represented by Cyprus Crustal Study Project drillholes Cy-2a and Cy-4. The composite section includes volcanics altered at low temperature, a subsurface sulfide mineralization, hydrothermally altered sheeted dikes, and underlying gabbros. The volcanics average 400 ppm S, and contain sulfide with $\delta^{34}S$ = +4‰, consistent with a possible primary arc-like sulfur source. The sulfide mineralization in CY-2a ranges up to 45 wt‰ S, and pyrites have uniform $\delta^{34}S$ values of 6.4 ± 0.4‰. The hydrothermally altered rocks in the underlying feeder zone contain an average of 3.5‰ S, and pyrites have $\delta^{34}S$ values averaging +5.9 ± 0.9‰. The $\delta^{34}S$ values of the mineralization and underlying rocks are consistent with mixtures of 15–35‰ Cretaceous seawater sulfur +17‰) plus sulfur remobilized from elsewhere in the ophiolite. Gabbros of hole Cy-4 contain 10–1270 ppm S, having both gained and lost S through hydrothermal alteration. $\delta^{34}S$ values of sulfide generally range from +0.2 to +7.7‰. The lower values are consistent with primary MORB-like sulfur, and the higher values reflect local incorporation of seawater-derived sulfur during hydrothermal alteration, and the possible presence of arc-like primary sulfur (+4 to +5‰).

Overall, processes affecting sulfur in the Troodos ophiolite are generally similar to those occurring in oceanic crust. The $\delta^{34}S$ values for the Troodos section average around +5‰ and are higher than those for oceanic crust, however, which average around 0 to +1‰. The ^{34}S-enrichment of ophiolitic crust is attributed to incorporation of greater amounts of seawater-derived S in the ophiolite, plus the possibility of a component of ^{34}S-enriched arc-like primary sulfur.

Tj. Peters et al. (Eds), Ophiolite Genesis and Evolution of the Oceanic Lithosphere, 427–443.

Introduction

Knowledge of the distribution and isotopic composition of sulfur in oceanic and ophiolitic crust is important in order to understand the role that submarine hydrothermal systems play in the cycling of sulfur in the oceans through geologic time (Edmond et al., 1979). If altered oceanic crust is enriched in ^{34}S through exchange with seawater, then subduction of altered crust ($\pm$ sediment) could contribute to ^{34}S-enrichment of island arc volcanics (Ueda and Sakai, 1984; Woodhead et al., 1987) and lead to S-isotopic heterogeneities in the mantle (Chaussidon et al., 1987; Harmon et al., 1987). Sources of sulfur in volcanogenic massive sulfide deposits on the seafloor and in ophiolites can also be determined through examination of the geochemistry of sulfur in the deposits and associated crust.

The sulfide mineralogy and S-contents of massive sulfide deposits and feeder zones in Cyprus are well-documented (Clark, 1971; Constantinou and Govett, 1973; Herzig and Friedrich, 1987); Pyrites from the sulfide deposits have $\delta^{34}S$ values ranging from -1 to $+7.5‰$, but most fall between $+4$ and $+7‰$ (Clark, 1971; Hutchinson and Searle, 1971; Johnson, 1972; Jamieson and Lydon, 1987), significantly higher values than those from modern MOR deposits, which average around $+3‰$ (Arnold and Sheppard, 1981; Styrt et al., 1981; Bluth and Ohmoto, 1987; Woodruff and Shanks, 1987). Despite this data for sulfide deposits, very little is known about the geochemistry of sulfur in other portions of ophiolitic crust, and a profile of sulfur in a complete section through ophiolitic crust is lacking.

This paper presents new data on the sulfide mineralogy, S contents, and S-isotopic compositions of a section through the Troodos Ophiolite, Cyprus, as represented by drillcore from the Cyprus Crustal Study Project (CCSP). The data presented here are preliminary results of an ongoing detailed study of sulfur in ophiolitic crust. The objectives of this work are to understand seawater-crustal reactions involving sulfur, and to obtain a preliminary mass balance for S in ophiolitic crust. The question of the ^{34}S-enrichment of massive sulfides in Troodos compared to those of modern MOR deposits is also addressed, and the results from Troodos are compared to recent data for oceanic crust.

Sampling and Methods

Studied samples are from a section through the Troodos Ophiolite as represented by CCSP drillcores CY-2a and CY-4 (Figure 1). Hole CY-2a penetrates 689 m, and passes through volcanics, a stockwork sulfide mineralization, and into underlying intrusive dikes. The rocks comprise basaltic andesites, andesites, and dacites, which formed by seafloor spreading in an arc-type setting above a subduction zone (Robinson et al., 1983; Rautenschlein et al., 1985; Bednarz et al., 1987). The core can be divided into three

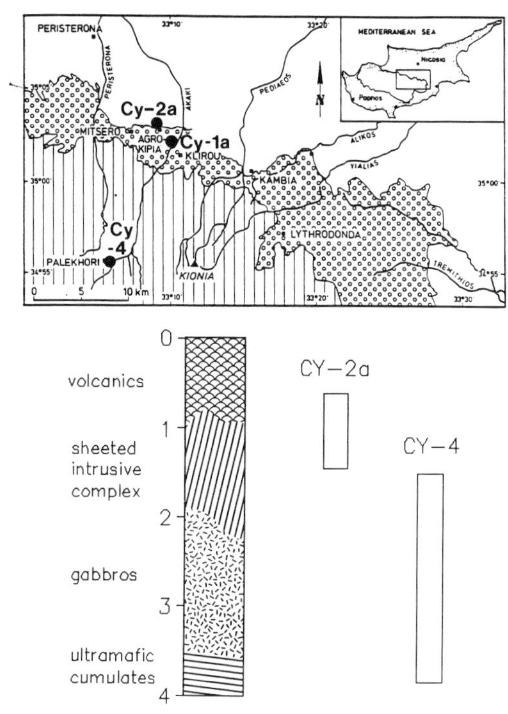

Figure 1. Top: Locations of CCSP drillholes in the Troodos ophiolite. Circles indicate volcanics, vertical lines indicate sheeted dikes. Bottom: Schematic ophiolite lithostratigraphy (depth in km) and approximate vertical locations of drillholes in the ophiolite.

different alteration zones (Figure 2A). The interval from 0 to 154 m comprises volcanics altered at low temperature (<150°C; Herzig and Friedrich, 1987; Sunkel et al., 1987), and which contain smectite, celadonite, zeolites, calcite, and Fe-oxyhydroxides (Cann et al., 1987). A highly mineralized zone occurs from 154 to 300m, where volcanics and dikes have been extensively altered to chlorite, quartz, illite and sulfide, with zones of massive sulfide, at temperatures of 250–300°C (Cann et al., 1987; Herzig and Friedrich, 1987; Sunkel et al., 1987). From 300 m to the bottom of the core, volcanics and dikes have been hydrothermally altered at temperatures of 230–300°C to assemblages of chlorite, albite, quartz, epidote, sulfide, sphene, and calcite (Herzig and Friedrich, 1987; Sunkel et al., 1987).

Hole CY-4 penetrated a total of 2263 m (Figure 1): 640 m of the lower sheeted dike complex, 1110 m of gabbros (plus sporadic dikes), and 513 m of ultramafic (websteritic) cumulates. The dikes range from basalt to dacite in composition, and are recrystallized to greenschist facies assemblages of mainly plagioclase plus actinolite, with minor quartz, chlorite, and epidote, plus later zeolites and calcite (Baragar et al., 1987). The gabbros contain secondary diopside, amphibole (hornblende, actinolite, and tremolite), plagi-

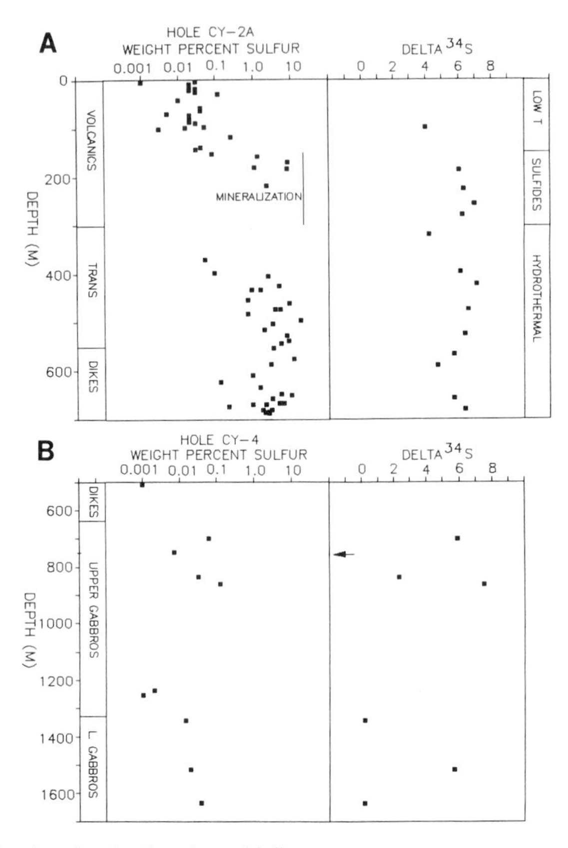

Figure 2. Sulfur data for the Troodos ophiolite.
(A) Sulfur contents and sulfur isotopic data for hole CY-2A. $\delta^{34}S$ for pyrite sulfur in ‰ CDT. Lithostratigraphy shown at left, and alteration zones at right. Data from Table 1, Auclair and Ludden (1987), Cann et al., 1987, and Jamieson and Lydon (1987). Vertical bar indicates location of sulfide mineralization with S contents of up to 45 wt%.
(B) Sulfur contents and sulfur isotopic data for hole CY-4. $\delta^{34}S$ in ‰ CDT. Same scale as in Figure 2, arrow indicates sample that plots off-scale. Data from Table 1. Lithostratigraphy shown at left.

oclase, epidote, quartz, sphene, prehnite, and later-stage, lower temperature zeolites and calcite (Barriga et al., 1985; Thy et al., 1989).

Sulfide mineralogy and sulfur contents have already been documented for the mineralized and hydrothermally altered zones in Hole CY-2a (Herzig and Friedrich, 1987; Jamieson and Lydon, 1987; Sunkel et al., 1987; Auclair and Ludden, 1987). Preliminary mineralogical and whole-rock sulfur content analyses were conducted on four samples of volcanics from the low-temperature alteration section of Hole CY-2a. Sulfur isotopic analyses were performed on sulfur extracted from these whole rocks, and on thirteen pyrite separates from the mineralized and hydrothermally altered portions of the core. Ten samples from Hole CY-4 were examined for sulfide mineralogy, and analyzed for whole-rock sulfur contents and sulfur isotopic compositions.

Samples were studied in polished thin sections, and pyrites for sulfur isotopic analysis were physically separated from massive sulfides, veins, and disseminations in the rocks. Monosulfide sulfur, soluble sulfate, and "pyrite" sulfur (actually pyrite + chalcopyrite) were extracted from whole rock powders for isotopic analysis using a stepwise extraction technique (see Alt et al., 1989, for details).

Results

Sulfide Petrology

The samples from the CY-2a volcanic section contain only traces of sulfide minerals. These typically consist of tiny, 1–6 μm grains of pyrite in interstitial areas with smectite, celadonite, or calcite. A coarse grained sample from 5.35 m contains chalcopyrite grains, up to 80 μm in size, in interstitial areas.

Dikes contain recrystallized igneous sulfides as globular grains and aggregates, 5–200 μm in size, consisting of pyrite ± chalcopyrite ± magnetite. Igneous pyrrhotite relics also occur as 5 μm inclusions in plagioclase. Porous grains of secondary pyrite, up to 200 μm in size, occur interstitially and replacing silicates, and secondary pyrite and chalcopyrite are present in and along veins filled with chlorite, actinolite, quartz, and epidote.

Gabbros ranging from unaltered (sample CY-4 1238.5) to slightly altered (with pyroxene partly replaced by amphibole and olivine partly altered to talc; sample CY-4 1637.22) contain recrystallized igneous sulfides as 1–20 μm inclusions of chalcopyrite and pyrite in plagioclase and pyroxene. Rare relics of igneous pyrrhotite also occur as inclusions in plagioclase. A 100 μm aggregate of pyrite, pyrrhotite and chalcopyrite interstitial to green amphibole in sample 1637.22 is another example of relict igneous sulfide. Secondary pyrite and chalcopyrite occur in these two gabbros as 10–40 μm grains in green and brown amphibole replacing pyroxene.

Other gabbro samples range from partly altered (samples CY-4 1345.65, 1517.67), where pyroxene and olivine are completely replaced by amphibole and talc, respectively, but plagioclase is mostly visibly unaltered, to samples that are completely recrystallized (all other samples). Sulfides in these samples consist primarily of pyrite and chalcopyrite, with lesser pyrrhotite and traces of other phases. Small (5–20 μm diameter) monomineralic grains of pyrite, chalcopyrite, pyrrhotite, and pyrite + chalcopyrite intergrowths in altered plagioclase and pyroxene, and in interstitial areas of these samples are recrystallized igneous sulfides. Clearly secondary pyrite replaces silicates in irregular to euhedral patches, generally <100 μm in size, but ranging up to 1 mm areas with chalcopyrite and pyrrhotite inclusions in sample CY-4 861.62. Secondary pyrite, chalcopyrite, pyrrhotite, and pyrite + chalcopyrite intergrowths ± magnetite are common as 1–10 μm grains in altered pyroxene, in green amphibole, and in talc replacing olivine. Pyrite and chalcopyrite also commonly occur with Ca-metasomatic assemblages: large pyrite grains are associated with epidote in samples CY-4 861.62 and 702.15, and disseminated pyrite and chalcopyrite occur in prehnite and zeolite in samples 861.62 and 1255.22.

Pyrite and chalcopyrite are common as 5–60 μm grains associated with late, low-temperature smectite and calcite replacing olivine and in interstitial areas. Low-temperature oxidation is indicated by partial replacement of pyrrhotite by pyrite and marcasite in samples CY-4 750.15 and 861.62, and rims of bornite and covellite chalcopyrite in sample 861.62. The replacement of Fe-Ti oxides by hematite and the presence of Fe- oxyhydroxides in sample 750.17 are also consistent with low temperature oxidation. Although both these samples exhibit low-temperature oxidation effects, sulfides in sample CY-4 861.62 are dominated by large hydrothermal sulfides described above, whereas CY-4 750.17 is characterized mainly by low-temperature sulfide assemblages.

Sulfur Contents

The analyzed volcanics from Hole CY-2a have relatively low sulfur contents (Table 1), ranging from 10 to 150 ppm and averaging 60 ppm total S. Sulfur is predominantly as pyrite-S, with only traces of monosulfide-S. Two samples contain small amounts of sulfate sulfur, comprising 30–40% of the total sulfur present.

Two diabase dikes from Hole CY-4 contain variable amounts of S, from 40 to 330 ppm (Table 1). Both have similar small amounts of sulfate sulfur (10–30 ppm). The gabbros have an even wider range of sulfur contents, from 10 to 1270 ppm, averaging 350 ppm (Table 1). Excluding one unusually high-S sample (Cy-4 861.62), however, sulfur contents of gabbros are less variable and average 220 ppm. As in the volcanic section, sulfur is present in the gabbros mostly as pyrite-S, with only traces of monosulfide-S. Sulfate sulfur contents are generally low, 10–40 ppm, comprising 0 to 50% of the total sulfur in the gabbros.

Sulfur Isotopes

Only one sample of the volcanics from Hole CY-2a yielded sufficient sulfur to analyze isotopically (Table 1). Pyrite S in sample 98.30 has $\delta^{34}S = +4.0‰$. Pyrites from the mineralized zone of Hole CY-2a have uniform $\delta^{34}S$ values, averaging $+6.4 \pm 0.4‰$ (Table 1). In the hydrothermally altered zone (>300 m depths), pyrites have $\delta^{34}S$ values generally similar to those in the mineralized zone, but range to lower $\delta^{34}S$ values (+4.2‰; Fig. 2A, Table 1). No consistent differences in $\delta^{34}S$ were observed among pyrites separated from massive sulfide zones, veins, or disseminated in the rocks.

Table 1. Sulfur data for Troodos Drillcore.

Sample	Rock/Vein Type + Alteration	Whole-Rock Sulfur Content (ppm)				$\delta^{34}S$	
		Mono-Sulfide	Pyrite	SO_4	Total	Whole-Rock Pyrite	Vein Mineral Pyrite
CY-2							
5.35	Vol, sl	tr	10	nd	10		
69.99	Vol, sl	tr	30	20	50		
98.30	Vol, sl	tr	150	nd	150	4.0	
101.38	Vol, sl	tr	20	10	30		
186.25	V						6.1
223.10	V						6.3
255.10	M						6.9
278.50	D						6.2
322.86	V						4.2
397.21	V						6.1
425.10	V						7.2
475.73	V						6.5
526.45	V						6.3
567.86	D						5.7
591.85	V						4.7
658.80	V						5.7
684.6	V						6.4
CY-4							
125.4	V						7.7
509.20	DI, ext	tr	10	30	40		
702.15	GB, ext	20	610	nd	630	5.9	
750.15	GB, ext	10	60	30	100	−16.3	
837.07	DI, sl	10	310	10	330	2.3	
861.62	GB, ext	10	1240	20	1270	7.5	
1238.5	GB, fresh	10	10	20	40		
1255.22	GB, ext	tr	10	nd	10		
1345.65	GB, part	tr	140	10	150	0.2	
1517.67	GB, part	tr	190		190	5.7	
1637.22	GB, sl	10	360	40	410	0.2	

Sample number corresponds to depth in meters in drillcores. Blanks, not analyzed; nd, none detected; tr, trace; $\delta^{34}S$ in ‰ CDT ± 0.1% one-sigma; VOL, volcanic; DI, diabase; GB, gabbro; sl, slightly altered; part, partly altered; ext, 100% altered (see text); V, vein; D, disseminated; M, massive.

Only one of the two dike samples from Hole CY-4 contained enough sulfur to obtain a sulfur isotopic analysis. Pyrite sulfur in sample 837.07 has $\delta^{34}S$ = + 2.3‰. Six of the eight gabbros studied contained sufficient pyrite sulfur for analysis of isotope ratios. Most $\delta^{34}S$ values are positive, ranging from + 0.2 to +7.5‰ (Table 1, Figure 2A). Pyrite sulfur in sample 750.15, however, has a strongly negative $\delta^{34}S$ value of −16.3‰. None of the samples analyzed from Cy-2a or CY-4 yielded enough sulfate or monosulfide sulfur for isotopic analysis.

Discussion

Primary Sulfur Contents

In order to discuss sulfur fluxes in ophiolitic crust, a knowledge of the primary sulfur contents of the rocks is required. The igneous sulfur contents of Troodos volcanics can be estimated from experimental work and from measurements of sulfur contents of volcanic glasses from Troodos and modern oceanic arcs. One of the main factors controlling solubility of sulfur in silicate melts is FeO content, with sulfur solubility increasing significantly as FeO content increases (Haughton et al., 1974). Basaltic to basaltic andesite volcanic glasses from Troodos contain about 7 to 11 weight per cent FeO^T (Robinson et al., 1983). At these iron contents, S solubility in anhydrous basaltic melts at 1 atmosphere total pressure ranges from 500 to 700 ppm S (Haughton et al., 1974). Consistent with this estimate are measured sulfur contents of unaltered basaltic glasses from Troodos, which contain 500 to 1000 ppm S (Muenow et al., 1990). Although the errors are quite large, extrapolation of higher pressure data for the solubility of S in hydrous dacite melts to P = 0.2 kb, corresponding to an eruption water depth of 2 km, gives saturated S-contents of about 250 ppm (for 11–12.5 wt% FeO^T at f_{O2} = QFM buffer; Carroll and Rutherford, 1985). Dacitic melts with lower FeO^T contents would presumably have correspondingly lower S contents. These data suggest a range of primary S contents for Troodos magmas, assuming saturation of the melts with sulfide. The presence of relict igneous sulfides included in silicates in many gabbro samples provides evidence for saturation of parent melts with sulfide at least at some point in their history. These data thus provide estimates of primary S contents of Troodos magmas ranging from around 200 up to 1000 ppm.

Besides the estimates given above, data from modern oceanic arcs suggest the possibility of relatively low primary sulfur contents for some of the ophiolitic section. For comparison, the sulfur contents of basaltic to andesitic submarine volcanic glasses from modern oceanic arcs are low, averaging around 100 ppm S (Garcia et al., 1979; Muenow et al., 1980; Alt and Shanks, 1989). Glass inclusions in phenocrysts in these lavas have similarly low S contents and have been interpreted to indicate a low-S mantle source for oceanic arc lavas (Ueda and Sakai, 1984; Garcia et al., 1979).

The chemistry of at least some of the Troodos volcanics is consistent with derivation from a sulfur-depleted mantle source. Troodos volcanics have been divided into a lower, arc-tholeiitic suite of basaltic andesites to dacites, and an upper group of more mafic lavas with boninitic affinities (high MgO, low Ti; Robinson et al., 1983; Bednarz et al., 1987). These two magmatic suites have also been identified in the dike complex and gabbros (Thyet al., 1989; Baragar et al., 1987). The change in rock type from arc tholeiitic to boninitic indicates a corresponding change in mantle source, from a MORB-type mantle that was metasomatized by a component from the subducting plate to a more depleted mantle from which basaltic melts had already been extracted. Such a refractory mantle would also be depleted in S, which behaves incompatibly during melting and would have been mostly extracted during prior melting (Hamlyn et al., 1985). Second-stage melting of depleted mantle could thus result in magnesian, low-Ti lavas of boninitic affinities with low S contents, similar to those of boninites (around 50 ppm S, Hamlyn et al., 1985). Sulfur contents of the least altered samples from Hole CY-4 range from 40 to 410 ppm, consistent with a range in primary sulfur contents down to low sulfur totals (40 ppm in sample CY-4 1238.5, Table 1). More data on the sulfur contents of unaltered rocks from Troodos are needed, however, in order to confirm that rocks with low primary S contents are present.

Primary Sulfur Isotopic Compositions

MORB glasses contain an average of 800 ppm S with $\delta^{34}S = +0.1 \pm 0.5‰$ (Sakai et al., 1984). In contrast, oceanic arc volcanics have much lower sulfur contents and generally higher $\delta^{34}S$ values, averaging 50–100 ppm and +4 to +5‰, respectively (Ueda and Sakai, 1984; Alt and Shanks, 1989). These differences could be due in part to degassing of H_2S from oxidized, SO_4-rich magmas, but may also reflect a low-S mantle source that is enriched in ^{34}S relative to MORB (Ueda and Sakai, 1984; Alt and Shanks, 1989). The supra-subduction zone origin of the Troodos ophiolite thus suggests the possibility that primary sulfur in some of the rocks had high $\delta^{34}S$ values similar to modern oceanic arcs (around +4‰). If the $\delta^{34}S$ of sulfur in slightly to partly altered rocks from hole CY-4 (+0.2 to +5.7‰, Table 1) reflects primary compositions, then sulfur ranged from MORB-like to arc-like in isotopic composition.

Volcanics Altered at Low Temperature in Hole CY-2a

The pyrite and chalcopyrite in samples from the CY-2a volcanics are typical of sulfides in oceanic volcanics altered at low temperatures (0–100°C, Alt et al., 1989). The analyzed volcanics average 40 ppm S (Table 1), significantly less than the mean of other reported analyses of CY-2a volcanics (mean = 400 ppm, range = <100 to 2600 ppm; Auclair and Ludden, 1987; Cann et al., 1987). The reason for the differences in S-contents is uncertain. The means of other analyses by coulometry (Cann et al., 1987) and x-ray fluorescence (Auclair and Ludden, 1987) agree, with both averaging 400 ppm S. The yields for the HCl-$CrCl_2$ extraction technique used in the present study average better than 95% of the yields from coulometric analyses on the same samples (Alt et al., 1989), and the extractions are more accurate than the other techniques at very low sulfur contents. The discrepancy between the different data sets can perhaps best be attributed to sampling bias. Low sulfur contents in the various data sets could reflect low primary values, degassing during eruption (Moore and Fabbi, 1971), or oxidation of sulfides and loss of sulfur during low temperature alteration. The rocks contain common Fe oxyhydroxides, and are oxidized relative to igneous values: Fe^{3+}/Fe^{T} of CY-2a volcanics averages 0.60 versus 0.14 for unaltered volcanic glass (Rautenschlein et al., 1985; Bednarz et al., 1987). Similar oxidation of sulfides and loss of S to seawater occurs during seafloor weathering of oceanic volcanics (Alt et al., 1989).

In seafloor rocks affected by low temperature alteration, sulfide sulfur, and particularly pyrite sulfur, has $\delta^{34}S$ values similar to or significantly lower than igneous values. Sulfides in MORBs altered at low temperature generally had $\delta^{34}S$ values that range from −20 to +2‰ (Krouse et al., 1977; Hubberten, 1983; Field et al., 1984): for example, see the Hole 504B volcanic section in Figure 3A. The negative $\delta^{34}S$ values are interpreted to be caused by isotopic fractionation during partial oxidation of igneous sulfides by seawater (Andrews, 1979). This produces sulfur species of intermediate oxidation state, which are unstable in solution and spontaneously disproportionate into oxidized sulfate and reduced sulfide components. Fractionation of sulfur isotopes between oxidized and reduced species leads to loss of ^{34}S-rich sulfate to solution, whereas ^{34}S-depleted sulfide can form secondary pyrite with low $\delta^{34}S$ values. Assuming that similar processes occurred during alteration of Troodos volcanics, the $\delta^{34}S$ of sulfide in the low-temperature section of hole CY-2a (+4‰, Table 1) is less than or equal to primary values, suggesting relatively high $\delta^{34}S$ for igneous sulfur in the rocks, i.e., around +4‰. This $\delta^{34}S$ value is consistent with a possible arc-like primary sulfur isotopic composition. More samples from the volcanic section are needed to confirm this result, however.

Sulfide Mineralization and Hydrothermally Altered Zones of CY-2a

The sulfide mineralogy of the CY-2a mineralized and hydrothermally altered zones (mainly pyrite, pyrrhotite, and chalcopyrite, with minor sphalerite, and trace galena; Herzig and Friedrich, 1987) is generally similar to that of sulfide deposits and metabasalts from the seafloor (e.g., Styrt et al., 1981; Alt et al., 1989). The CY-2a mineralization is similar in geological setting to the stockwork-like sulfide mineralization in DSDP Hole 504B: both formed below the seafloor, where upwelling, hot (250–350°C) hydrothermal fluids mixed with cooler seawater circulating in overlying rocks (Herzig and Friedrich, 1987; Alt et al., 1986; 1989). The main differences lie in the extent of mineralization: in hole CY-2a a 150 m thick section averages greater than 30 wt% S and ranges up to 45% S in the massive sulfide ore zones (Adamides, 1987), whereas in DSDP hole 504B a 20 m interval only ranges up to 3% S, with a mean of 0.4% (Alt et al., 1989). Beneath the CY-2a mineralization, rocks in the hydrothermally altered zone average 3.5% S (Fig. 2A).

The $\delta^{34}S$ values of pyrites from the mineralized and hydrothermally altered zones of hole CY-2a average +6‰ (Table 1). This is similar to other reported values for Cyprus pyrite mineralization, which range from −1.1 to + 7.5‰, with most values around +4 to +7‰ (Clark, 1971; Hutchinson and Searle, 1971; Johnson, 1971; Jamieson and Lydon, 1987). The coexistence of pyrite and pyrrhotite in the mineralized and hydrothermally altered zones suggests that conditions were sufficiently reducing that essentially all sulfur in circulating fluids was present as H_2S (or HS^-), and that the $\delta^{34}S$ values of sulfide minerals closely reflect the $\delta^{34}S$ of total sulfur in the hydrothermal fluids (Ohmoto and Rye, 1979). The isotopic data can thus be considered to represent a mixture of basaltic sulfide (MORB = + 0.1‰, Sakai et al., 1984) and sulfide derived from reduction of seawater sulfate (Cretaceous seawater SO_4 = + 17‰; Claypool et al., 1972; versus + 21‰ for present day seawater; Rees et al., 1978). Assuming (1) a basaltic component with $\delta^{34}S$ = + 0.1‰ and a seawater component of + 17‰, (2) that seawater sulfate was essentially totally reduced to sulfide (Janecky and Shanks, 1988; Alt et al., 1989), and (3) ignoring slight fractionation of sulfur isotopes between pyrite and fluid (about 1‰, Ohmoto and Rye, 1979), yields 27 to 40% (mean = 35%) seawater sulfur contribution to CY-2a pyrites, versus 5–30% (mean = 15%) for mineralization in DSDP hole 504B and modern MOR sulfide deposits. The Troodos ophiolite formed in an arc-related setting, however, and primary MORB-like $\delta^{34}S$ values may not be appropriate. If it is assumed that the proportions of seawater-derived and crustal sulfur in the CY-2a pyrites are similar to that in modern MOR deposits (15% seawater), then the calculated basaltic component has $\delta^{34}S$ =+ 4.1‰, identical to that of sulfur in modern oceanic arc volcanics (+ 4 to + 5‰: Ueda and Sakai, 1984; Alt and Shanks, 1989). Thus, the data are consistent with formation of sulfides in hole CY-2a from mixtures of remobilized crustal sulfur and reduced seawater sulfur. The proportion of seawater sulfur was either similar to or greater than that occurring in present-day seafloor sulfide deposits. The S-enrichments of CY-2a reflect addition of seawater sulfur to the crust, but also require transfer of crustal sulfur from elsewhere in the ophiolitic crust.

Gabbros and Dikes of Hole CY-4

The dikes in hole CY-4 were recrystallized under greenschist facies conditions. The sulfide minerals present (pyrite, chalcopyrite, and trace pyrrhotite) are stable at these conditions, and are identical to those occurring in altered sheeted dikes from oceanic crust (Alt et al., 1989).

Sulfide minerals are associated with all stages of silicate mineral paragenesis in the gabbro section. The replacement of olivine by talc + magnetite + pyrite + chalcopyrite may be related to early, high temperature (500–600°C) hydration reactions at very low water/rock ratios (Alt and Anderson, 1990). The small blebs of secondary magnetite in altered pyroxenes of CY-4 gabbros may be caused by similar oxidation reactions. The other occurrences of pyrite, chalcopyrite and pyrrhotite are consistent with the generally decreasing temperature hydrothermal regime (Barriga et al., 1985; Thy et al., 1990):

pyrite, pyrrhotite, and chalcopyrite with green amphibole; the same phases with epidote and Ca-silicates; and pyrite ± chalcopyrite with late calcite and smectite. Fe oxides are associated with smectite in some samples, whereas sulfides occur with smectite and calcite in others, indicating either that late, low temperature alteration varied or progressed from more reducing to more oxidizing.

No consistent trend in gain or loss of sulfur with alteration is apparent for the gabbros (Table 1). The least altered samples range from 40 to 410 ppm S, whilst the completely recrystallized samples exhibit an even wider range, from 10 to 1270 ppm, indicating both additions and losses of S due to hydrothermal metamorphism.

All of the $\delta^{34}S$ values for CY-4 are positive except for sample 750.15, which has a value of −16.3‰. The negative value reflects isotopic fractionation produced during partial oxidation of sulfides at low temperatures (<100°C), and subsequent formation of secondary pyrite (Andrews, 1979). Such negative values are typical of pyrites in basalts altered at low temperature (Krouse et al., 1977; Hubberten, 1983; Field et al., 1984; see previous discussion of CY-2A volcanics). Sample CY-4 750.15 contains abundant evidence of low-temperature oxidation (Fe-oxyhydroxides, hematite).

As in the CY-2a mineralized zone, the coexistence of pyrrhotite and pyrite in CY-4 rocks indicates that essentially all the sulfur in hydrothermal fluids was present as sulfide, and that the isotopic composition of sulfide minerals closely reflects that of sulfur in the hydrothermal fluids. No clear trend of $\delta^{34}S$ with sulfur content exists, but the highest $d^{34}S$ value (+ 7.5‰) occurs in the sample with the highest S content (1270 ppm, sample 861.62, Table 1). This sample is entirely recrystallized and has large secondary sulfides associated with epidote, indicating a hydrothermal origin of sulfides and that hydrothermal sulfur had high $\delta^{34}S$ values. As discussed in the preceding section, such high $\delta^{34}S$ values are consistent with mixtures of crustal and seawater-derived sulfur. The $\delta^{34}S$ of the least altered samples from CY-4 (+ 0.2 to + 2.3‰, Table 1) suggest a more MORB- like (+ 0.1‰) than arc-like (+ 4‰) primary S isotopic composition. Assuming mixtures of Cretaceous seawater (+ 17‰) and MORB-type primary S, the $\delta^{34}S$ of sulfur in CY-4 samples indicate that from less than 1 up to 44% of the S in the rocks is derived from seawater. If a ^{34}S-enriched arc-type sulfur composition is used for the crustal component (+4‰), then the proportion of seawater sulfur is less than 27%, and the low $\delta^{34}S$ values (<4‰) cannot be accounted for in this manner. These low values require MORB sulfur or that low-$\delta^{34}S$ sulfides are also present, e.g., from late, low temperature alteration.

Oceanic gabbros from the Cayman Rise and from ODP hole 735B have lost S and exhibit locally increased $\delta^{34}S$ values due to seawater interaction only in the upper 200–300 m of these sections (Fig. 3B; Alt and Anderson, 1990 and unpublished data). In contrast, sulfur isotopic data for hole CY-4 indicate incorporation of seawater-derived sulfur at depths of 877 m into the gabbro section (sample 1517.67, Table 1 and Figure 2B). Deeper penetration

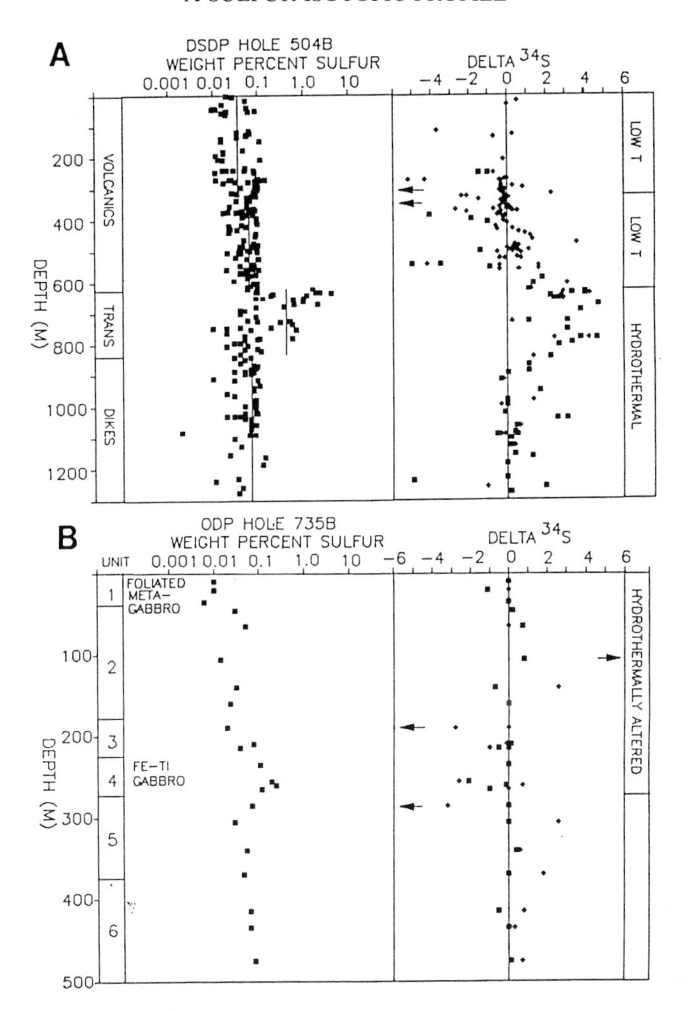

Figure 3. Sulfur data for oceanic crust.
(A) Whole-rock data for DSDP Hole 504B (from Alt et al., 1989). Squares, pyrite sulfur; diamonds, monosulfide sulfur. $\delta^{34}S$ in ‰ CDT; arrows indicate samples plotting off-scale; vertical line indicates MORB $\delta^{34}S$ value. Vertical lines in wt% S plot indicate mean S contents for the different alteration zones, which are listed at right. Lithostratigraphy given at left.
(B) Whole-rock data for gabbroic section of ODP Hole 735B (from Alt and Anderson, 1989). Lithologic unit divisions shown at left: the section is mostly olivine gabbros, but significantly different units are labelled. The foliated metagabbros of Unit 1 are the most recrystallized, but hydrothermally altered rocks occur in the upper 200 m. Symbols as in (A).

and incorporation of greater amounts of seawater-derived sulfur into Troodos crust are consistent with higher water/rock ratios and generally greater extents of recrystallization of the ophiolitic rocks compared to the drillcores from oceanic crust.

Table 2. A sulfur mass-balance for Troodos ophiolitic crust.

Rock type	S Content* (ppm)	Density**	S Content of 1 km^2 Section ($\times 10^{12}$ g S)	S loss# 1 km^2 Section (10^{12} g S)
1 km volcanics	400	2.34	0.94	0
1 km dikes	190	2.56	0.48	0.38
3 km gabbros	300	2.9	2.61	0.87
Total				1.25

*From Figure 2 and Table 1. Gabbro S content assumed intermediate between averages with and without sample 861.62. **From Smith and Vine, 1987. #Assuming 400 ppm primary S contents.

A Mass Balance for Sulfur in Troodos Ophiolitic Crust

A preliminary mass-balance for sulfur in Troodos ophiolitic crust is given in Table 2. Assuming 400 ppm S for the average primary sulfur content, the volcanic section, on average, has neither lost nor gained sulfur, whereas the dikes and gabbros have lost S overall. The sulfide mineralization in CY-2a, as well as the hydrothermally altered rocks in the lower portion of CY-2a represent significant sinks for sulfur in Troodos crust. From data given in Adamides (1987), it is estimated that the Agrokipia B sulfide deposit and associated alteration halo (which was penetrated by hole CY-2a) contain 30.7×10^{12}g S. If 15–35% of this is derived from Cretaceous seawater as indicated by the isotope mass-balance, then 20–26 $\times 10^{12}$g S must be transferred from elsewhere in Troodos crust. Given the data in Table 2, the excess S in Agrokipia B can be derived from about 15–20 km^2 of altered crust (with sulfur distribution as in Table 2, plus the seawater component). This estimate can be modified upward or downward, depending on assumptions about primary S contents of the crust, and requires refining by additional analyses of Troodos rocks. The net result, however, is redistribution of sulfur within the crust and incorporation of seawater-derived sulfur into the crust. The latter leads to a positive shift in the $\delta^{34}S$ of the crust.

Acknowledgements

This work was supported by NSF-EAR 8904788. The author thanks Janet Pariso and Peter Herzig for providing samples for this preliminary study, and Pat Shanks at the USGS and Jack Liu at the Illinois Geological Survey for use of their laboratories for the isotopic analyses. Helpful reviews were provided by Marc Chaussidon and A.E. Fallick.

References

Adamides, N.G., 1987. Diverse modes of occurrence of Cyprus sulphide deposits and comparison with recent analogues. In: P.T. Robinson, I.L. Gibson, and A. Panayiotou (Editors), Cyprus Crustal Study Project: Initial Report, Holes Cy-2 and 2a. Geol. Surv. Canada paper 85–29: 153–168.

Alt, J.C. and W.C. Shanks, 1989. Sources and cycling of sulfur in subduction zones: The Mariana island arc and back arc trough, Geol. Soc. Am. Ann. Mtg. Abstracts with programs (abstract).

Alt, J.C., T.F. Anderson, and L. Bonnell, 1989. The geochemistry of sulfur in a 1.3 km section of hydrothermally altered oceanic crust, DSDP Hole 504B., Geochim. Cosmochim. Acta, 53: 1011–1023.

Alt, J.C. and T.F. Anderson, 1990. The mineralogy and isotopic composition of sulfur in Layer 3 gabbros from the Indian Ocean, ODP Hole 735B. In: P.T. Robinson, R. VonHerzen, et al., Proc. Ocean Drilling Program, Scientific Results, 118: in press.

Andrews A.J., 1979. On the effect of low-temperature seawater-basalt interaction on the distribution of sulfur in oceanic crust, Layer 2., Earth Planet. Sci. Lett., 46: 68–80.

Arnold M. and Sheppard S.M.F., 1981. East Pacific Rise at latitude 21°N: Isotopic composition and origin of the hydrothermal sulphur., Earth Planet. Sci. Lett., 56: 148–156.

Auclair, F., and J.N. Ludden, 1987. Cyclic geochemical variation in the Troodos pillow lavas: evidence from the Cy-2a drill hole. In: P.T. Robinson, I.L. Gibson, and A. Panayiotou (Eds), Cyprus Crustal Study Project: Initial Report, Holes Cy-2 and 2a. Geol. Surv. Canada paper 85–29: 221–236.

Baragar, W.R., M.B.Lambert, N. Baglow, and I. Gibson, 1987. Sheeted dikes of the Troodos ophiolite. In: H.C. Halls and W.F. Fahrig (Eds), Mafic Dyke swarms., Geol. Assoc. Can. Spec. Paper, 34: 257–272.

Barriga, F.J.A.S, J. Munha, W.S. Fyfe and N.J.Vibetti, 1985. Extreme hydrothermal alteration in the intrusive layers of the Troodos ophiolite (Cyprus)., EOS Trans AGU, 66: 1128 (abstract).

Bednarz, U., G. Sunkel and H.-U. Schmincke, 1987a. The basaltic andesite- andesite and the andesite-dacite series from the ICRDG drill holes CY-2 and CY-2a. In: P.T. Robinson, I.L. Gibson, and A. Panayiotou (Eds), Cyprus Crustal Study Project: Initial Report, Holes Cy-2 and 2a., Geol. Surv. Canada paper 85–29: 183–204.

Bluth, G. and Ohmoto H. (1988) Sulfur isotope study of sulfide-sulfate chimneys on the East Pacific Rise, 11–13°N latitude., Can. Mineral., 26: 505–516.

Cann, J.R., P.J. Oakley, H.G. Richards, and C.J. Richardson, 1987. Geochemistry of hydrothermally altered rocks from Cyprus Drill Holes Cy-2 and Cy-2a compared with other Cyprus Stockworks. In: P.T. Robinson, I.L. Gibson, and A. Panayiotou (Eds), Cyprus Crustal Study Project: Initial Report, Holes Cy-2 and 2a., Geol. Surv. Canada paper, 85–29: 87–102.

Carroll, M.R. and M.J. Rutherford, 1985. Sulfide and sulfate saturation in hydrous silicate melts, Proc. 15th Lunar Planet. Sci. Conf., part 2: C601–C612.

Chaussidon M.F., Albarede F. and Sheppard S.M.F., 1987. Sulphur isotope heterogeneity in the mantle from ion microprobe measurements of sulphide inclusions in diamonds., Nature, 330: 242–244.

Clark, L.A., 1971. Volcanogenic ores: comparison of cupriferous pyrite deposits of Cyprus and Japanese Kuroko deposits., Soc. Mining Geol. Japan Spec. Issue, 3: 206–215.

Claypool, G.E., W.T. Holser, I.R. Kaplan, H. Sakai, and I. Zak, 1980. The age curves of sulfur and oxygen isotopes in marine sulfate and their mutual interpretation., Chem. Geol., 28: 199–260.

Constantinou, G. and G.J.S Govett, 1973. Geology, geochemistry and genesis of Cyprus sulfide deposits., Econ. Geol., 68: 843–858.

Edmond J.M., Measures C. McDuff R.E., Chan L.H., Collier R., Grant, B., Gordon, L.I. and Corliss J.B., 1979. Ridge crest hydrothermal activity and the balances of the major and minor elements in the ocean: the Galapagos data., Earth. Planet. Sci. Lett., 46: 1–18.

Field C.W., Sakai H. and Ueda A., 1984. Isotopic constraints on the origin of sulfur in oceanic rocks. In: A. Wauschkuhn, C. Kluth and R.A. Zimmermann (Eds), Syngenesis and epigenesis in the formation of mineral deposits: pp. 573–589.

Garcia, M.O., N.W.K. Liu and D.W. Muenow, 1979. Volatiles in submarine volcanic rocks from the Mariana Island arc and trough., Geochim Cosmochim Acta., 43: 305–312.

Hamlyn, P.R., R.R. Keays, W.E. Cameron, A.J. Crawford, and H.W.Waldron, 1985. Precious metals in magnesian low-Ti lavas: implications for metallogenesis and sulfur saturation in primary magmas., Geochim. Cosmochim. Acta, 49: 1797–1811.

Harmon, R.S., J. Hoefs, K.H. Wedepohl, 1987. Stable isotope relationships in Tertiary basalts and their mantle xenoliths from the Northern Hessian Depression, W. Germany., Contrib. Mineral. Petrol., 95: 350–369.

Haughton D.R., Roeder P. and Skinner B.J., 1974. Solubility of sulfur in mafic magmas., Econ. Geol., 69: 451–467.

Herzig, P.M.and G.H. Friedrich, 1987. Sulfide mineralization , hydrothermal alteration and chemistry in the drill hole CY-2a, Agrokipia, Cyprus. In: P.T. Robinson, I.L. Gibson, and A. Panayiotou (Eds), Cyprus Crustal Study Project: Initial Report, Holes Cy-2 and 2a., Geol. Surv. Canada paper, 85–29: 103–138.

Hubberten H.W., 1983. Sulfur content and sulfur isotopes of basalts from the Costa Rica Rift (Hole 504B, DSDP Legs 69 and 70). In: J. Honnorez, R.P. VonHerzen et al., Init. Repts. DSDP, Vol. 69: 629–635.

Hutchinson, R.W. and D.L. Searle, 1971. Stratabound pyrite deposits in Cyprus and relations to other sulfide ores., Soc. Mining Geol. Japan Spec Issue, 3: 198–205.

Jamieson, H.E. and J.W. Lydon, 1987. Geochemistry of a fossil ore solution aquifer: chemical exchange between rock and hydrothermal fluid recorded in the lower portio of research drill hole CY-2a, Agrokipia, Cyprus. In: P.T. Robinson, I.L. Gibson, and A. Panayiotou (Eds), Cyprus Crustal Study Project: Initial Report, Holes Cy-2 and 2a. Geol. Surv. Canada paper, 85–29: 139–152.

Janecky, D.S. and W.C. Shanks III, 1988. Computational modeling of chemical and sulfur isotopic reaction processes in seafloor hydrothermal systems: chimneys, massive sulfides, and subjacent alteration zones., Can Mineral., 26: 805–825.

Johnsón, A.E., 1972. Origin of Cyprus pyrite deposits. 24th IGC: pp. 291–298.

Ohmoto H. and Rye R.O., 1979. Isotopes of sulfur and carbon. In: H.L. Barnes (Ed), Geochemistry of hydrothermal ore deposits: pp. 509–567.

Rautenschlein, M., G.A. Genner, J. Hertogen, A.W.Hoffmann, R. Kerrich, H.U. Schminke and W.M. White, 1985. Isotopic and trace element composition of volcanic glass from the Akaki canyon, Cyprus: Implications for the origin of the Troodos Ophiolite., Earth Planet Sci Lett., 75: 369–383.

Rees C.E., Jenkins W.J. and Monster J., 1978. The sulphur isotopic composition of ocean water sulphate., Geochim. Cosmochim. Acta, 42: 377–381.

Robinson, P.T., W.G. Melson, T.O'Hearn and H.U. Schmincke, 1983. Volcanic glass compositions of the Troodos Ophiolite, Cyprus., Geology, 11: 400–404.

Sakai H., DesMarais D.J., Ueda A. and Moore, J.G., 1984. Concentrations and isotope ratios of carbon, nitrogen, and sulfur in ocean-floor basalts., Geochim. Cosmochim. Acta, 48: 2433–2441.

Smith, G.C. and F.J. Vine, 1987. Seismic veolocities in basalts from CCSP drill holes CY-2 and CY-2a at Agrokipia Mines, Cyprus. In: P.T. Robinson, I.L. Gibson, and A. Panayiotou (Eds), Cyprus Crustal Study Project: Initial Report, Holes Cy-2 and 2a., Geol. Surv. Canada paper, 85–29: 295–306.

Styrt M.M, Brackmann A.J., Holland H.D., Clark B.C., Pisutha-Arnold V.M., Eldridge C.S. and Ohmoto H., 1981. The mineralogy and isotopic composition of sulfur in hydrothermal sulfide/sulfate deposits on the East Pacific Rise, 21°N latitude., Earth Planet. Sci. Lett., 53: 382–390.

Sunkel, G., U. Bednarz and H.-U. Schmincke, 1987. The basaltic andesite-andesite and andesite-dacite series from the ICRDG drill holes Cy-2 and Cy-2a. II. Alteration. In: P.T.

Robinson, I.L. Gibson, and A. Panayiotou (Eds), Cyprus Crustal Study Project: Initial Report, Holes Cy-2 and 2a. Geol. Surv. Canada paper, 85–29: 205–220.

Thy, P., P. Schiffman and E.M. Moores, 1989. Igneous mineral stratigraphy and chemistry of the Cyprus Crustal Study Project drill core in the plutonic sequences of the Troodos ophiolite. In: I.L Gibson, J. Malpas, P.T. Robinson and C. Xenophontos (Eds), Cyprus Crustal Study Project: Initial Report, Hole CY-4., Geol. Surv. Canada paper, 88–9: 147–186.

Ueda, A. and H. Sakai, 1984. Sulfur isotope study of Quaternary volcanic rocks from the Japanese Islands Arc., Geochim. Cosmochim. Acta, 48: 1837–1848. Woodhead J.D., Harmon R.S. and Fraser D.G., 1987. O, S, Sr and Pb isotope variations in volcanic rocks from the Northern Mariana Islands: implications for crustal recycling in intra-oceanic arcs., Earth Planet. Sci. Lett., 83: 39–52.

Woodruff L.G. and Shanks W.C. III, 1988. Sulfur isotope study of chimney minerals and vent fluids from 21°N, East Pacific rise: Hydrothermal sulfur sources and disequilibrium sulfate reduction., J. Geophys. Res., 93: 4562–4572.

Part IV
Tectonics of Emplacement and Metamorphism

Obduction *Versus* Subduction and Collision in the Oman Case and Other Tethyan Settings

A. MICHARD[1], F. BOUDIER[2] and B. GOFFÉ[3]
[1] *Université de Paris Sud and Ecole Normale Supérieure, Laboratoire de Géologie, URA 1316 du C.N.R.S., 24 rue Lhomond 75231 Paris Cedex 05, France*
[2] *Université des Sciences et Techniques du Languedoc, Laboratoire de Tectonophysique, URA 1370 du C.N.R.S., place E. Bataillon, 34060 Montpellier Cedex, France*
[3] *Ecole Normale Supérieure, Laboratoire de Géologie, URA 1316 du C.N.R.S., 24 rue Lhomond, 75231 Paris Cedex, France*

Abstract

The Oman obduction can be considered as the type-obduction. It involved two main stages of thrusting of oceanic lithosphere: i) onto the adjacent oceanic crust and lower continental margin, without emergence; ii) onto the upper continental margin, with emergence of the ophiolite. This lithosphere shortening is a process distinct from subduction and collision. The first stage was accompanied in the upper plate by a differentiated magmatism that has previously been ascribed to subduction. By contrast, we argue that the magmatism must be ascribed to the fusion of an hydrated, thin mantle wedge at shallow depth during the lithospheric duplication, with partial anatexis in the HT-LP metamorphic sole at the base of the ophiolite. Such phenomenon might have occurred at a recently active mid-ocean ridge. At the end of this submarine stage, the oceanic slab moved onto the continental margin while bulldozing volcanic and sedimentary formations into a nappe complex. During the second, subaerial stage, a thick tectonic prism composed of continental cover and basement rocks formed under the thrust slab. The inner parts of the prism were affected by HP-LT recrystallization up to the eclogite facies in the deepest slices. The P-T evolution began with a stage of relatively high thermal gradient typified by counter-clockwise paths in the upper nappes of the prism. Such metamorphism is like that developed within collisional settings (Western Alps) and different from the subduction type (East-Pacific belts).

The Late Cretaceous Oman obduction affected an area of very recent continental crustal thinning and oceanic accretion (spreading changed to convergence within 5 Ma). Continental collision did not occur in Oman, while it affected the coeval Taurus obduction after a 40 Ma span of time. Other Tethyan obductions followed still different scenarios. In particular, Alpine and Himalayan obductions were broadly coeval with collision (respectively during Late Cretaceous and Eocene times) and involved rather old and deformed oceanic lithosphere. Rapid reversal of plate kinematics favor the largest, Oman-type obductions but other geodynamic circumstances in-

Tj. Peters et al. (Eds), Ophiolite Genesis and Evolution of the Oceanic Lithosphere, 447–467.

troduce various combinations of obduction, subduction and collision with different P-T-t signatures.

Introduction

"Obduction" corresponds to the emplacement of oceanic lithosphere onto a continental margin, with occasional blueschists development in the footwall of the ophiolitic overthrust (Coleman, 1971). It has frequently been considered as the end product or "blocking stage" of an intra-oceanic subduction process (Davies, 1971; Mattauer and Proust, 1976; Nicolas and Le Pichon, 1980). In the case of the Oman obduction that conception is supported by the arc-type volcanic rocks on top of the ophiolitic slab (Pearce et al., 1981) but it is contradicted by geometrical and chronological arguments (Boudier et al., 1985; Michard et al., 1985a). Obduction can also be compared to collision, since both processes create large thrust slabs of crystalline rocks (Armstrong and Dick, 1974). A collisional emplacement of the Western Alps ophiolites was advocated by Auzende et al. (1983).

The Oman obduction is probably the best obduction example available, due to lack of later continental collision, the excellent outcrops, and the intensive geological work begun by Allemann and Peters (1972) and Glennie et al. (1973, 1974) and continued until recent years (Coleman, ed., 1981; Lippard et al., 1986; Boudier and Nicolas, eds., 1988; Nicolas, 1989; Robertson, Searle and Ries, eds., 1990; this vol.). In the present paper, we intend to use this example and some of the other Tethyan ophiolites in addressing the semantic problem of obduction *versus* collision and subduction and in discussing the geodynamic circumstances that characterize obduction as a specific process of plate tectonics.

The Oman Obduction: Geological Setting and Stratigraphic Constraints

The Oman obduction developed within the Eurasian-African collisional zone in front of a re-entrant of the Arabian margin. This re-entrant has spared Oman the effects of further collision (Fig. 1). The Samail or Sumail ophiolitic nappe extends for more than 25,000 km^2 (Fig. 2) and, despite erosion, it is locally more than 10 km-thick (Shelton, 1984). This giant lithospheric slab overlies the Arabian margin and the Hawasina nappe complex of sedimentary rocks and associated alkaline basalts. The Hawasina complex is reduced to a discontinuous mélange at the trailing edge of the ophiolite nappe (Fig. 3). The Arabian sediments under the outer part of the thrust pile were only slightly deformed (Michard et al., 1984) and formed open ramp-folds that produced the Jabal Akhdar and Saih Hatat tectonic windows (Searle, 1985; Bernoulli and Weissert, 1987). These same continental rocks were strongly deformed and suffered a high-pressure, low-temperature (HP-LT) metamor-

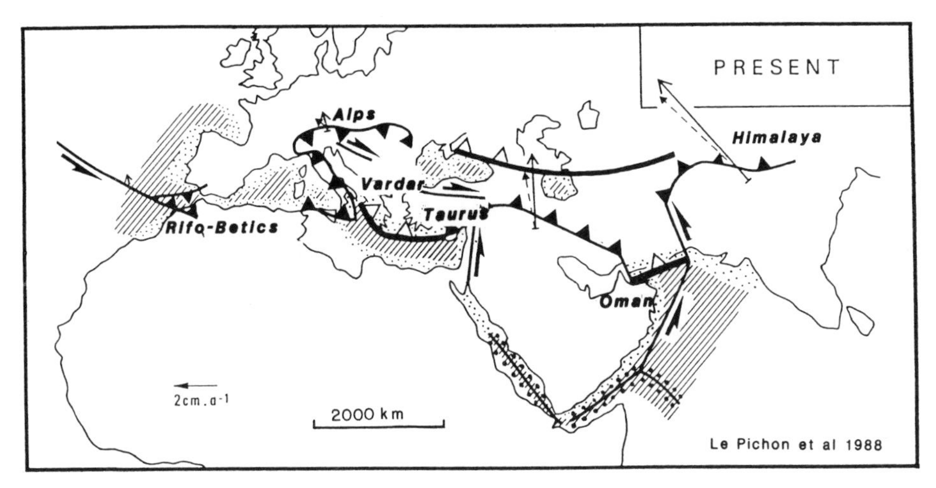

Figure 1. The Oman and other remarkable Tethyan obductions in the Eurasia-Gondwana collisional zone. Present kinematics after Le Pichon et al., 1988.

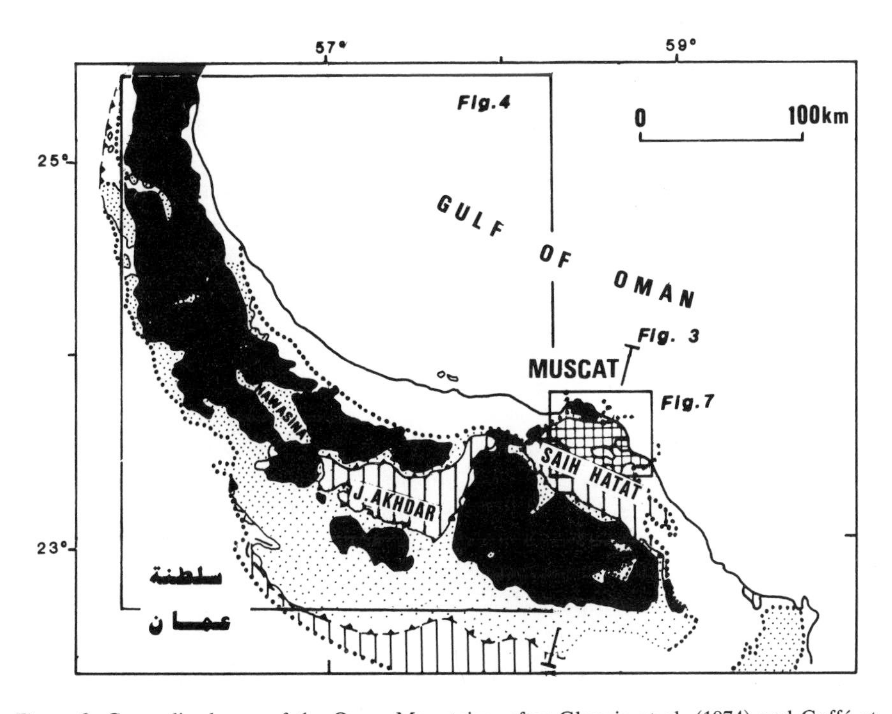

Figure 2. Generalized map of the Oman Mountains, after Glennie et al. (1974) and Goffé et al. (1988), with location of Figs. 3, 4 and 7. Black: Samail ophiolite; stippling: Hawasina nappes and mélanges; vertical shading: autochthonous (foreland) and parautochthonous (tectonic windows) terrains; square hatching: HP-LT metamorphics; dotted line: erosional limit of the Upper Maastrichtian – Tertiary onlap (white).

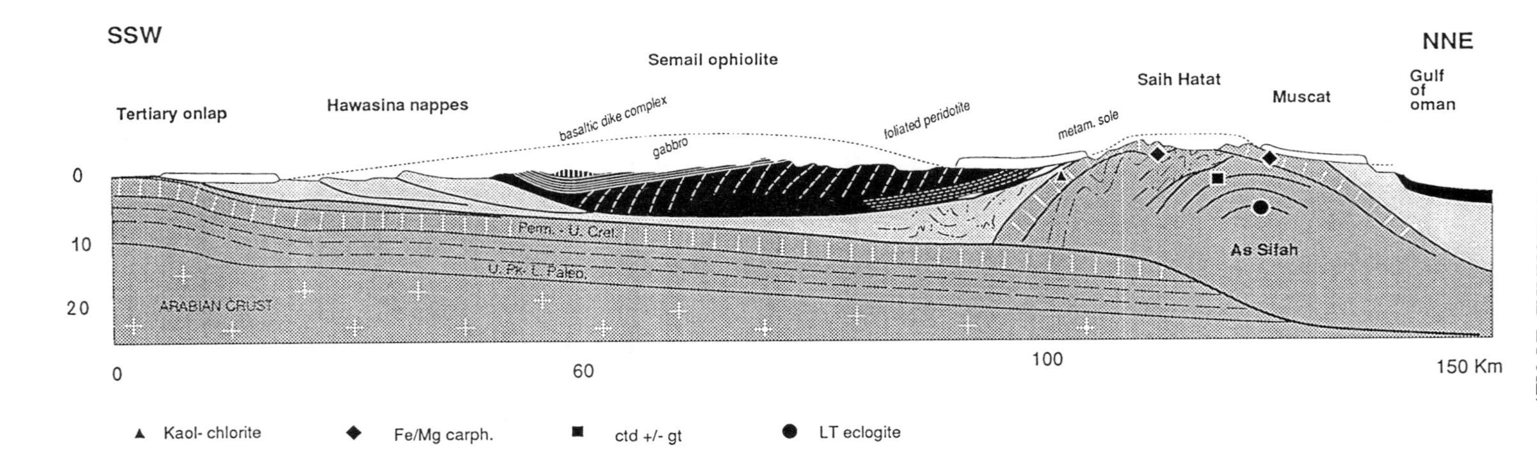

Figure 3. Schematic cross-section of the Oman Mountains (Fig. 2 for location) slightly modified after Goffé et al. (1988) and Nicolas (1989, p. 152). The early Late Cretaceous Moho is shown at the footwall of the gabbroic layer. The steeply dipping foliation in the underlying peridotite is related to mantle flow perturbations along transform faults, prior to the initiation of obduction (detachment of the future ophiolite).

phism in the internal, northeastern part of the latter window. The whole tectonic pile is unconformably overlain by a virtually undeformed, sedimentary onlap.

The continental sequences beneath the oceanic nappes record a passive margin setting throughout most of Mesozoic time and even as early as Late Permian (Béchennec et al., 1988; de Wever et al., 1988). Normal faulting sharply increased during Turonian-Campanian time, mainly as a result of flexural bending in front of the advancing nappes (Robertson, 1987; Michard et al., 1989). Campanian turbiditic marls are the youngest sediments observed under the nappes. Upper Campanian-Lower Maastrichtian marls in front of the nappes include boulders of Hawasina and Samail rocks. The earliest calcarenites of the post-nappe, mainly Tertiary cover sequence are dated as late Maastrichtian and follow an undated period of lateritic weathering (Coleman, 1981). Hence the time of progressive emplacement of the nappes onto the continental margin is rather accurately dated as Campanian-Lower Maastrichtian (85–80 to 70 Ma).

However, thrusting of the oceanic slab began earlier. The age of the Samail crust is essentially Early Cretaceous, the youngest MOR-basalts (Geotimes sequence) being dated as Albian-Cenomanian, about 98 Ma (Tilton et al., 1981; Mc Culloch et al., 1981; Tippit et al., 1981). The overlying, differentiated volcanics and associated radiolarian gave Turonian-Santonian and even Early Campanian ages (Schaaf and Thomas, 1986; Beurrier et al., 1987). By contrast, the youngest sediments dated in the underlying Hawasina nappes are Cenomanian pelagic marls. The Hawasina basin can be restored as a pre-oceanic, outer continental margin basin approximately 300 km-wide, flanked oceanward by carbonate buildups (Glennie et al., 1974; Lippard et al., 1986; Bernoulli and Weissert, 1987; Béchennec et al., 1988). It opened during Late Permian, at the same time as the inner continental margin began to subside, and closed as early as late Cenomanian, likely due to the Samail thrust encroachment upon it. Therefore two main periods can be isolated in the Samail nappe long-distance (about 400 km) travel: i) the Cenomanian-Turonian intra-oceanic, submarine detachment and early thrusting; ii) the Campanian-Lower Maastrichtian marginal thrusting, mostly at open air.

The Intra-Oceanic Stage

The submarine, early stage of obduction was accompanied in the Samail nappe by a transitional to calc-alkaline volcanism that has been considered as the signature of a classical subduction setting (Pearce et al., 1981; Lippard et al., 1986). Several lines of data used to recognize a different tectonic setting and define a model of initiation of the obduction by intra-oceanic thrusting at the ridge axis (Boudier and Coleman, 1981; Boudier et al., 1985, 1988) are recalled here.

The Metamorphic Sole

The ophiolite sole includes a continuous band (a few hundred meters thick) of fine grained porphyroclastic to mylonitic peridotites, deformed at high strain rate and temperatures of the order of 800 to 1000°C (Boudier et al., 1982; Lippard et al., 1986). The peridotites are underlain by discontinuous metamorphic lenses (five hundred meters maximum thickness) composed of metabasalts and metacherts. This metamorphic sole is characterized by a sharply downward decreasing metamorphic gradient, from garnet amphibolites (T ~ 870°C; P ~ 5 kbar) located in a meter-thick band at the peridotites contact (Searle and Malpas, 1980; Ghent and Stout, 1981), to greenschist facies formations overlying the Hawasina very low grade rocks. At the contact with the peridotites the amphibolites may develop anatexis; locally anatexis is also found in the metasediments. The steep inverted gradient observed throughout the whole sole results from both the stacking of slices developed at different P-T conditions in disparate areas, and an inverted thermal gradient within individual slices, observed at least in the highest grade rocks (Ghent and Stout, 1981; Searle and Malpas, 1982). The stacking of upper crustal oceanic slices of decreasing metamorphic grade during the initiation of obduction was progressive as supported by the K-Ar dating: 90–100 Ma on hornblende from the amphibolites to 85–70 Ma on muscovite from the greenschists (Lanphere, 1981; Montigny et al., 1988). These dates indicate a 5–15 Ma range of metamorphism recorded in a hundred-meter pile.

Structural and kinematic analysis in the metamorphic sole and basal peridotites (Boudier et al., 1982, 1985) has shown that two transport directions (E-W and NNW-SSE) were recorded in the basal peridotites and amphibolites (Fig. 4). In low grade rocks, the transport directions are more dispersed though a NE-SW direction is dominant. Systematic structural mapping in the mantle section revealed the presence of strike-slip, NNW-trending ductile faults developed at the same temperature conditions as those prevailing during the deformation of the peridotite sole. These shear zones were inferred to be related to the oceanic thrusting episode (Boudier et al., 1988). They are the favored site of gabbroic magma injections, forming locally sets of closely spaced gabbro dikes.

Thrusting at the Samail Ridge Axis

Several evidences have led to the proposal of a model of thrusting at the ridge axis: these evidences are discussed fully in Boudier et al. (1988) and summarized in the present paper.

i) chronological grounds: the ages of high grade amphibolites from the sole: 93 to 101 Ma (Lanphere, 1981; Montigny et al., 1988) overlap the ages of the sediments interbedded in the MORB-type volcanics of the Geot-

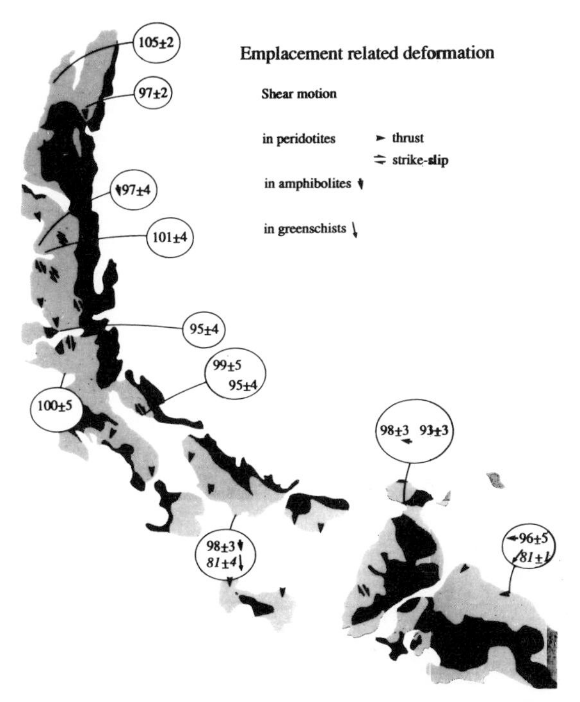

Figure 4. Directions of intra-oceanic thrusting of the Oman ophiolite deduced from kinematic analysis of the HT-LP metamorphic sole and basal peridotites (in Boudier et al., 1985), with K/Ar ages on amphiboles and muscovite from the sole (Montigny et al., 1988).

imes unit (Fig. 5): lower Cenomanian, 94 to 97.5 Ma (Tippit et al., 1981; Schaaf and Thomas, 1986; Beurrier et al., 1987). Considering age uncertainties, the fragment of oceanic lithosphere was less than 5 Ma old at the time of its detachment.

ii) geometrical grounds: the high-T thrust plane between the basal peridotites and the amphibolitic rocks is closely parallel to the paleo-Moho (Fig. 3). The thickness of the thrust lithosphere measured perpendicular to the thrust plane does not exceed 15 km in the external zone where the complete section is preserved. Evaluations of the nappe thickening across Bahla-Rustaq and Wadi Tayin massif sections (Fig. 4; Boudier et al., 1988), and gravity modeling (Shelton, 1984) show a wedge-shaped nappe with taper in the range 5–10°.

These criteria constrain a model of high-T decoupling of a young and thin lithosphere along a flat surface. The decoupling surface proposed is the 1000°C isotherm surface which is the base of the lithosphere at the spreading axis (Fig. 6). The attitude of this isothermal surface has been calculated up to 200 km from the ridge axis for a 5 cm/yr spreading rate (Boudier et al., 1988) using Kusznir's (1980), and Morton and Sleep's (1985) thermal models at the ridge vicinity, and transient equation of heat conduction for lithosphere

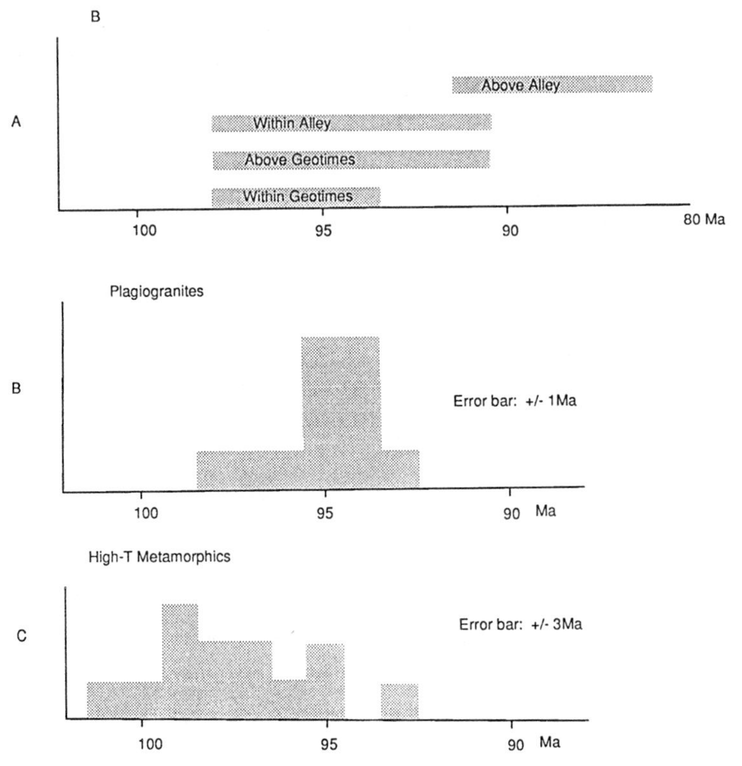

Figure 5. Chronological data for the intra-oceanic stage of the Oman obduction. – A: Biostratigraphic ages of sediments in the volcanics of Samail nappe (Tippit et al., 1981; Schaaf and Thomas, 1986; Beurrier et al., 1987). – B: U/Pb ages on zircon in plagiogranites intruding the plutonic section (Tilton et al., 1981). – C: K/Ar ages on hornblende from amphibolites of the metamorphic sole (Lanphere, 1981; Montigny et al., 1988).

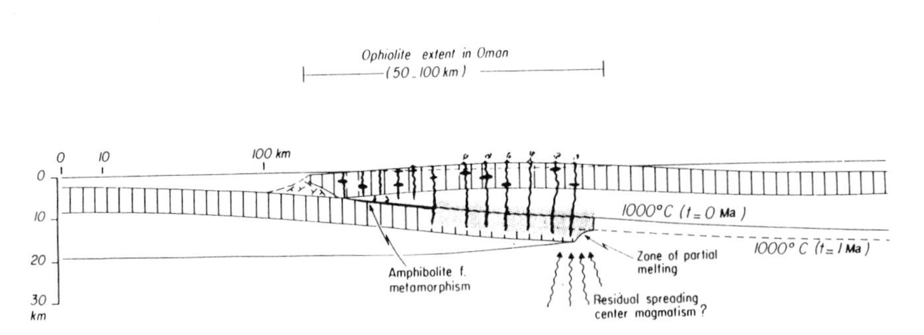

Figure 6. Model of intra-oceanic thrusting at the ridge for the Oman obduction, after Boudier et al. (1988). The decoupling surface is the 1000°C isotherm at the base of the lithosphere. The lithosphere structure is modelled for a 5 cm/year spreading rate. The dashed line shows the 1000°C isotherm after 1 Ma of static cooling.

older than 1 Ma. The resulting average dip for a 100 km wedge, corresponding to the transverse dimension of the Samail ophiolite, is about 6° (Boudier et al., 1988, Fig. 11). It fits the estimations derived from the field, and accounts for the model of Figure 6.

iii) thermal grounds: the deformation at about 900°C in the basal peridotites and 870°C in the immediately underlying garnet amphibolites implies that the maximum temperature was located at or above the peridotite/oceanic crust interface. The steeply decreasing metamorphism away from the contact and the presence of low angle thrusts faults separating the lithologic units are consistent with a model of piling of metamorphosed slices at decreasing temperature due to bulldozering of oceanic crust by the hot sole of the overriding lithospheric wedge: the "ironing effect". According to Hacker's (1990) simulation, the differential stress supported then by basaltic rocks in the fault zone was on order of 100 MPa (1 kbar). The consistent decrease of the metamorphic grade away from the sole precludes a large contribution of shear heating. Additionnally, the model of thrusting of hot lithosphere above a dying spreading center calls for some contribution of residual spreading heat to the heating of the lower plate. This residual heat could be traced by the multiple gabbro dikes injecting the detachment-related vertical shear zones above mentioned.

Intra-Oceanic Thrusting and Associated Magmatism

The occurrence in ophiolites of volcanics having the chemical characters of subduction environments supports the contention that many ophiolites formed in "immature island arcs". In Oman the volcanics stratigraphy is well documented (Lippard et al., 1986; Ernewein et al., 1988):

- The Geotimes formation, basaltic to andesitic N-MORB in composition, is co-genetic with the sheeted dikes and layered gabbros (plutonic section);
- the Lasail, Alley and Cpx-phyric formations (V2 in Ernewein et al., 1988) represent lavas differentiated from primitive Mg-rich magmas depleted in REE. They have a parental relationship with picritic to troctolitic dikes in the sheeted complex and with later mafic to ultramafic intrusions cross-cutting the plutonic section.
- the Salahi formation (V3), locally exposed, is represented by differentiated lavas strongly enriched in incompatible elements. Late magmatism is also represented by K mineral-bearing granitic intrusions intruding the mantle section.

The two last groups of volcanics are interpreted as seamounts capping the ophiolite; the V3 volcanics have bio-stratigraphic ages as recent as late Santonian, 83 Ma (Tippit et al., 1981), even Campanian, 81 Ma (Schaaf and Thomas, 1986). The process of intra-oceanic thrusting at the ridge accounts for the possibility of generating at least part of this "supra ophiolitic" volcanism. At some stage of the thrusting, hydrated oceanic crust will contact the HT base of the lithospheric wedge and promote hydrous melting in the mantle wedge at pressures of 1–5 kb, following the multi-stage process of Duncan and Green (1980) to produce picritic magmas. These magmas could be parents of the V2 volcanism, having the chemical signature of "supra-subduction". A second anatexis could develop at the expense of the components of the metamorphic sole (metabasites and metasediments), producing K- and incompatible elements-enriched magma represented by the K-bearing granitic intrusions cross-cutting the peridotite (Briqueu et al., this volume).

In conclusion, the major difference between the model of detachment at the ridge axis and arc-related models (Lippard et al., 1986) for genesis and detachment of the Oman ophiolite is that in the first model (Fig. 6), the ophiolite is created at an oceanic spreading center, and only the late magmatism (seamounts) is related to convergence process.

Crossing the Lower Continental Margin

The submarine transport of the oceanic slab was achieved by crossing the Hawasina basin, during the Turonian-Santonian span of time, corresponding to the age of the differentiated magmatism. Most of the lower margin sediments and underlying volcanics were then bulldozed by the leading edge of the ophiolitic slab, a process that evokes the formation of an accretionary prism in a subduction setting. However, the tectonics of the obduction-related Hawasina nappe complex differs from that of a classical accretionary prism in the occurrence of early, layer-parallel decollements in the sedimentary pile (Bernoulli and Weissert, 1987). This would reflect the stronger mechanical anisotropy of the Hawasina marginal sediments with respect to the unconsolidated oceanic oozes involved in the typical accretionary prisms.

The Marginal Stage

By contrast with the preceding stage, "climbing" onto the continental margin resulted in emergence of the nappe pile (except its leading edge?). HP-LT metamorphism of the loaded continental rocks has been regarded as the signature of this stage (Boudier and Michard, 1981; Michard, 1983; Lippard, 1983; Michard et al., 1984; Boudier et al., 1985; Lippard et al., 1986). However, it was ascribed to pre-obduction events by others: Le Métour et al. (1986) suggested an oceanward- dipping "subduction" dated as Campanian but older than the Samail obduction, while Montigny et al. (1988) and El-Shazly and Coleman (1990) argued for an Early Cretaceous crustal thickening. Several lines of data were used to recognize an obduction-related HP-LT metamorphism (Goffé et al., 1988; Michard et al., 1989): they are briefly recalled here.

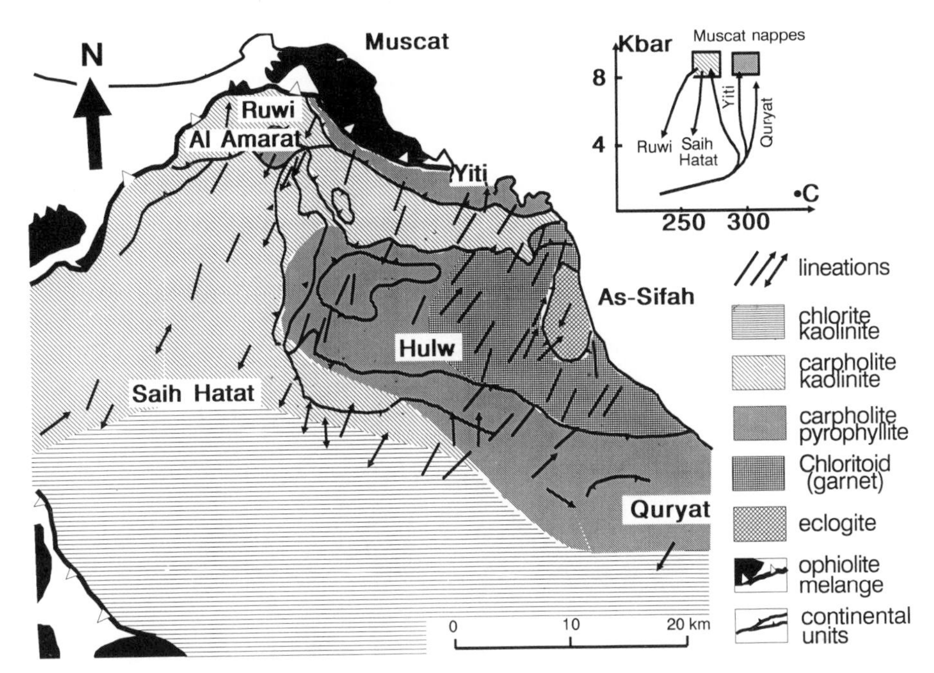

Figure 7. Simplified structural and metamorphic map of the Saih Hatat window, after Goffé et al. (1988) and Michard et al. (1989). Bars: stretching lineation; arrows: stretching lineation with clear kinematic indicators. *Insert*: P-T evolution of the low-grade HP-LT units.

The Metamorphic Tectonic Prism

In the most internal part of the Saih Hatat window (Figs. 3 and 7), the continental footwall of the oceanic thrust is intensely deformed into a pile of thrust slices that involve not only the Permo-Mesozoic carbonates but also their Proterozoic-Paleozoic basement. HP-LT assemblages occur in the highest units (Saih Hatat unit and Muscat nappes), including Fe/Mg-carpholite ± lawsonite associated with kaolinite or pyrophyllite in metapelites, lawsonite, crossitic glaucophane, and locally jadeite in metabasites. Peak conditions there are roughly 8 kbar, 270–300°C. Metamorphism progressively vanishes southwestward within the Saih Hatat unit itself. By contrast, high-grade rocks occur in the deepest units of the innermost part of the prism (As-Sifah units), including glaucophane-bearing eclogites with peak conditions close to 11 kbar, 400–500°C. Intermediate units (Hulw) show intermediate grade assemblages with chloritoid-garnet schists and garnet-glaucophane metabasites. This overall normal gradient precludes heating by the overlying oceanic slab ("ironing effect") during the HP-LT recrystallization of the continental rocks. Only locally are inverted gradients found in the upper nappes (pyrophyllite-bearing slice upon kaolinite-bearing), and these can be related to late metamorphic displacements. From the preceding data we must conclude that the overburden above the uppermost continental

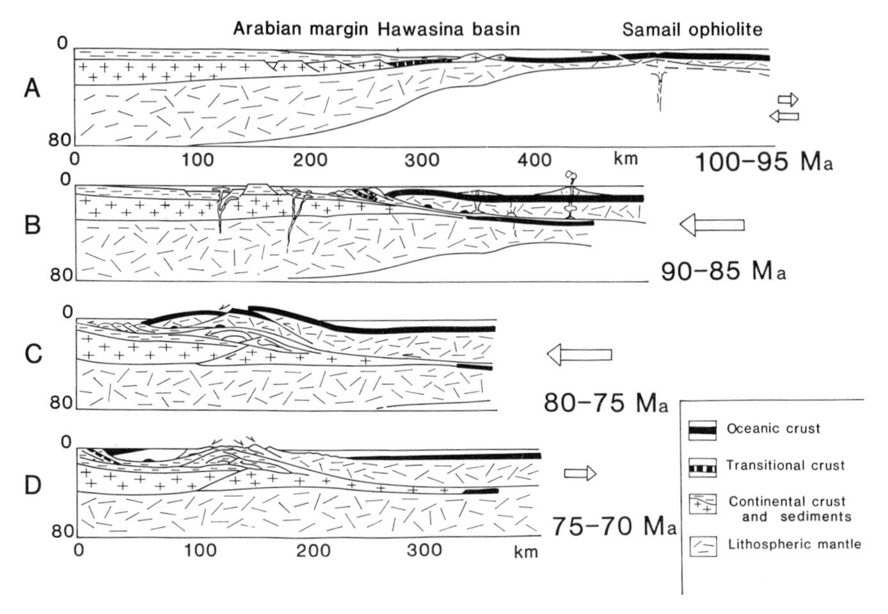

Figure 8. Tectonic evolution of the Oman obduction, after Boudier et al. (1988) and Goffé et al. (1988). A: Intra-oceanic thrusting at the ridge axis; initial setting. B: End of the submarine evolution; the oceanic slab encroaches upon the continental margin. C: Marginal stage; maximum thickening of the oceanic overburden and stacked continental slices underneath. D: Extensional collapse with uplift of the HP-LT metamorphics prior to the Late Maastrichtian-Tertiary onlap.

slices (Muscat nappes) was relatively cold and about 25 km-thick, which resulted in a 30 km-thick total overburden above the deepest exposed units (As-Sifah).

Structural and kinematic analysis in the whole metamorphic pile (Michard et al., 1984; Le Métour, 1987; Le Mer, 1988; Michard et al., 1989; Mann and Hanna, 1990) revealed stretching lineations trending consistently NNE-SSW, with a shear direction either to the south or the north in some place (Fig. 7). Both shear senses can be found associated with either the growth of the HP-minerals or their partial LP alteration.

An Obduction-Related HP-LT Metamorphism

Several lines of evidence support a model of obduction-related HP-LT metamorphism in the footwall of the Oman ophiolite (Fig. 8):

i) Chronological grounds: in the upper continental slices, Turonian-Santonian-(Campanian?) sediments are affected by low-grade, HP-LT metamorphism. Since the metamorphic and structural mapping supports the hypothesis of a single HP-LT event in this restricted area, this event (whatever its grade and the depth of the slice in the tectonic prism) appears to be broadly coeval with the ophiolite emplacement onto the upper continental margin. It is difficult to accept Le Métour et al.'s (1986) suggestion of a Late Cretaceous, "subduction"-related metamorphism prior to the Samail emplacement, since it would imply removal of a 25 km-thick overburden (responsible for the metamorphism) by the Hawasina-Samail nappes in a span of time restricted to part of the Campanian (5 to 10 Ma). By contrast, these chronological data are consistent with the idea that HP-LT metamorphism was triggered by the obduction encroachment onto the continental margin. This is supported by several K/Ar measurements on phengite, falling in the 80–100 Ma interval (Lippard, 1983; Montigny et al., 1988). Other results are scattered back to 240 Ma, with a 110–140 Ma cluster which was used to suggest an Early Cretaceous initiation of metamorphism (Montigny et al., 1988). We suggest that the discrepancy between these measurements and the stratigraphic data could result from the presence of extraneous argon, probably excess argon, i.e. argon introduced after crystallization, since 40Ar overpressures are common in HP-LT metamorphism (Chopin and Maluski, 1980).

ii) geometrical grounds: south-directed transport lineations in the HP-LT prism parallel the youngest lineations from the intra-oceanic metamorphic sole. This support the concept of a progressive displacement of the ophiolitic thrust from the oceanic and lower margin domain to the upper continental margin in the same geodynamic setting. Structures indicating N-directed transport are also present, but these can be ascribed, according to the specific case, either to compressional, discrete back-thrust zones (Fig. 8C), or to extensional, gravity-driven deformation of the thickened ductile prism (Fig. 8D).

iii) petrologic grounds: P-T-t data are provided in the low-grade units by crystallization of chloritoid prior to Fe/Mg-carpholite growth (Goffé et al., 1988). It may be shown that pressure increased syntectonically while temperature was stable or even decreased (Fig. 7, insert). P-T paths from the higher-grade units are not as well constrained and probably record a more complex evolution (El-Shazly and Coleman, 1990), as frequently observed in other belts (e.g. Hunziker and Martinotti, 1984). The lower-grade, blueschists P-T paths imply the progressive emplacement onto the upper continental margin of a low-temperature, wedge-shaped thrust complex. The cooled ophiolitic wedge, at the end of its intra-oceanic evolution, fits these constraints. The temperature of the ophiolitic thrust base must not have been significantly greater than 300°C, which is the maximum temperature reached by the continental rocks at their contact.

Prograde evolution in these rocks began with a rather "hot" gradient (30–35°C/km) which is likely to be related (at least partially) to the recent (up to early Late Cretaceous) thinning of the continental margin. Then, during the metamorphic climax, the mean geothermal gradient was about 15°C/km.

At that time, the 25 km-thick overburden above the Muscat nappes can be accounted for by the obducted slab itself, provided its sole thrust would have had a steeper dip in its trailing part than close to the leading edge, or that tectonic duplication would have occurred in the slab (Fig. 8C). Low temperature was maintained due to thrusting of the whole pile toward more external continental areas (Davy and Gillet, 1986; Gillet and Goffé, 1988). The retrograde evolution should have been controlled both by erosional and tectonic (likely as gravity spreading) unloading (Fig. 8D).

Discussion

The Oman Type-Obduction versus Subduction and Collision

Oman obduction appears as a lithospheric-scale thrust process at a converging plate margin (Figs. 1 and 8). With its two successive stages, it was active during about 20 Ma and the convergence accommodated there reached about 400 km. These time and length parameters are closer to that of a collision process than that of a typical subduction, which is virtually unlimited in time and space.

Magmatism developed in the upper plate during the early, intra-oceanic thrusting with characteristics similar to that of an immature island-arc, which was used as an argument to consider obduction as a starved subduction. However, the Oman obduction geometry is more similar to that of an Alpine-type collision than that of an island-arc subduction, at least at shallow and intermediate depth. The dip of the infra-ophiolitic sole thrust remained weak (less than 5°, taking into account the estimated taper of the thrust slab, see Figs. 3 and 6) except along discrete, intra-oceanic or marginal ramps. In the footwall of the lithospheric overthrust, shallow detachment horizons were active in the oceanic and outer continental margin domains, but the bulldozed nappe complex differs from an oceanic accretionary prism. The decoupling levels occurred at greater depths in the inner margin domain, allowing a thick tectonic prism to form, similar to that of a collision zone.

P-T-t data from the associated metamorphic rocks offer another way to characterize obduction with respect to subduction and collision. Metamorphism at the footwall of the Oman ophiolite changed from HT-LP to HP-LT conditions while thrusting proceeded from ocean to continent. During the earliest stage, inverted gradient developed in the metamorphic sole, as frequently observed in subduction settings (Peacock and Norris, 1989). During the later, marginal stage itself, normal gradient prevailed in the sub-ophiolitic continental slices and (at least in the upper slices) recrystallization followed counter-clockwise P-T paths. Such P-T paths are relatively scarce in metamorphic terranes (Bohlen, 1987; Ernst, 1988; Tournon et al., 1989). Collision or subduction-related metamorphisms should usually be characterized by rapid burial at low temperature followed by some heating at

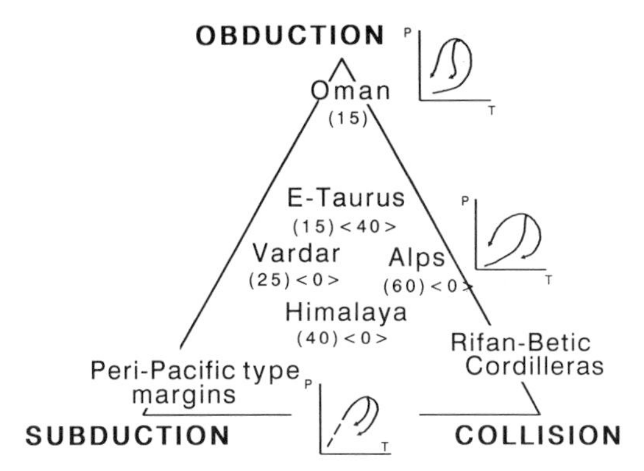

Figure 9. A schematic triangular diagram Obduction – Subduction – Collision with some characteristic P-T paths and selected Tethyan examples. (15): age of ophiolite (in Ma) when obducted on the outer continental margin; –40–: time span from obduction to collision (in Ma).

decreasing pressure, resulting in clockwise P-T paths (Thompson and Ridley, 1987). In the Alps, where collision is associated with obduction (next section, Fig. 9), some counter-clockwise P-T paths were found (Goffé and Velde, 1984; Goffé, 1984; Gillet and Goffé, 1988), but they have a steeper prograde path, reflecting a more rapid burial and/or cooler initial thermal structure. The specific form of the Oman P-T paths is likely related to an initial high thermal gradient produced by the youth of the thrust oceanic lithosphere and the recent stretching of the loaded continental margin.

The Oman Obduction versus Other Tethyan Obductions

The fascinating Samail thrust is worth to be taken as the type-example for large obductions such as the coeval, but more dissected obductions of Turkey and Cyprus or the Paleozoic obduction of Newfoundland (reviews in Gass et al., 1984; Nicolas, 1989). However, it would certainly be incorrect to equate every obduction zone with the Oman model, as can be shown in the Tethyan system (Figs. 1 and 9).

In the Alps and the Himalayas, obduction and collision are virtually contemporaneous. The Late Cretaceous tectonic phase of the Alps combined obduction of the Ligurian ophiolites with thrusting of the Austro-Alpine continental crust during the Apulian-European collision. By contrast with the Oman setting, the Ligurian ocean was old (about 60 Ma) when plate convergence began. Its lithosphere had been fragmented in Jurassic time by a dense array of normal and transform faults which were likely converted into (or cross-cut by) thrust faults during convergence (Auzende et al., 1983; Lagabrielle, 1987; Tricart and Lemoine, 1988). The deformed oceanic lithosphere was involved, together with the neighbouring, previously thinned continental lithosphere (Liassic paleomargins), in the collisional process. Hence, the Alpine obduction can be regarded as a sort of collision, involving not only two continents but an intervening, old oceanic lithosphere. The term "subduction" is frequently used here since it can describe the underthrusting of the Ligurian and European lithosphere beneath the Apulian leading edge. Evidently the process was much more limited than in the case of a true, Pacific-type subduction and did not give birth to any arc magmatism prior to the final (Oligocene) stage of plate convergence.

By contrast, for the Himalayan orogen, typical subduction is documented by the Trans-Himalayan batholith and accommodated the Cretaceous consumption of Neo-Tethyan lithosphere. The Spontang obduction is distinctly younger and ended during Lower Eocene, at the time of the Indian-Eurasian collision (Colchen et al., 1987). One can consider that both ophiolitic and further continental thrusts accommodated the ongoing plate convergence in place of the former subduction. However, although obduction and collision are broadly coeval in the Himalaya as in the Alps, flat-lying, high temperature shear zones were found in the Spontang ophiolite that would document a stage of intra-oceanic detachment prior to obduction, as in the Oman case (Reuber, 1986). Some blueschist facies metabasites and meta-radiolarites from the Indus-Tsangpo suture were dated at 70 to 100 Ma by the K/Ar and 39Ar/40Ar methods (Honneger et al., 1989). The Zanskar mélange sedimentation, coeval with the ophiolite encroachment upon the Indian margin, initiated as early as Late Santonian-Campanian time (Colchen et al., 1987).

Other Tethyan ophiolites offer still different association and timing of convergence mechanisms (Fig. 9). For example, in the Eastern Taurus belt, subduction and obduction occurred almost concomitantly during Late Cretaceous time, shortly after the latest accretion phase, while collision began only during Eocene time (Michard et al., 1985b). In the Vardar case, the oceanic closure followed a similar scheme (although it occurred during an earlier, Late Jurassic time), except that collision accompanied obduction (Sengör and Yilmaz, 1981; Mountrakis, 1986). Finally, in the westernmost part of the African-European collisional belt, i.e. the Rifan-Betic Cordilleras, ophiolites are quite scarce (Puga et al., 1989), and both obduction and subduction were limited by the narrowness of the Neo-Tethyan opening there.

A Triangular Diagram "Obduction-Subduction-Collision"

As far as we consider the Oman case, obduction appears as to be a specific process of lithospheric shortening along a convergent plate margin, different from both Pacific-type subduction and Alpine-type collision in several ways. However, various intermediate processes are possible between these end-member models, as can be observed within the single Tethyan belt. The proposed diagram (Fig. 9) was designed to visualize these different convergence scenarios.

Selected Tethyan obductions were plotted according to the relative importance of each of the three plate convergence processes involved. The largest obductions correspond to those which affected young oceanic lithosphere (Oman, Taurus: 15 Ma when obducted on the inner continental margin, less than 5 Ma at the beginning of the intra-oceanic shearing). Smaller obductions can develop during collisional events at the expense of old and deformed oceanic areas (Alps, Himalayas). Before the oceanic lithosphere was sampled by obduction, large parts of it disappeared by subduction (Eastern Taurus, Himalaya). The P-T paths for the associated HP-LT metamorphic rocks partly record the type of convergent margin, since they depend on the thermal status of both the upper and subducting lithospheres, on their geometry (upper plate wedge angle) and on kinematics (burial rate versus thermal equilibration). A typical obduction setting should favor counter-clockwise paths (Oman) by contrast with the most classical, subduction-collision related clockwise paths. Both types of P-T paths are observed in the obduction-collision setting of the Alps.

Conclusion

The Oman giant obduction is generally accepted as the type-obduction. Geometrical and petrological (P-T paths) criteria allow distinction between this type of convergent margin and both the subduction and collision types, although similarities and intermediate combinations exist. The question arises as to what circumstances favor obduction with respect to subduction in a convergence setting, allowing ophiolitic thrusts to form?

It has often been argued that young, low density oceanic lithosphere will be more easily obducted than old, denser lithosphere. This fits the chronological data for Oman, where the Samail ophiolite was 5 Ma-old when its thrusting began within its oceanic homeland, and about 15 Ma-old when it reached the continental margin. The rule should have been respected as early as Late Proterozoic, according to Sacchi and Cadoppi (1987). However, from other Tethyan examples and particularly the Alps and Himalayas, it appears that rather old oceanic lithosphere can also form ophiolitic nappes, although smaller, in collisional setting. This would mainly be controlled by early deformation in the oceanic basin.

Acknowledgements

Field works were supported by grants from the Institut National des Sciences de l'Univers – Centre National de la Recherche Scientifique. Facilities from the Oman Ministry of Petroleum and Minerals represented by M. H. Kassim and by the Bureau de Recherches Géologique et Minière represented by J. Caïa and M. Villey are also acknowledged. We thank R.G. Coleman, B.R. Hacker and A. Nicolas for their quite valuable criticism.

References

Allemann, F. and Peters, T., 1972. The ophiolite radiolarite belt of the North-Oman Mountains., Eclogae Geol. Helv., 65: 657–697.

Auzende J.M., R., Polino R., Lagabrielle, Y. and Olivet, J.L., 1983. Considérations sur l'origine et la mise en place des ophiolites des Alpes occidentales: apport de la connaissance des structures océaniques., C. R. Acad. Sci. Paris, 296: 1527–1532.

Armstrong, R.L. and Dick, H., 1974. A model for the development of thin overthrust sheets of crystalline rocks., Geology, 2: 35–40.

Béchennec, F., Le Métour, J., Rabu, D., Villey, M. and Beurrier, M., 1988. The Hawasina basin: a fragment of a starved passive continental margin, thrust over the Arabian platform during obduction of the Semail nappe., Tectonophysics, 151: 323–343.

Bernoulli, D. and Weissert, H., 1987. The upper Hawasina nappes in the Central Oman Mountains. Stratigraphy, palinspastics and sequence of nappes emplacement., Geodinamica Acta, Paris, 1: 47–58.

Beurrier, M., Bourdillon de Grissac, Ch., De Wever, P. and Lescuyer, J.L., 1987. Biostratigraphie des radiolarites associées aux volcanites ophiolitiques de la nappe de Samail, Oman., C.R. Acad. Sci., Paris, 304: 907–910.

Bohlen, S. R., 1987. Pressure-temperature-time paths and a tectonic model for the evolution of granulites., J. Geol., 95: 617–632.

Boudier, F. and Coleman, R.G., 1981. Cross section through the peridotite in the Semail ophiolite, Oman., J. Geophys. Res., 86: 2573–2592.

Boudier, F. and Michard, A., 1981. Oman ophiolite, the quiet obduction of oceanic crust., Terra Cognita, Cambridge, 1: 109–118.

Boudier, F., Nicolas, A., Bouchez, J.L., 1982. Kinematics of oceanic thrusting and subduction from basal sections of ophiolites., Nature, 296: 825–828.

Boudier, F., Bouchez, J.L., Nicolas, A., Cannat, M., Ceuleneer, G., Misseri, M. and Montigny, A., 1985. Kinematics of oceanic thrusting in the Oman ophiolite: model of plate convergence., Earth Planet. Sci. Lett., 75: 215–222.

Boudier, F. and Nicolas, A. (eds), 1988. The ophiolites of Oman., Tectonophysics, spec. issue, 401 p.

Boudier, F., Ceuleneer, G. and Nicolas A., 1988. Shear zones, thrusts and related magmatism in the Oman ophiolite: initiation of thrusting on an oceanic ridge., Tectonophysics, 151: 275–296.

Briqueu, L., Mevel, C. and Boudier, F., 1990. Calc-alkaline plutonic suite in Oman ophiolite related to the obduction process. Sr, Nd and Pb evidences. This volume.

Chopin, Ch. and Maluski, H., 1990. 40Ar/39Ar dating of high pressure metamorphic micas from the Grand-Paradis area (Western Alps): evidence against the blocking temperature concept., Contrib. Mineral. Petrol., 74: 109–122.

Coleman, R.G., 1971. Plate tectonic emplacement of upper mantle peridotites along continental edges., J. Geophys. Res., 76: 1212–1222.

Coleman, R.G., (ed.), 1981. Oman ophiolites. J. Geophys. Res. (spec. issue), 86: 2495–2782.

Coleman, R.G., 1981. Tectonic setting for ophiolite obduction in Oman., J. Geophys. Res. (spec. issue) 86: 2497–2508.

Colchen, M., Reuber, I., Bassoulet, J.P., Bellier, J.P., Blondeau, A., Lys, M. and De Wever, P., 1987. Données biostratigraphiques sur les mélanges ophiolitiques du Zanskar, Himalaya du Ladakh., C. R. Acad. Sci. Paris, 305 (II): 403–406.

Davies, H.L., 1971. Peridotite-gabbro-basalt complex in Eastern Papua: an overthrust plate of oceanic mantle and crust., Bur. Mineral. Res. Austral. Bull., 128 p.

Davy, Ph. and Gillet, Ph., 1986. The stacking of thrust slices in collision zones and its thermal consequences., Tectonics, 5: 309–320.

Duncan, R.A. and Green, D.H., 1980. Role of multistage melting in the formation of oceanic crust., Geology, 8: 22–26.

El Shazly, A. and Coleman, R.G., 1990. Metamorphism in the Oman mountains in relation to the Semail ophiolite emplacement., Geol. Soc. London (spec. publ.), 49: 475–495.

Ernewein, M., Pflumio, C. and Whitechurch, H., 1988. The death of an accretion zone as evidenced by the magmatic history of the Semail ophiolite (Oman)., Tectonophysics, 151: 247–274.

Ernst, W.G., 1988. Tectonic history of subduction zones inferred from retrograde blueschist P-T paths., Geology, 16: 1081–1084.

Gass, I.G., Lippard, S.J. and Shelton, A.W., eds., 1984. Ophiolites and oceanic lithosphere, Geol. Soc. London (spec. publ.) 13, 258 p.

Ghent, E.D. and Stout, M.Z., 1981. Metamorphism at the base of the Samail ophiolite, southeastern Oman Mountains., J. Geophys. Res., 86: 2557–2571.

Gillet, Ph. and Goffé, B., 1988. Significance of aragonite in the Western Alps., Contrib. Miner. Petr., 99: 70–81.

Goffé, B. and Velde, B., 1984. Contrasted metamorphic evolutions in thrusted cover units of the Briançonnais zone, French Alps: a model for the conservation of HP-LT metamorphic mineral assemblages., Earth Planet. Sci. Lett., 68: 351–360.

Goffé, B., 1984. Le faciès à carpholite-chloritoide dans la couverture briançonnaise des Alpes ligures: un témoin de l'histoire tectono-métamorphique régionale., Mem. Soc. Géol. It., 28: 461–479.

Goffé, B., Michard, A., Kienast, J.R. and Le Mer, O., 1988. A case of obduction-related high pressure, low temperature metamorphism in upper crustal nappes, Arabian continental margin, Oman; P-T paths and kinematic interpretation., Tectonophysics, 151: 363–386.

Glennie, K.W., Boeuf, M.G.A., Hughes-Clarke, M.W., Moody-Stuart, M., Pilaar, W.F.H. and Reinhardt, B.M., 1973. Late Cretaceous nappes in Oman mountains and their geologic evolution., Bull. Am. Ass. Petrol. Geol., 57: 5–27.

Glennie, K.W., Boeuf, M.G.A., Hughes-Clarke, M.W., Moody-Stuart, M., Pilaar, W.F.H. and Reinhardt, B.M., 1974. Geology of the Oman mountains., Verh. K. Nederl. Geol. Mijnb. Gen., 31, 423 pp.

Hacker, B.R., 1990. Simulation of the metamorphism and deformation history of the metamorphic sole of the Oman ophiolite., J. Geophys. Res., 95 (B4): 4895–4907.

Honneger, K., Le Fort, P., Mascle, G. and Zimmermann, J.L., 1989. The blueschists along the Indus suture zone in Ladakh, NW Himalaya., J. metam. Geol., 7: 57–72.

Hunziker, J.C. and Martinotti, G., 1984. Geochronology and evolution of the western Alps: a review., Mem. Soc. geol. It., 29: 43–56.

Kusznir, N.J., 1980. Thermal evolution of the oceanic crust, its dependance on spreading rate and effect on crustal structure., Geophys. J. R. Astr. Soc., 61: 167–181.

Lagabrielle, Y., 1987. Les ophiolites: marqueurs de l'histoire tectonique des domaines océaniques., Thèse Sci., Univ. Bretagne occ., Brest, 350 p.

Lanphere, M.A., 1981. K-Ar ages of metamorphic rocks at the base of the Samail ophiolite, Oman., J. Geophys. Res., 86: 2777–2782.

Le Métour, J., 1987. Géologie de l'autochtone des montagnes d'Oman dans la fenêtre du Saih Hatat., Mém. Sci. Terre Univ. P. et M. Curie, Thèse Sci. Paris. 87-13, 420 p.

Le Métour, J., Rabu, D., Tegyey, M., Béchennec, F., Beurrier, M. and Villey, M., 1986a. Le métamorphisme régional crétacé de faciès éclogites-schistes bleus sur la bordure omanaise de la plate-forme arabe: conséquence d'une tectogenèse précoce anté-obduction., C.R. Acad. Sci., Paris, 302: 905–910.

Le Pichon, X., Bergerat, F. and Roulet, M.J., 1988. Plate kinematics and tectonics leading to the Alpine belt formation; a new analysis., Geol. Soc. Amer. Bull. (spec. pap.) 218: 11–131.

Lippard, S.J., 1983. Cretaceous high pressure metamorphism in NE Oman and its relationship to subduction and ophiolite nappe emplacement., J. Geol. Soc., 140: 97–104.

Lippard, S.J., Shelton, A.W. and Gass, I.G., 1986. The ophiolite of Northern Oman., Geol. Soc. London Mem., 11, 178 p.

Mann, A. and Hanna, S., 1990. The tectonic evolution of pre-Permian rocks, Central Oman Mountains. Geol. Soc. London (spec. publ.), 49: 307–326.

Mattauer, M. and Proust, F., 1976. La Corse alpine: un modèle de genèse du métamorphisme haute-pression par subduction de croûte continentale sous du matériel océanique., C. R. Acad. Sci. Paris, 282: 1249–1252.

Mc Culloch, M.T., Gregory, R.T., Wasserburg, G.J. and Taylor, H.P.Jr., 1981. Sm-Nd, Rb-Sr, and $^{18}O/^{16}O$ isotopic systematics in an oceanic crustal section: evidence from the Samail ophiolite., J. Geophys. Res., 86: 2721–2735.

Michard, A., 1983. Les nappes de Mascate, Oman, rampe épicontinentale d'obduction à faciès schiste bleu, et la dualité apparente des ophiolites omanaises., Sci. Géol., Bull., Strasbourg, 36: 3–16.

Michard, A., Bouchez, J.L. and Ouazzani-Touhami, M., 1984. Obduction-related planar and linear fabrics in Oman., J. Struct. Geol., 6: 39–49.

Michard, A., Juteau, T. and Whitechurch, H., 1985a. L'obduction, revue des modèles et confrontation au cas de l'Oman., Bull. Soc. géol. France (8), 1: 189–198.

Michard, A., Whitechurch, H., Ricou, L.E., Montigny, R. and Yazgan, E., 1985b. Tauric subduction (Malatya - Elazig provinces) and its bearing on tectonics of the Tethyan realm in Turkey. Geol. Soc. London (spec. publ.), 17: 361–373.

Michard, A., Le Mer, O., Goffé, B. and Montigny, R., 1989. Mechanism of the Oman mountains obduction onto the Arabian continental margin, reviewed., Bull. Soc. Géol. France (8), 5: 241–252.

Montigny, R., Le Mer, O., Thuizat, R. and Whitechurch, H., 1988. K/Ar and 40Ar/39Ar study of metamorphic rocks associated with the Oman ophiolites: blueschist metamorphism prior to emplacement of the oceanic crust onto the Arabian platform., Tectonophysics, 151: 345–362.

Morton, J.L. and Sleep, N.H., 1985. A mid-oceanic ridge thermal model: constraints on the volume of axial hydrothermal heat flux., J. Geophys. Res., 90: 11345–11353.

Mountrakis, D., 1986. The Pelagonian zone in Greece: a polyphase-deformed fragment of the Cimmerian continent and its role in the geotectonic evolution of the Eastern Mediterranean., J. Geol., 94: 335–347.

Nicolas, A., 1989. Structures of ophiolites and dynamics of oceanic lithosphere., Kluwer Acad. Publ., Dordrecht, 368 p.

Nicolas, A. and Le Pichon, X., 1980. Thrusting of young lithosphere in subduction zones with special reference to structures in ophiolitic peridotites., Earth Planet. Sci. Lett., 46: 397–406.

Peacock, S.M. and Norris, P.J., 1989. Metamorphic evolution of the Central Metamorphic Belt, Klamath Province, California: an inverted metamorphic gradient beneath the Trinity ophiolite., J. metam. Geol., 7: 191–209.

Pearce, J.A., Alabaster, T., Shelton, A.W. and Searle, M.P., 1981. The Oman ophiolite as a Cretaceous arc-basin complex: evidence and implications., Phil. Transact. Roy. Soc., A 300: 299–317.

Puga, E., Diaz de Federico, A., Fedinkova, E., Bondi, M. and Morten, l., 1989. Petrology, geochemistry and metamorphic evolution of the ophiolitic eclogites and related rocks from the Sierra Nevada Betic Cordilleras, SE-Spain., Schweiz. Mineral. Petrogr. Mitt., 69: 435–455.

Reuber, I., 1986. Geometry of accretion and oceanic thrusting of the Spontang ophiolite, Ladakh-Himalaya., Nature, 321: 592–596.

Robertson, A.H., 1987. The transition from a passive margin to an Upper Cretaceous foreland basin related to ophiolite emplacement in the Oman Mountains., Geol. Soc. Amer. Bull., 99: 633–653.

Robertson, A.H.F., Searle, M.P. and Ries, A.C. eds, 1990. Geology and tectonics of the Oman region. Geol. Soc. London (spec. publ.) 49: 416–345

Sacchi, R. and Cadoppi, P., 1987. Ophiolite obduction today and yesterday., Ofioliti, 12: 393–402.

Schaaf, A. and Thomas, V., 1986. Les radiolaires campaniens du Wadi Ragmi (nappe de Semail, Oman): un nouveau repère chronologique de l'obduction omanaise., C.R. Acad. Sci. Paris, 303: 1593–1598.

Searle, M.P., 1985. Sequence of thrusting and origin of culminations in the northern and central Oman Mountains., J. Struct. Geol., 7: 129–143.

Searle, M.P. and Malpas, J., 1980. Structure and metamorphism of rocks beneath the Semail ophiolite of Oman and their significance in ophiolite obduction., Trans. R. Soc. Edinburgh, Earth Sci., 71: 247–262.

Searle, M.P. and Malpas, J., 1982. Petrochemistry and origin of sub-ophiolitic metamorphic and related rocks in the Oman Mountains., J. Geol. Soc., 139: 235–248.

Sengör, A.M.C. and Yilmaz, Y., 1981. Tethyan evolution of Turkey: a plate tectonic approach., Tectonophysics, 75: 181–241.

Shelton, A.W., 1984. Geophysical studies on the northern Oman ophiolite., Ph.D. thesis, Open Univ., Milton Keynes (U.K.), 323 p.

Thompson, A. B. and Ridley, J.R., 1987. Pressure-temperature-time histories of orogenic belts., Phil. Trans. R. Soc., London, A 321: 27–45

Tilton, G.R., Hopson, C.A. and Wright, J.E., 1981. Uranium-lead isotopic ages of the Samail ophiolite, Oman, with applications to Tethyan ridge tectonics., J. Geophys. Res., 86: 2763–2775.

Tippit, P.R., Pessagno, E.A. and Smewing, J.D., 1981. The biostratigraphy of sediments in the volcanic unit of the Samail ophiolite. J. Geophys. Res., 86, 2756–2762.

Tournon, J., Triboulet, C. and Azema, J., 1989. Amphibolites from Panama: anti-clockwise P-T paths from a Pre-upper Cretaceous metamorphic basement in Isthmian Central America., J. metam. Geol., 7: 539–546.

Tricart, P. and Lemoine, M., 1988. A l'origine de la structure des Schistes lustrés à ophiolites du Queyras Alpes françaises: un mode atypique d'obduction, conséquence de la structure particulière de la croûte océanique ligure., C. R. Acad. Sci. Paris, 306: 301–306.

De Wever, P., Bourdillon de Grissac, C. and Béchennec, F., 1988. Permian age from radiolarites of the Hawasina nappes, Oman Mountains., Geology, 16: 912–914.

The Obduction of the Northern Oman Ophiolite – Crustal Loading and Flexure

A.W. SHELTON [1] and S.S. EGAN [2]
[1] *Member of the Department of Earth Sciences at Sultan Qaboos University, in the Sultanate of Oman*
[2] *Member of the Department of Geology, University of Keele, Keele, Staffs, England*

Abstract

The unique opportunity to study a 20 km-deep section of oceanic lithosphere in the Oman Mountains has attracted a wealth of expertise and generated a great deal of geological data. Lippard et al. (1986) offered a solution to the problem of obduction at the former passive margin of Eastern Arabia which incorporated all age data known at that time and accounted for the broad spectrum of geological evidence presented in the mountains.

Their emplacement scheme involves the loading and flexure of the transitional zone of a passive margin. Indeed the 'isostatic rebound' of the margin is central to the gravity emplacement of the ophiolite. This paper explores the feasibility of such crustal flexure using a numerical model of lithosphere shortening with best-estimate deep crustal and mantle physical parameters.

Introduction

The emplacement scheme put forward in Lippard et al. (1986) for the obduction of the Semail Nappe was presented as a solution to the geological data gathered by Glennie et al. (1974), the USGS-Santa Barbara group headed by Prof R.G. Coleman and his team (Special issue of JGR 1981), and the Open University Group led by Prof I.G. Gass. It appeared to account for most of the structural, geochemical, geophysical and dating information established prior to 1984 but has not taken account of the more recent and detailed work carried out in Oman. The model as presented depended largely on the response of transitional lithosphere to loading; the separation and subsequent obduction of the ophiolite nappe were powered entirely by the flexural and isostatic rebound of a margin released from an expired subduction zone. It was important to test the response of lithosphere in such a setting to establish the validity of the mechanism because, as the model was first presented, it required what appeared to be unreasonable flexure in the

Tj. Peters et al. (Eds), Ophiolite Genesis and Evolution of the Oceanic Lithosphere, 469–483.

continental lithosphere. We here show that such effects are, for the most part, a predictable response of an ancient passive margin subjected to loading.

The depth-dependent numerical modelling we have used incorporates the geometric, thermal and flexural isostatic effects of lithospheric shortening. The complexity of the obduction scheme excludes the possibility of modelling the entire margin detail and history but various frames, or facets, of the sequence can be examined. In particular, the modelling is used to examine:

1. The development of a foreland basin;
2. Post-thrusting thermal and erosional uplift;
3. The effects of extensional reactivation in the later stages of the model.

The mathematical background to the modelling is given in Egan (1989). Essentially, numerical modelling of the lithosphere shortening uses the "Chevron Construction" modified to allow collapse of the hanging wall on to the footwall shoulder. It is assumed that major low-angle thrusts flatten into horizontal detachment horizons as properties change from brittle to ductile (Jackson and Mckenzie, 1983; Kusznir and Park, 1987). Below that detachment horizon, compression is achieved by pure shear in an amount that balances the thrust movement in the upper part of the model.

Both thrust movement (simple shear) and pure shear perturb the lithosphere temperature field. More precisely, thrusting causes a temperature discontinuity across the fault as warm hanging-wall material is aligned with a cooler environment in the footwall. Shortening by pure shear in the lower lithosphere deepens the lithosphere-asthenosphere thermal boundary and an overall cooling results. The associated thermal contraction induces surface subsidence. Following shortening the lithosphere temperature field re-equilibrates by conduction. This re-heating requires around 100 Ma and is accompanied by surface uplift. A two-dimensional form of the heat conduction equation has been used to quantify the re-equilibration with time. The equation has been solved numerically for small time increments using the finite difference method.

Loading of the lithosphere due to crustal thickening and thermal effects is assumed to be compensated using flexural theory. A simplified, laterally constant, elastic thickness has been used to define the flexural rigidity of the lithosphere.

Inertial Plate Model

The 'inertial' plate view merely stated that a system involving a 1000 km lithospheric cross-section would be unable to make any sudden acceleration

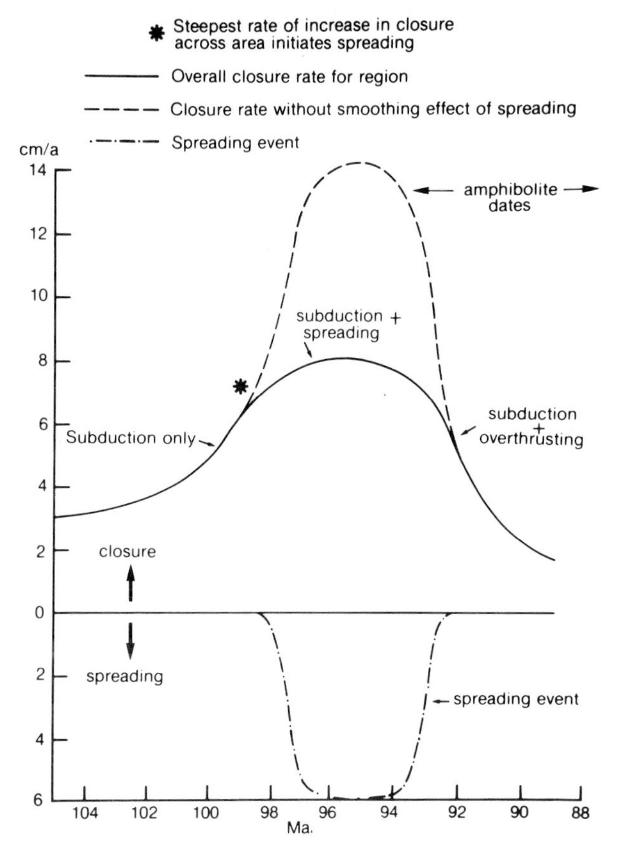

Figure 1. Postulated rate of consumption of crust at the southern Tethys margin assuming subduction initiated at 110 Ma. From Lippard et. al. (1986).

changes, the relative velocity changes of the components of the model must be described by a smooth function (Figure 1).

These velocities increase as the subduction zone develops and then diminish as it becomes progressively involved with the passive continental margin. In the postulated consumption-rate function used for this reconstruction, spreading was initiated at the maximum relative acceleration. This is given an age corresponding to the oldest ophiolite formation date. This spreading moderated the overall crustal consumption rate, keeping it within bounds observed as plate-tectonically feasible. The descending side of the function has little to constrain the amplitude and after 88 Ma the consumption of crust started to be transferred to new subduction zones at the northern margin of Tethys and out of this geologic reference frame.

Using these velocity changes the model was able to account for the main geological and geochemical data then available, while preserving volume and keeping vertical and horizontal movements compatible with the range of present day plate-motion measurements.

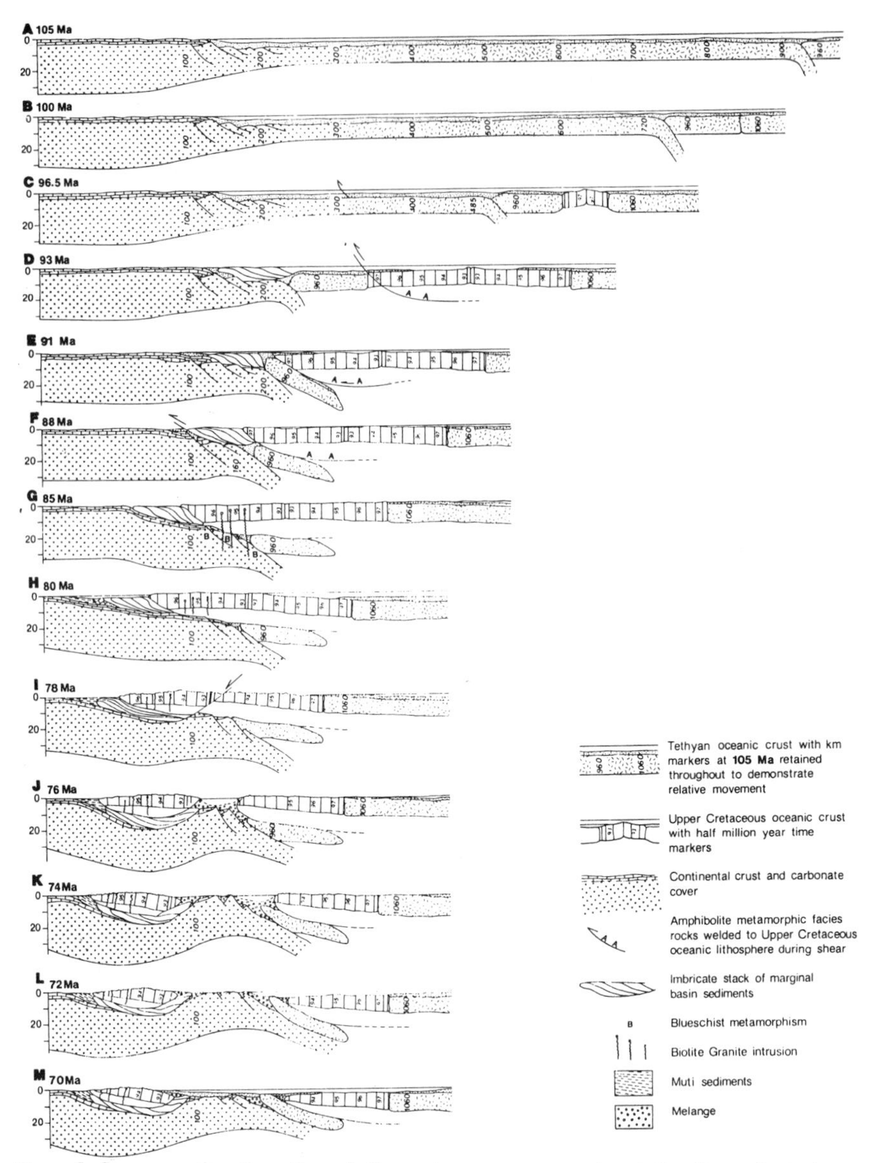

Figure 2. Sequence of sections through the southern Tethys margin relating the geological, age-dating and geochemical evidence via the rate curve in the previous figure. From Lippard et al. (1986).

Figure 2, from Lippard et al. (1986), shows generalized sections through the southern Tethys as it developed from an Atlantic-type margin into the thrust belt now observable in Northern Oman. The detailed geological evidence is discussed in that paper; only a brief summary of the most important stages leading to loading of the margin is given here.

Between 110–100 Ma a north to north-east dipping subduction zone developed some 800 km from the margin as a response to changing conditions in the closure of the Tethys Ocean (A).

The increasing subduction rate by 100 Ma as the descending slab length increased (B) resulted in an extensional regime above the subduction zone and the generation of new oceanic crust (C).

By 93 Ma the continental margin had been drawn into the subduction system. This both depressed the transitional crust and, in slowing the subduction rate, brought the system into compression. Spreading ceased and the stress concentration at the junction of old, cold Triassic crust and the Upper Cretaceous crust caused a dislocation (D). The marginal sediments had by then been subjected to thin-skinned thrusting which resulted in their thickening by c. 3:1.

Closure was then principally taken up by the new oceanic lithosphere overriding its older counterpart (E), the transitional continental lithosphere was still being drawn into the subduction system but was rapidly choking it. By 88 Ma that lithosphere had been loaded by the overthrust nappe pile (F) and shelf-edge sediments were becoming involved in the thrusting. Blueschist metamorphism was being impressed on parts of the subducted continental margin.

By 85 Ma (G) the overlying 15 to 25 km thickness of hot oceanic lithosphere (Lippard 1983) was locally sufficient to melt partially the continental crust and provide the biotite granite magmas. Overall closure was still slowing and the area was thus still under compression.

The combined effects of the, by then, extinct subduction zone and loading of the sedimentary and ophiolitic nappes brought the outer margin to its greatest depth. At 80 Ma overthrusting continued but that outer margin was attempting to re-equilibrate (H). Sub-aerial weathering on the upper surface of the ophiolite nappe attests to its uplift. Sediments were deposited in a foreland basin that migrated inland with time (Glennie et al. 1974).

Constraints on Numerical Modelling

The obduction stages at around 80 Ma can be simplified for computer modelling as one section of lithosphere being loaded upon another. Note that it is not claimed that the ophiolite has been thrust onto the margin. Its role is mostly passive and it does not have sufficient strength to perform otherwise. The dying subduction zone has done all the hard work in drawing down the marginal continental material beneath the new oceanic section. Upon complete cessation of subduction the system begins to re-equilibrate.

The margin is here represented by a *loading* of crust and mantle to a thickness of 23 km – the maximum preserved stratigraphic thickness of ophiolite to our knowledge. In order to account for the supra-subduction zone geochemistry and to conform to the minimum arc-SZ separation seen in present-day plate tectonics, approximately 100 km of marginward motion are required to position the ophiolite over the subduction zone. It is assumed that the passive margin, which displayed rifting, extension and heating in the Triassic had, by the Late Cretaceous, re-equilibrated or was in the last stages of very gentle subsidence. Certainly from modelling and examples of Steckler and Watts (1982) some 85% of the vertical movement is achieved within the first 100 Ma.

Most of the lithospheric physical properties can be estimated with some confidence from present-day geophysical measurements (e.g. Turcotte and Schubert, 1982). The parameter with the least certainty needed to model the lithospheric shortening is the flexural rigidity of the lithosphere at the time of emplacement. The effective flexural rigidity of oceanic lithosphere (early stages of emplacement) is strongly related to the age of the sea floor at the time of loading. Rheological models of such lithosphere (e.g. Bodine et al., 1981) indicate that the strength of the rocks and thus the equivalent elastic thickness (T_e) can be modelled as the depth to the 500°C isotherm. Additional evidence comes from interpretation of Free-Air gravity anomalies over loads forming at ridge crests (Watts et al., 1980); where the anomalies can be explained by a T_e of 5 km. Loads forming in off-ridge situations can be explained with elastic thicknesses of 25 km.

As for the flexural rigidity of marginal continental lithosphere, the calculations of Watts *et al.* (1982) provide a range of elastic thickness dependent on the tectonic setting. Rigidity is clearly linked to thermal age (i.e. time since rifting) with young continental lithosphere having a T_e of around 5 km increasing with age to above 30 km. The elastic thickness is again defined as the brittle-ductile transition depth, in this case at the 450°C isotherm.

These thicknesses represent extreme values. Even if restrained by age and tectonic type the range is large for predictive purposes because the geometrical outcome of crustal shortening is so sensitive to changes in rigidity. The constraints for this property are fortunately provided by the geological evidence for the geometry of the foredeep.

Foreland Basin Geometry and Flexural Rigidity

The width and more usefully the depth of the foredeep are controlled principally by the flexural rigidity and the amount of shortening (Beaumont, 1981; Jordan, 1981).

For a constant amount of shortening, the flexing of the lithosphere (as represented by width of foreland basin development) has an almost linear relationship with flexural rigidity (Figure 3). The initial width of the foredeep is estimated at 50 km for the north and central Oman Mountains (A.H.F. Robertson *pers comm*. 1990). The present gravity expression of the foredeep, though obscured on the eastern margin by the high gradients of the ophiolite nappe, predicts that the major expression of the foredeep does not exceed

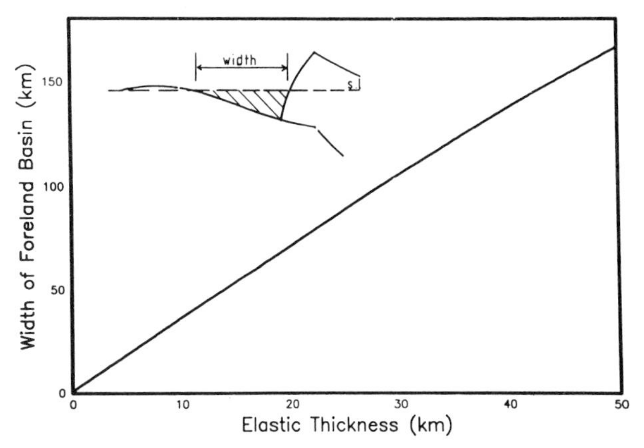

Figure 3. Width of foreland basin as a function of flexural rigidity after 100 km of shortening.

70 km in width. There is, however, evidence of minor subsidence extending over 110 km to the west of the mountains.

A foredeep width of 50 km after 100 km of shortening indicates an elastic thickness of about 14 km (Figure 3). That prediction can be cross-checked with the more sensitive relationship between flexural rigidity and depth of basin for the same loading. There is well-constrained geological evidence from Burruss et al. (1983) for the rapid deposition of at least 4 km of sediment in the northern foredeep during the Campanian. In Figure 4 flexural rigidity, as defined by elastic thickness, is plotted against the depth of the basin. At zero rigidity there is negligible basin but the depth increases rapidly to a maximum of around 7 km at a T_e of 30 km. Further increases in rigidity actually decrease the depth as the thrust load is increasingly supported by the flexural strength of the lithosphere rather than isostatic restoring forces. For 100 km of shortening a minimum elastic thickness of 8 km is predicted if the foredeep has a minimum original depth of 4 km.

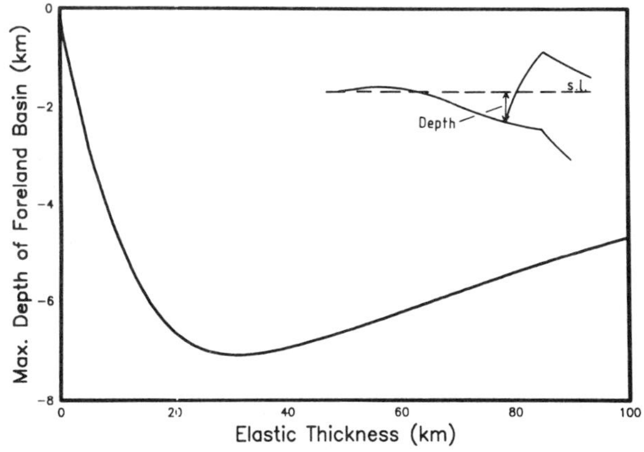

Figure 4. Depth of foreland basin as a function of flexural rigidity after 100 km of shortening.

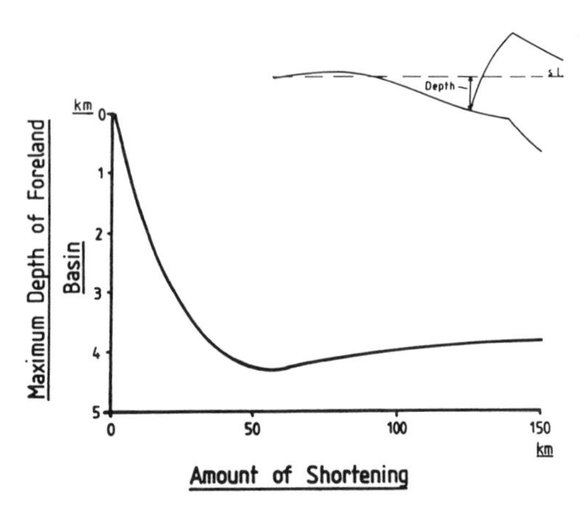

Figure 5. Depth of foreland basin as a function of shortening with elastic thickness fixed at 10 km.

A realistic elastic thickness lies somewhere between this minimum value and the 14 km more uncertainly predicted by estimates of original basin width. The sedimentary controls indicate that a $T_e = 10$ km is unlikely to have been exceeded at the time of emplacement.

The above calculations were made for a fixed amount of shortening and it might appear that any calculations based on the foredeep geometry would be invalidated if the amount of shortening were not accurately known. Fortunately for this argument, the basin geometry for a given T_e, turns out to be insensitive to loading of the lithosphere above a minimum amount of shortening that is most certainly exceeded in the case of the Oman margin.

Figure 5 shows that increasing the load by greater amounts of shortening (while keeping the elastic thickness fixed at the indicated 10 km) results in a maximum depth of 4.25 km beyond some 50 km. The depth stabilizes as the wavelength of the crustal load exceeds the flexing wavelength of the lithosphere.

The foredeep, insofar as it can be measured with the present over-thrusting on its seaward side, is now likely to be more extensive than it was in the Late Cretaceous due to the increased loading and the sediment infill. Figure 6 shows the flexural response to sediment load. In this calculation the addition of the sediment load increases the width of the basin to around 75 km and the depth by up to 2 km. In reality this depth increase will be compensated by the fact that the sediment infill is derived from the adjacent thrust sheets. This erosion results in unloading of the lithosphere and flexural uplift.

Lithospheric Response to Erosion

Having determined a reasonable value for T_e for the lithosphere at that time and place we can look at the effects of this and extreme values on the

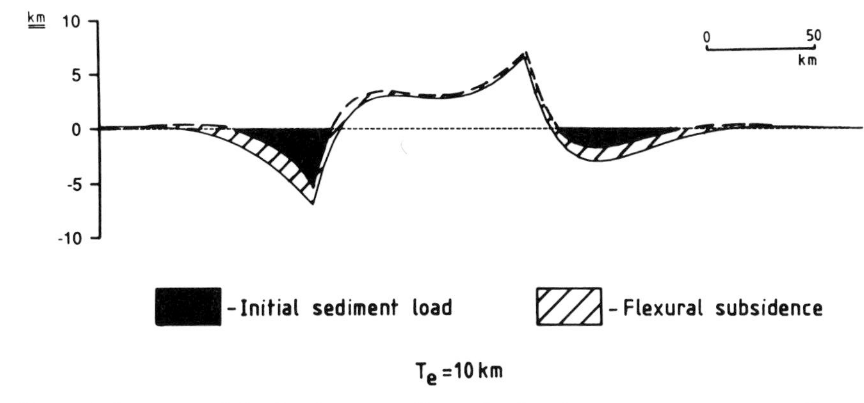

Figure 6. Response of the foredeep to sediment loading from an external source. Flexural rigidity is constrained to an effective elastic thickness of 10 km.

detailed geometry. Figure 7 shows the effects of increasing rigidity as defined by elastic thicknesses of 5, 10 and 20 km after 100 km of shortening.

It is evident that, even with very low flexural rigidity values, a considerable topographic expression will result from the shortening. The laterite first reported by Bailey and Coleman (1975) and interpreted in Coleman (1981) is indisputable evidence of the uplift of the new oceanic crust. The weathering product is up to 30 m thick and restricted to the upper surface of the ophiolite and is, on the field evidence, a post emplacement and pre-Maastrichtian

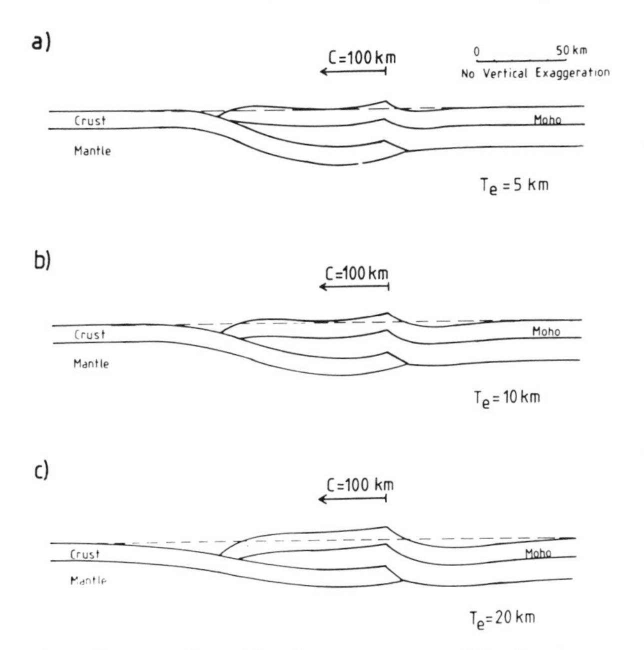

Figure 7. Comparison of topography and foredeep geometry resulting from extremes of flexural rigidity for the margin model.

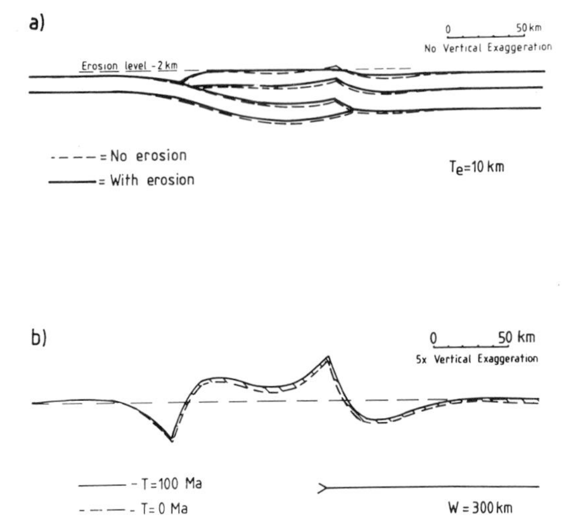

Figure 8.(*a*)·Isostatic and flexural response of the 10 km T_e case after 3 km of erosion from the highest sections. (b) Thermal uplift resulting from the re-equilibration of the temperature field after shortening.

process. The best-estimate elastic thickness results in approximately 5 km of relief. This value can be used in conjunction with the estimates of the amount of erosion that has taken place in the Oman mountains. Glennie et al. (1974) calculated that at least 3 km of erosion were required to account for the volume of flysch deposits comprising the western Gulf of Oman.

Figure 8(a) illustrates the result of the calculations for the erosion of only the elevated (those above 2 km) portions of the 10 km elastic-thickness loaded lithosphere. There is a resultant rebound of some 2 km over the entire history of the margin. A modest erosion rate of 0.23 mm/a would account for 3 km in 13 Ma, the time constrained by the age dating between loading of the lithosphere and the requirement for uplift. A significant proportion of the possible total can be achieved in those first 13 Ma.

As the greatest topography is over the outer margin the achievement of isostatic equilibrium results in a significant marginward slope and is proposed as a major contributor to the gravity-driven separation and final emplacement of the ophiolite nappe.

Modelling predictions suggest that the uplift will be increased by the effects of thermal re-equilibration. Following the relatively "instantaneous" shortening there is a regional uplift which could contribute as much as 500–1000 m up to the present day. Once again, as the thermal uplift follows an exponential rise, a significant proportion of this could be achieved in the early history of the margin. Figure 8(b) shows basement immediately following shortening and the situation 100 Ma later. The variation in uplift due to the re-equilibration across the margin is shown and it appears that this offers a further contribution to the required marginward slope.

The effects due to the pure shear thermal re-equilibration are more difficult to predict for this margin. It seems more likely that this region would be situated on the oceanic side of the zone where the lithosphere is thinnest, as such the subsequent uplift would only affect the sedimentation history of the Gulf of Oman and would not contribute to the emplacement mechanism.

Lithospheric Responses to Subduction

The Lippard et al. (1986) model asserts that the outer margin re- equilibration is the mechanism responsible for separating the western flank of the new oceanic lithosphere from its eastern counterpart by low-angle faulting at the palaeo-spreading axis (Figure 2, stage I). The uplift at the outer margin was

> "associated with a sympathetic down-flexure further inshore the development of a foredeep which migrated inland, ahead of the advancing nappes, is recorded in well-logged sedimentary sequences west of the present-day mountains".

That migrating fordeep was first defined by Glennie et al. (1974). It was considered by Lippard et al. as an unquantified flexural response to the isostatic rebound of marginal continental crust released from slab pull by thermal equilibration and possibly by break-up of the subducted oceanic lithosphere. It has been shown here to be a calculable response to loading of the lithosphere with an oceanic section. Moreover, the remaining necessity for an outer margin uplift to provide the gravity-driven emplacement and account for the inland migration of the basin has been provided by the erosion and thermal re-equilibration mechanisms outlined above.

The ill-defined "isostatic uplift" of the 1986 emplacement history has been quantified to an extent except that, because the modelling is limited to geologically simplistic situations, it is not yet possible to calculate the effects of the subduction zone.

The calculations of Molnar and Gray (1979) predicted that some 10 km of standard thickness continental material could be subducted before bringing the process to a standstill. A further 10 to 20 km of thinner marginal lithosphere are possibly involved in this consumption process.

The re-equilibration of the thermal field of the subducted slab is difficult to model with confidence and the resulting uplift will depend on the strength and uniformity of that slab. The tendency to neck or break-up increases unpredictably with depth. The uplift available from the subducted continental section is, if released in one burst, more than adequate to raise the ophiolite section above sea-level (Shelton 1984).

Royden and Karner (1984) observed that, for the Apennine and Carpathian thrust belts, the continental lithosphere was deflected by a much greater amount than could be caused by loading alone. An almost equivalent effect was considered to be due to the negative buoyancy of the subducted slab or possibly to thinning of the overriding slab by back-arc rifting. The analogue of this at the Oman margin would provide for the release of forces equal to the load upon the cessation of subduction.

The physical characteristics of an extinct subduction zone are extremely difficult to quantify. The cessation of slab-pull is governed by, among other factors, the thermal re-equilibration and depth reached by the subducted slab. Critical factors such as the subduction angle leave little evidence other than broad geochemical similarities to present day situations. The best-guess on depths reached by the lithosphere travelling at a moderate (45°) subduction angle would place it at the olivine/spinel phase change at around 97 Ma. The interaction with the phase-change boundary would act to accelerate crustal consumption at the time the ophiolite was being generated. As suggested in Lippard et al. (1986), initiation of spreading may be a direct result of this steepening in the crustal consumption rate curve (Figure 1). The combination of the extensional forces placed on the subducting slab and the resistance to subduction provided by the increasing thickness of buoyant continental crust at depth could have promoted the break-up of the slab which would have led to more rapid (and more easily modelled) isostatic rebound of the margin.

Lithospheric Response to Extension

In Figure 2, stages J-M, the ophiolite was shown as emplaced by a gravity glide from uplift generated by the rebound of the outer margin and into a foredeep controlled by lithospheric flexure. It was implied in the 1986 model that the rebound was assisted by the change to an extensional regime from ~ 76 Ma onwards as subduction gathered pace on the northern margins of the seaway under present-day Iran. This is on the downward slope of the consumption rate curve (Figure 1) and as such was very poorly constrained. Figure 9 shows the calculated response of the loaded lithosphere to extension of 30 km. The modelling considers extensional reactivation for about 10 Ma along a low-angle fault that surfaces at the palaeo-ridge axis. The geometrical predictions give the crustal section of the obducted ophiolite a realistic 30 degree gulfward dip. In response to the thinning of the lithosphere there has been a flexural rebound of nearly 3 km (3 mm/yr).

With the inclusion of realistic flexural rigidity in the model it becomes apparent that the stages following J in the emplacement scheme are too extreme; it is apparent that full restoration of the old margin sea-level surface will require considerably more than the indicated 100 km separation of the ophiolite sections. Further modelling is in progress to predict the response of the margin to 100 km and greater extension movements.

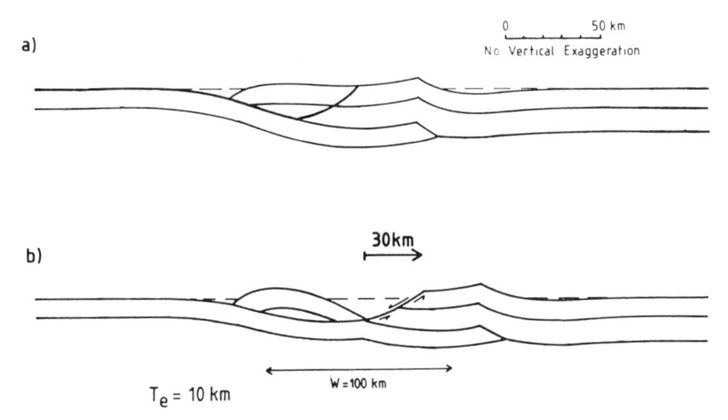

Figure 9. Response of the loaded lithosphere to the early stages of extensional reactivation. To be compared with the later stages (J to M) of Figure 2.

Conclusions

The emplacement scheme put forward in Lippard et al. (op. cit) has been tested by the modelling processes and quantified where possible by the inclusion of realistic lithospheric rheology.

The modelling has shown that the loading of a 23 km oceanic lithospheric section on to older oceanic or marginal continental crust will result in a foredeep that is consistent with the geological evidence. In particular, the depth and width of the foredeep are compatible with an elastic thickness of about 10 km – a value intermediate between that of newly-created and old lithosphere. The extent of the foredeep is likely to have been increased by the sediment infill, although the overall lithospheric response has been negligible because the sediments derive from the adjacent nappe.

Given the value of T_e derived from the foredeep geometry, the modelling predicts up to 6 km of topographic relief after shortening – an aspect only alluded to in the emplacement scheme. This relief would have been the object of substantial erosion during the Late Cretaceous. The early erosion provides a significant uplift as the lithosphere is unloaded – sufficient indeed with the contribution of the thermal uplift to generate the onshore slope necessary for gravity glide of the ophiolite nappe and foredeep migration.

There are still many uncertainties: most of these are associated with the subduction zone. It is held as evident that following ingestion of the continental lithosphere, subduction will cease and that, on re-equilibration of the cold oceanic slab, there will be sufficient isostatic response to lift the overlying oceanic lithosphere and emplace a section of it. What had not previously been considered was the timing of such processes.

It is here suggested that the subducted slab would have encountered the olivine/spinel phase-change boundary at the time that the first new oceanic crust was being created. It is further suggested that the resulting extensional forces set up in the slab combined with the resistance to subduction encountered only a few Ma later would lead to rapid break-up of the downgoing slab. In turn this would allow fast isostatic recovery of the margin (there are approximately 20 Ma allowed in the scheme between depression and uplift of the margin).

The final stimulus to uplift is the proposed extensional reactivation of the margin following re-establishment of subduction on the northern margin of Tethys. On this point the lack of rheological constraints in the emplacement model have lead to an exaggerated lithospheric response. The restoration of the old margin surface to sea-level required a separation of the obducted nappe and its parental slab by more than the 100 km indicated in the original scheme. Other than in this last aspect the emplacement scenarios put forward in Lippard et al. (1986) appear compatible with lithospheric behaviour.

References

Bailey, E.H. and Coleman, R.G., 1975. Mineral deposits in the Samail ophiolite of northern Oman., Geol. Soc. Amer. Abstr. Program, 7(3): 293.

Beaumont, C. 1981. Foreland Basins., Geophys. J.R. astr. Soc., 65: 291–329.

Bodine, J.H., Steckler, M.S. and Watts, A.B., 1981. Observations of flexure and the rheology of the oceanic lithosphere., J. Geophys. Res. 86: 3695–3707.

Burruss, R.C., Cercone, K.R. and Harris, P.M., 1983. Fluid intrusion petrography and tectonic-burial history of the Al Ali No 2 well: Evidence for the timing of diagenesis and oil migration, northern Oman foredeep., Geology 11: 567–570.

Coleman, R.G., 1981. Tectonic setting for ophiolite obduction in Oman., J. Geophys. Res. 86 (B4): 2497–2508.

Egan, S.S., 1989. Rheological, thermal and isostatic constraints on continental lithosphere and compression., PhD Thesis, Keele University, 155 p.

Glennie, K.W., Boeuf, M.G.A., Hughes Clark, M.W., Moody-Stuart, M., Pilaar, W.F.H. and Reinhardt, B.M., 1974. The geology of the Oman Mountains., Konin. Neder. Geol. Mijnbouw. Genoot. Verdh.31, part 1 (text) 423 p., part 2 (illustrations), part 3 (enclosures).

Jackson, J.A. and McKenzie, D.P., 1983. The geometrical evolution of normal fault systems., J. Struct. Geol., 5: 471–482.

Jordan, T.E., 1981. Thrust loads and foreland basin evolution, Cretaceous, Western United States., Bull. Am. Ass. Petrol. Geol., 65: 2506–2520.

Kusznir, N.J. and Park, R.G., 1987. The extensional strength of the continental lithosphere: its dependence on geothermal gradient, crustal composition and thickness. In: Coward, M.P., Dewey, J.F. and Hancock, P.L. (eds), Continental extensional tectonics., Geol. Soc. London Spec. Pub. 28 35–52.

Lippard, S.J., 1983. Cretaceous high pressure metamorphism in NE Oman and its relationship to subduction and ophiolite nappe emplacement., J. Geol. Soc. London, 140: 97–104.

Lippard, S.J., Shelton, A.W. and Gass, I.G. 1986. The ophiolite of Northern Oman., The Geological Society Memoir number 11, Blackwell Scientific Publications, Oxford, 178 p.

Molnar, P. and Gray, D., 1979. Subduction of continental lithosphere: Some constraints and uncertainties., Geology, 7: 58–62.

Royden, L. and Karner, G.D., 1984. Flexure of the continental lithosphere beneath the Apennine and Carpathian foredeep basins., Nature, 309: 142–144.

Shelton A.W., 1984. Geophysical studies on the Northern Oman Ophiolite., PhD thesis, Open University, 353 p.

Steckler, M.S. and Watts, A.B., 1982. Subsidence history and tectonic evolution of Atlantic-type continental margins. In: Scrutton, R.A. (ed), Dynamics of Passive Margins, Geodynamics Series Vol 6, AGU, Washington, pp. 184–196.

Turcotte, D.L. and Schubert, G., 1982. Geodynamics: Applications of Continuum Physics to Geological Problems. John Wiley and Sons, 450 p.

Watts, A.B., Bodine, J.H. and Ribe, N.M., 1980. Observations of flexure and the geological evolution of the Pacific ocean basin., Nature, 283: 532–537.

Watts, A.B., Karner, G.D. and Steckler, M.S., 1982. Lithospheric flexure and the evolution of sedimentary basins. In: Kent, P., Bott, M.H.P., McKenzie, D.P. and Williams, C.A. (eds.), The Evolution of Sedimentary Basins., Phil. Trans R. Soc., A305: 249–281.

Amagmatic Extension and Tectonic Denudation in the Kizildağ Ophiolite, Southern Turkey: Implications for the Evolution of Neotethyan Oceanic Crust

YILDIRIM DILEK *[1], ELDRIDGE M. MOORES [1],
MICHEL DELALOYE [2], and JEFFREY A. KARSON [3]
[1] Department of Geology, University of California, Davis, CA *95616*, USA
[2] *Mineralogy Department, The University, CH-1211 Geneva 4, Switzerland*
[3] *Department of Geology, Duke University, Durham, NC 27706, USA*
* *Current Address: Department of Geology & Geography, Vassar College, Poughkeepsie NY 12601, USA*

Abstract

The Cretaceous Kizildağ ophiolite in southern Turkey represents a remnant of oceanic lithosphere formed in a southern strand of the Mesozoic Neotethys. The northeast-southwest trending ophiolite complex consists, from bottom to top, of serpentinized peridotites, cumulate and isotropic gabbros, sheeted dikes, and massive and pillow lava flows. The well-preserved sheeted dike complex and underlying plutonic rocks are best exposed in a topographically and structurally defined narrow graben on the southeastern flank of a laterally extensive antiform the core of which is an axial high of serpentinized peridotites. High-angle normal faults associated with local horst-graben structures and low-angle normal faults associated with rotated blocks are common within the sheeted dike complex and are reminiscent of spreading related structures at slow-spreading ridge segments. The fault contact between sheared gabbros and serpentinized peridotites dips away from peridotites and beneath gabbros in a fashion similar to detachment surfaces documented in several other ophiolite complexes and in modern oceanic crust. The extrusive sequence of the ophiolite is commonly underlain by serpentinite and gabbro and includes highly tilted massive and pillow lavas locally intruded by sheeted and isolated dikes. These structures and contact relationships in the Kizildağ ophiolite are atypical of a characteristic ophiolite template and are interpreted to have resulted from a period of tectonic extension and denudation following magmatic construction of ophiolitic crust. A low-angle fault dipping southeast (in present coordinates) toward the axial graben (inferred spreading axis) is interpreted to have accommodated tectonic extension at crustal levels and uplift and exposure of upper mantle rocks. This postulated master normal fault is locally preserved on the northwest and southeast flanks of the antiform, which probably developed during isostatic rebound of upper mantle rocks as a result of excessive serpentinization and diapiric activity during and after displacement of oceanic crust from its original spreading environment. We envision a model of asymmetric exten-

Tj. Peters et al. (Eds), Ophiolite Genesis and Evolution of the Oceanic Lithosphere, 485–500.

sion, analogous to that suggested for magma starved segments of the Mid-Atlantic Ridge around 23°N latitude. The geometry of the inferred asymmetric extension was parallel to an extensional shear zone that may have cut through the entire lithosphere of the northern edge of Arabia resulting in continental break-up in Late Triassic time. The Kizildağ ophiolite represents a fragment of a Neotethyan oceanic spreading center that was subsequently emplaced onto the passive margin of Arabia possibly by conversion of a ridge-parallel normal fault to a thrust fault (incipient subduction zone?) as a result of a change in relative plate motion in late Cretaceous time.

Introduction

Ophiolites have been interpreted as on-land fragments of oceanic lithosphere developed as a result of seafloor spreading at oceanic spreading centers (Coleman, 1971; Dewey and Bird, 1971; Moores and Vine, 1971; Pallister and Hopson, 1981). The inferred "layer-cake" structure of oceanic lithosphere is thought to form from a steady-state magma chamber located beneath the spreading center (Cann, 1974; Dewey and Kidd, 1977). Periodic replenishments of such a magma chamber would result in production of lavas with limited ranges in composition (Pallister and Hopson, 1981). Geological, geophysical, and petrological observations from intermediate to fast-spreading ridge environments are consistent with this steady state magma chamber model (e.g., Detrick et al., 1987); recent findings from slow-spreading ridges indicate, however, the absence of laterally continuous and steady-state magma chambers beneath the spreading axes (Karson, 1990; Purdy and Detrick, 1986). Planar and listric normal faults stretching and thinning the existing oceanic crust are widespread in slow-spreading ridge environments; in places mafic and ultramafic plutonic rocks are exposed on the seafloor along and across slow-spreading ridge segments away from fracture zones. These structures are inferred to indicate the occurrence of amagmatic (tectonic) extensional processes separating phases of magmatic construction during evolution of oceanic crust (Karson, 1990; Tiezzi and Scott, 1980). Similar extensional structures have been observed in some ophiolite complexes (e.g., Troodos ophiolite in Cyprus, Varga and Moores, 1985; Josephine ophiolite in northern California, Harper, 1988) indicating that these ophiolites underwent large amounts of tectonic extension during their evolution and prior to their emplacement.

The nature of tectonic extension inferred to occur at slow-spreading ridge segments or to have been a significant part of ophiolite generation processes is problematic. Tectonic extension at slow-spreading ridges is commonly assumed to be a homogeneous, pure-shear style of deformation occurring along normal faults parallel to the ridge axis resulting in a thinned crust symmetric with respect to the spreading axis. Such interpretations require, however, very large magnitudes of extension relative to magmatic construc-

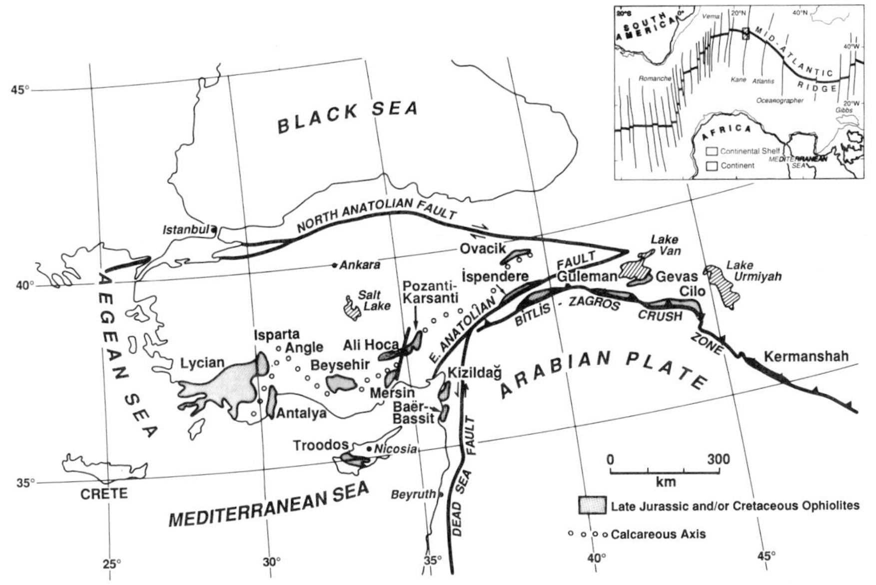

Figure 1. Outline map of the eastern Mediterranean region showing the distribution of ophiolites and other major tectonic elements (after Dilek and Moores, 1990). The calcareous axis represents discontinuous exposures of Mesozoic carbonate platforms underlain by a Paleozoic and older (?) basement of Gondwana-derived continental fragments. Inset map shows the location of the MARK area (the Mid-Atlantic Ridge near the Kane Transform) and other significant fracture zones along the Mid-Atlantic Ridge in the central Atlantic Ocean. Note that the inset map is rotated 90° clockwise to show the similarities between the distribution and spacing of spreading ridge-fracture zone segments northwest of Africa with that of the Cretaceous ophiolites around the northern edge of the Arabian promontory.

tion for the exposure of mafic-ultramafic rocks at the walls of the spreading cells. An alternative kinematic model suggests asymmetric extension with respect to the median valley that occurs along a system of detachment faults intersecting the seafloor near the crest of the median valley (Karson, 1990). This kind of asymmetric extension, accommodated by low-angle crustal shear zones, is well-documented in some domains of the Basin and Range province of the western U.S. (Wernicke, 1985), along some passive continental margins (Boillot et al., 1987; Lemoine et al., 1987), and in the MARK area located south of the Kane Transform on the Mid-Atlantic Ridge at about 23°N latitude (Figure 1; Karson, 1990), but has not been demonstrated in ophiolite complexes.

The Cretaceous Kizildağ ophiolite in southern Turkey (Figure 2) may be an example of a slow-spreading ridge fragment that underwent tectonic extension possibly along a low-angle shear zone following its magmatic construction. The Kizildağ ophiolite, a remnant of Neotethyan ocean crust, contains all subunits of an ophiolite stratigraphy and tectonically overlies relatively undeformed Lower Cretaceous shelf carbonates of the Arabian platform (Delaloye and Wagner, 1984; Selçuk, 1981). A northeast-southwest trending, laterally extensive exposure of serpentinized tectonites occupies an axial high in the ophiolite, whereas highly attenuated and thinned crustal sequence crops out within a small synform on the southeast flank of this axial high and as low-angle fault blocks over the peridotite in the northeastern and northwestern parts of the ophiolite. We suggest that these contact relationships and the internal structure of the ophiolite are pre-obduction features that developed due to tectonic denudation at a slow-spreading environment after crystallization of the magma chamber. The inferred slow-spreading origin and the regional geologic setting of the Kizildağ ophiolite suggest that this piece of oceanic crust may represent the edge of the Mesozoic Neotethys adjacent to the passive margin of Afro-Arabia, which evolved after a continental break-up in Late Triassic time.

Regional Geology

The Neotethys represents a Mesozoic ocean basin that evolved between the Afro-Arabian platform to the south and the Eurasian hinterland to the north as the Paleozoic-early Mesozoic Paleotethys was diminishing (Sengör and Yilmaz, 1981; Dilek and Moores, 1990). The inferred paleogeography of Neotethys, albeit controversial, suggests the existence of multiple strands of an ocean floored by oceanic crust and separated by continental fragments and volcanic arc terranes. Cretaceous ophiolites exposed in nearly east-west trending discontinuous belts around the Arabian promontory (*le croissant ophiolitique péri-arabe* of Ricou, 1971; Figure 1) are possible remnants of Neotethyan oceanic lithosphere. The ophiolites cropping out north of the microcontinents (the calcareous axis in Figure 1 as introduced by Ricou et al., 1975) are generally deformed and dismembered, associated with metamorphic soles and melanges, and commonly intruded and/or overlain by island-arc rocks (Dilek and Moores, 1990). The ophiolites in the southern belt south of the calcareous axis include relatively undeformed and stratigraphically complete complexes, namely the Antalya and Kizildağ ophiolites in southern Turkey (Dubertret, 1955; Çoğulu et al., 1975; Delaloye et al., 1977; Juteau et al., 1977), the Baër-Bassit ophiolite in Syria (Parrot, 1980), and the Troodos massif in Cyprus (Gass, 1967; Moores and Vine, 1971; Figure 1).

The tectonic position of the ophiolites in the southern belt between the Afro-Arabian platform on the south and the east-west trending microcontinents of possible Gondwana origin (Sengör and Yilmaz, 1981; Dilek and Moores, 1990) on the north suggests that Neotethyan oceanic crust in the southern strand developed after disruption and rifting-off of the northern margin of Afro-Arabia. The onset of this continental rifting is interpreted to have occurred around Late Triassic time as suggested by the existence of Upper Triassic alkalic lavas and rift-related terrigenous sedimentary rocks tectonically associated with the ophiolites (Delaloye and Wagner, 1984; Whitechurch et al., 1984). Cretaceous igneous ages of the ophiolites indicate, however, that there was a considerable time gap between inception of the continental break-up (Late Triassic) and development of oceanic floor (early Late Cretaceous) in southern Tethys. Mafic rocks from ophiolitic subunits in the Antalya complex yielded a wide range of ages clustering between 91.9 ± 2.3 to 55.5 ± 4.7 Ma (K/Ar whole rock, plagioclase, and hornblende ages from gabbro, diorite, and sheeted diabase rocks; Yilmaz, 1984). The sheeted dikes from the Kizildağ ophiolite revealed isochron ages between 73 to 99 Ma (K/Ar whole rock ages) although older ages (i.e., Late Jurassic?) cannot be ruled out as indicated by secondary amphiboles separated from gabbroic rocks (Delaloye and Wagner, 1984). The sheeted dikes and pillow lavas in the Troodos ophiolite provided 75 to 85 Ma K/Ar dates (Desmet et al., 1978; Delaloye and Desmet, 1979; Staudigel et al., 1986); however, recent U/Pb isotopic dates from plagiogranite rocks in the plutonic complex indicate much older (90.3 ± 0.7 to 92.4 ± 0.7 Ma) ages for the ophiolite (Mukasa and Ludden, 1987).

The ophiolites in the southern belt have well-developed sheeted dike complexes that imply extension and magmatic spreading. Some ophiolites (e.g., Troodos and Antalya ophiolites in Figure 1) display rock types, structures, and contact relationships typically observed at and/or near ridge-transform intersections (Reuber, 1984; Varga and Moores, 1985) and thus are thought to have developed in such tectonic environments (Dilek and Moores, 1990). The orientation of sheeted dike complexes, the inferred geometry of ridge-transform intersections, and the mantle flow structures in these ophiolites suggest nearly east-west trending paleo-spreading axes separated by north-south trending transform faults (Whitechurch et al., 1984; Dilek and Moores, 1990). This inferred geometry of the paleo-spreading centers is consistent with a regional north-south rifting event following the Late Triassic continental break-up.

Geology of the Kizildağ ophiolite

The Kizildağ ophiolite includes a core composed of serpentinized tectonites overlain to the southeast successively by gabbros and sheeted dikes preserved in a synform and to the northwest and northeast in fault-bounded blocks of gabbros, sheeted dikes, and lavas (Figure 2). The estimated minimum thickness of the serpentinized peridotite is around 2.5 km based on the elevation difference between the highest and lowest exposures. Most of the peridotite is made of harzburgite with local lenses and bands of dunite. Harzburgitic rocks display layering and foliation that are subsequently folded about northeast-southwest trending axial surfaces. Numerous pegmatitic gabbroic dikes and veins, which are oblique and/or perpendicular to the foliation planes, intrude tectonized peridotites and are commonly oriented parallel to the general northeast trend of the ophiolite.

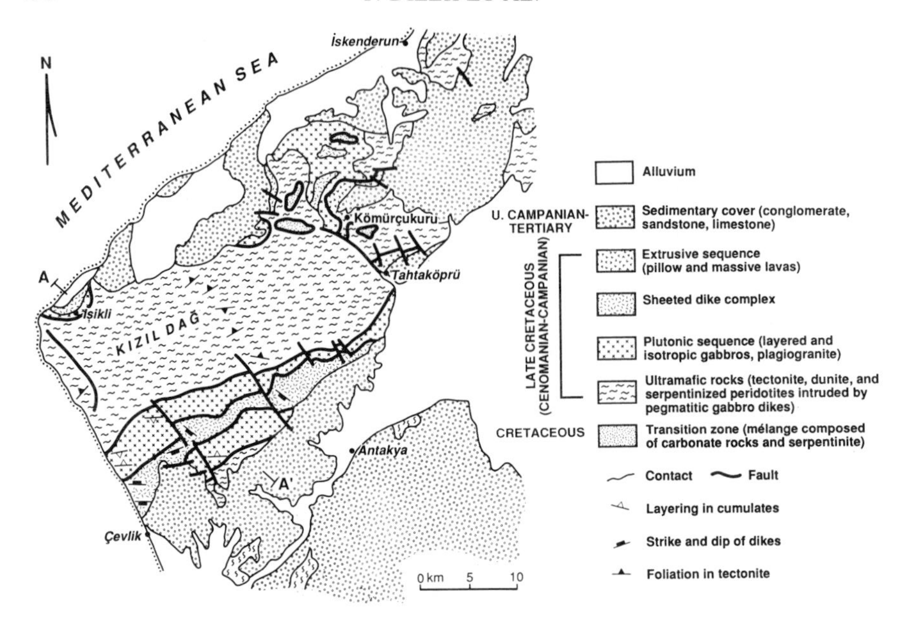

Figure 2. Simplified geologic map of the Kizildağ ophiolite. Data from Selçuk (1981), Delaloye and Wagner (1984), Erendil (1984), Dilek and Moores (1985), Piskin et al. (1986). A-A′ is the cross-section line in Figure 4A.

The contact between serpentinized peridotites and the plutonic sequence is faulted both on the north and the south (Selçuk, 1981; Dilek and Moores, 1985; Piskin et al., 1986). North of Çevlik along the coast (Figures 2 and 3A) the plutonic sequence is separated from underlying peridotites along a moderately to gently southeast dipping fault. Gabbroic rocks above this contact show well-preserved cumulate textures that are commonly cut by low-angle mylonitic shear zones. Boudinage structures and necking of layers are abundant at this level. Stratigraphically upward in the plutonic sequence are isotropic gabbros consisting mainly of clinopyroxene, amphibole, and plagioclase. Isotropic gabbros are intruded by numerous plagiogranite dikes that are in turn cross-cut by abundant diabase dikes. However, locally both isotropic gabbro and plagiogranite display chilled margins against each other, suggesting recurring and intermittent intrusive relationships (Dilek and Moores, 1985).

The well-developed sheeted dike complex is separated in places from the underlying plutonic sequence by a high- to low-angle fault zone, whereas in other places the base of the dike complex is intruded by small gabbroic plutons (Figures 3A and B). High-angle normal faults associated with horst and graben structures are common within the sheeted dike complex (see Figure 4; Tinkler et al., 1981; Dilek and Moores, 1985). Based on cross-cutting relationships and outcrop appearance, at least four generations of dike intrusions with one-sided and two-sided chilled margins can be distin-

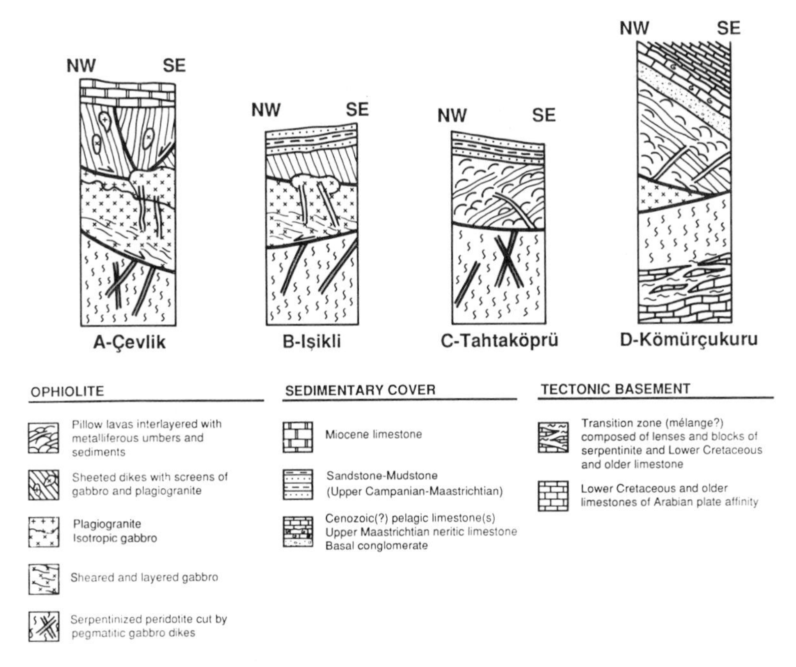

Figure 3. Schematic column sections depicting internal stratigraphy and structure of ophiolitic subunits, and various contact relationships between these units, sedimentary cover, and tectonic basement in Kizildağ. Location names refer to the villages that are shown in Figure 2.

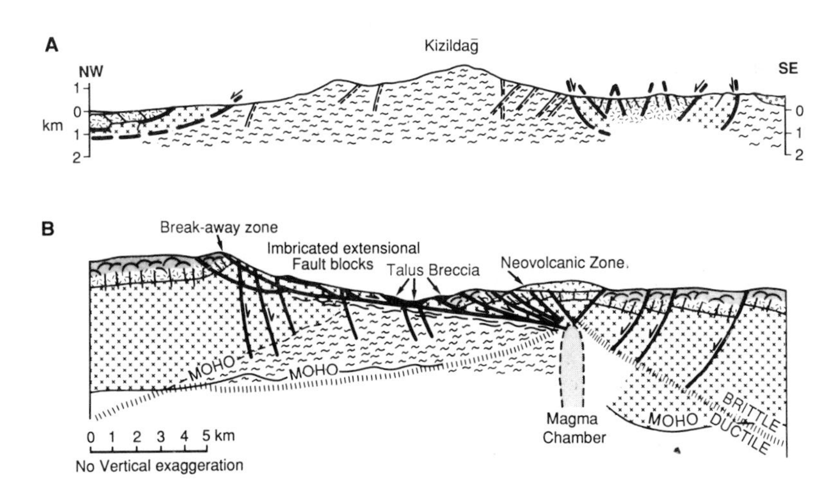

Figure 4. Geologic cross-sections of (A) the Kizildağ ophiolite and (B) the western wall of the Mid-Atlantic Ridge (MAR) south of the Kane Transform (MARK area; after Karson, 1990). See the inset in Figure 1 for the location of the MAR segment and Figure 2 for the legend.

guished in the dike complex (Dilek and Moores, 1985). The first and second generation dikes form subparallel dike swarms, have irregular chilled margins, and include screens of isotropic gabbro. These dikes are cut by thin mafic dikes that are in turn cross-cut locally by dark-colored and fine-grained mafic dikes with zig-zag outcrop patterns. The east-west orientation of sheeted dikes is oblique to the northeast-southwest trending foliation and lineation in tectonized peridotites and to the generally northeast trend of the ophiolite. On the northwestern flank of the peridotite exposure near Isikli (Figure 3B) sheeted and isolated dikes striking northeast and dipping to the southeast intrude a thin sliver of a sheared gabbro and/or are cut by a gentle fault surface separating them from the gabbro outcrop.

The extrusive sequence is poorly preserved in the ophiolite and is in tectonic contact with underlying mafic plutonic rocks and peridotites (Figure 2). Near Tahtaköprü a 200 to 300 meter thick extrusive sequence composed of pillow lavas, massive lava flows, and interflow sediments rest tectonically on serpentinized peridotites along a southeast-dipping fault surface (Figure 3C). Shape and size of pillow lavas vary considerably, and brecciated pillows are very common in this outcrop. Massive lava flows pass laterally and vertically into pillow lavas, and both pillow and massive lava flows are interlayered with thin red chert and brownish manganiferous mudstone beds (Erendil, 1984). The existence of dikes and sills subparallel to pillow margins indicates intrusive activity contemporaneous with extrusion of the lavas. Extrusive rocks in this locality are stratigraphically overlain by a sequence of interbedded coarse-grained sandstones and mudstones of a possible shallow sea environment.

Extrusive rocks near Kömürçukuru (Figures 2 and 3D) directly overlie gabbroic and ultramafic rocks along gently west-northwest dipping fault(s) and composed mainly of pillow lavas with steep to vertical inclinations. Dike swarms intruding these lavas dip moderately (30 to 40°) to the west-northwest. Characteristically pillow lavas are highly brecciated and include zones of extensive bleaching enriched in sulphide minerals, such as pyrite, chalcopyrite, and malachite; sediment horizons interbedded with pillow lavas include brownish silicified manganiferous mudstones containing manganite and pyrolusite (Erendil, 1984). These features suggest extensive hydrothermal alteration in the extrusive sequence. Farther northeast these lava flows are stratigraphically overlain by steeply east-dipping sedimentary rocks. This sedimentary sequence starts at the bottom with a conglomerate-breccia horizon passing upward into a neritic limestone with reworked Upper Maastrichtian rudists and lamellibranches, that in turn is overlain by a pelagic limestone of Paleogene(?) age.

A separate outcrop of extrusive rocks mainly composed of pillow lavas up to 300 meter thick, termed sakalavite by Dubertret (1955), crops out in a tectonic contact with a serpentinite exposure approximately 12 km south of Antakya (Figure 2). Individual pillows are embedded in a glassy rim with well-developed quench textures. Both pillows and the glassy rim contain egg-shaped micro-pillows. The lavas have high-Mg, boninitic composition (Laurent et al., 1980) and are interbedded with metalliferous sediments associated with turbiditic sandstones and siltstones. These clastic rocks contain material derived from a plutonic terrane of a possible ophiolite origin (Robertson, 1986). The fossils recovered from interlayered and overlying sedimentary rocks (Selçuk, 1981) indicate an Upper Campanian age for the high-Mg lavas which is similar to or slightly younger than the age of the Kizildağ ophiolite.

The autochthon beneath the Kizildağ ophiolite is exposed northwest of Kömürçukuru (Figure 2) where serpentinized ultramafic rocks rest tectonically on Lower Cretaceous and older carbonate rocks of possible Arabian platform affinity. Both serpentinites and carbonates are sheared and mutually include each other as tectonic blocks and/or sedimentary lenses in a transition zone beneath the contact. Pebbly and bituminous limestones beneath the transition zone locally contain shale, iron oxide, and reworked serpentinite layers and display conjugate sets of en échelon tension gashes and boudinage structures.

Implications for Tectonic Extension in the Ophiolite

Relatively undeformed Upper Campanian and younger sedimentary rocks overlying various ophiolitic subunits indicate that the Kizildağ ophiolite and its sedimentary cover did not experience any major post-emplacement deformation. Complicated crustal structures and contact relationships within the ophiolite are, therefore, likely to have resulted from syn- or pre-emplacement tectonic events. Well-preserved igneous textures and sedimentary contact relationships at upper crustal levels, undeformed pillow lavas and dikes, and the absence of thrust-related imbrication as seen in the Baër-Bassit ophiolite farther south in Syria (Parrot, 1980) indicate that internal deformation during emplacement of the ophiolite was slight. These relations suggest a pre-emplacement and possibly spreading-related origin of the structure in the Kizildağ ophiolite.

The northeast-southwest trending synform that includes the sheeted dike complex and plutonic rocks on the southeast flank of the peridotite exposure is geometrically and morphologically similar to median valleys of slow-spreading ridge segments (Macdonald, 1982). Horst and graben structures and planar to listric normal faults within the dike complex are commonly observed in such environments. The dike complex is separated from the underlying plutonic sequence by faults dipping toward each other and toward the center of the synform (Figure 4A). This structure defines a graben that resembles a spreading axis of a slow-spreading mid-ocean ridge. Similar graben structures are observed within the sheeted dike complex of the Troodos ophiolite (e.g., the Solea and Mitsero grabens) and are interpreted as abandoned axial valleys of spreading centers (Varga and Moores, 1985).

Thus the northeast-southwest trending synform may be a fossil spreading axis in the ophiolite (Dilek and Moores, 1985).

Several lines of evidence suggest that the Kizildağ ophiolite underwent tectonic extension following its magmatic construction. The plutonic sequence of the ophiolite is extremely thin compared to that of normal oceanic crust and is sheared along the fault contact that juxtaposes it against serpentinized peridotites (Figure 3A). Ultramafic cumulates commonly seen in a transition zone between layered gabbros and upper mantle peridotites in well-preserved ophiolites are either absent or insignificantly thin in the Kizildağ ophiolite. In places (e.g., Kömürçukuru and Tahtaköprü) highly brecciated and tilted pillow lavas directly overlie serpentinized peridotites and gabbroic rocks suggesting that upper crustal levels of the ophiolite (i.e., gabbros and sheeted dikes) were partly to entirely stripped away due to tectonic extension and that lower crustal levels and upper mantle rocks were exposed on the seafloor during and/or before the eruption of lavas.

The exposure of mafic-ultramafic rocks on the seafloor and thinning of the plutonic sequence imply large magnitudes of tectonic extension during generation of the ophiolite (Figure 5). A symmetrical, pure-shear extension model would require very large magnitudes of extension relative to magmatic construction to expose these rocks on the seafloor. Asymmetric extension along a low-angle shear zone (detachment surface) that intersects the seafloor near the crest of the spreading axis would explain, however, exposure of deep crustal rocks along and near the spreading center (Figure 5). A similar model is suggested by Karson (1990) for the geology of the western median valley wall along the Mid-Atlantic Ridge (MAR) south of the Kane Transform (MARK Area in Figure 1) where mafic and ultramafic rocks are exposed on the seafloor along low-angle detachment faults and shear zones dipping toward and beneath the median valley axis. Figure 4B shows a reconstructed schematic diagram, at the same scale as the geologic cross-section of the Kizildağ ophiolite, depicting asymmetrical geometry of extension across a MAR segment near the Kane Transform where the magma budget is very small. Comparison of the two diagrams in Figure 4 suggests an apparent structural asymmetry in both regions. The three structural domains, defined by the synformal upper plate composed of gabbros and sheeted dikes on the southeast, the antiformal uplift composed of serpentinized harzburgites in the lower plate, and the wedge-shaped upper plate composed of gabbros and sheeted dikes on the northwest, suggest the existence of an originally southeast dipping master low-angle fault beneath the spreading axis of the Kizildağ ophiolite. This inferred geometry is similar to that of the MARK area; however, there is a major difference between the two regions in that the mantle section beneath the break-away zone in the MARK area does not display a topographic high as do the mantle rocks in the Kizildağ ophiolite. It is likely, nonetheless, that highly extended terrane (the detachment surface) in the MARK area will be uplifted to the rift mountains as spreading continues during the next few thousand years. We think that the current domal

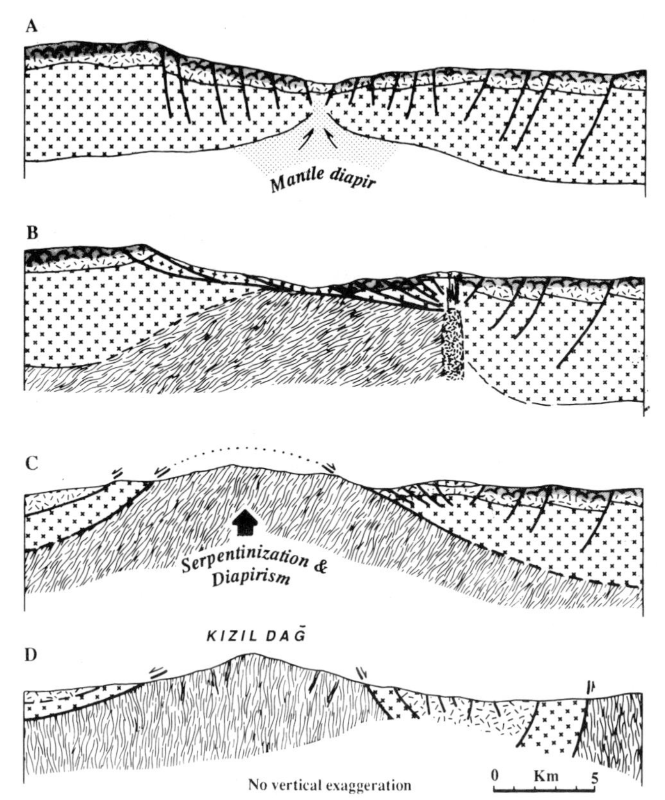

Figure 5. Interpretive geologic cross-sections depicting sequential stages during tectonomagmatic evolution of the Kizildağ ophiolite. Legend is the same as in Figure 2. (A) Seafloor spreading and magmatic extension developing ophiolitic crust. (B) Tectonic extension along a low-angle detachment surface and thinning of the ophiolitic crust. Eruption of high-Mg lavas occurs following the period of tectonic extension. Triangle-pattern designates talus and mafic breccia that has been subsequently removed by erosion. (C) Tectonic denudation, isostatic rebound and warping of the detachment surface, and uplifting of upper mantle rocks. Extensive serpentinization and associated diapirism occur at this stage. (D) Stripping away of crustal units, normal faulting, and further uplifting of serpentinized peridotites. This stage may have occurred during and/or after emplacement of the ophiolite onto the northern margin of Arabia.

shape of the postulated master normal fault in the Kizildağ ophiolite may have resulted from uplifting of the unloaded footwall analogous to metamorphic core complexes in the western U.S. Cordillera in addition to isostatic rebound and warping of the low-angle normal fault, and uplifting of upper mantle rocks as a result of extensive serpentinization and diapiric activity during and after displacement of the ophiolite from its original spreading environment (Figure 5).

The Upper Campanian age of high-Mg lavas overlying ultramafic rocks indicates that these primitive lavas were erupted following the generation of ophiolitic crust and possibly after the period of amagmatic extension. The presence of ophiolite derived sediments interlayered with the lavas supports

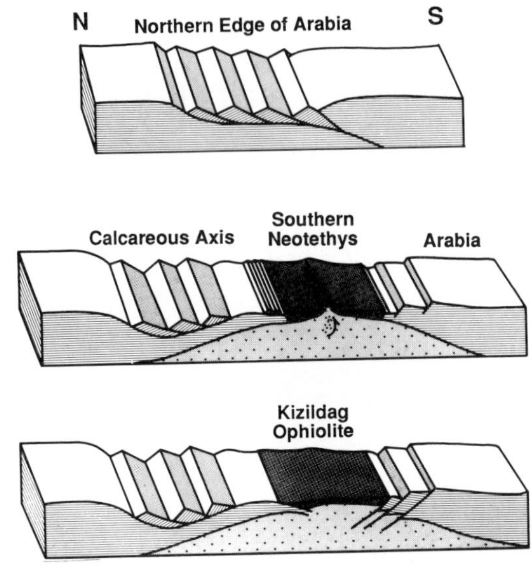

Figure 6. Schematic block diagrams depicting the inception of continental break-up in Late Triassic time at the northern edge of Arabia along a low-angle crustal shear zone (top diagram) and the evolution of the passive margin of Arabia and an oceanic spreading center in the southern strand of Neotethys in Cretaceous time (middle diagram). This model is reminiscent of the inferred low-angle crustal shear zones in continental extensional regimes as proposed by Wernicke (1985). The Kizildağ ophiolite, developed at this spreading center, is emplaced southward onto the Arabian passive margin due to a regional contraction in Late Cretaceous time (bottom diagram). See text for discussion.

this interpretation. These kinds of lavas are interpreted as products of partial melting of depleted upper mantle that are erupted directly on the seafloor in the absence of a magma chamber (see, for example, Harper, 1988), and this interpretation would suggest that the magma chamber beneath the Kizildağ spreading axis was already crystallized before their eruption. This implication is consistent with the episodic nature of magmatism and the absence of a steady-state magma chamber at slow-spreading ridge segments.

Evolution of the Arabian Passive Margin

The internal structure of the Kizildağ ophiolite shows evidence for amagmatic and asymmetric extension of Neotethyan ocean crust at a slow-spreading environment. We propose that the geometry of this extension may have been inherited from initial stages of the opening of the southern strand of Neotethys. The Late Triassic continental break-up along the northern edge of Arabia may have resulted from an asymmetric extension along a low-angle crustal shear zone dipping southeast and cutting through the entire lithosphere of northern Arabia (Figure 6A). This model is similar to the documented large-scale, uniform-sense normal simple shear of the continental lithosphere in parts of the Basin and Range province of the western U.S. (Wernicke, 1985) and to the inferred asymmetrical uncovering of the mantle by an oblique detachment fault during opening of the Ligurian Tethys in Jurassic time between the European and Apulian continental blocks (Lemoine et al., 1987). Continued extension along the postulated master detachment fault would eventually lead to continental break-up and to the formation of an ocean basin immediately north of Arabia around Late Triassic time (Figure 6B). This ocean basin may have evolved as an intra-cratonic rift or foredeep without oceanic substratum until Early to Late Cretaceous time, during which the first oceanic floor composed of mantle-derived peridotites appeared. The transition zone between the serpentinites and the platform carbonates at the basement of the ophiolite may characterize this stage of the ocean basin evolution (Tinkler et al., 1981; Erendil, 1984; Dilek and Moores, 1985; Robertson, 1986). Decompression and subsequent partial melting of mantle beneath the low-angle denudation zone may have developed a magma chamber beneath a slow-spreading ridge from which crustal rocks of the Kizildağ ophiolite (i.e., gabbros and sheeted dikes) were generated. A regional compression due to rapid northward convergence of Afro-Arabia around Late Cretaceous time (Dilek and Moores, 1990) would have ceased spreading and resulted in conversion of a ridge-parallel normal fault to a thrust fault (incipient subduction zone?) and/or reactivation of extensional low-angle faults as thrust faults that facilitated emplacement of the ophiolite southward onto the Arabian passive margin (Figure 6C).

The proposed asymmetric extension model for evolution of the Arabian passive margin and of Neotethyan ocean crust is consistent with observations from both ocean-continent transitions (Boillot et al., 1987) and slow-spreading mid-ocean ridge boundaries (Karson, 1990), where the magma budget is low. A similar model has been presented by Lemoine and others (1987) for ophiolites of the Ligurian Tethys in the Alps, Appennines, and Corsica. Thus we think that the outlined tectonic scenario for the Kizildağ ophiolite and the Arabian passive margin may provide an adequate explanation for the architecture of ophiolite laden passive margins and should be tested with additional data from the periphery of the Arabian promontory and other regions.

Acknowledgments

Field studies in the Kizildağ ophiolite are supported by the Geological Society of America Penrose Research Award and the Department of Geology, University of California, research awards to Y. Dilek and by NSF Grant 84–11854 to E. M. Moores. The logistic support provided by the Adana Division of the Mineral Research and Exploration Institute of Turkey during field studies is gratefully acknowledged. We thank Ö. Piskin, H. Selçuk, and J.-J. Wagner for their significant contribution to our understanding of the groundwork geology of the Kizildağ ophiolite. This study benefited from field discussions and communications with M. Erendil and O. Tekeli. R. J. Twiss and J. S. McClain critically read an earlier version of the paper and provided constructive comments for its improvement. Critical and thorough reviews by A. Stampfli and H. Zeck further improved the paper.

References

Boillot, G., Recq, M., Winterer, E. L., Meyer, A. W., Applegate, J., Baltuck, M., Bergen, J. A., Comas, M. C., Davies, T. A., Dunham, K., Evans, C. A., Girardeau, J., Goldberg, G., Haggerty, J., Jansa, L. F., Johnson, J. A., Kasahara, J., Loreau, J. P., Luna-Sierra, E., Moullade, M., Ogg, J., Sarti, M., Thurow, J., and Williamson, M., 1987. Tectonic denudation of the upper mantle along passive margins: a model based on drilling results (ODP leg 103, western Galicia margin, Spain)., Tectonophysics, 132: 335–342.

Cann, J. R., 1974. A model for oceanic crust developed., Roy. Astr. Soc. Geophys. Jour., 39: 169–187.

Çoğulu, E., Delaloye, M., Vuagnat, M., and Wagner, J.-J., 1975. Some geochemical, geochronological and petrophysical data on the ophiolitic massif from the Kizil Dağ, Hatay, Turkey., C. r. Séances Soc. Phys. Hist. nat. Genève, 10: 141–150.

Coleman, R. G., 1971. Plate tectonic emplacement of upper mantle peridotites along continental edges., Jour. Geophys. Res., 76: 1212–1222.

Delaloye, M., Vuagnat, M., and Wagner, J.-J., 1977. K-Ar ages from the Kizil Dağ ophiolite complex (Hatay, Turkey) and their interpretation. In: B. Biju-Duval and L. Montadert (Eds.), The Structural History of the Mediterranean Basins, Editions Technip, Paris, pp. 73–77.

Delaloye, M. and Desmet, A., 1979. Nouvelles données radiometriques sur les pillow-lavas du Troodos (Chypre). Paris, Académie des Sciences Comptes Rendus, 288: 461–464.

Delaloye, M. and Wagner, J.-J., 1984. Ophiolites and volcanic activity near the western edge of the Arabian plate. In: J.F. Dixon and A.H.F. Robertson (Eds), The Geological Evolution of the Eastern Mediterranean., Geol. Soc. London Spec. Pub. 17: 225–233.

Desmet, A., Lapierre, H., Rocci, G., Gagny, Cl., Parrot, J.F., and Delaloye, M., 1978. Constitution and significance of the Troodos sheeted complex., Nature, 273: 527–530.

Detrick, R., Buhl, P., Mutter, J., Vera, E., Orcutt, J., Madsen, J., and Brocher, T., 1987. Multichannel seismic imaging of magma chamber along the East Pacific Rise., Nature, 326: 35–41.

Dewey, J. F. and Bird, J. M., 1971. Origin and emplacement of the ophiolite suite: Appalachian ophiolites in Newfoundland. Jour. Geophys. Res., 76: 3179–3206.

Dewey, J. F. and Kidd, W. S. F., 1977. Geometry of plate accretion., Geol. Soc. America Bull., 88: 960–968.

Dilek, Y. and Moores, E. M., 1985. Structure, petrology, and origin of Kizil Dagh ophiolite, southern Turkey: Comparison with Troodos., EOS Transactions, American Geophys. Union, 66: 1129.

Dilek, Y. and Moores, E. M., 1990. Regional tectonics of the eastern Mediterranean ophiolites. In: J. Malpas, E. M. Moores, A. Panayiotou, and C. Xenophontos (Eds), Ophiolites, Oceanic Crustal Analogues, Proceedings of the Symposium "Troodos 1987", The Geological Survey Department, Nicosia, Cyprus, 295–309.

Dubertret, L., 1955. Géologie des roches vertes du nort-ouest de la Syrie et du Hatay (Turquie). Notes Mém. Moyen-Orient 6, 227 p.

Erendil, M., 1984. Petrology and structure of the upper crustal units of the Kizildağ ophiolite. In: O. Tekeli and M.C. Göncüoğlu (Eds), Geology of the Taurus Belt, Proceedings International Symposium, Turkey 1983, 269–284.

Gass, I.G., 1967. The ultramafic volcanic assemblage of the Troodos Massif, Cyprus. In: Ultramafic and Related rocks. John Wiley, New York, pp. 121–134.

Harper, G. D., 1988. Episodic magma chambers and amagmatic extension in the Josephine ophiolite., Geology, 16: 831–834.

Juteau, T., Nicolas, A., Dubessy, J., Fruchard, J.C., and Bouchez, J.L., 1977. Structural relationships in the Antalya ophiolite complex, Turkey: possible model for an oceanic ridge., Geol. Soc. America Bull., 88: 1740–1748.

Karson, J. A., 1990. Seafloor spreading on the Mid-Atlantic Ridge: implications for the structure of ophiolites and oceanic lithosphere produced in slow-spreading environments. In: J. Malpas, E.M. Moores, A. Panayiotou, and C. Xenophontos (Eds), Ophiolites, Oceanic Crustal Analogues, Proceedings of the Symposium "Troodos 1987", The Geological Survey Department, Nicosia, Cyprus, 547–556.

Laurent, R., Delaloye, M., Vuagnat, M., and Wagner, J.-J., 1980. Composition of parental basaltic magma in ophiolites. In: A. Panayiotou (Ed), Ophiolites, Proceedings International Symposium, Cyprus 1979, pp. 172–182.

Lemoine, M., Tricart, P., and Boillot, G., 1987. Ultramafic and gabbroic ocean floor of the Ligurian Tethys (Alps, Corsica, Appennines): In search of a genetic model., Geology, 15: 622–625.

Macdonald, K. C., 1982. Mid-Ocean Ridges: Fine scale tectonics, volcanic and hydrothermal processes within the plate boundary zone., Ann. Rev. Earth Planet. Sci., 10: 155–190.

Moores, E. M. and Vine, F. J., 1971. Troodos massif, Cyprus and other ophiolites as oceanic crust: Evaluation and implications., Phil. Trans. Roy. Soc. London, Ser. A, 268: 443–466.

Mukasa, S.B. and Ludden, J.N., 1987. Uranium-Lead isotopic ages of plagiogranites from the Troodos ophiolite, Cyprus, and their tectonic significance., Geology, 15: 825–828.

Pallister, J. S. and Hopson, C. A., 1981. Samail ophiolite plutonic suite: Field relations, phase variation, cryptic variation and layering, and a model of a spreading ridge magma chamber., Jour. Geophys. Res., 86: 2593–2644.

Parrot, J.-F., 1980, The Baër-Bassit (northwestern Syria) ophiolitic area., Ofioliti, 2: 279–295.

Piskin, O., Delaloye, M., Selçuk, H., and Wagner, J.-J., 1986. Guide to Hatay geology (SE-Turkey)., Ofioliti, 11: 87–104.

Purdy, G. M. and Detrick, R. S., 1986. Crustal structure of the Mid-Atlantic Ridge at 23°N from seismic refraction studies., Jour. Geophys. Res., 91: 3739–3762.

Reuber, I., 1984. Mylonitic ductile shear zones within tectonites and cumulates as evidence for an oceanic transform fault in the Antalya ophiolite, SW Turkey. In: J.F. Dixon and A.H.F. Robertson (Eds), The Geological Evolution of the Eastern Mediterranean., Geol. Soc. London Spec. Pub. 17: 319–334.

Ricou, L.E., 1971. Le croissant ophiolitique péri-arabe, une ceinture de nappes mises en place au Crétacé superieur., Rev. Géogr. phys. Géolo. dyn., 18: 327–349.

Ricou, L.E., Argyriadis, I., and Marcoux, J., 1975. L'axe calcaire du Taurus, un alignement de fenêtres arabo-africaines sous des nappes radiolaritiques, ophiolitiques et métamorphiques., Bull. Soc. géol. France, 17: 1024–1043.

Robertson, A. H. F., 1986. Geochemistry and tectonic implications of metalliferous and volcaniclastic sedimentary rocks associated with Late Cretaceous ophiolitic extrusives in the Hatay area, southern Turkey., Ofioliti, 11: 121–140.

Selçuk, H., 1981. Etude géologique de la partie méridionale du Hatay (Turquie). Published Ph.D. thesis (No. 1997), Université de Génève, 116 p.

Sengör, A. M. C. and Yilmaz, Y., 1981. Tethyan evolution of Turkey: a plate tectonic approach., Tectonophysics, 75: 181–241.

Staudigel, H., Gillis, K., and Duncan, R., 1986. K/Ar and Rb/Sr ages of celadonites from the Troodos ophiolite, Cyprus., Geology, 14: 72–75.

Tiezzi, L. J. and Scott, R. B., 1980. Crystal fractionation in a cumulate gabbro, Mid-Atlantic Ridge at 26°N., Jour. Geophys. Res., 85: 5483–5454.

Tinkler, C., Wagner, J.-J., Delaloye, M., and Selçuk, H., 1981. Tectonic history of the Hatay ophiolites (south Turkey) and their relation with the Dead Sea rift., Tectonophysics, 72: 23–41.

Varga, R. J. and Moores, E. M., 1985. Spreading structure of the Troodos ophiolite, Cyprus., Geology, 13: 846–850.

Wernicke, B., 1985. Uniform-sense normal simple shear of the continental lithosphere., Canad. Jour. Earth Sci., 22: 108–125.

Whitechurch, H., Juteau, T., and Montigny, R., 1984. Role of the eastern Mediterranean ophiolites (Turkey, Syria, Cyprus) in the history of the Neo-Tethys. In: J.F. Dixon and A.H.F. Robertson (Eds), The Geological Evolution of the Eastern Mediterranean., Geol. Soc. London Spec. Pub. 17: 301–317.

Yilmaz, P.O., 1984. Fossil and K-Ar data for the age of the Antalya complex. In: J.F. Dixon and A.H.F. Robertson (Eds), The Geological Evolution of the Eastern Mediterranean., Geol. Soc. London Spec. Pub. 17: 335–347.

Processes of Ophiolite Emplacement in Oman and Newfoundland

PETER A. CAWOOD
Centre for Earth Resources Research, Department of Earth Sciences, Memorial University of Newfoundland, St John's, Newfoundland, Canada A1B 3X5

Abstract

The Semail Ophiolite in the Sultanate of Oman, and the Bay of Islands and St. Anthony Ophiolite Complexes in western Newfoundland show a similar multistage history of ophiolite obduction. This involved both intra-oceanic detachment of the ophiolite slabs and their emplacement onto the continental margin.

Emplacement has generally been modelled as taking place through progressive foreland propagation of the thrust front (in-sequence or piggy-back thrusting). However, structural relations within the Oman and Newfoundland examples clearly indicate a more complex scenario of assembly which involved initial delamination between the drift and rift facies segments of the continental margin sequence followed by major break-back (out-of-sequence) thrusting. The main evidence for delamination comes from the presence of continental margin rift facies lithologies directly under the ophiolite slab. Rift facies lithologies occur both in the metamorphic sole welded to the ophiolite and in the major thrust sheets directly underlying the ophiolite. Continental margin drift facies and foreland basin sequences are generally restricted to the lower sheets of the thrust stack. The presence of out-of-sequence thrusts are indicated by the truncation of hanging-wall and footwall structures at the base of the ophiolite and rift-facies sheets.

Delamination and break-back thrusting are a consequence of ophiolite obduction over a continental margin. Delamination develops during the early phases of collision: drift facies sequences are bulldozed ahead of the advancing ophiolite, whereas the underlying rift facies lithologies are largely overridden by the ophiolite. Break-back thrusting is probably related to final emplacement of the thrust wedge over the continental shelf. Ramping of the thrust wedge over the shelf requires the wedge to thicken and increase its critical taper. This is achieved by internal deformation of the thrust wedge through break-back thrusting.

Tj. Peters et al. (Eds), Ophiolite Genesis and Evolution of the Oceanic Lithosphere, 501–516.

Introduction

Ophiolite obduction involves the thrusting of dense oceanic lithosphere over buoyant continental crust (Coleman, 1977). This is generally envisaged as the outcome of attempted subduction of a continental margin beneath an incipient island arc (e.g. Searle and Stevens, 1984; Leitch, 1984). Obduction is a multistage process involving both intra-oceanic detachment of the ophiolite slabs and their emplacement onto a continental margin. Detachment takes place through formation of an intra-oceanic high-temperature shear zone. This is preserved in the metamorphic sole welded to the base of the ophiolite (Williams and Smyth, 1973; Jamieson, 1986). Emplacement involves the incorporation of the ophiolite within an imbricate stack of continental margin strata and the emplacement of this assembled thrust stack over a collapsed shallow water passive margin sequence (Coleman, 1977; Glennie et al., 1974; Williams and Stevens, 1974). The ophiolite slabs lie at the top of the thrust stack and obduction has generally been modelled as taking place through progressive foreland propagation of the thrust front (in-sequence or piggy-back thrusting), such that the ophiolites represent the farthest travelled and first formed of the thrust sheets.

The aim of this paper is to look at the structural relations within the thrust stack below the ophiolite slab and discuss the implications for models of ophiolite emplacement. This study will concentrate on the thrust stacks in the Oman mountains of the Alpine Orogen and in the west Newfoundland segment of the Appalachian-Caledonian Orogen, which preserve the most complete obduction sequences in the world. Recent work in these regions has shown that relationships within the thrust stack are inconsistent with assembly through simple foreland propagation of the thrust front. In Oman, Bernoulli and Weissert (1987) have shown that the highest of the continental margin thrust sheets consists almost exclusively of rift facies lithologies. In Newfoundland, Lindholm and Casey (1989) have outlined a similar scenario for the highest thrust sheet below the Bay of Islands ophiolite. This paper establishes that such relationships are not anomalous features restricted to one specific area but rather reflect a general process of continental margin delamination and break-back (out-of-sequence) thrusting during emplacement of an ophiolite slab onto the margin.

Stratigraphic and Structural Relations

Both the Alpine and Appalachian orogens evolved through a Wilson cycle of ocean opening and closing (Fig. 1). In the Alpine Orogen this involved the generation and destruction of the Tethys Ocean and took place between the late Paleozoic and the Tertiary. The Appalachian cycle for initiation and destruction of the Iapetus Ocean occurred between the late Precambrian and the late Paleozoic.

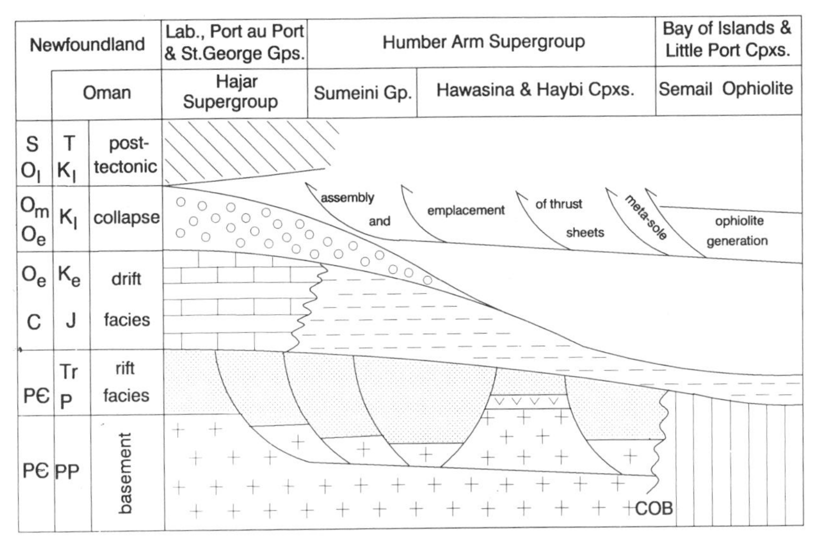

Figure 1. Paleogeographic setting and age range of rock units in Oman and the Humber Arm allochthon, west Newfoundland. Abbreviations: COB – continent-ocean boundary; LPC – Little Port Complex; meta-sole – metamorphic sole to ophiolites; T – Tertiary; K_L – Late Cretaceous; K_E – Early Cretaceous; J – Jurassic; Tr – Triassic; P – Permian; PP – Pre-Permian; D – Devonian; S – Silurian; O_L – Late Ordovician; O_M – Middle Ordovician; O_E – Early Ordovician; C – Cambrian; C_E – Early Cambrian; PC – Precambrian.

Oman Mountains

The Oman mountains consist of a thrust stack of Late Permian to Late Cretaceous continental margin and oceanic rocks structurally emplaced onto the time-equivalent shelf sequence of the Arabian margin and its unconformably underlying early Paleozoic and older metamorphic basement (Fig. 1). Emplacement of the thrust stack occurred in the Late Cretaceous in association with obduction of the Semail ophiolite (Glennie et al., 1974). The thrust stack is divisible into four main tectonostratigraphic slices (cf. Glennie et al., 1974; Searle and Malpas, 1980). From structurally lowest to highest these are (Figs. 2 and 3): 1) Sumeini Group shelf edge sediments; 2) Hawasina Complex base of slope and deep sea sediments; 3) Haybi Complex volcanic rock and reefal limestones, melanges and metamorphic sheet; 4) Semail ophiolite complex. Each of the slices is internally imbricated, and in particular the Hawasina Complex is divisible into a series of discrete thrust bounded continental margin assemblages. Figure 3 outlines the age range of the principal sheets within the Oman mountain thrust stack.

The overall age range of the rock units within the three lower slices is similar to that of the underlying Arabian margin shelf sequence and extends from the Late Permian to Late Cretaceous. These age relationships, along with the regular ordering of the thrust sheets lead Glennie et al. (1974) and most subsequent workers to conclude that the successively lower sheets lay

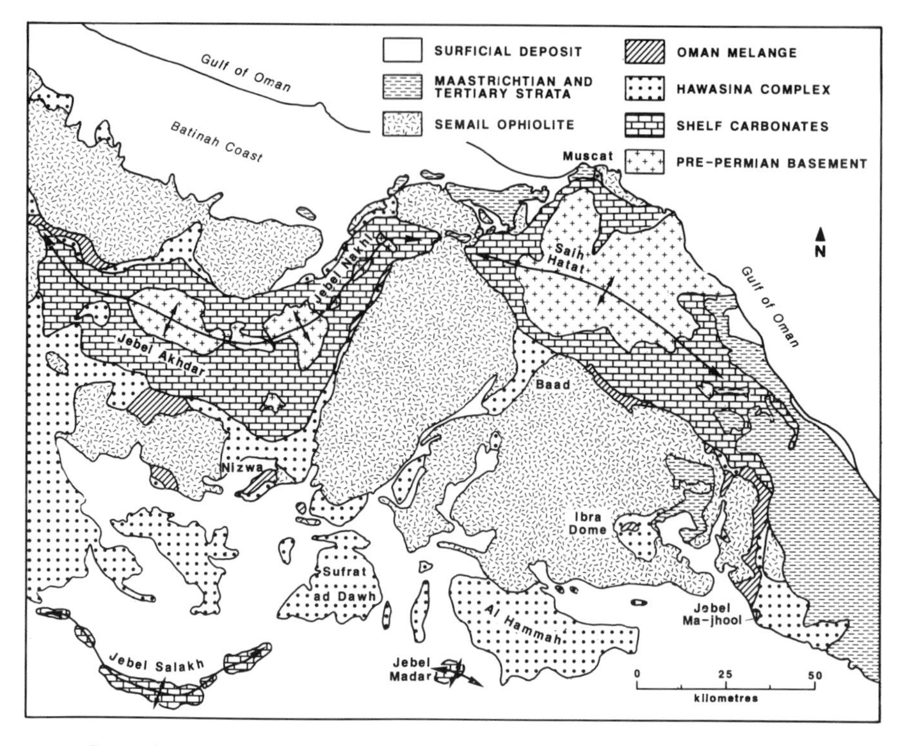

Figure 2. Simplified geologic map of the central and southern Oman Mountains.

at progressively more proximal sites along the margin. Reversals in the stacking order, both within and between, the four major tectonostratigraphic units have been recognized and related to late break-back (out-of-sequence) thrusting (Barrette and Calon, 1987; Cawood et al., 1990; Searle, 1985). These late-stage reversals have been interpreted as being imposed on a thrust stack originally assembled by a simple and regular propagation of the thrust front towards the foreland in a piggy-back manner.

Recent work by Bernoulli and Weissert (1987) and Bernoulli et al. (1990) has shown that individual tectonostratigraphic units within the Hawasina Complex are not time equivalent (Fig. 3). They demonstrate that the lower thrust sheets contain mainly Jurassic and Cretaceous strata representing drift cycle continental margin deposits, and that the upper thrust sheets contain mainly Triassic age strata which accumulated in rift basins during initiation of the Arabian margin. This observation indicates that assembly of the thrust stack did not take place by piggy-back thrusting and that higher slices are not the more distal time-equivalent facies of lower slices. Bernoulli and Weissert (1987) propose that the lower sheets originally formed the cover for

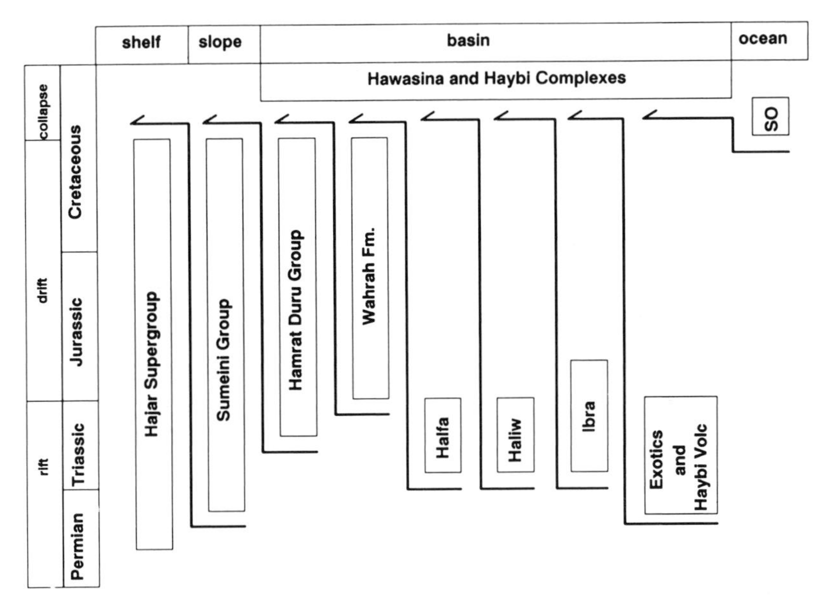

Figure 3. Age range of principal thrust sheets within the Oman Mountain thrust stack. Abbreviations: SO – Semail Ophiolite.

the upper thrust sheets and that the present stacking order is a consequence of multistage thrusting.

The Haybi Complex with its characteristic association of Permian and Triassic reefal limestones (Oman Exotics) and associated volcanic rocks also forms part of the structural trend outlined by Bernoulli and Weissert (1987) of older strata constituting the higher thrust sheets. The Oman Exotics developed on a basement of alkalic and transitional tholeiitic mafic igneous rocks (Haybi Volcanics), and are interpreted to represent a series of seamounts which developed during rifting of the Oman continental margin in the late Permian and Triassic (Searle and Graham, 1982). Similar mafic igneous rocks occur as flows and sills at the base of the Triassic rift-related strata in the upper thrust slices of the Hawasina Complex (Glennie et al., 1974; Searle, 1984; Cawood et al., 1990). Younger alkalic igneous rocks, at least in part of Jurassic and Cretaceous age, occur as intrusions and fault blocks within the Haybi Complex. Amphibolites and greenschists within the metamorphic sheet at the base of the Semail ophiolite were derived from a protolith of mafic igneous rock, chert, argillite and carbonate (Glennie et al., 1974; Searle and Malpas, 1980; Ghent and Stout, 1981). Whole-rock geochemistry on the amphibolites suggest derivation from a tholeiitic protolith, possibly similar to either the Haybi tholeiite series or transitional MORB compositions (Searle and Malpas, 1982). Carbonate-rich metasediments in the greenschists are similar to the underlying Oman Exotics and Hawasina Complex (Searle and Malpas, 1980).

West Newfoundland

During the late Precambrian to early Paleozoic, western Newfoundland lay along the ancient continental margin of eastern North America. Pre-Late Ordovician rocks within this region are divisible into a Precambrian crystalline basement, unconformably overlain by a latest Precambrian to Middle Ordovician continental margin cover sequence, structurally overlain by a thrust stack of sedimentary and igneous thrust slices (Fig. 4). The transported rocks were emplaced during the Early to Middle Ordovician Taconian Orogeny and are best exposed in the Humber Arm and Hare Bay allochthons (Williams and Cawood, 1989; Williams and Smyth, 1983).

The thrust stack in the *Humber Arm allochthon* consists of three internally imbricated structural slices, separated by melange. Lower and intermediate structural slices consist of a latest Precambrian to early Ordovician deep-water continental margin succession. This sequence is time-equivalent to, and the offshore facies equivalent of, the underlying shallow water, continental margin cover sequence. The upper structural slices of the allochthon consist of igneous rocks and are dominated by the Bay of Islands ophiolite complex (Fig. 5).

Age, lithologic and structural relations between the Bay of Islands ophiolite sheet, underlying structural slices, foredeep sediments and the shelf sequence, have, as in Oman, lead to the idea that successively higher sheets occupied progressively more outboard sites along the continental margin and that assembly of the thrust stack occurred through piggy-back thrusting (cf. Stevens, 1970; Williams, 1975).

The principal rock unit within the intermediate structural slices, the Blow-Me-Down Brook Formation, was until recently considered to be of early Ordovician age and represent foreland basin strata which accumulated ahead of, and was ultimately incorporated into, the advancing thrust stack. However, the lack of detritus derived from the thrust stack, and the discovery of Early Cambrian trace fossils within the unit (Lindhlom and Casey, 1989), indicate that it represents a rift facies association which accumulated at the initiation of the Appalachian cycle (Fig. 5; Stevens, 1983; Quinn, 1985; Waldron, 1985). A similar rift facies sedimentary sequence also occurs at the stratigraphic base of the lower structural slices and is termed the Summerside Formation (Williams and Cawood, 1989). The Early Cambrian and older age of the Blow-Me-Down Brook Formation and the lack of stratigraphically overlying younger continental margin rock units indicates that assembly of the thrust stack could not have taken place in a simple piggy-back manner.

A series of thrust bounded, igneous rock dominated, units occur locally

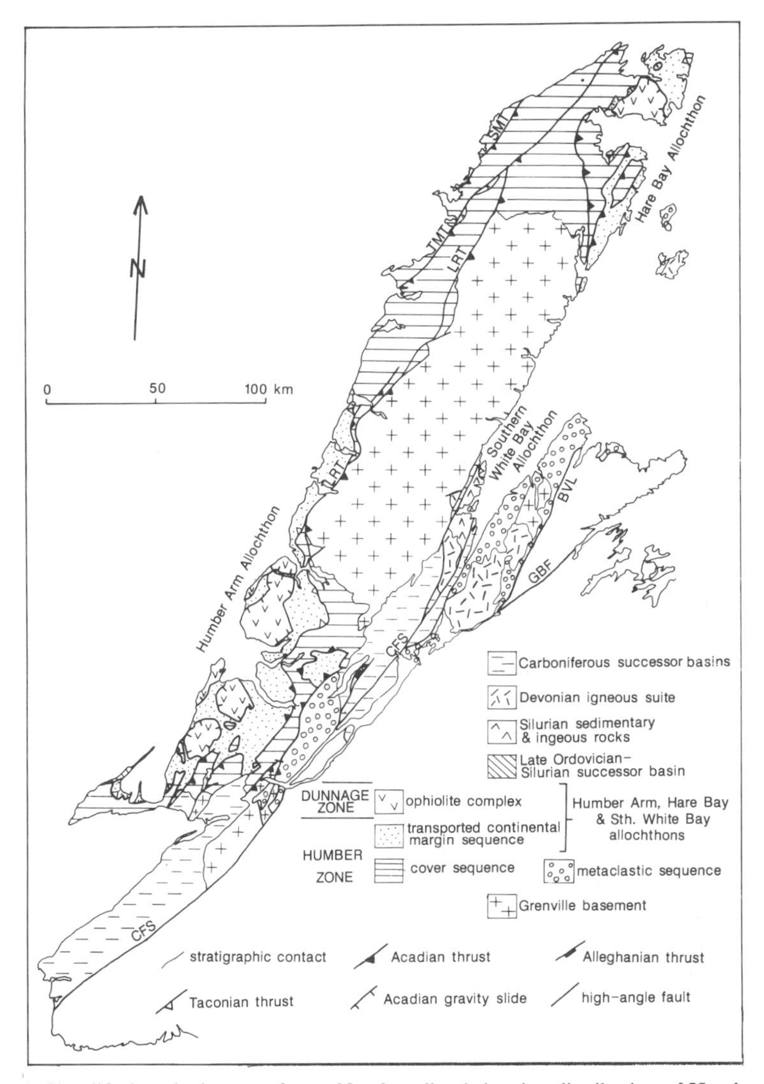

Figure 4. Simplified geologic map of west Newfoundland showing distribution of Humber Arm and Hare Bay allochthons. Abbreviations: LRT - Long Range Thrust; TMT - Ten Mile Lake Thrust; SMT - St. Margaret Bay Thrust; BVL - Baie Verte Lineament; CFS - Cabot Fault System.

within the intermediate structural slices (e.g. Skinner Cove Formation and equivalent units, Williams and Cawood, 1989). These igneous rocks are, at least in part, of alkalic affinities (Strong, 1974) and interpreted to represent seamounts developed at of near the continental margin, which were plucked and incorporated into the thrust stack during its assembly and emplacement. On Woods Island in Humber Arm, a base faulted sliver of these igneous rocks are in stratigraphic continuity with, and included within, the Blow-Me-

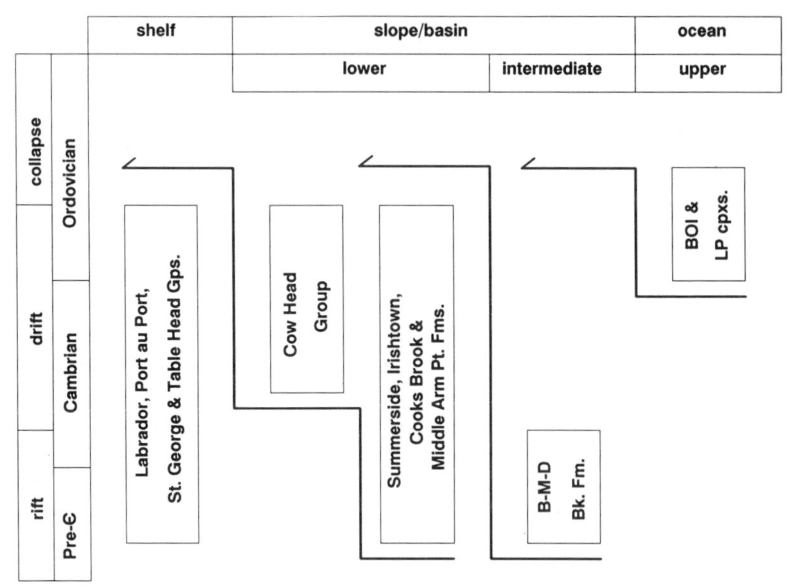

Figure 5. Age range of principal thrust sheets within the Humber Arm allochthon. Abbreviations: B-M-D Bk. Fm. Blow-Me-Down Brook Formation; BOI & LP cpxs – Bay of Islands and Little Port complexes.

Down Brook Formation (Stevens, 1970). They are interpreted to represent volcanism associated with the inception of rifting.

The metamorphic sole welded to the base of the Bay of Islands ophiolite consists predominantly of mafic igneous derived amphibolite and greenschist (Williams and Smyth, 1973; Malpas, 1979). Casey and Gultekin (1988) suggest that these rocks were derived from a MORB-type protolith similar to the Bay of Islands ophiolite. The lowest structural and metamorphic grade portions of the sole, incorporate a sedimentary protolith of arkose and shale derived from the Blow-Me-Down Brook Formation (Cawood unpub. data). Rocks of the metamorphic sole, including the sedimentary lithologies, are separated from the underlying Blow-Me-Down Brook Formation by a late stage post-metamorphic brittle fault (Cawood and Williams, 1988; Cawood, 1990).

The thrust stack of the *Hare Bay allochthon* at the northwest tip of Newfoundland contains lower slices of continental margin sedimentary and igneous rock, capped by the St. Anthony ophiolite complex which consists of the White Hills Peridotite and underlying metamorphic sole. The thrust stack at Hare Bay lacks the structural and stratigraphic coherency of the transported rocks in the Oman or Humber Arm regions. The thrust stack consists of two main slices, the upper Hare Bay slice (St. Anthony Complex) and the underlying Maiden Point slice, which locally are separated by a series of discontinuous and isolated volcanic and sedimentary slices and melange (Grandois slice, Milan Arm melange and Cape Onion slice; Williams and

Smyth, 1983). None of the sub-ophiolitic slices contain a complete and coherent continental margin sequence. The Maiden Point Formation consists of undated arkosic sandstones with interstratified alkalic volcanic rocks near the base. These are correlated with the latest Precambrian to Early Cambrian rift facies sandstones in the Humber Arm allochthon.

Sedimentary and volcanic rocks within the low grade and structurally lowest parts of the metamorphic sole to the St. Anthony ophiolite sheet are ascribable to a rift facies protolith similar to the Maiden Point Formation (Williams, 1975; Jamieson, 1977; Williams and Smyth, 1983). The lowest igneous rocks incorporated within the sole, the Ireland Point volcanics, are geochemically similar to, and could be derived from, the alkalic volcanic rocks in the Maiden Point Formation (Jamieson, 1977). Similarly, the sedimentary rocks incorporated in the sole on the Fishot islands are lithologically identical to the arkoses of the Maiden Point Formation (Williams, 1975; Cawood, unpub. data; Stevens, pers. commun., 1989). In addition, igneous rocks within the higher grade parts of the sole directly underlying the ophiolitic peridotite include an alkalic assemblage of jacupirangite and syenite, again of probable ocean island derivation (Jamieson and Talkington, 1980).

Discussion

Stratigraphic and structural observations in the thrust stacks of continental margin and ocean basin rocks below the Semail, Bay of Islands and St. Anthony ophiolite slabs demonstrate a series of common relationships which must be taken into account when modelling their emplacement onto continental margin lithosphere. Principal amongst these is that the upper thrust sheets below the ophiolite slabs do not represent the distal time equivalent facies of the lower sheets. Lithologic associations within the upper thrust sheets indicate formation during the rift cycle of continental margin development and/or in seamounts adjacent to the margin. In Oman, and possibly also in Newfoundland, it has been proposed that the rock units in the upper thrust slices originally underlay those in the lower thrust slices (Bernoulli and Weissert, 1987; Lindholm and Casey, 1989). Foreland basin lithologies, which accumulated during assembly and emplacement of the thrust stack, are also locally incorporated into the upper thrust sheets: Late Cretaceous Aruma Group sediments in the Haybi Complex of Oman; Cenomanian volcanics within the Haybi Complex; shale matrix developed in melanges directly underlying the Bay of Islands ophiolite may have accumulated in a foreland basin setting; Milan Arm melange below St. Anthony Complex. Passive continental margin drift cycle strata are conspicuously absent from the upper thrust sheets, including the metamorphic sole to the ophiolites. The presence of the stratigraphically lowest and oldest continental margin lithologies, as well as ocean basin rocks, in the metamorphic sole indicate that initial detachment of the hot ophiolite slab occurred adjacent to the continental margin (cf. Malpas, 1979; Jamieson, 1980). In addition, the presence of rift cycle lithologies and the exclusion of drift cycle lithologies, within the upper thrust sheets requires that the latter must have been selectively removed from this segment of the continental margin as it was overridden by the ophiolite.

The structural and stratigraphic relationships developed in the thrust stacks below the Semail, Bay of Islands and St. Anthony ophiolites are not isolated occurrences unique to these locations but may be a characteristic feature of thrust stacks emplaced during ocean closure in broad segments of the Alpine and Appalachian-Caledonian orogens. Whitechurch et al. (1984) have pointed out that Triassic alkalic volcanic rocks directly underlie the Taurus-type ophiolites and their metamorphic soles in the eastern Mediterranean. These volcanics are analogous to the time equivalent igneous rocks underlying the Semail ophiolite, and represent rift-related volcanism associated with initiation of the Tethys. In the Taconic allochthon of New York, the upper thrust sheets contain only the older latest Precambrian to Early Cambrian rift facies lithologies whereas lower slices contain a complete cycle of latest Precambrian to Ordovician rift and drift facies lithologies (Zen, 1967; Stanley and Ratcliff, 1985). In addition, Zen (1967) reasoned that emplacement of the thrust stack did not take place by piggy-back thrusting but by break-back thrusting with higher slices emplaced over lower slices.

Tectonic Model

Figure 6 presents a tectonic model for the assembly and emplacement of an ophiolite capped thrust stack onto a continental margin which is consistent with the stratigraphic and structural relations outlined above. The main features of this model are the delamination of rift and drift cycle lithologies during emplacement of the upper plate ophiolite over the lower plate continental margin followed by break-back thrusting of the delaminated margin.

Geochemical data constrains the site of ophiolite generation to a suprasubduction setting either within a back-arc basin or to a spreading zone formed at the inception of subduction (Pearce et al., 1984; Leitch, 1984). Obduction is a consequence of the attempted subduction of a continental margin beneath the supra-subduction zone ophiolite (Searle and Stevens, 1984).

Delamination commences at the inception of obduction and is a consequence of continental margin morphology. The contrasting character of rift and drift cycle lithologies results in their delamination as the ophiolite impacts against the buoyant continental margin (Fig. 6). Rift cycle sediments accumulate in fault bounded basins divisible into a series of horsts and grabens and are characterized by discontinuous rock units showing rapid lateral facies variation. Horsts may represent either attenuated continental basement highs or volcanic edifices built up through magmatism close to the eventual continent-ocean boundary. Drift cycle sediments accumulate during thermal sub-

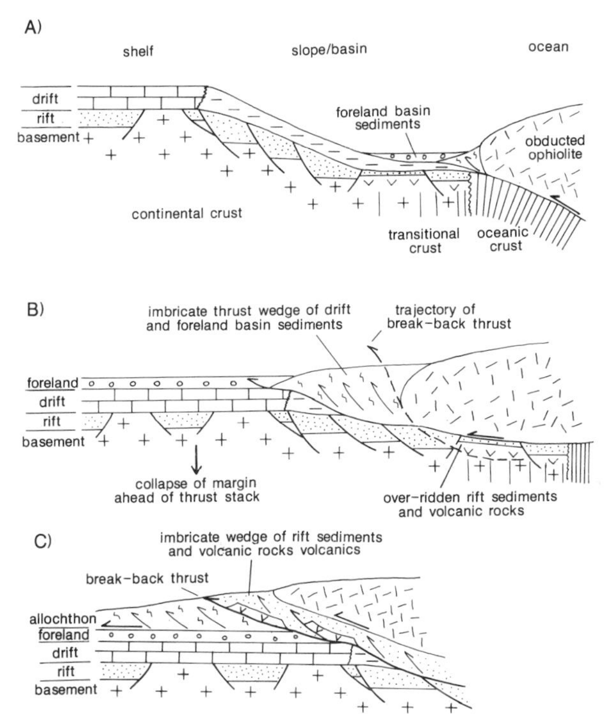

Figure 6. Tectonic model for the assembly and emplacement of an ophiolite capped thrust stack onto a continental margin: A) initial impingement of obducted ophiolite slab with continental margin; B) Delamination of drift and rift facies rocks. Drift facies rocks, with of without rift facies rocks, are bulldozed ahead of ophiolite slab whereas remaining rift facies rocks are over-ridden by the ophiolite slab. Delamination of rift and drift facies sequences is probably a function of the component nature of structural horsts within the rift facies sequence with the basal detachment of the thrust wedge ramping over the horst. C) Break-Back thrusting of within the thrust wedge allows rift facies rocks and their ophiolite cap to be structurally emplaced over the lower part of the thrust wedge.

sidence of the passive margin and are characterized by lateral continuity and relatively uniform character. During arc-continent collision sediments at the buoyant continental margin will probably be bulldozed ahead of the ophiolite slab forming a thrust wedge. Initially this may involve rift as well as drift cycle sediments. However, as the thrust front impinges against a rigid basement high, such as a buried horst, then the basal detachment of the thrust wedge is likely to climb above this rheologically competent structure. It will be confined to the upper part of the continental margin section, dominated by drift cycle sediments along with any foreland basin sediments accumulating ahead of the advancing thrust stack (Fig. 6). The horst block and any subsequent sedimentary sequences such as rift facies sediments which accumulated in grabens inboard of the horst will be subducted beneath the ophiolite slab. This process of delamination is analogous to that observed at normal convergent plate margins in which material lying in grabens on the block faulted oceanic plate is subducted beneath the arc (Hilde, 1983).

Delamination of the rift and drift facies continental margin sequences enables development of the basic structural relations seen in the upper levels of the thrust stacks in Oman and Newfoundland with ophiolite directly overlying rift facies sediments. However, further deformation is then required to place the ophiolite and rift facies associations over the drift facies dominated sequence bulldozed ahead of the ophiolite. This can be achieved through stepping back of the thrust front from the toe of the imbricating wedge to within the structural pile. This could lead to further deformation and imbrication of the rift facies sequence and the emplacement of this sequence along with its ophiolite cap over the wedge of imbricated drift and foreland basin strata (Fig. 6).

Evidence for break-back trusting is well developed in the Oman Mountains. Recent structural studies have emphasized that assembly of the Oman thrust stack was not a simple single event but required multiple thrusting phases resulting in re-imbrication and out-of-sequence relationships within the thrust stack (Searle, 1985, Hanna, 1986; Barrette and Calon, 1987, Bernoulli and Weissert, 1987; Cooper, 1988; Cawood et al., 1990). In particular, Cooper (1988) using a combined structural and stratigraphic data base suggested that the sequence of thrusting in the Oman Mountains required the continental margin sequence to be both bulldozed ahead of and partially subducted beneath the advancing ophiolite. Searle (1985) has pointed out that the base of the Semail ophiolite and Haybi Complex are characterized by break-back thrusts which truncate both footwall and hanging wall stratigraphy. Although the basal thrusts of both the Semail and Haybi Complexes were subsequently remobilized as extensional structures during gravitational driven collapse of late-stage structural culminations, they were probably initiated as break-back structures during assembly of the thrust stack (Hanna, 1986; Cawood, 1990). Mapping in the southern Oman Mountains around Al Hammah (Cawood unpub. data) indicates that the upper rift facies thrust sheets are separated from the lower drift facies thrust sheets by a major break-back thrust.

Evidence for break-back thrusting in the west Newfoundland thrust stacks is hindered by the poor level of exposure and the lack of detailed structural analysis of the Taconian structures within the stacks. However, Cawood (1989, 1990) has pointed out that the base of the Bay of Islands and St. Anthony ophiolite complexes are marked by post-metamorphic brittle fractures which truncate both hanging wall and footwall structures. Although these structures, like the Semail fault, now represent late-stage gravity driven extensional slide surfaces, they probably originated during assembly of the thrust stack as break-back structures. The strain regime during gravity driven extension is unlikely to have been sufficient for the initiation and propagation of these structures, rather they probably followed a pre-existing plane of weakness initiated during assembly of the thrust stack. No structural studies have been carried out on the basal detachment to the rift facies slices in the Humber Arm and Hare Bay allochthons (Blow-Me-Down Brook and Maiden Point formations, respectively). However, regional map patterns within at least the Humber Arm allochthon suggest a thinning and cutting out of footwall units below these sequences which is consistent with a break-back thrust at the base of the rift facies lithologies. In addition, bedding in both the hanging wall and footwall along the contact between the intermediate and lower structural slices in the allochthon are at a high-angle to, and probably trucated along, the contact (Williams and Cawood, 1989).

The driving mechanism for break-back thrusting is probably related to final emplacement of the thrust wedge up and on to the collapsed continental margin shelf succession. Analysis of the thrust stack as a coulomb wedge (Davis et al., 1983; Jamieson and Beaumont, 1989) indicates that for the stack to climb up over the shelf edge and onto the shelf it will have to increase the critical taper of the wedge. This is achieved through internal deformation of the wedge by stepping back of the thrust front leading to break-back thrusting in the wedge.

Conclusions

Thrust stacks in the Oman mountains and western Newfoundland show a consistent relationship in which passive continental margin drift facies lithologies are confined to the lower slices of the stack. Rift facies lithologies characterize the upper slices, directly below the ophiolite, but may also be present in stratigraphic continuity below drift cycle lithologies in the lower slices of the stack.

These relationships indicate that assembly and emplacement of the Oman and Newfoundland thrust stacks did not take place by simple footwall collapse and foreland propagation of the thrust front through a continental margin sequence beneath an obducted ophiolite slab, but required a multistage process of continental margin delamination followed by break-back thrusting. Delamination and break-back thrusting are probably an inherent consequence of ophiolite obduction over a continental margin. Delamination between rift and drift facies lithologies develops during the early phases of collision. Drift facies lithologies are bulldozed ahead of the ophiolite, whereas rift facies lithologies are carried beneath the overriding ophiolite. Delamination and the selective movement of the rift facies lithologies beneath the ophiolite is probably facilitated by the irregular and fault disrupted character of rift facies basins relative to the lateral continuity of drift cycle sequences. Thus, graben structures developed during rifting isolate and insulate the sediments deposited within these basins, allowing them to be overridden by the advancing ophiolite. Delamination of rift and drift facies sequences establishes the basic structural framework which is then exploited by break-back thrusting, resulting in the emplacement of the ophiolite and rift facies sheets over an imbricated sequence of drift facies rocks. Break-back thrusting is probably related to final emplacement of the thrust wedge over the continental shelf. Ramping of the thrust wedge over the shelf requires the wedge to thicken and increase its critical taper. This is achieved by internal deformation of the thrust wedge through break-back thrusting.

Acknowledgements

I thank S.S. Hanna, R.K. Stevens, and H. Williams for many enjoyable discussions on Oman and Newfoundland geology and A.W. Shelton and an anonymous reviewer for comments on the manuscript.

References

Barrette, P.D., and Calon, T.J., 1987. Re-imbrication of the Hawasina allochthons in the Sufrat ad Dawh Range, Oman Mountains., Journal of Structural Geology, 9: 859–867.

Bernoulli, D. and Weissert, H., 1987. The upper Hawasina nappes in the central Oman Mountains: stratigraphy, palinspastics and sequence of nappe emplacement., Geodynamica Acta, 1: 47–58.

Bernoulli, D., Weissert, H. and Blome, C.D., 1990. Evolution of the Triassic Hawasina Basin, Central Oman Mountains. In: A.H.F. Robertson, M.P. Searle and A.C. Ries (Eds), The Geology and Tectonics of the Oman Region. Geological Society Special Publication No 49: 189–202.

Coleman, R.G., 1977. Ophiolites, Ancient Oceanic Lithoshpere? Springer, New York, 229 p.

Casey, J.F. and Gultekin, S., 1988. Subduction-obduction history recorded in the Bay of Islands and Coastal Complex ophiolites of western Newfoundland. Geological Association of Canada, Program with Abstracts, 13: A18.

Cawood, P.A., 1989. Acadian remobilization of a Taconian ophiolite, western Newfoundland., Geology, 17: 257–260.

Cawood, P.A., 1990. Late-stage gravity sliding of ophiolite thrust sheets in Oman and western Newfoundland. In: Malpas, J., Moores, E.M., Panayioutou, A. and Xenophontos, C., (Eds), Ophiolites, Oceanic Crustal Analogues. Proceedings of the Symposium "TROODOS 87", Geological Survey Department, Ministry of Agriculture and Natural Resources, Nicosia, Cyprus, pp. 433–455.

Cawood P.A. and Williams, H., 1988. Acadian basement thrusting, crustal delamination and structural styles in and around the Humber Arm Allochthon, western Newfoundland., Geology 16: 370–373.

Cawood, P.A., Green, F.K. and Calon T.J., 1990. Origin of structural culminations within the southeast Oman Mountains at Jebel Ma-jhool and Ibra Dome. In: A.H.F. Robertson, M.P. Searle and A.C. Ries (Eds), The Geology and Tectonics of the Oman Region., Geological Society Special Publication No 49: 429–445.

Cooper, D.J.W., 1988. Structure and sequence of thrusting in deep-water sediments during ophiolite emplacement in the south-central Oman Mountains., Journal of Structural Geology, 10: 473–485.

Davis, D., Suppe, J. and Dahlen, F.A., 1983. Mechanics of fold-and-thrust belts and accretionary wedges., Journal of Geophysical Research, 88: 1153–1172.

Ghent, E.D. and Stout, M.Z., 1981. Metamorphism at the base of the Semail Ophiolite, southeastern Oman Mountains., Journal of Geophysical Research, 86: 2557–2571.

Glennie, K.W., Boeuf, M.G.A., Hughes-Clark, M.W., Moody-Stuart, M., Pilaar, W.F.H. and Reinhardt, B.M., 1974. Geology of the Oman Mountains. Verhandelingen van het Koninklijk Nederlands Geologisch Mijnbouwkundig Genootschap. Volume 31: Martinus Nijhoff, The Hague, 423 p.

Hanna, S.S., 1986. The Alpine (Late Cretaceous and Tertiary) tectonic evolution of the Oman Mountains: A thrust tectonic approach., Symposium on the hydrocarbon potential of intense thrust zones, 2, OAPEC, Kuwait, pp. 125–174.

Hanna, S.S., 1990. The Alpine deformation of the Central Oman Mountains. In: A.H.F. Robertson, M.P. Searle and A.C. Ries (Eds), The Geology and Tectonics of the Oman Region., Geological Society Special Publication No 49: 341–359.

Hilde, T.W.C., 1983. Sediment subduction versus accretion around the Pacific., Tectonophysics 99: 381–397.

Jamieson, R.A., 1977. A suite of alkalic basalts and gabbros associated with the Hare Bay Allochthon of western Newfoundland., Canadian Journal of Earth Sciences, 14: 346–356.

Jamieson, R.A., 1986. P-T paths from high temperature shear zones beneath ophiolites., Journal of Metamorphic Geology, 4: 3–22.

Jamieson, R.A. and Beaumont, C., 1989. Deformation and metamorphism in convergent orogens: a model for uplift and exhumation of metamorphic terrains. In J.S. Daly, R.A. Cliff B.W.D. Yardley (Eds), Evolution of Metamorphic Belts., Geological Society Special Publication No 43: 117–129.

Jamieson, R.A. and Talkington, R.W., 1980. A jacupirangite-synite assemblage beneath the White Hills Peridotite, northwestern Newfoundland., American Journal of Science, 280: 459–477.

Leitch, E.C., 1984. Island arc elements and arc-related ophiolites., Tectonophysics, 106: 177–203.

Lindholm, R.M. and Casey, J.F., 1989. Regional significance of the Blow Me Down Brook Formation, western Newfoundland: New fossil evidence for an Early Cambrian age., Geological Society of America Bulletin, 101: 1–13.

Malpas, J., 1979. Dynamothermal aureole beneath the Bay of Islands ophiolite in western Newfoundland., Canadian Journal of Earth Sciences, 16: 2086–2101.

Pearce, J.A., Lippard, S.J. and Roberts, S., 1984. Characteristics and tectonic significance of supra-subduction zone ophiolites. In B.P. Kokelaar and M.F., Howells (Eds), Marginal Basin Geology., Geological Society Special Publication No. 15: 77–94.

Quinn, L., 1985. The Humber Arm Allochthon at South Arm, Bonne Bay, with extension in the Lomond area west Newfoundland. MSc. thesis., Memorial University of Newfoundland, St. John's, Canada, 188 pp.

Searle, M.P., 1984. Alkaline peridotite, pyroxenite, and gabbroic intrusions in the Oman Mountains, Arabia., Canadian Journal of Earth Sciences, 21: 396–406.

Searle, M.P., 1985. Sequence of thrusting and origin of culminations in the northern and central Oman Mountains., Journal of Structural Geology, 7: 129–143.

Searle, M.P. and Graham, G.M., 1982. The "Oman Exotics": Oceanic carbonate build-ups associated with the early stages of continental rifting., Geology, 10, 43–49.

Searle, M.P. and Malpas, J., 1980. The structure and metamorphism of rocks beneath the Semail Ophiolite of Oman and their significance in ophiolite obduction. Transactions of the Royal Society of Edinburgh, Earth Science, 71: 247–262.

Searle, M.P. and Malpas, J., 1982. Petrochemistry and origin of sub-ophiolitic metamorphic and related rocks in the Oman Mountains., Journal of the Geological Society of London, 139: 235–248.

Searle, M.P. and Stevens, R.K., 1984. Obduction processes in ancient, modern and future ophiolites. In: I.G. Gass, S.J. Lippard, and A.W. Shelton (Eds), Ophiolites and oceanic lithosphere., Geological Society of London Special Publication No. 13: 303–319.

Stanley, R.S. and Ratcliffe, N.M., 1985. Tectonic synthesis of the Taconian orogeny in western New England., Geological Society of America Bulletin, 96: 1227–1250.

Stevens, R.K., 1970. Cambro-Ordovician flysch sedimentation and tectonics in west Newfoundland and their possible bearing on a Proto-Atlantic Ocean. In: J. Lajoie (Ed), Flysch Sedimentology in North America: Geological Association of Canada Special Paper No. 7: 165–177.

Stevens, R.K., 1983. The transported sedimentary rocks of western Newfoundland: evolution of the ancient continental margin of western Newfoundland. Geological Association of Canada, Newfoundland Section, Program with abstracts.

Strong, D.F., 1974. An off-axis alkali volcanic suite associated with the Bay of Islands ophiolites, Newfoundland., Earth and Planetary Science Letters, 21: 301–309.

Waldron, J.W.F., 1985. Structural history of continental margin sediments beneath the Bay of Islands Ophiolite, Newfoundland., Canadian Journal of Earth Sciences, 22: 1618–1632.

Williams, Harold, 1975. Structural succession, nomenclature, and interpretation of transported rocks in western Newfoundland., Canadian Journal of Earth Sciences, 12: 1874–1894.

Williams, H. and Cawood, P.A., 1989, Geology, Humber Arm Allochthon, Newfoundland. Geological Survey of Canada, Map 1678A, scale 1:250 000.

Williams, H. and Smyth, W.R., 1973. Metamorphic aureoles beneath ophiolite suites and Alpine peridotites: tectonic implications with west Newfoundland examples., American Journal of Science, 273: 594–621.

Williams, H., and Smyth, W.R., 1983, Geology of the Hare Bay Allochthon. In: Geology of the Strait of Belle Isle Area, Northwestern Insular Newfoundland, Southern Labrador, and adjacent Quebec., Geological Survey of Canada, Memoir 400: 109–141.

Williams, H. and Stevens, R.K., 1974. The ancient continental margin of eastern North America. In: C.A. Burk and C.L. Drake (Eds), The geology of continental margins. Springer-Verlag, New York, pp. 781–796.

Zen, E-An, 1967. Time and Space Relationships of the Taconic Allochthon and Autochthon., Geological Society of America Special Paper No. 97: 107 p.

Sr, Nd and Pb Isotopic Constraints in the Genesis of a Calc-Alkaline Plutonic Suite in Oman Ophiolite Related to the Obduction Process

L. BRIQUEU[1], C. MÉVEL[2], and F. BOUDIER[3]

[1]*Laboratoire de Géochimie Isotopique, URA1371, Département Sciences de la Terre, Université Montpellier II, 34095 Montpellier Cedex 5, France*

[2] *Laboratoire de Pétrographie, Université Paris VI, IPG, 4 place Jussieu, 75712 Paris, France*

[3] *Laboratoire de Tectonophysique, Département Sciences de la Terre, Université Montpellier II, 34095 Montpellier Cedex 5, France*

Abstract

K-mineral bearing felsic intrusions in the Oman ophiolite have been studied for incompatible elements, and Pb, Sr, Nd isotopes. The magmatic bodies (mainly from the Khawr Fakkan massif) represent differentiated tens of meter sized intrusions or metric dikes cross-cutting the ophiolite at mantle level. Metamorphic schists and quartzites having reached anatexis, belonging to the sole of the same massif are also studied for comparison.

Both incompatible element and isotopic data indicate that these intrusive bodies are fundamentally different from oceanic plagiogranites. Hygromagmaphile elements are globally depleted relative to oceanic magmas (basalts, gabbros and plagiogranites), but they present a relative enrichment in highly incompatible elements (Th, U and LREE). The least differentiated samples have a V positive anomaly indicative of high oxygen fugacity during genesis.

The lead isotopic compositions are characterized by radiogenic lead enrichment ($^{207}Pb/^{204}Pb$ and $^{206}Pb/^{204}Pb$ range respectively from 15.55 to 15.69, and from 18.48 to 18.76) suggesting the effect of a crustal component similar to that of the metasediments of the metamorphic soles. On the other hand, the initial Sr and Nd isotopic compositions are dominated by a mantle component ($^{87}Sr/^{86}Sr = 0.7049$ and $^{143}Nd/^{144}Nd = 0.5128$). Significant heterogeneities of the isotopic components are observed at the scale of the large body ($^{87}Sr/^{86}Sr$ range from 0.7049 to 0.7081). It is concluded that the K-mineral bearing felsic intrusions in the Oman ophiolite have a mixed (mantle and crustal) origin; melting of the metamorphic sole (amphibolites + quartz pelite sediments) would best account for their genesis.

Introduction

Late felsic intrusions crosscut the ophiolite section at any level in the Semail nappe. They split into two groups. The first group represents a minor, but widespread component of the plutonic section, occurring as differentiated

Tj. Peters et al. (Eds), Ophiolite Genesis and Evolution of the Oceanic Lithosphere, 517–542.

plutons which cover compositions from wehrlite to gabbro and trondjhemite, or as irregular masses or dikes of plagiogranite (Tilton et al., 1981; Lippard et al., 1986; Beurrier, 1987). Chemically this group is characterized by low potassium contents. On the basis of major and trace elements and isotopic chemistry these authors assign these intrusions to the late stages of the accretion process. The second group is represented by small intrusive bodies (a few tens to one hundred meter sized) or dikes of biotite granite or K-feldspar bearing aplite, intrusive in the mantle section of the Semail nappe. Such intrusions have been described in the Khawr Fakkan massif (UAE) (Browning and Sweming, 1981; Browning, 1982), and in the South Central Oman mountains: Sumail, Haylayn, Bahla and Sarami massifs (Beurrier, 1987). There, following this author, the granitic intrusions are concentrated along major pre-ophiolite-emplacement strike slip faults. From our observations, K-bearing granitic dikes are also present in the mantle section from the Fizh massif. Thus, these K-bearing granitic intrusions are characterized by a large dispersion all over the mantle section of the ophiolite, and their very small size is indicative of the involvement of small amounts of magma.

This study focus on the K-bearing granitic intrusions of the Khawr Fakkan massif which are numerous along the shore between Ras Dadnah and Khawr Fakkan, in the mantle section exposed just beneath the paleo-Moho. These Khawr Fakkan intrusions are exposed only two kilometers above the metamorphic sole of the ophiolite, due to the limited thickness of the mantle section in this massif (Nicolas et al., 1988). Interestingly, the metamorphic sole of the Khawr Fakkan massif, which is especially well exposed in the Hajar window, shows spectacular evidence of partial anatexis. Partial anatexis has been observed in other exposures of the metamorphic sole in the Semail ophiolite (Searle, 1980) and (Fig. 2C), limited and restricted to the high grade rocks which commonly are metabasalts. In the Hajar window, the anatexis affects metasediments rich in white mica, anatectic gneisses (Fig. 2E) produce dikes of acidic melt which reintrude the metamorphic section (Fig. 2F). These observations correlated with the relative vicinity of the K-bearing granitic intrusions (Fig. 1B) lead Boudier et al., (1988) to propose a possible parental relationship of these specific intrusions with the metasediments of the basal sole. A similar hypothesis had been proposed by Pedersen and Malpas (1984) for the Kamoy ophiolite of Norway, on the basis of REE modelling. On the other hand, Lippard et al. (1986) suggest on the basis of major and trace elements analysis that the K-bearing granitic intrusions in the Samail nappe were produced by the partial melting of the autochtonous Arabic shield. A few available isotopic dates do not contradict these hypotheses. K-Ar dating of 'retrograded' amphibolites from this area gives a minimum age of 85 ± 5 Ma (Alleman and Peters, 1972), which is slightly younger than the average in the metamorphic sole of the Semail ophiolite: 95 to 100 Ma (Lanphere, 1981; Montigny et al., 1988). K-Ar dating on biotite from the granite of Ra's Dadnah has provided ages of 85 ± 3 Ma (Rex, In Lippard et al., 1986).

The aim of the present paper is to infer through a complete isotopic study of the K-granitic rocks the characters of their source and to check possible relationships with the basal metamorphics.

Studied Areas

Except for two intrusive bodies from the Haylayn massif, all the studied areas concern the Khawr Fakkan massif.

Granitic Intrusions in the Moho Transition Zone

In the Haylayn massif, small pods or dikes of granitic rocks intrude the mantle peridotites and the Moho transition zone along the Wadi Haymilyah. One small dike of amphibole diorite, 80 cm thick, and a large body (20 m wide) of granite-biotite cross-cutting dunites and wehrlites, have been sampled (Table 1).

Differentiated Granitic Complexes

Two intrusions of this type were studied in the Khawr Fakkan massif at Dadnah and Wadi Zikht (Fig. 1A). Both intrusions occur in the mantle section, at the Moho level.

The Dadnah body, about 100 meters thick, forms a heterogeneous, roughly layered mass (Fig. 2B), with local magmatic breccias. Lenses of dark gabbro to diorite, five to ten centimeters sized, are scattered in a matrix progressively varying in color from greyish in the center to white toward the contacts with the enclosing mantle rocks, marked by bands of pyroxenite and wehrlite. The contacts between the lenses and the matrix never display chilled margins and are sometimes lobate, suggesting a coeval emplacement by injection of a late acidic magma disrupting an early, immiscible, more mafic mush. The lenses are generally homogeneous but locally display a spotted texture, corresponding to centimeter-sized aggregates of amphibole. Late dikes of K-feldspar bearing pegmatites, often with spectacular graphic textures, cross cut all the rock types.

Small masses of similar albeit deformed rocks crop out away from the coast, on a small plateau west of the village of Zikht. The dark facies form gneissic amphibolites, with deformed pebble-like amphibole aggregates. The light facies form biotite gneisses. Individual small dikes of biotite granite and muscovite-pegmatite also cross-cut the peridotite.

Granitic Dikes in Mantle Section

The two specimens studied belong to the mantle section of the Khawr Fakkan massif (Fig. 1A and 2A). The exposure of Ra's Dibba (Table 1) corresponds

Table 1. Rock types and mineral assemblages. Mineral between parenthesis are completely replaced by secondary products.

Granitic intrusions				
Sample no.	Location	Rock type	Magmatic assemblages	Secondary minerals
87OM155	Wadi Haymilyah	diorite	pl-amp-(oxide)	se-sph
87OM156	Wadi Haymilyah	granite	qz-pl-Kf-bi-amp-oxide	ep
87OM60	Dadnah	qz-gabbro	pl-amp-qz	scarce veins of ep-pr-chl
87OM61	Dadnah	qz-diorite	pl-amp-(bi)-qz-oxide	chl-ep-sph
87OM62	Dadnah	qz-diorite	pl-amp-(bi)-qz	chl-ep-se
87OM63	Dadnah	qz-diorite	pl-amp-(bi)-qz	chl-ep-se
87OM64	Dadnah	granite	qz-pl-(Kf)-(bi)	chl-ep-ab
87OM65	Dadnah	granite	qz-pl-(Kf)-(bi)-to	chl-ep-ab
87OM69	Dadnah	spotted qz-diorite	pl-amp-qz-(bi)-oxide	chl-ep-sph
87OM71	Zikht	qz-diorite	pl-amp-qz-bi-oxide	chl-Sph
87OM59	Wadi Fujayrah	granite	qz-pl-Kf-bi-to	chl-se
86OA62C	Ras Dibba	deformed granite	qz-pl-Kf-bi	

Metamorphic sole				
Sample no.	Location	Rock type	Typomorphic assemblage	Retromorphic minerals
86OA87U	Wadi Timarit	quartzite	alternating layers of qz-amp	cc veins
86OA87D	Wadi Timarit	quarztite	qz-mu-bi-pl	chl-ep-cc
86OA88D	Wadi Timarit	quartzite	qz-amp-pl-bi	chl-ep-se-cc
86OA98I	Wadi Adhan	amphibolite	amp-(pl)-cpx-gt-(oxide)	ep-ab-sph-cc

pl = plagioclase, amp = amphibole, se = sericite, sph = sphene, qz = quartz, ab = albite, bi = biotite, Kf = potassium feldspar, ep = epidote, pr = prehmite, chl = chlorite, to = tourmaline, cpx = clinopyroxene, cc = calcite, gt = garnet, mu = muscovite.

to a 3 meters thick dike, undeformed, having sharp contact with the enclosing harzburgite. The dike from the east side of Wadi Ham (Fig. 1A) is 1 meter thick, and is locally deformed.

Metamorphic Sole

The metamorphic sole has been sampled in two areas from the Khawr Fakkan massif: Wadi Limarit in the Hajar window, and Wadi Adhan in the Ayim-Sidr window.

The metamorphic sole of Hajar window is exposed over 10 km below the mylonitic sole of the mantle peridotite (Fig. 1 and 2). Along this contact the peridotite overlies directly metasediments, and not metabasalts; this is not the common situation and may explain the uncommon development of anatexis in this window. In this high grade metamorphic horizon, the metasediments are garnet-amphibole quartzites. A recent petrological survey in this

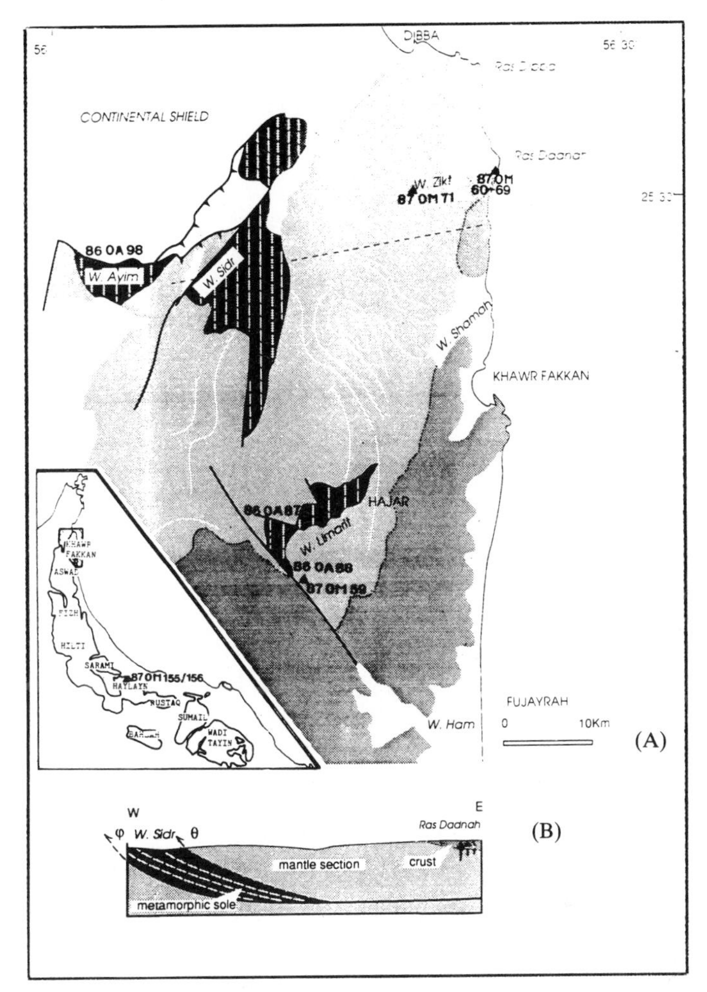

Figure 1. A, Map of the Khwar Fakkan massif in UAE, with location of the studied areas (see Table 1). Light lines in the mantle section are asthenospheric flow trajectories. B,Cross-section (located by dashed line on the map) with Dadnah intrusions; θ, high-T thrust contact of peridotite and metamorphics; ϕ, low-T emplacement thrust.

area reported assemblages of quartz-cordierite-enstatite and sapphirine (Kurz, 1990). The rest of the metamorphic sole, 300 meters thick, is composed of amphibole to epidote quartzites and micaschists. Occasionally they grade to anatectic biotite gneiss generating acidic melts which reintrude the metamorphic section during the course of the deformation (Fig. 2E). Three specimen of biotite-amphibole and biotite-epidote quartzites from this area have been studied (Table 1).

The metamorphic sole of Ayin-Sidr window shows the standard succession of an amphibolite band a few meters thick, in contact with the peridotite, reaching garnet amphibolite facies (Searle and Malpas, 1980; Lippard et al.,

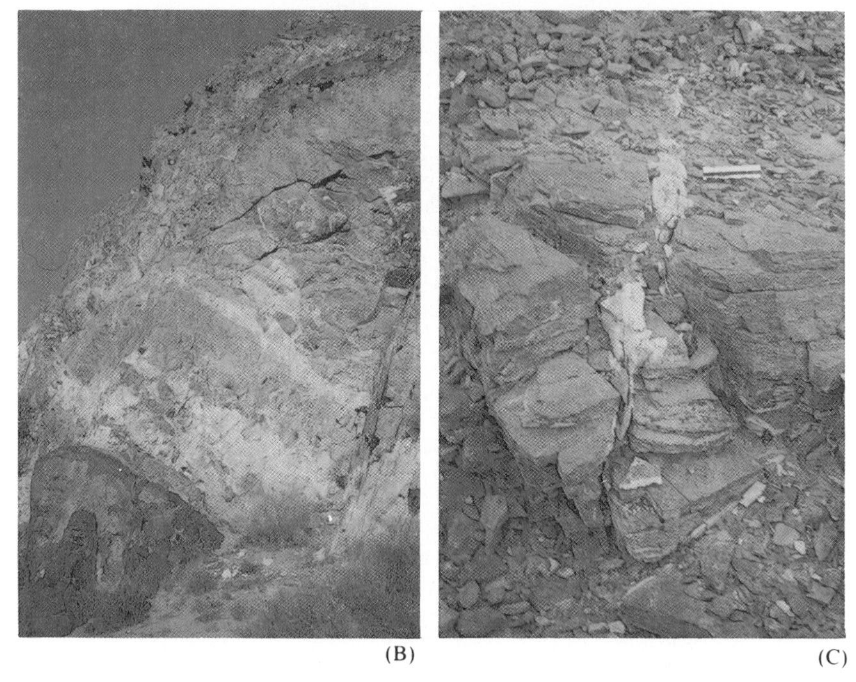

(B) (C)

Figure 2. Exposures of the specimens studied. A, two sets of granitic dikes cross-cutting the mantle section (86OA69), Ra's Dadna; B, granite-diorite relationships in the intrusive body of Ra's Dadnah (87OM60 to 87OM65); C, tension gash filled with quartz + felspar partial melt in amphibolites from Wadi Tayin massif; D,metamorphic sole of Wadi Limarit in Khawr Fakkan massif with basal peridotite to the right, metacherts to the left, and granitic melt in the central area (86OA87U); E, anatectic gneisses in metasediments from Wadi Limarit (86 OA87D); F, deformed felsic dike re-intruding metasediments in Wadi Limarit.

(D)

(E)

(F)

Fig.2(D),(E)(F)

1986). Recalculation of garnet-clinopyroxene geothermometry of Ziegler and Stoessel (1985) gives 810 ± 50°C (M. Bucher, pers. communication), elsewhere Ghent and Stout (1981) state a 875°C temperature for a similar facies in the Wadi Tayin massif of Oman. The amphibolites are underlain by lower grade metasediments. Aplitic dikes cross-cutting the amphibolite section have been described (Lippard et al., 1986; and our observations), suggesting possible anatexis in the underlying metasediments. One sample of amphibolite is studied (Table 1).

Petrographic Descriptions

Sixteen samples have been selected for geochemical studies: 12 from granitic intrusions and 4 from the metamorphic sole. Mineral assemblages are listed in Table 1.

Granitic Intrusions

The largest variety of samples comes from the Dadnah intrusion. One gabbro sample (87OM60), corresponding to the dark lenses described supra, is made up of zoned euhedral plagioclase, green amphibole and scarce interstitial quartz. In the quartz-diorites, corresponding either to a dark lens (87OM61) or to the greyish part of the intrusion (87OM62 and 63), amphibole is less abundant, but biotite occurs as well as iron-titanium oxides in some of the samples. The spotted diorite (87OM69) has the same mineral assemblage but amphibole crystals form small aggregates. In the granites, corresponding to the white part of the intrusion (87OM64 and 65), biotite is the only ferromagnesian mineral. Besides abundant quartz, two generations of plagioclase occur. The first one consists of zoned (An35 to An28) euhedral crystals. The second one corresponds to large anhedral crystals, An5 in sample 87OM64 and An15 in sample 87OM65, enclosing euhedral plagioclase, biotite and sometimes quartz grains. In this series of samples, biotite is always completely chloritized. In the granites, anhedral plagioclase mimics K-feldspar occurring in the non-altered granite dikes. We suggest that the granites have been albitized, resulting in the complete replacement of potassium feldspar by a sodic plagioclase. The geochemical effect of this albitization will be discussed in the next section. Scarce veins of epidote, chlorite and prehnite are also present, and iron-titanium oxides are replaced by sphene.

The quartz-diorite from Zikht (87OM71) is similar to that from Dadnah, although less altered: biotite is partly preserved and secondary chlorite and sphene are scarce.

The granite dike from Wadi Fujayrah (87OM59) is very fresh and contains only small flakes of secondary chlorite and sericite in feldspars. All the biotite is preserved. The granite dike from Ra's Dibba is also very fresh, but slightly deformed.

Finally, the granite from Wadi Haymilyah (87OM156) is very fresh, with only minor secondary epidote. It contains small amounts of magmatic amphibole. On the opposite, the diorite (87OM155) is largely altered, with abundant sericite, as well as sphene replacing oxides.

To conclude, the rocks belonging to the granitic intrusives are characterized by the widespread occurrence of quartz, even in the more mafic terms (gabbros) and the abundance of biotite in the diorites and the granites. Potassium feldspar is widespread in the granites, but has likely been albitized in the Dadnah intrusion. These mineralogical characters contrast with the rocks from the ophiolite sequence.

Metamorphic Sole

Three metasedimentary quartzites and one metabasite, have been selected. 86OA87U is a quartzite displaying alternating layers, at a millimetric scale, of quartz + amphibole and quartz + clinopyroxene + biotite + amphibole. Biotite is locally replaced by chlorite and calcite veins cross-cut the rock. 86OA87D is a quartzite with scarce plagioclase, partly chloritized biotite, and muscovite. 86OA88D is a more massive quartzite, containing plagioclase, biotite and green amphibole.

The amphibolite is made up of green hornblende, clinopyroxene, plagioclase, garnet and iron-titanium oxide. Plagioclase is altered to cloudy albite and oxides to sphene.

The mineral assemblages in the quartzites and in the amphibolite are consistent with amphibolite facies conditions.

Analytical Techniques

Major and some trace elements (Ba, Zr, Y and V) were measured by X-ray fluorescence. The accuracy is over 5%. Other trace element (Rb, Sr, Nd, Sm, U, Th and Pb) were analyzed using isotopic dilution and mass spectrometry techniques. The accuracy is better than 1% for the absolute concentration and 0.5% for the isotopic abundance ratios. After dissolution in a HF-HNO_3-$HClO_4$ mixture, Sr and the REE are extracted on cation exchange resin. Ca-Sr are separated using ammonium citrate as the complexing agent (Birck and Allègre, 1978). Nd is purified from Ba and other REE using a column of 'Teflon' powder coated with HDEHP (Richard et al., 1976). Pb is separated following the very efficient anionic exchange microprocedure described by Manhès et al. (1978). Total system blanks were found to be less than 70 ng, 20 ng and 100 ng for Sr, Nd and Pb respectively. Blanck are in all cases insignificant. Sr isotopic compositions are measured on a single tungsten filament previously covered with a tantalum activator. Nd isotopic compositions are measured on triple filament. Pb is analyzed using the silica-H_3PO_4 method on a single Re filament. Sr and Nd are corrected for mass discrimination assuming respectively $^{86}Sr/^{88}Sr = 0.1194$ and $^{146}Nd/^{144}Nd = 0.7219$. The Pb isotopic results are corrected with a mass discrimination factor of 0.13% per atomic mass unit difference. The errors are assumed to be less than 0.5% for all Pb isotopic ratios. Results for isotopic standards during the general period of this study are: Sr NBS 987: $^{87}Sr/^{86}Sr = 0.71024 \pm 0.00001$, Nd JMC 361 $^{143}Nd/^{144}Nd = 0.51114 \pm 0.00001$, Pb NBS981 $^{206}Pb/^{204}Pb = 16.947 \pm 0.007$, $^{207}Pb/^{204}Pb = 15.504 \pm 0.007$, $^{208}Pb/^{204}Pb = 36.746 \pm 0.019$.

Major and Trace Elements Geochemistry

Major element concentrations are listed in Table 2 in order of increasing silica content at each location. This magmatic evolution index varies from 54 to 77%. The geochemical characters of this sample suite is generally consistent with calc-alkaline plutonic series: high aluminum content, inverse correlation between Si and Al, Mg, Fe and Ca. However, the potassium content is somewhat surprising. Some of the analyzed granites (OM59 and OM156) as well the two granites analyzed by Searle et al. (1980) and Lippard et al. (1986) display 'normal' potassium contents with respect to the silica index, consistent with a calc-alkaline trend. However all the other samples, whatever their silica index, display very low potassium contents (K20 $> 0.1\%$). In an extended REE diagram (Fig. 3), this results in strong negative anomalies with respect to LREE and associated element (Th and U). These anomalies are ascribed to secondary processes, chloritization of biotite and albitization of potassium feldspars. As will be discussed later, Sr isotopes are also in favor of this assumption.

Trace Elements Geochemistry

The inconsistent behavior of potassium mentioned concerns also Rb and Ba. In an extended REE diagram (Fig. 3), K, Rb, Ba exhibit important negative anomalies uncorrelated with the magma differentiation trend. The most mafic terms of the Dadnah series are enriched in LREE and associated elements (U, Th), depleted in intermediate and HREE, with the K, Rb and Ba negative anomalies already mentioned. The average level of these REE patterns is lower than normally observed in oceanic crust components (Pallister and Knight, 1981; Lippard et al., 1986); they lie between the MORB and peridotite trends. These REE patterns are definitely different from the flat, slightly depleted in incompatible elements (Rb, Ba, K, Th and LREE) patterns normally observed in the oceanic plagiogranites.

Our representation (Fig. 3) stresses the homogeneous distribution of Ti, whereas V has a strong positive anomaly, suggesting high fluid pressure for the generation of the magma. Indeed in high fluid pressure conditions, V has its maximum valence (+5), and its partition coefficient is lowered (Bou-

Table 2. Major and trace element analysis of grano-dioritic whole-rock samples.

	87OM60	87OM61	87OM62	87OM69	87OM63	87OM64	87OM65	87OM71	87OM59	87OM156	87OM155
SiO_2	53.88	56.25	56.75	64.22	66.99	75.45	75.82	64.70	77.23	57.69	73.75
TiO_2	0.19	0.23	0.57	0.23	0.26	0.17	0.16	0.90	0.11	0.60	0.31
Al_2O_3	17.1	15.93	15.56	15.54	15.39	12.96	13.77	14.47	12.15	12.33	10.69
Fe_2O_3	1.37	1.52	2.13	1.32	0.63	0.37	0.25	1.49	0.13	0.29	0.88
FeO	6.99	7.14	5.95	4.96	4.81	1.56	1.80	4.46	0.61	2.51	2.20
MnO	0.19	0.18	0.14	0.13	0.12	0.05	0.06	0.18	0.08	0.12	0.15
MgO	5.90	5.65	6.18	3.58	2.08	0.90	0.87	3.81	0.65	2.08	3.31
CaO	10.07	9.12	9.85	7.68	5.89	2.77	2.64	7.45	0.99	5.89	2.90
Na_2O	1.78	1.33	2.26	1.89	2.06	4.58	4.05	1.33	3.04	5.96	2.28
K_2O	0.09	0.04	tr	tr	0.05	tr	0.04	tr	4.25	tr	2.44
P_2O_5	0.03	0.04	0.07	0.06	0.06	0.10	0.06	0.02	0.04	0.31	0.15
LOI	1.53	1.72	0.15	0.15	1.39	0.98	0.13	0.78	0.46	0.20	0.86
Σ	99.13	99.15	99.61	99.76	99.33	99.89	99.65	99.87	99.74	99.65	99.92
Ba	<10	57	52	43	28	44	41	<10	179	12	197
Zr	11	24	32	13	41	98	13	224	62	41	98
Y	8	10	11	11	12	14	12	22	22	20	16
V	313	348	294	43	170	45	43	97	<10	153	91
Co	16	16	15	12	10	<10	<10	11	<10	<10	<10
Cr	37	43	84	30	41	23	29	30	19	38	49
Ni	11	19	28	13	<10	11	<10	37	17	157	75
Cu	13	<10	93	49	39	<10	<10	25	13	<10	49
Zn	41	44	41	48	48	23	26	87	33	26	53

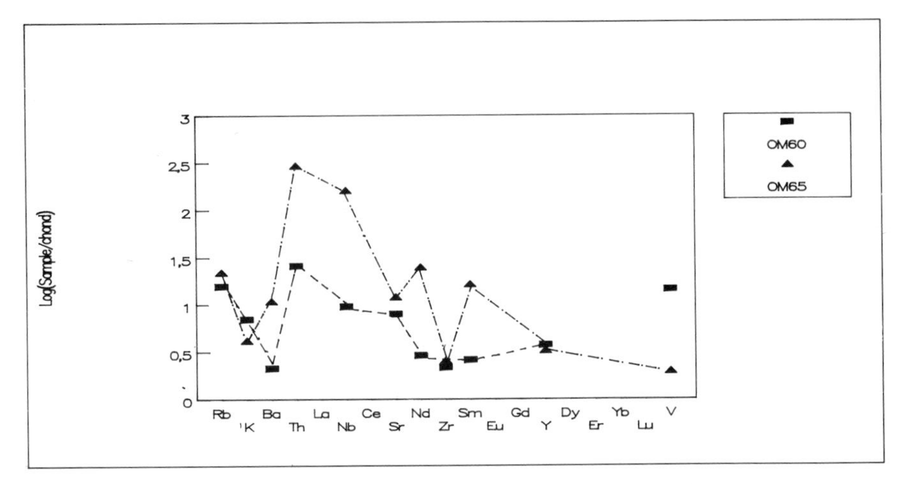

Figure 3. Comparison of normalized abundance patterns of incompatible elements from diorites and granites.

gault, 1980; Briqueu et al., 1984). In those conditions is a high incompatible element whose behaviors is directly comparable to La.

Magmatic Differentiation

The REE pattern in the granite is nearly parallel to that of diorite (Fig. 3), although granite is enriched in highly incompatible elements (Th/Sm: 2.64

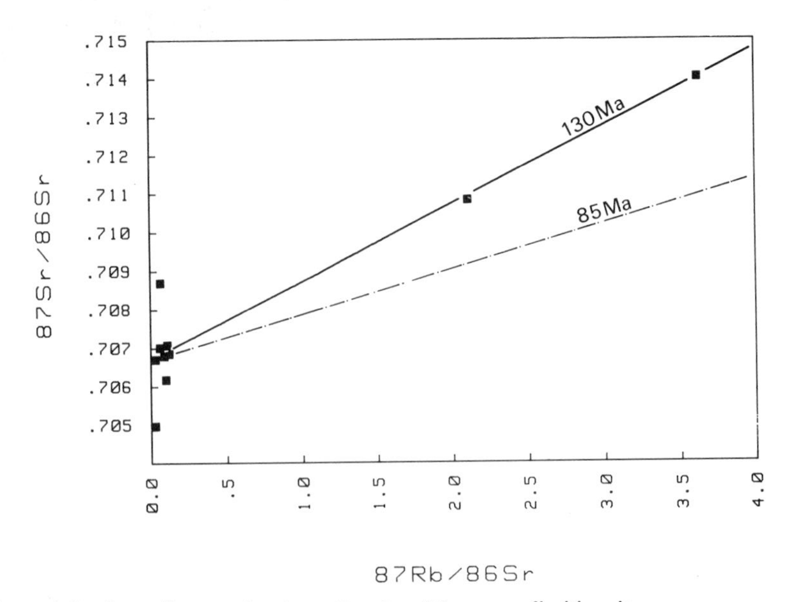

Figure 4. Isochron diagrams for the wall rocks of the grano-dioritic suite.

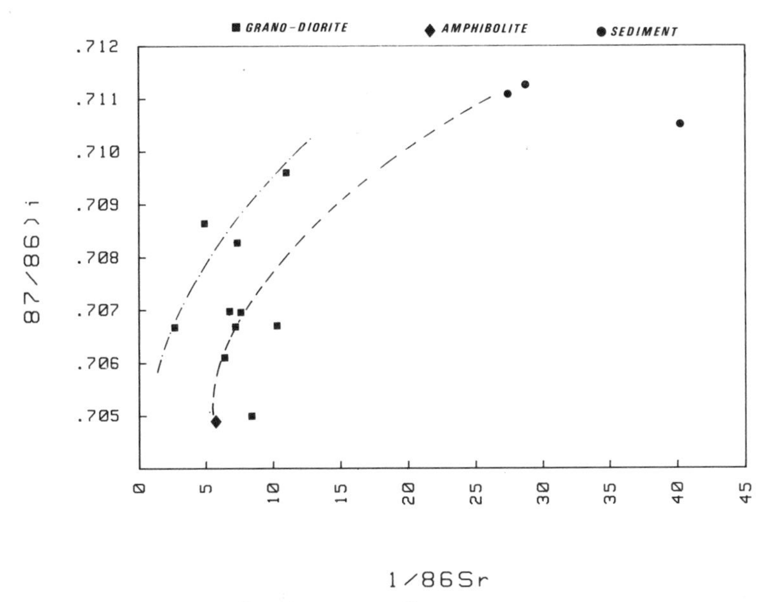

Figure 5. Variations of initial $^{87}Sr/^{86}Sr$ versus $1/^{86}Sr$ for all sample of the grano-dioritic suite.

for granite; Th/Sm = 1.53 for diorite), and presents several negative anomalies; low Ti and V contents result from rich Fe-Mg mineral fractionation, low P from apatite fractionation, low Sr from plagioclase removal. However, these geochemical characteristics cannot be simply interpreted as resulting from fractionation, since all members of the series have different isotopic compositions in Sr, Nd and Pb. Other processes like mixing or contamination have to be invoked.

Isotopic Geochemistry Sr, Nd and Pb

Sr, Nd and Pb isotopic ratios are reported in Tables 3, 4 and 5 respectively. These data concern the granodioritic series, and specimens from the metamorphic sole (amphibolites and metasediments), other units of the ophiolite (from peridotite to basalt) being already documented (Mc Culloch et al., 1980, 1981; Lanphere et al., 1981; Tilton et al., 1981; Chen and Pallister, 1981, Hamelin et al., 1984).

Rb/Sr isochron

In this diagram (Fig. 4), our data do not define a true isochron allowing to precisely determine the age of the granitic intrusions. For instance, all the samples from the Dadnah intrusion plot at the origin, although Sr isotopic ratios vary from 0.705 to 0.709. The extreme values correspond to the two

Table 3. Rb–Sr analytical results on the grano-dioritic suite and the metamorphic sole.

Sample	Rb ppm	Sr ppm	$^{87}Rb/^{86}Sr$	$^{87}Sr/^{86}Sr$	$^{87}Sr/^{86}Sr)i$	E.Sr)i
87OM59	109.2	81	3.892	0.71399 ± 2	0.70929	68.1
87OM60	3.9	86	0.131	0.70685 ± 1	0.70669	31.3
87OM61	3.0	168	0.052	0.70872 ± 3	0.70866	59.1
87OM62	0.5	106	0.013	0.70500 ± 2	0.70498	7.0
87OM63	21.0	141	0.430	0.70685 ± 1	0.70633	26.1
87OM64	28.1	174	0.465	0.70582 ± 2	0.70526	10.9
87OM65	5.5	131	0.121	0.70708 ± 3	0.70693	34.7
87OM69	13.2	120	0.313	0.70725 ± 2	0.70687	33.8
87OM71	12.0	381	0.091	0.70670 ± 3	0.70659	29.8
87OM155	16.0	159	0.291	0.70621 ± 2	0.70586	19.4
87OM156	95.4	122	2.261	0.71083 ± 1	0.70810	51.2
86OA62C	63.3	162	1.130	0.70669 ± 3	0.70533	11.8
biotite	343.1	39	25.65	0.71882 ± 7		
86OA87D	12.3	241	1.560	0.71225 ± 1	0.71037	83.4
86OA87U	6.0	32	0.537	0.71167 ± 2	0.71102	92.7
86OA88D	8.6	31	0.784	0.71216 ± 2	0.71121	95.4
86OA98I	4.0	155	0.075	0.70500 ± 1	0.70491	0.6

diorites with the same silica content (56%), while the most differentiated rocks display intermediate isotopic ratios. These isotopic variations cannot result only from the radioactive decay of ^{87}Rb. Only the two granites displaying normal alkali contents are enriched in radiogenic strontium. An isochron calculated using these two samples and the cluster of points close to the origin gives an age of 130 Ma. This geochronological information, however, probably has little geological significance: this age corresponds to the formation of the oceanic crust (Sm/Nd age of McCullogh et al., 1980, 1981) and not to the obduction 80–85 Ma). The age obtained from the two Wadi Haymilyah samples give an even older apparent age of 165 Ma. The Rb/Sr

Table 4. Sm/Nd analytical results and E.Nd)i parameter on the grano-dioritic suite and the metamorphic sole.

Sample	Nd ppm	Sm ppm	$^{147}Sm/^{144}Nd$	$^{143}Nd/^{144}Nd$	E.Nd)i
87OM60	1.72	0.49	0.1724	0.51252 ± 1	−2.1
87OM61	3.65	1.09	0.1808	0.51248 ± 4	−2.9
87OM63	9.71	2.26	0.1409	0.51248 ± 3	−2.5
87OM64	13.7	3.24	0.1432	0.51240 ± 1	−4.1
87OM65	14.9	3.12	0.1287	0.51254 ± 2	−1.2
87OM69	8.03	1.92	0.1447	0.51247 ± 1	−2.8
87OM71	7.69	1.96	0.1543	0.51277 ± 3	+3.0
86OA87D	10.9	2.05	0.1138	0.51228 ± 2	−6.2
86OA88D	17.7	4.12	0.1409	0.51188 ± 2	−14.2
86OA87I	13.1	3.62	0.1673	0.51287 ± 2	+4.7

Table 5. U, Th and Pb concentrations and isotopic parameters for the calc-alkaline suite and the metamorphic sole. The two last digits of the isotopic ratios are the corrected values for radiogenic Pb from decay of Th and U.

Sample	U ppm	Th ppm	Pb ppm	μ	k	$^{206}Pb/^{204}Pb$	$^{207}Pb/^{204}Pb$	$^{208}Pb/^{204}Pb$
87OM59	0.14	0.45	2.28	3.95	13.12	18.762(71)	15.687(68)	38.959(90)
87OM60	0.38	0.72	3.31	3.33	6.52	18.659(62)	15.672(67)	38.828(80)
87OM61	0.97	3.55	3.54	8.23	31.11	18.527(42)	15.591(59)	38.854(72)
87OM63	1.37	3.72	9.93	8.85	24.83	18.761(64)	15.640(63)	38.794(70)
87OM64	1.63	7.56	14.6	7.18	34.42	18.747(65)	15.680(68)	38.986(84)
87OM65	1.53	8.20	13.2	7.37	40.83	18.669(57)	15.549(54)	38.368(19)
87OM69	1.24	3.56	10.3	7.67	22.76	18.663(56)	15.664(66)	38.399(30)
87OM71	1.46	3.85	11.5	8.08	22.02	18.482(38)	15.616–61)	38.548(45)
86OA87D	0.31	0.85	4.62	4.31	12.21	18.649(59)	15.647(64)	39.027(97)
86OA87U	1.42	7.07	15.1	6.09	31.40	18.852(77)	15.692(69)	38.915(78)
86OA88D	1.46	5.44	17.0	5.53	21.28	18.746(67)	15.686(68)	39.028(93)
86OA98I	0.19	1.02	1.84	6.60	36.59	18.675(59)	15.538(53)	38.679(52)

system has been evidently strongly disturbed after the emplacement of the intrusions in the peridotite. The K, Rb and Ba negative anomalies of the most differentiated samples likely result from a leaching of these elements during late hydrothermal processes. Centimeter-sized biotite crystals separated from sample 86OA62C (Table 3) give an apparent age of 35 Ma. This age cannot be interpreted as the age of the perturbation but clearly demonstrates that the alkalis and likely the earth alkalis have been affected by secondary processes. Because of these perturbations, it is difficult to precisely correct the Sr isotopic ratios before petrogenetic discussions. However, given the low Rb/Sr ratios, the correction is small (in the order of 0.0003) for most of the analyzed samples. As for the ophiolite itself (Lanphere et al., 1981), the isotopic ratios corrected for 80 Ma will be considered as representative of initial magmas.

Sr-Nd Variations

Numerous studies (McCulloch et al., 1980, 1981) on the various components of ophiolite have shown that hydrothermalism and low-grade metamorphism strongly increases Sr isotopic ratio, but do not affect Nd, whose isotopic composition reflects the asthenospheric source state. The granodioritic series has Sr isotopic ratios higher than other ophiolite components, moreover the Nd isotopic ratios vary largely (Fig. 6); they trace definitely the contribution of a continental component in the genesis of the magma (initial ϵNd >0). Except one specimen off-set towards the oceanic crust field, all members of the intrusions are grouped in a same field intermediate between the oceanic crust array and that defined by the metasediments.

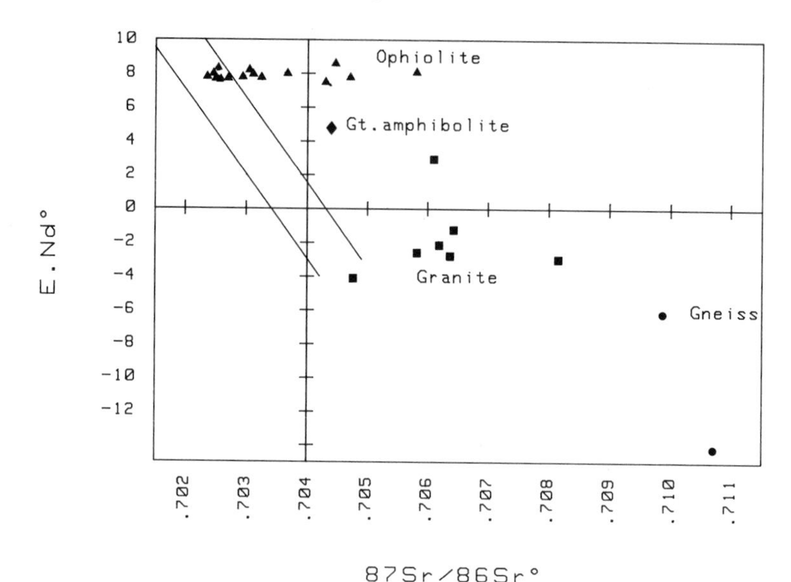

Figure 6. Comparison of initial ϵNd – $^{87}Sr/^{86}Sr$ values of the grano-diorite suite with the Oman ophiolite.

Pb Isotopes

When a contribution of a crustal component is expected in a magma genesis, lead isotopic ratios are the most sensitive tracers of this possible contribution genesis (Kay, 1980; Gill, 1981; White and Dupré, 1986; Sun, 1980; Davidson, 1986; Ben Othman et al., 1989). On both diagrams of Figure 7A and B, diorites and granites gather between MORBs and ophiolites field defined by Chen and Pallister, 1981 and Hamelin et al., 1984, and the general field of continental crust as defined by Sun (1980), Zartman and Doe (1981). The granodioritic field is tangent to that of continental crust. We observe that the metasediments, representing a possible source for continental contribution to the granodioritic magma genesis plot in the sub-field of present pelagic and detrital sediments.

Discussion

In all the isotopic diagrams, the granitic intrusions field is systematically intermediate between the ophiolite (oceanic crust and mantle) field and a crustal component. The studied suite likely results from a mixing between these two end-members. The problem is to precisely identify the two components of the mixing. Given the geodynamical context, there are two possible sources for both the crustal and mantle components. The crustal component can be either a) the autochtonous arabic shield or b) the metasediments of the metamorphic soles. The mantle component can be either c) the peridotite itself or d) the metabasalts from the metamorphic soles. Components b) and d) correspond to two members of the same geological unit.

Identification of the continental component

a) Arabic basement

Taking into account the petrologic, REE geochemical and isotopic characteristics, Lippard et al. (1986) have proposed partial anatexis of the autochtonous arabic shield as a possible source of crustal contamination for the calc-alkaline biotite granites, initiated during the ophiolite emplacement. The main objection to this hypothesis is based on thermal considerations: the temperature of the ophiolite nappe, when reaching the continental margin is below 400°C (Goffé et al., 1988; Michard et al., this volume), which is too low to promote crustal anatexis. Moreover the autochtonous shield is overlain by a thick (2–3 km) series of carbonates; their contribution should be traced in the magma composition. Finally, the initial isotopic compositions in Sr ($^{87}Sr/^{86}Sr = 0.705$) and Nd (eNd = + 3.0) of the dioritic part of the series record the contribution of a mantle component. This component is difficult to envisage in the hypothesis discussed, again for thermal reasons: felsic magmas are too cold to enhance partial melting and/or assimilation in a mantle slab whose temperature is below 400°C.

b) Metasediments

The other possible continental source which we favor is represented by the sediment blanket of the underthrust lithosphere. These sediments contained detritical component as suggested by the occurrence of phyllosilicates in the metasediments. The thermal conditions of the high grade metamorphism (about 900°C for Ghent and Stout, 1981) are capable to initiate partial melting, particularly when the high-T peridotite sole is in direct contact with the sediments. The isotopic compositions of the metasediments are quite consistent with those of the intrusive grano-dioritic series; particularly in Pb/Pb diagrams.

Two samples of granites have a Pb isotopic composition similar to that of the metasediments. One of them (87OM59) is normally enriched in potassium (K2O = 4.25%). Its high initial Sr isotopic composition ($^{87}Sr/^{86}Sr = 0.7093$) is directly comparable to that of the metamorphic sediments. This granite could be the pure product of partial melting of the sediments. On the contrary, the second one (87OM64) presents strong negative anomalies in alkali elements (K and Rb) and low initial Sr isotopic ratio ($^{87}Sr/^{86}Sr = 0.7058$). This granite cannot be the direct product of the mélting of sediment; a mantle component is required in the genesis of such sample.

Identification of the Mafic Component

Two sources carrying mantle signatures are potential parents for the mafic part of the calc-alkaline intrusions: the mantle peridotite itself, and the amphibolites representing the mafic part of the underthrust oceanic crust.

c) *Ultramafic rocks*

Mafic magmas could originate from hydrous partial melting of high-T peridotite due to dehydration of the underlying metamorphic sole during incipient obduction.

Ultrabasic rocks that are most probably directly involved in the genesis of the dioritic series can only be upper levels (10 to 20 km) of harzburgites and dunites below the Moho. Therefore, Chen and Pallister (1981) have analyzed Pb isotopic compositions of parts of these unit including hydrothermal galena. They show that the peridotites are enriched in radiogenic Pb. This enrichment is interpreted as due to contamination by crustal constituents of which hydrothermal galena is a direct evidence. The galena has the same Pb isotopic composition as the metasediments that we have analyzed (Fig. 7A and B). In the $^{207}Pb/^{204}Pb$ vs $^{206}Pb/^{204}Pb$ diagram, the isotopic composition of the ultramafics are intermediate between the field of ophiolites and that of metasediments. They are comparable to the members of the granodioritic series. On the other hand, the peridotites fall off the mixing line between the ophiolite and the metasediment on the $^{208}Pb/^{204}Pb$ vs $^{206}Pb/^{204}Pb$ diagram; they are lower in ^{208}Pb. These relationships could trace the possible contribution of the metasediments in hydrothermal fluids isotopic characters which have percolated at this upper mantle level. They could be responsible for an hydrous partial melting of the peridotite.

A further investigation would be to compare the geochemical characteristic of the grano-diorite suite with the secondary volcanism of calc-alkaline affinity which has been interpreted as resulting from hydrous melting during obduction of the peridotite (Ernewein et al., 1988; Boudier et al., 1988). The isotopic compositions of this magmatism should reflect those of the source and of the fluid inducing the melting. The secondary magmatisms have typically mantle Sr and Nd isotopic compositions (C. Pflumio and A. Michard, pers. comm.). The crustal contribution traced by isotopic contents, in the secondary volcanism is very low compared to that observed in the granodioritic series analyzed here. Thus, the petrogenetic processes generating the secondary volcanism and the granodioritic suite are different from the isotopic point of view.

d) *Metamorphic sole*

Partial melting could also involve the amphibolite of the metamorphic sole. Green and Ringwood (1968), Green (1973) have experimented melting of amphibolite and shown that at P = 10 kb and PH_2O, clinopyroxene and

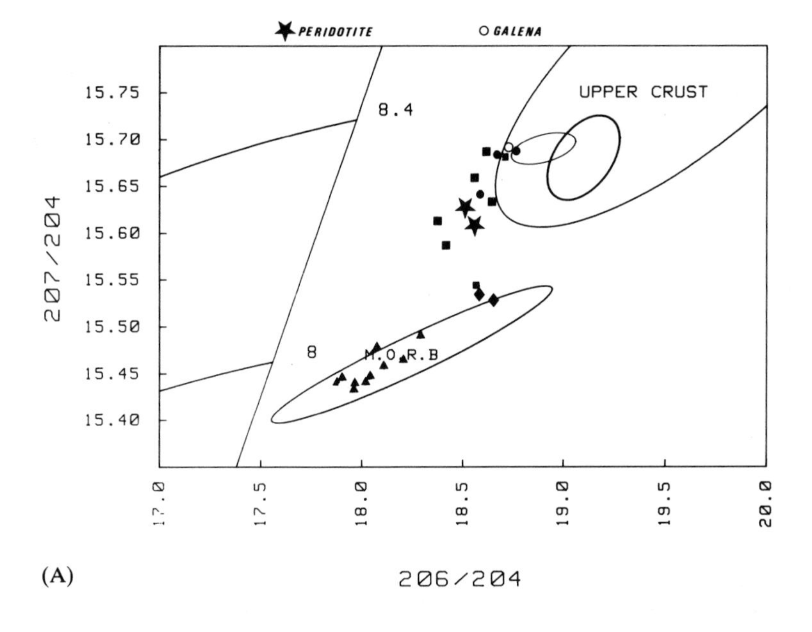

(A)

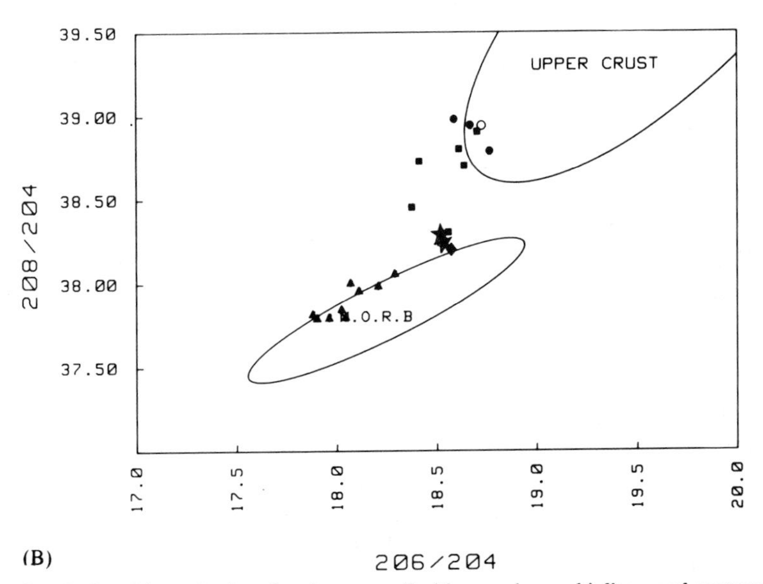

(B)

Figure 7*A*, *B*. Lead isotopic data for the grano-dioritic samples, ophiolites, and metamorphic sole. Data for ophiolite are from Chen and Pallister (1981) and Hamelin et al. (1982).

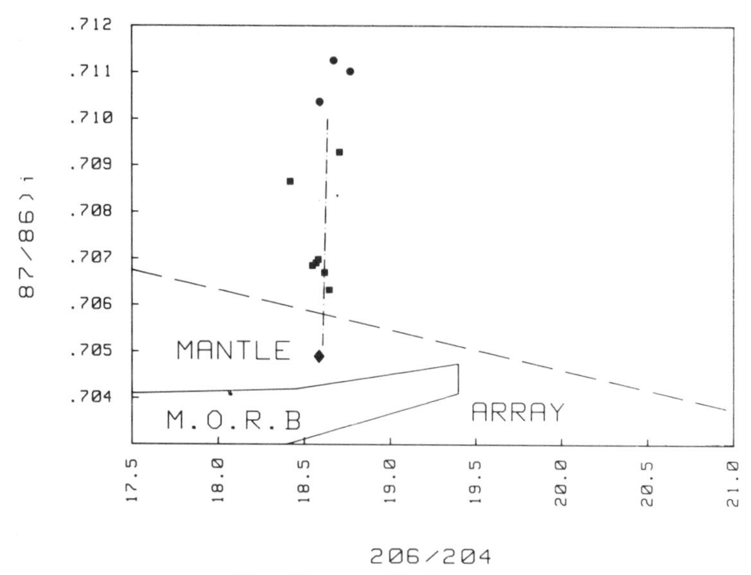

Figure 8. Relations between $^{206}Pb/^{204}Pb$ and $^{87}Sr/^{86}Sr$ initial ratios of diorites and granites, ophiolites and metasediments.

plagioclase are in equilibrium with andesitic magma at 940°C. At lower temperature (900°C) the residual assemblage is amphibole + clinopyroxene + garnet. These temperatures are in the range or just above those inferred in high grade amphibolites from Oman (Ghent and Stout, 1981). Concerning the trace elements, partial melting of a garnet amphibolite containing 10% garnet could produce calc-alkaline magmas presenting a U-shaped REE pattern (equilibrium with amphibole) and a depletion in HRE (equilibrium with garnet). Low degree of partial melting would generate rhyolitic magma, higher degree of melting would produce andesitic to basaltic liquid. Pb isotopic composition of the garnet amphibolite analyzed (Table 5) is identical to those of a metabasalt analyzed by Chen and Pallister (1981). They are enriched in radiogenic Pb compared to gabbros and ophiolite basalts, but lay within the general field for MORB (Fig. 7a and b). Sr and especially Nd isotopic compositions are significantly different from those of the ophiolites (ϵNd = + 4.7, Fig. 6).

In the ophiolite detachment occurred along the oceanic ridge (Nicolas, 1989), metamorphic sole amphibolites should have isotopic signatures similar to those of the volcanic pile from which they originate. According to Searle and Malpas (1982), the amphibolites actually show compositions consistent with MORB, particularly for their REE contents. However the sample we have studied differs significantly by its Nd isotopic composition (Table 4) from the ophiolitic basalts. But its much lower ϵNd (+ 4.7) could result from the metamorphic event due to the overthrusting of the ophiolite. The hydrothermal circulations and melting processes recorded in the metasediments intercaled between the basaltic layers were likely sufficient to significantly modify by impregnation the isotopic compositions of the metabasalts. It could explain the enrichment in incompatible elements also observed in the amphibolites by Searle and Malpas (1982). The identification of the amphibolite protolith would require a complete isotopic investigation of the amphibolites.

Magma Mixing or Differentiation

On the basis of major and trace element behavior (enrichment in incompatible elements and progressive depletion in compatible elements – Ti, V, P, Zr), the granodioritic series could be interpreted in terms of magmatic differentiation by fractional crystallization with superimposition of a variable rate of contamination. In such an assimilation-fractional crystallization process (AFC) the variation in isotopic composition should evolve progressively between the extreme end-members of the two components (Briqueu and Lancelot, 1978; De Paolo, 1981). It is not the case in the calc-alkaline series studied. The most basic dioritic members (87OM60 and 87OM61) do not show the lowest crustal contribution as expected in an AFC scheme. Such isotopic heterogeneities uncorrelated with a magmatic differentiation trend suggest that the granodioritic series results from a complex process of mixing between a differentiated fraction (diorite to granodiorite) produced by the melting of a mafic component, and a silica-rich fraction produced by the melting of metasediments. Field data fit with this hypothesis of magmatic hybridization.

However, quantification of the mixing processes is difficult if not impossible due to isotopic heterogeneities. For example, binary mixing between two unique components would result in mixing lines on the Pb-Pb diagram irrespective of the nature of the process (mass to mass mixing or selective contamination by fluid phases). For the granodioritic calc-alkaline series, some plots are significantly far from a single mixing line. It is clear that several components were involved.

From the petrogenetic point of view, the granodioritic series could be interpreted as the result of melting of the metamorphic sole induced by the obduction of the peridotite. Melting of amphibolites would produce a suite of calc-alkaline magmas ranging from andesitic basalts to rhyolites (depending on degree of melting) variably contaminated by fluids; granitic magmas would result from direct melting of quartzo-pelitic sediments. At this stage, the hybridization of these liquids of different nature and origin could still be incomplete. During their ascent and emplacement at higher structural levels, they interact with their peridotitic host.

Geodynamic Conditions of the Anatexis

The present study shows that the isotopic record of the calc-alkaline granitic rocks is consistent with an origin by anatexis of the metamorphic sole, in agreement with field observations. The metamorphic sole components could supply both the mantle and continental isotopic parents represented respectively by the metabasalts and metasediments. This conclusion raises the following questions and put some constraints on the system.

i) Representativity of the process at the scale of the Semail ophiolite
Although very limited in size, the calc-alkaline bodies or dikes are widespread along the chain. Moreover, Beurrier (1987) relates these intrusions in the South central Oman mountains to early shear zones, which is consistent with their emplacement during incipient obduction. Ductile strike-slip shear zones in the mantle have been already assigned to this episode (Boudier et al., 1988). The occurrence of these small intrusions all along the ophiolitic nappe implies that conditions of anatexis best observed in the metamorphic sole of the Hajar window were produced locally at other places, although the process appears to be more developed in the northern part of the chain. On the other hand, the small size of the intrusions suggests a punctual effect and a limited metamorphic sole melting.

ii) Conditions of the anatexis
The thermal conditions of the metamorphism in the ophiolite soles (Ghent and Stout, 1981) imply very high temperature conditions (about 900°C) at the initiation of the ophiolite nappe detachment, and a very steep thermal gradient emphasized by the tectonic piling of successive metamorphic slabs of lowering grade. The model accounting for these conditions is that of a high-T (1000°C) peridotite slab overriding the oceanic crust and bulldozering successive slices at decreasing T-conditions 'the iron model', as proposed in Omàn by Boudier and Coleman (1981), Ghent and Stout, (1981), Boudier et al. (1988). These thermal conditions lead the last authors to locate the decoupling plane at the ridge axis along the 1000°C isotherm. Thus the conditions are fulfilled to reach punctually anatexis conditions in the high grade metamorphics, especially when the peridotite sole rests directly on the metasediments, as corroborated by field observations. A second order evidence fits with the thrust-related melting: it is the observation of alignments of calc-alkaline granitic intrusions along early ductile strike-slip vertical shear zones (Beurrier, 1987). The kinematics of such shear zones have been studied in the mantle section (Ceuleneer, 1986; Boudier et al., 1988), and interpreted as the vertical wall of strike slip listric planes, acting as decoupling surfaces during the detachment. Such planes are the best candidates for concentrating detachment-related anatectic products.

iii) Timing

Isotopic ages on metamorphic minerals (Lanphere, 1981; Montigny et al., 1988) cluster around 95–100 Ma; they provide good time constraints on the metamorphic evolution, suggesting that high grade metamorphism occurred within a period of 5 My after dying accretion. This time gap is also consistent with the various heat sources expected to maintain the amphibolite facies thermal conditions: the 'iron effect' should participate for a 1–2 My duration (Boudier et al., 1988); shear heating is advocated, and estimated at about 200°C for a time thrusting of 5 My at a rate of the order of 8 cm yr^{-1} (Malpas, 1979); residual spreading heat could also participate for an unknown percent. Thus, 5 My is a reasonable duration for high grade thermal conditions, with the consequence that the oceanic crust involved, 5 My aged, had a sedimentary cover of sufficient thickness, including a flyschoïd component accounting for the mixed isotopic signature.

Conclusion

The following conclusions are derived from the study of K-mineral bearing felsic intrusions in the Semail ophiolite.

1. K-mineral felsic intrusions form hundred meters sized bodies intrusive in the mantle section, widespread along the chain but more developed in its northern part.
2. They belong to a K-poor calc-alkaline suite grading from diorite to K-bearing granite.
3. REE show an important negative K, Rb, Ba anomaly uncorrelated with the magma differentiation trend, but consistent with a mixing or contamination process.
4. Sr, Nd, Pb isotopic ratios plot in a field intermediate between oceanic and continental arrays. They appear to be unrelated with the volcanism capping the ophiolite. They are interpreted as derived from a mixed source represented by the metamorphic sole at the base of the ophiolite: the mica-bearing metasediments providing the continental source, and the metabasalts providing the mantle component. A partial contamination during percolation through the mantle wedge is not discarded.
5. Such partial anatexis of the metamorphic sole involves high-T conditions (above 700°C) maintained at the sole of the ophiolite during the initiation of the obduction. The 'iron-effect' of a young and hot lithosphere detached at the ridge accounts for the required temperature conditions.

Acknowledgments

This paper is indebted to A. Nicolas for his contribution in discussion of the data, J. Lancelot for lab. work support, and I. Reuber for field work contribution. The manuscript has been improved, thanks to constructive reviews by C. Pin and U. Ziegler. Field work has been partly supported by the ATP 'Blocs et Collision, 1987' from the CNRS-INSU.

References

Alleman, F., and Peters, T., 1972. The ophiolite-radiolarite belt of the North Oman mountains., Eclogae Geol. Helv., 65: 657–698.

Ben Othman, D., White, W.M., and Patchett, J., 1989. The geochemistry of marine sediments, island arc magma genesis, and crust-mantle recycling., Earth Planet. Sci. Lett., 94: 1–21.

Beurrier, M., 1987. Géologie de la nappe ophiolitique de Samail dans les parties orientale et centrale de l'Oman., Thèse Doc. Etat, Paris 6, 406 p.

Birck, J.L., and Allègre, C.J., 1978. Chronology and chemical history of the parent body of basaltic achondrites studies by the 87Sr/86Sr method., Earth Planet. Sci. Lett., 39: 37–51.

Boudier, F., and Coleman, R.G., 1981. Cross section through the peridotites in the Samail ophiolite. Southeastern Oman., J. Geophys. Res., 86: 2573–2592.

Boudier, F., Ceuleneer, G., and Nicolas, A., 1988. Shear zones, thrusts and related magmatism in the Oman ophiolite: intiation of thrusting on an oceanic ridge., Tectonophysics, 151: 275–296.

Bougault, H., 1980. Apport des éléments de transition à la compréhension des basaltes océaniques., Thèse, Univ. Paris VI, 178 p.

Briqueu, L., and Lancelot, J.R., 1978. Rb-Sr. systematics and crustal contaminations models for calc-alkaline igenous rocks., Earth Planet. Sci. Lett., 43: 385–396.

Briqueu, L., Bougault, H. and Joron, J.L., 1984. Quantification of Nd, Ta, Ti and V anomalies in magmas associated with subduction zones. Petrogenetic implications., Earth Planet. Sci. Lett., 68: 297–308.

Browning, P., 1982. The petrology, geochemistry and structure of the plutonic rocks of the Oman ophiolite., Ph. D. Thesis, The Open Univ., Milton Keynes, 404 p.

Browning, P., and Smewing, J.D., 1981. Processes in magma chambers beneath spreading axes: evidence for magmatic associations in the Oman ophiolite., J. Geol. Soc. London, 138: 279–280.

Chen, J.H., and Pallister, J.S., 1981. Lead isotopic studies of the Semail ophiolite, Oman., J. Geophys. Res., 86: 2699–2708.

Ceuleneer, G., 1986. Structure des ophiolites d'Oman: flux mantellaire sous un centre d'expansion océanique et charriage à la dorsale., Thèse Doc. Univ. Nantes, 152 p.

De Paolo, D.J., 1981. Trace elements and isotopic effects of combined wall rocks assimilation and fractional crystallization. Earth., Planet. Sci. Lett., 53, 189–195.

Davidson, J.P., 1986. Isotopic and trace element constraints on the petrogenesis of subduction related lavas from Martinique, Lesses Antilles., J. Geophys. Res., 91, 5943–5962.

Ernewein, M., Pflumio, C. and Whitechurch, H., 1988. The death of an accretion zone as evidence by the magmatic history of the Sumail ophiolite (Oman)., Tectonophysics, 151, 247–274.

Ghent, E.D., and Stout, M.Z., 1981. Metamorphism at the base of the Semail ophiolite, Southeastern Oman mountains., J. Geophys. Res., 86: 2557–2573.

Gill, J., 1981. Orogenic andesites and plate tectonics. Springer Verlag Ed., 385 p.

Goffé, B., Michard, A., Kienast, J.R. and Le Mer, O., 1988. A case of obduction-related

high-pressure, low-temperature metamorphism in upper crustal nappes. Arabian continental margin, Oman: p-T paths and kinematic interpretation., Tectonophysics, 151, 363–386.
Green, D.H., 1973. Experimental melting studies on a model upper mantle compositions at high pressure under water-saturated conditions., Earth Planet. Sci. Lett., 19: 37–53.
Green, T.H., Ringwood, A.E., 1968. Genesis of the calc-alkaline ignous rocks suite., Contrib. Mineral. Petrol., 18: 105–162.
Hamelin, B., Dupré, B. and Allègre, C.J., 1984. Lead-strontium isotopic variations along the East Pacific Rise and the Mid-Atlantic ridge: a comparative study., Earth Planet. Sci. Lett., 67: 340–350.
Kay, R.W., 1980. Volcanic arc magmas implications of a melting mixing model for element recycling in the crust-upper mantle system., J. Geol., 88: 497–522.
Kurz, D., 1990. Sapphivine-quartz assemblages in the high-grade metamorphic rocks below the Semail ophiolite in the northern Oman Mountains., Abst. Symp. on Ophiolite genesis and evolution of oceanic lithosphere. Muscat, 1990.
Lanphere, M.A., 1981. K-Ar ages of metamorphic rocks at the base of the Samail ophiolite, Oman., J. Geophys. Res., 86: 2777–2782.
Lanphere, M.A., Coleman, R.G. and Hopson, C.A., 1981. Sr isotopic tracer study of the Samail ophiolite, Oman., J. Geophys. Res., 86: 2709–2720.
Lippard, S.J., Shelton, A.W. and Gass, I.G., 1986. The ophiolite of Northern Oman., Geol. Soc. Memoir., v. 11, chap. 4, 178 p.
Malpas, J., 1979. The dynamo-metamorphic aureole of the Bay of Islands ophiolite suite., Cand. J. Earth Sci., 16: 2086–2101.
Manhés, G., Minster, J.F. and Allègre, C.J., 1978. Comparative Uranium-thorium-lead and rubidium strontium study of the Saint Severin amphoterites: consequences for early solar system chronology., Earth Planet. Sci. Lett., 39: 14–24.
McCulloch, M.T., Gregory, R.T., Wasserburg, G.J. and Taylor, H.P., 1980. A neodynum, strontium and oxygen isotope study of the Cretaceous Samail ophiolite and implication for the petrogenesis and seawater-hydrothermal alterations of ocean crust., Earth Planet. Sci. Lett, 46: 201–211.
McCulloch, M.T., Gregory, R.T., Wasserburg, G.J. and Taylor, H.P., 1981. Sm-Nd, Rb-Sr and 18O/16O isotopic systematics in an oceanic crustal section. Evidence from the Samail ophiolite., J. Geophys. Res., 86: 2721–2736.
Michard A., Boudier, F. and Goffé, B. Obudction versus subduction and collision: from the Oman case to other Tethyan settings., Tectonophysics, this volume.
Montigny, R., Le Mer, O, Thuizat, R. and Whitechurch, H., 1988. K-Ar and 40Ar/39Ar study of metamorphic rocks associated with the Oman ophiolite: tectonic implications., Tectonophysics, 151: 345–362.
Nicolas, A., 1989. Structures of ophiolites and dynamics of oceanic lithosphere. Kluwer Acad. Pub., 367 p.
Nicolas, A., Ceuleneer, G., Boudier, F. and Misseri, M., 1988. Structural mapping in the Oman ophiolites: mantle diapirism along an oceanic ridge., Tectonophysics, 151: 27–56.
Pedersen, R.B. and Malpas, J., 1984. The origin of oceanic plagiogranites from the Karmoy Ophiolite, Western Norway., Contr. Mineral. Petrol., 88: 36–52.
Pallister, J.S.and Knigt, R.J., 1981. Rare-Earth Element Geochemistry of the Samail Ophiolite near Ibra, Oman., Tectonophysics, 151: 2699–2708.
Richard, P., Shimizu, N., and Allegre, C.J., 1976. 143Nd/144Nd, a natural tracer: an application to oceanic basalt., Earth Planet. Sci. Lett., 31: 269–275.
Searle, M.P., 1980. The metamorphic sheet and underlying volcanic rocks beneath the Semail ophiolite in the Northern Oman mountains of Arabia., Unpubl. Ph. D. Thesis, Open Univ., p. 213. Searle, M.P., and Malpas, J., 1980. Petrochemistry and origin of sub-ophiolitic metamorphic and related rocks in the Oman Mountains., J. Geol. Soc., London, 139: 235–248.
Sun, S.S., 1980. Lead isotopic study of young volcanic rocks from mid-ocean ridges, ocean islands and island arcs., Ph. Los. Trans. R. Soc., London, 297: 409–444.

Tilton, G.R., Hopson, C.A. and Wright, J.E., 1981. Uranium lead isotopic ages of the Samail ophiolite, Oman, with applications to Tethyan Ocean Ridge Tectonics., J. Geophys. Res., 86: 2763–2775.

White, W.M. and Dupré, B., 1986. Sediment subduction and magma genesis in the lesser Antilles: isotopic and trace element constraints., J. Geophys. Res., 91: 5927–5941.

Zartman, R.E., and Doe, B.R., 1981. Plumbotectonics. The model., Tectonophysics, 75: 135–162.

Ziegler, U.R.F. and Stoessel, G.F.U., 1985. The metamorphic series associated with the Semail Ophiolite Nappe of the Oman Mountains in the United Arab Emirates., MS thesis, Univ. Bern, Switzerland, 293 p.

Mineral Equilibria in Metagabbros: Evidence for Polymetamorphic Evolution of the Asimah Window, Northern Oman Mountains, United Arab Emirates

MARTIN BUCHER
Min.-petr. Inst., University of Bern, Baltzerstr.1, CH – 3012 Bern

Abstract

The greenschist metamorphic areas of Wadi Dibba – Dahir – Wadi al Fay and Asimah are dominated isoclinally folded metacherts with less abundant intercalated metabasites, metacarbonates, and some metamorphic gabbros.

The gabbros are slightly alkaline rocks of transitional to tholeiitic affinity that intruded the Hawasina-like pelagic sediments, forming small bodies and dykes. The primary igneous phases are kaersutite, clinopyroxene, biotite, plagioclase (An_{60-80}) and ilmeno-hematite.

Two metamorphic events are recognized in the region. M_1 includes a typical upper greenschist- to epidote amphibolite-facies paragenesis which in the gabbro is:
Mg-hornblende – albite – quartz – chlorite$_1$ - epidote$_1$ (X_{Ps} 0.17–0.26) – sphene$_1$. M_2 overprints M_1, resulting in the lower greenschist-facies paragenesis:
actinolitic hornblende/actinolite – albite – chlorite$_2$ – epidote$_2$ (X_{Ps} 0.12–0.15) – sphene$_2$ (0.023–0.043 Fe^{3+} pfu) ± stilpnomelane.

M_2 amphiboles are characterized by lower edenite substitution ((Na^A+K), $Al^{IV} \Leftrightarrow [A]$, Si), minor Tschermak substitution ((Al^{VI},Fe^{3+},Ti), $Al^{IV} \Leftrightarrow (Fe^{2+},Mg,Mn)$, Si), some richterite substitution (Na^A, $Na^{M4} \Leftrightarrow [A]$, Ca^{M4}) and slightly higher X_{Mg} values compared to M_1 amphiboles. Chlorite in M_2 is lower in Tschermak substitution (Al^{VI}, $Al^{IV} \Leftrightarrow (Fe^{2+},Mg,Mn)$, Si) and generally richer in $X_{Fe^{2+}}$.

Mineral assemblages, phase compositions, trends of partitioning of $(Fe^{2+} + Mn)/Mg$ and $FeMg_{-1}$, and Tschermak substitution in amphibole and chlorite suggest that the Asimah rocks passed first through medium P/medium T (P = 500 ± 50 MPa; T = 450 – 500°C) conditions, then through medium P/low T (P = 400 ± 100 MPa; T = 340 – 380°C) conditions during their metamorphic evolution.

Tj. Peters et al. (Eds), Ophiolite Genesis and Evolution of the Oceanic Lithosphere, 543–571.

1. Introduction

Earlier studies dealing with metamorphic amphiboles in mafic systems have provided considerable information on the history of metamorphism in specific areas (e.g. Compton, 1958; Cooper and Lovering, 1970; Harte and Graham, 1975; Laird and Albee, 1981a, b). Changes in the mineral assemblage of metamorphosed mafic rocks have been used as indicators of metamorphic conditions. In particular, the applicability of the variation in amphibole composition as a guide to metamorphic pressures and temperatures is widely recognized. Amphiboles derived from high-pressure facies series are rich in glaucophane (Miyashiro, 1961), and even medium-pressure facies mafic rocks contain amphiboles richer in the glaucophane molecule than those of low pressure (Brown, 1977a, b). Other observations that have been used to make a distinction between low- and medium-pressure amphiboles are lower Al^{VI} (Leake, 1965; Raase, 1974; Laird and Albee, 1981 b), higher edenite substitution (Laird and Albee, 1981 b), notably higher Ti/Al ratios (Hynes, 1982), and a poorly developed compositional gap between actinolite and hornblende (Hynes, 1982).

This paper deals with amphiboles from metamorphic gabbros and greenschists s.str. in the greenschist-facies area of the Asimah window (Stoessel and Ziegler, 1985) or Tayybah anticline (Lippard et al., 1986), Dibba zone, Northern Oman Mountains. The amphibole compositions are compared to published amphibole data from other terrains in order to obtain a better understanding of the pressure and temperature conditions of metamorphism, and to relate these findings to the tectonic setting of the greenschist metamorphic thrust sheets in the Northern Oman Mountains.

2. Geologic Setting

2.1. *Geology of the Oman Mountains*

The Oman Mountains, located at the southeastern edge of the Arabian Peninsula, comprise a northwest to west facing system of oceanic nappes that were thrust over a Precambrian to Cretaceous autochthon in Mid- to Late Cretaceous time. During this interval from the Albian to the Campagnian (101–70 Ma), the allochthonous nappes were assembled and faulted onto the Arabian continental margin. By Maastrichtian time (70–65 Ma), all the nappes were partially eroded and subsequently transgressed by marine shallow water carbonate rocks. The autochthonous rocks include continental shelf rocks of the stable Arabian platform resting on a cratonal basement of Proterozoic age (Würsten et al., this volume). The nappes, from the structurally lowest in the west to the structurally highest in the east are: the Sumeini Group, the Hawasina Complex, the Haybi Complex (which was included by Bechennec et al. (1988) in the Umar Group of his Hawasina nappes), the

Semail ophiolite nappe, and the ophiolite-related Batinah Complex. The Sumeini, Hawasina, and Haybi represent shelf-edge, continental slope, partly intra-oceanic continental rise, and pelagic sedimentary and volcanic rocks. The history of deposition started in an intracontinental basin situated along the northeastern edge of Gondwana during Late Permian (Bechennec et al., 1988), and passed into a passive margin environment and ocean basin northeast of the Arabian basement during a period ranging from Middle/Late Triassic to Late Cretaceous (Glennie et al., 1974; Woodcock and Robertson, 1982; Lippard et al., 1986; Bechennec et al., 1988).

The Metamorphic sheet or sole is situated in the thrust zone at the base of the Semail ophiolite, and is inferred to show a facies-gradient from partially melted (Searle and Malpas, 1980, 1982; Boudier et al., 1988) lower granulite at the contact, grading downward into amphibolite, then into greenschist, and finally into unmetamorphosed Hawasina-like pelagic and volcanic rocks. The least disrupted exposures of metamorphic rocks reveal in the southern part of the mountain range a ≤ 250 m thickness of amphibolite-facies rocks and a ≤ 300 m thickness of greenschist-facies rocks. In the Northern Oman Mountains (UAE), exposures are generally larger and contain in the Bulaydah-Hajar-Shis area south of Masafi a ≤ 2000 m thickness of transitional amphibolite- to lower granulite-facies rocks; whereas in the Asimah window the amphibolite-facies is ≤ 500 m thick and the greenschist-facies rocks are ≥ 2000 m thick. The lower granulite- to amphibolite-facies rocks comprise a great variety of felses, gneisses and schists, and minerals exhibit crystal-plastic deformation features. The majority of the amphibolites are intermediate in composition between mid-ocean ridge basalts and within-plate tholeiites, comparable to the volcanic rocks in the underlying Haybi Complex (Searle, 1980; Lippard et al., 1986). Peak metamorphic temperature and pressure conditions are estimated at 765–875°C and 500 MPa, respectively (Ghent and Stout, 1981). Various high- and low-grade assemblages in different textural relationships in (1) metaquartzites with sapphirine – spinel – ilmeno-hematite – magnetite – quartz, quartz – cordierite ± enstatite – magnetite – ilmeno-hematite (D. Kurz, pers. commun., 1990); (2) metahemipelites with quartz – cordierite – phlogopite – ilmeno-hematite – magnetite, fibrous sillimanite – spinel – magnetite in cordierite grains, kyanite relics and secondary andalusite (D. Kurz, pers. commun., 1990); and (3) amphibolites with pargasitic hornblende – diopside – orthopyroxene (En_{75}) – plagioclase – ilmenite – magnetite are observed in the Hajar-Shis area in the Northern Oman Mountains (UAE). Preliminary determinations of temperature and pressure conditions suggest a more complex metamorphic history for the Metamorphic sheet than was previously assumed. The lower granulite- and amphibolite-facies rocks are overprinted by greenschist-facies assemblages which formed after a mylonitic deformation phase.

Greenschist-facies rocks tectonically underlying the amphibolite-facies rocks include metamorphosed shale, chert, limestone, slightly alkaline gabbros and volcanic rocks, perhaps derived from the Hawasina Complex (Bechennec et al., 1988) or the Haybi Complex (Searle, 1980; Lippard et al., 1986). Initial metamorphism of the greenschist-facies rocks developed during a deformation phase, producing tight isoclinal folds with a penetrative axial plane schistosity. Later cross-cutting mineral growth of a second metamorphic event in lower greenschist-facies can probably be attributed to large-scale open folding events.

The lower granulite-, amphibolite- and greenschist-facies rocks yield K-Ar and $^{40}Ar/^{39}Ar$ cooling ages ranging from ~101–72 Ma. High-grade metamorphism spans a period of ~12 Ma from 101–89 Ma. Most data, however, define a closer interval of ~7 Ma, from 101–94 Ma (including one-sigma standard deviation). Greenschist-facies rocks were formed in an interval of ~14 Ma from 86–72 Ma (Allemann and Peters, 1972; Searle et al., 1980; Lanphere, 1981; Montigny et al., 1988).

2.2. *Geology of the Greenschist Metamorphic Areas in the Asimah Window*

The greenschist metamorphic rocks (monometamorphic series of Allemann and Peters, 1972) crop out in the Dibba zone. This is a narrow elongate belt of complex geology comprising a sequence of six major imbricated and strongly folded, generally northwest to west facing thrust units. These are composed of (1) the Late Palaeozoic-Mesozoic Musandam carbonate platform (Hajar Supergroup); (2) Mesozoic shelf-edge and slope carbonates (Sumeini Group); (3) basinal facies carbonate platform slope, continental rise and abyssal plain sediments (Hawasina Complex); (4) sedimentary and tectonic mélange-type rocks, Oman Exotics and volcanics (Haybi Complex); (5) metamorphic rocks of different grade (Metamorphic sheet, included by Searle, 1980; Robertson et al., 1990 in the Haybi Complex); and (6) the Semail ophiolite nappe as the highest tectonic unit in the Dibba zone (Fig. 1). Particularly good exposures of greenschist metamorphic rocks are to be found in the Wadi Dibba – Dahir – Wadi al Fay and Asimah areas and in a narrow zone north of Masafi (Fig. 1). The south-eastern and south- to north-western borders of the Asimah window (Tayybah anticline) are in thrust contact with either amphibolite-facies rocks or the Semail ophiolite nappe (Fig. 2), whereas to the north and northeast the greenschist metamorphic rocks partly overlie either the Hamrat Duru Group of the Hawasina Complex (Glennie et al., 1974) or the Kub mélange of the Haybi Complex (Robertson et al., 1990).

The rock sequence of the Asimah window comprises intercalated quartzites, quartz-sericite(±piemontite) schists, quartz-white mica(±garnet) schists to phyllites, some metabasites and metacarbonates. The latter two lithologies are not very abundant, each comprising about 5–10% of the mapped area. The maximum structural thickness is about 6 km. This total thickness, however, may be the composite thickness of several individual thrust sheets. Initial metamorphism of the greenschist-facies rocks developed during a deformation phase, which produced chiefly tight isoclinal folds on a mm to

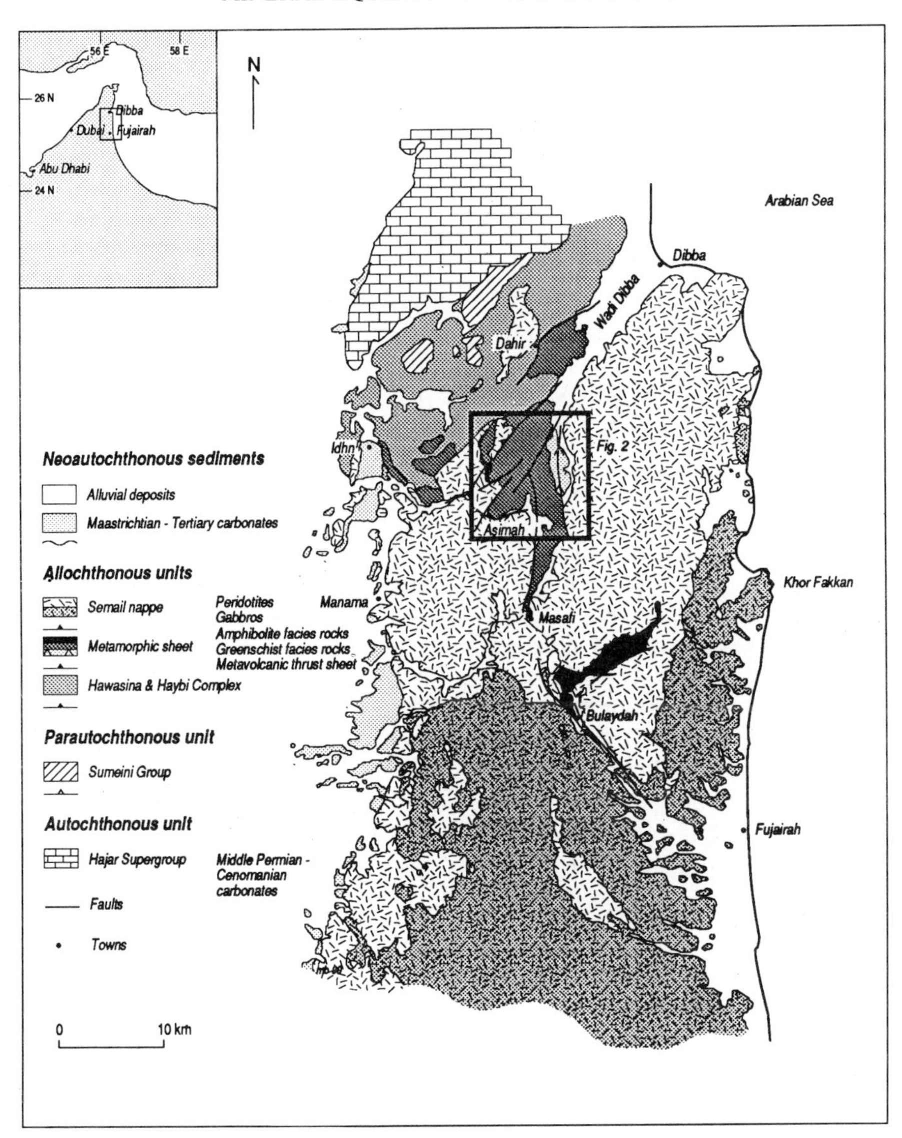

Figure 1. Regional tectonic map of the Northern Oman Mountains (modified after Allemann and Peters, 1972).

m scale, and an early, dominant, axial plane schistosity S_1, oriented generally parallel to the lithological boundaries. This penetrative foliation was in some places cut by mylonite fabrics belonging to a second folding phase. These fabrics are of a S_2 schistosity on which surfaces partly static diablastic growth of lower greenschist-facies minerals occurred. Later rather small scale features include open folds and kink bands.

A conspicuous feature of the Asimah window is the alkaline Metavolcanic

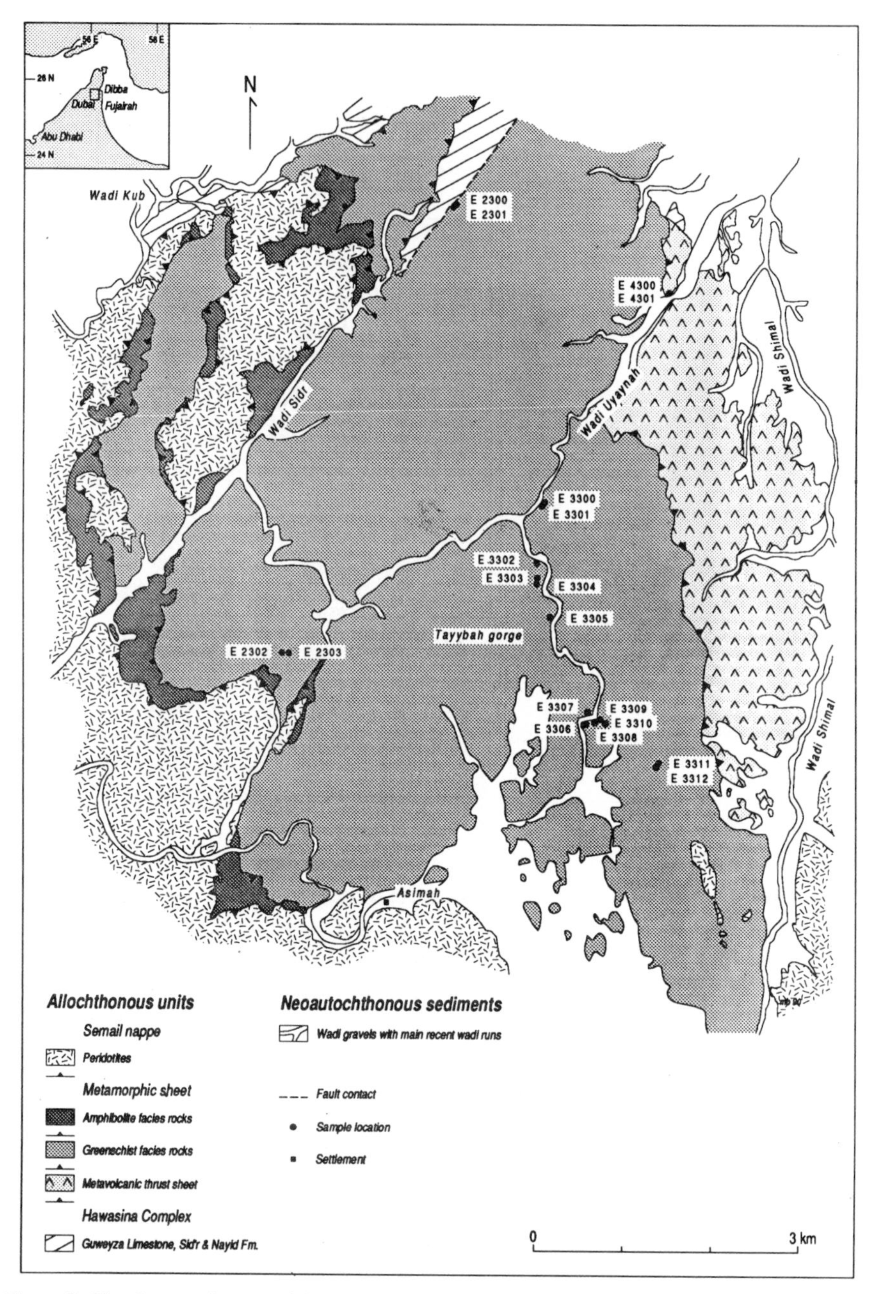

Figure 2. Sketch tectonic map of the Asimah window (Tayybah anticline) with sample locations of metagabbroic bodies and dykes.

thrust sheet (Stoessel and Ziegler, 1985), which is exposed between Wadi Uyaynah and Wadi Shimal (Fig. 2). In this unit are significant amounts of meta-pillowlavas, meta-tuffites, meta-hyaloclastites, meta-carbonatites, and some metacherts and metacarbonates. Metamorphic gabbros are less abundant and they can be traced in and along the Tayybah gorge, in Wadi Uyaynah, and in some outcrops E of Wadi Sidr (Fig. 2). These slightly alkaline rocks of transitional to tholeiitic affinity occur as small intrusive bodies and dykes in the greenschist-facies quartzitic, mafic and carbonatic rocks. The observed relict igneous phases are kaersutite, clinopyroxene and biotite. The composition of the plagioclase (An_{60-80}) and opaques (ilmeno-hematite) were calculated from bulk rock geochemistry (CIPW norm and least-square modeling). Table 1 lists the mineralogy of some representative metagabbros.

2.3. *Evidence of Polymetamorphism*

Mineral assemblages observed in metagabbroic rocks (and to a less extent in greenschists s.str. and quartz-mica schists) from the Asimah window reveal two episodes of metamorphic recrystallization. Evidence for polyphase metamorphism in these rocks includes textural relations among the phases, several generations of structures and minerals, relict phases and their replacements and overgrowths, composite grains with optically and chemically distinct cores and rims (amphibole and epidote), incomplete reactions between assemblages formed during successive events, and different phase compositions. The M_1 assemblages formed under medium P/medium T conditions in upper greenschist- to epidote amphibolite-facies, reflecting moderate crustal levels (or overload) and thermal gradients. During the M_2 event, metamorphic assemblages developed at lower temperatures but only slightly lowered pressures. Metamorphic conditions for the Asimah window are difficult to determine because suitable lithologies (e.g. metapelites) are lacking. However, quite reasonable constraints on P and T can be derived from experimentally studied phase equilibria, and from correlations of mineral compositions (amphibole, chlorite and epidote) with other metamorphic areas. Mafic metaigneous (metagabbros) and to a less extent metavolcanic rocks (greenschists s.str.) contain phase assemblages clearly identifying the two metamorphic events. Because of the rather small extent of mafic rocks in the area, and the continuous nature of the reactions that produce these assemblages in epidote amphibolite- and greenschist-facies conditions, no isograds could be mapped during the field work. It is however striking that the growth of stilpnomelane (assigned to M_2) in quartz-mica schists and metagabbros seems to be restricted to zones along, and east to southeast of the Tayybah gorge, and to quartzitic rocks in the alkaline Metavolcanic thrust sheet.

Table 1. Phase assemblages of representative metagabbros from the Asimah area (x = phase present in the sample).

Sample	E2301	E3303	E3306	E3308	E4301
Residual magmatic assemblage					
Clinopyroxene (Augite)	x				
Kaersutite	x	x	x	x	x
Biotite			x	x	
Apatite	x	x	x	x	x
M_1					
Mg-hornblende	x	x	x	x	x
Chlorite (Al^{IV}, $X_{Fe^{2+}}$)	1.15–1.17, 0.28–0.29	not analyzed	1.13–1.15 0.43 0.45	1.15–1.25 0.41–0.44	1.16–1.19 0.47–0.49
Epidote ($X_{Fe^{3+}}$)	not analyzed	0.25–0.26	0.18–0.25	≥0.17	0.18–0.22
Albite (coarse-grained)	x	x	x	x	x
Sphene	x	x	x	x	x
M_2					
Actinolite	x	x	x	x	x
Chlorite (Al^{IV}, $X_{Fe^{2+}}$)	1.06–1.1, 0.29–0.3	0.99–1.03, 0.52–0.54	not analyzed	0.83–1.09, 0.45–0.48	1.08–1.09, 0.49–0.50
Epidote ($X_{Fe^{3+}}$)	not analyzed	not analyzed	0.13–0.14	0.12–0.15	≤0.16
Albite (fine-grain., subgrains)	x	x	x	x	x
Sphene (recryst., Fe^{3+} cnt.)		x	x	x	x
Stilpnomelane			x		
Hematite (traces)	x	x	x	x	x
White mica (sericite ?)	x	x	x	x	

3. Analytical Techniques

Mineral analyses were carried out using an ARL® SEMQ microprobe at the University of Berne with wavelength- and energy-dispersive detector systems. Operating voltage and beam current were 15 kV and 20 nA respectively. All minerals were analyzed with a 10–μm beam. Counting times were 10 s and 20 s. Various well-characterized natural and synthetic silicate and oxide standards were used for calibration, some of which were regularly analyzed as internal standards to check for systematic errors.

All amphibole analyses were normalized on the basis of (Cations-Na-K-Ca) = 13 and 24 oxygen atoms. Chlorite was normalized to 14 anhydrous oxygens assuming all iron as ferrous. Mineral formulae for epidote and sphene were calculated on the basis of 12.5 and 5 oxygens respectively, with all iron considered as ferric.

4. Petrology and Mineral Chemistry

4.1. *Mineral Assemblages in Mafic Rocks*

Mafic metamorphic rocks contain in the greenschist- and epidote amphibolite-facies actinolitic hornblende/actinolite or Mg-hornblende, chlorite, albite, clinozoisite or epidote, usually sphene as a Ti-bearing phase, sometimes quartz, K-mica, a carbonate, and an iron oxide. Laird and Albee (1981b) refer to their 'common' mafic assemblage with amphibole + chlorite + epidote + plagioclase + quartz + Ti-phase ± carbonate ± K-mica ± Fe^{3+}-oxide. They say that the compositions of individual minerals in this 'common' assemblage are buffered by the assemblage and are therefore a function of P and T and not of whole rock geochemistry. In contrast to this assumption, Brown (1977b) notes that iron oxide buffers the amount of glaucophane in amphiboles so that in absence of an iron oxide, the maximum quantity of glaucophane cannot be realized. Furthermore, Harte and Graham (1975) and Hynes (1982) notice higher metamorphic temperatures for the Al substitution in amphiboles (actinolite-hornblende transition) for rocks with lower Mg/Fe, but they do not recognize an appreciable effect on the composition of the amphiboles in metabasic rocks.

4.2. M_1 *Medium-P/medium-T metamorphism*

4.2.1. *Mineral Assemblages*

Metagabbroic rocks – The M_1 upper greenschist- to epidote amphibolite-facies assemblages represent the first metamorphic event, because they replace and cut relict igneous phases, and because they are overgrown and/or replaced by M_2 assemblages. Parts of the M_1 assemblage are common in all metagabbros from the Asimah window and consist of Mg-hornblende – $chlorite_1$ – $epidote_1$ – albite – $sphene_1$. Mg-hornblende commonly occurs as relatively large (0.3–0.8 mm long) euhedral to anhedral elongate grains growing on, or totally replacing, magmatic clinopyroxene and kaersutite. Fine-grained, slightly greenish-yellow and generally very small aggregates of M_1 chlorite occur as inclusions in, or are partly surrounded, by Mg-hornblende. According to the classification of Hey (1954), $chlorite_1$ is of pycnochlorite composition. Feldspars from the M_1 assemblage are coarse- to medium-grained (0.1–0.8 mm diam.) anhedral porphyroblasts of almost pure albite. M_1 epidote is a ubiquitous phase. It occurs as (1) subhedral, mainly coarse-grained (0.1–0.3 mm diam.) crystals, sometimes oriented parallel to the main foliation; and (2) a very fine-grained anhedral phase adjacent to Mg-hornblende and $chlorite_1$. Sphene is a common rock-forming mineral and occurs as (1) fine-grained granular inclusions in Mg-hornblende and sometimes albite; and (2) coarse-grained (0.2–0.8 mm diam.) porphyroblastic crystals.

Greenschists s.str. – These are fine- to medium-grained rocks with some compositional layering parallel to the penetrative foliation. The layering consists of leucocratic quartz-albite and melanocratic amphibole-chlorite ± white mica layers, and is most probably inherited by original volcano-sedimentary features.

Metasedimentary rocks – With the exception of the initial structural style leading to the pervasive foliation S_1, quartzites and quartz-mica schists to phyllites contain almost no evidence of the M_1 metamorphic event. The observed compositional layering, although it is tightly isoclinally folded, might represent relict chert bedding of primary deposition but apart from muscovite flakes growing parallel S_1, no mineral phases were observed that can so far directly be related to M_1. Microstructures of quartz grains, however, reveal some indications of an older event that was preserved during M_2.

4.2.2. *Mineral Chemistry*

Amphiboles – M_1 amphiboles are Mg-hornblende in composition (Leake, 1978) with moderate $Mg/(Mg + Fe^{2+})$ of 0.42–0.68 and Si contents of 6.85–7.25 pfu. Average Mg-hornblende has the following general formula:

$$(K_{0.09}Na_{0.32})(Ca_{1.73}Na_{0.27})(Mg_{2.33}Fe^{2+}{}_{1.75}Mn_{0.02}Fe^{3+}{}_{0.21}Ti_{0.03}Al^{VI}{}_{0.66})$$
$$(Si_{7.01}Al^{IV}{}_{0.99})O_{22}(OH)_2.$$

Amphiboles from the M_1 assemblage characteristically have Al^{IV} values ranging from 0.7–1.2 pfu showing significant Tschermak substitution; and minor edenite substitution clearly demonstrated by analyses plotting below the 1:1 tremolite/actinolite – tschermakite join (Fig. 3C). Octahedral Ti is very low (≤ 0.04 pfu, Fig. 4). Al^{VI} ranges from a maximum value of 0.88 pfu to a minimum of 0.25 pfu and is connected with the systematic variation in Al^{IV},

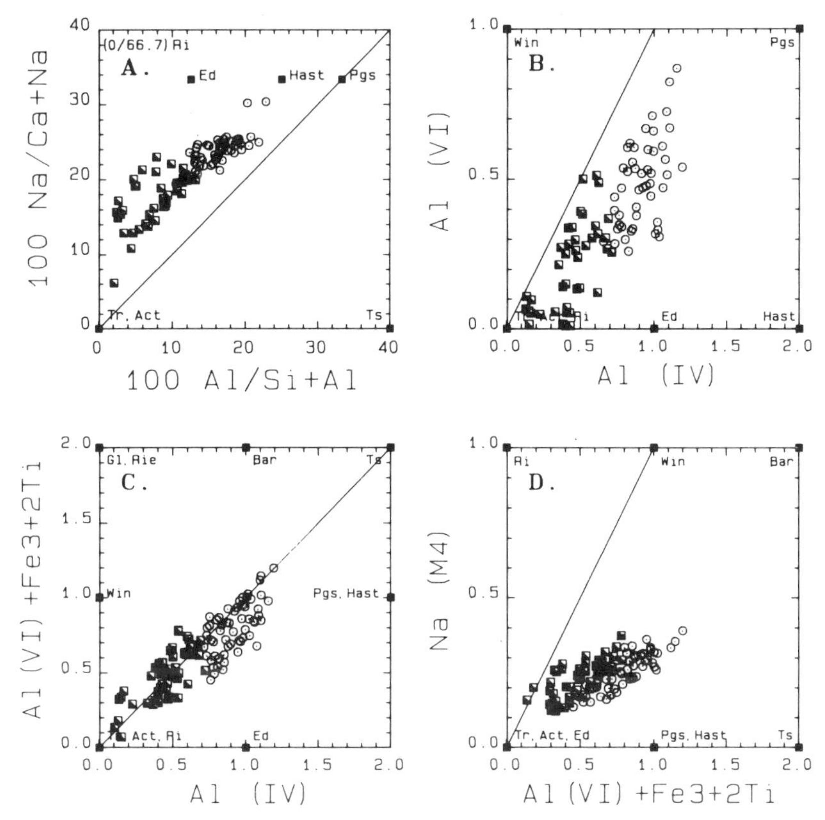

Figure 3. Plots of selected amphibole compositions for Asimah metagabbros (and greenschists s.str.) containing M_1 and M_2 assemblages (after Laird and Albee, 1981b). Endmember amphibole compositions are indicated by filled squares, circles = M_1 amphiboles, half-filled squares = M_2 amphiboles.

A) Plot of total Na/(Ca + Na) versus total Al/(Si + Al), this plot is independent of normalization method (1:1 line is for reference only).

B) Plot of Al^{VI} versus Al^{IV}. The line of slope 1 corresponds to a pure Al-Tschermak substitution from tremolite/actinolite at the origin.

C) Plot of (Al^{VI} + Fe^{3+} + 2Ti) versus Al^{IV} illustrates M2–site substitution in amphibole with increasing grade (Ferri-Tschermak and Al-Tschermak substitutions along a line of 1:1).

D) (Al^{VI} + Fe^{3+} + 2Ti) versus Na^{M4} shows glaucophane substitution. The line of slope 1 corresponds to a pure glaucophane substitution.

indicating a change in composition along a ± parallel line to the pure 1:1 Al-Tschermak substitution (Fig. 3B). The M_1 amphiboles show M4-site Na of 0.18–0.4 pfu (Fig. 3D), and have A-site occupancies of 0.35–0.55 pfu.

Chlorite – M_1 chlorites show a range in $Fe^{2+}/(Mg + Fe^{2+})$ from 0.42–0.50 and contain in some cases small amounts of Ti (≤0.01 pfu). Tetrahedral Al^{IV} varies between 1.15–1.25 pfu. Analyses of $chlorite_1$ plotting above the 1:1 line in Fig. 7 show dioctahedral substitution (Al_2Mg_{-3}) and have

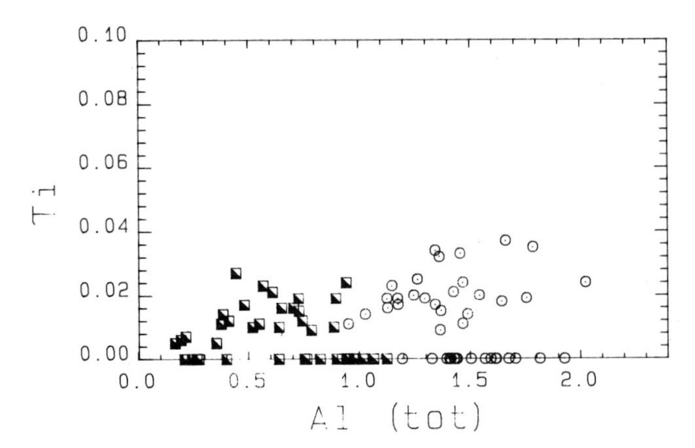

Figure 4. Total Ti versus total Al, after normalization of amphiboles to 23 oxygens with all iron considered as Fe^{2+} (Symbols as in Fig. 3).

octahedral vacancies. Average M_1 chlorite has the general formula:

$$(Mg_{2.7}Fe^{2+}{}_{2.12}Mn_{0.03}Al^{VI}{}_{1.15})(Si_{2.8}Al^{IV}{}_{1.2})O_{10}(OH)_8.$$

Plagioclase – Plagioclase from the M_1 assemblage is almost pure albite with the general formula $(Na_{0.99}K_{0.01})AlSi_3O_8$.

Epidote – M_1 epidotes are continuously zoned from cores with higher $X_{Fe^{3+}}$ toward slightly lower $X_{Fe^{3+}}$ rim compositions. $Fe^{3+}/(Fe^{3+}+Al)$ ratios ($= X_{Ps}$) in two populations (depending on whole rock geochemistry) range from 0.17–0.21 (Fe^{3+} cnt. = 0.5–0.65 pfu; Fig. 5) and 0.25–0.26 (Fe^{3+} cnt. = 0.75–0.76; Fig. 5), respectively. Average general formulas of rims and cores respectively are:

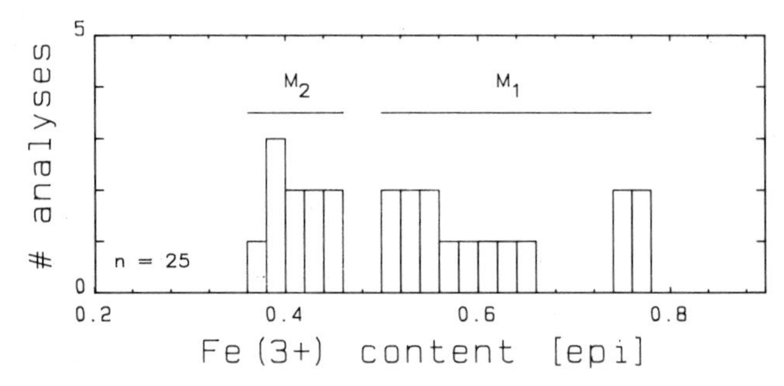

Figure 5. Plot of Fe^{3+} content in octahedral site of epidote (in per formula units based on 12.5 oxygens).

$$(Ca_{1.98})(Mn_{0.02}Fe^{3+}{}_{0.52}Ti_{0.01}Al^{VI}{}_{2.47})Si_{3.0}O_{12}(OH)$$

and

$$(Ca_{1.99})(Mn_{0.01}Fe^{3+}{}_{0.76}Ti_{0.01}Al^{VI}{}_{2.23})Si_{3.0}O_{12}(OH).$$

Sphene – M_1 sphene has the general average formula

$$Ca(Ti_{0.93}Al_{0.07})(SiO_4)(O, OH).$$

Representative analyses of M_1 phases are given in Table 2.

4.4. *M_2 Medium-P/low-T metamorphism*

4.4.1. *Mineral Assemblages*

Metagabbroic rocks and greenschists s.str. – All metagabbros in the Asimah area were affected by M_2 metamorphism. Actinolite/actinolitic hornblende – albite – chlorite$_2$ – epidote$_2$ – sphene$_2$ ± hematite typify M_2 lower greenschist-facies assemblages. Actinolitic hornblende/actinolites are generally unzoned, fine-grained and have a stretched fibrous habit (0.2–0.7 mm long) defining in some thin sections a nematoblastic foliation. Rim and zonar replacements of Mg-hornblende by ragged and fibrous crystals of actinolite are very common. Medium-grained, flaky aggregates of M_2 chlorite occur as prominent phases in the M_2 assemblage. They grow at the expense of Mg-hornblende and chlorite$_1$. According to the classification of Hey (1954), chlorite$_2$ is of ripidolite composition. Plagioclase from M_2 assemblages is very fine-grained pure albite, and it forms as an interstitial phase between actinolite and chlorite$_2$. M_2 epidote occurs most commonly as (1) a disseminated medium- to fine-grained (0.1–0.2 mm diam.) subhedral matrix phase; and (2) grows along rims of previous M_1 epidote. The fabrics formed by M_2 mineral assemblages vary between diablastic growth and rimmed M_1 phases, to a moderately developed S_2 schistosity, which is defined by a preferred orientation of actinolitic amphiboles growing across Mg-hornblende and albite. Good evidence for the second metamorphic event in greenschists s.str. is provided by the compositions, occasional zoning, and growth directions of amphiboles, together with diablastic intergrowth of M_2 white mica on M_1 white micas (Stoessel and Ziegler, 1985).

Metasedimentary rocks – Quartzites, and phyllitic quartzites and quartz-mica schists containing alternating laminations rich in fine-grained quartz and muscovite-chlorite associations are most abundant in the Asimah area. The laminations are parallel to S_1 and reflect the pervasive M_1 structural regime that was chiefly reequilibrated during M_2. Metasedimentary M_2 assemblages are characterized by quartz + muscovite + chlorite + epidote ± sphene ± stilpnomelane, in addition to feldspar (either albite or K-feldspar)

Table 2. Representative analyses of M_1 phases.

Sample	3306 am 15	3308 am 6	3308 am 12	3308 am 39	3308 am 61	4301 am 5	3308 chl 1	3308 chl 8	3308 chl 13	3303 ep 1	4301 ep 3	4301 sph 7
SiO_2	48.12	48.02	47.02	47.40	46.77	48.63	26.34	27.21	27.29	38.80	38.60	31.25
TiO_2	0.17	0.00	0.18	0.00	0.00	0.20	0.00	0.00	0.00	0.00	0.12	37.68
Al_2O_3	7.35	9.84	8.79	8.57	10.37	6.50	19.22	19.03	19.21	24.03	26.16	1.89
Fe_2O_3	4.31	2.23	2.92	1.89	1.80	2.59	0.00	0.00	0.00	12.89	10.67	0.00
FeO	16.20	14.38	14.59	15.14	15.44	15.30	25.42	23.60	23.84	0.00	0.00	0.00
MnO	0.32	0.00	0.33	0.00	0.00	0.32	0.00	0.28	0.54	0.00	0.09	0.00
MgO	9.29	10.51	10.64	10.85	9.83	10.65	16.42	17.43	17.82	0.00	0.00	0.00
CaO	10.63	10.91	11.20	11.37	10.92	10.95	0.00	0.00	0.00	23.86	23.94	28.63
Na_2O	1.64	1.90	2.07	1.96	2.08	1.67	0.00	0.00	0.00	0.00	0.00	0.00
K_2O	0.58	0.42	0.49	0.42	0.56	0.37	0.00	0.00	0.00	0.00	0.00	0.00
H_2O	2.02	2.05	2.03	2.02	2.03	2.01						
Total	100.62	100.25	100.25	99.63	99.80	99.20	87.40	87.55	88.70	99.59	99.58	99.45

	Cations calculated on the basis of 13 cations and 24 O						Cations based on 14 O, [1]			12.5 O, [2]		5 O, [2]
Si	7.124	7.015	6.941	7.022	6.916	7.239	2.785	2.842	2.820	3.028	2.990	1.020
Ti	0.019	0.000	0.020	0.000	0.000	0.023	0.000	0.000	0.000	0.000	0.007	0.925
Al	1.282	1.694	1.529	1.496	1.807	1.140	2.395	2.343	2.339	2.210	2.388	0.073
Fe^{3+}	0.480	0.245	0.325	0.211	0.200	0.290	0.000	0.000	0.000	0.757	0.622	0.000
Fe^{2+}	2.005	1.756	1.802	1.876	1.910	1.904	2.248	2.061	2.059	0.000	0.000	0.000
Mn	0.040	0.000	0.041	0.000	0.000	0.040	0.000	0.025	0.047	0.000	0.006	0.000
Mg	2.050	2.289	2.343	2.396	2.168	2.364	2.589	2.714	2.745	0.000	0.000	0.000
Ca	1.686	1.707	1.771	1.804	1.729	1.747	0.000	0.000	0.000	1.995	1.986	1.001
Na	0.471	0.537	0.592	0.563	0.597	0.482	0.000	0.000	0.000	0.000	0.000	0.000
K	0.110	0.078	0.091	0.080	0.106	0.071	0.000	0.000	0.000	0.000	0.000	0.000
OH	2.000	2.000	2.000	2.000	2.000	2.000						
	Site distribution and ratios											
X Mg/Mg + Fe^{2+}	0.505	0.566	0.565	0.561	0.532	0.554	0.535	0.568	0.571			
X $Mg/Fe_{(tot)}$	0.452	0.533	0.524	0.535	0.507	0.519						
Al (IV)	0.878	0.985	1.059	0.978	1.084	0.761	1.215	1.158	1.180	0.000	0.010	
Al (VI)	0.406	0.709	0.470	0.517	0.723	0.379	1.180	1.186	1.159	2.210	2.378	0.073
Na (M4)	0.314	0.293	0.229	0.196	0.271	0.253						
Na (A)	0.157	0.245	0.363	0.367	0.326	0.229						
Fe, Mg on (M4)	1.686	1.707	1.771	1.804	1.729	1.747						
Fe^{3+}/Fe^{3+} + Al (X Ps)										0.255	0.207	

[1] all Fe as FeO
[2] all Fe as Fe_2O_3
Cr_2O_3 and NiO below detection limit.

Table 3. Representative analyses of M_2 phases.

Sample	2303	2303	3303	3308	4301	4301	3308	3308	3308	3306	3308	3306
	am 2	am 15	am 8	am 9	am 3	am 13	chl 7	chl 15	chl 17	ep 5	ep 1	sph 2
SiO_2	55.56	53.99	51.32	52.09	50.98	51.63	30.08	29.30	29.37	38.60	39.37	29.86
TiO_2	0.00	0.00	0.00	0.24	0.09	0.17	0.10	0.00	0.00	0.13	0.12	36.15
Al_2O_3	1.44	2.37	4.74	2.56	5.12	5.20	15.58	15.72	15.04	28.31	28.93	0.88
Fe_2O_3	2.00	4.87	4.55	4.59	2.80	1.60	0.00	0.00	0.00	6.77	6.74	1.69
FeO	8.96	5.82	14.26	11.57	12.79	13.99	24.80	26.02	27.82	0.00	0.00	0.00
MnO	0.19	0.16	0.00	0.31	0.18	0.23	0.32	0.25	0.25	0.09	0.13	0.10
MgO	17.02	17.87	11.47	14.14	12.90	12.40	17.08	16.55	15.53	0.00	0.00	0.00
CaO	11.47	11.33	10.44	11.43	11.14	11.09	0.00	0.00	0.00	23.92	24.33	28.09
Na_2O	1.15	1.48	1.64	0.97	1.49	1.39	0.00	0.00	0.00	0.00	0.00	0.00
K_2O	0.07	0.07	0.24	0.15	0.27	0.26	0.00	0.00	0.00	0.00	0.00	0.00
H_2O	2.12	2.12	2.06	2.06	2.06	2.06						
Total	99.98	100.09	100.72	99.72	99.81	100.03	87.96	87.85	88.01	97.81	99.62	96.77

	Cations calculated on the basis of 13 cations and 24 O						Cations based on 14 O, [1]			12.5 O, [2]		5 O, [2]
Si	7.868	7.622	7.465	7.572	7.424	7.502	3.135	3.084	3.116	3.001	3.004	1.011
Ti	0.000	0.000	0.000	0.027	0.010	0.019	0.008	0.000	0.000	0.008	0.007	0.920
Al	0.240	0.394	0.813	0.438	0.879	0.891	1.914	1.950	1.881	2.594	2.601	0.035
Fe^{3+}	0.213	0.518	0.498	0.502	0.307	0.175	0.000	0.000	0.000	0.396	0.387	0.043
Fe^{2+}	1.062	0.688	1.735	1.358	1.557	1.700	2.161	2.289	2.468	0.000	0.000	0.000
Mn	0.023	0.019	0.000	0.039	0.023	0.028	0.028	0.023	0.022	0.006	0.009	0.003
Mg	3.593	3.760	2.488	3.064	2.801	2.686	2.654	2.597	2.456	0.000	0.000	0.000
Ca	1.741	1.714	1.626	1.780	1.738	1.727	0.000	0.000	0.000	1.992	1.988	1.019
Na	0.316	0.404	0.461	0.274	0.421	0.391	0.000	0.000	0.000	0.000	0.000	0.000
K	0.013	0.013	0.044	0.029	0.049	0.048	0.000	0.000	0.000	0.000	0.000	0.000
OH	2.000	2.000	2.000	2.000	2.000	2.000						
	Site distribution and ratios											
X Mg/Mg + Fe^{2+}	0.772	0.845	0.589	0.693	0.643	0.612	0.551	0.531	0.499			
X Mg/$Fe_{(tot)}$	0.738	0.757	0.527	0.622	0.600	0.589						
Al (IV)	0.132	0.378	0.535	0.428	0.576	0.498	0.865	0.917	0.884	0.000	0.000	
Al (VI)	0.109	0.016	0.278	0.010	0.303	0.393	1.049	1.033	0.997	2.594	2.601	0.035
Na (M4)	0.259	0.286	0.374	0.220	0.262	0.273						
Na (A)	0.056	0.118	0.087	0.054	0.159	0.118						
Fe, Mg on (M4)	1.741	1.714	1.626	1.780	1.738	1.727						
Fe^{3+}/Fe^{3+} + Al (X Ps)										0.132	0.130	

[1] all Fe as FeO
[2] all Fe as Fe_2O_3
Cr_2O_3 and NiO below detection limit.

and a number of accessory phases, including spessartine-rich garnet, piemontite and several Fe-Ti oxides.

4.4.2. *Mineral Chemistry*

Amphiboles – Amphiboles in M_2 parageneses are of calcic type and range in composition from actinolitic hornblende to actinolite (Leake, 1978). M_2 amphiboles contain high $Mg/(Mg + Fe^{2+})$ numbers (0.51–0.81) and Si contents in the tetrahedral site (7.3–7.87 pfu). Actinolite has the general average formula:

$$(K_{0.03}Na_{0.25})(Ca_{1.74}Na_{0.26})(Mg_{3.18}Fe^{2+}{}_{1.43}Mn_{0.03}Fe^{3+}{}_{0.09}Ti_{0.02}Al^{VI}{}_{0.25})$$
$$(Si_{7.6}Al^{IV}{}_{0.4})O_{22}(OH)_2.$$

Amphiboles from the M_2 assemblage show some Tschermak substitution; generally no or only very minor edenite substitution (Fig. 3C); and some richterite substitution (Na^A, Na^{M4} $\Leftrightarrow$ [A], Ca^{M4}) shown by analyses plotting above the 1:1 tremolite/actinolite – tschermakite join (Fig. 3C). Octahedral Ti is very low ($\leq$0.03 pfu, Fig. 4). Al^{VI} decreases from a maximum value of 0.5 pfu to a minimum of 0.02 pfu, and this is correlated with systematically lower contents of tetrahedral Al^{IV} (Fig. 3B). The M_2 amphiboles have M4-site Na of 0.16–0.37 pfu (Fig. 3D), and A-site occupancies of 0.07–0.30 pfu.

Chlorite – M_2 chlorites show a range in $Fe^{2+}/(Mg + Fe^{2+})$ from 0.44–0.54 and tetrahedral Al^{IV} varies between 0.83–1.1 pfu. Analyses of $chlorite_2$ plotting above the 1:1 line in Fig. 7 show dioctahedral substitution and octahedral vacancies. Average M_2 chlorite has the general formula:

$$(Mg_{2.46}Fe^{2+}{}_{2.47}Mn_{0.02}Al^{VI}{}_{1.0})(Si_{3.12}Al^{IV}{}_{0.88})O_{10}(OH)_8.$$

Plagioclase – Plagioclase from the M_2 assemblage is pure albite.

Epidote – Zonation patterns for M_2 epidote are not obvious and the nearly constant compositions reveal $X_{Fe^{3+}}$ values between 0.12–0.15 (Fe^{3+} cnt. = 0.36–0.44 pfu; Fig. 5). An average general formula is:

$$(Ca_{1.99})(Mn_{0.01}\,Fe^{3+}{}_{0.4}Ti_{0.01}Al^{VI}{}_{2.59})Si_{3.0}O_{12}(OH).$$

Sphene – M_2 sphenes form as very fine-grained, partly idiomorphic, almost colorless crystals along rims of $sphene_1$. Ferric iron contents (0.021–0.043 Fe^{3+} pfu) replacing Al^{VI} are very striking. The average general formula is:

$$Ca(Ti_{0.92}Al_{0.05}Fe^{3+}{}_{0.03})(SiO_4)(O, OH).$$

Representative analyses of M_2 phases are given in Table 3.

4.5. *Breakdown of the M_1 assemblage*

Textural relations and variations in mineral composition (see Figs. 3, 5 & 7) and abundance suggest that the metamorphic change from M_1 to M_2 assemblages took place via a continuous retrogressive reaction. The following weighted general hydrous reaction (cf. projection in Fig. 8) was calculated from the analyses of several sets of samples.

1 Ca-amp$_{1\,(Mg\text{-}hbl)}$ + 0.632 Epi$_1$ (Fe^{3+} cnt. = 0.56 pfu) + 2.176 Chl$_1$ + 1.544 Qtz + 0.97 H_2O = 0.644 Ca-amp$_{2(act)}$ + 0.953 Epi$_2$ (Fe^{3+} cnt. = 0.4 pfu) + 2.453 Chl$_2$ + 0.347 Ab$_2$

Except for the variation in the epidote composition, the same reaction was proposed by Laird (1980) for the change of a 'common' mafic schist during progressive medium-pressure metamorphism between the biotite-albite and the garnet-albite zone in Vermont. Compared to M_1 phases, amphibole (actinolite/actinolitic hornblende) belonging to M_2 show higher X_{Mg} values, less edenite and Tschermak substitutions, and some richterite substitution; chlorite$_2$ is lower in Tschermak substitution (Al^{VI}, Al^{IV} $\Leftrightarrow$ (Fe^{2+}, Mg, Mn), Si) and slightly richer in $X_{Fe}{}^{2+}$; epidote$_2$ is lower in $X_{Fe}{}^{3+}$; and quartz and water are consumed.

5. P-T Conditions

5.1. *Phase Assemblages*

Many workers (e.g. Eskola, 1939; Turner and Verhoogèn, 1960; Maruyama et al., 1983) have defined the greenschist- to amphibolite-facies boundary on the basis of (1) plagioclase (albite to oligoclase transition) and (2) amphibole composition (actinolite to hornblende transition). Fig. 11A shows these phase transitions from the experimental study of Maruyama et al. (1983). Thermal breakdown of chlorite (reversed experiment) and epidote (not reversed experiment) by the reactions chlorite + sphene + quartz + actinolite = hornblende + ilmenite + fluid and epidote + hornblende$_I$ + quartz = hornblende$_{II}$ + oligoclase + fluid were experimentally studied by Liou et al. (1974) and Apted and Liou (1983), respectively. The chlorite and epidote consuming ('out') reactions, however, depend on mineral compositions and are drawn in Fig. 11A for reference only.

Phase equilibria to constrain pressure conditions for the M_1 event are difficult to find. It is interesting to note that albite coexisting with Mg-hornblende does not show any chemical variation. This strongly suggests that the albite to oligoclase transition was never reached. Because of the generally concomitant low-grade appearance of more anorthitic plagioclase when actinolite changes to hornblende in low-pressure facies series (Miyashiro, 1961;

Laird and Albee, 1981b; Hynes, 1982; Maruyama et al., 1983), the lack of chemical variation may be considered evidence against low pressures of equilibration. However, an apparent relationship with data from medium-pressure terrains (Vermont, New Zealand) as can be seen from Fig. 10 is very striking.

Figure 11B displays relevant reaction equilibria for the M_2 assemblages in metagabbros. The pumpellyite-actinolite to greenschist-facies transition was identified by Nakajima et al. (1977) in a reversed equilibrium study with the reaction: pumpellyite + hematite + quartz = epidote + chlorite + actinolite + fluid. Compositions of epidote in the buffered assemblage epidote + pumpellyite + chlorite + actinolite + albite + chlorite + quartz are sensitive to temperature changes in this transition. In a NCFMASH system, the assemblage becomes trivariant and the compositions of the phases are dependent on P, T and bulk rock Fe^{2+}/Mg ratio (represented by the constant Fe_{tot}/Mg of chlorite). At given P and bulk rock Fe^{2+}/Mg ratio, the compositions of epidote and pumpellyite are a function of temperature. This function has been calibrated by thermodynamic calculations (Nakajima et al., 1977) and is drawn on the figure for a product assemblage containing epidote with X_{Ps} = 0.10. Also shown is the experimentally investigated reversed equilibrium: pumpellyite + chlorite + quartz = actinolite + epidote + fluid, described by Nitsch (1971) using natural Fe-bearing phases, and confirmed by Liou et al. (1985) for an NCMASH system. Introducing FeO and Fe_2O_3 into the system, the reaction becomes continuous and behaves like the one described above. The P-T location of the reaction is drawn for epidote with $X_{Ps} = 0.10$. Based on these equilibria, M_2 assemblages formed at temperatures high enough to crystallize actinolite and epidote, i. e. at T > ~ 330°C.

As the indicated equilibria are nearly independent of pressure, the empirical geobarometer from the experimentally studied reaction glaucophane + -clinozoisite + quartz + H_2O = tremolite + chlorite + albite (Maruyama et al., 1986) defining the greenschist-blueschist facies transition was used to estimate the pressure of metamorphism. One isopleth of the reaction is drawn on Fig. 11B with $X_{gl} = 0.30$ and $X_{Ps} = 0.37$ for reference only. This methods yields a pressure of approximately 400 MPa for M_2 amphiboles. Brown (1977b) suggested that according to the reaction crossite + epidote = actinolite + albite + chlorite + iron-oxide + H_2O (estimated from associated equilibria), the crossite content of Ca-amphibole coexisting with albite + iron oxide + chlorite can be used as a potential geobarometer. An empirical relationship between the Na^{M4} content of Ca-amphibole with pressure was established (Fig. 3.10 of Brown, 1977b). Na^{M4}-site values of M_2 actinolites ranging from 0.16–0.37 pfu (Fig. 3D) indicate moderate pressures of 300–500 MPa.

5.2. *Amphibole Composition*

The change in amphibole composition as a function of grade is a long known fact in metamorphic petrology (e.g. Compton, 1958; Cooper and Lovering,

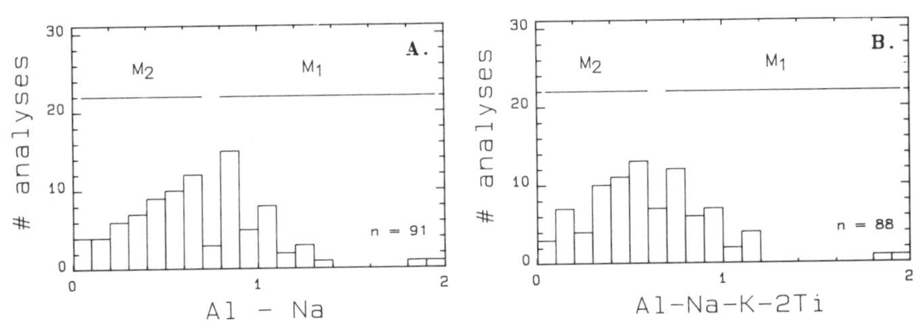

Figure 6. Histograms of (Al-Na) and (Al-Na-K-2Ti) for amphiboles from the Asimah area. All data were normalized to 23 oxygens with all iron considered as Fe^{2+} (see text for discussion).

1970). Laird and Albee (1981b) developed several formula proportion diagrams which associate amphibole compositions with metamorphic facies for metabasites containing a 'common' mafic assemblage. From their plots, it seems to be possible – at least for an approximate approach – to compare amphibole compositions from areas in which the metamorphic conditions are known, with those in which metamorphic pressure and temperature are difficult to determine. Amphibole compositions from metamorphic gabbros are plotted on a selection of these diagrams in Fig. 3. In general, the compositions of M_1 amphiboles are similar to those reported by Graham (1974) for Dalradien rocks, and by Laird and Albee (1981b) for medium-pressure, upper Biotite-Albite to lower Garnet-Albite zone metabasites in Vermont. In contrast, amphiboles belonging to M_2 assemblages lie in a field which is defined by the medium-pressure Biotite-Albite zone (Laird and Albee, 1981b). On Fig. 3A, which is independent of normalization, all analyzed amphiboles plot above the 1:1 reference line. This feature is generally attributed to amphiboles from medium- or higher-pressure areas (Laird and Albee, 1981b). Amphiboles from medium-pressure facies series have more glaucophane substitution than those from low-pressure terrains (Brown, 1977b; Laird and Albee, 1981b). On Fig. 3D, all Asimah samples lie in a medium-pressure field above a line 1:4 which approximately represents the oligoclase isograd of Laird and Albee (1981b).

Hynes (1982) shows that the clearest distinction between amphiboles from medium- and low-pressure areas is due to (1) lower Ti contents at the same $Al_{(tot)}$ level, and (2) a generally better developed compositional gap between actinolite and hornblende in medium-pressure metabasites. From Fig. 4 it is obvious that the Ti contents in all amphiboles from the Asimah area are very small, reaching values of at most 0.04 pfu. These contents are suggestive of medium-pressure conditions during the metamorphic history. The histograms in Figs. 6A and 6B show the extent of the compositional gap. The range of (Al-Na) is a measure of $Al_{(tot)}$ in amphiboles other than that directly associated with glaucophane and/or edenite substitutions; (Al-Na-K-2Ti) on the other hand is a value of the tschermakite-component in amphiboles (Hynes, 1982). In contrast to the compiled data of Hynes (1982), the distribu-

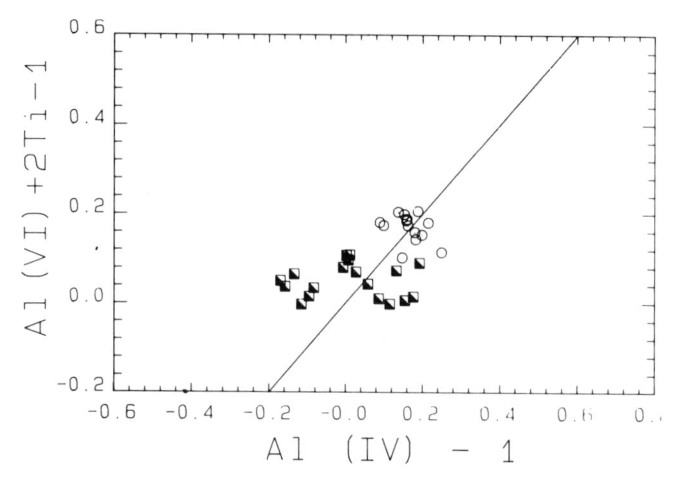

Figure 7. Chlorite analyses from Asimah metagabbros. Analyses above 1:1 line show dioctahedral substitution (Al_2Mg_{-3}) and have octahedral vacancies. Tschermak substitution parallels this line, the origin is clinoclore/chamosite. (Symbols: circles = M_1 chlorite, half-filled squares = M_2 chlorite).

tion shown on Fig. 4 is not that pronounced. However, a narrow but distinct gap between M_1 and M_2 amphiboles (indicating kind of a bimodal distribution for the Asimah area) can be seen on Fig. 6A, a fact which is somewhat blurred in Fig. 6B. The narrowing of the compositional gap was suggested to be a function of higher f_{O_2} during metamorphism, since increased f_{O_2} could significantly reduce the temperature range of the epidote amphibolite-facies, thereby leading to the appearance of albite – epidote – Mg-hornblende at lowered pressures (Liou et al., 1974).

5.3. *Amphibole and Chlorite Pairs*

The compositional variation of coexisting amphibole and chlorite from the Asimah area is systematic (Figs. 8 and 9). M_1 pairs have about the same values of $Mg/(Mg + Fe^{2+})$, and show an increase in Tschermak substitution compared to the M_2 phases. Tschermak exchange in the amphiboles is, however, larger than in the coexisting chlorites. Chlorite in the M_2 assemblage is more Fe-rich than coexisting actinolite/actinolitic hornblende. These findings seem to be a general relationship when mafic rocks pass from greenschist- to epidote amphibolite- and/or amphibolite-facies conditions (e.g. Laird, 1980). Potential geothermobarometers involve the $MgFe_{-1}$ exchange and are empirical (Laird, 1982).

The exchange reaction controlling the $(Fe^{2+} + Mn)/Mg$ partition between the two phases is: $(Fe^{2+} + Mn)Mg_{-1}$ [chl] = $(Fe^{2+} + Mn)Mg_{-1}$ [amp]. K_D $[(Fe^{2+} + Mn)/Mg]$chl/amp values change from 1.6 for lower grade M_2 pairs to 1.08 and 0.9 for M_1 compositions, respectively (Fig. 9). K_D and ln K_D numbers are very similar to data for the Garnet-Albite and Biotite-Albite

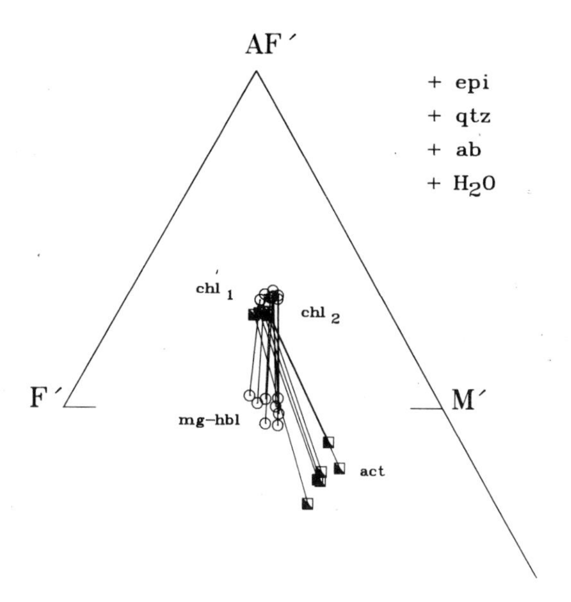

Figure 8. Projection of M_1 and M_2 amphibole/chlorite pairs in a NCFMASH system (Symbols: circles = M_1 amphibole/chlorite, half-filled squares = M_2 amphibole/chlorite). For simplicity, only coexisting phases of E3308 are plotted.

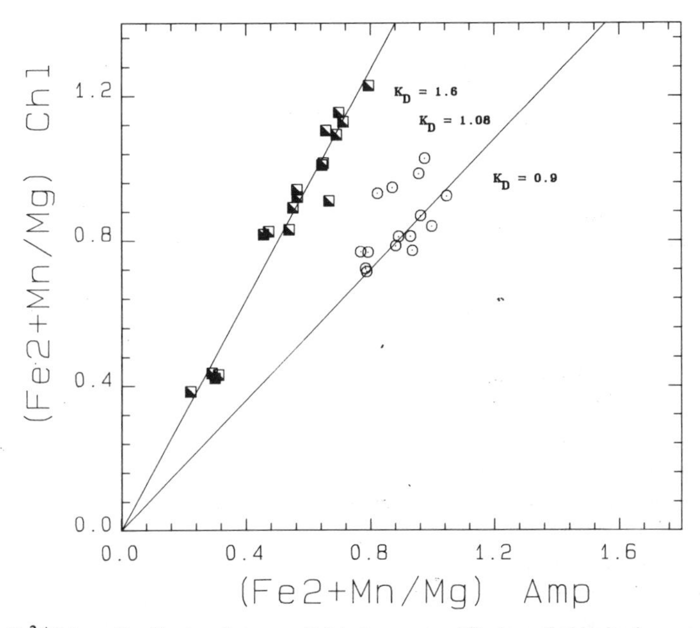

Figure 9. $Fe^{2+}Mg_{-1}$ distribution (atom units) between amphibole and chlorite in metagabbros from Asimah. K_D is $[(Fe^{2+} + Mn)/Mg]$chl/amp. Each point is representative of coexisting compositions (Symbols as in Fig. 7).

zones of Laird and Albee (1981b). The metamorphic temperatures for these zones were estimated using calcite-dolomite, muscovite-paragonite, garnet-clinopyroxene, garnet-biotite and stable isotope equilibria in the mafic schists and intercalated pelitic and calc-silicate schists. According to these calibrations, M_1 amphibole-chlorite pairs ($K_D^{chl/amp}$ 0.9–1.08) indicate upper greenschist- to epidote amphibolite-facies, with a temperature range of probably 450–500°C. Coexisting M_2 phases with a $K_D^{chl/amp}$ of 1.6 reflect significantly lower temperature conditions of about 350–400°C, typical for lower greenschist-facies.

The pressure dependence of $FeMg_{-1}$ between actinolite and chlorite was studied by Cho (pers. commun. in Laird, 1988). This study showed that the K_D (Mg/Fe)act/(Mg/Fe)chl may provide a geobarometer for low grade metabasites. Several M_2 pairs from the Asimah area were tested with this calibration and yielded pressures ranging from 340–520 MPa.

5.4. *Epidote Composition*

Compositional zoning of epidote is common in mafic metamorphic rocks (e.g. Raith, 1976; Laird and Albee, 1981a). Epidote compositions have been used as indicators of metamorphic grade (Raith, 1976; Nakajima et al., 1977; Maruyama et al., 1986). Other studies have shown that epidote can be zoned from high-$X_{Fe^{3+}}$ cores to rims with lower $X_{Fe^{3+}}$ during prograde metamorphism (Laird and Albee, 1981b). Two generations of epidote (and rare epidote compositional zoning at the same grains) are typical in metagabbros from the Asimah area, where M_2 grains (and rim compositions) are poorer in $X_{Fe^{3+}}$ compared to M_1 relict grains (and cores). The observed (although rare) inverse zonation, and the occurrence of epidote$_2$ with lower $X_{Fe^{3+}}$ is an interesting point. In general, zoning to Al-richer rims may either indicate increasing pressure (Maruyama et al., 1986) or decreasing temperature (Brown, 1986), or both. Epidote$_2$ compositions stable with the assemblage chlorite$_2$ – actinolite/actinolitic hornblende suggest that M_2 grains and rims formed under lower T (°C) conditions.

M_2 assemblages containing epidote + chlorite + actinolite are equivalent to the high temperature product assemblage formed from pumpellyite + hematite + quartz (Nakajima et al., 1977). This reaction is not univariant due to the compositional variation of the phases. Nakajima et al. (1977) calculated the dependence of the temperature of reaction on the basis of the variability of mole fraction of ferric iron in epidote and pumpellyite for a specific chlorite composition ($X_{Fe^{2+}} = 0.45–0.5$), and arbitrarily selected a temperature of 300°C where epidote in the reaction assemblage has $X_{Fe^{3+}} = 0.31$. In three samples, chlorite$_2$ stable with the M_2 assemblage has $X_{Fe^{2+}}$ values of 0.44–0.5 at epidote$_2$ compositions of $X_{Fe^{3+}} = 0.12–0.15$. This Fe partitioning indicates a fairly narrow temperature range of ~330–370°C. This estimate is considered a minimum value because no pumpellyite could be detected in the assemblage.

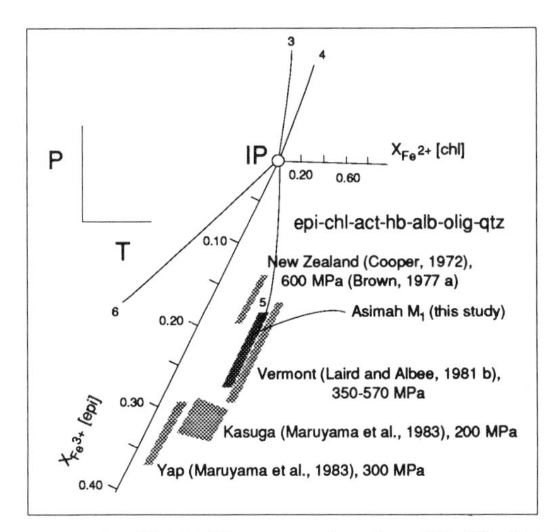

Figure 10. P-T relation in the NCMASH system after the addition of FeO and Fe_2O_3. The invariant assemblage, clinozoisite – chlorite – tremolite – hornblende – oligoclase – albite – quartz – fluid, in the model system becomes divariant. The quantitative effect of the addition of ferric and ferrous iron can be estimated by $X_{Fe^{3+}}$ (in epidote) and $X_{Fe^{2+}}$ (in chlorite) isopleths. Reactions are: (3) clz + tr + chl = hbl + fluid; (4) clz + chl = hbl + olig + fluid; (5) olig + chl + tr = hbl + fluid; (6) clz + chl = olig + tr + fluid (see Maruyama et al. (1983) for further discussion). The data for the M_1 assemblages from Asimah are concordant with moderate pressures.

6. Conclusions

The available petrologic data and phase equilibria allow the following conclusions:

1. The first metamorphic event, M_1, occurred between ~450–500°C, i.e. at temperatures typical for upper greenschist- to epidote amphibolite-facies. Metamorphic pressures during M_1 are poorly constrained. Amphibole and plagioclase compositions, however, indicate that M_1 reached medium-pressure conditions. Comparison with other data (e.g. Laird and Albee, 1981b; Fig. 10) allows an approximate estimate of 500 ± 50 MPa.
2. M_2 recrystallization in the Asimah area occurred at moderate metamorphic temperatures, ranging from ~340–380°C. This is in good agreement with peak metamorphic temperatures of greenschists in the Wadi Tayin area (400°C at 200 MPa) estimated by Ghent and Stout (1981). The M_2 assemblages do not permit direct determination of metamorphic pressures, but the Na^{M4} content of actinolite/actinolitic hornblende and the $FeMg_{-1}$ partitioning between actinolite and chlorite indicate pressures of around 400 ± 100 MPa. Such temperature and pressure conditions are representative of lower greenschist-facies.
3. The inferred metamorphic conditions for rocks from the Asimah window suggest the following:

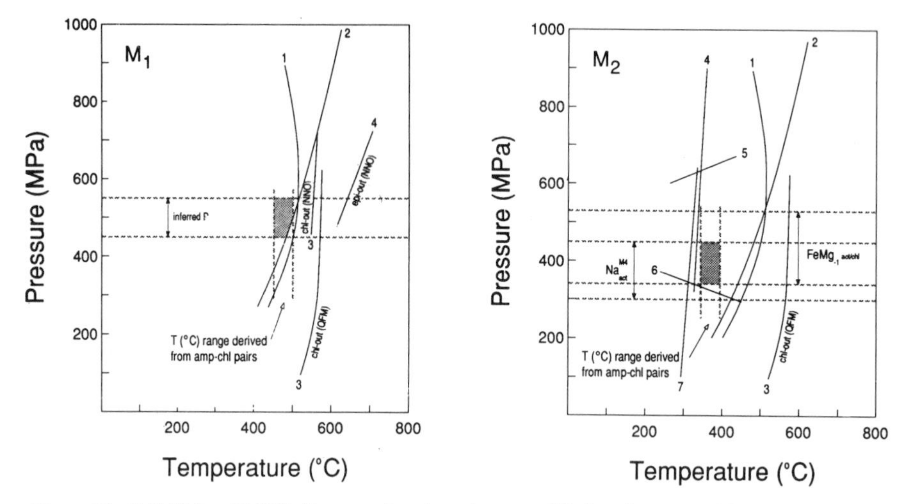

Figure 11. P (MPa) – T (°C) diagram showing phase equilibria relevant to M_1 and M_2 assemblages in metagabbros and greenschists s.str.; shaded regions show inferred P-T conditions for M_1 and M_2 assemblages.
A) M_1 assemblage:
Reactions are: (1) actinolite-hornblende transition (Maruyama et al., 1983); (2) albite-oligoclase transition (Maruyama et al., 1983); (3) chlorite + sphene + quartz + actinolite = hornblende + ilmenite + fluid (QFM & NNO buffers; Liou et al., 1974); (4) epidote + hornblende$_I$ + quartz = hornblende$_{II}$ + oligoclase + fluid (NNO buffer; Apted and Liou, 1983). Horizontal lines at 450/550 MPa show inferred pressure range. See text for discussion.
B) M_2 assemblage:
Reactions are: (1) actinolite-hornblende transition (Maruyama et al., 1983); (2) albite-oligoclase transition (Maruyama et al., 1983); (3) chlorite + sphene + quartz = hornblende + ilmenite + fluid (QFM buffer; Liou et al., 1974); (4) pumpellyite + chlorite + quartz = actinolite + epidote ($X_{Ps} = 0.10$) + fluid (Nitsch, 1970); (5) crossite + epidote = actinolite + albite + chlorite + hematite + fluid (Brown, 1977a); (6) glaucophane ($X_{gl} = 0.30$) + clinozoisite ($X_{Ps} = 0.37$) + quartz + fluid = tremolite + chlorite + albite (Maruyama et al., 1986); (7) pumpellyite + hematite + quartz = epidote ($X_{Ps} = 0.10$) + chlorite + actinolite + fluid (Nakajima et al., 1977).
Horizontal lines at 300/450 and 340/520 MPa show pressure constraints indicated by the Na^{M4} content of actinolites (Brown, 1977b) and $FeMg_{-1}$ partitioning between actinolite and chlorite (M. Cho, pers. commun. in Laird, 1988), respectively. See text for discussion.

i) The Asimah rocks underwent a retrograde metamorphic path characterized by a significant temperature decrease, but presumably with only a slight pressure drop. The form of this P-T path is deduced to have been concave towards the temperature axis.
ii) The metamorphic conditions of M_1 suggest that the 'Asimah thrust sheet' was part of a descending slab in a shallow dipping subduction zone and was accreted to the base of the hanging wall.
iii) The retrograde metamorphic path, M_2, could be the result of cooling of the 'Asimah thrust sheet' during obduction of the Semail ophiolite nappe onto the Arabian continental margin. M_2 pressures of 400 MPa require

that the overlying Semail nappe was 12–15 km thick. This agrees with a 16–20 km thickness at the time of detachment, as inferred by Hopson et al. (1981).

Acknowledgements

I thank Tj. Peters for his advice, encouragement, patience, and review of the manuscript. Constructive reviews by E. H. Brown and L. Diamond substantially improved both the scientific and stylistic aspects of the paper. I am indebted to the Ministry of Petroleum and Mineral Resources of the United Arab Emirates for permission to work in the Northern Oman Mountains. Long discussions with N. Waber, R. Oberhänsli, U. Raz, A. Feenstra, I. Mercolli, E. Gnos, F. Stössel and U. Ziegler were funny and stimulating. This study would not have been possible without the support of Daniel Kurz who was and still is the most reliable partner along our way to find a solution for these metamorphic rocks. Any remaining errors or misinterpretations remain to the responsibility of the author.

References

Allemann F. and Peters Tj., 1972. The ophiolite-radiolarite belt of the North-Oman Mountains., Eclog. Geol. Helv., 65/3: 657–697.

Apted, M.J. and Liou, J.G., 1983. Phase relation among greenschist, epidote-amphibolite, and amphibolite in a basaltic system., Am. J. Sci., 283–A (Orville vol.): 328–354.

Bechennec, F., Le Metour, J., Rabu, D., Villey, M. and Beurrier, M., 1988. The Hawasina basin: a fragment of a starved passive continental margin, thrust over the Arabian platform during obduction of the Semail nappe., Tectonophysics, 151: 323–343.

Boudier, F., Ceuleneer, G. and Nicolas, A., 1988. Shear zones, thrusts and related magmatism in the Oman ophiolite: Initiation of thrusting on an oceanic ridge., Tectonophysics, 151: 275–296.

Brown, E.H., 1977a. Phase equilibria among pumpellyite, lawsonite, epidote and associated minerals in low-grade metamorphic rocks., Contrib. Mineral. Petrol., 64: 123–136.

Brown, E.H., 1977b. The Crossite Content of Ca-Amphibole as a Guide to Pressure of Metamorphism., J. Petrol., 18: 53–72.

Brown, E.H., 1986. Geology of the Shuksan Suite, North Cascades, Washington, U.S.A.. In: B.W. Evans and E.H. Brown (Eds), Blueschists and eclogites., Geol. Soc. Am. Mem., 164: 143–154.

Compton, R.R., 1958. Significance of amphibole paragenesis in the Bidwall Bar region, California., Am. Mineralogist., 43: 890–907.

Cooper, A F., 1972. Progressive metamorphism in metabasic rocks from the Haast schist group of Southern New Zealand. Greenschist amphiboles from Haast River, New Zealand., J. Petrol., 13: 457–492.

Cooper, A.F. and Lovering, J.F., 1970. Greenschist amphiboles from Haast River, New Zealand., Contrib. Mineral. Petrol., 27: 11–24.

Eskola, P., 1939. In: Barth, T.F.W., Correns, C.W. and Eskola, P. (Eds), Die Entstehung der Gesteine. Springer, Berlin, 422 p.

Ghent, E.D. and Stout, M.Z., 1981. Metamorphism at the base of the Samail ophiolite, southeastern Oman mountains., J. Geophys. Res., 86: 2557–2571.

Glennie, K.W, Boeuf, M.G.A., Hughes-Clarke, M.W., Moody-Stuart, M., Pilaar, W.F.H. and Reinhardt, B.M., 1974. Geology of the Oman Mountains (2 vol.)., Verh. K. Ned. Geol. Mijnbouwkd. Genoot. Ged. Ser., 31: 1–423.

Graham, C.M., 1974. Metabasite amphiboles of the Scottish Dalradien., Contrib. Mineral. Petrol., 13: 269–294.

Harte, B. and Graham, C.M., 1975. The Graphical Analysis of Greenschist to Amphibolite Facies Mineral Assemblages in Metabasites., J. Petrol., 16: 347–370.

Hey, M.H., 1954. A new review of the chlorites., Mineral. Mag., 30: 277–293.

Hopson, C.A., Coleman, R.G., Gregory, R.T., Pallister, J.S. and Bailey, E.H., 1981. Geologic section through the Samail Ophiolite and associated rocks along a Muscat-Ibra transsect, Southeastern Oman Mountains., J. Geophys. Res., 86: 2527–2544.

Hynes, A., 1982. A Comparision of Amphiboles from Medium- and Low-pressure Metabasites., Contrib. Mineral. Petrol., 81: 119–125.

Laird, J., 1980. Phase Equilibria in Mafic Schist from Vermont., J. Petrol., 21: 1–37.

Laird, J., 1982. Amphiboles in metamorphosed basaltic rocks: greenschist to amphibolite facies., Reviews in Mineralogy, 9B: 113–135.

Laird, J., 1988. Chlorites: Metamorphic Petrology., Reviews in Mineralogy, 19: 405–453.

Laird, J. and Albee, A.L., 1981a. High-pressure metamorphism in mafic schist from Vermont., Am. Jour. Sci., 281: 97–126.

Laird, J. and Albee, A.L., 1981b. Pressure, temperature, and time indicators in mafic schist: their application to reconstructing the polymetamorphic history of Vermont., Am. Jour. Sci., 281: 127–175.

Lanphere, M.A., 1981. K-Ar ages of metamorphic rocks at the base of the Semail ophiolite, Oman., J. Geophys. Res., 86: 2777–2782.

Leake, B.E., 1965. The relationship between tetrahedral aluminium and the maximum possible octahedral aluminium in natural calciferous and subcalciferous amphiboles., Am. Mineral., 50: 843–851.

Leake, B.E., 1978. Nomenclature of Amphiboles., Am. Mineral., 63,: 1023–1052.

Liou, J.G., 1973. Synthesis and stability relations of epidote, $Ca_2Al_2FeSi_3O_{12}(OH)$., J. Petrol., 14: 381–413.

Liou, J.G., Kuniyoshi, S. and Ito, K., 1974. Experimental studies of the phase relations between greenschist and amphibolite in a basaltic system., Am. Jour. Sci., 274: 613–632.

Liou, J.G., Maruyama, S. and Cho, M., 1985. Phase equilibria and mineral parageneses of metabasites in low-grade metamorphism., Mineral. Mag., 49: 321–333.

Lippard, S.J., Shelton, A.W. and Gass, I.G., 1986. The ophiolites of Northern Oman., Geol. Soc. Lond. Mem., 11: 178 p.

Maruyama, S., Cho, M. and Liou, J.G., 1986. Experimental investigations of blueschist-greenschist transition equilibria: Pressure dependence of Al_2O_3 contents in sodic amphiboles – A new geobarometer. In: B.W. Evans and E.H. Brown (Eds), Blueschists and eclogites., Geol. Soc. Am. Mem., 164: 1–16.

Maruyama, S. and Liou, J.G., 1985. The stability of Ca-Na pyroxene in low-grade metabasalts of high-pressure intermediate facies series., Am. Mineralogist, 70: 16–29.

Maruyama, S., Suzuki, K. and Liou, J.G., 1983. Greenschist-amphibolite transition equilibria at low pressures., J. Petrol., 24: 583–604.

Miyashiro, A., 1961. Evolution of metamorphic belts., J. Petrol., 2: 277–311. Montigny, R., Le Mer, O., Thuziat, R., Whitechurch, H., 1988. K-Ar and $^{40}Ar/^{39}Ar$ study of the metamorphic rocks associated with the Oman ophiolite., Tectonophysics, 151: 345–362.

Nakajima, T., Banno, S. and Suzuki, T., 1977. Reactions leading to the disappearence of pumpellyite in low-grade matamorphic rocks of the Sanbagawa metamorphic belt in central Shikoku, Japan., J. Petrol., 18: 263–284.

Nitsch, K.H., 1970. Experimentelle Bestimmung der oberen Stabilitätsgrenze von Stilpnomelan., Fortschr. Miner., 47, Beiheft 1: 48–49.

Nitsch, K.H., 1971. Stabilitätsbeziehungen von Prehnit- und Pumpellyit-haltigen Paragenesen., Contrib. Mineral. Petrol., 30: 240–260.

Raase, P., 1974. Al and Ti contents of hornblende, indicators of pressure and temperature of regional metamorphism., Contrib. Mineral. Petrol., 45: 231–236.

Raith, M., 1976. The Al-Fe(III) Epidote miscibility gap in a metamorphic profile through the Penninic series of the Tauern Window, Austria., Contrib. Mineral. Petrol., 57: 99–117.

Robertson, A.H.F., Blome, C.D., Cooper, D.W.J., Kemp, A.E.S. and Searle, M.P., 1990. Evolution of the Arabian continental margin in the Dibba zone, Northern Oman Mountains. In: A.H.F. Robertson, M.P. Searle and A.C. Ries (Eds), The Geology and Tectonics of the Oman Region., Geol. Soc. London, Special Publication, 49: 251–284.

Searle, M.P., 1980. The metamophic sheet and underlying volcanic rocks beneath the Semail ophiolite in the northern Oman Mountains of Arabia. Unpubl. PhD thesis, Open University, 213 p.

Searle, M.P. and Malpas, J., 1980. Structure and metamorphism of rocks beneath the Semail ophiolite of Oman and their significance in ophiolite obduction., Trans. Roy. Soc. Edinburgh, Earth Sciences, 71: 247–262.

Searle, M.P. and Malpas, J., 1982. Petrochemistry and origin of subophiolite metamorphic and related rocks in the Oman mountains., J. geol. Soc. London, 139: 235–248.

Stoessel, F. and Ziegler, U., 1985. Metamorphic series associated with the Semail ophiolite nappe of the Oman Mountains in the United Arab Emirates., Unpubl. ms. thesis, University of Berne, 293 p.

Turner, F.J. and Verhoogen, J., 1960. Igneous and Metamorphic Petrology. McGraw-Hill Book Company, 250 p.

Woodcock, N.H. and Robertson, A.H.F., 1982. The upper Batinah Complex, Oman: allochthonous sediment sheets above the Semail ophiolite., Can. J. Earth Sci., 19: 1635–1656.

Part V
Paleogeographic Setting of the Oman Ophiolite

Al Aridh Formation, Oman: Stratigraphy and Palaeogeographic Significance

WOLFGANG BLENDINGER*
Petroleum Development Oman, XXG/43, P.O. Box 81, Muscat, Oman

Abstract

The Al Aridh Formation is a tectonic unit above the Hamrat Duru Group and below the "Oman Exotics" and the Haliw Formation, respectively. At its base, mafic volcanic rocks (?Permian) occur locally. The Triassic consists of radiolarite and a thick unit of platy limestone with thin-shelled bivalves. Jurassic oolitic turbidites are typically interbedded with radiolarite and locally fill erosional channels. The Lower Cretaceous consists of radiolarite. Carbonate breccia horizons are a distinctive feature of the Al Aridh Formation and are thickest in the Upper Triassic and the Upper Jurassic or Lower Cretaceous. They contain blocks of Permian reef limestone and Triassic reef- and backreef limestone. Erosion at the base of the megabreccias locally amounts to 200 m. Palaeocurrents measured from breccia beds and oolitic turbidites indicate a southern sediment source. Because particularly the Jurassic oolite turbidites record a bypass slope environment and the "Oman Exotics", the previously inferred source of the Al Aridh resediments, were drowned in the Jurassic, this slope can not have been part of the "Oman Exotics". The Al Aridh Formation is, therefore, here interpreted as slope deposits of the Arabian platform margin that was emplaced as an "out-of-sequence" thrust unit.

The Problem of the Al Aridh Formation

The Al Aridh Formation forms part of the Hawasina nappes that have been thrust faulted in the Late Cretaceous onto the Arabian continent along with the Semail ophiolite (e.g. Glennie et al., 1974; Robertson & Searle, 1990). It is a tectonic unit that is imbricated between the Hamrat Duru Group below and the "Oman Exotics" and the Haliw Formation (Umar Group of

*Present address: KSEPL Shell Research, RR/28, Volmerlaan 6, NL-2288GD Rijswijk.

Tj. Peters et al. (Eds), Ophiolite Genesis and Evolution of the Oceanic Lithosphere, 575–592.

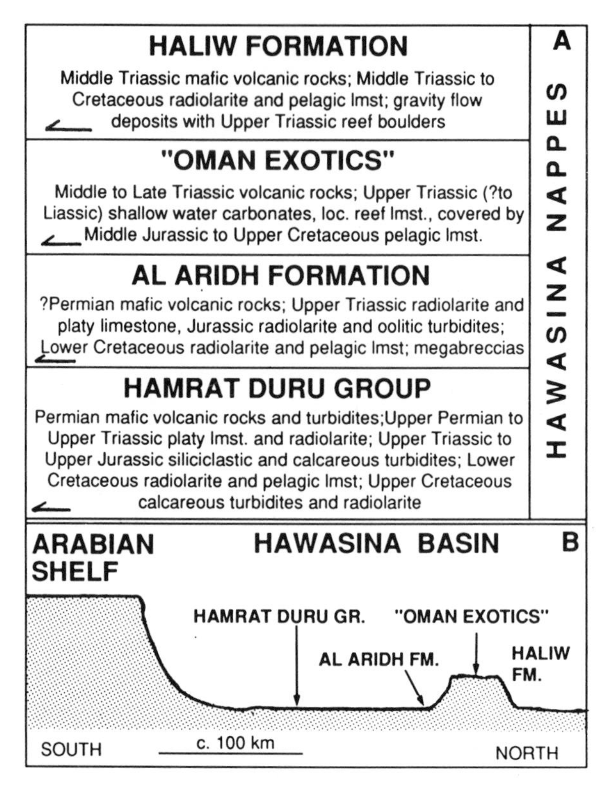

Figure 1. A: The tectonic units of the Hawasina nappes in the central Oman Mountains. B: Traditional palinspastic restoration of the Oman continental margin and position of the Al Aridh Formation (e.g., Bernoulli and Weissert, 1987).

Béchennec, 1988) above (Fig. 1A). Because of its tectonic position, it has traditionally been interpreted to have formed palaeogeographically between the Hamrat Duru Group and the "Oman Exotics" (first option in Glennie et al., 1974; Bernoulli and Weissert, 1987; Searle and Graham, 1982; Béchennec, 1988)(Fig. 1B). However, as yet no facies analysis of the Al Aridh Formation has been carried out to support this assumption. The purpose of this paper is to test this hypothesis through a facies analysis of the Al Aridh Formation.

Previous Work and Regional Occurrence of the Al Aridh Formation

Glennie et al. (1974) mapped the Al Aridh Formation as a Jurassic formation containing abundant sedimentary breccias occurring in the area between the Dibba zone in the northern Oman Mountains and the Batain coast of northeastern Oman. Bernoulli and Weissert (1987) studied the Al Aridh

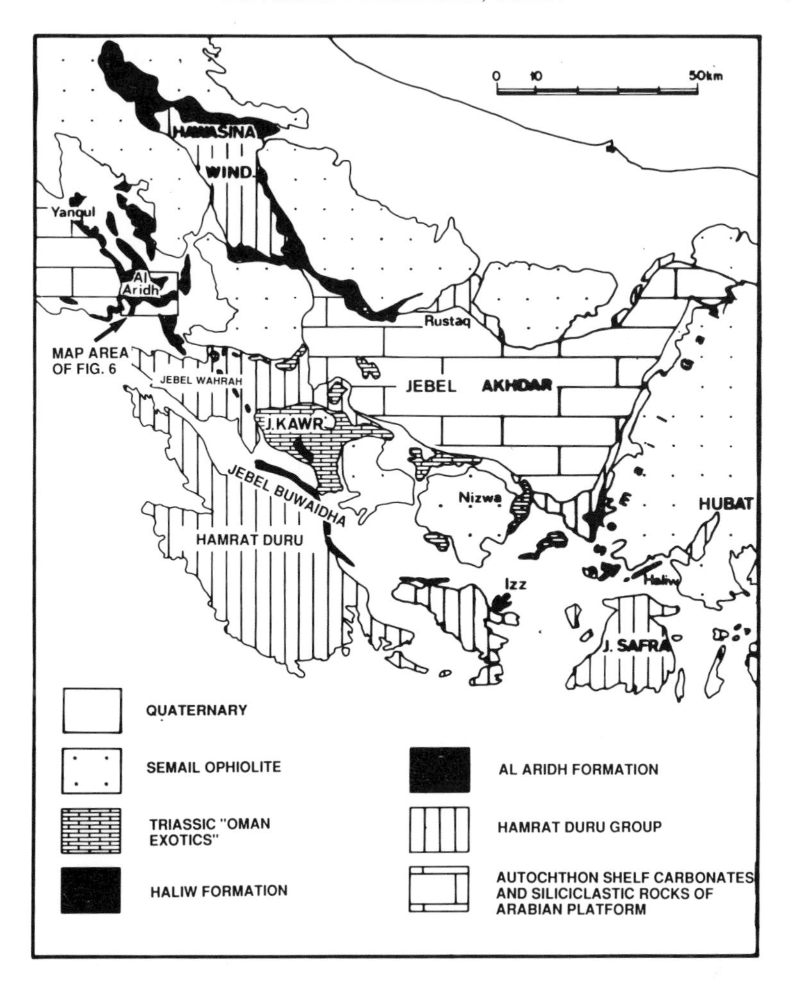

Figure 2. Simplified geological map of the central Oman Mountains, modified after Béchennec (1988, pl. 1).

Formation in Jebel Buwaidha (Fig. 2) and the type area and recognized that it ranges from the Upper Triassic to the Lower Cretaceous. They called it the Al Aridh *Group*, but without establishing different Formations. Béchennec (1988) established an Al Aridh *Group*, distinguished different *Formations*, and mapped it between Yanqul in the north to the Samad area east of the Semail Gap. Béchennec's *Formations* of the Al Aridh Group, however, are entirely based on dating of radiolarite, whereas the stratigraphic significance of microfossils in calcareous sediments was largely ignored. The stratigraphy proposed in this paper differs considerably from that of Béchennec (1988), and the rank as a Formation, as originally defined by Glennie et al. (1974), will be kept.

The Al Aridh Formation consists of volcanic rocks, pelagic sediments (platy limestone, radiolarite), oolitic gravity flows and calcareous breccias. These facies occur only between Yanqul and Izz (Fig. 2). The eastern limit of the Al Aridh Formation thus coincides with the southern extension of the Semail Gap, the northern roughly with the southern limit of the Sumeini Group, a slope unit of the northern Oman Mountains (Watts and Garrison, 1986). All other occurrences mapped as Al Aridh Formation north and east of this area (Glennie et al., 1974; Béchennec, 1988, pl. 1), respectively, are essentially radiolarite with calcareous breccias and have affinities with the Haliw Formation.

Stratigraphy

Volcanic rocks locally form the base of the Al Aridh Formation such as at Jebel Buwaidha (Fig. 3, 4). They are vesicular, alkaline basalts with geochemical characteristics similar to the Permian volcanic rocks at the base of the proximal part of the Hamrat Duru Group and have an intraplate character (Béchennec, 1988, Fig. 94–99, 107, 109, 110). The pillow basalts are locally overlain by massive volcaniclastic rocks with poorly sorted grains of sand to boulder size, pillow breccias and, locally, reef boulders such as in the western part of Jebel Buwaidha. There, the volcanic unit is approximately 250 m thick, but its base is marked by a thrust fault. At Yanqul, sheared volcanic rocks occur at the base of Carnian radiolarite (Fig. 5).

Age. The volcanic rocks have not been dated. They have been interpreted as Upper Jurassic to Lower Cretaceous by Béchennec (1988, p. 196), but they are certainly Triassic or older (Fig. 3) and could be as old as Permian (see section *Palaeogeographic position of the Al Aridh Formation*).

Platy limestones form a significant part of the Al Aridh Formation in the type area (Fig. 5, 6). They are centimetre- to decimetre bedded, light grey to pinkish grey lime wackestones to graded grainstones with shale partings and contain abundant thin-shelled bivalves of the *Halobia*-type, peloids, and calcified radiolarians (Fig. 7A, B). Interbeds of green and white radiolarite occur (Fig. 5) as well as breccia beds with Permian reef boulders, and oolitic turbidites, particularly west of Al Aridh (cf. Glennie et al., 1974). The platy limestone unit grades upward into radiolarite (Fig. 3, 5) and varies in thickness from about 70 m in the type area to 15 m in Jebel Buwaidha (Fig. 3, 5).

Age. The platy limestone unit has been mapped as Ar3 by Béchennec (1988) and interpreted as of Late Jurassic to Early Cretaceous age. The abundance of thin-shelled bivalves, however, is a typical feature of Triassic basinal sediments (Bernoulli et al., 1990), and conodonts of radiolarites below and above the platy limestones show that it is a unit of Late Triassic age spanning the Carnian (base of the upper thrust slice at Yanqul, Fig. 5) and the Norian (Jebel Buwaidha, Fig. 3).

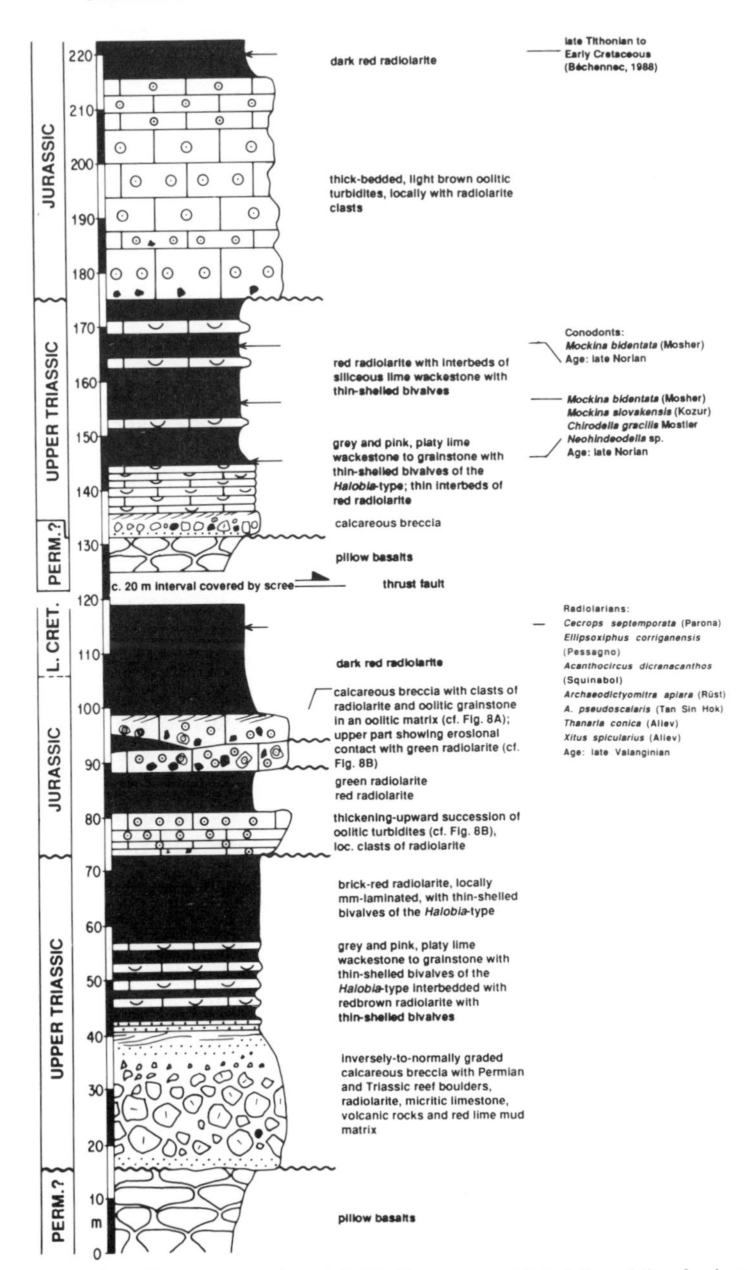

Figure 3. Stratigraphic column of the Al Aridh Formation of Jebel Buwaidha; for location see Fig. 4.

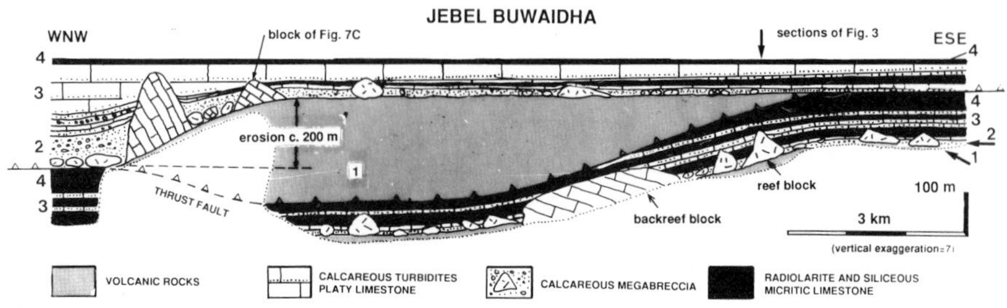

Figure 4. Facies profile of Jebel Buwaidha, approximately perpendicular to the palaeocurrents (Fig. 9) and the thrust direction. Each thrust slice shows volcanic rocks at the base (?Permian) (1) overlain by an Upper Triassic megabreccia (2) that cuts for about 200 m into the volcanic rocks in the western part of the upper thrust slice. Upward follow Upper Triassic platy limestones interbedded with radiolarite, and Jurassic oolitic turbidites (3) that are interbedded with radiolarite forming lensoid bodies in the lower thrust slice indicating a channellized bypass slope. Note westward thickening of the Jurassic trubidites at the expense of the underlying Norian radiolarite in the upper thrust slice indicating an erosional base and a large channel fill. Upper Jurassic to Lower Cretaceous radiolarites are preserved in both thrust slices (4). Thicknesses partly after Béchennec (1988).

Radiolarites of the Al Aridh Formation typically consist of centimetre- to decimetre-thick chert with millimetre-thick shale partings. Most radiolarites show no sedimentary structures, but millimetre-laminations and normal grading of radiolarians are occasionally observed. Radiolarians and sponge spicules make up 20–30% of the rock volume. Siliceous lime wackestones with radiolarians are occasionally interbedded.

Age. Late Triassic radiolarite occurs at the base of thrust slices in the Al Aridh type area and Yanqul, but conodont fragments in the lower thrust slice of the Yanqul section could be reworked (cf. Fig. 5). Late Triassic radiolarite also occurs at Jebel Buwaidha (Fig. 3). Jurassic and Early Cretaceous radiolarites have been dated at Jebel Buwaidha (Béchennec, 1988, Fig. 67), and at Al Aridh and Yanqul (Fig. 5). A Late Cretaceous age (Turonian to Santonian) has been obtained from an isolated radiolarite occurrence at Jebel Kawr (Béchennec, 1988, p. 206), but the tectonic position of these radiolarites is unclear.

Oolitic turbidite facies. Calcareous turbidites derived from a neritic source typically contain oolite grains with diametres of 0.2 to 1 mm, peloids, aggregate grains, bioclasts, and a few volcanic grains. Decimetre-sized lithoclasts of radiolarite and oolitic grainstone are locally abundant (Fig. 8A). The beds are decimetre- to metre-thick, show normal size grading and usually are capped by cross-bedded intervals that can be up to several decimetres thick. The oolite turbidites occur as discrete sequences, up to 100 m thick in the upper thrust slice of Jebel Buwaidha (Fig. 4). In the lower thrust slice, they occur as two to four separate, lenticular sequences, separated by radiolarite

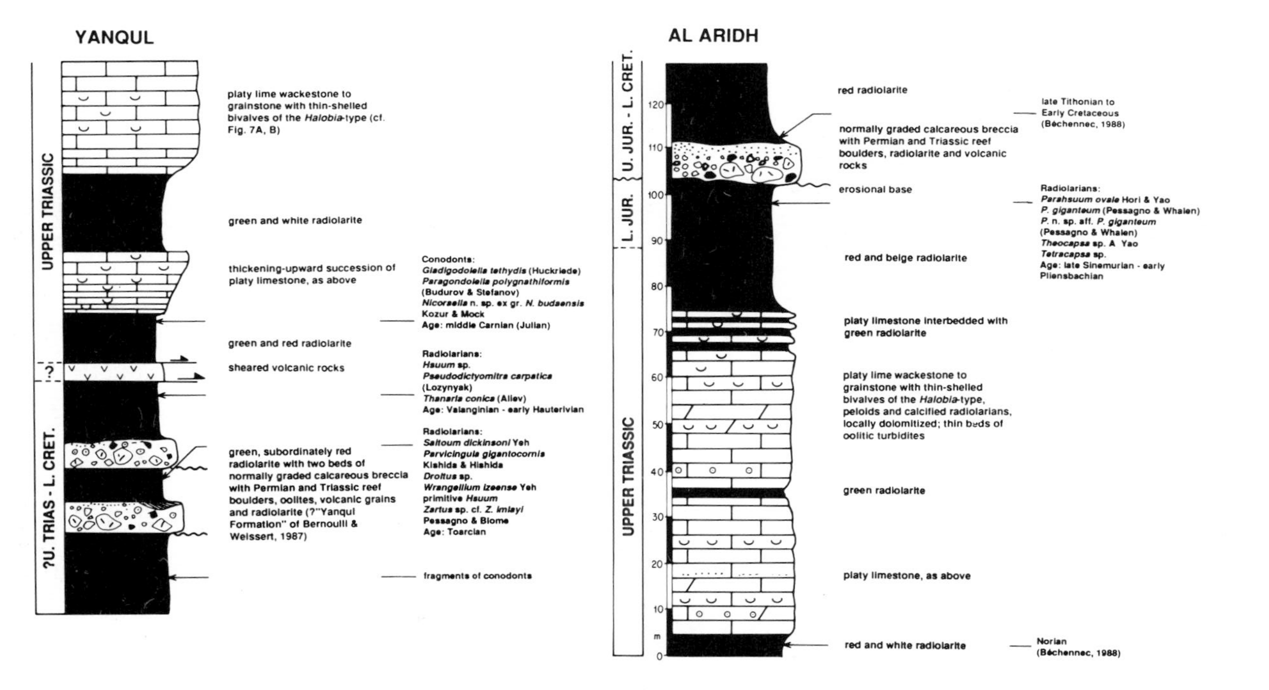

Figure 5. Stratigraphic columns of the Al Aridh Formation in the type area (for location, see Fig. 6) and at Yanqul. The Yanqul section is probably identical to that mentioned by Bernoulli and Weissert (1987), although they gave no details of their locality. The section corresponds to the foreground in Glennie et al. (1974, Fig. 5.40.4).

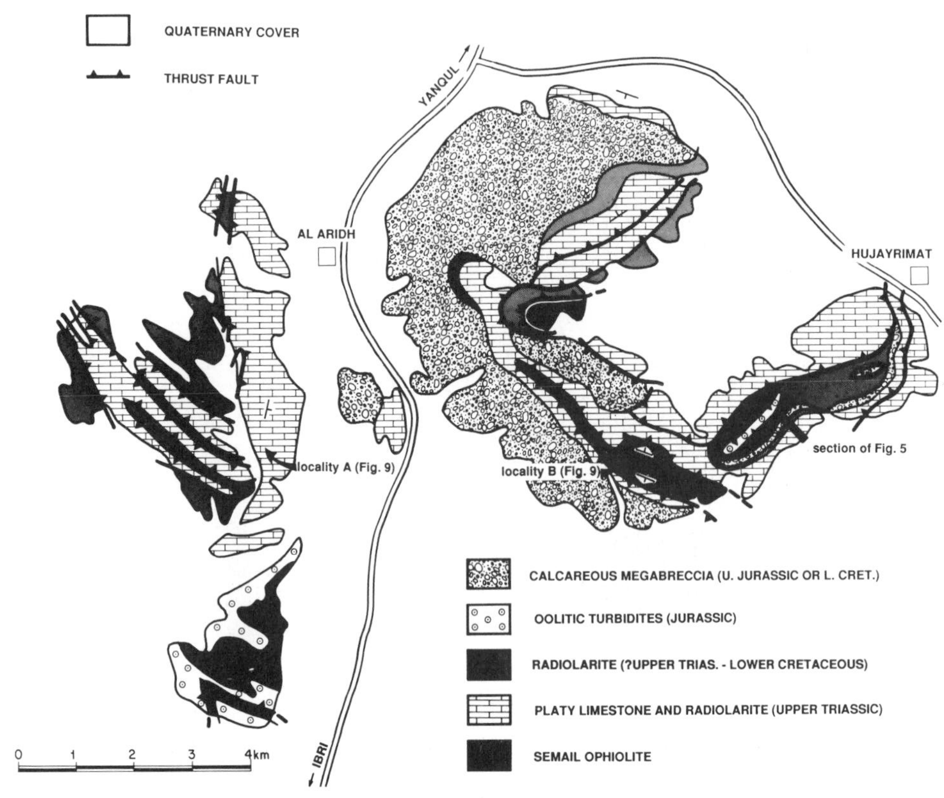

Figure 6. Geological map of the Al Aridh area (for location, see Fig. 2), modified after Minoux and Janjou (1986), who mapped the entire part of the Al Aridh Formation wets of the Yanqul-Ibri road as Wahrah Formation. The tectonic position of the Al Aridh Formation is not typical here as it overlies the Semail ophiolite according to Minoux and Janjou.

(Fig. 8B). Each sequence is between 8 and 10 m thick showing fining-and-thinning-upward and coarsening-and-thickening-upward successions, respectively. A facies association similar to the lower thrust slice of Jebel Buwaidha has been called the "Yanqul Formation" by Bernoulli and Weissert (1987)(Fig. 5).

Age. The oolitic turbidites contain Jurassic foraminifers (Béchennec, 1988, p. 204, 205; Glennie et al., 1974, Fig. 4.4.28, 4.4.29; M. Septfontaine, pers. comm., 1990). Despite this, Béchennec (1988) inferred an Early Cretaceous age for the oolitic turbidites he called Ar4. He based his assumption on Lower Cretaceous radiolarite below the thrust fault at Jebel Buwaidha (Fig. 3, 4) which remained unrecognized in his profiles. Triassic radiolarite below (Fig. 3), interbedded Jurassic radiolarites (e.g., Yanqul, Fig. 5), and Upper Jurassic to Lower Cretaceous radiolarite above (Fig. 3) prove the Early to Late Jurassic age of the oolite turbidites.

Figure 7. A: Upper Triassic thin-bedded and platy limestones, about 30 m thick, at Yanqul (cf. Fig. 5). Arrow marks the stratigraphic way-up. The platy limestones indicate a period of periplatform ooze deposition which can be correlated with the *Halobia*-Limestone Member of the Hamrat Duru Group. Greater thickness than in the Hamrat Duru Group indicates closer position to the shelf source area. B:Negative print of a thin section of the platy limestone at Yanqul showing thin-shelled bivalves (arrow) set in a micritic matrix (light coloured). C: Block derived from Triassic back-reef sources occurring in the Upper Triassic megabreccia at Jebel Buwaidha (for location, see Fig. 4). The breccia is overlain by Norian radiolarites (cf. Fig. 3) which in turn are covered by Jurassic oolite turbidites. Poorly exposed ?Permian volcanics occur below the breccia. Structural dip is about 45° into the plane of the photograph. The megabreccia indicates a collapse of the shelf margin and relative proximity to it. D: Negative print of a thin section of the Triassic shallow water facies of block of Fig. 7C. The grainstone consists of dasycladaceans (*Andrusoporella* sp.) (1), aggregate grains (2) and blue-green algal thalli (3).

Calcareous megabreccia facies. A typical feature of the Al Aridh Formation is the presence of thick, coarse calcareous conglomerates and breccias. They form sequences up to 100 m thick (western part of Jebel Buwaidha, Fig. 4) that can be traced laterally for several kilometres. These breccias contain blocks of limestone, up to a few hundred metres across, which were generally mapped as "Oman Exotics" by Glennie et al. (1974) (Fig. 7C). Breccia components are, in decreasing order of abundance:

Figure 8. A: Lower part of a Jurassic gravity flow deposit in the lower thrust slice of Jebel Buwaidha. The coarse conglomerate contains clasts of radiolarite (1) and lighter coloured clasts of oolitic grainstone (2) in an oolitic matrix. The clasts indicate erosion of slope deposits associated with the advancing gravity flow. B: Coarsening-and-thickening-upward succession of oolitic turbidites, about 8 m thick (white arrow), breccia (cf. Fig. 8A), and radiolarites in the lower thrust slice of Jebel Buwaidha. The radiolarite horizon above the breccia has been completely eroded by the overlying oolite turbidites about 100 m to the right of the photograph (cf. Fig. 3) evidencing a channellized bypass slope.

1. Permian reef limestone. These are typically sponge-*Archaeolithoporella*-coral-*Tubiphytes* boundstones with abundant crinoid ossicles and laminated, red lime mud filling primary voids;
2. Lime wackestones and packstones with radiolarians and thin-shelled bivalves;
3. Triassic reef limestone. Boundstones with colonial corals and solenoporacean red algae are most common, but *Tubiphytes*-calcareous sponge boundstones are common as well;
4. Triassic bedded shallow water limestones consisting of white aggregate grainstones with abundant dasycladaceans (*Andrusoporella* sp., Fig. 7C, D);
5. Radiolarite;
6. Volcanic grains and boulders of mafic volcanic rocks.

The clasts are generally angular to moderately rounded and show stylolitic contacts. They are roughly bimodally sorted and show average diametres of 10–30 cm and coarse sand size. The fabric is characterized by clast-support; a red lime mud matrix is occasionally present.

The breccia beds consist of a basal, roughly inversely graded unit up to a metre thick. Upward follows a massive or crudely normally graded unit of boulders and sand-size grains which may be 10 m thick. The top of a typical breccia bed is commonly made up of a metre-thick unit with low-angle planar cross-bedding and parallel lamination.

Boulders project up to 50 m above the top of the breccia units and are draped by radiolarite. Other blocks apparently foundered into the underlying units such as the volcaniclastic rocks of Jebel Buwaidha.

Age. The megabreccia of Jebel Buwaidha contains Triassic shallow water limestones (Fig. 7D) and is covered by Norian radiolarite. It is, therefore, Late Triassic in age. The megabreccia of the Al Aridh area (Fig. 6) overlies Lower Jurassic radiolarite (Fig. 5) and is overlain by Tithonian-Lower Cretaceous radiolarite. It is, therefore, of Late Jurassic to Early Cretaceous age.

Erosional Features

Erosion associated with gravity flow deposits is indicated by clasts of shallow and deep water origin, volcanic rocks (Fig. 8A), and by channel fill deposits. In the western part of the upper thrust slice of Jebel Buwaidha, the Triassic megabreccia cuts about 200 m into the volcanic rock substrate (Fig. 4) and is covered by a thick succession of platy limestone and breccia beds (cf. Béchennec, 1988, Fig. 67). Similarly, the Jurassic oolite turbidites are thickest in this area suggesting that it persisted as a morphological trough into the Jurassic. Along the upper thrust slice of Jebel Buwaidha, the Jurassic oolite turbidites thicken westward at the expense of the underlying Norian radiolarite that is totally absent above the breccia channel. In the lower thrust slice

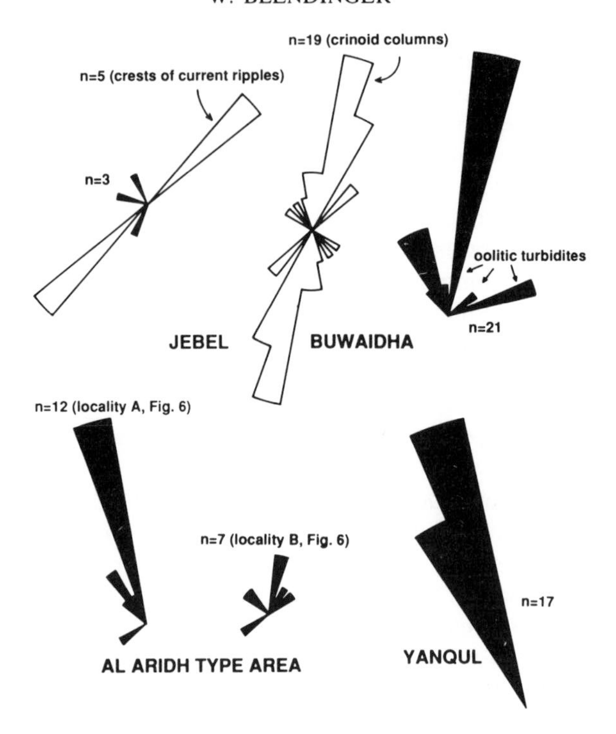

Figure 9. Palaeocurrents in the Al Aridh Formation. Black rose diagrams were obtained from measurements of cross bedding and flute casts in megabreccias and oolite turbidites, blank diagrams show linear features.

of Jebel Buwaidha, lensoid oolite turbidite sequences are cut into radiolarite (Fig. 3, 8). At Al Aridh (Fig. 5), most parts of the Jurassic are missing below the megabreccia. All this points to more or less significant erosion induced by the sediment gravity flows.

Palaeocurrents

Palaeocurrents of the Al Aridh Formation were measured from flute casts at the base of calcareous breccias, from cross-bedded successions that typically cap such deposits, and from cross bedding developed in oolitic turbidites. Some current ripple orientations and the orientation of centimetre-long crinoid stems in turbidite beds were included as non-vectoral data (Fig. 9).

The palaeocurrents clearly indicate that the calciclastic sediments were transported mainly northward. This is in agreement with the few palaeocurrents measured by Glennie et al. (1974, Fig. 8.2.1).

Interpretation and Discussion

Depositional Environment: Bypass Slope

The association of radiolarite and sediment gravity flows indicates a deep marine environment for the Al Aridh Formation. The Triassic *platy limestones* can be interpreted as periplatform ooze (cf. Schlager and James, 1978). The *oolite turbidites* indicate the penecontemporaneous existence of oolite shoals in an adjacent shallow shelf located in the south. Boulders of Permian and Triassic reef limestone were derived from a typical shelf edge facies and indicate that the deep water environment of the Al Aridh Formation has existed since the Permian. Sedimentary structures similar to those observed in the *calcareous megabreccias* either suggest extremely thick turbidites (e.g., Skaberne, 1987) or debris avalanches (Watts and Garrison, 1986). The megabreccias probably derived from erosion of the adjacent shelf edge and may have been triggered by a variety of causes such as oversteepening of the slope and/or earthquake shocks.

Although there are no palinspastic handholds to determine the distance of the Al Aridh Formation from the shelf edge, its relative position can be established. The association of submarine erosion phenomena, sediment gravity flows with a proximal aspect, and pelagic rocks indicates a channelized bypass slope. In such a setting, shallow water derived sediment is transported downslope in gullies or canyons cut into fine-grained, pelagic sediment (Mullins et al., 1984; Watts and Garrison, 1986). The dominance of radiolarite (as compared to calcareous ooze in the Bahamian example: Mullins et al., 1984) in the Al Aridh Formation indicates deposition below the calcite compensation depth (CCD) and/or elevated surface productivity of radiolarians in the proximity of the shelf edge due to upwelling, as will be discussed below. The environment of the Al Aridh Formation has been calculated to be about 40 km wide (Béchennec, 1988, p. 358).

Palaeogeographic Position of the Al Aridh Formation

Distal origin hypothesis. Two hypotheses have been put forward for the palaeogeographic position of the Al Aridh Formation. The first and most popular one is the distal origin hypothesis proposed by Glennie et al. (1974) and later adopted by Searle and Graham (1982), Bernoulli and Weissert (1987) and Béchennec (1988)(Fig. 1B). It is based on simple unstacking of the Hawasina nappe pile. According to this hypothesis, the Al Aridh Formation would have been positioned between the distal part of the Hamrat Duru Group (Wahrah Formation) and the Triassic carbonate platforms of the "Oman Exotics".

This model has the advantage of simplicity, but it is challenged by sedimentological data. *Palaeocurrents* show that the calciclastic gravity flows were derived from a source located in the south, whereas a northern source would have to be expected according to the above hypothesis. *Permian reef boulders* can not be derived from the "Oman Exotics" carbonate platforms, as these show only Upper Triassic shallow water carbonates above several hundred metres of volcanic rocks (Béchennec, 1988). The strongest argument against the above hypothesis comes from the *Jurassic oolite turbidites and the slope environment they record*. The oolite turbidites can not be derived from the "Oman Exotics" because these were drowned since at least the Middle Jurassic (Glennie et al., 1974, Fig. 4.4.46; Baud et al., 1990) excluding that the slope of the Al Aridh Formation was located along the "Oman Exotics". The "Yanqul Formation" of Bernoulli and Weissert (1987), recently interpreted as derived from the north of the "Oman Exotics" (Bernoulli et al., 1990), contains Permian reef boulders and oolite turbidites in the Jurassic and is similar to the lower thrust slice of Jebel Buwaidha. It is here included in the Al Aridh Formation.

The Al Aridh Formation can, therefore, not be considered as a facies link between the Hamrat Duru Group and the "Oman Exotics"/Haliw Formation ("Umar Group"). The most distal facies of the Hamrat Duru Group, the Wahrah Formation, shows neither a facies transition into nor southward directed palaeocurrents (Glennie et al., 1974, Fig. 8.2.1) expected from the "Oman Exotics". Between the "Exotic" offshore carbonate platforms and the Wahrah Formation, a considerable amount of sea floor must have been entirely subducted.

Proximal origin hypothesis. Another hypothesis, offered as an alternative option by Glennie et al. (1974), regards the Al Aridh Formation as derived from the block-faulted margin ("borderland") of the Oman shelf. Although block faulting has not yet been demonstrated, this hypothesis is in agreement with the palaeocurrents. The Al Aridh Formation were then an "out-of-sequence" thrust unit and located palaeogeographically between the Hamrat Duru Group and the shelf edge (Fig. 10). This would help explain the Al Aridh Formation for three reasons.

First, the undated mafic volcanic rocks of Jebel Buwaidha can, due to the erosional base of the overlying breccia, be interpreted as Permian. Permian volcanic rocks are known to underlie large parts of the Hamrat Duru Group sediments (Béchennec, 1988) and extend onto the shelf area as more acidic rocks exposed on the Saiq Plateau in Jebel Akhdar, and in Saih Hatat (Béchennec et al., 1989, Fig. 5). In a bypass slope environment, where much non-deposition and erosion occur, subsea exposures of Permian (rift-related) volcanic rocks could be the source for the volcanic grains and boulders in younger sediments of the Al Aridh Formation.

Second, the inferred bypass slope setting of the Al Aridh Formation can provide an explanation for the lack of siliciclastic rocks in the Triassic and Lower Jurassic part of the Al Aridh Formation as compared with the Hamrat Duru Group. Siliciclastic debris either never reached the shelf edge in this sector of the continental margin, or bypassed the slope environment effec-

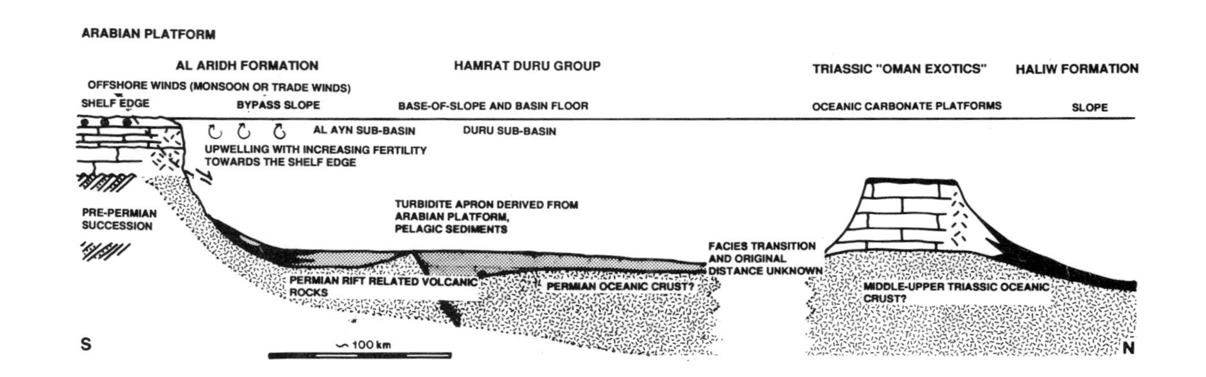

Figure 10. Conclusion: Cross section of the Arabian continental margin in the Jurassic for the central parts of the Oman Mountains inferring the Al Aridh Formation as a slope unit between the Hamrat Duru Group (modified after Cooper, 1990) and the shelf edge of the Arabian platform. Horizontal scale according to the palinspastic values of Béchennec (1988, p. 358); vertically extremely exaggerated. Distance between the shelfbreak and the Al Aridh Formation speculative.

tively and/or was funnelled into the Hamrat Duru Group environment *via* widely spaced channels or canyons (cf. Cooper, 1990) not preserved in the Al Aridh Formation.

Third, the abundance of radiolarite throughout the Al Aridh Formation, another major argument for its interpreted position far from the Arabian shelf (Bernoulli & Weissert, 1987) can be explained with the palaeowind model of Cooper (1990). Palaeocurrents suggest that the slope of the Al Aridh Formation showed a strike around E-W. In the palaeowind model proposed by Cooper (1990), such a shelf edge would have had a leeward position. The leeward position could have promoted upwelling and associated high productivity of radiolarians in surface waters close to the shelf edge. This could have been caused either by trade winds (as suggested by Cooper, 1990) or by the monsoon (as on the present southeastern Oman coast: Prell et al., 1990). By contrast, the established slope facies of the Sumeini Group were apparently deposited on an east-facing, windward slope (Watts, 1990), and are accordingly more calcareous including prolific reef growth recorded in the Jebel Wasa Formation (Watts and Garrison, 1986).

Conclusions

1. The Al Aridh Formation is a tectonic unit that typically overlies the Hamrat Duru Group in the central part of the Oman Mountains. Its base consists locally of undated, probably Permian mafic volcanic rocks, elsewhere it shows Upper Triassic deep water sediments above the basal *décollement*. The sedimentary rocks of the Al Aridh Formation consist of Upper Triassic platy limestones and radiolarites, Jurassic oolitic gravity flows with varying amount of interbedded radiolarite, and Lower Cretaceous radiolarite and pelagic limestone. Megabreccias with Permian and Triassic shallow water limestone blocks can be dated as Late Triassic and Late Jurassic or Early Cretaceous.
2. Erosional features indicate a bypass slope environment for the Al Aridh Formation. Palaeocurrents show that the calciclastic gravity flows of the Al Aridh Formation are derived from a southern source. As particularly the oolite turbidites can not be derived from the "Oman Exotics", the Arabian shelf must have been the source for all gravity flows.
3. The absence of siliciclastic rocks distinguishes the Upper Triassic and Lower Jurassic of the Al Aridh Formation from coeval deposits in the Hamrat Duru Group and can be explained if one assumes that a wide spacing of feeder canyons allowed effective bypassing of the slope environment. The dominance of radiolarite, particularly in the Jurassic, can be explained as the result of upwelling and associated high productivity of radiolarians in surface water close to the north-facing, leeward shelfbreak.
4. The Al Aridh Formation can be explained as an "out-of-sequence" thrust unit derived from a position between the edge of the Arabian shelf and

the Hamrat Duru Group. It can be considered as a lateral equivalent of the Sumeini Group in an area between Yanqul and the Semail Gap. This is, in principle, the alternative option of Glennie et al. (1974).

Acknowledgments

The author is particularly indebted to Dr H. Kozur (Budapest) for the extraction and identification of conodonts and radiolarians from radiolarite samples. Dr M. Septfontaine (Lausanne) kindly inspected thin sections of oolite turbidites. The quality of the paper benefited from comments by Drs K.F. Watts (Fairbanks, Alaska) and A.H.F. Robertson (Edinburgh, U.K.). The paper appears with the approval of Petroleum Development Oman and the Oman Ministry of Petroleum and Minerals.

References

Baud, A., Marcoux, J. and Stampfli, G., 1990. Evolution of the Oman margin, from rifting to passive margin stage (Permian to early Mesozoic). In: Symposium on ophiolite genesis and evolution of the lithosphere, abstracts (Ed. by Ministry of Petroleum and Minerals, Muscat) (no page numbers).

Béchennec, F., 1988. Géologie des nappes Hawasina dans les parties orientale et centrale des montagnes d'Oman., Documents du BRGM, 127: 474 p.

Béchennec, F., Le Métour, J., Rabu, D., Beurrier, M., Bourdillon-Jeudy-de-Grissac, C., De Wever, P., Tégyey, M. and Villey, M., 1989. Géologie d'une chaîne issue de la Téthys: les montagnes d'Oman. Societée géologique de la France Bulletin, 8: 167–188.

Bernoulli, D. and Weissert, H., 1987. The upper Hawasina nappes in the central Oman Mountains: stratigraphy, palinspastics and sequence of nappe emplacement. Geodinamica Acta, 1: 47–58.

Bernoulli, D., Weissert, H., and Blome, C.D., 1990. Evolution of the Triassic Hawasina Basin, Central Oman Mountains. In: The Geology and Tectonics of the Oman Region (Ed. by A.H.F. Robertson, M.P. Searle and A.C. Ries), Geological Society Special Publication, 49: 189–202.

Cooper, D.J.W., 1990. Sedimentary evolution and palaeogeographical reconstruction of the Mesozoic continental rise in Oman: evidence from the Hamrat Duru Group. In: The Geology and Tectonics of the Oman Region (Ed. by A.H.F. Robertson, M.P. Searle and A.C. Ries), Geological Society Special Publication, 49: 161–187.

Glennie, K.W., Boeuf, M.G.A., Hughes Clarke, M.W., Moody-Stuart, M., Pilaar, W.H.F. and Reinhart, B.M., 1974. Geology of the Oman Mountains. Koninklijk Nederlands Geologisch Mijnbouwkundig Genootschap Verhandelingen, 31: 423 p.

Minoux, L. and Janjou, D., 1986. Geological map of Ibri 1:100.000. Explanatory notes. Ministry of Petroleum and Minerals, Directorate General of Minerals, Muscat.

Mullins, H.T., Heath, K.C., van Buren, M. and Newton, C.R., 1984. Anatomy of a modern open-ocean carbonate slope: northern Little Bahama Bank. Sedimentology, 31: 141–168.

Prell, W.L., and shipboard party of ODP Leg 117, 1990. Neogene tectonics and sedimentation of the SE Oman continental margin: results from ODP Leg 117. In: The Geology and Tectonics of the Oman Region (ed. by A.H.F. Robertson, M.P. Searle and A.C. Ries), Geological Society Special Publication, 49: 745–758.

Robertson, A.H.F. and Searle, M.P., 1990. The northern Oman continental margin: stratigraphy, structure, concepts and controversies. In: The Geology and Tectonics of the Oman Region (ed. by A.H.F. Robertson, M.P. Searle and A.C. Ries), Geological Society Special Publication, 49: 3–25.

Schlager, W. and James, N.P., 1978. Low-magnesian calcite limestones forming at the deep-sea floor, Tongue of the Ocean, Bahamas. Sedimentology, 25: 675–702.

Searle, M.P. and Graham, G.M., 1982. "Oman Exotics" – Oceanic carbonate build-ups associated with the early stages of continental rifting. Geology, 10: 43–49.

Skaberne, D., 1987. Megaturbidites in the Paleogene Flysch in the region of Anhovo (W Slovenia, Yugoslavia). Societa Geologica Italiana Memorie, 40: 231–239.

Watts, K.F., 1990. Mesozoic carbonate slope facies marking the Arabian platform margin in Oman: depositional history, morphology and palaeogeography. In: The Geology and Tectonics of the Oman Region (ed. by A.H.F. Robertson, M.P. Searle and A.C. Ries), Geological Society Special Publication, 49: 139–159.

Watts, K.F. and Garrison, R.E., 1986. Sumeini Group, Oman – evolution of a Mesozoic carbonate slope on a South Tethyan continental margin. Sedimentary Geology, 48: 107–168.

Igneous Rocks in the Hawasina Nappes and the Hajar Supergroup, Oman Mountains: Their Significance in the Birth and Evolution of the Composite Extensional Margin of Eastern Tethys

F. BÉCHENNEC*, M. TEGYEY, J. LE MÉTOUR, B. LEMIÈRE, J.L. LESCUYER, D. RABU and J.P. MILÉSI
Service Géologique National, BRGM, B.P. 6009, 45060 Orleans Cedex 2, France

Abstract

The main phases in the structural evolution of the Arabian continental margin were accompanied by significant magmatism the geochemistry of which gives data on its geodynamic setting.

The history of the margin began in the Late Permian with a phase of extension and rifting within Gondwana that gave rise to the Hamrat Duru basin, separating the Arabian platform to the south from the Baid platform to the north.

This phase was characterized by considerable magmatic activity, represented in the basin by the Al Jil Formation and along the edges of the two bordering platforms by the Saiq and Baid formations. The volcanic rocks, mainly basaltic pillow lavas, are dominantly alkaline with minor transitional components and are characteristic of an intracontinental basin. However, MORB tholeiites are encountered locally.

Renewed extension and rifting in the Middle to Late Triassic, affecting mainly the Baid platform, led to the formation of the Al Aridh trough, the Misfah horst and the Umar basin, on the internal side of the Hamrat Duru basin. Important magmatic activity in the new structures marked this phase. Volcanic rocks, mainly basalts and andesites, are in general of within-plate alkaline type, characteristic of continental rift zones. However, in the most distal part,the Umar basin, the volcanic sequences evolve towards transitional and locally to MORB tholeiitic types. This suggests strong crustal thinning in this area, which was probably continuous with the Neo-Tethyan sea floor.

During the Jurassic and Cretaceous, magmatic activity was less important on the Arabian continental margin, and was characterized mainly by alkaline to peralkaline sills and dykes intruding a variety of rock types. However a small amount of extrusive rocks were poured out in the Hamrat Duru basin in the Jurassic and along the border of the Arabian Platform in the Late Cretaceous (Muti Formation) concomitantly with the onset of the Eo-Alpine tectogenesis.

Tj. Peters et al. (Eds), Ophiolite Genesis and Evolution of the Oceanic Lithosphere, 593–611.

Regional Setting

The Oman Mountains form an arc extending for 700 km along the northeast edge of the Arabian Peninsula, their high relief being due to Pliocene-Quaternary uplift. They are composed of four main geological units (Allemann and Peters, 1972; Glennie et al., 1974) which are, from bottom to top:

1. The crystalline and sedimentary basement of Proterozoic and Palaeozoic age, unconformably overlain by arabian platform sediments of Permian to Late Cretaceous age.
2. The Sumeini and Hawasina Nappes, consisting mainly of deep-marine sedimentary and volcanic rocks, thrust onto the Arabian platform.
3. The Samail Ophiolite Nappe, in thrust contact with either the Hawasina Nappes or the Arabian platform sediments.
4. Sedimentary rocks of end Cretaceous, Tertiary and Quaternary age unconformably overlying all the above units.

The work done by numerous investigators (Reinhart, 1969; Allemann and Peters, 1972; Glennie et al., 1974; Gealey, 1977; Searle et al., 1980; Coleman, 1981; Lippard et al., 1986; Beurrier, 1987; Nicolas and Boudier, 1988) has shown that the ophiolites of the samail nappe are a fragment of neo-tethyan oceanic lithosphere that was thrust, during the Late Cretaceous, onto the southern passive margin of tethys.

This margin is represented, in a now classic scheme after the recent works (Graham, 1980; Searle and Graham, 1982; Searle et al., 1983; Watts and Garrison, 1986; Béchennec 1987; Le Métour, 1987; Rabu 1987; Bernoulli and Weissert 1988; Béchennec et al., 1988; Béchennec et al., 1990; Cooper, 1990) by the Hawasina Nappes the Sumeini Nappes and the northeastern part of the Arabian Platform.

Our own work based on mapping at 1:100 000 and 1:50 000 scales over the Oman Mountains between the Batain coast in the South-East and the Hatta area in the North-West (Fig. 1) leads to a new paleogeographical reconstruction of the margin. Indeed the numerous new lithostratigraphic, biostratigraphic and structural data has led to determine six main structural units which correspond to six palaeogeographical domains: the Arabian platform, the Sumeini continental slope, the Hamrat Duru basin with the Baid relict horst, the Al Aridh trough, the Misfah horst, and the Umar basin the last four domains belonging to the Hawasina basin (Béchennec, 1987; Le Métour, 1987; Rabu, 1987). The main stages of evolution of the margin has been also determined, based on sedimentological, paleontological and structural features. Nevertheless these main stages can be also characterized through the petrographic and chemical variations in the magmatic rocks that were emplaced in the different palaeogeographic domains between the Permian and the Late Cretaceous.

Part of the magmatic rocks of the margin, associated with the so-called

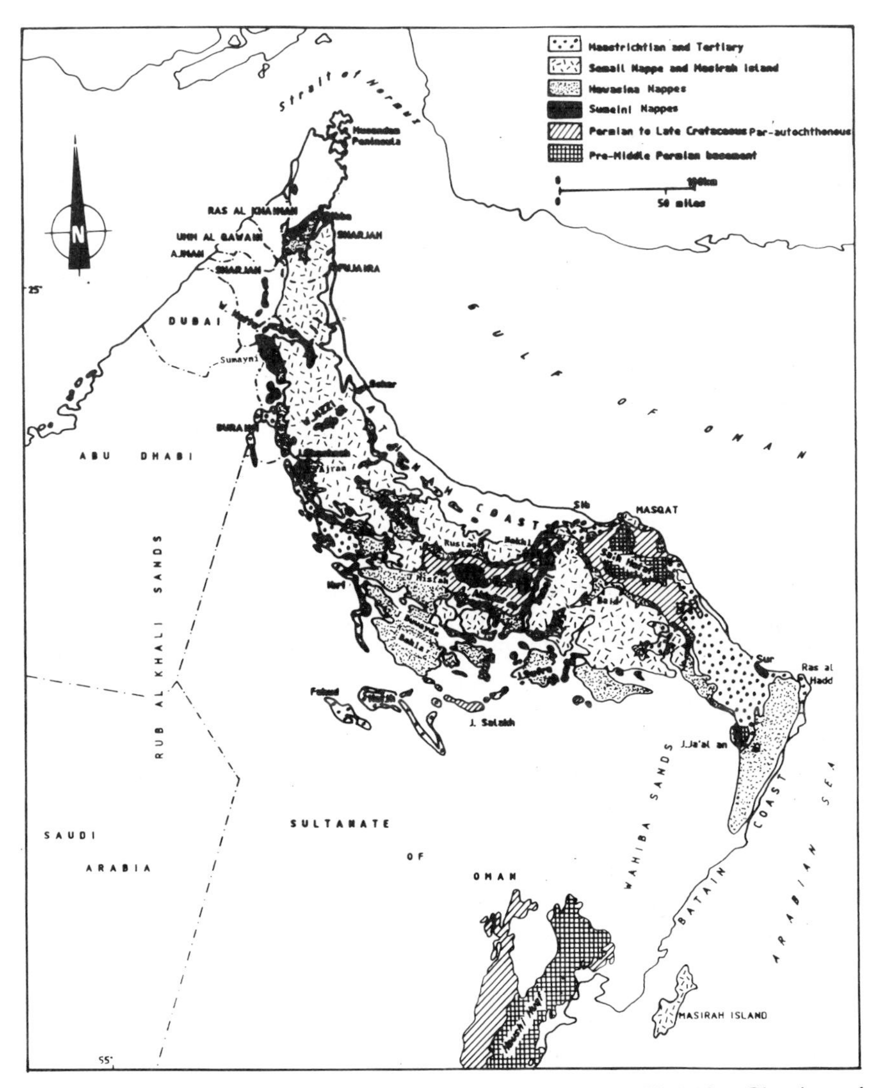

Figure 1. Simplified Geological map of the Oman Mountains (modified after Glennie et al. 1974).

Exotics, and the so-called Haliw, Halfa, Ibra, Al Aridh, Dhera, and Al Ayn Fms. were previously mapped or pointed out by Glennie et al. (1974) Later, the "Haybi Complex" was defined by Searle et al. (1980) in the Hawasina Window at the upper part of the Hawasina allochthon. Its major component is a thick unit of mainly Triassic volcanic rocks, the "Haybi volcanics", associated with limestone exotics and radiolarian cherts. In the "Ḥaybi volcanics" these authors recognized alkaline volcanic rocks at the base, dominantly tholeiitic lavas in the upper part, and late (Cretaceous) differentiated sills intruding the volcanic rocks.

However, part of the "Haybi volcanics" have been recognized of Late Permian age in the Hawasina window (Villey et al., 1986; Béchennec, 1987; De wever et al., 1988). Furthermore it appears that a lot of the margin volcanic rocks, mapped in the whole Oman Mountains, are not related to the so-called "Haybi volcanics", due to their different ages (from Late Permian to Late Cretaceous) and their completely different structural and palaeogeographical settings (edge of the Arabian platform, Hamrat Duru basin, Baid horst, Al Aridh trough, Misfah horst, Umar basin). We thus propose to characterize the margin magmatism connected to his palaeogeographical and structural evolution between the Permian and the Late Cretaceous.

Late Permian Magmatism

The history of the southern passive margin of Tethys began in the Late Permian (Béchennec, 1987; De Wever et al., 1988) with a broad marine transgression across the northern margin of Gondwana. This region was at the same time affected by an important phase of extension and rifting, with the formation of the Hamrat Duru basin separating the Arabian platform to the south from the Baid platform to the north (Fig. 2A). The first sediments deposited in this basin were Murghabian radiolarites, followed by carbonate turbidites, forming the Lower Member of the Al Jil Formation in the Hamrat Duru Group (Béchennec, 1987).

Important igneous activity occurred during this phase, both on the borders of the two platforms and in the basin (Béchennec et al., 1986; Le Métour et al, 1986a, b; Béchennec, 1987; Le Métour, 1987).

Description of the Late Permian Igneous Rocks

The edge of the Arabian platform: In northern Saih Hatat, shallow marine carbonates of Late Permian age (the Saiq Fm.) enclose two volcanic sequences, still well recognisable despite the metamorphism that affected them in the Late Cretaceous, during subduction/obduction related tectonics in the Eo-Alpine orogenic episode. The first sequence (Sq1V), at the base of the Saiq Fm., is represented essentially by tuffs and tuffites, with subordinate andesitic to rhyodacitic lavas, laid down during an important episode of explosive submarine volcanic activity. This basal sequence was laid down in the northeast quadrant of Saih Hatat, in a foundered shelf environment in the Hulw half-graben. At the western edge of the graben it ranges from 0 to 100 m thick, but in the east it is more than 500 m thick at its center (Le Métour, 1987).

The second sequence (Sq2V), the thickness of which ranges from 0 to

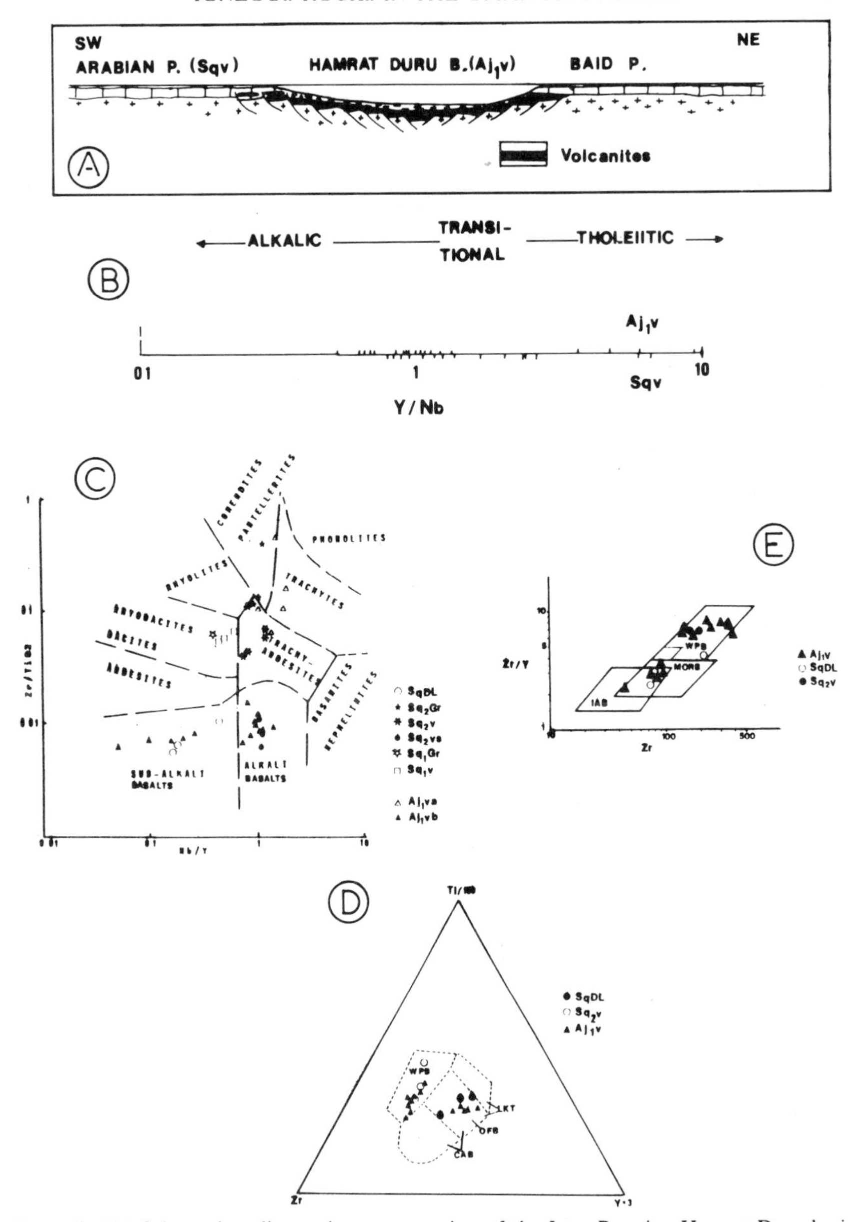

Figure 2. (A) Schematic palinspastic reconstruction of the Late Permian Hamrat Duru basin. (B) Determination of petrologic character (after Pearce and Cann, 1973) of the Late Permian volcanites. Aj1v: magmatic rocks of the Hamrat Duru basin, sqv: magmatic rocks of the Arabian platform. (C) Zr/TiO_2 – Nb/Y diagram (after Winchester and Floyd, 1977) showing distribution of the Late Permian magmatic rocks. Arabian platform, Saiq Fm: SqDl doleritic sill, Sq_2Gr microgranite, Sq_2V basalt, Sq_2Vs doleritic sill, Sq_1Gr microgranite, Sq_1v tuf. Hamrat Duru basin, Al Jil Fm: Ajlva acid hypabyssal dyke and plug, Aj_1vb basalt. (D) Zr/Y – Zr diagram (after Pearce and Norry, 1979) and (E) Ti/100 – Zr-YYx_3 diagram (after Pearce and Cann, 1973) discriminant for tectonic setting.

150 m, was deposited in the north of Saih Hatat, in a shallow-marine carbonate shelf environment that succeeded the early horst and graben topography. Indeed this is interstratified with the carbonates of the middle part of the Saiq Fm. and represents essentially effusive volcanism. It begins with a ten-meter thick tuffite unit enclosing blocks of acidic lavas and fine-grained crystalline rocks, which is succeeded by 30 m of dolomite. Then follows a series of basalt and trachyandesite porphyritic lavas, in small pillows, with flow textured rhyodacitic lava flows, with which are associated intervals composed of lava blocks, hyaloclastites and tuffites. The volcanic rocks are cut by doleritic sills and plugs of rhyodacite with microgranite cores (SqGr).

Numerous dykes, sills and laccoliths of dolerite (SqDl) that cut the pre-Permian rocks and the Saiq Fm. up to the level of the lavas, have also been attributed to the Permian magmatic activity (Le Métour, 1987).

The edge of the Baid platform: The Baid platform is only known as small relics among the Hawasina Nappes, so that information on its igneous activity is also fragmentary. There is however, to the northeast of Baid (Villey et al. 1986), and near Nakhl (Rabu et al., 1986) a volcanic succession several hundred meters thick composed of pillowed basalt and basaltic andesite beneath Late Permian shallow marine carbonate (the Baid Fm.).

The Hamrat Duru basin: Important igneous activity during the Late Permian has been identified at the bases of several tectonic units among the Hawasina Nappes that are composed of members of the Hamrat Duru Group (Béchennec 1987). The principal representatives of this activity are along the Batain coast, in the Baid area, northeast of Jabal Akhdar, and to the north of the Hawasina Window, where they underlie pelagic and turbidite sedimentary sequences of Late Permian age (Béchennec, 1987; De Wever et al., 1988).

Along the Batain coast, the Late Permian magmatic rocks consist mainly of pillowed basalt often associated with tuffite enclosing blocks of volcanite and blocks and boulders of shallow-marine carbonate with fusulinids and corals and of grey limestone with Permian brachiopods. Locally it appears also thick massive trachy-andesitic lava.

In the Baid area, in the east of the Oman Mountains, this activity was particularly diverse, being represented by basalts and andesites in tubular pillows, hyaloclastites, locally enclosing blocks of andesitic and trachytic lavas, tuffites with subordinate flows of sodic trachyte lava with well-defined flow structure, and keratophyre, Intercalations several meters thick of coarse breccia enclose clasts of reworked volcanic rocks and blocks and boulders of Late Permian shallow-marine carbonates. The entire succession is intruded by dykes and sills several meters thick, of dolerite and porphyritic andesite.

On the northeast border of Jabal Akhdar the Permian igneous rocks consist essentially of basalt in thick tubular pillows, locally cemented by a carbonate matrix. Intercalations of hyaloclastite enclosing blocks of lava, and tuffs including rounded and flattened bombs can be seen locally. The latter demonstrate that the essentially effusive, submarine volcanism of this episode in places became subaerial and explosive. This thick succession is cut by meter-thick dykes of andesite and sills, locally abundant, of sodic andesite. In some place (near Rustaq) at the top of the volcanic sequence appears brownish limestone with intervals containing abundant crinoids and ammonoids, Permian in age for some of them. (J. Marcoux pers. comm. 1990)

To the north of the Hawasina Window, in the Buday'ah area, Permian magmatism is represented by a very thick, monotonous sequence of pillow basalts, in tubes 20 cm to 1 m in diameter, with locally variolitic cortexes, which tend to become autobrecciated at the flow fronts. At the top of this sequence are thin intercalations of shales and radiolarites of Murghabian age (Béchennec, 1987; De Wever et al., 1988).

Geochemical Characterization of the Late Permian Magmatism

Eighteen samples of the range of Late Permian igneous rocks from the Arabian platform and nineteen from the Hamrat Duru basin were analyzed for major and trace elements (Tables 1 and 2). Large variations in major element contents, in particular SiO2, Al2O3, MgO, FeO and Fe2O3, show the variety of rock types present, ranging from basalt to rhyolite. The rocks have in general been strongly altered (high L.O.I.) and in places metamorphosed (as in Saih Hatat), so the geochemical interpretation is based essentially on the contents of elements known to be immobile, such as Zr, Nb, Ti and Y (Floyd and Winchester, 1978; Pearce and Norry, 1979).

The Y/Nb ratio, plotted on the diagram of Pearce and Norry (1973), shows that most of the Permian igneous rocks have the chemistry of alkaline to transitional basalt (Fig. 2B). Eight samples, however, have tholeiitic composition; these samples are of dolerites (SqDl) from the Arabian platform and basalts from the Hamrat Duru basin collected in the north of the Hawasina Window. This distribution is confirmed by the variation in Zr/TiO2 as a function of Nb/Y (Fig. 2C) plotted on the diagram of Winchester and Floyd (1977), where the same eight samples fall in the field of subalkaline basalt. This diagram also shows that the intermediate to acidic rocks of the first volcanic sequence (Sq1V) on the Arabian platform fall in the field of subalkaline rhyodacites. In contrast, the rocks of the second sequence, like most of the rocks of the Hawasina basin, extend from the alkali basalt through the trachyandesite and trachyte fields into the field of comendite-pantellerite, indicating that they probably belong to a single differentiated series.

The tectonic setting of formation of the Permian basaltic rocks can be identified using the log Zr/Y vs log Zr and Ti/100 – Zr-Y × 3 discriminant diagrams respectively of Pearce and Norry (1979) and Pearce and Cann (1973) given in Figures 2D and 2E. These plots show that most of these rocks, both on the Arabian platform and in the Hamrat Duru basin are within-plate basalts. The emission of transitional to alkaline basalt indicates

Table 1. Major (%) and trace (ppm) element compositions of the Late Permian arabian platform (Saih Hatat area) volcanic and hypabyssal rocks (Saiq Fm., Sq_1V: metatuf and metatuffite – Sq_1Gr: microgranite – Sq_2V: basic and intermediate lava – Sq_2Gr: microgranite – SqDL: dolerite). [Major elements by X-Ray fluorescence at BRGM, and Lyon University (*). Trace elements by X-Ray fluorescence at Lyon University (*) and by ICP at BRGM, Orléans.]

Unité	N°Ech.	SiO_2	Al_2O_3	Fe_2O_3	FeO	MgO	CaO	Na_2O	K_2O	TiO_2	P_2O_5	MnO	P.feu	Total	Zr*	Y*	Nb*	Sr*	Rb*	Cr'	Ni'
	1 JPM 18	73.30	13.90	0.44	1.20	0.20	0.56	3.55	5.00	0.21	0.27	0.02	0.57	99.22	150	37	21	43	150	12	8
Sq_1V	2 JM 2046	71.30	14.00	2.15	–	0.48	0.62	2.70	6.00	0.26	0.25	0.02	1.20	98.98	158	40	19	49	195	17	35
	3 JM 2046 A	74.50	13.50	2.25	–	0.39	0.56	4.00	4.30	0.25	0.21	0.03	0.90	100.89	144	38	18	28	107	27	40
Sq_1Gr	4 JLL 50	75.40	12.70	0.57	1.05	0.37	0.63	2.25	5.40	0.22	0.31	0.01	0.87	99.78	146	46	17	63	207	8	7
	5 JM 1081*	61.21	15.70	7.76	–	0.48	3.40	6.36	1.00	1.04	0.45	0.04	1.58	99.02	407	57	45	316	15	55	14
	6 JM 1082*	68.53	11.60	7.31	–	1.07	2.44	4.58	0.81	0.84	0.31	0.08	1.49	99.06	318	49	34	196	11	84	17
	7 JPM 5	68.50	13.80	5.20	1.50	0.27	0.61	4.85	4.30	0.55	0.08	0.04	0.13	99.83	583	66	57	89	116	21	12
Sq_2V	8 JM 2045 A	63.80	14.40	9.70	–	0.57	2.45	4.10	3.25	0.81	0.22	0.07	1.50	100.87	518	49	57	280	61	25	39
	9 JM 2045 B	62.50	15.40	7.45	–	1.90	3.15	4.45	2.50	0.89	0.19	0.10	1.90	100.43	582	52	63	548	35	18	37
	10 JM 2045 C	50.00	16.20	11.60	–	3.65	7.90	2.05	2.10	2.45	0.57	0.16	3.15	99.83	225	31	30	384	36	69	58
	11 JPM 12	47.20	14.40	6.20	5.65	7.15	9.45	3.15	0.46	2.15	0.29	0.15	3.35	99.80	155	23	27	693	8	195	114
	12 JPM 6	47.40	14.20	7.80	5.65	4.55	8.50	3.65	0.79	3.75	0.57	0.22	3.10	100.18	193	28	30	515	13	43	42
	13 JM 2044 A	69.10	13.10	5.80	–	0.10	1.15	1.60	7.50	0.43	0.07	0.06	0.84	99.75	511	55	50	93	92	44	39
Sq_2Gr	14 JM 2044 B	64.00	16.10	7.25	–	0.32	0.58	3.45	7.45	0.52	0.07	0.06	0.54	100.34	674	68	67	62	120	34	39
	15 JLL 78	69.10	12.20	2.65	1.25	0.31	2.85	1.80	0.26	0.06	0.04	2.05	100.02	1035	89	106	270	60	20	13	
	16 JM 625 A	46.30	14.50	12.80	–	3.70	8.75	4.35	0.62	2.10	0.26	0.33	5.45	99.16	211	50	21	152	21	29	50
SqDl	17 JM 2047	46.70	15.60	11.90	–	6.15	11.40	1.70	0.10	1.25	0.10	0.24	5.55	99.69	68	29	4	139	2	168	86
	18 JM 2048	48.50	14.00	12.60	–	6.80	10.60	2.20	0.08	1.25	0.09	0.17	3.15	99.44	79	27	4	132	2	186	85

Table 2. Major (%) and trace (ppm) element compositions of the Late Permian Hamrat Duru basin volcanic and hypabyssal rocks (Aj_1V: Al Jil Fm.). (Major elements by X-Ray fluorescence at BRGM, and Lyon University (*). Trace elements by X-Ray fluorescence at Lyon University (*) and by ICP at BRGM, Orléans).

	NECH	SiO_2	Al_2O_3	Fe_2O_3	FeO	MgO	CaO	Na_2O	K_2O	TiO_2	P_2O_5	MnO	H_2OP	H_2OM	P.few	Total	Zr*	Y*	Nb*	Sr*	Rb*	Cr	V
L	3	50.60	13.40	5.50	5.45	3.95	5.75	5.75	0.18	2.75	0.71	0.15	0.00	0.00	4.85	99.04	345	45	46	278	4	11	101
L	7	48.70	14.60	4.70	6.95	5.20	4.60	5.45	0.17	3.10	0.70	0.15	0.00	0.00	4.75	99.07	349	48	47	438	2	11	131
L	54	46.50	14.00	6.05	1.50	3.75	13.50	3.80	0.10	1.70	0.39	0.10	0.00	0.00	9.30	100.69	158	21	24	456	1	137	126
L	82	50.00	12.90	7.60	1.80	2.85	8.90	3.80	1.20	2.45	1.10	0.21	0.00	0.00	7.15	99.96	397	67	53	121	19	6	53
L	84	45.00	18.00	8.50	1.75	4.80	10.60	2.65	0.82	1.90	0.27	0.21	0.00	0.00	5.55	100.05	149	23	20	230	10	62	161
L	96	45.70	16.00	4.65	6.85	5.65	5.85	3.85	0.53	2.70	0.32	0.12	0.00	0.00	6.40	98.62	181	30	22	623	8	43	242
M	30	41.00	15.20	3.75	9.70	7.65	6.35	2.60	0.20	2.85	0.55	0.10	0.00	0.00	9.20	99.15	251	34	36	200	4	156	189
M	36	43.40	13.70	5.95	3.10	2.50	13.10	5.10	0.51	2.45	0.50	0.13	0.00	0.00	9.40	99.84	226	28	40	353	9	46	176
M	26	47.30	18.10	8.10	3.30	4.30	3.20	3.40	3.05	2.90	0.41	0.45	0.00	0.00	5.25	99.76	299	38	36	164	52	15	299
M	28	56.40	14.90	2.15	8.25	1.90	3.10	3.80	1.70	0.66	0.20	0.16	0.00	0.00	5.75	98.97	745	58	104	171	38	6	8
M	35	58.00	17.10	3.25	4.80	1.10	3.15	6.35	3.05	0.80	0.23	0.22	0.00	0.00	1.25	99.30	531	58	80	306	23	17	8
M	24	62.20	14.50	8.50	1.50	0.18	1.10	5.95	4.15	0.63	0.10	0.25	0.00	0.00	0.21	99.27	693	86	89	83	52	26	7
L	80	63.70	16.30	6.95	1.00	0.20	0.68	7.85	2.45	0.43	0.05	0.08	0.00	0.00	0.18	99.87	727	53	100	73	16	16	38
M	47	69.10	16.00	3.05	0.55	0.37	0.60	9.55	0.17	0.34	0.07	0.02	0.00	0.00	0.14	99.96	1659	93	140	53	4	40	15
L	135*	46.11	16.80	2.60	5.32	7.32	12.00	2.54	0.61	0.70	0.03	0.14	4.22	0.49	4.71	98.88	181	30	22	623	8	43	242
L	136*	50.30	15.20	3.26	6.36	7.68	9.65	3.35	0.18	1.14	0.11	0.17	1.80	0.34	2.14	99.54	82	31	3	226	3	275	241
L	138*	43.43	14.80	6.33	2.68	7.01	11.53	4.17	0.17	1.12	0.12	0.17	7.86	0.16	8.02	99.57	92	27	7	334	0	411	254
L	113*	44.93	14.20	7.68	2.84	5.60	11.43	3.52	1.08	1.33	0.15	0.14	6.04	0.50	6.54	99.44	94	32	5	176	13	226	278
M	33	47.50	14.10	3.25	6.45	8.50	12.40	1.95	0.10	0.94	0.07	0.20	0.00	0.00	3.95	99.41	69	25	5	90	1	285	213

a continental crust under extension, in our case this crust was the edge of Gondwanaland. Some of these rocks however, in particular all those from the western part of the Hamrat Duru basin, fall in the field of MORB. This distribution can be interpreted as the result of pronounced crustal thinning, apparently with incipient oceanization during the Late Permian in the western part of the Hamrat Duru basin, which is now exposed in the Hawasina Window, thinning which was arrested before the development of any true oceanic crust.

Middle-Late Triassic Magmatism

A second phase of extension and rifting affected the South Tethyan passive continental margin during the Middle to Late Triassic. This was little felt on the Arabian platform, where the regressive episode that began during the Djulfian continued during the Triassic with inter- to supratidal carbonate deposition. On the continental slope and in the Hamrat Duru basin, the extensional phase is mainly represented by radiolaritan cherts related in part to the deepening of the basin, from the upper Anisian-Ladinian (Member E of the Maqam Fm., Watts and Garrison, 1986, and Upper Member of the Al Jil Fm., Béchennec, 1987).

The extension was most strongly expressed on the internal border of the Hamrat Duru basin, where, from the Early-middle Triassic, the Baid platform broke up, with the formation of the Al Aridh trough, the Misfah horst and the Umar basin (Fig. 3A), which henceforward, with the Hamrat Duru basin, together formed the Hawasina basin. The Triassic sedimentary rocks in the Al Aridh trough and the Umar basin are radiolarian cherts and micritic limestones containing radiolaria and pelagic bivalves, of Ladinian to Carnian -Norian age (the Al Ghafat and Aqil formations, Béchennec, 1987). At the same time as these were being deposited, a very thick sequence of platform carbonates developed on the Misfah horst (Misfah Fm., Béchennec, 1987).

Magmatic activity also occurred during this phase of extension. While little in evidence in the Hamrat Duru basin, it was important in the new fault-controlled structures, where igneous rocks underlie all the Triassic sedimentary successions (Searle et al. 1980; Béchennec, 1987).

Description of the Middle-Late Triassic Igneous Rocks

The Hamrat Duru Basin: Triassic igneous activity in the Hamrat Duru basin was relatively insignificant and very localized. Thus, a volcanic sequence a few hundred meters thick in the Upper Member of the Al Jil Fm. has been recognized within radiolarites of Middle to Late Triassic age in the Wadi Ajran area, (Dank 1:100 000 map area, sheet NF 40–2B). This includes

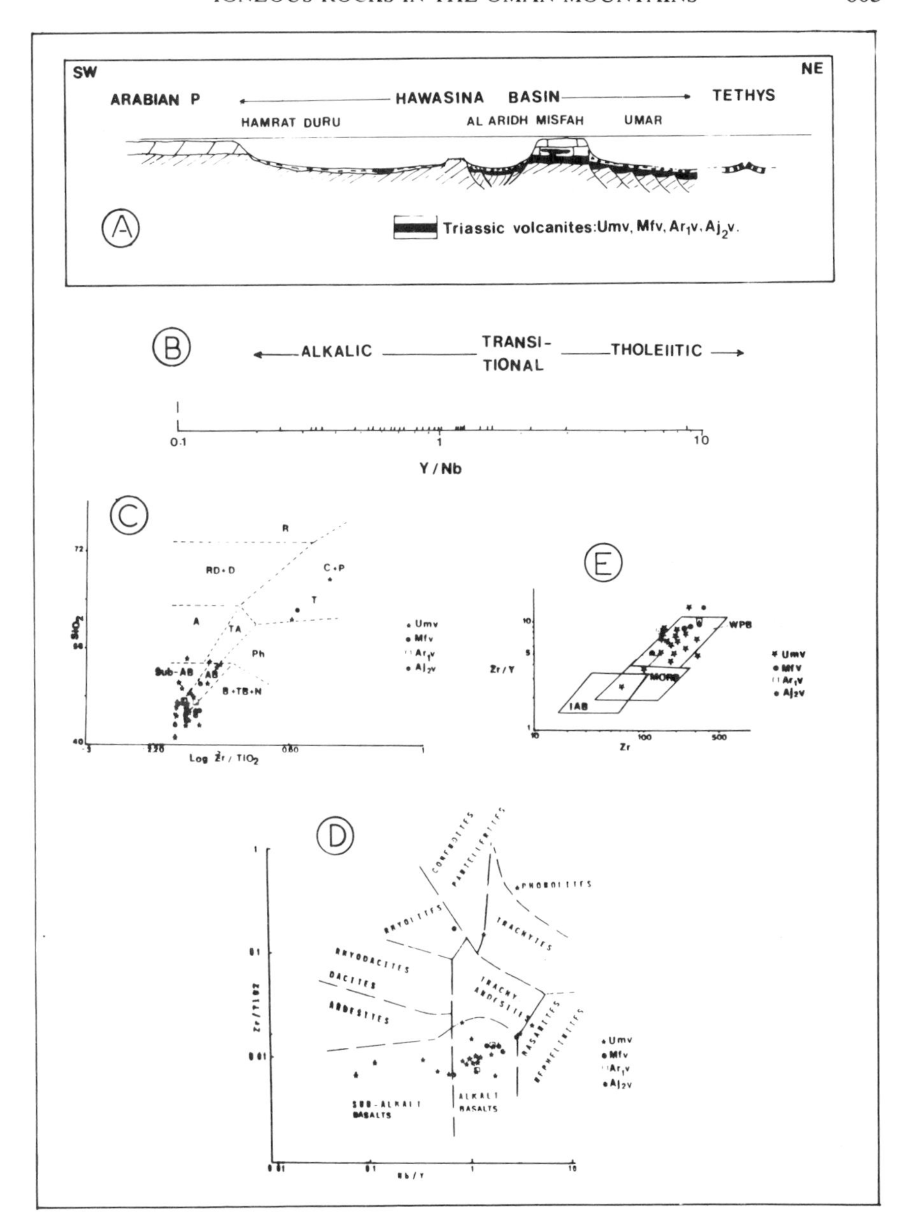

Figure 3. (A) Schematic palinspastic reconstruction of the Triassic arabian continental margin. (B) Determination of petrologic character (after Pearce and Cann, 1973) of the Middle-Late Triassic volcanic rocks. (C) SiO_2 – Zr/Tio$_2$ diagram (after Floyd and Winchester, 1978) showing the bimodal distribution of the various Middle-Late Triassic magmatic rocks. Hamrat Duru basin, Aljil Fm.: Aj_2v basaltic and trachytic lava; Al Aridh trough, Al Ghafat Fm.: Ar_1v basalt; Misfah horst, Misfah Fm.: Mfv basaltic lava and trachytic sill; Umar basin, Sinni Fm.: Umv basalt, basaltic andesite, trachytic dyke. (D) Zr/TiO_2 – Nb/Y diagram (after Winchester and Floyd, 1977) showing the distribution of the various Middle-Late Triassic magmatic rocks.
E) Zr/Y – Zr diagram (after Pearce and Norry, 1979) discriminant for tectonic setting.

small-pillow basalts, tuffs, keratophyres in large tubular pillows and massive trachytes.

The Al Aridh trough: The Middle-Late Triassic igneous rocks (Ar1V) (Al Ghafat Fm., Béchennec et al., 1986; Béchennec, 1987), best represented in Jabal Buwaydah (Bahla 1:100 000 map area, Sheet NF 40–07A), consist of up to 400 m of basaltic andesite, either massive or in tubular pillows, with tuffs, fine-grained hyaloclastites enclosing blocks of porphyritic andesite, and in places intervals of agglomerate composed essentially of flattened bombs. These agglomerates show that the essentially submarine volcanism at times became subaerial. Microdolerite sills in places cut the volcanic sequence.

The Misfah horst: In the Triassic Misfah Fm. (Béchennec, 1987) platform carbonates overlie a thick volcanic sequence (Mfv) that is exposed in jabals Misfah, Ghull and Kawr. (Rustaq 1:100 000 map area, sheet NF 40–3D). This consists of pillowed and massive basalts and basaltic andesites, massive trachyandesites, tuffs and hyaloclastites, which in places enclose blocks of lava and of platform carbonates. The formation is cut by dykes and sills of dolerite, andesite and sodic trachyte. Towards the top, the volcanic rocks become interstratified with Norian platform carbonates. Pipe breccia can be seen locally, enclosing rounded blocks, some tens of centimeters across, of diorite, gabbro and carbonatized peridotite and several-meters-wide blocks of platform carbonate torn up from the underlying rocks.

The Umar basin: The lowest exposed deposits from the Umar basin are the volcanic rocks of the Sinni Fm. (Umv, Béchennec, 1987), several hundred meters thick, that outcrop at the base of the topmost tectonic units of the Hawasina throughout the length of the Oman Mountains.

These rocks are particularly well exposed in the Sinni area on the northwest flank of Jabal Akhdar (Rustaq 1:100000 map area, Sheet NF 40–3D), in the north of the Hawasina Window (Yanqul 1:100 000 map area, Sheet NF 40–2C) and southeast of Sumayni (Shinas 1:100 000 map area, Sheet NE 40–14B). The formation is composed of two main units; the lower (Umv1) consists of several hundred meters of basalt, in small tubular pillows, with variolitic borders, in places surrounded by fine-grained hyaloclastite.

The upper unit (Umv2) consists essentially of sodic andesite pillow lavas, in large pillows, but is more varied than the lower unit and includes intervals of fine-grained hyaloclastite enclosing blocks of lava. Towards the top, discontinuous intervals of radiolarite and shale, with breccias composed of clasts of radiolarite and volcanic rocks appear, with block and boulders of shallow-marine carbonate. The upper unit is cut by common dykes and sills, several meters thick, of sodic trachyte, keratophyre and granophyre, and in places by plugs some hundreds of meters across of quartz diorite (in Wadi Tifili, in the Hawasina Window) and of gabbro with associated microgabbro, quartz-gabbro and diorite (at Hayl village, in the Samail Gap).

Geochemistry of the Middle-Late Triassic Igneous Rocks

Major and trace elements analyses have been made of samples of the various igneous rocks from the Hamrat Duru basin (2), the Al Aridh trough (2), the Misfah horst (5) and the Umar basin (22), the results of which are given in Table 3. The wide range in major elements composition (in particular SiO2 and TiO2 Fig. 3C) of these volcanic rocks shows a bimodal distribution, with basaltic and acid trachytic components, but no representatives of intermediate compositions. The rocks are commonly strongly altered (high L.o.I.) so they have been geochemically characterized using elements known to be immobile during alteration – Nb, Y, TiO2 and Zr (Floyd and Winchester, 1978; Pearce and Norry, 1979).

The Y/Nb ratios (Fig. 3B) show that 22 of the samples have alkaline chemistry, 6 have the chemistry of transitional basalts and 3 that of tholeiitic basalts. The tholeiites are all from the Sinni Fm. in the Umar basin, while the transitional basalts are some of the samples from the upper volcanic assemblage (Umv2) in the same basin, and from the Misfah horst (Mfv). The Zr/TiO2 vs Nb/Y diagram (Fig. 3D; Winchester and Floyd, 1977) also shows that most of the basic rocks fall in the field of alkali basalts, but two fall on the trachyandesite/basanite boundary and five in the field of subalkaline basalts. The acid rocks fall in the trachyte and phonolite fields. The basalts showing subalkaline chemistry are thus all from the Umar basin, in a very distal situation on the Arabian continental margin.

Most of the Triassic lavas (the alkali and transitional basalts) are within-plate basalts, whereas MORB are represented only by two of the tholeiitic basalts (Fig. 3E).

The emission of thick sequences of alkaline to transitional basalts, associated with alkaline acid rocks, indicates that major crustal extension took place in the most distal parts of the Arabian passive continental margin during the Middle to Late Triassic. This extension affected mainly the pre-existing Permian Baid platform, which was broken up at this time. In the Umar basin, the magmatic evolution from early alkaline to transitional and tholeiitic terms would appear to indicate progressive crustal thinning during extension. This could have led to true oceanization in the most internal parts of the basin, which would then have been in continuity with Neo-Tethys, which also began to develop at this time (Whitechurch et al., 1984).

Jurassic and Cretaceous Volcanism

Volcanic activity was rare and sporadic during the Jurassic and Cretaceous on the continental Arabian margin.

In the Hamrat Duru basin, however, weak volcanic activity was manifested during the Lias and Bathonian – Kimmeridgian. Indeed it has been pointed out, interbedded in the liassic turbiditic sandstone and shale of the upper

Table 3. Major (%) and trace (ppm) element compositions of the Middle to Late Triassic Hawasina basin volcanic and hypabyssal rocks. (Umv: Sinni Fm., Umar basin – Mfv: Misfah Fm., Misfah horst – Ar_1V: Al Ghafat Fm., Al Aridh trough – Aj_2V: Al Jil Fm., Hamrat Duru basin). (Major elements by X-Ray fluorescence at BRGM, and Lyon University (*) – Trace elements by X-Ray fluorescence at Lyon University (*) and by ICP at BRGM, Orléans).

		NECH	SiO_2	Al_2O_3	Fe_2O_3	FeO	MgO	CaO	Na_2O	K_2O	TiO_2	P_2O_5	MnO	H_2OP	H_2OM	P.few	Total	Zr*	Y*	Nb*	Sr*	Rb*	Cr	V
	L	139*	49.95	14.22	6.71	2.42	1.75	8.50	6.34	1.06	1.94	0.81	0.07	5.14	0.17	5.31	99.08	307	43	r 44	361	22	23	125
	L	140*	45.23	14.00	5.12	3.31	1.45	11.49	5.53	0.79	2.43	0.71	0.14	8.65	0.22	8.87	99.07	236	31	38	241	15	13	190
	L	141*	46.68	13.60	1.89	3.11	3.70	11.91	5.67	0.33	1.82	0.65	0.10	9.06	0.20	9.26	98.72	154	22	22	198	2	88	241
	L	12	48.60	14.60	5.15	5.95	4.85	9.65	4.55	0.89	2.55	0.50	0.14	0.00	0.00	2.70	100.13	252	32	35	244	19	35	190
	L	20	46.70	16.70	4.60	4.65	4.10	11.40	3.40	1.30	1.60	0.31	0.00	0.00	4.70	99.59	148	19	22	646	17	67	131	
	L	22	46.00	12.20	10.10	1.90	3.75	11.60	4.65	0.79	2.05	0.32	0.17	0.00	0.00	6.70	100.23	184	30	27	354	10	26	226
	L	137*	47.90*	15.87	5.74	6.20	2.20	7.59	6.16	0.15	2.43	0.55	0.12	4.37	0.12	4.49	99.40	261	25	40	449	3	23	196
	L	107*	54.25	15.50	0.82	6.00	1.30	6.40	6.38	0.08	1.08	0.22	0.22	6.00	0.09	6.09	98.34	99	27	3	221	1	490	206
	L	109*	52.81	17.65	3.14	2.97	2.14	4.74	5.75	2.71	1.67	0.54	0.16	4.49	0.35	4.84	99.12	347	24	99	1618	33	9	44
	L	110*	41.18	9.35	3.62	5.90	8.48	18.50	1.55	0.75	2.00	0.26	0.17	6.96	0.34	7.30	99.06	134	15	26	401	10	954	248
(Umv)	L	111*	53.82	17.27	1.38	4.97	3.40	4.00	6.14	1.99	1.88	0.56	0.13	3.33	0.17	3.50	99.04	318	27	82	679	16	23	92
	L	112*	50.08	17.00	6.26	4.21	5.00	3.86	4.68	1.99	2.01	0.24	0.10	3.47	0.68	4.15	99.58	149	28	13	553	35	65	265
	L	151*	49.29	16.22	4.01	4.97	3.06	5.81	5.75	0.81	2.53	0.41	0.12	6.44	0.20	6.64	99.62	201	30	26	193	12	171	322
	L	152*	47.00	14.43	5.49	4.15	5.83	12.92	3.36	0.15	1.01	0.09	0.15	4.44	0.19	4.63	99.21	70	28	2	107	4	349	277
	L	153*	43.26	15.10	3.73	2.62	1.85	14.90	4.97	0.73	1.58	0.45	0.11	9.32	0.14	9.46	98.76	204	24	40	402	14	147	194
	L	155*	53.00	15.43	3.21	1.89	0.72	8.66	7.78	0.51	1.45	0.83	0.07	5.73	0.08	5.81	99.36	316	65	54	286	6	170	59
	L	157*	46.80	13.75	4.58	2.81	3.33	12.10	5.35	0.84	1.57	0.26	0.11	7.05	0.19	7.24	99.74	146	22	18	375	7	164	219
	L	158*	46.26	13.35	10.26	1.85	1.61	11.78	4.68	0.46	2.58	0.95	0.09	5.43	0.16	5.59	99.46	180	35	23	355	7	95	283
	L	159*	43.87	15.75	2.57	5.74	4.33	11.44	4.32	0.41	1.92	0.40	0.12	7.43	0.41	7.84	98.71	187	25	24	464	4	273	194
	L	171*	60.60	14.30	7.36	0.00	0.73	3.20	4.35	4.47	0.40	0.07	0.18	3.59	0.21	3.80	99.46	608	58	80	57	74	123	11
	L	172*	67.06	14.05	2.53	0.00	0.41	3.37	6.00	2.87	0.11	0.01	0.16	2.96	0.20	3.16	99.73	481	67	200	42	75	63	16
	M	22	43.90	12.80	9.05	3.70	3.35	15.30	3.10	0.40	1.80	0.26	0.18	0.00	0.00	5.15	98.99	171	39	13	162	10	92	264
	L	24	62.20	15.00	8.40	0.10	0.81	1.75	8.50	0.10	0.63	0.09	0.08	0.00	0.00	1.80	99.46	1134	110	78	111	2	6	7
	L	26	43.20	15.20	4.15	3.70	3.55	15.10	3.45	0.84	1.80	0.21	0.18	0.00	0.00	8.60	99.98	122	24	15	671	15	60	169
(Mfv)	L	67	45.60	15.50	6.15	4.55	7.40	8.55	2.65	1.40	2.05	0.67	0.14	0.00	0.00	5.15	99.81	265	29	54	1005	32	154	112
	L	69	44.80	15.20	8.85	1.75	6.40	9.15	4.00	1.05	2.00	0.70	0.19	0.00	0.00	4.95	99.04	236	28	58	854	16	146	111
	L	71	50.00	14.90	8.75	3.00	3.25	4.95	6.45	0.25	2.45	0.98	0.15	0.00	0.00	3.95	99.08	321	33	48	672	3	15	137
(Ar_1V)	L	44	46.80	16.50	6.15	2.55	6.30	9.60	4.25	0.55	1.80	0.22	0.11	0.00	0.00	5.05	99.88	132	16	18	1007	5	102	147
	L	62	46.80	15.20	9.50	1.40	3.65	10.90	3.60	1.65	2.70	0.65	0.13	0.00	0.00	3.60	99.78	296	30	44	691	31	58	190
(Aj_2V)	F	543 A	60.90	16.40	6.30	0.90	0.20	1.80	6.10	5.05	0.98	0.44	0.06	0.00	0.00	0.36	99.49	–	–	–	–	–	–	–
	F	543 B	45.70	14.60	4.80	4.25	2.95	10.30	5.10	0.92	2.25	0.57	0.16	0.00	0.00	8.05	99.65	361	25	70.9	–	21	–	–

Member of the Matbat Fm., some volcanic rocks (Béchennec 1987). In the southeast of the Hawasina Window and in the Sinni area (Yanqul 1/100 000 map area, sheet NF 40–2C and Rustaq 1/100000 map area sheet NF 40–3D) the volcanic activity is manifested most commonly as blocks of reworked basalt and andesite several tens of centimeters across in sedimentary calcirudite, more rarely as thin intervals of pillowed basalt. In the Sumayni area in the same member, pillowed and massive basalt form thicker units and are commonly cut by trachyandesite sills.

Volcanic rocks have also been found interbedded in the Bathonian – Kimmeridgian turbiditic calcarenite of the Guwayza Fm., near Al Fay in the Wadi Hatta area; but only about ten meters of tuffs and scoriaceous trachyandesitic lavas are present here.

In the Umar basin there was some volcanic activity in the lower Jurassic. This is demonstrated in the Buraymi area by rare and thin intervals of pillowed basalt and tuffites associated with thin beds of radiolarian cherts, dated as Upper Lias to Early Dogger, and with thick unit of sedimentary breccia and megabreccia.

On the Arabian platform (north flank of Jabal Akhdar and Saih Hatat) Late Cretaceous volcanic rocks have been observed within the sedimentary rocks of the Muti Fm. (Rabu, 1987; Le Métour, 1987). The rocks are vesicular trachyandesites and tuffs and pillowed spilitic basalts which, in Saih Hatat, were metamorphosed during the Late Cretaceous Eo-Alpine orogenic episode.

Alkaline and Peralkaline Plutonic Intrusions

Alkaline and peralkaline plutonic intrusions have been recognized both in the sedimentary succession of the Hamrat Duru basin and in the volcanic rocks of the Al Aridh trough and Umar basin (Searle et al., 1980; Searle, 1984; Béchennec, 1987).

Sills of wehrlite several meters thick, with chilled margins, have been observed in the Triassic volcanic rocks of the Al Aridh trough (Al Ghafat Fm., Jabal Buwaydah; Béchennec, 1987) and the Umar basin (Haybi Complex and Sinni Fm., Hawasina Window and Ajran area; Searle, 1984; Béchennec, 1987). Wehrlite sills have also been seen in Liassic turbidites deposited in the Hamrat Duru basin (Upper Member of the Matbat Fm., Béchennec, 1987), in the southeast of the Hawasina Window, where they are associated with sills of jacupirangite and of kaersutite gabbro (Searle, 1984).

Jacupirangite sills several meters thick have also been recognized in the east of Jabal Safra (Birkat al Mawz 1:100000 map area, Sheet NF 40–7B), where they cut and metamorphose limestones enclosing pelagic bivalves of Late Triassic age from the Hamrat Duru basin (Lower Member of the Matbat Fm; Béchennec, 1987). The jacupirangites consist of subhedral plagioclase, titaniferous augite, kaersutite, biotite, sphene, apatite, Fe-Ti oxide and carbonate. They have also been recognized in the Hawasina Window, where they cut Liassic sedimentary rocks of the Hamrat Duru basin (Matbat Fm., Béchennec, 1987) in the Ajran area,in the Wadi Kub (UAE) and Jabal Ghawil (Searle, 1984), where they cut Triassic volcanic rocks of the Umar basin.

Camptonite sills, composed of labradorite, titanaugite, kaersutite, hastingsite, analcime, apatite and Fe-Ti oxides, have been seen cutting Late Triassic limestone of the Hamrat Duru basin in the east of Jabal Ja'alan and of Jabal Safra (Ja'alan 1:100000 and Birkat al Mawz map areas, Sheets NF 40–8B and NF 40–7B), and cutting Triassic volcanic rocks of the Umar basin in the Hawasina Window (Searle, 1984).

Ouachitite sills have been observed in Tithonian-Valanginian radiolarites of the Hamrat Duru basin (Wahrah Fm; Béchennec, 1987) in the east of Jabal Ja'alan and southwest of Bahla. They are composed of titaniferous biotite, augite, hastingsitic hornblende, abundant apatite and carbonate, and Fe-Ti oxides.

Finally, teschenite sills cutting limestone turbidites of Late Permian-Early Triassic age have been recognized near Rustaq. They consist of labradorite, augite, red-brown hornblende, Fe-Ti oxides, apatite and analcime.

The ages of these alkaline to peralkaline intrusions range from Early Liassic to Late Cretaceous. The Rustaq teschenites have been dated at 194 ± 6 Ma (sinemurian-Toarcian; K-Ar on hornblende; Béchennec, 1987); the jacupirangites intruding the Matbat Formation in the Hawasina Window have given ages of 162 ± 6 and 159 ± 6 Ma (Bathonian-Oxfordian; K-Ar on biotite; Lippard and Rex, 1982); the jacupirangites in the Ajran area, intruding the volcanic rocks of the Umar basin, have been dated at 129 ± 5 Ma (Tithonian-Berriasian; K-Ar on biotite; Lippard and Rex, 1982), and the jacupirangites in the Jabal Ghawil area, intrusive into the volcanic rocks of the Umar basin, have yielded ages of 93 ± 5 and 92 ± 4 Ma (Cenomanian-Coniacian; K-Ar on biotite; Searle, 1980).

Conclusions

It is thus clear that the main stages in the development of the Arabian continental margin correspond to the two periods of major volcanic activity, in the Late Permian and then again in the Middle to Late Triassic. This activity took place in the zones of maximum crustal stretching.

During the Late Permian this was essentially in, and on the margins of the Hamrat Duru basin – in the northeast of the Arabian platform on its southern side and in the south of the Baid platform on its northern side, whereas in the Middle-Late Triassic volcanic activity was very limited in the Hamrat Duru basin but extensive in the newly formed Al Aridh trough, Misfah horst and Umar basin.

The geochemistry of the rocks formed during these two magmatic episodes indicates that the Late Permian activity produced a single differentiated series with basic, intermediate and acid representatives, whereas the Middle-Late Triassic episode was bimodal, with basic and acid representatives only. During both these episodes, however, the rocks produced were essentially alkaline to transitional, indicating extension within a continental plate. During the Late Permian this was the edge of the Gondwana macroplate, and during the Middle-Late Triassic it was the Baid platform, situated distally on the Arabian continental margin.

However, the appearance of thick MORB tholeiite pillow basalts during the Late Permian could indicate the initial stages of oceanisation in the western part of the Hamrat Duru basin. Similarly, in the Late Triassic, the evolution, in the Umar basin, of the early alkaline magmatism to later transitional and locally tholeiitic compositions indicates pronounced crustal thinning which could, in the most internal zones, have led to true oceanization (formation of Neo-Tethys?).

The contrast between this Permo-Triassic evolution, which produced the main features of the continental margin, and that which took place during the Jurassic and Cretaceous, is striking. During the whole of the latter long period of time, including the Tithonian-Berriasian, when the sedimentary succession reveals that a 250 km retreat of the continental slope of the Arabian Platform took place, with general foundering of the margin, igneous activity on the continental margin was very limited. Volcanism was very sporadic, in the Hamrat Duru basin during the Jurassic and on the Arabian platform during the Late Cretaceous.

Furthermore, the magmatic activity in the Hawasina basin took the form essentially of the emplacement of small intrusion of medium to fine-grained crystalline rocks of alkaline to peralkaline composition, this underlines the stability and the continental character of the Arabian continent margin, which, from the Late Triassic onwards, reacted to the various stresses as a single block up to the time of the obduction that took place during the Late Cretaceous.

Acknowledgements

The authors express their gratitude to Mohamed Bin Hussain Bin Kassim, Director General of Minerals and Dr. Hilal Al Azri, Director of Geological Surveys, Ministry of Petroleum and Minerals, Sultanate of Oman for their warm reception and the constant help throughout the mapping project. John Kemp, Senior geologist, service Géologique National, France, translated the text.

References

Alleman, F. and Peters, T., 1972. The ophiolite-radiolarite belt of the North Oman Mountains. Eclogae Geol. Helv. 65 (3): 657–697.

Béchennec, F., Beurrier, M., Rabu, D., and Hutin, G., 1986a. Geological map of Barka. Sheet NF 40–3B. Scale, 1:100000. Explanatory notes. Dir. Gen. Miner., Oman Minist. Pet. Miner., 33 p.

Béchennec, F., Beurrier, M., Rabu, D., and Hutin, G., 1986b. Geological map of Bahla. Sheet NF 40–7A. Scale, 1:100000. Explanatory notes. Dir. Gen. Miner., Oman Minist. Pet. Miner., 44 p.

Béchennec, F., 1987. Geologie des Nappes Hawasina dans les parties orientale et centrale des Montagnes d'Oman. These Doct. d'Etat, Univ. P. et M. Curie, Paris 6. (Doc. Bur. Rech. Geol. Min. 127, 474 p.)

Béchennec, F., Le Métour, J., Rabu, D., Villey, M., and Beurrier, M., 1988. The Hawasina Basin: A fragment of a starved passive continental margin, thrust over the Arabian Platform during obduction of the Sumail Nappe., Tectonophysics, 151: 323–343.

Béchennec, F., Le Métour, J., Rabu, D., Bourdillon de Grissac, Ch., De wever, P., Beurrier, M., and Villey, M., 1990. The Hawasina Nappes: Stratigraphy, palaegeography and structural evolution of a fragment of the South-Tethyan passive continental margin. In: A.H.F. Robertson, M.p. Searle and, A.C. and Ries (Eds). The Geology and Tectonics of the Oman Region. Geol. Soc. London, Spec. publ., 49: 215–225.

Bernoulli, D. and Weissert, H., 1987. The upper Hawasina nappes in the central Oman Mountains: Stratigraphy, palinspastics and sequence of nappe emplacement., Geodin. Acta, 1: 47–58.

Beurrier, M., 1987. Geologie de la Nappe ophiolitique de Samail dans les parties orientale et centrale des Montagnes d'Oman. These Doct. d'Etat, Univ. P. et M. Curie, Paris 6. (Doc. Bur. Rech. Geol. Min., 128: 412 p).

Coleman, R.G. (Ed), 1981. Oman Ophiolites., J. Geophys. Res., 86: 2495–2782

Cooper, D.J.W, 1990. Sedimentary evolution and palaeogeographical reconstruction of the Mesozoic continental rise in Oman: Evidence from the Hamrat Duru Group. In: A.H.F. Robertson, M.P. Searle and A.C. Ries, (Eds), The Geology and Tectonics of the Oman Region. Geol. Soc. London, Spec. Publ., 49: 161–187.

De wever, P., Bourdillon de Grissac, Ch. and Béchennec F., 1988. Permiam age from radiolarites of the Hawasina nappes, Oman Mountains., Geology, 16: 912–914.

Floyd, P.A. and Winchester, J.A., 1978. Identification and discrimination of altered and metamorphosed volcanic rocks using immobile elements., Chemical Geol., 21: 291–306.

Gealey, W.K., 1977. Ophiolite obduction and geologic evolution of the Oman Mountains and adjacent areas., Geol. Soc. Am. Bull., 88: 1183–1191.

Glennie, K.W., Boeuf, M.G.A., Hughes Clarke, M.W., Moody-Stuart, M., Pilaar, W.F., and Reinhardt, B.M., 1974. Geology of the Oman Mountains.. Verh. K. Ned. geol. Mijnbouwkd. Genoot. Ged. Ser., 31 (1–3): 423 p.

Graham, G.M., 1980. Evolution of a passive margin and nappe emplacement in the Oman Mountains. In: A. Panayiotou (Ed), Ophiolites, Proc. Int. Ophiolite Symp. (Cyprus, 1979) Nicosia, pp 414–423.

Le Métour, J., Villey, M. and de Grammont, X. 1986a. Geological map of Quryat, Sheet NF 40–4D. Scale, 1:100 000. Explanatory notes. Dir. Gen. Miner., Oman Minist. Pet. Miner., 72 p.

Le Métour, J., Villey, M., and de Grammont, X. 1986 b. Geological map of Masqat. Sheet NF 40–4A. Scale 1:100 000. Explanatory notes. Dir. Gen. Miner. Oman Minist. Pet. Miner., 45 p.

Le Métour, J., 1987. Geologie de l'autochtone des Montagnes d'Oman: La fenetre du Saih Hatat. These Doct. d'Etat, Univ. P. et M. Curie, Paris 6. (Doc. Bur. Rech. Geol. Min., 129, 466 p.)

Lippard, S.J. and Rex, D.C., 1982. K-Ar ages of alkaline igneous rocks in the northern Oman

Mountains, N.E. Arabia, and their relations to rifting, passive margin development and destruction of the Oman Tethys., Geol. Mag., 119 (5): 497–503.
Lippard, S.J., Shelton, A.W. and Gass I.G., 1986. The Ophiolite of Northern Oman. Geol. Soc., London, Memoir 11, 178 p.
Nicolas, A. and Boudier, F. (Eds), 1988. The Ophiolites of Oman., Tectonophysics, 151, 401 pp.
Pearce, J.A. and Cann, J.R., 1973. Tectonic setting of basic volcanic rocks dertermined using trace element analysis., Earth Planet. Sc. lett., 19: 290–300.
Pearce, J.A. and Norry, M.J., 1979. Petrogenetic implications of Ti, Zr, Y and Nb Variations in Volcanic rocks., Contrib. Miner. Petrol., 69: 33–47.
Rabu, D., Béchennec, F., Beurrier, M. and Hutin, G., 1986. Geological map of Nakhl. Sheet NF 40–3E Scale 1:100 000. Explanatory notes. Dir. Gen. Minr., Oman Minist. Pet. Miner., 76 p.
Rabu, D., 1987. Geologie de l'autochtone des Montagnes d'Oman: la fenetre du Jabal Akhdar – La semelle metamorphique de la nappe ophiolitique de Samail dans les parties orientale et centrale des Montagnes d'Oman: une revue. These Doct. d'Etat, Univ. P. M. Curie, Paris 6. (Doc. Bur. Rech. Geol. Min., 130, 613 p).
Reinhardt, B.M., 1969. On the genesis and emplacement of ophiolites in the Oman Mountains geosyncline. Schweizer Min. Pet. Mitt. 49: 1–30.
Searle, M.P., Lippard, S.J., Smewing, J.D. and Rex, D.C., 1980. Volcanic rocks beneath the Samail Ophiolite nappe in the northern Oman mountains and their significance in the Mesozoic evolution of Tethys., J. Geol. Soc. London, 137: 589–604.
Searle, M.P. and Graham, G.M., 1982. "Oman exotics": oceanic carbonate build-ups associated with the early stages of continental rifting., Geology, 10: 43–49.
Searle, M.P., James, N.P., Calon, T.J. and Smewing, J.D., 1983. Sedimentological and structural evolution of the arabian continental margin in the Musandam Mountains and Dibba zone, United Arab Emirates., Geol. Soc. Am. Bull., 94: 1381–1400.
Searle, M.P., 1984. Alkaline peridotite, pyroxenite and gabbroic intrusions in the Oman Mountains, Arabia., Canadian J. Earth Sci., 21: 396–406.
Watts, K.F. and Garrison, R.E., 1986. Sumeini Group, Oman. Evolution of a Mesozoic carbonate slope on a South Tethyan continental margin., Sediment. Geol., 48: 107–168.
Whitechurch, H., Juteau, Th. and Montigny, A., 1984. Role of the Eastern Mediterranean ophiolites (Turkey, Syria, Cyprus) in the history of the Neo-Tethys. In: J.E. Dixon and A.H.F. Robertson, (Eds), The Geologi- cal Evolution of the Eastern Mediterranean. Geol. Soc. London, Spec. publ. 17: 301–317.
Villey, M., Béchennec, F., Beurrier, M., Le Métour, J. and Rabu, D., 1986. Geological map of Yanqul. Sheet NF 40–2C. Scale 1:100 000. Explanatory notes. Direct. Gen. Miner., Oman Minist. Pet. Miner., 59 p.
Winchester, J.A. and Floyd, P.A., 1977. Geochemical discrimination of different magma series and their differentiation products using immobile elements., Chemical Geol. 20: 325–343.

The Uplift History of the Precambrian Crystalline Basement of the Jabal J'alan (Sur Area)

FELIX WÜRSTEN[1], MARKUS FLISCH[1], INGRID MICHALSKI[1], JOËL LE MÉTOUR[2], IVAN MERCOLLI[1], UWE MATTHÄUS[1] and TJERK PETERS[1]

[1] *Institute of Mineralogy and Petrology, University of Bern, Switzerland*

[2] *Service Géologique National, BRGM, Nantes, France*

Abstract

Field relations and geochronological data from the Precambrian crystalline complex of the Jabal J'alan area allow the reconstruction of its uplift history. K/Ar cooling ages on biotites date the uplift of the whole complex (metamorphics and plutonics) through the 300°C isotherm at 820 Ma. Sedimentary rocks on top of the crystalline complex in the Northern Jabal J'alan area indicate uplift during Late Proterozoic and Jurassic times. Fission track data on apatite indicate the complex reached the 100°C isotherm during Upper Jurassic times (146 Ma). After Upper Cretaceous denudation of the exposed crystalline rocks, Maastrichtian and Tertiary sediments were deposited. Tertiary burial to depths of 2–3 km partially reset the apatite ages. Final uplift of the crystalline block took place in the Upper Oligocene.

Introduction

Proterozoic crystalline basement is well exposed in a window through the Upper Cretaceous/Tertiary cover in the Jabal J'alan (Sur area). Because it is the nearest crystalline basement to the Oman ophiolite and Hawasina nappes and the northeasternmost basement of the Arabian platform, it is of major importance as an indicator of tectonic activity and as a source of detritus for the sedimentation in the authochtonous and Hawasina basins.

The Southern Jabal J'alan area was mapped in detail by F. Würsten and U. Matthäus and a preliminary description of the geological relationships is given in Mercolli et al. (1989). The whole Jabal J'alan area was mapped in 1988/1989 at 1:100.000 scale by the BRGM team (Roger et al. 1990).

The aim of the present note is to discuss the uplift history of the basement as determined from geochronological data (K/Ar, apatite fission tracks).

Tj. Peters et al. (Eds), Ophiolite Genesis and Evolution of the Oceanic Lithosphere, 613–626.

Geological Setting

The Southern Jabal J'alan area is characterized by five lithological units:

- metamorphic rocks: metapelites, amphibolites, migmatites
- dioritic intrusive bodies: fine- grained diorite, "Schollen"-diorite
- granitoid intrusive bodies: porphyric granite, amphibole granite, fine-grained granite, granodiorite, garnet granite
- Dykes: aplites, pegmatites, basaltic and minor andesitic to rhyolitic dykes, dolerites
- Cretaceous/Tertiary sedimentary cover.

The distribution of the different lithologic units is outlined in Fig. 1.

Since the different plutonic rocks intrude the metamorphic rocks, the latter represent the oldest rocks exposed in the area. The metamorphic complex is mainly composed of high-grade pelitic gneisses. In the northern and north-eastern region, these gneisses gradually pass into migmatites. Amphibolitic layers are intercalated in the metapelites. Some rare calcsilicate lenses are interbedded within the pelitic gneisses. Many veins and some small bodies of a garnet-bearing leucogranite cut through the gneisses. Their origin might be closely related to the anatexis of the metapelites. The metamorphic conditions (Table 1) reached upper amphibolite facies, which is demonstrated by the migmatites and the K-feldspar – sillimanite-cordierite-garnet – biotite paragenesis of the pelitic rocks. A subsequent greenschist-facies overprint partly obliterated the higher grade paragenesis so that sillimanite, cordierite and K-feldspar can only be found as relics or pseudomorphs.

The oldest plutonic rocks (as indicated by intrusive relations) (Fig. 2) are fine- grained diorites and "Schollen"-diorites (Table 1).

The fine- grained diorite forms large (1–50 m) slabs within "Schollen"-diorite and granites. "Schollen"-diorite and granites always clearly intrude the fine- grained diorite, demonstrating that this is the oldest magmatic rock.

The "Schollen"-diorite is a very inhomogeneous rock type, characterized by the random distribution of fine- grained dioritic "Schollen" in a medium-grained quartz-dioritic to granodioritic matrix (Table 1). The "Schollen" are highly variable in aspect ratio, shape (angular, round, ellipsoidal, lobate) and size (cm to 1 m). The normal "Schollen" type is a few dm long, lenticular in shape and has an aspect ratio of 1:3. The matrix can be roughly divided into three main types: hornblenditic, quartz-dioritic and granodioritic. The lithologic transition from one type to another is always gradual and the geometrical relationships between different types do not show any mappable systematics on a regional scale. However, the southern area is dominated by quartz-dioritic to granodioritic types, whereas the eastern area is characterized by the hornblenditic to quartz-dioritic types. These three rock types display similar geochemical trends that suggest close genetic relationships (Fig. 2).

Table 1. Mineralogical composition of the crystalline rocks from the Jabal J'alan.

Magmatic rocks	Mineralogy	Description
Dolerite	2% qz, 45% plag, 38% cpx, 5% amph, 2% bi, 3% sap, 5% opm, 1% chl; apa.	Fine- to medium grained; intense weathering.
Fine-grained granite	25–35% qz, 30–40% kf, 27–36% plag, 1–6% amph, 2–6% bi; tit, apa, epi, chl, zr.	Fine-grained, equigranular; may show slight deformation (alignment of the mafic minerals); colour: grey to pink.
Amphibole-granite	30% qz; 30–45% kf, 30–40% plag, 0–4% cpx, 7–15% amph; tit, apa, epi, chl, zr, (orth).	Medium-grained, equigranular (locally porphyric); stalked, up to 3 mm large amph; colour: pink to red.
Porphyric granite	27–32% qz, 20–30% kf, 30–40% plag, 0–2% amph, 5–15% bi; tit, apa, epi, zr, (orth).	Fine- to medium-grained with porphyric kf; colour varies between white, pink and red.
Granodioritic "Schollen"-diorite	13–28% qz, 0–31% kf, 44–62% plag, 0–19% amph, 7–30% bi; (tit), apa, epi, zr.	Medium-grained; extremely inhomogeneous distribution of the minerals; often schlieric texture; continuous transition between the different types.
Qz-dioritic "Schollen"-diorite	2–12% qz, 30–60% plag, 0–4% cpx, 10–50% amph, 10–23% bi; (tit), apa, epi, zr.	"Schollen": dioritic mineralogy; fine-grained highly variable in size and shape.
Hornblenditic "Schollen"-diorite	0–2% qz, 2–8% plag, 3–25% cpx, 60–75% amph, 10% bi; (apa), epi, pyt.	Coarse-grained, may show pegmatitic zones with large amph and plag.
Fine-grained diorite	3–5% qz, 50–65% plag, 17–32% amph, 8–15% bi; tit, apa, zr.	Fine-grained, dark diorite, which forms large slabs in the "Schollen"-diorite and the granite.

Metamophic rocks	1. metamorphic event upper amphibolite facies	2. metamorphic event upper greenschist to amphibolite facies	late low-grade alteration
garnet granite	qz, micl, plag, gar, ±bi max. 5% mafic minerals		gar + bi → chl Alteration: weak
migmatites	qz, micl, plag, bi		bi → chl rarely
pelitic gneisses	kf, cor, sill, gar, bi	cor → bi + mu/chl + mu kf, sill → mu, qz Overprint: strong	gar → chl + mu + qz bi → chl/mu + illm Alteration: weak
amphibolites (sedimentary)	amph, plag, ±bi, ±gar, ±qz	partly recristallized	amph → chl + clzoi + cc Alteration: weak
amphibolites (magmatic)	porphiric plag; plag, amph, bi, tit, ±kf		
calcsilicate rocks	cc, di, trem, ±micl, ±plag, ±tit; cc, di, qz, ±gar	clzoi, plag, cc, hydrogar Overprint: medium	

qz = quartz, kf = kalifeldspar, micl = microcline, plag = plagioclase, cpx = clinopyroxene, amph = amphibole, bi = biotite, tit = titanite, apa = apatite, epi = epidote, zr = zircone, pyt = pyrite, cc = calcite, clzoi = clinozoisite, cor = cordierite, sill = sillimanite, trem = tremolite, di = diopsid, chl = chlorite, hydrogar = hydrogarnet, gar = garnet, mu = muscovite, illm = illmenite, sap = sapporite, opm = opaque minerals.

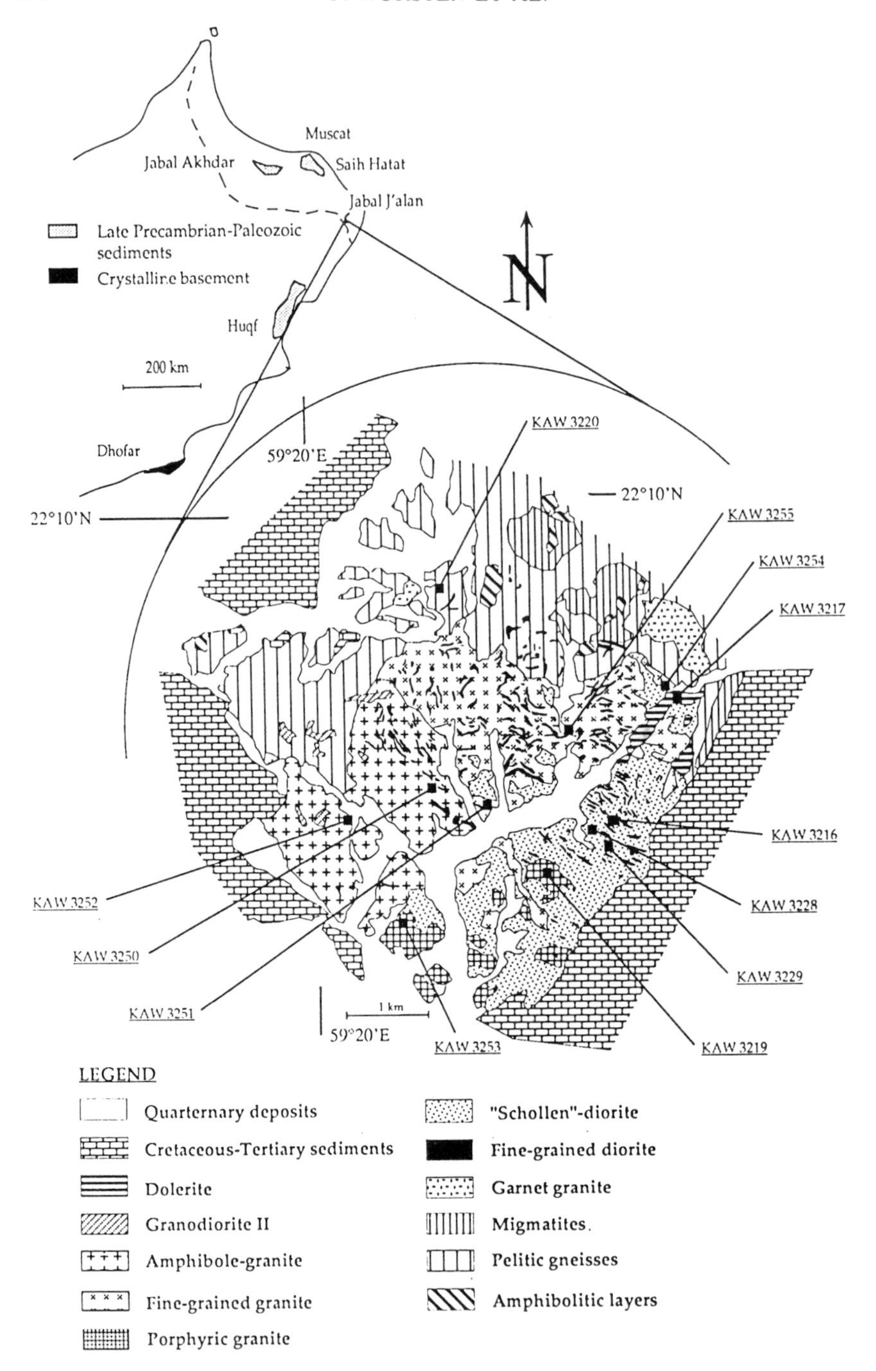

Figure 1. Geological map of the Jabal J'alan area with the location of the samples used for geochronology.

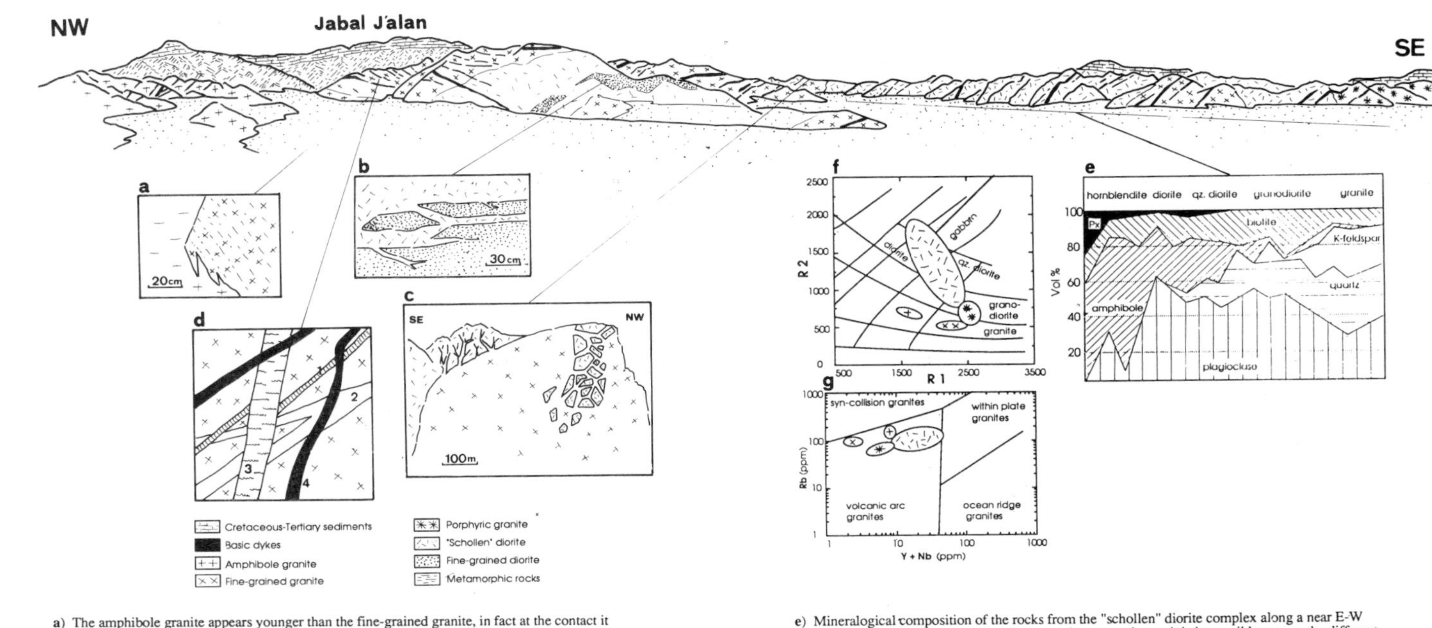

a) The amphibole granite appears younger than the fine-grained granite, in fact at the contact it clearly cuts across a weak foliation in the fine-grained granite.
b) As demonstrated by the mutual contact relationships, the "schollen" diorite is youger then the fine-grained diorite.
c) Stoping phenomena at the intrusive contact between fine-grained granite and "schollen" diorite.
d) Schematic crosscuting relationships between various basic dykes in a selected area of ca. 100 m^2. 1- Dark, fine-grained basaltic dykes (lamprophyres). 2- and 3- Basaltic to basaltic-andesitic dykes with cpx, plag and Kfsp phenocrysts. 4- Trachy-andesitic dykes.
Appart from few selected areas, the relative timing of the dykes is difficult to recognize, owing to their almost subparallel attitudes. All these dykes are younger than the aplites and pegmatites (not shown) related to the intrusion of the granites, and older than the large dolerite dyke (fig. 1).

e) Mineralogical composition of the rocks from the "schollen" diorite complex along a near E-W profile. As the changes in mineralogy are almost continous, it is impossible to map the different rock types, exept the porphyric granite, at the 1:20000 scale (fig.1).
f) The De La Roche diagram shows clearly two distinct magmatic series. The first series includes the fine-grained diorite, the "schollen" diorite and the porphyric granite, and represents the first magmatic event. The second series includes the fine-grained granite and the amphibole granite.
g) The Rb/Y+Nb diagram indicates for the whole igneous activity a possible link with a subduction environment.

Figure 2. Schematic representation of some important geological observations in the Jabal J'alan area.

The porphyric granite shows characteristic centimeter-sized K-feldspars. It is quite variable in grain-size, mineralogy and color. Continuous transitions from the "Schollen"-diorite to the porphyric granite (dioritic "Schollen" in porphyric granite near the contact) suggest that the porphyric granite could be the end-member of the dioritic suite. Geochemical data (Table 2, Fig. 2) also favor a genetic relationship to the diorites.

The amphibole granite and fine- grained granite comprise the large intrusive complex in the central part of the Jabal J'alan area (Fig. 1). The two granites show clear mineralogical and geochemical differences (Table 1 and 2, Fig. 2) outlining a probable different origin of the melts. The mutual contact relationships of these two bodies are unclear, but some field observations seem to demonstrate that the amphibole granite intrudes the fine-grained granite (Fig. 2).

The contact between these two granites and the "Schollen"-diorite is clearly intrusive. Different contact types can be distinguished: sharp boundary surfaces; granitic and pegmatitic apophyses injected into the "Schollen"-diorite; brecciated contacts with pieces of "Schollen"-diorite in the granite; big blocks of "Schollen"-diorite in the granite (roof pendants); intergranular infiltration of granitic melt into the "Schollen"-diorite. These various contact phenomena show clearly that the two granites are younger than the "Schollen"-diorite.

A large number of subparallel dykes (mainly basaltic and in smaller amounts of andesitic to rhyolitic composition) cut through the plutonic and metamorphic rocks. They make up about 25% of the mapped surface. These dykes are 1–10 m thick. Crosscutting relationships between the different dykes and geochemical investigations show a relative chronological evolution from basaltic to rhyolitic composition.

The large doleritic dykes (Table 1 and 2), outcropping in the eastern region, cut through all other rock types and, therefore, represent the latest intrusive bodies.

The crystalline complex is overlain in the Southern part of the Jabal J'alan area by Cretaceous-Tertiary sedimentary rocks. These rocks were deposited on a weathered surface. The basal unit is a fluviatile conglomerate with subrounded radiolarian cherts of the Hawasina formations and local crystalline blocks (Qahlah Fm). It is a fining-upward sequence with intercalations of coarse Cretaceous lime conglomerate. This fluviatile sequence was followed by a marine transgression represented by alternating nodular shallow-water limestones and laminated marls (Simsima Fm). These layers are very rich in fossils of Maastrichtian age, e.g. rudista, other bivalves, gastropods, corals, sponges, echinites and loftusies of Maastrichtian age.

In the Northern Jabal J'alan area, near Yistin, the BRGM team discovered two sedimentary series on top of the crystalline basement (Roger et al., 1990). At the base is 250–300 m of probable Late Proterozoic conglomerate, arkosic sandstone and shale, with a thin basaltic intercalation (Abu Mahara

	Fine-grained diorite	Horn-blendite	"Schollen"-diorite qz-diorite	"Schollen"-diorite granodiorite	Porphyric granite	Amphibole granite	Fine-grained granite	Dolerite
SiO_2	54.12	46.42	54.08	64.29	71.08	66.23	71.30	48.13
TiO_2	1.51	1.16	1.19	1.06	0.37	0.36	0.19	3.02
Al_2O_3	16.22	9.42	18.17	15.37	15.05	16.82	17.16	12.81
Fe_2O_3	11.20	10.22	8.07	8.58	2.52	2.49	1.30	16.84
MnO	0.19	0.15	0.12	0.15	0.04	0.04	0.02	0.27
CaO	7.09	11.69	6.16	3.26	2.13	2.57	1.64	9.08
Na_2O	3.30	1.84	4.29	3.47	3.94	4.47	4.98	2.69
K_2O	1.35	0.66	2.21	2.95	4.00	5.47	4.57	0.97
P_2O_5	0.27	0.20	0.47	0.48	0.14	0.19	0.09	0.52
LOI	0.85	5.81	1.54	0.67	0.82	0.70	0.54	0.56
Total	100.42	98.83	100.62	101.49	101.20	100.33	102.71	100.22
F	417	1067	929	538	433	565	486	643
Ba	371	223	850	1066	1271	3053	2307	559
Rb	27	6	44	68	66	105	116	15
Sr	295	389	1238	246	640	1974	1799	288
Pb	<5	<5	<5	11	5	17	40	<5
Th	<5	<5	<5	<5	<5	<5	<5	<5
U	<5	<5	<5	<5	<5	<5	<5	<5
Nb	<4	<4	<4	18	<4	<4	<4	<4
La	<20	<20	<20	39	<20	61	<20	33
Ce	32	66	66	75	45	97	33	68
Nd	30	42	40	35	<10	21	<10	38
Y	49	21	21	41	5	10	<2	43
Zr	202	85	195	285	152	269	145	232
V	259	181	218	74	47	47	23	448
Cr	53	323	62	6	8	14	<6	60
Ni	36	263	44	<3	5	9	5	35
Co	33	62	25	19	12	13	10	53
Cu	39	<3	51	<3	<3	<3	<3	35
Zn	110	100	92	119	43	33	24	139
Hf	<1	<1	1	<1	<1	5	<1	<1
Sc	35	41	21	29	3	4	<2	48

Fm.). This serie is overlain by 50 m of Jurassic shelf carbonate of the Sahtan Group.

The plutonic suite was not affected by strong deformation since its emplacement in the Precambrian. Neither foliations nor folding (particularly of the dykes) have been observed, but an intense jointing overprinted the complex. A steep strike-slip fault crosses the entire area in N-S direction. Sinistral displacement along this fault, shown mainly by offset dykes as well as by vertical displacement at the crystalline-sediment contact, does not exceed 100 m. A strong hydrothermal alteration (ca. 100 m thick) of the crystalline rocks (particularly the granites), characterized by calcite and barite veins, is related to this fault. As this fault also cuts the sedimentary rocks, it is clearly of Tertiary age.

Geochronology

Rubidium- strontium, potassium- argon and fission track data have been used to answer the following questions:

- when did the granodioritic to granitic plutons intrude?
- when did the basement complex of the Jabal J'alan cool down to about 300°C?
- when did the complex cool down to below 100°C for the last time?

Rb/Sr Whole Rock Data

An attempt to use Rb/Sr to date the intrusion age of the amphibole granite and the fine- grained granite failed. The requirements for the Rb/Sr whole rock dating technique were not fulfilled, as Sr isotope homogenization was not reached in the two types of granites. This inhomogeneity is also reflected by mineralogical and geochemical observations, which suggest different sources for the two magmas. The number of analyzed samples (three samples of each granite) was insufficient to date the two granite types separately.

K/Ar Data on Micas

One muscovite and five biotite K/Ar model ages have been calculated from six different rock types of the Jabal J'alan (Table 3a). Judging from the agreement between the mica K/Ar model ages from different rock types, the ages between 828 Ma and 815 Ma can be interpreted as cooling ages of the whole crystalline complex to the temperature range of about 350–300°C.

Muscovite has a slightly higher blocking temperature (350 ± 50°C) than the biotite (300 ± 50°C). Therefore the muscovite should show a slightly higher age. It does not, possibly because of the resolution limits of the K/Ar dating technique in the high age range (errors of about ±12 Ma for a single

Table 3a. K/Ar isotopic data.

Sample KAW Nr.	Rock type	Mineral	Fraction mesh	K wt%	40 K x1e-8 mol/g	40 Ar rad xle-9 mol/g	40 Ar rad %	40/36 Ar xle + 4	40 Ar/ 36 Ar xle + 5	Age Ma ±2 sigma
3216	fine-grained diorite	biotite	>130	6.54	19.52	11.70	99.6	8.094	13.458	816 ± 11
3218	"Schollen"-diorite	biotite	60–90	6.26	18.68	11.41	99.7	10.260	16.750	828 ± 11
3219	porphyric granite	biotite	60–90	5.89	17.58	10.63	99.5	6.202	10.208	822 ± 12
3220	metapelite	biotite	40–80	6.36	18.98	11.36	99.5	5.470	9.905	815 ± 11
3229	muscovite pegmatite	muscov.	40–60	9.13	27.25	16.32	99.7	10.078	16.780	815 ± 12
3217	dolerite	biotite	>100	4.68	13.97	7.51	99.0	2.872	5.285	747 ± 11

Table 3b. Analytical data of the fission tracks in apatite.

Sample	Mineral ρ_s (N_s) crystals	Spontaneous ρ_i (N_i)	Induced or	s'%	Irrad	Glass	Dosimeter ρ_d (N_d)	Age Ma	Mean length ±st. deviation (number of tracks)
KAW 3250	apatite (200/200)	16.1 (1021)	25.4 (1605)	3	BeH16	612	5.18 (3771)	73.1 ± 5.3	11.2 ± 2.2 (60)
KAW 3253	apatite (48/48)	13.0 (198)	10.15 (154)	12	BeH16	612	5.16 (3771)	146 ± 18	10.8 ± 3.0 (81)
KAW 3254	apatite (150/150)	4.9 (932)	18.2 (3462)	5	BeH16	612	5.14 (3771)	30.8 ± 3	12.0 ± 2.6 (49)

Notes:
–track densities (ρ) are as measured and ($\times 10^5$ tr cm^{-2});
–all ages calculated with zeta-612 = 223
–for population method analyses, relative standard error of mean track count (s') is shown.

age determination). A slight disturbance of the K/Ar system is also possible and could be explained by the moderate chloritization of the separated biotite. Chloritization is also responsible for the relatively low K content of the biotite concentrates and it could result in ages that are slightly too old or slightly too young for cooling to the blocking temperature.

The biotite sample from the dolerite KAW3217 shows a significantly lower age of 747 ± 11Ma. The dolerite is the youngest magmatic rock in this area on the basis of intrusive relations. The K content of this biotite concentrate is also very low because of the strong chloritization. It can not be determined if the younger biotite age is a true cooling age or if it is related to the strong chloritization and represents a partly rejuvenated age.

Fission Track (FT) Data on Apatite

The FT count data for 3 apatite samples are presented in Table 3b. All errors are 1 standard deviation and typically ± 5–10%. The ages vary between 30–146 Ma. The mean of track length measurements ranges from 10.8–12.0 μm with standard deviations of 2.2–3.0 μm. The length distributions are typical of mixed ages.

Because all FT apatite ages are mixed ages, the three calculated ages must be interpreted only as limiting ages. This means that the crystalline basement has suffered renewed burial down to the 60° and 100° isotherms (2 to 4 km sedimentary cover) in the time between the Jurassic (146 Ma.) and the Oligocene (30 Ma.).

The interpretation of the regional variation of the ages in terms of differential uplift of the basement is not conclusive, because of the small number of dated samples. Nevertheless, an E-W tilting of the crystalline block could be in good agreement with the N-S to NNW-SSE trending Tertiary structures (Filbrand et al., 1990).

Radiometric Ages from the Literature

Glennie et al. (1974) report a K/Ar age of 872 ± 17 Ma on biotite from a large dyke of biotite granite and an Rb/Sr biotite and whole rock age of 848 ± 15 Ma from the same sample. The fact that the K/Ar cooling age (300°C) is older than the Rb/Sr age (higher closing temperature) make it difficult to interpret these data.

Pallister et al. (1990) report two $^{207}Pb/^{206}Pb$ model ages of 803 and 823 Ma for zircons from a biotite tonalite. These two zircon populations give an U-Pb upper intercept of the concordia at 834 ± 6 Ma. Judging from its field location (Pallister, pers. comm. 1990), the dated biotite tonalite belong to our dioritic complex. The BRGM team (Le Métour pers. com.) has found similar ages around 815 Ma for zircons from different magmatic lithologies. The tendency of these zircon ages and our K/Ar ages to group around 820 Ma, within an error of ±10 Ma, can be interpreted as the time of the cooling of the whole crystalline complex to temperatures between 600°C and 300°C, while the 834 Ma U-Pb age (Pallister et al., 1990) could represent the intrusion age of the diorites.

Much more difficult to interpret are the ages given by Gass et al. (1990). The Rb/Sr whole rock isochron for the "igneous suite I", defined by Gass et al., (1990, p. 593) as "basic horblende-rich rocks, diorites, granodiorites, biotite granites and pink granites", yields an age of 850 ± 27 Ma, could represent the crystallization age of the igneous complex as interpreted by the authors. Unfortunately, many of the rocks used to construct the isochron have intrusive contacts which clearly demonstrate a relative age succession of the intrusions. Even if the time lapse between the different intrusions is very small (e.g. a few million years), it seems questionable to assume an overall Rb/Sr homogenization for rocks of probably different magmatic origin, as indicated by our problematic results for the Rb/Sr system. In this sense, the age of 850 Ma could be the result of a pseudo-isochron.

Gass et al. (1990) report another Rb/Sr whole rock isochron for a "suite II" defined as follows: "six samples from loose boulders in the wadi fill of strongly banded and deformed igneous rocks" (Gass et al., 1990, p. 593). This isochron yields an age of 773 ± 34 Ma and is interpreted by the authors as "the age of the deformation and amphibolite-facies metamorphism which produced the banding in the rocks."(Gass et al., 1990, p. 593) From the Jabal J'alan rocks, to our knowledge, only the migmatites fit the given description of the "suite II" rocks. The migmatites, together with the gneisses and the amphibolites, form the country rocks in which the diorites and granites intruded. Therefore, the "igneous suite I" cannot be older than the metamorphic "suite II", in which it is supposed to be intrusive. The proposed interpretation is also in contradiction with the fact that the "igneous suite I" is not affected by the 100 Ma younger amphibolite-facies metamorphism dated by "rock suite II" and that this metamorphic event does not reset the K/Ar ages.

The Uplift History

As discussed above, the preliminary Rb/Sr data on the granitic rocks do not allow the definition of the age of the intrusion. From the geological and petrographic relations we tend to relate the peak of metamorphism and the slightly younger magmatism to the same orogenic event active before 820 Ma.

The concordant K/Ar cooling ages on biotites of around 820 Ma (Table 3a) for all rock types of the Jabal J'alan area (metamorphics and plutonics), indicate that the whole complex reached the 300°C isotherm at the same time. This means further that the crystalline basement has not experienced an extensive reheating since 820 Ma. This fact has an important consequence. The controversial "Hercynian" metamorphism in the pre-Permian series of the Saih Hatat and Jabal Akhdar (Glennie et al.1974, Michard 1982, Rabu 1987, Le Métour 1987, Man and Hanna, 1988) is not detectable in the Jabal J'alan. Assuming that the Jabal J'alan belongs to the same basement as Saih Hatat and Jabal Akhdar, this result seems to support the position of Rabu (1987) and Le Métour (1987) who dispute the existence of this event. Otherwise one would have to accept that this metamorphism was very weak (<300°C), at least in this area. The lack of obvious penetrative deformation of the plutonic suite and of the dykes after their emplacement in the Precambrian, corroborates the hypothesis of a weak "Hercynian" orogenic event.

After the uplift through the 300°C isotherm at 820 Ma, the crystalline rocks reached the surface in Late Proterozoic times and were covered by continental, mainly detritic, sediments.

An attempt to define the uplift through the 200°C isotherm with fission tracks on zircons has failed due to the intense metamictization of the zircon grains (a further indication that the granites never again reached the cooling temperature of the zircons after their emplacement).

The fission tracks on apatites from the granitoids yield mixed ages. Nevertheless, the 146 Ma age sets a lower limit for the cooling down to 100°C. This means that at least in the Upper Jurassic the presently exposed crystalline rocks were at 3–4 km depth, but it is impossible to say if cooling to 100°C had already occurred in the lower Mesozoic or even in the Paleozoic. The Jurassic age agrees with the occurrence of Jurassic carbonates on top of the crystalline rocks and, on the regional scale, with the marine transgression on the Huqf and adjacent areas in Bathonian times (Morton, 1959; Murris, 1980). From the K/Ar cooling ages and from the fission tracks on apatite, it can at least be observed that, from the Late Proterozoic to the Upper Jurassic, the thickness of the sedimentary cover must have varied between a minimum of 3 to a maximum of 10 km, assuming an average geothermal gradient of 30°C/km.

During the Upper Cretaceous the Hawasina nappes were thrusted over the crystalline basement and its Late Proterozoic and Jurassic cover. This tectonic thickening may have caused a first partial annealing of the fission tracks in apatite.

The basal sedimentary unit unconformably overlying the weathered crystalline rocks in the Southern Jabal J'alan area is a fluviatile conglomerate with radiolarian cherts of probable Hawasina origin, indicating that shortly after the emplacement of the Hawasina nappes the Jabal J'alan block began to subside. A marine transgression followed immediately with deposition of the Maastrichtian units.

From this geological evidence one would expect the apatite fission track ages to reflect the Maastrichtian event and show coherent values around 80–75 Ma. This is not the case, and therefore, it is necessary to call on a second burial of the crystalline rocks in order to reset the ages. The incomplete resetting of the fission tracks indicates that the temperatures did not exceed

60–100°C for a long period of time, which implies a burial within the range of 2–3 km. In "Report on geological survey of the Sultanate of Oman, Sur area" of the Metal Mining Agency of Japan (1982) the thickness of the Maastrichtian-Tertiary sequence in this area is estimated at 2000 m. Similar thicknesses (2500 to 3000 m) are reported by Roger at al. (1990) for the Northern Jabal J'alan area. However, South of the North Jabal J'alan Fault (Filbrand et al., 1990) the Maastichtian is only about 200 m thick, with an eroded top, and the Tertiary (Lower Eocene) is reduced to 100 m (Roger et al., 1990). The Middle-Upper Eocene and the Oligocene seem to have been eroded subsequently. Under these conditions it is difficult to estimate the thickness of the Tertiary cover; nevertheless, taking into account the Miocene erosion, the required thickness of 2–3 km (60–100°C) could be realistic. The youngest apatite age (31 Ma) is an indication of the definitive uplift through the 100°C isotherm during the Oligocene, which is consistent with the mentioned erosion of the Upper Eocene and Oligocene sediments.

The last stage of this history consists of the exhumation of the crystalline complex during the Miocene and Pliocene, leading to the present day configuration.

Acknowledgements

Mohamed Kassim, director general of Minerals and Dr. Hilal al Azri, director of Geological Surveys from the Ministry of Petroleum and Minerals, Muscat, Sultanate of Oman, are thanked for their permission to publish and for their co-operation and assistance. J.S. Pallister and A. Kröner are thanked for their constructive criticism of the paper. Furthermore, we are indebted to the "Stiftung zur Förderung der wissenschaftlichen Forschung an der Universität Bern" for their financial support.

References

Filbrand, J.B., Nolan, S.C. and Ries, A.C., 1990. Late Cretaceous and early Tertiary evolution of the Jebel Ja'alan and adjacent areas, NE Oman. In: A.H.F. Robertson, M.P. Searl and A.C. Ries (Eds), The Geology and Tectonic of the Oman Region., Geol. Soc. Spec. Publ. No. 49: 585–599.

Gass, I.G., Ries, A.C., Shackleton, R.M. and Smewing J.D., 1990. Tectonics, geochronology and geochemistry of the Precambrian rocks of Oman., pp. 697–714. In: A.H.F. Robertson, M.P. Searl and A.C. Ries (Eds), The Geology and Tectonic of the Oman Region. Geol. Soc. Spec. Publ. No. 49.

Glennie, K.W., Beouf, M.G.A., Hughes Clark, M.W., Moody-Stuart, M., Pilaar, W.F.H. and Reinhard, B.M., 1974. The geology of the Oman mountains., Konin. Neder. Geol. Mijnbouw. Genoot. Verdh. 31.

Le Métour J., 1987. Géologie de l'Autochtone des Montagnes d'Oman: La fenêtre du Saih Hatat. Thèse Doct. d'Etat, Université P. et M. Curie, Paris VI, Documents du B.R.G.M. No. 129.

Mann, A. and Hanna, S., 1988. The tectonic evolution of pre-Permian rocks, central Oman mountains., Intern. Oman Conf. Edinburgh, March 1988, abstr. vol., p. 31.

Mercolli, I., Peters, T., Mattäus, U. and Würsten, F., 1989. Oman as a test region for tectonic and petrologic models: the crystalline basement of the Jabal J'alan (Sur area)., Schweiz. mineral. petrogr. Mitt. 69: 135–139.

Michard A., 1982. Contribution à la connaissance de la marge nord du Gondwana: une chaîne plissée paléozoic, vraisemblablement hercynienne, en Oman., C. R. Acad. Sci., Paris, 295: 1031–1036.

Morton, D. M., (1959). The geology of Oman. Proc. 5th World Pet. Congr., New York, Section I, Paper 14: 227–280.

Murris, R. J., 1980. Middle East: stratigraphic evolution and oil habitat., A.A.P.G. Bull. 64: 597–618.

Pallister, J.S, Cole, J.C., Stoesser, D. B. and Quick, J. E., 1990. Use and abuse of crustal accretion calculations., Geology, 18: 35–39.

Rabu, D., 1987. Géologie de l'Autochtone des Montagnes d'Oman: La fenêtre du Jabal Akhdar. Thèse Doct. d'Etat, Université P. et M. Curie, Paris VI, Documents du B.R.G.M. No.130.

Roger, J., Bechennec, F., Janjoud, D., Le Métour, J., Wyns, R and Beurrier M., 1990. Geological map of J'alan, sheet NF 40–8E, scale 1:100.000, Explanatory Notes: Directorate General of Minerals, Oman Ministry of Petroleum and Minerals.

Meta-Carbonatites in the Metamorphic Series Below the Semail Ophiolite in the Dibba Zone, Northern Oman Mountains

U. ZIEGLER, F. STÖSSEL and TJ. PETERS
Min.-petr. Institute, Bern University, Baltzerstr. 1, CH-3012 Bern, Switzerland

Abstract

Within the metamorphic series below the Semail Ophiolite, meta-carbonatites occur in a meta-volcanic unit in the Dibba region of the United Arab Emirates (UAE). The meta-carbonatites occur as flows, sills or constituents of meta-tuffites. An intimate association with alkaline basaltic meta-pillow lavas, meta-hyaloclastites, meta-pillow breccias, meta-tuffites, meta-radiolarian cherts and meta-siliceous carbonates can be observed. The succession represents a suite of subaerial and subaquatic volcanic rocks deposited together with deep sea sediments.

The meta-carbonatites are geochemically characterized by high P_2O_5 contents and Ce, La and Nd contents between 500 ppm and 15000 ppm. Compared to N-type MORB, these rocks are strongly enriched in LIL-elements. Chondrite normalized REE patterns show a strong enrichment in light REE compared to heavy REE. With the exception of the meta-tuffites, where orthite is the major REE carrying phase, the REE are favouredly enriched in apatite. The initial $^{87}Sr/^{86}Sr$ ratios between 0.7037 and 0.7043 confirm a mantle source for the analyzed meta-carbonatites. The Sm-Nd isotope data reflect $\epsilon_{Nd(T)}$ values below those commonly observed in magmatic rocks deriving from depleted mantle sources and thus imply a source in a low ϵ_{Nd} subcontinental mantle. A comparative sedimentary carbonate sample yields $\epsilon_{Sr} - \epsilon_{Nd}$ characteristics strongly differing from the meta-carbonatites and, furthermore, indicates the presence of a 1.68 Ga old source region.

The meta-volcanics associated with the meta-carbonatites are considered to be the metamorphic equivalents of the mid-Triassic non-metamorphic alkaline volcanic rocks from the Haybi complex and the Umar group. They probably represent magmas of volcanism related to the Triassic rifting of the Neotethys along the passive eastern margins of the Arabian platform. They possibly formed in a transition zone of continental to oceanic crust similar to the situation of the Canary Islands, where carbonatites are encountered at the margin of the African continent.

Tj. Peters et al. (Eds), Ophiolite Genesis and Evolution of the Oceanic Lithosphere, 627–645.

Introduction

The presence of ankaramites and other volcanic rocks of alkaline affinity in the Middle Triassic lavas of the Umar Group (Béchennec, 1987) or in the Haybi volcanics (Searle et al., 1980) was used as an argument for a Permo-Triassic rifting of the Neotethys east of the Arabian continent. During an exploration campaign in the Dibba area (Allemann and Peters, 1972) large occurrences of apatite- and magnetite-rich carbonatitic rock associated with greenschist facies meta-cherts and meta-volcanics were detected.

Detailed fieldwork and geochemical investigations confirmed the existence of meta-carbonatites in this area and revealed that the whole rock sequence was deposited partly in a subaquatic, oceanic environment, but partly also in a subaerial environment. The occurrence of carbonatites within an oceanic environment, however, is rather seldom and has been described before only by Allègre et al. (1971) for the Cape Verde and Canary Islands. It implies that carbonatites can also form outside intracontinental rift zones.

The objective of this study is firstly, to provide new information about the chemical and isotopic characteristics of these carbonatites and their source region and, secondly, to derive their geotectonic position.

Geology

The meta-carbonatites occur between Wadi Uyaynah and Wadi Shimal in a meta-alkaline volcanic rock unit which itself is part of the larger greenschist facies metamorphic complex in the Oman mountains (Fig. 1). This meta-alkaline volcanic rock unit is separated from this complex to the west by an eastward dipping thrust plane and to the east by a major fault belonging to the large Dibba fault system. This fault brings the Semail ophiolite side by side with the meta-volcanic rock unit (Fig. 1).

Some meta-cherts occur within the meta-alkaline volcanic rock unit, whereas the other metamorphic unit mainly consists of meta-cherts with only subordinate meta-volcanic rocks and meta-carbonates. In the meta-alkaline volcanic rock unit, the meta-pillow lavas, meta-pillow breccias, meta-hyaloclastites and meta-tuffites show well preserved volcanic textures alternating with red and green quartzites (meta-radiolarian cherts) and occasional marly metacarbonates. Interbedded in this sequence are four stringers of dark-brown, weathered meta-carbonatite bodies, standing out due to their greater resistance to weathering. Some of them can be followed over several hundreds of meters (Fig. 1). A cross-section E of Wadi Uyaynah (Fig. 2) depicts the close relationship of the meta-carbonatites with meta-tuffites and meta-hyaloclastites. Some of the meta-carbonatites are tuffaceous, others, however, seem to be flows. The major part of the extrusive meta-carbonatites seems to have formed under subaerial conditions. This is also confirmed by the observations of Woolley et al. (1990).

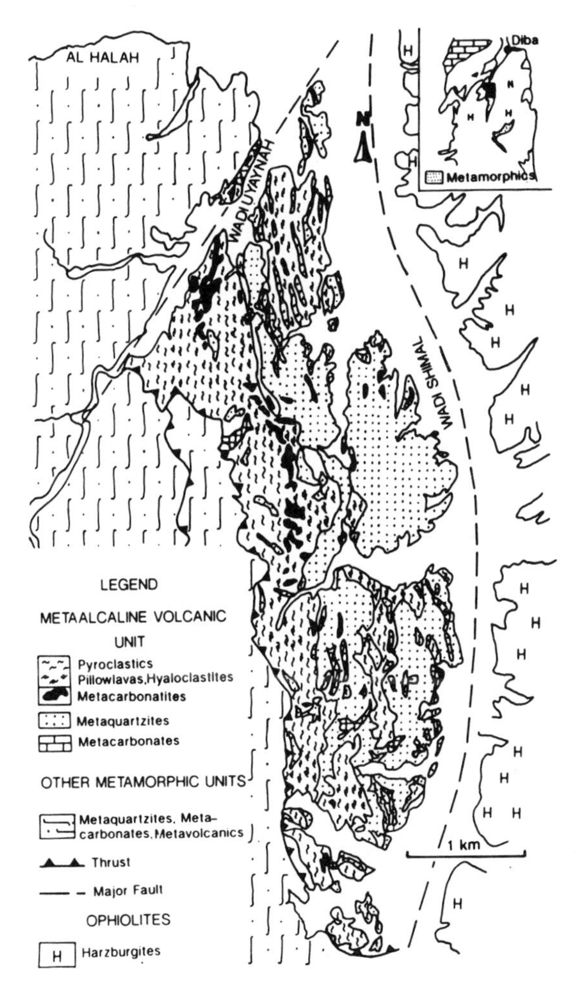

Figure 1. Geologic sketch map of the meta-alkaline volcanic unit with the distribution of the metacarbonatite occurrences.

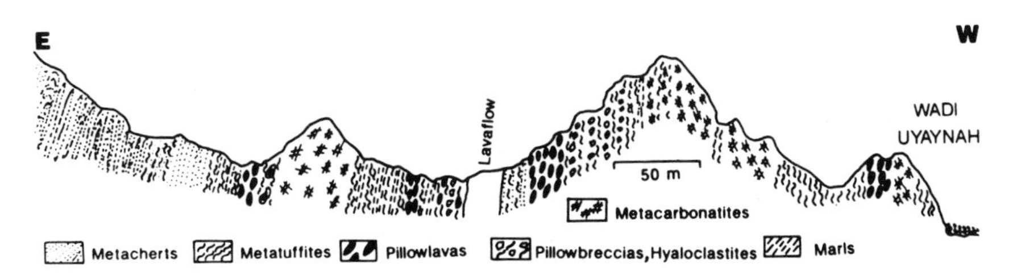

Figure 2. Cross section through the meta-alkaline volcanic unit East of Wadhi Uyaynah.

Petrology and Mineralogy

Metacarbonatites

The more massive meta-carbonatites display a porphyric texture with phenocrysts of apatite and magnetite in a fine-grained calcite matrix. Apatite occurs in the shape of euhedral prisms of up to 1 cm length and is normally broken with cracks healed by calcite and chlorite. Other primary minerals in the massive carbonatites are orthite, green amphibole and sphene (?). In the more tuffaceous carbonatites Ti-rich sodic amphiboles, plagioclase and biotite are common. Chlorite, crossite, talc, stilpnomelane, epidote and sphene are secondary products of the metamorphism. Although a major part of the carbonate material has been recrystallized during metamorphism, it can be inferred from the distribution of the iron hydroxide pigment that the original carbonatite was rather fine-grained. Apart from the more tuffaceous type, modal analyses of the carbonatites yielded the following composition: calcite 50–65%, apatite 15–30%, magnetite 10–15%, silicates 4–15%. The more calcitic carbonatites continuously grade into the more silicate-rich tuffites. In one case, remnants of brown magnesio chromite were observed, similar to those described by Bailey (1989) in the Rufunsa carbonatite volcanoes.

Mineral Chemistry

Three silicate-poor and two silicate-rich carbonatites were investigated using the electron microprobe in order to determine their mineral composition and mainly to see which minerals were the bearers of the rare earth elements (REE).

The *calcites* of the carbonate-rich carbonatites that texturally look like primary magmatic phases show a limited range of solid solution: $Ca_{0.94-0.98}Mg_{0.01-0.02}Fe_{0.05-0.02}Mn_{0.0-0.02}Sr_{0.0-0.02}CO_3$. The recrystallized calcites which probably formed during metamorphism are either 100% pure calcite or also display a limited range of solid solution: $Ca_{0.92-0.98}Mg_{0.0-0.06}Fe_{0.005-0.02}Mn_{0.0-0.02}Sr_{0.0-0.004}CO_3$.

Among the *Mg-calcites* the one that was analyzed yielded a composition of $Ca_{0.48}Mg_{0.41}Fe_{0.01}Sr_{0.10}CO_3$, which either indicates an extremely high Sr solid solution or an intergrowth with extremely fine-grained *strontianite*.

In all the analyzed carbonates the REE contents were below the detection limit of our Electron Microprobe (200–300 ppm) but also below the detection limit (a few ppm) of the cathode luminescence spectrometry.

The *magnetites* show a wide range of solid solution for ulvo-spinel ranging from 0.1% to 36 mol.% and a limited range of 0 to 9 mol.% for jacobsite. The microscopically observed martitizations at the borders and cracks of the grains chemically appear as somewhat more magnetite-rich solid solutions. No REE were detected in the magnetites. In the almost pure fluor-*apatites* the REE are concentrated as could be seen under the cathode luminescence microscope with attached spectrometer.

The brown *magnesio chromite* in one sample has a similar composition to those described by Bailey (1989) and was interpreted to be indicative of a mantele origin for the carbonatites.

Orthite is the main bearer of the REE in the tuffaceous meta-carbonatites. The structural formula has been calculated on the base of 25 oxygen from the mean chemical analyses: $(Ca_{2.05}Ce_{0.87}La_{0.54}Nd_{0.18})_{3.64}$ $(Fe_{3.00}Mg_{1.03})_{4.03}Al_{2.09}(Si_{5.77}Al_{0.23})_{6.00}$ $O_{24}(OH)$ is only remarkable with respect to its huge Mg content. The following range of REE concentrations have been determined: Ce_2O_3 11.8–14.2%, La_2O 6.4–9.9%, Nd_2O_3 2.4–3.2% and Sm_2O_3 0–0.38%.

The one *monazite* specimen shows the following REE concentrations: Ce_2O_3 27%, La_2O 15% and Nd 7.3%.

Some *sphenes* contained appreciable amounts of Ce. However, they seem to have formed during metamorphism.

Metavolcanic Rocks

Due to different cooling rates, the primary textures of the pillow lavas, breccias and hyaloclastites are extremely variable. Plagioclase, Ti-rich alkali-amphibole and zoned sodium-rich clinopyroxene were observed as phenocrysts. In the groundmass only primary plagioclase, Ti-magnetite and ilmenite are preserved. Albite, chlorite, muscovite, epidote, calcite, sphene, green biotite and Mg-crossite occur as secondary minerals. They indicate the quartz-albite-biotite subfacies of the greenschist facies metamorphism.

Metacherts, Metacarbonates and Metapelites

The meta-cherts are composed of almost monomineralic, fine-grained granoblastic quartz alternating with layers rich in muscovite and chlorite. Continuous transitions to more pelitic protoliths are characterized by an increase in white mica and chlorite content. In some layers actinolitic hornblende appears. Stilpnomelane enrichments occur mainly along shear zones. Epidotes, grossular-rich garnet, tourmaline, magnetite and sphene are common accessories. Apart from talc formation no other mineral reactions could be observed in the carbonate-rich layers.

Major and Trace Element Geochemistry

Major and trace elements were analyzed by X-ray fluorescence spectrometry (XRF) according to Nesbitt et al. (1979). CO_2 was determined coulometrically. The REE La, Ce and Nd have been analyzed by XRF, Sm, Er, Tb and Yb by neutron activation techniques (Krähenbühl, 1985). The results are presented in Tables 1 and 2. Due to the very high trace element contents in these rocks, some of the major element totals are rather low. The samples richest in carbonate only contain about 2 wt.% SiO_2. Some of the tuffs contain up to 40 wt.% SiO_2. The very low alkaline content of the massive carbonatites indicates that originally they did not contain alkali-rich silicates such as aegirine augites. The tuffaceous types, however, still contain phlogopite and sodium amphiboles. Among the trace elements, Ba and Sr contents are very high. Ba reaches up to 1 wt.%. Ba and Sr concentrations vary considerably, but no relation with either phosphate or carbonate content could be detected, nor is there any inverse relationship as described by Barber (1974). The analyses were normalized to N-type MORB and presented in a spider diagram (Pearce, 1982, 1983; Fig. 3). The trace element patterns are almost identical in shape, indicating a common magmatic evolution for the carbonatites and the more siliceous tuffs. They display similar "humps" as within plate basalts originating from sub-continental lithosphere (Pearce, 1983). A similar plot of average carbonatite composition from data of Heinrich (1966) is congruent with the UAE carbonatites.

The very high REE concentrations as well as the strong light REE enrichment (Fig. 4) are also characteristic for carbonatitic rocks (Sun and Nesbitt, 1977; Nelson et al., 1988).

Rb-Sr and Sm-Nd Isotope Systematics

Rb-Sr and Sm-Nd studies were carried out on four meta-carbonatites and one metasedimentary carbonate sample of a mass of 1–2 kgs. An aliquot of each of the samples was completely dissolved in a mixture of hydrofluoric and perchloric acid. In addition, the carbonaceous material of every sample was leached from another aliquot using 0.1 normal hydrochloric acid. The Rb-Sr concentrations and the Sr isotopic compositions were determined using the isotope dilution method according to Jäger (1979). The Sm-Nd concentrations and the Nd isotopic compositions were determined according to the method described by Stössel and Ziegler (1989) using a two column element separation procedure. The Rb analyses were carried out on an "Ion Instruments" solid source mass spectrometer while elemental Sr, Sm and Nd measurements were carried out on a "VG Sector" thermal ionization mass spectrometer by single and triple filament modes, respectively. The $^{87}Sr/^{86}Sr$ ratios are normalized to $^{88}Sr/^{86}Sr = 0.1194$. The mean $^{87}Sr/^{86}Sr$ ratio for 7 measurements of the NBS SRM-987 standard is 0.71022 ± 3 (ϵ). The $^{143}Nd/^{144}Nd$ ratios are normalized to a $^{146}Nd/^{144}Nd$ ratio of 0.7219. The mean value for 14 analyses of the La Jolla isotopic standard was 0.511843 ± 15 (ϵ). Total blanks were below 1 ng for Sm, 0.01–0.1 ng for Nd and below 10 ng for Sr. The results of the Sr analyses are listed in Table 3 while those of the Sm-Nd analyses are given in Table 4.

The Rb concentration of the meta-sedimentary carbonate rock is 0.22

Major and Trace Elements	F 94	F 24	F 67–87c	UAE 4759	F 23	UAE 4776	F 93	UAE 4748	UAE 4020	UAE 4714	UAE 4757
SiO_2	1.84	2.23	2.72	3.84	4.67	5.93	6.38	6.69	7.97	9.41	10.49
TiO_2	0.74	0.89	0.82	0.34	0.89	0.28	1.29	1.04	0.72	1.33	1.00
Al_2O_3	1.04	1.22	1.26	0.83	1.71	0.81	2.12	2.50	2.09	2.02	2.27
Fe_2O_3	9.46	10.29	10.82	1.66	13.30	2.14	16.36	10.80	7.40	10.37	7.84
FeO	n.d.	n.d.	n.d.	4.37	n.d.	1.88	n.d.	2.40	2.50	2.15	2.16
MnO	0.39	0.34	0.38	0.78	0.52	0.49	0.66	0.44	0.57	0.41	0.81
MgO	1.21	1.69	2.53	12.78	2.96	3.16	5.26	2.61	2.74	5.32	3.96
CaO	43.66	42.33	40.95	32.03	39.14	44.76	34.96	39.77	40.80	36.66	34.28
Na_2O	0.01	0.01	0.01	0.48	0.02	0.54	0.01	0.83	0.66	0.47	0.49
K_2O	0.01	0.01	0.01	0.05	0.03	0.03	0.02	0.08	0.09	0.03	0.12
P_2O_5	6.09	6.15	6.73	3.29	8.45	4.01	8.61	7.03	6.34	7.17	4.90
H_2O	n.d.	n.d.	n.d.	2.05	n.d.	1.03	n.d.	1.27	1.73	1.67	0.99
CO_2	29.60	28.84	27.53	36.70	23.11	33.00	20.03	23.70	25.40	21.80	23.60
Total	94.05	94.00	93.76	99.20	94.80	98.04	95.70	99.16	99.01	98.80	92.91
Nb (3)	291	245	315	149	243	377	332	309	209	491	744
Zr (4)	934	965	1175	682	1170	768	1503	1038	874	1230	1151
Y (4)	119	104	119	118	141	77	169	132	109	135	257
Sr (4)	1597	3072	3300	2490	2822	9350	3367	1699	2670	3660	8524
U (7)	22	32	27	37	42	61	44	17	20	33	119
Rb (4)	9	11	14	10	12	16	14	12	9	11	15
Th (6)	41	39	38	12	46	7	64	23	18	24	103
Pb (6)	<5	<5	<5	7	<5	18	<5	<6	<6	13	31
Ga (2)	<2	<2	<2	<2	<2	<2	5	7	<2	10	6
Zn (5)	243	236	262	216	324	104	425	354	253	361	656
Cu (7)	<3	<3	<3	<7	<3	25	<3	<7	<7	9	27
Ni (7)	42	26	34	64	46	42	46	71	67	45	65
Co (6)	32	33	41	43	44	42	39	68	61	59	45
Cr (10)	<6	6	21	44	37	<10	35	59	43	<10	67
V (3)	29	26	56	15	35	30	48	56	70	170	62
Ce (23)	807	700	938	1218	1110	545	1108	1032	727	1217	8859
Nd (12)	390	337	446	615	534	287	572	615	434	624	3098
Ba (2)	63	263	573	126	315	195	225	966	426	291	3523
La (12)	467	340	504	664	506	270	572	451	330	638	4400
Sc (1)	13	9	14	11	17	14	16	25	23	20	3

Table 1. Continued.

Major and Trace Elements	UAE 4792	UAE 4729	UAE 4713	UAE 4712	UAE 4742	UAE 4716	UAE 4743	UAE 4725	UAE 4761	UAE 4790	UAE 4744
SiO_2	11.29	12.09	14.48	14.43	15.32	16.87	16.98	17.07	17.36	19.77	42.95
TiO_2	0.83	1.64	1.09	1.44	0.18	1.41	0.23	0.92	1.99	2.12	0.93
Al_2O_3	1.91	2.35	2.07	3.00	1.41	2.93	1.78	2.22	4.44	3.99	3.22
Fe_2O_3	4.77	14.59	11.16	5.63	3.05	15.62	3.59	3.66	10.66	9.42	4.49
FeO	2.39	2.46	2.43	3.28	1.93	3.02	2.11	3.64	2.72	2.43	6.18
MnO	0.67	0.86	0.54	0.52	0.64	0.61	0.52	0.52	1.51	0.46	0.69
MgO	3.05	11.69	10.73	9.87	5.28	15.06	5.50	4.66	4.91	4.90	12.43
CaO	40.47	26.27	29.04	29.24	33.47	22.11	34.65	35.88	25.97	29.58	14.41
Na_2O	0.71	0.38	0.35	1.34	0.81	0.32	0.77	1.47	0.44	0.45	1.73
K_2O	0.50	0.03	0.03	0.07	0.06	0.04	0.06	0.17	1.12	0.28	0.09
P_2O_5	3.46	7.62	8.10	3.52	5.03	11.95	3.42	3.43	5.13	3.80	4.82
H_2O	0.60	2.22	2.18	2.39	1.00	5.04	0.52	1.52	2.17	3.01	2.07
CO_2	28.70	16.90	16.80	24.40	21.80	4.00	22.80	23.80	16.20	19.60	0.50
Total	99.35	99.10	99.00	99.13	89.98	98.98	92.93	98.96	94.62	99.81	94.51
Nb (3)	336	260	388	446	1023	112	775	319	753	206	922
Zr (4)	837	1551	1308	378	199	1591	251	1162	1315	831	466
Y (4)	136	147	156	115	453	222	365	136	223	79	440
Sr (4)	1579	2183	3429	2418	3229	2563	2645	2682	4601	1037	1964
U (7)	26	28	34	26	<7	29	11	27	80	22	41
Rb (4)	23	12	11	10	<4	13	5	17	35	24	<4
Th (6)	23	35	34	39	263	44	183	23	90	11	238
Pb (6)	15	<6	<6	10	73	<6	64	<6	27	<6	57
Ga (2)	<2	11	12	6	<2	16	<2	4	11	10	5
Zn (5)	252	381	343	286	375	440	369	168	586	225	622
Cu (7)	<7	<7	8	26	18	13	15	48	17	11	21
Ni (7)	147	59	47	310	295	57	327	50	93	79	371
Co (6)	48	50	45	62	43	57	44	52	57	69	54
Cr (10)	82	45	33	320	407	57	386	25	95	78	510
V (3)	47	24	43	154	43	38	45	125	135	172	96
Ce (23)	900	1275	1362	1757	12525	2306	8291	827	8004	606	1005
Nd (12)	423	706	769	747	3554	1270	3123	433	2259	328	3434
Ba (2)	185	505	363	1112	10789	86	10349	530	3474	893	1652
La (12)	384	452	674	792	7615	1024	4242	383	5095	301	7141
Sc (1)	11	17	18	16	9	21	11	12	8	17	11

Table 2. Rare Earth Element concentrations in ppm of the metacarbonatites compiled from neutron activation (*), XRF (**) and mass spectrometry (***) data.

Sample	La	Ce	Nd	Sm	Eu	Tb	Yb
UAE 4714	638**	1217**	624**	72*	20*	8.7*	19*
UAE 4716	1024**	2306**	1270**	115*	32*	15*	19*
UAE 4742	7615**	12525**	4935***	631**	58*	55*	78*
UAE 4744	7147**	11005**	3434**	395*	76*	61*	78*
UAE 4761	5095**	8004**	2128***	249***	40*	19*	118*

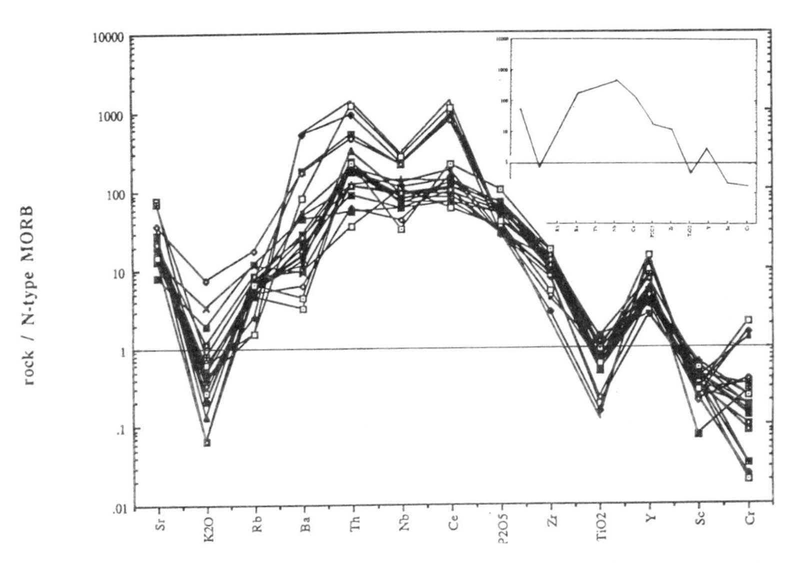

Figure 3. N-type MORB normalized diagram of metacarbonatites and metacarbonatitic tuffs. For comparison (insert) mean carbonatite composition by Heinrich (1966).

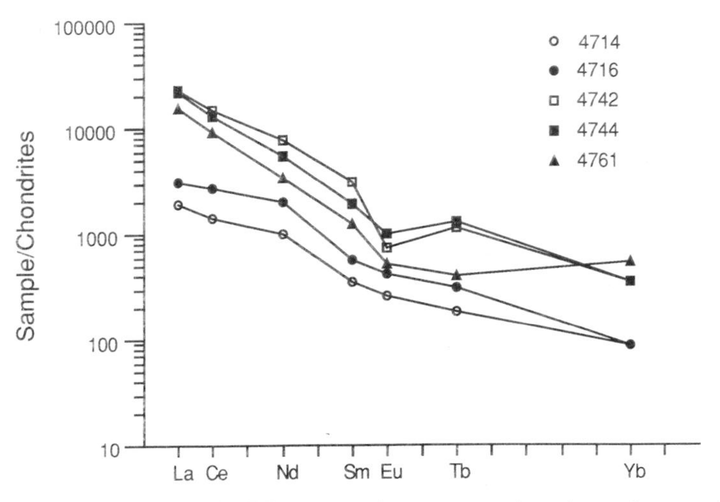

Figure 4. Chondrite normalised REE patterns of some metacarbonatites and metacarbonatitic tuffs. Chondrite normalizing values are those of Nakamura (1974) and Haskin et al. (Tb, 1968).

Table 3. Tabulation of the Rb and Sr results. Constants used are: ^{87}Rb decay constant = $1.42*10^{-11}a^{-1}$ (Steiger and Jaeger, 1977), $^{85}Rb/^{87}Rb$ = 2.59265, $^{88}Sr/^{86}Sr$ = 0.1194, $^{84}Sr/^{86}Sr$ = 0.056584 (Steiger and Jaeger, 1977), $^{87}Sr/^{86}Sr_{UR}$ = 0.7045 (e.g. De Paolo, 1988), $^{87}Rb/^{86}Sr_{UR}$ = 0.0827 (e.g. De Paolo, 1988).

Sample	Remark	$^{87}Sr/^{86}Sr$ (today)	+/− 2σ	$^{87}Rb/^{86}Sr$	+/− 2σ	Sr ppm	Rb ppm	$^{87}Sr/^{86}Sr$ (T = 250 Ma)	Epsilon Sr (today)	Epsilon Sr (T = 250 Ma)
Metacarbonatites										
KAW 3192/UAE 4020	leachate	0.703950	30	0.000104	0.000001	3621	0.13	0.703950	−7.31	−3.64
KAW 3192/UAE 4020	whole rock	0.703947	13	0.000308	0.000003	2857	0.22	0.703946	−7.85	−3.69
KAW 3194/UAE 4742	leachate	0.704100	23	0.000035	0.000001	6555	0.08	0.704100	−5.68	−1.51
KAW 3194/UAE 4742	whole rock	0.704068	16	0.000099	0.000001	4058	0.14	0.704068	−6.13	−1.96
KAW 3195/UAE 4761	leachate	0.704345	204	0.017621	0.000176	5941	36.20	0.704282	−2.20	1.09
KAW 3195/UAE 4761	whole rock	0.703758	19	0.015955	0.000160	5233	28.87	0.703701	−10.53	−7.17
KAW 3196/UAE 4776	leachate	0.703765	57	0.000034	0.000001	11775	0.14	0.703765	−10.43	−6.26
KAW 3196/UAE 4776	whole rock	0.703731	7	0.000048	0.000001	10664	0.18	0.703731	−10.92	−6.75
Metamorphic Limestone										
KAW 3193/UAE 4206	leachate	0.712021	26	0.000606	0.000006	574	0.12	0.712019	106.76	110.95
KAW 3193/UAE 4206	whole rock	0.712154	100	0.001229	0.000012	515	0.22	0.712150	108.64	112.80

Table 4. Tabulation of the Sm and Nd results, Constants used are: ^{147}Sm decay constant = 0.00654 b.y.$^{-1}$, $^{143}Nd/^{144}Nd_{CHUR(0)}$ = 0.51264, $^{143}Nd/^{144}Nd_{DM(0)}$ = 0.51316, $^{147}Sm/^{144}Nd_{CHUR}$ = 0.1967. $^{147}Sm/^{144}Nd_{DM}$ = 0.2136.

Sample	Nd ppm	Sm ppm	$^{143}Nd/^{144}Nd$	+/− 2σ	$^{147}Sm/^{144}Nd$	E Nd(0)	E Nd(250)
Metacarbonites							
KAW 3192/UAE 4020	1943	68.04	0.512690	5	0.02117	0.90	6.50
KAW 3194/UAE 4742	4935	630.8	0.512683	3	0.07727	0.76	4.58
KAW 3195/UAE 4761	2128	249.4	0.512675	5	0.07088	0.60	4.62
KAW 3196/UAE 4776	250.1	39.14	0.512696	10	0.09462	1.01	4.28
Metamorphic Limestone							
KAW 3193/UAE 4206	7.18	1.12	0.512067	30	0.09439	−11.26	−7.99

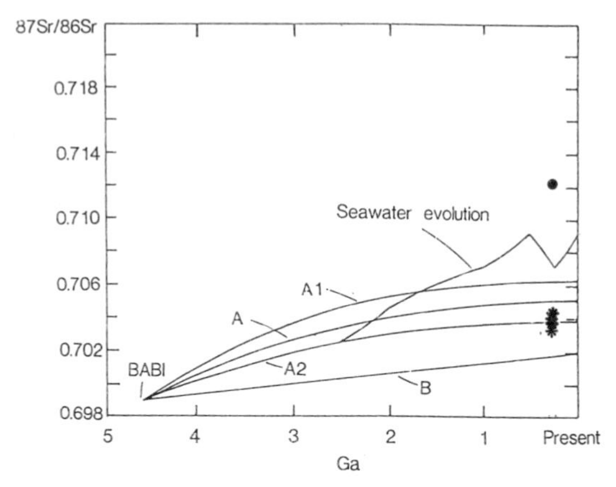

Figure 5. Position of the carbonatite samples (stars) and the carbonate sample (circle) in relation to the isotopic evolution of terrestrial strontium according to Faure (1986). A, A1, A2 = hypothetical evolutionary paths of Sr in the sub-continental mantle (see text for explanation). B = strontium evolution of Rb depleted mantle regions S = simplified seawater evolution curve

ppm while that of the leachate is 0.12 ppm. The measured $^{87}Sr/^{86}Sr$ ratio of 0.71202 of the leachate is somewhat below the value of 0.71215 obtained for the whole rock sample and thus indicates the presence of some detrital material in sample KAW3193. The initial $^{87}Sr/^{86}Sr$ ratio of the meta-sedimentary carbonate calculated for an assumed maximum age of 250 Ma plots well above the maximum $^{87}Sr/^{86}Sr$ value of seawater during its Phanerozoic evolution (Fig. 5). This elevated initial $^{87}Sr/^{86}Sr$ ratio can be explained by the combination of two processes. Firstly, this marine carbonate was contaminated with some radiogenic detrital material which increased the initial whole rock $^{87}Sr/^{86}Sr$ ratio of this sample above the value of seawater at its time of formation, and, secondly, it has to be assumed that some radiogenic Sr was possibly added to the sample in question during diagenetic and metamorphic processes.

The Sm and Nd concentrations of 1.12 and 7.18 ppm of the meta-sediment are within a normal range for marine carbonates (e.g. Keto and Jacobsen, 1988). The analyzed meta-sedimentary carbonate plots far off the mantle array into the field of crustal rocks in the ϵ_{Sr}-ϵ_{Nd} diagram (Fig. 6). As according to considerations of Keto and Jacobsen (1988) the ϵ_{Nd} value of seawater is directly related to the mean ages of the continental sources and because of the short residence time of Nd in seawater (approx. 200 a according to De Paolo, 1988) it should thus be possible to calculate a model age for the hinterland of the analyzed meta-sediment. Since Sm and Nd are commonly fractionated relative to their continental sources at their age of deposition (Keto and Jacobsen, 1987), it was appropriate to calculate a two stage model age for the meta-sediment instead of calculating a single stage Nd model age as it is usually used for clastic sediments. This calculation, assuming a maximum sedimentary age of 250 Ma, yields a two stage model

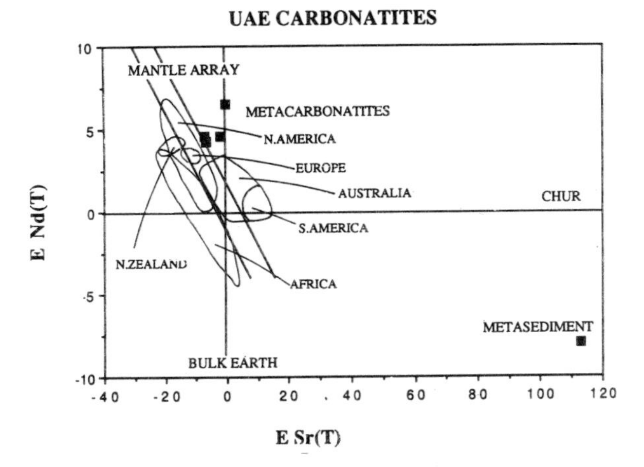

Figure 6. Plot of $\epsilon_{Sr(T)}$ versus $\epsilon_{Nd(T)}$ for the analysed metacarbonatites and the analysed metasediment of the UAE. Fields show $\epsilon_{Nd(T)}$ and $\epsilon_{Sr(T)}$ carbonatite data from different continents according to the compilation of Bell and Blenkinsop (1989).

age of 1678 Ma with a mean continental crustal residence time of the eroded material of 1428 Ma. This model age of the analyzed meta-sediment indicates that an influx of sediment from a Proterozoic source region controlled the Sm-Nd characteristics of the paleocean in question at the time of sedimentation. This in turn indicates that in the source regions neighboring the ocean basin in question (e.g. the Arabian shield), crustal material with an average age of approximately 1.7 Ga was available for erosion during the Mesozoic.

With the exception of KAW3195, whose elevated content is due to its intermediate mineralogical composition comprising the presence of amphiboles and micas, the Rb concentrations of all the analyzed meta-carbonatitic samples is below 1 ppm. The whole rock Sr concentrations of the meta-carbonatitic samples are ranging between 2857 ppm and 10667 ppm which clearly distinguishes them from the meta-sedimentary carbonate with an Sr concentration of 515 ppm. The leaching experiment shows that the Sr contents of the leachates is generally higher than the whole rock Sr concentration, while, with the exception of KAW3195, the Rb contents of the leachates are generally lower than those of the whole rock aliquots. This confirms the fact that Sr is enriched in the soluble carbonate phases while Rb is enriched in the insoluble residual minerals. The measured $^{87}Sr/^{86}Sr$ ratios of the analyzed meta-carbonatites range between 0.70373 and 0.70435. The differences of the obtained $^{87}Sr/^{86}Sr$ ratios between the whole rock aliquots and the leachates is, with the exception of KAW3195 (where the difference is 1 per mil of the measured value), within analytical error. This indicates that no significant amounts of radiogenic Sr were added to the high Sr concentrations of the analyzed meta-carbonatites by post formation hydrothermal and metamorphic processes. Due to the very high Sr concentrations and the very low Rb concentrations the initial $^{87}Sr/^{86}Sr$ ratios of the meta-carbonatites (calculated for an assumed age of 250 Ma) are almost identical with the $^{87}Sr/^{86}Sr$ ratios measured today (see Table 3). As the $^{87}Sr/^{86}Sr$ ratio of seawater ranges between 0.7065 and 0.7090 since the Precambrian (Faure, 1986), the very low initial $^{87}Sr/^{86}Sr$ ratios of the analyzed meta-carbonatites again confirm the magmatic origin of these metamorphic rocks. The initial $^{87}Sr/^{86}Sr$ ratios of the meta-carbonatites plot slightly above the strontium evolution line of an Rb depleted mantle which connects the Basaltic Achondritic Best Initial value (BABI) to a present value of 0.702 (Fig. 5). Nonetheless, they lie within the range of the hypothetical evolutionary paths for Sr in a subcontinental mantle postulated by Faure (1986). As significant post formation contamination with non-magmatic Sr can be excluded for the analyzed meta-carbonatites (see above), it may be assumed that they derive from a mantle source which had a slightly increased Rb/Sr ratio (similar to curve A2 in Fig. 5) in comparison to an Rb depleted mantle.

The meta-carbonatites show very high Sm and Nd concentrations ranging between 68 ppm and 630 ppm and between 250 ppm and 4935 ppm, respectively. The very low $^{147}Sm/^{144}Nd$ ratios (0.021 to 0.095) reflect the strong LREE enrichment in the analyzed meta-carbonatites. Due to the very small spread of the $^{147}Sm/^{144}Nd$ ratios it was not possible to calculate an Sm-Nd whole rock isochron age for the UAE meta-carbonatites. In Fig. 6 their $\epsilon_{Sr(T)}$ values are plotted versus their $\epsilon_{Nd(T)}$ values assuming a formation age T of 250 Ma. All the carbonatitic samples plot in a distinct field with somewhat elevated ϵ_{Nd} values compared to the Chondritic Uniform Reservoir (CHUR) and slightly lower ϵ_{Sr} values than the bulk earth. So far, no other carbonatites have been described to plot into this same field of the $\epsilon_{Nd} - \epsilon_{Sr}$ diagram. A comparison with literature data (e.g. De Paolo, 1988; De Paolo and Johnson, 1979; Staudigel et al., 1984) shows that the meta-carbonatites nonetheless plot within the area commonly occupied by mantle derived rocks. It is of interest, however, that a comparison with data of e.g. Carmichael and De Paolo (cit. De Paolo, 1988), Hawkesworth et al. (1984), De Paolo (1988), Menzies et al. (1983) and Perry et al. (1987) shows that especially magmatic rocks with continental affinities are found to plot in the same area as the analyzed meta-carbonatites. A comparison of the ϵ_{Nd} values of the UAE meta-carbonatites with the ϵ_{Nd} values of the Cretaceous Semail ophiolite, which has ϵ_{Nd} values of approximately $+8$ (McCulloch et al., 1980) and with a model depleted mantle source, which at T = 250 Ma had an ϵ_{Nd} value of approximately $+9$, shows that the ϵ_{Nd} values of the meta-carbonatites are significantly lower than those of average depleted upper mantle sources. This observation is in good agreement with literature data on the Sm-Nd isotope systematics of carbonatites (e.g. Bell and Blenkinsop, 1989; Nelson et al., 1988). The commonly low initial ϵ_{Nd} values of carbonatites and their relatively large variation (see Fig. 6) thus lead to the hypothesis that the subcontinental mantle is heterogeneous with respect to its Sm-Nd characteristics (e.g. Bell and Blenkinsop, op. cit; Nelson et al., 1988). From the present data it may furthermore be deduced that this on a global scale heterogeneous subcontinental mantle, which serves as source region for carbonatitic magmas, has low ϵ_{Nd} values compared to model depleted upper mantle areas and present day MOR-basalts and ophiolites. This fits well into the observations of Jacobsen and Wasserburg (1978), Menzies et al. (1983), Semken (1984) and Perry et al. (1987) who studied magmatic rocks of intracontinental rift zones which led them to the idea of low ϵ_{Nd} subcontinental mantle. Due to the scarcity of data there is much discussion on the characteristics of the mantle source of carbonatites. Bell and Blenkinsop (1989) favor a lithosphere depleted by crustal extraction while Barreiro and Cooper (1987) and Meen et al. (1989) favor a metasomatized depleted lithosphere. Nelson et al. (1988) assume a depleted mantle source originating in asthenospheric mantle plumes as sources for carbonatite magmas while Perry et al. (1987) interpreted the low ϵ_{Nd} values which they obtained for basalts of the Rio Grande rift (USA) as representatives of upwelling low ϵ_{Nd} mantle which means that alkali basalts with low ϵ_{Nd} values originate in deeper mantle areas than high ϵ_{Nd} tholeiitic basalts. In spite of this large uncertainty on the origins of carbonatite magmas it can be said that the $\epsilon_{Nd} - \epsilon_{Sr}$ characteristics of the UAE meta-carbonatites add a new facet to the diversity of low ϵ_{Nd} subcontinental mantle compositions. Further Sm-Nd work on the meta-volcanic sequence accompanying the analyzed meta-carbonatites, and on the peridotites of the Semail nappe overlying the metamorphic sheets of the UAE is expected to give more evidence on the evolution and structure of the upper mantle in the UAE area and on the location of carbonatite genesis within this mantle.

Geotectonic Implications

Carbonatites usually occur closely associated with alkaline rocks in intracontinental rift zones (e.g. King and Sutherland, 1966; Woolley, 1989). The only known exceptions so far are carbonatites from the Cape Verde and Canary Islands which are found in oceanic environments (Allègre et al., 1971). According to Le Bas (1984), these occurrences are of purely intraoceanic nature contrasting the carbonatites of continental origin. The alkaline volcanic rocks of these islands may, however, also be interpreted within the context of a transition from continental to oceanic crust, as discussed by Emery and Uchapé (1984) on the basis of the sediment record.

The Haybi volcanic series consists of non-metamorphic, alkaline volcanic rocks (ankaramites, nephelinites and trachytes) and intrusives (alkaline gabbros and peridotites) (Searle et al., 1980). Based on geochemical and radiometric age data, they were partly interpreted as within-plate magmas of Triassic age erupting from volcanic islands at the sites where initial rifting of the Tethys occurred. The alkaline meta-volcanic rocks associated with the meta-carbonatites can be considered to be the metamorphic equivalents of the non-metamorphic alkaline volcanic rocks of Searle's Haybi Complex

(1980) or the Umar Group (Ziegler and Stössel, 1985). As there seems to be a general agreement with respect to the opening of the Neotethys NE of the Arabian continental block during Triassic times (Stöcklin, 1974; Sengör, 1985; Descourt et al., 1986), it can be inferred that the meta-carbonatites represent magmas related to the volcanism associated with the Triassic rifting. A relationship with the movements of the Paleo-Tethys of Sengör (op. cit.) or Tethys 1 of Dewey et al. (1983) is ruled out based on the observations by Bèchenec (1987) who clearly assigned the alkaline volcanic rocks to the aforementioned Neotethys. Glennie et al. (1973, 1974), Lippard et al. (1986) and Bèchenec (1987) place the origin of the allochthonous sediments NE of the Arabian platform at the passive margin of the Neotethys. According to this model, the meta-carbonatites can be explained as mantle magmas marking the site where the deep reaching Dibba fault zone crosscuts the Triassic SE-NW running rift faults. In order to explain the measured paleocurrent data of the Exotics and the Umar Group, Blendinger (in press) assumes an original position further south and movements along a transform faulting adjacent to an SW directed thrust to bring them into their present position. Using this model, the connection with the Dibba fault zone becomes arbitrary. In both models, however, the meta-carbonatites would have formed along the passive margin of the Arabian continent in a transition zone from continental to oceanic crust in an oceanic island setting similar to the situation of the Canary Islands at the margin of the African continent.

Acknowledgements

Prof. U. Krähenbühl, Dr. M. Flisch, Dr. K. Ramseyer and Prof. R. Oberhänsli are thanked for analytical help, and Messrs. M. Bucher, D. Kurz and M. Giger for field and laboratory assistance. We are indebted to the Ministry of Petroleum and Mineral Resources of the United Arab Emirates for permission to work in the U.A.E. and for their logistic support. The study was partly financed by the "Swiss National Foundation". We are grateful to Prof. P. Stille who critically reviewed and improved a first version of the manuscript.

References

Allègre, C., Pineau, F., Bernat, M. and Javoy, M., 1971. Evidence for the occurrence of carbonatites on the Cape Verde and Canary Islands., Nature Phys. Sci., 233: 103–104.

Allemann, F. and Peters, Tj., 1972. The Ophiolite-radiolarite belt of the North Oman Mountains., Eclogae Geol. Helv., 65 (3): 657–697.

Bailey, D.K., 1989. Carbonate melts from the mantle in the volcanoes of south-east Zambia., Nature, 338: 415–418.

Barber, C., 1974. The Geochemistry of Carbonatites and related rocks from two Carbonatite Complexes, South Nyanza, Kenya., Lithos, 7: 53–63.

Barreiro, B.A. and Cooper, A.F., 1987. A Sr, Nd and Pb isotope study of alkaline lamprophyres and related rocks from Westland and Otago, South Island, New Zealand. In: E.M. Morris, and J.D. Pasteris (Eds), Spec. Pap. Geol. Soc. Amer, 215: 115–125.

Bechenec, F., 1987. Géologie des nappes Hawasina dans les parties orientales et centrales des montagnes d'Oman., BRGM, Orléans, 127: 474 p.

Bell, K., 1989. Carbonatites, genesis and evolution., Unwyn Hyman, London, 618 p.

Bell, K. and Blenkinsop, J., 1989. Neodymium and strontium isotope geochemistry of carbonatites. In: K. Bell (Ed), Carbonatites, genesis and evolution. Unwyn Hyman, London, pp. 278–300.

Blendinger, W., in press. The Upper Hawasina Nappes, Oman: Remnants of offshore carbonate highs or fragments of a south Tethyan continental margin?.

De Paolo, D.J., 1988. Neodymium Isotope Geochemistry. An Introduction. Springer, Berlin, 187 pp.

De Paolo, D.J. and Johnson, R.W., 1979. Magma genesis in the New Britain island arc: constraints from Nd and Sr isotopes and trace element patterns., Contrib. Mineral. Petrol., 70: 367–379.

Descourt, J., Zonenshain, L.P., Ricou, L.E., Kazmin, V.G., Le Pichon, X., Knipper, A.L., Grandjacquet, C., Sbortshikov, I.M., Geyssant, J., Lepvrier, C., Pechersky, D.H., Boulin, J., Sibuet, J.C., Savostin, L.A., Sorokhtin, O., Westphal, M., Bazhenov, M.L., Lauer, J.P. and Bijou-Duval, B., 1986. Geological evolution of the Tethys belt from the Atlantic to the Pamirs since the Lias., Tectonophysics, 123: 241–315.

Dewey, J.F., Pitman, W.C., Ryan, W.B.F. and Bonin, J., 1973. Plate tectonics and the evolution of the Alpine system., Bull. Geol. Soc. Amer., 84: 3137–3180.

Emery, K.O. and Uchupi, E., 1984. The geology of the Atlantic Ocean. Springer, New York, 1050 p.

Faure, G., 1986. Principles of Isotope Geology. Second Edition. Wiley, New York, 589 p.

Glennie, K.W., Boeuf, M.G.A., Hughes-Clarke, M.W., Moody-Stuart, M., Pillar, W.F.H., and Reinhardt, B.M., 1973. Late Cretaceous nappes in the Oman Mountains and their geologic evolution., Bull. Amer. Assoc. Petroleum Geol., 57 (1): 5–27.

Glennie, K.W., Boeuf, M.G.A., Hughes-Clarke, M.W., Moody-Stuart, M., Pillar, W.F.H. and Reinhard, B.M., 1974. Geology of the Oman Mountains (2 volumes). Koninkl. Nederlands. Geol. Mijnbouwkundig Genootschap.

Greenwood, J.E.G.W. and Loney, P.E., 1968. Geology and Mineral Resources of the Trucial Oman Range., Inst. Geol. Sci. London, 108 p.

Haskin, L.A., Haskin, A., Frey, F.A. and Wildemann, T.R., 1968. Relative and absolute terrestrial abundances of the rare earths. In: L.H. Ahrens (Ed), Origin and Distribution of the Elements, 1. Pergamon, Oxford, pp. 889–911.

Hawkesworth, C.J. and van Calsteren, P.W.C., 1984. Radiogenic Isotopes – Some Geological Applications. In: P. Henderson, (Ed), Rare Earth Element Geochemistry. Elsevier, Amsterdam, pp. 375–421.

Heinrich, E.W., 1966. The geology of carbonatites. Rand McNally, Chicago, 555 p.

Jacobsen, S.B. and Wasserburg, G.J., 1978. Nd and Sr isotopic study of the Permian Oslo Rift. US Geol. Surv. Open File Rep., 78–701: 194–196.

Keto, L.S. and Jacobsen, S.B., 1987. Nd and Sr isotopic variations of early Paleozoic oceans. Earth and Planet. Sci. Lett., 84: 27–41.

Keto, L.S. and Jacobsen, S.B., 1988. Nd isotopic variations of Phanerozoic paleoceans. Earth and Planet. Sci. Lett., 90: 395–410.

Krähenbühl, U., 1985. Möglichkeiten und Grenzen der Neutronenaktivierungsanalyse. Swiss. Chem, 7 (10): 55–56.

Jaeger, E., 1979. The Rb-Sr Method, 13–26. In: E. Jaeger and J.C. Hunziker (Eds), Lectures in Isotope Geology. Springer, Berlin, 329 p.

King, B.C. and Sutherland, D.S., 1966. The carbonatite complexes of eastern Uganda. In: O.F. Tuttle and J. Gittins (Eds), Carbonatites. Interscience, New York, pp. 73–126.

Le Bas, M.J.,1984. Oceanic carbonatites. In: J. Kornprobst (Ed), Kimberlites I. Kimberlites and related rocks. Elsevier, Amsterdam, pp. 169–178.
Lippard, S.J., Shelton, A.W. and Gass, I.G., 1986. The ophiolite of Northern Oman. Mem. Geol. Soc. London, 11. Blackwell Scientific Publications, Oxford-London.
McCulloch, M.T., Gregory, R.T., Wasserburg, G.J. and Taylor, H.P., 1980. A neodymium, strontium and oxygen isotopic study of the Cretaceous Semail Ophiolite and implications for the petrogenesis and seawater-hydrothermal alteration of oceanic crust. Earth and Planet. Sci. Lett., 46: 201–211.
Meen, J.K., Ayers, J.C. and Fregeau, E.J., 1989. A model of mantle metasomatism by carbonated alkaline melts: trace element and isotopic compositions of mantle source regions of carbonatite and other continental igneous rocks. In: K. Bell (Ed), Carbonatites, genesis and evolution. London, Unwyn, Hyman, pp. 448–461.
Menzies, M., Leeman, W.P. and Hawkesworth, C.J., 1983. Isotope geochemistry of Cenozoic volcanic rocks reveals mantle heterogeneity below western USA., Nature, 303: 205–207.
Nakamura, N., 1974. Determination of REE, Ba, Mg, Na and K in carbonaceous and ordinary chondrites., Geochim. cosmochim. Acta, 38: 757–775.
Nelson, D.R., Chivas, A.R., Chappell, B.W. and McCulloch, M.T., 1988. Geochemical and isotope systematics in carbonatites and implications for the evolution of ocean-island sources. Geochim. cosmochim., Acta, 52: 1–17.
Nesbitt, E.B., Dietrich, V. and Esenwein, A., 1979. Routine trace element determination in silicate minerals and rocks by X-ray fluorescence., Fortschr. Mineral., 57, 264–279.
Pearce, J.A., 1982. Trace element characteristics of lavas from destructive plate boundaries. In: R.S. Thorpe (Ed), Andesites. J. Wiley & Sons, New York, pp. 525–548.
Pearce, J.A., 1983. Role of the Sub-continental Lithosphere in Magma Genesis at Active Continental Margins. In: C.J. Hawkesworth and M.J. Norry (Eds), Continental Basalts and Mantle Xenoliths. Shiva Publ. Ltd., Norwich, pp. 230–249.
Perry, F.V., Baldridge, W.S. and De Paolo, D.J., 1987. Role of astenosphere and lithosphere in the genesis of Late Cenozoic basaltic rocks from the Rio Grande Rift and adjacent regions of the southwestern United States., J. Geophys. Res., 92: 9193–9213.
Searle, M.P., 1980. The metamorphic sheet and underlying volcanic rocks beneath the Semail Ophiolite in the northern Oman Mountains of Arabia. PhD thesis, The Open University, Department of Earth Sciences, Great Britain. 213 p.
Searle, M.P., Lippard, S.J., Smewing, J.D. and Rex, D.C., 1980. Volcanic rocks beneath the Semail Ophiolite nappe in the northern Oman mountains and their significance in the Mesozoic evolution of Tethys., J. geol. Soc. London, 137: 589–604.
Semken, S.C., 1984. A neodymium and strontium isotopic study of late Cenozoic basaltic volcanism in the southwestern Basin and Range province. MS thesis, University of California, Los Angeles, 68 p.
Sengör, A.M.C., 1984. Die Alpiden und die Kimmeriden: Die verdoppelte Geschichte der Tethys., Geol. Rdsch., 74: 181–213.
Staudigel, H., Zindler, A., Hart S.R., Leslie, T., Chen, C-Y., Clague, D., 1984. The isotope systematics of a juvenile intraplate volcano: Pb, Nd, and Sr isotope ratios of basalts from Loihi seamount, Hawaii., Earth and Planet. Sci. Lett., 69: 13–29.
Steiger, R.H. and Jäger, E., 1977. Subcommission on geochronology: convention on the use of decay constants in geo- and cosmochronology., Earth and Planet. Sci. Lett., 36: 358–362.
Stöcklin, J., 1974. Possible ancient continental margins in Iran. In: C. Burk and C.L. Drake (Eds), The Geology of Continental Margins. Springer, Berlin, pp. 873–878.
Stössel, G.F.U. and Ziegler, U.R.F., 1989. Age determinations in the Rehoboth Basement Inlier, SWA/Namibia. Doctoral thesis, Univ. Bern, Switzerland, 250 p.
Sun, S. and Nesbitt, R.W., 1977. Chemical heterogeneity of Archaean mantle, composition of the earth and mantle evolution., Earth and Planet. Sci. Lett., 35: 429–448.
Woolley, A.R., 1989. The spatial and temporal distribution of carbonatites In: K. Bell (Ed), Carbonatites, genesis and evolution. Unwyn, Hyman, London, pp. 15–37.

Woolley, A.R., Barr, M.W.C., Din, V.K., Jones, G.C., Wall, F. and Williams, L.T., 1990. Extrusive Carbonatites from the United Arab Emirates. Abstract, IAVCEI Meeting, Mainz 1990.

Wyllie, P., 1966. Experimental studies of carbonatite problems: The origin and differentiation of carbonatite magmas. In: O.F. Tuttle and J. Gittins (Eds), Carbonatites. Interscience, New York, pp. 311–352.

Ziegler, U.R.F. and Stoessel, G.F.U., 1985. The Metamorphic Series associated with the Semail Ophiolite Nappe of the Oman Mountains in the United Arab Emirates. MS thesis Univ. Bern, Switzerland, 293 p.

Chert-Hosted Manganese Deposits in the Wahrah Formation: A Depositional Model

W. KICKMAIER and TJ. PETERS
Mineralogisch-Petrographisches Institut, Baltzerstrasse 1, CH-3012 Bern, Switzerland

Abstract

The non-metamorphic, stratiform manganese deposits in the Wahrah Formation (Al Hammah Range, 125 km south of Muscat) are of upper Berriasian to Hauterivian age. The manganese enriched horizons are continuously exposed over several 100 km^2. Sedimentological and geochemical features prove that the host rock of the manganese deposits exclusively consisting of red cherts are of biogenic, non-hydrothermal origin, deposited in a distal part of the Hawasina basin. A depositional (geochemical) environment comparable to that of a siliceous ooze facies is highly likely. Geochemically the manganese layered cherts and nodules, with their extremely high Mn/Fe ratio and the low minor element content, are similar to hydrothermal deposits associated with volcanic activity or to diagenetic manganese enrichments in hemipelagic regions in sediments with relatively high organic-C contents. Diagenetic shallow water deposits are geochemical similar to the Wahrah Formation enrichment also. Regional geological considerations, mineralogical, sedimentological and REE-patterns evidences in the Wahrah Formation are, however in conflict with these models. Therefore, we propose a alternative manganese transport and accumulation model. "Continentally" derived manganese is introduced to the sea in suspended and/or dissolved form and deposited on the continental margin and -slope. Upward migrating pore fluids, diagenetically enriched in manganese, are released to the water column and horizontally transported through the Oxygen-minimum zone towards the open basin. During selective element scavenging onto different sinking particles and the formation of manganese-rich microparticles, element separation takes place. Sedimentary features indicate that manganese microparticles, precipitated in the water column and the sea water sediment interface, are mechanically enriched to fine laminae and thicker layers. Hydrodynamically controlled separation processes during transport of clay minerals, manganese-rich microparticles and biogenic particles (main element carrier phases) explain the geochemical composition. Geochemical differences and similarities of nodular and layered ore types are attributed to later minor element redistributions.

Tj. Peters et al. (Eds), Ophiolite Genesis and Evolution of the Oceanic Lithosphere, 647–674.

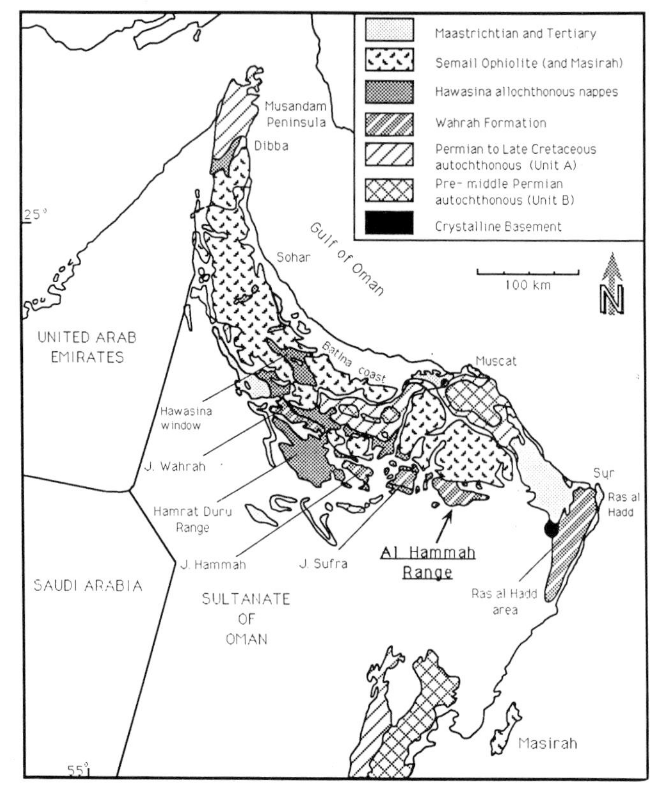

Figure 1. Geological sketch map of Oman. (Simplified after Glennie et al. 1974)

Introduction and Geological Situation

The non-metamorphic manganese deposits of the Al Hammah Range occur in red radiolarian cherts of the Wahrah Formation, which occurs in one of the lowermost Hawasina nappes (Fig. 1).

The stratigraphic sequence in the Al Hammah Range starts with fine-grained turbiditic limestones followed by a transition zone of variegated mudstones and cherts which continuously grade into almost $CaCO_3$-free red radiolarian cherts, covered by 1–5 m of (strongly) silicified limestones (Fig. 2).

The variegated mudstones and red radiolarian cherts were deposited during late early Tithonian to Aptian time.

Although more fine grained and silicified, the lithologic sequence of the Wahrah Formation in the Al Hammah Range can be correlated with the Guwayza and Sid'r Formations belonging to the Hamrat Duru Group. Glennie et al. (1974) regarded the Wahrah Formation as "... equivalent of the Hamrat Duru Group, deposited in a greater distance from a common source area " Glennie et al. p. 217. Béchennec et al. (1988) included the Wahrah

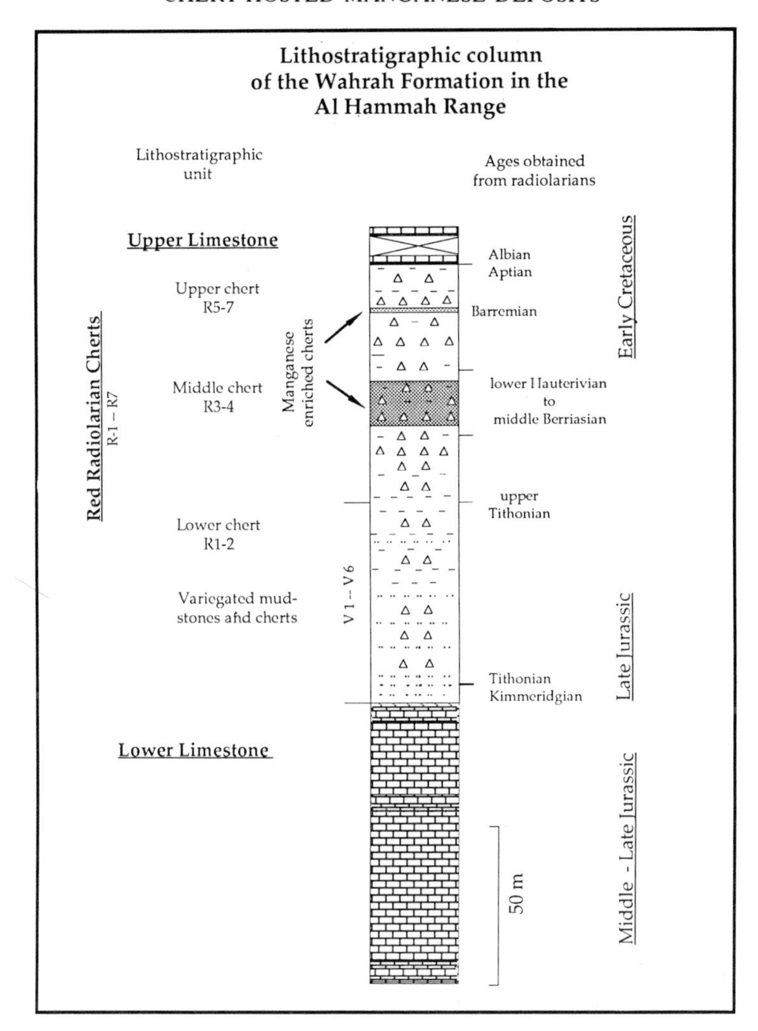

Figure 2. Schematic lithostratigraphic column of the Wahrah Formation in the Al Hammah Range. The main phase of manganese enrichment has been dated by radiolaria as Valanginian (middle Berriasian to lower Hauterivian). During the Barremian a second, minor manganese enrichment took occurred as well. This manganese enriched cherts of this age are relatively rare in the Al Hammah Range.

Formation within the Hamrat Duru Group. They concluded that the eastern to central part of the Hamrat Duru Basin is developed in a more siliceous facies of both, chert and limestones. The lower limestones of the Wahrah Formation in the Al Hammah Range should be regarded as the Guwayza limestone. The cherts however should belong to the Wahrah Formation. (Béchennec pers. commun. 1990). To the authors it seems more likely *not* to separate the cherts from the limestones in different formations (Fig. 2). In this case the Wahrah Formation with the limestone-chert-limestone sequence is the distal more fine grained equivalent of the Hamrat Duru Group. The top of the Guwayza, the base of the Sid`r Formation respectively was dated by De Wever et al. (1990) as Kimmeridgian/Tithonian. The base of the red cherts in the Al Hammah Range, has been dated by radiolaria as Tithonian so that the siliceous facies of the Sid`r and red bedded cherts of the Wahrah Formation are time equivalent. According to Béchennec (1987) the Nayid Formation is of Cenomanian to Turonian age. Thus the silicified upper limestones in the Al Hammah Range (top red cherts is Aptian) has to be correlated with the silicified calcareous parts within the Sid'r Formation. Unfortunately these horizons yielded no palaeontological data (De Wever et al. 1990). In the Ras al Hadd area the lower limestones are thinner and more silicified; the upper ones have not been found. The chert facies with the variegated mudstones is developed like in the Al Hammah Range. This general trend of more silicified and fine grained (distal) facies towards the south has been reported already by Glennie et al. (1974) and Béchennec (1988).

Based on detailed stratigraphic work by Biaggi and Steinmann (1990 in prep.) the red cherts can be grouped into three members which are characteristic for the whole Al Hammah Range. The middle member is characterized by manganese enriched horizons of up to 10 m, which can be subdivided in several zones (Kickmaier & Peters 1990).

The main phase of manganese enrichment could be dated by radiolaria as Valanginian but began in the late Berriasian. In the Barremian a second manganese occurred as well (preliminary dating by radiolaria by R.Jud).

Although the thickness and the ore enrichment types (brown cherts; laminated to layered cherts, black cherts and nodular ore) are variable, the manganese horizons are more or less continuous over large distances. The main difficulties to correlate the manganese enriched cherts are the intense folding and thrusting of the chert sequence.

Analytical Methods

With the exception of Mn and Fe major and trace elements were determined by an automatic XRF spectrometer. Manganese and iron were analyzed by AAS. The main error in the major element analyzes is SiO_2 content since the high SiO_2 content is above the normal standardization limit. The mineralogy was determined in thin section, by x-ray diffractometer (XDR) and X-ray diffraction photographs using a Guinier camera. In addition REM and TEM together with X-ray spectra was used for mineral identification.

Sedimentological and Geochemical Characteristics of the Red Cherts

Generally marine manganese deposits have been geochemically classified in hydrothermal – hydrogenetic – diagenetic or mixed type enrichments (Bonatti et al. 1972, Roy 1981, Halbach & Puteanus 1988). Considering the geological situation of the Wahrah Formation and the geochemical composition of the manganese ore, the manganese enrichment cannot be explained by one of these "classic" depositional models.

As the depositional environment of the red cherts is of special interest for the deposition and accumulation model, the following part summarizes the main lithologic and geochemical characteristics of the middle chert member.

The red chert sequence consists of alternating layers of impure red to brown radiolarian cherts (mean bed thickness 7–10 cm) and siliceous shales (1–3 cm). The cherts are often triple layered (Iijima and Utada 1983) with a regular increase in clay content towards the siliceous shale interbeds, but asymmetrically chert beds are common as well. The most dominant sedimentary features are the subparallel and flaser bedding and the intense bioturbation. Sedimentary features, like graded bedding and ripple marks which would unambiguously indicate a turbiditic origin (as one of a possible depositional model) are missing. But nevertheless, in rare cases vague cross lamination and the fine rhythmic bedding of the cherts can be attributed to benthic currents or low density turbidity currents reworking the radiolarian debris. It is to note that one single chert layer is built up of several sedimentation units, each composed of a cherty and more argillaceous part. These sedimentation, units are not comparable to the sedimentation intervals described by Nisbet and Price (1974). The siliceous shale interbeds represent the normal pelagic sedimentation supplying the fine clay fraction. The absence of more pronounced current induced sedimentary features is possible due to the homogeneous particle size, which does not allow any clearly visible grading and the intense bioturbation, destroying the primary sedimentary features.

Mineralogically the red cherts consist of chalcedonic and microcrystalline quartz, radiolaria, and (mainly) detritically derived clay minerals (smectite, illite and kaolinite). Some hematite/goethite and calcite is also present. No opal or a coarse detrital fraction have been determined.

Geochemically the cherts are characterized by high SiO_2 contents of 84–93%. The main geochemical variations are due to the variable influence of detritic clay minerals supplying mainly Al_2O_3, Fe_2O_3, TiO_2 and Zr. The element distribution patterns are similar over the whole stratigraphic sequence including the cherts directly associated with the manganese enrichment. Using discrimination diagrams like the Al, Fe, Mn diagram (Fig. 3) or by calculating various element ratios such as Al/Al + Fe + Mn (Yamamoto 1987, Varanavas and Punganos 1983)) the cherts of the Wahrah Formation point to typical biogenic deep sea cherts without any significant hydrothermal influence.

The siliceous shale interbeds are geochemically comparable to the average pelagic deep sea clays (Turekian and Wedepohl 1961), although they are diluted by radiolarians (relatively high SiO_2 content). The Al/Si plot (Fig. 4) indicates that the different lithologies, ranging from pure chert to siliceous

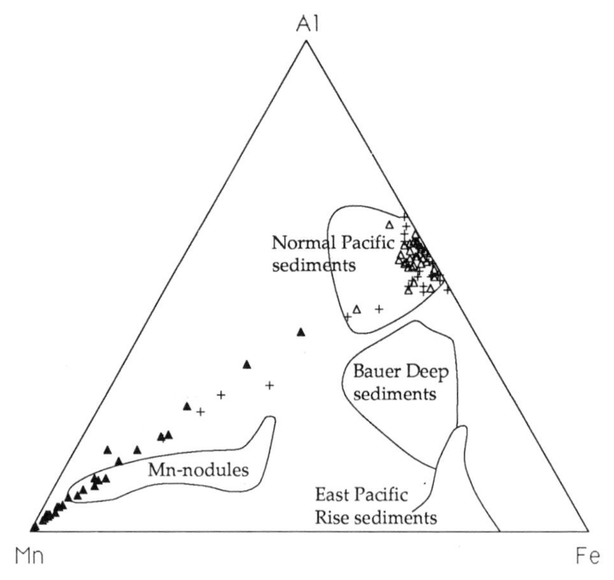

Figure 3. Ternary Al-Fe-Mn-plot (after Varnavas & Pungnos 1983). Open triangles = red cherts; crosses = siliceous shale interbeds including manganese rich shales; solid triangles = black cherts, layered and nodular ore.

shale, are the result of a variable supply of biogenically derived SiO_2 and detritic clay fraction. This is also supported by the almost identical element distribution patterns of cherts and shales.

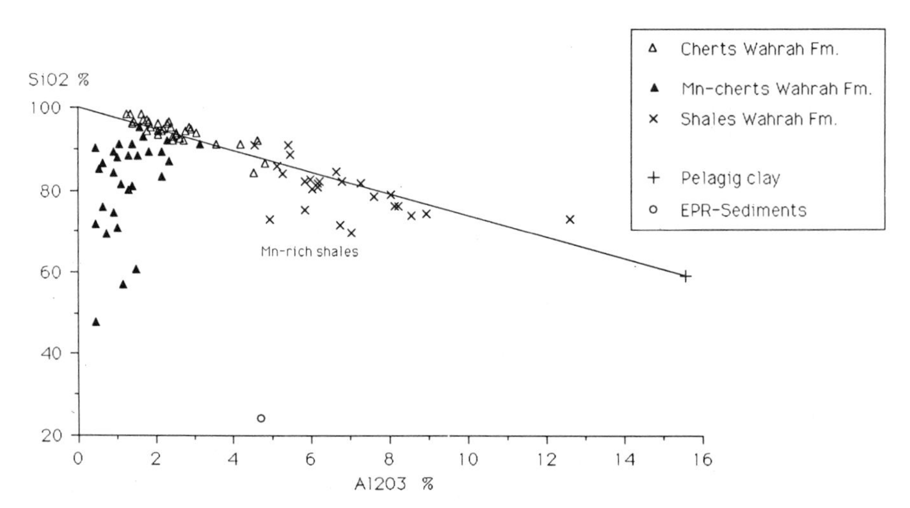

Figure 4. Correlation diagram SiO_2-$Al2O_3$. (Pelagic clay and East Pacific rise sediment composition in: Matsumoto & Iijima 1983).

Discussion

The stratigraphic sequence, sedimentary structures, mineralogical and geochemical composition as well as the regional distribution of the Wahrah Formation indicate a deposition of the cherts in a distal position of the Hawasina basin. The influence of turbidity currents contributing to the deposition of the radiolarian cherts is doubtful, but rare sedimentary structures show the influence of weak (benthic) currents.

The low $CaCO_3$-content supports a deposition below the CCD, the intense bioturbation an O_2-rich depositional environment.

The sedimentation rates of $\approx$ 2–4 mm/1000 y (sedimentation rates estimated from stratigraphic thickness and radiolaria dating) and the geochemical rock composition are comparable to the pelagic marine sedimentation found in present oceans (Andreyev and Kulikov 1987).

For the cherts of the Wahrah Formation a geochemical sedimentary environment analogue to that of the siliceous ooze facies in recent oceans is highly likely.

Manganese-Enriched Cherts

The middle chert member is characterized by stratiform manganese horizons which are built up of several manganese zones. In these zones one ore enrichment type, (e.g. brown cherts, laminated to layered cherts) is dominating, although all ore types may occur together in one chert layer (for further details see Kickmaier and Peters, 1990). The horizontal and vertical distribution of the manganese enriched cherts, showed that they form large, lense shaped ore rich bodies with extreme length to thickness ratios (kilometer by deca-meters). Thus, the zones seem to be completely parallel to each other. This form of manganese enrichment is possibly controlled by the relief of the basin with a preferential manganese deposition in small basins. Beside this large scale manganese distribution patterns thinning out of manganese enriched cherts have been observed over distances smaller than 10 m.

Brown Laminated and Layered Manganese Cherts

The brown and finely laminated cherts occur as individual chert units (manganese zones) or as interlayers between the manganese layered cherts.

Field and thin section observations show all transitional stages from simple brown cherts, with homogeneously distributed manganese, over finely laminated to layered cherts. The laminated brown cherts are characterized by small, manganese-rich laminae running almost parallel to the bedding plane. Both thickness and spacings of the laminae are quite variable but in typical laminated cherts it is about 0.3 to 1 mm. The main ore minerals are pyrolusite and cryptomelane but also traces of todorokite?, vernandite and psilomelane have been determined. The iron is bound to clay minerals and fine hematite/goethite pigment.

The manganese in the laminae is concentrated to isolated microparticles of variable size ranging from 0.01–0.05 mm. This particle size is much smaller than the average size of radiolarian tests, so that it is possible to distinguish between structureless radiolaria, diagenetically replaced by manganese and microparticles. Further manganese occurs as thin coatings on biogenic remains (mainly radiolaria) so that shell structures are well preserved. Larger manganese particles are missing. Beside these manganese enrichments the brown cherts show structures similar to the manganese free cherts.

The microparticles may be concentrated to layers of 1 cm. Even in the case of these thick layers the microparticles are still visible supporting an analogue accumulation process for both, layered and brown finely laminated cherts. Internally the manganese layers show parallel layering or they are graded with respect to the abundance of the microparticles. Diagenetic manganese redistribution and chert impregnation leads to thicker, macroscopically homogeneous black layers of maximal 3 cm in thickness in the case of more closely spaced primary enrichments.

Erosional contacts of manganese-rich layers (vague cross bedding) favor a current derived manganese enrichment to fine laminae. An early formation and consolidation of the manganese-rich layers is indicated by the brittle behavior of manganese layers during dehydration and compaction, whereas the chert interlayers still show ductile deformation textures.

In most cases however, the primary deposition features are overprinted by intense recrystallizations and mobilization processes which will be the subject of further investigations.

Nodular Ore

The second type of ore enrichment discussed here are the manganese nodules which can be grouped into three classes:

I. Oval nodules slightly elongated parallel to the bedding plane with generally smooth surfaces. Nodules of this type occur isolated in red cherts and can reach up to 8 cm in length.
II. Disk-like forms which dominate in the siliceous shale interbeds and are characterized by extreme width/length ratio.
III. Oval to elliptical nodules with uneven surfaces elongated to the bedding plane often associated with manganese layers.

The nodules occur isolated or closely spaced in the pale red or brown radiolarian cherts and show generally sharp contacts with the cherts. In most cases no primary contact with the upper or lower manganese-rich chert is visible. With the exception of the extreme small microparticles all nodules are built up of two zones. The largest central part is macroscopically homogeneous fine grained and of dull black color (zone 1). No internal concentric structures nor a nucleus are visible even on polished surfaces. This central part is (completely) surrounded by a dense hard black rim (zone 2) with a higher reflectance and small white dots (quartz filled radiolarian tests). Generally the contact between the core, the outer rim is sharp.

Mineralogically the nodules consist of pyrolusite, microcrystalline/chalcedonic quartz and minor amounts of cryptomelane. Pyrolusite may reach up to 95 vol%. In this homogeneous pyrolusite matrix, numerous radiolarian tests are scattered randomly. The radiolaria show a variable mode of preservation. The best preserved radiolaria are filled with chalcedonic and/or microcrystalline quartz, whereby the (porous) structure of the shell is visible by fine pyrolusite coating (or infills). Starting from this coating or from the fine pores, pyrolusite grows towards the center of the test (open space filling structures). Some calcite is found in the red cherts, preferably near the margin of the nodule.

In some cases the continuously layered manganese enrichment changes into a series of parallel aligned lense-shaped nodular ores, which are often connected with brown cryptomelane and calcite-"rich" mm-thick layers. According to the density of nodules in the chert layer the optical impression is a more nodular or a slab like one.

Regional Manganese Distribution

On regional scale manganese enriched cherts of the Wahrah Formation are found in the Ras al Hadd area east of the Al Hammah Range, towards the west-northwest in the Sufrad ad Dawh Range and in Jabal Wahrah.

In the Ras al Hadd area the manganese enrichment types, the mineralogical and geochemical composition is equivalent to those found in the Al Hammah Range. Kickmaier and Peters (1990) showed that the scale and continuity of the manganese enrichment is tectonically controlled also. In the Ras al Hadd area several m^3 large, tectonically formed manganese-rich blocks have been found. Thus an exact evaluation of the total (primary) manganese enrichment without detailed structural work or drilling is impossible. But within these limitations no major differences in the amount of manganese enrichment between the Al Hammah Range and the Ras al Hadd area seems to exist.

Towards the west-northwest of the Al Hammah Range a decrease of manganese enrichment in the red cherts is obvious. This regional variation of the manganese enriched chert facies correlates with the facies development in the Wahrah Formation which becomes more fine grained and siliceous towards the east. This indicates that only the distal, siliceous chert facies in the Wahrah Formation is a suitable environment for the enrichment.

Additionally it is remarkable that manganese bearing cherts are exclusively bound to the Wahrah Formation. In other chert dominated facies, such as

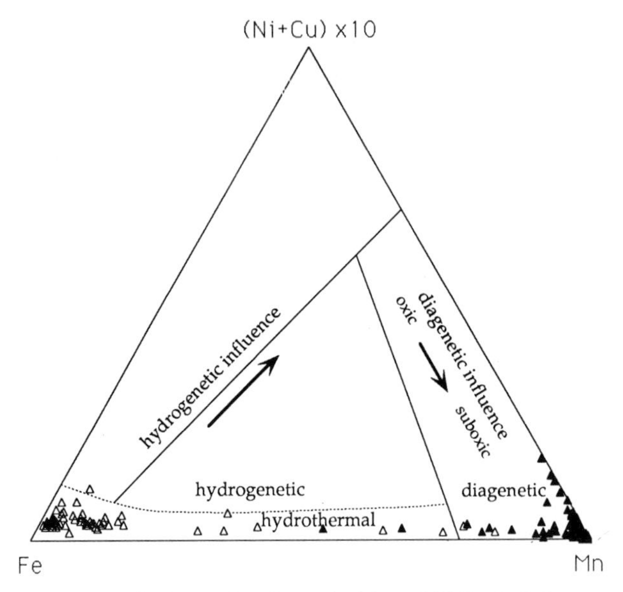

Figure 5. Mn-Fe-(Ni + Cu) diagram (after Bonatti 1971 and Halbach & Puteanus 1988) open triangles = red cherts; filled = black cherts nodular and layered ore of the Wahrah Formation.

parts of the Umar Group, neither manganese nor iron enrichments have been found.

Geochemistry

With respect to the genetic type of manganese enrichments a geochemical comparison with recent and fossil marine manganese deposits does not allow an unambiguous answer. The extreme Mn/Fe ratio and the low minor element content (basic classification parameters) point to a hydrothermal precipitation associated with volcanic activity or to a diagenetic enrichment in shallow water or hemipelagic sediments (Fig. 5) Both models however, contrast with REE-patterns and field evidence as well as with regional geological considerations.

Beside a set of 32 whole rock analyses of manganese cherts, 33 single nodules and manganese-rich layers were analyzed. For comparison and to calculate enrichment factors the average chert composition (based on 31 samples) was used (Table 1a–d). REE analyses from all typical lithologies were also available.

Based on statistical analyses and geological considerations four main element groups can be established.

1. Radiolaria -SiO_2
2. Clay minerals (mainly detritic), hematite -Al_2O_3, Fe_2O_3, TiO_2, Zr

Table 1a. Major and minor element composition of siliceous shales. Major elements in (weight %); Trace elements in (ppm).

Siliceous shales																								
SiO_2	TiO_2	Al_2O_3	Fe_2O_3	MnO	MgO	CaO	Na_2O	K_2O	P_2O_5	Ba	Co	Cr	Cu	Ga	Nb	Ni	Pb	Rb	Sr	Th	V	Y	Zn	Zr
80.91	0.52	6.81	4.00	0.43	0.03	1.96	0.01	1.14	0.12	89	31	58	35	6	15	49	<6	40	143	<13	60	25	58	89
82.31	0.39	6.00	3.88	0.52	1.58	1.12	0.59	1.09	0.09	115	22	45	61	6	<11	43	<6	38	123	<13	41	19	42	62
89.29	0.43	5.86	4.56	0.05	1.58	1.33	0.01	0.98	0.11	89	23	53	44	6	<11	41	<6	40	127	<13	91	18	48	65
82.10	0.46	6.26	4.20	0.14	1.51	1.32	0.05	0.97	0.14	73	23	47	58	6	11	38	<6	35	122	<13	49	26	48	80
83.93	0.35	5.27	3.10	0.04	1.49	1.56	0.07	0.99	0.11	136	19	51	31	6	<11	41	<6	36	130	<13	51	24	47	69
80.45	0.45	6.07	4.23	0.46	1.76	1.98	0.08	1.02	0.16	133	36	54	53	6	11	44	<6	39	146	<13	58	32	47	84
81.50	0.44	6.18	3.85	0.33	1.61	1.36	0.46	1.07	0.11	97	31	40	49	<6	15	46	<6	40	138	<13	54	22	47	97
71.58	0.41	6.74	3.78	9.78	1.37	1.13	0.18	0.99	0.08	176	n.d.	61	147	<6	<11	18	<6	44	314	<13	83	30	66	80
85.86	0.31	5.12	2.38	0.17	1.34	1.06	0.38	0.88	0.06	103	15	40	37	6	<11	33	<6	28	152	<13	60	11	31	67
78.27	0.50	7.64	4.81	0.09	1.80	1.39	0.20	1.55	0.15	129	n.d.	61	61	8	13	53	6	56	169	<13	62	27	62	86
76.15	0.62	8.13	5.54	0.26	1.90	1.56	0.31	1.79	0.17	166	50	74	73	9	12	60	<6	65	174	<13	69	35	71	105
74.01	0.69	8.57	6.18	0.15	1.97	1.95	1.06	1.98	0.24	153	55	74	68	10	20	55	<6	64	170	<13	86	45	72	125
76.18	0.66	8.23	5.71	0.08	1.96	1.61	0.16	1.82	0.24	175	32	77	216	9	18	57	<6	61	159	<13	60	45	80	119
74.41	0.66	8.96	6.20	0.05	2.11	1.38	0.54	2.00	0.15	125	42	77	74	10	14	72	<6	70	134	<13	104	28	79	108
75.05	0.47	5.86	4.21	6.55	1.35	1.22	0.11	1.01	0.15	131	50	55	83	<6	<11	83	<6	43	408	<13	62	21	63	80
73.04	0.87	12.60	7.70	0.21	2.41	1.64	0.39	2.31	0.19	173	82	68	128	18	16	91	<6	99	214	<13	128	45	115	155
69.65	0.47	7.03	4.02	12.80	1.42	1.56	0.71	1.37	0.15	197	32	65	238	7	<11	91	<6	64	247	<13	329	26	91	78
78.89	0.65	8.03	4.91	3.01	1.83	1.58	0.35	1.50	0.09	226	67	80	99	10	13	71	<6	60	196	<13	165	25	66	101
88.38	0.34	5.46	2.65	0.24	1.39	0.96	0.07	1.12	0.13	127	19	37	29	<6	<11	38	<6	34	140	<13	46	18	42	64
91.90	0.22	4.67	2.24	0.04	0.94	0.66	0.16	0.54	0.04	69	19	27	43	<6	<11	21	<6	17	123	<13	60	9	24	43
90.97	0.26	4.56	2.33	0.08	1.13	0.77	0.02	0.70	0.12	97	17	18	35	6	<11	28	<6	23	108	<13	31	22	32	47
90.58	0.27	5.41	2.26	0.06	1.04	0.69	0.42	0.70	0.07	81	22	25	37	<6	<11	24	<6	22	104	<13	40	13	27	47
84.52	0.45	6.69	4.40	0.74	1.73	1.36	0.22	1.13	0.11	123	26	37	68	<6	11	47	<6	44	138	<13	105	21	46	69
81.41	0.52	7.30	4.90	1.86	1.85	1.69	0.61	1.28	0.18	169	31	48	52	<6	<11	45	<6	47	322	<13	137	22	51	79
72.95	0.36	4.94	2.89	13.36	1.41	1.87	0.19	0.85	0.12	1077	50	46	137	<6	<11	76	<6	26	200	<13	433	14	58	54
82.28	0.53	6.79	4.92	0.56	2.02	2.33	0.12	1.21	0.14	145	43	39	38	<6	11	51	<6	42	248	<13	82	21	45	81
90.89	0.23	4.24	2.25	0.05	1.15	0.98	0.01	0.68	0.14	76	19	19	33	<6	<11	17	<6	<14	86	<13	10	9	17	22
84.04	0.27	4.58	2.74	0.16	1.51	2.40	0.40	0.73	0.07	238	19	54	45	6	<11	43	7	25	292	<13	55	16	39	58

Table 1b. Major and minor element composition of red cherts. Major elements in (weight %); Trace elements in (ppm).

Chert SiO_2	TiO_2	Al_2O_3	Fe_2O_3	MnO	MgO	CaO	Na_2O	K_2O	P_2O_5	Ba	Co	Cr	Cu	Ga	Nb	Nl	Pb	Rb	Sr	Th	V	Y	Zn	Zr
91.86	0.18	2.78	1.62	0.19	0.66	0.46	0.18	0.51	0.04	82	20	25	32	<6	<11	19	<6	16	73	<13	16	<5	15	<10
92.64	0.16	2.50	1.64	0.02	0.60	0.48	0.01	0.41	0.04	62	n.d.	13	28	<6	<11	15	<6	14	83	<13	38	7	17	27
93.27	0.13	2.09	1.19	0.06	0.49	0.54	0.10	0.31	0.05	46	27	10	28	<6	<11	17	<6	<14	83	<13	16	7	15	23
94.08	0.11	1.79	0.99	0.03	0.47	0.43	0.14	0.30	0.05	64	32	13	27	<6	<11	14	<6	<14	76	<13	21	8	12	21
91.95	0.15	2.46	1.32	0.20	0.66	0.72	0.68	0.39	0.05	100	24	17	24	<6	<11	17	<6	<14	86	<13	23	10	14	30
93.04	0.17	2.57	1.53	0.15	0.65	0.49	0.80	0.51	0.06	79	24	17	21	<6	<11	20	<6	15	72	<13	20	9	19	34
92.00	0.16	2.66	1.33	0.11	0.74	0.60	0.55	0.43	0.04	47	26	19	22	<6	<11	14	<6	16	124	<13	22	7	15	33
90.58	0.21	3.63	1.77	0.03	0.85	0.62	0.35	0.61	0.04	73	14	27	29	<6	<11	25	<6	20	114	<13	37	8	27	43
93.55	0.15	2.55	1.33	0.03	0.55	0.41	0.48	0.43	0.05	65	15	19	26	<6	<11	17	<6	13	89	<13	28	18	10	28
94.03	0.12	2.10	1.09	0.13	0.58	0.35	0.12	0.37	0.04	55	27	21	25	<6	<11	17	<6	<14	72	<13	22	7	13	22
94.50	0.14	2.25	1.14	0.04	0.41	0.33	0.93	0.45	0.05	57	28	22	25	<6	<11	14	<6	<14	74	<13	30	6	14	28
94.20	0.12	2.08	1.07	0.02	0.50	0.40	0.18	0.35	0.05	65	25	21	50	<6	<11	15	<6	<14	79	<13	15	9	13	23
93.83	0.14	2.19	1.15	0.02	0.52	0.51	0.82	0.40	0.04	47	15	18	24	<6	<11	16	<6	<14	59	<13	51	7	13	24
95.89	0.10	1.43	1.04	0.07	0.27	0.25	0.89	0.27	0.04	52	31	17	26	<6	<11	11	<6	<14	49	<13	23	6	9	17
96.80	0.08	1.72	0.70	0.33	0.41	0.99	0.45	0.26	0.03	46	14	10	27	<6	<11	14	<6	<14	125	15	13	6	13	18
96.05	0.13	2.11	1.24	0.04	0.51	0.40	0.03	0.41	0.06	65	23	24	36	<6	<11	18	<6	<14	68	<13	33	8	17	24
95.95	0.15	2.35	1.38	0.11	0.55	0.43	0.01	0.43	0.05	85	20	17	23	<6	<11	20	<6	<14	67	<13	17	6	18	27
94.22	0.18	2.79	1.66	0.15	0.83	0.68	1.09	0.58	0.08	81	25	10	31	<6	<11	19	<6	14	81	<13	23	10	24	31
96.24	0.12	2.36	1.18	0.07	0.42	0.38	0.70	0.40	0.04	394	18	103	36	21	14	65	35	178	352	18	133	27	111	141
98.16	0.07	1.38	0.73	0.05	0.24	0.22	0.17	0.21	0.02	43	25	9	23	<6	<11	8	<6	<14	59	<13	22	<5	7	14
98.02	0.12	1.69	0.92	0.01	0.36	0.21	0.05	0.27	0.03	54	25	19	15	<6	<11	12	<6	<14	50	<13	22	<5	14	24
94.70	0.18	2.97	1.64	0.16	0.77	0.63	0.06	0.49	0.05	78	14	22	27	<6	<11	19	<6	<16	157	<13	23	9	20	34
95.02	0.19	2.89	1.62	0.19	0.63	0.51	0.50	0.52	0.04	61	24	13	19	<6	<11	23	<6	<16	278	12	26	8	19	40
93.55	0.22	3.11	2.05	0.23	0.75	0.67	0.15	0.54	0.04	105	37	22	13	<6	<11	24	<6	16	179	<13	30	9	22	41
94.85	0.17	2.45	1.49	0.71	0.52	0.43	0.09	0.42	0.03	306	32	16	15	<6	<11	18	<6	14	92	<13	36	5	17	31
96.17	0.13	1.85	1.25	0.19	0.38	0.56	0.94	0.38	0.06	42	18	12	40	<6	<11	14	<6	<14	63	<13	24	8	12	23
98.19	0.08	1.28	0.71	0.08	0.18	0.20	0.77	0.25	0.04	42	29	13	11	<6	<11	9	<6	<14	35	<13	11	<5	7	13
95.17	0.11	1.90	1.31	0.04	0.53	1.53	0.11	0.32	0.06	48	15	23	37	<6	<11	15	<6	<14	84	<13	38	9	13	23
96.36	0.08	1.47	0.90	0.03	0.32	0.32	2.01	0.53	0.03	19	8	16	14	<6	<11	4	<6	<14	45	<13	14	<5	6	<10
94.93	0.12	2.43	1.09	0.05	0.63	0.53	0.08	0.37	0.07	104	22	25	41	<6	<11	27	<6	21	104	<13	28	19	33	46
96.89	0.11	1.82	1.03	0.01	0.45	0.37	0.07	0.28	0.05	28	13	10	16	<6	<11	12	<6	<14	60	<13	20	5	10	18

Table 1c. Major and minor element composition of manganese-rich cherts. Major elements in (weight %); Trace elements in (ppm).

Mn-cherts

SiO_2	TiO_2	Al_2O_3	Fe_2O_3	MnO	MgO	CaO	Na_2O	K_2O	P_2O_5	Ba	Co	Cr	Cu	Ga	Nb	Ni	Pb	Rb	Sr	Th	V	Y	Zn	Zr
80.58	0.08	1.45	0.53	12.54	0.39	0.49	0.67	0.23	0.04	1062	72	45	61	<6	<11	23	<6	<14	117	<13	201	<5	24	<10
75.73	0.04	0.69	0.28	17.78	0.25	0.65	0.54	0.16	0.05	1249	115	74	63	<6	<11	33	<6	<14	140	<13	328	<5	30	<10
85.05	0.03	0.55	0.19	10.36	0.31	0.53	0.57	0.12	0.03	749	86	31	40	<6	<11	17	<6	<14	141	<13	165	<5	19	<10
89.82	0.03	0.46	0.14	8.67	0.14	0.12	0.96	0.24	0.03	3237	97	15	33	<6	<11	31	<6	<14	377	<13	352	<5	23	<10
69.34	0.05	0.78	0.30	21.70	0.34	0.64	0.45	0.31	0.06	2170	132	53	25	<6	<11	55	<6	<14	598	<13	520	<5	33	<10
70.42	0.07	1.07	0.42	18.89	0.32	0.85	0.22	0.32	0.04	1849	107	28	36	<6	<11	42	<6	<14	495	<13	996	<5	22	<10
90.83	0.09	1.43	0.71	4.24	0.30	0.28	0.37	0.23	0.03	452	30	21	53	<6	<11	22	<6	<14	66	<13	172	7	14	12
83.74	0.06	0.97	0.45	10.85	0.26	0.51	0.28	0.17	0.04	1016	85	21	71	<6	<11	41	<6	<14	133	<13	328	<5	35	<10
91.69	0.14	2.35	1.25	1.24	0.54	0.54	0.18	0.37	0.05	126	18	11	26	<6	<11	17	<6	<14	99	<13	30	8	14	23
86.63	0.15	2.37	1.39	5.38	0.54	0.63	0.13	0.39	0.04	1011	38	26	43	<6	<11	28	<6	<14	144	<13	85	8	21	28
74.17	0.06	0.96	0.37	18.87	0.21	0.59	0.15	014	0.04	1173	131	37	84	<6	<11	50	<6	<14	127	<13	181	<5	36	<10
89.10	0.13	2.20	0.58	7.01	0.52	0.73	0.09	0.33	0.03	511	37	19	41	<6	<11	21	<6	<14	155	<13	54	5	16	22
71.41	0.02	0.49	0.12	25.00	0.13	0.39	0.07	0.34	0.08	2886	74	46	31	<6	<11	60	<6	<14	363	<13	396	<5	46	<10
47.72	0.03	0.50	0.11	39.60	0.12	1.01	0.05	0.72	0.11	8439	113	337	609	<6	<11	169	<6	<14	655	<13	1153	<5	51	<10
87.97	0.08	1.34	0.71	7.33	0.27	0.38	0.20	0.22	0.03	158	36	39	13	<6	<11	18	<6	<14	180	<13	99	<5	18	17
87.59	0.07	1.04	0.50	7.38	0.22	0.20	0.31	0.22	0.03	2430	93	29	80	<6	<11	33	<6	<14	154	<13	130	<5	16	10
89.00	0.06	0.94	0.48	7.73	0.15	0.18	0.44	0.16	0.04	163	34	20	152	<6	<11	15	<6	<14	77	<13	104	<5	12	<10
94.73	0.10	1.62	0.75	2.54	0.29	0.36	0.24	0.30	0.04	142	24	17	62	<6	<11	16	<6	<14	92	<13	93	5	15	16
n.d.	n.d.	n.d.	0.23	25.82	n.d.	n.d.	n.d.	n.d.	n.d.	247	45	170	619	8	<11	388	<6	<14	315	<13	695	<5	93	<10
n.d.	n.d.	n.d.	0.44	8.51	n.d.	n.d.	n.d.	n.d.	n.d.	254	45	63	420	<6	<11	122	21	<14	98	<13	80	<5	31	<10
n.d.	n.d.	n.d.	0.97	15.71	n.d.	n.d.	n.d.	n.d.	n.d.	457	38	50	172	<6	<11	31	<6	<14	201	<13	208	<5	37	17
60.52	0.10	1.54	0.72	32.02	0.35	0.50	0.11	0.27	0.11	1167	59	157	480	<6	<11	50	<6	<14	264	<13	440	<5	59	<10
90.93	0.21	3.20	1.57	2.86	0.70	0.50	0.07	0.54	0.06	470	31	12	50	<6	<11	28	<6	16	115	<13	105	8	28	38
82.98	0.16	2.19	1.25	11.11	0.43	0.35	0.39	0.40	0.05	2495	72	30	72	<6	<11	40	<6	<14	175	<13	224	5	40	20
92.59	0.11	1.72	0.95	4.11	0.33	0.36	0.54	0.36	0.04	689	51	15	63	<6	<11	32	<6	<14	119	<13	54	<5	21	17
88.88	0.12	1.88	0.89	7.00	0.47	0.75	0.04	0.32	0.03	1875	32	1429	2453	<6	<11	235	<6	<14	333	<13	2002	<5	319	<10
86.13	0.06	0.69	0.36	11.80	0.18	0.41	0.72	0.18	0.04	501	132	12	12	<6	<11	34	<6	<14	184	<13	132	<5	25	<10
90.79	0.08	1.09	0.62	6.72	0.21	0.25	0.01	0.19	0.03	452	85	18	66	<6	<11	23	<6	<14	89	<13	82	<5	17	10
88.11	0.09	1.58	0.71	8.44	0.32	0.40	0.12	0.27	0.04	548	41	23	243	<6	<11	21	<6	<14	152	<13	67	6	17	18
81.28	0.06	1.16	0.41	14.48	0.18	0.57	0.26	0.21	0.05	286	29	44	276	<6	<11	35	<6	<14	106	<13	215	<5	28	<10
79.57	0.08	1.34	0.53	16.54	0.26	0.33	1.04	0.28	0.09	253	48	41	163	<6	<11	35	<6	<14	91	<13	220	<5	26	10
56.70	0.07	1.17	0.45	36.28	0.22	0.27	0.02	0.17	0.10	601		84	1390	<6	<11	54	<6	<14	155	<13	716	<5	56	<10

Table 1d. Major and minor element composition of nodular and layered ore. Major elements in (weight %); Trace elements in (ppm).

Layered ore Fe_2O_3	MnO	Ba	Co	Cr	Cu	Ga	Nb	Ni	Pb	Rb	Sr	Th	V	Y	Zn	Zr
0.12	17.78	8756	111	47	141	<6	<11	46	<6	<4	495	<13	626	<5	31	<10
0.94	79.30	663	57	709	2491	<6	<11	140	<6	<14	1411	<13	693	15	199	<10
0.64	60.50	660	58	1376	2829	<6	<11	143	<6	<14	1443	<13	713	15	215	<10
0.20	65.40	3691	173	195	460	<6	<11	69	<6	<14	207	<13	486	<5	66	<10
0.28	73.80	4928	184	95	282	<6	<11	48	<6	<14	249	<13	442	<5	53	<10
0.45	32.00	4912	115	96	203	<6	<11	48	c6	<14	252	<13	424	<5	203	<10
0.44	30.90	1578	83	12	20	<6	<11	21	<6	<14	119	<13	107	<5	19	16
0.20	65.40	664	43	513	4068	<6	<11	153	<6	<14	624	<13	1046	<5	214	<10
0.33	41.40	14161	87	72	127	<6	<11	65	<6	<14	1374	<13	861	<5	49	<10
n.d.	n.d.	7646	110	28	132	<6	<11	48	<6	<14	484	<13	608	484?	27	<10
1.00	60.68	3399	103	1	1518	<6	<11	161	<6	<14	329	<13	2114	<5	181	<10
0.36	66.88	2266	100	1	1679	<6	<11	208	<6	<14	275	<13	2208	<5	235	<10
0.51	31.90	5626	123	142	265	57	<11	55	<6	<14	314	<13	450	<5	65	<10
0.26	68.30	636	45	34	115	<6	<11	22	<6	<14	138	<13	110	<5	24	18
0.19	74.34	2033	181	1	386	<6	<11	194	<6	<14	610	<13	2033	<5	299	<10
0.23	69.33	710	94	1	162	<6	<11	153	<6	<14	307	<13	850	<5	138	<10
0.38	31.76	13757	742	127	40	15	<11	148	<6	<14	1695	<13	413	<5	54	<10
0.27	70.24	384	200	1	95	<6	<11	162	<6	<14	225	<13	712	<5	147	<10
0.40	33.57	6857	362	106	17	399	<11	93	<6	<14	825	<13	414	<5	59	<10
0.36	57.84	867	142	787	41	<6	<11	154	<6	<14	255	<13	516	<5	99	<10
0.03	51.26	583	40	1	1	<6	<11	74	<6	<14	526	<13	2293	<5	115	<10

Table 1d (continued)

Nodular ore

Fe_2O_3	MnO	Ba	CO	Cr	Cu	Ga	Nb	Ni	Pb	Rb	Sr	Th	V	Y	Zn	Zr
0.93	61.20	10254	77	1	951	<6	<11	219	<6	<14	732	<13	1375	<5	185	<10
0.12	74.89	13513	43	1	2461	<6	<11	113	<6	<14	709	<13	1436	<5	233	<10
0.11	63.01	298	51	1	1167	<6	<11	193	<6	<14	670	<13	2598	<5	195	<10
0.45	76.90	583	47	1	8639	<6	<11	165	<6	<14	256	<13	1138	<5	205	<10
0.37	75.40	496	61	1	7551	<6	<11	149	<6	<14	224	<13	1138	<5	202	<10
0.17	70.80	35465	224	1	3073	<6	<11	38	<6	<14	1675	<13	2812	<5	83	<10
0.11	73.30	68	10	1	1891	<6	<11	262	<6	<14	1418	<13	696	<5	287	<10
0.83	60.30	2779	411	754	1602	<6	<11	92	<6	<14	876	<13	1041	<5	241	<10
0.12	73.34	68	43	1	4544	<6	<11	191	<6	<14	184	<13	1765	<5	235	<10
0.12	71.91	12633	37	1	2464	<6	<11	107	<6	<14	683	<13	712	<5	236	<10
0.21	74.88	1824	291	1	2310	<6	<11	132	<6	<14	607	<13	1348	<5	255	<10
0.43	56.56	971	81	1	4396	<6	<11	156	<6	<14	307	<13	2045	<5	186	36
0.46	75.02	1160	136	1	11253	<6	<11	161	<6	<14	420	<13	884	<5	214	35
0.51	75.41	1293	193	1	8413	<6	<11	167	<6	<14	421	<13	982	<5	212	39

3. Carbonate particles including plankton -$CaCO_3$, P_2O_5, Sr, (Ba)
4. Manganese particles -Mn, Cu, Ni, Co, (Ba, K_2O)

The different sources and carrier phases of Mn and Fe partly explain the strong fractionation of Mn and Fe (up to Mn/Fe 690).

Beside the strong positive correlation of 0.98 of Al with Fe in cherts and shales and 0.90 in manganese cherts (implying detritically derived Fe), the positive correlation between elements of group 3 (the biogenic carbonate factor) and Mn is remarkable (Table 2). CaO shows only a weak correlation of 0.63 because $CaCO_3$ is mobilized during late diagenesis and tectonic events. But manganese replaced calcareous foraminifers and the preferential occurrence of calcite together with manganese nodules and layers verify the statistical results.

The constant Al/Fe ratio in cherts, shales, and manganese cherts (Fig. 3) indicates that composition of the clay fraction is almost constant and Fe is not mobilized over large distances. The negative correlation between the detritically derived elements (group 2) and the manganese enrichment supports that manganese is concentrated during times of relative reduced detrital sedimentation.

Using ternary discrimination diagrams the separation of nodules and layered ore is geochemically justifiable. As shown in Fig. 6 the nodular ores are dominated by a relative copper enrichment with only minor Ba contents; the layered ores are characterized by a wide spectrum of Mn/Ba ratios. The third field reflects a transition zone between nodular and layered ore types, already seen in the field.

The Ni-Cu-Co diagram (Fig. 7) again shows the geochemical differences between "nodules" and manganese layers, but more important is the clear separation between recent marine manganese deposits and the Wahrah ore. Analyses used for comparison include hydrogenetic, diagenetic and hydrothermal ores from several localities (cf. data compiled by Baturin 1988 and Halbach et al. 1988).

Fig. 8 shows the element distribution patterns of nodules, manganese layers and associated cherts in a normalized plot. As a normalizing factor the average Wahrah-chert composition was used. Based on enrichment factors (ef) three element groups can be established:

I. Mn, Cu, V with ef-values > 40
II. Zn, Ni, Sr, Co, Ba with ef-values between 5–15
III. Fe, Zr, Cr are depleted ($ef < 0$)

Generally, the element distribution patterns are the same in nodular and layered ores implying similar primary deposition history for both ore types.

The cherts associated with the nodules do not show any significant difference to the normal red cherts. The higher manganese content in these cherts

Table 2a. Correlation matrix of siliceous shales and red cherts, nodular and layered ore.

n = 60	SiO_2	TiO_2	Al_2O_3	Fe_2O_3	MnO	MgO	CaO	Na_2O	K_2O	P_2O_5	Ba	Co	Cr	Cu	Ga	Nb	Ni	Rb	Sr	V	Y	Zn	Zr
SiO_2	1.00																						
TiO_2	−0.92	1.00																					
Al_2O_3	−0.91	0.98	1.00																				
Fe_2O_3	−0.91	0.99	0.98	1.00																			
MnO	0.52	0.23	0.22	0.20	1.00																		
MgO	−0.85	0.88	0.91	0.90	0.21	1.00																	
CaO	−0.80	0.75	0.75	0.76	0.29	0.77	1.00																
Na_2O	0.04	−0.09	−0.07	−0.07	−0.02	−0.03	0.07	1.00															
K_2O	−0.90	0.98	0.97	0.97	0.20	0.88	0.71	0.02	1.00														
P_2O_5	−0.84	0.88	0.85	0.88	0.26	0.81	0.69	−0.13	0.88	1.00													
Ba	−0.43	0.25	0.24	0.22	0.69	0.28	0.35	−0.09	0.22	0.24	1.00												
Co	−0.63	0.71	0.65	0.68	0.33	0.53	0.37	−0.18	0.66	0.57	0.37	1.00											
Cr	−0.78	0.79	0.77	0.78	0.27	0.68	0.62	−0.06	0.79	0.69	0.40	0.54	1.00										
Cu	−0.75	0.64	0.63	0.62	0.65	0.56	0.47	−0.08	0.65	0.66	0.43	0.49	0.62	1.00									
Ga	−0.44	0.56	0.59	0.55	−0.04	0.46	0.28	−0.01	0.59	0.45	0.18	0.46	0.79	0.39	1.00								
Nb	−0.76	0.87	0.83	0.85	0.08	0.70	0.59	−0.15	0.84	0.80	0.22	0.60	0.85	0.53	0.72	1.00							
Ni	−0.89	0.84	0.83	0.82	0.57	0.74	0.64	−0.15	0.81	0.74	0.53	0.70	0.87	0.76	0.65	0.78	1.00						
Rb	−0.55	0.61	0.62	0.61	0.16	0.52	0.37	−0.04	0.62	0.52	0.35	0.46	0.89	0.48	0.92	0.77	0.78	1.00					
Sr	−0.56	0.51	0.51	0.51	0.42	0.50	0.53	−0.06	0.46	0.45	0.40	0.39	0.64	0.41	0.51	0.54	0.73	0.65	1.00				
V	−0.65	0.46	0.46	0.44	0.87	0.43	0.46	−0.03	0.44	0.41	0.81	0.46	0.54	0.71	0.27	0.35	0.73	0.46	0.48	1.00			
Y	−0.80	0.88	0.86	0.87	0.14	0.76	0.64	−0.13	0.88	0.88	0.24	0.60	0.85	0.65	0.73	0.92	0.80	0.75	0.50	0.39	1.00		
Zn	−0.79	0.82	0.82	0.80	0.36	0.69	0.57	−0.11	0.82	0.74	0.44	0.62	0.93	0.71	0.83	0.86	0.94	0.92	0.67	0.62	0.88	1.00	
Zr	−0.78	0.86	0.85	0.85	0.16	0.74	0.60	−0.11	0.85	0.77	0.32	0.63	0.92	0.58	0.83	0.95	0.87	0.89	0.64	0.45	0.92	0.95	1.00

Table 2b. Correlation matrix of maganese rich cherts, nodular and layered ore.

n = 60	SiO_2	TiO_2	Al_2O_3	Fe_2O_3	MnO	MgO	CaO	Na_2O	K_2O	P_2O_5	Ba	Co	Cr	Cu	Ni	Rb	Sr	V	Zn	Zr
SiO_2	1.00																			
TiO_2	0.22	1.00																		
Al_2O_3	0.36	0.96	1.00																	
Fe_2O_3	0.25	0.92	0.90	1.00																
MnO	−0.99	−0.27	−0.40	−0.27	1.00															
MgO	0.17	0.84	0.88	0.83	−0.20	1.00														
CaO	−0.58	−0.08	−0.11	0.13	0.63	0.15	1.00													
Na_2O	−0.27	−0.30	−0.34	−0.38	0.22	−0.28	0.05	1.00												
K_2O	−0.39	0.26	0.14	0.27	0.42	0.22	0.44	−0.26	1.00											
P_2O_5	−0.87	−0.22	−0.27	−0.15	0.88	−0.09	0.67	0.28	0.17	1.00										
Ba	−0.34	−0.09	−0.23	−0.22	0.33	−0.22	−0.07	−0.07	0.73	−0.03	1.00									
Co	−0.01	−0.45	−0.55	−0.49	0.03	−0.41	−0.07	0.10	0.10	−0.24	0.39	1.00								
Cr	−0.27	0.03	0.02	0.08	0.27	0.16	0.46	0.04	0.18	0.29	0.01	−0.15	1.00							
Cu	−0.73	−0.05	−0.14	−0.07	0.70	0.01	0.56	0.41	0.09	0.75	−0.03	−0.27	0.72	1.00						
Ni	−0.75	−0.07	−0.20	−0.13	0.75	−0.06	0.35	0.00	0.31	0.65	0.31	−0.11	0.59	0.75	1.00					
Rb	0.36	0.77	0.85	0.76	−0.40	0.78	−0.10	−0.27	0.10	−0.20	−0.24	−0.48	−0.19	−0.23	−0.30	1.00				
Sr	−0.84	−0.28	−0.37	−0.25	0.86	−0.15	0.56	0.07	0.55	0.77	0.51	0.06	0.27	0.53	0.67	−0.31	1.00			
V	−0.57	−0.08	−0.21	−0.18	0.55	−0.05	0.24	0.03	0.42	0.33	0.51	0.11	0.71	0.63	0.83	−0.39	0.56	1.00		
Zn	−0.69	0.00	−0.11	−0.02	0.69	0.07	0.44	0.11	0.17	0.69	0.09	−0.26	0.72	0.87	0.93	−0.25	0.61	0.76	1.00	
Zr	0.47	0.82	0.89	0.82	−0.50	0.75	−0.17	−0.32	0.01	−0.32	−0.35	−0.48	−0.31	−0.36	−0.44	0.90	−0.46	−0.52	−0.39	1.00

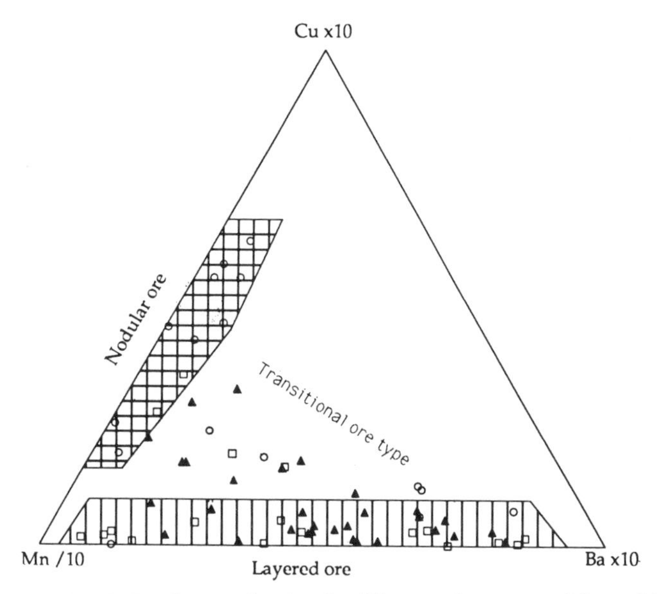

Figure 6. Ternary Cu-Mn-Ba diagram showing the differences between nodular and layered ore types of the Wahrah Formation. The transitional ore type includes nodules as well as layered ores. Circles = nodular ore, squares = layered ores, solid triangles = whole rock composition of manganese enriched cherts.

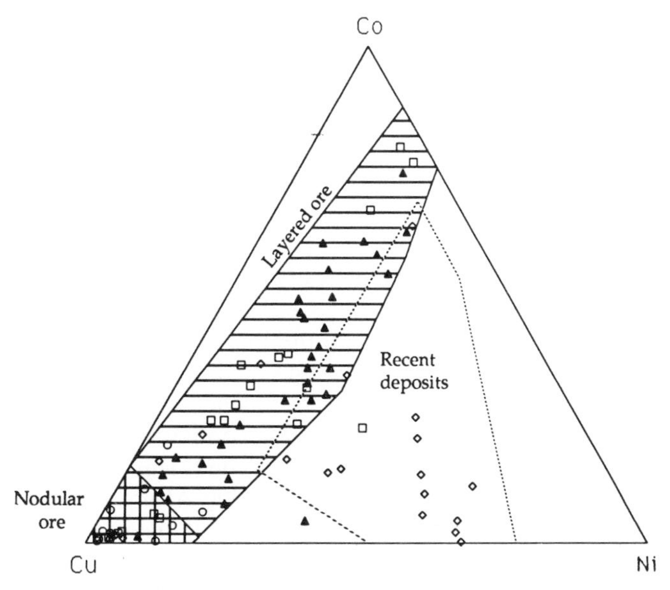

Figure 7. Co-Ni-Cu triangle showing the differences between recent marine ferromanganese deposits and the Wahrah Formation ores. The Wahrah deposits are compared with hydrogenetic, diagenetic and hydrothermal ore enrichments from several localities (data compiled by Baturin 1988, Halbach 1988, and Toth 1980). Circles = nodular ore, squares = layered ores, solid triangles = whole rock composition of manganese enriched cherts, open diamonds = recent deposits.

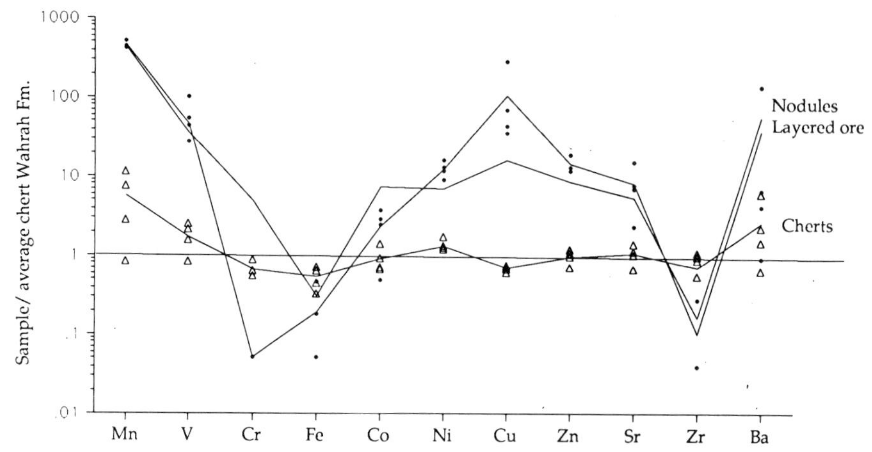

Figure 8. Element distribution patterns of nodular ore and associated cherts in a normalized plot. As normalizing factor the average Wahrah chert composition was used. For comparison the element distribution pattern of the layered ore is also plotted.

is due to minor amounts of microparticles, but it is to keep in mind that enrichment factors of 10 are caused by manganese contents of only 1% MnO.

Assuming a diagenetic growth of manganese nodules with a metal supply from the underlying sediments one would expect more pronounced differences between the normal cherts and the manganese nodule bearing cherts, especially in the case of highly enriched elements (such as Mn,Cu,V). A relative decrease of detrital supply during times of manganese accumulation is indicated by slight negative concentration values of Fe and Zr representing the detrital influence in this diagram.

Numerous papers dealing with REE distribution patterns in marine ferromanganese nodules and sediments have been published over the last years (Glasby 1987 et al. and references therein). By normalizing the REE data to an average shale composition (NASC) it should be possible to distinguish between hydrothermally and hydrogenetically enriched ore deposits.

All authors reported a strong negative Ce anomaly and a minor negative Eu anomaly in fast accumulating hydrothermal deposits near spreading ridges. Generally, hydrothermal deposits show similar REE distribution patterns as sea water, whereas the hydrogenetic deposits show an " .mirror-image relationship compared to the sea water composition" (Piper 1974). Fig. 9 shows the REE patterns of manganese enriched samples in the Wahrah Formation compared to recent manganese deposits. Although lithologically and geochemically different, no significant difference between the nodular and layered ore types in the normalized plot is visible.

Both are characterized by a positive Ce anomaly and an obvious enrichment of Lu compared to Yb. Recent (hydrogenetic) ferromanganese nodules have analogue REE patterns but the Wahrah deposits are strongly depleted in total REE content with values of 0.1–1 and 5–20 respectively (Toth 1980).

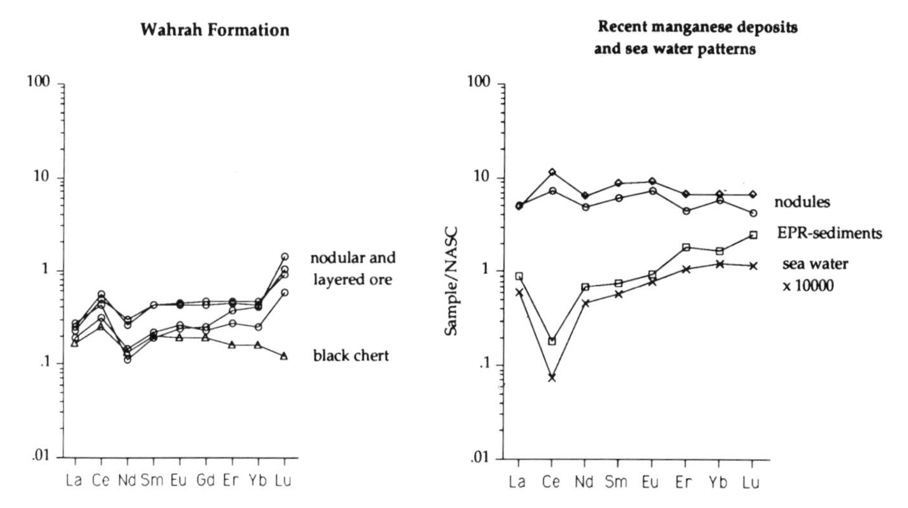

Figure 9. REE patterns of the Wahrah Formation ores (right) compared with East Pacific Rise Sediments (hydrothermal), seawater and recent manganese nodule REE distribution patterns (left); (data from Baturin 1988, Henderson 1983).

In a study of ferromanganese nodules Elderfield et al. (1981) found a strong positive correlation between increasing Mn/Fe ratios and a decreasing total REE content. He proposed that ferrooxyhydroxide phases play an important role in the accumulation processes of REE, although the "nature of the particles carrying REE" is not quite understood. Therefore, the low total REE content can be explained by the extreme Mn/Fe ratio. As suggested by Toth (1980) an La/Ce ratio < 1 (0.25 in North Pacific nodules) for recent nodules supports a hydrogenetically controlled formation process of the Wahrah deposits with values about 0.25–0.30.

Discussion

Sedimentary structures in the manganese enriched cherts point to a manganese enrichment process at the sediment water interface, with a relatively early consolidation of the manganese-rich layers. The generally well preserved biogenic remains also support an early accumulation process. Otherwise the fine structures would have been destroyed. The primary ore are small manganese-rich microparticles transported and deposited by benthic currents. Hayes (1988) showed the variability of benthic current systems with respect to their velocity, frequency and direction in the Pacific. By varying the supply of microparticles or the current intensity the microparticles are concentrated to finer laminae or to thicker layers.

Sinking radiolaria contribute to the "growth" of layered and nodular ore.

The outer rim of the nodules is due to an early diagenetic impregnation of the chert.

The fractionation of Mn from Fe is the result of their different sources and their incorporation to different carrier phases (detrital clay minerals and microparticles, respectively). Similar element distribution patterns of layered and nodular ore, the simultaneous occurrences of both in one chert layer and transitional lithologic appearances, support a similar primary element source. The geochemical differences can be attributed to minor postdepositional element mobilization processes. However, comparisons of normal ("manganese-free") red cherts with cherts directly associated with the manganese ore do not support element redistribution processes over distances larger than a few cm.

Manganese Accumulation Models

Diagenetic Enrichment

Geochemical and sedimentological characteristics of the red cherts (the host rock of the manganese deposits) indicate a depositional environment comparable to recent radiolarian oozes. The radiolarian ooze facies is characterized by slow sedimentation rates and a low organic C-content resulting in an entirely oxic sedimentary milieu. According to Müller and Mangini (1980) the organic C-content in the siliceous clay facies is below 1% at the sediment water interface and rapidly decreases to a constant value of 0.1–0.2% after a few cm. Oxic pore waters up to a depth of over 10m are found in pelagic sediments in the pacific ocean. According to Bender (1971) and Elderfield (1976) and considering the oxic environment of the radiolarian ooze facies a diagenetic metal supply from the underlying sediments over large distances is not likely (cf. oxic diagenesis Dymond et al. 1984 and deep diagenesis Müller et al. 1988, respectively). Only the uppermost cm in pelagic oxidizing sediments can contribute to the growth of manganese nodules (surficial diagenesis). But geochemical comparisons of the nodule bearing cherts and the manganese free cherts directly underlying the manganese layered cherts with the average Wahrah-chert composition do not favor a diagenetic enrichment process.

In contrast to the oxic environment in the siliceous clay and radiolarian ooze facies, continental-margin slope and hemipelagic regions are characterized by a thin oxidized upper sediment layer which overlays sediments with reducing conditions.

The reducing environment is the result of high sedimentation rates and the oxygen consumption by the decay of buried organic material. In such cases a mobilization of manganese and associated elements followed by an upward migration of metal enriched pore fluids, is a well known process (e.g. Bonatti et al. 1971).

Hydrogenetic Manganese Precipitation

A hydrogenetic (sensu stricto) manganese enrichment with slow precipitation of manganese from sea water is not favored, because of the low minor element content and the extreme fractionation of Mn/Fe. The REE patterns in the Wahraḥ Formation deposits are similar to hydrogenetically precipitated ore. But the total REE content is too low (factor 10) for typical hydrogenetic enrichments. Based on biostratigraphic dating the time span during which the manganese cherts of both horizons (including the manganese free cherts between them) are deposited is 20 my at the most. With growth rates of 1–7 mm/my proposed for recent hydrogenetic deposits (Baturin 1988), a hydrogenetic ore enrichment process is impossible considering that a single layered chert bed can contain up to 10 cm manganese-rich layers. The resulting accretion rates would exceed any known hydrogenetic accumulation process. Therefore, a "normal" hydrogenetic metal enrichment, analogous to the ferromanganese enrichments in recent oceans, is considered here as an unlikely model for the Wahrah Formation.

Hydrothermal Enrichment

A hydrothermal origin of the manganese enrichment has to be discussed here as well. The geochemical data could imply a hydrothermal manganese source far away from the "Wahrah-basin", allowing the strong fractionation of iron from manganese and the almost complete overprint of the hydrothermal element distribution patterns (including REE) by hydrogenetic and detrital influences. A recent analogue of this metal enrichment process would be the metalliferous sediments of the Bauer Deep in the Pacific. Metal enriched solutions, contributing to the sediments, are thought to originate at the East Pacific Rise and are transported by currents into the Bauer Deep (Heath and Dymond 1977). The higher pelagic/detrical influence in the Bauer Deep compared to the East Pacific Rise sediments is represented by higher Al values in Fig. 3. Because of the strong overprinting processes during transport, deposition and diagenetic alterations, it has to be emphasized that an unambiguous genetic answer based on geochemical evidence is not possible.

If a primary hydrothermal element source is assumed several problems arise. The Semail ophiolite with its Albian-Cenomanian age can not be the source for the Neocomian manganese enrichment in the cherts. Furthermore, no hydrothermal/magmatic activity during the relevant time span is known from the Hamrat Duru Basin or the more distal Umar Basin.

Therefore a hydrothermal source would have to be located "outside" the Hawasina Basin.

The ophiolites of Masirah are the only other known source of magmatic activity in the region. Only one of the two radiolarian dated sediments associated with the Masirah volcanics gave a Neocomian age (Beurrier 1988).

If the hydrothermal activity is the source for the manganese enrichment, also iron-rich formations must have formed which are not known to the authors. Furthermore the manganese enriched cherts in the Ras al Hadd area, located nearer to the hydrothermal metal source, would not show the almost identical geochemical and lithologic characteristics like the deposits in the Al Hammah Range. (In recent oceans always geochemical variations of the ferromanganese enrichments are reported with increasing distance from active hydrothermal centers). A manganese supply originating from the Masirah area would lead to a more widespread distribution of metalliferous sediments (e.g. also in the Umar Group) and cannot explain the unique occurrence of manganese-rich cherts in the Wahrah basin.

Thus considering the geochemical, sedimentological features, field relations as well as the history of the Hawasina Basin we propose an alternative model of manganese enrichment.

Proposed Manganese Transport and Accumulation Model

The model proposed here should explain:

- the sedimentary structures supporting a transport and deposition of manganese microparticles by benthic currents and a sedimentary formation of manganese-rich layers of variable thickness;
- the regional distribution patterns of manganese enriched cherts in the Al Hammah Range;
- the relatively high accumulation rates of manganese-rich layers in the red cherts;
- the contribution of biogenic particles especially radiolaria to the growth of layered and nodular ore;
- the constant Al/Fe ratio and the extreme separation of Mn/Fe;
- the low minor element content;
- the geochemical differences and similarities of nodular and layered ores and REE patterns.

The model (Fig. 10) assumes a "primary" manganese supply in dissolved and particulate form the continent similar to the manganese input in recent oceans. Calculations of the marine manganese balance e.g. by Bender et al. (1977) or Sunbey et al. (1981) proved that the supply from continents can explain the total manganese content in nodules and crusts. Without this continental source (only by hydrothermal input to the ocean water) the excess of manganese in pelagic sediments cannot be explained.

After a relatively short distance manganese is deposited on the continental margin together with terrigenous particles, iron and organic material (autochthonous sediments of the Arabian platform, Sathan and Kahmah Group). High sedimentation rates together with the oxygen consuming decay of bu-

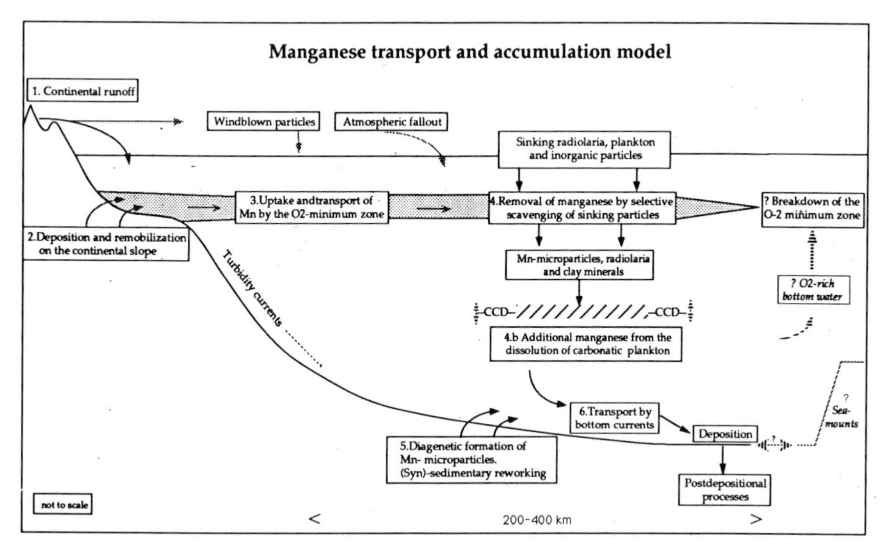

Figure 10. Proposed manganese transport and accumulation model.

ried organic matter lead to reducing sedimentary environments analogue to conditions found in recent oceans. This environment favors the mobilization of manganese and the upward migration of the diagenetically enriched pore fluids, which is a well known, generally accepted process in the continental slope sediments. Due to the different geochemical behavior of iron and manganese (Krauskopf 1957) Fe remains mainly fixed in the sediment column. Sunbey et al. (1981) showed that after complex recycling processes manganese is "exported" towards the open ocean in significant amounts.

The oxygen-minimum zone may act as a medium for the horizontal transport of manganese towards the open sea. Geochemical seawater studies of Klinkhammer and Bender (1980) showed that manganese is especially enriched in this zone and can be transported over large distances. Martin and Knauer (1984) calculated that about 70% of the dissolved manganese in the oxygen-minimum zone is supplied through adjective, diffusive processes from the continental slope region. Further manganese is attributed to the oxygen-minimum zone by the breakdown of organic material. The dissolution of $CaCO_3$ with an OH- release favors the precipitation of manganese onto carbonatic particles. Thus, ph values of ≈ 9 are found in the micro-environment of globigerina tests, (P. Halbach pers. commun., 1989) allowing the oxidation of dissolved Mn^{2+} to hydrated manganese oxides (manganese microparticles). During (selective) element scavenging onto sinking particles and the formation of manganese microparticles during oxidation processes below the oxygen-minimum zone a further element separation takes place. Element distribution patterns in grain size fractions of the manganese nodule field in the Pacific (Halbach et al. 1979) demonstrate that Fe is dominantly bound to the finest sediment fraction between 1–4 μm. Mn,Cu and Ni show a pronounced maximum in the grain size class between 4–20 μm; Si is enriched in the coarsest fraction.

Reaching the sediment water interface the particles are further transported together with microparticles formed in the uppermost sedimentary layer or the sediment-water interface by benthic currents. Due to the different hydrodynamic behavior of the element carrier phases a separation and enrichment of elements takes place during transport. Controlled by the relief of the sea bottom and the intensity of the currents, manganese microparticles are enriched to laminae and thicker layers in small basins. Additionally to this current derived material, sinking radiolaria contribute to the formation of the manganese-rich layers. This sedimentary enrichment of "hydrogenetically formed" manganese microparticles explains the fractionation of manganese from iron which is bound to the finer clay fraction.

The exact process of nodule formation is not yet clear, but the nodular ore type is possibly due to locally accumulated microparticles. This, as well as recrystallization processes, could explain the missing concentric layering and nucleus. As opal A, organic matter and calcite are the main carrier phases for (biogenic) copper, sufficient amounts of copper (and minor manganese) are mobilized during their early decomposition and can contribute to the nodules preferentially "growing" after the deposition of the microparticle-rich layers. Small scale diagenetic mobilization of manganese and copper have been proven by Hartmann (1979) and Boyle et al. (1977), respectively. Sedimentary structures indicate that the nodule formation process must take place at the sediment water interface or in the uppermost unconsolidated layer of the radiolarian ooze.

Postdepositional processes such as recrystallization of the hydrous manganese oxides to pyrolusite and cryptomelane, the dehydration and chertification of the radiolarian ooze and tectonic stress lead to several structural modifications in the manganese enriched cherts, which will be discussed in a later paper.

Acknowledgements

Mohamed Bin Hussain Bin Kassim, Director General of Minerals and Dr. Hilal Al Azry, Director of Geological Survey, Ministry of Petroleum and Minerals, Sultanate of Oman are thanked for giving us the possibility to work in Oman and their kind help. Mr. Cherian Zachariah put his maps and knowledge of the Al Hammah Range at our disposal, for which we are grateful. The radiolaria dating was done by R. Jud. Our colleagues I. Mercolli, D. Biaggi and Ph. Steinmann contributed to the work and the discussions. We thank F. Béchennec and W. Blendinger for the carefully review of the first manuscript. The work was financially supported by the Swiss National Science Foundation and the "Hochschulstiftung".

References

Andreyev, S.N. and Kulikov, A.N. 1987. Sedimentation rates in areas of manganese-nodule formation in the Pacific. In: Transactions (Doklady) of the USSR, Academy of sciences: earth science section, 297 (6): 75–77.

Baturin, G.N. 1988. The Geochemistry of Manganese nodules in the Ocean. D. Reidel Publishing Company, 243 pp.

Béchennec, F. 1987. Géologie des nappes Hawasina dans les parties orientale et centrale des montangnes de'Oman. Thése University Tierre et Marie Currie, Paris. Documents du BRGM 127, 474 pp.

Béchennec, F. Le Métour, J. Rabu, D., Villey, M. and Beurrier, M. 1988. The Hawasina Basin: a fragment of a starved passive continental margin, thrust over the Arabian Platform during obduction of the Semail Ophiolite., Tectonophysics, 151 (1–4): 223–243.

Beurrier, M. 1988. Geologie de la nappe ophiolitque de Samail dans les parties orientale et centrale des montagnes d'Oman. Thése University Tierre et Marie Currie, Paris. Documents du BRGM 128, 412 pp.

Bender, M.L., 1971. Does upward diffusion supply the excess manganese in pelagic sediments., Journal of Geophysical Research, 76: 4214–4215.

Bender, M.L., Klinkhammer, G.P. and Spencer, D.W., 1977. Manganese in sea water and the marine manganese balance., Deep Sea Research, 24: 799–812.

Bonatti, E., Fischer, D.E., Joensuu, V. and Rydell, M.S., 1971. Postdepositional mobility of some transition elements, phosphorous, uranium and thorium in deep sea sediments., Geochimica et Cosmochimica Acta, 35: 189–201.

Bonatti, E., Kraemer, T. and Rydell, H., 1972. Classification and genesis of submarine iron-manganese deposits. In: Horn, D.R. (Ed), Ferromanganese Deposits on the Pacific Floor. National Science Foundation, Washington D. C., pp. 149–166.

Boyle, E.A., Sclater, E.R. and Edmund, J.M., 1977. The distribution of dissolved copper in the Pacific., Earth and Planetary Science Letters, 37: 38–54.

De Wever, P., Bourdillon des Grissac, Ch. and Béchennec, F., 1990. Permian to Cretaceous radiolarian biostratigraphic data from the Hawasina complex, Oman Mountains. In: Robertson, A.H.F., Searle, M.P., Ries, A.C., (Eds). The Geology and tectonics of the Oman region. The Geological Society, Special Publication, 49: 225–238.

Dymond, J. Lyle, M., Finney, B., Piper, D.Z., Murphy, K., Conrad, R. and Pisas, N., 1984. Ferromanganese nodules from MANOP sites H,S, and R – Control of mineralogical and geochemical composition by multiple accretionary processes., Geochimica et Cosmochimica Acta, 48: 931–949.

Elderfield, H., 1976. Manganese fluxes to the oceans., Marine Chemistry, 4: 103–132.

Elderfield, H. Hakesworth, C.J., Greaves, M.J., 1981. Rare earth element geochemistry of the oceanic ferromanganese nodules and associated sediments., Geochimica et Cosmochimica Acta, 45: 531–528.

Glasby, G.P., Gwozdz, R., Kunzendorf, H., Friedrich, G.T., Thijssen, T., 1987. The distribution of rare earth and minor elements in manganese nodules and sediments from the equatorial and S.W. Pacific., Lithos, 20: 97–113.

Glennie, K.W., Boeuf, M.G.A., Hughes Clarke, M.W., Moody Stuart, M. Pilaar, W.F.H. and Reinhardt, B.M., 1974. Geology of Oman Mountains. Verhandelingen van het Koninklijk Nederlands geologisch mijnbouwkundig Genootschap, 31, 423 pp.

Halbach, P., Friederich, G. and Stockelberg, U., (Eds) 1988. The manganese nodule belt of the Pacific Ocean, Enke, Stuttgart, 254 pp.

Halbach, P. and Puteanus, D. 1988. Geochemical trends of different genetic types of nodules and crusts. In: Halbach, P., Friederich, G. and Stockelberg, U. (Editors) The manganese nodule belt of the Pacific Ocean, Enke, Stuttgart, 61–67.

Halbach, P., Rehm, E. and Marching, V., 1979. Distribution of Si, Mn, Fe, Ni, Cu, Co, Zn, Pb, Mg and Cu in grain-size fractions of sediment samples from a manganese nodule field in the Central Pacific Ocean., Marine Geology, 29: 237–253.

Hartmann, M., 1979. Evidence for early diagenetic mobilization of trace elements from discolorations of pelagic sediments., Chemical Geology, 26: 277–293.

Hayes, S.P., 1988. Benthic currents in the deep ocean. In: Halbach, P., Friederich, G. and Stockelberg, U. (Eds) The manganese nodule belt of the Pacific Ocean, Enke, Stuttgart, pp. 90–102.

Heath, G.R. and Dymond, J., 1977. Genesis and transformation of metalliferous sediments from the East Pacific Rise, Bauer Deep and central Basis, North-West Nazca Plate., Geological Society of America Bulletin, 88: 723–733.

Iijima A. and Utada, M., 1983. Recent developments in sedimentology of siliceous deposits in Japan. In: Iijima A., Hein, J.R., Siever, P., (Eds), Siliceous deposits in the Pacific region. Developments in Sedimentology, 36, Elsevier, Amsterdam, pp. 45–64.

Kickmaier, W. and Peters, Tj., 1990. Manganese occurrences in the Al Hammah Range- Wahrah Formation. In: Robertson, A.H.F., Searle, M.P., Ries, A.C., (Eds). The Geology and tectonics of the Oman region., The Geological Society, Special Publication, 49: 239–249.

Klinkhammer, G.P. and Bender, L.M., 1980. Distribution of manganese in the Pacific Ocean., Earth and Planetary Science Letters, 46: 361–384.

Krauskopf, K.B., 1957. Separation of manganese from iron in sedimentary processes., Geochimica et Cosmochimica Acta, 12: 61–84.

Martin, J.H. and Knauer, G.A., 1984. VERTEX: manganese transport through oxygen minima., Earth and Planetary Science Letters, 67: 35–47.

Matsumoto, R. and Iijima, A., 1983. Chemical sedimentology of some Permo-Jurassic and Tertiary bedded cherts in central Honshu, Japan. In: Iijima A., Hein, J.R., Siever, P., (Eds), Siliceous deposits in the Pacific region. Developments in Sedimentology, 36: 175–191, Elsevier, Amsterdam.

Müller, P.J.and Mangini A., 1980. Organic C-carbon decomposition rates in sediments of the Pacific manganese nodule belt dated by 230 Th and 231Pa., Earth and Planetary Science Letters, 51: 94–114.

Müller, P.J., Hartmann, M. and Suess, E., 1988. In: Halbach, P., Friederich, G. and Stockelberg, U. (Eds) The manganese nodule belt of the Pacific Ocean, pp. 71–90, Enke, Stuttgart.

Nisbet, E.G.and Price, J., 1974. Turbidite derived cherts. In: Hsü, K.J. and Jenkyns, H.C. (Eds) Pelagic Sediments: on Land and under Sea. Special Publication of the International Association of Sedimentologists, 1: 351–366.

Piper, D.Z., 1974. Rare earth elements in the sedimentary cycle: A summary., Chemical Geology, 14: 285–304.

Roy, S., 1981. Manganese Deposits, Academic Press, London, 456 p.

Sunby, B., Silverberg and Chesslet, R. 1981. Pathways of manganese in the open esdurine system. Geochimica et Cosmochimica Acta, 45: 239–307.

Toth, J.R., 1980. Deposition of submarine crusts rich in manganese and iron., Geological Society of America Bulletin, I (91): 44–54.

Turekian, K. K. and Wedepohl, K. H., 1961. Distribution of the elements in some major units of the earth's crust., Geological Society of America, Bulletin, 72: 175–192.

Varanavas, S. P. and Pungnos, A. G., 1983. The use of trace metals in elucidating the genesis of some Greek and Cyprus manganese and ferromanganese deposits. In: Augustihis, S.S. (Ed). The significance of trace elements solving petrogenetic problems and controversies. Theophrastus publications, Athenes, pp. 819–860.

Yamamoto, K., 1987. Geochemical characteristics and depositional environments of cherts and associated rocks in the Franciscan and Shimanto terranes., Sedimentary Geology, 52: 65–108.

Tertiary Basaltic Intrusions in the Central Oman Mountains

M.S. AL-HARTHY[1], R.G. COLEMAN[2], M.W. HUGHES- CLARKE[3] and S.S. HANNA[4]
[1] *Exploration Department, Petroleum Development Oman, PO Box 81, Muscat, Oman*
[2] *Dept. of Geology, Stanford University, California 94305*
[3] *Consultant; Manningtree, Essex, UK*
[4] *Dept. of Earth Sciences, Sultan Qaboos University, PO Box 32486, Al-Khod, Oman*

Abstract

Basalt intrusions cutting Tertiary strata have been newly discovered north of the central Oman Mountains. The intrusions are of small size but their presence has considerable implications. The intrusives are undersaturated alkali olivine basalts whose depth of origin is generally considered to be circa 50 km. This occurrence implies an extensional regime at the time of their emplacement and that the intrusion could be via deep seated fault planes. They are associated with deep mantle melting and ponding of these lavas in the crust could develop hydrothermal circulation related to mineralization in this area.

Introduction

A map summarizing the geology of the central Oman Mountains is given as Fig. 1. From mid Permian to mid Cretaceous, SE Arabia formed part of a wide carbonate shelf on the southern side of Neo-Tethys, a remnant of which now constitutes the Gulf of Oman. In North Oman carbonate shelf sequence were deposited unconformably upon various early Palaeozoic and Proterozoic sediments and on crystalline basement. Contraction of Neo-Tethys involved development of a north easterly dipping subduction zone during the Cretaceous (Pearce et al. 1981). Consumption of the Arabian Plate in this subduction led to the obduction of a slab of newly formed oceanic lithosphere onto the Oman segment of the Arabian continental margin to form the Semail Ophiolite. This emplacement was completed late in the Cretaceous (Coleman 1981, Lippard et al. 1986) and during the process a syn-tectonic foreland basin developed in which thick sediments of the Late Cretaceous Aruma Group were deposited (Glennie et al. 1974).

The tectonically emplaced Semail Ophiolites and associated Hawasina oceanic sediments are overlain unconformably by a Maastrichtian and lower

Tj. Peters et al. (Eds), Ophiolite Genesis and Evolution of the Oceanic Lithosphere, 675–682.

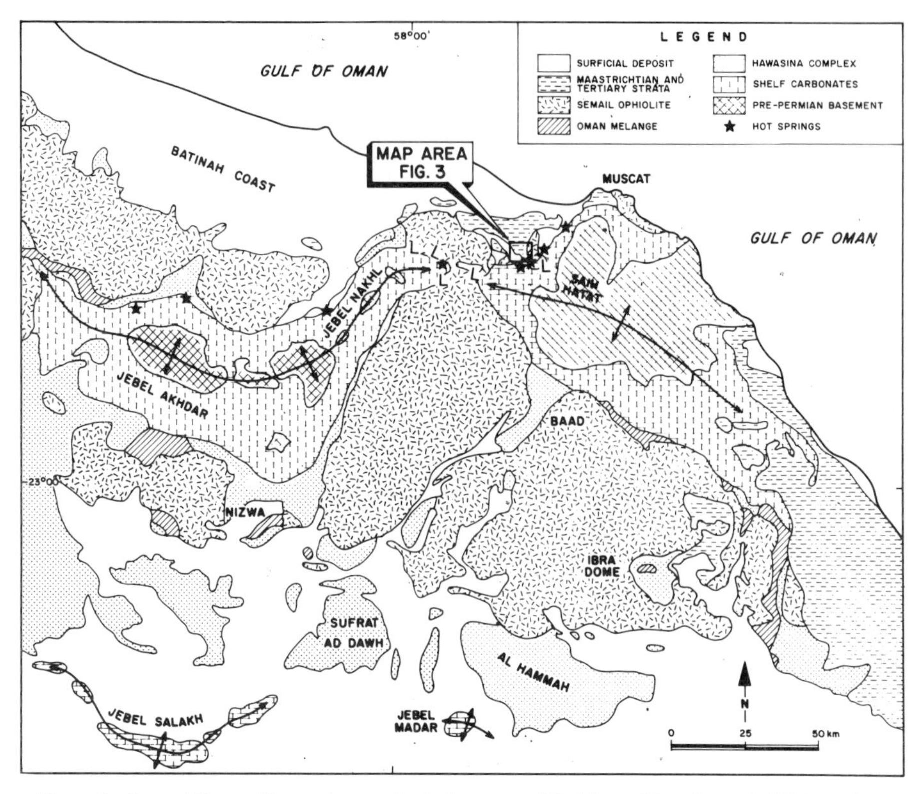

Figure 1. Central Oman Mountains: geological map modified from Glennie et al. 1974, to show the area detailed in Fig. 3, location of hot springs and the location of some Lisvenite outcrops 'L' near the base of the ophiolite.

Tertiary shelf sequence. This cover sequence has been moderately deformed by mid to late Tertiary movements associated with the major uplift of the Oman Mountain range.

The Tertiary deformation and uplift of the range was initially assigned to compressional events (Glennie et al. 1974) but later work, using reasoning based on structural-stratigraphic arguments, has developed an alternative extensional model (Hanna 1986, 1990, Mann et al. 1990). The only previous evidence of Tertiary intrusives and volcanicity is confined to the extreme eastern corner of Oman around the Jebel Ja'Alan basement block (map in Shackleton et al, 1990); but this occurrence has been linked by us and by others to extensional stresses resulting from strike-slip faulting along the Masira Line (Fig. 4). The presence of the basalt intrusions in the central Oman Mountains, however, implies that an extensional component must have been involved at some time in the Tertiary movements that gave rise to the mountain range.

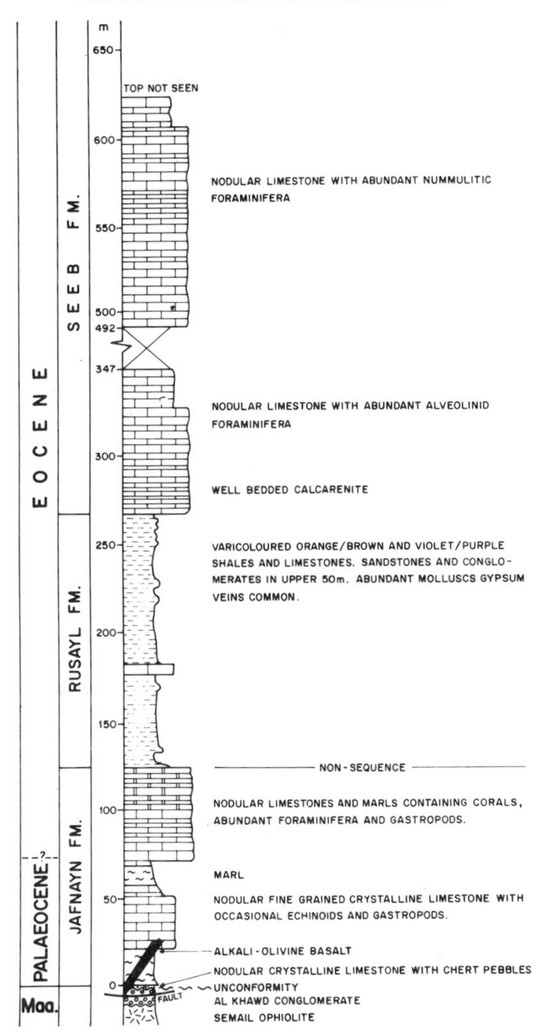

Figure 2. Tertiary Stratigraphy (modified from Nolal et al., 1990).

The Intrusions

The stratigraphy of the Lower Tertiary around the Oman Mountains is described by Nolan et al, 1990 (Fig. 2). Detailed field studies of the Palaeogene sequence in outcrops south-west of Muscat carried out by one of us (M.S. A-H) brought to light a small cluster of intrusions cutting the lower part of the marine Paleocene sequence (Jafnayn Formation). This marine limestone sequence is here underlain by post emplacement Maastrichtian polymict conglomerates (Al-Khod Formation) containing fossil wood and, nearby Dinosaur (Ornithischian) and Turtle bones. The conglomerates lie on a weathered surface of the Semail Ophiolite. Further small cluster of

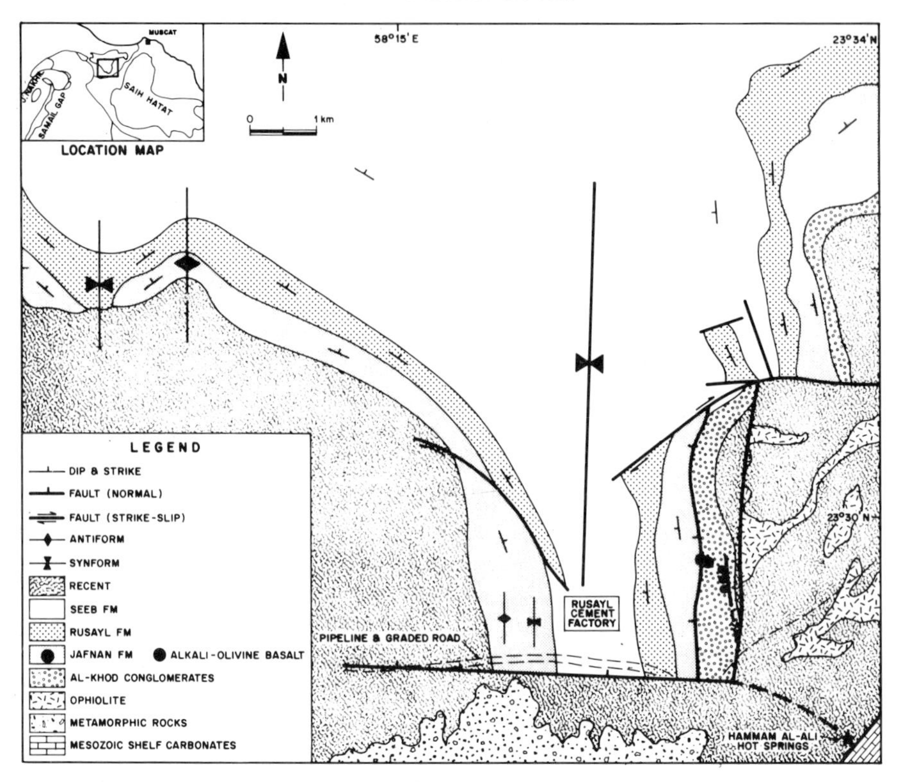

Figure 3. Simplified geological map of the area of Tertiary outcrops south west of Muscat to indicate the location of the basaltic intrusions.

intrusions were recently found cutting these Al-Khod conglomerates (Fig. 3).

The regional dip of the Tertiary strata is northward off the mountain range but this is complicated by a series of smaller anticlines and synclines on axes close to north-south. The intrusions occur on the eastern limb of a broad synform where the contact of the Jafnayn limestones and the Al-Khod conglomerates is a N-S fault. Figure 3 is a simplified geological map of the area where it can be seen that the intrusions are also close to the basal contact of the ophiolite suite upon the Mesozoic shelf carbonates.

The largest intrusion (maximum outcrop dimensions some 50 m) is a small irregular dyke updoming the overlying beds which bakes and recrystallizes the limestone country rock to give a reddened aureole around the periphery of the intrusive body. The smaller intrusions are similar but less clearly seen at outcrop.

Under the microscope, the intrusive is a basalt with large euhedral phenocrysts of Ti-augite having the typical mauve color in plain light and showing zoning under crossed nicols. Extremely fresh euhedral to anhedral grains of Olivine are less common; many have chromian spinel inclusions but all show only minor alteration to serpentine. The groundmass is fine grained and patchy with areas of plagioclase showing abrupt change to zeolites or nephline (analcite). Fine grained titaniferous clinopyroxine predominates and is accompanied by dark brown barkevikite, magnetite and biotite. In small areas, zeolite and analcite concentrations appear to be recrystallization products of glass. Carbonate is present throughout the rock and appears to be primary but is always interstitial.

Importantly, the rock also contains a large amount of xenocrystic and xenolithic fragments up to 2 cm (most around 2–5 mm) in dimensions. Bright green fragments of chromium diopside are associated with coarse grained xenoliths (up to 2 cm) of dunite, harzburgite and lherzolite. These xenoliths and xenocrysts compare well with alkali basalts of known provenance and imply that they are fragments of the upper mantle Green and Ringwood (1967).

The observations made so far indicate that both zeolites and analcite are widely present in the groundmass suggesting that these rocks are very undersaturated with regard to silica.

The presence of biotite, barkevikite and aegerine is also evidence of an alkali-rich magma. In sum, the rock could be called a camptonite, because of the large amount of barkevikite in the groundmass and the lack of plagioclase, but, here, we use alkali olivine basalt as a more suitable rock name as such rocks characteristically contain mantle xenoliths and are generally undersaturated with respect to silica.

Although no such work has been attempted by us, the depth of origin of rocks of this type could well be estimated using co-existing mineral pairs.

Discussion

The intrusives documented here are the first evidence of Tertiary igneous activity in this north-facing margin of Oman. The intrusives on the eastern corner, noted above, seem related to the strike-slip structural history of the south-east margin and have been dated at 38–42 Ma.

This new occurrence of basaltic intrusions lies close to the line of outcrop of the basal thrust zone on which the Semail and Hawasina nappes were emplaced. This thrust zone has been proposed by one of us (Hanna 1983 (in Lippard et al. 1986, p. 159), 1986, 1990) to be part of a master decollement fault under the nappe zones of north Oman which roots to the north. Such master obductive faults have been postulated to root in the mantle. Should the inversion that took place with the Tertiary uplift of the Oman Mountain range have involved any phases of extension, then these deep rooted faults would have provided the most facile route for intrusion from the 50 km depths necessary for these undersaturated alkali olivine basalts magma to form.

The age of the youngest country rock hosting these basalt intrusives is

Paleocene or earliest Eocene and this provides a minimum age for the intrusions. Absolute dating of the basalts has not yet been carried out but is under consideration; as is an attempt at geothermometry and geobarometry of the minerals within the xenoliths which could provide some guide to depth of melting. Nevertheless, the field relationships are so clearly unambiguous in proving a Tertiary age that they are sufficient to allow this short paper to consider the implications.

The factual establishment of mid Tertiary igneous activity in the central sector of the Oman Mountains has pertinence in the explanation of other geological factors peculiar to this area. The northern mountain front of the central Oman Mountain range has especially well developed silica carbonate (lisvenite) in the periodotites of the Semail. Additionally, this same sector contains the only significantly hot springs (issue temperatures 40 to over 60 degrees C) known in Oman. The spring line effectively follows the outcrop of the basal thrust zone of the Semail (Fig. 1). It, perhaps, should not be considered only coincidental that the hottest spring recorded (Hammam Al-Ali, 62°C. see Fig. 3) is less than 3 km away from the intrusives described here. Although not the only explanations, present day residual heat flow from Tertiary intrusives and associated longer-term hydrothermal circulation could adequately explain both the restricted belt of hot springs and the

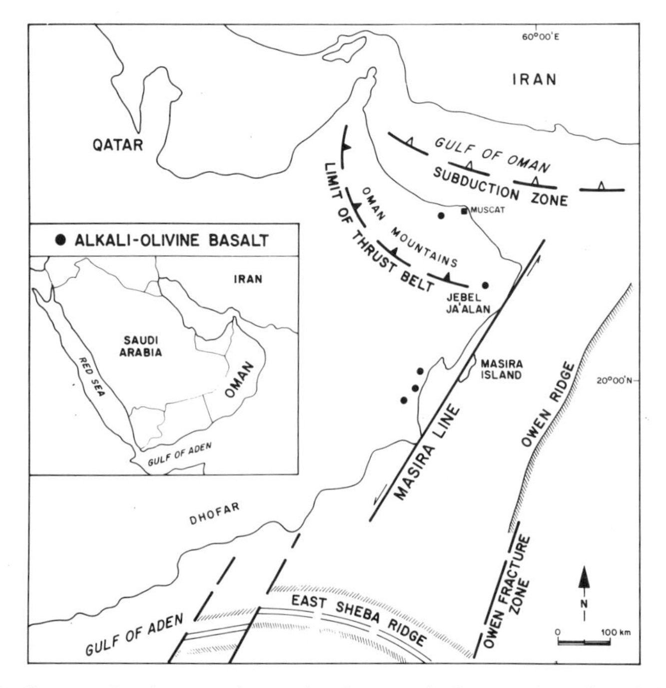

Figure 4. Outline map drawing attention to the plate margin features that affected the Tertiary movements in north Oman. The Masirah line, although inactive at present, was a locus of strike slip movement during the mid Tertiary.

localized areas of enhanced alteration of the ultrabasics. If enhanced hydrothermal circulation is suspected, then the belts of possible extensional faulting could deserve further examination with regard to mineralization of potential economic value.

The presence of deep seated intrusives in the Tertiary sequence seems inescapable evidence of structural extension during the tectonism associated with the Eocene and younger Oman Mountains uplift. Whilst several possibilities exist for the origin of the faults on which the extension occurred. It must be considered that they could have been controlled by the original extensional faults formed during the Permian and Triassic rifting stage which defined the north Oman Mesozoic continental margin. These same faults would have acted as useful lines of weakness during the Cretaceous convergence and emplacement event (positive inversion) and retained their facility as deep seated failure lines during any Tertiary extension (negative inversion). Whilst the overall stress field in which the major uplift of the Oman Mountains range occurred (seemingly mainly in the mid Tertiary) is as yet unclear, it cannot be ignored that the today nearby subduction beneath the Makran coast north of the Gulf of Oman could be engendering a tensional regime in north Oman. The phasing of the stresses in the Makran subduction that could be transferred to the Oman margin are probably best understood by inter-relating all knowledge of plate movements affecting southern Arabia, in particular the better documented extentional history of the Red Sea and Gulf of Aden (Fig. 4).

Acknowledgement

M.S. Al-Harthy wishes to thank Petroleum Development of Oman for the support he received during his work in the area. The authors would like to thank the two referees Professors Tj. Peters and J. Mercolli for their constructive comments which improved the manuscript.

References

Coleman R.G., 1981. Tectonic setting for ophiolite obduction in Oman., Journal of Geophysical Research, 86: 2497–2508.

Glennie K.W., Boeuf M.G.A., Hughes Clarke M.W., Moody-Stuart M., Pilaar W.F.H. and Reinhardt B.M., 1974. Geology of the Oman Mountains., Verhandelingen v.d. Koninglijk Nederlands geologisch mijnbouwkundig Genootschsap, 31, 433 p.

Green D.H and Ringwood, A.E., 1967. The genesis of basaltic magmas., Contrib. Mineral Petrol. 15: 103–190.

Hanna S.S., 1986. The Alpine (Late Cretaceous and Tertiary) tecontic evolution of the Oman mountains: A thrust tectonic approach. Symposium on the hydrocarbon potential of intense thrust zones, 2 OPAEC, Kuwait, pp. 125–174.

Hanna S.S., 1990. The Alpine deformation of the Central Oman Mountains. In: Robertson

A.F.H., Searle M.P. and Ries A.C. (Eds), The Geology and Tectonics of the Oman region. Geological Society of London Special Publication 49: 341–352.

Lippard S.J., Shelton A.W. and Gass I.G., 1986. The Ophiolite of Northern Oman. Geological Society London, Memoir No. 11., 178 p.

Mann A., Hanna S.S. and Nolan S.C., 1990. The post-Campanian structural evolution of the Central Oman Mountains: Tertiary extension of the east Arabian margin. In: Robertson et al. (Eds), ibid., pp. 549–563.

Nolan S.C., Clissold B.P., Smewing J.D. and Skelton P.W., 1990. Late Campanian to Tertiry palaeogeography of the central and northern Oman Mountains. In: Robertson et al. (Eds) ibid., pp. 495–519.

Pearce J.A., Alabaster T., Shelton A.W. and Searle M.P., 1981. The Oman Ophiolite as a Cretaceous arc-basin complex: evidence and implications. Philosophical Transactions, Royal Society of London, A3, pp. 299–317.

Shackleton R.M., Ries A.C., Bird P.R., Filbrand J.B., Lee C.W. and Cunningham G.C., 1990. The Batain Melange of NE Oman. In: Robertson et al. (Eds) ibid., pp. 673–696.

Part VI
Paleoenvironment of Other Ophiolites

Geology, Geochemistry, and the Evolution of an Oceanic Crustal Rift at Sithonia, NE Greece

K. MUSSALLAM
D-2000 Hamburg 26, Lohhof 9, Germany

Abstract

At Sithonia (NE Greece), an approximately 3 km thick pseudostratified volcanic arc consisting of abyssal volcanics intruded by granophyres and alternating with turbidites and platform limestones is exposed. During the Late Jurassic, the volcanic arc was subjected to short-termed rifting that resulted in the formation of an oceanic crustal sequence comprising a sheeted dyke complex, submarine volcanics and a coeval Upper Jurassic shallow-water sedimentary cover. Field relations indicate that this sequence was created within a wall-to-wall, NE-trending rift that is at most 30 km wide and 10 km long. Rifting commenced without passing through phases of extensional faulting, subsidence and sedimentation. The sheeted dykes immediately invaded the previously deformed arc volcanics. Rift-related magmatism is meager and is restricted to microdioritic dykes of island-arc affinity. After a spreading period of at most 1.5 Ma, the rift closed by NE-directed overthrusting of first its axial sequence along a former transform onto the adjacent arc volcanics prior to the SW transport of the Serbomacedonian massif.

The arc volcanics range from boninite-like to low-K and transitional tholeiites. The order and chemistry of their eruptions correspond to those in some modern fore-arc regions. The transition from arc to spreading volcanism is chemically gradational. The sheeted dykes and overlying volcanics have compositions closest to marginal basin tholeiites. Off-axis volcanics are more varied and comprise alkaline basalt, low-K andesite and potassic rhyolite.

The shallow-water sedimentation and the absence of metamorphic peridotites are suggestive of ophiolite generation prior to the break-up of the lithosphere. The tip of a spreading axis that failed to develop into an oceanic basin is envisaged as the most appropriate setting. The nearby submarine volcanics at Kassandra probably represent the abyssal segment of this rift.

Tj. Peters et al. (Eds), Ophiolite Genesis and Evolution of the Oceanic Lithosphere, 685–704.

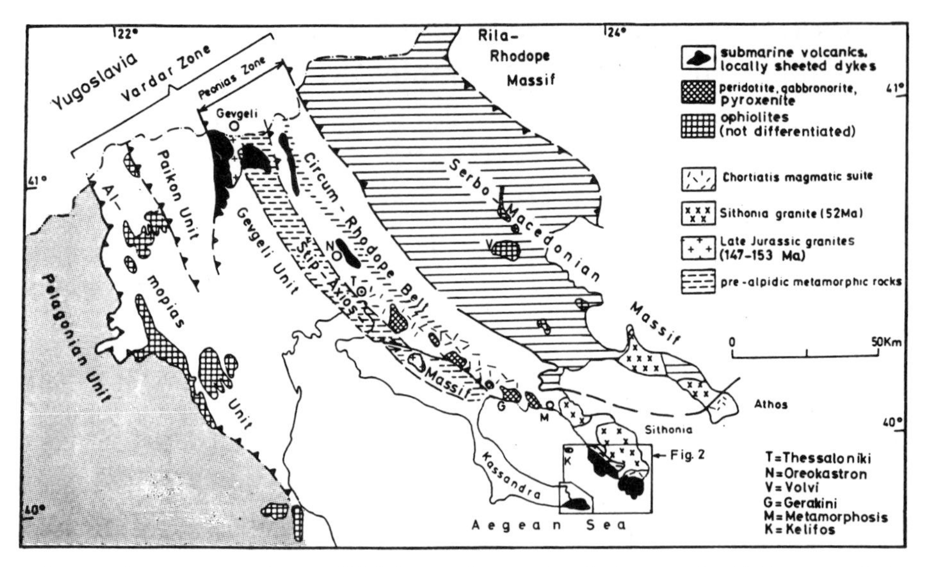

Figure 1. Structural sketch map of NE Greece showing the distribution of ophiolites (modified after Kockel 1986).

Introduction

The ophiolites of the eastern Vardar zone of Greece (Fig. 1) are intimately associated with and preceded by Na-dominant salic rocks of the Chortiatis magmatic suite (Kockel and Mollat, 1977). This suite is believed to represent an island-arc complex formed above a subduction zone (Schünemann, 1985; Mussallam and Jung, 1986a). Several authors (e.g. Vergely, 1984; Schünemann, 1985; Mussallam and Jung, 1986a) suggested that subduction was NE directed. Recent data obtained from this study rather favor S and SW subduction of a Tethys branch in the Vardar zone by the Middle to Upper Jurassic. The island arc and ophiolite association is sandwiched between the Middle to probably Upper Jurassic autochthonous Svoula Formation of the continental slope and rise in the E (Kauffmann et al., 1976) and the locally exposed crystalline basement of the Paikon (Stip-Axios massif of Kockel, 1986) in the W. Prior to the Upper Jurassic transgression, the ophiolites were locally intruded by 147 to 153 Ma old granites (Borsi et al., 1966; Kockel, 1986).

The most conspicuous feature of the arc/ophiolite association is the lateral covariation in lithology and tectonic setting (Mussallam and Jung, 1986a). The arc rocks change from tonalites and trondhjemites intrusive in Triassic limestones and narrowly interbanded by intermediate and acidic sills at central Chalkidiki and Athos peninsula to abyssal volcanics in SE at Sithonia and probably also in NW. The complementary variations displayed by the ophiolite belt are such that plutonic mafic and ultramafic rocks occur in its middle part, meanwhile volcanics and subvolcanics occur at the NW and SE parts (Fig. 1). This correspondence points at a cause and effect relationship. Variations within the island-arc complex were attributed by the present author (Mussallam and Jung, 1986a) to a change of the tectonic setting from active continental margin to oceanic island-arc, whereas variations within the ophiolites relate to changes in the plate boundary from conservative at central Chalkidiki to constructive at Sithonia.

The auto- to parautochthonous mafic-ultramafic association comprises five independent protrusions which consist of cumulate peridotites displaying tectonite fabrics and intruded by pyroxenitic and gabbroic rocks (Mussallam et al., 1981). The absence of sheeted dykes and submarine volcanics is suggestive of origin at a tectonic setting other than a spreading center. Jung and Mussallam (1985) and Mussallam and Jung (1986a and b) proposed therefore emplacement by a transcurrent-related process referred to by Page et al. (1979) as transduction. On the other hand, Bébien et al. (1986) believe in the presence of volcanics and of a particularly thin doleritic crust between these and the plutonic formation. They interpret this association as an ophiolite and inspired by the above model invoke uplift of mantle and crustal rocks along an oceanic fracture zone.

In the present paper, the results of a detailed geological and petrological study of the southeasternmost exposed part of the Vardar zone ophiolite belt at Sithonia are summarized. Remapping of this area revealed that the ophiolite there is intrusive into abyssal arc volcanics and that it represents a nearly completely preserved rift referred to here as the Sithonia rift. These findings make a re-evaluation of earlier data inevitable and allow to elucidate some processes related to the initiation of spreading and the genesis of the Vardar-zone ophiolites.

The Sithonia Rift

The Sithonia rift is discontinuously exposed along the western coast of Sithonia peninsula for about 30 km (Fig. 2). Judging from the prominent direction of the sheeted dykes, the rift trends NE and forms a high angle with the NW-SE striking stratigraphical boundaries. A well exposed SE wall, an axial trough and parts of the NW wall are identified. These are differentially rotated with corresponding reversals of dip directions. They behave in this regard as independent blocks that were separated from each other by ridge-parallel faults (Fig. 4).

The southeastern rift wall comprises the area between Mavrochoma and Toroni with an exposed width of about 9 km. A complete section from the arc volcanics via intrusive sheeted dykes, submarine volcanics and overlying reefal limestone is exposed. The whole sequence is overturned with steep NE dips. At Toroni, both the sheeted and volcanic units are tectonically removed and the reefal limestones are juxtaposed to the arc volcanics (Figs. 2 and 4).

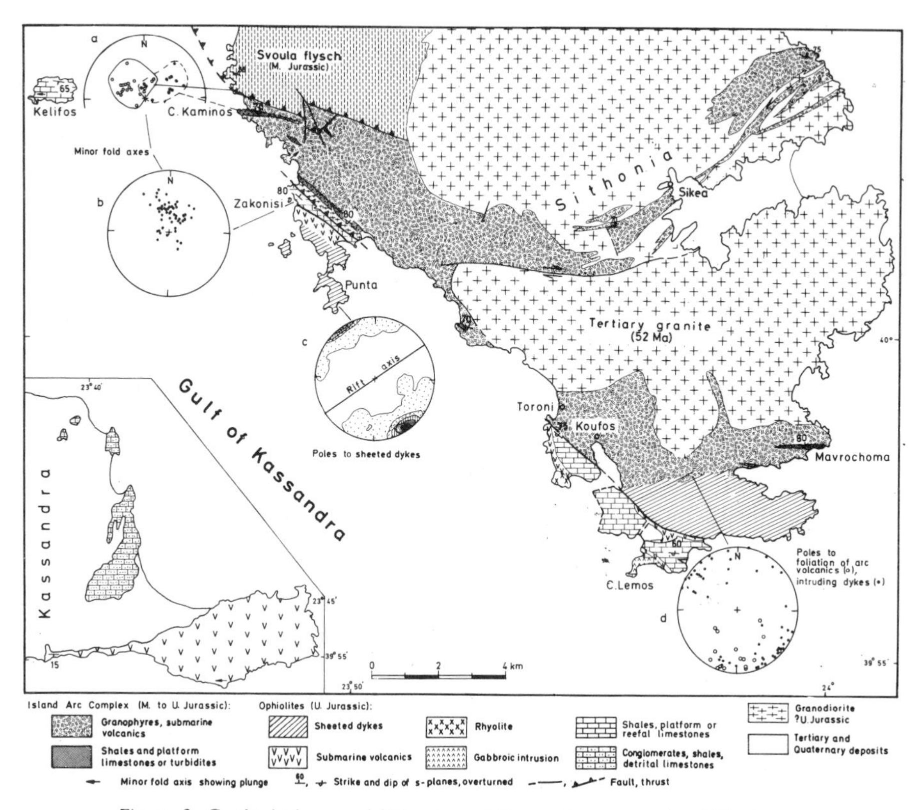

Figure 2. Geological map of Sithonia and Kassandra peninsulas, NE Greece.

Only the innermost part of the 15 km wide rift trough is exposed between Punta and Zakonisi. The lowest outcropping stratigraphic unit is the sheeted complex which achieves a thickness of 1.2 km before it submerges beneath the sea. A pronounced magnetic anomaly in this area (Makris and Röwer, 1986) is suggestive of the possible presence of mafic or ultramafic rocks at depth. The ophiolite has normal NE dips and is tectonically bounded against the overturned and NE-dipping arc volcanics by a high-angle thrust. The volcanics are devoid of dykes and other rift-related structures. The bounding thrust has apparently acted as a ridge transform which truncated the rift axis against the arc volcanics. Later it was the site of ophiolite thrusting and subsequent intrusion of granodiorite prior to the SW-directed transport of the Serbomacedonian massif. Post-emplacement reactivation of this fault resulted in the above mentioned tectonic removal at Toroni.

From the northwestern rift wall which encompasses the area around Kelifos, only the sedimentary unit and interfingered pillowed basalt are exposed. The similarity in structure and sedimentary facies and rare dyklets in the arc volcanics along the opposing Sithonia coast are suggestive of similar relations as at the SE wall. In contrary to the latter, the sedimentary sequence at

Kelifos moderately dips toward the south. Locally it is folded with nearly horizontal E-W trending fold axes.

The continuation of the rift axis towards SW is determined by the occurrence of 1.2 km thick abyssal volcanics and interbedded radiolarian chert at Kassandra peninsula lying some 12 km west of Sithonia (Fig. 2). These gently southward tilted volcanics substantially differ from the shallow-water spilites at Sithonia that a direct connection between both can with certainty be excluded. Tentatively, it is proposed that the Kassandra volcanics may represent the abyssal segment of the Sithonia rift from which they were separated by a ridge transform. Their different style of deformation, the concordantly overlying transgressive sediments which contain individual gravity-slided pillows are suggestive of contemporaneous uplift along this transform.

1. *The Basement*

The volcanic arc constituting the basement to the Sithonia ophiolite is represented by a 3 km thick pseudostratified sequence that consists of diorites and granophyres interbanded with aphanitic, several m thick basic to acidic sills in the lower part and of submarine volcanics intercalated with sedimentary rocks in the upper part. The volcanics are developed as pillowed or massive lavas, less common as pyroclastics and are frequently intruded by granophyres. The rocks were subjected to hydrothermal and subsequent regional metamorphism. Primary mafic minerals are replaced by chlorite and actinolitic or tschermakitic hornblende. Mn-rich garnet was locally encountered.

The volcanic arc is bounded against the Svoula flysch by a high-angle thrust, across which the fold axes change in trend and plunge from NW in the stronger deformed flysch into NE in the volcanic arc (Fig. 2, stereonet a). The Svoula flysch is here interfingered with several m thick horizons of extremely altered tholeiites. The arc volcanics are overturned with steep N and NE dips. The change in strike from N90°E in NW and SE parts to N120°E in between probably points at a domal structure.

Aside from thin radiolarian chert and argillaceous intercalations, sedimentary rocks form two 130 and 175 m thick successions. These have a limited lateral extent and apparently accumulated in isolated troughs between the volcanic piles during periods of volcanic quiescence (Fig. 3).

The lower succession consists of shales with minor limestone interbeds overlain either by platform limestone in SE at Mavrochoma or by turbidites in NW at Cap Kaminos. In SE, the shales contain Fe- and Mn-rich lenses and a 1 m thick pyrite horizon and are intruded by 1 to 40 m thick basic sills. The turbidites at C. Kaminos locally contain up to 1 m large lenticles of jaspilite interbanded with Fe-Ti-oxides and partly incrusted with malachite.

The upper succession is best exposed north of Zakonisi, where it formed the tectonic base to the ophiolite thrusting prior to the intrusion of both by granodiorite. The facies is most similar to that of the lower succession except for the predominance of turbidites and conglomeratic sandstones.

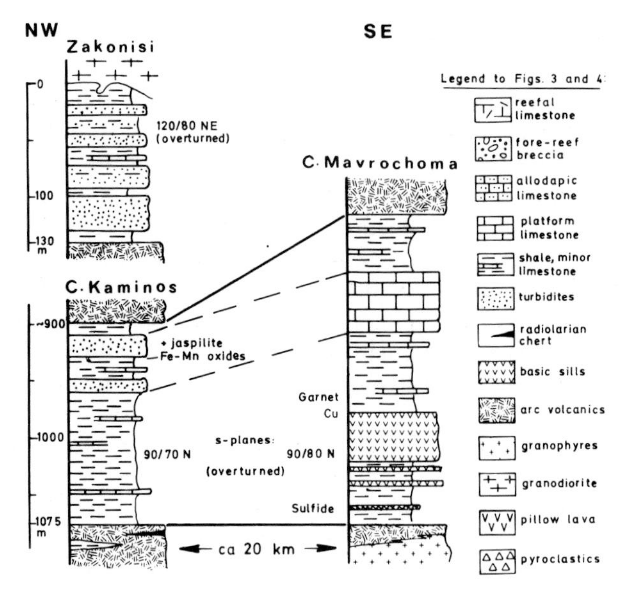

Figure 3. Columnar sections of the two sedimentary successions in the arc volcanics at Sithonia.

The lateral variation of the sedimentary facies points at increasing water depth towards NW which is consistent with the increased abundance of radiolarian chert. This variation roughly coincides with the pinch-out termination of the Svoula flysch, which supplied most of the detrital material in the turbidites.

2. *The Sheeted Dyke Complex*

The base of the sheeted complex is well exposed at the SE rift wall, where individual dykes extend for some hundred m through foliated arc volcanics. The dykes are devoid of foliation and range in thickness from a few dm to more than 20 m. On the scale of an outcrop, the dykes are discordant to the N- to NW-dipping foliation of their hosts. On a large scale however, they, make only a small angle with this foliation which roughly coincides with the s-planes of the arc volcanics (Fig. 2, stereonet d). Some dykes can be traced to their roots where they gradually terminate indicating that they were fed from an overlying magma chamber floored by the arc volcanics. The geological relations strongly resemble those in the Tihama Asir ophiolite along the eastern coast of the Red Sea which was formed prior to the formation of oceanic crust in the axial valley (Coleman et al., 1979). They are also similar to those of the Tortuga and Sarmiento ophiolites of southern Chile (Bruhn et al., 1978). At Chalkidiki, the crystalline basement (Stip-Axios massif) of the Gevgeli ophiolite is intruded by dolerite dykes (Leube and Kockel, 1962; Bébien, 1982; own observations), which probably belong to this ophiolite. Moreover, Dixon and Dimitriadis (1984) pointed out possible intrusive relations between the sheeted dykes of the Volvi ophiolite and pre-existing country rocks.

Vertically, the dykes increase rapidly in abundance to constitute an over 3 km thick unit consisting of 100% subparallel dykes. Across the rift, systematic variations in dip direction of the dykes are not present (Fig. 2, stereonets c and d) probably because the time lapse between the opening and closure of the rift was short to allow for large-scale rotations. Nor is there a preferred orientation of their chilled margins; the majority of the dykes were evidently emplaced along the margins of previous ones.

In terms of mineralogical composition, the sheeted complex is essentially made of doleritic dykes with clinopyroxene and plagioclase as the main constituents. Leucocratic rocks amount about 3%. Doleritic dykes are fine to medium grained and phaneritic, a few are phyric with cm-large plagioclase phenocrysts (microgabbro). Leucocratic dykes include hornblende diorite, albitite, and trondhjemite. Hornblende diorite occurs as several m thick, synmagmatically brecciated dykes. Albitite dykes are weak oligoclase-phyric and up to 3 m thick. Trondhjemite is mostly confined to the chilled margins of the most basic microgabbroic dykes where it occurs as a network of cm to dm thin veinlets or as irregular, dm large nodules. Ample field evidences are suggestive of crystallization from an immiscible liquid.

In terms of temporal and field relations, Mussallam and Jung (1986b) distinguished between ridge and off-ridge dykes. Both consist of doleritic and leucocratic rocks in the above ratio.

Ridge dykes are those making up the framework of the sheeted complex, contributing with more than 95% to its volume and ranging in thickness between 1 and 13 m (average 4.2 m). Mussallam and Jung (1986b) pointed out the periodicity of these dykes and defined an ideal injection period to start with a plagioclase-phyric dyke and to terminate with albitite. The spatial and temporal relations of these dykes point at a spreading axis that oscillates within a range of a fraction to a few multiples of average dyke width.

Off-ridge dykes are on average 0.4 m thick and include finer-grained clinopyroxene-phyric dolerite and rare trondhjemite. They display various kinds of dislocations that are attributable to active tectonics operating during their injection.

3. *The volcanic Unit*

The base of the volcanic unit is not well defined and is essentially made up of brecciated dolerite intruded by irregular plagiogranitic and basic veins. The overlying some 750 m thick sequence can be divided into three parts. The lower part consists of massive flows with minor pillowed basalt and pyroclastic horizons and is intruded by 40 to 50 m thick keratophyre at its top. In the middle part, pillow lava intercalated by two a few m thick shale horizons predominates. The pillows are highly vesicular and change upsequence in composition and diameter from intermediate with an average diameter of 15 cm to basic measuring more than 1 m in diameter. The upper part is developed as pillow lavas with jaspilite filling their interpillows. These pass laterally into boulder-sized autobrecciated pyroclastics.

Unlike the sheeted dykes which have mostly retained their primary minerals, the volcanics are invariably present as spilites. These display weak plagioclase-phyric textures and consist of albite, chlorite and calcite. Epidote and quartz are not uncommon and occur also as vesicle-infillings.

4. *The sedimentary Unit*

Across the rift, the facies and thickness of the sedimentary and associated magmatic rocks vary widely (Fig. 4). Both rift walls are covered by reefal limestones which extend at the SE wall for more than 5 km with an exposed thickness of approximately 1.5 km. At Cap Lemos the reef limestones are underlain by violet phyllites which commonly constitute the substratum of fossil reefs. At Kelifos, a more than 0.9 km thick reef complex is exposed. This complex consists of a repetitive alternation of reefal limestone, fore-reef breccia and shales intercalated by partly allodapic limestones. The abundant micro- and macrofauna obtained from the Kelifos section yield a late Kimmeridgian age (Mussallam and Jung, 1986b). The Sithonia ophiolite seems therefore to be the youngest among the Vardar zone ophiolites which are transgressively overlain by Upper Jurassic sediments. Towards the rift trough, the reefs are replaced by shales with thin limestone interbeds. Higher in the sequence, platform limestones locally make up the major part of the sedimentary column of the rift trough. The facies distribution points at a rather rigid topography also reflected in the lateral lithological variations of the subjacent volcanics.

5. *Off-axis Magmatic Activity*

The rift walls which stand according to the sedimentological data as topographic highs, were the locus of intense magmatic activity. This activity is most pronounced at the SE rift wall (Fig. 2 and 4), where the reefal limestones at its rear flank are intruded by an approximately 350 m thick gabbroic sheet, whereas in its innermost part they are overlain by 200 m thick pillow lava and by probably more than 300 m rhyolite. In between no magmatic rocks are present. At the rift trough, magmatic rocks are except for a 4 m thick volcanic lens virtually absent.

The off-axis volcanics occur at several levels within the sedimentary unit but they do not constitute a continuous, laterally traceable horizon. The number and thickness of the flows as well as their chemistry and rock association vary markedly. At Toroni, pillowed intermediate spilites are intruded by partly brecciated or flow-banded rhyolite which extends as 3 to 4 m thick sills in them. The rhyolite is plagioclase-phyric. As sills it contains in addition abundant sanidine and quartz phenocrysts. At Kelifos, two pillowed basalt

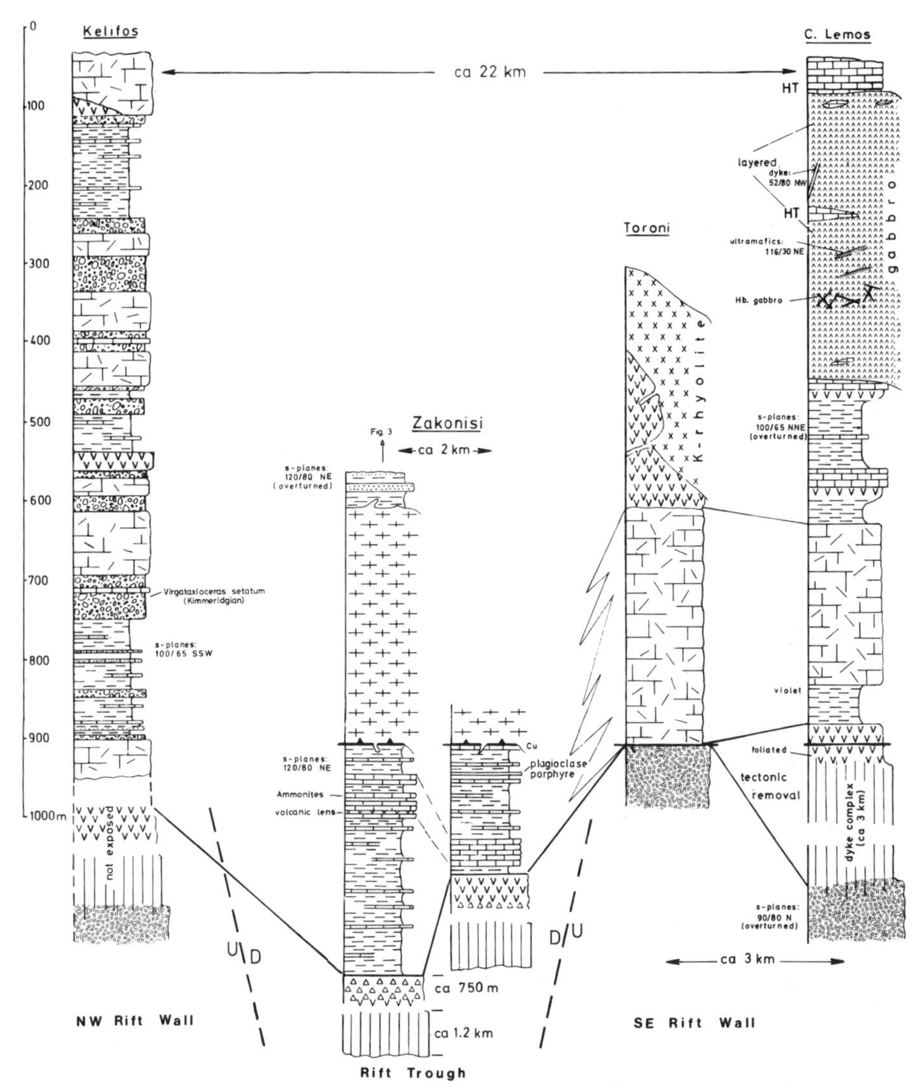

Figure 4. Columnar sections across the Sithonia ophiolite showing the facies variations of the sedimentary and associated magmatic rocks. Heay dashed lines: inferred ridge-parallel faults, U and D indicate the up- and down-thrown sides respectively. HT = hight-temperature contact metamorphism with garnet and diopside (for legend see Fig. 3).

horizons are present. The more than 1 m measuring pillows are highly vesicular with calcite filling both the vesicles and interpillows. The pillow lavas at Kelifos have exceptionally retained their clinopyroxene, which occurs as 0.6 mm large phenocrysts set in a Fe-rich glassy to cryptocrystalline groundmass. Traces of biotite were also encountered.

Geochemistry

The autochthonous setting of the Sithonia ophiolite and its foundation on a volcanic arc provide the opportunity to highlight some aspects related to the pre- or syn-rift magmatism, the chemical characteristics of the arc volcanism and possible changes in the magma composition accompanying the change from arc to spreading magmatism. .The following discussion will therefore focus on these points. Post-spreading volcanism and the relationship of the Kassandra volcanics are outlined.

1. *Rift-Related Magmatism*

The only evidence for a magmatic activity that may be related to the initiation of spreading and formation of the Sithonia ophiolite comes from local occurrence of hornblende microdiorite and biotite andesite dykes that are concentrated in a narrow zone along the southwestern border of the mafic-ultramafic association in the Gerakini-Metamorphosis area (Fig. 1). At Gerakini, the dykes trend mostly N to NNE and dip S to ESE and are frequently succeeded by perigranitic pegmatites. At Metamorphosis they trend E to ENE and dip in northern directions. The dykes were evidently introduced subsequent to the emplacement of their mafic and ultramafic hosts. Chemically, they are quartz and hypersthene normative and correspond to the low-K basaltic andesite to andesite series rocks.

In Fig. 5, the dykes plot mostly within and partly to the left of or above the field of island-arc tholeiites. A comparison with Fig. 6 reveals that they strongly resemble the arc volcanics constituting the basement to the Sithonia ophiolite. This similarity is also reflected in major and other minor element abundances. However, the dykes are considerably younger than the regionally metamorphosed arc volcanics and were emplaced long after the closure of an assumed Middle Jurassic ocean in this part of the Vardar zone. Their island-arc affinity implies therefore the presence of a subducting slab of that vanished ocean. Anatexis or contamination of continental crustal material can in view of the exceedingly low K and Sr contents and concomitant depletion in HFSE be excluded. The island-arc affinity of the dykes and their restricted occurrence set the Sithonia rift apart from continental rifts which are mostly characterized by extensive predominantly alkali volcanism. Nevertheless, slightly alkaline or calc-alkaline magmatism is widespread in subduction-related rifts. Voluminous silicic and minor mafic volcanics of calc-alkalic character were erupted prior to the initiation of spreading and the formation of the Chilean Tortuga and Sarmiento back-arc basin ophiolites (Bruhn et al., 1978). Similarly, several km thick calc-alkaline and subordinate alkaline volcanics were erupted in the Mesozoic marginal basins of central Peru without formation of an oceanic crust (Atherton et al., 1985).

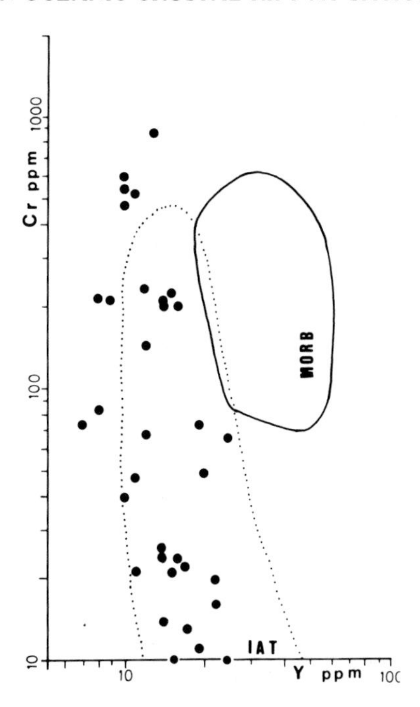

Figure 5. Cr-Y plot for microdioritic dykes from the Gerakini Metamorphosis are (after Pearce et al., 1984). Analyses by Jung & Mussallam in Schünemann (1985) are considered.

2. *The Transition from Arc to Spreading Magmatism*

We concern here with the chemical characteristics of the arc volcanics and emphasize possible changes in magma chemistry accompanying the change from arc to spreading magmatism. To elucidate this, the sheeted dykes intruding the arc volcanics are envisaged to represent the earliest spreading-related magmatic rocks. Analyses from these and from the arc volcanics are portrayed in Figs. 6 and 7.

In Fig. 6 the arc volcanics show a large scatter of points reflecting their heterogeneity and the presence of probably more than one magma type. The majority of them classifies as IAT, a few as MORB and a third group plots above and to the left of IAT. Samples classifying as IAT and MORB have similar average MgO content (7.3%) and FeO^*/MgO ratio (1.5). The third group is characterized by higher MgO contents (more than 10%) and lower FeO^*/MgO ratio (less than 1) by an average SiO_2 of 54.2% and displays other points of similarity to boninite or Mg-andesite. The dykes join with a small overlap the arc volcanics and plot, as with the sheeted dyke unit within and below MORB. The sheeted unit ($n = 220$) was sampled at different localities across the rift and classifies in a Cr-Ti diagram as low-K tholeiites (Jung and Mussallam, 1985). Spatial chemical variations are not present and the wide compositional range is attributable to fractional crystallization.

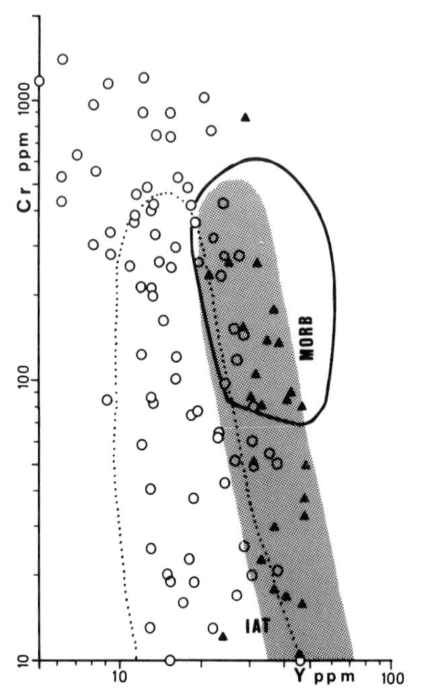

Figure 6. Cr-Y plot for the island arc volcanics at Sithonia (open circles) and the sheeted dykes (closed triangles) intruding them. Shaded area: Field of the sheeted dyke unit from Jung and Mussallam (1985). After Pearce et al., 1984.

Fig. 7 shows a continuous increase of Ti and Zr contents at a fairly constant rate. Again, the dykes lie at and overlap with the variation trend of the arc volcanics, simulating thus some kind of consanguinity. The arc volcanics displaying boninitic affinity in Fig. 6 have the lowest Zr and Ti

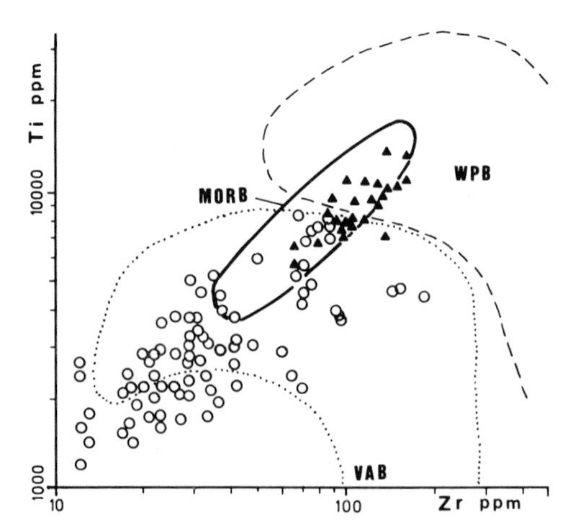

Figure 7. Ti-Zr plot for the island arc volcanic and the sheeted dykes intruding them at Sithonia (after Pearce et al., 1984). Symboles as in Fig. 8.

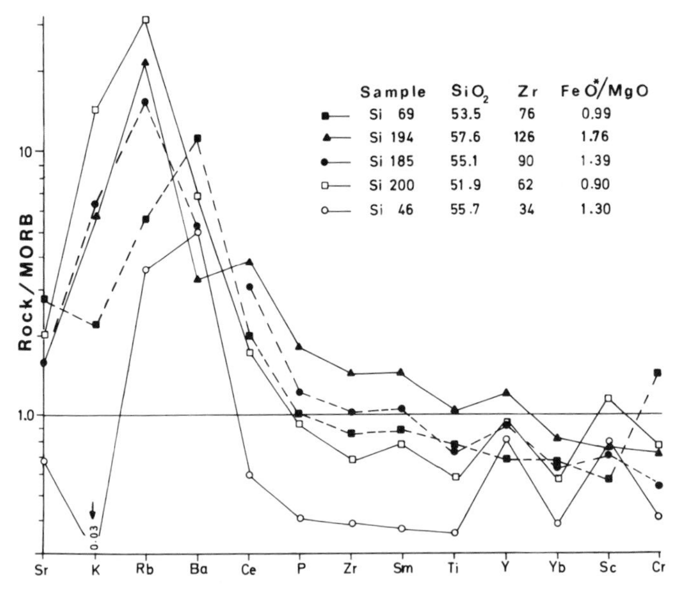

Figure 8. MORB-normalized patterns for: island arc volcanics (Si 46), sheeted dykes (Si 200), axis volcanics (Si 185, 194), and an off-axis volcanic sample (Si 69) from Sithonia (normalizing values after Pearce, 1982).

contents and fall below the VAB field. They pass gradually to those having IAT features. Samples plotting as MORB come from the uppermost part of the volcanic arc, where they occur as screens between the intruding sheeted dykes. The trend in Fig. 7 corresponds to the vertical variation trend in passing from the base of the arc volcanics to the sheeted dykes. The change in the chemical affinity of the volcanic arc eruptions roughly coincides with quiescent intervals which permitted the deposition of the two sedimentary successions in Fig. 3. The boninite-like volcanics belong to the earliest and short-termed eruption phase that took place prior to the deposition of the lower sedimentary sequence. The next phase brought about the outpouring of extensive low-K tholeiites. Arc volcanism concluded after renewed sedimentation with minor extrusion of lavas having MORB compositions. This stratigraphy favorably compares to that of some modern fore-arc regions (e.g. Hawkins et al., 1984). The rather gradational transition from arc to spreading magmatism is striking in view of the contrasted mineralogies and metamorphic histories of the arc volcanics and sheeted dykes. Additional features of this transition are illustrated by the following spiderdiagrams.

In Fig. 8, the volcanic arc sample (Si 46) displays a strong depletion in the elements Ce to Cr consistent with a classification as IAT according to Figs. 6 and 7. As already mentioned, the K- and Sr-depletion is an overall feature of the island-arc complex and of the younger subduction-related microdioritic dykes discussed above regardless their varied modes of occurrence and degrees of alteration. The patterns for the ophiolite samples are

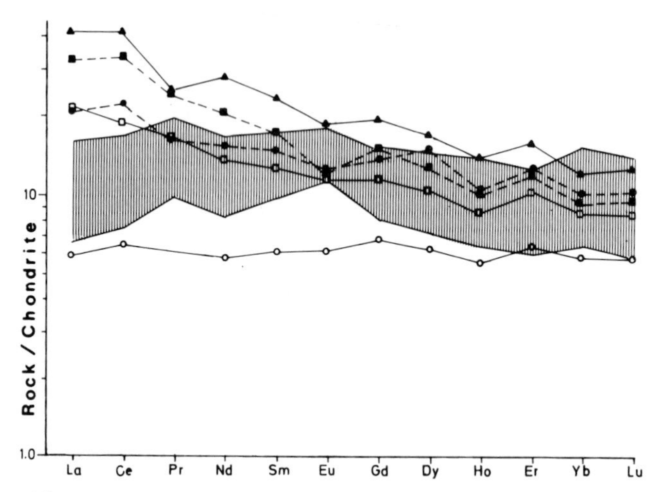

Figure 9. Chondrite-normalized patterns for the same samples as in Fig. 8 (analyses with ICP-MS). Ruled area shows the patterns for the Kassandra volcanics (4 analyses with ICP-AES). Normalizing values after Haskin et al. (1968).

broadly similar. Compared with the volcanic arc sample and starting with the sheeted dyke pattern (Si 200), the extent of depletion in the HFS elements progressively decreases from the sheeted dykes to the ophiolite volcanics. In the most silicic ophiolite volcanic sample (Si 194), this depletion is confined to Yb. The ophiolite rocks are in general more enriched in the LIL elements.

Fig. 9 illustrates chondrite-normalized REE data for the same samples of Fig. 8. The pattern of the volcanic arc sample is nearly horizontal at 6 times chondrite. The flat pattern and low total REE (20.09 ppm) resemble those of island-arc and marginal basin tholeiites (Kay and Hubbard, 1978). The ophiolite patterns are mostly parallel and display clear heavy- to light-REE fractionation with La_N/Yb_N increasing from 1.95 to 3.4. The total REE content (49.11 to 96.61 ppm) increases with SiO_2, Zr, and FeO^*/MgO as to be expected by fractional crystallization. The enrichment in the light REE sets the ophiolite apart from N-type MORB. The REE patterns and the major and minor element abundances are closest to the volcanics from the Bransfield Strait which formed during the initial stage of back-arc spreading (Weaver et al., 1979). They are also comparable to those of the ensialic Sarmiento (Stern, 1980) and Gevgeli (Haenel-Remy and Bébien, 1987) ophiolites.

3. *Off-Axis Volcanism*

Axis and off-axis volcanics are plotted in Fig. 10. Whilst the axis volcanics uniformly display transitional features, the off-axis volcanics are more varied. The two basalt samples from the NW rift wall at Kelifos plot within or to the left of IAT. Both samples are highly nepheline and olivine normative

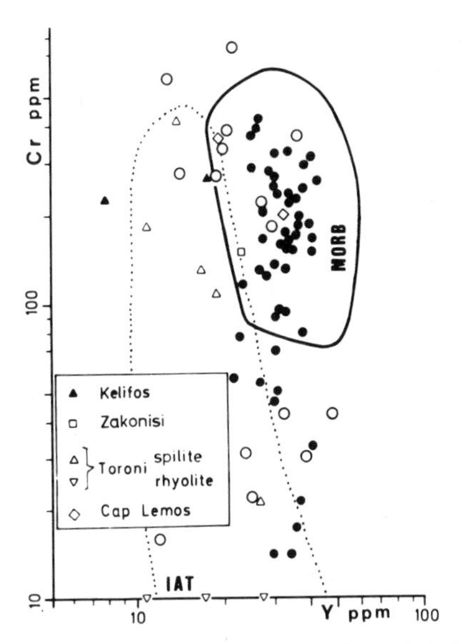

Figure 10. Cr-Y plot for the axis volcanics of the Sithonia (closed circles) and Kassandra (open circles) ophiolites. Off-axis Sithonia volcanics are also shown. After Pearce et al., 1984.

and are in addition strongly enriched in Ce (225, 252 ppm), Zr (297, 348 ppm), P_2O_5 (0.94, 1.1%) and in alkaline rare earths. These properties are evidence for an alkalic signature. By contrast, the volcanics from the outermost SE rift wall (C. Lemos, two analyses) and from the rift trough (Zakonisi, one analysis) rather classify as MORB. Lastly, the intermediate spilites from the innermost SE rift wall at Toroni (five samples) classify as IAT and chemically correspond to low-K andesites. MORB- and chondrite-normalized patterns for one of these samples (Si 69 in Figs. 8 and 9) show a stronger depletion in the HFS and heavy REE when compared with those of the axis volcanics indicating that they are in part less fractionated. The rhyolites are as opposed to the adjacent low-K andesites calc-alkalic and contain in less altered samples up to 4.5% K_2O. This bimodality along with the high amount of rhyolite exclude a comagmatic origin. Since the ophiolite sequence here is underlain by some km thick K-poor arc volcanics, contamination by these rocks cannot be responsible for the high K content in the rhyolite. The diversity of off-axis volcanism is well known from present-day spreading centers. Dick (1982) described interdigitated but unrelated Ti-rich alkalic and tholeiitic sills intruding pelagic sediments from the northern Philippine Sea. In the Gulf of California, off-axis dolerite sills show alkaline to subalkaline affinities (Saunders et al., 1982). On the other hand, intermediate and acidic volcanics are with the exception of plume-related settings virtually absent both in major and in marginal ocean basins. Pearce et al. (1984) reported two fractionation series of basaltic to rhyolitic lavas from the Oman ophiolite which they similarly relate to the marginal basin setting of this ophiolite.

4. *Relationship of the Kassandra Volcanics*

It was suggested earlier that the pillowed basalts at Kassandra may represent the abyssal segment of the Sithonia rift. To verify such a relation and to point out possible along-axis chemical variations, a comparison of these volcanics with those from Sithonia is made. In Fig. 10, the Kassandra volcanics plot to a greater part outside MORB. Nevertheless, the Ti/Zr, Ti/V, Zr/Y, and Rb/Sr ratios are similar to those of the Sithonia volcanics. Both are depleted in some HFS elements. Notwithstanding the broad similarity in major and minor element abundances, the REE patterns for the Kassandra volcanics (ruled area in Fig. 9) considerably differ from those of the Sithonia. Firstly, the total REE (25.6 to 49.77 ppm) is at a similar FeO^*/MgO ratio half as high as in the Sithonia volcanics. Secondly, they display a weak but significant LREE depletion with Sm_N/La_N and Ce_N/La_N more than 1 and with La_N/Yb_N fluctuating around unity (0.95 to 1.36). Thirdly, the Kassandra patterns invariably show a positive Eu anomaly as opposed to a negative one in the Sithonia patterns. These differences can hardly be explained in terms of fractional crystallization. The patterns and low REE abundances of the Kassandra volcanics are similar to N-type MORB. This similarity does not conflict with the proposal that the Kassandra volcanics represent the abyssal segment of the Sithonia rift. The enriched features of the volcanics from this rift may be attributed to a lower degree of partial melting of the same source or to stronger involvement of a subduction component.

Discussion and Conclusions

The presented geological and chemical data enables us to reconstruct the events leading to the opening and closure of the Sithonia rift and to put them forward in an evolutionary scenario. Starting with the pre-spreading conditions, these events can be summarized as follows.

The arc volcanics which were subjected to rifting correspond to some modern forearc regions. The derivation of the lithic clasts in the turbidite horizons from the formerly deformed Svoula flysch indicates that this flysch was early scrapped off and accreted to the forearc prior to or contemporaneous with the onset of arc volcanism. A second phase of deformation and uplift affected the volcanics resulting in their penetrative deformation and greenschist-facies metamorphism.

The first step of rifting was uplift and probable arching of the volcanic arc as inferred from its domal structure and substantiated by the change in the sedimentary facies from abyssal during the development of the volcanic arc to neritic during spreading and ophiolite formation. Subaerial erosion, subsidence and sedimentation did not take place; the sheeted dykes immediately invade the arc volcanics indicating that spreading commenced soon after their uplift. The paucity of rift-related volcanism may be attributed to insufficient crustal attenuation and lack of extensional faulting which would facilitate partial melting or alternatively to the absence of a major heat source. These features are in marked contrast with common models of rifted continental margins which envisage a rifting period of several tens of million years prior to the development of an ocean.

The rift opened along a NE direction that is oblique to subduction. At Chalkidiki, there are several examples of oblique rifting and obliquely oriented extension during the Mesozoic. At Gevgeli, the sheeted dykes strike predominantly NE (Bébien, 1982). Dixon and Dimitriadis (1984) described a NE trending sheeted complex in the Volvi ophiolite. The microdioritic dykes crosscutting the southeastern protrusions of the peridotite-gabbro association trend mostly NE. Aside from extensional structures, most fold axes associated with the Vardar-zone ophiolites have NE trends. It is worth noting that oblique spreading is presently taking place in some marginal basins as well as in plate-boundary related or transform dominated spreading systems (Curray et al., 1978; Weissel et al., 1982). Because of the instability of this direction and because a readjustment of plate motion direction did not take place, the rift soon closed at an almost right angle by NE-directed overthrusting of first the axial trough along a former transform.

Assuming a least half spreading rate of 1 cm/a, spreading lasted at most 1.5 Ma to account for a 30 km crustal accretion. The parallelism of the sheeted dykes and the periodicity of their injection point at a highly regular spreading. Geophysical studies by Makris and Röwer (1986) reveal the presence of a 30 to 32 km thick continental crust and, except for a local magnetic anomaly at the rift trough, provide no evidences for the occurrence of ultramafic rocks at depth in the Sithonia-Kassandra area. We conclude from these data that a break-up of the lithosphere was not achieved and that the Sithonia ophiolite most likely originated at the tip of a spreading axis which failed to develop into an ocean. This ophiolite may be compared with an early stage in the evolution of the transform dominated Gulf of California which obliquely opened along short segments several million years after the cessation of subduction (Curray et al., 1982). Unlike the Sithonia volcanics, those recovered from the Gulf of California exhibit N-type MORB chemistries attributed to the presence of slab-free window and a Pacific Ocean-type mantle beneath the Gulf (Saunders et al., 1982).

The nearly orthogonal configuration of the Sithonia and Kassandra ophiolites in SE and those of Oreokastron and Gevgeli in NW with respect to the non-ophiolitic peridotite – gabbro association of central Chalkidiki led to the proposal, that these ophiolites may relate by a transcurrent fault along which the peridotite – gabbro association was emplaced (Jung and Mussallam, 1985; Mussallam and Jung, 1986a, b; Bébien et al., 1986, 1987). Interestingly, Dixon and Dimitriadis (1984) proposed that the Volvi-?Therma-?

Gomati rift some 40 km to the west of the Vardar-zone ophiolite belt (see Fig. 1) may be part of an intra-Serbomacedonian massif extensional basin, which they relate to "Late Jurassic-Early Cretaceous pull-apart tectonics along and within an oblique subduction-zone margin". Though our model sounds convincing, evidences for lateral displacement and structures distinctive of wrench faulting are virtually absent (Mussallam and Jung, 1986a). The boundaries of the peridotite-gabbro association against the adjacent older formations are clear-cut and sharp. Local NW-verging folds within and near the western borders of this association may indeed be evidence of some NW-directed lateral movement, but this movement clearly postdates the emplacement of the peridotites and associated gabbroic and pyroxenitic rocks. Available paleontological and radiometric data reveal that these ophiolites were created at different times during the Middle and Upper Jurassic implying that spreading was diachronous. The crystallization age obtained from the Gevgeli ophiolite (Spray et al., 1984) and the cross-cutting relations of its sheeted dykes (Bébien, 1982) indicate that spreading diffusely starts during the latest Middle to earliest Upper Jurassic. Failing there to develop into an ocean (Bébien et al., 1986, 1987), spreading recommenced in the Sithonia-Kassandra area by the Late Jurassic where it experienced despite its regularity the same fate.

Acknowledgements

I am indebted to J. Lodziak, H. Raschka, B. Stütze, W. Vogel for the XRF and to U. Siewers for the REE analyses. K. Burgath and M. Mohr made REE analyses on some samples possible. Discussions with F. Kockel are acknowledged. Thanks are due to Mrs. Barbara Cornelisen for drawing some figures.

References

Atherton, M.P., Warden, V. and Sanderson, L.M., 1985. The Mesozoic marginal basin of Central Peru: a geochemical study of within-plate-edge volcanism. In: W.S. Pitcher, M.P. Atherton, E.J. Cobbing and R.D. Beckinsale (Eds), Magmatism at a Plate Edge: The Peruvian Andes, Blackie, Glasgow and London, pp. 47–58.

Bébien, J., 1982. L'Association ignée de Guévguéli (Macédoine Grecque): Expression d'un magmatisme ophiolithique dans une déchirure continentale. D. S. thesis, Nancy, p. 467.

Bébien, J., Dubois, R. and Gauthier, A., 1986. Example of ensialic ophiolites emplaced in a wrench zone: Innermost Hellenic ophiolite belt (Greek Macedonia)., Geology, 14: 1016–1019.

Bébien, J., Baroz, F., Capedri, S. and Venturelli, G., 1987. Magmatismes basiques associes a l'ouverture d'un bassin marginal dans les Hellénides internes au jurassique., Ofioliti, 12: 53–70.

Borsi, S., Ferrara, G. and Mercier, J., 1966. Age stratigraphique et radiométrique jurassique supérieur d'un granite des zones internes des Hellénides (granite de Fanos, Macédoine, Grèce)., Rev. Géogr. Phys. Géol. Dyn., 8: 279–287.

Bruhn, L.R., Stern, C.R. and de Wit, M.J., 1978. Field and geochemical data bearing on the development of a Mesozoic Volcano-tectonic rift zone and back-arc basin in southernmost South America., Earth Planet. Sci. Lett., 41: 32–46.

Coleman, R.G., Hadley, D.G., Fleck, R.G., Hedge, C.T. and Donato, M.M., 1979. The Miocene Tihama Asir ophiolite and its bearing on the opening of the Red Sea. In: A.M.S. Al-Shanti (Ed), Evolution and mineralization of the Arabian-Nubian shield, vol., 1: 173–186.

Curray, J.R., Moore, D.G., Lawer, L.A., Emmel, F.J., Raitt, R.W., Henry, M. and Kieckhefer, K., 1978. Tectonics of the Andaman Sea and Burma. In: J. Watkins and L. Montadert (Eds), Geological and geophysical investigations of continental margins. Amer. Assoc. Petrol. Geol. Mem., 29: 189–198.

Curray, J.R., Moore, D.G., Kelts, K. and Einsele, G., 1982. Tectonics and geological history of the passive continental margin at the tip of Baja California. In: J. Blakeslee, L.W. Platt and L.N. Stout (Eds), Initial reports of the Deep Sea Drillin Project, vol., 64: 1089–1116.

Dick, H.J.B., 1982. The petrology of two back-arc basins of the Northern Philippine Sea., Am. Jour. Sci., 282: 644–700.

Dixon, J.E. and Dimitriadis, S., 1984. Metamorphosed ophiolitic rocks from the Serbo-Macedonian Massif near Lake Volvi, north-east Greece. In: J.E. Dixon and A.H.F. Robertson (Eds), The Geological Evolution of the Eastern Mediterranean., Spec. Publ. Geol. Soc. Lond., 17: 603–618.

Haenel-Remy, S. and Bébien, J., 1987. Basaltes et dolerites riches en magnésium dans l'association ignée de Guévguéli (Macédoine Grecque): Les temoins d'une évolution dépuis des tholéiites abyssales jusqu' à des basaltes continentaux?, Ofioliti, 12: 91–106.

Haskin, L.A., Haskin, M.A., Frey, F.A. and Wildeman, T.R., 1968. Relative and absolute terrestrial abundances of the rare earths. In: L.H. Ahrens (Ed), Origin and Distribution of the Elements, I. Peramon, Oxford, pp. 889–911.

Hawkins, J.W., Bloomer, S.H., Evans, C.A. and Melchior, J.T., 1984. Evolution of intra-oceanic arc-trench systems., Tectonophysics, 102: 175–205.

Jung, D. and Mussallam, K., 1985. The Sithonia ophiolite: A fossil oceanic crust., Ofioliti, 10: 329–342.

Kauffmann, G., Kockel, F. and Mollat, H., 1976. Notes on the stratigraphic and paleogeographic position of the Svoula Formation in the innermost zone of the Hellenides., Bull. Soc. Geol. France, 18: 225–230.

Kay, R.W. and Hubbard, N.J., 1978. Trace elements in ocean ridge basalts., Earth Planet. Sci. Lett., 38: 95–116.

Kockel, F., 1986. Die Vardar- (Axios-) Zone. In: V. Jacobshagen (Ed), Geologie von Griechenland. Gebrüder Borntraeger, Berlin, Stuttgart, pp. 150–168.

Kockel, F. and Mollat, H., 1977. Erläuterung zur geologischen Karte der Chalkidiki und angrenzenden Gebiete 1:100.000 (Nord-Griechenland). Federal Institute of Geosciences and Natural Resources, Hannover, 119 p.

Leube, A. and Kockel, F., 1962. Bericht über die Untersuchung der Molybdänerz-Lagerstätte Mavro-Dendron. Federal Institute of Geosciences and Natural resources, Hannover, Unpubl. Report.

Makris, J. and Röwer, P., 1986. Struktur und heutige Dynamik der Lithosphäre in der Ägäis. In: V. Jacobshagen (Ed), Geologie von Griechenland. Gebrüder Borntraeger, Berlin, Stuttgart, pp. 241–256.

Mussallam, K. and Jung, D., 1986a. Petrology and geotectonic significance of salic rocks preceding ophiolites in the eastern Vardar Zone, Greece., Tschermaks Min Petr Mitt, 35: 217–242.

Mussallam, K. and Jung, D., 1986b. Geologie und Bau des Sithonia-Ophioliths (Chalkidiki, NE-Griechenland): Anmerkungen zur Bildung ozeanischer Krusten., Geol. Rdsch., 75: 383–409.

Mussallam, K., Jung, D. and Burgath, K., 1981. Textural features and chemical characteristics of chromites in ultramafic rocks, Chalkidiki complex (northeastern Greece)., Tschermaks Min Petr Mitt, 29: 75–101.

Page, B.G.N., Bennet, J.D., Cameron, N.R., Bridge, D.McC., Jeffery, D.H., Kreats, W. and Thaib, J., 1979. A review of the main structural and magmatic feature of northern Sumatra., J. Geol. Soc. Lond., 136: 569–579.

Pearce, J.A., 1982. Trace element characteristics of lavas from destructive plate boundaries. In: R.S. Thorpe (Ed), Andesites. J. Wiley and Sons, Chichester, pp. 525–547.

Pearce, J.A., Lippard, S.J. and Roberts, S., 1984. Characteristics and tectonic significance of supra-subduction zone ophiolites. In: B.P. Kokelaar and M.F. Howells (Eds), Marginal basin geology., Spec. Publ. Geol. Soc. Lond., 14: 77–94.

Saunders, A.D., Fornari, D.J., Joron, J.-L., Tarney, J. and Truil, M., 1982. Geochemistry of basic igneous rocks, Gulf of California, Deep Sea Drilling Project Leg 64. In: J. Blakeslee, L.W. Platt and L.N. Stout (Eds), Initial reports of the Deep Sea Drillin Project, vol., 64: 595–642.

Schünemann, M., 1985. Contributions to the geology, geochemistry and tectonics of the Chortiatis Series metamorhic calc-alkaline suite, Chalkidiki, Northern Greece. Unpubl. Ph.D. thesis, University of Hamburg, 181 p.

Spray, J.G., Bébien, J., Rex, D.C. and Roddick, J.C., 1984. Age constaints on the igneous and metamorphic evolution of the Hellenic-Dinaric ophiolites. In: J.E. Dixon and A.H.F. Robertson (Eds), The Geological Evolution of the Eastern Mediterranean., Spec. Publ. Geol. Soc. Lond., 17: 619–627.

Stern, C.R., 1980. Geochemistry of Chilean ophiolites: Evidence for the compositional evolution of the mantle source of back-arc basin basalts., J. Geopghys. Res., 85: 955–966.

Vergely, P., 1984. Tectonique des ophiolites dans la Hellénides internes (déformations, métamorphismes et phénomènes sédimentaires). Conséquences sur l'évolution des régions téthysiennes occidentales. D. S. thesis, Orsay, Paris, p. 661.

Weissel, J.K., Taylor, B. and Karner, G.D., 1982. The opening of the Woodlark Basin, subduction of the Woodlark spreading system, and the evolution of the northern Melanesia since mid-Pliocene time., Tectonophysics, 87: 253–277.

Weaver, S.D., Saunders, A.D., Pankhurst, R.J. and Tarney, J., 1979. A geochemical study of magmatism associated with the initial stages of back-arc spreading: The Quaternary volcanics of Bransfield Strait, form South Shetland Islands., Contrib. Mineral. Petrol., 68: 151–169.

Upper Triassic-Early Jurassic Sedimentary Breccias in the Ophiolitic Suite of the Lesser Caucasus

A.L. KNIPPER
Geological Institute of the USSR Academy of Sciences, Moscow, USSR

Abstract

Sedimentary breccias, consisting of the clasts of gabbro, diabase, greenschist, and carbonate rocks altered to a different degree, have been found inside the ophiolitic suite of the Sevan-Akera zone of the Lesser Caucasus. The breccias are of Late Triassic to Early Jurassic age, having been formed during the break-up of the continental crust, related to the destruction of the Gondwana northern margin, according to an asymmetric rift model. Later on, the crustal fragment was transported northward and, in the middle of the Coniacian, obducted on the Mesotethys active margin.

Short Description of the Ophiolite Sequence

The ophiolites in the Sevan-Akera zone of the Lesser Caucasus belong to the northern ophiolitic branch of the Alpine foldbelt, stretching from the Apennines, through the Alps, into the Vardar zone of Yugoslavia and Greece and farther eastward, along the southern boundary of the Pontides, into Armenia and Azerbaijan. There these rocks compose two major tectonic nappes thrusted on the Mesotethys active margin in the middle of the Coniacian (Knipper and Sokolov, 1974; Knipper, 1980). The lower, Vagazin nappe is situated within an autochthonous ophiolite – clastic olistostrome of the Upper Cenomanian to Lower Coniacian, and the upper, Ipiak nappe is structurally uppermost in the allochthonous sequence. The structure of the ophiolitic suite of the Ipiak nappe is shown in Fig. 1. The following points should be noted:

1. The column represents the composite sequence which is compiled from many repeating sections with the correctly established relations between the various members of the ophiolitic suite.
2. The K-Ar age (on micas) of plagiogranite, intruding ultrabasic tectonites and gabbro on the northeastern coast of Lake Sevan, is 168 ± 8 Ma

Tj. Peters et al. (Eds), Ophiolite Genesis and Evolution of the Oceanic Lithosphere, 705–713.

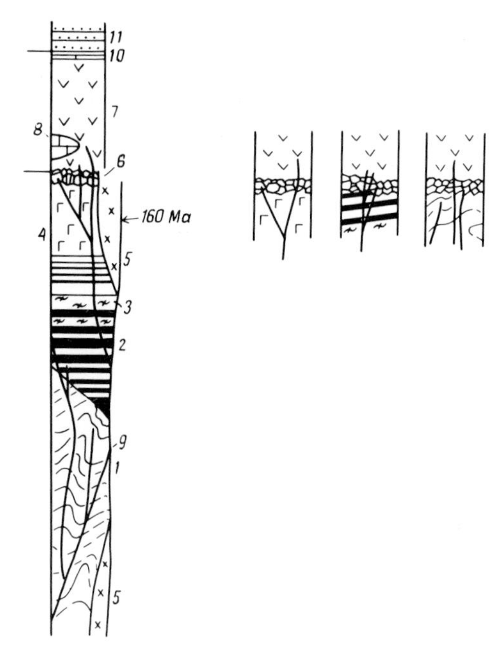

Figure 1. Summary section of the ophiolitic suite of the Ipiak nappe, Sevan-Akera zone. 1) harzburgite tectonite, 2) layered gabbro, 3) flaser gabbro, 4) isotropic gabbro, 5) plagiogranite, 6) breccia, 7) volcanic rocks (upper Trias-Cenomanian), 8) exotic blocks of limestone, 9) diabase dykes, 10) radiolarite, 11) flysch (Upper Cenomanian).

(Morkovkina and Arutiunian, 1971), its U-Pb (on zircons) yield 160 ± 4 Ma (Zakariadze et al., 1990). Therefore, the age of the lower part of the ophiolite sequence is, at least, pre-Late Jurassic.

3. The crystallization of gabbro occurred under conditions of plastic flow accompanied by the formation of a flaser to amphibolite fabric on the rocks of cumulus complex and isotropic gabbro.
4. The complex of sheeted dykes is absent. Diabase dykes are represented by individual bodies or dyke swarms, which are not localized just in the roof of the gabbro, but intrude both harzburgite tectonites and gabbro.
5. The formation of gabbro and ultrabasic tectonites is separated from the volcanic stage by the gap recorded by the presence of peculiar breccias, which resemble the Ligurian breccias of the Apennines (Knipper, 1978). These breccias will be described below.
6. The sequence of the ophiolitic volcanic rocks of the Lesser Caucasus will not be discussed in detail in this paper, since it was done earlier (Zakariadze et al., 1983, 1986). Only two points should be noticed. First, the wide range of volcanic rocks, formed in different geodynamic environments from the Carnian (Upper Triassic) to the Cenomanian (Middle Cretaceous): tholeiitic basalts with intraplate affinity (Upper Triassic-Jurassic), alkaline and subalkaline differentiated basaltic-andesite suites

of a seamount type (Neocomian), and differentiated basaltic–andesite suites and boninites of immature ensimatic island arcs (Albian-Cenomanian). These groups of rocks compose different incomplete sequences, whose parts, mutually overlying one another, have a direct stratigraphic contact with the gabbro-peridotite basal sequence of the Ipiak nappe. Second, the volcanic sequence in some places contains a great number of limestone and marble blocks. Famenian corals (Gasanov, 1985), Upper Paleozoic conodonts (Kariakin and Aristov, in press), Upper Triassic bivalves, and Upper Jurassic rudists have been collected in some of these blocks.

Breccias at the Base of the Volcanic Sequence

Three main types of breccias may be identified, differing in structure and composition of clastic material.

Zod Type. The type section is located at the old Zod Pass (Armenia and Azerbaijan frontier). The section consists of three parts and contains three levels of breccias (Fig. 2). The lower part of the lowermost layer of breccias consists of compact diabase scree. The rock fabric and degree of crystallization indicate its dyke nature. The clasts are absolutely unrounded and unsorted, being of very different shape. Their size varies from some mm to 30–40 cm. Sometimes, unclear lenses of flour-like rock, composed of finest clasts of the same diabase and its minerals, are found among the monolithic breccias. The scree composition changes upward along the section: clasts of cumulative and isotropic gabbro, gabbro-pegmatite and flaser gabbro appear in it. Occasionally, such clasts are of a rounded or semirounded shape. A single quartz porphyry pebble, 5 cm in diameter, and a single marble pebble, approximately of the same size, have been found in the lenses of flour-like rock. To all appearance this characteristic indicates the sedimentary origin of the breccia suite as a whole. The breccia is overlain by basalt and basaltic andesite flows, which in their lower part include breccia fragments. These aphyric and amygdaloid lavas have no traces of brecciation. A second horizon of breccia, overlying the lavas, does not differ in composition of clasts from those of the lower layer. The gabbro blocks reach 1 m in diameter. However, this part of the sequence contains, as a whole, the greater amount of fine-grained matrix, which shows more clearly its sedimentary origin (unclear bedding and some elements of sorting).

The breccias are overlain by a volcano-sedimentary member consisting of pelites, radiolarites, contourites, and distal turbidites of graywacke composition with interbedded layers of basaltic andesite. Radiolarians of the Upper Carnian (*Canoptum cf. verrucosum*, Bragin, *Capnuchosphaera tricornis*, De Wever, *C. lea*, De Wever, *C. cf. triassica* De Wever, *Pentaspongodiscus cf. dercourti*, De Wever, *Triassocampe nova*, Yao, and *T.* sp.) occur in the

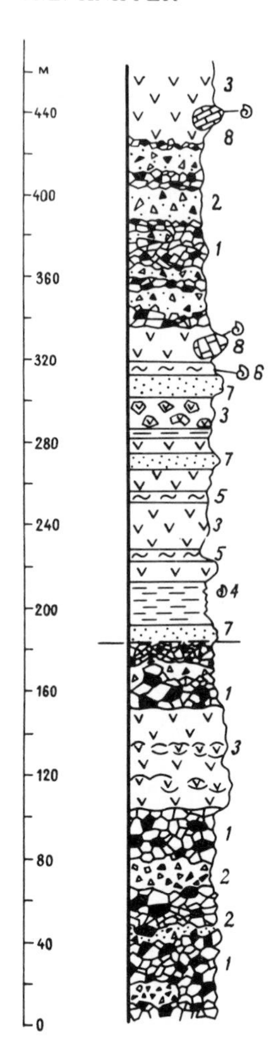

Figure 2. Section of the Zod type breccias (old Zod Pass).
1) gabbro-diabase breccia, 2) sandstone layers, 3) basalt and basaltic andesite, 4) pelite with Carnian radiolarians, 5) radiolarite, 6) radiolarite with Toarcian radiolarians, 7) distal turbidite and contourites of graywacke composition, 8) blocks of limestone with Carnian bivalves.

pelites of the lower part of the sequence (Fig. 2). Upward, Toarcian radiolarians *Crubus wilsonensis*, Carter, *Hsuum cf. minoratum*, Sashida, *Paronaella variabilis*, Carter, *Parvicingula*(?) *gigantocornis*, Kishida and Hisada, *Trillus cf. elkhornensis*, Pessagno and Blome, *Acanthocircus* sp., and *Bernoullius* sp. have been found in radiolarite beds. The roof of the volcano-sedimentary member contains blocks of nodular limestone with middle Norian bivalves Halobia norica MJS. and H.(?) sp. juv. (H. ex gr. salinarum BRONN.) and the Middle to Upper Norian conodonts *Neogondolella navicula* (Huckriede) and *Epigondolella* cf. *postera* (Kozur and Mostler). (Determinations by N.A. Bragin – radiolarians and conodonts and I.V. Polubotko – bivalves.)

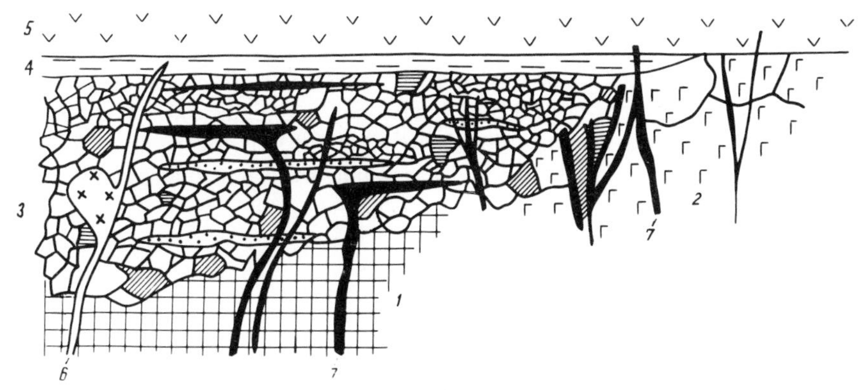

Figure 3. Schematic picture of the structure of Geydara type breccias (Geydara).
1) harzburgite tectonite, 2) gabbro, 3) predominantly diabase breccia with greenschist blocks (hatched) and sandstone layers (dotted), 4) black shale, 5) basalt, 6) quartz porphyry dyke, 7) diabase dykes grading into sills and lava sheets.

It should be noted, that the layers of contourites, in the lower part of this member, contain numerous very fine clasts of serpentinite and the turbidites of the upper part abundant fragments of quartz and muscovite.

The third level of breccias differs only slightly from the underlain breccias, however, together with diabase and gabbro scree, numerous clasts of aphyric and amygdaloid basalt and basaltic andesite were found.

Geydara type. The type section is situated on the western slope of the Geydara watershade. Breccias of this type are composed of gigantic blocks, reaching 2000 or even 3000 m in volume. The blocks are represented by greenschist and phyllite (protholith-mudstone, fine-grained graywacke, with rare and thin layers of limestone), enveloped by breccias of sharply angular scree of diabase, dolerite and gabbro-diabase. The K-Ar age of greenschists from Kylychly Village yields 207 Ma (Wh.rock, Gasanov, 1985), corresponding to the Triassic - Jurassic boundary. It is clearly seen, that the crushed basic rocks, which display different fabrics, are derived from several generations of dykes. Rather frequently these large blocks include relics of brecciated and pull-apart dikes, which partly preserve their original shape and can be traced for a distance up to 10 m. Sometimes, horizons and lenses of sandstone with unclear bedding and sorting occur in the matrix, occasionally together with rounded to flattened pebbles of marble, limestone, and garnet-rich skarn (after carbonate rocks). The whole section, including its sedimentary components, is cut by dykes of almost non-brecciated diabase, which does not differ in composition and fabric from the same rock, composing breccias. These dykes display chilled margins and sometimes grade into sills and lava sheets (Fig. 3).

Tekiakaya type. The type section is situated on the right bank of Tekiakaya River, above the village of the same name. The breccia cement is represented by a fine, sometimes flour-like matrix, consisting of clasts of diabase, greenschist, marble, re-crystallized limestone, and limestone. The cement contains huge blocks (up to 100–150 m in length) of the same carbonate rocks. In some places the blocks of carbonate rocks are injected by dykes and basalts.

Discussion

The above data suggest the following conclusions:

1. The breccias under consideration are sedimentary formations originated as a result of the redeposition of metamorphic rock of a continental affinity (greenschist, marble, limestone and skarn) and ophiolitic suite rocks (various gabbros and diabase, sometimes serpentinite).
2. The breccia formation processes, dykes intrusion and redeposition occurred almost simultaneously; consequently, the dykes of early generation have been crushed and redeposited, whereas those of later generation, preserved the original shape and chilled contacts, cut the redeposited rocks as feeders of sills and lava flows.
3. The breccias of Geydara and Tekiakaya type are collapse formations accumulating along the feet of steep scarps. The Zod type breccias have been relatively more removed from the erosion source.
4. The structure of blocks in the Geydara type breccias suggests that their formation has resulted not only from the simple mixing of various clasts during their deposition. In this case, already brecciated rocks have been redeposited.

Let us notice one further circumstance. It was already mentioned, that the single occurrences and swarms of diabase dykes, cross-cutting peridotites and gabbros, are present in the ophiolite sequence of the Sevan-Akera zone. These bodies, not differing in their rock composition and fabric from the diabases which are the parts of breccias, are traceable along the strike for many hundreds of meters and can be easily mapped. They cross-cut successively ultrabasic rocks, cumulate gabbro, and then isotropic gabbro, too. They are not brecciated and have chilled contacts. Then, suddenly, the zone of breccias begins. An impression is created, that the area of brecciation has been confined to some zone along the boundary between the peridotite-gabbro and metamorphic complexes. Just here most dykes ended their existence, undergoing brecciation. This evidence suggests, that a continental

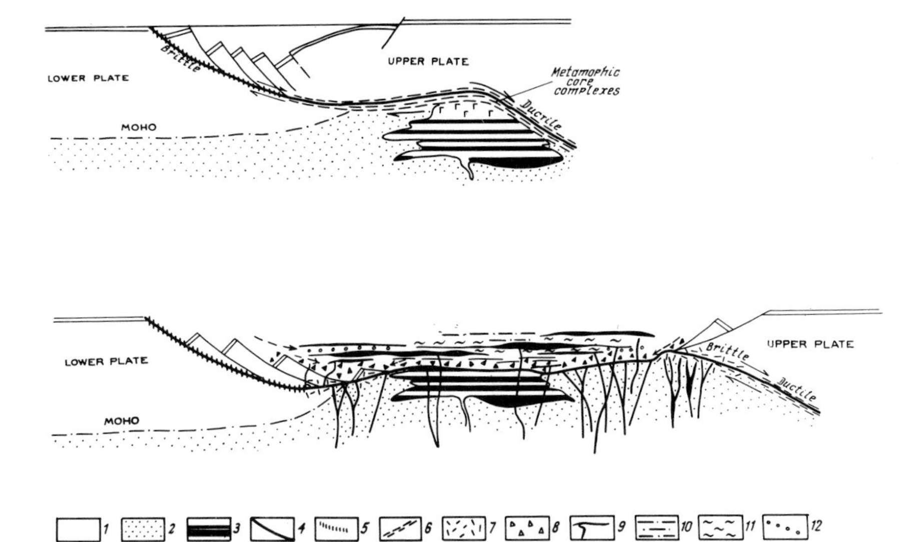

Figure 4. Schematic illustration of the Gondwana northern margin destruction in the Late Triassic – Early Jurassic (detailed explanations in the text).
1) continental crust, 2) upper mantle, 3) gabbro, 4) detachment fault, 5) area of brittle deformation, 6) area of ductile deformation, 7) area of brittle deformation overlapped by ductile deformation, 8) breccia, 9) diabase dykes and basalt lava flows, 10) black shale (Lias), 11) radiolarite (Upper Triassic-Lias), 12) turbidite and contourites of graywacke composition.

sequence, which subsequently has mysteriously disappeared, once had been situated above peridotites and gabbros.

My ideas about the formation history of the breccias are shown on Fig. 4, which requires some additional elucidations.

Some evidences of the Late Triassic destruction of the northern side of Godwana seem to be recorded by the ophiolite sequence of the Lesser Caucasus. This process, which related to the opening of the Mesotethys oceanic basin, was accompanied by the detachment of continental blocks from the African-Arabian craton. At that time, a series of listric faults, related to a deep detachment fault were formed, according to the Wernicke's model (1981, 1985), supplemented and developed by Lister et al. (1986). Proceeding from the data above, the succession of interlinked events can be presented in the following way: the rise of a mantle diapir, the melting of gabbro and its crystallization under a directional tensile stress (appearance of flaser gabbro), the simultaneous thinning of continental crust, the brecciation of rocks in the area of brittle deformation, and their metamorphism in the area of ductile deformation along the detachment fault at the boundary between an upper plate and lower one. Further on: the intrusion of early generations of diabase dykes, cutting through both the mantle diapir rocks and continental crust.

These dykes are brecciated along the detachment fault, which at that time, as a result of the rise of "metamorphic core" complexes, partly passed from the area of ductile deformation into the area of brittle deformation. Then: the break-up of continental crust, the bringing of mantle rocks and breccias formed along the detachment fault onto the sea-basin floor, and their subsequent erosion, collapse, and redeposition (Geydara type breccias). Simultaneously, the carbonate rocks of the Gondwana platform cover are eroded and transported (Tekiakaya type breccias). Diabase dykes continue to intrude and are no more brecciated in the area of sea-basin floor; submarine extrusions of basalt and basaltic andesite begin, together with the deposition of sedimentary strata with radiolarite, micritic limestone, and distal turbidite, characteristic of the Zod type breccia sequences.

According to the faunal datings, the break-up of the continent has already occurred in late Carnian time. The source areas of ophiolite-clastic and sialic material were situated not far from the basin, where the Zod type sequence accumulated till the Toarcian. Therefore, the disjoining process of upper and lower plates may be envisaged as non-catastrophic, with a duration of at least 35–40 Ma (if the destruction had begun in the early Carnian).

Later on, the Gondwana fragment began its northward journey, during which it was entering various paleogeodynamic environments, reconstructed by geochemical affinity of volcanic rocks. This fragment, like Tom Thumb, marking its way by white stones (the marble and limestone blocks now buried in the lavas of the Middle-Upper Jurassic and Lower Cretaceous age), reached the northern Thetys margin in the beginning of the Coniacian stage.

References

Gasanov, G.A., 1985. Ophiolity Malogo Kavkaza. Moskwa: Nedra (in Russian), 240 p.

Kariakin, Yu.V., and V.A. Aristov. In press. O vozraste i geologicheskoy pozicii "ekzoticheskikh porod" Touragaichaiskoy zony (Maliy Kavkaz). DAN SSSR (in Russian).

Knipper, A.L., 1978. Ophicaltziti i nekotoriye drugie tipy brecchiy, soprovogdajustchie doorogenoje obrazovaniye ophiolitovogo kompleksa., Geotectonika No. 2: 50–66 (in Russian).

Knipper, A.L., 1980. The tectonic position of ophiolites of the Lesser Caucasus. In: Ophiolites, Proc. Int. Sym., Nicosia, Cyprus, 1979, pp. 372–376.

Knipper, A.L., and S.D. Sokolov, 1974. Predverkhnesenonskiye tectonicheskiye pokrovy Malogo Kavkaza., Geotectonika No. 6: 74–80 (in Russian).

Lister, G.S., M.A. Etheridge, and P.A. Symonds, 1986. Detachment faulting and the evolution of passive continental margins., Geology 14: 246–250.

Morkovkina, V.F., and G.S. Arutiunian, 1971. O radiometriches- kom vozraste ultraosnovnykh porod Sevanskogo khrebta. Izv. AN SSSR. Ser. geol. No. 11: 133–137 (in Russian).

Wernicke, B., 1981. Low-angle normal faults in the Basin and Range province: Nappe tectonics in an extending orogen., Nature 291: 645–648.

Wernicke, B., 1985. Uniform-sense normal simple shear of the continental lithosphere., Can. Journ. of Earth Sciences 22: 108–125.

Zakariadze, G.S., A.L. Knipper, A.V. Sobolev, O.P. Tsamerian, L.V. Dmitriev, V.S. Vishnevskaya, and G.M. Kolesov, 1983. The ophiolite volcanic series of the Lesser Caucasus., Ofioliti 8 (3):439–466.

Zakariadze, G.S., A.L. Knipper, A.V. Sobolev, O.P. Tsamerian, L.V. Dmitriev, V.S. Vishnevskaya, and G. Kolesov, 1986. Osnovniye cherty structurnogo pologenia i sostava vulkanicheskikh tolsch ophiolitovoy serii Malogo Kavkaza. In: Bogatikov, O.A. (Ed), Oceanic magmatism. Moscow: Nauka, pp. 218–241 (in Russian).

Zakariadze, G.S., A.L. Knipper, Ye.V. Bibikova, S.A. Silantyev, S.K. Zlobin, G.V. Grachov, S.A. Makarov, and G.M. Kolesov, 1990. Istoriya formirovaniya i vozrast plutonicheskoy chasti ophiolitovogo kompleksa severo-vostochnogo poberejia ozera Sevan., Izv. AN SSSR 3: 17–30 (in Russian).

Origin of Volcanics in the Tethyan Suture Zone of Pakistan

GEORGE R. MCCORMICK
Professor of Geology, The University of Iowa, Iowa City, Iowa, 52242 USA
Fulbright Professor, Centre of Excellence in Mineralogy, University of Baluchistan, Quetta, Pakistan

Abstract

The south Tethyan suture zone in Baluchistan has two distinct tectonic segments: (1) a north-south segment along the fold belt from Karachi to the Pamir Mountains and (2) an east-west segment west of the fold belt in the Makran region. Along the fold belt ophiolites are obducted over folded Jurassic and Cretaceous rocks. The Maastrichtian age Parh Formation on which ophiolites are obducted contains volcanic agglomerates, flows and breccia. Preliminary petrographic and chemical analyses as well as field evidence indicates these volcanic units are oceanic island type basalts.

These volcanic units may represent passage of Tethyan ocean floor over the Reunion Island "hot spot" prior to the passage of the Indian continent over it which produced the Deccan rocks and prior to the passage of the Indian Ocean floor over it which has produced the Chaggos, Maldive and Laccadive islands, movement similar to that proposed for the Ninetyeast Ridge as it passed over the Kerguelen Island "hot spot." Preliminary dates for basalts are: 72 my near Uthal, north of Karachi; 68 my in the Deccan; 32 my at site 706 in the Indian Ocean; 10 my at Mauritius and 0–2 my at Reunion. The oceanic islands formed by the movement over the "hot spot" north of the Deccan are now in the collision zone between the Indian plate and the Afghan micro-continent.

Introduction and Setting

The south Tethyan suture zone in Baluchistan has two distinct tectonic segments: (1) a north-south segment along the fold belt from Karachi to the Pamir Mountains and (2) an east-west segment (Ras Koh and Chagai Hills) west of the Chaman-Ornach-Nal fault in the Makran region (Fig. 1).

Coleman (1981) described the ophiolites in the Iranian and Pakistani Makran as being of the same tectonic origin and age as the ophiolites in Oman; these ophiolites originated at a paleo spreading center located in the

Tj. Peters et al. (Eds), Ophiolite Genesis and Evolution of the Oceanic Lithosphere, 715–722.

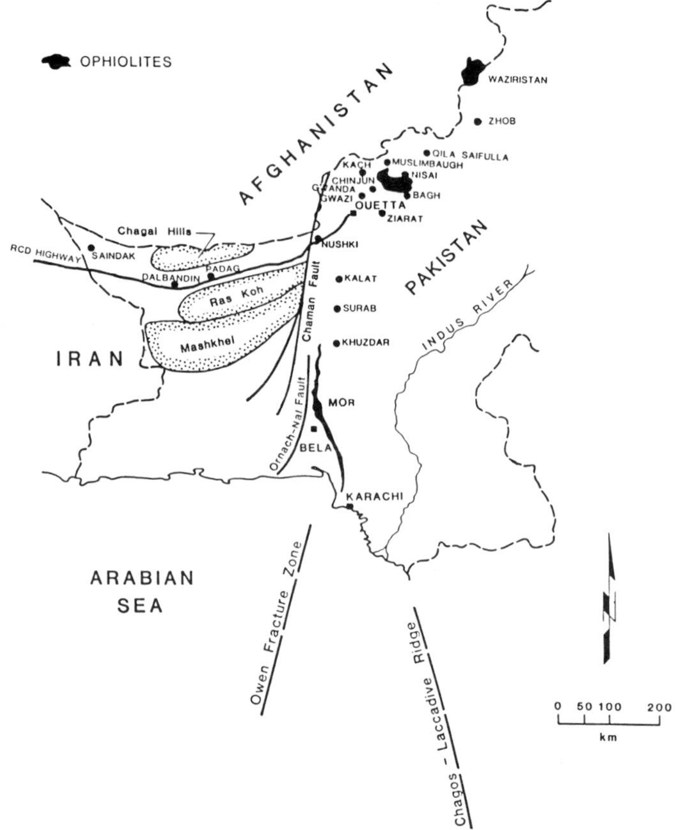

Figure 1. Map of Pakistan showing location of major topographic features, structural features and ophiolite massifs in Baluchistan.

present Gulf of Oman about 95 my. Closure and detachment began about 90 my and obduction of the ophiolite in Oman and Makran occurred in the Turonian. Subduction of the Arabian plate under the Makran began in the Eocene.

The origin of the ophiolites and associated volcanic rocks in the north-south axial zone appear to be of a different origin. Numerous investigators have described the western margin of the Indo-Pakistan continent to be an Atlantic-type margin throughout almost all of its geologic history.

It is known that the axial zone ophiolites (Fig. 1) were obducted in the Paleocene, however, the origin of the ophiolites has been in doubt. Most investigators have assumed the origin to be a spreading center followed by closure of the Tethyan ocean as that described for Oman by Coleman (1981). If they were formed by a spreading center it could not be the same one from which the Oman ophiolites originated because paleo-geographic reconstructions, such as that of Powell (1979) located the Indo-Pakistani continent nowhere near the proposed Omani spreading center at 95 my.

Axial Zone Volcanics

Volcanic flows, breccias and pillow lavas are associated with melanges and with the Parh formation all along the axial zone from Bela to Waziristan (Fig. 1); the Bibai formation (Kazmi, 1979) and Porali conglomerates (De Jong, 1979) are thought to be equivalents of the Parh formation. Kazmi interpreted the Bibai formation as a remnant island arc. A Japanese-Pakistani research group chaired by Y. Okimura of Hiroshima University and A.N. Fatmi of the Geological Survey of Pakistan has recently concluded that the axial zone ophiolites were formed by a spreading center followed by subduction of the Tethyan crust eastward under the Indo-Pakistan continent. Apparently this eastward subduction zone was based wholly on the Bibai volcanics being arc-like; there are no chemical analyses to substantiate this.

The Parh formation in the axial zone has classically been dated as Maastrichtian on the basis of its stratigraphic position. The Japanese-Pakistani research group has recently completed a thorough paleontological study of the Parh formation in the axial zone and determined it to be Triassic, just slightly younger than the Permo-Triassic Alozai formation in the axial zone. The research group has called the Triassic Parh in the axial zone the false "Parh" group. They still consider the true Parh group in the calcareous zone over which the axial zone is thrust to be Cretaceous; however, they admit to finding no fossil evidence to justify this conclusion.

Petrochemistry of Axial Zone Volcanics

McCormick (1989) analyzed pyroxenes from flows at Gwanda Gwazi (Fig. 1) and determined they were within-plate basalts. Khan (1986) analyzed 15 samples of volcanic rocks from the Parh formation near Chinjun (Fig. 1) by XRF and identified both olivine tholeiites and quartz tholeiites. He interpreted these volcanics to be of island-arc affinity, however, he did not examine the immobile element relationships. In recent months the Geological Survey of Pakistan has obtained 39 XRF and Neutron Activation analyses of the Bela volcanics and intrussives (Fig. 1) and McCormick has obtained 11 new XRF analyses including samples from Muslimbagh and Zhob (Fig. 1).

The analyses indicate the volcanic rocks in the axial zone are within-plate basalts. All specimens plot as tholeiites on the FeO/MgO vs. SiO_2 diagram

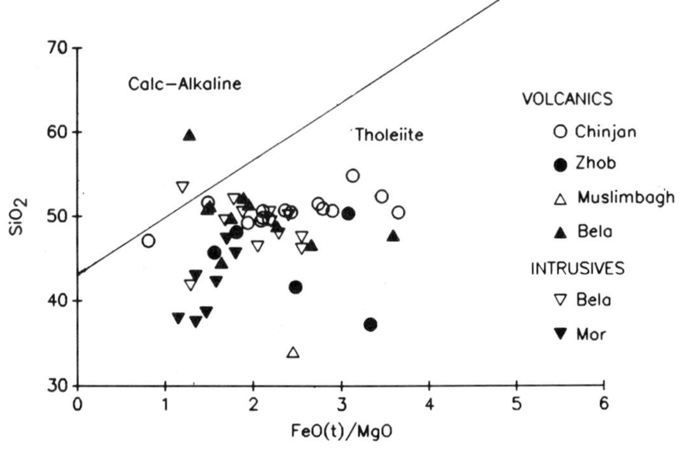

Figure 2. SiO_2 vs FeO/MgO plot for volcanics from the axial zone, after Miyashro (1974).

(Fig. 2) and the majority are sub-alkaline on the $Na_2O + K_2O$ vs. SiO_2 diagram (Fig. 3); most of the samples are in the tholeiite field in the AFM diagram (Fig. 4).

All but two samples plot in the field of within-plate basalts in the Zr vs. TiO_2 diagram (Fig. 5) and also plot in or near the field of within-plate basalt of the Zr/Y vs. Zr diagram (Fig. 6). An average value for the Reunion Island basalts (R) is shown in Figure 7 to be similar to the axial zone samples. Selected samples for which Nb values were available plot in or near the within-plate basalt field in the Zr vs. Nb diagram (Fig. 7).

Wood et al. (1979) distinguished between N- and E-type MORB on the basis of (La/Ta), (Hf/Ta) and (Th/U) values (Table 1). These values for the

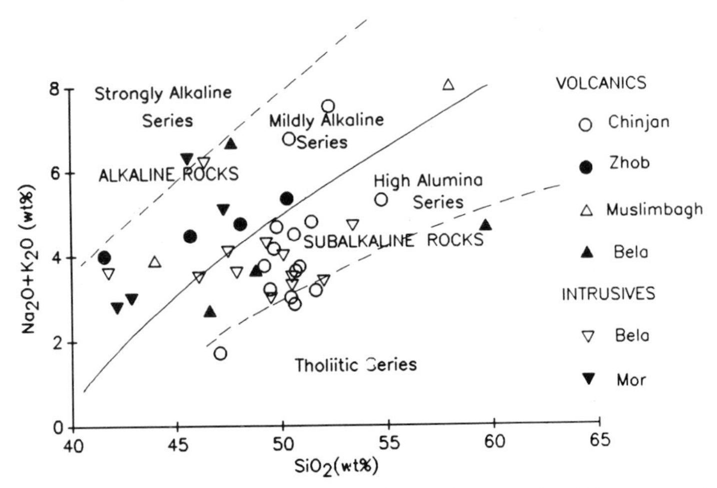

Figure 3. Alkali vs SiO_2 plot for volcanics from the axial zone, after Schwarzer and Roger (1974).

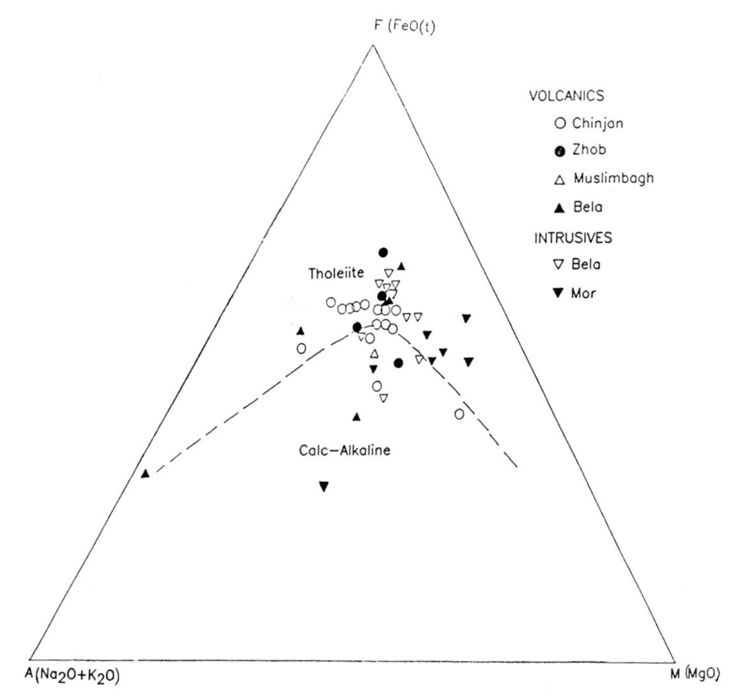

Figure 4. AFM plot for the axial zone volcanics, after Irvine and Barager (1971).

axial zone volcanics, Reunion Island basalts and Deccan trap rocks are all similar (Table 1). All of the volcanics are identified as ocean-island type or within-plate basalts by the (Hf/Ta) values but the axial zone and Reunion Island volcanics have N-type MORB (La/Ta) values; lanthanum values were not available for Deccan plateau volcanics.

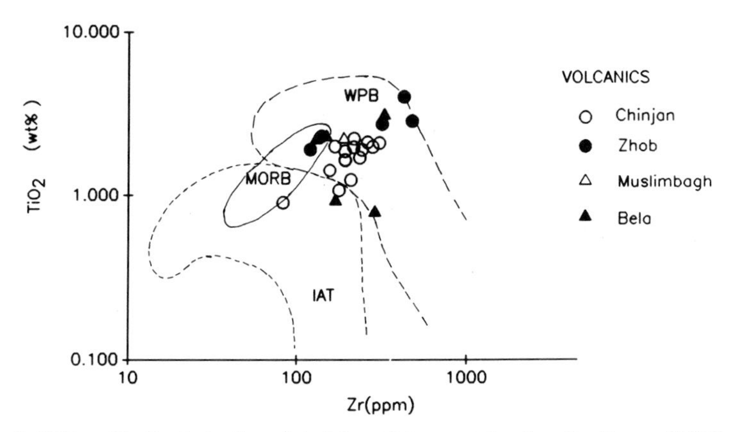

Figure 5. TiO_2 vs Zr discrimination plot of the axial zone volcanics after Pearce (1980). WPB = within-plate basalt.

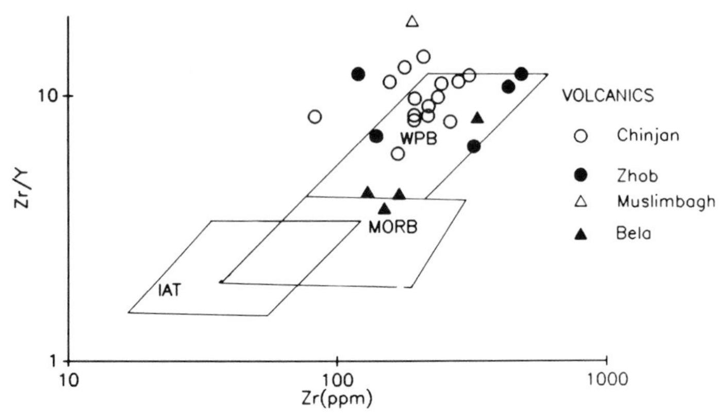

Figure 6. Zr/Y vs Zr discrimination plot of the axial zone volcanics after Pearce and Norry (1979). WPB = within-plate basalt and IAT = Island arc tholeiite.

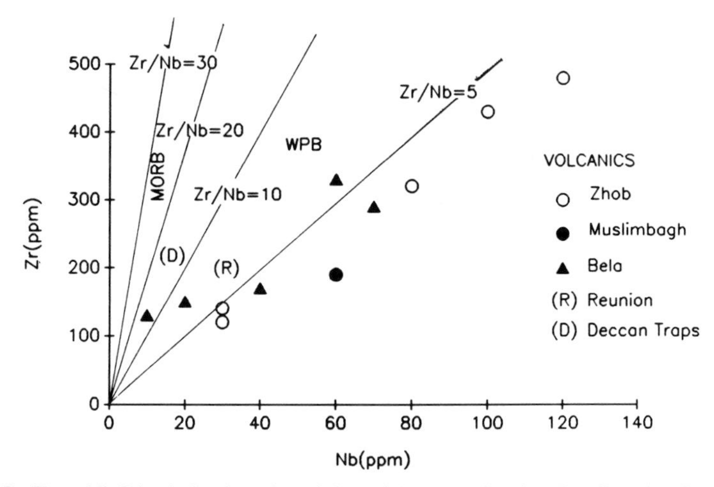

Figure 7:. Zr vs Nb Discrimination plot of the axial zone volcanics. R = Reunion Island basalt average and D = Deccan basalt average.

Table 1.

	(La/Ta)	(Hf/Ta)	(Th/U)
N-type MORB (Wood, 1979)	~15	>7	~2
E-type MORB (Wood, 1979)	~10	<7	~4
Ocean Island E-type MORB (Wood, 1979)	~10	2–7	~4
Axial zone volcanics	10–30	2–5	1–7
Deccan plateau volcanics		4–6	
Reunion island volcanics	14–18	3–4	

Interpretation

Sclater and Fisher (1974), Duncan and Pyle (1988), Courtillot et al (1988) and Fisk, Duncan et al. (1989) have interpreted the volcanic islands of the Chagos-Laccadive chain and the Deccan traps of India to have formed by passage over the Reunion Island hot spot. Morgan (1981) proposed that the Deccan flood basalts were the first manifestation of the Reunion Island hot spot. Duncan and Pyle (1988) summarized several of the models involving the hot spot origin with the Cretaceous/Tertiary boundary: meteorite impact and volcanic catastrophe. Duncan and Pyle (1988) did determine that the Deccan lavas can be correlated with the Cretaceous/Tertiary boundary.

Proposed Model

I propose that the Reunion Island hot spot was present at least in the middle Triassic and the axial zone volcanics were formed by passage of the Indo-Pakistan plate over the hot spot prior to the passage of the Indian continent over the hot spot at Cretaceous/Tertiary time when the Deccan traps were formed. As the plate continued its northward move over the hot spot the Chagos and Laccadive islands were formed and at the present time Reunion Island is over the hot spot. The ophiolites and volcanics were obducted onto the Indo-Pakistan continent in Paleocene time when the plate motion changed to northwest from a north direction.

Acknowledgements

Wazir Khan Baluch of the Geological Survey of Pakistan and the Department of Geology at the University of Iowa compiled the chemical data and constructed and drafted the diagram. I am very appreciative of his efforts on the diagrams and his discussions concerning the interpretation of the data.

References

Coleman, Robert G., 1981. Tectonic setting for ophiolite obduction in Oman., Journal of Geophysical Research, 86 B4: 2497–2508.

Courtillot et al., 1988. Deccan flood basalts and the Cretaceous/Tertiary boundary., Nature, 333: 843–846.

De Jong, Kees A. and Subhani, A.M., 1979. Note on the Bela ophiolites with special reference to the Kanar Area in Geodynamics of Pakistan, Abul Farah and Kees DeJong, (Ed), Geological Survey of Pakistan, pp. 263–269.

Duncan, R.A. and Pyle, D.G., 1988. Rapid eruption of the Deccan flood basalts at the Cretaceous/Tertiary boundary., Nature, 333: 841–843.

Fisk, Martin R., Duncan, Robert A., Baxter, Alistair N., Greenough, John D., Hargraves,

Robert B., and Tatsumi, Yoshiyuki,, 1989. Reunion hotspot magma chemistry over the past 65 m.y.: Results from Leg 115 of the Ocean Drilling Program., Geology, 17: 934–937.

Irvin, T.N. and Barger, W.R.A., 1971. A guide to the classification of the common volcanic rocks., Canadian Journal of Earth Science, 8: 523–549.

Kazmi, Ali Hamza, 1979. The Bibai and Gogai nappes in the Kach-Ziarat area of northeast Baluchistan in Geodynamics of Pakistan, Abul Farah and Kees De Jong, (Ed), Geological Survey of Pakistan, pp. 333–339.

Khan, Wazir, 1986. Geology and petrochemistry of a part of the Parh Group volcanics near Chinjun, Loralai District, Baluchistan, Pakistan. M. Phil. thesis, Centre of Excellence in Mineralogy, University of Baluchistan, Quetta, Pakistan, 149 p.

McCormick, George R., 1989. Geology of the Baluchistan (Pakistan) portion of the southern margin of the Tethys sea in Tectonic Evolution of the Tethyan Region, A.M.C. Sengor (Ed), NATO ASI Series C: Mathematical and Physical Sciences, 259: 277–288.

Miyashero, A., 1974. Volcanic rock series in island arcs and active continental margins., American Journal of Science, 274: 321–355.

Morgan, W.J., 1981. Hot spot tracks and the opening of the Atlantic and Indian oceans in the oceanic lithosphere, Cesare Emiliani, (Ed), in the collection *The Sea*, John Wiley and Sons, pp. 443–487.

Pearce, J.A., 1980. Geochemical evidence for the genesis and eruptive setting of lavas from Tethyan ophiolites, in Panayioutou, A. (Ed), Ophiolites, Proceedings of International Ophiolite Symposium, Cyprus, 1979; Nicosia, Cyprus Geological Survey Department, pp. 261–272.

Pearce, J.A. and Norry, M.J., 1979. Petrogenetic implications of Ti, Zr, Y and Nb variations in volcanic rocks. Contributions to Mineralogy and Petrology, 69: 33–47.

Powell, C. McA., 1979. A speculative tectonic history of Pakistan and surroundings: Some constraints from the Indian Ocean in Geodynamics of Pakistan, Abul Farah and Kees De Jong, (Ed), Geological Survey of Pakistan, pp. 5–24.

Sclater, J.G. and Fisher, R.L., 1974. Evolution of the east-central Indian Ocean with emphasis on the tectonic setting of the ninetyeast ridge., Geological Society of America, Bulletin, 85: 683–702.

Schwarzer, R.R. and Rogers, J.J.W., 1974. A world-wide comparison of alkali olivine basalts and their differentiation trends., Earth and Planetary Science Letters, 23: 286–296.

Wood, D.A., Joron, J.L. and Treuil, M., 1979a. A re-appraisal of the use of trace elements to classify and discriminate between magma series in different tectonic settings. Earth and Planetary Science Letters, 45: 326–336.

Time-Space Distribution and Petrologic Diversity of Japanese Ophiolites

AKIRA ISHIWATARI
Department of Earth Science, Faculty of Science, Kanazawa University, Marunouchi 1-1, Kanazawa 920 Japan

Abstract

The Japanese ophiolites occur as nappes and melanges. Nappe-type ophiolites formed and were emplaced within 20–30 Ma in the Ordovician, Permian, and Jurassic-Cretaceous periods, corresponding to the circum-Pacific ophiolite pulses. The Paleozoic ophiolite nappes (Yakuno, Oeyama, Miyamori and others) are distributed in Honshu, while the Mesozoic ones (Horokanai, Poroshiri and others) are in Hokkaido. In southwestern Honshu, the Jurassic accretional complex is overthrust by the Permian Yakuno ophiolite which in turn is overridden by the Ordovician Oeyama ophiolite. This relationship is a mirror image of the superposing ophiolite nappes of corresponding ages in the Klamath Mountains. Downward (oceanward) younging of ophiolite nappes is a common feature on both sides of the Pacific.

The fragments in a melange-type ophiolite may span more than 100 Ma in age. Paleozoic-Mesozoic ophiolite melanges in Honshu (Omi, Kurosegawa, Mikabu, and Motai) and Hokkaido (Kamuikotan and Tokoro) are all affected by high-pressure blueschist metamorphism, and some of them are overlain by nappe-type ophiolites. Cenozoic ophiolite melanges (Setogawa and Mineoka) occur in front of the Izu island arc which has collided with Japan since Miocene, but do not contain high-pressure metamorphic rocks.

The Japanese ophiolites show wide petrologic diversity. The residual mantle peridotite ranges from fertile lherzolite to highly depleted harzburgite. The mafic-ultramafic cumulates include plagioclase-type, clinopyroxene-type, and orthopyroxene-type. MORBs are dominant among Japanese ophiolitic basalts, but alkali basalts and picrites are common in melange-type ophiolites. The occurrence of highly depleted harzburgite which can not coexist with MORB, phlogopite-bearing harzburgite, and gabbros of arc tholeiite mineralogy indicates island-arc origin of some ophiolites. Most of Japanese nappe-type ophiolites may be fragments of ancient oceanic crust-mantle forming island arc-marginal basin systems, and the melange-type ophiolites may represent underlying subduction zone, in which subducting MORB crust-mantle, seamounts and oceanic arc volcanoes were crushed and mixed-up.

Tj. Peters et al. (Eds), Ophiolite Genesis and Evolution of the Oceanic Lithosphere, 723–743.

Introduction

Ten years after the worldwide rise of ophiolite boom, the first complete ophiolite suite was reported from Japan in 1978, and several other findings followed in the subsequent years (Ishiwatari 1978, Asahina and Komatsu 1979, Arai 1980, Ishizuka 1980, Ozawa 1983, Miyashita 1983, Kurokawa 1985). Much data on their geology, petrology and isotopic age have accumulated in 1980's, in which geological framework of the Japanese Islands has been completely re-interpreted from the viewpoint of allochthonism on the basis of successful radiolarian biostratigraphy (Saito and Hashimoto 1982, Hara 1982, Charvet et al. 1985, Watanabe et al. 1987). An up-to-date version of the old Japanese nappe tectonics (Huzimoto 1937, Kobayashi 1941) is now on construction by many Japanese workers, who were mostly autochthonists before 80's. This paper provides a short comprehensive overview on the Japanese ophiolites with special reference to their time-space distribution and petrologic diversity.

The term "ophiolite" is used in this paper in broad sense including dismembered, metamorphosed mafic-ultramafic complexes of "Cordilleran type" occurrences (Coleman 1986). A rudely stratified association of mafic volcanic-hypabyssal rocks, mafic-ultramafic cumulate rocks, and residual peridotite is hereafter called "complete ophiolite" regardless of the absence of sheeted dike complex.

The plate-like geologic bodies bounded by thrust faults are variously denoted as "terranes", "slabs" or "plates" in U.S.A., and "they would probably be treated as nappes in European Alpine terminology" (Coleman et al. 1988, p. 1063). The term "nappe" is used throughout this paper in preference of its definite structural and geometrical meaning.

Time-Space Distribution of Ophiolites

Tethyan ophiolites like those in Cyprus and Oman formed in a very short geologic time and were emplaced on continental margins as huge nappes shortly after their birth. The age of Tethyan ophiolites is mostly Cretaceous (Cyprus, Oman, and others) and partly Jurassic (Vourinos, Liguria, and others), and all of them are of Mesozoic age. The uniformity of age is also observed in the Appalachian-Caledonian chain and Ural Mountains, in which majority of ophiolites formed in early Paleozoic (mainly Ordovician).

Abbate et al. (1985) compiled the known ages of the ophiolites in the world, and concluded that ophiolite age is not randomly distributed through Phanerozoic time but concentrated in relatively short periods, namely Cretaceous/Jurassic and Ordovician (Fig. 1). These apparent "ophiolite pulses" may have resulted, at least in part, from the uniformity of ophiolite age in the well-studied, ophiolite-rich orogenic belts as mentioned above. In the circum-Pacific orogenic belts, however, the ophiolite ages are not so uniform

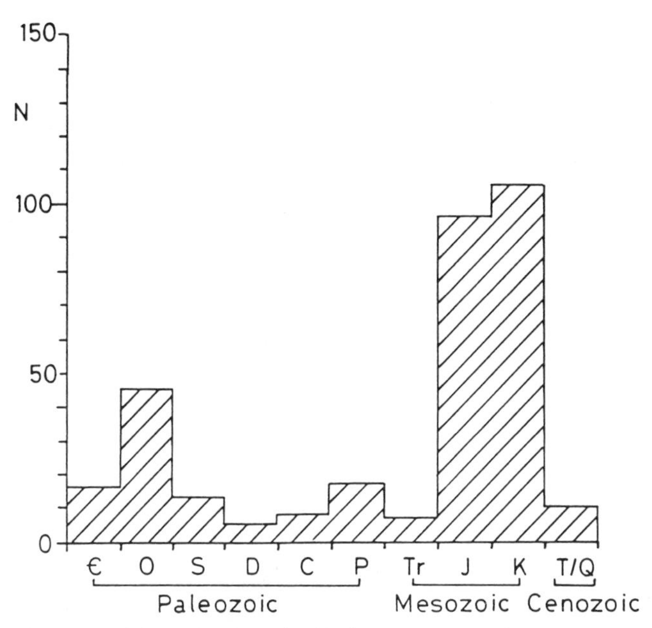

Figure 1. Histogram of formation age of ophiolites in Phanerozoic orogenic belts (based on the compilation of Abbate et al. 1985). The ages are indicated by the initial letters of geologic periods from Cambrian (€) to Quaternary (Q). The ophiolite pulses in Jurassic/Cretaceous and Ordovician are apparent, and another small pulse may be in Permian.

as in the other orogenic belts, and are scattered through the whole Phanerozoic periods.

The excellent example of the multiple ophiolite belts is reported from Klamath Mountains (Irvine 1977, Saleeby et al. 1982, Coleman 1986, Coleman et al. 1988), western U.S.A., in which late Jurassic Josephine ophiolite (Harper 1980) on the west is tectonically overlain by dismembered ophiolite bodies such as the Preston Peak complex (Snoke 1977) and Seiad complex (Cannat 1985) of the Western Paleozoic and Triassic Belt (actually also including early Jurassic rocks), which is in turn overridden by Devonian high-pressure metamorphic rocks and Ordovician Trinity ophiolite on the east (Lindsley-Griffin 1977, Quick 1981) (Fig. 2). The oceanward and downward younging of the ophiolites through the nappe pile is remarkably shown by this example.

The recent geologic studies have revealed essentially the same geologic structure in the Japanese Islands, which are composed of the pile of nappes with or without ophiolitic basement. The stacking relationship of these ophiolites forms a mirror image of the Klamath Mountains as described below.

Petrologic Diversity of Ophiolites

Ophiolitic residual peridotite has been divided into two major groups: lherzolite and harzburgite (Nicolas and Jackson 1972, Coleman 1977). Ophiolitic

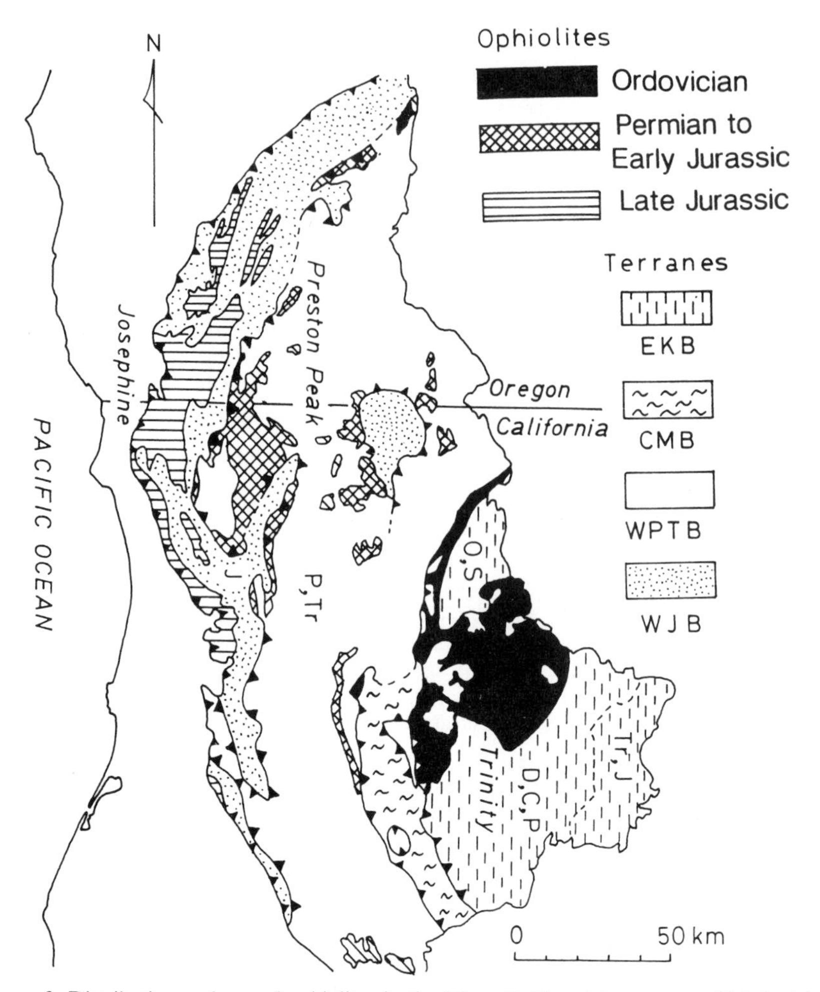

Figure 2. Distribution and age of ophiolites in the Klamath Mountains, western U.S.A. (simplified after Snoke 1977, Fig. 1). Oceanward (downward) younging of the ophiolite age through the nappe pile from early Paleozoic to Mesozoic is evident. The name of the terranes are as follows (Irwin 1977): EKB – Eastern Klamath Belt (Paleozoic), CMB – Central (high-pressure) Metamorphic Belt (Paleozoic), WPTB – Western Paleozoic and Triassic Belt (actually including lower Jurassic rocks: Coleman et al. 1988), WJB – Western Jurassic Belt. The O, S, D, etc. are initial letters of geologic ages from Ordovician to Jurassic.

cumulates have been classified into several groups on the basis of crystallization sequence of the cumulus minerals (Church and Riccio 1977). The diversity of ophiolitic volcanics and hypabyssal rocks has once been the focus of dispute against the mid-oceanic ridge origin of ophiolites (Miyashiro 1973, 1975), and is now an important measure to establish the geotectonic setting of their birth place in comparison with the present day volcanic rocks.

Ishiwatari (1985b) found regular and reasonable correspondence in pet-

Japan. Ophiolite	Poroshiri	Yakuno	Horokanai
Petrologic Type	Liguria	Yakuno	Papua
Sediment. Cover	Cret.? Mudstone	Permian Mudstone	Jurassic Chert
Basaltic Volc.	Evolved MORB	Primitive MORB	Primitive MORB
Sheeted Dikes	Absent	Absent	Absent
Cumulate Type Cpx TiO_2 in maf.	Plagioclase 0.8 wt.%	Clinopyroxene 0.4 wt.%	Orthopyroxene 0.1 wt.%
Crystallization Sequence of Cumulates (Mafic) (Seismic Moho) (Ultramafic)	Ol Pl Cpx Opx	Ol Pl Cpx Opx	Ol Pl Cpx Opx
Residual Mantle Peridotite Bulk Al_2O_3+CaO Cr# of spinel Al_2O_3 of opx Olivine Fo	Lherzolite and Cpx-rich Harzburgite 3-5 wt.% 30-50 2-4 wt.% 90	Cpx-bearing (common) Harzburgite 1-2 wt.% 50-70 1-2 wt.% 90-91	Cpx-free (pure) Harzburgite 0-1 wt.% 70-90 0-1 wt.% 92
(Top) Internal Metamorphism of Mafic Rocks (Moho)	Greenschist Epi. Amphibolite Amphibolite Hornbl. Granulite	Prehnite-Pumpelly. Greenschist Ab-Epi.Amphibolite Amphibolite Hornbl. Granulite Pyroxene Granulite	Zeolite Greenschist (Plagio.-actinol.) Amphibolite Hornbl. Granulite

Other Examples (igneous aspect)	Alps Trinity	Bay of Islands Semail(Oman)	Troodos Miyamori	Vourinos Mariana	Khan Taishir Adamsfield

Figure 3. Lithologic and petrologic characteristics of the three complete ophiolites in Japan. References: Poroshiri (Miyashita 1983, Miyashita and Yoshida 1988), Yakuno (Ishiwatari 1985a,b, Koide et al. 1987), Horokanai (Ishizuka, 1980, 1981, 1985, 1987). Other examples (Ishiwatari 1985b,c, Quick 1981, Varne and Brown 1978). They represent increasing degrees of mantle melting from left to right. Cr# is 100 Cr/(Al + Cr).

rologic nature among the three ophiolite members (Fig. 3). He divided the harzburgite into two types: clinopyroxene-bearing common harzburgite and clinopyroxene-free pure harzburgite. The chemical difference between the two harzburgites is equal to or greater than that between the lherzolite and common harzburgite. The majority of ophiolitic residual peridotite is the common harzburgite, while lherzolite and pure harzburgite are rather rare.

He classified the ophiolitic cumulates into three types, which are named after the mineral crystallized next to olivine; plagioclase-type, clinopyroxene-type, and orthopyroxene-type (Fig. 3). The plagioclase-type is characterized by the early appearance of plagioclase to form troctolite or plagioclase dunite and by the scarcity of orthopyroxene throughout the cumulates. The clinopyroxene-type is characterized by wehrlite and clinopyroxenite as well as by moderate abundance of orthopyroxene. The orthopyroxene-type is characterized by poikilitic harzburgite or orthopyroxenite layers overlying dunite and by the abundance of orthopyroxene through the cumulates.

The ophiolitic volcanics and hypabyssals are too diverse to be simply grouped, but nepheline-normative alkalic basalt, common high-alumina olivine tholeiite, and low-alumina quartz tholeiite with boninitic tendency may be distinguished as major groups.

Among the ophiolites in the world, almost without exception, the lherzolitic residual peridotite is associated with plagioclase-type cumulates, the common harzburgite with clinopyroxene-type cumulates, and the pure harzburgite with orthopyroxene-type cumulates. The associated basaltic volcanics may change from alkalic basalt through olivine tholeiite to quartz tholeiite in this order, though exceptional associations are not rare. These associations are named after typical examples from Liguria, Yakuno and Papua, respectively, and are reasonably explained as the result of increasing degrees of partial melting of primary mantle peridotite at depths and subsequent low-pressure crystallization of the extracted melt (Ishiwatari 1985b).

The ophiolites with clinopyroxene-type cumulates and clinopyroxene-bearing harzburgite residue are the commonest on the earth (ibid.), and the other types are rather rare. It seems likely that a wide area in Tethyan belt is dominated by a single type of ophiolite in view of the dominance of lherzolite in the West Mediterranean area and the abundance of harzburgite (clinopyroxene-bearing type) in the East Mediterranean-Middle East area (Nicolas and Jackson 1972). In the circum-Pacific orogenic belts, however, petrologic nature of ophiolites is far more diverse even in a small area as shown below.

Occurrence of the Japanese Ophiolites

Distribution, occurrence, and age of the Japanese ophiolites are shown in Fig. 4. The occurrences of the Japanese ophiolites are classified into two major types: nappe and melange. An ophiolite nappe occurs as a thrust sheet of 10 km or larger size, more or less retaining its original igneous stratigraphy. An ophiolite melange occurs as a sheared breccia zone or a conglomerate bed (olistostrome) including various ophiolite fragments from 1 cm to 1 km in size buried in the matrix of serpentinite or mudstone.

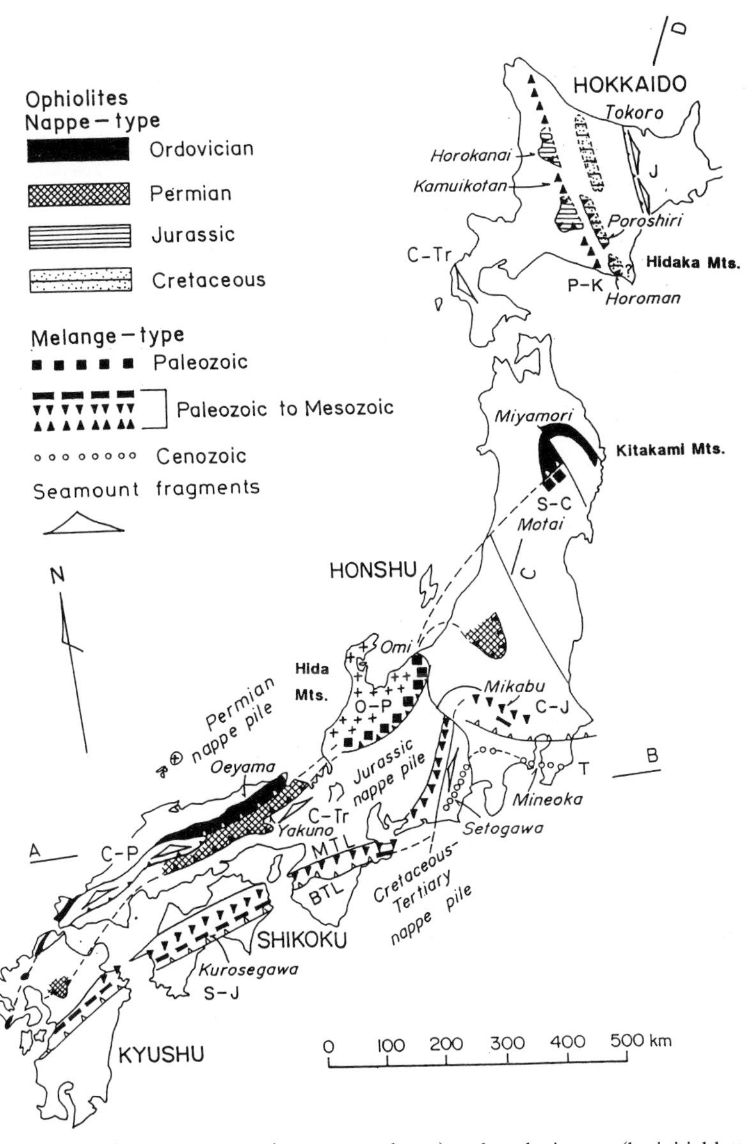

Figure 4. Distribution, occurrences (nappe or melange) and geologic ages (by initial letters) of the Japanese ophiolites. The schematic cross sections along the lines A-B and C-D are shown in Fig. 5. The Northeast Japan and Southwest Japan are divided by Tanakura Fault. MTL (Median Tectonic Line) divides SW Japan into Inner Zone (north) and Outer Zone (south), and BTL (Butsuzo Tectonic Line) is the boundary between northern Jurassic nappe pile and southern Cretaceous nappe pile.

Table 1. Age and occurrence of Japanese ophiolites.

Age	Southwest Japan	Northest Japan
Tertiary	{Mineoka} {Setogawa}	----------
Cretaceous	[Shimanto]	(**Poroshiri**) {Kamuikotan}
Jurassic	[Chichibu] [Tamba]	(**Horokanai**) {Tokoro}
	{Kurosegawa} {Omi} {Mikabu}	[N Kitakami] [SW Hokkaido]
Permian	(**Yakuno**) [Akiyoshi]	{Motai}
Ordovician	(**Oeyama**)	(**Miyamori**)

Explanations
( ) Intact ophiolite nappe, complete or partial.
{ } Ophiolite melange.
[ ] Accreted seamounts with limestone caps in fore-arc sediments.
The melanges and seamounts include older rocks.

In addition to these ophiolites, a considerable amount of basaltic volcanics (mainly pillow lavas of alkalic and tholeiitic affinity) separately occurs in the trench-fill or fore-arc deposits forming a major part of the Japanese Islands. These volcanics are often capped with limestone, but are generally in tectonic contact with the adjacent arc-derived sediments. The volcanics are interpreted to be fragments of the topmost part of seamounts, which were scraped off and incorporated in the arc-derived sediments during subduction. It should be noted, however, that some gabbro bodies of MORB affinity were intruded into the fore-arc deposits with clear chilled margins (Miyake 1985, Jakes and Miyake 1984), and some MORB-like basaltic rocks are interlayered with fore-arc sediments (Miyashita and Yoshida 1988). The occurrences of ophiolite nappes, ophiolite melanges, and the accretional complexes with seamount bodies are summarized in Table 1.

The Ophiolitic Nappe Pile in Southwest Japan

Southwest Japan is divided into the Inner Zone and Outer Zone by the Median Tectonic Line (MTL). Although MTL is a vivid, linear topographic feature when viewed from a satellite, an equally important geologic boundary may be Butsuzo Tectonic Line (BTL) to the south running parallel with MTL (Figs. 4 and 5). The BTL is the boundary between the southern Cretaceous- Tertiary nappe pile and the overlying, northern Jurassic nappe pile, which spreads on both sides of the MTL.

(1) *Inner Zone*

The Jurassic nappe pile to the north of MTL is-called Tamba (or Mino) zone, which consists of the upper (T2) and lower (T1) nappes (Ishiga 1983). The upper nappe includes seamount fragments and chert of Carboniferous to Triassic age, mudstone of early-middle Jurassic age, and sandstone and

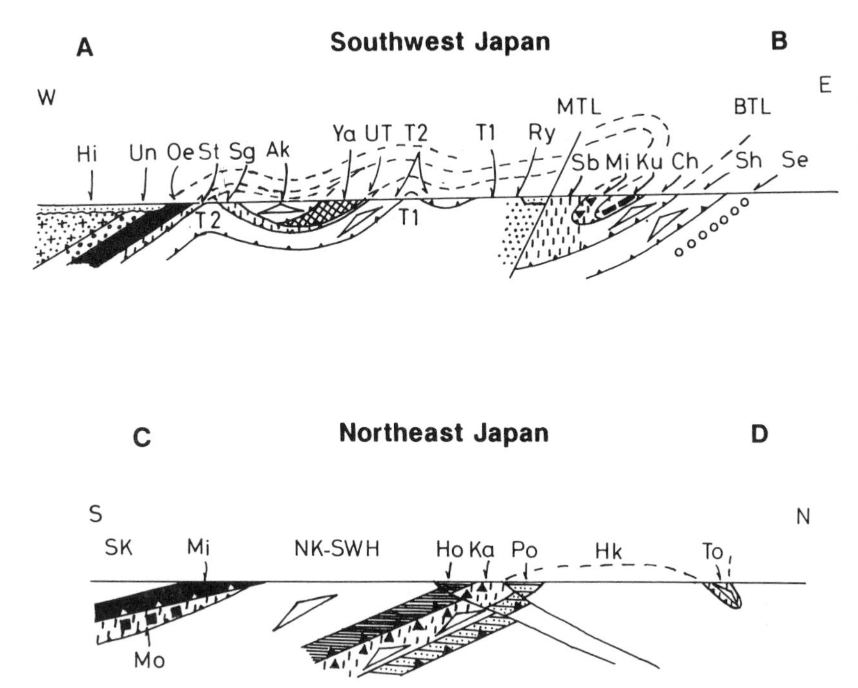

Figure 5. Schematic and highly speculative cross sections of the ophiolitic nappe piles of Japan. The cutting positions are shown in Fig. 4. The Kurosegawa zone in the Outer Zone of Southwest Japan is speculated to be fragmental klippes of the Omi melange (including the elements of Akiyoshi, Shitani, and Oeyama nappes) with the elements of the Unazuki and Hida nappes in the Inner Zone. The Tokoro melange of Northeast Japan is also speculated to be a klippe of the Kamuikotan melange. Thus the downward (but not exactly oceanward) younging in the nappe pile becomes regular and clear. The nappe pile of NE Japan is complicated by the Miocene backthrusting caused by the collision of Chishima (Kurile) arc. High pressure metamorphic belt is dashed, and low or medium pressure belt is dotted. Syn- or post-orogenic Cretaceous sedimentary cover in SW Japan is strippled.

The name of the nappes (or terranes) are listed below. Geologic ages are shown in abbreviations, and a slash indicates significant unconformity. The sections are not to scale.

Southwest Japan (from left to right) Hi – Hida low-P metamorphic belt (Pre-Cambrian-P/J), Un – Unazuki medium-P metamorphic belt (C-P/J), Oe – Oeyama ophiolite (O), St – Shitani high-P metamorphic belt (C-P), Sg – Sangun high-P metamorphic belt (Tr-J), Ak – Akiyoshi seamount-limestone complex (C-P/Tr), Ya – Yakuno ophiolite and its cover (Maizuru Zone) (P/Tr), UT – Ultra-Tamba Nappe (sediments) (P), T2 – Upper Tamba Nappe (seamounts and sediments) (C-J/K), T1 – Lower Tamba Nappe (sediments) (Tr-J/K), Ry – Ryoke low-P metamorphic belt (K), MTL – Median Tectonic Line, Sb – Sanbagawa high-P metamorphic belt (?-J/K), Mi – Mikabu ophiolitic melange (C-J), Ku – Kurosegawa melange (S-J/K), Ch – Chichibu (C-J/K), BTL – Butsuzo tectonic Line, Sh – Shimanto Belt (seamounts and sediments) (K-T), Se – Setogawa Belt (ophiolite melange and sediments) (T).

Northeast Japan (from left to right): SK – Southern Kitakami belt (S-J/K), Mi – Miyamori ophiolite (O/K), Mo – Motai melange (S-C), NK-SWH – Northern Kitakami-Southwestern Hokkaido (C-J/K), Ho – Horokanai ophiolite and its cover (Sorachi Zone) (J/K), Ka – Kamuikotan ophiolitic melange (P-K/T)I Po – Poroshiri ophiolite (K?), Hk – Hidaka Zone (K/T), To – Tokoro melange (seamounts in high-P metamorphic belt) (J-K).

olistostrome of late Jurassic age. The stratigraphy of the lower nappe is slightly different in that the chert sedimentation continued through early Jurassic, and sandstone appeared only in the latest Jurassic time. The T2 nappe also appears in some windows behind the overlying Permian nappe pile to the north (Fig. 5, A-B). The southern part of the Tamba zone provided protoliths for the Cretaceous low-pressure Ryoke metamorphism.

The Jurassic nappe pile is in turn overlain by the Permian nappe pile. The nappe directly resting upon the Tamba zone is Ultra-Tamba zone, which consists of Upper Permian chert and sandstone (Ishiga 1986). The next nappe is the Permian Yakuno ophiolite (Ishiwatari 1985a,b) and its sedimentary cover composed of Permian mudstone, sandstone, and Triassic coal-bearing conglomerates. The nappe is called Maizuru zone as a whole. The Yakuno ophiolite is interpreted to be generated in an island arc-marginal basin system, because ophiolite bodies of MORB-type and arc tholeiite-type petrology coexist in the Maizuru zone (Ishiwatari et al. 1990). The Maizuru zone was then overthrust by Akiyoshi seamount-limestone bodies (Permo-Carboniferous), the high-pressure Shitani metamorphic rocks of Carboniferous metamorphic age and the Ordovician Oeyama ophiolite (Kurokawa 1985, Arai 1980, Nishimura and Shibata 1989, Ishiwatari 1990). The Oeyama ophiolite, the Shitani metamorphic rocks, the Akiyoshi limestone bodies, and the Yakuno ophiolite may converge into Omi ophiolitic melange in the Hida marginal zone to the east (Fig. 4), in which limestone of Siluro-Devonian age and Jurassic-Cretaceous conglomerates are also incorporated. The Unazuki zone made of the upper Paleozoic medium-pressure metamorphic rocks (Hiroi 1981) and the Hida zone made of Proterozoic-Paleozoic low-pressure gneiss with Triassic-Jurassic granite intrusions occupy the structurally highest position in the nappe pile of the Inner Zone (Komatsu et al. 1985).

(2) *Outer Zone*

The ophiolites in the Outer Zone are more fragmental, and their chronological and structural position is a matter of dispute. The most conspicuous is the Mikabu ophiolite melange, a narrow belt of basaltic volcanic rocks and cumulus mafic-ultramafic rocks stretching from the vicinity of Tokyo to the eastern Kyushu for nearly 1,000 km. Residual peridotite has never been reported from this melange. The Mikabu melange occurs between the Sanbagawa zone, a Jurassic nappe pile affected by the latest Jurassic high-pressure metamorphism, and the Chichibu zone, a highly fragmented, non-metamorphic Jurassic nappe pile with chert and seamount-limestone bodies of Carboniferous to Triassic ages. Ernst (1972) first described the Mikabu complex, allegedly of Permian age, as an ophiolite marking a suture zone. Sugisaki et al. (1972) clearly showed that the Mikabu greenstones are depleted in LREE in contrast to the LREE-enriched seamount fragments in the Tamba and Chichibu zones, although their rift-zone model is geologically unrealistic.

Iwasaki (1979) and Saito et al. (1979) have found that most of the Mikabu "ophiolite" is composed of sedimentary rocks (basaltic, gabbroic, and dunite-wehrlite breccia), in which gabbro-wehrlite and pillow lava bodies of up to 1 km size are buried. The fossil age determined in this zone ranges from Carboniferous to late Jurassic. Nakamura (1971) carried out detailed petrologic work on the mafic-ultramafic plutonics near Toba, and concluded that they were formed through fractional crystallization of a hydrous picritic magma. In fact, they contain much "igneous" hornblende. The Mikabu melange is affected by the low-grade Sanbagawa metamorphism.

The next ophiolitic belt to the south of Mikabu is the Kurosegawa melange, which is composed of various blocks such as garnet amphibolite, gneiss and granite of early Paleozoic metamorphic age, and the fossiliferous sediments and metabasites (some with glaucophane) of Silurian to Jurassic ages buried in the serpentinite matrix (Maruyama et al. 1984). They are free from the Cretaceous Sanbagawa metamorphism, and are interpreted to rest on the Sanbagawa-Mikabu zone as a superficial nappe (Suzuki et al. 1990). This interpretation is contrary to the conventional idea regarding the Kurosegawa melange as an upheaved fragment of a collided microcontinent beneath the Sanbagawa zone (Maruyama et al. 1984, Charvet et al. 1985). If Suzuki et al. (1990) are right, the Kurosegawa zone may correspond to the Omi ophiolitic melange of the Inner Zone in terms of its structural position and age-lithology of the constituent blocks.

It is possible that the Pre-Cretaceous nappe pile of the Outer Zone may represent "tectonic outlier" of the Inner Zone, in the manner like the Prealps representing the pile of outpoured nappes coming from the main Alpine orogen. The idea assuming long-way displacement of nappes from the Inner Zone to Outer Zone was first proposed by Huzimoto (1937) in Kwanto (Kanto) Mountains to the northwest of Tokyo, and has recently been revived in Shikoku by Faure and Charvet (1984), who have postulated that a part of the Chichibu zone came from the Tamba zone or Akiyoshi zone on the basis of deformation history and lithology. The cross section in Fig. 5 A-B follows this idea and my speculation that the Kurosegawa zone is also a chain of klippes came from the Oeyama ophiolite-Hida marginal zone (Omi melange) and the structurally overlying nappes of the Inner Zone. In fact, the protoliths of serpentinite in the Kurosegawa zone (Yokoyama 1987) petrologically resemble lherzolitic residual peridotite of the Oeyama ophiolite.

The mafic-ultramafic bodies in the Sanbagawa metamorphic belt may mostly be the fragments of an island arc lower crust-wedge mantle section, which was originally equilibrated in garnet and spinel lherzolite facies (Mori and Banno 1973, Yokoyama 1980) and then tectonically emplaced in the deeper part of the subduction zone. Some of the mafic-ultramafic bodies gave "high-pressure contact metamorphism" to form eclogite aureole in the surrounding albite-epidote amphibolite due to hot emplacement (Takasu 1984, Kunugiza et al. 1986). Such mafic-ultramafic bodies can not be called ophiolites, but some other peridotite bodies formed by dehydration of serpentinite due to Sanbagawa metamorphism could have been dismembered ophiolite bodies.

Seamount fragments and ophiolitic melanges are also present in the Cretaceous and Tertiary nappe pile on the south of BTL. In the Shimanto belt, many pillow lava bodies of MORB-type chemistry (Sugisaki et al. 1979) are tectonically included in the turbidite formations. Very rare serpentinite melange is also present. The youngest ophiolitic rocks in Japan occur in the Tertiary Setogawa and Mineoka melange distributed in front of the Izu island arc, which has collided against Japan since Miocene. The radiometric ages of the ophiolitic rocks give early Tertiary age (35–50 Ma) (Taniguchi and Ogawa, 1990). This is one of the youngest ophiolite on the earth only next to the Miocene Lichi Melange of East Taiwan (Jahn 1986).

The Ophiolitic Nappe Pile in Northeast Japan

Major ophiolites in Northeast Japan also occur as nappes. In the Kitakami Mountains, the Ordovician Miyamori ophiolite (Ozawa 1983, 1988, Ozawa et al. 1988) occurs as nappes overlying the Paleozoic Motai ophiolitic melange affected by blueschist metamorphism (Maekawa 1981) and other late Paleozoic strata. On the east and north of the Miyamori-Hayachine ophiolite belt, a vast extent of Jurassic accretional complex with occasional Carboniferous to Triassic chert and seamount-limestone bodies occupies the northern Kitakami Mountains and the southwestern Hokkaido (Ishiga and Ishiyama 1987). The lithostratigraphy closely resembles that of the upper Tamba nappe and Chichibu zone of Southwest Japan. Saito and Hashimoto (1982) argued that the southern Kitakami block came from "Pacifica" continent farther to the south, then collided, and thrust over the northern Kitakami block by the Cretaceous time. Tazawa (1988) presented a clear cross section showing the southern Kitakami nappe with the early Paleozoic ophiolitic (not continental) basement overthrust upon the northern Kitakami nappe.

The Jurassic ophiolites (e.g. Horokanai ophiolite; Ishizuka 1985, 1987) in central Hokkaido rest upon the Kamuikotan ophiolitic melange affected by Cretaceous blueschist metamorphism (Watanabe and Maekawa 1985, Maekawa 1989). Jolivet (1986) considered that this thrusting happened from west to east at the Jurassic/Cretaceous boundary due to collision between Eurasia and Okhotsk block. The Japanese geologists, however, insist upon a late Cretaceous collision (Kiminami et al. 1985, Niida and Kito 1986). The Cretaceous ophiolites in the Hidaka zone (s.l.) to the east may have originally occupied the structural positions lower than the Jurassic ophiolite (Fig. 5 C-D). In the southern Hokkaido, however, the structural relationship is very complicated due to backthrusting (east to west) caused by the Miocene collision of the Chishima (Kurile) arc against Hokkaido. In fact, the Poroshiri ophiolite was overturned and overridden by an island-arc crustal section of the Hidaka main zone on the east (Miyashita 1983, Komatsu 1985).

The Horoman plagioclase lherzolite-layered gabbro body was once reported as a high temperature peridotite intrusion with contact aureole (Onuki 1965). However, Niida (1985) mapped out that the body is completely surrounded by faults, and is not related to the adjacent granulite-facies rocks, which may represent lower-crustal section of the collided Hidaka island arc. It is possible that the Horoman body also represents upper mantle-lower crust portion of the Cretaceous oceanic lithosphere (Komatsu, 1985).

The high-pressure metamorphic rocks of the Tokoro belt originated in accreted seamount bodies (Sakakibara 1986) have been used as an evidence of eastward subduction beneath the Okhotsk block with the support of turbiditic paleocurrent from the east (Niida and Kito 1986). However, it is also possible that the Tokoro belt, which is exposed in a synclinal manner, represents a klippe of the Kamuikotan melange as an analogous pair in Southwest Japan (Kurosegawa-Omi).

Thus, the ages and stacking relationship of the ophiolites in Southwest and Northeast Japan resemble with each other, and roughly correspond to those in Klamath Mountains. The Ordovician ophiolite occupies the highest structural position in the nappe pile, and is tectonically underlain by Permian or Jurassic/Cretaceous ophiolite. The downward younging of the ophiolites through the nappe pile is a common feature on both sides of the Pacific, though the presence of tectonic outliers complicates their oceanward younging pattern in Japan.

Igneous Petrologic Diversity of Japanese Ophiolites

Only three complete ophiolites have so far been reported from Japan, namely Yakuno (Permian), Horokanai (Jurassic), and Poroshiri (Cretaceous?). The lithologic and petrologic features of the three ophiolites are summarized in Fig. 3. All of them lack sheeted dike complex, and were thoroughly metamorphosed from the top to the Moho.

Igneous petrologic nature of the three ophiolites are very diverse. The residual peridotite is pure harzburgite in the Horokanai ophiolite, clinopyroxene-bearing harzburgite in the Yakuno ophiolite, and lherzolite or clinopyroxene-rich harzburgite in the Poroshiri ophiolite. The mafic-ultramafic cumulates are of orthopyroxene-type in Horokanai, clinopyroxene-type in Yakuno, and plagioclase-type in Poroshiri. The associated volcanic rocks are all MORB-type basalt among the three ophiolites, while those in Poroshiri is relatively evolved and richer in alkalic elements.

The Ordovician ophiolites lack volcanic member, and are composed of dominant residual peridotite and some cumulates. The residual peridotite of the Oeyama ophiolite is lherzolitic harzburgite closely resembling abyssal peridotite (Ishiwatari 1990), and commonly bears "dancing spinel", a sinuous aggregate of minute chromian spinel (Cr/(Al + Cr) = 0.5 or so) occurring with clinopyroxene and orthopyroxene. The aggregate sometimes forms a pseudomorph after an equant mineral, possibly garnet. This may be a partial melt texture of the former garnet peridotite ascended to the spinel lherzolite facies level.

The residual peridotite of the Miyamori ophiolite is clinopyroxene-bearing harzburgite which also invariably bears hornblende and sometimes phlogopite. The common occurrence of these hydrous, potassic minerals forced Ozawa (1988) to conclude that the residual peridotite is a fragment of the sub-arc wedge mantle, which has been impregnated with hydrous, potassic fluid emanated from the subducting oceanic crust below during the partial melting event. The "unconformity" between harzburgite tectonite and wehrlitic cumulates has been well documented in this ophiolite (Ozawa 1983), while the two members seems to be conformable in the other Japanese ophiolites (e.g. Ishiwatari 1985b).

Igneous petrologic nature of the melange-type ophiolites is far more diverse. As pointed out by Miyashiro (1975), ophiolites in high pressure metamorphic belt (mostly melange-type) bears alkalic basalt along with tholeiite. Picrite is also common in the Mikabu and Setogawa-Mineoka melange, and curious rocks such as high-magnesian dunite (Fo_{94}) (Arai and Uchida 1978) and plagioclase (An_{100}) lherzolite (Takasawa 1976) were reported from the latter.

Internal Metamorphism of Japanese Ophiolites

Every ophiolite experienced hydrous recrystallization under the temperatures steeply increasing downward soon after its birth in the birthplace (Internal metamorphism), and sometimes affected by later regional metamorphism in the subduction zone (external metamorphism) (Coleman, 1977). Ophiolite may also be metamorphosed during detachment from the oceanic lithosphere (Nicolas 1989), and may cause metamorphism in the underlying sediments during its hot obductive emplacement, but such cases are not yet known in Japan.

In most ophiolites, the internal metamorphism is restricted to the volcanics and sheeted dikes, and sometimes extends to the upper part of the cumulates. Among the Japanese ophiolites, however, the cumulate section is also thoroughly metamorphosed in hornblende granulite and pyroxene granulite facies. Ishiwatari (1985a) concluded that the mafic-ultramafic boundary (seismic Moho) was as deep as 15 to 30 km on the basis of olivine-plagioclase reaction and spinel lherzolite-facies mineral chemistry of the metacumulates at the Moho level. Analogous, higher-pressure, mafic-ultramafic metacumulates bearing garnet have recently been reported from Alaska (Tonsina complex: DeBari and Coleman 1989).

Metamorphic facies series of the Yakuno ophiolite includes prehnite-pumpellyite and albite-epidote amphibolite facies, suggesting less steep thermal gradient than the other ophiolitic metamorphism. Ishizuka (1985) presented detailed metamorphic petrology of the Horokanai ophiolite, in which zeolite, greenschist, amphibolite, and hornblende granulite facies were recognized. Metamorphic sequence of the Poroshiri ophiolite includes greenschist, epidote amphibolite, amphibolite, and hornblende granulite facies, and the recrystallization of the metacumulates is so incomplete as to preserve igneous mineral assemblage in part (Miyashita 1983). The thoroughly metamorphosed ophiolites with high-pressure metacumulates in Japan and other circum-Pacific areas may represent a tectonically reduced section of the relatively thick crust beneath an island arc-marginal basin system.

The residual mantle peridotite is invariably tectonized through high-temperature solid flow. Spinel and orthopyroxene lineation is well developed in the strongly deformed Miyamori tectonite (Ozawa 1983), while even kink-band in olivine is rather rare in the weakly deformed Oeyama tectonite, in which partial melting texture is preserved as mentioned above. A peculiar metasomatism to form monticellite took place in the highly depleted mantle peridotite of the Horokanai ophiolite during its solid flow (Nagata 1982). The residual peridotite is often cut by later gabbroic dikes (Ishizuka 1987, Kurokawa 1985, etc.).

External Metamorphism of Japanese Ophiolites

The ophiolite melanges in Omi, Mikabu, Kurosegawa, Motai, Kamuikotan and Tokoro all experienced the external high-pressure metamorphism in glaucophane schist facies. However, the ophiolite nappes were generally escaped from the high-pressure metamorphism, and only experienced very low-grade external metamorphism in prehnite-pumpellyite or lower grade. The only exception may be the Oeyama ophiolite, which suffered medium pressure metamorphism to form kyanite and staurolite in amphibolite (Kuroda et al. 1976, Kurokawa 1985). The medium-pressure type metamorphic belt is rare in Japan, and only Unazuki metamorphic belt of late Paleozoic age has so far been reported from the eastern Hida Mountains (Hiroi 1982). The occurrence of kyanite-staurolite assemblage in the Oeyama ophiolite suggests common metamorphic history with the Unazuki belt.

The Ordovician Oeyama and Miyamori ophiolites commonly bear "cleavable olivine", an olivine with well-developed cleavages in three directions (Kuroda and Shimoda 1967). This may be produced by the low-grade contact metamorphism caused by nearby granite intrusions, and may not be a primary feature (Uda 1984).

Conclusion

The Japanese ophiolites form a multiple ophiolite belt, in which a younger ophiolite (as young as Tertiary) was emplaced beneath the older ophiolite (as old as Ordovician) possibly through the process analogous to "underplating by duplex accretion" (Sample and Fisher 1986). Such a multiple ophiolite belt with full-Phanerozoic time span is characteristic of the circum-Pacific areas such as Klamath Mountains, Alaska, Koryak Range (A.P. Stavsky and S.D. Sokolov, personal communication), Chilean Andes (Herve et al. 1987; Saunders et al. 1979), and Tasman Orogen (E. Australia), and is in contrast with the ophiolites of uniform age in the Tethyan, Appalachian-Caledonian and Ural orogenic belts. Among the circum-Pacific orogenic belts, the oldest ophiolite may be Cambrian ophiolites in Tasmania, e.g. Papua-type Adamsfield ophiolite (Varne and Brown 1978), and the youngest ophiolite (on land) may be Miocene East Taiwan ophiolite (Jahn 1986). The age data of Japanese and other circum-Pacific ophiolites support the idea of "ophiolite pulses", i.e. most of preserved ophiolites were formed in Jurassic/Cretaceous and Ordovician. The "ophiolite pulses" may not have resulted from faulty sampling, but reflect global tectonic events in fact. It should be noted, however, that Permian ophiolites are not rare in the circum-Pacific regions (e.g. Canyon Mountain (Oregon), Preston Peak (California), Dun Mountain (New Zealand), Yakuno (Japan), and some Alaskan ophiolites).

The Japanese ophiolites are petrologically very diverse. From the aspect of igneous petrology, every type of ophiolite representing varying degrees of partial melting in the mantle is present. This implies their origin in island arc-marginal basin systems with intense magmatic heterogeneity rather than relatively homogeneous mid-ocean ridge systems.

The three complete ophiolites in Japan all lack sheeted dikes, and are thoroughly metamorphosed down to the Moho. The amphibolite xenoliths occurring with mantle peridotite xenoliths in Ichinome-gata volcano on the Japan Sea coast (Aoki 1971) suggest that a metamorphosed, relatively thick, mafic crust is present in a back-arc setting. The marginal-basin origin of these ophiolites are also consistent with the fact that the basaltic rocks are commonly intercalated with mudstone (Yakuno and Poroshiri). Moreover, some ophiolites include arc-related rocks such as highly depleted harzburgite which can not coexist with MORB (Horokanai), amphibole and phlogopite-bearing harzburgite (Miyamori), and gabbro with arc tholeiite mineralogy (Yakuno). It seems likely that most, if not all, Japanese ophiolite nappes originated in island arc-marginal basin system.

All ophiolitic melanges in Japan, except Setogawa-Mineoka, were affected by blueschist metamorphism, and some ophiolitic nappes (Horokanai, Miyamori, Oeyama) emplaced over these melanges. The ophiolitic melanges often bear alkalic basalt and picrite, suggesting incorporation of seamount fragments. This suggests that the ophiolitic melanges represent subduction zone itself, and the ophiolitic nappes represent hanging wall of the subduction zone, i.e. crust-mantle section of the island arc-marginal basin system.

Acknowledgements

I am grateful to Professor R. G. Coleman of Stanford University for his critical review of the manuscript with many valuable suggestions and proper corrections. Dr. S. Miyashita of Niigata University is thanked for fruitful discussion about the ophiolites in Hokkaido. The Grant-in-Aid for Encouragement of Young Scientists Nos. 62740480 and 63740463 offered by Japanese Ministry of Education, Science and Culture (Monbusho) is also acknowledged.

References

Abbate, E., Bortolotti, V., Passerini, P. and Principi, G., 1985. The rhythm of Phanerozoic ophiolites. Ofioliti, 10: 109–138.

Aoki, K., 1971. Petrology of mafic inclusions from Ichinomegata, Japan. Contrib. Mineral. Petrol., 30: 314–331.

Arai, S., 1980. Dunite-harzburgite-chromitite complexes as refractory residue in the Sangun-Yamaguchi zone, western Japan. J. Petrology, 21: 141–165.

Arai, S. and Uchida, T., 1978. Highly magnesian dunite from the Mineoka belt, central Japan. J. Japan. Assoc. Min. Petr. Econ. Geol., 73: 176–179.

Asahina, T. and Komatsu, M., 1979. The Horokanai ophiolitic complex in the Kamuikotan tectonic belt, Hokkaido, Japan. J. Geol. Soc. Japan, 85: 317–330.

Cannat, M., 1985. Tectonics of the Seiad massif, northern Klamath Mountains, California. Geol. Soc. Amer. Bull, 96: 15–26.

Charvet, J., Faure, M., Caridroit, M. and Guidi, A., 1985. Some tectonic and tectogenetic aspects of SW Japan: an alpine-type orogen in an island-arc position. In: N. Nasu et al. (Eds), "Formation of Active Ocean Margins", pp. 791–817, Terra Sci. Publ., Tokyo.

Church, W.R. and Riccio, L., 1977. Fractionation trend of the Bay of Islands ophiolite of Newfoundland: polycyclic cumulate sequences in ophiolites and their classification. Can. J. Earth Sci., 14: 1156–1165.

Coleman, R.G., 1977. Ophiolites ancient oceanic lithosphere?. Springer Verlag, Berlin; 229 pp.

Coleman, R.G., 1986. Ophiolites and accretion of the North American Cordillera. Bull Soc. geol. France, 1986: 961–968.

Coleman, R.G., Manning, C.E., Donato, M.M., Mortimer, N. and Hill, L.B., 1988. Tectonic and regional metamorphic framework of the Klamath Mountains and adjacent Coast Ranges, California and Oregon. In W.G. Ernst (Ed), "Metamorphism and Crustal Evolution of the Western United States", (Rubey Volume) pp. 1061–1097, Prentice-Hall, London.

DeBari, S.M. and Coleman, R.G. (1989): Examination of the deep levels of an island arc: evidence from the Tonsina ultra-mafic-mafic assemblage, Tonsina, Alaska. J. Geophys Res., 94: 4373–4391.

Ernst, W.G., 1972. Possible Permian oceanic crust and plate junction in central Shikoku, Japan. Tectonophsics, 15: 233–239.

Faure, M. and Charvet, J., 1984. Mesozoic nappe structures in SW Japan, from the example of eastern Shikoku and Kinki area. Sci. Geol. Bull. (Strasbourg), 37: 51–63.

Hara, I., 1982. Evolutional processes of paired metamorphic belts: Hida belt and Sangun belt. Mem. Geol. Soc. Japan. No. 21, pp. 71–89.

Harper, G.D., 1980. The Josephine ophiolite: remains of a late Jurassic marginal basin in northwestern California. Geology, 8: 333–337.

Herve, F., Godoy, E., Parada, M.A., Ramos, V., Rapela, C., Mpodozis, C. and Davidson, J., 1987. A general view on the Chilean-Argentine Andes, with emphasis on their early history. In: J.W.H. Monger and J. Francheteau (Eds), "Circum-Pacific orogenic belts and evolution of the Pacific ocean basin", Geodynamics Ser. Vol. 18, pp. 97–113, AGU-GSA.

Hiroi, Y., 1981. Subdivision of the Hida metamorphic complex, central Japan, and its bearing on the geology of the Far East in pre-Sea of Japan time. Tectonophysics, 76: 317–333.

Huzimoto, H., 1937. The nappe-theory with reference to the northeastern part of the Kwanto Mountainland. Sci. Rep. Tokyo Bunrika Univ., Sec. C, No. 6, 215–244.

Irwin, W.P., 1977. Ophiolitic terranes of California, Oregon, and Nevada. In: R.G. Coleman and W.P. Irwin (Eds), "North American Ophiolites", Oregon State Dept. Geol. Min. Indst. Bull., No. 95, 75–92.

Ishiga, H., 1983. Two suites of stratigraphic succession within the Tamba Group in the western part of the Tamba belt, Southwest Japan. J. Geol. Soc. Japan, 89: 443–454 (in Japanese with English abstract).

Ishiga, H., 1986. Ultra-Tamba zone of Southwest Japan. J. Geosci. Osaka City Univ., 29, 45–88.

Ishiga, H. and Ishiyama, D., 1987. Jurassic accretionary complex in Kaminokuni terrane, Southwestern Hokkaido, Japan. Mining Geology (Tokyo), 37: 381–394.

Ishiwatari, A., 1978. A preliminary report on the Yakuno ophiolite in the Maizuru zone, Inner Southwest Japan. Earth Sci. (Chikyu Kagaku), 32, 301–310 (in Japanese with English abstract).

Ishiwatari, A., 1985a. Granulite-facies metacumulates of the Yakuno ophiolite, Japan: evidence for unusually thick oceanic crust. J. Petrology, 26, 1–30.

Ishiwatari, A., 1985b. Igneous petrogenesis of the Yakuno ophiolite (Japan) in the context of the diversity of ophiolites. Contrib. Mineral Petrol., 89: 155–167.

Ishiwatari, A., 1985c. Alpine ophiolites: product of low-degree mantle melting in a Mesozoic transcurrent rift zone. Earth Planet. Sci. Lett., 76: 93–108.

Ishiwatari, A., 1990. Yakuno ophiolite and related rocks in the Maizuru Terrane. In Ichikawa K. et al. (Eds), "Pre-Cretaceous Terranes of Japan" (Publ. IGCP Prbject 224), pp. 109–120, Osaka.

Ishiwatari, A., Ikeda, Y. and Koide, Y., 1990. The Yakuno ophiolite, Japan: fragments of Permian island arc and marginal basin crust with a hot spot. Proceedings of the Troodos 87 Symposium, pp. 497–506.

Ishizuka, H., 1980. Geology of the Horokanai ophiolite in the Kamuikotan tectonic belt, Hokkaido. J. Geol. Soc. Japan, 86: 119–134 (in Japanese with English abstract).

Ishizuka, H., 1981. Geochemistry of the Horokanai ophiolite in the Kamuikotan tectonic belt, Hokkaido, Japan. J. Geol. Soc. Japan, 87: 17–34.

Ishizuka, H., 1985. Prograde metamorphism of the Horokanai ophiolite in the Kamuikotan Zone, Hokkaido, Japan. J. Petrology, 26: 391–417.

Ishizuka, H., 1987. Igneous and metamorphic petrology of the Horokanai ophiolite in the Kamuikotan zone, Hokkaido, Japan: a synthetic thesis. Mem. Fac. Sci., Kochi Univ., Ser. E. Geology, 8: 1–70.

Iwasaki, M., 1979. Gabbroic breccia (olistostrome) in the Mikabu green stone belt of the eastern Shikoku. J. Geol. Soc. Japan, 85: 481–487.

Jahn, B.-M., 1986. Mid-ocean ridge or marginal basin origin of the East Taiwan ophiolite: chemical and isotopic evidence. Contrib. Mineral. Petrol., 92: 194–206.

Jakes, P. and Miyake, Y.. 1984. Magma in forearcs: implication for ophiolite generation. Tectonophysics, 106: 349–358.

Jolivet, M., 1986. A tectonic model for the evolution of the Hokkaido central belt: Late Jurassic collision of the Okhotsk with Eurasia. Monogr. Assoc. Geol. Collab. Japan, No. 31, 355–377.

Kiminami, K., Kito, N. and Tajika, J. (1985): Mesozoic Group in Hokkaido: stratigraphy and age, and their significance. Earth Sci. (Chikyu Kagaku), 39, 1–17 (in Japanese with English abstract).

Kobayashi, T., 1941. The Sakawa orogenic cycle and its bearing on the origin of the Japanese Islands. J. Fac. Sci.. Univ. Tokyo, Sec. 2: 219–578.

Koide, Y., Sano, S., Ishiwatari, A. and Kagami, H., 1987. Geochemistry of the Yakuno ophiolite in Southwest Japan. J. Fac. Sci.. Hokkaido. Univ., Ser. IV, 22: 297–312.

Komatsu, M., 1985. Structural framework of the axial zone in Hokkaido: its composition, characters and tectonics. Mem. Geol. Soc. Japan, 25: 137–155 (in Japanese with English abstract).

Komatsu, M., Ujihara, M. and Chihara, K., 1985. Pre-Tertiary basement structure in the Inner Zone of Honshu and the North Fossa Magna region. Sci. Rept. Niigata Univ., Ser. E, No. 6, 17–35 (in Japanese with English abstract).

Kunugiza, K., Takasu, A. and Banno, S., 1986. The origin and metamorphic history of the ultramafic and metagabbro bodies in the Sanbagawa metamorphic belt. Geol. Soc. Amer. Mem., 164: 375–385.

Kuroda, Y., Kurokawa, K., Uruno, K., Kinugawa, T., Kano, H., and Yamada, T., 1976. Staurolite and kyanite from epidote-hornblende rock in the Oeyama (Komori) ultramafic mass, Kyoto Prefecture, Japan. Earth Sci. (Chikyu Kagaku), 30: 331–335.

Kuroda, Y. and Shimoda, S., 1967. Olivine with well-developed cleavages: its geological and mineralogical meanings. J. Geol. Soc. Japan, 73: 377–388.

Kurokawa, K., 1985. Petrology of the Oeyama ophiolitic complex in the Inner Zone of Southwest Japan. Sci. Rept. Niigata Univ. Ser. E, No. 6, 37–113.

Lindsley-Griffin, N., 1977. The Trinity ophiolite, Klamath Mountains, California. In: R.G. Coleman and W.P. Irwin (Eds), 'North American Ohiolites", Oregon State Dept. Geol. Min. Indst. Bull. 95: 107–120.

Maekawa, H., 1981. Geology of the Motai Group in the southwestern part of the Kitakami Mountains. J. Geol. Soc. Japan, 87: 543–554 (in Japanese with English abstract).

Maekawa, H., 1989. Two modes of mixing of Biei ophiolitic melange, Kamuikotan blueschist belt, Japan. J. Geol., 87: 543–554.

Maruyama, S., Banno, S., Matsuda, T. and Nakajima, T., 1984. Kurosegawa zone and its bearing on the development of the Japanese Islands. Tectonophysics, 110: 47–60.

Miyake, Y., 1985. MORB-like tholeiites formed within the Miocene forearc basin, Southwest Japan. Lithos, 18: 23–34.

Miyashiro, A., 1973. The Troodos ophiolitic complex was probably formed in an island arc. Earth Planet. Sci. Lett., 19: 218–224.

Miyashiro, A., 1975. Classification, characteristics. and origin of ophiolites. J. Geology, 83: 249–281.

Miyashiro, A., 1986. Hot regions and the origin of marginal basins in the Western Pacific. Tectonophysics, 122: 195–216.

Miyashita, S., 1983. Reconstruction of the ophiolite succession in the western zone of the Hidaka metamorphic belt, Hokkaido. J. Geol. Soc. Japan, 89: 69–86 (in Japanese with English abstract).

Miyashita, S. and Yoshida, A., 1988. Pre-Cretaceous and Cretaceous ophiolites in Hokkaido, Japan. Bull. Soc. Geol. France, 1988: 251–260.

Mori, T. and Banno, S., 1973. Petrology of peridotite and garnet clinopyroxenite of the Mt. Higasi-Akaisi mass, central Shikoku, Japan: subsolidus relation of anhydrous phases. Contrib. Mineral. Petrol., 41: 301–323.

Nagata, J., 1982. Magnesioferrite-olivine rock and monticellite-bearing dunite from the Iwanaidake alpine-type peridotite mass in the Kamuikotan structural belt, Hokkaido, Japan. J. Japan. Assoc. Min. Petr. Econ. Geol., 77: 23–31.

Nicolas, A., 1989. Structures of ophiolites and dynamics of oceanic lithosphere. Kluwer Academic Publishers, Dordrecht, 367pp.

Nicolas, A. and Jackson, E.D., 1972. Repartition en deux provinces des péridotites des châines alpines longeant la Méditerranée: implications géotectoniques. Schweiz. Min. Petrogr. Mitt., 52: 479–495.

Niida, K., 1984. Petrology of the Horoman ultramafic rocks in the Hidaka metamorphic belt, Hokkaido, Japan. Jour. Fac. Sci., Hokkaido Univ., Ser IV, 21: 197–250.

Niida, K. and Kito, N., 1986. Cretaceous arc-trench systems in Hokkaido. Monogr. Assoc. Geol. Collab. Japan, 31: 379–402 (in Japanese with English abstract).

Nishimura, Y. and Shibata, K., 1989. Modes of occurrence and K-Ar ages of metagabbroic rocks in the "Sangun metamorphic belt", Southwest Japan. Mem. Geol. Soc. Japan, 33: 343–357 (in Japanese with English abstract).

Onuki, H., 1965. Petrochemical research on the Horoman and Miyamori ultramafic intrusives, northern Japan. Sci Rept. Tohoku Univ., Ser. III, 9: 217–276.

Ozawa, K., 1983. Relationships between tectonite and cumulate in ophiolites: the Miyamori ultramafic complex, Kitakami Mountains, northeast Japan. Lithos, 16: 1–16.

Ozawa, K., 1984. Geology of the Miyamori ultramafic complex in the Kitakami Mountains, northeast Japan. J. Geol. Soc. Japan, 90: 697–716.

Ozawa, K., 1988. Ultramafic tectonite of the Miyamori ophiolitic complex in the Kitakami Mountains, Northeast Japan: hydrous upper mantle in an island arc. Contrib. Mineral Petrol., 99: 159–175.

Ozawa, K., Shibata, K. and Uchiumi, S., 1988. K-Ar ages of hornblende in gabbroic rocks from the Miyamori ultramafic complex of the Kitakami Mountains. J. Japan. Assoc. Petr. Min. Econ. Geol., 83: 150–159. (in Japanese with English abstract)

Quick, J.E., 1981. Petrology and petrogenesis of the Trinity peridotite, an upper mantle diapir in the eastern Klamath Mountains, northern California. J. Geophys. Res., 86: 11837–11863.

Saito, Y. and Hashimoto, M., 1982. South Kitakami region: an allochthonous terrane in Japan. J. Geophys. Res., 87: 3691–3696.

Saito, Y., Tiba, T. and Matsubara, S., 1979. Ultramafic complex and its mechanical sedimentary derivatives in the Tonmakuyama area, north of Hamana-ko, Central Japan. Mem. Nation. Sci. Mus., Tokyo, 12: 29–44.

Sakakibara, M., 1986. A newly discovered high-pressure terrane in eastern Hokkaido, Japan. J. metamorphic Geol., 4.

Saleeby, J.B., Harper, G.D., Snoke, A.W. and Sharp, W.D., 1982. Time relations and structural-stratigraphic patterns in ophiolite accretion, West Central Klamath Mountains, California. J. Geophys. Res., 87: 3831–3848.

Sample, J.C. and Fisher, D.M., 1986. Duplex accretion and underplating in an ancient accretionary complex, Kodiak Islands, Alaska. Geology 14: 160–163.

Saunders, A.D., Tarney, J., Stern, C.R. and Dalziel, I.W.D., 1979. Geochemistry of Mesozoic marginal basin floor igneous rocks from southern Chile. Geol. Soc. Amer. Bull., 90: 237–258.

Shibata, K. and Nishimura, Y., 1989. Isotopic ages of the Sangun crystalline schists, Southwest Japan. Mem. Geol. Soc. Japan, 33: 317–341. (in Japanese with English abstract)

Snoke, A.W., 1977. A thrust plate of ophiolitic rocks in the Preston Peak area, Klamath Mountains, California. Geol. Soc. Am. Bull., 88: 1641–1659.

Sugisaki, R., Mizutani, S., Hattori, H., Adachi, M. and Tanaka, T., 1972. Late Paleozoic geosynclinal basalt and tectonism in the Japanese Islands. Tectonophysics, 14: 35–56.

Sugisaki, R., Suzuki, T., Kanmera, K., Sakai, T. and Sano, H., 1979. Chemical compositions of green rocks in the Shimanto belt, Southwest Japan. J. Geol. Soc. Japan, 85: 455–466.

Suzuki, H., Isozaki, Y. and Itaya, T., 1990. Tectonic superposition of the Kurosegawa terrane upon the Sanbagawa metamorphic belt in eastern Shikoku, Southwest Japan: K-Ar ages of weakly metamorphosed rocks in northeastern Kamikatsu Town, Tokushima Prefecture. J. Geol. Soc. Japan, 96: 143–153 (in Japanese with English abstract).

Takasawa, K., 1976. Anorthite in peridotites from the Setogawa Group, Shizuoka Prefecture, Central Japan. Earth Sci. (Chikyu Kagaku), 30: 163–169.

Takasu, A., 1984. Prograde and retrograde eclogites in the Sanbagawa metamorphic belt, Besshi district, Japan. J. Petrol, 25: 619–643.

Taniguchi, H. and Ogawa, Y., 1990. Occurrence, chemistry and tectonic significance of alkali basaltic rocks in the Miura Peninsula, Central Japan. J. Geol. Soc. Japan, 96: 101–116. (in Japanese with English abstract)

Tazawa, J., 1988. Paleozoic-Mesozoic stratigraphy and tectonics of the Kitakami Mountains, northeast Japan. Earth Sciene (Chikyu Kagaku), 42: 165–178 (in Japanese with English abstract).

Uda, S., 1984. The contact metamorphism of the Oeyama ultrabasic mass and the genesis of the "cleavable olivine". J. Geol. Soc. Japan, 90: 393–410 (in Japanese with English abstract).

Varne, R. and Brown, A.V., 1978. The geology and petrology of the Adamsfield ultramafic complex, Tasmania. Contrib. Mineral. Petrol. 67: 195–207.

Watanabe, T. and Maekawa, H., 1985. Early Cretaceous dual subduction system in and around the Kamuikotan tectonic belt, Hokkaido, Japan. In: N. Nasu et al. (Eds), "Formation of Active Ocean Margins", pp. 677–699, Terra Sci. Publ., Tokyo.

Watanabe, T., Tokuoka, T. and Naka, T., 1987. Complex fragmentation of Permo-Triassic and Jurassic accreted terranes in the Chugoku region, Southwest Japan and the formation of the Sangun metamorphic rocks. In: E.C. Leitch and E. Scheibner (Eds) "Terrane Accretion and Orogenic Belts" (Geodynamic Series Vol. 19), pp. 275–289.

Yokoyama, K., 1990. Nikubuchi peridotite body in the Sanbagawa metamorphic belt: thermal history of the Al-pyroxene-rich suite' peridotite body in high-pressure metamorphic terrane. Contrib. Mineral. Petrol., 73: 1–13.

Yokoyama, K., 1987. Ultramafic rocks in Kurosegawa tectonic zone, Southwest Japan. J. Japan. Assoc. Min. Petr. Econ. Geol., 87: 319–335.

Tectonostratigraphic Relationships and Obduction Histories of Scandinavian Ophiolitic Terranes

BRIAN A. STURT and DAVID ROBERTS
Geological Survey of Norway, P.O. Box 3006 – Lade, N-7002 Trondheim, Norway

Abstract

The Scandinavian Caledonides represents a now linear montage of nappes and thrust-sheets derived by imbrication of the Baltoscandian continental rise prism and miogeocline and from the accretion of far-travelled, outboard, oceanic terranes. The originally near-continuous pattern of such geotectonic features was destroyed during the episodic Caledonian, contractional and strike-slip, orogenic processes that eventually led to juxtaposition of tectonic units which, in many cases, represent terranes of considerable, though unknown, geographical separation.

The earliest recorded, subduction-related, ensimatic island arc construction (Tremadoc), subduction of the Baltoscandian miogeocline (ca. 505 Ma), and the thrusting of these elements together with obduction of the Group I ophiolites onto the margin of Baltica or a drifted microcontinental block occurred during the Finnmarkian orogeny. These ophiolites were obducted during early (Central Norway) and latest (SW Norway) Arenig times, and the difference in timing may relate to separation and segmentation by transform faults. Both the early and the later ophiolites and island arc complexes can be related to subduction processes off the margin of Baltica and we see no reason for invoking westward obduction onto Laurentia prior to an eventual overthrusting onto Baltica during the Scandian collision. The orientations, dips and polarity of the subduction zones were probably complex, and it is difficult to view the developing orogen in terms of belt-parallel subduction. The problem of faunal provincialism in the Scandinavian Caledonides, epitomized by the anomalous Arenig fauna of mainly North American affinity in the western Trondheim region can perhaps be better explained by the combination of an ephemeral physical barrier and a belt-parallel warm-water ocean current, than by separation across a wide Iapetus Ocean.

Tj. Peters et al. (Eds), Ophiolite Genesis and Evolution of the Oceanic Lithosphere, 745–769.

Introduction

The Caledonide orogen of western Scandinavia extends over a strike distance of almost 2,000 km, and is characterized structurally by a series of nappes or thrust-sheets transported eastwards onto the Baltoscandian platform in Early Palaeozoic time. In its simplest terms the tectonostratigraphy comprises imbricated miogeoclinal assemblages in the east, indigenous to Baltica; and orogen-internal, mainly oceanic sequences of more exotic character and suspect provenance in the west, in the higher tectonic units. Geochemical studies have shown that many of the volcanite sequences in the higher, exotic nappe-complexes, can be recognized as island arc products and, in some cases, as fragmented ophiolites (e.g. Gale and Roberts 1974, Furnes et al. 1980), considered to represent vestiges of the oceanic crust of the Early Palaeozoic Iapetus Ocean (Roberts and Gale 1978). Later work has revealed a network of primitive and mature arc/forearc associations, rifted arc sequences and marginal basin products, a complex mosaic which brings to mind the present-day scenario in the SE Asia/SW Pacific region (Hamilton 1979). The complexity of palaeoenvironments in the Caledonides is, however, compounded by the polyphase tectonometamorphic evolution from Late Precambrian to Devonian time, involving repeated thrust and strike-slip dissection of the original rock associations and culminating in collision between the continents Baltica and Laurentia. The Caledonide arcs, fragmented ophiolites and associated basinal deposits thus form the major component of a "broad tract of tectonic flotsam crunched between collided continents" (Hamilton 1988).

The Scandinavian Caledonides today depict an essentially rectilinear foldbelt resulting from the *Scandian* collisional orogeny in Mid Silurian to Early Devonian times. As a consequence, the palaeogeographic and tectonic patterns which existed during the constructive and destructive cycles of Iapetus Ocean development can only be appreciated in an approximate manner. The stratigraphic and igneous record is considerably telescoped within the nappe stack and major parts of the Iapetus system are missing due to subduction processes and tectonic erosion.

The Ophiolitic Complexes of Southwestern Norway

The ophiolitic rocks of this region are now well-known (Fig. 1) from the literature (see Sturt et al. 1984, Pedersen et al. 1988), and an extensive review will not be attempted here. Instead, a number of points bearing on the development of the oceanic crust and the subsequent timing of obduction will be emphasized.

The Karmøy Ophiolite is the best known and contains the most complete record of oceanic crustal development of these ophiolite fragments, in this case in a supra-subduction zone setting (Pearce et al. 1984). Three major

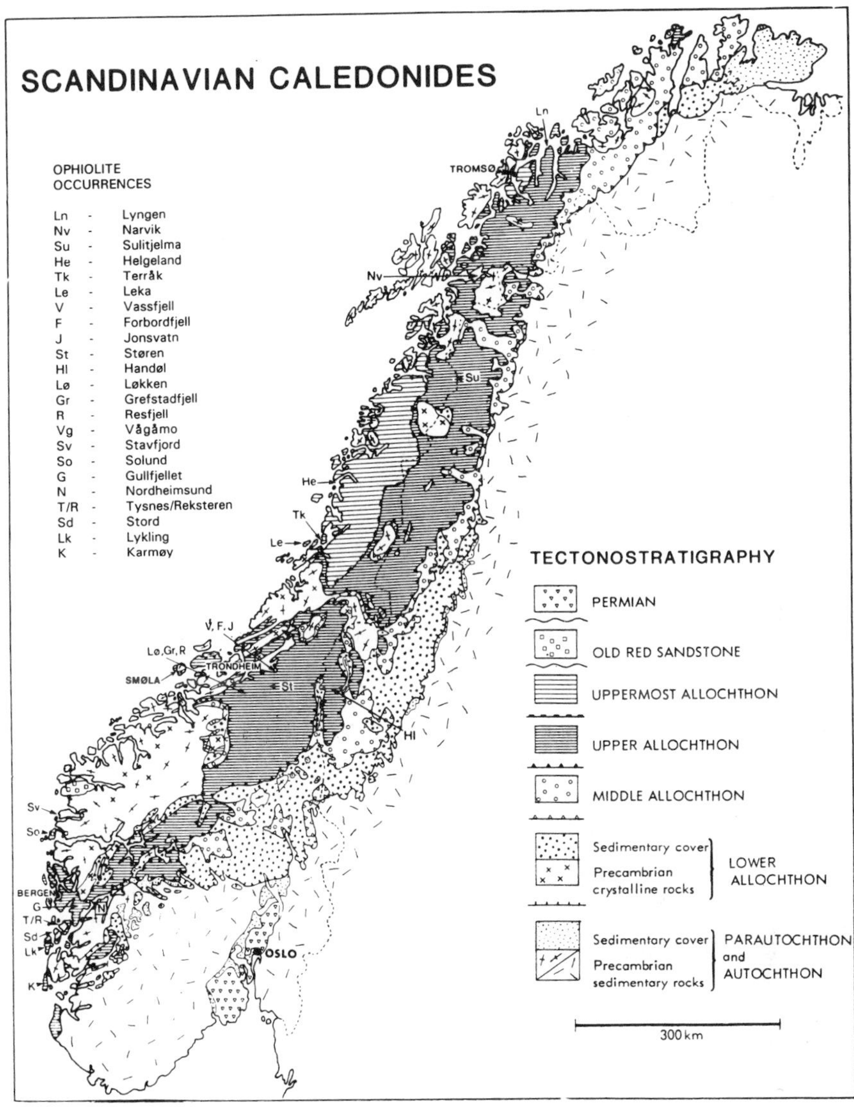

Figure 1. Principal tectonostratigraphic subdivisions of the Scandinavian Caledonides showing the locations of the main ophiolite complexes and ophiolitic associations. Modified from Sturt et al. (1984).

stages in the construction of the oceanic crust (Pedersen et al. 1988) have been reported:

i) Formation of an axis sequence by spreading-related magmatism around 493 +7/−4 Ma.

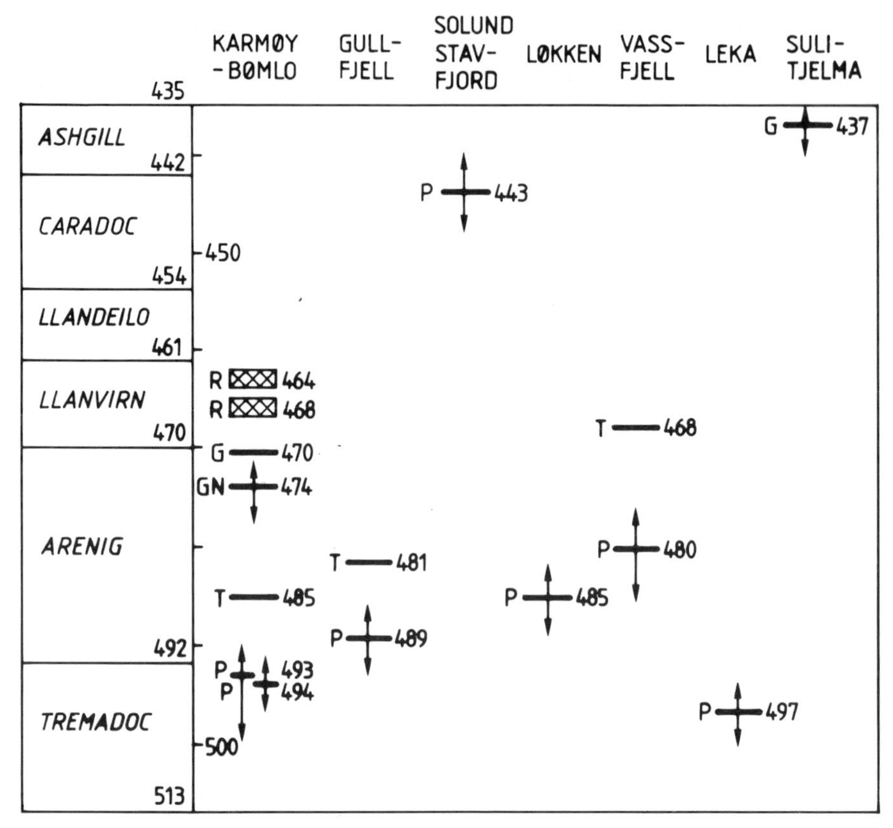

Figure 2. Isotopic dates for Norwegian Caledonian ophiolites; compiled from Dunning (1987), Dunning and Pedersen (1988), Pedersen et al. (1988 and in press) and Pedersen (pers. comm.). The Rb-Sr dates are from Furnes et al. (1986). The arrows on the bars indicate the error range. With the exception of R (rhyolite – Rb/Sr), all the dates are U-Pb on zircons. P –plagiogranite; T – trondhjemite; G – gabbro; GN – granite.

ii) Construction of an island arc upon this crust, as evidenced by the widespread presence of basaltic rocks with an IAT geochemistry.

iii) Splitting of this arc and formation of a new basin, the latter constrained as having occurred between 485 ± 2 and 470 +9/−5 Ma.

The age determinations quoted here are all U/Pb zircon ages (Dunning et al. 1988), and show a spread of ages through the Early Ordovician to around the base of the Llanvirn (Fig. 2). One of the interesting features implied by these studies is that the Torvastad Group (Sturt et al. 1980) sediments and volcanites are Arenig in age, and that their deformation and metamorphism (upper greenschist facies) probably related to obduction of the ophiolite prior to emplacement of the granitic rocks of the West Karmøy Igneous Complex (WKIC) (Sturt and Thon 1978, Ledru 1980).

The timing of obduction of the Karmøy Ophiolite is constrained by the emplacement of the WKIC (Ledru 1980). The WKIC was clearly derived from a continental crustal source as indicated by its calc-alkaline character and the initial $^{87}Sr/^{86}Sr$ ratios in the order 0.7220 (Ledru 1980). Compelling evidence for a continental crustal source is also seen in the multitude of xenoliths of continental crustal rocks which the granitoids have brought up through the ophiolite (Ledru 1980). Recent U-Pb geochronology (R.B. Pedersen pers. comm. 1989) provides an age of 474 + 3/−2 Ma for one of the granites, and this overlaps in error-bar with the dating of the latest stage of oceanic crustal construction. This indicates rapid obduction of the Karmøy Ophiolite, close to the Arenig-Llanvirn boundary, and implies that the WKIC may have originated via a disturbance in thermal regime consequent upon crustal loading by the Karmøy Ophiolite (Ledru 1980).

The *Lykling (Bømlo) Ophiolite Complex* shows many features similar to the Karmøy Ophiolite in having an early axis sequence upon which developed an island arc complex. An early stage in this evolution, on Geitung, has been dated by U-Pb on zircons at 494 ± 2 Ma (R.B. Pedersen, pers. comm. 1989). The minimum age of obduction is given by the bimodal volcanites of the unconformably overlying Siggjø Group (Llanvirn) which have yielded Rb-Sr whole-rock isochrons at 468 ± 23 Ma (meta-andesites) and 464 ± 16Ma (metarhyolites) (Fig. 2). The initial $^{87}Sr/^{86}Sr$ ratios, in the range 0.70740–0.70755, are taken by Furnes et al. (1986) to be comparable with continental volcanites of Andean type, in contrast with the arc-type volcanic rocks of the Lykling Ophiolite where quartz keratophyres record a $^{87}Sr/^{86}Sr$ ratio at 0.70452 ± 0.0002, (Furnes et al. (1986)). The trace-element geochemistry of the Siggjø volcanites is considered by Furnes et al. (1986) to be comparable with volcanites of the 'Basin-and-Range' type. It will be noted that the initial $^{87}Sr/^{86}Sr$ ratios of the Siggjø volcanites differ considerably from those of the WKIC, probably as the result of crustal contamination of mantle sourced magmas in the former case (Furnes et al. 1986) and magmas derived via crustal anatexis in the latter (Ledru 1980).

The ophiolite fragments of the *Major and Minor Bergen Arcs* are considerably fragmented and have been dated by U-Pb on zircons at 489 ± 3 (Pedersen et al. 1988) in the case of Gullfjellet and 485 ± 2 (R.B. Pedersen pers. comm. 1989) for a gabbro pegmatite from the Minor Bergen Arc.

The *Solund-Stavfjord Ophiolite Complex* (SSOC), previously assumed to be coeval with those discussed above (Sturt et al. 1984, Furnes et al. 1985), has been shown to be considerably younger (Dunning and Pedersen 1988). A pegmatitic diorite cut by two generations of dyke swarms, the latter with MORB geochemistry, has yielded a U-Pb zircon date of 443 ± 3 Ma (Dunning and Pedersen 1988). The ophiolite was generated in an areally restricted marginal basin (Pedersen et al. 1988). The SSOC is apparently separated from underlying Lower to Middle Silurian continental margin deposits by an obduction mélange (Andersen et al. 1990). The ophiolitic rocks of Southwest Norway thus fall into two main groups:

i) Early Ordovician complexes developed in a supra-subduction zone setting. The Karmøy Ophiolite demonstrates the most complex oceanic crust construction, and isotopic and field data indicate two distinct spreading events separated by island arc magmatism, the latest during Late Arenig time. As Pedersen et al. (1988) state, the Iapetus Ocean was probably floored by 60–80 Ma old oceanic crust at the time that the oldest parts of the preserved ophiolites developed in arc-related basins. The obduction of the Karmøy Ophiolite onto the continental margin occurred very rapidly, close to the Arenig-Llanvirn boundary.

ii) The Late Ordovician-Early Silurian complex of Solund-Stavfjord. This was developed in some sort of marginal basin after the ophiolites (i) had accreted onto the continental crust, and its obduction almost certainly relates to the Scandian collisional event.

The Ophiolitic Complexes of Central and Northern Norway

Although this segment covers a strike length of more than 900 km, the majority of the fragmented ophiolites (Fig. 1) occur in the southern half within or adjacent to the Trondheim Region of Central Norway (Prestvik 1974, Furnes et al. 1980, 1985, Roberts et al. 1984, Grenne 1989, and references therein). One small tectonic lens has also been reported from part of the Jämtland district of Sweden (Gee and Sjöström 1984).

The *Vågåmo Ophiolite* (Fig. 1) lies with thrust contact upon a psammite-dominated succession, the Heidal Series. Both these units are overlain unconformably by the Sel Series, which carries the Otta Conglomerate at its base (Sturt et al. in press). Cumulate and isotropic gabbros, sheeted dolerite dykes and pillow lavas are the main units in the ophiolite, with the dykes showing an N-MORB composition. The clast population in the Otta Conglomerate varies from mainly serpentinite in the type locality, to a polymict admixture derived from different parts of the ophiolite pseudostratigraphy. The matrix, at one locality, contains a varied fauna of mixed Baltic and North American affinity, of Late Arenig-Early Llanvirn age (Bruton and Harper 1981), thus placing a minimum age, Late Arenig, on ophiolite obduction.

In the southwestern Trondheim Region, ophiolite complexes displaying all members of the typical pseudostratigraphy, except for the ultramafic layer, have been described from *Vassfjell* (Grenne et al. 1980), *Grefstadfjell* (Ryan et al. 1980), *Løkken* (Grenne 1986, 1989) and *Resfjell* (Heim et al. 1987). Both the sheeted dyke and the basaltic lava units generally exceed 1 km in thickness. Field relationships, lithological character and geochemical traits are such that it would appear that these now separate, faulted and, in part, tectonically inverted complexes are correlatable. The geochemistry of these fragmented ophiolites denotes generation in some form of extensional, marginal basin setting, with a transitional MORB/IAT component appearing in the higher lava units at Løkken.

Even though fossils, mainly but not exclusively of North American affinity, are comparatively common in the sedimentary sequences of this district, there are ambiguous sediment/ophiolite relationships which call for further detailed field study. Faunas of Late Arenig to Early Llanvirn age occur in Lower Hovin Group (LHG) sediments above, and locally intercalated with, the uppermost lavas of the Løkken and Grefstadfjell Ophiolites. Graptolites of Mid Arenig (Castlemain 2) age are present in one shaly formation, believed by Ryan et al. (1980) to be located stratigraphically beneath the Grefstadfjell Ophiolite, but the true relationships of this shale to the ophiolite are contentious (cf. Heim et al. 1987).

Preliminary isotopic dating (U-Pb) on zircons from plagiogranite dykes has yielded ages of 493 ± 10 and 487 ± 5 Ma for Løkken and 480 ± 4 Ma for Vassfjell (Dunning 1987 and pers. comm. 1990) (Fig. 2), approximately Mid Arenig, which broadly coincides with the age assessment from the faunas. An interesting and anomalous feature of this group of fragmented ophiolites is that they do not show the usual signs of having been obducted directly after their formation. Shear zones occur within the Løkken Ophiolite which appear to pre-date the Hovin sediments, and there is a marked primary unconformity within the thick Resfjell lava sequence denoting local uplift and erosion of the ophiolite during its actual formation (Heim et al. 1987). These features may relate to some form of abortive obduction in an unstable basinal setting during the Late Arenig (Roberts et al. 1984).

The *Støren Ophiolite* unit is considered to have been obducted along with deep-water pelagic sediments, upon the subjacent Gula Complex. A tectonic mélange developed during the obduction process (Horne 1979) and the entire sequence of ophiolite, pelagics, mélange and Gula was deformed, weakly metamorphosed and then intruded by abundant trondhjemite dykes and sheets prior to uplift, erosion and sedimentation of the LHG. The basal polymict conglomerate of the LHG is composed largely of ophiolitic (and trondhjemitic) material, and is directly succeeded by the Krokstad Formation, which has yielded just one trilobite of Arenig age and European affinity (Spjeldnæs 1978). Obduction and tectonic deformation of the Støren Ophiolite is thus considered to be syn- to pre-Middle Arenig and pre-LHG. This equates with the 'Trondheim disturbance' of Holtedahl (1920) and the Early Ordovician *Finnmarkian* orogeny, and should be viewed in a more regional context. The Gula Complex and, in the east, the Tremadoc, rifted-arc Fundsjø Group (Grenne 1987) with *Dictyonema* of European affinity (Spjeldnæs 1985) were also affected by this same tectonothermal event. Thus, two phases of obduction, the later one probably abortive, may have obtained in this region, separated by some 15–20 million years.

Metabasalt/gabbro units of probable ophiolitic affinity are known from other parts of the western Trondheim Region, within equivalents of the Lower Hovin Group, e.g. *Forbordfjell*, *Jonsvatn*, *Frosta* (Roberts et al. 1984). These are of extensional basin origin, developed in a thickened continental margin setting. Age constraints are poor, but within the range Llanvirn to Early Caradoc.

To the north of the Trondheim Region, the *Leka Ophiolite Complex* (LOC) exposes undoubtedly the best preserved harzburgite tectonite of all the Caledonian ophiolites, with ultramafic cumulates lying directly above the tectonite and being succeeded by metagabbro layers (Furnes et al. 1988). One of a number of minor acidic (keratophyre) intrusions within the gabbro/dyke unit has yielded a U-Pb zircon date of 497 ± 2 Ma (Dunning and Pedersen 1987). Geochemical studies indicate a supra-subduction zone setting for the initial phases of the LOC, with a more arc-remote, WPB basinal milieu taking over higher up, locally with ocean island development (Furnes et al. 1988).

As yet, there is no good minimum constraint on the age of obduction of the LOC. Deformation, metamorphism and uplift of the ophiolite were succeeded by deposition of the initially continental sediments of the Skei Group (Sturt et al. 1985). No fossils have yet been recovered from these sediments. On the mainland to the east of Leka, the higher Helgeland Nappe Complex (HNC) contains several imbricate lenses of fragmented and dismembered ophiolite with unconformably overlying sediments (Furnes et al. 1985). A granite (Heilhornet) cutting one of a series of mylonites close to the base of the HNC has given an age of 444 ± 11 Ma (Nordgulen and Schouenborg, 1990), thus providing a minimum age (Ashgill) for *initial* HNC translation above the Skei Group sediments. Since the Seve Nappes, which were high-P metamorphosed and rapidly uplifted in Early Ordovician time (490–480 Ma) (Dallmeyer et al. 1985), are considered to occur at a level below the HNC, and below the ophiolites in general, then obduction of the LOC may have occurred at or around this time.

At *Sulitjelma* (Fig. 1), an inverted and otherwise tectonized association of gabbro, sheeted basic dykes and pillow lavas has been inferred to represent an ophiolite sequence (Boyle 1980, 1989). Petrochemical studies denote a back-arc E-MORB character. This, together with the fact that the gabbros intrude, and contain, rafts of metasediments, seems to signify magma generation by back-arc spreading in an ensialic marginal basin (Boyle 1980). U-Pb dating of zircon and sphene from a pegmatitic portion of the gabbro has given a crystallization age of 437 ± 2 Ma (Pedersen et al. in press) (Fig. 2), which providing a maximum age for accretion of the arc/basin terrane upon the continental margin sequences.

In northern Norway, the tectonized and layered *Lyngen Gabbro Complex* (Fig. 1), with bodies of serpentinite, is structurally underlain by a sequence of sheared amphibolitic lavas with profuse mafic and felsic dykes, the Kjosen Formation. The complex has been interpreted as ophiolitic (Furnes et al. 1980, 1985). Tectonic lenses of comparable amphibolitic lavas, metagabbros and dyke units occur at the same tectonostratigraphic level further south (Boyd 1983, Barker 1986). Geochemical studies on the Lyngen complex have indicated a MORB to IAT situation (Furnes et al. 1985, Fuller 1986), interpreted by Fuller as a probable supra-subduction zone, back-arc basin. No isotopic dating is yet available from this complex, but the ophiolite is believed to have been obducted, polyphasally deformed, overturned and eroded during the Finnmarkian orogenic phase, in Early Ordovician time (Furnes et al. 1985, Minsaas and Sturt 1985). Lying unconformably upon the ophiolite is a clastic-dominated sedimentary succession of Late Ordovician to Early Silurian age, the Balsfjord Group. The clast petrography of the conglomerates indicates derivation from both the ophiolite complex and Finnmarkian-deformed metasediments, as well as from an exotic magmatic arc complex (Minsaas and Sturt 1985) not known to occur anywhere in northern Norway.

An additional element in the Early Caledonian magmatic evolution of Central Norway is that of island arcs. The Tremadoc, ensimatic, rifted-arc setting of the Fundsjø Group (Grenne and Lagerblad 1985, Grenne 1987) has already been mentioned, and a comparable situation exists at Stekenjokk, in Sweden, to the east of the HNC and below the level of the Gjersvik Nappe. There, a U-Pb zircon age of 492 ± 3 Ma for a trondhjemite (Claesson et al. 1987) indicates a minimum age, at the Tremadoc-Arenig boundary, for the ensimatic arc volcanism. The magmatic complex of the Gjersvik Nappe is also of ensimatic arc character (Gale and Roberts 1974, Lutro 1979), with subsequent arc rifting (Grenne and Reinsbakken 1981). A different situation exists, however, on the island of Smøla (Fig. 1) where a mature, calc-alkaline arc (Roberts 1980) is spatially related to, and succeeds, limestones of Late Arenig to Llanvirn age (Bruton and Bockelie 1979); plutonic rocks of the arc range up into the Early Silurian (Gautneb and Roberts 1989). Thus, as in Southwest Norway, there appear to be two main groups of ophiolites or rocks of ophiolitic affinity:

i) Early Ordovician, pre-Mid Arenig ophiolites, partly developed in a supra-subduction zone setting. Associated with these in some cases are Tremadoc ensimatic island arc complexes, which have not been identified in the Southwest Norwegian areas. On the basis of faunal constraints the obduction of these ophiolites appears to be somewhat earlier than in SW Norway, i.e. pre-Mid Arenig in the case of Storen and pre-Late Arenig for Løkken, Vassfjell and Vågåmo. The history of obduction in the Early Ordovician may thus involve two stages.

ii) Late, or possibly Mid-Ordovician to Early Silurian spreading in ensialic marginal basins, as indicated by the complexes at Frosta, Forbordfjell, Jonsvatn and Sulitjelma. Of these, only the Sulitjelma unit has so far been isotopically dated, 437 ± 2 Ma. Other mafic-ultramafic bodies of a similar age are known from the central/ northern Caledonides: the Råna Intrusion, at 437 + 1/-2 Ma (Boyd et al. 1988), Artfjäll 435 ± 5 (Otten 1983), Umbukta pre-447 ± 7 Ma (Claesson 1979), all of which are considered to represent spreading-related magmatism (Stephens et al. 1985). In all these cases, deformation and thrusting occurred during the Scandian orogeny.

The existence of spreading-related, extensional, marginal basins throughout the Ordovician and Early Silurian, as recorded in the suspect terranes of the Upper Allochthon, is now well documented. As pointed out by Grenne and Roberts (1980), there is no reason to expect such ephemeral spreading centres to show belt-length contemporaneity. Development of these spreading centres, along different segments of the continental margin, may have been regulated by varying rates and azimuths of plate convergence and changing rates and angles of subduction within the transform-bounded segments.

Tectonostratigraphic setting

The fragmented ophiolites in the Scandinavian Caledonides occur in a comparatively high tectonostratigraphic position within the orogen. The metamorphic allochthon is divided into four main complexes, the Lower, Middle, Upper and Uppermost Allochthons; in addition, there is the Autochthon/Parautochthon in the eastern external zone (Roberts and Gee 1985). The telescoped continental margin of Baltica and associated miogeoclinal sequences are represented by the Parautochthon and Lower and Middle Allochthons. Outboard of these arkosic sandstone nappes is a transitional continent-to-ocean assemblage (the Seve Nappes) composing the lowest part of the Upper Allochthon (Andreasson and Gee 1989). Most of the Upper Allochthon comprises the orogen-internal, exotic, suspect terranes (Sturt 1984, Roberts 1988, Stephens 1988) (the Köli Nappes) derived from the oceanic/eugeoclinal realm of Iapetus, including most of the preserved ophiolites and island arc assemblages (Fig. 1). The highest and most outboard Uppermost Allochthon is a heterogeneous complex of thrust-sheets and granitoid plutons, and which also includes fragmented ophiolites. This unit has generally been considered to derive from the Laurentian side of Iapetus (Roberts et al. 1985, Stephens and Gee 1985, 1989). Of the two principal groups of ophiolite complexes recognised in the Norwegian Caledonides, Group I ophiolites are believed to have originated in arc-related, intra-oceanic basins, and been involved in Finnmarkian orogenic deformation and thrusting/obduction in Early Ordovician time. Group II ophiolite associations post-date this early Caledonian orogenesis; they formed in small marginal basins along a continental margin, and are considered to be of Mid to Late Ordovician age. In recent years, U-Pb (zircon and sphene) isotopic dating from some of these ophiolite fragments (Fig. 2) have largely confirmed this basic division, and led to revisions in our understanding of early Caledonide orogenic evolution (Dunning and Pedersen 1988, Pedersen et al. 1988, 1980). In the preceding section the Scandinavian Caledonian ophiolites have been discussed in terms of two separated geographical regions, a division which appears to reveal certain significant differences in oceanic crustal development and the age of obduction, metamorphism and uplift of the early Group I ophiolites.

i) The construction of the oceanic crust and associated arc rocks in the ophiolites of SW Norway appears to extend to younger chronostratigraphic levels than in the central region. In the case of Karmøy, oceanic crust and arc development apparently extends to or even straddles the Arenig-Llanvirn boundary, and its obduction occurred almost immediately at this time.
ii) In the above region, the post-obduction Llanvirn igneous activity included intrusion of crustally derived calc-alkaline granitoids generated by partial melting of the continental crust (WKIC) and the effusion of 'Basin-and-Range' type bimodal continental volcanites (Siggjø Volcanic Complex).
iii) In the Central Norwegian ophiolites, initial obduction and mélange formation can be stratigraphically constrained to pre-Mid Arenig for Støren and pre-Late Arenig for a number of the other complexes. In this some region, and in Sweden, the Tremadoc ensimatic island arc complexes occur in three separate nappes.
iv) On Smøla, a mature island arc developed on 'continental' crust, was initiated in Late Arenig time or earlier, presumably after the Group I ophiolites of this region had already accreted onto the continental margin, or the margin of a microcontinent.

These differences between the two segments may have arisen from subtle changes in palaeotectonic regime with resultant differences in histories of collision and obduction. Based on recent collisional events at active continental margins where major transform faults separate segments of oceanic crust, slight differences, measured in terms of a few millions of years, in the rates and vectors of collision/obduction with a continental margin may occur on either side of such a bounding transform (Fig. 3), thus resulting in disparate oceanic crustal character and development. Such a configuration could possibly explain the differences in timing between the two sectors under discussion; the Group I ophiolites and their related ensimatic arcs could have collided with and been obducted onto a continental margin during the Early to Middle Arenig whilst oceanic crustal development with suprajacent arc build-up was still occurring to the south of the theoretical transform. Consequently, collision/obduction could have been retarded to a time close to the Arenig-Llanvirn boundary in the southwestern region (Fig. 3).

In the Central segment, the ophiolitic rocks of Vassfjellet, Grefstadfjell, Løkken and Resfjell represent somewhat of an anomaly in this picture as they appear to have developed subsequent to the emplacement onto the continental margin of the Støren Ophiolite and the Tremadocian ensimatic island arc complexes of Fundsjø and Stekenjokk. This emplacement has been shown to be associated with deformation and weak metamorphism (Grenne

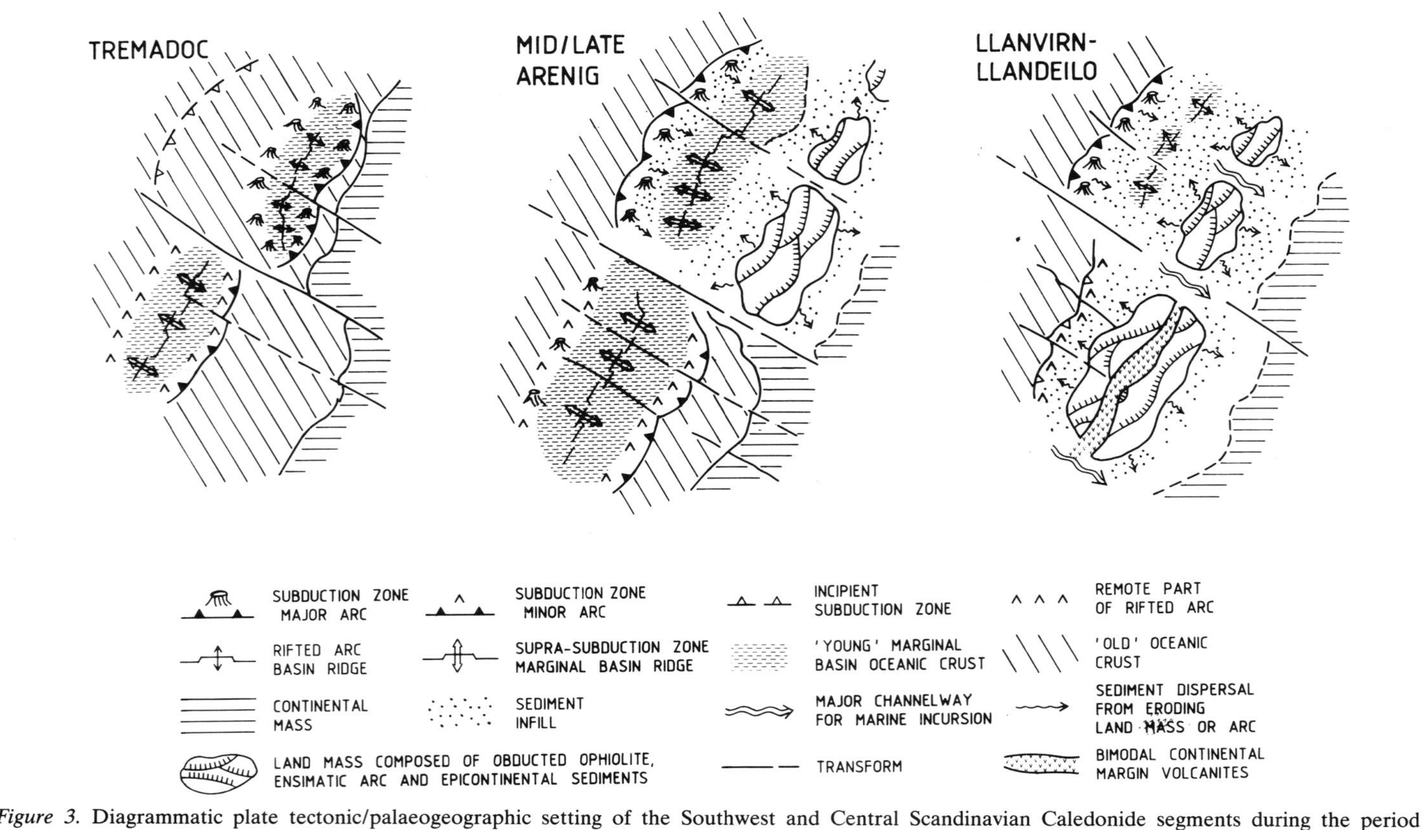

Figure 3. Diagrammatic plate tectonic/palaeogeographic setting of the Southwest and Central Scandinavian Caledonide segments during the period Tremadoc to Llandeilo. The upper half of each diagram (NE of the main NW-SE transform) is meant to portray the Central segment; and the lower half the Southwestern segment. For full explanation, compare the legend and the main text. Tremadoc – the pre-obduction situation. Mid/Late Arenig –obduction has already accurred in the Central segment, represented by the eroding landmass composed of thrust slices. Llanvirn-Llandeilo – in the Central Segment, a mature arc is present in the northwest and a marginal basin occurs between this
arc and the eroding, dimishing landmass. In the Southwest segment, obduction has occurred and the uplifted, eroding landmass is affected by rifting with the production of bimodal volcanic rocks.

and Lagerblad 1985), and in the case of the Støren Ophiolite and Fundsjø Complex to pre-date deposition of the Hovin Group and equivalent sediments in the east (Roberts et al. 1984).

The anomalous ophiolite complexes form an intricate part of the architecture of an extensional back-arc basin and interdigitate within the lithostratigraphy of this basin. In time, they coincide with the effusion of the distinctly calc-alkaline mature arc volcanites of Smøla (Late Arenig-Llanvirn). It is tempting to consider that these slightly later ophiolites, and the ophiolitic rocks of Forbordfjell, Jonsvatn and Frosta, were formed in a marginal basin, or basins, related to a more westerly situated Smøla Arc (Fig. 3). The marginal basin ophiolites were apparently formed within a fault-controlled shallow marine basin (Bruton and Bockelie 1980) which appears to have deepened into Late Ordovician times. There is also clear evidence of a major input of continentally derived detritus from a southeasterly source, and arc-derived material from the northwest (Roberts et al. 1984). This situation, involving the development of a fault-controlled extensional basin consequent upon collapse of a crust thickened by orogenic contraction, is a fairly normal one immediately succeeding collision/obduction.

The site of derivation of the early Group I ophiolites within the Iapetus Ocean is a matter of speculation must relate to the siting and geometry of the subduction zones which controlled the formation of volcanic arcs and associated back-arc/marginal basins. The consumption of enormous volumes of oceanic crust by subduction is also a major factor which affected the former palaeogeographical location of such features, and this considerable though unknown telescoping makes reconstruction profoundly difficult. An allied problem in deciphering the provenance of our oceanic suspect terranes is that of continental margin-parallel and trench-linked megafractures (Hutton 1987, Roberts 1988). Concerning the question of the obduction polarity of the Group I ophiolites, there are essentially three possible models:

i) That the ophiolites and primitive arc rocks were all obducted onto a former, more westward extension of the continent Baltica (e.g. Sturt 1984).
ii) That they were obducted first onto the Laurentian continent and then thrust over the Baltic margin during the Scandian continent/continent collision in Mid Silurian to Early Devonian times (e.g. Stephens and Gee 1985, Pedersen et al. 1988).
iii) That they were obducted first onto an intervening microcontinent and subsequently translated across the Baltic margin during the Scandian collisional event (e.g. Roberts 1980).

It is difficult to find unequivocal proof for these various models in relation to each individual Group I ophiolite, but a number of significant lines of evidence can be observed via terrane-linking features:

a) In the case of the Mid to Late Ordovician marginal basin ophiolitic complexes of the Trondheim Region there is good evidence for the presence of a continental block along the E or SE margin of the basin with the Smøla Arc or a comparable arc occurring along the western margin of the basin (Fig. 3), this arc being constructed above easterly subducting oceanic lithosphere (Roberts et al. 1984). The continental block would represent either a former westward extension of Baltica or a microcontinent (Roberts 1980), thus supporting either models (i) or (iii). An anomalous element here is that of the mainly North American faunas in the Early Ordovician basinal sediments, though there are two examples of fossils of European affinity, in the one case directly above the Støren Ophiolite and the other associated with the Fundsjø arc volcanites.
b) The Vågåmo Ophiolite was emplaced onto the 'sparagmitic' rocks of the Heidal Series which, in turn, rests with stratigraphic unconformity upon continental gneissic rocks. Above the Vågåmo Ophiolite the unconformably overlying basal Otta Conglomerate, with its mixed Baltic-North American fauna, oversteps the ophiolite onto the underlying miogeoclinal rocks of the Heidal Series, thus providing an elegant example of a *terrane-linking unconformity*. This terrane linkage is further reinforced by the petrography of conglomerates in the Sel Series which clearly demonstrate contemporaneous derivation from both the ophiolite and its miogeoclinal substrate. The previous continuity of the miogeocline subsequent to its telescoping by thrusting would thus favor model (i).
c) The relationships described earlier from Lyngen show a terrane-linkage via conglomerate petrography in the unconformably overlying clastics between the ophiolite and subjacent nappes of metamorphosed miogeoclinal rocks. Although the terrane-linking nature of the actual unconformity has not been established, the evidence from the conglomerates would favor model (i). The conglomerates also contain debris derived from an ensialic arc complex. Minsaas and Sturt (1985) take this to imply provenance from a westerly sited mature arc, i.e. reminiscent of the Smøla situation.
d) The Karmøy Ophiolite, which lies directly upon continental crustal rocks of Baltic Shield affinity, is pierced by the terrane-linking West Karmøy Igneous Complex. The terrane-linking relationship, however, is not unambiguous as the rocks of the WKIC have not been observed to directly cut Baltic Shield basement.
e) In the epicontinental Oslo Basin there is a clearly marked positive geochemical anomaly of Fe, Mg, Cr, Ni and V in shales of Llanvirn and Llandeilo age. These shales, which were derived from an uplifted area in the northwest, also have an anomalous content of detrital high-Mg chlorites and sporadic clastic grains of chromite (Bjørlykke and Englund 1979). These features clearly imply a provenance from a mafic/ultramafic source, a source which was likely to be represented by ophiolites already obducted onto the western margin of Baltica (Sturt 1984).

Although the evidence is as yet incomplete, and in the case of faunal affinities to a large degree contradictory or ambiguous, we would submit that the balance would tend to favour an eastward obduction polarity onto a formerly more westward extension of Baltica. This would be commensurate with westward subduction of part of the Baltic continental crust and the Baltoscandian miogeocline as indicated by high-pressure metamorphism of the Seve and equivalent units. Eclogites from the Seve have been dated at around 505 Ma (U-Pb zircon) (Mørk et al. 1988) and their retrogression/uplift/ cooling dated by Ar^{40}/Ar^{39} hornblende plateau ages to around 490 Ma (Dallmeyer and Gee 1986). The primary age of eclogite formation coincides well with the construction of the Tremadoc ensimatic island arcs during subduction of oceanic crust (Stephens and Gee 1985). The Group I ophiolites, in the Central Norwegian Caledonides, could thus be viewed in a scenario of westward subduction of oceanic crust producing the primitive arcs and allowing generation of the 'young' oceanic crust of the ophiolites in a suprasubduction zone setting. This was broadly coeval with, and partly followed by subduction of continental crust and the Baltoscandian rise prism/miogeocline complexes, producing the high-pressure metamorphic assemblages in the accretionary wedge. This situation, with subduction of magmatosedimentary successions down to sufficient depths for the formation of eclogites occurring at the same time as ophiolite generation was going on, is known e.g. from Oman. The bouyancy effect of the subducting continental lithosphere also contributed to the arc/continent collision, with the upward imbrication of parts of the descending continental slab containing the relics of high-P assemblages. This would also allow for the initial obduction of the ophiolites during Early to Mid Arenig times. The model presented here has many analogues, both in the processes and in their timing related to ophiolite generation and obduction as, e.g. in Oman (Lippard et al. 1988, Coleman and El Shazly 1990). Subsequent to this obduction it would appear that eastward-directed subduction became operative or that a former eastward-directed subduction zone approached the margin of Baltica, as a result of which the Smøla Arc was formed together with the development of a back-arc marginal basin generated by rifting in an already accreted and then extending crust.

Let us now address the apparently contradictory evidence provided by faunal provinciality in the Early Ordovician, which has in fact been one of the principal arguments for a wide Iapetus Ocean gradually closing during the later part of the Ordovician. Indeed, the largely North American affinity of the Hølonda fauna has been taken by most authors to imply that the Lower Hovin Group, which bears this fauna, was definitively deposited along the Laurentian margin of Iapetus and subsequently shunted over to Baltica during the Scandian continental collision (Bruton and Bockelie 1980, Cocks and Fortey 1982, Stephens and Gee 1985, 1989, Bruton 1986). The mixed North American and Baltic fauna at Otta posed a problem in this regard and Neuman (1984) conceived an innovative theory to explain this. Neuman considered the Otta fauna in a context of oceanic islands moving from west to east across Iapetus providing a transport mechanism for the North American fauna, thus producing a migration route for this fauna and enabling its eventual mixing with that of Baltica. The interpretation of the stratigraphic relationships of the Lower Ordovician rocks of the Otta-Vågåmo region is obviously critical to the resolution of this problem. Sturt et al. (in press) have described how the sediments of the Sel Series, containing the Otta fauna, rest unconformably upon both the Vågåmo Ophiolite and sediments of the subjacent Heidal Series with its continental basement plinth. This unconformity would thus be of a terrane-linking nature, stitching together the obducted ophiolite, the miogeoclinal sediments and the basement of Baltica. Thus it would be inconceivable that the sediments of the Sel Series, and hence the Otta fauna, could have been deposited on the Laurentian side of the Iapetus Ocean and subsequently moved into place during the Scandian collision. The sparse record of a solitary European trilobite in slates of the Krokstad Formation, and also *Dictyonema* in phyllites associated with the Fundsjø rifted arc (Spjeldnæs 1978, 1985), would thus not be an anomaly in this pattern.

The Hølonda fauna, however, still remains a problem owing to its predominantly North American affinities. As noted previously, the sediments of the Lower Hovin Group, in which the Hølonda fauna is found, represent an integral part of a marginal basin sequence to the east of an evolving island arc sited on thickened crust, and with a continental block lying further to the east of this basin (Roberts et al. 1984). The models involving a considerable geographic separation of the Baltic and North American faunas during the Early Ordovician depend on the existence of a wide Iapetus Ocean. The faunas, however, really represent differences in the water temperature regime and there are alternative explanations for such differences. One such model involves a combination of separation by physical barriers and longshore oceanic currents bringing waters from different, geographically controlled, temperature regimes (Whittington 1966, Williams 1969a, 1969b, Whittington and Hughes 1972). The fact that ocean currents must have played a significant role in the regional distribution of faunas in the Ordovician has recently been emphasized (Bergström 1990, Burrett et al. 1990, Finney and Xu 1990, Rickards et al. 1990, Burrett et al. 1990). Good examples of these are seen today e.g. in the North Atlantic, in the warm-water Gulf Stream and the cold-water Labrador Current. These observations could be taken as a basis for proposing that the difference between the cold-water Baltic fauna, developed in an essentially epicontinental sea, was separated from a warm-water longshore current, bearing the North American fauna, by a mountain range composed of the obducted Group I ophiolites and their eastward-thrusted substrate (Fig. 4). The situation during the early part of the Early Ordovician of western Baltica would have not been unlike that of Oman at the present time, with the ophiolite and its thrusted substrate

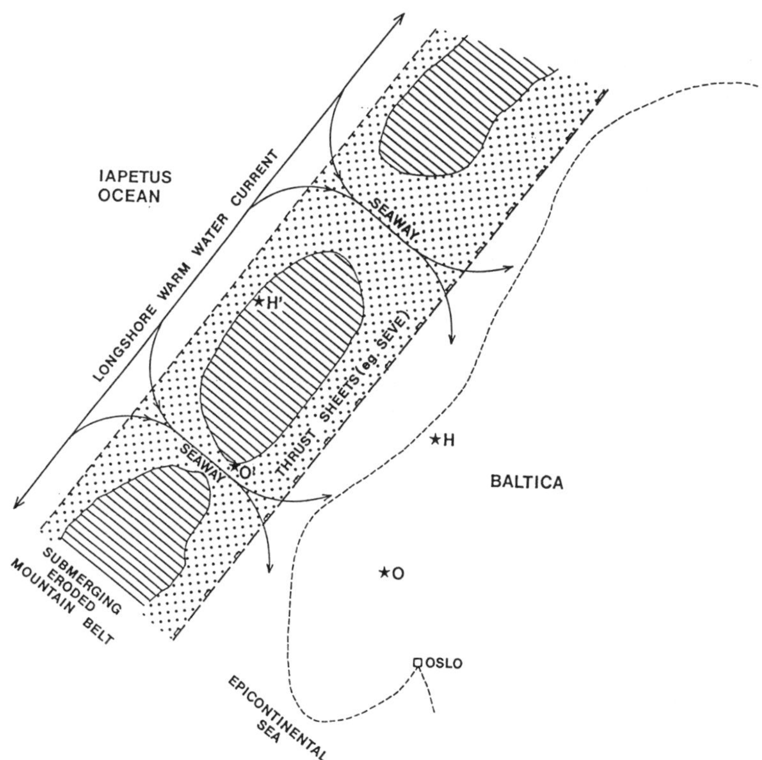

Figure 4. Possible palaeogeographical relationships of ophiolite(s) in the Scandinavian Caledonides during Late Arenig/Early Llanvirn times, indicating seaway(s) connecting Iapetus with the epicontinental sea of Baltica.

forming the Northern Oman Mountains indented by 'corridors' exposing the subjacent rocks (Lippard et al. 1986). During a subsequent submergence, such corridors would have formed seaways allowing for the ingress of waters transported by longshore currents (Fig. 4). Such a scenario would allow for the Hølonda fauna to be controlled by a warm-water current flowing broadly parallel to the continental margin with little or no mixing with the waters of the epicontinental sea, and intermixing with the cold waters via a newly established seaway to the Otta Region (Fig. 4). It must be remembered here that a considerable contraction of the Cambrian Iapetus must have occurred during the subduction cycle responsible for the formation of the Tremadoc ensimatic island arcs, the construction of the supra-subduction zone ophiolites and the obduction of these onto western Baltica during the early part of the Lower Ordovician. As Dean (1985) commented, in his review of Cambro-Ordovician faunas, there are many problems relating to a wide Iapetus in the Lower Ordovician and "it is possible that Iapetus was narrower than has been postulated". The erosion and submergence of the mountain belt would, in time, effectively remove the physical barrier and allow faunal mixing to occur progressively during the Ordovician.

Concluding Remarks

In this contribution we have endeavoured to show that the first major stage in the gradual destruction of the Iapetus Ocean occurred at a time close to the Cambrian-Ordovician boundary, and culminated in the obduction of Group I ophiolites and rifted-arc volcanites onto a westward extension of the Baltoscandian margin and miogeocline during the time interval Early Arenig to possibly earliest Llanvirn. This initial closure coincided with the consumption of a considerable, though unknown, volume of oceanic crust. The subduction would appear to have had a westerly vector commensurate with the development of medium- to high-pressure metamorphic assemblages in the down-going slab, which also involved rocks of the Baltoscandian miogeocline and continental rise prism (e.g. Seve Nappes). This also initiated the formation of the Tremadoc ensimatic arcs, fostered the construction of supra-subduction zone ophiolites, and led to their subsequent rapid obduction onto the miogeocline and continental margin. Slight differences in the apparent timing of the oceanic crustal development and eventual obduction of the Group I ophiolites, e.g. between southwestern and central Norwegian areas, is considered to be the result of regional segmentation and separation by a series of transform faults (Fig. 3). The character of the sedimentation which succeeded the obduction of the Group I ophiolites, with thick clastic deposits dominated by both ophiolitic and continentally derived detritus, attests to the rapid erosion of a mountainous area. This situation is very similar to that seen in Oman between the Semail Ophiolite and the overlying Late Maastrichtian and Lower Tertiary cover sequence (Skelton et al. 1990). We consider that when the Group I ophiolites, and their metamorphic substrate, were thrust eastwards onto the margin of Baltica, or a microcontinental block which had rifted and drifted away from Baltica, they formed an uplifted mountain range. We submit that, in Early Ordovician times, this could have formed an effective physical barrier which prevented mixing of the cold waters of the Baltic epicontinental sea with those of the Iapetus oceanic system. In the latter we conceive of a warm-water longshore current providing the basis for the predominantly 'North American' fauna of the western Trondheim Region; and that faunal mingling, producing the mixed Otta fauna, occurred via the drowning of a seaway during the Late Arenig (Fig. 4). This would in fact be convenient, as an available supply of ophiolite-derived material is also demanded by the geochemical anomalies in the shales of Llanvirn and Llandeilo age in the Oslo area. The general mixing of the Baltic and North American faunas could then have been achieved by the progressive erosion, breaching and submergence of the mountain barrier (Fig. 4).

The resolving of exact palaeogeographies in the Scandinavian Caledonides, especially for Cambrian and Early Ordovician time, is not yet possible, and indeed may never get beyond the modelling stage owing to the considerable inherent difficulties in this highly telescoped tectonostratigraphy. The great variety of palaeo-environments proposed and inferred in the literature implies a complex pattern of subduction systems with related sedimentary basins, island arcs, spreading centres and rifted-arc basins, and even continental blocks. In addition, there is the complicating factor of major strike-slip displacement of suspect terranes, which is a key feature in interpreting the accretionary history of e.g. the British Caledonides (Hutton 1987). In a complex package such as this, the loss of both oceanic and in part continental crust by a combination of subduction and tectonic erosion, together with highly variable strain regimes during deformation, make for difficulties in attempting even approximate connections between juxtaposed nappes. It is, in fact, virtually impossible to lay out a map-plan, in more than cartoon fashion, for any particular time interval. However, the indications in the tectonostratigraphic and magmatic record allow for the type of reconstructions made by Roberts (1988) and Stephens and Gee (1989) where the patterns and pieces are viewed in terms of terrane accretion. For the Silurian to Early Devonian period there is, however, good evidence for subduction of continental Baltic crust during the Scandian collisional event. This is indicated principally by the high-pressure eclogite and granulite assemblages developed in the northwest gneiss region (Cuthbert et al. 1983, Bryhni and Sturt 1985, Griffin et al. 1985). The peak of the high-pressure metamorphism has been dated at about 425 Ma (Griffin and Brueckner 1980), and this is considered to represent the acme of continental subduction. As the high-P assemblages show a progressive temperature gradient more or less orthogonal to the present coastline (Griffin et al. 1985), we assume that the down-going slab had an approximately NE-SW strike. It would thus appear that two of the subduction stages, which were approximately belt-parallel, involved the continental miogeocline prism in latest Cambrian to earliest Ordovican times and the Baltic margin during the Mid to Late Silurian Scandian collisional orogeny. The orientation and polarities of subduction zones during the intervening period are essentially unknown, although they may have had considerable complexity (Fig. 5).

It would appear that the orogenic history recorded in the various miogeoclinal and eugeoclinal (outboard) terranes cannot be classified into simple belt-long events as has been previously suggested (e.g. Roberts and Sturt 1980). Accretion onto the Baltic margin seems to have been of a rather more progressive nature with collisions occurring within the Iapetus system often at a considerable distance from their present siting (cf. Dallmeyer and Gee 1986). The collisional Scandian orogeny had the unifying effect of welding these disparate elements, including the fragmented and imbricated ophiolites,

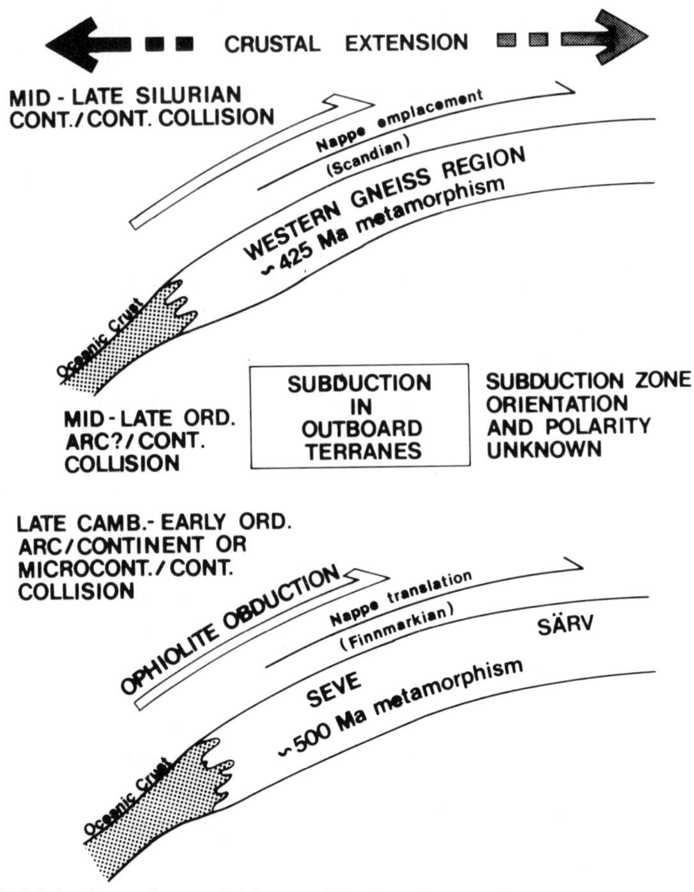

Figure 5. Subduction and tectonic history of the Scandinavian Caledonides.

into an essentially linear belt of overthrust nappes. This orogenic contraction gradually produced a thickened crust which was succeeded by a phase of gravitational collapse involving the development of fault-controlled extensional and transtensional Devonian basins.

Acknowledgements

The authors thank the Geological Survey of Norway for funding this research, and are grateful to a number of colleagues for helpful and stimulating discussions. Contribution no. 140 to the Norwegian ILP project.

References

Andersen, T.B., Skjerlie, K.P. and Furnes, H., 1990. The Sunnfjord Melange, evidence of Silurian ophiolite accretion in the West Norwegian Caledonides. Journal Geological Society, London, v. 147: 59–68.

Andreasson, P.G. and Gee, D.G., 1989. Baltoscandia's outer margin (the Seve Nappe Complex) in the Kebnekaise-Singis area of Norrbotten, Swedish Caledonides. Geologiska Förenings i Stockholm, Förhandlingar, v. 11: 378–381.

Barker, A.J., 1986. The geology between Salangsdalen and Gratangenfjord, Troms, Norway. Norges geologiske undersøkelse Bulletin 405: 41–56.

Bergström, S.M., 1990. Relations between conodont provincialism and the changing palaeogeography during the Early Palaeozoic. In: McKerrow, W.S. and Scotese, C.R., (Eds),: Palaeozoic Palaeogeography and Biogeography. Geological Society of London, Memoir, 12: 105–122.

Bjørlykke, K. and Englund, H., 1979. Geochemical response to Upper Precambrian rift basin sedimentation and Lower Palaeozoic epicontinental sedimentation in South Norway. Chemical Geology, v. 27: 271–295.

Boyd, R., 1983. The Lillevik dyke complex, Narvik: Geochemical and tectonic implications of a probable ophiolite fragment in the Caledonides of the Ofoten region, North Norway. Norsk Geologisk Tidsskrift, v. 63: 39–55.

Boyd, R., Tucker, R.D. and Barnes, S.J., 1988. A U-Pb zircon age for the Råna mafic-ultramafic intrusion and its implications for the timing of deformation and metamorphism in the Narvik *Nappe Complex, Ofoten, North Norway. Abstract, 18. Nordiske Geologiske Vintermøde, København 1988, p. 61.

Boyle, A.P., 1980. The Sulitjelma amphibolites, Norway: Part of a Lower Palaeozoic ophiolite complex? Proceedings of the International ophiolite Symposium, Nicosia 1979, pp. 567–575.

Boyle, A.P., 1989. The geochemistry of the Sulitjelma ophiolite and associated basic volcanics: tectonic implications. In: The Caledonide Geology of Scandinavia (Ed R.A. Gayer), pp. 153–163, Graham and Trotman, London.

Bruton, D.L., 1986. Recognition and significance of fossil-bearing terranes in the Scandinavian Caledonides (extended abstract). Geologiska Föreningens i Stockholm Förhandlinger, v. 108: 272–273.

Bruton, D.L. and Bockerlie, J.F., 1980. Geology and palaeontology of the Hølonda area, western Norway – a fragment of North America? In: Wones, D.R. (Ed),: The Caledonides in the U.S.A. Virginia Polytectonic Institute. Memoir 2: 41–47.

Bruton, D.L. and Harper, D.A.T., 1981. Brachiopods and trilobites of the Early Ordovician serpentinite Otta Conglomerate, south central Norway. Norsk Geologisk Tidsskrift, v. 61:3–18.

Bryhni, I. and Sturt, B.A., 1985. Caledonides of southwestern Norway. In: Gee, D.G. and Sturt, B.A. (Eds), The Caledonide Orogen – Scandinavia and related areas. John Wiley and Sons, Chichester, pp. 89–108.

Burrett, C. Long, J. and Stait, B., 1990. Early-Middle Palaeozoic biogeography of Asian terranes derived from Gondwana. In: McKerrow,W.S. and Scotese, C.R. (Eds),: Palaeozoic Palaeogeography and Biogeography. Geological Soviety of London, Memoir, 12: 105–122.

Claesson, S., 1979. Pre-Silurian orogenic deformation in the north-central Scandinavian Caledonides. Geologiska Föreningens i Stockholm Förhandlingar, v.101: 353–356.

Claesson, S., Stephens, M.B. and Klingspor, I., 1987. U-Pb zircon dating of felsic intrusions, Middle Köli Nappes, central Scandinavian Caledonides. Norsk Geologisk Tidsskrift, v. 67: 89–97.

Cocks, L.R.M. and Fortey, R.A., 1982. Faunal evidence for oceanic separations in the Palaeozoic of Britain. Journal of the Geological Society of London, v. 139: 465–478.

Coleman, R.G. and El-Sharzly, A.K., 1990. Metamorphism and emplacement of the Oman Ophiolite. Symposium on ophiolite genesis and evolution of oceanic lithosphere, UNESCO – Sultan Qaboos University, Muscat. Abstract.

Cuthbert, S., Harvey, M.A. and Carswell, D.A., 1983. A tectonic model for the metamorphic evolution of the Basal Gneiss Complex, western south Norway. Journal of Metamorphic Geology, v. 1: 63–90.

Dallmeyer, R.D., Gee, D.G. and Beckholmen, M., 1985. $^{40}Ar/^{39}Ar$ mineral age record of early Caledonian tectonothermal activity in the Baltoscandian miogeocline, central Scandinavia. American Journal of Science, v. 285: 532–568.

Dallmeyer, R.D. and Gee, D.G., 1986. $^{40}Ar/^{39}$ mineral dates from retrogressed eclogites within the Baltoscandian miogeocline. Implications for a polyphase Caledonian evolution. Geological Society of America Bulletin, v. 97: 26–34.

Dean, W.T., 1985. Relationships of Cambro Ordovician faunas in the Caledonide-Application region, with particular reference to Trilobites. In: Gayer, R.A. (Ed), The tectonic evolution of the Caledonide-Appalachian orogen, Vieweg, Braunschweig/Wisbaden, pp. 17–47.

Dunning, G.R., 1987. U-Pb zircon ages of Caledonian ophiolites and arc sequences: implications for tectonic setting. Abstract, EUG IV, Strasbourg 1987, p. 179.

Dunning, G.R. and Pedersen, R.B., 1988. U/Pb ages of ophiolites and arc-related plutons of the Norwegian Caledonides: Implications for the development of Iapetus. Contributions to Mineralogy and Petrology, v. 98: 13–23.

El-Sharzly, A.K. and Coleman, R.G., 1990. Metamorphism in the Oman Mountains in relation to the Semail Ophiolite emplacement. In: Robertson, A.H.F., Searle, M.P. and A. Ries (Eds), The geology and tectonics of the Oman region. Geological Society of London Special Publication 49: 473–494.

Finney, S.C.f and Xu, C., 1990. The relationship of Ordovician graptolite provincialism to palaeogeogrphy. In: McKerrow, W.S. and Scotese, C.R. (Eds),: Palaeozoic Palaeogeography and Biogeography. Geological Society of London, Memoir, 12: 123–128.

Fuller, B.A., 1986. Geochemistry of the Kjosen Formation, Arctic Norway. Unpubl. B.A. thesis, Middlebury College, Vermont, USA, 56 p.

Furnes, H. Roberts, D., Sturt, B.A., Thon, A. and Gale, G.H., 1980. Ophiolite fragments in the Scandinavian Caledonides. In: Panayioutou, A. (Ed), Ophiolites – Proceedings of the International Ophiolite Symposium, Cyprus 1979, pp. 582–600.

Furnes, H., Ryan, P.D. Grenne, T., Roberts, D., Sturt B.A. and Prestvik, T., 1985. Geological and geochemical classification of the ophiolite fragments in the Scandinavian Caledonides. In: Gee, D.G. and Sturt, B.A. (Eds), The Caledonide Orogen – Scandinavia and related areas. John Wiley and Sons, Chichester, pp. 657–670.

Furnes, H., Brekke, H., Nordås, J. and Hertogen, J., 1986. Lower Palaeozoic convergent plate margin volcanism on Bømlo, southwest Norwegian Caledonides. Geological Magazine, v. 123: 123–142.

Furnes, H., Pedersen, R.B. and Stillman, C.J., 1988. The Leka ophiolite complex, central Norwegian Caledonides field characteristics and geotectonic significance. Journal of the Geological Society of London, v. 145: 401–412.

Gale, G.H. and Roberts, D., 1974. Trace element geochemistry of Norwegian Lower Palaeozoic basic volcanics and its tectonic implications. Earth and Planetary Science Letters, v. 22: 380–390.

Gautneb, H. and Roberts, D., 1989. Geology and petrochemistry of the Smøla-Hitra Batholith, Central Norway. Norges geologiske undersøkelse 416: 1–24.

Gee, D.G. and Sjöström, H., 1984. Early Caledonian obduction of the Handöl ophiolite (abs.) In: Armands, G. and Schager, S. (Eds), Abstracts 16th Nordic Geological Winter Meeting 1984: Meddelanden från Stockholms Universitets Geologiska Institution, v. 255: 72.

Grenne, T., 1986. Ophiolite-hosted Cu-Zn deposits at Løkken and Høydal, Trondheim Nappe Complex, Upper Allochthon. In: Stephens, M.B. (Ed), Stratabound mineralizations in the Central Scandinavian Caledonides. Sveriges Geologiske Undersökning, Ser.Ca. v. 60: 55–65.

Grenne, T., 1987. Marginal basin type metavolcanites of the Hersjø Formation, eastern Trondheim District, Central Norwegian Caledonides. Norges geologiske undersøkelse Bulletin 412: 29–42.

Grenne, T., 1989. Magmatic evolution of the Løkken SSZ Ophiolite, Norwegian Caledonides:

relationships between anomalous lavas and high-level intrusions. Geological Journal v. 24, p. 251–273.

Grenne, T., Grammeltvedt, G. and Vokes, F.M., 1980. Cyprus-type sulphide deposits in the western Trondheim district, central Norwegian Caledonides. In: Panayioutou, A. (Ed), Ophiolites – Proceedings of the International Ophiolite Symposium, Cyprus 1974, pp. 582–600.

Grenne, T. and Reinsbakken, A., 1981. Possible correlations of island arc greenstone belts and related deposits from the Grong and eastern Trondheim districts of the Central Norwegian Caledonides. Abstract, Transactions of the Inst. of Mining and Metallurgy, B v. 90: 59.

Grenne, T. and Lagerblad, B., 1985. The Fundsjø Group, central Norway – A Lower Palaeozoic island arc sequence: Geochemistry and regional implications. In: Gee, D.G. and Sturt, B.A. (Eds), The Caledonide Orogen – Scandinavia and related areas. John Wiley and Sons, Chichester, pp. 745–760.

Griffin, W.L. and Brueckner, H.K., 1980. Caledonian Sm-Nd ages and a crustal origin for Norwegian eclogites. Nature, v. 285: 319–321.

Griffin, W.L., Austrheim, H., Brastad, K., Bryhni, I., Krill, A.G., Krogh, E.J., Mørk, M.B.E., Qvale, H. and Tørudbakken, B., 1985. High pressure metamorphism in the Scandinavian Caledonides. In: Gee, D.G. and Sturt, B.A. (Eds), The Caledonide Orogen – Scandinavia and related areas. John Wiley and Sons, Chichester, pp. 783–801.

Hamilton, W.B., 1979. Tectonics of the Indonesian Region. United States Geological Survey. Professional Paper, v. 1078, 345 p. Hamilton, W., 1988. Plate tectonics and island arcs. Bulletin of the Geological Society of America, v. 100: 1503–1527.

Heim, M., Grenne, T. and Prestvik, T., 1987. The Resfjell Ophiolite fragment, Soutwest Trondheim Region, Central Norwegian Caledonides. Norges geologiske undersøkelse Bulletin, v. 409: 49–71.

Holtedahl, O., 1920. Palaeogeography and diastrophism in the Atlantic-Arctic region during Palaeozoic time. American Journal of Science, v. 49: 1–25.

Hutton, D., 1987. Strike-slip terranes and a model for the evolution of the British and Irish Caledonides. Geological Magazine, v. 124: 405–525.

Ledru, P., 1980. Evolution structurale et magmatique du complexe plutonique de Karmøy. Societe Geologique et Mineralique du Bretagne, v. 12: 1–106.

Lippard, S.J., Shelton, A.W. and Gass, I.G., 1986. The Ophiolite of Northern Oman. Geological Society of London, Memoir, p. 11.

Lutro, O., 1974. The geology of the Gjersvik area, Nord-Trøndelag, Central Norway. Norges geologiske undersøkelse 354: 53–100.

Minsaas, O. and Sturt, B.A., 1985. The Ordovician – Silurian clastic sequence overlying the Lyngen Gabbro Complex, and its environment significations. In: Gee, D.G. and Sturt, B.A. (Eds), The Caledonide Orogen – Scandinavia and related areas. John Wiley and Sons, Chichester, pp. 379–393.

Mørk, M.B.E., Kullerud, K. and Stabel, A., 1988. Sm-Nd dating of Seve eclogites, Norbotten, Sweden – Evidence for early Caledonian (505 Ma) subduction. Contributions to Mineralogy and Petrology, v. 99: 344–351.

Neumann, R.B., 1984. Geology and paleobiology of islands in the Ordovician Iapetus Ocean: review and implications. Geological Society of America Bulletin, v. 95: 1188–1201.

Nordgulen, Ø. and Schouenborg, B., 1990. The Caledonian Heilhornet Pluton, north-central Norway: geological setting, radiometric age and geochemistry. Geological Society of London (in press).

Otten, M.T., 1983. The magmatic and subsolidus evolution of the Artfjället gabbro, Central Swedish Caledonides. Unpubl. Ph.D. thesis, Univ. of Utrecht, 186 p.

Pearce, J.A., Lippard, S.J. and Roberts, S., 1984. Characteristics and tectonic significance of supra-subduction zone ophiolites. In. Kokelaar, B.P. and Howells, M.F. (Eds), Margin Basin Geology. Volcanic and tectonic processes in modern and ancient marginal basins. Geological Society of London Special Paper, v. 16: 77–94.

Pedersen, R.B., Furnes, H. and Dunning, G.R., 1988. Some Norwegian ophiolite complexes reconsidered Norges geologiske undersøkelse, Special Publication 3: 80–85.

Pedersen, R.B., Furnes, H. and Dunning, G.R., 1990. The U/Pb age of the Sulitjelma Gabbro, north Norway: Further evidence of a Caledonian marginal basin in Ashgillian/Llandoverian time. Geological Magazine – In Press.

Rickards, R.B., Rigby, S. and Harris, J.H., 1990. Graptoloid biogeography: recent progress, future hopes. In: McKerrow, W.S. and Scotese, C.R. (Eds),: Palaeozoic Palaeogeography and Biogeography. Geological Society of London, Memoir, v. 12: 134–146.

Roberts, D., 1980. Petrochemistry and palaeogeographic setting of Ordovician volcanic rocks of Smøla, Central Norway. Norges geologiske undersøkelse 359: 43–60. Roberts, D., 1988. The terrane concept and the Scandinavian Caledonides: a synthesis. Norges geologiske undersøkelse. Bulletin v. 413: 93–99.

Roberts, D. and Gale, G.H., 1978. The Caledonian – Appalachian Iapetus Ocean. In: Tarling, D.H. (Ed), Evolution of the Earth's crust. Academic Press, London and New York, pp. 255–342.

Roberts, D. and Gee, D.G., 1985. An introduction to the structure of the Scandinavian Caledonides. In: Gee, D.G. and Sturt, B.A. In: Gee, D.G. and Sturt, B.A. (Eds), The Caledonie Orogen – Scandinavia and related areas. John Wiley and Sons, Chichester, pp. 55–68.

Roberts, D. and Grenne, T. and Ryan, P.D., 1984. Ordovician marginal basin development in the central Norwegian Caledonides. In: Kokelaar, B.P. and Howells, M.F. (Eds), Marginal Basin Geology. Volcanic and tectonic processes in modern and ancient marginal basins. Geological Society of London Special Paper, v. 16: 233–244.

Roberts, D., Sturt, B.A. and Furnes, H., 1985. Volcanite assemblages and environments in the Scandinavian Caledonides and the sequential development history of the mountain belt. In: Gee, D.G. and Sturt, B.A. (Eds), The Caledonide orogen – Scandinavia and related areas. John Wiley and Sons, Chichester, pp. 919–930.

Ryan, P.D. Williams, D.M. and Skevington, D., 1980. A revised interpretation of the Ordovician stratigraphy of Sør-Trøndelag, and its implications for the evolution of the Scandinavian Caledonides. In: Wones, D.R. (Ed),: The Caledonides in the USA. Virginia Polytectnic, Geological Sciences Memoir, v. 2: 99–105.

Skelton, P.W. Nolan, S.C. and Scott, R.W., 1990. The Maastrichtian transgression onto the northwestern flank of the Proto-Oman Mountainhs: sequences of rudite-bearing beach to open shelf facies. In: Robertson, A.F.H., Searle, M.P. and Ries, A.C. (Eds), The Geology and Tectonics of the Oman Region. Geological Society of London Special Publication, v. 49: 521–547.

Spjeldnæs, N., 1978. Faunal provinces and the Proto-Atlantic. In: Bowes, D.R. and Leake, B.E. (Eds), Crustal evolution in northwestern Britain and adjacent regions. Geological Journal, Special Issue, v. 10: 139–150.

Spjeldnæs, N., 1985. Biostratigraphy of the Scandinavian Caledonides. In: Gee, D.G. and Sturt, B.A. (Eds), The Caledonide Orogen – Scandinavia and related areas. John Wiley and Sons, Chichester, pp. 317–329.

Stephens, M.B., 1988. The Scandinavian Caledonides; A complexity of collisions. Geology Today, v. 4: 20–26.

Stephens, M.B. and Gee, D.G., 1985. A tectonic model for the evolution for the eugeoclinal terranes in the central Scandinavia Caledonides. In: Gee, D.G. and Sturt, B.A. (Eds), The Caledonide Orogen – Scandinavia and related areas: John Wiley and Sons, Chichester, pp. 953–978.

Stephens, M.B. and Gee, D.G., 1989. Terranes and polyphase accretionary history in the Scandinavian Caledonides. Geological Society of America Special Paper 230: 17– 30.

Sturt, B.A., 1984. The accretion of ophiolitic terranes in the Scandinavian Caledonides. Geologie en Mijnbouw, v. 63: 201–212.

Sturt, B.A., Andersen, T.B. and Furnes, H., 1985. The Skei Group, Leka, an unconformable sequence overlying the Leka Ophiolite. In: Gee, D.G. and Sturt, B.A. (Eds), The Caledonide orogen – Scandinavia and related areas: John Wiley and Sons, Chichester, pp. 395–406.

Sturt, B.A., Ramsay, D.M. and Neumann, R.B., 1990. The Otta Conglomerate and the Vågåmo Ophiolite – further indications of Early Ordovician orogenesis in the Scandinavian Caledonides. Norsk Geologisk Tidsskrift (in press).

Sturt, B.A., Roberts, D. and Furnes, H. 1984: A conspectus of Scandinavian Caledonian

ophiolites. In: Gass, I.G., Lippard, S.J. and Shelton, A.W. (Eds), Geological Society of London Special Publications, v. 13: 381–391.
Sturt, B.A., and Thon, A., 1978. An ophiolite complex of probable early Caledonian age discovered on Karmøy, v. 275: 538–539.
Sturt, B.A., Thon, A. and Furnes, H., 1980. The geology and preliminary geochemistry of the Karmøy ophiolite, southwest Norway. In: Panayioutou, A. (Ed), Ophiolites – Proceedings of the International Ophiolite Sympsium, Cyprus 1979, pp. 538–554.
Thon, A., 1985. The Gullfjellet ophiolite complex and the structural evolution of the major Bergen Arc, west Norwegian Caledonides. In: Gee, D.G. and Sturt, B.A. (Eds), The Caledonide orogen – Scandinavia and related areas. John Wiley and Sons, Chichester, pp. 671–678.
Whittington, H.B. and Hughes, C.P., 1972. Ordovician palaeogeography and faunal provinces deduced from trilobite distributions. Philosophical Transactions of the Royal Society of London, B. 263: 235–278.
Williams, A., 1969. Ordovician of the British Isles. In: Marshall Kay (Ed), North Atlantic – Geology and Continental Drift. American Association of Petroleum Geologists Memoir, v. 12: 236–264.

Genesis and Emplacement of the Supra-Subduction Zone Pindos Ophiolite, Northwestern Greece

G. JONES[1], A.H.F. ROBERTSON[1] and J.R. CANN[2]
[1]*Grant Institute of Geology and Geophysics, West Mains Road, Edinburgh, Scotland, EH9 3JW*
[2]*Department of Earth Sciences, University of Leeds, Leeds, LS2 9JT, UK*

Abstract

The north Pindos Mountains of mainland Greece expose a series of thrust sheets, which were emplaced towards the southwest during the Early Tertiary. The Jurassic Pindos ophiolite forms the upper part of this assemblage, and reveals a complex history, interpreted as mainly the result of supra-subduction zone spreading processes. Ophiolitic thrust sheets are additionally found structurally beneath the main peridotite body, and contain variable volcanic sequences, including units of island arc tholeiitic and boninitic affinities. The metamorphic sole formed in response to overthrusting of the young, hot, ophiolite, followed by emplacement onto the Pelagonian continental margin to the east. During the Cretaceous, the Pindos ocean existed as a remnant basin to the west of the emplaced ophiolite, and was infilled by deep marine sediments. During final Late Eocene collision, the Cretaceous sediments and Tertiary foreland basin deposits, were overthrust by the Pindos ophiolite and underlying thrust sheets, then emplaced onto Apulia to the southwest.

Introduction

The purpose of this paper is to summarize field observations and geochemical data, that have a bearing on the processes of formation, intra-oceanic displacement and subsequent emplacement of the Pindos ophiolite in northwestern Greece. The Pindos ophiolite preserves a complex history of sea-floor spreading and subduction-accretion events. An overall description and interpretation of the Pindos ophiolite and related units has been presented elsewhere (Jones and Robertson, 1990; Jones, 1990); here we focus on the genesis and emplacement of the Pindos ophiolite in its regional geological context.

Tj. Peters et al. (Eds), Ophiolite Genesis and Evolution of the Oceanic Lithosphere, 771–799.

Previous Studies

Several aspects of the north Pindos Mountains were considered by previous workers; in particular, the igneous geochemical evolution (Parrot, 1967, 1969; Montigny et al., 1973; Parrot and Verdoni, 1974; Terry, 1974, 1979; Capedri et al., 1980, 1981, 1982; Dupuy et al., 1984; Kostopoulos, 1989), also the metamorphic sole (Whitechurch and Parrot, 1978; Spray and Roddick, 1980), and the sedimentary and tectonic history (Brunn, 1956; Terry, 1971, 1975; Terry and Mercier, 1971; Kemp and McCaig, 1984). Many of these studies, however, considered only relatively small geographical areas, or did not provide sufficient information regarding the localities or field relationships, to allow the data to be usefully assimilated into a regional interpretation. Recent work in the area has involved study of the structure and sedimentology (Jones, 1990), the deformation of the peridotite sequence and chromite exploration (A. Rassios, pers. comm., 1989), geochemical analysis (Kostopoulos, 1989; Jones, 1990), and investigation of hydrothermal alteration processes (Valsami, 1990). Together, this has allowed an integrated tectonic model to be developed (Jones, 1990; Jones and Robertson, 1990).

Regional Tectonic Setting

The Pindos ophiolite of northwestern Greece (Fig. 1), is one of a series of emplaced oceanic crustal and mantle fragments, which form the Tethyan ophiolite belts of the Eastern Mediterranean region. These ophiolites represent tectonic sutures formed during progressive closure of the Mesozoic Neotethyan ocean, and include the Cretaceous Semail (Oman), Troodos (Cyprus), and Antalya (S.W. Turkey) ophiolites, together with the Jurassic Vourinos and Othris ophiolites of northern and central mainland Greece. The Pindos ophiolite is believed to have formed during the Jurassic, based on radiometric dating of the metamorphic sole (165 ± 3 Ma.; Spray et al., 1984). The ophiolite is considered to be continuous with the Vourinos ophiolite (Moores, 1969), beneath the Mid-Tertiary "molasse" sediments of the Meso-Hellenic trough (Fig. 1; Smith, 1979). Recent work favours an origin for the Pindos and Vourinos ophiolites within a single oceanic basin, the Pindos ocean (Jones and Robertson, 1990), which was sited within the present day Pindos-Olonos and Sub-Pelagonian isopic zones of Aubouin et al. (1970).

The western Greek ophiolites, which include Othris (Smith et al., 1979), Vourinos (Moores, 1969; Rassios et al., 1983; Wright, 1986), Kastoria (Mountrakis, 1984; 1986), Koziakas (Ferriere, 1982; Capedri et al., 1985; Lekkas, 1988) and Pindos, are here interpreted as having been emplaced generally northeastwards (Naylor and Harle, 1976; Smith, 1979; Jones and Robertson, 1990) onto the margin of the Pelagonian Zone, which is interpreted as a microcontinent. One alternative view agrees that Othris was em-

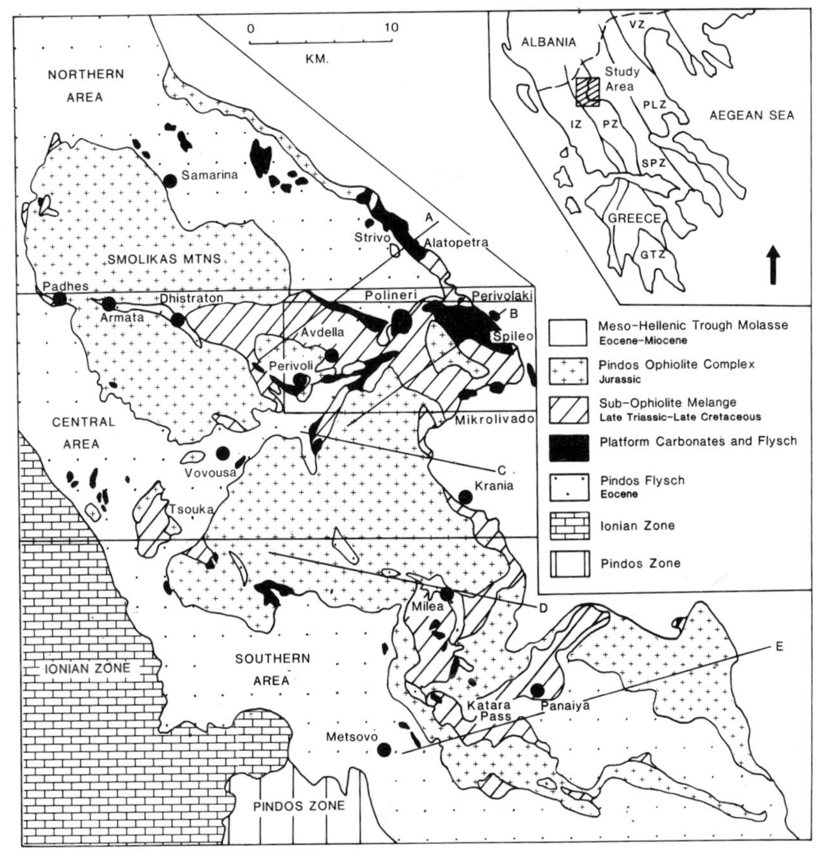

Figure 1. Outline geological and locality map of the northern Pindos Mountains. Modified from Jones and Robertson, (1990).

placed from southwest to northeast, but argues that Pindos, Vourinos, Kastoria and Koziakas were emplaced towards the southwest, across the Pelagonian Zone (e.g. Vergely, 1976, 1977, 1982). Another view is that all ophiolites originated in the northeast (Jacobshagen et al., 1978). The timing of emplacement of the Vourinos ophiolite is constrained by Late Jurassic (Kimmeridgian-Tithonian) transgressive sediments (Mavrides et al., 1979), that are exposed in the western Vourinos area (Krapa Hills). We consider these as having accumulated on the Vourinos ophiolite following its regional emplacement onto the Pelagonian margin, whilst it was still submerged. Significantly, similar sediments (of Tithonian age) have recently been shown to locally transgress the Pindos Avdella Melange (see below; Jones and Robertson, 1990).

Following initial Jurassic emplacement, the Pindos thrust sheets were further deformed by Early Tertiary collision of Apulia with the Pelagonian microcontinent, within the broad zone of collision between the African and

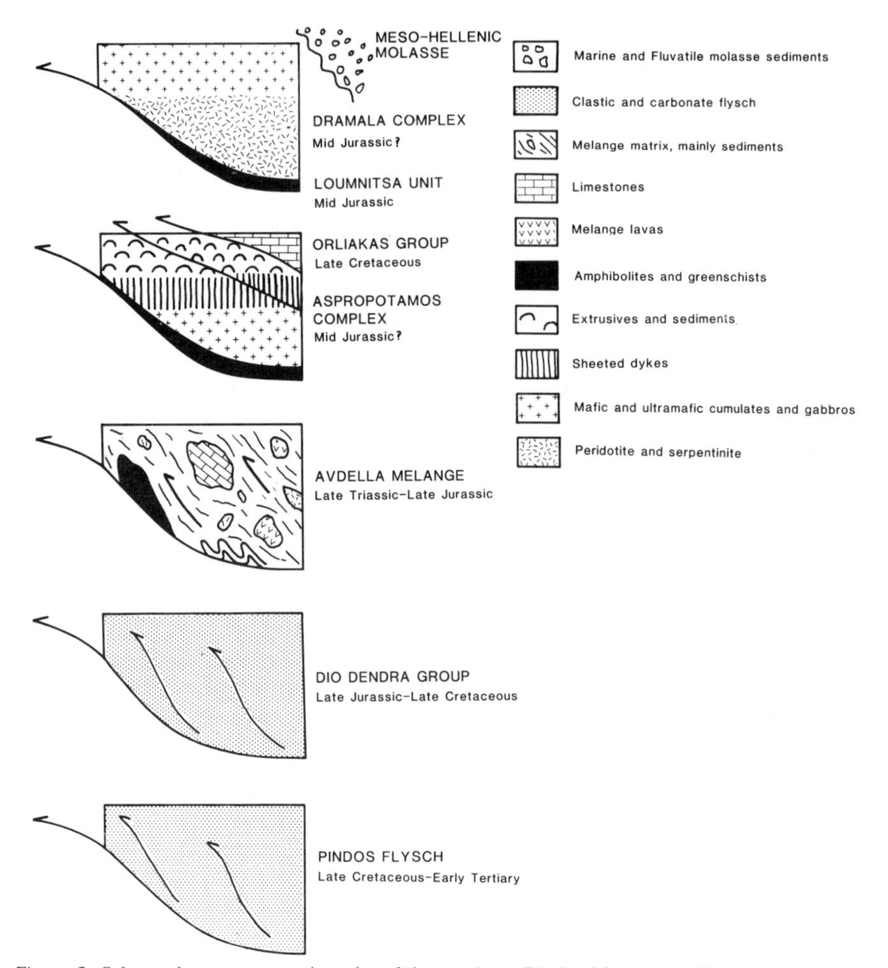

Figure 2. Schematic tectono-stratigraphy of the northern Pindos Mountains. See text for details.

Eurasian plates. This collision telescoped the ophiolite and associated sedimentary sequences, and emplaced them southwestwards onto the marginal sediments of the Ionian platform (Fig. 1), as a conventional foreland-propagating thrust stack (Kemp and McCaig, 1984; Jones and Robertson, 1990).

Tectono-stratigraphy

The tectono-stratigraphy of the north Pindos Mountains has been set out elsewhere (Jones and Robertson, 1990), and only a brief summary is given here. The Pindos thrust stack involves six major thrust sheets, which despite folding, occur in a regionally consistent structural order (Fig. 2). The *Pindos Ophiolite Group* consists of three main thrust sheets: the *Dramala Complex*, the *Aspropotamos Complex* and the *Loumnitsa Unit*. The Dramala Complex consists of a thick (>5 km) mantle peridotite sheet, with a locally preserved cumulate sequence; higher parts of the ophiolite were apparently completely eroded. The peridotite and cumulate rocks of the Dramala Complex overlie the remaining thrust sheets along a major out-of-sequence thrust, the *Achladi thrust* (Jones and Robertson, 1990). The Loumnitsa Unit represents the basal metamorphic sole of the Pindos ophiolite, and consists of amphibolite to greenschist facies meta-igneous and meta-sedimentary rocks. These metamorphic rocks are locally still attached to the base of the Pindos ophiolite, but elsewhere are found as detached slabs of metre-to hundreds-of-metre scale, within the Avdella Melange (see below).

Beneath the Achladi thrust is the Aspropotamos Complex, which consists of ophiolitic thrust slices, up to 2.5 km thick. The Complex displays oceanic crustal sequences, consisting of ultramafic and mafic cumulates, gabbros, sheeted dykes and extrusives. Associated sediments include radiolarites, metalliferous mudstones and pelagic limestones. These occur as thin thrust units, or conformably overlie the ophiolitic extrusives. The Aspropotamos Complex is extensively deformed by thrusting, but some important igneous intrusive relationships are still preserved. The Aspropotamos Complex is locally overthrust by the Late Cretaceous *Orliakas Group* platform carbonates, seen mainly on the eastern margin of the region.

The ophiolitic sheets (Pindos Ophiolite Group) are, in turn, structurally underlain by a tectono-sedimentary melange unit, the *Avdella Melange*. This is generally of block-in-matrix-type, and is more than 1 km thick. The melange is pervasively deformed, but is mainly unmetamorphosed, based on illite crystallinity studies (Jones, 1990). In some areas, single lithologies, or distinctive lithological associations occur, and can be mapped as thrust sheets. The most common lithologies are basalt, serpentinite, radiolarite and sandstone-shale sequences, together with pelagic and redeposited carbonates. The melange is, in turn, structurally underlain by the *Dio Dendra Group*, which consists of Cretaceous deep-marine sediments, and finally, all these sheets are underlain by the *Pindos Flysch*, the lowest exposed unit in the thrust stack.

Tectonic Evolution

Late Triassic – Early Jurassic

a) Rift Phase

Continental rifting and break-up in the Greek area occurred along the northern margin of Gondwana during the Mid-Late Triassic (Smith et al., 1979). The Pindos ocean basin was sited between the Ionian (southwesterly) and Pelagonian (northeasterly) continental basement areas, and is today represented by the Pindos and Sub-Pelagonian Zones of the "external" Hellenides.

The Ionian and Pindos Zone sequences of western and central Greece represent the Mesozoic to Tertiary carbonate platform and passive margin of Apulia (Green, 1983), situated to the southwest of the Pindos ocean basin.

In the studied area, Mid-Late Triassic rifting is documented by the extrusion of a range of within-plate (WPB) and transitional to mid-ocean ridge (MORB) basic volcanic rocks. These units are now found within the Avdella Melange (Table 1; Kostopoulos, 1989; Jones and Robertson, 1990). Locally, at Avdella village (Fig. 1), vesicular, alkalic within-plate basalts in the melange are dated as Late Triassic by the presence of *Halobia* (Terry, 1971). Southwest of Vovousa (Fig. 1), igneous activity associated with rifting is evidenced by basic volcanism of WPB, and WPB-MORB transitional type (Table 1; Fig. 3). These volcanics have a pelagic sedimentary cover, which includes limestones and radiolarites, dated as Late Triassic (P. De Wever, pers. comm., 1990). The volcanics and sediments are intruded by transitional to MORB dykes, with well-preserved chilled margins.

Geochemically similar Late Triassic volcanics of this type are known elsewhere in Greece (e.g. Othris; Hynes, 1972; Peloponesse; Pe-Piper, 1982) and are also known from other Tethyan ophiolites in the Eastern Mediterranean and Middle Eastern areas (e.g. Haybi Complex, Oman; Searle and Malpas, 1982). These extrusives are interpreted to represent magmas erupted during rifting and the earliest stage of spreading, to form a new, Late Triassic Pindos ocean basin.

Basinal sediments were also deposited at an early stage within the developing Pindos ocean. Pelagic limestones containing a Mid Triassic (Anisian-Carnian) ammonite fauna (M.K. Howarth, pers. comm., 1990) are the oldest sediments so far identified. Other widespread deposits include pelagic carbonates, carbonate and siliciclastic turbidites and radiolarites of Late Triassic age (P. De Wever, pers. comm., 1990).

Evidence of subsidence, following continental break-up to form the Pindos ocean basin, is preserved within sedimentary sequences overlying Late Triassic extrusives, and also within the Avdella Melange. At Alatopetra (Fig. 1; Jones and Robertson, 1990), WPB-type vesicular basalts (trachybasalts) are conformably overlain by Late Triassic-Early Jurassic thick-bedded and nodular carbonates, followed by pelagic marls, shallow-water redeposited calcarenites, and then by manganiferous and ferruginous mudstones. Above come thinly-bedded red shales, and minor volcaniclastic arenites, which pass transitionally into ribbon radiolarites of inferred early Middle Jurassic age. This conformable sequence documents the subsidence of a "high", or seamount through the carbonate compensation depth (CCD), into a deeper-water basinal setting. This seamount was located within the Pindos ocean basin, probably adjacent to the passive Pelagonian microcontinental margin.

b) Early Spreading Stage

By the Late Triassic (Carnian-Norian?), a spreading ridge was established in the Pindos basin. The main evidence is the occurrence within the Avdella

Table 1. Representative chemical analyses of basic rocks from the Pindos ophiolite and associated units. Samples 7SP13 to 66/89 from the ophiolitic Aspropotamos Complex; 45/89 to 86/89 from the sub ophiolite Avdella Melange, and 40/89 to 183/89 from the Loumnitsa Unit metamorphic sole. Refer also to Figs. 3 and 9. All rocks analysed at Edinburgh University. See Jones (1990) for details of analytical techniques. 7SP13 = high Ti-MORB sheeted dyke; 53/89 = N-MORB pillowed basalt (both from W of Spileo, Fig. 1); 68/89 = IAT pillowed basalt (from Krania, Fig. 1); 95/89 = boninitic sheeted dyke; 7AD1 = low-Zr plagiogranite (from Avdella, Fig. 1); 66/89 = higher Zr plagiogranite (from W of Spileo, Fig. 1). 45/89 = WPB massive basalt; 46/89 = WPB/MORB dyke from dyke swarm; 25/89 = N-MORB pillowed basalt; 86/89 = depleted MORB massive basalt (see Fig. 3 for localities). 40/89 = WPB greenschist; 126/88 = MORB amphibolite; 124/88 = depleted MORB amphibolite; 33/89 = IAT amphibolite; 183/88 = boninitic amphibolite (see Fig. 9 for localities).

	Aspropotamos Complex						Avdella Melange				Loumnitsa Unit				
	7SP13	53/89	68/89	95/89	7AD1	66/89	45/89	46/89	25/89	86/89	40/89	126/88	124/88	33/89	183/88
SiO_2	51.92	48.02	50.92	54.54	73.89	70.61	45.02	44.84	37.04	48.53	46.56	47.65	47.66	54.51	44.10
Al_2O_3	11.80	15.56	14.92	13.03	12.53	12.78	15.68	15.21	14.72	13.81	15.85	16.47	14.66	12.75	15.94
Fe_2O_3	14.09	8.48	9.60	7.23	1.63	4.58	5.19	8.08	10.94	10.43	12.13	11.19	7.66	8.11	5.25
MgO	6.69	8.69	4.32	11.08	1.47	0.97	4.92	6.70	5.32	7.46	5.92	8.21	7.77	8.92	12.05
CaO	6.17	8.84	7.00	6.39	2.64	2.76	10.62	10.69	12.86	12.58	8.62	12.47	16.51	7.36	18.70
Na_2O	5.20	3.92	4.27	4.67	5.26	5.42	5.41	3.31	3.56	3.11	3.35	2.39	2.23	5.24	0.48
K_2O	0.03	0.55	1.89	0.05	0.48	0.26	1.05	2.20	0.15	0.25	1.08	0.35	0.30	0.12	0.01
TiO_2	2.37	1.11	0.61	0.30	0.40	0.52	1.98	1.93	1.53	0.98	2.40	1.29	0.73	0.66	0.25
MnO	0.22	0.14	0.18	0.18	0.03	0.04	0.17	0.12	1.60	0.13	0.20	0.16	0.14	0.15	0.11
P_2O_5	0.25	0.10	0.08	0.03	0.06	0.12	0.48	0.37	0.13	0.11	0.34	0.14	0.07	0.05	0.00
LOI	1.48	4.50	6.20	3.12	1.48	1.10	9.37	6.81	11.72	2.77	2.73	2.64	2.54	1.92	3.88
Total	100.22	99.91	99.99	100.62	99.87	99.16	99.89	100.26	99.57	100.16	99.18	102.96	100.27	99.79	100.77

Table 1. Continued

	Aspropotamos Complex						Avdella Melange				Loumnitsa Unit				
	7SP13	53/89	68/89	95/89	7AD1	66/89	45/89	46/89	25/89	86/89	40/89	126/88	124/88	33/89	183/88
Ni	79	132	25	176	12	11	136	120	87	96	88	119	134	144	227
Cr	230	334	15	530	21	5	110	159	293	236	122	390	435	483	705
V	393	231	270	205	23	58	181	207	363	324	251	319	229	216	166
Sc	56	41	46	43	6	16	36	26	57	55	33	43	34	41	38
Cu	10	67	105	1	0	1	33	40	58	0	48	30	34	55	17
Zn	127	60	63	60	5	15	69	71	93	30	140	88	57	62	27
Sr	25	263	106	100	144	93	175	243	120	344	560	137	93	122	74
Rb	0	1	20	0	4	1	7	10	1	2	14	3	4	1	0
Zr	161	72	39	22	27	112	306	202	87	31	177	80	45	28	12
Nb	4	1	2	2	3	4	12	10	1	3	22	5	3	3	1
Ba	0	37	199	15	11	45	84	100	160	49	294	29	71	39	21
Pb	0	1	0	1	3	2	3	0	4	0	3	2	3	1	3
Th	0	0	0	0	1	1	0	0	0	0	0.6	0	0	0	0
La	3	2	5	2	0	5	19	8	606	3	21	0	0	13	0
Ce	13	3	4	0	12	3	48	31	17	4	48	1	0	23	0
Nd	14	6	3	0	3	4	27	20	8	5	24	11	5	9	0
Y	53	24	15	9	16	25	36	33	31	21	24	29	18	19	7

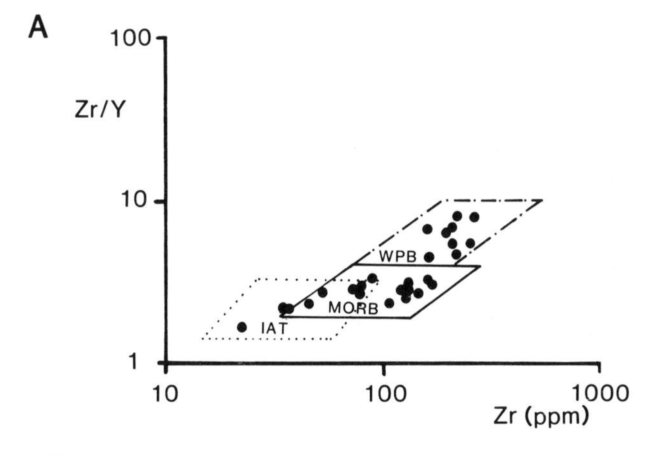

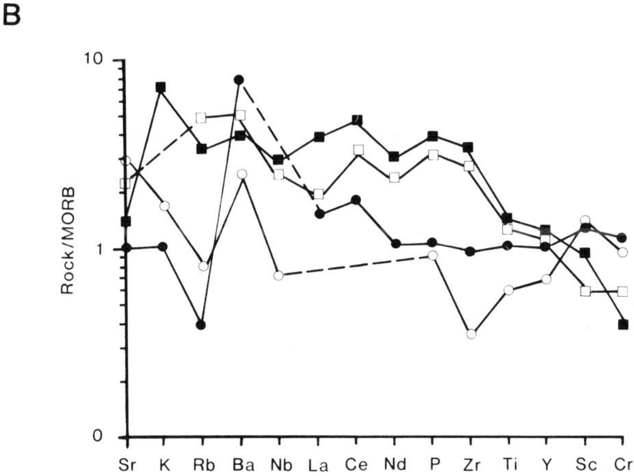

Figure 3. A) Zr/Y-Zr discrimination diagram (fields after Pearce, 1980) for Avdella Melange basic rocks. Data from Jones (1990). B) MORB-normalised multi-element plots for Avdella Melange basic rocks. Refer to Table 1 for original data. Filled square = within plate basalt (specimen 45/89) from Tsouka (Fig. 1); Open square = transitional to MORB dyke (46/89) intruding the above basalt; Filled circle = MORB-type basalt (25/89) from the western Smolikas Mountains (NE of Padhes, Fig. 1); Open circle = transitional MORB/IAT-type basalt (86/89), depleted in Ti, Zr, Y and Nb relative to normal MORB (from E of Avdella, Fig. 1). Normalising values (partly after Pearce, 1980) Sr = 120 ppm, K_2O = 0.15%, Rb = 2 ppm, Ba = 20 ppm, Nb = 4 ppm, La = 4.5 ppm, Ce = 10 ppm, Nd = 8 ppm, P_2O = 0.12%, Zr = 90 ppm, TiO_2 = 1.5%, Y = 30 ppm, Sc = 40 ppm, Cr = 250 ppm.

Melange of extrusives of geochemically normal (N) MORB-type (Table 1; Fig. 3), dated as Late Triassic, based on conodonts from conformably overlying pelagic sediments (Kostopoulos, 1989).

Abundant high-Ti MORB-type sheeted dykes and extrusives (Table 1) are also found at the basal part of the preserved ophiolite slices of the Aspropotamos Complex (see below; Capedri et al., 1980; Kostopoulos, 1989), but their age is uncertain. These basal units could therefore represent either: i) Late Triassic MORB-type crust, which was subsequently intruded by the products of later subduction-related magmatism (see below); or ii) younger Jurassic crust formed immediately prior to, or during, subduction-influenced spreading.

The Pindos oceanic spreading ridge was probably well established by the Early Jurassic. Most of the oceanic crust presumed to be of this age has not been preserved (i.e. it was subducted). Spreading took place to the northeast of a passive margin, now represented by the present day Pindos Zone. The maximum size of the Pindos ocean basin is not known, but was perhaps in the order of 500–1000 km, comparable with modern day basins such as the Gulf of Aden (Cochran, 1981) or the southern Red Sea (Bonatti, 1985), where the extrusives are transitional between WPB and MORB (Girdler and Underwood, 1985).

Early-Middle Jurassic

a) Supra-Subduction Zone Spreading

Based on recent studies of the Oman ophiolite, some workers suggest that the features of ophiolites previously attributed to a supra-subduction zone origin (SSZ), could instead have been generated at a fast spreading mid ocean ridge (e.g. Nicolas et al., 1988). In the case of the Pindos ophiolite, Dupuy et al. (1984) suggested that partial melting of a rising mantle diapir could also explain the complex geochemistry, without the need for a subduction influence. However, much evidence has now accumulated that the Pindos ophiolite, at least, developed above a subduction zone (e.g. Capedri et al., 1981), as will now be discussed.

The Dramala Complex is considered to be of supra-subduction-type, mainly because of its lithologic nature as a harzburgite restite peridotite sequence, with abundant podiform dunite (Pearce et al., 1984a). However, in detail, the overlying layered crustal cumulate sequence shows a complex magmatic history, including cumulates typical only of MORB-type crust (e.g. plagioclase dunite-troctolite-anorthosite-gabbro cumulates; Kostopoulos, 1989). Other cumulate sequences are interlayered with wehrlites. This contrasts with the intrusive wehrlite bodies that are reported to cross-cut the layered sequence in Oman (Juteau et al., 1988). Also, chromite bodies are extremely rare within the Dramala Complex, in contrast to the harzburgitic Vourinos ophiolite (Rassios and Roberts, 1986).

Locally at the base of the Dramala Complex, at Smolikas Mountain (Fig. 1), breccias of harzburgite and jasper are cemented by ophicalcite, indicating exposure of ultramafic rocks at the ocean floor, as at rifted ridges or transform faults (Casey and Dewey, 1984), a point which could be relevant to metamorphic sole formation.

In summary, the Dramala Complex shows some evidence of both MORB and supra-subduction features, but the latter appears to dominate.

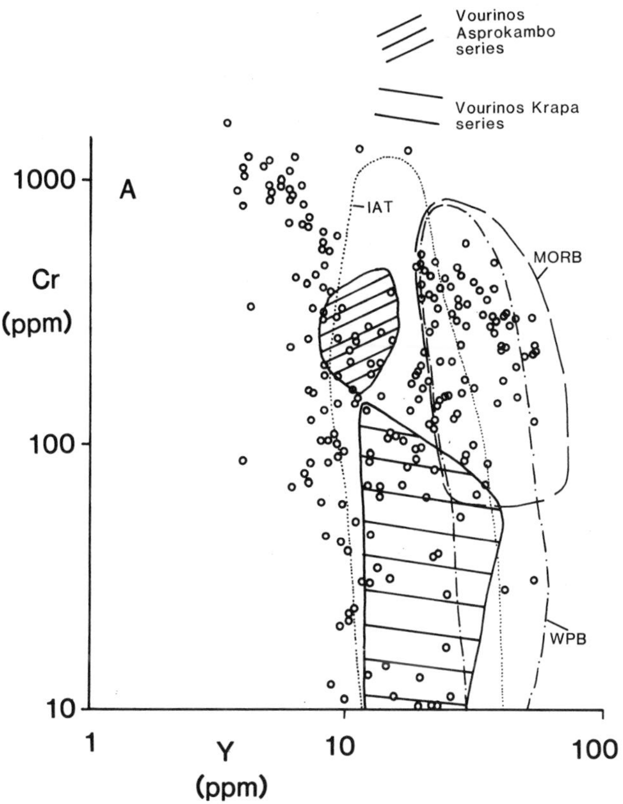

Figure 4. Cr-Y discrimination diagram (after Pearce, 1982) for the Aspropotamos Complex dykes and extrusive volcanic units (data from Jones, 1990, and Kostopoulos, 1989), with fields for the Vourinos intrusive (Asprokambo series) and extrusive (Krapa series) units superimposed (Vourinos data from Noiret et al., 1981; Beccaluva et al., 1984; this study). IAT = island arc tholeiite; MORB = mid ocean ridge basalt; WPB = within plate basalt.

The Aspropotamos Complex demonstrates the influence of subduction most clearly. A series of intrusive and volcanic events can be reconstructed, in which a general pattern of the progressive eruption of more high field strength (HFS) element-depleted, and more evolved volcanics can be recognized (N-MORB, MORB/IAT, IAT, Boninite Series Volcanics (BSV); Kostopoulos, 1989). Cr-Y and Zr/Y-Zr discrimination diagrams for the Aspropotamos Complex are presented in Figs. 4 and 5. Based on detailed geochemical work on the lower Aspropotamos section, the relationships shown in Fig. 6 were determined by Kostopoulos (op. cit.). For example, at this locality NE-trending IAT- and boninitic-type dykes are observed to cross-cut MORB and other more HFS element-depleted dykes (MORB-IAT and IAT; Kostopoulos, 1989). Also, at Avdella (Fig. 1), an isolated kilometer-thick succession, shows cumulate gabbros passing into sheeted dykes and extrusives,

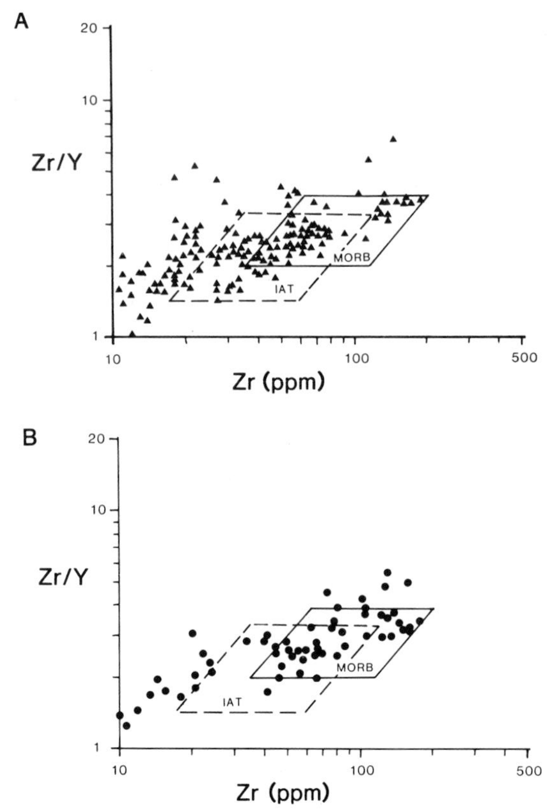

Figure 5. Zr/Y-Zr discrimination diagrams (after Pearce, 1980) for the Aspropotamos Complex dykes and extrusive volcanic units. A) Data from Kostopoulos, (1989) collected from the central and lower Aspropotamos River valley; B) Data from Jones, (1990), collected from localities elsewhere in the north Pindos Mountains (see Jones, 1990 for details)

which include pillow lavas and pillow breccias. All of the volcanic units and dykes display boninitic affinities at this locality (Table 1; Kostopoulos, 1989).

Some additional points from our study are as follows: southwest of the village of Spileo (Fig. 1), high level gabbros with anastomosing plagiogranite bodies pass conformably into extensive sheeted dykes and extrusives. These sheeted dykes trend NW-SE (Fig. 7), and are of MORB-type (Table 1). They are intruded by both parallel, and obliquely-trending dykes of boninite-type (Fig. 7). Near the village of Perivolaki (Fig. 1), high level gabbros with sheeted dykes of IAT- and boninite-type, pass locally upwards into boninitic pillow lavas, with abundant weathered glassy rims.

Plagiogranites occur within the Aspropotamos Complex in two known settings: the first, near Avdella, is associated entirely with HFS element-depleted extrusives and dykes, mainly boninites. These plagiogranites are relatively depleted in Zr (Table 1). Secondly, plagiogranites collected west of Spileo are associated with more heterogeneous assemblages (see above);

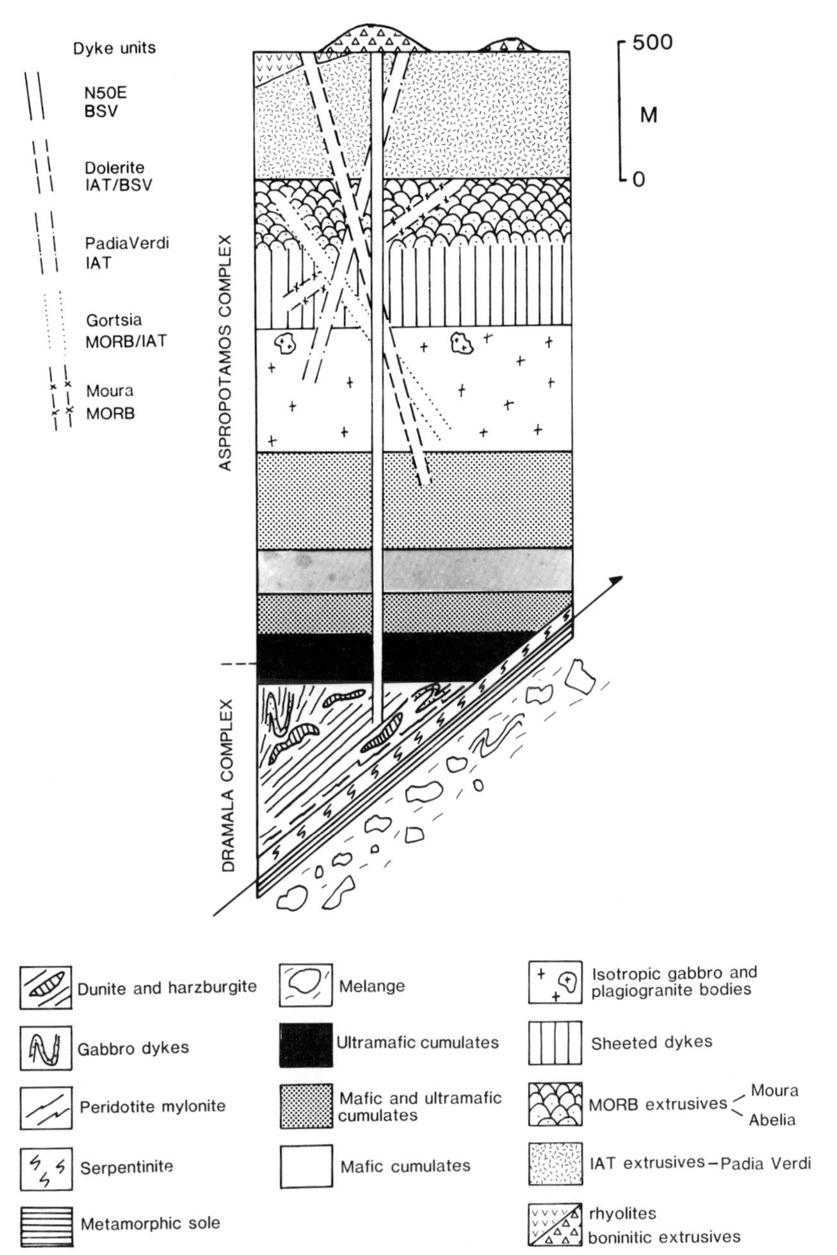

Figure 6. Summary reconstructed stratigraphy of the Pindos ophiolite (modified after Kostopoulos, 1989). Scale bar refers only to the crustal succession; the mantle sequence is approximately 5 km-thick at maximum development.

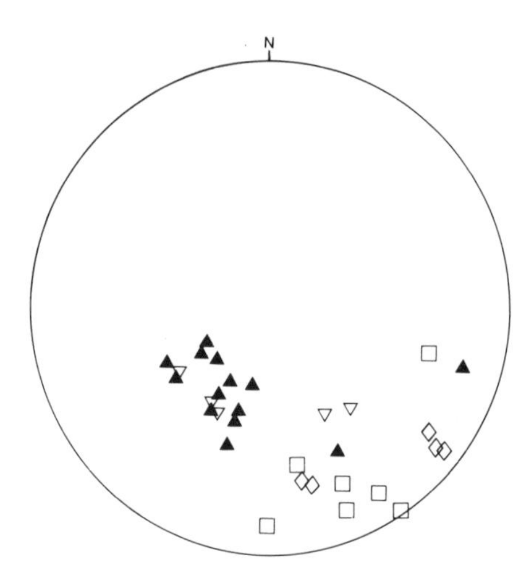

Figure 7. Lower hemisphere stereographic projection of dyke trends in the Aspropotamos Complex. Key: Closed triangles = sheeted MORB dykes (W of Spileo, Fig. 1); Inverted open triangles = individual and sheeted IAT and boninitic dykes (W of Spileo, Fig. 1); Open diamonds = sheeted dykes of IAT-boninitic type (from Perivolaki; Fig. 1); Open squares = sheeted IAT-boninitic dykes (from 2 km S of Perivolaki, Fig. 1).

these have higher Zr. In general, the plagiogranites all plot in the Volcanic Arc Granite field of Pearce et al. (1984b; Fig. 8). These magmas were presumably derived from basaltic parents, on average lower in HFS elements (e.g. Zr, Nb), and thus show arc-like affinities in evolved as well as primitive magmas. Three of the Pindos plagiogranites plot in the Vourinos field, further suggesting a close genetic relationship between these two ophiolites.

The Aspropotamos Complex has now been recognized in other areas of the Pindos Mountains (e.g. Vovousa, Armata, Dhistraton, Katara Pass, Fig. 1; Jones, 1990), proving that it forms a regionally extensive thrust sheet. These ophiolitic units usually comprise tectonized cumulates, intrusives and extrusives, all of which show comparable geochemical sequences to the type area, based on regional geochemical reconnaissance (Jones, 1990). However, not all chemical types are necessarily present at any one locality.

The volcanic stratigraphy of the Aspropotamos Complex (Fig. 6) shows general similarities to other Tethyan ophiolites, for example Oman (Alabaster et al., 1982), Troodos (Pearce, 1975; Murton, 1989) and Hatay (Piskin et al., 1986), in that all show more HFS element-depleted extrusives, including boninites, towards the top of the succession. However in detail, Troodos (and Hatay) comprise lower arc-like lavas, overlain by boninitic-type extrusives. In Oman, lavas of MORB-IAT transitional type (low Cr Geotimes Lavas) are overlain by arc-type extrusives (Lasail and Alley Lavas).

The most similar known modern day settings to the Pindos and Vourinos ophiolites relate to the initiation of the southwest Pacific volcanic arcs (e.g. Marianas), during the Eocene (Fujiaka et al., 1989). In principle, MORB-

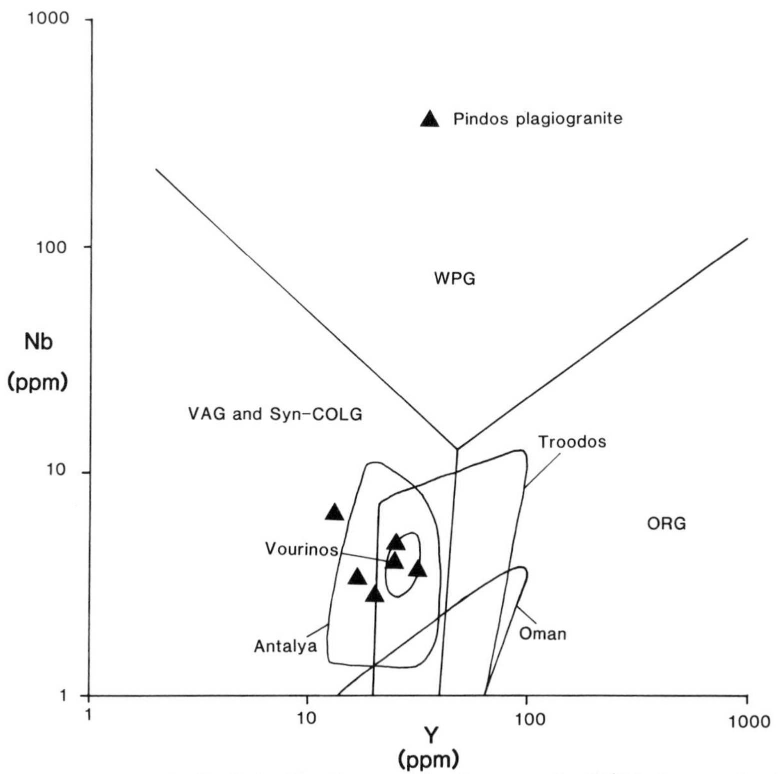

Figure 8. Nb-Y tectonic discrimination diagram (after Pearce et al., 1984b), for oceanic plagiogranites of the Pindos ophiolite. Syn-COLG = syn-collision granite; VAG = volcanic arc granite; ORG = ocean ridge granite; WPG = within plate granite. Fields for Troodos, Oman, Antalya and Vourinos after Pearce et al. (1984b).

type intrusives and extrusives could represent either old stranded remnants of earlier crust (i.e. ?Late Triassic-Early Jurassic), younger remnant crust formed immediately prior to the initiation of subduction (Early Jurassic), or related to later back-arc basin formation, where MORB-type crust is also found (e.g. Saunders and Tarney, 1979). In Pindos, the MORB-type lavas and extrusives appear to pre-date the boninitic extrusives, and thus formation in a back arc basin seems least likely. The MORB units are therefore more likely to represent either older crustal remnants, or perhaps the youngest crust formed before a gradual evolution to subduction influenced melting. The second option would seem to be suggested by the gradual evolution of the chemistry, and the complex interlayering of lithologies in the cumulate sequence (Kostopoulos, 1989).

b) Intra-Oceanic Displacement and Metamorphic Sole Formation

In the late Middle Jurassic (165 ± 3 Ma; Spray et al., 1984; Bathonian on the Harland et al., 1989 time scale), a high temperature metamorphic sole

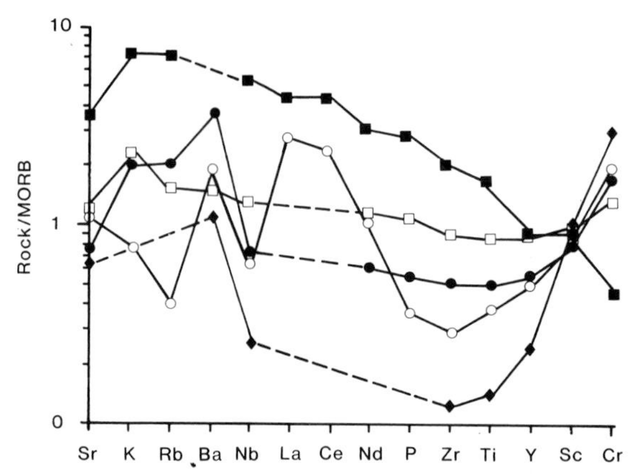

Figure 9. Representative MORB-normalised multi-element plots for the Loumnitsa Unit metamorphic sole rocks of the Pindos ophiolite. Normalising values as in Fig. 3. Analyses are presented in Table 1. Filled squares = Within plate-type greenschist facies metabasite (sample 40/89), eastern Smolikas Mountains (see Fig. 1); Open squares = MORB-type amphibolite (126/88), from east of Perivoli (Fig. 1); Filled circles = depleted MORB-type amphibolite (124/88, east of Perivoli); Open circles = IAT-type amphibolite (33/89), from Padhes (Fig. 1). Filled diamonds = boninitic-type amphibolite, east of Perivoli.

(Loumnitsa Unit) was formed at the base of the Pindos ophiolite (Whitechurch and Parrot, 1978; Spray and Roddick, 1980), similar to many other Tethyan ophiolites, including Othris and Vourinos (Woodcock and Robertson, 1977). The formation of a basal metamorphic sole is believed to represent the initial displacement of young, hot, ocean crust and mantle within an oceanic setting (e.g. Searle and Malpas, 1980; Spray, 1984), probably in response to regional compression. The metamorphic sole of the Pindos ophiolite contains lower amphibolite and greenschist facies metabasic and metasedimentary rocks, and is structurally underlain by unmetamorphosed units of the Avdella Melange.

A critical observation from the Pindos ophiolite is that the metamorphic sole is not cross-cut by the latest magmatic event (the boninitic dykes), which therefore must have formed prior to initial displacement. A few amphibolites analyzed from the sole have IAT and boninitic chemical signatures (Fig. 9; Table 1), and must therefore postdate subduction. Notably in the Pindos Mountains, the metamorphic sole rocks are found in original thrust contact, not only with peridotites, but also cumulates of *both* the Dramala *and* the Aspropotamos Complexes. South of Samarina (Fig. 1), well exposed examples of banded amphibolites and greenschists of the metamorphic sole are found in thrust contact with a 3 km-thick section of the Dramala Complex. The Dramala Complex is represented there by cumulate dunites and intrusive gabbros, with a thin (15 m) basal serpentinized unit, the sole being mainly composed of sheared metabasic rocks and pelagic sediments. Locally, greenschist-facies pillow basalts in the metamorphic sole have conformably overlying metalliferous-oxide sediments preserved.

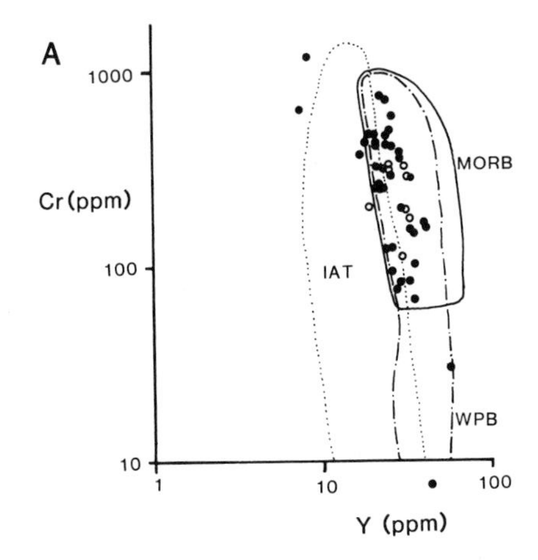

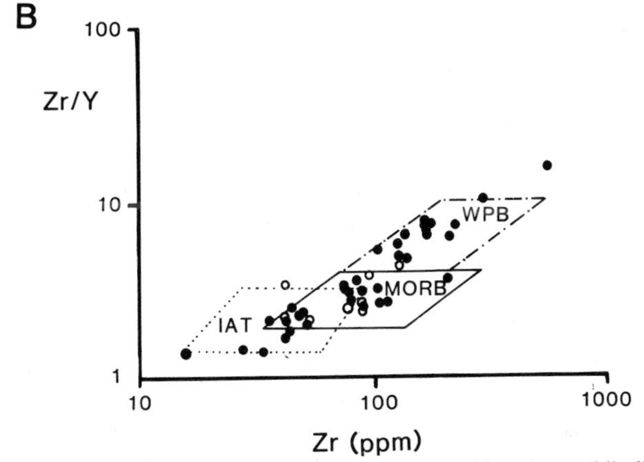

Figure 10. A) Cr-Y diagram for the Loumnitsa Unit metamorphic sole amphibolites and greenschists (closed circles), and Vourinos melange greenschist facies basalts (open circles). Key as for Fig. 4. B) Zr/Y-Zr diagram for the Loumnitsa Unit metamorphic sole amphibolites and Vourinos melange greenschist facies basalts. Key as for A.

Despite the occasional presence of IAT and boninitic-derived amphibolites described above, the metabasic lithologies of both the amphibolite and greenschist facies are mostly of WPB and MORB compositions (Fig. 10; Jones and Robertson, 1990). The structurally underlying Avdella Melange contains rocks of similar lithology and composition, which therefore form the obvious protoliths for most of the metamorphic sole (Jones, 1990). The amphibolites analyzed from the Oman sole (Searle and Malpas, 1982), show WPB, MORB and depleted MORB compositions.

The metamorphic sole occurs at the base of the Aspropotamos Complex at several localities, including west of Zakas village (Fig. 1). There, the protoliths are mainly sedimentary, and are found in thrust contact with cumulate plagioclase dunites, troctolites and gabbros (ca. 2 km thick). At this locality, late-stage boninitic dykes are seen to cross-cut the cumulate rocks, but not the metamorphic sole. In the lower Aspropotamos valley (SE of Avdella, Fig. 1), a thin amphibolitic sole unit is structurally overlain by a sequence of cumulates, intrusives and extrusives. Elsewhere, in the Venetikos valley (N of Krania, Fig. 1), amphibolite to greenschist facies metabasalts and metasediments appear to be welded onto high-level gabbros, sheeted dykes and even basalts, but further mapping of this complex tectonic contact is required to confirm if these are early metamorphic tectonic contacts, or the result of later thrusting.

Study of high-temperature folds within the Pindos sole (Loumnitsa Unit), has unfortunately not yielded conclusive data as to the direction of emplacement, due to Tertiary re-thrusting. We believe it was towards the present day northeast, based on the recognition of northeast verging mylonite structures in both the Dramala Complex and the Vourinos peridotites (Ross et al., 1980; Ross and Zimmerman, 1982; A. Rassios, pers. comm., 1989), and also considering the regional emplacement vectors for Vourinos and Othris (Smith and Woodcock, 1976; Wright, 1986).

The remnants of subsided off-margin highs, similar to those described from Alatopetra (see above), can also be recognized within the greenschist part of the metamorphic sole. At Padhes (Fig. 1), the Dramala Complex basal serpentinized peridotite passes structurally downwards into banded amphibolites, and then into a variety of greenschist facies metabasalts, marbles and metalliferous sediments (Fig. 11). The amphibolites analysed from this locality are of IAT-type, showing depletions of HFS elements and enrichment in light rare earths (Fig. 9; Table 1). Within the greenschist facies part of the sole, a coherent "seamount-type" sequence also occurs (Fig. 11). Structurally downwards this comprises deep-water cherts and shales, passing into nodular 'ammonitico rosso' pelagic carbonates, and then into a 300–m thick shallow-water carbonate sequence, of Early Jurassic age (I.G.M.E. Konitsa sheet, 1983). The carbonates are thick-bedded and, at the top of the sequence, contain replacement chert nodules.

Alternative settings for sole formation are: i) during the initiation of subduction, before supra-subduction zone (SSZ) spreading (Casey and Dewey, 1984); ii) during "displacement" following SSZ spreading, prior to tectonic emplacement (Searle and Malpas, 1980, 1982); iii) during ocean basin closure, unrelated to subduction (e.g. Oman: Nicolas and Le Pichon, 1980). From the above data, it is clear that model ii); i.e. post-subduction sole formation, applies best to the Pindos ophiolite.

Comparison with the Vourinos Ophiolite

We believe the Vourinos and Pindos ophiolites originated as essentially one area of oceanic crust and mantle (Jones and Robertson, 1990), that is now continuous beneath the Tertiary Meso-Hellenic molasse basin. However,

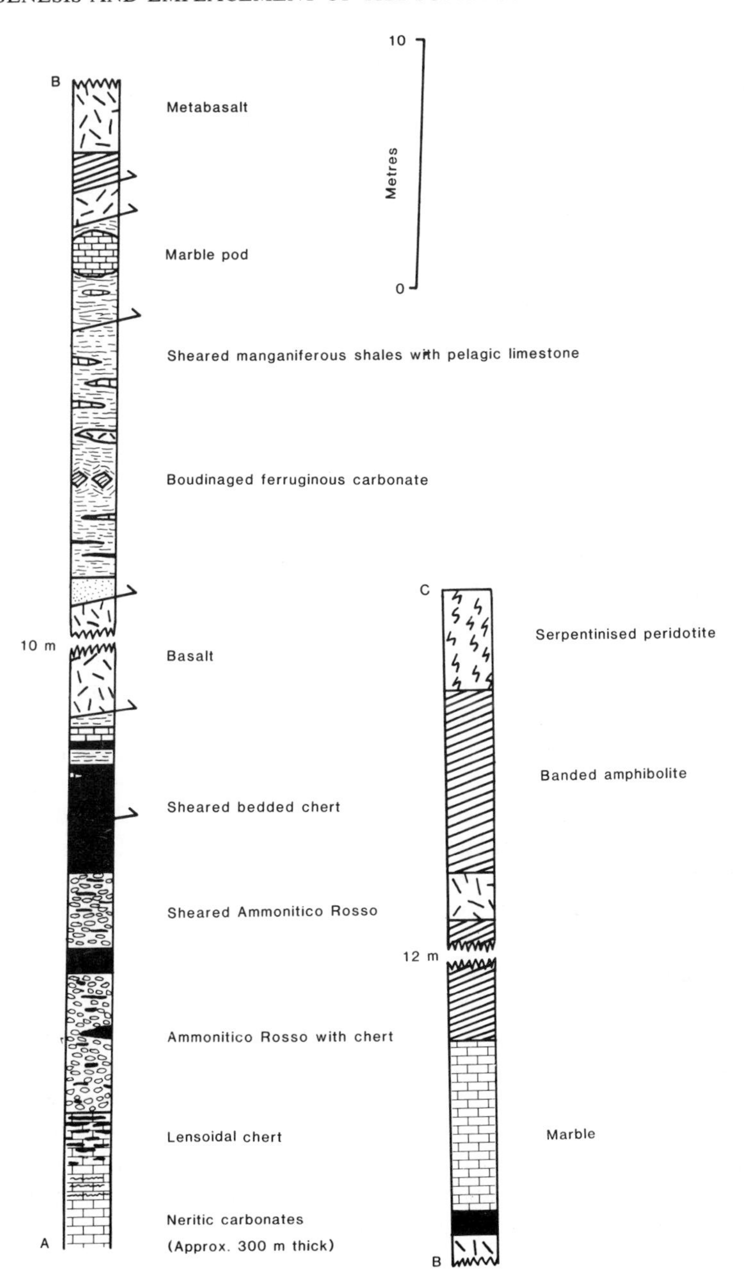

Figure 11. Measured section through the metamorphic sole section observed at Padhes village. See text for discussion.

since there is no continuous outcrop, the possibility clearly exists that Pindos and Vourinos could have originated in two separate supra-subduction zone settings within the Pindos ocean. However, we assume an origin as remnants of the same supra-subduction zone slab, as it is the simplest hypothesis.

The Vourinos ophiolite is a harzburgite-dunite dominated peridotite, with a partly preserved overlying crustal succession (Moores, 1969; Rassios et al., 1983). Many authors have argued in favour of a spreading ridge origin. Dunite bodies present within the harzburgite are either deformed into the dominant peridotite foliation, or cross-cut the foliation and therefore undeformed (Harkins et al., 1980). Orbicular, layered and disseminated chromite ores are found within the undeformed dunite pods, and elongate oribicular, boudinaged and folded chromite is found in the deformed pods (Harkins et al., 1980). The dunites were interpreted as originating as magmatic segregations related to multiple intrusion at a constructive margin, as has been inferred for the Masquad area of the Oman ophiolite (Nicolas et al., 1988), based on similar evidence. Structural studies of olivine fabrics within the peridotites, together with pyroxene thermobarometry and paleopiezometry, have further led to interpretation of Vourinos as the remnant of a mantle diapir and overlying magma chamber, located at a spreading ridge (Ross et al., 1980). Peridotite foliations (formed at an estimated 60 km depth) are transposed towards the horizontal in a zone of rotation within the peridotite, as seen also in the Semail peridotites of Oman (Ceuleneer et al., 1988). The conformably overlying crustal succession contains cyclic cumulates, considered to have originated by magmatic crystallization in a large magma chamber located at a mid-ocean ridge (Jackson et al., 1975). However, we believe these spreading-related features could equally well have developed above an active subduction zone.

In this context, the locally preserved extrusives (Krapa series) within the Vourinos ophiolite, are entirely of IAT geochemical type, and are cross-cut by transitional to boninitic-type intrusive dykes (Asprokambo series; Beccaluva et al., 1984), similar to those found within the Aspropotamos Complex. Isotopic and trace element geochemical studies of the Vourinos ophiolite, led Noiret et al. (1981) to postulate an origin as an "island arc" ophiolite. On the Cr-Y geochemical discrimination diagram presented (Fig. 4), the Vourinos volcanics and dykes are shown to correlate with the more HFS element-depleted rocks of the Pindos suite, although the Pindos boninitic rocks show the greatest overall depletion. The intrusion of the Asprokambo dykes into the Krapa series volcanics is significant, as it provides independent evidence for the timing of the intrusion of the boninitic magmas, already suggested to be a late stage event in the Aspropotamos Complex. Further south, in the Koziakas area, a boninitic dyke has also been discovered intruding harzburgite, further evidence of late-stage occurrence (Capedri et al., 1985).

An important point from Vourinos, is that the earliest displacement related fabrics developed whilst the peridotite was still hot (770–823°; Ross et al., 1980). This subsequent mylonitic deformation was superimposed on the earlier spreading-related fabric (1200–900°; Ross et al., op. cit.). These mylonites formed at an estimated 30 km depth. The sense of overthrusting determined by shear sense within olivine crystals is towards the N-NNE (Ross et al., op. cit.). A similar relationship was inferred for the Dramala Complex (Ross and Zimmerman, 1982; A. Rassios, pers. comm., 1989)

In summary, the accumulated evidence favors an interpretation of the Vourinos ophiolite as representing crust formed at a supra-subduction zone spreading axis, in an oceanic setting, which produced magmas of only IAT and boninitic affinities. Subsequently, the ophiolite was thrust towards the present N-NNE, whilst still within the oceanic environment, producing a metamorphic sole similar to that of the Dramala Complex. The young, hot, Vourinos ophiolite was then emplaced northeastwards, metamorphosing the underlying Pelagonian passive margin sediments to greenschist facies.

Tectonic Setting for the Genesis and Emplacement of the Pindos and Vourinos Ophiolites

Following regional rifting and break-up during the Mid-Late Triassic, the Pindos basin developed oceanic crust during the Latest Triassic and Early Jurassic. The crust formed during these extensional stages is now preserved within the Avdella Melange, and probably also within the basal MORB-type sections of the Aspropotamos Complex. Subduction was then initiated, possibly near the spreading ridge (Dixon and Robertson, 1985), or perhaps at a ridge-transform intersection (Casey and Dewey, 1984).

In our preferred model, the Dramala Complex, the Aspropotamos Complex and the Vourinos ophiolite formed part of an originally continuous supra-subduction zone slab. The Dramala Complex formed the most westerly part of the slab, the Vourinos ophiolite the central part, and the Aspropotamos Complex the eastern part, nearest the subduction zone (Fig. 12). When the Pelagonian passive margin collided with the trench, the hot supra-subduction zone slab was first disrupted, generating the metamorphic sole. Then the Dramala Complex and Vourinos ophiolite combined, over-rode the Aspropotamos Complex, and went on to be finally emplaced over the Pelagonian margin.

In justification of this model, a number of points can be made. The Pindos ophiolitic units are inferred to represent mainly mantle (Dramala Complex), and crustal (Aspropotamos Complex) sections, formed by supra-subduction zone spreading (Fig. 12). In common with many Tethyan ophiolites, development was probably by a "pre-arc" spreading process (Pearce et al., 1984a), without the development of any true volcanic arc (e.g. without arc-derived volcaniclastic sediments). The Dramala Complex possibly represents early supra-subduction spreading, since it also contains MORB-type crustal units. By the time the Vourinos ophiolite was forming, the Dramala Complex

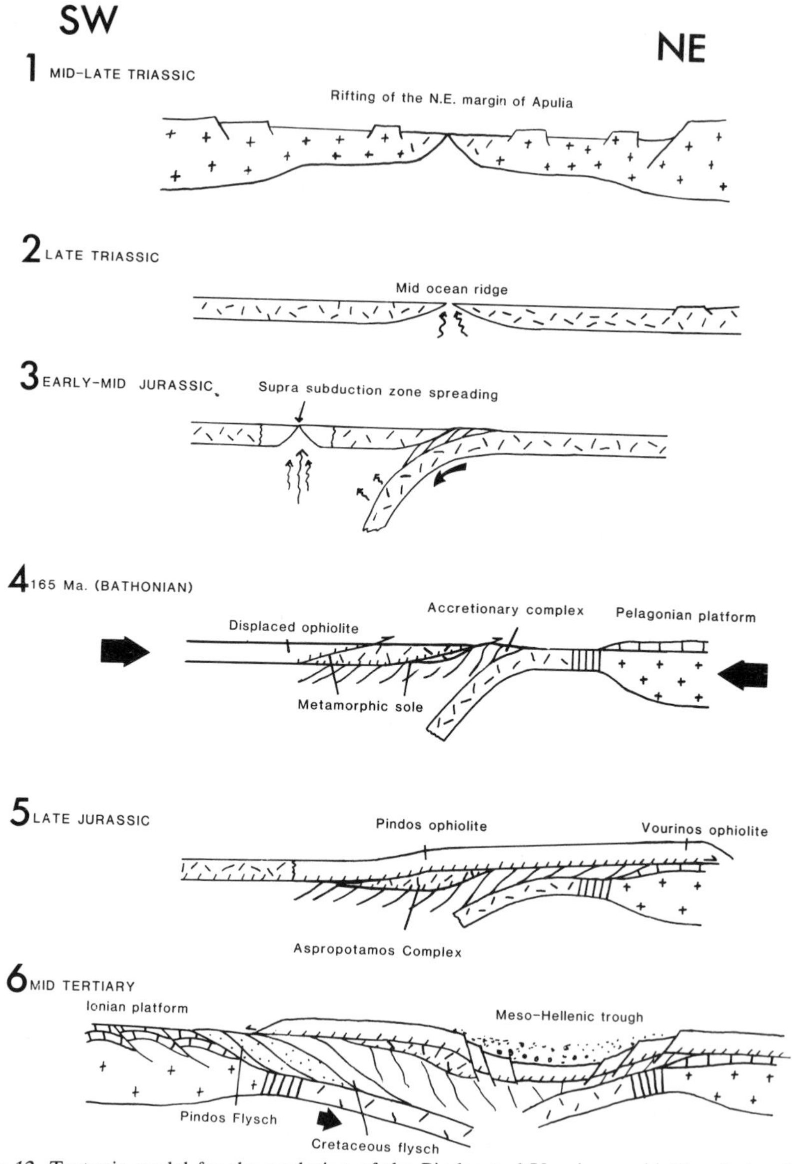

Figure 12. Tectonic model for the evolution of the Pindos and Vourinos ophiolites during the Mesozoic and Tertiary. See text for details.

mantle and crust had already spread away some distance (i.e. westwards of Vourinos).

The Pindos mantle sequence lacks abundant massive chromites: those present are highly elongate and thin, as are the dunite bodies. One explanation is that the mantle flow away from the ridge had the effect of streaking out any original chromite diapiric pods, as envisaged by Culeneer et al. (1988) for Oman. By contrast, the Vourinos peridotites and crustal sequences contain abundant podiform chromite, some of which is undeformed. This is explicable if the spreading axis was still active immediately prior to initial displacement. The Vourinos ophiolite is interpreted to represent a lateral (i.e. eastwards) equivalent of the Dramala Complex, located nearer to, or at, a spreading center above the subduction-zone at the time of emplacement (Jones and Robertson, 1990).

In the model, the Aspropotamos Complex preserves the crust formed nearest the subduction zone. Any more distal crust, (i.e. the "feather edge") was subducted, or perhaps accreted into the metamorphic sole. The original crustal thickness of the Aspropotamos Complex appears to have been anomalously thin (approx. 2 km; Kostopoulos, 1989), consistent with this setting. The MORB dykes are potentially rifted remnants of the former spreading axis, which was apparently orientated northwest-southeast, assuming no subsequent rotation. From the trends of the boninite dykes, we infer that the trend of supra-subduction zone spreading was both parallel, and oblique to, the earlier extension direction.

Possible tectonic scenarios for boninite development are as follows: i) remelting of a previously depleted mantle source during "roll back" of a subducting slab (Smith and Spray, 1984); ii) generation at various stages during supra-subduction zone crust development; iii) generation at a late stage by the "roll-forward" and disruption of the "fore-arc slab" during the initial stages of intra-oceanic displacement (i.e. at the trench and/or along transform faults). Further work is required to establish which, if any, of these processes could apply to the Aspropotamos Complex. However, early-stage genesis seems unlikely, since the boninites appear to mainly occur late in the volcanic sequence.

The Avdella Melange formed by subduction-accretion processes, as an accumulation of oceanic sediments and crust scraped from the descending slab. These accretionary processes, and subsequent Tertiary re-thrusting, were responsible for the pervasive tectonic deformation and mixing of the Avdella Melange.

The metamorphic sole (Loumnitsa Unit) formed during the earliest stages of displacement within the ocean, after supra-subduction zone spreading effectively had ended. The general WPB and MORB nature of the Pindos sole, suggests that initial detachment took place some distance from the locus of SSZ spreading, possibly near the boundary between the old, mainly MORB-type crust and newly created SSZ crust (Lippard et al., 1986). Subsequent foreland-directed thrusting occurred towards and/or at the trench, accreting basalts and sediments to the base of the Aspropotamos Complex, as proposed by Whitechurch and Parrot (1978). However, detachment along ocean fracture-zone-like segments is also hinted at by the occurrence of cemented ultramafic breccias. Metamorphic rocks could be welded onto cumulate sequences either at transform faults, and/or near the trench (Fig.

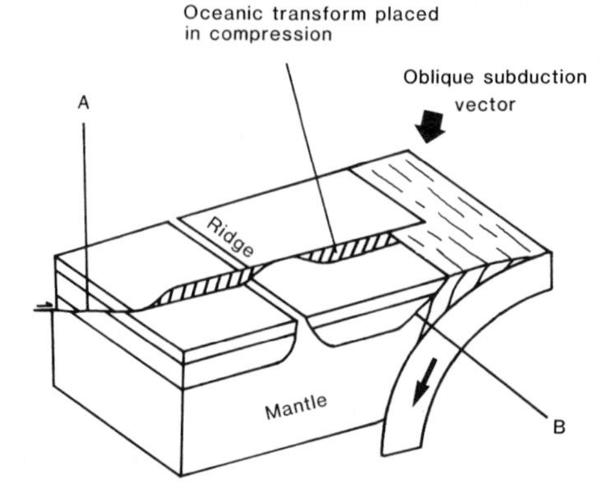

Figure 13. Block diagram to show two possible oceanic environments for the formation of metamorphic soles on ophiolitic cumulates. A) Oceanic transform placed in compression, causing local overthrusting of the crustal succession. B) Thrusting occurs along a basal detachment (basal thrust of the subduction zone?), cutting up-section through the crustal sequence during initial intra-oceanic detachment (e.g. Nicolas and Le Pichon, 1980).

13). In summary, the metamorphic sole preserves a record of at least two major intra-oceanic thrusts cutting up-section through oceanic mantle and crust: firstly during the early stages of Jurassic emplacement (i.e. initial break along MORB/SSZ join); and secondly, towards the trench.

The driving force in metamorphic sole formation and intra-oceanic thrusting, appears to have been the collision of the subduction trench with the Pelagonian microcontinental passive margin to the east. A similar mechanism of emplacement is inferred for Oman (Lippard et al., 1986). During convergence, the leading edge, represented by the Aspropotamos Complex, was detached and overridden by the young, hot, crust and mantle, represented by the Vourinos ophiolite.

The Pindos ocean was sutured further north in Albania and Yugoslavia (Karamata, 1988) during the Late Jurassic collisional event, and subduction ceased in northern Greece. As a result, the Dramala Complex and the Aspropotamos Complex remained stranded in the eastern part of the remnant Pindos ocean during the Cretaceous and Early Tertiary. These units were then emplaced onto Apulia during Early Tertiary convergence, and final closure of the remnant Pindos ocean. This two phase emplacement (Jurassic and Early Tertiary) has contributed to the preservation of the Aspropotamos Complex fore-arc sliver, unlike the comparable section in Oman.

Conclusions

The north Pindos Mountains preserve the complete record of the genesis, initial displacement and two-stage emplacement of oceanic mantle, crust and sediments formed within the Mesozoic Pindos ocean basin. The Triassic rift and break-up history is preserved in sequences found within an accretionary complex (Avdella Melange). After genesis of a ca. 1000 km-wide ocean, intra-oceanic subduction was initiated during the Early to Mid Jurassic, leading to supra-subduction zone spreading, which created the Pindos (Dramala and Aspropotamos Complexes), and the Vourinos ophiolite. Boninite-type magmatism took place in an inferred "fore-arc" setting. Due to the onset of regional compression during the Late Mid Jurassic (ca. 165 Ma), the supra-subduction zone crust was displaced within the ocean, and as a result, a basal metamorphic sole was accreted to the hot overriding slab. This thrusting may have taken place at the boundary between old and new lithosphere within the overriding plate, and/or near the trench. Further compression resulted in the hot and thin "fore-arc" crust (Aspropotamos Complex) being completely overridden by the remainder of the ophiolite, which was subsequently emplaced as a relatively hot slab onto the Pelagonian microcontinental margin to the east. This crust is now represented by the Vourinos ophiolite, whilst the trailing edge is represented by the Dramala Complex (Pindos ophiolite). During the Early Tertiary, eastward-dipping subduction was initiated, leading to collision of the Pindos ophiolitic units and accretionary complex with passive margin and platformal sequences of the Apulian passive margin to the southwest. The "fore-arc" crust with its attached metamorphic sole (Aspropotamos Complex), was then exhumed as thin thrust slices beneath the main peridotite. Presently, no more complete record of a supra-subduction zone ophiolite is known anywhere.

Acknowledgements

The authors would like to thank the University of Edinburgh (AHFR), the Royal Society (GJ and JRC) and the NERC (GJ) for financial support. Helpful discussions were had with Ann Rassios and John Dixon.

References

Alabaster, T., Pearce, J.A. and Malpas, J., 1982. The volcanic stratigraphy and petrogenesis of the Oman ophiolite complex. Contributions to Mineralogy and Petrology, 81: 168–183.

Aubouin, J., Bonneau, M., Celet, P., Charvet, J., Clement, B., Degardin, J.M., Dercourt, J., Ferriere, J., Fleury, J.J., Guernet, C., Maillot, H., Mania, J.H., Mansy, J.L., Terry, J., Thiebault, P., Tsoflias, P. and Verriex, J.J., 1970. Contribution à la géologie des Hellénides: le Gavrovo, le Pinde et la zone ophiolitique subpélagonienne. Annales Société géologique du Nord, 90: 277–306.

Beccaluva, L., Ohnenstetter, D., Ohnenstetter, M. and Paupy, A., 1984. Two magmatic series with island arc affinities within the Vourinos Ophiolite. Contributions to Mineralogy and Petrology, 85: 253–271.

Bonatti, E., 1985. Punctiform initiation of seafloor spreading in the Red Sea, during transition from a continental to an oceanic rift. Nature, 316: 33–37.

Brunn, J.H., 1956. Contribution a l'étude géologique du Pinde septentrional et d'une partie de la Macédonie occidentale. Annales géologique des pays hélléniques, 7: 358 p.
Capedri, S., Venturelli, G., Bocchi, G., Dostal, J., Garuti, G. and Rossi, A., 1980. The geochemistry and petrogenesis of an ophiolitic sequence from Pindos, Greece. Contributions to Mineralogy and Petrology, 74: 189–200.
Capedri, S., Venturelli, G., Bebien, J. and Toscani, L., 1981. Low and high Ti ophiolites in northern Pindos: petrological and geological constraints. Bulletin Volcanologique, 44 (3): 439–449.
Capedri, S., Venturelli, G. and Toscani, L., 1982. Petrology of an ophiolitic cumulate sequence from Pindos, Greece. Journal of Geology, 17: 223–242.
Capedri, S., Lekkas, E., Papanikolaou, D., Skarpelis, N., Venturelli, G. and Gallo, F., 1985. The ophiolite of the Koziakas Range, Western Thessaly, (Greece). Nues Jahrbuch Mineralogie, 152 (1): 45–64.
Casey, J.F. and Dewey, J.F., 1984. Initiation of subduction zones along transforms and accreting plate boundaries, triple junction evolution, and fore-arc spreading centres-implications for ophiolitic geology and obduction. In: Gass, I.G., Lippard, S.J. and Shelton, A.W. (Eds), Ophiolites and oceanic lithosphere. Geological Society of London Special Publication, 13: 269–290.
Cochran, J.R., 1981. The Gulf of Aden: structure of a young ocean basin and continental margin. Journal of Geophysical Research, 86: 263–288.
Céuleneer, G., Nicolas, A. and Boudier, F., 1988. Mantle flow patterns at an oceanic spreading centre: the Oman peridotites record. Tectonophysics, 151: 1–26.
Dixon, J.E. and Robertson, A.H.F., 1985. Asymmetrical ridge-collapse model for genesis of arc-type magmas in Ophiolites. Terra Cognita, 5: 127.
Dupuy, C., Dostal, J., Capedri, S. and Venturelli, G., 1984. Geochemistry and petrogenesis of ophiolites from northern Pindos (Greece). Bulletin Volcanologique, 47(1): 39–46.
Ferriere, J., 1982. Palaéogeographie et tectoniques superposées dans les Héllenides internes: les massifs de l'Othrys et du Pelion (Grèce continental). Société de la Géologie de Nord, Lille Publication, 8: 1–970.
Fujiaka, A., and leg 126 shipboard scientific party, 1989. Arc volcanism and rifting. Nature, 342: 18–20.
Girdler, R.W. and Underwood, M., 1985. The evolution of early oceanic lithosphere in the Red Sea. Tectonophysics, 116, 95–108.
Green, T.J., 1983. The sedimentology and structure of the Pindos zone in southern mainland Greece. Unpublished Ph.D thesis: University of Cambridge.
Harkins, M.E., Green, H.W. II, and Moores, E.M., 1980. Multiple intrusive events documented from the Vourinos ophiolite complex, northern Greece. American Journal of Science, 280(A): 284–295.
Harland, W.B., Armstrong, R.L., Cox, A.V., Craig, L.E., Smith, A.G. and Smith, D.G., 1989. A Geologic time scale. Cambridge University Press.
Hynes, A.J., 1974. Igneous activity at the birth of an ocean basin in Eastern Greece. Canadian Journal of Earth Sciences, 11: 842–853.
Institute of Geological and Mineral Exploration, 1987. Konitsa sheet, 1:50,000 scale. Athens.
Jackson, E.D., Green, H.W. and Moores, E.M., 1975. The Vourinos ophiolite, Greece: Cyclic units of lineated cumulates overlying harzburgite tectonite. Geological Society of America Bulletin, 86: 390–398.
Jacobshagen, V., St. Durr, F., Kockel, K.O. and Kowalczyk, G., 1978. Structure and geodynamic evolution of the Aegean region. In: Closs, J., Roeder, D. and Schmidt, K. (Eds), Alps, Appenines, Hellenides. Schweizerbart, Stuttgart, pp. 537–564.
Jones, G., 1990. Tectono-stratigraphy and evolution of the Pindos ophiolite and associated units, northwest Greece. Unpublished Ph.D thesis, University of Edinburgh, U.K. 397 p.
Jones, G., and Robertson, A.H.F., 1990. Tectono-stratigraphy and evolution of the Pindos ophiolite and related units, northwest Greece. Journal of the Geological Society of London, 148(2), 267–268.

Juteau, T. Ernewein, M., Reuber, I., Whitechurch, H. and Dahl, R., 1988. Duality of magmatism in the plutonic sequence of the Sumail nappe, Oman. Tectonophysics, 151: 107–135.

Karamata, S., 1988. The "Diabase-Chert Formation". Some genetic aspects. Bulletin T XCV de l'Academie Serbe des Science et des Arts (Sciences naturelles), 28: 1–11.

Kemp, A.E.S., and McCaig, A., 1984. Origins and significance of rocks in an imbricate thrust zone beneath the Pindos ophiolite, northwestern Greece. In: Robertson, A.H.F. and Dixon, J.E. (Eds), The Geological Evolution of the Eastern Mediterranean. Geological Society of London, Special Publication, 17: 569–580.

Kostopoulos, D., 1989. Geochemistry and tectonic setting of the Pindos ophiolite, northwestern Greece. Unpublished PhD thesis, University of Newcastle-upon-Tyne, U.K.

Lekkas, E., 1988. Geological structure and geodynamic evolution of the Koziakas Mountain Range, Western Thessaly. University of Athens Geological Monographs, 1, Athens. 281 pp.

Lippard, S.J., Shelton, A.W., and Gass, I.G., 1986. The ophiolite of northern Oman. Geological Society of London, Memoir 11.

Mavrides, A., Skourtsis-Coroneou, V. and Tsalia-Monopolis, S., 1979. Contribution to the Geology of the Subpelagonian Zone (Vourinos area, West Macedonia). 6th Colloquium on the Geology of the Aegean Region. Institute of Geological and Mineral Exploration, Athens, pp. 175–195.

Montigny, R., Bougault, H., Bottinga, Y. and Allegre, C.J., 1973. Trace element geochemistry and genesis of the Pindos Ophiolite suite. Geochemica et Cosmochemica Acta, 37: 2135–2147.

Moores, E.M., 1969. Petrology and structure of the Vourinos Ophiolitic complex of Northern Greece. Geological Society of America Special Paper, 118, 74 pp.

Mountrakis, D., 1984. Evolution of the Pelagonian Zone in N.W. Macedonia. In: Dixon, J.E. and Robertson, A.H.F. The Geological Evolution of the Eastern Mediterranean. Geological Society of London Special Publication, 17: 581–590.

Mountrakis, D., 1986. The Pelagonian Zone in Greece: a polyphase deformed fragment of the Cimmerian continent and its role in the geotectonic evolution of the eastern Mediterranean. Journal of Geology, 94: 335–347.

Murton, B.J., 1989. Tectonic controls on boninite genesis. In: Saunders, A.D., and Norry, M.J. (Eds), Magmatism in the ocean basins. Geological Society of London Special Publication, 42: 347–377.

Naylor, M.A. and Harle, T.J., 1976. Palaeogeographic significance of rocks and structures beneath the Vourinos Ophiolite, northern Greece. Journal of the Geological Society of London, 132: 667–676.

Nicolas, A. and Le Pichon, X., 1980. Thrusting of young lithosphere in subduction zones with special reference to structures in ophiolitic peridotites. Earth and Planetary Science Letters, 46: 397–406.

Nicolas, A., Ceuleneer, G., Boudier, F. and Misseri, M., 1988. Structural mapping in the Oman ophiolites: mantle diapirism along an oceanic ridge. Tectonophysics 151: 27–56.

Noiret, G., Montigny, R. and Allegre, C.J., 1981. Is the Vourinos Complex an island arc ophiolite? Earth and Planetary Science Letters, 56: 375–386.

Parrot, J.F., 1967. Le cortege ophiolithique du Pinde Septentrional (Grèce). Cahiers ORSTOM Séries Géologie, 114 p.

Parrot, J.F., 1969. Etude d'une coupe de réference dans le cortège ophiolithique du Pinde Septentrional: la valee de l'Aspropotamos. Cahiers ORSTOM Séries Géologie, 1, 35–59.

Parrot, J.F., and Verdoni, P-A., 1976. Conditions de formations de deux assemblages ophiolitiques Mediterraneens (Pinde et Hatay) d'apres l'etude des mineraux constitutifs. Cahiers ORSTOM, Séries Géologie, 8(1): 69–94.

Pearce, J.A., 1975. Basalt geochemistry used to investigate past tectonic environments on Cyprus. Tectonophysics, 25: 41–67.

Pearce, J.A., 1980. Geochemical evidence for the genesis and eruptive setting of lavas from Tethyan ophiolites. In: Panayiotou, A. (Ed), Ophiolites; proceedings of the International Ophiolites Symposium, Cyprus, 1979, pp. 261–272.

Pearce, J.A., 1982. Trace element characteristics of lavas from destructive plate boundaries. In: Andesites (Ed R.S. Thorpe), pp. 525–547. Chichester: J. Wiley and Sons.

Pearce, J.A., Lippard, S.J. and Roberts, S., 1984a. Characteristics and tectonic significance of supra-subduction zone ophiolites. In: Kokelaar, B.P. and Howells, M.F. (Eds), Marginal Basin Geology. Geological Society of London Special Publication, 16: 77–89.

Pearce, J.A., Harris, N.B.W. and Tindle, A.G., 1984b. Trace element discrimination diagrams for the tectonic interpretation of granitic rocks. Journal of Petrology, 25 (4): 956–983.

Pe-Piper, G., 1982. Geochemistry, tectonic setting and metamorphism of mid-Triassic volcanic rocks of Greece. Tectonophysics, 85: 253–272.

Piskin, O., Delaloye, M., Selcuk, H. and Wagner, J-J., 1986. A guide to Hatay Geology. Ofioliti, 11: 87–104.

Rassios, A., Beccaluva, L., Bortolotti, V., Mavrides, A. and Moores, E.M., 1983. The Vourinos Ophiolitic complex. Ofioliti, 8 (3): 275–292.

Rassios, A., and Roberts, S., 1986. Results of the chromite research program, Vourinos, Greece (1984–1986): a synthesis. In: The application of a multidisciplinary concept for chromite exploration in the Vourinos Complex, northern Greece. European Economic Community Report, part 1: 1–96.

Ross, J.V., Mercier, J-C. C., Ave Lallement, H.G., Carter, N.L. and Zimmerman, J., 1980. The Vourinos ophiolite complex, Greece: the tectonite suite. Tectonophysics, 70: 63–83.

Ross, J.V., and Zimmerman, J., 1982. The Pindos Ophiolite complex, Northern Greece: evolution and significance of the tectonic suite. Abstract in: Dixon, J.E. and Robertson, A.H.F. (Eds), 1984. The Geological Evolution of the Eastern Mediterranean. Conference Abstracts, Edinburgh, 1982, p. 95.

Saunders, A.D. and Tarney, J., 1979. The geochemistry of basalts from a back-arc spreading centre in the Scotia Sea. Geochemica et Cosmochemica Acta, 43: 555–572.

Searle, M.P. and Malpas, J., 1980. Structure and metamorphism of rocks beneath the Semail Ophiolite of northern Oman and their significance in ophiolite obduction. Transactions of the Royal Society of Edinburgh (Earth Sciences), 71: 247–262.

Searle, M.P., and Malpas, J., 1982. Petrochemistry and origin and sub-ophiolite metamorphic and related rocks in the Oman Mountains. Journal of the Geological Society of London, 139, 5–24.

Smith, A.G., 1979. Othris, Pindos and Vourinos Ophiolites and the Pelagonian Zone. 6th Colloquium on the Geology of the Aegean Region. Institute of Geological and Mineralogical Exploration, Athens, pp. 1369–1373.

Smith, A.G., and Woodcock, N.H., 1976. The earliest Mesozoic structures in the Othris region, Eastern Central Greece. Bulletin de la Société Géologique de France, 18: 245–251.

Smith, A.G., Woodcock, N.H. and Naylor, M.A., 1979. The structural evolution of a Mesozoic continental margin, Othris Mountains, Greece. Journal of the Geological Society of London, 146: 589–603.

Smith, A.G., and Spray, J.G., 1984. A half-ridge transform model for the Hellenic-Dinaric ophiolites. In: Dixon, J.E. and Robertson, A.H.F. The Geological Evolution of the Eastern Mediterranean. Geological Society of London Special Publication, 17: 629–644.

Spray, J.G., 1984. Possible causes and consequences of upper mantle decoupling and ophiolite displacement. In: Dixon, J.E. and Robertson, A.H.F. The Geological Evolution of the Eastern Mediterranean. Geological Society of London Special Publication, 17: 255–269.

Spray, J.G., and Roddick, J.C., 1980. Petrology and $^{40}Ar/^{39}Ar$ geochronology of some Hellenic sub-ophiolite metamorphic rocks. Contributions to Mineralogy and Petrology, 72: 43–55.

Spray, J.G., Bebien, J., Rex, D.C. and Roddick, J.C., 1984. Age constraints on the igneous and metamorphic evolution of the Hellenic-Dinaric ophiolites. In: Dixon, J.E. and Robertson, A.H.F. The Geological Evolution of the Eastern Mediterranean. Geological Society of London Special Publication, 17: 619–628.

Terry, J., 1971. Sur l'age Triassique de laves associées a la nappe ophiolitique du Pinde septentrional (Epire et Macedonie, Grèce). Comptes rendus sommaires de la Société géologique de France, pp. 384–385.

Terry, J., 1974. Ensembles lithologiques et structures internes du cortège ophiolitique du Pinde séptentrionale (Grèce), construction d'un model pétrogenetique. Bulletin de la Société Géologique de France, 16: 204–213.

Terry, J., 1975. Echo d'une téctonique jurassique: les phenomènes de résedimentation dans le secteur de la nappe des ophiolites du Pinde septentrional (Grèce). Comptes rendus sommaires de la Societé géologique de France, pp. 49–51.

Terry, J., 1979. Distinction géochemique de plusieurs groupes dans les ensembles volcaniques de la nappe ophiolitique du Pinde septentrional (Grèce). Bulletin de la Société Géologique de France, 21(6): 727–735.

Terry, J., and Mercier, M., 1971. Sur l'éxistence d'une serie détritique bèirrasienne intercalée entre la nappe des ophiolites et le flysch Eocene de la nappe du Pinde (Pinde septentrional, Grèce). Comptes rendus sommaires de la Société géologique de France, pp. 71–73.

Valsami, E., 1990. Hydrothermal alteration processes in the Pindos ophiolite, northwest Greece. Unpublished Ph.D thesis, University of Newcastle-Upon-Tyne, U.K.

Vergely, P., 1976. Origine "vardarienne" chevauchement vers l'ouest et rétrocharriage vers l'est des ophiolites de Macedonie (Grèce) au cours du Jurassique supérieur-Eocrétace. Comptes Rendus de la Academie de Science, Paris, D280: 1063–1066.

Vergely, P., 1977. Discussion of the palaeogeographic significance of rocks beneath the Vourinos ophiolite, Northern Greece. Journal of the Geological Society of London, 133: 505–507.

Vergely, P., 1984. Téctonique des ophiolites dans les Hellenides internes. Consequences sur l'évolutions des regions Tethysiennes occidentales. Thèse de l'Université de Paris-Sud, Orsay.

Whitechurch, H. and Parrot, J.F., 1978. Ecailles métamorphiques infraperidotitiques dans le Pinde septentrional (Grèce): croûte oceanique, métamorphisme et subduction. Comptes Rendus de la Academie de Sciénce, Paris t.286: 1491–1494.

Woodcock, N.H. and Robertson, A.H.F., 1977. Origins of some ophiolite-related metamorphic rocks of the "Tethyan" belt. Geology, 5: 373–76.

Wright, L., 1986. The effect of deformation on the Vourinos Ophiolite. In: The application of a multi-disciplinary concept for chromite exploration in the Vourinos Complex, N Greece. European Economic Community report, part 1, Athens, 1986, pp. 155–216.

The Circum-Izu Massif Peridotite, Central Japan, as Back-Arc Mantle Fragments of the Izu-Bonin Arc System

SHOJI ARAI
Department of Earth Sciences, Faculty of Science, Kanazawa University, Kanazawa 920, Japan

Abstract

The Circum-Izu Massif peridotite, central Japan, is an upper mantle fragment of the Shikoku Basin, a back-arc or an inter-arc basin of the Izu-Bonin arc system. The Circum-Izu Massif peridotite had been uplifted along a transcurrent plate boundary during the opening of the Shikoku Basin. The Circum-Izu Masslf peridotite was investigated in order to understand petrological characteristics of the upper mantle beneath the back-arc (or inter-arc) basin. It is characterized by harzburgite or clinopyroxene-poor lherzolite (Fo_{90-92}) frequently with calcic plagioclase (An_{88-96}). Chromian spinel (Cr/(Cr + Al) ratio, 0.4–0.6) often has orbicular primary inclusions of pargasite +K- (or Na-) phlogopite + orthopyroxene.

The back-arc mantle peridotite resembles some refractory ocean-floor peridotite but generally has lower-Ti mineral chemistry. Unlike the ocean-floor peridotite the back-arc mantle peridotite could have primary hydrous minerals.

Introduction

The origin and tectonic setting of the alpine-type peridotite mass has been controversial (Dick and Bullen, 1984; Arai, 1989a, 1990). One of the main obstacles to this problem is that the ages of some peridotite masses are too old to deal within the framework of the plate interaction working now on the earth. It is less difficult, however, to interpret a provenance of the peridotite masses which are young enough to be included within the present-day plate tectonic framework. In order to understand the petrological characteristics of the upper mantle peridotites of a given tectonic setting, we should work much more on such young peridotite masses. In this article I describe young peridotite masses derived from a back-arc basin.

The Circum-Izu Massif peridotites (or serpentinites) (Arai and Ishida, 1987) occur in the Oligo-Miocene sedimentary rocks around the Izu Massif,

Tj. Peters et al. (Eds), Ophiolite Genesis and Evolution of the Oceanic Lithosphere, 801–816.

central Japan (Fig. 1). The Circum-Izu Massif peridotite masses are the youngest of all ophiolitic fragments exposed on the Japan arcs (Ishiwatari, 1989). The Circum-Izu Massif serpentine belt is almost parallel with the present-day plate boundary between the Philippine Sea and the Eurasia plates (Fig. 1). Considering the Tertiary tectonic history of this region (e.g., Kobayashi and Nakada, 1978; Ogawa et al., 1985; Taira et al., 1989) the Circum-Izu Massif peridotites are interpreted to be fragments of the upper mantle materials of the northern tip of the Philippine Sea plate (= the Shikoku Basin) of the middle Miocene (Arai and Takahashi, 1988). The Shikoku Basin at that time was an inter-arc basin or a back-arc basin of the Izu-Bonin arc system (e.g., Kobayashi and Nakada, 1978). Petrology of the Circum-Izu Massif peridotites, therefore, may give us clues to understanding of otherwise unknown petrological characteristics of the upper mantle peridotite beneath the inter-arc (or back-arc) basin.

Geological Background

The Circum-Izu Massif peridotites crop out mainly in four regions, Setogawa, Kobotoke, Miura and Mineoka, from west to east, forming the Circum-Izu Massif serpentine belt (Arai and Ishida, 1987) (Fig. 1). They are usually brecciated or comminuted (Fig. 2) and make very small masses in Oligo-Miocene sedimentary rocks, most of which belong to the Shimanto belt in a broad sense. The Circum-Izu Massif peridotite masses in the Setogawa region (or belt) are of syn-sedimentary origin, having supplied detrital particles to the surrounding sediments of the Setogawa Group of Oligo-Miocene (Arai et al., 1978). In contrast, those in the Mineoka region (or belt) are probably fault-bounded with the Oligo-Miocene sediments of the Mineoka Group, which are free from serpentine and related particles (Arai, 1981). Young (possibly Miocene) serpentine sandstone blocks are included in the comminuted Circum-Izu Massif peridotites of the Mineoka region (Arai et al., 1983). Kano et al. (1978) reported that the Circum-Izu Massif peridotite masses in the Miura region are gigantic bounders in the Miocene Hayama Group. Small serpentinite masses in the Kobotoke region are probably boulders in sediments of the Oligo-Miocene Kobotoke Group (Ishida, 1987; Ishida and Arai, in prep.).

The Circum-Izu Massif peridotites occur as small masses in the Oligocene-Miocene sediments with fault or sedimentary contacts. They are strongly serpentinized and crushed (Fig. 2). In the Mineoka region the serpentinite is sometimes clayey. Clinopyroxenites and cumulative peridotites were commonly found in the Setogawa region. Gabbroic to dioritic rocks are commonly associated with the serpentinite. In the Mineoka region gabbroic to dioritic rocks occur as small (usually <10 m) blocks or thin (<1 m) dikes within the serpentinite. In the Setogawa region they occur as boulders or pebbles in conglomerates (pebbly mudstones) or breccias (Arai et al., 1978).

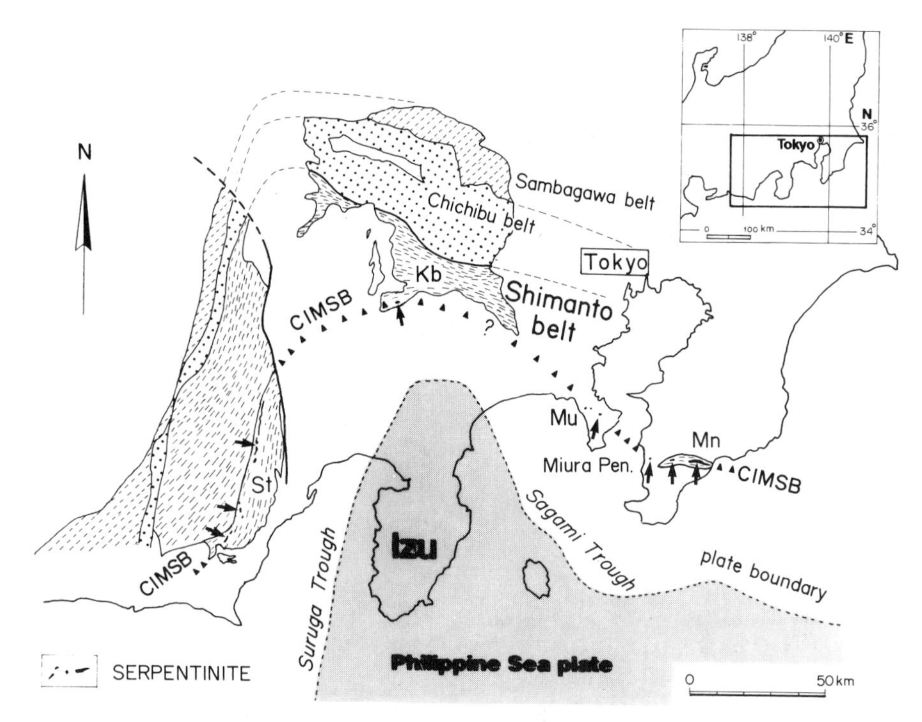

Figure 1. Locality map of the Circum-Izu Massif peridotite masses, central Japan. CIMSB, the Circum-Izu Massif serpentine belt. St, Kb, Mu & Mn; Setogawa, Kobotoke, Miura & Mineoka regions, respectively. Main localities are shown by arrows. A possible trace of the serpentine belt is indicated by solid triangle lines. Note that the Circum-Izu Massif serpentine belt is almost parallel with the present plate boundary.

The gabbroic rocks are varied in lithology, from olivine gabbro, noritic gabbro to hornblende-bearing varieties. Basalts (mainly tholeiitic, sometimes picritic and rarely alkaline) are often pillowed and are commonly associated with the serpentinites (Sameshima, 1960; Kanehira, 1976; Tazaki and Inomata, 1980; Kanie et al., 1987; Ishida et al., 1988; Taniguchi and Ogawa, 1990).

Tectonic Development of the Circum-Izu Massif Serpentine Belt

The Circum-Izu Massif serpentine belt is northward convex around the northern margin of the Izu Massif (or Izu Block) (Fig. 1). It is located near the southern (oceanward) rim of the Shimanto Supergroup (or Group) (cf., Taira et al., 1989). The Izu Massif had collided against the Eurasia plate at Pliocene (5 Ma) after Taira et al. (1989) or at Pleistocene (0.5 Ma) after Niitsuma and Akiba (1985). The Circum-Izu Massif serpentine belt, therefore, had been almost straight before the collision. The Circum-Izu Massif serpentine belt

Figure 2. An outcrop of the Circum-Izu Massif peridotite (Mineoka region). Note the comminuted appearance.

had begun to be produced at Oligo-Miocene because the peridotite (or serpentinite) masses had been protruded onto the sea floor and had supplied detrital particles during the sedimentation of the Setogawa and the Kobotoke Groups (Oligocene to Miocene) (Fig. 3).

In the Miocene the northern end of the Philippine Sea plate had ceased to subduct and opened to form the Shikoku Basin (Kobayashi and Nakada, 1978). The opening axis was almost north to south, and therefore the east–west trending transcurrent fault had worked between the Eurasia plate and the Philippine Sea plate in that time (Fig. 3). It is possible that initial uplift of the mantle peridotites had been performed during the Shikoku Basin opening along the transcurrent plate boundary by analogy with the oceanic fracture zones where serpentinite protrusion is ubiquitous (Bonatti, 1978; Hamlyn and Bonatti, 1980). Subsequent protrusion and accretion of the peridotite masses had been promoted by the opening of the Sea of Japan (=

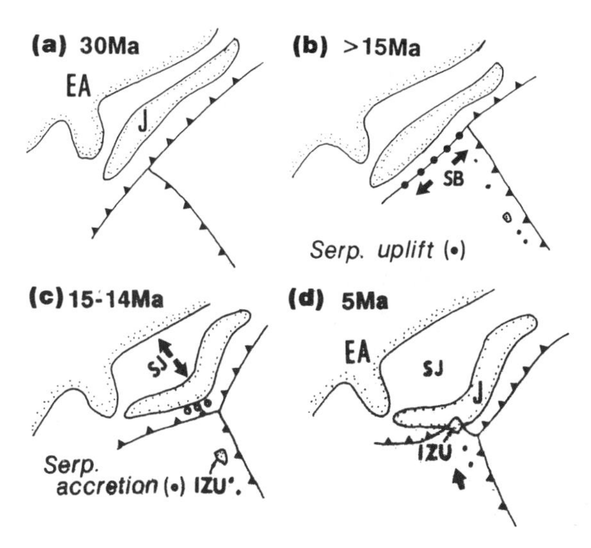

Figure 3. A tectonic model of formation of the Circum-Izu Massif serpentine belt. EA, Eurasia continent. J, Japan arcs. SB, Shikoku Basin. SJ, Sea of Japan. The Circum-Izu Massif peridotite masses had been uplifted along a transcurrent plate boundary during the opening of the Shikoku Basin (b) and had been accreted to the Japan arc during the opening of the Sea of Japan (c). Later the Circum-Izu Massif serpentine belt had been bent northwards by the collision of the Izu Massif (d).

oceanward progression of the Japan arcs) during the subsequent Miocene time (14 to 16 Ma) (Otofuji et al., 1985).

Petrography of the Circum-Izu Massif Periodotites

Harzburgite is predominant over dunite in abundance (Arai and Uchida, 1978; Uchida and Arai, 1978; Arai, 1981). There is a definite modal gap between harzburgite and dunite; harzburgite and dunite contain more than 10 volume % and on average less than 4 volume % of total pyroxenes, respectively (Fig. 4). A few harzburgite samples have more than 5 volume % of clinopyroxene (= lherzolite in a strict terminology). Chromian spinel is euhedral to subhedral in dunite and anhedral in harzburgite (Fig. 5ab). The harzburgite often has plagioclase (or its saussuritized equivalent) (Fig. 5c). The modal amount of plagioclase (less than 10%) varies in harmony with that of pyroxenes (especially clinopyroxene) (Arai and Takahashi, 1988). Small orbicular inclusions composed of phlogopite, pargasite and orthopyroxene are frequently found within chromian spinel grains both in harzburgite and in dunite (Fig. 5ab). Similar inclusions are sometimes found in chromian spinel from ophiolitic chromitites (e.g., Talkington et al., 1986; Augé, 1987). Distribution of the inclusions within the chromian spinel grain is irregular (Fig. 5a). Amounts of discrete grains of hydrous minerals (amphi-

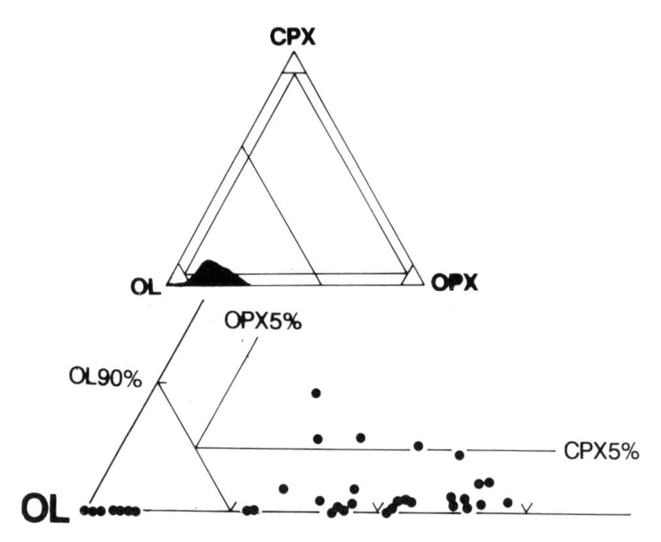

Figure 4. Modal amounts of olivine, orthopyroxene and clinopyroxene in the Circum-Izu Massif peridotites. OL, olivine. OPX, orthopyroxene. CPX, clinopyroxene.

bole and mica) in harzburgite vary from region to region (Arai and Ishida, 1987). The hydrous minerals are rare in peridotites from the Mineoka and the Kobotoke regions and are sometimes abundant in those from the Setogawa and the Miura regions. Fine acicular tremolitic amphiboles are ubiquitous as low-temperature alteration products.

The serpentine sandstones are mainly composed of serpentinite particles with calcite or muddy matrix (Fig. 5d) (Arai et al., 1983). Particles of mafic rocks (basalt and dolerite) and basic schists are common but subordinate in amount. Mineral particles in the serpentine sandstones are dominated by clinopyroxene, hornblende, plagioclase and chromian spinel. The chromian spinel grains with various color in thin section are sometimes concentrated to make thin layers (Fig. 5d).

Mineral Chemistry

Constituent minerals were analyzed with microprobe at the University of Tokyo (JXA-5), the University of Tsukuba (JXA-50A) and the Geological Survey of Japan (JCXA-733). Selected analyses are listed in Table 1.

Harzburgite

The Fo content of olivine varies from 90 to 92 (Fig. 6) and is negatively correlated both with clinopyroxene/pyroxenes volume ratio and with plagioclase volume in harzburgite (Arai and Takahashi, 1988). The Cr/(Cr + Al)

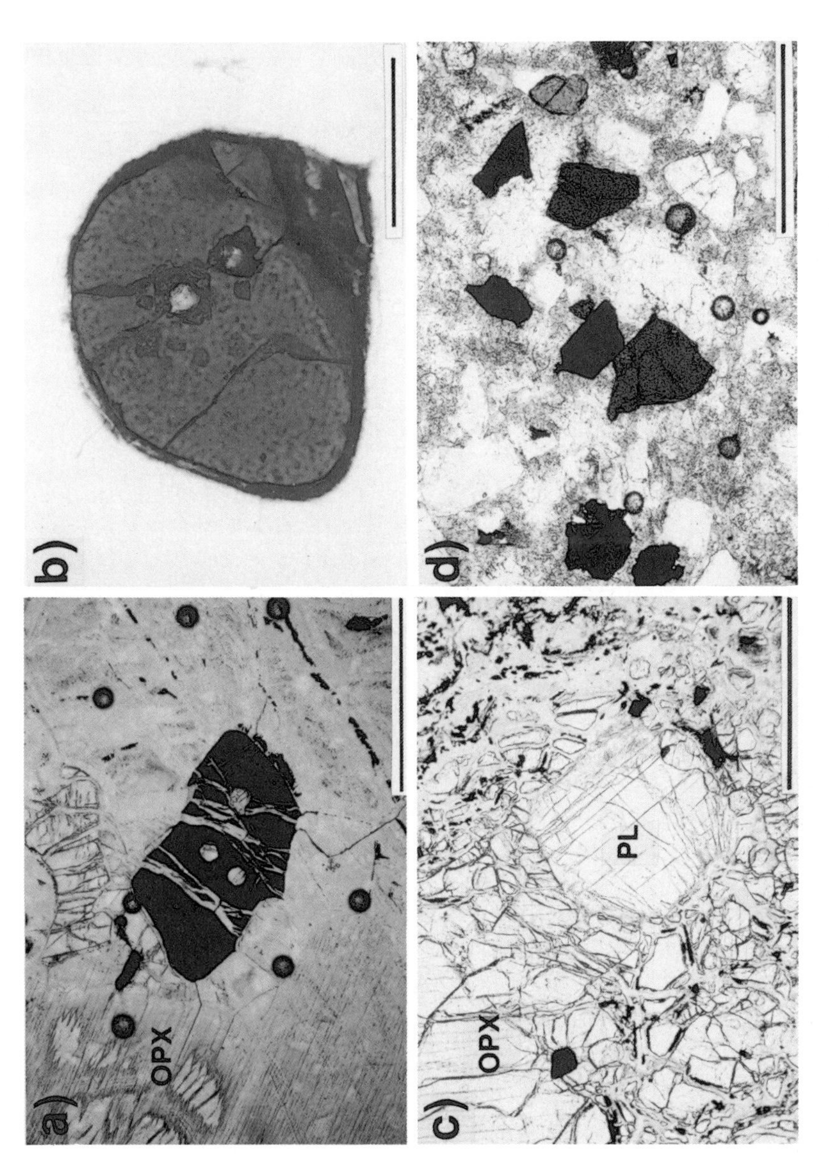

Figure 5. Photomicrographs of the Circum-Izu Massif peridotites. Plane-polarized light. OPX, partially serpentinized orthopyroxene. PL, plagioclase. Scale bar, 0.5 mm for (a)(c) & (d). 0.2 mm for (b). (a) Spherical silicate inclusions in chromian spinel of harzburgite. Miura region (Fig. 1). (b) Spherical inclusions in chromian spinel of dunite Mineoka region (Fig. 1). (c) Plagioclase-bearing harzburgite. Setogawa region (Fig. 1). (d) Serpentine sandstone with a spinel–rich seam from the Mineoka region. Note the difference of the color of spinel.

Table 1. Selected microprobe analyses of minerals in the Circum-Izu Massif peridotites (Mineoka region). OL, OPX, CPX, SP & PL; olivine, orthopyroxene, clinopyroxene and plagioclase, respectively. FeO*, total iron as FeO. Cationic fractions in spinel were calculated assuming spinel stoichiometry. Mg#, Mg/(Mg + total Fe) atomic ratio except spinel, for which Mg/(Mg + Fe^{2+}) atomic ratio. Ca#, Ca/(Ca + Na) atomic ratio. Cr#, Cr/(Cr + Al) atomic ratio. Mg, Fe*, Ca; Mg, total Fe and Ca atomic fractions to (Mg + total Fe + Ca), respectively. Cr, Al, Fe^{3+}; Cr, Al and Fe^{3+} atomic fractions to (Cr + Al + Fe^{3+}), respectively.

	PL-bearing harzburgite (1611)**					Harzburgite (11143i)				Dunite (273-3)		Dunite (273-6)	
	OL	OPX	CPX	SP	PL	OL	OPX	CPX	SP	OL	SP	OL	SP
SiO_2	40.84	55.56	51.86	0.07	46.45	40.54	57.00	53.65	nd	40.53	0.05	40.38	0.02
TiO_2	0.02	0.02	0.12	0.19	0.00	0.00	0.00	0.11	0.05	0.00	0.15	0.00	0.16
Al_2O_3	0.01	3.22	4.35	24.56	33.30	0.03	2.71	3.65	31.35	0.02	17.95	0.01	27.05
Cr_2O_3	0.01	0.82	1.44	41.97	0.03	0.05	1.05	1.41	39.43	0.00	51.67	0.01	41.62
FeO*	8.88	5.95	2.30	17.75	0.32	7.99	5.07	2.33	15.51	6.29	14.28	8.68	17.07
NiO	0.41	0.05	0.11	0.26	0.00	0.44	0.15	0.15	nd	0.51	0.06	0.36	nd
MnO	0.04	0.01	0.10	0.12	0.00	0.05	0.11	0.13	0.18	0.09	0.49	0.08	0.52
MgO	49.04	32.53	16.41	13.47	0.05	50.28	33.42	17.19	15.11	51.61	14.52	50.01	14.28
CaO	0.04	1.22	23.22	0.02	17.65	0.03	1.72	22.07	nd	0.13	0.00	0.06	0.00
Na_2O	0.00	0.03	0.33	0.02	1.30	nd	nd	0.72	nd	nd	nd	0.02	0.00
K_2O	0.00	0.00	0.01	0.00	0.00	nd	nd	0.00	nd	nd	nd	nd	nd
Total	99.29	99.41	100.25	98.43	99.10	99.41	101.23	101.41	101.62	99.18	99.17	99.60	100.72
Mg#(Ca#)	0.908	0.907	0.927	0.622	(0.882)	0.918	0.922	0.923	0.652	0.936	0.681	0.911	0.636
Cr#				0.534					0.457		0.659		0.508
Mg(Cr)		0.885	0.477	(0.514)			0.891	0.500	(0.451)		(0.643)		(0.492)
Fe*(Al)		0.091	0.038	(0.448)			0.076	0.038	(0.535)		(0.333)		(0.476)
$Ca(Fe^{3+})$		0.024	0.485	(0.037)			0.033	0.462	(0.014)		(0.025)		(0.032)

**After Arai and Takahashi (1988).

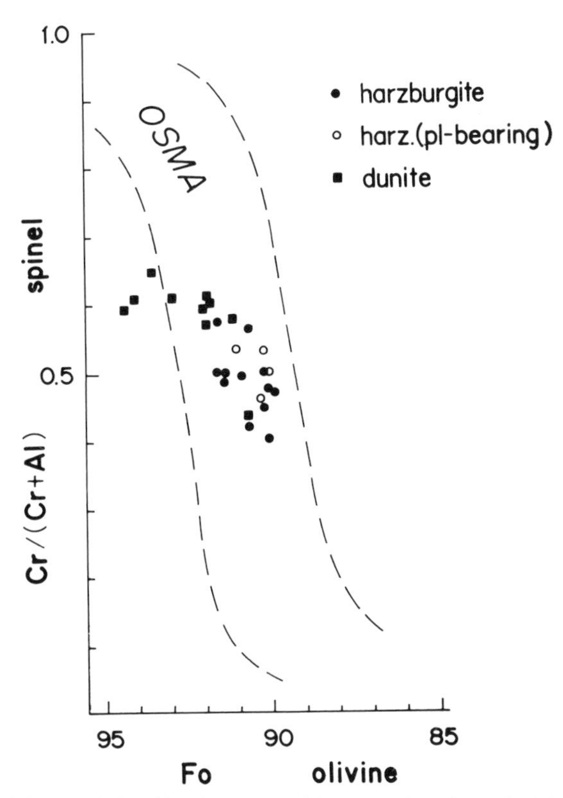

Figure 6. Compositional relationships between olivine and chromian spinel in the Circum-Izu Massif peridotites. OSMA, olivine-spinel mantle array (Arai, 1987, 1989a, 1990a), in which most of mantle-derived spinel peridotites are included.

atomic ratio (= Cr#) of chromian spinel ranges from 0.2 to 0.7 (most frequently 0.4 to 0.6) (Arai and Abe, 1987; Arai and Ishida, 1987). Plagioclase is very calcic, An_{88-96} (Takasawa, 1976; Arai and Takahashi, 1988). Discrete phlogopite is extremely Ti-poor. Discrete amphiboles are Na-rich, Ti-poor and are classified as pargasites (Arai and Ishida, 1987; Arai et al., 1990).

Dunite

Olivine is more enriched with Fo and NiO than in harzburgite (Fig. 6). Fo content varies from 91 to 94 (Fig. 6). CaO content of olivine is slightly but definitely higher in dunite than that in harzburgite. The Cr# of chromian spinel varies positively with the Fo content of coexisting olivine from 0.4 to 0.6 (Fig. 6). The Cr# of chromian spinel is not so very high even in the highly magnesian dunite (Fo_{94}). Some magnesian dunites ($> Fo_{93}$) lie in the Fo-rich part clearly off the olivine-spinel mantle array on the Fo-Cr# diagram (Arai, 1987) (Fig. 6). Such dunites are very rare; dunites of any derivations are usually plotted within or off the olivine-spinel mantle array towards the Fo-poor direction on the Fo-Cr* diagram (Arai, 1989b).

Solid Inclusions in Chromian Spinel

The orthopyroxene is low in CaO, Al_2O_3 and Cr_2O_3. The amphibole is Na-rich pargasite. The mica is either K-phlogopite or its Na-analogue. These mineralogical characteristics are almost equivalent to those of solid inclusions sometimes found in chromian spinel of chromitites from some ophiolites such as Oman or Bay of Islands (e.g., Talkington et al., 1986; Augé, 1987).

Serpentine Sandstone

Among detrital mineral particles clinopyroxene and chromian spinel are important to give some constraints on the tectonic history of the serpentinite masses (Arai et al., 1983; Arai and Okada, in prep.). Serpentine sandstones from the Mineoka region were examined. The clinopyroxene grains are classified into two categories in terms of chemistry; chromian diopside and augite (or salite) (Arai et al., 1983). The chromian diopside group had been derived from peridotites whereas the augite (or salite) group from mafic igneous rocks (mainly gabbros). The detrital chromian spinel in the Mineoka serpentine sandstone has a much wider compositional spread (the Cr#, from 0.17 to 0.82) than the chromian spinel in harzburgite and dunite from the Mineoka belt (Fig. 7). This fact may indicate that serpentinite masses with much wider lithology than the present ones had been exposed probably at the Miocene time in the Mineoka region (Arai et al., 1983; Arai and Okada, 1991).

Discussion

In terms of the Fo-Cr# relationship the Circum-Izu Massif harzburgite and clinopyroxene-poor lherzolite are included in the refractory part of the whole spread of ocean-floor peridotites (e.g., Dick and Bullen, 1984; Shibata and Thompson, 1986) (Fig. 6). The TiO_2 content of chromian spinels from the Circum-Izu Massif peridotites is low and is included in the range of ocean-floor spinel peridotites (Dick and Bullen, 1984) (Fig. 7). It is especially noteworthy that the chromian spinel in the Circum-Izu Massif plagioclase-bearing peridotites is low in TiO_2 (0.1 to 0.3 wt%) while the oceanic equivalents often have high-TiO_2 (>1 wt%) spinels (Dick and Bullen, 1984) (Fig. 7). Hamlyn and Bonatti (1980) reported highly calcic plagioclase (An_{91-96}) in lherzolites (Fo_{89} and Cr#, 0.4) dredged from the Owen fracture zone. Plagioclase in other documented ocean-floor peridotites (Melson et al., 1972; Prinz et al., 1976; Roden et al., 1984) is less calcic (An_{58-80}) than that in the Circum-Izu Massif peridotites. Na_2O content of chromian diopside in the

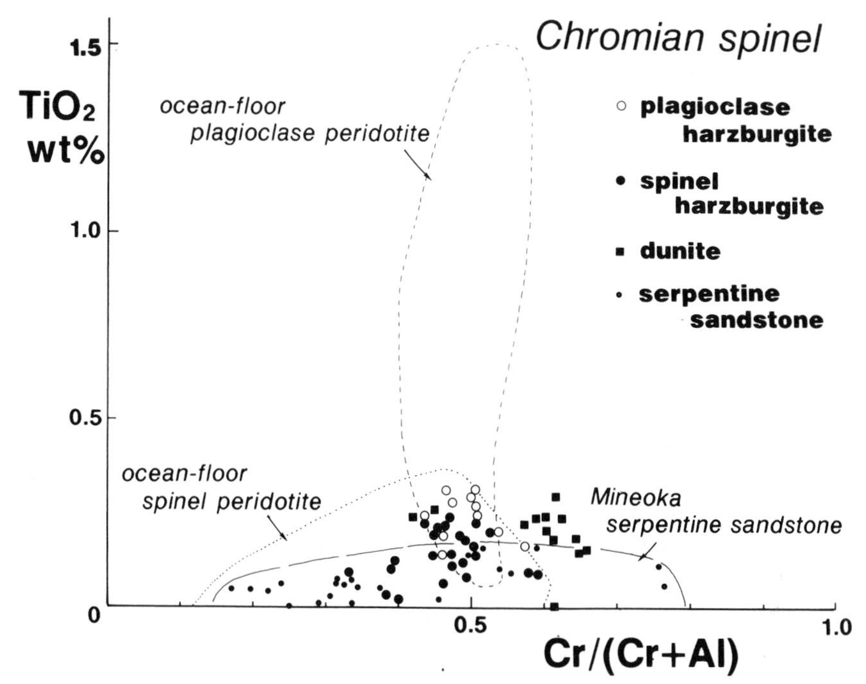

Figure 7. TiO_2 wt % vs. Cr/(Cr + Al) atomic ratio (Cr#) of the Circum-Izu Massif chromian spinels. The ranges of the ocean-floor peridotites (Dick and Bullen, 1984) are also shown for comparison.

Circum-Izu Massif harzburgite and lherzolite is low but usually higher than that in the ocean-floor peridotites (Fig. 8). In terms of a combination of the Na_2O content in diopside and the Cr# of chromian spinel the Circum-Izu Massif peridotites are intermediate between the ocean-floor peridotites and the Japan-arc peridotites (xenoliths in the Cenozoic magmas of the Japan arcs) (Fig. 8), which are interpreted as arc-magma restites (Arai, 1989b).

The Cr# of chromian spinel in the Shikoku Basin basalts, 0.4 to 0.6 (Dick and Bullen, 1984), is almost equivalent to that of the most common type of Circum-Izu Massif peridotites (Fig. 6). Considering the variation of the Cr# of spinel on magmatic differentiation (Arai, 1990ab), the restites for the Shikoku Basin basalts should resemble the Circum-Izu Massif peridotites. The tholeiitic pillow basalts in the Mineoka-Miura region are similar in chemistry to the MORB (Tazaki and Inomata, 1980; Kanie et al., 1987). Their high alkali contents relative to MgO and FeO (total iron), however, indicate their similarity to the Shikoku Basin tholeiites (Dick, 1982).

Dioritic rock (and some hornblende gabbro) blocks commonly found in the Setogawa and the Mineoka regions are possibly fragments of crust of the Paleo-Izu–Bonin arc which had been split into two fragments, the present Izu-Bonin arc and the Kyushu-Palau ridge, during the opening of the Shikoku Basin (Kobayashi and Nakada, 1978). The highly refractory peridotites which

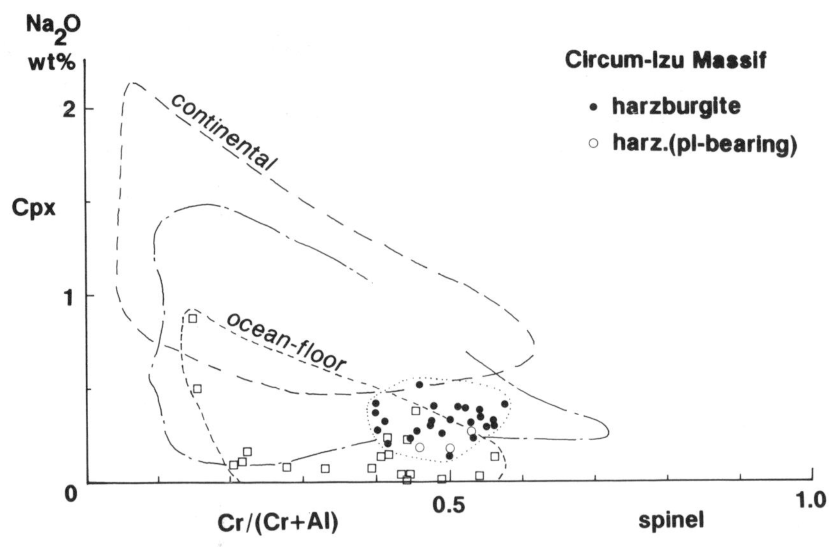

Figure 8. The relationships between Na_2O wt % of clinopyroxene and Cr/(Cr + Al) atomic ratio (Cr#) of chromian spinel in the Circum-Izu Massif peridotites. The regions of continental, Japan-arc and ocean-floor mantle peridotites are also shown.

are deduced from the serpentine sandstone mineralogy (Fig. 7) may have been protruded at the Miocene prior to the presently exposed Circum-Izu Massif peridotites. They are possibly fragments of the uppermost mantle of the Paleo-Izu-Bonin arc.

Hydrous silicate inclusions in chromian spinel (Fig. 5ab) are absent in the ocean-floor peridotites, whereas they are common in some ophiolitic chromitites (Talkington et al., 1986; Augé, 1987). This kind of inclusion is usually interpreted as primary one based on the textural features (Talkington et al., 1986; Augé, 1987). This may in turn lead to a conclusion that the chromite with such hydrous silicate inclusions was formed from some hydrous magma (Augé, 1987) or some hydrothermal fluid at high temperatures (Zohan et al., 1983). The hydrous silicate inclusions in spinel may indicate that the Circum-Izu Massif harzburgite was a residue of a hydrous magma. Alternatively the hydrous silicate inclusions may be reaction products between anhydrous peridotite and later impregnated hydrous melt. The hydrous magma or melt may be most easily available in a supra-subduction zone mantle.

The characteristic mineral assemblage, olivine + orthopyroxene + clinopyroxene + chromian spinel (Cr# = ca. 0.5) + plagioclase (An_{90}), in the Circum-Izu Massif peridotites indicates a low-pressure (~5 kb) origin. The assemblage magnesian olivine + calcic plagioclase itself denotes an equilibrium pressure less than 10 kb (Kushiro and Yoder, 1966; Green and Hibberson, 1970). Jaques and Green (1980) demonstrated that the assemblage olivine (Fo_{90}) + orthopyroxene + clinopyroxene + plagioclase + chromian spinel (Cr#, probably ca. 0.5) could be equilibrated with a basaltic melt at 1,200°C and 5 kb; that is, calcic plagioclase can be one of the residual minerals in the shallow mantle. Alternatively the plagioclase could be precipitated from trapped impregnated melt (e.g., Dick, 1989), which was hydrous and was also due to precipitation of hydrous silicate inclusions in spinel. The similar on-land low-pressure peridotites with calcic plagioclase ($An_{\sim 90}$) and chromian spinel (Cr#, ~0.5) have been reported only from the Trinity ophiolite, California (Quick, 1981).

Summary and Conclusions

The Circum-Izu Massif peridotites are likely upper mantle fragments of the Shikoku Basin (the northernmost tip of the Philippine Sea plate), a back-arc or inter-arc basin of the Izu-Bonin arc system. They had been uplifted during the opening of the Shikoku Basin. The Circum-Izu Massif peridotites have an intermediate petrological character between the ocean-floor peridotites and the sub-arc (Japan arc) peridotites. They are similar to some refractory oceanic peridotites in terms of the Fo-Cr# relationship. Chromian spinel in the plagioclase-bearing Circum-Izu Massif peridotites is usually lower in TiO_2 (0.1–0.3 wt%) than in the oceanic equivalents (TiO_2, 0.1–1.5 wt%) (Dick and Bullen, 1984). The Na_2O content of chromian diopside in the Circum–Izu Massif peridotites is intermediate between the ocean-floor peridotites and the Japan arc peridotites at a fixed Cr# of coexisting chromian spinel. Presence of primary hydrous solid inclusions in chromian spinel, which have not been found in the oceanic peridotites, is consistent with a supra-subduction zone origin.

The back-arc mantle peridotite is similar in lithology to the ocean-floor peridotite but generally has a low-Ti mineral chemistry. The back-arc mantle peridotite, unlike the ocean-floor peridotite, could have primary or impregnated hydrous minerals.

Acknowledgements

The author wishes to acknowledge many people, especially T. Uchida, H. Okada, K. Shimokawa, T. Ito, T. Shibue, N. Abe, T. Ishida and N. Takahashi, for their cooperation both in field and laboratory works. K. Nakamura helped to make figures.

References

Arai, S., 1981. Igneous and ultramafic rocks in the Mineoka Belt, the Boso Peninsula. Excursion Guidebook, 88th Ann. Meet. Geol. Soc. Japan, pp. 59–72. (in Japanese)

Arai, S., 1987. An estimation of the least depleted spinel peridotite on the basis of olivine-spinel mantle array. N. Jb. Mineral. Mh., 1987, pp. 347–354
Arai, S., 1989a. Origin of ophiolitic peridotites. Jour. Geogr. (Tokyo), 98: 231–240. (in Japanese with English abstract)
Arai, S., 1989b: Upper mantle peridotites beneath Japan island arcs. Kaiyo Monthly, 21: 47–54. (in Japanese)
Arai, S., 1990a. Origin of upper mantle spinel peridotites. Kagaku, 60: 103–112. (in Japanese)
Arai, S., 1990b. What kind magmas could be equilibrated with ophiolitic peridotites? Proceed. Intern. Ophiolite Conf. TROODOS 87, Nicosia, Cyprus. (in press)
Arai, S. and Abe, N., 1987. Ultramafic rocks from the Mineoka belt. Abstr. 94th Ann. Meet. Geol. Soc. Japan, p. 483. (in Japanese)
Arai, S. and Ishida, T., 1987. Petrological characteristics of serpentinites in the Kobotoke Group, the Sasago area, Yamanashi Prefecture: A comparison with other Circum-Izu Massif serpentinites. Jour. Mineral. Petrol. Econ. Geol., 84: 336–344. (in Japanese with English abstract)
Arai, S. and Okada, H., 1991. Petrology of serpentine sandstone as a key to tectonic development of the serpentine belt. Tectonophys.
Arai, S. and Takahashi, N., 1988. Relic plagioclase-bearing harzburgite from the Mineoka belt, Boso Peninsula, central Japan. Jour. Mineral. Petrol. Econ. Geol., 83: 210–214. (in Japanese with English abstract)
Arai, S. and Uchida, T., 1978. Highly magnesian dunite from the Mineoka belt, central Japan. Jour. Japan. Assoc. Mineral. Petrol. Econ. Geol., 73: 176–179.
Arai, S., Ito, M., Nakayama, N. and Masuda, F., 1990. A suspect serpentinite mass in the Tokyo Bay area: petrology and provenance of serpentinite pebbles in the upper Cenozoic system in the Boso Peninsula, central Japan. Jour. Geol. Soc. Japan, 96: 171–179. (in Japanese with English abstract)
Arai, S., Ito, T. and Ozawa, K., 1983. Ultramafic-mafic clastic rocks from the Mineoka belt, central Japan. Jour. Geol. Soc. Japan, 89: 287–297. (in Japanese with English abstract)
Arai, S., Shimokawa, K. and Takahashi, T., 1978. On the mode of emplacement of ultramafic-mafic rocks in the Setogawa belt, Shizuoka Prefecture. Jour. Geol. Soc. Japan, 84: 691–603. (in Japanese)
Augé T., 1987. Chromite deposits in the northern Oman ophiolite: Mineralogical constraints. Mineral. Deposita, 22: 1–10.
Bonatti, E., 1978. Vertical tectonism in oceanic fracture zones. Earth Planet. Sci. Lett., 37: 369–379.
Dick, H.J.B., 1982. The petrology of two back-arc basins of the northern Philippine Sea. Amer. Jour. Sci., 282: 644–700.
Dick, H.J.B., 1989. Abyssal peridotites, very slow spreading ridges and ocean ridge magmatism. In: A.D. Saunders and M.J. Norry (Editors), Magmatism in the Ocean Basins. Geol. Soc. Spec. Pub. 42: 71–105.
Dick, H.J.B. and Bullen, T., 1984. Chromian spinel as a petrogenitic indicator in abyssal and alpine-type peridotites and spatially associated lavas. Contrib. Mineral. Petrol., 86: 54–76.
Green, D.H. and Hibberson, W., 1970. The instability of plagioclase in peridotite at high pressure. Lithos, 3: 209– 221
Hamlyn, P.R. and Bonatti, E., 1980. Petrology of mantle-derived ultramafics from the Owen fracture zone, northwest Indian Ocean: implications for the nature of the oceanic upper mantle. Earth Platet. Sci. Lett., 48: 65–79.
Ishida, T., 1987. Serpentinite from the Kobotoke Group in the Sasago district, Yamanashi Prefecture, central Japan. Jour. Geol. Soc. Japan, 93: 233–236. (in Japanese)
Ishida, T. and Arai, S. (in prep.) Talc-amphibole rocks in the Kobotoke Group, central Japan.
Ishida, T., Arai, S. and Takahashi, N., 1988. The occurrence of picrite basalt in the Kobotoke Group, the Hatsukari area, Yamanashi Prefecture, central Japan. Jour. Pet. Miner. Econ. Geol., 83: 43–50. (in Japanese with English abstract)

Ishiwatari, A., 1989. Ophiolites of Japan. Jour. Geogr. (Tokyo), 98: 290–303. (in Japanese with English abstract)

Jaques, A.L. and Green, D.H., 1980. Anhydrous melting of peridotite at 0–15 Kb pressure and the genesis of tholeiitic basalts. Contrib. Mineral. Petrol., 73: 287–310.

Johan, Z., Dunlop, H., Le Bel, L., Robert, J.L. and Volfinger, M., 1983. Origin of chromite deposits in ophiolitic complexes for a volatile- and sodium-rich reducing fluid phase. Fortsch. Mineral., 61: 105–107.

Kanehira, K., 1976. Modes of occurrence of serpentinite and basalt In the Mineoka district, southern Boso Peninsula. Mem. Geol. Soc. Japan, 13: 43–50. (in Japanese with English abstract)

Kanie, Y., Fujioka, K., Koga, K. and Taniguchi, H., 1987. Paleogene pillow lava from the Miura Peninsula, south-central Japan. Sci. Rep. Yokosuka City Museum, 35: 23–28. (in Japanese with English abstract)

Kano, K., Ito, T. and Masuda, T., 1975. Sedimentary serpentinite near Kinugasa in the Miura Peninsula, Japan. Jour. Geol. Soc. Japan, 81: 641–644. (in Japanese)

Kobayashi, K. and Nakada, M., 1978 . Magnetic anomalies and tectonic evolution of the Shikoku Inter-arc Basin. Jour. Phys. Earth, 26: Suppl., S391–S402.

Kushiro, I. and Yoder, H.S., Jr., 1966. Anorthite-forsterite and anorthite-enstatite relations and their bearing on the basalt-eclogite transformation. Jour. Petrol., 7: 337–362.

Melson, W.G., Hart, S.R. and Thompson, G., 1972. St. Paul's Rocks, equatorial Atlantic, radiometric ages, and implications on sea-floor spreading. Geol. Soc. Amer. Mem., 132: 241–272.

Niitsuma, N. and Akiba, F., 1985. Neogene tectonic evolution and plate subduction in the Japanese island arcs. In: N. Nasu et al. (Eds), Formation of Active Ocean Margins, Terra Sci. Pub., Tokyo, pp. 75–108.

Ogawa, Y., Horiuchi, K., Taniguchi, H. and Naka, J., 1985. Collision of the Izu Arc with Honshu and the effects of oblique subduction in the Miura-Boso Peninsulas. Tectonophys., 119: 349–379

Otofuji, Y., Matsuda, T. and Nohda, S., 1985. Paleomagnetic evidence for the Miocene counter-clockwise rotation of Northeast Japan – rifting process of the Japan Arc. Earth Planet. Sci. Lett., 75: 265–277.

Prinz, M., Keil, K., Green, J.A., Reid, A.M., Bonatti, E. and Honnorez, J., 1976. Ultramafic and mafic dredge samples from the equatorial Mid-Atlantic Ridge and fracture zone. Jour. Geophys. Res., 81: 4087–4103.

Quick, J.E., 1981. Petrology and petrogenesis of the Trinity peridotite, an upper mantle diapir in the eastern Klamath Mountains, northern California. Jour. Geophys. Res., 86: 11837–11863.

Roden, M.K. Hart, S.R., Frey, F.A. and Melson, W.G., 1984. Sr, Nd and Pb isotopic and REE geochemistry of St. Paul's Rocks: the metamorphic and metasomatic development of an alkali basalt mantle source. Contrib. Mineral. Petrol., 85: 376–390.

Sameshima, T., 1960. Picrite basalt dikes in the Paleogene formation in Central Japan. Rept. Liberal Art Sci. Fac. Shizuoka Univ., Sec. Nat. Sci., 3: 77–80.

Shibata, T. and Thompson, G., 1986. Peridotites from the Mid-Atlantic Ridge at 43°N and their petrogenetic relation to abyssal tholeiites. Contrib. Mineral. Petrol., 93: 144–159.

Taira, A., Tokuyama, H. and Soh, W., 1989. Accretion tectonics and evolution of Japan. In: Z. Ben-Avraham (Ed), The Evolution of the Pacific Ocean Margins, Oxford Univ. Press, Oxford, pp. 100–123.

Takasawa, K., 1976. Anorthite in peridotite from the Setogawa Group, Shizuoka Prefecture, central Japan. Earth Sci., 30: 163–169.

Talkington, R.W., Watkinson, D.H., Whittaker, P.J. and Jones, P.C., 1986. Platinum group element-bearing minerals and other solid inclusions in chromite of mafic and ultramafic complexes: chemical compositions and comparisons. In: B. Cater et al. (Eds), Metallogeny of Basic and Ultrabasic Rocks (Regional Presentations), Theophrastus Publications, Athens, pp. 223–249.

Taniguchi, H. and Ogawa, Y., 1990. Occurrence, chemistry and tectonic significance of alkali basaltic rocks in the Miura Peninsula, central Japan. Jour. Geol. Soc. Japan, 96: 101–116. (in Japanese with English abstract)

Tazaki, K. and Inomata, M., 1980. Picritic basalts and tholeiitic basalts from Mineoka tectonic belt, Central Japan. Jour. Geol. Soc. Japan, 86: 653–671. (in Japanese with English abstract)

Uchida, T. and Arai, S., 1978. Petrology of ultramafic rocks from the Boso Peninsula and the Miura Peninsula. Jour. Geol. Soc. Japan, 84: 561–570.

Geology and Chemistry of the Early Proterozoic Purtuniq Ophiolite, Cape Smith Belt, Northern Quebec, Canada

D.J. SCOTT[1,2], M.R. St-ONGE[3], S.B. LUCAS[3], and H. HELMSTAEDT[1]

[1]*Department of Geological Sciences, Queen's University, Kingston, Ont., K7L 3N6, Canada*
[2]*Present address: GEOTOP,. Université du Québec à Montréal, C.P. 8888, Succursale A, Montréal, Québec, H3C 3P8, Canada*
[3]*Geological Survey of Canada, 601 Booth St., Ottawa, Ont., K1A 0E8, Canada*

Abstract

The two-billion year-old Purtuniq ophiolite comprises pillowed mafic flows, sheeted mafic dykes, gabbros, and minor plagiogranites, and an extensive suite of layered mafic and ultramafic cumulate rocks; depleted mantle rocks have not been observed. The tectonically dismembered ophiolite is similar in most physical and chemical respects to Phanerozoic ophiolites, and represents direct evidence for modern-style plate tectonic processes in the early Proterozoic. The preserved *crustal* thickness is of the order of 7.5–8 km; but may originally have been as much as 9–10 km thick. The pillowed volcanic rocks and sheeted dykes are tholeiitic, with rare earth and other trace element abundances most similar to modern MORBs. The cumulate rocks also follow a tholeiitic trend in major element composition. The mafic nature of the volcanic pile, the absence of extensive pyroclastic rocks, and the presence of a well developed sheeted dyke complex stratigraphically underlying the volcanic rocks supports an oceanic spreading center origin for the ophiolite. Modally graded layers and adcumulate textures observed in thin section suggest that the layered mafic and ultramafic rocks are dominantly the result of cumulus crystallization. Cryptic compositional variations in relict igneous minerals, such as forsterite content in olivine, and chromium and titanium content of clinopyroxene do not vary systematically with stratigraphic height in the cumulate pile. This is thought to record the periodic input of fresh batches of primitive magma into the magma chamber. Nd-isotopic data suggest that two time-integrated depleted mantle sources were responsible for the generation of the ophiolite, one highly depleted (ϵ_{Nd} + 4.6 to +5.3), the other less so (ϵ_{Nd} +2.5 to +3.6). Each source produced a suite of cumulate rocks and sheeted mafic dykes. The mutually intrusive nature of sheeted dykes from the two suites suggests that the two sources operated simultaneously, and in close proximity to one another. The physical extent of the ocean basin in which the ophiolite was generated is not well constrained.

Tj. Peters et al. (Eds), Ophiolite Genesis and Evolution of the Oceanic Lithosphere, 817–849.

Introduction

The significance of ophiolitic rocks in plate tectonic interpretations of Phanerozoic orogenic belts has long been recognized (e.g. Gass, 1968; Moores and Vine, 1969; 1971; Dewey and Bird, 1971; Coleman, 1971). The presence of ophiolites in modern mountain belts is taken as compelling evidence of the existence of a former ocean basin, and the nearly complete subduction of its oceanic crust (e.g. Coleman, 1977). In contrast, the apparent absence of ophiolites from ancient mountain belts has prompted speculation on the nature of Proterozoic (e.g. Moores, 1986) and Archean oceanic crust (e.g. Sleep and Windley, 1982; Arndt, 1983; Nisbet and Fowler, 1983; Hargraves, 1986). In addition, this has lead to speculation on when plate tectonic processes involving modern-style subduction of oceanic crust may have begun (e.g. Condie, 1976; Kröner, 1983). Whereas some well-studied ancient mountain belts are thought to have formed by processes compatible with modern plate tectonics (e.g. the early Proterozoic Wopmay Orogen, Hoffman and Bowring, 1984; St-Onge and King, 1987), undisputed oceanic crust has rarely been identified. To date, only the Jormua complex, northeastern Finland, has been interpreted as a dismembered but complete example of oceanic crust and upper mantle of early Proterozoic age (Kontinen, 1987) which resembles a modern ophiolite. Recent work (St-Onge et al., 1987; 1988a) in the Cape Smith Belt, an early Proterozoic collisional thrust belt in northern Quebec, Canada, has lead to the interpretation of a package of pillowed mafic volcanic rocks, sheeted mafic dykes, and layered gabbroic and ultramafic rocks as ancient oceanic crust. As will be shown in the sections that follow, this package of rocks is compelling evidence of the existence of a former ocean basin in northern Quebec in the early Proterozoic. As such, it provides an additional opportunity to examine plate tectonic and oceanic magmatic processes in the early Proterozoic.

The Cape Smith Belt (CSB) is a thrust belt (Hynes and Francis, 1982; Lucas, 1989) which comprises a deformed, north-facing continental margin (Francis et al., 1981; 1983; Picard, 1990; St-Onge et al., 1989), and the suite of rocks which are interpreted as a 2.00 Ga ophiolite (Scott et al., 1988; St-Onge et al., 1989; St-Onge and Lucas, 1990a, and references within). The thrust belt is the result of southward-directed imbrication of this margin and ophiolite during the early Proterozoic Trans-Hudson Orogen (See Hoffman, 1988 for an overview). The belt is allochthonous with respect to the granitic gneisses of the Superior province upon which it now sits (St-Onge et al., 1986; 1987; 1988a; St-Onge and Lucas, 1990a). The development of the continental margin and formation of the ophiolitic rocks, and their subsequent deformation, can be explained in terms of modern plate tectonics (St-Onge and Lucas, 1990a).

The present contribution provides an introduction to the geology and chemistry of the newly recognized ophiolite in the CSB. In the first section of the paper, a general overview of the tectonostratigraphy and deformation

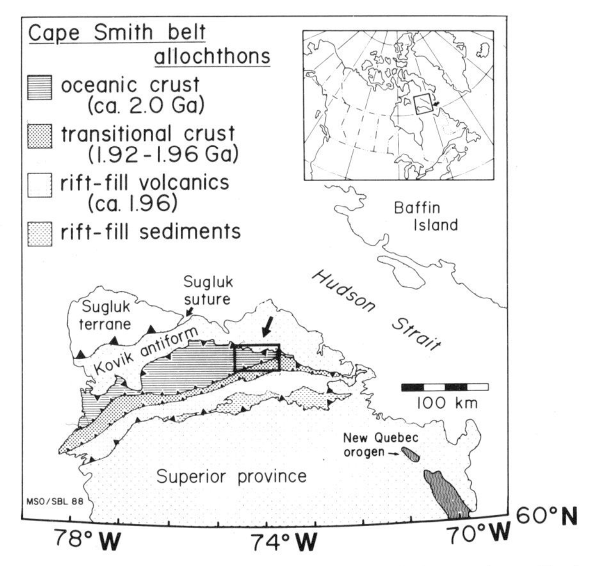

Figure 1. Map showing the location of the Cape Smith Belt, northern Quebec, Canada, with simplified tectonostratigraphic units shown. Inset (upper right-hand corner) locates the main figure in northern Canada. The eastern portion of the ophiolite, as described in the text, is outlined in the main figure. The proposed suture between the underthrust Superior province and the overriding Sugluk block is indicated. After St-Onge and Lucas, 1990a.

history of the CSB is given, together with an outline of the units interpreted as oceanic crust. In the main body of the paper, field and petrographic aspects of the individual igneous components of the ophiolite and their whole-rock chemistry are described. Well-constrained cross-sections through various parts of the ophiolite are presented, and are used to generate a composite section of the ophiolite sequence. Finally, possible environments of formation of the ophiolite are discussed, and related to the tectonic evolution of the CSB.

The Cape Smith Belt

The Cape Smith Belt is located on the northern margin of the Superior structural province of the Canadian Shield (Figure 1). It consists, from south to north, of five distinct tectonostratigraphic packages: (1) a fluvio-deltaic sedimentary package (St-Onge et al., 1988a); (2) sedimentary and volcanic remnants of a *ca*. 1960 Ma north-facing continental rift margin (Povungnituk Group, Hynes and Francis, 1982; Picard et al., 1990); (3) *ca*. 1920 Ma transitional- to MORB (mid-ocean ridge basalt)-like volcanic rocks (Chukotat Group) (Francis et al., 1981, 1983; Picard et al., 1986, 1990); (4) deep-water sedimentary rocks (Spartan Group) (Lamothe et al., 1984; Lamothe, 1986) and (5) older, 1998 Ma oceanic crust (Watts Group) (Scott et al., 1988; 1989; St-Onge et al., 1989, St-Onge and Lucas, 1990a). All mineral age determinations are from Parrish (1989). All rocks have been metamorphosed at grades ranging from greenschist to upper-amphibolite facies (Bégin and Carmichael, 1987; Bégin, 1989a; 1989b). The prefix "meta-"applies thus to all rock-types subsequently described, but is omitted. Field-based research and geological mapping by St-Onge et al. (1986; 1987; 1988a; 1988b) in the eastern half of the belt has shown that packages (2) through (5) are allochthonous. They were tectonically imbricated during southward transport (D_1) onto Archean tonalitic gneisses of the Superior province. An overview of the deformation and metamorphic history is presented below.

Detailed mapping of the eastern portion of the CSB (St-Onge et al., 1986; 1987; 1988a) led to the recognition of two generations of D_1 thrusts; 1) an early set of regular- (piggyback) stacking sequence faults, and 2) a later, out-of-sequence set. The earlier set of thrusts developed above a basal décollement, prior to the thermal peak of metamorphism, in a sequence in which the faults young in the direction of propagation (i.e. towards the foreland) (Lucas, 1989). The later set of thrusts formed during and after peak metamorphic conditions, and re-imbricated the earlier, regular-sequence thrust stack and basal décollement (*ibid*). This out-of-sequence generation of faults incorporates basement into the thrust stack (Lucas, 1989; St-Onge and Lucas, 1990a). Various criteria were used to identify thrust faults in the field, including the recognition of repeated stratigraphic units separated by narrow zones of intense high strain (Lucas, 1989; St-Onge et al., 1986; 1987; and 1988a). The re-imbricated thrust stack was folded (D_2) together with the underlying Superior province basement into an east-west trending synclinorium (Lucas, 1989; St-Onge and Lucas, 1990a). At a regional scale, the D_1 and D_2 folds are statistically coaxial (St-Onge et al., 1986; Lucas, 1989) and their initial plunges are thought to have been principally subhorizontal. The D_2 synclinorium was refolded by a northwest-trending set of folds (D_3), causing a dome and basin-type fold-interference pattern which preserves D_1 thrust sheets as a series of tectonic klippen (Hoffman, 1985; St-Onge et al., 1989; St-Onge and Lucas, 1990a). The D_3 cross-folding event has caused a variation in plunge from east to west, and has provided exposure of more than 18 km of structural relief on the present erosion surface (Lucas, 1989).

Metamorphic isograds have overprinted and transected regular-sequence thrusts, whereas out-of-sequence fault zones are characterized by growth of thermal-peak or retrograde mineral assemblages (Lucas, 1989). Regional metamorphism resulting from crustal thickening during early, regular-sequence thrusting reached mid-amphibolite facies (Bégin, 1989a; Lucas, 1989; St-Onge and Lucas, 1989a). Out-of-sequence thrusting-related retrograde assemblages crystallized at greenschist-facies conditions (Bégin, 1989a).

The sequence of deformation events has been bracketed by U-Pb zircon dating of intrusions (Parrish, 1989) with various relationships to the D_1 thrusts (St-Onge and Lucas, 1990a). The D_1 piggyback-sequence thrusting may have started as early as 1918 $+9/-7$ Ma, the age of the youngest dated tectonostratigraphic unit. Tonalitic plutons with ages in the range 1880–1870

Ma have been observed to cut piggyback-sequence thrusting and thermal peak out-of-sequence thrusts, thus providing a minimum age constraint on this phase of the deformation. A tonalite pluton dated at 1876 ± 1.5 Ma is carried in the hanging-wall of a post-thermal peak out-of sequence thrust, and is thus interpreted to post-date movement on the thrust; this provides a maximum age for this latest phase of D_1 thrusting. All generations of D_1 thrust faults are deformed by D_2 folding. The D_1 and D_2 deformation events occurred during the Trans-Hudson orogeny.

The emplacement and subsequent deformation of the CSB are interpreted to have resulted from a collision between the Superior province and an overriding, exotic terrane (Sugluk terrane, see Figure 1) to the north (Hoffman, 1985; St-Onge and Lucas, 1990b). The collisional suture lies to the north of the Kovik antiform (Figure 1), which separates the synformal CSB from the Sugluk terrane. Subsequent discussion will be limited to the two northernmost tectonostratigraphic packages of the CSB, the Watts Group igneous rocks, and the Spartan Group sedimentary rocks.

Geology and Chemistry of the Ophiolitic Units

The northern margin of the Cape Smith Belt consists of a package of layered mafic and ultramafic rocks, sheeted mafic dykes and gabbros, and pillowed mafic volcanic rocks. These units comprise the Watts Group (see Figure 2), and are believed to represent the igneous crustal components of an ophiolite sequence. Field and chemical characteristics of these units are discussed in the following sections, based on observations made on the eastern portion of the ophiolite, the area shown in Figure 2.

Geology of the Igneous Units

Volcanic Sequence

The volcanic sequence is dominated by fine-grained pillowed flows, although 1–2 m thick massive basalt flows and 2–3 m thick mafic sills are also present. Pillows are generally less than one meter in diameter, and are commonly defined by epidote-rich selvages which are interpreted as representing former chilled margins. The facing direction of the pillows indicates that the sequence is upright. Primary igneous phenocrysts or their metamorphosed pseudomorphs are not observed in the volcanic rocks. Metamorphic mineral assemblages of the basalts correspond to the upper-greenschist- and amphibolite-facies, with a greenschist-facies retrograde overprint in the vicinity of post-thermal peak out-of-sequence faults (Bégin, 1989a). Deposits of interflow sediment are only rarely observed.

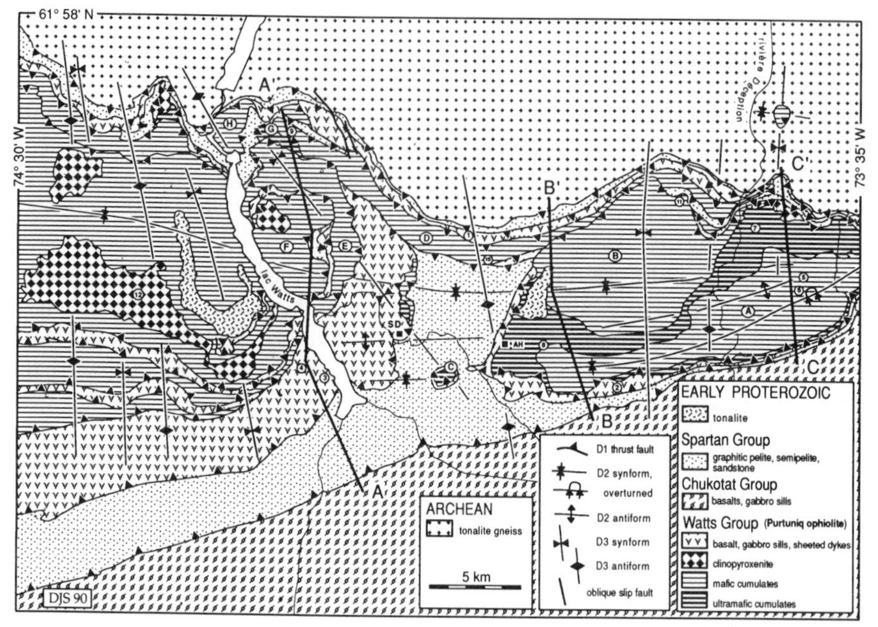

Figure 2. Generalized geology of the eastern portion of the Purtuniq ophiolite, Cape Smith Belt, northern Quebec. Location of this figure is outlined in Figure 1. Sections AA′, BB′, and CC′ are shown in Figure 5. "AH" denotes former Asbestos Hill mine at Purtuniq. "SD" indicates location of sheeted dykes referred to in text. Circled letters A-H identify cumulate packages referred to in text. Locations of samples listed in Table 1 are shown as circled numbers 1–12. Modified from St-Onge et al., 1988b.

Sheeted Dykes

A dense swarm of mafic dykes occurs at the base of the principal mafic thrust sheet. This feature is best developed at a location approximately 5 km due east of lac Watts (Figure 2, location "SD"), where the outcrop is composed of virtually 100% dykes. All samples analyzed for this study were collected at this location, which has an along-strike exposure of approximately 700 m, and exposed width of 150 m. Individual dykes range in width from less than 20 cm up to 50–60 cm, and can be traced along strike for tens of meters. The dykes strike NNE, and dip steeply toward the ESE. Dykes which show only one chilled margin ("half-dykes") are present, and make up approximately half of the population. The total number of half-dykes observed is however insufficient to determine a statistically valid spreading polarity based on "younging" direction. No primary igneous minerals were recognized in the dykes; the upper greenschist assemblage is similar to that in adjacent volcanic rocks (Bégin, 1989a). Imbrication of the Watts Group has preserved only the upper-most part of the sheeted dyke complex; along strike from location "SD" (Figure 2), the underlying thrust fault cuts upsection in its hanging-wall, thereby eliminating the sheeted dykes from the base of the thrust-bound volcanic sequence.

The transition from the sheeted dykes into the overlying volcanic sequence is well-exposed, and occurs gradually over a thickness of approximately 300–400 meters. Directly overlying the zone of 100% dykes is foliated, fine-grained, mafic material and rare screens of pillowed basalt cut by sheeted dykes. The fine-grained mafic material is thought to represent screens of dykes with unrecognizable margins, or thin massive sills or flows. Pillowed flows steadily become more common up-section, and individual dykes become correspondingly less frequent. Dykes are extremely rare in the volcanic sequence above this transition zone. Individual dykes have been observed in the volcanic section at the base of this thrust sheet along strike to the west of location SD. As the mafic sequence (south end of lac Watts, Figure 2) is thrust-bound at top and bottom, the original contacts with the overlying sedimentary rocks and underlying crustal cumulates are not observed.

Layered Gabbros

Layered mafic rocks are volumetrically the most extensive unit preserved in the ophiolite (Figure 2). Compositional layering is dominantly centimeter- to meter-scale, and is defined by modal variations of the metamorphic products of primary plagioclase and clinopyroxene. The thinnest layers are rarely continuous laterally for more than several meters, but thicker bands may be traced for several tens of meters. Individual layers range in composition from anorthosite to clinopyroxenite, with gabbroic compositions most common. During regional metamorphism, primary calcic plagioclase recrystallized to albite and clinozoisite, whereas the primary mafic phases were replaced by hornblende-actinolite. The metamorphic minerals define a schistosity (S_1) which is in general parallel to primary compositional layering. This coincident orientation is due to transposition of the primary layering into the S_1 plane during regional deformation and metamorphism (St-Onge et al., 1987; 1988a; St-Onge and Lucas, 1990a).

Primary, fine-scale textures are preserved locally in areas of lower strain. Examples of centimeter-scale modally graded, rhythmically repeated gabbroic layers are observed, with sharply defined mafic bases and felsic tops. Although they are rare, centimeter-scale mafic dykes are observed cross-cutting the dominant compositional layering, indicating an intrusive origin for some of the gabbroic material. Concordant, centimeter-scale layers of homogeneous gabbro are common, and may represent intrusive sills or in extremely deformed cases transposed dykes. Irregularly shaped, decimeter-scale lozenges of gabbroic pegmatite are rare, but have been observed throughout the layered gabbro section.

In addition to the centimeter-scale layers which characterize individual outcrops, macroscopic meter- to decameter-scale bands are recognized. These bands may be traced laterally for at least tens of meters. Laterally discontinuous, decameter thick lenses of layered ultramafic rocks are found throughout the layered gabbros.

Two samples of layered gabbro from the lac Watts area (Figure 2) have been dated using the U-Pb system in zircon. The zircons of the first sample are interpreted as being igneous in origin, and yielded an age of 1998 ± 2 Ma (Parrish, 1989). This determines the age of formation of the Watts Group (*ibid*). Zircons of the second sample have cloudy igneous cores dated at 1995–2000 Ma, and clearer overgrowths dated at 1977 ± 3 Ma (*ibid*). Neither the igneous ages nor the overgrowth age have been obtained elsewhere in the CSB, outside of the Watts Group; the significance of this will be discussed in a subsequent section.

Layered Ultramafic Rocks

Layered ultramafic rocks are the lowest preserved pseudostratigraphic unit in the ophiolite. They occur as decameter- to kilometer-thick bodies within the mafic cumulates, or as smaller, thrust-bound lozenges at the base of mafic cumulate thrust sheets. Centimeter-scale layering is defined by modal variations of relict primary clinopyroxene and olivine, and their metamorphic recrystallization products. Individual layers are dominantly millimeters to tens of centimeters thick. Packages of individual layers, and rare single layers, are up to tens of meters thick. The lateral extent of layers ranges from meters to tens of meters depending on layer thickness. Decameter-thick layers can be traced for hundreds of meters.

Modally graded layers observed in the field, and textures observed in thin-section (both described below) suggest a cumulate origin for these rocks. Thin (1–10 cm), modally graded layers have dunitic bases grading abruptly into wehrlitic-clinopyroxenitic tops. Adcumulate textures are commonly preserved in the wehrlitic and clinopyroxenitic rocks. Rhythmic repetitions of graded layers are observable from hand-specimen- to outcrop-scale. Homogeneous, fine-grained clinopyroxenitic layers with sharply defined contacts form centimeter-scale concordant bands within the layered ultramafic rocks. Decameter-scale units defined by systematic variations of individual layer thickness have been observed. At the base of each unit, primary compositional layers are thin (1–3 cm) and grade abruptly from dunitic to clinopyroxenitic compositions. Individual layers gradually increase in thickness upsection; near the top of each unit, layers are several meters thick, and are dominated by wehrlitic to clinopyroxenitic compositions. At a location studied in detail (thrust sheet G, Figure 2), eight such units were observed in a 400 m thick fault-bound lozenge.

Clinopyroxenitic layers contain varying amounts of hornblende and/or tremolite-actinolite, serpentine and magnetite, whereas dunitic layers are dominantly serpentine and magnetite, with minor amounts of brucite, chlorite, and calcite. Two texturally distinct olivine populations are recognized in thin sections of the ultramafic cumulates; the first population consists of relict igneous grains, the second of metamorphic grains.

Igneous grains are recognized by their generally fresh appearance, relatively coarse size (up to 1–2 mm in diameter), and relict adcumulate texture. Serpentinization of primary olivine-rich layers is extensive; the freshest dunitic rocks rarely contain more than 5–10% relict igneous olivine Individual grains are compositionally homogeneous, as is the population in any one thin-section. The observed range of compositions, based on microprobe analyses of primary olivines from 8 thin-sections is Fo_{92} to Fo_{81}. This range is believed to reflect variation in primary magma composition. Forsterite content in olivine does not systematically decrease with increased height in the cumulate pile, as would be expected had the cumulates crystallized from a single batch of magma in a closed chamber. This suggests that the magma chamber from which these rocks crystallized was periodically recharged with fresh, more primitive magma.

Layers in which primary olivine is absent locally contain metamorphic olivine. This olivine population consists of very fine neoblastic grains, which range in composition from Fo_{99} to Fo_{96}. They are observed only in the northern portion of the ophiolite, which, at amphibolite facies, represents the area of highest regional metamorphic grade (Bégin, 1989a). An early serpentinization event could explain the magnesium-rich composition of the metamorphic olivine, and the observed presence of magnetite. Hydration of igneous olivine lead to the formation of magnesium-rich serpentine and magnetite. Subsequently, during regional prograde metamorphism, this early serpentine is dehydrated to produce the magnesium-rich metamorphic olivine. Alternatively, the neoblastic nature of the metamorphic olivine and the mutually exclusive occurrence of the two olivine populations suggests that the metamorphic grains were derived directly at the expense of the igneous grains during prograde metamorphism. The loss of iron during this transition can be explained by the presence of magnetite (Scott, 1990).

Relict chromium-rich spinels are a common accessory phase in the ultramafic cumulates, and are subsequently referred to as chromites. Grain size is generally less than 1–2 mm, although aggregates of such grains may reach 5 mm in diameter. A full range of grain shapes, from equant to elongate, and euhedral to anhedral is present. Thin, concordant seams of chromite up to tens of centimeters in length are only rarely observed in the ultramafic cumulates. Individual grains commonly display a core overgrown by an optically distinct rim. Typically, the grains have chromium-rich, alumimum-poor cores (Cr^{3+} : 0.70–0.90; Al^{3+} : 0.02–0.20; Fe^{3+} : 0.01–0.25), and iron-rich overgrowths (Cr^{3+} : 0.00–0.16; Al^{3+} : 0.00–0.02; Fe^{3+} : 0.83–0.99). Individual cores are compositionally homogeneous. The core-overgrowth boundaries are abrupt. Chromite grains are locally surrounded by, and occasionally include chromium-bearing chlorite.

The chromium-rich, alumimum-poor core compositions of chromite grains from the CSB ophiolite are at one extreme of the compositional range reported for fresh igneous grains from other ophiolites and oceanic igneous rocks (e.g. Smith and Elthon, 1988; Dick and Bullen, 1984). Compared to examples in the literature, many of the chromites from the CSB ophiolite are distinctly aluminum-poor. This may be due in part to a combination of alteration and subsequent metamorphism. In these cases, cores may not preserve primary igneous compositions. The chlorite spatially associated with chromite is a likely sink for aluminum removed from the igneous cores.

Relict clinopyroxene is common in the layered ultramafic rocks. It is generally fresh, with only minor recrystallization to actinolite at grain boundaries. Adcumulate textures are commonly observed in clinopyroxene-rich rocks. Individual grains are usually 3–5 mm in diameter, but range up to 1–2 cm in extreme cases. Poikilitic grains enclosing olivine crystals, or their serpentinized pseudomorphs, and/or chromite are common. Clinopyroxene is diopsidic in composition and is not zoned chemically. Grains are of constant composition in a single thin-section. Clinopyroxene grains in modally graded layers show no compositional differences compared with those in more homogeneous layers. This suggests that both types of layers were derived from chemically similar magmas. Chemical homogenization is unlikely, as the samples studied are from areas that did not reach hornblende-clinopyroxene facies (Bégin, 1989a). The ratio $Mg^{2+}/(Mg^{2+} + Fe^{2+})$ in clinopyroxene ranges from 0.84 to 0.96, but does not vary systematically with position in the cumulate pile. The Cr_2O_3 content of clinopyroxene, which ranges between samples from < 0.5–1.5 weight percent, similarly does not correlate with stratigraphic height. These observations are not accounted for by simple fractional crystallization of a melt without periodic recharge of the melt with more primitive liquid. One possible explanation is that the magma chamber from which these rocks crystallized may have behaved as an open system, periodically being recharged with fresh, more primitive magma as material left the system as cumulates and erupted volcanics. The present data are compatible with such an hypothesis, as modelled by O'Hara (1977) and O'Hara and Mathews (1981) (Scott, 1990).

Primary igneous minerals and textures, such as modally graded layering, are more common in the ultramafic cumulates than in the mafic cumulates. Compositional layering (S_0) in the layered mafic rocks is generally parallel to a metamorphic schistosity (S_1). In contrast, where S_1 is developed in ultramafic rocks, it is generally at a high angle to compositional layering. These observations suggest that the layered mafic rocks are more highly recrystallized, and may record a larger amount of strain than do the ultramafic rocks (Lucas, 1990).

The stratigraphic transition from ultramafic- to mafic cumulates is well exposed east of rivière Déception (see Figure 2). It occurs gradationally, over an interval approximately 100 meters thick. The base of the transition is marked by the appearance of plagioclase-bearing layers, which increase in abundance upsection. The absence of dunitic layers defines the upper limit of the transition. This transitional relationship between the mafic cumulates and the underlying ultramafic cumulates (e.g. section CC′, Figure 5) supports their interpreted consanguineous origin (Scott, 1990). The crystallization sequence inferred from gross cumulate stratigraphy, is olivine + Cr-spinel followed by clinopyroxene then plagioclase. Orthopyroxene is not observed in the cumulate sequence.

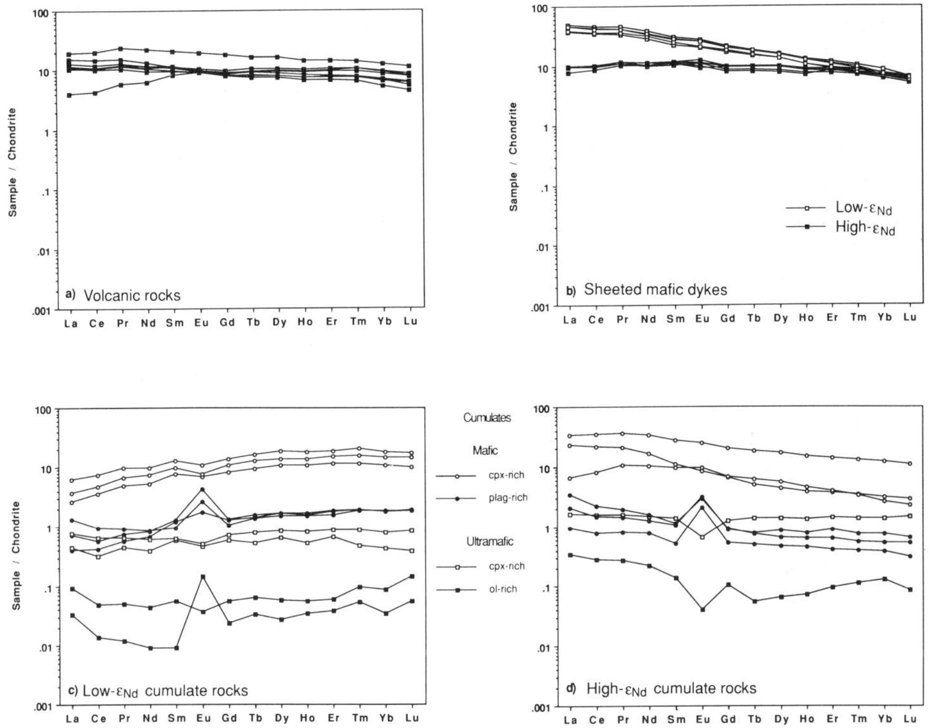

Figure 3. Chondrite normalized plots of: a) Watts Group volcanic rocks; b) Watts group sheeted dykes; c) low-ϵ_{Nd} cumulates, and; d) high-ϵ_{Nd} cumulates. Normalizing values used (ppm): La = 0.367; Ce = 0.957; Pr = 0.137; Nd = 0.711; Sm = 0.231; Eu = 0.087; Gd = 0.306; Tb = 0.058; Dy = 0.381; Ho = 0.0851; Er = 0.249; Tm = 0.0356; Yb = 0.248; Lu = 0.0381; from Taylor and McLennan, (1985, Appendix 2).

Plagiogranites

Plagiogranites are relatively rare in the Watts Group, but are observed in both the mafic volcanic pile and the cumulate sequence. In the volcanic pile, individual intrusive bodies occur as thin (<5 m thick) sills which are laterally extensive for tens of meters, locally observed being fed by cross-cutting dykes. Intrusions in the cumulate rocks tend to be somewhat larger, and more irregular in shape. The plagiogranites are poor in ferromagnesian minerals.

Non-Layered Clinopyroxenite Bodies

Numerous irregularly-shaped bodies of clinopyroxenite without compositional layering are found in the mafic and ultramafic cumulates, at all stratigraphic levels. These coarse-grained rocks show no signs of fine-scale modal- or size-graded layering, but vary laterally in their primary mafic mineralogy. Generally fresh diopside regularly comprises ≥ 90% of the rock, the balance being olivine (Fo_{85-80}) commonly variably serpentinized, and accessory Cr-

Fe spinels. Secondary growth of amphibole (actinolite) is observed, but is restricted to the boundaries of igneous grains. The samples studied are from areas of upper greenschist facies regional metamorphism (Bégin, 1989a), below the grade at which metamorphic hornblende would be stable. The bodies range in size from tens of meters to several kilometers in longest dimension; the largest bodies are shown in Figure 2 (diamond pattern, see especially the west side of lac Watts). The presence of layered gabbro xenoliths along the margins of this body demonstrates that it is intrusive into the layered gabbros. The small body on the northern margin of the belt, east of rivière Déception (see Figure 2) cross-cuts primary compositional layering in the ultramafic cumulates, and is also interpreted as intrusive. Whereas host-rock xenoliths have not been observed in many of the smaller bodies, their similar field, petrographic and chemical nature suggests that they share a common origin with the lac Watts and rivière Déception bodies described above. They have been deformed and metamorphosed in a manner similar to their hosts, which suggests that they are close in age to, and thus consanguineous with the CSB ophiolitic suite of rocks. This suggestion remains to be tested by radiometric age determination.

Wehrlite intrusions similar to those described above have been documented in the crustal section of the Oman ophiolite (Pallister and Hopson, 1981; Benn et al., 1988; Juteau et al., 1988), where they are believed to originate in the crust-mantle transition zone as late-stage melts. As the crust-mantle transition is not preserved in the CSB ophiolite, such an origin cannot be demonstrated based on field criteria. Nevertheless, it is clear that the massive clinopyroxenites postdate part of the main suite of layered rocks.

Chemistry of the Igneous Units

Volcanic Sequence

Chemical compositions of flows in the Watts Group correspond to those of tholeiitic basalt (Jensen, 1976) (see Table 1, Locations 1–4). Chondrite-normalized rare earth element (REE) patterns are flat at 10× chondritic abundance (Scott et al., 1988; 1989) (see Figure 3a), a pattern which resembles modern N-MORB (e.g. Sun et al., 1979). The Nd isotopic composition of two samples from the volcanic sequence ($\epsilon_{Nd(2.0Ga)}$ + 3.3 and + 4.0) (all Nd isotope data Hegner and Bevier, 1989) supports this interpretation, suggesting no influence of continental material during generation and eruption of these basalts (Hegner and Bevier, 1989). On the Ti/Cr-Ni diagram of Beccaluva et al. (1979), the volcanics fall within the ocean floor tholeiite (OFT) field near the statistically derived line which separates them from Island Arc Tholeiites (IAT) (Figure 4a). On the Ti-V diagram of Shervais (1982), the volcanics analyzed plot dominantly within the MORB field toward lower values of Ti/V (Figure 4b). Shervais (1982) suggested that decreasing Ti/V ratios correspond to increasing fO_2, and that those rocks in the MORB

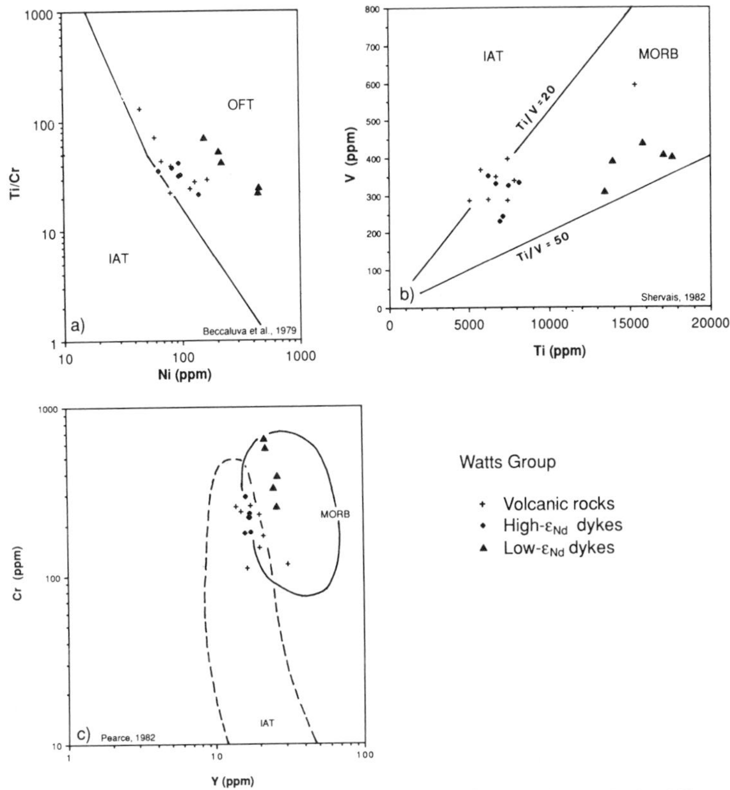

Figure 4. a) Ti/Cr – Ni diagram of Beccaluva et al., (1979). Statistically derived line separates Ocean Floor Tholeiites (OFT) from Island Arc Tholeiites (IAT). Watts Group volcanics and dykes plot in the OFT field, statistically similar to modern MORB (Beccaluva et al., (1983). b) Ti/V discrimination diagram of Shervais (1982). All Watts Group dykes plot in the field bounded by Ti/V = 17.5 to 50, as do most Watts Group volcanics analysed. Shervais (1982) has defined the field bounded by Ti/V = 20–50 as the range of modern MORB compositions. c) The Cr-Y diagram of Pearce (1982) does not clearly show the Watts Group in the MORB field, but rather as MORB- to MORB-IAT transitional.

field (Figure 4b) represent melts produced by 20–30% partial melting under relatively reducing conditions ($fO_2 \sim 10^{-11}-10^{-12}$). Figure 4b suggests that the Watts Group volcanic rocks may have been derived at the more oxidized end of this range. The volcanics are intermediate between MORB and IAT as defined on the Cr-Y diagram (Pearce, 1982) (Figure 4c). Based on whole-rock trace element and isotopic data, it is clear that the Watts Group volcanics formed in an oceanic environment. The rocks of the Watts Group are more similar to modern MORBs than to modern IATs.

Sheeted Dykes

Although all of the analysed dykes are tholeiitic (Jensen, 1976) (see Table 1, Location SD), they can be subdivided into two distinct populations on the

Table 1. Representative analyses of Watts group rock types. Locations 1–4-mafic volcanics; SD – sheeted dykes; Locations 5 & 9a-anorthositic layered gabbro; Locations 5 & 10-mafic layered gabbro; Locations 7 & 9b-clinopyroxenitic ultramafic cumulate; Locations 8 & 11-dunitic ultramafic cumulate; Location 12-intrusive clinopyroxenite. 1) Major element oxides (weight%), XRF, O. Mudroch, McMaster University. 2) LOI – Loss On Ignition. 3) Cr, V, Ni, Zr, (ppm), XRF, O. Mudroch, McMaster University. 4) Li-U (ppm), ICP-MS, S. Jackson, Memorial University of Newfoundland. 5) n/d – not detected.

	Volcanic rocks				High-ϵ_{Nd} Dykes			Low-ϵ_{Nd} Dykes		
Sample #	D247A	B367	S198	S199	B1	C5	E7	D3	E4	E5
location	1	2	3	4	SD	SD	SD	SD	SD	SD
SiO_2	45.50	52.70	45.70	49.10	49.80	46.33	47.60	43.95	42.40	41.25
Al_2O_3	14.90	12.80	15.20	11.70	12.10	13.15	12.30	12.85	9.54	9.97
CaO	11.00	7.10	9.96	10.12	10.84	11.01	10.12	11.54	11.84	11.66
MgO	8.72	7.28	8.58	6.27	7.29	8.47	9.03	7.90	11.64	11.22
Na_2O	2.31	3.13	2.07	2.11	2.53	2.18	2.13	0.96	0.41	0.38
K_2O	0.22	0.31	0.24	0.08	0.11	0.10	0.24	0.31	0.11	0.11
Fe_2O_3	13.70	13.03	12.29	14.28	13.28	15.18	13.24	16.52	17.19	19.51
MnO	0.19	0.16	0.17	0.19	0.17	0.23	0.18	0.22	0.22	0.27
TiO_2	1.12	1.24	1.24	0.96	1.12	1.19	1.04	2.95	2.32	2.33
P_2O_5	0.09	0.08	0.09	0.07	0.08	0.06	0.07	0.17	0.30	0.09
LOi	1.00	2.32	3.04	3.43	1.62	2.10	2.41	2.63	3.33	3.21
Total	98.80	100.15	98.58	98.13	98.94	100.00	98.36	100.00	99.30	100.00
cation mg#	0.56	0.53	0.58	0.47	0.52	0.52	0.57	0.49	0.57	0.53
Cr	240	172	257	147	179	220	294	254	644	560
Ni	130	68	166	82	84	99	142	158	447	460
V	349	397	286	367	330	245	353	400	389	307
Zr	80.4	69.4	91.5	70.3	72.7	80.3	72.9	201.6	144.2	136
Li	10.68	9.52	11.15	5.02	6.36	9.01	12.10	12.87	17.71	17.40
Be	0.39	0.18	0.37	0.42	0.40	0.35	0.60	1.49	1.01	1.20
Sc	35.57	36.37	31.57	40.57	42.62	40.38	35.46	26.18	32.00	28.09
Rb	1.33	3.99	4.20	0.51	0.53	0.69	2.79	6.91	0.96	1.12
Sr	195.20	96.02	229.54	172.30	156.07	158.99	157.60	538.15	257.52	137.28
Y	14.81	21.14	13.93	19.96	15.90	17.17	16.10	25.95	21.86	21.97
Nb	4.03	3.55	5.61	3.81	3.24	3.56	3.29	13.95	14.64	14.46
Cs	0.02	0.05	0.03	0.03	0.03	0.31	0.03	0.29	0.05	0.07
Ba	35.71	45.28	58.17	25.96	13.59	18.83	112.45	87.16	23.44	22.16
La	4.10	3.82	5.63	4.00	3.55	3.66	3.45	17.90	13.69	14.16
Ce	10.10	9.87	14.23	9.56	9.48	9.85	9.16	43.25	32.99	34.47
Pr	1.57	1.60	2.09	1.44	1.58	1.61	1.48	6.18	4.64	4.97
Nd	7.50	7.82	9.43	6.75	7.31	8.00	7.10	27.42	19.90	22.36
Sm	2.29	2.60	2.62	2.19	2.57	2.68	2.36	7.03	5.15	5.70
Eu	0.81	0.91	0.89	0.79	0.86	1.07	0.95	2.37	1.78	1.80
Gd	2.44	2.99	2.48	2.68	2.47	3.03	2.68	6.58	5.17	5.51
Tb	0.48	0.61	0.44	0.54	0.47	0.58	0.51	1.06	0.85	0.89
Dy	3.07	4.03	2.82	3.68	2.93	3.68	3.17	6.13	5.06	5.05
Ho	0.63	0.87	0.56	0.80	0.60	0.74	0.64	1.12	0.91	0.91
Er	1.83	2.63	1.66	2.42	2.08	2.05	1.87	2.89	2.20	2.32
Tm	0.26	0.37	0.23	0.37	0.26	0.28	0.25	0.36	0.28	0.30
Yb	1.61	2.27	1.33	2.33	1.60	1.76	1.53	2.11	1.65	1.68
Lu	0.23	0.30	0.17	0.33	0.21	0.23	0.20	0.25	0.22	0.22
Hf	0.28	0.53	0.63	0.64	0.80	0.66	0.72	2.37	2.52	2.26
Ta	0.35	0.28	0.35	0.55	0.42	0.77	0.29	0.81	1.42	1.00
T.	0.02	0.12	0.04	0.20	0.01	0.22	0.02	0.10	0.03	0.06
Pb	3.52	0.62	2.82	1.06	1.75	1.72	1.85	5.07	3.64	1.79
Bi	0.04	0.04	0.04	0.04	0.02	0.05	0.02	0.07	0.05	0.07
Th	0.36	0.33	0.46	0.31	0.25	0.26	0.21	1.37	0.98	1.09
U	0.10	0.10	0.13	0.11	0.08	0.09	0.08	0.33	0.29	0.40

High-ϵ_{Nd} Cumulates				Low-ϵ_{Nd} Cumulates				Cpx'ite
D144A 9a	D248A 10	D146D 9b	B387A 11	S318 5	S317 6	S297B 7	D251B 8	B357 12
48.00	47.10	50.10	42.16	46.90	44.10	45.25	41.39	45.20
10.30	28.80	1.11	0.31	13.60	20.10	1.58	3.22	3.11
13.20	16.50	15.30	0.06	13.38	12.50	12.12	1.12	14.34
12.70	1.47	21.10	37.13	8.02	4.84	21.78	34.05	21.75
0.63	2.44	0.01	0.06	0.96	1.72	0.17	0.04	0.16
0.10	0.13	0.03	0.01	0.07	0.02	0.01	0.01	0.01
11.30	1.63	6.06	9.01	13.71	11.83	14.21	10.35	9.60
0.21	0.03	0.15	0.15	0.20	0.18	0.15	0.14	0.15
0.57	0.11	0.05	0.03	0.80	0.72	0.11	0.10	0.20
0.19	0.02	0.01	0.02	0.06	0.00	0.03	0.01	0.00
1.77	2.39	5.39	10.93	1.88	3.10	4.43	9.07	3.96
99.20	100.30	99.70	99.87	99.58	99.11	99.84	99.50	98.33
0.69	0.64	0.87	0.89	0.54	0.45	0.75	0.87	0.82
1160	110	2750	8211	146	29	2408	3628	2211
210	20	340	1470	93	15	505	1361	429
230	42	130	26	366	248	343	96	270
62.6	47.1	–	9.4	27.9	21.3	5.2	4.4	10.9
3.59	5.00	0.87	0.43	2.65	2.03	2.37	0.30	0.96
1.00	0.04	0.33	0.12	0.05	0.21	n/d	n/d	n/d
25.33	5.12	41.74	2.62	59.28	44.54	55.40	13.97	61.78
0.37	1.75	0.11	0.08	0.59	0.17	0.13	0.18	0.27
170.11	314.26	63.65	8.45	90.32	164.90	6.95	6.70	14.19
8.61	1.29	1.71	0.14	16.47	3.62	1.02	1.86	3.67
7.68	0.23	0.24	0.17	2.35	0.22	1.32	0.13	0.07
0.05	0.12	0.08	0.03	0.05	0.02	0.05	0.07	0.04
12.25	40.09	2.27	5.61	20.42	8.70	42.79	4.02	3.93
8.50	1.26	1.11	0.13	0.65	0.27	0.16	0.13	0.07
20.43	2.15	1.48	0.27	2.57	0.54	0.30	0.33	0.29
2.88	0.27	0.19	0.04	0.57	0.10	0.06	0.07	0.08
11.86	1.14	0.75	0.16	3.59	0.59	0.27	0.35	0.59
2.51	0.26	0.17	0.03	1.68	0.30	0.14	0.16	0.34
0.74	0.25	0.09	0.00	0.65	0.37	0.04	0.04	0.13
2.06	0.28	0.16	0.03	2.59	0.41	0.18	0.23	0.51
0.29	0.04	0.03	0.00	0.56	0.09	0.03	0.04	0.11
1.65	0.25	0.23	0.03	3.98	0.65	0.25	0.33	0.72
0.32	0.05	0.06	0.01	0.88	0.14	0.04	0.07	0.15
0.91	0.16	0.22	0.02	2.67	0.46	0.17	0.23	0.45
0.12	0.02	0.03	0.00	0.40	0.07	0.02	0.06	0.06
0.78	0.13	0.27	0.03	2.39	0.46	0.11	0.23	0.37
0.11	0.02	0.05	0.00	0.39	0.07	0.01	0.04	0.05
0.30	0.06	8.39	0.11	0.68	0.15	0.08	0.10	0.15
0.78	0.31	0.24	0.02	1.98	0.20	0.35	0.02	0.08
0.05	0.03	n/d	0.03	0.02	0.02	0.01	0.02	0.01
1.47	1.06	1.55	1.91	1.22	0.83	0.75	2.29	0.72
0.02	0.03	n/d	0.04	0.00	0.05	0.01	0.05	0.01
0.40	0.05	0.15	0.03	0.02	0.06	0.01	0.06	0.01
0.10	0.02	0.17	0.04	n/d	0.01	0.00	0.02	0.00

basis of their major element covariations, trace element contents, and Nd isotopic composition. The first population is characterized by decreasing weight percent Al_2O_3 and increasing FeO* with decreasing MgO, flat chondrite-normalized REE patterns that are 10x chondrite (Figure 3b), and $\epsilon_{Nd(2.0Ga)}$ from + 4.0 to + 4.7. These characteristics are similar to modern N-MORB (e.g. Sun et al., 1979; BVSP, 1981). These are subsequently referred to as "high-ϵ_{Nd}" dykes. The second population shows increasing weight percent Al_2O_3 and slightly decreasing FeO* with decreasing MgO, a relatively strong enrichment in light REE (LREE) (Figure 3b) and other trace elements (Zr, Y, Th, U) (Table 1), and $\epsilon_{Nd(2.0Ga)}$ from + 3.0 to + 3.4. These are subsequently referred to as "low-ϵ_{Nd}" dykes. The major element covariations, and Cr, Sc, Zr, Nb, Y, Th, and REE contents of the low-ϵ_{Nd} dykes are very similar to Kilauea tholeiites (eg Basaltic Volcanism Study Project, 1981). The low-ϵ_{Nd} dykes have LREE contents approximately twice that of modern "plume" or enriched MORB (Sun et al., 1979; Basaltic Volcanism Study Project, 1981). These characteristics suggest that the low-ϵ_{Nd} Watts Group dykes may reflect the influence of a mantle plume.

All dykes analyzed plot within the OFT field on the Ti/Cr-Ni plot (Beccaluva et al., 1979) (Figure 4a). Both populations of dykes plot within the MORB field on the Ti-V plot (Figure 4b) (Shervais, 1982). The high-ϵ_{Nd} population clusters at lower Ti values than does the low-ϵ_{Nd} population. The low-ϵ_{Nd} group dykes plot in the MORB field on the Cr-Y diagram (Figure 4c) (Pearce, 1982); the high-ϵ_{Nd} dykes, however, lie just outside the MORB range in the IAT field.

At locality "SD" (Figure 2), the two dyke populations are indistinguishable in outcrop; they do not show consistent cross-cutting relationships, suggesting that emplacement was coeval. Of the samples analysed, the two populations occur in roughly equal proportions. Compositional variations between roughly coeval dykes have also been documented in sheeted dyke swarms of Phanerozoic ophiolites (e.g. Oman, Pearce et al., 1981; Troodos, Cyprus, Baragar et al., 1987; Lewis Hills, SW Newfoundland, Elthon et al., 1986), and in *in situ* oceanic crust (DSDP hole 504B, Emmermann, 1985; Tual et al., 1985). The two populations of dykes in the Watts Group are thought to represent the products of two discrete magma chambers; this suggestion will be discussed further in a subsequent section.

The overlying mafic volcanic rocks are compositionally more similar to the high-ϵ_{Nd} dykes than the low-ϵ_{Nd} dykes, based on major and trace element abundances (Table 1). This similarity is expressed on the REE plots (Figure 3a and b) and the trace element discrimination diagrams (Figure 4), supporting their interpreted genetic relationship. The neodymium isotopic compositions of two samples of the volcanic rocks ($\epsilon_{Nd(2.0Ga)}$ + 3.3 and + 4.0) are intermediate between those of the two dyke populations ($\epsilon_{Nd(2.0Ga)}$ + 3.0 to + 3.4 and + 4.0 to + 4.7). Preliminary sampling of the full exposed thickness of Watts Group volcanic rocks in the study area has not identified volcanic rocks which correspond chemically to the low-ϵ_{Nd} dyke population.

There are several possible explanations for this absence; (1) the flows fed by the low-ϵ_{Nd} dykes are not preserved; (2) such flows were not deposited in the eastern portion of the ophiolite; (3) the low-ϵ_{Nd} dykes did not feed volcanic flows. The first possibility may be the result of imbrication of the ophiolite; the upper limit of the preserved volcanic sequence is a fault contact (Figure 2). However, the complete removal of such flows would seem fortuitous. The paucity of dykes in the upper portion of the volcanic section supports the suggestion that low-ϵ_{Nd} volcanic rocks were not erupted in this area. Volcanic rocks related to the low-ϵ_{Nd} dykes may yet be identified elsewhere in the ophiolite, specifically to the west of the present study area.

Layered Mafic and Ultramafic Rocks

Rare earth element analyses of the compositionally layered gabbros and ultramafic rocks show patterns consistent with their interpreted igneous mineralogy. Plagioclase-rich gabbros show distinct positive Eu anomalies, consistent with the notion of plagioclase accumulation (Figure 3c and d). Clinopyroxene-rich gabbro samples in general have higher overall REE abundances than plagioclase-rich samples, with only weak development of negative Eu anomalies (Figure 3c and d). These observations support the interpretation that the layered rocks are cumulate in origin. The cumulate rocks have been found to be either LREE-depleted, or slightly LREE-enriched (compare Figure 3c and d). Within any one thrust sheet, LREE-depletion or enrichment is independent of bulk chemistry and modal mineralogy. This feature is readily observed when cumulates of similar bulk composition found in different thrust sheets are compared (see Table 1, compare Location 5 with 9a; 6 with 10; 7 with 9b; 8 with 11; Figure 3c and d). Cumulates from thrust-sheets A, C, E and G (Figure 2; Table 1, Locations 5–8) consistently show LREE-depleted compositions, whereas those in thrust-sheets B, D, F and H are slightly LREE-enriched (Table 1, locations 9–11). Within a single thrust-sheet, whole-rock abundances of trace elements do not vary systematically with stratigraphic height, further supporting the open-system fractionation hypothesis. The LREE-depleted population has $\epsilon_{Nd(2.0Ga)}$ values which range from $+2.5$ to $+3.6$, whereas the LREE-enriched samples range from $+4.6$ to $+5.3$ (E. Hegner, pers. comm., 1989).

The neodymium isotope data show that the high-ϵ_{Nd} population of sheeted dykes ($\epsilon_{Nd(2.0Ga)}$ $+4.0$ to $+4.7$) is isotopically similar to the LREE-*enriched* cumulate population ($\epsilon_{Nd(2.0Ga)}$ $+4.6$ to $+5.3$) whereas the low-ϵ_{Nd} dykes ($\epsilon_{Nd(2.0Ga)}$ $+3.0$ to $+3.4$) are isotopically within the range of the LREE-*depleted* cumulates ($\epsilon_{Nd(2.0Ga)}$ $+2.5$ to $+3.6$). This suggests a genetic relationship between each pair of dyke-cumulate populations. The polarity of ϵ_{Nd} populations is thought to indicate that two time-integrated depleted mantle sources were responsible for the generation of the Watts Group (Scott and Hegner, 1990). One source was highly depleted ($\epsilon_{Nd(2.0Ga)}$ $+4.0$ to $+5.3$), the other less so ($\epsilon_{Nd(2.0Ga)}$ $+2.5$ to $+3.6$). This suggestion requires at least two physically separate chambers to produce the observed chemical and isotopic variations. The mutually intrusive nature of the two dyke populations observed at location "SD" (Figure 2) indicates that the two mantle sources were active synchronously, and in close proximity to one another. The bimodality of the trace element abundance and neodymium isotopic data indicates that the sources remained physically independent; the observed juxtaposition reflects the tectonic imbrication of the formerly independent chambers. The synchroneity of the two populations is testable with further U-Pb geochronology; the two U-Pb age determinations discussed previously are both from LREE-enriched high-ϵ_{Nd} cumulate rocks. That high-ϵ_{Nd} cumulate rocks presently have LREE-*enriched* abundances strongly suggests that their source was enriched in LREE immediately prior to melt generation. This enrichment is thought to be the result of a mantle metasomatic event (Scott and Hegner, 1990).

Non-Layered Clinopyroxenite Bodies

Major-element analyses of the intrusive clinopyroxenite bodies are similar to concordant, clinopyroxene-rich layers in the cumulate ultramafic rocks (see Table 1, Location 12). Neodymium isotopic analyses of three samples range from $\epsilon_{Nd(2.0Ga)}$ + 1.9 to + 3.7. This range coincides with that obtained for the LREE-depleted cumulate rocks ($\epsilon_{Nd(2.0Ga)}$ + 2.5 to + 3.6), and suggests they have a common mantle source. These bodies are intrusive into both populations of cumulate rocks, further supporting the interpretation of a close spatial association of the two mantle sources and their respective products.

The Purtuniq Ophiolite

The units of the Watts Group, described above, represent all of the igneous crustal members of an ophiolite suite according to the Penrose Conference definition (Anonymous, 1972). They are subsequently referred to as the *Purtuniq* ophiolite (Figure 2) (St-Onge et al., 1987; 1988a; Scott et al., 1988). Tectonized, depleted mantle rocks have not yet been identified in the CSB. A simple "ophiolitic sequence" of mafic volcanic rocks, sheeted dykes, gabbros, and layered mafic and ultramafic cumulates is not immediately evident from the map pattern shown in Figure 2 due to dismemberment and imbrication during D_1 thrusting, and subsequent folding and cross-folding during D_2 and D_3 respectively. As presently preserved, the thrust sheets containing the mafic and ultramafic cumulates overlie, and are imbricated with, thrust sheets containing high-level gabbros, sheeted dykes and mafic flows. Similarly, thrust sheets with dykes and flows structurally overlie deep-water sedimentary rocks of the Spartan Group.

Spartan Group

The Spartan Group consists of sedimentary rocks which are tectonically interleaved between the underlying Chukotat (transitional- to MORB-like volcanic rocks) and overlying Watts Groups (ophiolitic rocks) (Figure 2). In the south, the stratigraphically lowest exposed part of the group comprises thinly bedded (2–3 mm to 1–2 cm) graphitic pelites interbedded with semipelite. The sequence coarsens up stratigraphic section into more quartz-rich semipelites. Individual semipelite beds commonly grade upward into graphite-rich pelite. The stratigraphically highest exposed portion of the group comprises sandstone turbidites. The graded pelite-semipelite beds and sandstone turbidites are interpreted as deep-water and distal fan deposits respectively (St-Onge and Lucas, 1990a)

Where the Spartan Group is exposed along the more outboard northern margin of the belt (Figure 2), it is dominantly composed of semipelite, and it contains a higher proportion of sandstone than in the southern exposures of the Group. It also includes a substantial thickness of mafic sedimentary schist consisting of the assemblage quartz-plagioclase-amphibole. Rare mafic flows, and gabbroic and peridotitic sills are also present in the northern exposures of the group, but primary textural features are not observed, due to their highly deformed state and amphibolite-grade metamorphism. Similar mafic sedimentary rocks are noted within the group along strike to the west and north of the present study area (St-Onge and Lucas, 1990b). These significant differences between northern and southern exposures of the Spartan Group are not fully understood in terms of their paleo-environmental implications.

The primary tectonostratigraphic relationship of the Spartan Group sedimentary rocks to the igneous rocks of the ophiolite (Watts Group) is uncertain, as the two Groups are not observed in stratigraphic contact, and the depositional age of the Spartan Group is unknown. Future U-Pb age studies of detrital zircons in the Spartan Group (R. Parrish, Geological Survey of Canada) will help to constrain both the age and provenance of the Group. If the rocks at the base of the Spartan Group contain detrital minerals which are significantly *younger* than the age of ophiolite, it would seem more reasonable to interpret the Spartan Group as having been deposited on the younger transitional crust which comprises the Chukotat Group, rather than on the older crust of the Watts Group.

Composite Crustal Section

Tectonic dismemberment by thrusting is such that no continuous, complete stratigraphic section through the early Proterozoic oceanic crust is preserved. Thus individual, dismembered segments have to be re-assembled to produce a composite section through the Purtuniq ophiolite crustal sequence. Structural (and to a lesser extent topographic) relief was used to construct a series of down-plunge constrained cross-sections projected onto vertical planes. Sections, presented in Figure 5, were drawn through various parts of the ophiolite, in order to determine individual segment thicknesses.

Each cross-section in Figure 5 was constructed by projecting surface geological data onto a vertical plane, after the method of Stockwell (1950). This method assumes that the geologic information illustrated on the map is predictable (and therefore projectable) at the scale of the structural domain. This assumption has been shown to be valid for the eastern Cape Smith Belt (Lucas, 1989). The cross-sections were constructed with no vertical exaggeration. Section CC' represents data from a single domain of uniform plunge, whereas sections AA' and BB' were generated using data from three and two adjacent plunge domains respectively. Each straight segment of the map traces of sections AA' and BB' (Figure 2) is the trace of the projection surface for a single structural domain. Taking into account the differences between the apparent thicknesses shown in vertical section (Figure 5) and true thicknesses in down-plunge fold-profile, the thicknesses of individual segments used to create the composite ophiolite section shown in Figure 6 can be determined directly from Figure 5. Individual segments used to construct Figure 6 are described below.

The maximum continuous thickness of ultramafic cumulates observed is approximately 2200m, as shown in section CC', on the north limb of a regional D_2 synformal structure (segment 1, section CC', Figure 5). The base of the ultramafic cumulate sequence is not preserved, as the lower end of segment 1 is a thrust fault. The top of segment 1 is the transition into mafic cumulates, which occurs over an approximately 100m thick interval. The base of segment 2 is taken as the ultramafic-mafic cumulate transition, on the southern limb of the same synform (section CC', Figure 5); its top is the map trace of the (overturned) synformal axis. This gives a total thickness of mafic cumulates of approximately 2100m. Segment 3 (section AA', Figure 5) represents the principal mafic thrust sheet, with some sheeted dykes preserved at its base. The total thickness of this thrust sheet is approximately 2800m. The thickness of the sedimentary package (segment 4) is estimated

Figure 5. Down-plunge constrained vertical cross-sections constructed along lines of section shown in Figure 2. The 1:50 000-scale maps of the eastern portion of the Purtuniq ophiolite (St-Onge et al., 1988b; St-Onge and Lucas, 1989b, c) which are summarized here in Figure 2, were divided into domains of statistically uniform plunge of D_1 and D_2 fold axes, and a mean plunge orientation was determined for each domain (Lucas, 1989). The systematic variations in mean plunge are due to variations in the amplitude of D_2 folds, and D_3 cross folding about axes oriented approximately perpendicular to the D_1 and D_2 structures (*ibid*). Data from individual structural domains are projected using the method of Stockwell (1950). The present sections are drawn with no vertical exaggeration. Southernmost portions of Sections AA′ and BB′ are constrained by surface dip data. Sections AA′ and CC′ are projected from the east, BB′ from the west. Line segments 2–4 are used to construct Figure 6. Circled numbers indicate locations of samples listed in Table 1 which are projected into the cross-sections. Circled letters identify cumulate bodies referred to in text. "SD" on section AA′ denotes the location of the sheeted dykes described in text. "SL" indicates sea level.

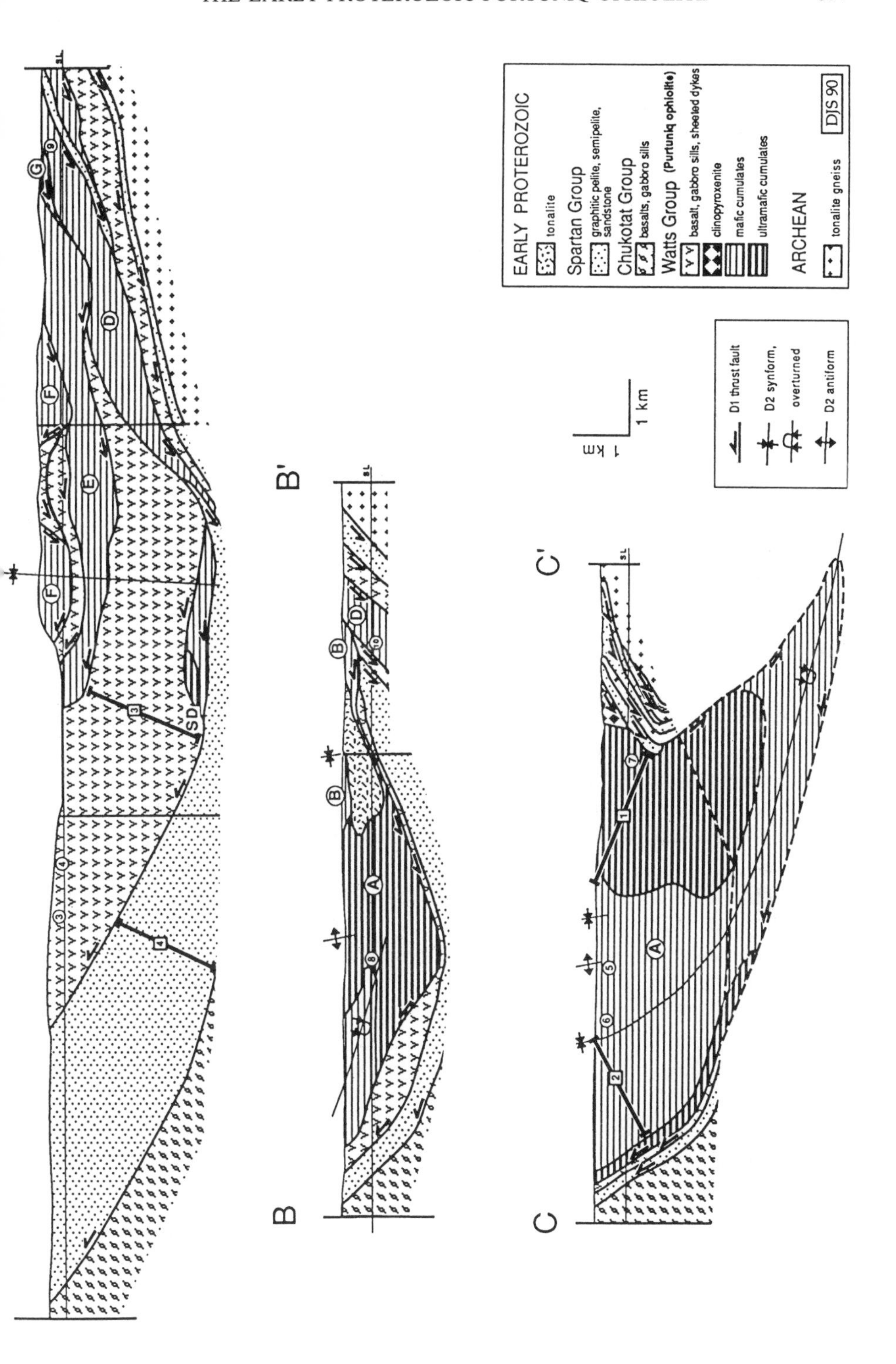

B
B'
C
C'
SL
SD
1 km
1 km
D1 thrust fault
D2 synform,
overturned
D2 antiform
EARLY PROTEROZOIC
tonalite
Spartan Group
graphitic pelite, semipelite, sandstone
Chukotat Group
basalts, gabbro sills
Watts Group (Purtuniq ophiolite)
basalt, gabbro sills, sheeted dykes
clinopyroxenite
mafic cumulates
ultramafic cumulates
ARCHEAN
tonalite gneiss
DJS 90

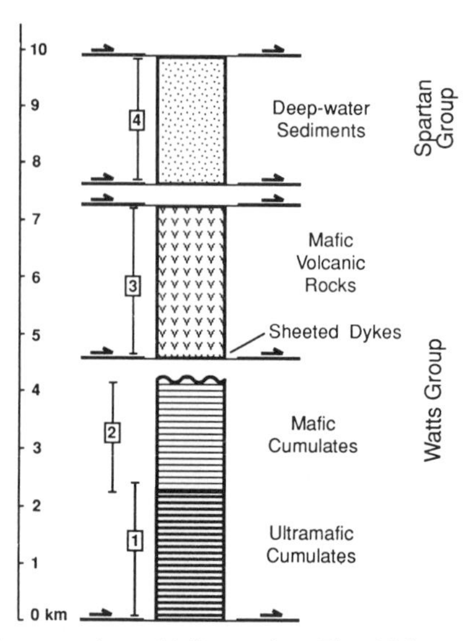

Figure 6. Reconstructed, composite ophiolite section. The thickness of segments 1–4 are corrected to account for differences between true thicknesses of segments and those shown in vertical sections in Figure 5. The locations of the segments used to construct this figure are indicated on sections AA′ and CC′ in Figure 5. Thicknesses of missing portions of stratigraphy (e.g. cumulate-dyke transition) are shown as minima.

from Section AA' (Figure 5) to be approximately 2000m; it is fault-bound at both top and bottom.

The thicknesses of the individual segments shown in Figure 6 should be considered well-constrained estimates. Structural repetitions along smaller thrusts and folds of S_0 that were not recognized may be present, resulting in over-estimation of primary thickness. The thickness of the sedimentary package (segment 4) is almost certainly inflated by meso- to microscopic scale folds (D_2) of bedding-parallel schistosity (S_1). Thicknesses of missing portions of the ophiolite section (e.g. cumulate-dyke transition) are shown as minimum estimates. The total thickness of igneous crustal rocks in the composite section (Figure 6) is greater than 7 km. The additional thickness of the mafic cumulate-sheeted dyke transition could reasonably increase this figure to 7.5–8 km. Similarly, the tectonic nature of the lower contact of the ultramafic cumulates (segment 1) and the upper contact of the volcanic pile (segment 3) allows the possibility of additional crustal thickness, possibly up to 9–10 km. This figure includes neither the overlying blanket of clastic sedimentary rocks, nor residual mantle material which has not been identified in the CSB.

The thickness of igneous crustal components preserved in the eastern portion of the Purtuniq ophiolite is somewhat greater than the thickest crustal sections reported from Phanerozoic ophiolites. Crustal thicknesses between 6 and 7 km have been reported from the Samail ophiolite (Hopson and

Pallister, 1980; Hopson et al. 1981; Christensen and Smewing, 1981); 4–5 km are reported for the Bay of Islands complex, Newfoundland (Malpas and Stevens, 1977; Suen et al., 1979); and up to 6 km from the Karmøy ophiolite, southwest Norway (Sturt et al., 1980).

Environment of Formation of the Purtuniq Ophiolite

Numerous lines of chemical and field evidence suggest that the Purtuniq ophiolite igneous suite is crust produced in an oceanic spreading environment. Relationships observed in the field, such as the stratigraphic transition from the sheeted dykes upwards into pillow basalts, and their association with mafic-ultramafic cumulates, are consistent with an oceanic interpretation. The chemistry of the Watts Group is indistinguishable from that of modern oceanic rocks. Concomitant Nd-Sm isotopic studies of all members of the igneous suite (Hegner and Bevier, 1989) support an oceanic interpretation. Due to the essential continuum of chemical compositions from various oceanic environments and the effects of alteration and metamorphism, it does not appear to be possible here to distinguish between a large, oceanic setting, a restricted back-arc setting, or an incipient arc solely on the basis of chemistry (e.g. Hawkins, 1980). The relative merits of the various oceanic settings are discussed below.

Several attributes of the Watts Group volcanic sequence observable in the field support an ocean-ridge rather than a subduction-related origin. The overall composition of the Watts Group volcanic pile is tholeiitic basalt; intermediate and silicic volcanics are not present. Island arc tholeiites rarely comprise more than 20% of the total of all volcanics in an arc (Basaltic Volcanism Study Project, 1981, p. 207). Plagiogranitic/dioritic intrusions are only rarely observed in the Watts Group, whereas they are common in subduction-related ophiolites (Pearce et al., 1984). The transition from sheeted dykes into volcanic flows observed in the Purtuniq ophiolite is similar to those described in various other ophiolites (Annieopsquotch, SW Newfoundland, Dunning, 1987; Bay of Islands, Newfoundland, Rosencrantz, 1983; Samail, below "Geotimes" unit, Pearce et al., 1981) interpreted as having formed at oceanic spreading ridges.

Mid-ocean ridge basalts are chemically similar to island arc tholeiites (IAT) in abundances of major elements, REE's, and certain trace elements. However, some significant differences do exist. As reported by numerous authors (e.g. Basaltic Volcanism Study Project, 1981), the Al_2O_3 content of IATs is somewhat higher, and TiO_2 lower than MORB at a given SiO_2 content (e.g. Sun et al., 1979). In this respect, the volcanics and dykes of the Watts Group more closely resemble MORB than IAT (Table 1). The values reported here for K and Ba are not considered meaningful due to the high mobility of these elements during metamorphism, and thus are not useful as environmental discriminators. The Watts Group mafic volcanics and sheeted dykes are not enriched in large-ion lithophile elements relative to high field strength elements as would be expected for an arc setting (e.g. Saunders et al., 1980). Trace element discrimination diagrams, such as those shown in Figure 4, suggest that the Watts Group volcanics and sheeted dykes share more similarities with MORBs than with IATs.

The compositions of relict primary igneous silicates in the cumulates of the Purtuniq ophiolite are comparable to those reported for the Bay of Islands complex of western Newfoundland (Smith and Elthon, 1988), thought to have formed at a mid-ocean ridge (Pearce et al., 1984, and references therein). The crystallization sequence in the cumulates of the Purtuniq ophiolite appears to be olivine and spinel, followed by clinopyroxene and then plagioclase. Arc cumulate sequences documented in the literature are dominated by orthopyroxene (e.g. the Tonsina assemblage, Alaska; DeBari and Coleman, 1989). The absence of orthopyroxene from the Purtuniq ophiolite cumulate sequence argues against an island arc tectonic setting for the origin of the cumulates (Pearce et al., 1984).

Tectonic Interpretation

The Purtuniq ophiolite section is somewhat thicker than that of Phanerozoic ophiolites. A greater thickness of oceanic lithosphere in the Proterozoic due to a higher degree of partial melting of a more fertile mantle has been predicted by Moores (1986), and in the Archean due to a hotter mantle by Sleep and Windley (1982), and Bickle (1986); the Purtuniq ophiolite may represent a positive test of these predictions. Moores (1986) suggested that Proterozoic ophiolites, if they existed at all, would have had crustal sections two to three times as thick as the 6 km generally accepted for modern oceans.

Numerous studies have shown that in general, Phanerozoic ophiolites are obducted soon after their formation (e.g. Spray, 1984). In contrast, regional U-Pb geochronological constraints suggest that the Purtuniq ophiolite may have been relatively older when it was obducted. As previously stated, the age of formation of the Purtuniq ophiolite is 1998 Ma (Parrish, 1989), and that of a high-temperature metamorphic event is 1977 Ma (*ibid*). If obduction of the ophiolite was not initiated until development of D_1 deformation as seen in the rest of the belt (as early as 1918 Ma), the Purtuniq oceanic crust would have been of the order of 80 Ma when obducted. Alternatively, the high-temperature zircon growth event which occurred at 1977 Ma (not recognized elsewhere in the CSB) may record the initiation of early, obduction-related, intra-oceanic thrusting. In this case, crustal age at obduction would have been approximately 20 Ma. It is unlikely that this metamorphic event represents intra-oceanic spreading-related deformation (e.g. Helmstaedt and Allen, 1977; Mével et al., 1990), unless renewed magmatic activity affected the rocks 20 Ma after their formation. Further U-Pb geochronology is necessary to resolve these questions.

The opportunity to directly date the obduction event using the metamorphic minerals from an obduction-related dynamothermal contact aureole is not available in the Purtuniq ophiolite, as such an aureole is not observed below the lowest exposed level of the ophiolite. The lowest exposed level of the ophiolite is presently in the hanging-wall of a post-thermal peak out-of-sequence thrust fault (Lucas, 1989). Whether this fault is the original obduction surface which was subsequently reactivated during post-thermal peak out-of-sequence thrusting or is a fault initiated during the out-of-sequence phase of thrusting is at present unknown. In either case, the documented late movement of this thrust has juxtaposed the hanging-wall against its present foot-wall rocks at a late stage in the development of the thrust belt. The out-of-sequence thrust which carries the ophiolite is younger than 1876 Ma (Parrish, 1989; St-Onge and Lucas, 1990a), *ca.* 140 Ma *younger* than the age of the ophiolite. Thus it is unlikely that an obduction-related metamorphic aureole would be presently observed directly below the lowest exposed units of the ophiolite. The role of out-of-sequence thrusting in ophiolite obduction has been documented elsewhere, both in Oman and western Newfoundland (e.g. Searle, 1985; Cawood and Williams, 1988; 1990).

Depleted mantle material is commonly obducted in Phanerozoic ophiolites, but has not been observed in the Purtuniq ophiolite. As mentioned previously, the uppermost part of the depleted mantle may originally have been obducted along with the crustal sequence, the crustal sequence subsequently being separated from the mantle during out-of-sequence faulting. Alternatively, a possible mechanism to explain this feature is delamination of the lithosphere prior to obduction, occurring at or above the ultramafic cumulate-mantle boundary. Such a model has been proposed by Hoffman and Ranalli (1988), wherein a thicker oceanic crust develops a strong-weak-strong rheological layering due to a nominally higher oceanic geothermal gradient and its greater thickness. The weak interface is postulated here to have been located within the lower stratigraphic levels of the crustal cumulates, allowing delamination to occur, thus preventing the uppermost part of the mantle from being incorporated into the accretionary assemblage.

The older age of the ophiolite (1998 Ma, Parrish, 1989) with respect to the rift sequence (*ca.*1960 Ma, Parrish, 1989) suggests that it is not part of a south- to north progression from rift to oceanic crust. Two possible explanations of the older age of the ophiolite are suggested; (1) it may have developed in a separate basin, outboard of the rift margin, or (2) it may be an older, along-strike equivalent segment of the same rift-ocean margin, subsequently juxtaposed against a younger portion of the margin by transpression (St-Onge et al., 1989; Scott et al., 1989).

The first (two-basin) model implies the existence of crust to separate the ophiolite from the younger rift-margin. No record of such crust has been identified within the CSB. The depositional basement to the sedimentary rocks of the Spartan Group which separate the Chukotat Group (part of the younger basin) from the igneous rocks of the ophiolite is not observed. The modern Red Sea rift serves as an actualistic model for the second scenario.

There, within a single basin, continental rifting and mid-ocean ridge-like spreading are occurring synchronously, several hundreds of kilometers along-strike from one another (Cochran and Martinez, 1988) The earliest truly oceanic rocks in the Red Sea are approximately 5 million years *older* than the youngest continental rifting (*ibid*). If subsequently deformed by large-scale transcurrent faulting, this would produce a sequence of rocks with the same relative age relationship observed in the CSB.

The age of the continental rifting event in the CSB (Povungnituk Group) is constrained by the 1958 +3.1/-2.7 Ma age of a rhyolitic body (Parrish, 1989) near the top of the rift sequence, indicating that the rifting event was initiated well before this time. Thus the 40 Ma age difference between the continental rift and ophiolite should be considered a maximum, as the age of initiation of rifting could in fact be the same age or older than the 1998 Ma age of the ophiolite. This possibility can only be tested by direct dating of erupted material at the base of the rift-related sequence. The Red Sea analogy implies an along-strike diachroneity within units of the Povungnituk, Chukotat and Watts Groups. Additional radiometric age determinations are required to further clarify the tectonic evolution of the basin in which the Purtuniq ophiolite formed, and of the entire northern Quebec peninsula.

The Purtuniq ophiolite in the CSB may be of particular interest to workers in other old mountain belts, for example the "greenstone" (greenschist-grade volcano-sedimentary) belts characteristic of Archean cratons. The lithostratigraphy of CSB, which is dominated by pillowed mafic volcanics and siliciclastic sedimentary rocks, is broadly similar to that typically found in Archean greenstone belts (e.g. Condie, 1981, and references within). The CSB, however, has not been as extensively modified by pervasive late faulting and granitoid intrusions (*ibid*), and thus better preserves important primary lithologic and structural relationships, allowing its tectonomagmatic and subsequent deformation history to be more readily studied. The CSB contains distinct volcanic suites from a variety of tectonic environments (rift basin, transitional crust and oceanic basin; St-Onge et al., 1989, and references within) in demonstrably tectonic juxtapositions. The CSB may thus serve as a useful conceptual model to explain similar lithologic and structural associations in Archean greenstone belts (e.g. Dimroth *et al.*, 1982; Jensen, 1985; Davis et al., 1988). The intermediate age (early Proterozoic) of the CSB establishes it as an important link in attempts to compare "ophiolites" in Archean greenstone belts (e.g. Helmstaedt et al., 1986; de Wit et al., 1987) with their Phanerozoic counterparts.

Summary

1. The rocks of the northeastern portion of the Cape Smith Belt represent the igneous crustal components of an ophiolite suite. This is direct evidence of

the involvement of modern-style plate tectonic processes in the early Proterozoic Trans-Hudson orogen in northern Quebec. Indirect evidence for subduction is the fact that only a small remnant of this oceanic crust is preserved.

2. The two-billion year-old oceanic crust preserved in the Cape Smith Belt is similar in most physical and chemical respects to more recent examples exposed as ophiolites in Phanerozoic orogenic belts. A principal difference is the greater crustal thickness preserved in the Purtuniq ophiolite. The preserved crustal thickness is of the order of 7.5–8 km, but the complete igneous crustal section may originally have been up to 9–10 km thick. This greater thickness is thought to be consistent with higher oceanic heat flow in the early Proterozoic.
3. Field relations and chemical data for the layered mafic and ultramafic rocks suggest that these units represent the products of dominantly cumulus crystallization in an open-system magma chamber. The systematic repetition of cyclical ultramafic units, and cryptic compositional variations observed in relict mafic minerals are thought to reflect the input of fresh batches of more primitive magma into the chamber.
4. Two physically separate, geochemically distinct mantle sources were simultaneously involved in the generation of the Purtuniq ophiolite. Each source produced a suite of cumulate rocks and sheeted mafic dykes. The long-term, more highly depleted of the two appears to have been enriched metasomatically immediately prior to melt generation.
5. The chemistry of the Purtuniq rocks is most similar to modern MORB-type ophiolites and oceanic crust. Field evidence supports the oceanic spreading center hypothesis to explain the origin of the Purtuniq ophiolite. The physical extent of the ocean basin in which the ophiolite was generated is not well constrained.

Acknowledgements

Normand Bégin (University of Calgary) is thanked for contributions during the summers of field work and winters in Kingston. Discussions with S. J. Edwards and J. Malpas of Memorial University of Newfoundland (MUN), and D. C. Hall and M. J. Van Kranendonk of Queen's University contributed to numerous ideas presented here. P. L. Roeder (Queen's) is thanked for helpful discussions regarding chromite, and for some of the chromite analyses. The dedicated efforts of P. L. Roeder and D. Kempson facilitated the expeditious use of the electron microprobe at Queen's. R. H. McNutt and O. Mudroch (McMaster University, XRF), and MUN's ICP-MS team are thanked for rapid, high-quality whole-rock data. Logistical support for fieldwork was provided by the Geological Survey of Canada (GSC), during three seasons of 1:50 000–scale mapping in the eastern half of the CSB, lead by

M. R. St-Onge. Whole-rock geochemistry was paid for by the GSC. Labwork at Queen's was supported by NSERC Operating Grant A8375 to H. Helmstaedt. Helmstaedt's 1987 field visit to Cape Smith, which resulted in his prediction and subsequent discovery of the sheeted dykes in the Watts Group, was covered by NSERC A8375. The first author gratefully acknowledges the financial support of an NSERC Canada scholarship and additional funds from the Department of Geological Sciences, Queen's University. A School of Graduate Studies and Research (Queen's) grant toward travel expenses facilitated the first author's participation in this Symposium. W. R. A. Baragar is thanked for his comments on an earlier version of this manuscript.

References

Anonymous, 1972. Penrose conference field report. Geotimes, v. 17: 24–25.

Arndt, N. T., 1983. Role of a thin, komatiite-rich oceanic crust in Archean plate tectonic processes. Geology, v. 11: 372–275.

Baragar, W. R. A., M. B. Lambert, N. Baglow, and I. Gibson, 1987. Sheeted dykes of the Troodos ophiolite, Cyprus, In: Mafic Dyke Swarms. H. C.Halls, and W. F. Fahrig, (Eds), Geological Association of Canada, Special Paper 34, pp. 257–272.

Basaltic Volcanism Study Project (BVSP), 1981. Basaltic volcanism on the terrestrial planets. Permagon Press, Inc., New York, 1286 p.

Beccaluva, L., D. Ohnenstetter and M. Ohnenstetter, 1979. Geochemical discrimination between ocean floor and island arc tholeiites- Application to some ophiolites. Canadian Journal of Earth Sciences, v. 16: 1874–1882.

Beccaluva, L., P. Di Girolamo, G. Macciotta, and V. Morra, 1983. Magma affinities and fractionation trends in ophiolites. Ofioliti, v. 8: 307–324.

Bégin, N. J., 1989a. P-T conditions of metamorphism inferred from the metabasites of the Cape Smith Belt, northern Quebec. Geoscience Canada, v. 16: 151–154.

Bégin, N. J., 1989b. Metamorphic zonation, mineral chemistry and thermobarometry, in metabasites of the Cape Smith Thrust-Fold Belt, northern Quebec: Implications for its thermotectonic evolution. unpublished Ph. D. thesis, Queen's University, 313 p.

Bégin, N. J. and D. M. Carmichael, 1987. Metabasites in the eastern Cape Smith Thrust-Fold Belt, northern Quebec: Metamorphic facies, mineral reactions and P-T-XCO_2 estimates. Geological Association of Canada, Program with Abstracts, v. 12, p. 24

Benn, K., A. Nicolas and I. Reuber, 1988. Mantle-crust transition and origin of wehrlitic magmas: Evidence from the Oman ophiolite. Tectonophysics, v. 151: 75–85.

Bickle, M. J., 1986. Implications of melting for stabilization of the lithosphere and heat loss in the Archean. Earth and Planetary Science Letters, v. 80: 314–324.

Cawood, P. A., and H. Williams, 1988. Acadian basement thrusting, crustal delamination, and structural styles in and around the Humber Arm allochthon, western Newfoundland. Geology, v. 16: 370–373.

Cawood, P. A., and H. Williams, 1990. Processes of ophiolite obduction in Oman and Newfoundland. Abstracts, Symposium on Ophiolite Genesis and Evolution of the Oceanic Lithosphere, Muscat, Oman. January, 1990.

Christensen, N. I., and J. D. Smewing, 1981. Geology and seismic structure of the northern section of the Oman ophiolite. Journal of Geophysical Research, v. 86: 2545–2555.

Cochran, J. R., and F. Martinez, 1988. Evidence from the northern Red Sea on the transition from continental to oceanic rifting. Tectonophysics, v 153: 25–53.
Coleman, R. G., 1971. Plate tectonic emplacement of upper mantle peridotites along continental edges. Journal of Geophysical Research, v. 76: 1212–1222.
Coleman, R. G., 1977. Ophiolites. Springer-Verlag, New York. 229 p.
Condie, K. C., 1976. Plate tectonics and crustal evolution. Permagon Press, Inc., New York, 288 p.
Condie, K. C., 1981. Archean greenstone belts. Elsevier Scientific Publishing Co., New York, 434 p.
Davis, D. W., R. H. Sutcliffe, and N. F. Trowell, 1988 Geochronological constraints on the tectonic evolution of a late Archean greenstone belt, Wabigoon subprovince, northwest Ontario, Canada. Precambrian Research, v. 39: 171–191.
Debari, S. M., and R. G. Coleman, 1989. Examination of the deep levels of an island arc: Evidence from the Tonsina ultramafic-mafic assemblage, Tonsina, Alaska. Journal of Geophysical Research, v. 94: 4373–4391.
Dewey, J. F. and J. M. Bird, 1971. Origin and emplacement of the ophiolite suite: Appalachian ophiolites in Newfoundland. Journal of Geophysical Research, v. 76: 3179–3206.
deWit, M.J., R. A. Hart and R. J. Hart, 1987. The Jamestown ophiolite complex, Barberton mountain belt: a section through 3.5 Ga oceanic crust. Journal of African Earth Sciences, v. 6: 681–730.
Dick, H. J. B., and T. Bullen, 1984. Chromian spinel as a petrogenetic indicator in abyssal and alpine-type peridotites and spatially associated lavas. Contributions to Mineralogy and Petrology, v. 86: 54–76.
Dimroth, E., L. Imreh, M. Rocheleau, and N. Goulet, 1982. Evolution of the south-central part of the Archean Abitibi Belt, Quebec. Part I: Stratigraphy and paleogeographic model. Canadian Journal of Earth Sciences, v. 19: 1729–1758.
Dunning, G. R., 1987. Geology of the Annieopsquotch Complex, southwest Newfoundland. Canadian Journal of Earth Sciences, v. 24: 1162–1174.
Elthon, D., J. A. Karson, J. F. Casey, J. Sullivan, and F. X. Siroky, 1986. Geochemistry of diabase dykes from the Lewis Hills Massif, Bay of Islands ophiolite: evidence for partial melting of oceanic crust in transform faults. Earth and Planetary Science Letters, v. 78: 89–103.
Emmermann, R., 1985. Basement geochemistry, hole 504B. In: Initial Reports of the Deep Sea Drilling Project, v. 83, Anderson, R. N., J. Honnorez, K. Becker et al., U. S. Government Printing Office, Washington: 183–199.
Francis, D. M., A. J. Hynes, J. N. Ludden, and J. Bédard, 1981. Crystal fractionation and partial melting in in the petrogenesis of a Proterozoic high-MgO volcanic suite, Ungava Quebec. Contributions to Mineralogy and Petrology, v. 78: 27–36.
Francis, D., J. Ludden, and A. Hynes, 1983. Magma evolution in a Proterozoic rifting environment. Journal of Petrology, v. 24: 556–582.
Gass, I. G., 1968. Is the Troodos massif of Cyprus a fragment of Mesozoic ocean floor? Nature, v. 220: 39–42.
Hargraves, R. B., 1986. Faster spreading or greater ridge length in the Archean? Geology, v. 14: 750–752.
Hawkins, J. W. Jr., 1980. Petrology of back-arc basins and island arcs: Their possible role in the origin of ophiolites. In: Proceedings of the International Ophiolite Symposium, Cyprus, 1979, Panayioutou, A., (Ed), pp. 244–254.
Hegner, E. and M.L. Bevier, 1989. Geochemical constraints on the origin of mafic rocks from the Cape Smith Belt. Geoscience Canada, v. 16: 148–151.
Helmstaedt, H., and J. M. Allen, 1977. Metagabbronorite from DSDP hole 334: an example of high-temperature deformation and recrystallization near the Mid-Atlantic Ridge. Canadian Journal of Earth Sciences, v. 14: 886–898.
Helmstaedt, H., W. A. Padgham, and J. A. Brophy, 1986. Multiple dikes in the Lower Kam

Group, Yellowknife greenstone belt: Evidence for Archean sea-floor spreading? Geology, v. 14: 562–566.
Hoffman, P. F., 1985. Is the Cape Smith Belt (northern Quebec) a klippe? Canadian Journal of Earth Sciences, v. 22: 1361–1369.
Hoffman, P. F., 1988. United plates of America, the birth of a craton: Early Proterozoic assembly of Laurentia. Annual Reviews of Earth and Planetary Sciences, v. 16: 543–603.
Hoffman, P. F. and S. A. Bowring, 1984. Short-lived 1.9 Ga continental margin and its destruction, Wopmay Orogen, northwest Canada. Geology, v. 12: 68–72.
Hoffman, P. F. and G. Ranalli, 1988. Archean oceanic flake tectonics. Geophysical Research Letters, v. 15: 1077–1080.
Hopson, C. A., and J. S. Pallister, 1980. Semail ophiolite magma chamber: I. Evidence from gabbro phase variation, internal structure and layering. In: Proceedings of the International Ophiolite Symposium, Cyprus, 1979, Panayioutou, A., (Ed): 402–404.
Hopson, C. A., R. G. Coleman, R. T. Gregory, J. S. Pallister, and E. H. Bailey, 1981. Geologic section through the Samail ophiolite and associated rocks along a Muscat-Ibra transect, southeastern Oman mountains. Journal of Geophysical Research, v. 86: 2527–2544.
Hynes, A., and D. M. Francis, 1982. A transect of the early Proterozoic Cape Smith Foldbelt, New Quebec. Tectonophysics, v. 62: 251–278.
Jensen, L.S., 1976. A new cation plot for classifying subalkalic volcanic rocks. Ontario Division of Mines, Miscellaneous Paper 66, 22 p.
Jensen, L.S., 1985. Stratigraphy and petrogenesis of Archean metavolcanic sequences, southwestern Abitibi subprovince, Ontario. In: Evolution of Archean Supracrustal Sequences, Ayers, L. D., P. C. Thurston, K. D. Card, and W. Weber, (Eds), Geological Association of Canada Special Paper 28: 65–87.
Juteau, T., M. Ernewein, I. Reuber, H. Whitechurch, and R. Dahl, 1988. Duality of magmatism in the plutonic sequence of the Sumail Nappe, Oman. Tectonophysics, v. 151: 107–135.
Kontinen, A., 1987. An early Proterozoic ophiolite- the Jormua mafic-ultramafic complex, northeast Finland. Precambrian Research, v. 35: 313–341.
Kröner, A., 1983. Proterozoic mobile belts compatible with the plate tectonic concept. In: Proterozoic Geology: Selected papers from an international symposium, Medaris, L. G., Jr., C. W. Byers, D. M. Mickelson and W. C. Shanks, (Eds), Geological Society of America, Memoir 161: 59–74.
Lamothe, D., 1986. Développements récents dans la Fosse de l'Ungava. In: Exploration en Ungava: données récentes sur la géologie et gîtologie. Lamothe, D., R. Gagnon, and T. Clark, (Eds), Ministère de l'Energie et des Ressources du Quebéc, DV 86–16: 1–6.
Lamothe, D., C. Picard, and J. Moorhead, 1984. Région du lac Beauparlant, bande de Cap Smith-Maricourt, Nouveau-Québec. Ministère de l'Energie et des Ressources du Québec, DP 84–39.
Lucas, S. B., 1989. Structural evolution of the Cape Smith Thrust Belt and the role of out-of-sequence faulting in the thickening of mountain belts. Tectonics, v. 8: 655–676.
Lucas, S. B., 1990. Relations between thrust belt evolution, grain-scale deformation, and metamorphic processes: Cape Smith Belt, northern Canada. Tectonophysics, v. 178: 151–182.
Malpas, J. and R. K. Stevens, 1977. The origin and emplacement of the ophiolite suite with examples from western Newfoundland. Geotectonics (English translation), v. 11: 453–466.
Mével, C., M. Cannat, and D. S. Stakes, 1990. Influence of lithospheric stretching on hydrothermal processes in gabbros from slow-spreading ridges. Abstracts, Symposium on Ophiolite Genesis and Evolution of the Oceanic Lithosphere, Muscat, Oman. January, 1990.
Moores, E., M., and F. J. Vine, 1969. Troodos massif, Cyprus, a deep ocean floor: Preliminary structural and petrologic evidence. EOS, Transactions of the American Geophysical Union, v. 50: 333
Moores, E., M., and F. J. Vine, 1971. The Troodos massif, Cyprus, and other ophiolites as

oceanic crust: Evaluation and implications. Transactions of the Royal Society of London, v. A268: 443–466.

Moores, E., M., 1986. The Proterozoic ophiolite problem, continental emergence, and the Venus connection. Science, v. 234: 65–68.

Nisbet, E. G., and C. M. R. Fowler, 1983. Model for Archean plate tectonics. Geology, v. 11: 376–379.

O'Hara, M. J., 1977. Geochemical evolution during fractional crystallization of a periodically refilled magma chamber. Nature, v. 266: 503–507.

O'Hara, M. J. and R. E. Mathews, 1981. Geochemical evolution in an advancing, periodically replenished, periodically tapped, continuously fractionated magma chamber. Journal of the Geological Society of London, v. 138: 237–277.

Pallister, J. S., and C. A. Hopson, 1981. Samail ophiolite plutonic suite: Field relations, phase variation, cryptic variation and layering, and a model of a spreading ridge magma chamber. Journal of Geophysical Research, v. 86: 2593–2644.

Parrish, R. R., 1989. U-Pb geochronology of the Cape Smith Belt and Sugluk block, northern Quebec. Geoscience Canada, v. 16: 126–130.

Pearce, J. A., 1982. Traces element characteristics of lavas from destructive plate margins. In: Andesites, Thorpe, R. S., (Ed), John Wiley and Sons: 525–548.

Pearce, J. A., T. Alabaster, A. W. Shelton, and M. P. Searle, 1981. The Oman ophiolite as a Cretaceous arc-basin complex: evidence and implications. Philosophical Transactions of the Royal Society of London, A300: 299–317.

Pearce, J. A., S. J. Lippard, and S. Roberts, 1984. Characteristics and tectonic significance of supra-subduction zone ophiolites, In: Marginal Basin Geology. Kokelaar, B. P., and Howells, M. F., (Eds), Geological Society Special Publication 16: 77–94.

Picard, C., 1986. Lithogéochimie de la partie centrale de la Fosse de l'Ungava. In: Exploration en Ungava: données récentes sur la géologie et gîtologie. Lamothe, D., R. Gagnon, and T. Clark, (Eds), Ministère de l'Energie et des Ressources du Québec, DV 86–16: 57–72.

Picard, C., D. Lamothe, M. Piboule, and R. Oliver, 1990. Magmatic and geotectonic evolution of a Proterozoic ocean basin system: The Ungava Trough (New Quebec). Precambrian Research, v. 47: 223–249.

Rosencrantz, E., 1983. The structure of sheeted dykes and associated rocks in North Arm massif, Bay of Islands ophiolite complex, and the intrusive process at oceanic spreading centers. Canadian Journal of Earth Sciences, v. 20: 787–801.

Saunders, A. D., J. Tarney, N. G. Marsh, and D. A. Wood, 1980. Ophiolites as ocean crust or marginal basin crust: A geochemical approach. *in* Proceedings of the International Ophiolite Symposium, Cyprus, 1979, Panayioutou, A., (Ed): 193–204.

Scott, D. J., 1990. Geology and geochemistry of the early proterozoic Purtuniq ophiolite, Cape Smith Belt, northern Quebec, Canada. Unpublished Ph.D. thesis, Queen's University, 289 p.

Scott, D. J., M. R. St-Onge, S. B. Lucas, and H. Helmstaedt, 1988. The 1999 Ma Purtuniq Ophiolite: Imbricated oceanic crust obliquely exposed in the Cape Smith Thrust-Fold Belt, northern Quebec, Canada. Geological Society of America, Abstracts with Program, v. 20: A158.

Scott, D. J., M. R. St-Onge, S. B. Lucas, and H. Helmstaedt, 1989. The 1998 Ma Purtuniq ophiolite: imbricated and metamorphosed oceanic crust in the Cape Smith Thrust Belt, northern Quebec. Geoscience Canada, v. 16: 144–148.

Scott, D. J. and E. Hegner, 1990. Two mantle sources for the two-billion year-old Purtuniq ophiolite, Cape Smith Belt, northern Quebec. Geological Association of Canada, Program with Abstracts, v. 15: A118.

Searle, M. P., 1985. Sequence of thrusting and origin of culminations in the northern and central Oman Mountains. Journal of Structural Geology, v. 7: 129–143.

Shervais, J. W., 1982. Ti-V plots and the petrogenesis of modern and ophiolitic lavas. Earth and Planetary Sciences Letters, v. 59: 101–118.

Sleep, N. H., and B. F. Windley, 1982. Archean plate tectonics: Constraints and inferences. Journal of Geology, v. 90: 363–379.

Smith, S. E., and D. Elthon, 1988. Mineral compositions of plutonic rocks from the Lewis Hills massif, Bay of Islands complex. Journal of Geophysical Research, v. 93: 3450–3468.

Spray, J. G., 1984. Possible causes and consequences of upper mantle decoupling and ophiolite displacement. In: Ophiolites and Oceanic Lithosphere. Gass, I. G., S. J. Lippard, and A. W. Shelton, (Eds), Geological Society Special Publication 13: 255–268.

Stockwell, C. H., 1950. The use of plunge in the construction of cross-sections of folds. Proceedings of the Geological Association of Canada, v. 3: 97–121.

St-Onge, M. R., and J. E. King, 1987. Thermo-tectonic evolution of a metamorphic internal zone documented by axial projections and petrological P-T paths, Wopmay Orogen, northwest Canada. Geology, v. 15: 155–158.

St-Onge, M. R., and S. B. Lucas, 1989a. Tectonic controls on the thermal evolution of the Cape Smith Thrust Belt. Geoscience Canada, v. 16: 154–158.

St-Onge, M. R., and S. B. Lucas, 1989b. Geology, lac Watts, Quebec, Geological Survey of Canada, Map 1721A, scale 1:50 000.

St-Onge, M. R., and S. B. Lucas, 1989c. Geology, Purtuniq, Quebec, Geological Survey of Canada, Map 1722A, scale 1:50 000.

St-Onge, M. R., and S. B. Lucas, 1990a. Evolution of the Cape Smith Belt: early Proterozoic continental underthrusting, ophiolite obduction and thick-skinned folding. In: The Early Proterozoic Trans-Hudson Orogen of North America. J. F. Lewry and M. E. Stauffer, (Eds), Geological Association of Canada, Special Paper, 37: 313–351.

St-Onge, M. R., and S. B. Lucas, 1990b. Early Proterozoic collisional tectonics in the internal zone of the Ungava (Trans-Hudson) orogen: Lacs Nuvilik and Sugluk map areas, Québec. In: Current Research, Part C, Geological Survey of Canada, Paper 90–1C: 119–132.

St-Onge, M. R., S. B. Lucas, D. J. Scott, and N. J. Bégin, 1986. Eastern Cape Smith Belt: an early Proterozoic thrust-fold belt and basal shear zone exposed in oblique section, Wakeham Bay and Cratere du Nouveau Quebec map areas, northern Quebec. In: Current Research, Part A, Geological Survey of Canada, Paper 86–1A: 1–14.

St-Onge, M. R., S. B. Lucas, D. J. Scott, and N. J. Bégin, 1987. Tectonostratigraphy and structure of the lac Watts - lac Cross - rivière Deception area, central Cape Smith Belt, northern Quebec. In: Current Research, Part A, Geological Survey of Canada, Paper 87–1A: 619–632.

St-Onge, M. R., S. B. Lucas, D. J. Scott, N. J. Bégin, H. Helmstaedt, and D. M. Carmichael, 1988a. Thin-skinned imbrication and subsequent thick-skinned folding of rift-fill, transitional-crust, and ophiolite suites in the 1.9 Ga Cape Smith Belt, northern Quebec. In: Current Research, Part C, Geological Survey of Canada, Paper 88–1C: 1–18.

St-Onge, M. R., S. B. Lucas, D. J. Scott, and N. J. Bégin, 1988b. Geology, eastern portion of the Cape Smith Belt, parts of the Wakeham Bay, Cratère du Nouveau Quebec, and Nuvilik Lakes map areas, northern Quebec. Geological Survey of Canada, Open File 1730, 16 maps, 1:50,000 scale.

St-Onge, M. R., S. B. Lucas, D. J. Scott, and N. J. Bégin, 1989. Evidence for the development of oceanic crust and for continental rifting in the tectonostratigraphy of the Early Proterozoic Cape Smith Belt. Geoscience Canada, v. 16: 119–122.

Sturt, B. A., A. Thon, and H. Furnes, 1980. The geology and preliminary geochemistry of the Karmøy ophiolite, S. W. Norway. *in* Proceedings of the International Ophiolite Symposium, Cyprus, 1979, Panayioutou, A., (Ed): 538–553.

Suen, C. J., F. A. Frey, and J. Malpas, 1979. Bay of Islands ophiolite suite, Newfoundland: Petrological and geochemical characteristics with emphasis on rare earth element geochemistry. Earth and Planetary Science Letters, v. 45: 337–348.

Sun, S. S., R. W. Nesbit, and A. Y. Sharaskin, 1979. Geochemical characteristics of mid-ocean ridge basalts. Earth and Planetary Science Letters, v. 44: 119–138.

Taylor, S. R., and S. M. McLennan, 1985. The continental crust: its composition and evolution. Blackwell Scientific Publications, 312 p.

Tual, E., B. M. Jahn, H. Bougault, and J. L. Joron, 1985. Geochemistry of basalts from hole 504B, Leg 83, Costa Rica rift. In: Initial Reports of the Deep Sea Drilling Project, v. 83, Anderson, R. N., J. Honnorez, K. Becker et al., U. S. Government Printing Office, Washington: 201–214.

Part VII
Mapping Ophiolites

The Use of Digitally-Processed Spot Data in the Geological Mapping of the Ophiolite of Northern Oman

S. CHEVREL[1], P. CHEVREMONT[1], R. WYNS[1], J. LE MÉTOUR[1], A. AL TOBA[2] and M. BEURRIER[1]
[1]*BRGM, BP 6009, 45060 Orleans, CEDEX, France*
[2]*Ministry of Petroleum and Minerals, P.O. Box 551, Muscat, Oman*

Abstract

Detailed geological mapping of the Shinas area in the Northern Oman mountains, close to the border with the United Arab Emirates, covered the Samail ophiolite sequence that is fully exposed in this mountainous and poorly accessible region.

Remote sensing, by means of aerial photographs and satellite imagery, was particularly well suited in the mapping of the sequence in this semi-arid zone with its excellent outcrop conditions.

The digital data from the SPOT satellite were subjected to geometric and radiometric digital processing, to obtain spectral difference enhanced images most suitable for the mapping purposes. Furthermore, the synoptic coverage of satellite images helps in understanding the regional extension and relationship of the ophiolitic units and their structural features.

The boundary between crustal and mantle sequences (petrological Moho) is clearly shown on the images and can be easily mapped. At the base of the mantle sequence, the Banded Unit is well seen on the image in the west, and a vertical zonation seems related to serpentinization gradient according to mylonitization.

Within the crustal sequence, only the major lithological units can be recognized. The sheeted dyke complex is well differentiated from gabbros in the west and volcanic rocks in the east. The highly reflectant trondhjemites are easy to map. Through image processing, some of the late-intrusive peridotite bodies can be seen within the gabbros, as well as late-intrusive uralitic gabbros within the sheeted dyke complex. Several gossans can be clearly identified, although their spectral response is close to that of the vegetation.

Finally, the geometrically and radiometrically enhanced image, combined with aerial photographs and field results, will be used as a base for drawing the maps.

Tj. Peters et al. (Eds), Ophiolite Genesis and Evolution of the Oceanic Lithosphere, 853–870.

1. Introduction

On behalf of the Ministry of Petroleum and Minerals of Oman, B.R.G.M. geologists have been intimately involved in the program of detailed geological mapping of the ophiolites of Northern Oman. Eight 1:50,000 scale maps, each covering 15′ × 15′, centered around the Shinas, Sumayni and Buraymi towns, are to be mapped these days.

Despite the better spectral range of Landsat Thematic Mapper data, the use of SPOT data for such a mapping is clearly justified by their excellent ground resolution (20 m in multispectral mode), which allows printing of excellent multispectral images at the 1:50,000 mapping scale. Their synoptic coverage (60 × 60 km, corresponding roughly to 30′ × 30′) allows coverage of one map with a single scene and thus simplifies the handling of numerous aerial photographs for interpretation and drafting of geological contacts.

Furthermore, the digital format of the data allows accurate geometric correction. Image interpretation is thus directly transferable to the topo maps and can also be used for the final map drawing. Digital data can be processed in different ways to get enhanced images for a specific geological purpose.

2. Geographic Setting

Among the areas mentioned above, the Shinas region has been chosen, as it appear representative for the developed methodology. The four 1:50,000 sheets of this area are bordered by latitudes 24° 30′ to 25° 00′ N and longitudes 56° 00′ to 56° 34′ E (Fig. 1).

The continental area is bordered to the east by the Gulf of Oman. The at most 10–km-wide Al Batinah coastal plain gently dips to the sea, Numerous settlements and farms can be found along it. Much of the western half of the area is mountainous, with a steep relief in the up to 1500 m high Al Hajar Al Gharbi mountains. Access is poor, apart from several dirt roads, coming from the coastal highway.

3. Geologic Setting

According to the work done by Glennie et al. (1974) (Fig. 1) and by Smewing et al. (1979), the major part of the area is covered by the allochthonous ophiolitic rocks of the Samail nappe, ranging from peridotites in the west to pillow lavas in the east, representing the whole ophiolite sequence (displayed in Fig. 2).

Beneath the ophiolitic rocks, allochthonous units of the metamorphic sole and the Hawasina nappes occur in the western central part of the Shinas map area.

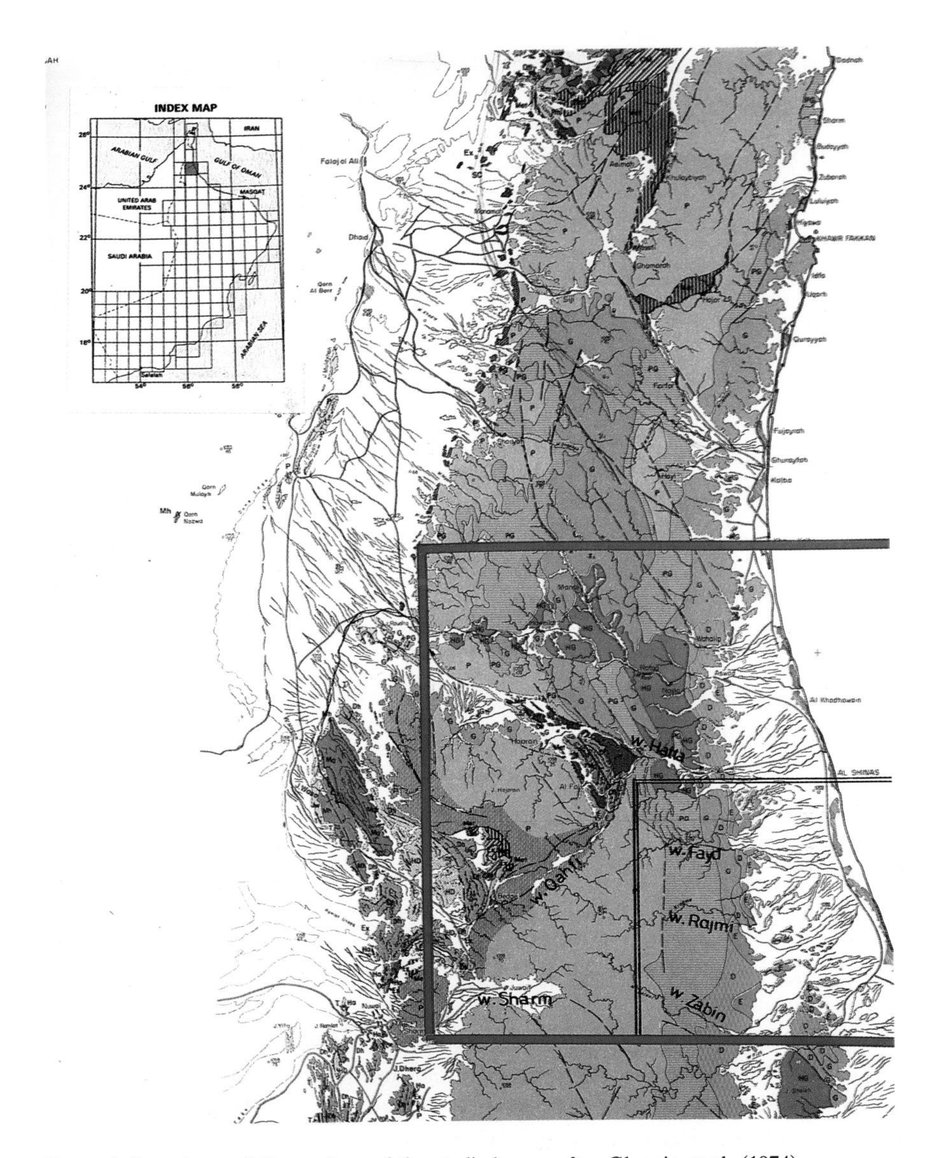

Figure 1. Location and the geology of the studied area, after Glennie et al. (1974).

Post-emplacement late Tertiary to Quaternary deposits occur as wadi terraces in the mountains and cover entirely the Batinah plain, where they form several levels of terraces and alluvial fans.

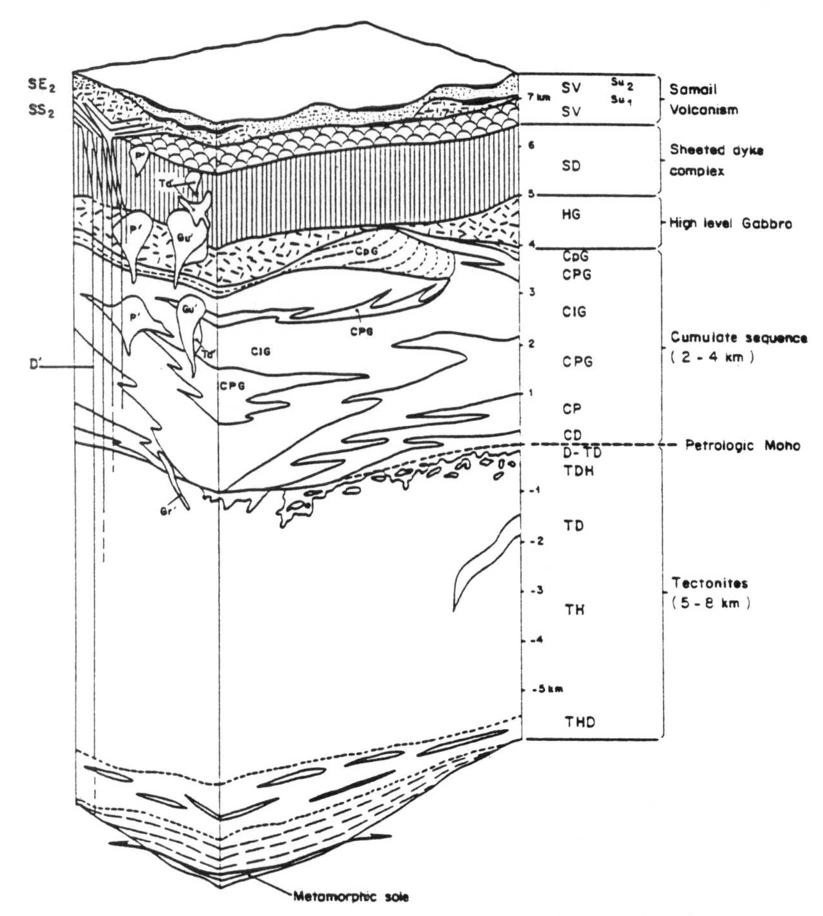

Figure 2. Sketch diagram of the ophiolite sequence, after M. Beurrier (1988).

4. Data Acquisition and Preparation

4.1. *Review on Spot Multispectral Data*

The SPOT satellite acquires data in digital format with a ground resolution cell of 20×20 m. in the multispectral mode used here.

The instrument is fitted with sensors operating in three spectral regions of the electromagnetic spectrum, defining three spectral bands as follows:

band XS1: 0.50 to 0.59 micrometer (green),
band XS2: 0.61 to 0.68 micrometer (red),
band XS3: 0.79 to 0.89 micrometer (near infrared).

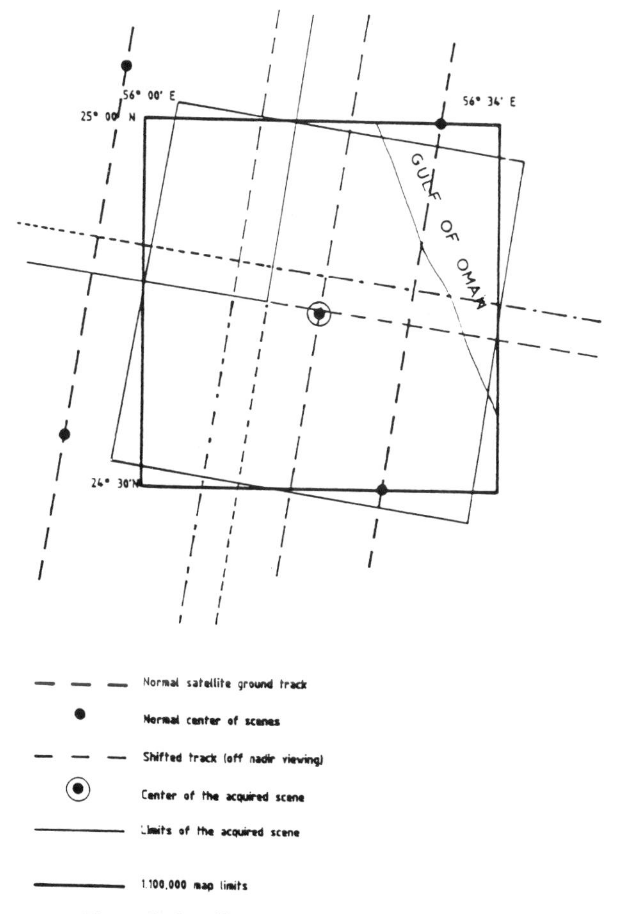

Figure 3. Specific programmation of the satellite.

4.2. *Specific Programming of Spot for Data Acquisition*

Specific programming of the SPOT satellite uses its off-nadir viewing capability.

Fig. 3 shows the limits of the Shinas area together with the normal path of the satellite. The coverage of the whole area requires acquisition of four SPOT scenes. Using the off-nadir viewing capability of the satellite and the data shift along the track facility offered by SPOT-IMAGE, it is possible to reduce the number of scenes to be acquired to only one. Despite the fact that the price of such a specifically programmed scene is much higher than that of one normal scene, it remains notably less than purchasing four scenes. Moreover, the processing cost is obviously minimized.

4.3. *Geometric Projection*

For mapping purposes, a very accurate geometric projection of the data is required. This projection is done with a cartographic accuracy (i.e. < 1 mm), using ground control points from the reference 1:100,000 topographic map.

Advantages of such a projection reside in the possibility of obtaining an image that is geometrically perfectly conformable with the reference topographic map, so that any interpretation done on it is directly transferable to the map.

4.4. *Enhancement*

A standard false-color composite image (i.e. band XS1 in blue, XS2 in green, XS3 in red) is produced (see Fig. 4), at a scale of 1:100,000, which is used as a reference image. For the mapping work itself, specific documents at a scale of 1:50,000 are produced, each one covering the corresponding 1:50,000 map (see Fig. 5). A specific software, particularly well suited for this semi-arid environment, was used. This software, so-called "mobile regional stretching", stretches each part of the image with respect to its environment, instead of processing the image as a whole. It is particularly suited for enhancing the subtle spectral differences, and allows visualization and interpretation of morpholithologic details that are more or less hidden on a standard image. It is thus useful in the drafting of the maps.

5. Specific Image Processings for Lithologic Differentiation

The above mentioned processings are not specifically adapted to the ophiolite sequence, and are too "systematic" to take into account the requirements of geological mapping.

Specific digital processings have thus been carried out, to enhance the lithologic differentiation within the ophiolites.

5.1. *Specific Stretching of the Image Dynamics*

Fig. 6 displays a linear mode stretching of the dynamic range that takes into account the statistical parameters related to the ophiolite sequence only, without consideration for the coastal plain and the Hawasina units.

The resulting image clearly displays the main lithologic units of the ophiolitic sequence: the upper volcanic rocks appear pale green, while the lower volcanic rocks are darker green; the sheeted-dyke complex is outlined by a medium green hue. The petrological Moho is easily distinguishable between green-brown cumulate gabbro of the crustal sequence in the east and yellow to brown harzburgite of the mantle sequence in the west. Light yellow

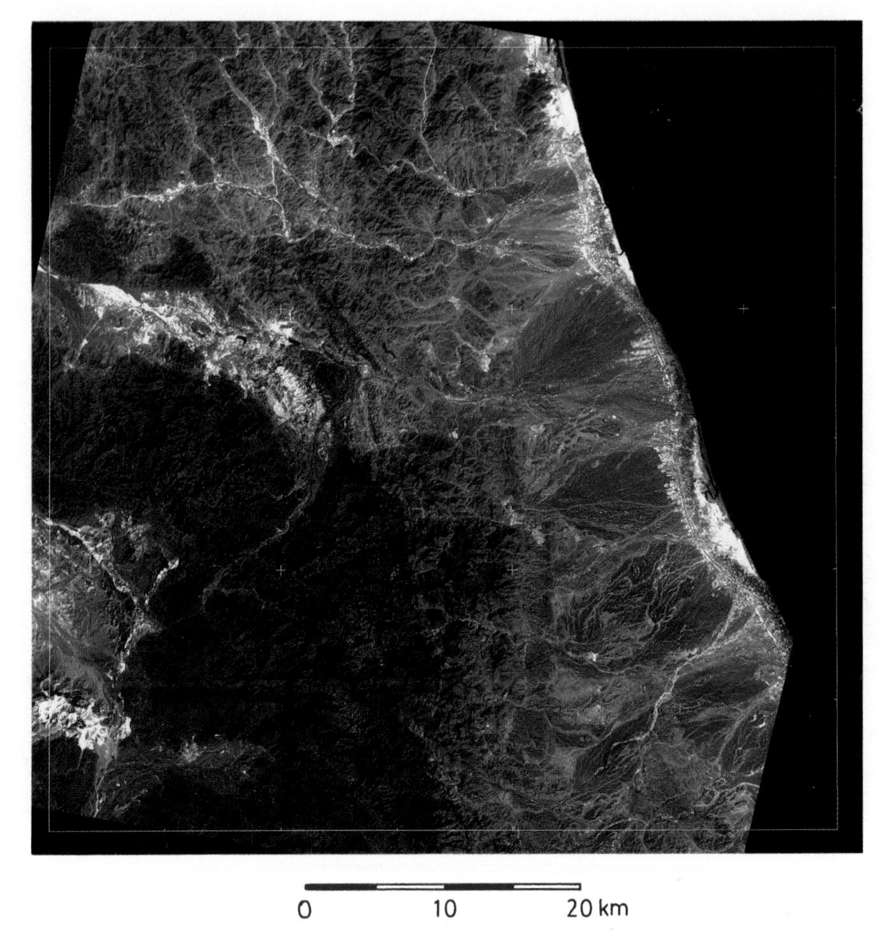

Figure 4. Standard false colour composite image. Geocoded SPOT image of Shinas area, 1:100,000 sheet NG 40-14 B.

corresponds to the late-intrusive peridotite bodies within the cumulate gabbro, while a light brown hue inside the sheeted-dyke complex is related to late-intrusive uralitic gabbro. Pale blue to light gray colors along the petrological Moho represent sheared gabbros. At the base of the mantle sequence, the banded unit (Searle, 1980) appears dark green. The highly reflectant trondhjemite bodies that intruded the crustal sequence can be recognized through their whitish color.

Sine stretching shown in Fig. 7 is useful in lithologic discrimination within the mantle sequence: a bluish-white color delineates extension of the harzburgite, and the banded unit at the base is clearly distinguishable with a light pink color.

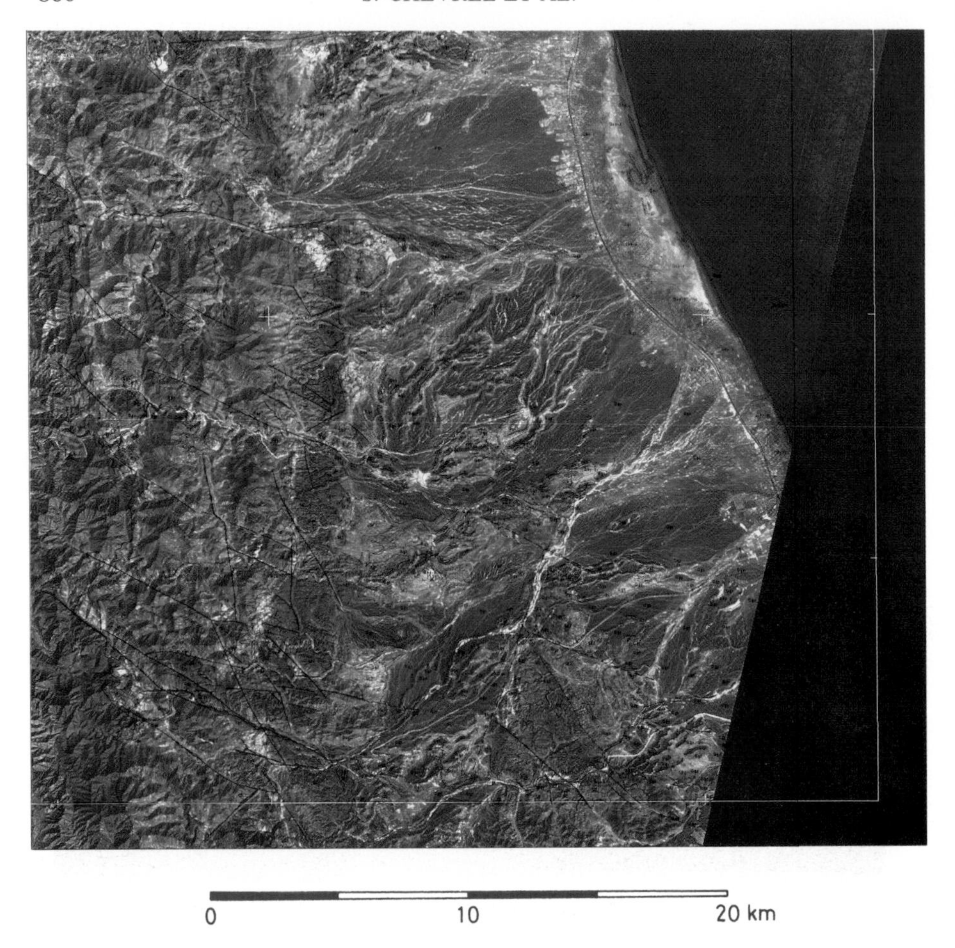

Figure 5. Specifically processed extract of the previous image prepared for morpho-lithologic interpretation at 1:50,000 scale.

5.2. *Image Classification*

The satisfactory results of the stretchings described above suggest a fairly good relationship between lithology and spectral response on the SPOT bands. Image classification procedures have thus been undertaken.

Digital image classification may be considered as an assigning of pixel into classes defined through decision rules (based on pixel values), such as a pixel belongs to one class if, and only if, it fits with the decision rules of the corresponding class for all the considered bands. The result of a classification is a new image in which all the pixels either have been assigned into a class, or rejected as unclassified. A color chart is then established to visualize the different classes. The newly obtained image is a "skeleton" of the standard image and can be of use in the visualization of the extension of the lithological units.

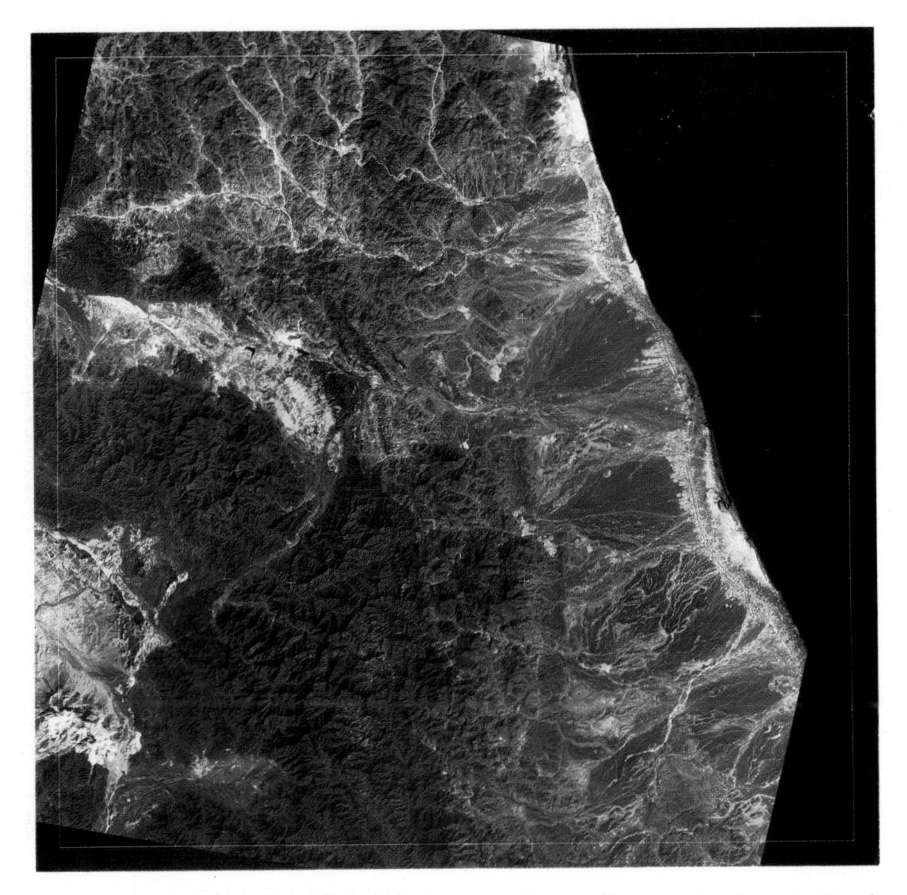

Figure 6. Specific thematic stretching of the dynamic. (Scale and band assignation as in Fig. 4).

5.2.1. *Classification from the three Spot Bands*

A total of 24 geologically known training areas were selected corresponding to the 11 following classes.

- **upper volcanic** rocks
- **lower volcanic** rocks
- **sheeted-dyke** complex
- late-intrusive **trondhjemite**
- **late-intrusive gabbro**
- late-intrusive **wehrlite**
- **late-intrusive dunite**
- **cumulate**-layered **gabbro**
- **sheared gabbro**
- **harzburgite**
- **banded unit**

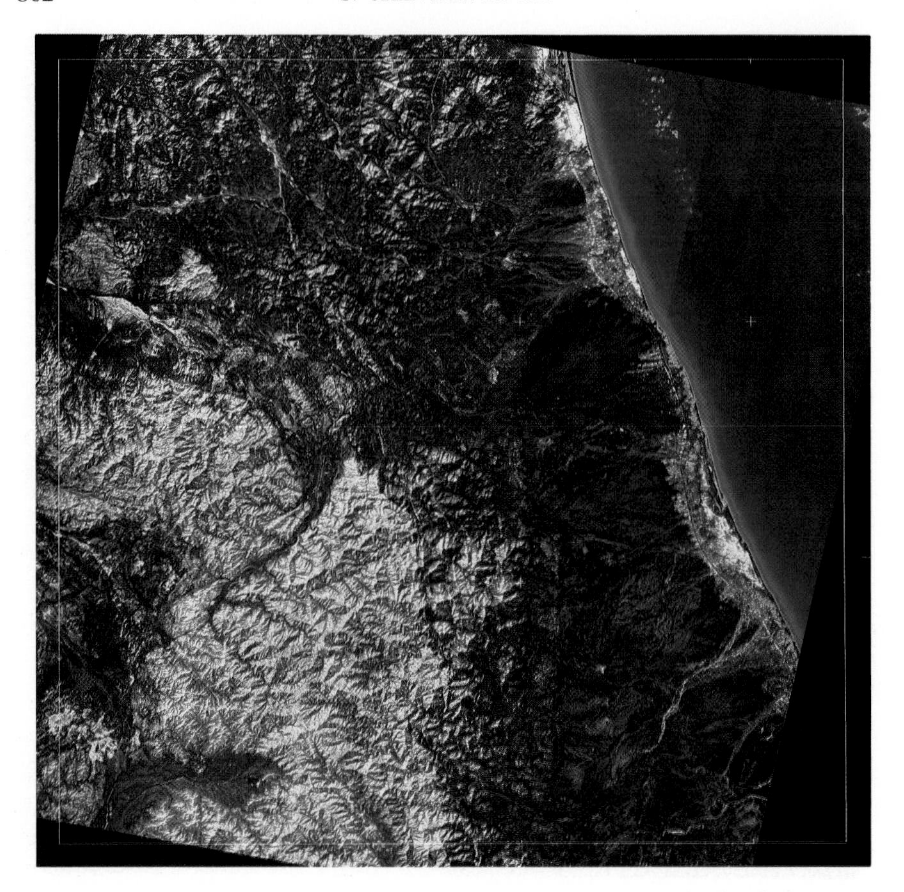

Figure 7. Sine stretching of the dynamic. (Scale and band assignation as in Fig. 4).

The training areas are large enough to be statistically representative and were chosen on sunlit slopes, while shadowed zones were avoided.

Table 1 displays the statistical parameters (mean and standard deviation) of the training areas for all the classes. Such a table allows a fast visualization of overlapping problems between classes, due to reflectance-convergence phenomena. For example, sheared gabbro and lower volcanic rocks, cumulate-layered gabbro and late-intrusive gabbro, all have very similar average reflectance in the 3 SPOT bands, and large reflectance overlapping exists between wehrlite bodies and the banded unit. Cumulate-layered gabbro and late-intrusive dunite are differentiated only in SPOT band 1.

The matrix of Table 2 displays, for each class, the percentage of pixels of the training areas classified in the related class, as well as the percentage of misclassified (i.e. classified as if belonging to another training area) or rejected pixels. For example, among the 803 pixels of the banded unit training areas, 57.5% are classified as banded unit, 38.4% as wehrlite, 1.6% as sheared gabbro, 1.9% as lower volcanic rocks and 0.6% rejected as unclassi-

Table 1. Statistical parameters of the training areas.

Class	Band XS3		Band XS2		Band XS1	
	Mean	S.D.	Mean	S.D.	Mean	S.D.
Trondhjemite	64.763	3.775	74.668	4.024	86.644	4.199
Sheeted dyke	47.121	3.011	56.421	2.886	66.930	2.704
Cumulate gabbro	52.378	2.783	58.460	3.265	67.521	3.617
Wehrlite	36.908	2.502	44.458	2.723	55.051	2.580
Late-int. gabbro	53.984	2.141	61.065	2.822	68.465	3.201
Late-int. dunite	52.491	3.184	56.439	3.235	61.903	3.288
Harzburgite	51.469	2.676	54.218	2.547	56.958	1.985
Banded unit	38.922	2.351	46.473	2.169	55.700	2.292
Sheared gabbro	44.763	2.446	52.370	2.977	62.905	3.851
Upper volcanics	49.578	1.774	59.709	2.022	70.315	3.125
Lower volcanics	43.897	3.212	52.404	2.792	62.538	3.355

fied. One can see that, apart from the very reflectant trondhjemite bodies, the percentage of misclassified pixels is generally very high, due to reflectance overlapping between the rock outcrops, or insufficiently representative training areas.

The result of the classification process applied to the whole image, using a maximum-likelihood algorithm, is presented in Fig. 8. Color assignation is given in Fig. 9.

Detailed examination reveals numerous misclassifications and confusion between different lithologic classes. A systematic confusion occurs between cumulate-layered gabbro, late-intrusive gabbro and sheared gabbro classes. Pixels classified as cumulate-layered gabbro can be found quite everywhere, except in harzburgite areas. Sunlit harzburgite is perfectly classified, but shadowed harzburgite is classified as banded unit. Sheeted dyke complex and volcanic rocks are commonly mixed.

Apart from the reflectance convergence phenomena, one of the reasons for misclassification is related to the steep relief, giving rise to reflectance differences between sunlit and shadowed slopes within the same lithology. Processing that minimizes slope effects must thus be applied.

5.2.2. *Classification Including Additional Computed Bands*

One way to minimize these unwanted slope effects is to proceed with band-to-band ratioing, which consists in dividing each pixel of one band by the corresponding pixel of one or more other bands. Following Rothery (1984), each SPOT band has been divided by the sum of the three bands, i.e. the new computed bands are pixel-by-pixel divisions corresponding to XS3/XS1 + XS2 + XS3, XS2/XS1 + XS2 + XS3 and XS1/XS1 + XS2 + XS3. A new six-band data set was then created by gathering the three SPOT bands with these three computed bands. After this, the same classification procedure, using the same training areas, was applied to this new six-band data.

Table 2. Confusion matrix between the classes for the training areas.

Class	1	2	3	4	5	6	7	8	9	10	11	Rej
Trondhjemite	99.5	0	0	0	0	0	0	0	0	0	0	0.5
Sheeted dyke	0	44.2	8.6	0	4.7	0	0	0.3	12.7	16.5	13.0	0
Cumulate gabbro	1.0	9.3	30.4	0	37.3	15.3	0	0	2.0	0.2	4.5	0
Wehrlite	0	0	0	68.2	0	0	0	26.5	0.2	0	5.1	0
Late-int. gabbro	1	1.6	21.9	0	65.5	8.4	0	0	1	0.6	0	0
Late-int. dunite	0	6.4	7.9	0	2.1	69.7	8.2	0	2.1	2.1	1.5	0
Harzburgite	0	0	0	0.2	0	28.2	64.1	0.5	7	0	0	0
Banded unit	0	0	0	38.4	0	0	0	57.5	1.6	0	1.9	0.6
Sheared gabbro	0	16.1	1	0	0.5	0	0	6.6	36.5	4.8	34.5	0
Upper volcanics	0.2	24.5	9.8	0	11.2	0	0	0	1.2	52.9	0.2	0
Lower volcanics	0	9.9	0	0.3	0	7.5	1	2.7	34.9	0	43.5	0

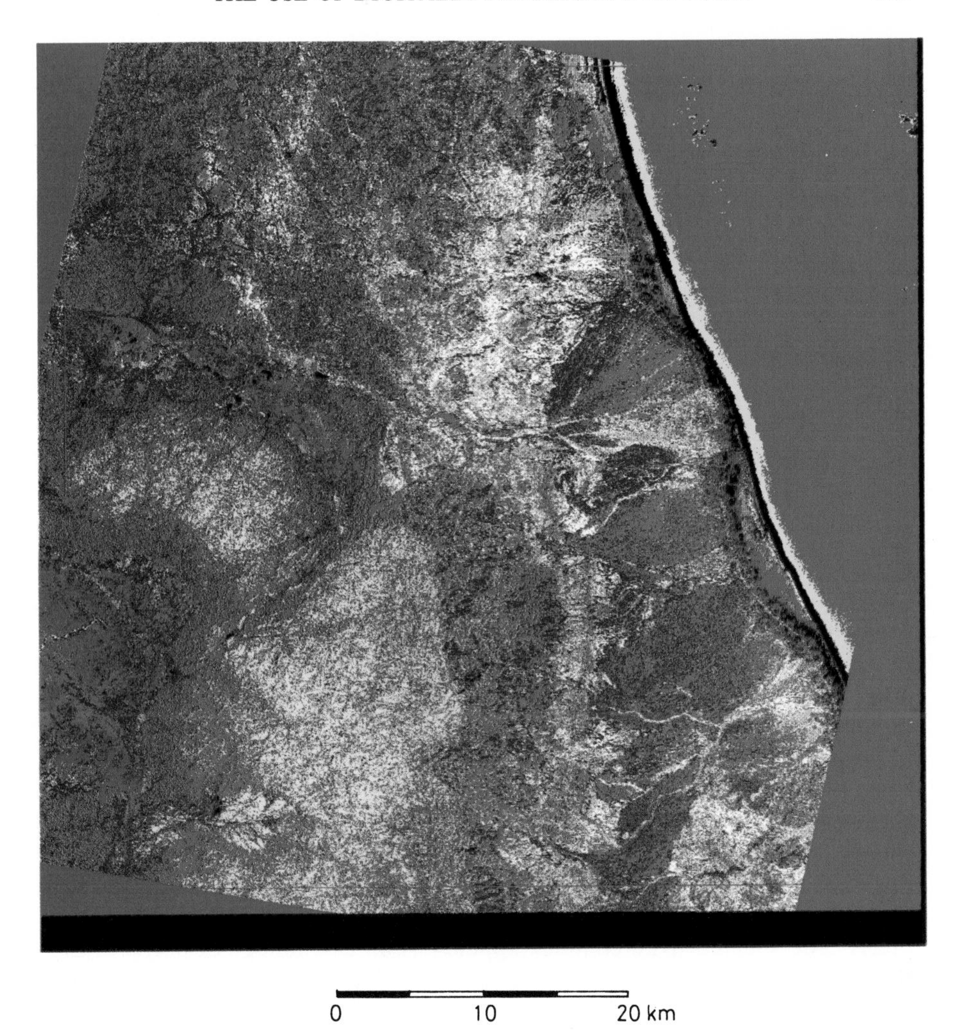

Figure 8. Image classification from 3 SPOT bands.

Statistical parameters for the new classification are listed in Tables 3 and 4.

Compared with the matrix of Table 2, the general dispersion between the classes is notably reduced (increasing number of zeros and decreasing percentages) and classification is improved for cumulate gabbro, late-intrusive dunite, banded unit and lower volcanic rocks classes. On the other hand, a decreasing percentage of correctly classified pixels for the other classes and a general shift of the confusion to the lower volcanic rocks class can be noted.

The result of the classification is presented in Fig. 10 and color assignation is the same as in Fig. 8 and 9. For the purpose of classification the result

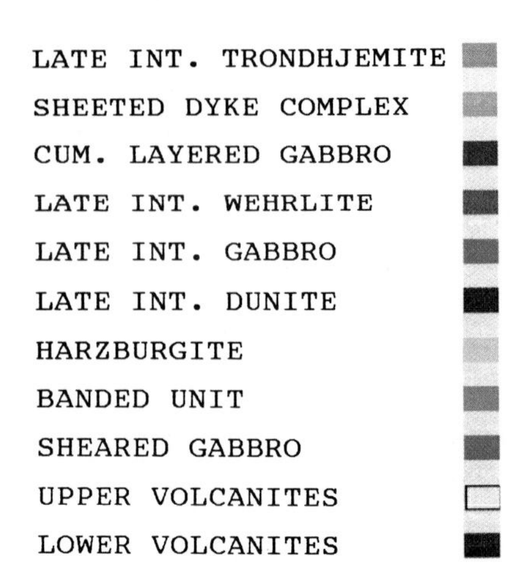

Figure 9. Colour assignation for the classification.

cannot be considered as successful. Indeed, apart from the harzburgite, which is correctly classified, confusion and misclassification are still present. Differences with the previous classification concern only the classes late-intrusive dunite, banded unit and lower volcanic rocks, which have nearly disappeared, and the late-intrusive wehrlite.

However, a quick view of the classified image outlines homogeneous assemblies, each of them composed of several classes, which are closely related to the main lithologic units. For example, the cumulate sequence is clearly delineated by an overall constant assemblage of pixels belonging to the sheared gabbro, wehrlite and cumulate-layered gabbro classes, while the banded unit is represented by an assemblage of wehrlite and sheared gabbro pixels. The sheeted dyke complex appears as an assemblage of pixels of the sheeted dykes, upper volcanic rocks and wehrlite classes. The classification is thus useful for the geologist, for a rapid view of the geographic distribution and relationships of the main lithologic units. The geologist is thus provided with a synthetic and synoptic document, useful in the understanding of the regional geological and structural context, which helps in completing the field investigations. In this manner it is possible to sketch regional geological maps.

Furthermore, when seen in detail, the classification outlines several intrusives within the crustal sequence, such as trondhjemites, late-intrusive uralitic gabbro in the sheeted-dyke complex, and late-intrusive dunites in the cumulate sequence. Larger amounts of smaller intrusions are outlined by a specific assembly of classes; for example, cumulate gabbro intruded by numerous trondhjemites, peridotites and uralitic gabbro can be distinguished from the rest of the cumulate sequence by an association of pixels classified as trondhjemite, wehrlite, sheeted dyke and sheared gabbro.

Another interest of the classification is related to the lithologic components

Table 3. Statistical parameters of the training areas.

Class	Band XS3		Band XS2		Band XS1		XS3/ADD		XS2/ADD		XS1/ADD	
	Mean	S.D.	Mean	S.D.	Mean	S.D.	Mean	S.D.	Mean	S.D.	Mean	S.D.
Trondhjemite	64.763	3.775	74.668	4.024	86.644	4.199	123.39	2.883	142.25	2.356	165.24	3.758
Sheeted dyke	47.121	3.011	56.421	2.886	66.930	2.704	119.28	2.711	143.06	2.435	169.86	3.827
Cumulate gabbro	52.378	2.783	58.460	3.265	67.521	3.617	127.05	2.585	141.68	2.409	163.73	3.591
Wehrlite	36.908	2.502	44.458	2.723	55.051	2.580	117.46	2.984	141.64	2.912	175.60	3.819
Late-int. gabbro	53.984	2.141	61.065	2.822	68.465	3.201	126.93	2.727	143.41	2.417	160.86	3.068
Late-int. dunite	52.491	3.184	56.439	3.235	61.903	3.288	119.69	26.900	135.73	17.901	151.82	6.698
Harzburgite	51.469	2.676	54.218	2.547	56.958	1.985	136.95	2.546	144.38	2.624	151.77	3.145
Banded unit	38.922	2.351	46.473	2.169	55.700	2.292	118.38	10.014	142.37	4.753	170.84	6.335
Sheared gabbro	44.763	2.446	52.370	2.977	62.905	3.851	121.08	2.927	141.61	3.129	170.10	3.543
Upper volcanics	49.578	1.774	59.709	2.022	70.315	3.125	119.14	2.792	143.56	2.120	169.09	3.338
Lower volcanics	43.897	3.212	52.404	2.792	62.538	3.355	118.59	10.392	142.30	4.883	170.12	7.145

Table 4. Confusion matrix between the classes for the training areas (3 bands + 3 ratios).

Class	1	2	3	4	5	6	7	8	9	10	11	Rej
Trondhjemite	98.7	0	0	0	0	0	0	0	0	0	0	1.3
Sheeted dyke	0	16.8	6.7	0	3.0	0.1	0	0.1	0	14.1	59.1	0
Cumulate gabbro	1.1	0.5	41.8	0	23.0	11.1	0	0	0	1.6	20.9	0
Wehrlite	0	0	0	43.9	0	0	0	48.4	0	0	7.5	0
Late-int. gabbro	1	0	17.4	0	59.0	18.7	0	0	0	1	3	0
Late-int. dunite	0	0	1.2	0	0.6	95.5	0	0	0	0	2.8	0
Harzburgite	0	0	0	0	0	45.8	48.6	0	0	0	5.6	0
Banded unit	0	0	0	7.6	0	0	0	71.6	0	0	18.4	2.4
Sheared gabbro	0	3.8	0	0	0	0	0	2.8	1.9	2.8	88.6	0
Upper volcanics	0.5	12.6	6.5	0	2.1	0	0	0	0	46.6	31.7	0
Lower volcanics	0	0.7	0	0	0	2.1	0	2.1	0.3	0	92.5	2.4

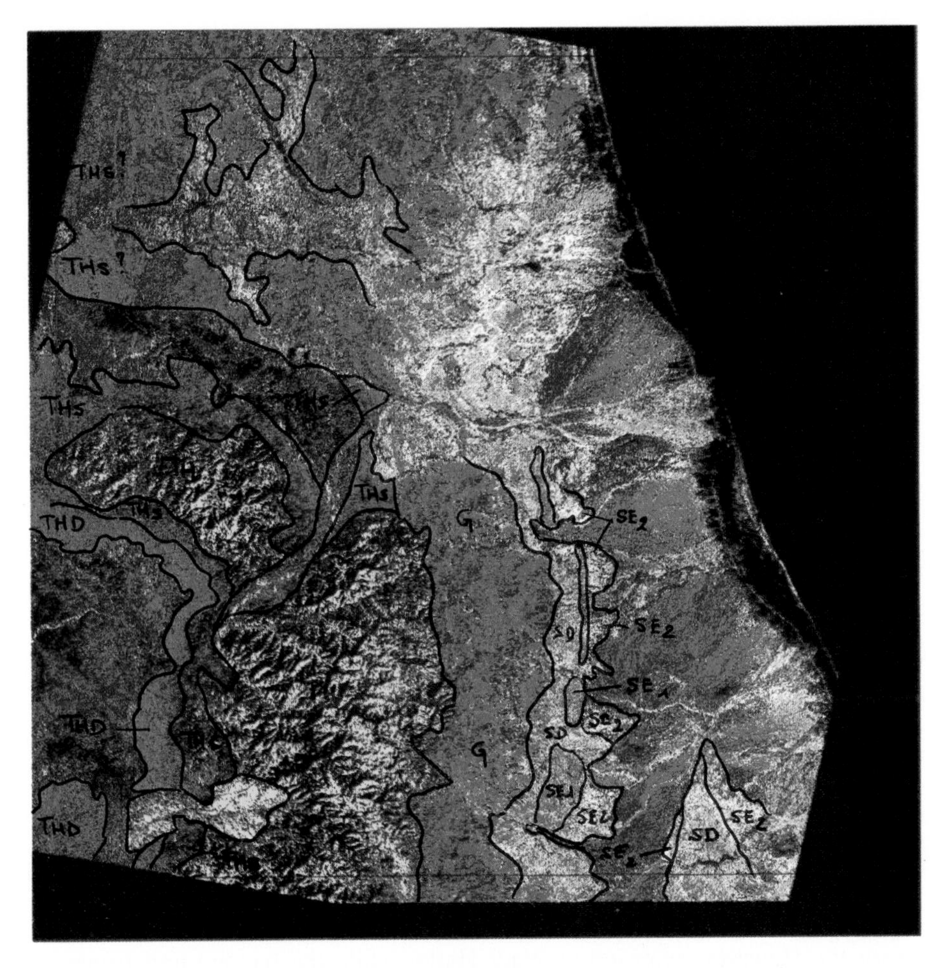

Figure 10. Classification after addition of 3 ratio. (Scale as in Fig. 8).

found in the late Tertiary to Quaternary terraces, which seem composed mainly of crustal rocks transported by east-flowing wadis (Wadi Zabin, Wadi Rajmi, Wadi Fayd), and of harzburgites in Wadi Sharm and the upper reaches of Wadi al Qahfi.

5.2.3. *Geological Interpretation of the Classified Image*

South of wadi Hatta, image classification clearly outlines the main lithologies of the ophiolite, which are from base to top:

- banded unit,
- harzburgite,
- cumulate sequence,

- sheeted dyke complex,
- lower and upper volcanic rocks,
- trondhjemite bodies,

as well as late Tertiary to Quaternary terraces. As indicated earlier, the spectral signature of these terraces is directly related to that of the ophiolite unit from which they derive. By the means of this well-contrasted classified image, geological contacts between these ophiolite units can be easily drawn.

North of wadi Hatta, the classified image is more confused and contacts appear fuzzy. An explanation can be found on the existing geological maps (Glennie, 1974) where the following features can be seen:

- south of wadi Hatta, the ophiolite sequence is tilted east, along a north-south trend, and the various rock units are seen in a cross-section,
- north of wadi Hatta, the sequence is almost horizontal, and the units are tangential to the topographic surface. Contacts are thus irregular and less perceptible.

South of wadi Hatta, two yellow zones of the classification correspond to harzburgite, and are bordered to the south and west, in contact with the Hawasina nappes, by a green and blue pixel fringe corresponding to the banded unit (alternating of sheared dunites and harzburgite). Between the banded unit and harzburgite, a fringe of blue, green and black pixels corresponds to harzburgite. Such a fringe can be seen , to the north, at the contact with the Hawasina exposed in Wadi Hatta, while the banded unit fringe is lacking. Field investigations here have revealed the absence of the banded unit over the Hawasina nappes; Harzburgite appears, on the classified image, with the same blue, green and black pixel association as the harzburgite overlying the banded unit in the south. It is thus possible that the most basal harzburgite has a specific spectral response (translated on the classified image by blue, green and black pixels), close to that of the dunite, due to a higher level of serpentinization near the tectonic contact. The absence of the banded unit to the north suggests that the sole "thrust" at the base of Samail nappe cuts up-section through the ophiolite and is close to the petrological Moho north of wadi Hatta. This particular structural feature suggests a gravity-induced nappe, accompanied by a higher serpentinization of the harzburgite slice. The partial serpentinization would explain the specific spectral response of the harzburgite north of wadi Hatta, as well as the smoother relief in this zone, where altitudes do not exceed 400–700 m, instead of 700–1500 m that are found further south.

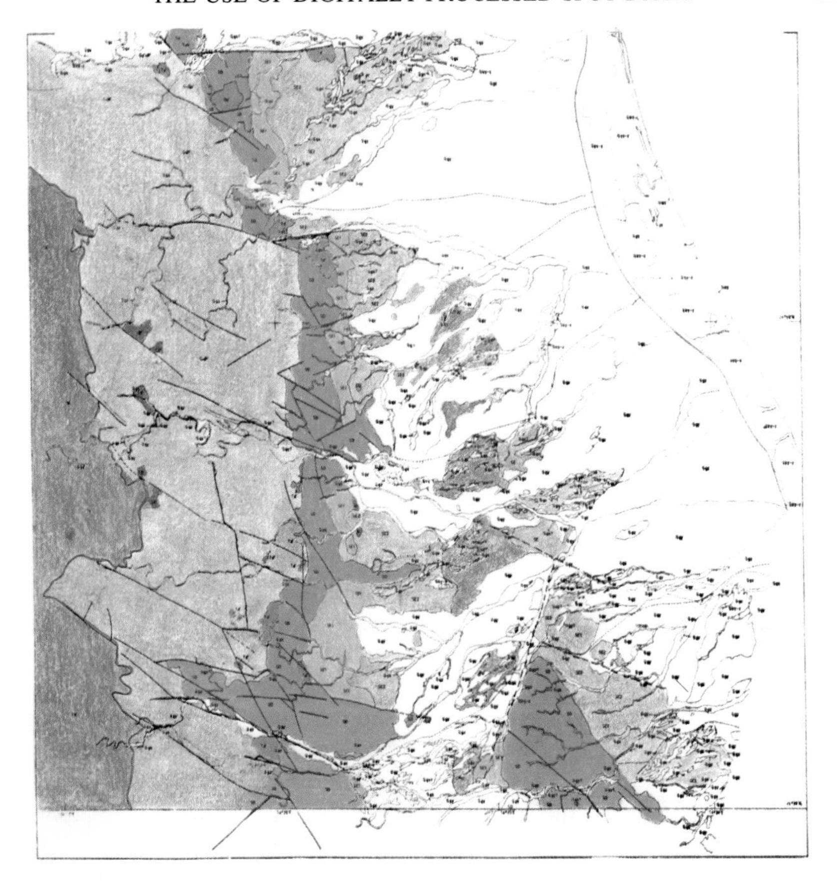

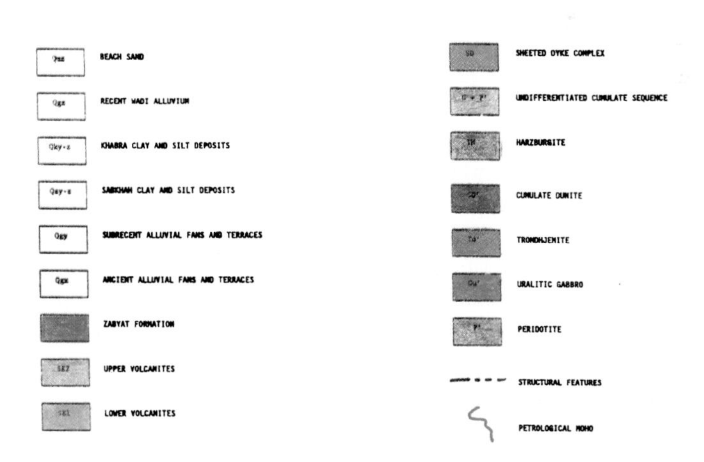

Figure 11. First rough draft of the geology. (Scale as in Fig. 5).

6. Image Interpretation

All the above-mentioned processings must be seen as intermediate steps to prepare and help the final image interpretation, which will be done on the specifically-processed document of Fig. 5. A first rough draft of the south eastern quadrangle is presented in Fig. 11. A close collaboration with the field geologists has led to the delimitation of the extension of the main lithologies: harzburgite, cumulate sequence, sheeted-dyke complex, lower and upper volcanic rocks as well as late Tertiary to Quaternary terraces and recent alluvium. Late intrusive bodies, such as trondhjemite, uralitic gabbro and dunite have also been mapped.

7. Conclusions

The study has proven that SPOT data, together with field work and aerial photographs, are a useful complementary tool for the mapping geologist.

Advantages reside in an excellent ground resolution, which allows production of documents up to 1:50,000 scale, a synoptic coverage, and multispectral digital data that are easily processed in various geometric and radiometric combinations. Another advantage, not used here, is the possibility to obtain stereoscopic pairs which can be of major interest for the geologist.

Specific radiometric processing improves lithologic differentiation through sometimes subtle reflectance differences.

Despite the fact that image classification strictly seen is unsatisfactory, mainly due to sharp relief inducing shadows, it provides the geologist with a synthetic document of use in the understanding of the geometric distribution of the major lithologic features. In the southern half of the mapping area, it allows drafting of geological contacts precise enough for 1:250,000-scale mapping. Classification steps should take place after field investigations to identify the main rock types for the selection of the training areas. After classification processing and interpretation, field control would be useful for checking the homogeneity of the classification, and for defining and resolving any uncertainties and discrepancies in the most complex zones.

Visual interpretations of the various processing steps can be drawn on specifically enhanced documents, and will complete field investigations and aerial photographs in the final drafting of the map.

References

Beurrier M., 1988. Géologie de la nappe ophiolitique de Samail dans les parties orientale et centrale des montagnes d'Oman. Document du B.R.G.M. No. 128.

Glennie, K.W., Boeuf, M.G.A., Hughes Clark, M.W., Moody-Stuart M., Pilaar, W.F.H.,

Reinhardt B.M., 1974. Geology of the Oman Mountains. Verh. Kon. Ned. Geol. Mijnbouwk. Genoot., 31, Part 1,2 and 3.

Rothery, D.A., 1984. Refelectances of ophiolite rocks in the Landsat MSS bands: relevance to lithological mapping by remote sensing. J. geol. Soc. London, Vol. 141: 933–939.

Searle, M.P., Malpas, J., 1980. Structure and metamorphism of rocks beneath the Samail ophiolite of the Oman and their significance in ophiolite obduction. Trans. Roy. Soc. Edinburg, Earth Sciences, vol. 71: 247–262.

Smewing, J.D., Brown, M.A., Searle, M.P., 1979. The Ophiolite Project, Sumeini-Shinas area, Map 1: Open University, Milton Keynes, UK, and Dept. of Minerals, Muscat, Sultanate of Oman, scale 1:100,000, 1 sheet.

Comparison Between Mapping at 1:25000 Scale and Decorrelation Stretched Landsat Thematic Mapper Images in the Wuqbah Block (Oman Mountains)

EDWIN GNOS[1], RENATO WYDER[1] and DAVID A. ROTHERY[2]
[1]*Mineralogisch-Petrographisches Institut, Baltzerstr. 1, 3012 Bern, Switzerland*
[2]*Department of Earth Sciences, The Open University, Milton Keynes MK7 6AA, England*

Abstract

Field mapping at 1:25000 scale of a 250 km^2 area in the western part of the Wuqbah Block (Oman Mountains) has enabled us to revise the distribution of rocks of the Semail Ophiolite Nappe, the Metamorphic Sheet, Haybi Volcanics (Umar Group), Exotics (Misfah Group) and Hawasina Sediments (Hamrat Duru Group). The large Exotic block (Hawrat al Asan) was recognized as mainly Triassic in age.

Comparison of the resulting geological map with enhanced decorrelation stretched Landsat Thematic Mapper (TM) false-color images, using channels within the short wavelength infrared part of the spectrum, has shown a very strong correlation between the two mapping methods.

Mapped areas of layered/laminated gabbros, the shape and distribution of the sheeted dyke complex and the associated isotropic gabbros, altered regions in the mantle and crustal sequence, and even newly mapped outcrops of sheeted dykes and metamorphic rocks, are all visible on the TM images.

Introduction

The Oman (Semail) Ophiolite is the largest and most intact example of ocean floor rocks exposed on land. It was formed by sea-floor-spreading at about 100 Ma and was emplaced over the Arabian continental margin before 70 Ma, together with sedimentary and volcanic rocks of different origins. The Ophiolite covers an area of about 20,000 km^2 and forms large parts of the Oman mountains.

A complete Ophiolite sequence is preserved (Penrose, 1972). Tectonized and partially strongly serpentinized harzburgite (the mantle sequence) with layers, lenses, pods or bodies of concordant or discordant spinel-dunites, pass upwards sharply or diffusely into other peridodites and layered, laminated and homogeneous gabbros. Isotropic gabbros, associated with sheeted

Tj. Peters et al. (Eds), Ophiolite Genesis and Evolution of the Oceanic Lithosphere, 875–885.

dykes and pillow lavas, and sometimes with more differentiated rocks, are believed to represent the undeformed rocks of the axial magma chamber below an oceanic spreading zone and its extrusive products. Late extrusive and intrusive rocks of different chemistry indicate off-axis magmatism.

Since the climate in Oman is arid, virtually no vegetation covers the rocks and exposures are excellent. Gravels are concentrated in wadi floors, intramontane plains and beyond the mountains in alluvial fans.

The first reasonably detailed map of the Oman ophiolite was by Glennie et al. (1974). It was based on interpretation of black and white photographs and spot checks and shows the geology of the whole of the Oman mountains at 1:500000 scale. Within the Open University Oman Ophiolite Project a new map at 1:100000 (Lippard and Rothery 1981, Rothery 1982), shows our area of interest (Figure 1) in more detail.

Our study area lies in the western end of the Wuqbah Block in the Wadi al Uqaybah and Wadi Harim area, north-west of Yanqul. It comprises parts of the Ophiolite Nappe from the base of the mantle sequence up to gabbros and sheeted dykes in the crustal sequence and as well parts of the Metamorphic Sheet, Exotics, Haybi Volcanics and Hawasina Sediments.

Our principal aims were to establish the distribution of metamorphic rocks, different gabbro types and sheeted dyke areas, the rocks of the mantle-crust transition zone, the grade of serpentinization in ultramafic rocks and to map any gossans or ancient slags.

The map (shown at a reduced scale in Figure 2) was produced by two of us (E.G, R.W) during three months of field work in the winter of 1988/89. 1:25000 color air photographs were used as a topographic base. The map is based on photo-interpretation only in the sedimentary part.

Our new mapping suggests that previous interpretation of the area (Lippard & Rothery 1981, Rothery 1982) is broadly correct.

Mapping Results

Mantle Sequence

In comparison to the 1:100000 Open University map (Lippard and Rothery, 1981) the shape and extent of the mantle sequence are generally the same. However, in parts of the contact with the crustal sequence in the west there are sometimes large areas with foliated or unfoliated spinel-, spinel-plagioclase- to plagioclase-dunites and wehrlites of intrusive (cumulate peridodites) or semi-intrusive (trapped melts in restitic harzburgite) origin. As the serpentinization is strong and there is a continuous transition to serpentinized and tectonized harzburgites, the distinction was sometimes impossible. Apart from their stratigraphic position, these intrusive rocks are similar in composition to the peridodites in the layered gabbro and to late intrusive, mainly undeformed ultramafic rocks in the crustal part, that were previously

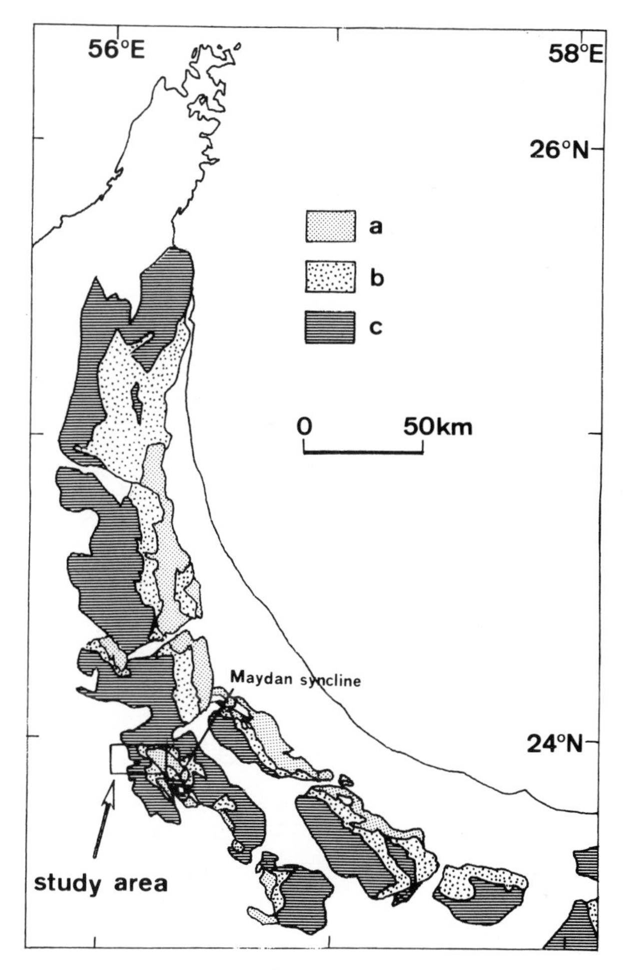

Figure 1. Simplified map of the northern part of the Oman Ophiolite. The study are is outlined. a) Sheeted dykes and extrusives (crustal sequence); b) Gabbros and cumulate peridodites (crustal sequence); c) Tectonized peridodite (mantle sequence).

recognized because of their relationship with late-stage cross-cutting dykes (Rothery, 1983; Browning and Smewing, 1981). It is difficult to decide in the field, whether serpentinization is 70% or 100%, so the strongly serpentinized zones shown on the map must therefore be regarded as incomplete.

Crustal Sequence

Within the gabbros (gabbros, troctolites and norites) layered and laminated types are shown as one unit on the map, as we have not been able to distinguish them with certainty over the entire area.

Isotropic gabbros with prismatic or sometimes dendritic-shaped amphi-

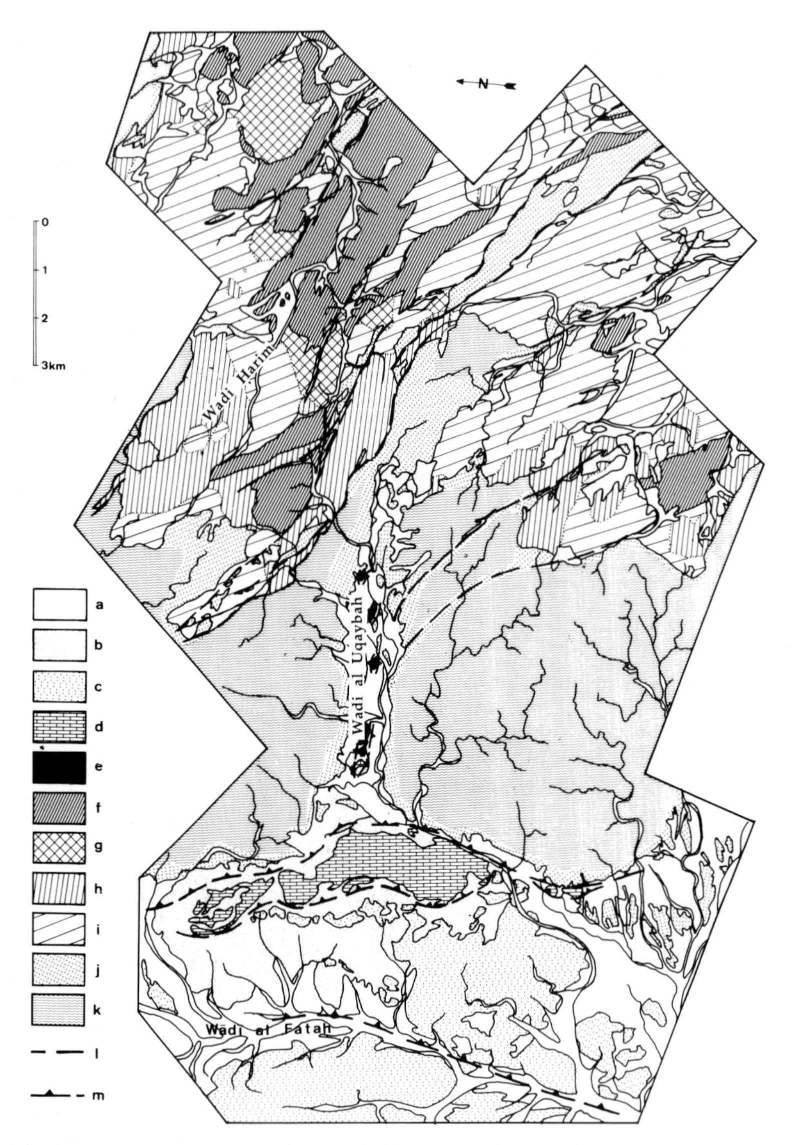

Figure 2. Reduced and slightly simplified geological map of the study area (see figure 1 for location). a) old and recent gravel; b) scree; c) Hawasina Sediments; d) Exotic limestone and autochtoneous cover; e) Metamorphic Sheet and associated volcanic and nonmetamorphic rocks; f) Sheeted dykes (>70% dykes); g) isotropic gabbros, sometimes with dykes; h) homogeneous gabbros, sometimes with dykes; i) layered and laminated gabbros, sometimes with dykes; j) strongly serpentinized ultramafic rocks (harzburgites, dunites, wehrlites and melatroctolites), sometimes with dykes; k) tectonized harzburgites, with concordant and discordant dunites; l) fault; m) thrust.

Figure 3. Decorrelation stretched TM 7, 5, 4 image of the study area (see figure 1 for location). For geological units see figure 2.

boles were always found in association with sheeted dykes. Small leuco-quartz-dioritic or leuco-tonalitic (trondhjemite) outcrops occur in the same position. Altered rocks are widespread, spatially related to the dykes and linked to an increased water content in these parts of the magma chamber. The water was either directly intruded from the sea floor or derived from older sheeted-dyke-complexes, where it was trapped in joints or in hydrated minerals (altered parts).

The sheeted dyke outcrops drawn on the map are areas where more than 70% dykes were observed (in accordance with the Open University map convention). Often the separation from host rocks is very sharp, but in some areas the transition from gabbros to more than 70% dykes is gradual. Observed areas with a mixture of gabbros or peridodites with dykes are not indicated on Figure 2. In the north-western area of the crustal sequence a new outcrop of sheeted dyke complex was mapped (Rothery et al. 1990), which is hosted by layered gabbro. In this area the peridodites of the mantle-crust transition zone contain a lot of doleritic, gabbroic and dioritic dykes as well, and the rocks are often brecciated by small plagiogranitic intrusions. Doleritic dykes were also observed within the upper part of the tectonized harzburgite. The formation of this sheeted dyke complex was off-axis.

Metamorphic Sheet

Outcrops of metamorphic rocks (amphibolites and greenschists) together with slices of volcanics (Umar group or Haybi volcanics) and non-metamorphic rocks occur within serpentinites along an east-west line in the upper part of Wadi al Uqaybah. The most easterly outcrops were newly found in the present study.

Sediments

Megalodonts observed in the Exotic limestones show that they (or large parts of them) are upper Triassic in age. The S_0-planes are nearly vertical or steeply dipping to the east. Volcanic rocks occur along the eastern thrust-contact to the Ophiolite (Umar-group or Haybi volcanics). On the eastern slope of the Triassic Hawrat al Asan-Exotic, remnants of autochthonous sedimentary cover were observed. A red, condensed layer was found in a tectonically preserved situation. It contains belemnites and therefore indicates a Jurassic or younger age. It is not clear if some volcanics, interfingered with sediments and above the condensed layer, are in the original stratigraphic position.

The Hawasina sediments (Hamrat Duru) west of the exotics are composed of green and red cherts or radiolarites alternating with calciturbiditic layers. On the eastern edge, they have been tectonically overridden by the mass of Exotics. The eastern part of the Hawasina Sediments, showing more distal facies, is folded, whereas the western section (west of Wadi al Fatah) is imbricated.

Scree and Gravel

The surfaces of scree or gravel terraces are covered with a red desert varnish of alteration minerals (ferric-iron-minerals). Older gravel terraces show darker weathering cortices. Only the recent or subrecent gravels in the wadi bottoms show fresh surfaces.

Carbonate (mostly calcitic) cements or lenses of carbonate, reflecting ancient aragonitic or calcitic deposits from oversaturated water (both detected in recent deposits) are widespread in the gravel terraces.

Serpentinized peridodites or illitic gabbros (hydrothermally altered) are typical below gravel terraces.

A description of the mineralogical composition of the rocks from the Ophiolite Nappe in this area was given by Rothery (1982).

TM Data and Processing

After our detailed mapping, we examined enhanced satellite images to determine how well these data can discriminate between rock units at this scale. Our remote sensing data are from the Thematic Mapper (TM) scanner, carried by Landsat 5 (Baker, 1985). From 7 possible spectral bands a triplet of channels (7, 5 and 4) in the short wavelength infrared were chosen, where spectral reflectance differences between varieties of natural rocks are greatest (Rothery, 1987, Abrams et al, 1988). Each pixel of the image corresponds to the information from 30 × 30 m of the earth's surface. We consider that the spectral discrimination provided by the short wavelength infrared bands of the Landsat TM more than compensates for the lower resolution compared to the SPOT satellite, which has 20 × 20 m pixels but is limited to the 0.5–0.9 μm region of the spectrum.

Our data were "decorrelation stretched". The advantage of this procedure has been discussed by Soha and Schwartz (1978), Gillespie et al (1986), Rothery (1987), and Abrams et al. (1988). Decorrelation stretching makes it possible to distinguish very small spectral reflectance differences while maintaining easily and consistently recognizable colors. Because the image is not broken into discrete classes (as in digital classification), the textural and contextural information of the image is available to guide the interpreter in addition to the spectral information. Band 7 (2.08–2.35 μm) contains information about the presence of hydroxyl-bearing minerals (Al-OH, Mg-OH in clays, micas, amphiboles and serpentine) and CO_3-bearing minerals (carbonates). Band 5 (1.55–1.75 μm) serves to characterize the general albedo of the materials. Altered rocks with ferric iron have higher reflectances in this band, whereas ferrous iron causes a depression. Band 4 (0.76–0.90 μm) contains information relating to the presence of ferric iron.

The most effective way of displaying these channels after decorrelation stretching is to show band 7 in red, band 5 in green and band 4 in blue.

Interpretation of the Decorrelation Stretched Infrared Image

Two distinct colors appear in the mantle sequence (see Figure 3). Dark purple colors show the basal part, areas around gravel terraces, fault zones and the upper part of the mantle sequence. The same colors occur in the peridodites of the transition zone and in peridodites of the Maydan syncline. A similar color with a more irregular pattern occurs in some parts of the sheeted dykes and metamorphic rocks. The purple appearance of all those rocks can be explained by their content of fine dispersed magnetite or magnetite/ilmenite which depresses absorption bands of other minerals. In the ultramafic rocks magnetite was formed by serpentinization process.

The purple color correlates with nearly 100%-serpentinized peridotitic rocks, wehrlites and some melatroctolites. The intrusive ultramafic rocks in the upper part of the tectonized harzburgite, in the transition zone and in the crustal sequence, seem to get serpentinized more readily than the tectonized harzburgite (owing to their porosity, foliation and distance to the ancient sea floor). However, as the olivines are also typically more iron-rich compared with those in the mantle harzburgites (Lippard et al., 1986), even a moderate degree of serpentinization may result in the formation of significant amounts of magnetite to quench the rock spectra.
Serpentinization at an early stage (subseafloor) could be a possible explanation; Stanger (1986) described high chloride concentrations in serpentinites. A second kind of serpentinization in the basal part of the ophiolite nappe is likely to relate to its emplacement, where water, derived from the underlying rocks or from a subduction related process (Smith et al., 1976) induced or intensified the growth of serpentine minerals (including antigorite). A third and certainly important reason is serpentinization by meteoric water at low temperatures (Dunlop and Fouillac, 1984). Our field study showed that covering of ultramafic rocks by gravels and water circulation through the gravels has encouraged alteration. Ancient gravel terraces found in the upper part of Wadi al Uqaybah show that large parts of the mapped area once were covered by gravels as a result of the more humid climate during the late Pleistocene. Today, this old gravel plain is being eroded from the west.

Dark-green to yellow colors are typical of the internal parts of the tectonized mantle sequence, where rocks are less altered. The color is largely controlled by the spectral response of the harzburgite in the visible-near infrared (NIR) wavelengths, which is a combination of the olivine, orthopyroxene and serpentine spectra. However, the variation from green to yellow on the image may be due to variations in the proportions of different weathering surface types. Normal reddish weathering surfaces have little influence on the rock spectra in the NIR (Pontual, 1990) but the presence of large proportions of dark red deeply weathered surfaces in a pixel area may produce a more yellow color due to the ferric oxide component in the weathering cortices (Pontual, op. cit.). On the TM images it is not possible to discriminate between harzburgite and dunite. This is largely due to the small sizes of most dunite pods relative to the TM pixel size in the lower part of the harzburgite. In addition to this, the spectra of partially serpentinized dunites and harzburgites, as in the upper part of the mantle sequence, are virtually indistinguishable in the visible and NIR wavelengths (Hunt and Evarts 1981, Pontual 1990).

Cyan colors in the crustal sequence show a similar distribution to the mapped layered/laminated gabbros. Homogeneous gabbros show green colors, whereas isotropic gabbros (often strongly altered), and areas with gabbros and dykes, are yellowish-orange on the infrared image. Differences in the content of mafic minerals (Rothery, 1990) and differences in alteration (increasing with increasing level in the ophiolite sequence) produce this color pattern (Pontual, 1989, 1990). The sheeted dykes show orange to red colors, in some parts even purple colors like serpentinized ultramafic rocks. Purple colors occur if the dykes contain dispersed, intergrown magnetite-ilmenite. A new ultramafic body was postulated in the Maydan syncline on the basis of this evidence (Rothery, 1987), but was later recognized as an error (Rothery, 1987, 1990). Rothery et al. (1990) show that the purple color is generally typical of the sheeted dykes in Oman, and that the orange to red colors that are common in the Wuqbah Block are a consequence of widespread epidotic alteration in this region.

The shape of the sheeted-dyke outcrops visible on the enhanced infrared image and on the map are exactly the same. Metamorphic rocks, greenschists to green quartzites and amphibolites in the Wadi al Uqaybah area, show similar yellow and orange (but never red) colors on the infrared image but are easy to distinguish from sheeted dykes because of their geotectonic position beneath the ophiolite. The most eastern outcrops of the metamorphic rocks are not visible on the TM-image, either because they are too small in size or in contrast, or because quartzites and surrounding serpentinites both contain magnetite, resulting in purple colors for both.

Ancient gravel terraces in the mapped area, with high amounts of altered gabbroic and doleritic rocks, typically show smooth orange-greenish colors. Fresh washed gravels in the wadi bottoms have cyan or purple colors depending on their origin.

Carbonate deposits in the upper Wadi al Uqaybah, and a slag-field are not recognizable on the infrared image. Their size is about that of one pixel, and is therefore critical. A single, pink pixel was identified as plagiogranite.

Generally, such plagiogranites are too small in the mapped area to be detected by TM.

An interpretation of minerals producing the observed absorption pattern in the different rocks of the Semail Ophiolite is given in Pontual (1988, 1989, 1990) and Rothery (1990).

Conclusions

Even though the reasons for different absorption colors in some parts are still uncertain, the correspondence between the results of 1:25000 scale mapping and enhanced reflected infrared images is remarkable. Images using visible wavelengths do not have this property.

These results demonstrate that, with previous experience of a terrain and its lithologies, it is possible to produce maps with much information regarding the extent, composition and condition (fresh or altered) of rocks and gravels. For future field-based research in Oman, we recommend the use of the decorrelation stretched Landsat Thematic Mapper images as additional information to the published maps, particularly when searching for areas with a particular rock type, rock association or rock condition.

Acknowledgments

Mohammed Bin Hussain Bin Kassim, Director of Minerals and Dr. Hilal Al Azry, Director of Geological Surveys are thanked for permission to work in Oman and for their kind help. The manuscript was improved by comments of Sasha Pontual and an anonymous reviewer.

References

Abrams, M.J., Rothery, D.A. and Pontual, A., 1988. Mapping in the Oman ophiolite using enhanced Landsat Thematic Mapper images. Tectonophysics, 151: 387–402.

Baker, J.L. (Ed), 1985. Landsat-4 Science Characteristics, Early Results. NASA, Washington, D.C., NASA CP-2355, 4 Volumes.

Browning, P. and Smewing, J. S., 1981. Processes in magma chambers beneath spreading axes: evidence from magmatic association in the Oman Ophiolite. Jour. Geol. Soc. London, 138: 279–280.

Dunlop, H.M. and Fouillac, C., 1984. O, H, Sr, Nd isotope systematics of the Oman Ophiolite. Ophiolites "Through Time" Conf. Nantes, abstr., p. 25.

Glennie, K.W., Boeuf, M.G.A., Hughes-Clarke, M.W., Mody-Stuart, M., Pilaar, W.H.F. and Reinhardt, B.M., 1974. Geology of the Oman Mountains. Verh. K. Ned. Geol. Mijnbouwkd. Gennot., Geol. Ser., 31, 423 p.

Hunt, G.R. and Evarts, 1981. The use of NIR spectroscopy to determine the degree of serpentinisation of ultramafic rocks. Geophysics, 46(3): 316–321.

Lippard, S.J. and Rothery, D.A (Eds), 1981. Wadi Ahin-Yanqul. Oman Geological Ophiolite Project map 3. Dir. Overseas Surv., Open Univ., Milton Keynes.

Lippard, S.J., Shelton, A.W. and Gass, I.G.,1986. The Ophiolite of Northern Oman. Geol. Soc. Mem., 11, 178 p.

Penrose Conference Participiants, 1972. Penrose field conference on Ophiolites. Geotimes 17: 24–25.

Pontual, A., 1988. Application of Landsat Thematic Mapper imagery for lithological mapping of poorly accesible semi-arid regions. In Proceedings, 6th Thematic Conference on Remote Sensing for Exploration Geology, May16–19 1988, Houston, Texas (publ. ERIM): pp. 339–348.

Pontual, A., 1989. Lithological information in enhanced Landsat Thematic Mapper images of arid regions. In Proceedings, 7th Thematic Conference on Remote Sensing for Exploration Geology, October 2–6 1989, Calgary, Alberta, Canada (publ. ERIM): pp. 379–393.

Pontual, A., 1990. Lithological information in Remotely Sensed Images and Surface Weathering in Arid Regions. Unpubl. PhD Thesis, The Open University.

Rothery, D.A., 1982. The evolution of the Wuqbah block and the application of remote sensing in the Oman ophiolite. Open University Ph.D.Thesis: 414 pp.

Rothery, D.A., 1983. The base of a sheeted dyke complex, Oman ophiolite: implications for magma chamber configuration at oceanic spreading axes. Jour. Geol. Soc. London 140: 287–296.

Rothery, D.A., 1987. Improved discrimination of rock units using Landsat Thematic Mapper imagery of the Oman ophiolite. Jour. Geol. Soc. London, vol. 44: 587–597.

Rothery D.A., Abrams, M.A. and Pontual A., 1990. Subdivision of mapped units in the Oman ophiolite using enhanced Landsat Thematic Mapper images. Proceeding, Troodos 87 Ophiolite and Ocean Lithospheres, Nicosia, Cyprus, October 4–10 1987.

Smith, A.G. and Woodcock, N,H., 1976. The earliest Mesozoic structures in the Othris region, Eastern Central Greece. Bull. Soc. Geol. France,.2: 245–251.

Soha, J.M. and Schwartz, A.A., 1978. Multispectral histogram normalization and contrast enhancement. In: Proc, Can. Symp. Remote Sensing, 5th (Victoria, B.C.), pp. 86–93.

Stanger, G., 1986: The hydrogeology of the Oman Mountains. Open University Ph.D.Thesis, 572 p.

Index

Tj. Peters et al. (Eds). Ophiolite Genesis and Evolution of the Oceanic Lithosphere, 887–903.

Petrology and Structural Geology

1. J. P. Bard: *Microtextures of Igneous and Metamorphic Rocks.* 1986
ISBN Hb: 90-277-2220-X; ISBN Pb: 90-277-2313-3
2. A. Nicolas: *Principles of Rock Deformation.* 1987.
ISBN Hb: 90-277-2368-0; ISBN Pb: 90-277-2369-9
3. J. D. Macdougall (ed.): *Continental Flood Basalts.* 1988
ISBN 90-277-2806-2
4. A. Nicolas: *Structures of Ophiolites and Dynamics of Oceanic Lithosphere.* 1989
ISBN 0-7923-0255-9
5. Tj. Peters, A. Nicolas and R.G. Coleman (eds.): *Ophiolite Genesis and Evolution of the Oceanic Lithosphere.* Proceedings of the Ophiolite Conference (Muscat, Oman, January 1990). 1991
ISBN 0-7923-1176-0

KLUWER ACADEMIC PUBLISHERS – DORDRECHT / BOSTON / LONDON